Advanced Manufacturing and Surface Technology

Advanced Manufacturing and Surface Technology

Topic Editors

Dingding Xiang
Junying Hao
Xudong Sui
Kaiming Wang

MDPI

Basel • Beijing • Wuhan • Barcelona • Belgrade • Novi Sad • Cluj • Manchester

Topic Editors
Dingding Xiang
School of Mechanical
Engineering and Automation
Northeastern University
Shenyang
China

Junying Hao
Lanzhou Institute of Chemical
Physics
Chinese Academy of Sciences
Lanzhou
China

Xudong Sui
E307-05 Research Unit of
Tribology
TU Wien
Vienna
Austria

Kaiming Wang
College of Automobile and
Mechanical Engineering
Changsha University of
Science and Technology
Changsha
China

Editorial Office
MDPI AG
Grosspeteranlage 5
4052 Basel, Switzerland

This is a reprint of the Topic, published open access by the journals *Journal of Functional Biomaterials* (ISSN 2079-4983), *Journal of Manufacturing and Materials Processing* (ISSN 2504-4494), *Lubricants* (ISSN 2075-4442), *Machines* (ISSN 2075-1702) and *Materials* (ISSN 1996-1944), freely accessible at: https://www.mdpi.com/topics/T4O9HU7809.

For citation purposes, cite each article independently as indicated on the article page online and as indicated below:

Lastname, A.A.; Lastname, B.B. Article Title. *Journal Name* **Year**, *Volume Number*, Page Range.

ISBN 978-3-7258-4651-1 (Hbk)
ISBN 978-3-7258-4652-8 (PDF)
https://doi.org/10.3390/books978-3-7258-4652-8

Contents

About the Editors

Dingding Xiang

Dingding Xiang holds a Ph.D from Tsinghua University. He is an Associate Professor at Northeastern University. In 2019, he worked as a research assistant at the Singapore Centre for 3D Printing, School of Mechanical and Aerospace Engineering, Nanyang Technological University, Singapore. In 2023, he was a visiting scholar at the Research Institute for Advanced Manufacturing, The Hong Kong Polytechnic University. He serves as a council member and committee member of five national-level societies, including the Tribology Branch, Surface Engineering Branch, Additive Manufacturing Technology (3D Printing) Branch, and Extreme Manufacturing Branch of the Chinese Society of Mechanical Engineering, and the Additive Manufacturing Technology Specialized Committee of the Chinese Society for Nonferrous Metals. He is a guest editor for eight international SCI journals such as "Journal of Functional Biomaterials" and "Lubricants", and a Special Issue editor, editorial board member, and young editorial board member for over ten EI journals such as "China Surface Engineering", "Surface Technology", and "Materials Engineering". He has published over 30 SCI papers and has led over 20 projects including the National Natural Science Foundation.

Junying Hao

Dr Junying Hao is currently appointed as a full professor of State Key Laboratory of Solid Lubrication (SKSL), Lanzhou Institute of Chemical Physics (LICP), Chinese Academy of Sciences (CAS). She completed his PhD degree at LICP, CAS in Material Science and Engineering in 2006. Dr Hao is the group leader of Lubrication and Protection Films Group. Her major research interests include solid lubricant films, solid–liquid composite lubrication, anti-corrosion films, CVD films, PVD films, etc. To date, she has published over 150 SCI journal papers, including Advanced Materials, Applied Surface Science, Surface & Coatings Technology, Tribology International, etc. In addition, she has declared 60 China invention patents, and 20 of which are authorized. She received two honors on Science and Technology Progress Awards of Guofang (Class I) and Natural Sciences Awards of Gansu Province (Class II). She headed over 10 research projects of the National Natural Science Foundation, National key Research and Development Program Sub-project, CAS Focus on Sub-projects, Provincial and Ministerial Levels Projects, and Industry Projects, etc.

Xudong Sui

Dr. Sui received his PhD from the Northeastern University in China in 2017 where he studied advanced hard coatings for high-speed cutting application under Prof. Dr. Qiang Wang and Prof. Dr. Guojian Li. After graduation, Dr. Sui joined Prof. Weimin Liu's group in Lanzhou Institute of Chemical Physics, Chinese Academy of Sciences. There, he conducted research on the design and fabrication of solid lubrication coatings, with a focus on tribological mechanisms, and was promoted to Associate Professor in 2018. Since January 2024, he has been working as a full-time guest professor at TU Wien, with research focused on tribochemical and sustainable lubrication. Dr. Sui has authored 49 peer-reviewed publications in high-impact journals such as *Advanced Materials, Matter, Advanced Science, Applied Surface Science, Tribology International, and Surface and Coatings Technology.* His research has received over 1327 citations (Google Scholar), with an h-index of 20 and an i10-index of 30, reflecting a sustained and broad academic impact in the areas of tribology and advanced coating materials. In addition to his academic output, he is also the inventor of four granted invention patents, demonstrating strong engagement in technological innovation and practical applications. Additionally, he is a member of the Chinese Mechanical Engineering Society and serves on the editorial board of the *Journal of Industrial Lubrication and Tribology*. He also serves as a reviewer for several SCI journals, including *Tribology International, Advanced Engineering Materials, Langmuir, Surface & Coatings Technology, and Vacuum.*

Kaiming Wang

Associate Professor Kaiming Wang received his PhD in Mechanical Engineering from the Department of Mechanical Engineering ,Tsinghua University, Beijing, China, in 2021. He is currently working at the Changsha University of Science and Technology in China. He has authored more than 40 publications in the areas of his research interests, which include intelligent manufacturing, additive manufacturing and laser cladding. Professor Wang is the recipient of the 2025 Hunan Science Foundation for Excellent Young Scholars.

materials

Article

Microstructure and Microhardness Research of Steel 304 After Forming Partially Regular Reliefs by Ball Burnishing Operation

Stoyan Dimitrov Slavov [1], Diyan Minkov Dimitrov [2], Desislava Yordanova Mincheva [3], Volodymyr Dzyura [4], Pavlo Maruschak [4,*] and Volodymyr Semehen [4]

1 Department of Manufacturing Technologies and Machine Tools, Technical University of Varna, Studentska Str. 1, 9010 Varna, Bulgaria; sdslavov@tu-varna.bg
2 Department of Mechanics and Machine Elements, Technical University of Varna, Studentska Str. 1, 9010 Varna, Bulgaria; dm_dimitrov@tu-varna.bg
3 Department of Materials Science and Technology, Technical University of Varna, Studentska Str. 1, 9010 Varna, Bulgaria; dmincheva@tu-varna.bg
4 Department of Motor Vehicles, Ternopil Ivan Puluj National Technical University, 46001 Ternopil, Ukraine; volodymyrdzyura@gmail.com (V.D.); semegen5@ukr.net (V.S.)
* Correspondence: maruschak.tu.edu@gmail.com

Abstract: The influence of regular relief formation modes on the microhardness of the formed groove surface near the apex at the bottom of the groove has been studied. It has been established that the rate of plastic deformation, expressed as the feed rate of the deforming element, has a significant impact on the plastic deformation mechanism, and the microstructure of the formed subsurface layer, as well as on the microhardness of the groove surface. The influence of the type of partially regular reliefs on the degree of plastic deformation was also investigated. It was found that the third type of partially regular relief, which has the highest groove density, provides higher microhardness values than the first and second types. This is explained by the significantly greater density of these type of partially regular relief grooves, which exert a mutual strengthening effect on the surface during formation. The experimental study conducted enabled the derivation of regression equations describing the influence of the feed rate of the deforming element and the type of partially regular relief created on the surface microhardness beneath the lateral ridges and the bottoms of the plastically deformed traces.

Keywords: surface microhardness parameters; microstructure; partially regular relief; ball burnishing; ANOVA; regression analyses

1. Introduction

The study of surfaces with different types of regular microreliefs is primarily aimed at determining the influence of technological factors on the physical and mechanical properties of the surface to improve its durability [1–3] and wear resistance [4,5].

The main patterns of changes in the physical and mechanical properties of metals after the formation of regular and partially regular so-called "microreliefs" are summarized in [6]. It has been established that the surface ball burnishing process with the formation of a regular microrelief affects the magnitude of compressive stresses. Ground samples with a fatigue strength limit of σ_{-1} = 255 MPa were studied. The application of regular microreliefs increased their fatigue strength limit to 334 MPa, which is a 20% increase. To determine the influence of compressive stress, the samples with a regular microrelief were subjected to heat treatment, namely, tempering in argon at 650 °C for 1.5 h. As a result, the fatigue strength limit decreased to 285 MPa, i.e., an 11% reduction. Thus, although heat treatment reduces the strength limit, it ensures a uniform distribution of physical and mechanical properties across the entire surface with a regular microrelief.

An increase in surface hardness after laser texturing was identified in [7]. The authors investigated the effect of laser surface hardening and ultrasonic impact treatment on the surface's roughness and microhardness of AISI D2 high-chromium tool steel. The study concluded that analyzing the hardness magnitude, depth, and width of the laser-hardened area allowed for selecting the optimal laser surface hardening (LSH) regime, which provided a ~150% increase in hardness due to transformation hardening with a minimal reduction in surface roughness. The optimal ultrasonic impact treatment (UIT) regime was determined based on the lowest achieved surface roughness. The optimal UIT led to severe plastic deformation, reducing surface roughness by ~80% and increasing hardness by ~50%.

The creation of an entirely new regular microrelief on the surface of a cylindrical shaft alters its operational properties [8]. The microrelief was applied using a vibrating tool with a radius of R = 2 mm under a load of P = 550 N. As a result of treatment, the surface roughness based on the arithmetic mean deviation was Ra = 0.63 μm, which is 7.6 times lower than the initial value. Experimental studies confirmed that different regimes of treatment significantly alter surface roughness, hardness, and microhardness. It was found that comprehensive treatment technology increased surface hardness by 1.2 to 1.6 times compared to the base metal hardness.

The microhardness of textured surfaces with grooves of different shapes, created using a laser method, was analyzed in [9]. A comparison was conducted between surfaces without the formed microstructures, surfaces with non-intersecting straight grooves, and surfaces with straight grooves intersecting at right angles. It was established that a surface with intersecting grooves provided a hardness 100 units HV higher than a surface without formed regular microstructures. Additionally, this surface exhibited a hardness 50 units HV higher than that of non-intersecting grooves. This indicates that a greater number of deformation-affected locations results in higher surface hardness.

The influence of the formed microrelief and its shape on the friction coefficient and surface wear of the Ti-6Al-4V alloy is described in [10]. The study also generalized the impact of surface texturing on its microhardness. Several configurations of the microrelief elements, in the form of hemispherical dimples, were analyzed for their effect on microhardness. It was found that a microgrid configuration provided the highest microhardness, reaching 400–405 units HV, which is 35–40 units HV higher than other microrelief configurations. Additionally, this configuration also ensured the lowest friction coefficient. Another important finding was that an increase in the relative area of the microrelief beyond 51% significantly reduced the friction coefficient due to the decrease in the contact surface area.

A new technological process has been considered, which additionally includes turning operations with or without cryogenic pre-cooling of the workpiece and surface plastic deformation operations by ball burnishing [11]. The method for performing cryogenic pre-cooling, rigid turning, and ball polishing to achieve satisfactory surface integrity and operational properties has been described. Surface roughness was evaluated in 2D and 3D dimensions, microstructure changes were analyzed using SEM/BSE methods, and microhardness distribution was assessed. Some geometric, mechanical, and physical improvements in the properties of the surface layer were identified.

During surface plastic deformation using a deforming element, a gradient of hardness surface is formed across the depth of the surface. This gradient depends on the treatment regime parameters and the geometric characteristics of the deforming element. The dependence of microhardness on the distance from the surface subjected to surface plastic deformation is described in [12]. It was found that its value initially increases in the subsurface layer before gradually decreasing with increasing depth.

The effect of burnishing parameters on microhardness changes was investigated in [13]. It was established that burnishing cylindrical surfaces with a spherical burnishing tool and increasing the feed rate within the range of small values (0.01–0.05 mm/rev) increases microhardness by 12%. It was found that the feed rate had the most significant impact on microhardness changes, whereas workpiece rotational speed had the least influence among all studied parameters.

Research results on the microhardness of surface layers on cylindrical surfaces with a formed regular microrelief using vibratory burnishing with a spherical deforming element under a deformation force of 350 N are known [14]. It was established that microhardness gradually decreases with increasing distance from the surface with the formed microrelief and stabilizes at approximately 0.32 mm depth.

An increase in the microhardness of the processed surface by 28% after burnishing with roller-type deforming elements compared to the initial state was found in [15]. It was determined that microhardness depended on the spindle speed of the machine tool, with values in the range of 1000–2500 rpm, causing a significant increase in surface microhardness. However, increasing the feed rate within the range of 200 to 450 mm/min led to a decrease in microhardness.

A contradictory result regarding the effect of the feed rate on microhardness changes was obtained in [16]. It was found that increasing the feed rate from 0.2 to 0.6 mm/rev for cylindrical surfaces led to a 10–15% increase in microhardness. Additionally, increasing the workpiece rotational speed from 355 to 710 rpm also resulted in an 8–10% increase in microhardness.

In [17], the dynamic plastic deformation of 304L stainless steel was investigated. A significant influence of the microstructure of the formed surface on the rate of its deformation was established. An improvement in mechanical properties due to the increase in dislocations α'-martensite phases was also noted.

In [18], the effect of different thickness reductions by cold-rolling on the microstructure and the mechanical properties of AISI 304 austenitic stainless steel were investigated. The results of the research indicated that the formation of strain-induced martensite resulted in a significant strengthening of the steel.

Thus, summarizing the effect of plastic deformation on microhardness changes, it should be noted that microhardness increases due to the strengthening of surface and subsurface layers. However, the effect of the feed rate of the deforming element on surface microhardness requires further research, as the literature presents contradictory findings. Additionally, the change in microhardness of the surface layer around the formed groove and at various distances from the groove requires further investigation. This will enable us to determine the optimal relative placement of microrelief grooves on the surface of the processed parts.

The main objective of this study is to assess the influence of some ball-burnishing operation regime parameters and three types of partially regular relief formatted onto AISI 304 steel flat specimens on the microhardness in the depth of the surface layer and to provide a physical–mechanical interpretation of these effects. The second objective is to derive appropriate mathematical models, which give the relation between the regime parameter "feedrate", and the type of the researched partially regular reliefs and resulting microhardness beneath the lateral ridges and bottoms of the plastically deformed traces after ball burnishing operation applying.

Due to the obvious discrepancy between the originally introduced terms "regular-microreliefs" and "partially-regular-microreliefs" in [6] and the actual sizes of the groove's patterns that are obtained, in the present work, the terms "regular relief" (RR) and "partially regular relief" (PRR) will be used instead of the original ones. We believe that using these proposed terms could be more appropriate considering the actual scales of the obtained plastically deformed groove's patterns, which are about 10^3 times larger than the micrometer unit scale and can reach up to several millimeters.

2. Materials and Methods

2.1. Preparing the Test Specimens

According to the main goal of the present research, three types of partial RR (PRR) are investigated, which are obtained by applying ball burnishing process on the top surface of a prismatic plate, made of AISI 304 stainless steel cold-rolled sheet. The types of PRR differ by how plastically deformed traces are patterned onto the specimen's surface (see Figure 1a–c). Depending on it, they can be classified as follows [6]:

(a) Pattern with traces that do not touch each other (type I);
(b) Pattern with traces that touch each other (type II);
(c) Pattern with traces that are crossing between each other (type III).

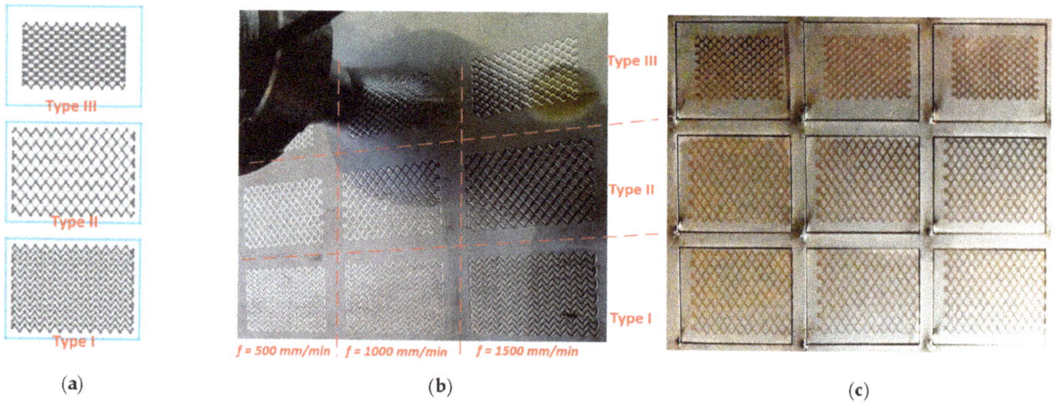

Figure 1. Diagram of the three types of partial regular reliefs (**a**); forming the PRR onto the top surface of the plate using BB process (**b**); the cut-out smaller specimens from the plate for microhardness measurements and structure analyses (**c**).

The three types of PRR with three different feed rates are formed onto a single plate using HAAS (Philadelphia, PA, USA) TM-1 CNC milling machine and specially developed deforming tool [19] (see Figure 1b). After that, the plate is cut out into smaller size specimens with dimensions of 30 × 40 × 4 mm (see Figure 1c), using a fiber laser CNC-machine, type GN NCF 3015 (Wuhan Gnlaser Equipment Manufacturing CO., LTD, Wuhan, China). An austenitic stainless-steel cold-rolled plate (grade AISI 304) with 4 mm thickness was utilized, which has the mechanical properties shown in Table 1, according to the certificate issued by the sheets' manufacturer. The chemical composition of the material is measured by using a model S1 Sorter XRF analyzer (Bruker Corp., Billerica, MA, USA). The resulting chemical composition in wt% is given in Table 2.

To keep the surface with PRR while cutting the pieces for metallography cross-sections, a thin copper film coating was applied to the upper surface of the specimens, by the "brush plating" technology. The laying of the coating is carried out as follows: The sample is caught in a handle that is connected to the negative pole of the power source. The graphite electrode wrapped in a textile braid connects to the positive pole of the power source. With the current source on, the electrode is immersed in the appropriate solution and by rapid movements, performs the application of the coating on a local part of the specimen surface.

Table 1. Mechanical properties of austenitic stainless-steel AISI 304 cold-rolled sheet.

Material	Mechanical Properties			
	Yield Strength, MPa	Ultimate Tensile Strength, MPa	Elongation A, %	$HV_{0.01}$
AISI 304	372 ± 10	710 ± 12	31.2 ± 0.8	205 ± 12

Table 2. Chemical composition of austenitic stainless-steel AISI 304 cold-rolled sheet.

Material	Cr	Mn	Mo	Ni	Si	Fe
AISI 304	18.23	1.76	0.27	8.33	0.38	71.03

One of the lateral surfaces of the kerfs, which pass normally through the deformed traces as sections, is additionally mechanically processed using a metallographic grinding and polishing machine, type MoPao 260E (Opassy Industry Group Co., Ltd., Dongguan, China), for the purpose of making suitable specimens for metallographic analyses and microhardness measurements and removing the material layer that is changed after the laser cutting operation. The cut-out smaller metallographic specimens are subjected to thermo-hardening plastic pressing first, using an XQ-2B metallographic specimen inlaying machine, and then to further polishing to reach the needed surface smoothness for microhardness measuring. For further microstructure analysis, the cross-sections are etched with a reagent suitable for austenitic steel composed of a 1:3 volume ratio mixture of concentrated nitric acid (HNO_3) and concentrated hydrochloric acid (HCl).

2.2. *Initial Microstructure and Microhardness Measurement of the PRR*

To prepare a substantiated experimental plan, a preliminary microstructure and microhardness measurements of specimens that have PRR from Type I were conducted.

The microhardness of the deformed structure was measured by the Vickers method with a load of 50 g, according to ISO6507-1 [20]. An HVS-1000 Digital Micro Hardness Tester (Opassy Industry Group Co., Ltd., Dongguan, China) was used.

The microhardness measurements in the depth of the surface layer and along the PMM trace section are shown in Figure 2a, and the diagrams of microhardness variation are shown in Figure 2b,c.

Measurements taken very close to the surface level are in the range of 340–380 $HV_{0.05}$ and only one indentation is above 380, which can be local structure fluctuation due to the strain-induced martensitic area, as shown in Figure 2b. There is no tendency of an increase in the values going from the ridge to the bottom of the groove.

The lower indentation traces indicate the varying of microhardness values in depth. The summarized graph in Figure 2c shows a sharp decrease in values to the depth of about 500 μm similar for both traces. Deeper, the bottom trace is higher than the ridge trace with microhardness value 225 $HV_{0.05}$. This trace reaches the steel microhardness 215 $HV_{0.05}$ below a 1000 μm depth. It can be summarized that even the hardness near the surface of both traces is similar, and the difference between the measured microhardness near the trace's bottoms and near the ridges around 30–50 $HV_{0.05}$ is also observed, which confirms the distribution of the degree of plastic deformation during the BB finishing operations derived in their classical application (i.e., smoothing the surface) [21].

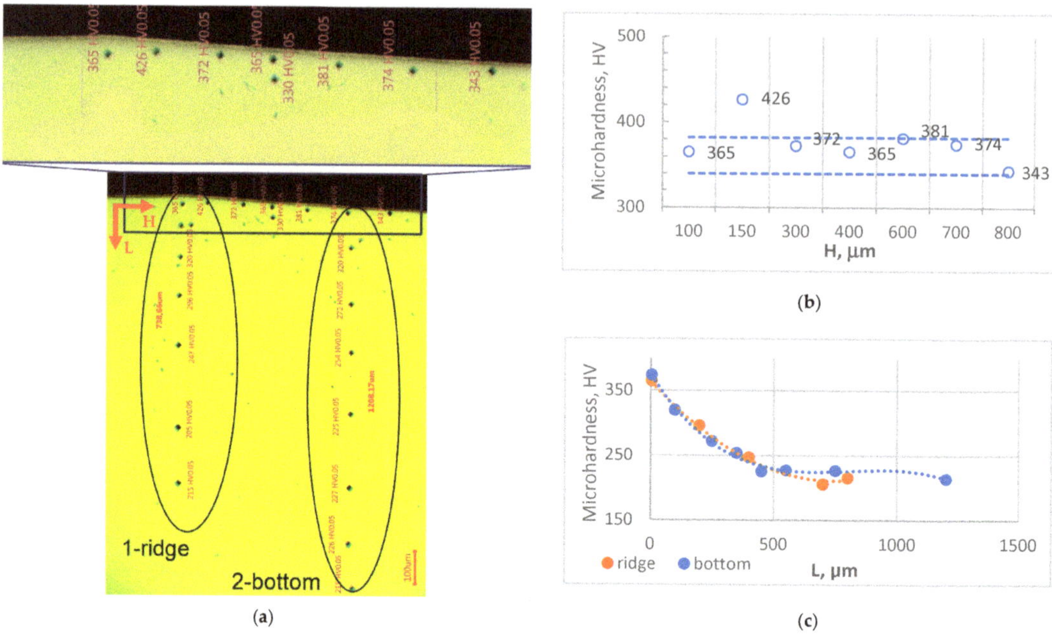

Figure 2. (**a**) The cross-section of measured microhardness for PRR from type I; (**b**) diagrams of changing the microhardness in the depth of the material beneath the traces ridge (1) and bottom (2); (**c**) diagram of the microhardness changing along the trace section.

2.3. Experimental Design Description

The BB operations are conducted with three different feed rates of the ball tool: f = 500, 1000, and 1500, mm/min for every of the three types of PRRs. Therefore, the main factors of the current experimental investigation are the three types of patterns and three feed rates of the BB operation. The deforming force F, N is kept constant for all formed PRRs and has a magnitude of 400 N.

After processing the PRRs, the test specimens are obtained, as described in Section 2.2. Microhardness measuring is performed in two main areas: near the top of the plastically deformed trace ridges (area 1), and near the bottom of the traces (area 2) (see Figure 3a).

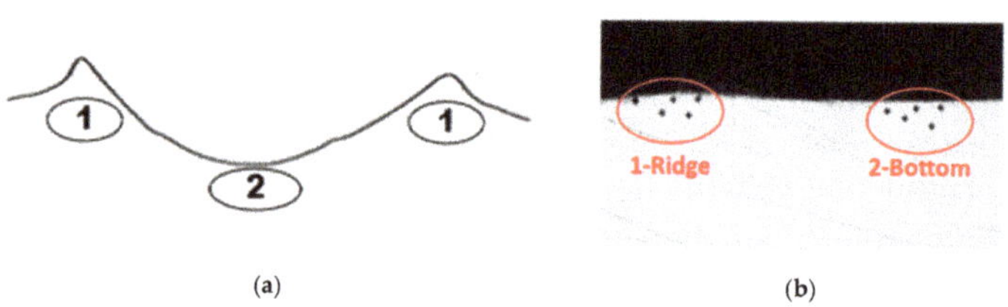

Figure 3. Diagram of the measurement areas of the PRR's traces cross-sections (**a**); image of microhardness measurements at the ridges $HV_{0.01}^{R}$ (area 1) and bottoms $HV_{0.01}^{B}$ (area 2) of PRR that are of Type I to III (**b**).

Because the HV measurement repetitions should not be too close to each other to avoid interference between them, and, at the same time, the area of measurements at the ridges section is comparatively narrow, the applied force of the test device indenter was chosen to be 10 g instead of 50 g. Therefore, the outcome parameters from the experimental investigation are $HV_{0.01}^{R}$ and $HV_{0.01}^{B}$. To compensate for the microstructure-dependent hardness values and so-called "indentation size effect" [22] on lower-load indents, measurements were carried out in five-time repetitions for every combination of the design factors levels to obtain more reliable results for $HV^R_{0.01}$ and $HV^B_{0.01}$. The experimental design and the results obtained for hardness measurements are shown in Table 3.

Table 3. Experimental design and results for microhardness values of $HV^R_{0.01}$ and $HV^B_{0.01}$.

NO.	Type of PRR	Feed Rate, $f_{in.i}$, [mm/min]	$HV^R_{0.01}$, [kgf/mm^2]	Mean $HV^R_{0.01}$, [kgf/mm^2]	$HV^B_{0.01}$, [kgf/mm^2]	Mean $HV^B_{0.01}$, [kgf/mm^2]
1	I	500	244		289	
2	I	500	256		279	
3	I	500	254	252.4	285	285.8
4	I	500	256		291	
5	I	500	252		285	
6	I	1000	265		311	
7	I	1000	263		312	
8	I	1000	257	261.4	308	311.4
9	I	1000	258		315	
10	I	1000	264		311	
11	I	1500	278		319	
12	I	1500	281		326	
13	I	1500	274	275.8	329	324.6
14	I	1500	271		326	
15	I	1500	275		323	
16	II	500	248		292	
17	II	500	258		286	
18	II	500	257	255.2	291	291.0
19	II	500	258		292	
20	II	500	255		294	
21	II	1000	278		324	
22	II	1000	275		328	
23	II	1000	272	272.2	330	327.2
24	II	1000	267		325	
25	II	1000	269		329	
26	II	1500	312		341	
27	II	1500	317		341	
28	II	1500	309	310.2	345	343.4
29	II	1500	305		344	
30	II	1500	308		346	
31	III	500	253		312	
32	III	500	266		313	
33	III	500	262	261.6	304	309.0
34	III	500	264		305	
35	III	500	263		311	
36	III	1000	268		334	
37	III	1000	278		340	
38	III	1000	276	274.4	337	333.8
39	III	1000	271		327	
40	III	1000	279		331	
41	III	1500	315		350	
42	III	1500	321		357	
43	III	1500	312	316.0	354	353.0
44	III	1500	319		351	
45	III	1500	313		353	

3. Results

3.1. *Results for Microhardness Change*

According to modern concepts of deformation processes, the application of RR caused localized plastic deformations on the surface of an austenitic stainless-steel plate (grade AISI 304). From a physical perspective, the plastic deformation process led to the accumulation of volumetric damage in the surface layer, where defects and damage were introduced into the steel's microstructure. It was found that, for all types of PRRs studied, an increase in the feed rate, and consequently in the deformation speed, resulted in material strengthening at the micro level for all PRRs types investigated.

This was confirmed by the increase in microhardness values of $HV^R_{0.01}$ and $HV^B_{0.01}$, as shown in Table 3 and the 3D scatter plots in Figure 4a,b. As can be seen, an overall increase in surface hardness occurred in comparison with the initial material hardness $HV_{0.01} = 205$ (see Table 1).

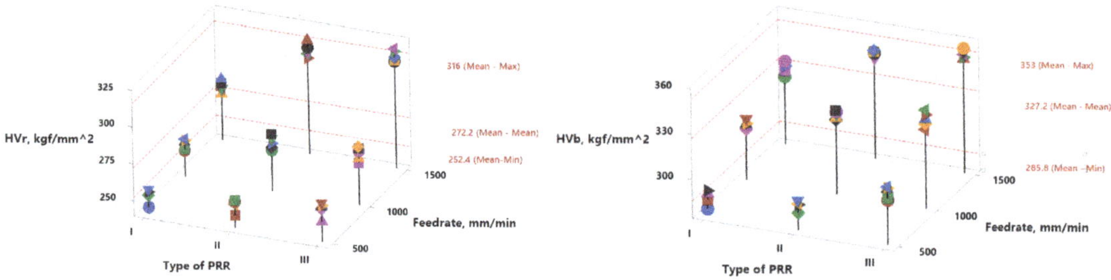

(a) (b)

Figure 4. Three-dimensional scatter and trend plots that visualize HVs' amendment, based on the values given in Table 2: (**a**) at the area of ridges of the PRRs; (**b**) beneath the bottoms of the PRRs' plastically deformed traces.

3.2. *Microstructure Analysis of the Deformed Surface Layer*

The microstructure of the steel in its initial state consists predominantly of austenite grains, with some areas of strain-induced martensite due to rolling. During the formation of the regular micro-relief, the bottom of the groove remains smooth, free from macro- and micro-defects. The traces of the deformation tool's movement are uniform, indicating the stability of the deformation effect on the material. At the bottom of the lubrication groove, a highly deformed microstructure is formed, characterized by a high dislocation density [23]. The strengthening capability of austenitic steel is higher than that of steels of other classes, as reflected in the distinctly plastically deformed layer. This layer is barely noticeable at a feed rate of $f_{in.i}$. = 500 mm/min, distinctly pronounced at $f_{in.i}$. = 1000 mm/min, and it exhibits clear boundaries with the base metal at $f_{in.i}$. = 1000 mm/min. It is formed by the subdivision of austenite grains into finer elements, followed by the amorphization of the surface layer, making individual grains indistinguishable and developing strain-induced martensite.

The subsurface layers are formed through plastic deformation and displacement of austenite grains (see Figure 5). The deformation twin lines and areas transformed to strain-induced martensite are visible. This occurs because the groove profile is created by the plastic indentation of the metal both forward and laterally. Beyond the deformed surface, the microstructure remains almost unchanged.

III-1000

III-1500

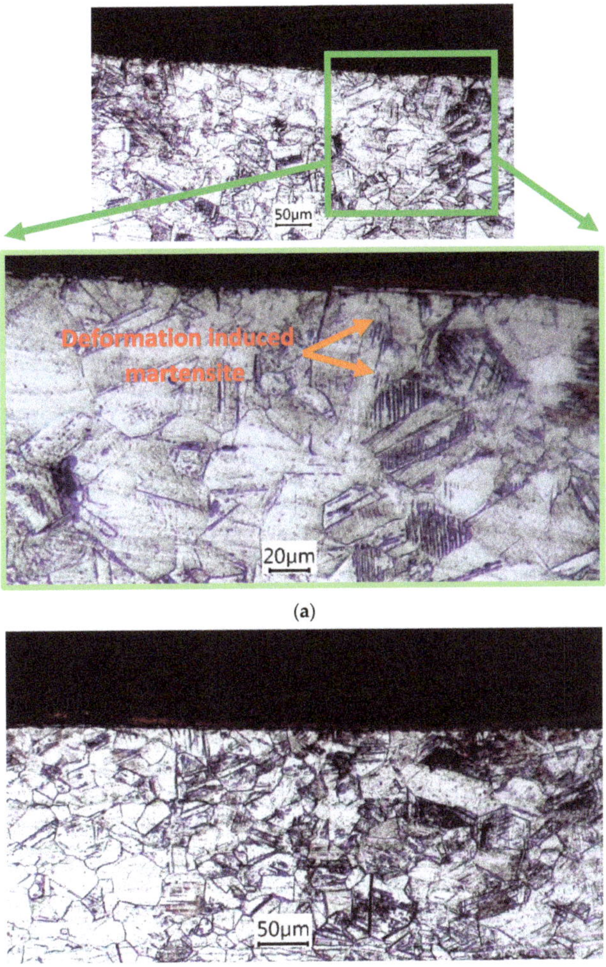

(**a**)

(**b**)

Figure 5. Optical micrographs of microstructures of austenitic stainless steel (grade AISI 304) obtained for different values of the feed rate, [mm/min]: (**a**) 1000; (**b**) 1500 for Type III of the relief.

With an increase in the feedrate, a greater deformation depth was observed. The maximum microhardness for the investigated processing conditions was similar, ranging from 314 to 377 $HV_{0.05}$, indicating the maximum material strengthening under the given temperature–force processing conditions.

4. Derived Regressions Models and Its Statistical Analyses

Thus, the intensity of microhardness increases around PRRs formation area, depending on:

- The type and parameters of the deforming tool.
- The ball indentation force.
- The feed rate, as demonstrated above.

Therefore, structural and parametric optimization of existing technological modes, as well as the development of some new ones for the formation of an ordered PRR, is essential.

The results for $HV^R_{0.01}$ and $HV^B_{0.01}$ that are shown in Table 1 are used to analyze the impact of the type of PRR and feed rate factors of the ball burnishing operation on the basis of derived regressions equations and statistical analyses on it carried out by using Minitab 2022 (Minitab, LLC, State College, PA, USA) software [24]. They are as shown below.

4.1. *Influence on the Microhardness Obtained at the Ridges Area of the PRRs Traces*

The obtained coefficients of the main effects and interactions, along with the full regression equation, which is related to the impact of different types of PRRs and feed rates over the microhardness at the ridges $HV^R_{0.01}$, are as follows:

$$HV^R{}_{0.01}(PRR,f) = 275.467 - 12.267{\cdot}PRR_I + 3.733{\cdot}PRR_{II} + 8.533{\cdot}PRR_{III} - 19.067{\cdot}f_{500} - 6.133{\cdot}f_{1000} + 25.200{\cdot}f_{1500} + 8.270{\cdot}PRR_I{\cdot}F_{500} + 4.330{\cdot}PRR_I{\cdot}f_{1000} - 12.600{\cdot}PRR_I{\cdot}f_{1500} - 4.930{\cdot}PRR_{II}{\cdot}f_{500} - 0.870{\cdot}PRR_{II}{\cdot}f_{1000} + 5.800{\cdot}PRR_{II}{\cdot}f_{1500} - 3.330{\cdot}PRR_{III}{\cdot}f_{500} - 3.470{\cdot}PRR_{III}{\cdot}f_{1000} + 6.800{\cdot}PRR_{III}{\cdot}f_{1500} \quad (1)$$

The statistical assessment of the regression Equation (1) shows a standard deviation of σ = 4.391, and an R-square = 96.8%, which means that the model has a very good fit against the experimentally obtained results. To assess the statistical significance of the regression predictors, using the null hypothesis theorem, an Analysis of Variance (ANOVA) was conducted. The results of it are shown in Table 4.

Table 4. Results from ANOVA analysis of the regression model (1).

Source	**DF**	**Seq SS**	**Contribution**	**Adj SS**	**Adj MS**	**F-Value**	***p*-Value**
Model	8	20,971.2	96.80%	20,971.2	2621.40	135.98	0.000
Linear	4	19,101.3	88.17%	19,101.3	4775.33	247.71	0.000
PRR_type	2	3558.4	16.42%	3558.4	1779.20	92.29	0.000
Feed rate	2	15,542.9	71.74%	15,542.9	7771.47	403.13	0.000
Two-Way Interactions	4	1869.9	8.63%	1869.9	467.47	24.25	0.000
PRR_type · Feed rate	4	1869.9	8.63%	1869.9	467.47	24.25	0.000
Error	36	694.0	3.20%	694.0	19.28		
Total	44	21,665.2	100.00%				

As seen from Table 4, all predictors and interaction between them from the regression model (1) are statistically significant due to the close-to-zero *p*-values calculated, which means that the null hypothesis theorem is rejected for all of them. Considering the percentages of contribution (see Table 4), the feed rate has a higher impact (71.74%), followed by the type of PRR (16.42%) and the interaction between them (8.63%) over parameter $HV^R_{0.01}$.

4.2. *Influence on the Microhardness Obtained at the Bottom Area of the PRRs' Traces*

The full regression equation that describes the influence of PRRs' types and feed rates over the microhardness at the bottom $HV^B_{0.01}$ is as follows:

$$HV^B{}_{0.01}(PRR,f) = 319.911 - 12.644{\cdot}PRR_I + 0.622{\cdot}PRR_{II} + 12.022{\cdot}PRR_{III} - 24.644{\cdot}f_{500} + 4.222{\cdot}f_{1000} + 20.422{\cdot}f_{1500} + 3.180{\cdot}PRR_I{\cdot}F_{500} - 0.090{\cdot}PRR_I{\cdot}f_{1000} - 3.090{\cdot}PRR_I{\cdot}f_{1500} - 4.890{\cdot}PRR_{II}{\cdot}f_{500} + 2.440{\cdot}PRR_{II}{\cdot}f_{1000} + + 2.440{\cdot}PRR_{II}{\cdot}f_{1500} + 1.710{\cdot}PRR_{III}{\cdot}f_{500} - 2.360{\cdot}PRR_{III}{\cdot}f_{1000} + 0.640{\cdot}PRR_{III}{\cdot}f_{1500} \quad (2)$$

The regression Equation (2) has a standard deviation of σ = 3.551, and an R-square = 97.84%, which means that the model has a very good fit to the experimentally obtained results. To assess the statistical significance of the regression predictors, using the null hypothesis theorem, an Analysis of Variance (ANOVA) was conducted, and its results are shown in Table 5.

Table 5. Results from ANOVA analysis of the regression model (2).

Source	DF	Seq SS	Contribution	Adj SS	Adj MS	F-Value	*p*-Value
Model	8	20,527.6	97.84%	20,527.6	2565.96	203.47	0.000
Linear	4	20,205.7	96.30%	20,205.7	5051.42	400.55	0.000
PRR_type	2	4572.0	21.79%	4572.0	2286.02	181.27	0.000
Feed rate	2	15,633.6	74.51%	15,633.6	7816.82	619.84	0.000
Two-Way Interactions	4	322.0	1.53%	322.0	80.49	6.38	0.001
PRR_type · Feed rate	4	322.0	1.53%	322.0	80.49	6.38	0.001
Error	36	454.0	2.16%	454.0	12.61		
Total	44	20,981.6	100.00%				

As seen from Table 5, all predictors and interaction between them from the regression model (2) are also statistically significant because the *p*-value statistics calculated for them are lower than 0.05 (i.e., 5%). This means that the null hypothesis theorem is rejected for all of them. The percentages of contribution in the model (see Table 5) are as follows: the feed rate has the highest impact (74.51%), followed by the type of PRR (21.79%). The interaction between them has contribution of just 1.53% over parameter $HV^{B}_{0.01}$ in model (2).

4.3. Main Effects and Interactions Diagrams

Based on the derived regression models (1) and (2), a graphical representation of the main effects and interactions of the regression models' factors is shown in Figure 6.

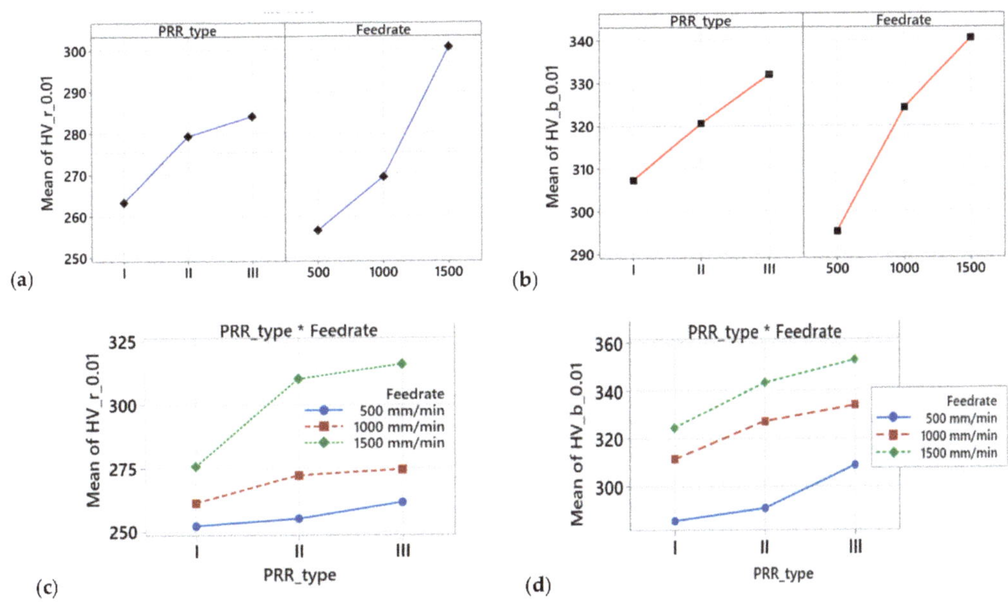

Figure 6. Main effects plot and interactions plot for: (**a**,**c**) microhardness $HV_{0.01}$ at the ridges of PRRs; (**b**,**d**) microhardness $HV_{0.01}$ at the bottoms of PRRs.

The diagrams of main effects (see Figure 6a,b) show that there is an increase in $HV_{0.01}$ for both areas of measurement—beneath the ridges and beneath the bottom of PRR traces, with an increase in the feed rate of the BB operation. A similar tendency occurs for different types of PRR, starting with PRR of Type I, where the traces do not touch each other, and reaching PRR of Type III, where they cross each other. As can be seen from the interaction plots shown in Figure 6c,d, the tendencies of the impact are very close to those observed for the main effects, both for microhardness $HV_{0.01}$ at the grooves' bottom and at the ridge areas.

5. Conclusions

Summarizing the research conducted, it can be stated that the type of PRR and the deformation speed significantly influence the microhardness variation in the subsurface layers of PRR's grooves. As can be seen from the Min and Max mean values on the scatter plots (see Figure 4), this increase varies from 23.1% to 54.1% around the ridges (see pos. 1 in Figure 3a) and from 39,4% to 72.2% beneath the bottom of the plastically deformed traces (see pos. 2 in Figure 3a). It depends both on the BB operation regime parameter, feed rate and on the type of PPR formed onto the surface. In fact, the type of PRR essentially characterizes the groove density on the surface area and, consequently, determines their mutual influence on microhardness during formation. The highest microhardness (both at the edges and at the bottom of the grooves) was observed for Type III PRR, where the groove density was the highest, while the lowest microhardness was recorded for Type I grooves, where the density was the lowest.

Thus, understanding the influence of PRR type and deformation speed on surface microhardness allows for controlled variation of this parameter within 23.1 to 72.2%, depending on the selected PRR type and its formation speed (i.e., BB operation feed rate). The obtained results indicate the possibility of regulating the microstructural state of the bottom and edges of the ordered PRR grooves by adjusting the deformation intensity in the area under treatment. The primary influencing parameter is the feed rate, which controls the deformation speed and determines the structural-phase state and microhardness of the surface and subsurface layers of AISI 304 stainless steel.

Although the obtained regression models (see Equations (1) and (2)) are empirical in nature since they were derived for a specific study under certain conditions, the statistical analysis shows that they correctly describe the influence of the operating mode parameters on the resulting microhardness. They can be used to plan further studies to search for optimal settings of the operating mode parameters of the BB operation outside the limits of variation used in the present study, which will be the future work of our team.

Author Contributions: Conceptualization, V.D. and S.D.S.; methodology, V.D., P.M., S.D.S. and D.M.D.; software, S.D.S.; validation, D.M.D., D.Y.M. and P.M.; formal analysis, V.D. and V.S.; investigation, P.M., S.D.S. and D.Y.M.; resources, D.M.D. and D.Y.M.; data curation D.Y.M. and S.D.S.; writing—original draft preparation, V.D. and S.D.S.; writing—review and editing, V.D. and V.S.; visualization, S.D.S. and V.S.; supervision, V.D., S.D.S. and P.M.; project administration, V.D., S.D.S. and P.M.; funding acquisition, S.D.S. All authors have read and agreed to the published version of the manuscript.

Funding: This research was funded by the Bulgarian National Science Fund (BNSF), grant number КП-06-Н57/6, and the APC was funded by grant contract КП-06-Н57/6, entitled "Theoretical and experimental research of models and algorithms for formation and control of specific relief textures on different types of functional surfaces".

Institutional Review Board Statement: Not applicable.

Informed Consent Statement: Not applicable.

Data Availability Statement: The original contributions presented in this study are included in the article. Further inquiries can be directed at the corresponding author.

Acknowledgments: The authors would like to thank Georgi Petrov from CERATIZIT Bulgaria AG for providing computing and specialized cutting tooling.

Conflicts of Interest: The authors declare no conflicts of interest.

References

1. Gurey, V.; Hurey, I.; Hurey, T.; Wojtowicz, W. Fatigue Strength of Steel Samples After Friction Treatment. In Proceedings of the Advanced Manufacturing Processes IV. Inter Partner 2022. Lecture Notes in Mechanical Engineering, Odessa, Ukraine, 6–9 September 2022; Tonkonogyi, V., Ivanov, V., Trojanowska, J., Oborskyi, G., Pavlenko, I., Eds.; Springer: Cham, Switzerland, 2023. [CrossRef]
2. Hurey, I.; Hurey, T.; Gurey, V. Wear Resistance of Hardened Nanocrystalline Structures in the Course of Friction of Steel-Grey Cast Iron Pair in Oil-Abrasive Medium. In Proceedings of Advances in Design, Simulation and Manufacturing II. DSMIE 2019, Sumy, Ukraine, 11–14 June 2019; Ivanov, V., Trojanowska, J., Pavlenko, I., Zajac, J., Peraković, D., Eds.; Lecture Notes in Mechanical Engineering. Springer: Cham, Switzerland, 2020. [CrossRef]
3. Swirad, S. Influence of Ball Burnishing on Lubricated Fretting of the Titanium Alloy Ti6Al4V. *Lubricants* **2023**, *11*, 341. [CrossRef]
4. Mezghani, S.; Demirci, I.; Zahouani, H.; El Mansori, M. The effect of groove texture patterns on piston-ring pack friction. *Precis. Eng.* **2012**, *36*, 210–217. [CrossRef]
5. Pawlus, P.; Reizer, R.; Wieczorowski, M. Reverse Problem in Surface Texture Analysis—One-Process Profile Modeling on the Basis of Measured Two-Process Profile after Machining or Wear. *Materials* **2019**, *12*, 4169. [CrossRef] [PubMed]
6. Schneider, Y.G. Service Properties of Parts with Regular Microrelief. Available online: https://books.ifmo.ru/file/pdf/147.pdf (accessed on 23 December 2024).
7. Lesyk, D.A.; Martinez, S.; Dzhemelinskyy, V.; Lamikiz, A.; Mordyuk, B.; Prokopenko, G. Surface microrelief and hardness of laser hardened and ultrasonically peened AISI D2 tool steel. *Surf. Coat. Technol.* **2015**, *278*, 108–120. [CrossRef]
8. Zyhulia, S.; Barauskiene, O. Effect of the integrated treatment on the manufacturing of printing cylinders. *East.-Eur. J. Enterp. Technol.* **2019**, *3*, 22–28. [CrossRef]
9. Menezes, P.L.; Kishore; Kailas, S.V.; Lovell, M.R. Role of Surface Texture, Roughness, and Hardness on Friction During Unidirectional Sliding. *Tribol. Lett.* **2011**, *41*, 1–15. [CrossRef]
10. Pratap, T.; Patra, K. Tribological performances of symmetrically micro-textured Ti-6Al-4V alloy for hip joint. *Int. J. Mech. Sci.* **2020**, *182*, 105736. [CrossRef]
11. Grzesik, W.; Zak, K. Producing high quality hardened parts using sequential hard turning and ball burnishing operations. *Precis. Eng.* **2013**, *37*, 849–855. [CrossRef]
12. Maximov, J.T.; Anchev, A.P.; Duncheva, G.V.; Ganev, N.; Selimov, K.F. Influence of the process parameters on the surface roughness, mi-cro-hardness, and residual stresses in slide burnishing of high-strength aluminum alloys. *J. Braz. Soc. Mech. Sci. Eng.* **2017**, *39*, 3067–3078. [CrossRef]
13. Luo, H.; Liu, J.; Wang, L.; Zhong, Q. The effect of burnishing parameters on burnishing force and surface microhardness. *Int. J. Adv. Manuf. Technol.* **2006**, *28*, 707–713. [CrossRef]
14. Nagîț, G.; Slătineanu, L.; Dodun, O.; Rîpanu, M.I.; Mihalache, A.M. Surface layer microhardness and roughness after applying a vibroburnishing process. *J. Mater. Res. Technol.* **2019**, *8*, 4333–4346. [CrossRef]
15. Yuan, X.L.; Sun, Y.W.; Gao, L.S.; Jiang, S.L. Effect of roller burnishing process parameters on the surface roughness and microhardness for TA2 alloy. *Int. J. Adv. Manuf. Technol.* **2016**, *85*, 1373–1383. [CrossRef]
16. Cobanoglu, T.; Ozturk, S. Effect of burnishing parameters on the surface quality and hardness. *Proc. Inst. Mech. Eng. Part B J. Eng. Manuf.* **2015**, *229*, 286–294. [CrossRef]
17. Lee, W.-S.; Lin, C.-F. Impact properties and microstructure evolution of 304L stainless steel. *Mater. Sci. Eng. A* **2001**, *308*, 124–135. [CrossRef]
18. Hedayati, A.; Najafizadeh, A.; Kermanpur, A.; Forouzan, F. The effect of cold rolling regime on microstructure and mechanical properties of AISI 304L stainless steel. *J. Mater. Process. Technol.* **2010**, *210*, 1017–1022. [CrossRef]
19. Slavov, S.; Markov, O. A Tool for Ball Burnishing with Ability for Wire and Wireless Monitoring of the Deforming Force Values. Acta Technica Na-pocensis-Series: APPLIED MATHEMATICS, MECHANICS, and ENGINEERING 2023, vol. 65, no. 4S. Available online: https://atna-mam.utcluj.ro/index.php/Acta/article/view/2066 (accessed on 5 January 2025).
20. *ISO 6507-1:2023*; "Metallic materials — Vickers hardness test — Part 1: Test method". Technical Committee ISO/TC 164/SC 3 of ISO; ISO: Geneva, Switzerland, 2023; p. 35.
21. Slavov, S.D.; Dimitrov, D.M.; Konsulova-Bakalova, M.I. Advances in burnishing technology. In *Advanced Machining and Finishing*; Elsevier: Amsterdam, The Netherlands, 2021; pp. 481–525. [CrossRef]
22. Nix, W.D.; Gao, H. Indentation size effects in crystalline materials: A law for strain gradient plasticity. *J. Mech. Phys. Solids* **1998**, *46*, 411–425. [CrossRef]

23. Chausov, M.; Maruschak, P.; Pylypenko, A.; Shmanenko, V.; Brezinová, J.; Lisnichuk, M. The effects of cryogenic treatment and impact-oscillatory loading on changes in the mechanical properties and structural condition of stainless steel 12Kh18N10T. *Mater. Sci. Eng. A* **2024**, *913*, 147007. [CrossRef]
24. Minitab 22 Support-Minitab. Available online: https://support.minitab.com/en-us/minitab/21/ (accessed on 7 January 2025).

Journal of
Manufacturing and Materials Processing

MDPI

Article

Alternative Real-Time Part Quality Monitoring Method by Using Stamping Force in Progressive Stamping Process

Juras Skardžius [1,2] and Justinas Gargasas [1,*]

1 Department of Mechanical and Materials Engineering, Faculty of Mechanics, Vilnius Gediminas Technical University, Saulėtekio al. 11, LT-10223 Vilnius, Lithuania; juras.skardzius@vilniustech.lt or jsk@stansefabrikken.lt
2 LLC Stansefabrikken Automotive, Pašilės g. 10, LT-20194 Ukmergė, Lithuania
* Correspondence: justinas.gargasas@vilniustech.lt; Tel.: +370-61651275

Abstract: The manufacture of automotive parts using progressive stamping tools demands high precision and efficiency to meet industry standards. This research explores integrating quality monitoring techniques, focusing on load sensors and tonnage monitoring, to enhance the production process. Progressive tools, which perform multiple operations within a single press cycle, are critical for maintaining dimensional accuracy and minimizing defects. This research examines the correlation between real-time load sensor data and the tonnage applied during stamping, aiming to detect anomalies and deviations that may define part quality. By analyzing variations in tool loads and press tonnage, this research identifies patterns that allow the user to connect the applied force with the observed part quality and could help to determine potential issues such as instability of force or part dimensions, tool wear or improper alignment. The results of this research demonstrate that incorporating advanced monitoring systems into progressive stamping processes does not only improve part quality but also extends tool life and reduces downtime. The proposed approach provides a robust framework for ensuring reliability and efficiency in the production of automotive components, aligning with the industry demand for high-quality, cost-effective manufacturing solutions.

Keywords: manufacture of automotive parts; stamping; progressive tool; quality monitoring; load sensor; tonnage monitoring

1. Introduction

In progressive stamping, the main factors like speed, force, tool design, and quality of production play a crucial role in the material flow and tool interactions. This directly affects the final part dimensions. Several studies have shown that increased press speed can lead to undesirable dimensional deviations. Hu et al. [1] noted that faster stamping rates can result in increased dynamic forces, which can cause material deformation and tool misalignment, which can alter component geometry. Higher stamping rates reduce material flow control, leading to problems like spring-back and feature misalignment on the stamped item [2]. In progressive stamping, press force has a significant effect on part quality. Even when sufficient force is needed for material deformation, excessive press force can cause undesirable dimensional deviations such as spring-back, thinning, and tool damage. Excessive press force during stamping can entail a variety of dimensional problems. Research shows that more force increases thinning and pushes strain over the material's elastic limit, resulting in irreversible deformation. The authors of [3] claim that excessive force causes an uneven distribution of stress, which leads to warping in high-strength steel components. Additionally, overstressing aluminum or other alloys results in spring-back, which reduces dimensional accuracy by forcing parts to partially return to their original shape [2].

High press forces make spring-back, a frequent dimensional problem in steel stamping, worse. As press forces increase, the steel deviates from the intended shape because of its elastic rebound during unloading. In the literature, the authors of [2] conclude in their analysis of advanced high-strength steels (AHSSs) that excessive press force increases residual stresses, which in turn results in increased spring-back effects. These variations are particularly noticeable in designs that are lighter and use thinner steel plates. Regarding unalignment and warping, the uneven distribution of stress during the stamping process is the cause of the warping and misalignment of steel plates under high-force circumstances. Localized deformation brought on by high loads results in bending surfaces and an uneven assembly fit [4–10]. This is extremely risky since cumulative mistakes have the potential to spread across subsequent stamping lines. This is particularly problematic in progressive stamping lines, where cumulative errors may span multiple steps. Excessive press forces accelerate punch and die wear, which leads to tool geometry problems. By creating burrs and irregular cuts on steel plates, excessive force usage reduces tool life and compromises dimensional accuracy [5–7,9]. The research emphasized the need for wear-resistant tool coatings and advanced lubricants in mitigating these effects [11]. In scientific articles, it was found that high press forces cause an uneven distribution of stress, which results in misalignments and out-of-plane distortions in multi-stage assemblies [12,13]. If these are increased forces, the likelihood of residual stresses, which worsen warping and lower assembly quality, increase [6]. Sophisticated monitoring systems and careful force optimization are required to lessen these issues [6,9,10].

It is crucial to recognize that tool design plays a key role in minimizing the impact of speed. Advanced progressive tools are more resilient to the effects of high-speed operations because they have characteristics like improved die clearance and sophisticated coatings [13]. It was noted that using optimized die geometries to reduce stress concentrations at faster speeds can improve the uniformity of parts [3,14–20]. Additionally, servo-driven presses allow for more precise control over stamping parameters, improving the ability to maintain dimensional tolerances.

A review of the literature suggests that the introduction of sensor technology into the stamping tool allows for independent or semi-independent monitoring of the process and part quality. The main aim of this research is to investigate the possibility of using load sensor technology to autonomously monitor part quality in a continuous process.

A couple hypotheses were formulated:

1. There is an existing relationship between the applied stamping force and the geometric dimensions of the part;

2. Stamping force could be used to monitor part bending angles and the overall width of stamping parts.

Both hypotheses are based on the principle of the force–distance measurement, where the force on the material grows exponentially with the tool top plate being lowered.

The primary objective of this research is to explore the application of load sensor technology for autonomous monitoring of part quality in a continuous progressive stamping process. The research aims to establish a direct relationship between stamping force and the geometric dimensions of stamped components.

The novelty of this research lies in real-time monitoring using load sensors. Unlike traditional quality control methods, this research investigates the feasibility of using load sensors to continuously monitor and assess the part quality without interrupting the manufacturing process.

Regarding force–dimension correlation, this research hypothesizes that applied stamping force directly influences part dimensions, particularly bending angles and overall width. This relationship has not been extensively researched for real-time quality control in progressive stamping.

The next section of this research reveals the methodology, simulations, and numerical setup of this research.

2. Experimental Research Methodology

In metal forming operations, stamping force is an essential characteristic since it has a direct impact on the quality of the finished product. By altering the press slide height, this parameter can be changed, giving operators more control over the force used during stamping. Under controlled circumstances, stamping tests were conducted to determine the relationship between the stamping force and the bending angle. The data during these tests were acquired from the flywheel angle sensor and the press load sensors (JZT1 and JZT2). On the diagonally opposite columns of the press, the JZT1 and JZT2 sensors were positioned strategically, as seen in Figure 1.

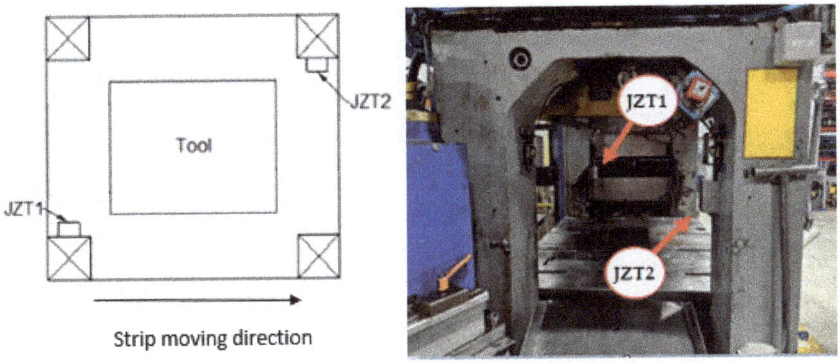

(**a**) (**b**)

Figure 1. Position of JZT1 and JZT2 sensors in two perspectives: (**a**) a sketch view and (**b**) a real-world view.

The total force exerted by the press is the combined result of forces generated during each operational phase and the resistance from the mechanical springs within the system. To better understand the specific contribution of the bending phase to the overall force, a mathematical model of the bending process was developed and analyzed through simulation. This approach offered deeper insights into the distribution of forces and their influence on the resulting bending behavior.

The experimental and simulated processes began with DC01 steel sheets having a nominal thickness of 1 mm. The press was operated at a speed of 40 cycles per minute, with lubrication levels set according to the specifications outlined in the tool instruction. These conditions ensured consistency and reliability for comparative analysis.

To measure the press load of the tool, two load sensors, JZT1 and JZT2, were installed on diagonally opposite columns of the press, as seen in Figure 2. These sensors were calibrated to ensure their combined output accurately reflected the total force exerted by the press. The technical specifications of the sensors are provided in Table 1.

Figure 2. The position of the eddy current sensor EC1 shown in two perspectives: (**a**) a model view and (**b**) a real-world view.

Table 1. Technical specification of the load sensor.

Characteristic	Value
Manufacturer	TR systems GmbH, Division Unidor, Trossingen, Germany
Model	JZT 127/S–MS

A shaft position sensor, model CEV58S–00117, was integrated into the press to monitor its operation. The technical specifications of this sensor are detailed in Table 2.

Table 2. Technical data of the shaft position sensor.

Characteristic	Value
Manufacturer	TR systems GmbH, Division Unidor
Model	CEV58S–00117
Resolution	≤33 bits
Measurement count	≤256,000

The sensors were linked to the data processing and control system, the compact PRESS smart LINE. An intelligent controller that continuously monitors and analyzes the signals from the connected sensors, ensuring the protection of both the press and the installed tooling, while optimizing and controlling the production process, is central to this system. Additional components of this system are outlined in Table 3, and their integration and practical application within the stamping line are illustrated in Figure 3.

Table 3. Technical data of the data processing system.

Characteristic	Value
Manufacturer	TR systems GmbH, Division Unidor
Model	UNiDOR compact PRESS smart LINE
Controller	Ether CAT Master RT
Press angle setting module	CEVV58S–00117
Analog signal input module	Ether CAT Box

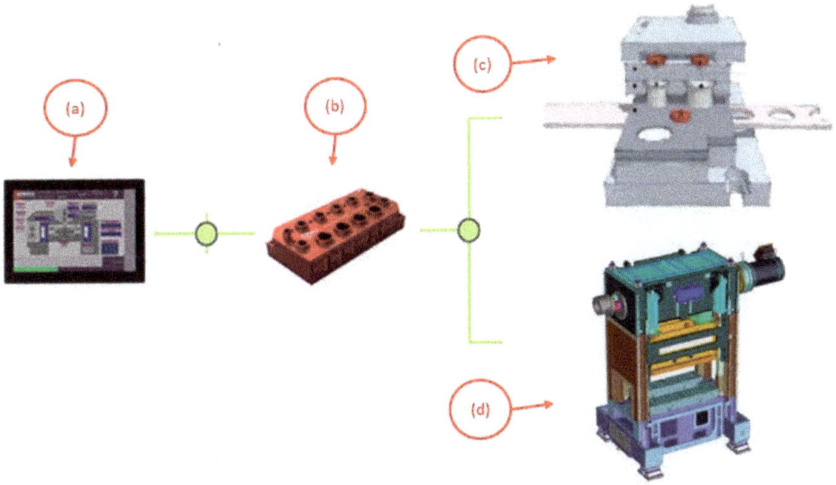

Figure 3. The communication structure of the sensor system is organized as follows: (**a**) display screen paired with a control unit, (**b**) connection adapter for the press and tool sensors, (**c**) the tool sensors, and (**d**) the press sensors.

An eddy current sensor was placed to compare the determined press slide with the real press slide. The eddy current sensor was selected because of its measuring range and good resolution. To provide precise, localized readings, it was set up on the backing plate next to the bending step. The locations of the eddy current sensor installation are illustrated in Figure 3. Technical specifications can be found in Table 4.

Table 4. Technical specification of the eddy current sensor.

Characteristic	Value
Manufacturer	TR systems GmbH, Division Unidor
Model	WSD S2/10 MF
Measurement range	0–2 mm
Linearity	±0.12 mm
Repeatability	0.02 mm
Resolution	<0.001 mm
Response time	<1 ms
Linearity deviation	+/−12 µm

In parallel, a mathematical model was developed based on the existing progressive tool design. This model was utilized to perform simulations under the same set-up parameters as those used in real production, allowing for a comparative analysis of theoretical and actual outcomes.

The Transvalor Forge, a non-linear solver program, was used to simulate the mathematical model. To isolate the effects of other stations and produce more lucid findings, the main goal of these simulations was to duplicate the bending process steps carried out in the tool with the same process settings. The oil lubrication between the sheet and bending punches was simulated using a Tresca-limited Coulomb friction model, with a friction factor of 0.1. Based on weak exchange conditions, the default heat transfer rate between the tools and the sheet was calculated to be 2000 W/m^2K. All model components were initially set to a temperature of 20 °C.

The mechanical press model used for the simulations had a nominal speed of 40 strokes per minute (SPM), with an operating contact length ranging from 9.6 to 11.2 mm. It took about 0.15 s to complete the entire simulated bending cycle. Both sheet and tool components were meshed using automatic remeshing. In the deformation and shearing zones, a fine 0.2 mm volumetric tetrahedral mesh was selected. A coarser mesh was used in places that were not in contact. A 0.6 mm trigger size was used to initiate remeshing in the deformation zones. Figure 4 displays a completely configured model with the mesh included.

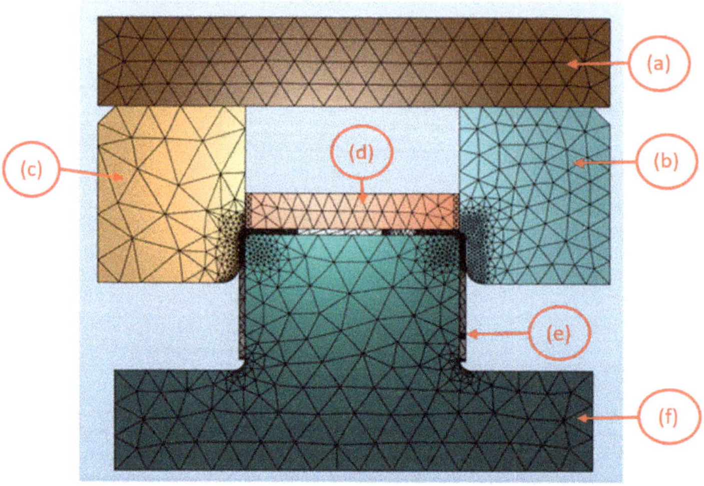

Figure 4. The meshed model includes the following components: (**a**) upper die, (**b**) right bending punch, (**c**) left bending punch, (**d**) stripper, (**e**) workpiece, and (**f**) bending matrix.

3. Results

As soon as the punches engage the workpiece, cutting and bending occur, resulting in a sudden rise in force. When the punches pierce the material, the load temporarily drops before suddenly increasing as the die descends, further bending and pushing the workpiece. This cycle phase ends when the rotor reaches around 180° and starts to move upwards. During this phase, the workpiece is fully formed; this explains our hypothesis that a relationship between the applied stamping force and the geometric dimensions of the part exists. At the end of the forming cycle, the stripper begins to rise, and the load on the columns rapidly declines until it reaches the free-running values seen at the beginning. In addition to monitoring press load and flywheel angle sensor readings, the data from the EC1 sensor, which measures stripper position, were also recorded. This additional eddy current sensor has been chosen to assess whether there are any discrepancies between the set-up and actual stripper positions. Moreover, the goal was to determine which of these process parameters serves as a better indicator of the force acting on the part.

Before the tests started, the slide of the press, using 1.0 mm thickness material, was determined. For this determination, common practice in production, the essence of which is to select the position of the press slide that produces the highest quality part, was used. In this case, the main criteria were the bending angles of the part, its width obtained after bending, and the imprints left on the surface of the part, indicating the level of compression of the part. The optimal slide position was identified and marked with an asterisk in the results seen in Table 5.

Table 5. Summary of test results of the parts measured using a coordinate measuring machine (CMM).

Determined Slide Position, mm	Sample No.	Measured Press Slide, µm	Maximum Force, kN	Angle RH, Degrees	Angle LH, Degrees	The Width of the Part, mm
382	1	1620.911	14.997	92.102	92.313	38.647
	2	1620.483	15.416	92.062	92.335	38.621
	3	1621.704	14.997	92.09	92.34	38.633
	4	1620.422	15.416	92.078	92.306	38.63
	5	1622.009	14.997	92.074	92.304	38.65
	Average	1621.106	15.165	92.081	92.304	38.636
381.8	1	1440.125	14.997	92.084	92.084	38.629
	2	1440.369	15.206	92.084	92.332	38.631
	3	1439.636	15.416	92.088	92.334	38.629
	4	1439.819	14.997	92.021	92.296	38.622
	5	1440.735	13.742	92.016	92.316	38.598
	Average	1400.137	14.872	92.059	92.272	38.622
381.6	1	1215.271	19.671	91.729	91.921	38.461
	2	1213.928	19.915	91.703	91.972	38.464
	3	1215.088	19.915	91.694	91.986	38.462
	4	1214.966	19.915	91.662	91.988	38.454
	5	1215.332	19.217	91.678	91.977	38.46
	Average	1214.917	19.727	91.693	91.969	38.46
381.4	1	1080.322	28.634	91.267	91.298	38.23
	2	1080.444	28.39	91.279	91.295	38.246
	3	1079.59	28.809	91.296	91.36	38.197
	4	1080.139	28.809	91.29	91.258	38.244
	5	1079.59	28.809	91.281	91.286	38.234
	Average	1080.017	28.69	91.283	91.299	38.23
381.2	1	915.1	39.028	91.009	90.72	38.122
	2	915.405	39.411	90.977	90.724	38.126
	3	916.504	39.446	90.966	90.637	38.119
	4	916.992	39.865	90.915	90.621	38.094
	5	916.138	39.446	90.948	90.617	38.098
	Average	916.028	39.439	90.963	90.664	38.112
381.1 *	-	836.243	52.769	-	-	-
381	1	813.965	60.512	89.629	88.704	37.421
	2	814.27	61.14	89.629	88.687	37.422
	3	814.514	61.349	89.6	88.72	37.42
	4	814.636	61.349	89.599	88.667	37.414
	5	814.026	61.523	89.608	88.678	37.416
	Average	814.282	61.175	89.613	88.691	37.419
380.9	1	792.297	74.533	89.131	88.587	37.286
	2	792.053	73.835	89.158	88.594	37.305
	3	792.542	74.254	89.147	88.516	37.271
	4	791.87	74.079	89.136	88.554	37.27
	5	791.565	73.835	89.114	88.556	37.275
	Average	792.065	74.107	89.137	88.561	37.281
380.8	1	736.145	103.62	88.32	88.198	37.027
	2	735.718	102.783	88.295	88.198	37.019
	3	735.901	103.62	88.334	88.197	37.031
	4	733.337	104.457	88.29	88.188	37.031
	5	733.826	106.166	88.319	88.162	37.023
	Average	734.985	104.129	88.312	88.189	37.026

Table 5. *Cont.*

Determined Slide Position, mm	Sample No.	Measured Press Slide, µm	Maximum Force, kN	Angle RH, Degrees	Angle LH, Degrees	The Width of the Part, mm
380.7	1	706.299	125.244	87.997	88.04	36.913
	2	705.75	125.035	88.033	88.06	36.937
	3	705.322	125.663	88.025	88.083	36.913
	4	705.139	124.442	88.023	88.081	36.933
	5	704.773	124.826	88.044	88.087	36.92
	Average	705.457	125.042	88.024	88.07	36.923
380.6	1	685.181	139.718	87.723	87.936	36.85
	2	685.791	139.718	87.745	87.968	36.85
	3	686.89	140.381	87.728	87.955	36.863
	4	687.256	140.381	87.799	87.998	36.877
	5	686.951	140.799	87.813	88.002	36.871
	Average	686.414	140.199	87.762	87.972	36.862
380.5	1	665.649	153.774	87.594	87.914	36.806
	2	665.344	153.948	87.571	87.919	36.799
	3	665.588	154.192	87.591	87.922	36.819
	4	665.161	153.53	87.601	87.957	36.828
	5	664.612	153.774	87.602	87.907	36.814
	Average	665.271	153.844	87.592	87.924	36.813
380.4	1	647.88	169.503	87.465	87.941	36.765
	2	648.315	168.666	87.478	87.951	36.74
	3	648.193	169.085	87.506	87.965	36.773
	4	648.473	168.666	87.477	87.978	36.766
	5	648.621	168.422	87.473	87.943	36.761
	Average	648.352	168.868	87.48	87.956	36.761

* Optimally injected tool.

The testing procedure began with the press slide raised to its highest position (382.0 mm) and then gradually lowered in increments of 0.2 mm. Once the full injection point was reached, the increments were reduced to 0.1 mm. At each increment, between 30 and 40 parts were produced at a speed of 40 strokes per minute, and the last five parts produced at each increment were labeled and set aside for measurements.

Labeled parts were measured using a coordinate measuring machine (CMM), and the results are presented in Table 5. A fully formed part with features and alignment in the CMM is presented in Figure 5. Plane Datum A was measured by 25 points, both circle Datum B and slot Datum C were measured by 15 points, and each plane (B and C) was measured by 20 points. Angle characteristic was evaluated as 3D angle between two planes in order to acquire a result without deviations, and part width evaluations were acquired by plane B and C average values. It is noticeable that the part-bending lines are of different lengths. The length of the left side's bending line (LH) was 34.3 mm, and that of the right side (RH) was 25.4 mm. Overall, 60 samples were measured. The results table also includes the determined and actual slide positions, as well as the stamping forces recorded by the JZT1 and JZT2 force sensors.

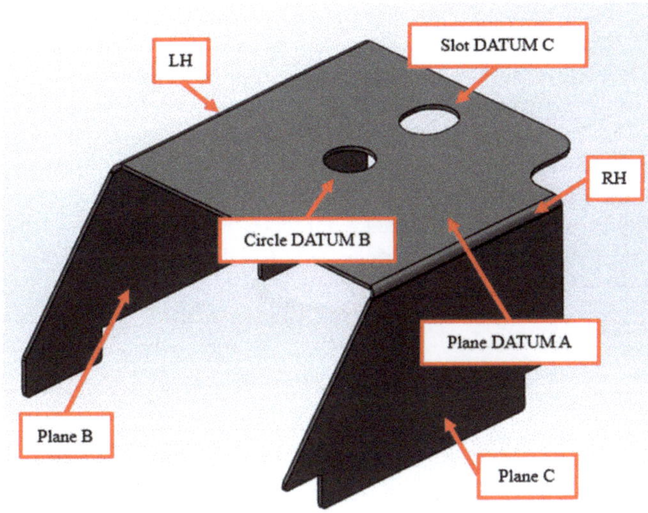

Figure 5. Fully formed part with marked left and right-side angles.

The results (Table 5) show that the relationship between the determined position of the angle (the primary independent variable) and its measured position is non-linear. As shown in Figure 6, the curve exhibits a nearly linear trend at the beginning but starts to deviate (bend) with the press slide lowering. This alteration occurs near the point of complete injection. Until this moment, the curve had a nearly perfect linear trajectory with a high coefficient of determination (R^2 = 0.9935). However, beyond the full injection point, the variance in measured position slows and eventually stabilizes. In the following phase, the coefficient of variation decreases to R^2 = 0.9615, indicating that the die is fully injected and the upper and lower sections rest on each other. As the slide lowers, the upper plate position becomes almost static, with any additional motion absorbed by the press and die elements.

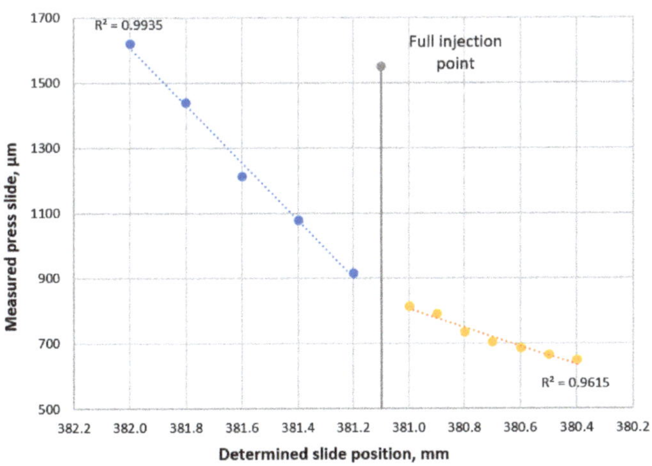

Figure 6. Dependence between the determined and measured slide position.

The next scientific observation is the dependence between the maximum generated force and the measured slide position. As shown in Figure 7a, the force initially increases gradually with the slide being lowered. However, the rate of increase accelerates before slowing down again, resulting in an S-shaped curve. The initial portion of the curve, or the left "tail", can be attributed to the stripper being too high. In this position, the die is not fully injected, allowing the rising pressure to be easily absorbed, which leads to only a slight change in the load on the press columns. When the slide approaches the ideal "bottom" position, the clearance between tool components and the workpiece diminishes, resulting in increased resistance and column load. The flattening of the curve at the end indicates that the stamping force has reached the point where the workpiece begins to experience plastic deformation of material, absorbing part of the load from the columns.

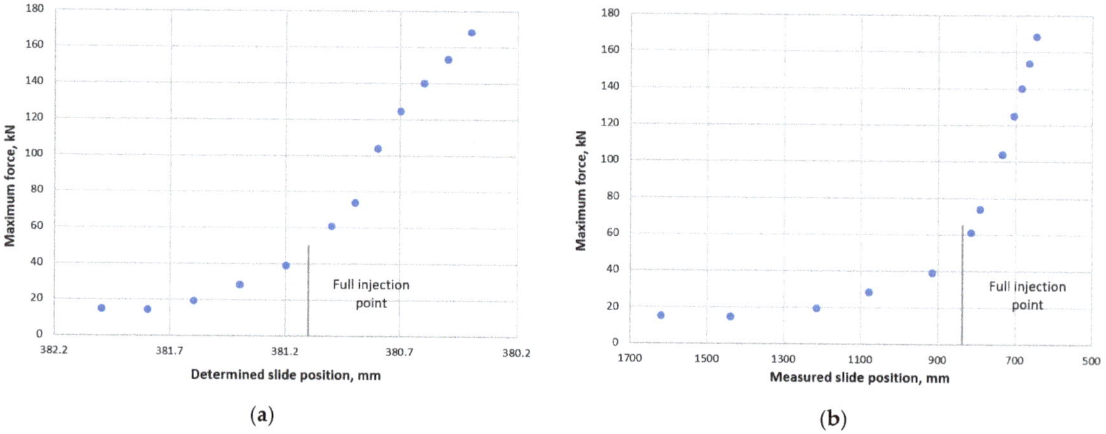

Figure 7. Real-life tests on the dependence of the maximum force of the tool. (**a**) The relationship between the maximum generated force and determined slide position; (**b**) the relationship between the maximum generated force and measured slide position.

The relationship between force and measured stripper position (Figure 7b) is simpler. Initially, the force gently rises until it accelerates sharply. In contrast to the fixed threshold position, there is no deceleration at the end of the curve. This phenomenon can be explained by the fact that, once the die is fully injected and the slide continues to fall, the actual location of the slide changes more slowly, as seen in Figure 6, and this is consistent with the deceleration of column elastic deformation previously discussed. These displacement decelerations appear proportional to one another, resulting in a curve that looks more like an L-shape than an S-shape, with no obvious flattening. Real-life test results are shown in (Figure 7), while (Figure 8) shows simulated data results of only the bending step.

The results of the simulated force influence show that the research area observed in the simulated bending step (Figure 8) closely mirrors the experimental findings. Before the optimal tool injection point, the force remains relatively constant. This can be explained by the force required to overcome the yield strength of the material during the initial stages. When the tool is fully injected, a significant increase in force is observed, marking the beginning of the bending phase. This rapid rise in force occurs because the bending punches start applying pressure to compress the material. With the upper plate moving downwards, the material begins to deform, and the force reflects the growing resistance against the punch motion. This resistance increases as the material is progressively shaped and compressed, requiring more force to continue the movement. At the first two increments following tool optimum injection, force remains constant with negligible growth due to material thickness reduction along the bending line.

This process aligns with the known theories of material deformation under pressure, where the force needed to deform a material increases as it resists being reshaped because it will change the inner structure of the material. For instance, thicker or stronger materials require more force to bend because they resist deformation more than thinner or softer materials [21].

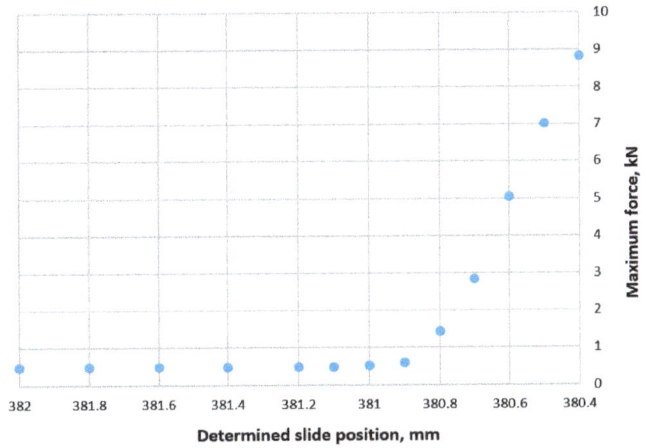

Figure 8. Simulated tests dependency of the tool's maximum force on a specific slide position.

An important observation when comparing the curves in Figure 7a,b is the force spike that occurs around the ideal injection point for the observed slide position. This spike is less pronounced in the case of the determined slide position, suggesting that combining the force generated by the press with the measured stripper position allows for a more accurate determination of full injection point. This would help achieve optimal part quality, maintain a stable production process, and prevent potential overloads on both the press and the die.

When comparing the effects of the measured and determined slide positions on the right-hand (RH) and left-hand (LH) angles of the part, a clear discrepancy emerges (see Figure 5). Both angles exhibit an S-shaped relationship with the stripper position, as shown in Figure 9. Initially, a small reduction in the stripper's position results in minimal changes to the angles, especially when the position is still far from the ideal. At this stage, the bending punch primarily bends the workpiece downward without significant pressure being applied to it, as there is still considerable space between the bending punch workpiece and the die (see Figure 4). However, as the stripper position approaches the optimal injection point, the workpiece begins to press against the edge of the matrix, which locks the bend and minimizes elastic recovery. This highlights the importance of achieving precise control over the stripper's position to ensure optimal part geometry and process stability. This behavior is in line with previous research findings that emphasize how important it is to align and place tools properly to get dependable stamping process results [5].

As the slide is lowered further, this impact intensifies, resulting in an almost linear influence on the angles. Toward the end, the change slows down once more before coming to a complete halt. This is because, despite the stripper's growing pressure, the part's sides rest against the support and are unable to bend any further. Furthermore, the graphs show that the component flap angles fall below the 90-degree mark. This is explained by the fact that the material is compressed and, due to elastic deformation, the workpiece can bend at an angle greater than 90 degrees.

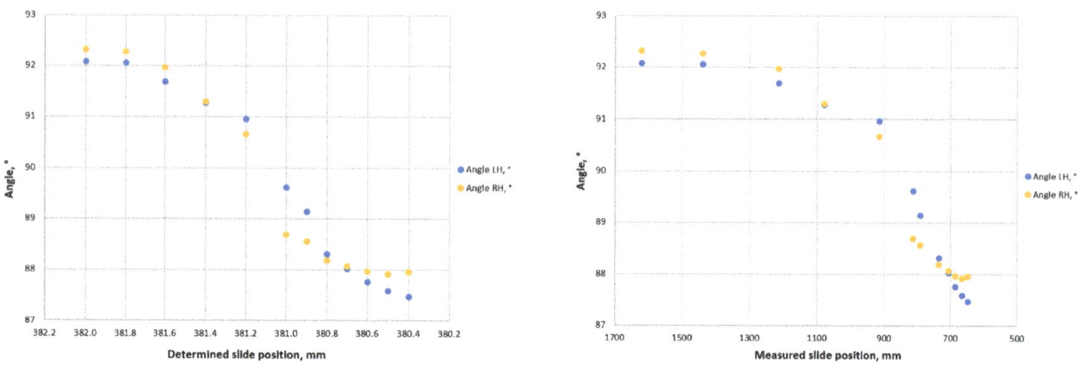

(a) (b)

Figure 9. Dependence of part bending angles on: (**a**) determined and (**b**) measured slide position.

In the case of the measured slide position, the curves (see Figure 9b) also exhibit an S-shape, but they are more convex and only slightly curve toward the end. This behavior can be explained by the fact that, as the stripper is lowered further, the actual position of the slide remains relatively unchanged because the die is already fully injected. The slowing of position change coincides with a reduction in the bending angle, as the workpiece begins to press against the matrix's side. This results in a less pronounced flattening of the curve compared to the determined slide position.

Given the observed differences, the question arises as to which of the following dependencies, seen in Figure 9a,b is more realistic and could be more useful for practical application. On the one hand, the slide's measured location should more accurately depict the dynamics of the bending angle since it displays the actual progress of the bending punch. However, as we know from the results discussed above, when the slide is low, even a slight change in its measured position results in a significant increase in force (see Figures 7b and 8). This force determines the plastic deformation of the part as well as the tool and press overload. It is simpler to forecast when these detrimental processes would start. Based on the angle's dependence on the slide's identified position, it is easier to forecast when these detrimental activities would start because curve bending is more noticeable there.

The results of simulated part angles are presented in Table 6 and Figure 10, with a comparison between the part angles derived from the measured samples and the simulated values. The disparity between the two datasets is the first obvious finding. For every determined slide position, the measured part angles show steadily higher values. This is caused by the CMM (coordinate measuring machine) measurement technique, which calculates angles by measuring the intersection of two planes constructed from user-defined points. The simulated part geometry has smaller plane deflections, and even small changes could significantly affect the results. However, the simulated samples only consider spring-back at the bending lines and exclude any plane deflection, which could be accountable for the differences between the measured and simulated angles.

Table 6. Results of angles derived from the observed samples and the calculated values.

Determined Slide Position, mm	382.0	381.8	381.6	381.4	381.2	381.1	381.0	380.9	380.8	380.7	380.6	380.5	380.4
Angle LH determined from simulation	91.310	91.320	91.090	90.890	90.610	90.060	89.820	89.630	89.260	89.050	88.900	88.880	88.800
Angle RH determined from simulation	91.340	91.330	91.110	90.850	90.650	90.040	89.800	89.700	89.310	89.100	88.980	88.930	88.850

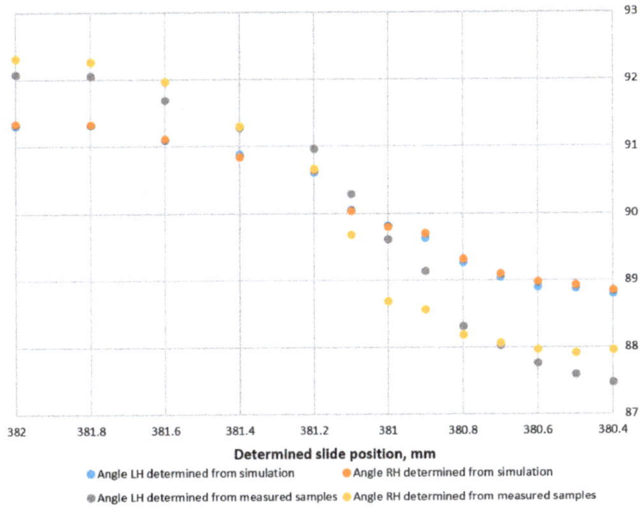

Figure 10. Disparities between part angles derived from simulation results and measurements.

Analyzing the impact of the measured and determined slide position on the part's width presents similar results. It is not surprising that the dependencies in Figure 11a,b mirror the corresponding curves in Figure 9a,b since this dimension represents the cumulative consequence of the side bend angles. Although the measured slide position is more convex and only starts to tilt slightly at the very end, the dependence on the determined slide position is still S-shaped.

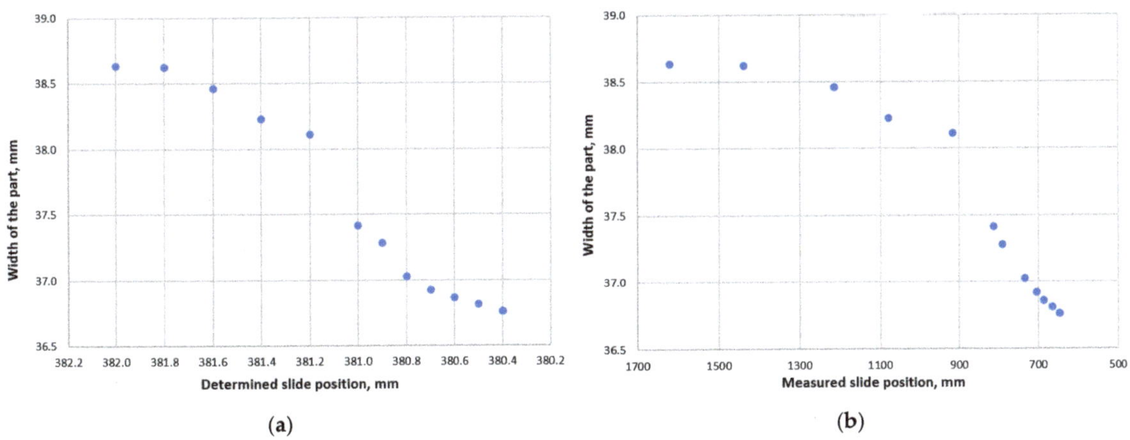

Figure 11. Dependence of the width of the part obtained by: (**a**) determined and (**b**) measured slide position.

Analysis of simulated and measured angles in the progressive stamping process shows that almost perfect matches in part widths were achieved/taken (Table 7 and Figure 12). Also, it was observed that variation in part angles between measured samples and simulated results offers significant details on the mechanics of the bending process and the precision of the simulation model. The consistent part width measurements across the experimental and simulated data indicate that the mathematical model used in simulations accurately predicts the deformation behavior along the bending line.

Table 7. Comparison of part width determined from simulations and measured samples.

Slide, mm	382.0	381.8	381.6	381.4	381.2	381.1	381.0	380.9	380.8	380.7	380.6	380.5	380.4
Part width determined from simulation, mm	38.628	38.630	38.455	38.272	38.086	37.636	37.455	37.338	37.046	36.886	36.776	36.749	36.691
Part width determined from measured samples, mm	38.636	38.622	38.460	38.230	38.112	37.609	37.419	37.281	37.026	36.923	36.852	36.813	36.761

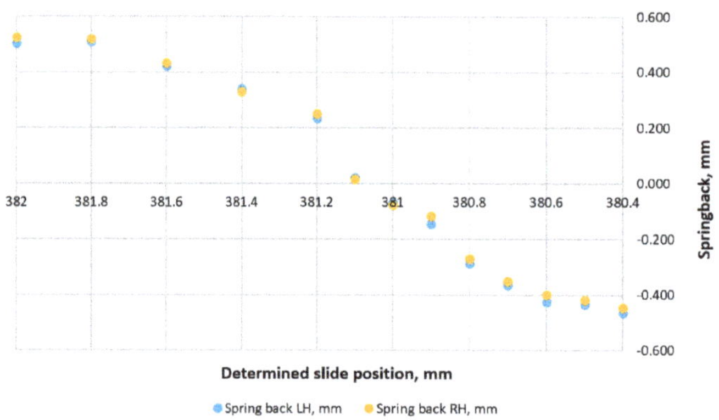

Figure 12. Differences between part width obtained from measurement and simulation results.

The deviation divergence in part angles can be attributed to deflection in the planes of the post-bending of measured samples. In real life, slight plane misalignments may be driven by minimal minute flaws, irregular substances, or elastic recovery effects. These are recorded during measurements by the coordinate measuring machine (CMM), which uses user-defined points to create planes. Deviations from the simulated results may result from minimal deflections in these planes, which may cause variability in the recorded angles. As a result, by assuming idealized conditions and calculating spring-back effects only along the bending lines without factoring potential plane deflections into account, the simulation streamlines the bending process.

The consistency in part widths between the measured and simulated samples suggests that, while plane deflection impacts angle measurements, it does not significantly affect the overall material flow or deformation in the width direction. This underlines the importance of considering plane deflections in experimental setups and supports the simulation model's capabilities to capture crucial aspects of the bending process. These findings highlight the value of sophisticated modeling methods in estimating process results and finding potential causes of measurement discrepancies in practical applications.

The results of the research support the hypothesis that the stamping force used to bend parts to form angles can be used to monitor the overall width of parts. This is confirmed by the data in Figure 13. The curves shown previously with the bending angle are first significantly affected by the stamping force, but at certain points, the angles basically cease to change; in this case, the break occurs after the force rises from 60 to 80 kN.

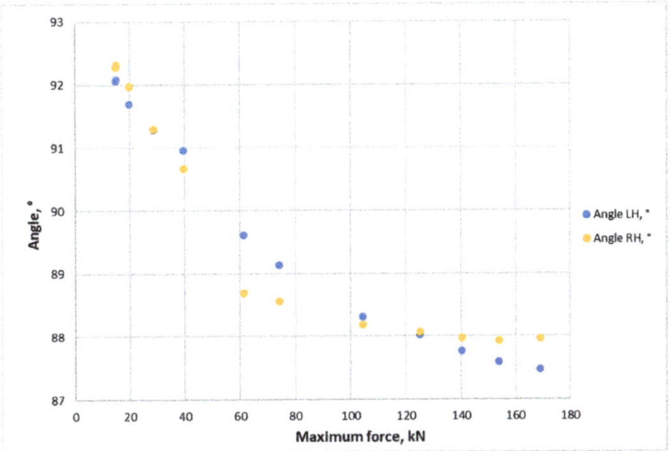

Figure 13. The influence of the stamping force on the bending angles of the part.

The correlation between stamping force and part width showed an analogous pattern (Figure 14). These results indicate that there may be a correlation between the stamping force and the indicators of the quality of components, though it is not linear and takes the shape of a L or sphere instead.

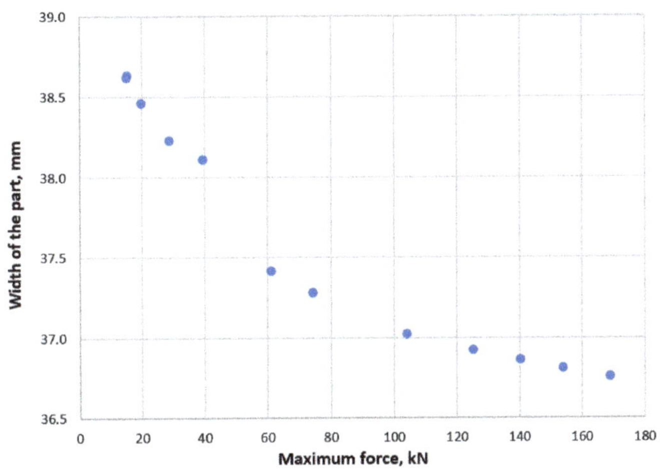

Figure 14. The influence of stamping force on the width of the part.

This can be explained by the previously indicated fact that the bending die wall limits the capacity of the workpiece to bend. In line with this, the L-shaped curve is extremely useful because, if a slowdown in the angle's decrease occurs over process modification, it could be assumed that the tool is already optimally injected and that lowering the slide further is not required. After all, performing so will only overload the die and the press and cause undesirable surface imprints throughout the parts.

4. Conclusions

This research confirms a strong relationship between stamping force, slide position, and part quality. After conducting research, the following conclusions were formulated:

1. The main findings include non-linear force-displacement behavior, deviations between experimental and simulated results due to the plane deflections, and the importance of optimizing tool injection for consistent part quality. These results validate the hypothesis that stamping force can monitor critical geometric parameters, aiding in process stability and defect prevention.

2. The relationship between the slide position and the stamping force is non-linear, showing S-shaped and L-shaped curves.

3. The R^2 values indicate strong linearity until full injection ($R^2 = 0.9935$) followed by reduced linearity ($R^2 = 0.9615$) during the die compression.

4. Measured angles exhibit a steady increase compared to simulated values, attributed to deflections in planes caused by elastic recovery and material irregularities.

5. Clear S-shaped dependency between the slide position and bending angles, emphasizing optimal tool injection role in minimizing elastic recovery.

6. Experimental and simulated part widths showed close alignment, confirming that width is less sensitive to the plane deflection compared to bending angles.

7. Incremental slide lowering during testing process revealed the influence of material compression on force dynamics and bending quality.

8. Deviations between measured and simulated angles were caused by the simulation assumption of ideal spring-back conditions without accounting for the plane deflections.

Author Contributions: Conceptualization, J.S. and J.G.; Methodology, J.S. and J.G.; Investigation, J.S. and J.G. All authors have read and agreed to the published version of the manuscript.

Funding: This research received no external funding.

Data Availability Statement: The article was prepared using the limited accessibility data provided by company LLC Stansefabrikken Automotive. For more detail, please contact directly by this e-mail: juras.skardzius@vilniustech.lt.

Acknowledgments: I would like to express my sincere gratitude to LLC "Stansefabrikken AUTOMOTIVE" for providing access to tools and equipment used in this research. Their contribution has been invaluable in facilitating the data collection process and enabling the analysis conducted in this research process. Without their support and cooperation with VILNIUS TECH–Vilnius Gediminas Technical University, this research would not have been possible.

Conflicts of Interest: Author Juras Skardžius was employed by the company LLC Stansefabrikken Automotive. The remaining author declare that the research was conducted in the absence of any commercial or financial relationships that could be construed as a potential conflict of interest. The company had no role in the design of the study; in the collection, analyses, or interpretation of data; in the writing of the manuscript, or in the decision to publish the results.

References

1. Hu, Y.; Qiu, Y.; Mei, M.; Chen, S.; Zhao, Y. Influence and improvement of stamping process parameters on V-shaped clamp forming quality. *Adv. Mech. Eng.* **2024**, *16*, 16878132241235076. [CrossRef]
2. Li, J.; Wang, Y.; Zhao, Q. Influence of stamping speed on high-strength steel parts in progressive tools. *J. Mater. Process. Technol.* **2021**, *284*, 116423.
3. Zhang, Y.; Liu, F.; Hu, Z. Optimizing stamping speed for part quality and production efficiency. *Procedia CIRP* **2020**, *84*, 553–558.
4. Yang, H.; Zhang, F.; Lin, Y. Material behavior and dimensional control in high-speed aluminum stamping. *Int. J. Adv. Manuf. Technol.* **2023**, *119*, 1523–1535.
5. Yang, H.; Lee, J. The impact of high-speed stamping on the dimensional accuracy of automotive components. *Int. J. Adv. Manuf. Technol.* **2017**, *91*, 703–710.
6. Zhang, Y.; Hu, Z. Effect of press force on warping in high-strength steel stamping. *J. Mater. Process. Technol.* **2021**, *288*, 116432.
7. Li, J.; Wang, Q.; Zhao, L. Residual stress and springback in aluminum alloy stamping under varying forces. *Int. J. Adv. Manuf. Technol.* **2022**, *119*, 1523–1535.

8. Li, J.; Zhao, Q.; Wang, L. Springback effects in advanced high-strength steels under varying press forces. *Int. J. Adv. Manuf. Technol.* **2022**, *120*, 1456–1472.
9. Singh, P.; Gupta, K.; Rao, A. Tool wear and dimensional consistency in high-force progressive stamping. *Procedia CIRP* **2023**, *118*, 245–252.
10. Singh, P.; Gupta, K.; Rao, A. Effects of tool wear on dimensional consistency in high-speed stamping processes. *Procedia CIRP* **2021**, *96*, 275–280.
11. Chen, X.; Lee, J.; Kumar, M. Real-time force monitoring and its effects on tool wear in progressive stamping. *J. Manuf. Syst.* **2023**, *74*, 325–334.
12. Otsu, M.; Uchimura, T.; Okada, M.; Yoshimura, H.; Matsumoto, R.; Muranaka, T. Friction Stir Incremental Forming of Preformed Sheets with Improving Bending Stiffness. *Procedia Eng.* **2017**, *183*, 131–136. [CrossRef]
13. Kumar, P.; Rao, D. Material behavior under high-speed stamping: An experimental study. *Mater. Sci. Eng. A* **2016**, *672*, 74–82.
14. Morel, X.; Mercier, S.; Bouscaud, D.; Martiny, M.; Sarre, B.; Sturel, T.; Berveiller, S. Residual stresses in the new press-hardening steels: Experiments and numerical simulations of V-bendings. *Mater. Today Commun.* **2025**, *44*, 111903.
15. Yang, H.; Kim, D. Quantification of springback effects in AHSS with increased press force. *Wear* **2023**, *501*, 203045.
16. Altan, T.; Tekkaya, A.E. *Sheet Metal Forming: Processes and Applications*; ASM International: Almere, The Netherlands, 2012.
17. Singh, R.; Kumar, M.; Gupta, V. Influence of tool wear on dimensional accuracy in high-speed stamping. *Wear* **2019**, *428–429*, 107–115.
18. Magliaro, J.; Cui, Z.; Shirzadian, S.; Green, D.E.; Altenhof, W.; Alpas, A.T. Evaluating die wear-induced edge quality degradation in trimmed DP980 steel sheets from in situ force response monitoring. *Wear* **2023**, *524–525*, 204792.
19. Hsu, K.; Lee, T. Advanced press force monitoring for dimensional accuracy in progressive stamping. *J. Manuf. Sci. Eng.* **2024**, *146*, 234–248.
20. Zhang, L.; Sun, Q. Advanced tooling solutions for high-speed progressive stamping. *Wear* **2022**, *510–511*, 203054.
21. Lorenzoni, B.R.; da Silva, F.P. Bending techniques for flat materials using cut patterns: A review. *Glob. J. Eng. Technol. Adv.* **2021**, *7*, 91–102. [CrossRef]

materials

MDPI

Communication

Effect of Electrode Surface Microstructuring on Electrochemical Biosensor Performance

Amal Kabalan * and Maliheh Azimi Roueini

Department of Electrical and Computer Engineering, Bucknell University, Lewisburg, PA 17837, USA; mar059@bucknell.edu
* Correspondence: aak012@bucknell.edu

Abstract: Electrode surface microstructuring involves the engineering of the topographical features of an electrode to enhance its performance in electrochemical sensing applications. By creating controlled micro- or nano-scale patterns, the active surface area can significantly increase, which leads to improved electron transfer and enhanced sensitivity to target analytes in devices such as biosensors. Geometrical parameters such as diameter, height, pitch, and position of the patterns can be optimized to enhance sensor detection. This paper introduces an electrochemical biosensor designed to detect *Moraxella catarrhalis*, a respiratory pathogen affecting young children. This paper investigates the effects of the radius of micropillars on adsorption in the electrochemical biosensor using COMSOL Multiphysics (Version: 6.0). The model demonstrates that the rate of surface adsorption depends on the position of the micropillars on the electrode. The paper also presents the effects of analyte concentration on the detection current of the biosensor using Cottrell's equation.

Keywords: electrochemical sensors; microstructuring; adsorption

1. Introduction

Electrochemical biosensors are analytical devices that convert a biological event—such as the binding of an enzyme, antibody, or other biorecognition element to a target molecule—into an electrical signal [1]. This signal can be measured in terms of current, voltage, or impedance and provides valuable information about the presence or concentration of specific analytes [1]. These sensors are characterized by high sensitivity and rapid response, which makes them attractive to applications in a wide variety of fields [2,3]. In medical diagnostics, they enable rapid point-of-care testing for biomarkers of diseases such as diabetes, cancer, and infectious pathogens, thereby facilitating timely clinical decisions [4]. In food safety, these sensors detect contaminants like pathogens, allergens, and chemical residues, ensuring product quality and consumer health. Environmental monitoring is another key area where electrochemical biosensors are used to measure pollutants, heavy metals, and other hazardous substances in water and soil, helping maintain ecological balance [5]. Advancements in electrode fabrication, such as micro- and nano-structuring, further enhance these sensors' performance by increasing the surface area, improving mass transport, and facilitating targeted surface functionalization [6].

In this study, we introduce an electrochemical biosensor that aims to detect M. Catarrhalis, which is a pathogen that causes ear infections, especially in young children [7–9]. Specifically, this paper investigates the effects of the radius of micropillars on the adsorption rate in the electrochemical biosensor using COMSOL Multiphysics (Version: 6.0) [10]. It also analyzes the adsorption rate with respect to the position of the micropillar on the electrode surface.

2. Materials and Methods

The sensing platform of the device is encapsulated in a microfluidic channel, as shown in Figure 1, and is composed of three electrodes (reference, working, and counter electrodes). The adsorption of the pathogen takes place at the working electrode, which is patterned by micropillars and shown in Figure 2. The surface of the working electrode is functionalized with antibodies to bond with the pathogen in a given fluid. The shape and the position of the micropillars impact the adsorption rate, which is analyzed in this study.

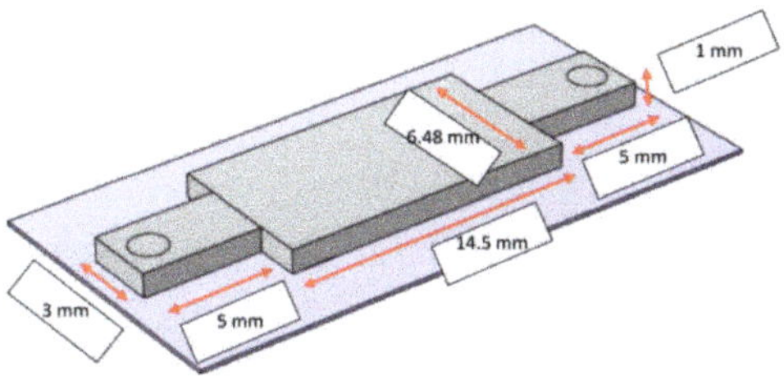

Figure 1. The sensing platform of the device [1].

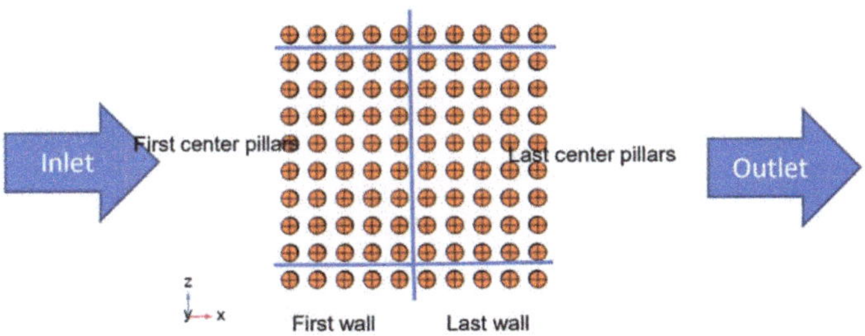

Figure 2. Patterned working electrode in the biosensor.

Analyte molecules (P—in this case, *Moraxella catarrhalis*) can adsorb and desorb from surface sites (S) on the micropillar's surfaces according to Equation (1):

$$p + s \overset{k_{ads}}{\leftrightarrow} ps \tag{1}$$

The adsorbed analyte (*Moraxella catarrhalis*) of Equation (1) can transform into a quenched state (QS) that does not contribute to the sensor signal, as shown in Equation (2).

$$PS \underset{K_2}{\overset{K_1}{\leftrightarrow}} QS \tag{2}$$

The rate of adsorption depends on the concentration of the analyte species in its stream. The equations describe the surface reactions that are coupled to free analyte flow. The movement of analyte molecules in a biosensor is due to the lateral flow in the device [11].

3. Results

The rate of adsorption with respect to the position of the pillars in the microfluidic channel was studied using COMSOL Multiphysics. Figure 3 shows the rate of adsorption at four different locations of the pillars for three different radii (0.3 mm, 0.4 mm, and 0.5 mm). The following references were consulted when choosing the dimensions of the pillars [11,12]. Adsorption starts 25 s after the fluid enters the channel. It starts at the first pillar in the center and wall rows. It peaks at 33% after 35 s in the center row, and it peaks at 16% after 42 s in the wall row. Adsorption starts at 43 s in the last pillars of the wall and first rows. The adsorption peaks at 24% and 10% after 70 s in the last pillars of the wall and center rows, respectively.

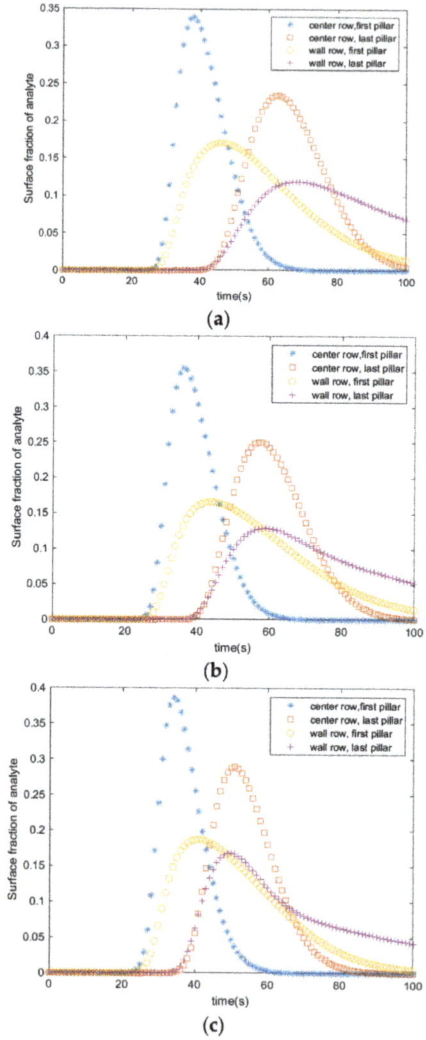

Figure 3. Average fractional surface coverage of adsorbed analyte in pillars with radii of (**a**) 0.3 mm, (**b**) 0.4 mm, and (**c**) 0.5 mm.

Figure 4a–c show the velocity (left) and adsorption fraction (right) in three micropillar radii (0.3 mm, 0.4 mm, and 0.5 mm) based on the location of the micropillar in the microfluidic channel. The data were obtained at t = 70 s (where t = 0 is the time the fluid enters the microfluidic channel). It was observed that as the radius of the pillars increased the adsorption fraction increased among the pillars in the channel. In Figure 4a, micropillars located at coordinates (0, −5), (0, −3), (1, −5), (1, −4), and (1, −2) have the highest absorption at t = 70 s. This pattern continues in Figure 4b, where the radius is set to 0.4 mm.

Figure 4c shows that the highest adsorption occurs in the center pillars at coordinates (3, −6), (4, −6), and (4, −4). Figure 5 depicts the concentration distribution in the analyte (*Moraxella catarrhalis*) stream and the surface coverage of the adsorbed species in various radii. The comparison of Figure 5a,b show that pillars located at the center of the electrode (Figure 5a) have higher adsorption than center micropillars (Figure 5b).

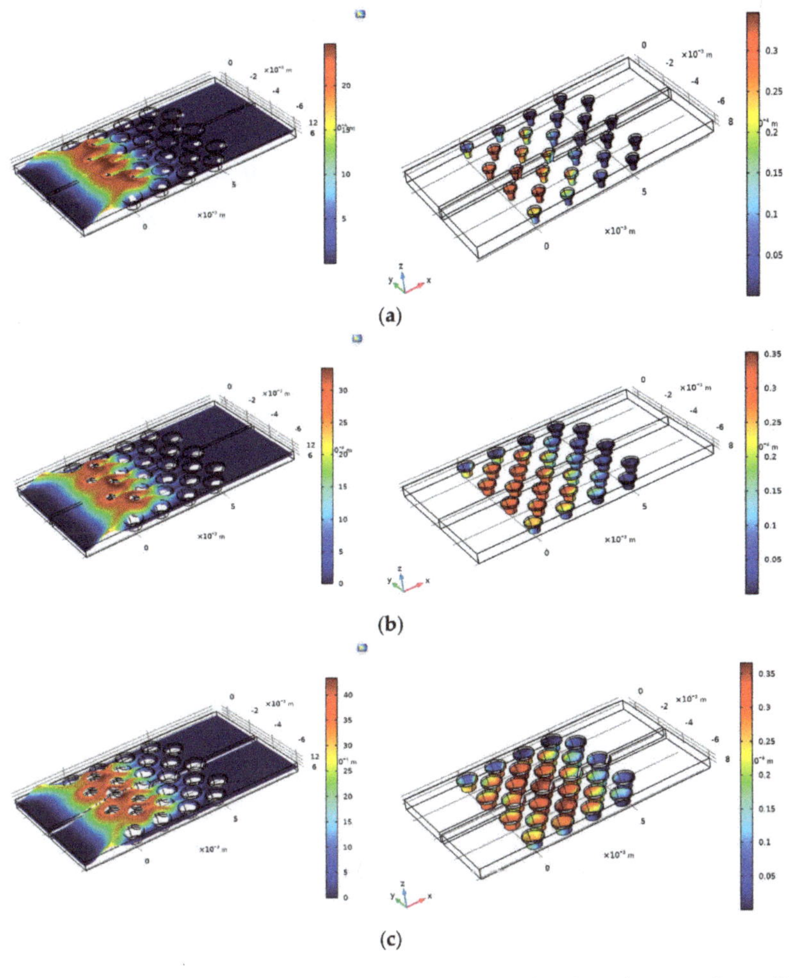

Figure 4. The average fractional surface coverage of adsorbed analytes shown on the gradient scale in pillars with a 0.3 mm radius: (**a**) 0.4 mm radius (**b**), and 0.5 mm radius (**c**). The left-hand images indicate the laminar flow in each channel.

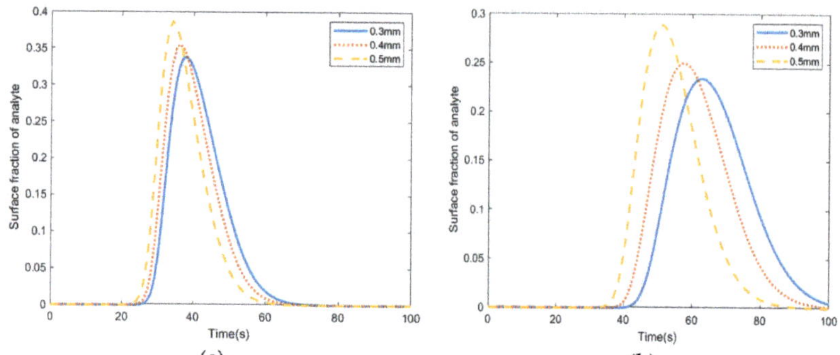

(**a**) (**b**)

Figure 5. The average fractional surface coverage of the adsorbed analyte in pillars (0.3 mm, 0.4 mm, and 0.5 mm radii) located in the (**a**) center of the electrode and (**b**) pillars located at the edges of the electrode.

Figure 6 shows the fraction of analyte absorption with varying radii of pillars in the channel. Based on the results in Figure 4, at t = 25 s, the sum of the adsorption of the analyte on pillars with a radius of 0.5 mm is higher than the other sizes; however, after 58 s, the biosensor that has pillars with the largest radius shows the lowest amount of adsorption. In other words, the largest pillars have the most adsorption and they desorb analytes quicker than other pillars.

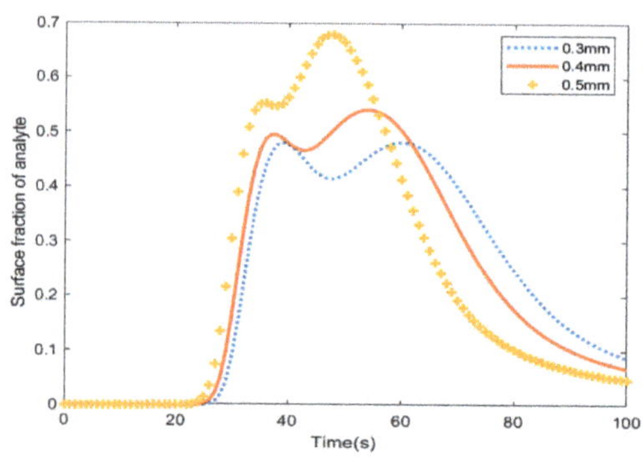

Figure 6. The total surface adsorption of the micropillars in the microfluidic channel.

We also investigated the changes in the current by varying the concentrations of the analyte. The current in the electrochemical biosensors can be calculated using the Cottrell equation [13]:

$$i = nFcA\sqrt{D/\pi t} \tag{3}$$

where i is the current density, n is the number of electrons transferred per molecule of the analyte, c is the bulk concentration of the analyte, D is its diffusion coefficient, A is the area, and F is Faraday's constant [14]. We varied the bulk concentration between 10 and 80 mM and investigated its effects on the current. Figure 7 shows that the absolute value of the response current increased proportionally with the increase in the analyte. The value of the current varied from 0.5 mA to 1 mA.

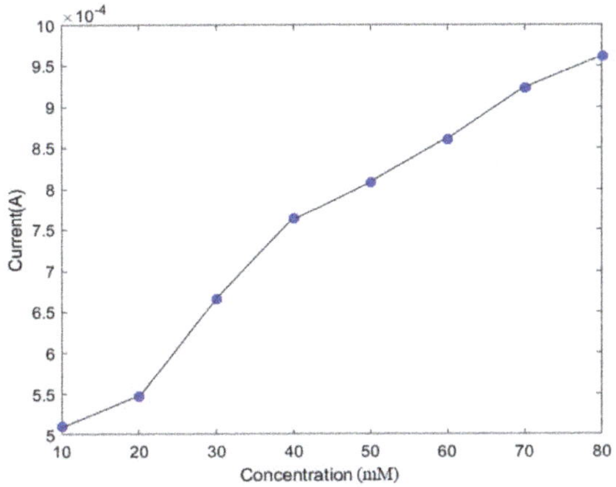

Figure 7. The effect of the pathogen concentration on the current.

4. Conclusions

In this study, a Multiphysics computational model was developed to study the effects of the micropillar radius on the absorption of the pathogen M. Catarrhalis under laminar flow. It was observed that the rate of adsorption varies based on the location of the micropillar within the microfluidic channel. Micropillars in the center of the channel exhibit the highest absorption rates. The increased radius size also contributes to an increase in the surface area, which increases absorption. In addition, the effects of the concentration of the analyte on the current over time were analyzed based on the Cottrell equation. Understanding the effects of the concentration of the analyte on the electric current response plays a critical role in the design parameters of electrochemical biosensors. The next step is to further optimize the parameters of the microfluidic channel and consider the effect of turbulence flow on the device.

Author Contributions: Conceptualization, A.K.; methodology, A.K.; software, M.A.R.; validation, A.K. and M.A.R.; formal analysis, M.A.R.; investigation, M.A.R.; resources, A.K.; data curation, M.A.R.; writing—original draft preparation, M.A.R.; writing—review and editing, A.K.; visualization, M.A.R.; supervision, A.K.; project administration, A.K.; funding acquisition, A.K. All authors have read and agreed to the published version of the manuscript.

Funding: This work was funded by a grant from the Bucknell Geisinger Research Initiative—Stage 2, under the direction of the Principal Investigator Amal Kabalan.

Institutional Review Board Statement: Not applicable.

Informed Consent Statement: Not applicable.

Data Availability Statement: The original contributions presented in this study are included in the article. Further inquiries can be directed to the corresponding authors.

Acknowledgments: We thank our collaborators at the Geisinger Medical Center, especially, Jonathan Spahr for the invaluable input.

Conflicts of Interest: The authors declare no conflicts of interest.

References

1. Bhalla, N.; Jolly, P.; Formisano, N.; Estrela, P. Introduction to biosensors. *Essays Biochem.* **2016**, *60*, 1–8. [CrossRef] [PubMed]
2. Tan, S.S.M.; Xing, W. Biosensors for food quality and safety. *J. Food Qual.* **2017**, *2017*, 1–15.

3. Bounegru, A.V.; Dinu Iacob, A.; Iticescu, C.; Georgescu, P.L. Electrochemical Sensors and Biosensors for the Detection of Pharmaceutical Contaminants in Natural Waters—A Comprehensive Review. *Chemosensors* **2025**, *13*, 65. [CrossRef]
4. Wang, J. Electrochemical biosensors: Towards point-of-care cancer diagnostics. *Biosens. Bioelectron.* **2006**, *21*, 1887–1892. [CrossRef] [PubMed]
5. Zhang, H.; Li, X.; Wang, J. Recent advances on electrochemical sensors for detection of contaminants of emerging concern (CECs). *Molecules* **2023**, *28*, 7916. [CrossRef] [PubMed]
6. Malhotra, B.D.; Ali, M.A. Nanomaterials in biosensors: Fundamentals and applications. In *Nanomaterials for Biosensors*; Elsevier: Amsterdam, The Netherlands, 2017; pp. 1–22.
7. Yang, D.K.; Kim, H.H.; Yoo, J.Y.; Lim, S.K.; Yoon, S.S.; Cho, I.S. Isolation and identification of Moraxella cuniculi from a rabbit with keratoconjunctivitis. *Korean J. Vet. Res.* **2017**, *57*, 201–204. [CrossRef]
8. Ruckdeschel, E.A.; Brauer, A.L.; Johnson, A.; Murphy, T.F. Characterization of proteins Msp22 and Msp75 as vaccine antigens of Moraxella catarrhalis. *Vaccine* **2009**, *27*, 7065–7072. [CrossRef] [PubMed]
9. Al-Ahmer, S.D.; Al-Amili, W.A.; Al-Asady, Z.H.; Al-Baiyati, N.I. Molecular Detection of Moraxella Catarrhalis Isolated from Children Infection with Otitis Media. *Int. J. Sci. Nat.* **2017**, *8*, 197–201.
10. Comsol. Biosensor Design. Available online: https://www.comsol.jp/trademarks (accessed on 3 January 2022).
11. Liu, B.; Lv, C.; Chen, C.; Ran, B.; Lan, M.; Chen, H.; Zhu, Y. Electrochemical Performance of Micropillar Array Electrodes in Microflows. *Micromachines* **2020**, *11*, 858. [CrossRef] [PubMed]
12. Chen, C.; Ran, B.; Wang, Z.; Zhao, H.; Lan, M.; Chen, H.; Zhu, Y. Development of micropillar array electrodes for highly sensitive detection of biomarkers. *RSC Adv.* **2020**, *10*, 41110–41119. [CrossRef] [PubMed]
13. Bard, A.J.; Faulkner, L.R. *Electrochemical Methods: Fundamentals and Applications*; John Wiley & Sons: Hoboken, NJ, USA, 2000.
14. Cottrell, A.H. The analysis of current transients following a potential step. *Trans. Faraday Soc.* **1903**, *1*, 293–310.

machines

Article

Prediction of Residual Life of Rolling Bearings Based on Multi-Scale Improved Temporal Convolutional Network (MITCN) Model

Keru Xia [1], Qi Li [2], Luyuan Han [3], Zhaohui Ren [2,*] and Hengfa Luo [2]

[1] School of Mechanical and Power Engineering, Shenyang University of Chemical Technology, Shenyang 110142, China; abc514863@126.com
[2] School of Mechanical Engineering and Automation, Northeastern University, Shenyang 110819, China; qli@mail.neu.edu.cn (Q.L.); lhf1456428108@163.com (H.L.)
[3] School of International Exchange, Guangzhou Maritime University, Guangzhou 510725, China; m17745196673@163.com
* Correspondence: zhhren_neu@126.com

Abstract: The method based on convolution neural networks (CNNs) has been widely developed and applied to residual life prediction, and many excellent results have been achieved. However, CNN models can only learn feature information relative to size, and it is difficult to extract complex time series features from data of long time series. In addition, the existing models still have some problems, such as capturing the correlation of each time series and generating a large amount of redundant information. In order to alleviate the above problems, this study proposes a residual life prediction method of rolling bearings based on a multi-scale improved temporal convolutional network (MITCN) model. It is used to solve problems such as the low accuracy of bearing life prediction and the difficulty of the temporal convolutional network (TCN) model to capture the correlation of each time series. The model adopts the framework of a time convolution network and has good ability to extract time series information. By introducing a multi-scale expanded causal convolution residual structure, improved temporal convolutional network (ITCN) modules with different expansion factors capture information on different time scales and combine soft threshold functions and channel attention mechanisms to adaptively generate thresholds and eliminate redundant information. Finally, the carbon border adjustment mechanism (CBAM) is an attention mechanism used to enhance useful features and suppress useless features, so as to realize the effective fusion of multi-scale features. The IEEE PHM 2012 challenge data set is hereby used to verify the proposed method, which can effectively solve the problem of the low prediction accuracy of the remaining life of bearings.

Keywords: convolution neural network; residual life prediction; MITCN model; attention mechanism; residual structure

1. Introduction

Rolling bearing is an indispensable component of rotating machinery, and the estimation of its remaining service life plays a vital role in the stable operation and production safety of equipment [1]. Actual data show that about 50% of bearing damage is attributed to insufficient lubrication, which leads to a significant rise in internal bearing temperature and vibration [2,3]. In order to avoid danger, it is necessary to accurately predict the remaining service life of rolling bearings [4]. Traditional fault diagnosis methods, such as the physical model [5], dynamic model prediction [6], and the Bayesian method [7], rely too much on manual experience and a theoretical basis and have a limited application scope and weak generalization abilities [8]. At present, remaining useful life (RUL) prediction technology is mainly divided into two methods: model-driven and data-driven methods [9].

The model-driven prediction method relies on the knowledge of a mechanical system's operation rule and fault mechanism and simulates the equipment's degradation process by mathematical and statistical models so as to achieve the effective estimation of bearing RUL [10–12]. The key premise is to select the mathematical model matching with the target system [13]. Yakout et al. [14] selected bearings with the same model and size and analyzed their dynamic characteristics. The experimental results show that the internal radial clearance has a significant impact on the dynamic characteristics of bearings. Cui et al. [15] proposed a comprehensive nonlinear dynamic model and simulated the bearing degradation process based on similarity theory. The degradation process was clearly divided into four stages: the steady-state stage, the defect formation stage, the defect diffusion stage, and the final damage stage. Wang et al. [16] proposed a nonlinear random coefficient regression (RCR) model, which updated the above information regarding random coefficients in real time under the Bayesian framework and calculated the RUL probability density function under specific failure threshold constraints, which effectively improved the RUL prediction accuracy compared to traditional Bayesian methods. Lei et al. [17] used the weighted minimum quantization error to fuse multi-feature information and accurately map it to the degradation process of bearings and then used the maximum likelihood estimation method to improve the accuracy of bearing RUL prediction. Although the model-driven method is very convincing in theory in predicting the RUL of mechanical equipment, the key to its success lies in whether it can build an accurate model reflecting the actual working state of the equipment. The limitation of this method is mainly reflected in its high sensitivity to the complex dynamic characteristics of rotating machinery, changeable operating environment, and external disturbances.

The prediction data-driven method provides an alternative scheme for predicting the RUL of rotating machinery and other equipment without relying on complex mathematical models. In RUL prediction, deep learning networks represented by CNNs and recurrent neural networks (RNNs) are most widely used [18–20]. Sloukia et al. [21] used the support vector machine (SVM) model to establish the correlation between the current and future state of bearings to predict their remaining life. Yan et al. [22] evaluated the bearing degradation stage through the dimensionless measurement index and trained the SVM classifier based on the measured values fitted by the generalized degradation model. Then, according to the classification results, a hybrid degradation tracking model was used to track the degradation process of bearings, thus optimizing RUL prediction. Rohani et al. [23] proposed a method to find a suitable parent wavelet, extract the required features from the decomposed wavelet coefficients, use the Weibull risk function and the exponential function for smooth fitting, and then feed this into the artificial neural network for RUL prediction. Zhang et al. [24] extracted spatial and temporal information from historical data with the help of 1D-CNN and the Bi-gated recurrent unit (BiGRU) hybrid model. The experimental results showed that the proposed parallel hybrid network could effectively predict the RUL. Guo et al. [25] proposed an RUL prediction method based on CNN-Bi-directional long short-term memory (BiLSTM)-Bootstrap model, which adaptively extracted vibration signal features and quantified the RUL prediction interval of rolling bearings by CNN and BiLSTM. Wang et al. [26] used a multi-scale convolution neural network to learn different receptive field features by expanding convolution blocks and improve the RUL prediction accuracy. The results showed that it was superior to the traditional one-dimensional convolution method. Dong et al. [27] proposed an RUL prediction method based on the

MSGCNN-TR model, which effectively extracted the spatio-temporal characteristics of rolling bearing life data through the depth convolution path with residual error and the gated cycle unit path. Jiang et al. [28] fused signal features using the Wasserstein distance, extracted weighted features through the double-attention mechanism and multi-scale convolution, and achieved RUL prediction, although the CNN was limited in terms of the insufficient capture of dynamic characteristics in time series data processing. To solve this problem, Bai et al. [29] proposed TCN, which used dilated causal convolution and residual learning to avoid future information leakage, and achieved remarkable results in the RUL prediction of rotating machinery through strict time constraints. He et al. [30] proposed a new RUL prediction framework based on TCN, which integrated a self-encoder and a self-calibration module, improved the ability of feature fusion and local information processing, and significantly improved the prediction accuracy and efficiency. Deng et al. [31] designed a multi-scale extended causal convolution residual block MsDCCRB and embedded SE units to adaptively calibrate feature responses, enhance feature learning ability, and directly use historical state monitoring data to achieve accurate RUL prediction.

With the development and progress of digital-twin technology, relevant enterprises and researchers began to consider the gradual integration of digital-twin technology and predictive maintenance [32]. Using virtual and real interactions and the real-time fusion characteristics of digital-twin technology, the prediction, early warning, and online diagnosis of the running state of bearings are of great significance for the efficient and reliable operation of equipment.

In summary, it is very common to predict the residual life of rolling bearings by introducing an improved CNN. Nevertheless, there are still some challenges in the existing research. Traditional CNNs have a fixed size of the convolution kernel. Convolution kernels with different sizes can only learn feature information relative to their size, so it is difficult to extract complex time series features from data derived from long time series. In addition, causal dilation convolution uses causality for filling, which produces a lot of redundant information.

In order to solve these challenges, this study proposes a residual life prediction method for rolling bearings based on the MITCN model. Firstly, time convolution, which has a better ability to extract time series information than other forms, is adopted as the network framework. Then, a new multi-scale extended causal convolution residual structure is proposed, which captures information on different time scales and adds a soft threshold function and a channel attention-adaptive threshold to eliminate redundant information. Finally, the CBAM form of the attention mechanism is introduced to enhance useful feature information and suppress useless feature information. The specific contributions of this study are as follows:

- A new multi-scale extended causal convolution residual structure is proposed, which uses several ITCN modules with different expansion factors to capture information on different time scales;
- We constructed an MITCN to predict the residual life of rotating machinery, adding a soft threshold function and a channel attention-adaptive generative threshold to eliminate redundant information, alongside the CBAM form of the attention mechanism to enhance useful feature information and suppress useless feature information;
- The effectiveness and superiority of the proposed method are hereby verified using bearing data sets.

The rest of this paper is structured as follows: The Section 2 introduces the specific details of the proposed method in detail. The Section 3 introduces the data processing and experimental setup. In the Section 4, the performance of the proposed method is verified using bearing data sets. The Section 5 draws the conclusion.

2. Methodology

This section describes the details of the proposed method. The overall residual life prediction scheme is shown in Figure 1. ITCN, CBAM, and MITCN were introduced. ICTN was the backbone of MITCN, and CBAM was the attention mechanism. Multiple ITCN modules with different size expansion factors were used to extract time series features of different scales in parallel. After feature fusion, the CBAM form of the attention mechanism was used to enhance the model's attention to key features and suppress invalid features to effectively predict the remaining life of bearings.

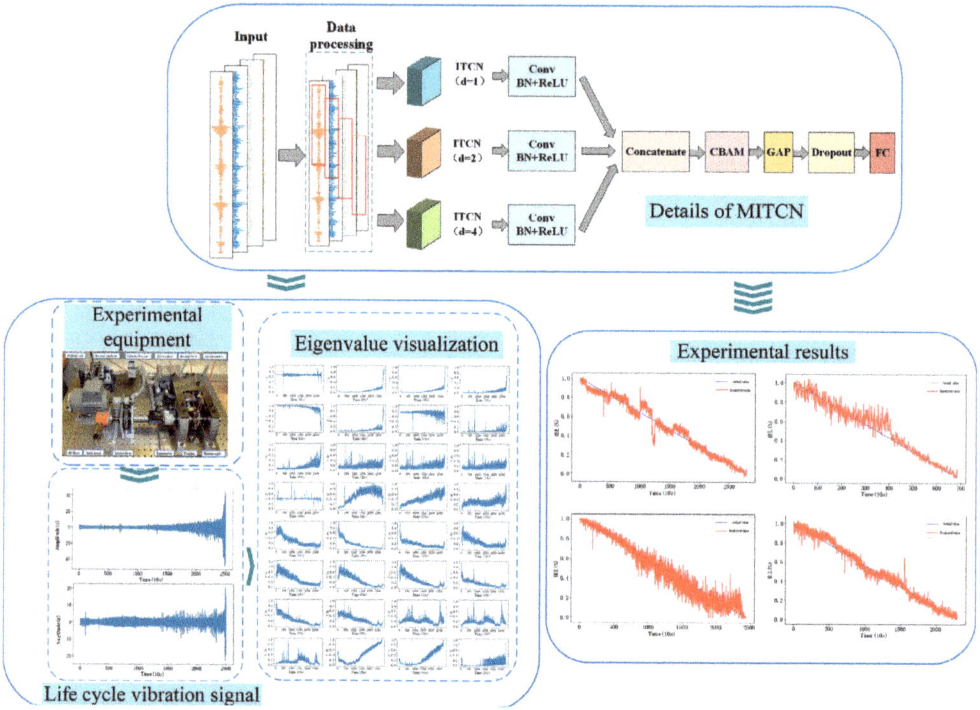

Figure 1. Frame of residual life prediction based on the MITCN model.

2.1. Details of ITCN

The ITCN module consisted of an extended causal convolution, a WeightNorm layer, an attention soft threshold layer, an ReLU activation function, and Dropout. Through residual connection and 1 × 1 convolution, the input and output dimensions were consistent, and the model's performance and feature learning ability were improved. The deep ITCN module was constructed by continuously stacking two layers of extended causal convolution. The network structure combination of extended causal convolution and residual connection could effectively improve the robustness and feature learning ability of the ITCN module. At the same time, a soft threshold was introduced into the network structure as a nonlinear transformation layer to effectively eliminate redundant information related to noise. Soft thresholds were dynamically generated by a specific subnetwork that adaptively calculated a set of thresholds for each vibration signal. In the ITCN module, the signal after two layers of extended causal convolution processing did not directly enter the next layer but first entered a specific subspace network. This subspace network was essentially a channel attention mechanism, which assigned different weights to each feature channel to determine their respective noise reduction thresholds. The processed soft threshold signal was fused with the primary feature before it was sent to the following layer network. Adaptive noise reduction in the soft threshold layer and the channel attention mechanism were used to distribute weights in the subspace network. The WeightNorm layer and the ReLU activation function were placed before the convolution to stabilize the network and accelerate training, while the Dropout layer reduced the risk of over-fitting. This design was helpful to learn deep features, prevent over-fitting, and ensure training stability. Figure 2 shows the proposed ITCN module.

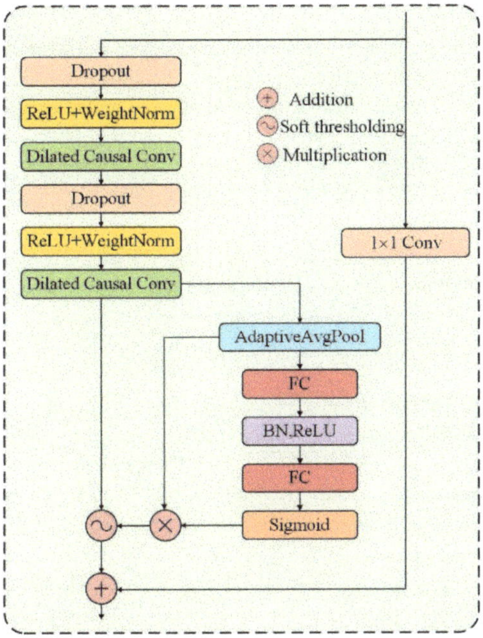

Figure 2. Details of ITCN.

2.2. Details of CBAM

CBAM was a kind of attention module combining channel and space, capable of enhancing useful features and suppressing useless features in the feature map, thus improving the ability of the network to extract implicit important features [33]. At the same time, CBAM was a plug-and-play simple attention block. Figure 3 is a schematic diagram of the CBAM type of attention modules. The two-channel attention module calculated the attention weight map through the channel and spatial attention modules and redefined the feature map. The output feature was activated by a sigmoid function, thus generating the final output result, via the following calculation process [34]:

$$\begin{aligned} M_c(F) &= \sigma(\mathrm{MLP}(\mathrm{AvgPool}(F)) + \mathrm{MLP}(\mathrm{MaxPool}(F))) \\ &= \sigma\left(W_1\left(W_0\left(F^c_{\mathrm{avg}}\right)\right) + W_1(W_0(F^c_{\mathrm{max}}))\right) \end{aligned} \quad (1)$$

where F represents the input feature, $M_c(F)$ represents the output, AvgPool and MaxPool indicate global average pooling and maximum pooling operations, MLP represents the multi-layer perceptron, and $\sigma(\cdot)$ represents the sigmoid activation function.

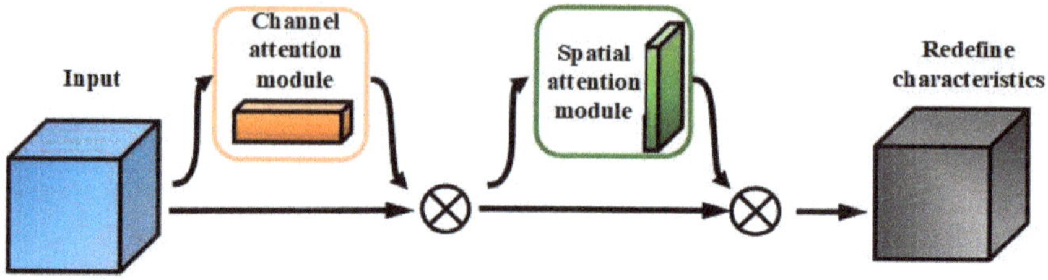

Figure 3. CBAM type of attention modules.

The spatial attention module took the output of the channel attention module as its own input and generated a maximum/average pooled feature using the maximum/average pooled operation. After this mosaic process, the spatial attention feature map was obtained by the convolution layer. The expression is as follows [34]:

$$\sigma\left(f^{7\times7}([\mathrm{AvgPool}(F);\mathrm{MaxPool}(F)])\right)=\sigma\left(f^{7\times7}\left(\left[F^{s}_{\mathrm{avg}};F^{s}_{\mathrm{max}}\right]\right)\right) \quad (2)$$

where $f^{7\times7}$ represents a convolution operation with a convolution kernel of 7 and [AvgPool(*F*); MaxPool(*F*)] represents the fact that the average and maximum pooling results are concatenated along the channel axis.

2.3. Details of MITCN

As shown in Figure 4, the MITCN module used several ITCN modules with different size expansion factors to extract time series features of different scales in parallel. After feature fusion, the CBAM type of attention mechanism was adopted to enhance the attention paid by the model to key features and suppress invalid features. The expansion coefficient d was usually set to an exponential power of 2, and the receptive field of the network expanded exponentially with the increase in the number of layers of the network. A branch of the ITCN module with multiple expansion factors (d = 1, 2, 4) was designed in the model. The expansion factor of d = 1 corresponded to the most basic time correlation, and the expansion rate of d = 4 enabled the model to explore time dependence over a longer range. Each ITCN module was followed by convolution layers of 3 and 1, and padding was set to 1 for further refinement of feature extraction. The activation function was ReLU. By fusing ITCN modules from different expansion factors and convolving the output, the model could integrate information across time scales. This cascading approach ensured that features at different time scales would be taken into account, providing a comprehensive representation of the features for subsequent attention mechanisms and decision levels. CBAM could assign different attention weights by learning the importance of each feature. Applying CBAM after feature fusion could effectively enhance the model's attention to key features and suppress unimportant information. The local average pooling layer (GAP) converted multi-dimensional features into a single-dimensional feature vector, which effectively reduced the dimension of data and prepared for the input of the full connection layer. Then, the Dropout layer randomly "discarded" (temporarily removed) some neurons in the neural network to prevent over-fitting of the network, and the full connection layer transformed the extracted features into the final prediction results. The parameters of MITCN model are shown in Table 1.

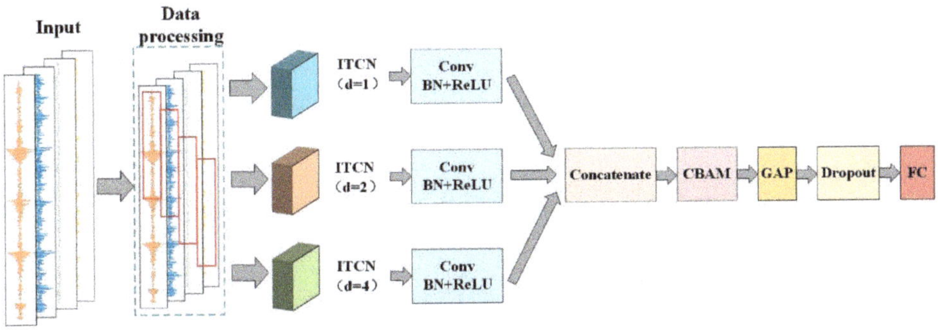

Figure 4. Details of MITCN.

Table 1. The parameters of the MITCN model.

Network Layer	Convolution Kernel	Step Size	Number of Channels	Output Layer
Convolution layer	3 × 3	1	16, 32, 64	(32, 16, 400), (32, 32, 400), (32, 64, 400)
Feature fusion layer	1 × 1	1	64	(32, 64, 400)
RELU	——	——	64	(32, 64, 400)
CBAM	——	——	64	(32, 64, 400)
Global mean pooling layer	1 × 1	1	64	(32, 64, 1)
Dropout layer	——	——	64	(32, 64)
Fully connected layer	——	——	1	(32, 1)
SoftMax classification layer	——	——	1	(32, 1)

3. Experimental Setting

This section describes the details for experimentation and data processing.

3.1. Description of Data Sets

The IEEE PHM 2012 Bearing Challenge data set [35] from the PRONOSTIA testbed was selected to verify the MITCN model-based prediction of rolling bearings' residual life proposed in this paper. PRONOSTIA is an experimental platform specifically designed to test, validate, and confirm methods related to bearing health assessment, diagnosis, and prediction [28]. The accelerated degradation residual life test platform is shown in Figure 5. Sensors were arranged in the vertical and horizontal directions of the bearing for collecting vibration signals. Temperature and acceleration sensors were installed in the vertical and horizontal directions of the bearing to record the relevant data of bearing degradation. The sampling frequencies of the acceleration sensor and temperature sensor were 25.6 kHz and 10 kHz, respectively. The bearing type used in the PRONOSTIA test bench (Besançon, France) was a single volleyball bearing; the specific model was NSK 6804DD (Saint-Cyr-sur-Loire, France). The testbed was developed by the FEMTO-ST Institute (Besançon, France) specifically to test and validate bearing fault detection, diagnosis, and prediction methods. The working condition information related to the experimental bearing is shown in Table 2. The Bearing 1_1 and Bearing 2_1 life cycle horizontal vibration signals are shown in Figure 6. The trend of the Bearing 1_1 acceleration amplitude over its whole life cycle showed that the bearing had experienced a smooth running-in period from 1 s to 15,000 s, during which the bearing could operate normally, usually including the normal operation period and the slight degradation period. Between 15,000 s and 23,000 s, the bearing entered the initial degradation stage, and although the performance was reduced, it could still meet the needs of normal operation. After more than 23,000 s, the bearing entered a rapid descent stage, until it could not work at all.

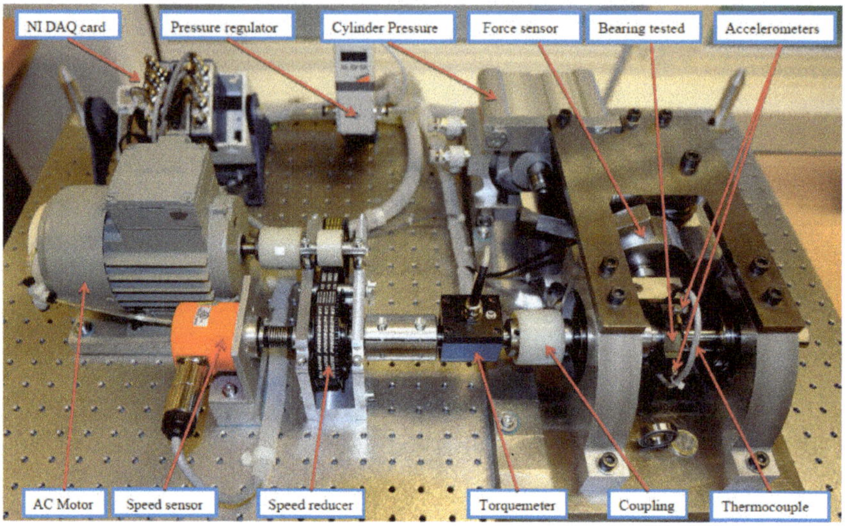

Figure 5. PRONOSTIA bearing life test platform.

Table 2. Test bearing condition information.

Condition	Load (N)	Rotational Speed (r/min)	Training Set	RUL (s)	Testing Set	RUL (s)
A	4000	1800	Bearing 1_1 Bearing 1_2	28,030 8710	Bearing 1_3 Bearing 1_4 Bearing 1_5 Bearing 1_6 Bearing 1_7	23,750 14,280 24,630 24,480 22,590
B	4200	1650	Bearing 2_1 Bearing 2_2	9110 7970	Bearing 2_3 Bearing 2_4 Bearing 2_5 Bearing 2_6 Bearing 2_7	19,550 7510 23,110 7010 2300
C	5000	1500	Bearing 3_1 Bearing 3_2	5150 16,370	Bearing 3_3	4340

3.2. Data Preparation

In order to control the input data, the collected vibration data were preprocessed. A total of 32 key features were extracted from the original data, as shown in Table 3. The horizontal feature visualization of bearing 1_1 is partly shown in Figure 7. F1 through F32 represent visualizations of the 32 different feature indicators in Table 3. In order to optimize the training speed, the features were normalized. Their value range was uniformly adjusted to [0,1] to effectively eliminate the amplitude difference between different features. Different time domain features had different sensitivities to various faults or their degradation degree. For example, kurtosis usually easily revealed early faults, while the root mean square value could reflect the long-term degradation degree. The use of time–frequency domain features allowed us to comprehensively consider information over the whole spectrum range, rather than relying only on a few specific frequency bands to avoid information loss. Trigonometric features provided more accurate trend analyses by converting input signals to different scales. Because the frequency resolution of vibration signal was low, this method did not involve the extraction of frequency domain features.

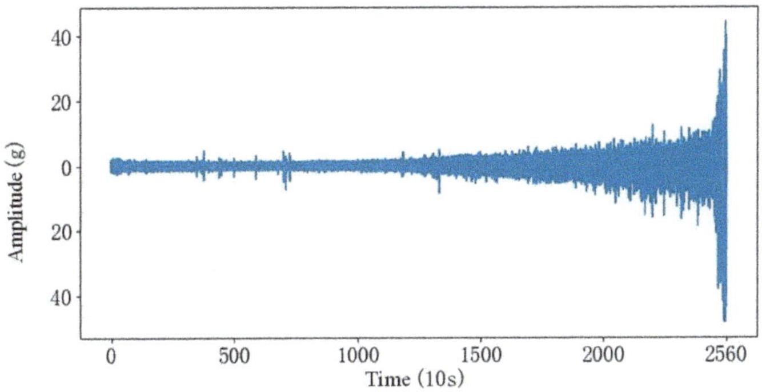

(a)

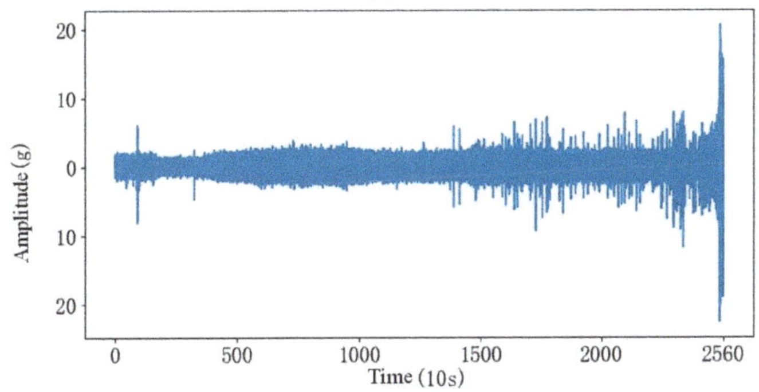

(b)

Figure 6. PRONOSTIA bearing life test platform: (**a**) Bearing 1_1 and (**b**) Bearing 1_1 horizontal life cycle vibration signals.

Table 3. Feature indicator.

Type	Characteristic		
Time domain feature	F1: Average F4: Maximum value F7: Skewness F10: Crest factor F13: Coefficient of variation	F2: Root mean square value F5: Minimum value F8: Kurtosis F11: Pulse factor	F3: Standard deviation F6: Peak–peak value F9: Waveform factor F12: Margin factor
Time–frequency domain characteristics	F14–F29: Sixteen-band energy		
Trigonometric function characteristics	F30: Inverse hyperbolic sine standard deviation	F31: Inverse hyperbolic cosine standard deviation	F32: Standard deviation of inverse hyperbolic tangent

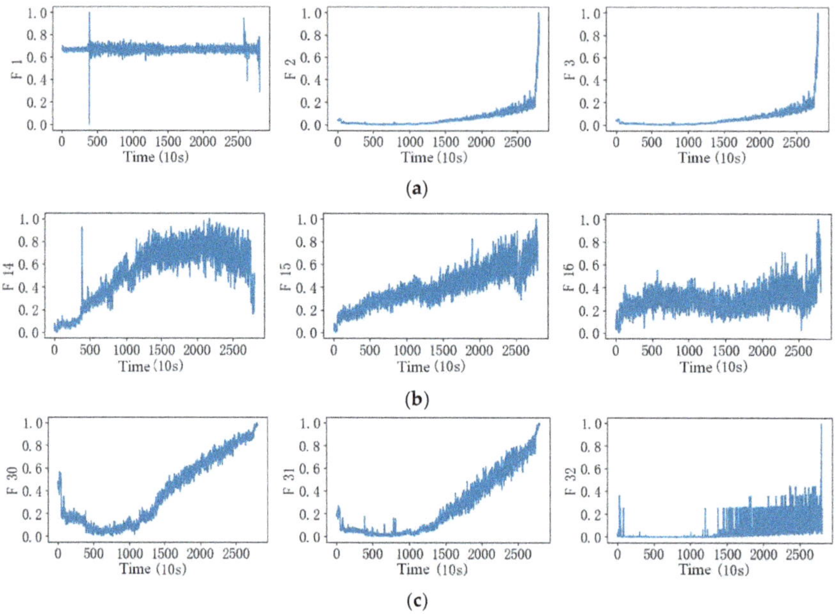

Figure 7. Bearing 1_1 character visualization chart in the horizontal direction: (**a**) time domain features; (**b**) time–frequency domain characteristics; and (**c**) trigonometric function characteristics.

Due to the small number of experimental data samples for testing bearings under Working Condition C in Table 2, this experiment selected bearing vibration data under two different operating conditions, Working Condition A and Working Condition B, dividing them into a training set, a verification set, and a test set. The specific division is shown in Table 4.

Table 4. Data set partitioning.

Experimental Group	Training Set	Verification Set	Test Set
S1	Bearing 1_1, Bearing 1_2, Bearing 1_3, Bearing 1_4, Bearing 1_6	Bearing 1_7	Bearing 1_5
S2	Bearing 2_1, Bearing 2_2, Bearing 2_3, Bearing 2_4, Bearing2_6	Bearing 2_7	Bearing 2_5

3.3. Data Preprocessing

The experiment was implemented using a Windows system with i7 11800H CPU and GTX 3060 GPU. The optimizer for model training was Adam, the learning rate was 0.001, and the loss function was MSE. The epoch was set to 30.

4. Experimental Results and Discussion

4.1. Prediction Effect of the Proposed Method

Figure 8 shows the remaining useful life prediction results for training sets Bearing 1_1 and Bearing 2_6 and test sets Bearing 1_5 and Bearing 2_5 in two experimental groups. The red curve in the figure represents the RUL prediction results of these bearings, and the blue line represents the actual RUL values. Because the accelerated degradation vibration data of these bearings were normalized to the range of 0 to 1 during network training, when the bearings started to run, their RUL was marked as 1; then, when the bearings ran to the end of failure, their RUL was marked as 0, and the RUL during operation was divided equidistantly. Therefore, the actual RUL value on the prediction result graph below appears as a diagonal line from 1 to 0. It can be observed from the figure that the predicted RUL value fluctuates around the actual RUL value, which has good monotonicity, fits the degradation trend of bearings well, and meets the basic requirements of prediction as a whole.

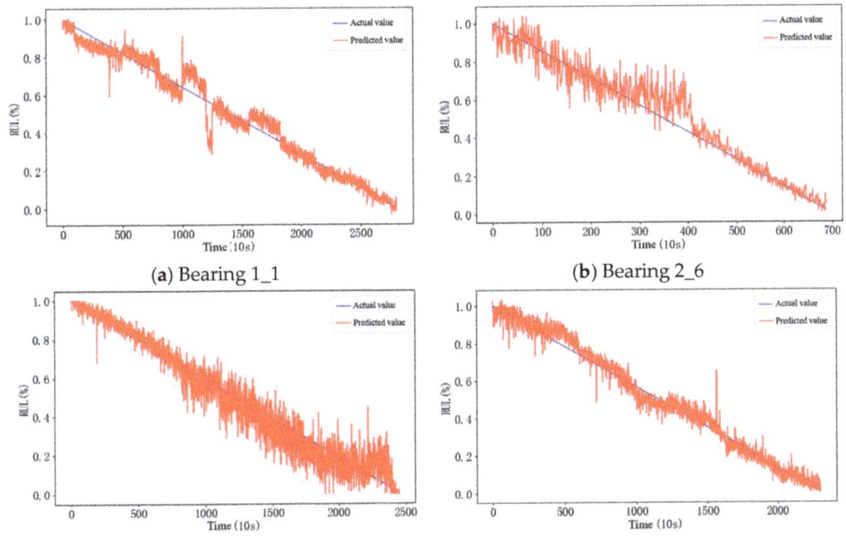

(**a**) Bearing 1_1 (**b**) Bearing 2_6 (**c**) Bearing 1_5 (**d**) Bearing 2_5

Figure 8. MITCN model's RUL prediction results.

In this study, the following indexes were used to evaluate the model quantitatively: root mean square error (RMSE), mean absolute error (MAE) and coefficient of determination (R). These indexes were helpful to quantitatively analyze the model's prediction accuracy of remaining service life. The results of the output test set's RMSE error, mean absolute error, and determination coefficient are shown in Table 5.

Table 5. Prediction results.

Experimental Group	RMSE	MAE	R^2
S1	0.0834	0.0542	0.9613
S2	0.1253	0.0917	0.9165

The experimental results showed that MITCN model performed well in the test set. In the S1 test set, the values of RMSE and MAE were small, and R was close to 1, which indicated that the model had a high degree of curve fitting. In the S1 experimental group, because the bearing ran for a long time under the actual load of 4000 N, more training data could be collected, which significantly enhanced the training effect of the model. On the other hand, the actual load of bearings in the S2 experimental group increased to 4200 N, which led to the decrease in the speed and running time, thus reducing the amount of data collected. Although the performance of the model in the S1 experimental group was reduced, the results in Table 4 show that the MITCN model trained using the S2 experimental group still had better test results.

4.2. Comparison Experiments with Other Methods

In order to further demonstrate the superiority of this model, it was compared with the TCN, MTCN (multi-scale TCN without the ITCN module), long short-term memory (LSTM) CNN, and transformer models. RMSE, MAE, and R2 were used as evaluation indexes, and Bearing1_5 of the S1 experimental group test set was used for comparative analysis among models. The other parameters were consistent, as shown in Table 6, outlining the model comparison results.

Table 6. Comparison results of different models.

Model	RMSE	MAE	R^2
TCN	0.1678	0.1058	0.9421
MTCN	0.1165	0.0989	0.9498
LSTM	0.2465	0.1963	0.9254
CNN-LSTM	0.2134	0.1574	0.9375
Transformer	0.1007	0.0859	0.8797
MITCN	**0.0834**	**0.0542**	**0.9613**

The MITCN model was the best under every performance index: the MAE and RMSE values of MITCN were the lowest, and the R^2 of MITCN was 0.9% higher than that of the transformer network, which fully shows that this model has certain advantages in predicting the remaining life of bearings.

5. Conclusions

In this study, the authors proposed a residual life prediction method called MITCN. MITCN could not only capture information at different time scales using a multi-scale extended causal convolution residual structure but also adaptively enhance useful feature information and suppress useless feature information through a soft threshold function and an attention mechanism. By paying attention to the correlation of time series, MITCN could deeply mine feature information in the whole life cycle signal and had excellent residual life prediction performance. The key features, including 13 time domain features, 16 time–frequency domain features, and 3 trigonometric function features, were extracted from the original data as inputs. The RUL prediction curve of the trained model on the test set fit well with the actual life curve in the degradation stage. The experimental results showed that the overall effect of MITCN on RMSE, MAE, and R2 was better than that of other comparative models, which proves the feasibility and superiority of MITCN in predicting the remaining service life of rolling bearings. Although MITCN achieved good results, further analysis is still needed in the case of large data sets and lifetime prediction of different research objects. In future research, the authors will adopt the method of data enhancement to solve the problem of residual life prediction with few data samples.

Author Contributions: Formal analysis, Q.L.; Investigation, Z.R.; Data curation, L.H.; Writing—original draft, H.L.; Writing—review & editing, K.X. All authors have read and agreed to the published version of the manuscript.

Funding: The project is supported by the Natural Science Foundation of China (No. 52275091), Natural Science Foundation of Liaoning Province (No. 2022-MS-125), Shenyang Natural Science Foundation (No. 23-503-6-02) and Fundamental Research Funds for the Central Universities (No. N2403005, No. N2303011).

Data Availability Statement: The data presented in this study are included in the article. Further inquiries can be directed to the corresponding author.

Conflicts of Interest: The authors declare no conflict of interest.

References

1. Tang, Y.F.; Xu, F.; Xu, D.L. Life Prediction of Rolling Bearing Based on Stacked Denoising Self-encoder. *Autom. Instrum.* **2024**, *39*, 124–130.
2. Eleonora, D.; Aleksandar, A.; Ljiljana, R.; Palinkaš, I.; Kiss, I. Selection, Dimensioning and Maintenance of Roller Bearings. In Proceedings of the 31st International Conference on Organization and Technology of Maintenance (OTO 2022), Osijek, Croatia, 12 December 2022; pp. 133–142.
3. Alexander, P.; Evgeny, T.; Damjan, S. Temperature Conditions and Diagnostics of Bearings. *Appl. Eng. Lett.* **2023**, *8*, 45–51.
4. Zhou, Y.G.; Zhang, J.C.; Sun, Y.P.; Yu, C.; Zhou, L. Prediction of Remaining Service Life of Rolling Bearings Based on TET and DSRNet-AttBiLSTM. *J. Vib. Shock* **2024**, *43*, 163–173.
5. Yuan, X.; Zhu, Y.S.; Zhang, Y.Y. Fault Diagnosis of Rolling Bearing Based on Model Identification. *Vib. Test Diagn.* **2013**, *33*, 12–17.
6. Liang, Y.; Jia, L.M.; Cai, G.Q. Research on Nonlinear Dynamic Fault Model of Rolling Bearing. *Chin. Railw. Sci.* **2014**, *35*, 98–104.
7. Tong, Z.Y.; Li, J.X.; Qiao, Z.R. Fault Diagnosis of Motor Bearing Based on Improved Bayesian Network. *Electron. Meas. Technol.* **2022**, *45*, 48–55.
8. Hoang, D.T.; Kang, H.J. A survey on deep learning based bearing fault diagnosis. *Neurocomputing* **2019**, *335*, 327–335. [CrossRef]
9. Cheng, C.; Ma, G.; Zhang, Y.; Sun, M.; Teng, F.; Ding, H.; Yuan, Y. A Deep Learning-Based Remaining Useful Life Prediction Approach for Bearings. *IEEE/ASME Trans. Mechatron.* **2020**, *25*, 1243–1254. [CrossRef]
10. Yu, W.; Kim, I.Y.; Mechefske, C. An Improved Similarity-Based Prognostic Algorithm for RUL Estimation Using an RNN Autoen Coder Scheme. *Reliab. Eng. Syst. Saf.* **2020**, *199*, 106926. [CrossRef]
11. Liu, J.; Lei, F.; Pan, C.; Hu, D.; Zuo, H. Prediction of Remaining Useful Life of Multi-Stage Aero-Engine Based on Clustering and LSTM Fusion. *Reliab. Eng. Syst. Saf.* **2021**, *214*, 107807. [CrossRef]
12. Li, X.; Jiang, H.; Liu, Y.; Wang, T.; Li, Z. An Integrated Deep Multiscale Feature Fusion Network for Aeroengine Remaining Useful Life Prediction with Multi-Sensor Data. *Knowl.-Based Syst.* **2022**, *235*, 107652. [CrossRef]
13. Yang, X.; Zheng, Y.; Zhang, Y.; Wong, D.S.-H.; Yang, W. Bearing Remaining Useful Life Prediction Based on Regression Shapalet and Graph Neural Network. *IEEE Trans. Instrum. Meas.* **2022**, *71*, 3505712. [CrossRef]
14. Yakout, M.; Nassef, M.G.A.; Backar, S. Effect of Clearances in Rolling Element Bearings on Their Dynamic Performance, Quality and Operating Life. *J. Mech. Sci. Technol.* **2019**, *33*, 2037–2042. [CrossRef]
15. Cui, L.L.; Wang, X.; Wang, H.Q.; Jiang, H. Remaining Useful Life Prediction of Rolling Element Bearings Based on Simulated Performance Degradation Dictionary. *Mech. Mach. Theory* **2020**, *153*, 103967. [CrossRef]
16. Wang, F.; Tang, S.; Sun, X.; Li, L.; Yu, C.; Si, X. Remaining Useful Life Prediction Based on Nonlinear Random Coefficient Regression Model with Fusing Failure Time Data. *J. Syst. Eng. Electron.* **2023**, *34*, 247–258. [CrossRef]
17. Lei, Y.; Li, N.; Gontarz, S.; Lin, J.; Radkowski, S.; Dybala, J. A Model-Based Method for Remaining Useful life Prediction of Machinery. *IEEE Trans. Reliab.* **2016**, *65*, 1314–1326. [CrossRef]
18. Guo, Y.R.; Mao, J.; Zhao, M. Prediction Method of Bearing Remaining Useful Life Based on CNN and Attention BiLSTM. *J. Shanghai Univ. Eng. Sci.* **2023**, *37*, 96–104.
19. Azyus, A.F.; Wijaya, S.K.; Naved, M. Prediction of Remaining Useful Life Using the CNN-GRU Network: A study on Maintenance Management. *Softw. Impacts* **2023**, *17*, 100535. [CrossRef]
20. Qiu, P.Z.; Wu, D.R.; Xia, Y. Review on The Application of Recurrent Neural Networks in The Prediction of Remaining Service Life. *Mod. Inf. Technol.* **2023**, *7*, 61–66.
21. Sloukia, F.; El Aroussi, M.; Medromi, H.; Wahbi, M. Bearings Prognostic Using Mixture of Gaussians Hidden Markov Model and Support Vector Machine. In Proceedings of the 2013 ACS International Conference on Computer Systems and Applications (AICCSA), Ifrane, Morocco, 27–30 May 2013; pp. 1–4.
22. Yan, M.; Wang, X.; Wang, B.; Chang, M.; Muhammad, I. Bearing Remaining Useful Life Prediction Using Support Vector Machine and Hybrid Degradation Tracking Model. *ISA Trans.* **2020**, *98*, 471–482. [CrossRef] [PubMed]
23. Rohani, B.A.; Aasi, A.; Arghand, H.A. Estimation of Remaining Useful Life of Rolling Element Bearings Using Wavelet Packet Decomposition and Artificial Neural Network. *Iran. J. Sci. Technol. Trans. Electr. Eng.* **2019**, *43*, 233–245. [CrossRef]
24. Zhang, J.; Tian, J.; Li, M.; Leon, J.I.; Franquelo, L.G.; Luo, H.; Yin, S. A Parallel Hybrid Neural Network with Integration of Spatial and Temporal Features for Remaining Useful Life Prediction in Prognostics. *IEEE Trans. Instrum. Meas.* **2022**, *72*, 3501112. [CrossRef]
25. Guo, J.; Wang, J.; Wang, Z.; Gong, Y.; Qi, J.; Wang, G.; Tang, C. A CNN-BiLSTM-Bootstrap Integrated Method for Remaining Useful Life Prediction of Rolling Bearings. *Qual. Reliab. Eng. Int.* **2023**, *39*, 1796–1813. [CrossRef]
26. Wang, R.; Shi, R.; Hu, X.; Shen, C. Remaining Useful Life Prediction of Rolling Bearings Based on Multiscale Convolutional Neural Network with Integrated Dilated Convolution Blocks. *Shock Vib.* **2021**, *2021*, 6616861. [CrossRef]
27. Guo, D.; Cao, Z.; Fu, H.; Li, Z. Remaining Useful Life Estimation for Rolling Bearings Using MSGCNN-TR. *IEEE Sens. J.* **2022**, *22*, 24333–24343. [CrossRef]

28. Jiang, F.; Ding, K.; He, G.; Lin, H.; Chen, Z.; Li, W. Dual-attention-based Multiscale Convolutional Neural Network with Stage Division for Remaining Useful Life Prediction of Rolling Bearings. *IEEE Trans. Instrum. Meas.* **2022**, *71*, 3525410. [CrossRef]
29. Bai, S.; Kolter, J.Z.; Koltun, V. An Empirical Evaluation of Generic Convolutional and Recurrent Networks for Sequence Modeling. *arXiv* **2018**, arXiv:1803.01271.
30. He, K.; Su, Z.Q.; Tian, X.Q.; Yu, H.; Luo, M. RUL Prediction of Wind Turbine Gearbox Bearings Based on Self-Calibration Temporal Convolutional Network. *IEEE Trans. Instrum. Meas.* **2022**, *71*, 3501912. [CrossRef]
31. Deng, F.; Bi, Y.; Liu, Y.; Yang, S. Remaining Useful Life Prediction of Machinery: A New Multiscale Temporal Convolutional Network Framework. *IEEE Trans. Instrum. Meas.* **2022**, *72*, 2516913. [CrossRef]
32. Li, C.B.; Sun, X.; Hou, X.B. On-line monitoring method of tool wear in CNC milling driven by digital twins. *China Mech. Eng.* **2022**, *33*, 78–87. [CrossRef]
33. Yu, Y.; Ma, J.; Wang, X.D.; Zhu, J.; Liu, G. Improved CNN Fault Diagnosis Model Based on Depth Separable Small Convolution Kernel and CBAM. *Electron. Meas. Technol.* **2022**, *45*, 171–178.
34. Woo, S.H.; Park, J.; Lee, J.Y.; Kweon, I.S. CBAM: Convolutional Block Attention Module. In Proceedings of the European Conference on Computer Vision (ECCV) 2018, Munich, Germany, 8–14 September 2018; Volume 11211, pp. 3–19.
35. Benkedjouh, T.; Medjaher, K.; Zerhouni, N.; Rechak, S. Remaining Useful Life Estimation Based on Nonlinear Feature Reduction and Support Vector Regression. *Eng. Appl. Artif. Intell.* **2013**, *26*, 1751–1760. [CrossRef]

Journal of
Manufacturing and Materials Processing

MDPI

Article

Plasma Dielectric Etching with $C_4H_2F_6$ Isomers of Low Global-Warming Potential

Minsu Choi [1], Youngseok Lee [2], Chulhee Cho [1], Wonnyoung Jeong [1], Inho Seong [1], Jami Md Ehsanul Haque [1], Byeongyeop Choi [1], Seonghyun Seo [1], Sijun Kim [2], Shinjae You [1,2,*] and Geun Young Yeom [3,4]

1 Department of Physics, Chungnam National University, Daejeon 34134, Republic of Korea; bss125576@naver.com (M.C.); paulati@o.cnu.ac.kr (C.C.); jwn1748@o.cnu.ac.kr (W.J.); showing123@o.cnu.ac.kr (I.S.); jami@o.cnu.ac.kr (J.M.E.H.); choibyeongyeop@o.cnu.ac.kr (B.C.); tnqmsqnzo11@o.cnu.ac.kr (S.S.)
2 Institute of Quantum Systems (IQS), Chungnam National University, Daejeon 34134, Republic of Korea; yslee92@cnu.ac.kr (Y.L.); sj_kim@cnu.ac.kr (S.K.)
3 Department of Advanced Materials Science and Engineering, Sungkyunkwan University, Suwon 16419, Republic of Korea; gyyeom@skku.edu
4 SKKU Advanced Institute of Nano Technology (SAINT), Sungkyunkwan University, Suwon 16419, Republic of Korea
* Correspondence: sjyou@cnu.ac.kr

Abstract: This paper presents the observed changes when replacing the widely used CHF_3 gas in semiconductor processing with two isomeric gases, $C_4H_2F_6$-iso and $C_4H_2F_6$-Z. This study investigates the etching process results of SiO_2 and Si_3N_4 by varying the ratios of CHF_3, $C_4H_2F_6$ gases. The process outcomes were analyzed using ellipsometer and X-ray photoelectron spectroscopy, while plasma radical densities were examined through quadrupole mass spectrometry. The results are compared across five conditions, with substitution gas ratios ranging from 0% to 100%. The process results indicated that the selectivity increased at certain gas ratios. The diagnostic results provided the ratios of various etchants within the plasma. This research advances the development of alternative precursors designed to mitigate the effects of global warming.

Keywords: plasma etching; global-warming potential; alternative precursors; plasma diagnostics; surface chemistry analysis

1. Introduction

The semiconductor industry has widely adopted plasma technology for the manufacturing of semiconductor devices [1,2]. Plasma has become indispensable in the etching process for nanoscale material processing, enabling precise and deep etching profiles to accommodate the shrinking dimensions of transistor features [3]. Consequently, plasma processing has become crucial for fabricating next-generation semiconductor devices, particularly for creating high-aspect-ratio contact (HARC) holes.

Plasma etching employs a variety of gases or gas mixtures customized for specific target materials, including semiconductors [4,5], metals [6,7], and dielectrics [8–10]. Fluorocarbon-based molecules like C_4F_8 and CHF_3 are commonly utilized for dielectrics such as SiO_2 and Si_3N_4 due to their polymerizing characteristics, which are especially advantageous for achieving anisotropic etch profiles [11–14]. During the etching process, fluorocarbon films form a passivation layer on the sidewalls of the dielectric trench, shielding them from reactive etchants. In contrast, the energetic bombardment of plasma ions removes the fluorocarbon film from the trench bottom, enabling etching to proceed and resulting in an anisotropic etch profile [15–17]. Given the importance of anisotropy in semiconductor manufacturing, the semiconductor industry extensively employs significant amounts of perfluorocarbon (PFC) and hydrofluorocarbon (HFC) precursors, such as C_4F_8 and CHF_3.

However, the use of PFC and HFC precursors is strictly regulated due to their substantial contribution to greenhouse gas emissions [18,19]. Global-warming potential (GWP) is a metric that quantifies the relative warming effect of greenhouse gases (GHGs) on Earth's climate over a defined time horizon, typically 20, 100, or 500 years, in comparison to CO_2 [20]. The GWP100 (global-warming potential over a 100-year period) for C_4F_8 and CHF3 is reported to be 10,592 and 12,400, respectively, highlighting that these precursors possess a greenhouse effect more than 10,000 times greater than that of CO_2 [21]. Amid significant global efforts to combat global warming in recent decades [22,23], the emission of these high-GWP gases is increasingly subject to stringent industry regulations.

In this context, researchers worldwide have been actively engaged in developing alternative plasma etching precursors with low global-warming potential (GWP) that deliver an etch performance comparable to that of conventional precursors [24]. Sung et al. investigated the etching characteristics of SiO_2 using a $C_6F_6/Ar/O_2$ gas mixture in both capacitively coupled and inductively coupled plasma systems, comparing the etching outcomes between the two plasma configurations [25]. Kim et al. examined the etching of SiO_2 using $C_4H_3F_7O$ isomers and compared their performance to that of C_4F_8, employing X-ray photoelectron spectroscopy, quadrupole mass spectroscopy, and scanning electron microscopy for analysis [21].

In general, the etching capability of certain fluorocarbon (FC) precursors can be simply predicted by the ratio of fluorine atoms to carbon atoms in their chemical structure, commonly referred to as the F/C ratio. A high F/C ratio typically indicates that the precursor exhibits a rapid etch rate and a reduced deposition rate, whereas a low F/C ratio suggests the reverse, characterized by a slower etch rate and an increased deposition rate. However, it has been observed that isomers of some FC precursors can exhibit different deposition or etch rates depending on their chemical structures [26]. Isomers, which are stoichiometrically identical, have variations in their molecular structures, such as double bonds, triple bonds, or rings. These differences result in varied patterns of electron–impact dissociation, leading to distinct radical compositions in the plasma or surface chemistry of the deposited FC films. Therefore, even a single type of FC precursor with multiple isomers can produce diverse etching outcomes.

In this study, we explore the etching capabilities of $C_4H_2F_6$ isomers, specifically $C_4H_2F_6$-iso and $C_4H_2F_6$-Z, for SiO_2 and Si_3N_4, as alternatives to the traditional precursor CHF_3, which is a leading etching gas for dielectrics [3]. We investigate the feasibility of substituting CHF_3 with $C_4H_2F_6$ precursors, known for their lower global-warming potential (GWP), by blending an isomer with CHF_3. The etch rate and selectivity are evaluated based on the mixing ratio of CHF_3 to $C_4H_2F_6$ isomers. Ellipsometry and plasma diagnostics are employed to assess the etching outcomes. The results present a comprehensive analysis of etch performance, focusing on etch rate and selectivity. The experimental setup and procedures are described in detail in the following sections.

2. Experiment

2.1. Plasma Processing

Figure 1 shows a schematic representation of the plasma chamber used for etching SiO_2 and Si_3N_4. A radiofrequency (RF) power of 400 W is delivered to the electrode, which has a diameter of 208 mm and is coated with a 2 mm thick ceramic plate, through an impedance matching box. Argon (Ar) is introduced into the chamber through a 1/4"(diameter of

42.7 mm) stainless steel (SUS) tube, with the flow rate controlled by a mass flow controller (MFC) (1179A611522K, MKS Instruments Inc., Andover, MA, USA), while the flow rate of HFC precursors is managed by a different MFC model (M3030VA, Line-Tech, Daejeon, Korea). The Ar (99.999%) flow rate is set at 20 sccm, while the total flow rate of HFC precursors (exceeds 99.5%) is maintained at 10 sccm across all gas mixing ratios, while the chamber pressure is kept at 20 mTorr by the throttle valve. The gas mixing ratio is defined as the flow rate of a $C_4H_2F_6$ isomer divided by the combined flow rates of CHF_3 and the $C_4H_2F_6$ isomer; thus, the ratio is 0% when no $C_4H_2F_6$ isomer is used and 100% when CHF_3 is absent. The chamber is evacuated to a base pressure below 10^{-4} Torr using a turbomolecular pump backed by an oil rotary pump. Notably, the canisters containing C4H2F6 isomers are heated to 130 °C to ensure their vaporization, as these compounds remain in a liquid state at room temperature.

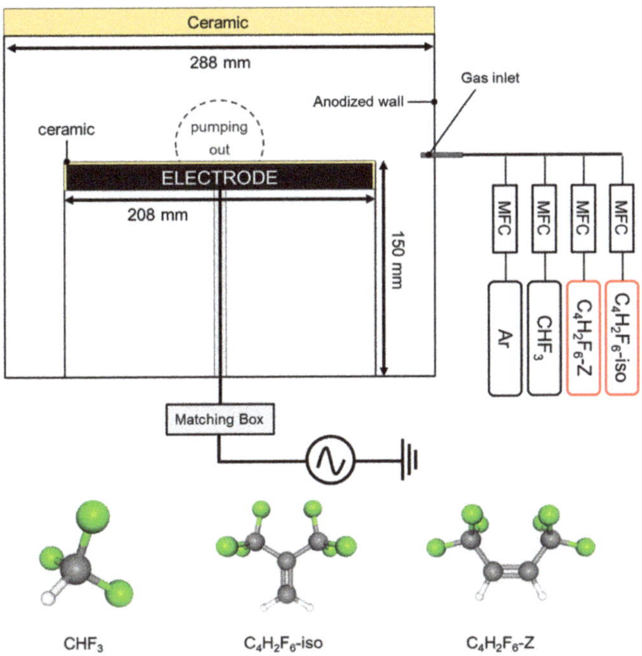

Name	Molecular Formula	CAS	Atmospheric Lifetime (Years)	GWP_{100}
Trifluoromethane	CHF_3	75-46-7	222	12,400
Hexafluoro-**iso**butylene	$C_4H_2F_6$	382-10-5	0.03	0.377
Z-1,1,1,4,4,4-Hexafluorobu-2-ene	$C_4H_2F_6$	692-49-9	0.07	2.08

Figure 1. Schematic of the experimental setup and gas information.

2.2. Plasma Diagnostics

2.2.1. Wave Cutoff Method

Microwave-based plasma diagnostic techniques have been extensively employed, particularly for reactive plasmas in semiconductor-manufacturing processes. A key advantage of microwave diagnostics is their minimal susceptibility to contamination from polymer deposition in processing plasmas, as well as their resistance to RF noise, which often affects other electrical diagnostic tools like the Langmuir probe. Among the various microwave diagnostic tools developed over the years, the cutoff probe was chosen for this study because of its simplicity in both fabrication and operation.

The cutoff probe operates based on the principle of wave cutoff in plasma, which dictates that microwaves with frequencies lower than the plasma frequency are attenuated, whereas those with frequencies higher than the plasma frequency can propagate through the plasma [27]. Because the plasma frequency is dependent on electron density, an S_{21} spectrum exhibiting cutoff behavior across a wide range of microwave frequencies can be used to determine the electron density from the cutoff frequency [28–30].

A cutoff probe typically consists of two coaxial cables and a signal generator, excluding the vacuum components. These two cables function as the transmitting and receiving antennas. In this study, the antennas were fabricated by stripping approximately 10 mm from one end of two sub-miniature version A (SMA) cables without connectors. These cables were spaced about 3 mm apart to allow plasma to fill the gap between them. A vector network analyzer (S33601B, SALUKI TECHNOLOGY, Taipei, Taiwan) was employed as a microwave signal generator, capable of varying frequencies from hundreds of kHz to 8.5 GHz.

2.2.2. Quadrupole Mass Spectroscopy

Quadrupole mass spectrometers (QMSs), (RGA-200, SRS Instruments Inc., Sunnyvale, CA, USA) are commonly used tools for plasma diagnostics, providing insights into the gas-phase species present in plasmas. In a QMS, gas-phase atoms or molecules are ionized through collisions with 70 eV energetic electrons. The resulting ions are then separated based on their mass as they traverse a quadrupole mass filter operating. This separation process enables the QMS to generate a signal intensity profile corresponding to the atomic mass of the ions [31–33].

The QMS is mounted on a chamber port and features its own dedicated vacuum space, separated from the main chamber by an orifice approximately 100 μm in diameter. This configuration maintains the QMS pressure at roughly two orders of magnitude lower than that of the main chamber. By minimizing collisions with background gas, this dedicated vacuum system ensures that gas-phase atoms or molecules pass through the QMS with minimal interference, thereby improving the reliability of the data.

2.3. *Sample Measurement Method and Analysis*

2.3.1. Dielectric Thickness Measurement

To measure the variation in dielectric thickness, an ellipsometer (FS-1, Film-Sense Instruments Inc., Lincoln, NE, USA) was utilized. An ellipsometer was employed to measure changes in dielectric thickness due to its high precision and non-destructive nature. This optical technique allows for accurate thickness determination at the nanoscale, making it ideal for sensitive material analysis. By measuring changes in the polarization of reflected light, it provides detailed information about thin film properties [34].

2.3.2. Dielectric Surface Analysis

X-ray photoelectron spectroscopy (XPS) (NEXSA G2, Thermo Hisher Scientific Instruments Inc., Waltham, MA, USA) was employed for surface analysis of the dielectric materials. This technique offers high surface sensitivity, allowing for the identification of elemental composition and chemical states. Additionally, XPS provides quantitative information, which is essential for understanding the material properties and surface modifications [35].

3. Results and Discussion

3.1. *SiO_2 and Si_3N_4 Etching with CHF_3 and $C_4H_2F_6$-Iso Gas Mixtures*

Figure 2 displays the etching results achieved with various gas mixing ratios of CHF_3 and $C_4H_2F_6$-iso at a constant 100 W RF power under 20 mTorr. Figure 2a shows the thickness of SiO_2 before and after etching with different gas mixing ratios, with error bars indicating the standard deviation from measurements of five different samples etched simultaneously. Figure 2b illustrates the thickness changes for Si3N4. When comparing the etched thicknesses of SiO_2 and Si_3N_4, it is evident that Si_3N_4 is etched significantly more than SiO_2, which aligns with previous findings. However, it is noteworthy that at a gas mixing ratio of 100%, where only $C_4H_2F_6$-iso is used, the etched amounts of SiO_2 and Si_3N_4 are nearly identical, indicating poor etch selectivity of SiO_2 over Si_3N_4.

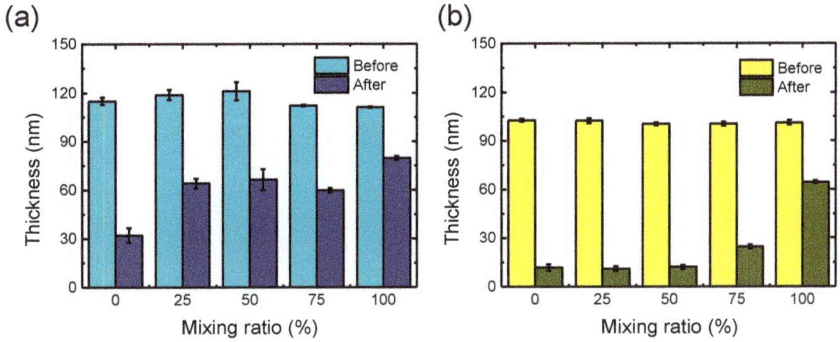

Figure 2. Sample thicknesses before and after etching with different gas mixing ratios between CHF_3 and $C_4H_2F_6$-iso for (**a**) SiO_2 and (**b**) Si_3N_4.

The data in Figure 2a,b were processed to offer a clearer understanding of the etching outcomes using general etching parameters such as etch rate and selectivity, as depicted in Figure 3. The etch rate was determined by dividing the change in thickness of SiO_2 or Si_3N_4, measured after etching with different gas mixing ratios, by the process duration of 1 min. The error bars indicate the standard deviation of the etch rates calculated from five different samples etched simultaneously. Selectivity is defined as the etch rate ratio of Si_3N_4 to SiO_2 at the given gas mixing ratio, assessing the etch performance of H-containing etch precursors typically used for Si_3N_4 etching. This selectivity can be altered by varying conditions, such as by incorporating oxygen [36,37].

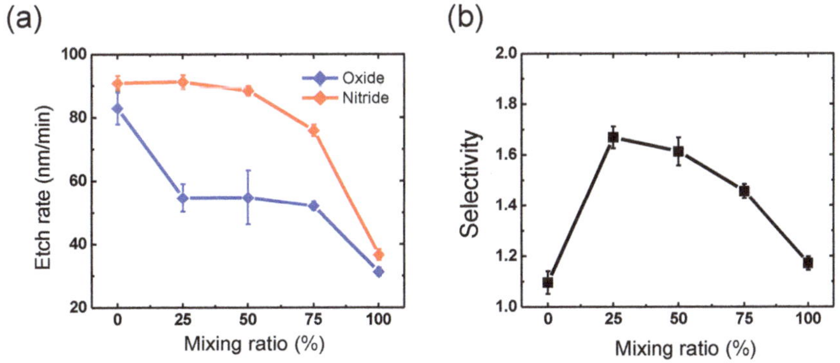

Figure 3. (**a**) Etch rate of SiO_2 and Si_3N_4 and (**b**) etch selectivity of Si_3N_4 over SiO_2 with different gas mixing ratios between CHF_3 and $C_4H_2F_6$-iso.

Figure 3a illustrates the etch rates of SiO_2 and Si_3N_4 across different gas mixing ratios of CHF_3 and $C_4H_2F_6$-iso. When only CHF_3 is used (0% mixing ratio), the etch rates of SiO_2 and Si_3N_4 are quite similar. However, as the mixing ratio increases to 75%, the etch rate of

SiO_2 significantly decreases at the 25% mixing ratio and then stabilizes, while the etch rate of Si_3N_4 declines more gradually. At a 100% mixing ratio, where CHF_3 is fully replaced by $C_4H_2F_6$-iso, the etch rates of both SiO_2 and Si_3N_4 decrease to similarly low levels.

Figure 3b depicts the relative dependence of the SiO_2 and Si_3N_4 etch rates on the mixing ratio. With only CHF_3, selectivity is slightly above one, indicating that the etch rates of SiO_2 and Si_3N_4 are very similar. As shown in Figure 3a, the SiO_2 etch rate changes more dynamically with varying mixing ratios compared to Si_3N_4. This is reflected in Figure 3b as a spike in selectivity with a small replacement of CHF_3 by $C_4H_2F_6$-iso, followed by a gradual decrease as the mixing ratio increases. When CHF_3 is fully replaced by $C_4H_2F_6$-iso, selectivity decreases further, reaching a level similar to the CHF_3-only case. These findings suggest that even a small replacement of traditional precursors with low GWP alternatives could significantly enhance the etching process and greatly reduce GHG emissions. The underlying mechanisms for such improvements will be discussed later with plasma diagnostics and surface analysis results.

Changes in plasma parameters with varying gas mixing ratios were monitored using a cutoff probe and a QMS, which are used to measure electron density and radical species density, respectively. As depicted in Figure 4a, the electron density slightly increases as the mixing ratio rises to 75%, and then jumps to approximately three times its original value at a 100% mixing ratio. Given the etching results show a rapid decrease in the etch rates of both SiO_2 and Si_3N_4 at this mixing ratio, the increase in electron density likely causes intense dissociation of precursors, resulting in low etch rates for both SiO_2 and Si_3N_4.

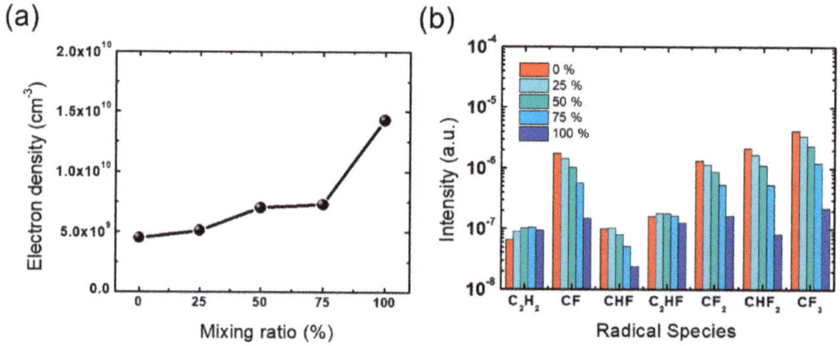

Figure 4. (**a**) Electron density and (**b**) radical species densities with different gas mixing ratios between CHF_3 and $C_4H_2F_6$-iso.

Radical density measurements, shown in Figure 4b, support this hypothesis. As the mixing ratio increases, the densities of major radical species such as CF, CF_2, and CHF_3 gradually decrease. However, it is noteworthy that certain radical species containing two carbon atoms in their molecular structure initially increase and then decrease as the mixing ratio rises. This suggests that these carbon-rich radical species may not significantly contribute to dielectric etching.

We performed XPS analysis on dielectric surfaces to examine changes in surface chemistry related to the fluorocarbon (FC) films deposited on the surfaces after etching, aiming to gain insights into the varying etch rate and selectivity of SiO_2 and Si_3N_4 depending on the gas mixing ratio. Figures 5a and 5b display the XPS spectra of C1s obtained from SiO_2 and Si_3N_4 surfaces, respectively, across different mixing ratios. For comparison, XPS analysis was also carried out on unetched reference samples, and these results are included alongside those from etched samples, labeled "ref." in Figure 5. The C1s XPS spectrum includes the CF peak (289 eV) [38], the CFx peak (286.2 eV) [38], and the C-C/C-F peak (284.8 eV) [39].

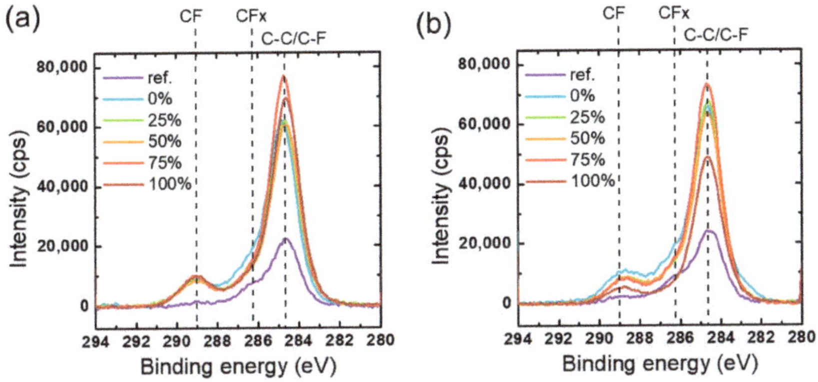

Figure 5. XPS analysis results of C1s spectrum on (**a**) SiO_2 and (**b**) Si_3N_4 surfaces with different gas mixing ratios between CHF_3 and $C_4H_2F_6$-iso.

It is evident from Figure 5a,b that post-etching, regardless of the mixing ratio, there is a substantial increase in the intensity of the CF and C-C/C-F peaks. However, while the increases in peak intensities appear similar across different mixing ratios for SiO_2, they show noticeable variation for Si_3N_4 under different experimental conditions. This observation aligns with the etch rate changes shown in Figure 3, where the etch rate of SiO_2 quickly altered and then stabilized, whereas that of Si_3N_4 declined more gradually and slowly. This could be attributed to the termination of deposited FC films by hydrogen atoms, which may impede the saturation of FC film formation. Further detailed analysis of the relationship between surface chemistry and etching outcomes is planned for future research.

3.2. SiO_2 and Si_3N_4 Etching with CHF_3 and $C_4H_2F_6$-Z Gas Mixtures

Figure 6 illustrates the etching results obtained with various gas mixing ratios between CHF_3 and $C_4H_2F_6$-Z, maintaining a constant RF power of 100 W at a pressure of 20 mTorr. Figure 6a displays the thickness of SiO_2 before and after etching with different gas mixing ratios, while Figure 6b shows the thickness changes for Si_3N_4. When comparing the etched thickness of SiO_2 and Si_3N_4, it is evident that the overall amount etched is significantly greater for Si_3N_4 than for SiO_2, except at a mixing ratio of 25%, where the etched amount of Si_3N_4 is notably lower compared to other conditions. This could be due to an abrupt and unexpected change in the plasma conditions.

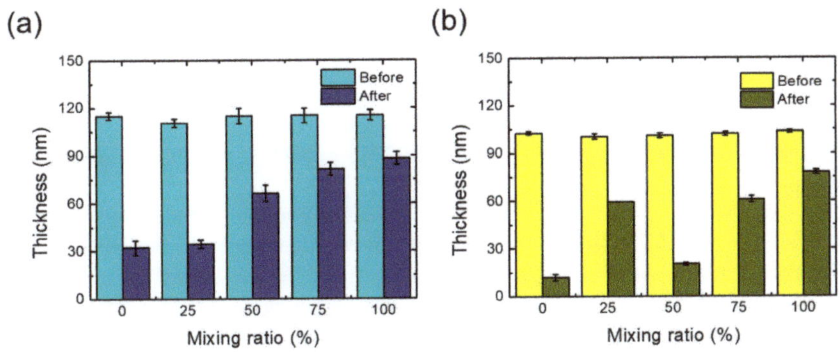

Figure 6. Sample thicknesses before and after etching with different gas mixing ratios between CHF_3 and $C_4H_2F_6$-Z for (**a**) SiO_2 and (**b**) Si_3N_4.

Figure 7a illustrates the varying etch rates of SiO_2 and Si_3N_4 based on the gas mixing ratio between CHF_3 and $C_4H_2F_6$-Z. When only CHF_3 is used, corresponding to a gas mixing ratio of 0%, the etch rates for SiO_2 and Si_3N_4 are quite similar. However, as the mixing ratio increases up to 100%, the etch rate of SiO_2 gradually declines, whereas the etch rate of Si_3N_4 undergoes significant changes, with a notable drop at a 25% mixing ratio, as observed in Figure 6b. When CHF_3 is completely replaced by $C_4H_2F_6$-Z, corresponding to a 100% mixing ratio, the etch rates for both SiO_2 and Si_3N_4 become significantly low and nearly equal.

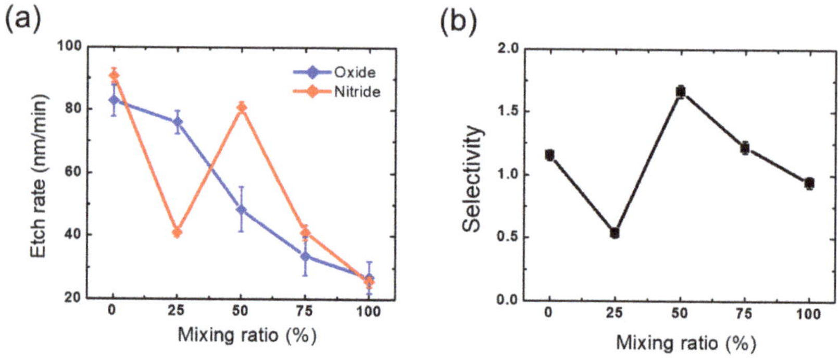

Figure 7. (**a**) Etch rate of SiO_2 and Si_3N_4 and (**b**) etch selectivity of Si_3N_4 over SiO_2 with different gas mixing ratios between CHF_3 and $C_4H_2F_6$-Z.

These relative dependencies of the etch rates of SiO_2 and Si_3N_4 on the mixing ratio are clearly depicted in Figure 7b. With only CHF_3, the selectivity is slightly above unity, indicating that the etch rates for SiO_2 and Si_3N_4 are very similar. As shown in Figure 7a, the change in SiO_2 etch rates with varying mixing ratios is more pronounced than that of Si_3N_4. This is reflected in Figure 7b, where the selectivity initially increases and then decreases towards unity as CHF_3 is progressively replaced with $C_4H_2F_6$-Z. It should be noted that the unusual behavior of the Si_3N_4 etch rate and the corresponding peculiar selectivity at the 25% mixing ratio were excluded from this discussion, as this result seems to require more thorough experiments to verify its reproducibility. The authors intend to address this analysis in future work.

Figure 8 illustrates the changes in plasma parameters with variations in the gas mixing ratio, as measured by the cutoff probe and QMS. As depicted in Figure 8a, the electron density increases slightly with a rise in the mixing ratio up to 75%, then jumps to approximately three times its original value at a 100% mixing ratio, similar to the behavior observed with $C_4H_2F_6$-iso. Notably, at a 25% mixing ratio, no plausible explanation for the abrupt decrease in the Si_3N_4 etch rate can be found in the electron density measurements.

Radical density measurements, shown in Figure 8b, indicate a steady decline in the densities of major radical species such as CF, CF_2, and CHF_3. Notably, the density of carbon-rich radicals remains high at a 100% mixing ratio, likely due to the absence of CHF_3 and the abundance of $C_4H_2F_6$-Z, which serves as a source of carbon-rich radicals. As observed in the etching results in Section 3.1, carbon-rich radicals do not appear to be effective in etching either SiO_2 or Si_3N_4.

As depicted in Figure 9a,b, after etching with any mixing ratio, there is a substantial increase in the intensity of the CF peak and the C-C/C-F peak. Notably, the spectrum for Si_3N_4 at a 25% mixing ratio is significantly different from the other spectra, which aligns with the abnormal etching results observed in Figures 6b and 7a,b. This suggests that the etch rate may be primarily influenced by surface chemistry rather than the plasma itself.

This finding is consistent with previous research indicating that surface chemistry plays a more dominant role in etching than gas-phase radical atoms or molecules [26].

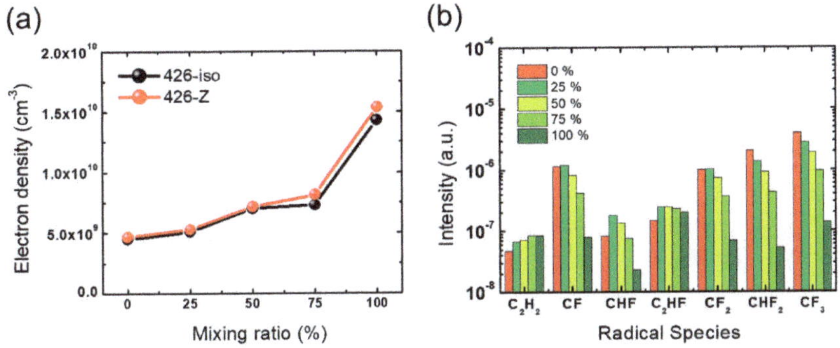

Figure 8. (**a**) Electron density and (**b**) radical species densities with different gas mixing ratios between CHF_3 and $C_4H_2F_6$-Z.

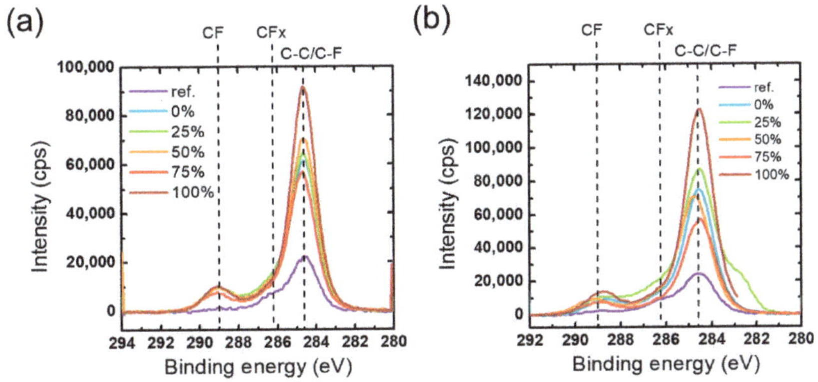

Figure 9. XPS analysis results of C1s spectrum on (**a**) SiO_2 and (**b**) Si_3N_4 surfaces with different gas mixing ratios between CHF_3 and $C_4H_2F_6$-Z.

4. Concluding Remarks

In this study, we investigated the feasibility of replacing the conventional precursor, CHF_3, with alternative low-GWP precursors, specifically $C_4H_2F_6$ isomers, in dielectric plasma etching. Considering the substantially lower GWP of these alternative gases, it is essential to determine whether they can deliver an etch performance comparable to that of conventional precursors. To evaluate their performance, we carried out etching experiments using gas mixtures of both conventional and alternative gases under fixed RF power and pressure conditions. The findings of this study demonstrated that the isomers, $C_4H_2F_6$-iso and $C_4H_2F_6$-Z, exhibited similar etch performance for SiO_2 and Si_3N_4, although their etch rates were lower than those of CHF_3. Notably, small amounts of mixtures of the isomers with CHF_3 resulted in increased etch selectivity, which is crucial for advanced semiconductor device fabrication, such as HARC processes. These findings are anticipated to promote the broader adoption of low-GWP gases in semiconductor manufacturing. Integrating these alternative gases into etching processes offers a promising pathway for the industry to significantly reduce its environmental impact while maintaining the required etch performance. Additionally, such adoption aligns with global efforts to achieve sustainability goals within high-tech industries. Future research should prioritize optimizing the operational parameters for low-GWP gases and addressing technical challenges to facilitate their seamless integration into commercial semiconductor fabrication processes on a larger scale.

5. Discussion

In this study, the results of the etching process were obtained using an ellipsometer, as previously described. The ellipsometer measures refractive index and thickness based on specific models; however, the accuracy of these models can vary depending on the surface structure, morphology, and elemental composition of the etched dielectric layer. For example, in Figure 7, the etching rate of nitride at a 25% mixing ratio showed a noticeable decline, which we attribute to potential model inaccuracies in the ellipsometer [40]. We found that under these conditions, the accuracy dropped significantly. This means that the accuracy of the model is reduced due to changes in the chemical composition of the surface. In addition, this observation may suggest that C4H2F6-Z and C4H2F6-iso are not entirely identical. This was confirmed through chemical measurement results in previous studies [37]. Therefore, further verification using more precise measurement equipment would be beneficial in future studies.

Author Contributions: Conceptualization, M.C., Y.L., C.C., W.J., I.S., J.M.E.H., B.C., S.S. and S.K.; validation, M.C. and Y.L.; formal analysis, M.C. and Y.L., investigation, M.C.; resources, G.Y.Y.; writing—original draft preparation, Y.L.; writing—review and editing, S.Y.; supervision, S.Y.; funding acquisition, M.C. All authors have read and agreed to the published version of the manuscript.

Funding: This research was supported by Basic Science Research Program through the National Research Foundation of Korea (NRF) funded by the Ministry of Education (RS-2020-NR049597), by a Korea Institute of Machinery & Materials (KIMM) institutional program (NK254B), by the SEMES research program [Research of semiconductor plasma arcing mechanism], by the Technology Innovation Program (or Industrial Strategic Technology Development Program-Public-private joint investment semiconductor R&D program (K-CHIPS) to foster high-quality human resources) ("RS-2023-00236642", Development of etching process and diagnostic technology using precursor with low GWP for next generation semiconductor process) funded By the Ministry of Trade, Industry & Energy (MOTIE, Korea) (1415187459) by the Technology Innovation Program (or Industrial Strategic Technology Development Program) (RS-2023-00235339, Development of Critical Dimension Control technology using ion beam) funded By the Ministry of Trade, Industry & Energy (MOTIE, Korea) (1415187783), by K-CHIPS (Korea Collaborative & High-tech Initiative for Prospective Semiconductor Research) (23023-15FC) funded by the Ministry of Trade, Industry & Energy (MOTIE, Korea).

Institutional Review Board Statement: Not applicable.

Informed Consent Statement: Not applicable.

Data Availability Statement: The data presented in this study are available on request from the corresponding author.

Conflicts of Interest: The authors declare no conflicts of interest.

References

1. Samukawa, S.; Hori, M.; Rauf, S.; Tachibana, K.; Bruggeman, P.; Kroesen, G.; Whitehead, J.C.; Murphy, A.B.; Gutsol, A.F.; Starikovskaia, S.; et al. The 2012 Plasma Roadmap. *J. Phys. D Appl. Phys.* **2012**, *45*, 253001. [CrossRef]
2. Adamovich, I.; Murphy, A.B.; Bruggeman, P.J.; Turner, M.M.; Hamaguchi, S.; Favia, P.; Oehrlein, G.S.; Pitchford, L.C.; Starikovskaia, S.; Bogaerts, A.; et al. The 2017 Plasma Roadmap: Low Temperature Plasma Science and Technology. *J. Phys. D Appl. Phys.* **2017**, *50*, 323001. [CrossRef]
3. Donnelly, V.M.; Kornblit, A. Plasma Etching: Yesterday, Today, and Tomorrow. *J. Vac. Sci. Technol. A* **2013**, *31*, 050825. [CrossRef]

4. Constantine, C.; Johnson, D.; Pearton, S.J.; Chakrabarti, U.K.; Emerson, A.B.; Hobson, W.S.; Kinsella, A.P. Plasma Etching of III–V Semiconductors in $CH_4/H_2/Ar$ Electron Cyclotron Resonance Discharges. *J. Vac. Sci. Technol. B Microelectron. Process. Phenom.* **1990**, *8*, 596–606. [CrossRef]
5. Vartuli, C.B.; Pearton, S.J.; Lee, J.W.; Hong, J.; MacKenzie, J.D.; Abernathy, C.R.; Shul, R.J. ICI/Ar Electron Cyclotron Resonance Plasma Etching of III-V Nitrides. *Appl. Phys. Lett.* **1996**, *69*, 1426–1428. [CrossRef]
6. Tang, C.C.; Hess, D.W. Tungsten Etching in CF_4 and SF_6 Discharges. *J. Electrochem. Soc.* **1984**, *131*, 115–120. [CrossRef]
7. Osipov, A.A.; Andrianov, N.A.; Speshilova, A.B.; Gagaeva, A.E.; Risquez, S.; Vorobyev, A.; Alexandrov, S.E. Highly Selective Plasma Etching Technique for Molybdenum. *Plasma Chem. Plasma Process.* **2023**, *43*, 697–707. [CrossRef]
8. Gasvoda, R.J.; Zhang, Z.; Wang, S.; Hudson, E.A.; Agarwal, S. Etch Selectivity during Plasma-Assisted Etching of SiO_2 and SiN_x: Transitioning from Reactive Ion Etching to Atomic Layer Etching. *J. Vac. Sci. Technol. A* **2020**, *38*, 050803. [CrossRef]
9. Schaepkens, M.; Oehrlein, G.S. A Review of SiO_2 Etching Studies in Inductively Coupled Fluorocarbon Plasmas. *J. Electrochem. Soc.* **2001**, *148*, 211–221. [CrossRef]
10. Standaert, T.E.F.M.; Hedlund, C.; Joseph, E.A.; Oehrlein, G.S.; Dalton, T.J. Role of Fluorocarbon Film Formation in the Etching of Silicon, Silicon Dioxide, Silicon Nitride, and Amorphous Hydrogenated Silicon Carbide. *J. Vac. Sci. Technol. A Vac. Surf. Film.* **2004**, *22*, 53–60. [CrossRef]
11. Jeong, W.N.; Lee, Y.S.; Cho, C.H.; Seong, I.H.; You, S.J. Investigation into SiO_2 Etching Characteristics Using Fluorocarbon Capacitively Coupled Plasmas: Etching with Radical/Ion Flux-Controlled. *Nanomaterials* **2022**, *12*, 4457. [CrossRef] [PubMed]
12. Seong, I.; Lee, J.; Kim, S.; Lee, Y.; Cho, C.; Lee, J.; Jeong, W.; You, Y.; You, S. Characterization of an Etch Profile at a Wafer Edge in Capacitively Coupled Plasma. *Nanomaterials* **2022**, *12*, 3963. [CrossRef] [PubMed]
13. Seong, I.H.; Lee, J.J.; Cho, C.H.; Lee, Y.S.; Kim, S.J.; You, S.J. Characterization of SiO_2 over Poly-Si Mask Etching in Ar/C_4F_8 Capacitively Coupled Plasma. *Appl. Sci. Converg. Technol.* **2021**, *30*, 176–182. [CrossRef]
14. Cho, C.; You, K.; Kim, S.; Lee, Y.; Lee, J.; You, S. Characterization of SiO_2 Etching Profiles in Pulse-Modulated Capacitively Coupled Plasmas. *Materials* **2021**, *14*, 5036. [CrossRef]
15. Rueger, N.R.; Beulens, J.J.; Schaepkens, M.; Doemling, M.F.; Mirza, J.M.; Standaert, T.E.F.M.; Oehrlein, G.S. Role of Steady State Fluorocarbon Films in the Etching of Silicon Dioxide Using CHF3 in an Inductively Coupled Plasma Reactor. *J. Vac. Sci. Technol. A Vac. Surf. Film.* **1997**, *15*, 1881–1889. [CrossRef]
16. Schaepkens, M.; Standaert, T.E.F.M.; Rueger, N.R.; Sebel, P.G.M.; Oehrlein, G.S.; Cook, J.M. Study of the SiO_2-to-Si_3N_4 Etch Selectivity Mechanism in Inductively Coupled Fluorocarbon Plasmas and a Comparison with the SiO_2-to-Si Mechanism. *J. Vac. Sci. Technol. A Vac. Surf. Film.* **1999**, *17*, 26–37. [CrossRef]
17. Schaepkens, M.; Oehrlein, G.S.; Hedlund, C.; Jonsson, L.B.; Blom, H.-O. Selective SiO_2-to-Si_3N_4 Etching in Inductively Coupled Fluorocarbon Plasmas: Angular Dependence of SiO_2 and Si_3N_4 Etching Rates. *J. Vac. Sci. Technol. A Vac. Surf. Film.* **1998**, *16*, 3281–3286. [CrossRef]
18. Roskova, Z.; Schneider, J.; Stengel, M. Predicted Hydrofluorocarbon (HFC) and Perfluorocarbon (PFC) Emissions for the Years 2010–2050 in the Czech Republic. *Atmosphere* **2023**, *14*, 111. [CrossRef]
19. Mikhaylov, A.; Moiseev, N.; Aleshin, K.; Burkhardt, T. Global Climate Change and Greenhouse Effect. *Entrep. Sustain. Issues* **2020**, *7*, 2897–2913. [CrossRef]
20. Available online: https://www.epa.gov/ghgemissions/understanding-global-warming-potentials (accessed on 6 August 2024).
21. Kim, Y.; Kim, S.; Kang, H.; You, S.; Kim, C.; Chae, H. Low Global Warming $C_4H_3F_7O$ Isomers for Plasma Etching of SiO_2 and Si_3N_4 Films. *ACS Sustain. Chem. Eng.* **2022**, *10*, 10537–10546. [CrossRef]
22. Kweku, D.; Bismark, O.; Maxwell, A.; Desmond, K.; Danso, K.; Oti-Mensah, E.; Quachie, A.; Adormaa, B. Greenhouse Effect: Greenhouse Gases and Their Impact on Global Warming. *J. Sci. Res. Rep.* **2018**, *17*, 1–9. [CrossRef]
23. Yoro, K.O.; Daramola, M.O. CO_2 Emission Sources, Greenhouse Gases, and the Global Warming Effect. In *Advances in Carbon Capture*; Woodhead Publishing: Cambridge, UK, 2020. [CrossRef]
24. Yoo, S.W.; Cho, C.; Kim, K.; Lee, H.; You, S. Characteristics of SiO_2 Etching by Capacitively Coupled Plasma with Different Fluorocarbon Liquids (C_7F_{14}, C_7F_8) and Fluorocarbon Gas (C_4F_8). *Appl. Sci. Converg. Technol.* **2021**, *30*, 102–106. [CrossRef]
25. Sung, D.; Wen, L.; Tak, H.; Lee, H.; Kim, D.; Yeom, G. Investigation of SiO_2 Etch Characteristics by $C_6F_6/Ar/O_2$ Plasmas Generated Using Inductively Coupled Plasma and Capacitively Coupled Plasma. *Materials* **2022**, *15*, 1300. [CrossRef] [PubMed]
26. Li, C.; Gupta, R.; Pallem, V.; Oehrlein, G.S. Impact of Hydrofluorocarbon Molecular Structure Parameters on Plasma Etching of Ultra-Low-K Dielectric. *J. Vac. Sci. Technol. A Vac. Surf. Film.* **2016**, *34*, 031306. [CrossRef]
27. Lieberman, M.A. *Principles of Plasma Discharges and Materials Processing*, 2nd ed.; John Wiley & Sons, Inc.: Hoboken, NJ, USA, 2015; ISBN 9786468600.
28. Kim, J.H.; Choi, S.C.; Shin, Y.H.; Chung, K.H. Wave Cutoff Method to Measure Absolute Electron Density in Cold Plasma. *Rev. Sci. Instrum.* **2004**, *75*, 2706–2710. [CrossRef]
29. Kim, J.H.; You, S.J.; Seong, D.J.; Shin, Y.H. Electron temperature measurements in plasmas with surface wave absorption and wave cutoff frequency. *Appl. Phys. Letts.* **2007**, *91*, 201502. [CrossRef]

30. Kim, D.W.; You, S.J.; Na, B.K.; Kim, J.H.; Chang, H.Y. An analysis on transmission microwave frequency spectrum of cut-off probe. *Appl. Phys. Letts.* **2011**, *99*, 131502. [CrossRef]
31. Lee, Y.S.; Oh, S.H.; Lee, J.J.; Cho, C.H.; Kim, S.J.; You, S.J. A Quantification Method in Quadrupole Mass Spectrometer Measurement. *Appl. Sci. Converg. Technol.* **2021**, *30*, 50–53. [CrossRef]
32. Cho, C.; Kim, S.; Lee, Y.; Jeong, W.; Seong, I.; Lee, J.; Choi, M.; You, Y.; Lee, S.; Lee, J.; et al. Refined Appearance Potential Mass Spectrometry for High Precision Radical Density Quantification in Plasma. *Sensors* **2022**, *22*, 6589. [CrossRef]
33. Yeom, H.J.; Lee, J.J.; Kim, S.J.; Lee, Y.S.; Cho, C.H.; Kim, J.H.; Lee, H.C.; You, S. Mass Filter Characteristic and Design Role of Quadrupole Mass Spectrometer for Radical Measurement in Low-Pressure Plasmas. *Appl. Sci. Converg. Technol.* **2020**, *29*, 77–81. [CrossRef]
34. Available online: https://film-sense.com/film-sense-products/ (accessed on 6 August 2024).
35. Available online: https://www.thermofisher.com/order/catalog/product/kr/ko/IQLAADGAAFFACVMAHV (accessed on 6 August 2024).
36. Lee, H.; Tak, H.; Kim, S.; Kim, S.K.; Park, T.H.; Kim, J.Y.; Sung, D.; Lee, W.; Lee, S.B.; Kim, K.; et al. Characteristics of high aspect ratio SiO_2 etching using $C_4H_2F_6$ isomers. *Appl. Surf. Sci.* **2023**, *639*, 158190. [CrossRef]
37. Cho, N.I.; Hong, J.W.; Yoo, H.J.; Eoh, H.J.; Kim, C.H.; Jeong, J.W.; Kim, K.L.; Kwak, J.H.; Cho, Y.J.; Kim, D.W.; et al. Etch characteristics of maskless oxide/nitride/oxide/nitride (ONON) stacked structure using $C_4H_2F_6$-based gas. *Sci. Rep.* **2024**, *14*, 22900. [CrossRef] [PubMed]
38. Metzler, D.; Bruce, R.L.; Engelmann, S.; Joseph, E.A.; Oehrlein, G.S. Fluorocarbon Assisted Atomic Layer Etching of SiO_2 Using Cyclic Ar/C_4F_8 Plasma. *J. Vac. Sci. Technol. A* **2014**, *32*, 020603. [CrossRef]
39. Lin, K.-Y.; Li, C.; Engelmann, S.; Bruce, R.L.; Joseph, E.A.; Metzler, D.; Oehrlein, G.S. Achieving Ultrahigh Etching Selectivity of SiO_2 over Si_3N_4 and Si in Atomic Layer Etching by Exploiting Chemistry of Complex Hydrofluorocarbon Precursors. *J. Vac. Sci. Technol. A* **2018**, *36*, 040601. [CrossRef]
40. Lysenko, G.A.; Krioukov, A.V.; Kachurin, Y.Y.; Pogodaev, V.V. Accurate measurements at interferomet-ric ellipsometer. *Opt. Eng.* **2006**, *45*, 023605. [CrossRef]

Journal of
Manufacturing and
Materials Processing

MDPI

Article

Effect of Al/Cu Ratio on Microstructure and High-Temperature Oxidation Resistance of $Al_xCoCrCu_yFeNi$ High-Entropy Alloy Coatings

Ling Zhou, Hongxi Liu *, Qinghua Zhang, Jiazhu Liang, Yuanrun Peng, Xuanhong Hao, Chen Yang, Yaxia Liu and Yueyi Wang

Faculty of Materials Science and Engineering, Kunming University of Science and Technology, Kunming 650093, China; 17869418798@163.com (L.Z.); zqhzam1001@163.com (Q.Z.); 20222130076@stu.kust.edu.cn (J.L.); pyr6798@163.com (Y.P.); hxh_instinct@163.com (X.H.); 20233130007@stu.kust.edu.cn (C.Y.); 20212130082@stu.kust.edu.cn (Y.L.); wyy_0710@126.com (Y.W.)

* Correspondence: piiiliuhx@sina.com or liuhx@kust.edu.cn

Abstract: To improve high-temperature oxidation resistance for Ti6Al4V alloy, $Al_xCoCrCu_y$-FeNi (x = 0, 0.3, 0.5, 0.7, 1.0; y = 1.0, 0.7, 0.5, 0.3, 0, x + y = 1.0) high-entropy alloy (HEA) coatings were prepared on the Ti6Al4V alloy substrate by a laser cladding technique. The results show that the coatings were mainly composed of FCC, BCC, and Ti-rich phases. Severe segregation of the Cu element occurred in the CoCrCuFeNi HEA coatings as a Cu-rich phase (FCC2). The Cu-rich phases decreased with a decreasing Cu content and completely disappeared until the Al content reached 1.0. The microhardnesses of the $Cu_{1.0}$, $Cu_{0.7}Al_{0.3}$, $Cu_{0.5}Al_{0.5}$, $Cu_{0.3}Al_{0.7}$, and $Al_{1.0}$ HEA coatings were 2.01, 2.06, 2.08, 2.09, and 2.11 times that of the substrate, and compared with those of a Ti6Al4V alloy substrate, the oxidation rates of the HEA coatings decreased by 55%, 51%, 47%, 42%, and 35%, respectively. The surface oxides of the five coatings were mainly composed of CuO, TiO_2, Fe_3O_4, Cr_2O_3, and Al_2O_3. The increase in the Al content promoted the generation of Al_2O_3 film and Cr_2O_3 on the surfaces of the coatings, which significantly improved the high-temperature antioxidant performance of the high-entropy alloy coatings for 50 h at 800 °C. When x = 1.0, the coating showed the best high-temperature antioxidant performance.

Keywords: high-entropy alloy; laser cladding; high-temperature oxidation resistance; microstructure; Ti6Al4V alloy

1. Introduction

Titanium alloys are widely used as important structural materials in fields such as aerospace, petrochemical, automobile, biomedicine, and architectural decoration due to their low density, high specific strength, excellent corrosion resistance, and biocompatibility [1]. Ti6Al4V alloy, as an important titanium alloy material, has attracted much attention. However, in some specific application scenarios, the relatively weak hardness and high-temperature oxidation resistance make Ti6Al4V alloy fail easily during service in extreme environments. Therefore, in order to solve this problem, many researchers are committed to exploring new surface treatment technologies [2–4].

Nowadays, the main surface modification methods include thermal spraying technology, vapor deposition technology, electrochemical surface modification, surface nanocrystallization, ion implantation, and laser cladding technology [5–12]. As an advanced material surface modification technology, laser cladding (LC) has the advantages of a high bonding strength between the coating and the substrate, a small heat-affected zone, and precise control of the composition of the coating, which can effectively reduce the impact on the properties of the substrate [13–15]. In recent years, many researchers have improved the surface properties of titanium alloys using laser cladding technology, and the material systems that have been investigated include Ni-based alloy systems [16,17], metal–ceramic composite systems [18–21], and rare earth element modification systems [22,23]. Unfortunately, in the material system described above, due to the large differences in the coating and the substrate's physical and chemical properties, the combination of the coating and the substrate is often not strong enough, and these materials are less elemental, easily forming a large number of intermetallic brittle phases, resulting in a reduction in toughness. Even some of the material preparation costs are high, and the process is complex and difficult to mass produce. Additionally, there are other factors affecting the effectiveness of the surface engineering. And it is difficult to meet the practical application requirements.

High-entropy alloys (HEAs), as a new type of alloy material, have a unique composition that imparts a high entropy effect, a hysteresis diffusion effect, a lattice distortion effect, and a "cocktail" effect, and it possesses high thermal stability [24–27]. The high entropy of mixing encourages the alloys to tend to form simple solid solutions, such as face-centered cubic (FCC), body-centered cubic (BCC), and hexagonal close-packed (HCP) [28,29]. Due to their unique microstructure and the synergistic effect between multiple elements, HEAs have excellent properties, such as high strength, corrosion resistance, and high-temperature resistance. CoCrFeNi high-entropy alloy has attracted extensive research interest due to its good toughness, which is mainly attributed to its single face-centered cubic (FCC) structure [30]. This structure can form a uniform, dense coating on the surface of the substrate and form a good bond, avoiding the brittle fracture that tends to occur in laser cladding coating materials. Compared with the traditional alloy coatings consisting of only one or two elements, the unique multi-element mixed structure of a CoCrFeNi high-entropy alloy coating makes dislocation movement more difficult. This unique microstructure impedes dislocation movement to a greater extent, thereby minimizing wear and deformation and enhancing the mechanical properties of the coating. It tends to form a simple multi-component solid-solution structure with a relatively single-phase composition and uniform element distribution, ensuring the consistency and stability of its performance over an extended period of use. Through the modulation of the types and amounts of elements, a diverse range of performance characteristics can be achieved, catering to the specific and individualized demands of various engineering applications in terms of material performance. The synergistic interaction among its multiple elements frequently endows it with performance superiority over traditional alloy coatings, consequently attracting more extensive attention and research efforts [31–33]. However, the lack of hardness and oxidation resistance of the CoCrFeNi high-entropy alloy poses a challenge for further use in high-temperature environments. It has been shown that the element Al can have a significant effect on the hardness and high-temperature oxidation behavior of HEAs, and increasing the amount of Al can improve the oxidation resistance of HEAs through the formation of an Al_2O_3 oxide layer [34–36]. Huang [37] prepared a TiVCrAlSi HEA coating on a Ti6Al4V alloy by laser cladding and found that the microstructure of the coating consisted of (Ti, V)$_5$Si$_3$ and BCC solid solutions. The SiO_2, Cr_2O_3, TiO_2, and Al_2O_3 and a small quantity of V_2O_5 produced during the oxidation process of the laser-clad TiVCrAlSi coating could effectively enhance the oxidation resistance of a Ti6Al4V alloy in the air at 800 °C. Zhang [38] prepared a TiAlNiSiV HEA coating with a BCC phase structure on the surface of a Ti6Al4V alloy by laser cladding, and the results showed that the hardness of the TiAlNiSiV HEA coating was 1151∼1357 $HV_{1.0}$, which was 4∼6 times that of the substrate, and the wear resistance of the TiAlNiSiV HEA coating at 800 °C was four times that of the substrate. Geng [39] prepared $(AlCu)_{3.5}CoCrNiFe$ and $(AlCu)_{3.5}CoCrNiTi$ HEA coatings on an AZ91 magnesium alloy using a laser cladding technique; both coatings were BCC and FCC phases. Their microhardnesses were 642.1 $HV_{1.0}$ and 656.2 $HV_{1.0}$, respectively, which were about 9 times higher than that of the AZ91 substrate (71.9 $HV_{1.0}$). The TiO_2 produced in $(AlCu)_{3.5}CoCrNiTi$ formed an oxide film during the oxidation process, resulting in excellent oxidation resistance of the substrate. It can be seen that the application of HEAs makes the surface properties of materials significantly improved. However, an excess of Al may cause the alloy to become too hard and brittle, thereby reducing its toughness and plasticity, which in turn negatively impacts its mechanical properties. The composition of the phases can be controlled to increase the toughness of the alloy by the addition of elements with good plasticity (such as Cu) and by adjusting the ratio of the elements.

In this study, Al and Cu elements were added to the CoCrFeNi alloy system. The addition of the Al element can effectively improve the microhardness and oxidation resistance of the coating but also make the coating brittle. In contrast, the introduction of the Cu element can enhance the plasticity of the coating. This paper innovatively adjusted the ratio of Al/Cu at the same time, allowing for the optimization of the coating's morphology and performance. The $Al_xCoCrCu_yFeNi$ (x = 0, 0.3, 0.5, 0.7, 1.0; y = 1.0, 0.7, 0.5, 0.3, 0, x + y = 1) HEA coatings were prepared on the surface of a Ti6Al4V alloy by laser cladding technology. The effects of varying the Al and Cu contents on the microstructure, microhardness, and high-temperature oxidation resistance were systematically analyzed to further elucidate the underlying principles.

2. Materials and Methods

2.1. Experimental Materials and Preparation of Coating

The substrate in this study is a Ti6Al4V alloy in a rolling state with a dual-phase $\alpha + \beta$ structure (sample size 150 mm × 15 mm × 10 mm). One side of each substrate sample was polished with 240-grit and 400-grit SiC abrasive paper to remove surface oxides and contaminants, which also helped reduce laser reflection. The samples were then thoroughly cleaned in absolute ethanol to eliminate oil, rust, and other impurities from the surface. Before laser cladding, the sample was dried in an electric box at 80 °C for 2 h.

Commercially available Al, Cu, CoCrFeNi alloy spherical powder (Beijing Yanbang Metal Materials Co., Ltd., China) with a purity of more than 99.9% (particle size from 15 μm to 53 μm) was selected as the raw material. The mixed powders were weighed according to the atomic ratio of $Al_xCoCrCu_yFeNi$ (x = 0, 0.3, 0.5, 0.7, 1.0; y = 1.0, 0.7, 0.5, 0.3, 0, x + y = 1). The specific alloy powder ratios are shown in Table 1. An MSK-SFM-1S-type high-flux planetary ball mill was used to mix the powders in dry grinding mode. To prevent damage to the sphericity of the initial powder and avoid oxidation during the ball milling process. We used the following parameters: a speed of 120 rpm, milling times of 8 h, and a ball-to-power ratio of 4: 1. After ball milling, the mixed powders were dried at 60~80 °C for 1~2 h. The SEM images and EDS elemental distribution of the raw powders are shown in Figure 1a-g. SEM and EDS analysis results indicate that the mixed powders are relatively uniform, with good sphericity, show no obvious fragmentation, and do not exhibit agglomeration or oxidation. Figure 1h shows the XRD pattern of the $Al_{0.5}CoCrCu_{0.5}FeNi$ power, which reveals no formation of new phases.

The mixture powder was preset on the Ti6Al4V alloy plate surface with CH_3CH_2OH organic binder (thickness 1 mm) and then dried in an electric oven at 80 °C for 5 h. A fiber laser (XL-F1500) with a power output of 1500 W was used to prepare $Al_xCoCrCu_yFeNi$ HEA coatings on the Ti6Al4V alloy surface. After conducting several orthogonal experiments, the optimal processing parameters for this study were determined as follows: laser power of 1100 W, laser scanning speed of 300 $mm \cdot min^{-1}$, circular spot diameter of 4 mm, and defocus of 13 mm. To prevent oxidation, Ar was used as the protection gas (99.99% purity, blowing to one side of the molten pool), with a gas flow rate of 15 L·min^{-1}.

Table 1. Composition of the mixed powder (at. %).

Groups	Ratios	Al/ Cu	Simplified
1	CoCrCuFeNi	0:1	$Cu_{1.0}$
2	$Al_{0.3}CoCrCu_{0.7}FeNi$	0.3:0.7	$Cu_{0.7}Al_{0.3}$
3	$Al_{0.5}CoCrCu_{0.5}FeNi$	0.5:0.5	$Cu_{0.5}Al_{0.5}$
4	$Al_{0.7}CoCrCu_{0.3}FeNi$	0.7:0.3	$Cu_{0.3}Al_{0.7}$
5	AlCoCrFeNi	1:0	$Al_{1.0}$

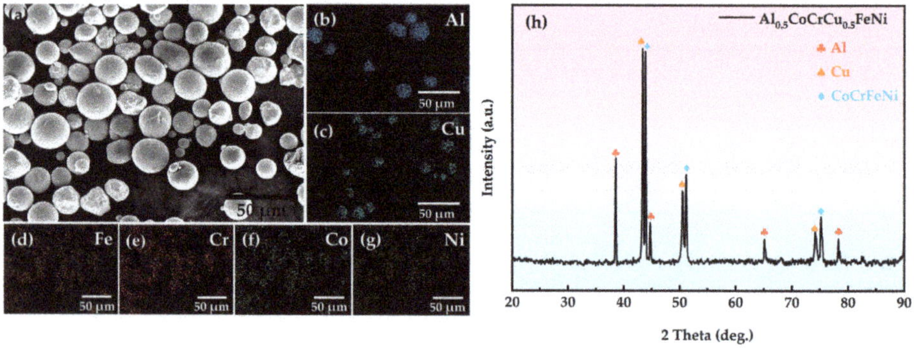

Figure 1. SEM image and EDS elemental distribution of the raw powder: (**a**) SEM image of the mixed powder; (**b–g**) EDS elemental distribution of the powder; (**h**) XRD pattern of $Al_{0.5}CoCrCu_{0.5}FeNi$ power.

2.2. Microstructure Characterization

The prepared high-entropy alloy coatings were cut into rectangular samples with the size of 20 mm × 10 mm × 10 mm using electric discharge wire cutting technology. The coating section of each sample was polished with 240, 600, 800, 1200, and 2000 mesh sandpapers to remove oil stains and oxides on the surface until smooth. The samples were then polished to a mirror finish using a 0.12 µm silica polishing liquid, ensuring the surface had no obvious scratches, and finally with anhydrous ethanol ultrasonic cleaning.

The microstructure of the laser cladding coating was analyzed by an EMPYREAN-type X-ray diffractometer (XRD) from Malvern-Panaco, Netherlands, with a Cu cathode target and $K\alpha$ line wavelength of 1.5406 Å, working voltage of 40 kV, and working current of 30 mA. The scan was conducted with a step size of 5°·min^{-1}over a 2θ range of 20° to 90° (CSM, continuous scan mode). Scanning electron microscopy (SEM; TESCAN VEGA 3., Czech Republic; HV: 20.0 kV) and its energy dispersive X-ray spectrometer (EDS) were used to analyze the phase microstructure, micro-morphology, and elemental distribution of the cross-sectional coating samples.

2.3. Performance Tests

The surface of the laser cladding coating sample was initially pre-treated by sanding and polishing until free of scratches and ultrasonically cleaned with alcohol. The microhardness of the laser cladding coating sample was tested using a 3S-1000QZD-type automatic Vickers hardness tester (Zhejiang Sansi Yongheng Technology Co., Ltd., China), with a load of 5 N and a dot placed at intervals of 0.2 mm from the surface of the laser cladding coating towards the substrate for 15 s.

The high-temperature oxidation resistance of the coatings and the substrate was evaluated using a GSL-1700X-type tube resistance furnace (Shenyang Kejing Automation Equipment Co., Ltd., China). The oxidation process followed a cyclic pattern at a constant temperature of 800 °C, with a total oxidation duration of 50 h. The coatings and the Ti6Al4V alloy substrate were taken out at 5 h, 10 h, 15 h, 30 h, and 50 h. After natural cooling, the oxidation weight gain of both the coatings and the Ti6Al4V alloy substrate was measured. The oxidation kinetics curve was subsequently plotted. The oxidation morphology and the oxides formed on the oxidized surfaces of the coatings and the Ti6Al4V alloy substrate were examined using a scanning electron microscope (SEM) equipped with energy dispersive spectrometer (EDS), and the oxidation mechanism of the coatings was analyzed.

3. Results and Discussion

3.1. XRD Phase Analysis and Microstructure Analysis

Figure 2 presents the X-ray diffraction (XRD) patterns of the $Al_xCoCrCu_yFeNi$ (x = 0, 0.3, 0.5, 0.7, 1.0; y = 1.0, 0.7, 0.5, 0.3, 0) HEA coatings with varying Al and Cu ratios. The microstructure of the coatings was composed of FCC, BCC, and a small amount of Laves phases. The FCC phase includes the matrix phase (FCC1) and the Cu-rich phase (FCC2), while the BCC phase is primarily Fe-Cr-rich and the hard Laves phase is rich in Ti. It can be seen that three prominent diffraction peaks appear at 43.12°, 63.32°, and 80.10°, corresponding to the main FCC solid solution phase, while a second set of peaks at 44.15°, 64.39°, and 81.80° corresponds to the BCC phase. This is attributed to the more constituent elements and the close atomic percentage of each element in the HEA, resulting in a higher mixing entropy and a lower Gibbs free energy in the system, which promotes the formation of random disordered solid solution phases with simple and stable microstructures [40]. It can be clearly seen that in addition to the main peaks, several smaller peaks are observed in the XRD patterns. According to the literature [41], and matching with PDF card data, these are attributed to Ti-rich Laves-reinforced phases. Furthermore, when the Cu content is 1.0, Cu-rich phases are detected at diffraction peaks of 43.12°, 50.08°, and 89.29°, but these peaks disappear when the Cu content is reduced to 0.3. This may be due to the low Cu content at which the X-ray diffraction cannot detect these phases. With the decrease in the Cu content, the intensity of the FCC peaks gradually diminishes. Simultaneously, with the increase in the Al content and the decrease in the Cu content, the intensity of the Laves phase diffraction peaks also decreases.

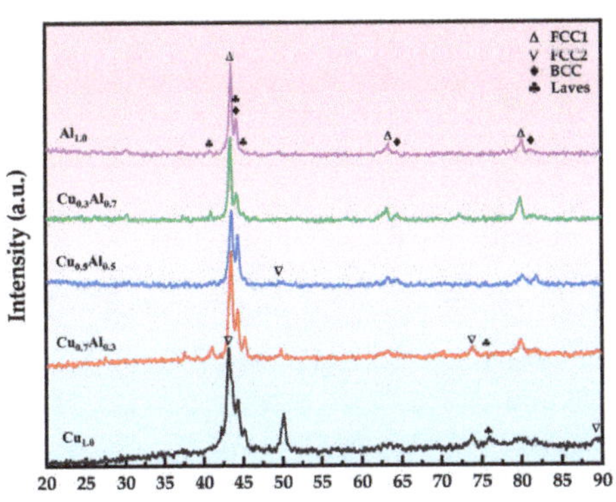

Figure 2. XRD patterns of the $Al_xCoCrCu_yFeNi$ HEA coatings.

The cross-sectional microscopic morphology and the EDS line scanning distribution curves of different elements along the cross-section of the $Al_{0.7}CoCrCu_{0.3}FeNi$ coating are shown in Figure 3a. According to the literature [42], the dilution rate is calculated using the following equation (Equation (1)):

$$\lambda = \frac{h}{H+h} \times 100\% \quad (1)$$

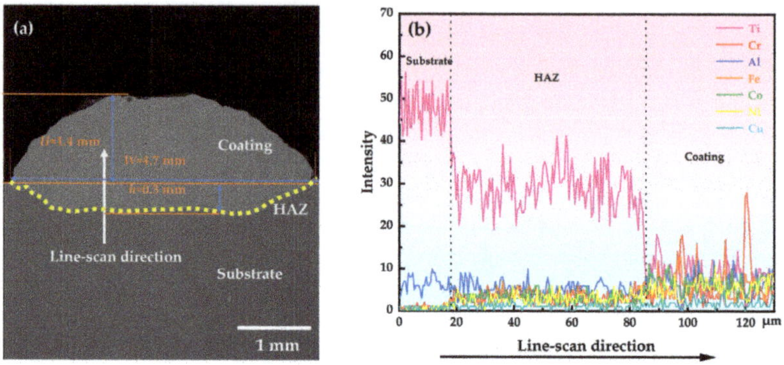

Figure 3. Macroscopic cross-section morphology of the $Al_{0.7}CoCrCu_{0.3}FeNi$ HEA coating: (**a**) The cross-sectional morphology; (**b**) the EDS line scanning distribution curves of the different elements along the cross-section.

According to the data shown in Figure 3a, the dilution rate of the coating is determined to be 17.6%. As indicated by the EDS line scanning results in Figure 3b, element diffusion is observed between the coating and the substrate. This demonstrates a strong metallurgical bond between the coating and the substrate. Generally, dilution is a critical parameter in laser cladding HEA coating. A low dilution rate in HEA coating may lead to insufficient melting and poor formability, while an excessively high dilution rate can alter the microstructure and properties of the coating. An optimal dilution rate enhances the bond strength between the coating and the substrate, results in a more uniform coating structure, and better preserves the original characteristics of the HEA, along with the stability of its structure [43,44].

Figure 4 shows the cross-sectional SEM images of the upper and middle regions of the coatings. It can be observed that the coating exhibits a typical dendrite microstructure. When the Cu content is 1.0, a light-gray microstructure is present in the inter-dendritic regions at the top of the coating, while a butterfly-like dark-gray microstructure is observed within the dendrites. With the decreases in the Cu content and the increases in the Al content, the grains evolve from a cellular structure to a chrysanthemum-like shape and ultimately to petal-like equiaxed crystals, and gradually decrease with a small amount of dispersed black petal-like or granular precipitates. The dark-grey microstructure in the dendrites and the light-grey microstructure in the inter-dendrites show that the grain size decreases from the top to the bottom of the coating, accompanied by a reduction in the size of the black granular precipitates. In addition, as shown in Figure 4a–d, white precipitates accumulate at the grain boundaries, with their content decreasing progressively. Due to the high mixing entropy effect [45], no intermetallic compounds were found in the middle region of the coatings, which promoted the dissolution of the main components and inhibited the precipitation of the intermetallic compounds.

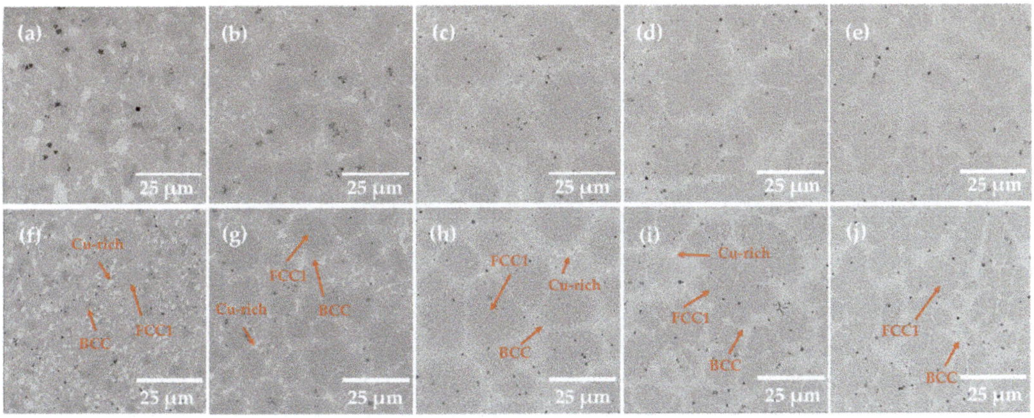

Figure 4. Microstructure of top and central coatings of $Al_xCoCrCu_yFeNi$ HEA coatings: (**a**,**f**) $Cu_{1.0}$; (**b**,**g**) $Cu_{0.7}Al_{0.3}$; (**c**,**h**) $Cu_{0.5}Al_{0.5}$; (**d**,**i**) $Cu_{0.3}Al_{0.7}$; (**e**,**j**) $Al_{1.0}$; (**a**–**e**) top of coating; (**f**–**j**) center of coating.

To further analyze the distribution of different elements, EDS scanning was performed, and the results are shown in Figure 5. Combined with the previously discussed XRD results, it is evident that the dark-gray microstructure in the dendrites corresponds to the FCC phase, while the light-gray microstructure in the inter-dendritic regions corresponds to the BCC phase. Due to the mutual dilution between the coating and the substrate, the Ti element from the Ti6Al4V alloy substrate is continuously diffused into the coating during the laser cladding process, leading to an increased Ti element in the coating. This promotes the formation of the FCC phase enriched in Al-Co-Ni-Ti, which grows into equiaxed crystals. During the solidification process, the content of Fe and Cr elements in the liquid molten pool gradually increases until it reaches the maximum solubility limit of the liquid metal, at which point these elements precipitate as an alloy phase. Since Cr has the highest melting point, it is preferred to precipitate to form a Fe-Cr-rich BCC phase. At the same time, the atomic radius of Ti is significantly larger than that of other elements, and the mixing enthalpy is low, resulting in the segregation of Ti elements in the coating and the formation of dispersed Laves phases. The EDS analysis results at different locations in Figure 5 are provided in Table 2. The black petal-like precipitates and the black particles dispersed in the coating are Ti-rich Laves phases.

Table 2. EDS results of areas in Figure 5.

Points	Elements (at. %)						
	Al	Co	Cr	Cu	Fe	Ni	Ti
1	-	14.5	10.6	10.4	11.5	18.0	35.0
2	8.1	10.8	3.9	3.9	5.8	10.3	57.3
3	14.0	14.9	7.4	3.3	9.0	15.1	36.4
4	29.2	11.0	6.9	2.1	7.7	10.2	32.8
5	22.7	13.6	6.0	-	8.0	12.2	37.4

Based on the above-mentioned analysis, it can be seen that the light-gray area of the $Cu_{1.0}$ coating corresponds to the Co-Ni-Ti-rich FCC phase, while the BCC phase in the inter-dendritic regions is Fe-Cr-rich and the white microstructure at the grain boundaries is the Cu-rich phase. During the process of cladding, the Ti atom, with its large atomic radius and slow diffusion rate, exhibits negative mixing enthalpy with Co and Ni elements, promoting the formation of Co_3Ti or Ni_3Ti phases. This, in turn, facilitates the aggregation of Fe and Cr elements into the dendrites. The mixing enthalpy values for different atomic pairs between the elements are listed in Table 3. The Cu element has a positive mixing enthalpy and a lower melting point, resulting in a weak affinity with other elements and low bonding strength. As a result, the Cu element does not readily form solid solutions with other elements and is extruded to the grain boundaries, forming Cu-rich phases. This process leads to a reduction in the grain size. In the $Cu_{0.7}Al_{0.3}$, $Cu_{0.5}Al_{0.5}$, $Cu_{0.3}Al_{0.7}$, and $Al_{1.0}$ coatings, chrysanthemum-like dendrites and petal-like equiaxed dendrites are observed in the FCC phase, which is rich in Al-Co-Ni-Ti, while the light-gray microstructure in the BCC phase is rich in Fe-Cr. The white monomer at the grain boundaries in $Cu_{0.7}Al_{0.3}$, $Cu_{0.5}Al_{0.5}$, and $Cu_{0.3}Al_{0.7}$ are also Cu-rich phases. With the decrease in the Cu content, the Cu-rich phase at the grain boundaries significantly decreases and eventually disappears when the Al content reaches 1.0, along with the disappearance of the FCC2 phase. With the increase in the Al content, the negative mixing enthalpy between Ti and Al causes the FCC1 phase to transition from being Co-Ni-Ti-rich to $AlNi_2Ti$- or $AlCo_2Ti$-rich.

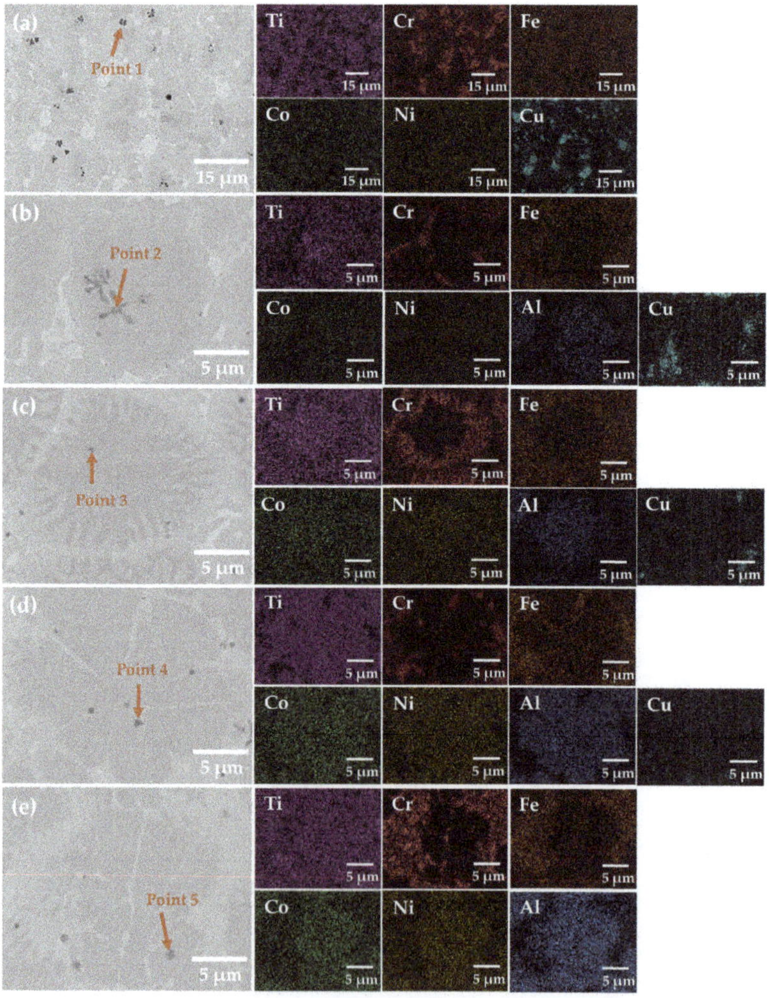

Figure 5. Microstructure of $Al_xCoCrFeCu_yNi$ HEA coatings: **(a)** $Cu_{1.0}$; **(b)** $Cu_{0.7}Al_{0.3}$; **(c)** $Cu_{0.5}Al_{0.5}$; **(d)** $Cu_{0.3}Al_{0.7}$; **(e)** $Al_{1.0}$.

Table 3. ΔH_{mix} enthalpy of mixing of different atomic pairs.

Elements	Enthalpy (kJ·mol^{-1})						
	Al	Co	Cr	Fe	Ni	Cu	Ti
Al	-	−19	−10	−11	−22	−1	−30
Co	-	-	−4	−1	0	6	−28
Cr	-	-	-	−1	−7	12	−7
Fe	-	-	-	-	−2	13	−17
Ni	-	-	-	-	-	4	−35
Cu	-	-	-	-	-	-	4
Ti	-	-	-	-	-	-	-

3.2. Microhardness of the Coating and the Ti6Al4V Alloy Substrate

The microhardness distribution curves and the average microhardness values of the HEA coatings and the Ti6Al4V alloy substrate are shown in Figure 6. As shown in Figure 6a, the microhardness increases significantly from the substrate to the surface of the coating. Figure 6b displays the average microhardness values of the Ti6Al4V alloy substrate and the $Al_xCoCrCu_yFeNi$ (x = 0, 0.3, 0.5, 0.7, 1.0; y = 1.0, 0.7, 0.5, 0.3, 0, x + y = 1) HEA coatings, which are as follows: 344.76 $HV_{0.5}$, 694.70 $HV_{0.5}$, 711.48 $HV_{0.5}$, 716.19 $HV_{0.5}$, 720.71 $HV_{0.5}$, and 727.61 $HV_{0.5}$. The average microhardness of the HEA coatings is 2.01, 2.06, 2.08, 2.09, and 2.11 times that of the Ti6Al4V alloy substrate, respectively. During the process of laser cladding, due to dilution between the Ti6Al4V alloy substrate and the coating, the Ti element from the substrate diffuses into the coating, resulting in an increase in the Ti element content in the coating. The large atomic radius of Ti enhances the lattice distortion effect within the coating, which promotes the segregation of BCC phases (Cr-Fe phase) and contributes to solid solution strengthening. In addition, the low mixing enthalpy of Ti with other elements further promotes the formation of the reinforced phase. Ti combines with other elements in the coating to form hard Laves phase and Ti-rich particles, which are dispersed throughout the coating. As a result of precipitation strengthening, the coating is strengthened, and the hardness is significantly increased. In addition, the large atomic radius of Al, combined with the increasing Al content, leads to greater lattice distortion, which impedes dislocation movement, resulting in a stronger solid solution strengthening effect. In summary, the microhardness of the coating is significantly enhanced due to the lattice distortion caused by the incorporation of elements with large atomic radii, which induces solid solution strengthening. The diffusion of the Ti element from the substrate into the coating also promotes the precipitation of the hard phases, contributing to the precipitation-strengthening effect.

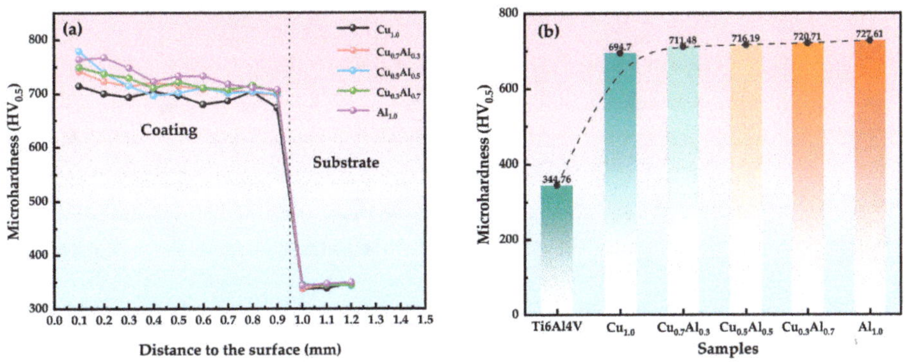

Figure 6. Microhardness results of the coatings and Ti6Al4V alloy: (**a**) microhardness distribution; (**b**) average microhardness.

3.3. Oxidation Behavior

The oxidation resistance of the Ti6Al4V alloy and HEA coatings was evaluated at 800 °C for a total duration of 50 h, with measurements taken at intervals of 5 h, 10 h, 15 h, 30 h, and 50 h. Figure 7a shows the oxidation mass gain per unit area for the Ti6Al4V alloy substrate and the $Al_xCoCrCu_yFeNi$ (x = 0, 0.3, 0.5, 0.7, 1.0; y = 1.0, 0.7, 0.5, 0.3, 0, x + y = 1) HEA coatings. The results indicate that the oxidation mass gain for the HEA coatings and Ti6Al4V alloy substrate are 0.156 mg·mm^{-2}, 0.143 mg·mm^{-2}, 0.133 mg·mm^{-2}, 0.118 mg·mm^{-2}, 0.098 mg·mm^{-2}, and 0.283 mg·mm^{-2}, respectively. Compared to the Ti6Al4V alloy, the oxidation rates of the $Cu_{1.0}$, $Cu_{0.7}Al_{0.3}$, $Cu_{0.5}Al_{0.5}$, $Cu_{0.3}Al_{0.7}$, and $Al_{1.0}$ HEA coatings decreased by 55%, 51%, 47%, 42%, and 35%, respectively. The oxidation mass gain of the HEA coatings is significantly lower than that of the Ti6Al4V alloy substrate. In addition, the addition of Al content improves the oxidation resistance of the $Al_xCoCrCu_yFeNi$ HEA coatings. Before 15 h, all of the HEA coatings exhibit a rapid initial increase in oxidation mass gain, which is attributed to the rapid oxide growth at the early stages of the oxidation process [46]. During this stage, the initial oxide layer nucleates and grows on the surface, primarily consisting of less protective oxides. Over time, the oxidation mass gain of the HEA coating increases slowly, whereas the oxidation mass gain of the Ti6Al4V alloy substrate increases rapidly. This difference in behavior suggests that the formation of the oxide layer is controlled by the outward diffusion of the constituent elements with high diffusion coefficients. To better understand the oxidation kinetics of the $Al_xCoCrCu_yFeNi$ HEA coatings, a power function [47,48] (Equation (2)) was used to fit the oxidation mass gain as a function of time:

$$\left(\frac{\Delta m}{A}\right)^2 = k_p \times t \qquad (2)$$

where Δm—the oxidation mass gain (mg), A—the surface area (mm^2), k_p—the parabolic oxidation rate constant, and t—the oxidation time (h) at high temperatures.

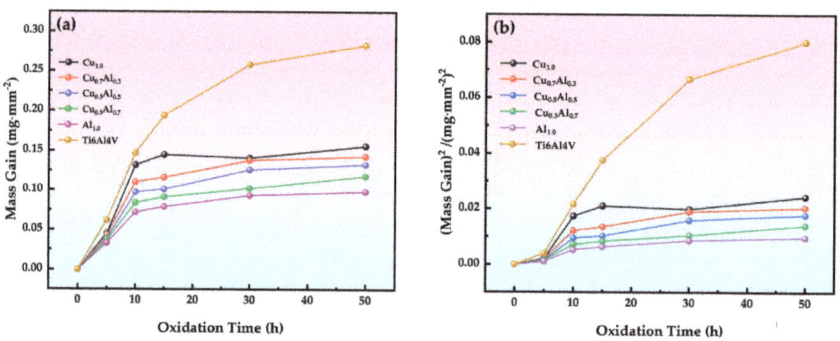

Figure 7. Oxidation kinetics curves of the $Al_xCoCrCu_yFeNi$ HEA coatings and Ti6Al4V alloy at 800 °C: (**a**) original curves; (**b**) linear fit curves.

Figure 7b shows the relationship between the square of the mass gain per unit area and oxidation time. In addition, the oxidation rate constant k_p values were calculated according to Equation (2). The oxidation rate constants k_p of each coating and the Ti6Al4V alloy substrate after oxidation at 800 °C for 50 h are 4.8672 × 10^{-4} mg^2·mm^{-4}·h^{-1}, 4.0898 × 10^{-4} mg^2·mm^{-4}·h^{-1}, 3.5378 × 10^{-4} mg^2·mm^{-4}·h^{-1}, 2.7848 × 10^{-4} mg^2·mm^{-4}·h^{-1}, 1.9208 × 10^{-4} mg^2·mm^{-4}·h^{-1}, and 16.0185 × 10^{-4} mg^2·mm^{-4}·h^{-1}, respectively. It can be seen that the AlCoCrFeNi HEA coating exhibited the best oxidation resistance. This shows that the high-entropy alloy coatings can significantly improve the high-temperature oxidation resistance of the Ti6Al4V alloy.

The XRD patterns of the oxidized surface of the coatings are shown in Figure 8. The results indicate that all phases on the surface of the HEA coatings are composed of various types of oxides. For the CoCrCuFeNi HEA coating, the surface oxides include CuO, TiO_2, Fe_3O_4, and Cr_2O_3. As the Al content increases, the intensity of the CuO peak decreases, and the Al_2O_3 phase appears on the oxidized surface of the coating. Moreover, the peak intensity of Al_2O_3 and Cr_2O_3 increases with a higher Al content. For the $Cu_{1.0}$, $Cu_{0.7}Al_{0.3}$, and $Cu_{0.5}Al_{0.5}$, the surface oxides are primarily CuO, Fe_3O_4, and TiO_2, while the surface oxides of the $Cu_{0.3}Al_{0.7}$ and $Al_{1.0}$ coatings are mainly Cr_2O_3 and Al_2O_3. With the increase in the Al content, the peak intensity of Al_2O_3 and Cr_2O_3 is gradually enhanced. This suggests that the addition of Al elements promotes the formation of Al_2O_3 and Cr_2O_3 on the surface of the HEA coatings. The presence of Al_2O_3 and Cr_2O_3 in the oxide layer increases the surface density of the HEA coatings, effectively preventing the inward diffusion of oxygen and slowing down the oxidation process.

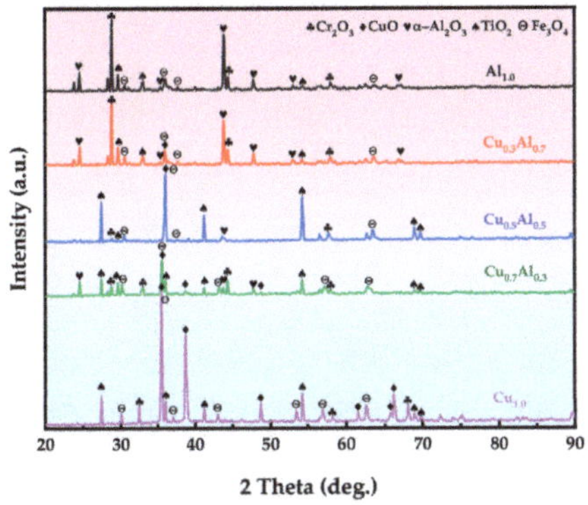

Figure 8. XRD patterns of $Al_xCoCrCu_yFeNi$ HEA coatings post-oxidation test.

Figure 9 shows the surface morphology of the coatings after 50 h of oxidation at 800 °C, while Figure 10 presents the cross-sectional morphology and elemental distribution of the coatings. The strong bonding between the oxide layer and the substrate, without any peeling phenomenon observed, indicates a robust bonding strength at the interface. In contrast, the Ti6Al4V alloy substrate experiences severe surface oxidation, leading to a very loose and flaky oxide film, which will not be discussed further. The results show that the upper layer of the surface oxide is rich in Fe, Ti, and Cr, while the lower layer has a horizontal boundary rich in Ni and Co. During the oxidation process, Ni and Co in the oxide film diffuse and concentrate internally, forming a transition zone with a high concentration of these elements. The preferentially oxidized elements diffuse to the surface, where they combine with oxygen to form an oxide layer, while the remaining elements, such as Ni and Co, diffuse inward, aggregating to form a transition layer [2]. The diffusion rates of elements from highest to lowest are as follows: Ti, Cr, Co, and Ni [49,50]. Based on these diffusion rates, it can be inferred that Ti and Cr will diffuse more quickly than Co and Ni, leading to their preferential reaction with oxygen to form oxides. On the other hand, Co and Ni are slower to oxidize and, therefore, accumulate inward, contributing to the formation of the transition layer. This transition layer not only hinders the diffusion of elements from the coating to the surface but also prevents the inward diffusion of oxygen atoms, thereby enhancing the coating's high-temperature oxidation resistance and protecting the Ti6Al4V alloy substrate.

The likelihood of each element reacting with oxygen can be determined by analyzing the change in Gibbs free energy with temperature. From a thermodynamic perspective, the more negative the value of the Gibbs free energy, the more the reaction is thermodynamically favored, and thus, the more likely the reaction will occur [51], Figure 11 shows the Gibbs free energy of each oxidation product in these high-entropy alloys at different temperatures. The surface oxides of the $Cu_{1.0}$ coating (Figures 9a and 10a) are mainly composed of CuO, Fe_3O_4, and TiO_2. As can be seen from Figure 9a, the primary reason for this phenomenon is the distribution of TiO_2 within the oxide layer, which creates significant gaps between the layers. These gaps provide channels for the rapid diffusion of oxygen, adversely affecting the oxidation resistance of the coating at high temperatures. In addition, the oxide layer is loosely arranged, and several diffused voids can be seen in the oxidation cross-section. This loose structure aggravates the oxidation process and results in the formation of a thicker oxidation layer. Compared to the $Cu_{1.0}$ coating, the oxidized surface of the $Cu_{0.7}Al_{0.3}$ coating shows significant changes following the addition of Al. The oxide morphology transitions from larger blocks to smaller plate-like structures (Figure 9b), and the oxide layer becomes denser and thinner. As can be seen from Figure 10b, TiO_2 is mainly concentrated in the outermost layer of the oxide layer, while Ni and Co are enriched in the innermost layer to form a horizontal transition zone. Other elements are evenly distributed, and the contents of CuO and Fe_3O_4 decrease, which aligns with XRD results. Based on the XRD analysis, it can be seen that the addition of Al leads to the formation of a certain amount of Al_2O_3 on the oxidized surface. Al easily reacts with oxygen at high temperatures to form a continuous and dense Al_2O_3 film, which effectively prevents high-temperature oxidation of the Ti6Al4V alloy substrate. However, there are still some voids in the oxide layer, resulting in a relatively thick oxide film. Compared with the $Cu_{1.0}$ and $Cu_{0.7}Al_{0.3}$ coatings, the $Cu_{0.5}Al_{0.5}$ and $Cu_{0.3}Al_{0.7}$ coatings have a more dense surface oxide film distribution and no obvious porosity (Figure 9c,d). The cross-section of the oxide layer is uniform and dense, showing no significant voids or holes, and the thickness of the oxide layer decreased significantly (Figure 10c,d). It is worth mentioning that for the $Cu_{0.3}Al_{0.7}$ coating, which contains a higher Al content, it can be seen that the concentration of Al_2O_3 and Cr_2O_3 increases. Al_2O_3 is primarily concentrated in the innermost layer. This can be attributed to the high diffusion coefficient of Al and the relatively low bond energy between Al and oxygen compared to other elements, making Al preferentially oxidize at the surface of the coating and form a dense Al_2O_3 film. This Al_2O_3 layer provides effective protection for the coating at an early stage [50]. As oxidation progresses, Cr diffuses outward through the pores at the interface between the Al_2O_3 layer and the coating–oxide boundary, and the Cr_2O_3-rich oxide protective film formed in the steady state phase further impedes the diffusion of oxygen, preventing the formation of new oxides with other elements. For the $Al_{1.0}$ coating, which has the highest Al content, the Al_2O_3 film generated during the oxidation process is denser and more concentrated. At the same time, the higher content of Cr_2O_3 oxides also provides better protection for the coating. As can be seen from Figure 10e, the oxide layer is the thinnest and densest, offering superior protection compared to the Ti6Al4V alloy substrate, which further improves the oxidation resistance at high temperatures.

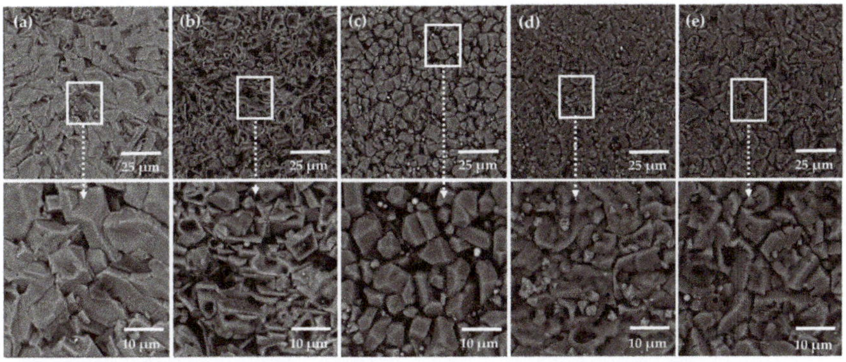

Figure 9. The SEM images of the oxidized surface of $Al_xCoCrCu_yFeNi$ HEA coatings post-oxidation test at 800 °C for 50 h: (**a**) $Cu_{1.0}$; (**b**) $Cu_{0.3}Al_{0.7}$; (**c**) $Cu_{0.5}Al_{0.5}$; (**d**) $Cu_{0.3}Al_{0.7}$; (**e**) $Al_{1.0}$.

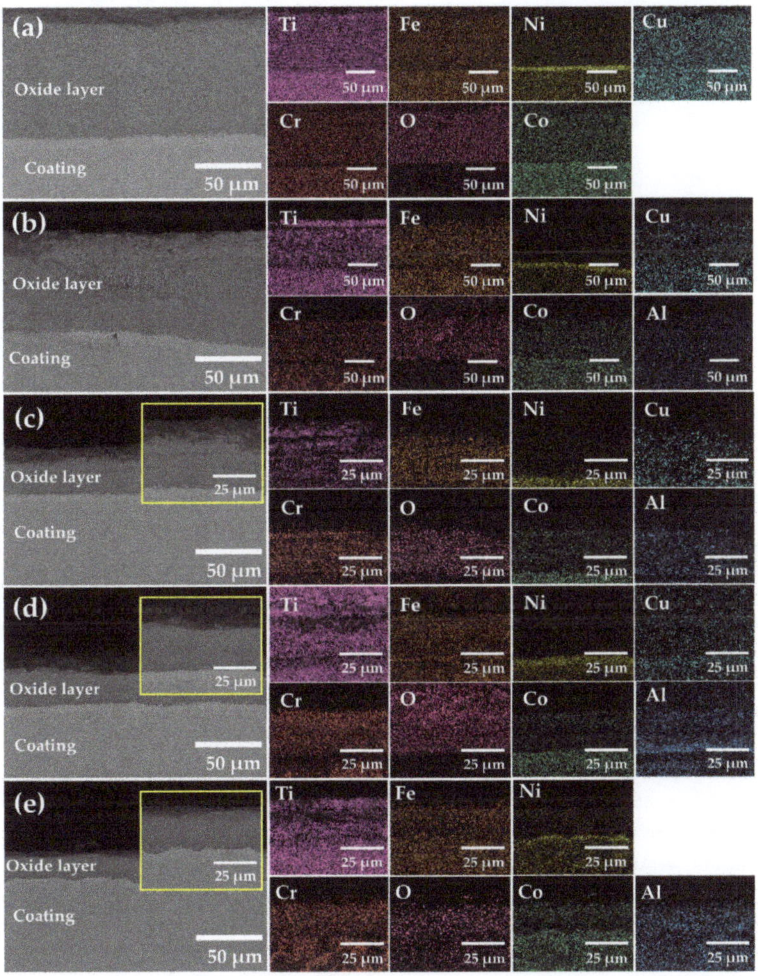

Figure 10. The SEM images and corresponding EDS mappings of the oxidation scale of $Al_xCoCrCu_yFeNi$ HEA coatings at 800 °C for 50 h: (**a**) $Cu_{1.0}$; (**b**) $Cu_{0.7}Al_{0.3}$; (**c**) $Cu_{0.5}Al_{0.5}$; (**d**) $Cu_{0.3}Al_{0.7}$; (**e**) $Al_{1.0}$.

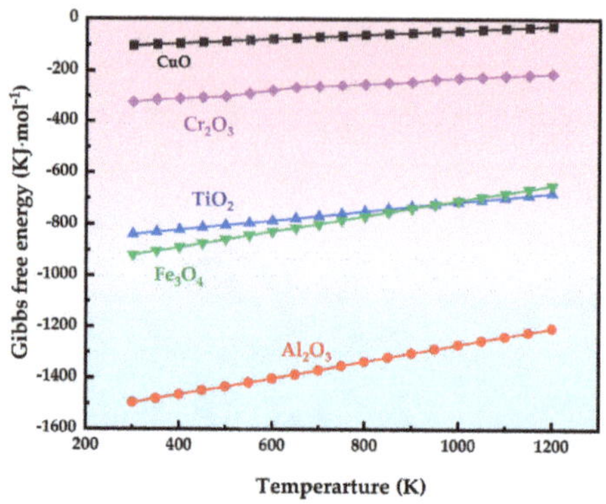

Figure 11. Gibbs free energy of CuO, Cr_2O_3, TiO_2, Fe_3O_4, and Al_2O_3 with temperature.

The oxidation mechanism of the coatings for $Cu_{1.0}$, $Cu_{0.5}Al_{0.5}$, and $Al_{1.0}$ is illustrated in Figure 12. During the initial stage of oxidation, the oxidation reaction is mainly dominated by the reaction rate of the coating surface, where the alloying elements and oxygen are more readily available for the oxidation process. According to the thermodynamic principle, the larger the absolute value of the Gibbs free energy, the more significant the tendency of the elements to undergo spontaneous reaction. As the oxidation process progresses, after a certain length of time, the elements involved in the oxidation at the surface of the coating become increasingly depleted. During this period, the oxidation behavior is largely controlled by the diffusion rates of alloying elements outward from the coating and the rate of oxygen diffusion inward. Therefore, the formation of the final oxidation products is predominantly attributed to the composition of the alloying elements in the coating during the slower oxidation phase, as well as their oxidation kinetics. In this process, oxygen and metal cations within the coating interact to form oxides, with the higher the content of alloying elements and the faster the diffusion rates. The higher the content of alloying elements and the faster their diffusion rates, the more likely the elements are to preferentially react with oxygen and form an oxide layer on the surface of the coating. This oxide layer plays a critical role in protecting the substrate, significantly influencing the overall performance of the coating as well as the evolution of the subsequent oxidation process.

At the early stage of oxidation, which is characterized by rapid oxidation, Fe has high reactivity with O. During this period, Fe fully reacts with O, and as oxidation proceeds, Fe_3O_4 oxide is generated. However, the structure of this oxide is extremely loose, which allows for relatively easy atomic diffusion and does not effectively prevent the inward diffusion of oxygen. Ti atoms, due to their relatively large atomic radius and strong binding affinity with other elements, require the surmounting of a certain energy barrier for lattice diffusion. Consequently, their diffusion is somewhat inhibited, leading to the formation of TiO_2 on the coating surface in a more dispersed, lamellar distribution. The diffusion of Cr contributes to the formation of a continuous and dense oxide film, which can play a certain protective effect on the substrate. The Cr-Fe phase composed of the BCC phase structure at the grain boundary is enriched with a large amount of Cr, attaining a high concentration. Moreover, the atomic arrangement at the grain boundaries is relatively disordered, and the migratory activation energy of atoms is relatively low, which favors the diffusion of Cr atoms and thus promotes the production of a relatively dense Cr_2O_3. In addition, in the alloy with a higher Cu content (Figure 12a, b), the phenomenon that Cu tends to aggregate at the grain boundaries leads to the formation of more CuO. However, this oxide exists in a loose arrangement and diffuse distribution, which cannot effectively limit the mutual diffusion of the atoms, resulting in a relatively thick oxide layer. Compared with other elements, the diffusion rate of Co and Ni is relatively low. As a result, during the oxidation process, they are unable to react with oxygen on time and instead aggregate in the inner layer to form a transition layer. Al has the lowest Gibbs free energy and the fastest diffusion coefficient, so it can combine with oxygen at a very fast rate and preferentially form an Al_2O_3 oxidized layer on the surface of the alloy. When the content of Al is relatively low (Figure 12b), the formed Al_2O_3 oxide film is discontinuous and low in density, which has a limited inhibitory effect in inhibiting atomic diffusion. With the increase in the Al content (Figure 12c), a layer with a uniform and continuous distribution of Al_2O_3 film is formed on the surface of the alloy. This continuous oxide layer serves as a barrier, effectively isolating the substrate from external oxygen atoms and preventing further reactions between oxygen and other elements. This enhances the antioxidant performance and reduces the thickness of the oxidized layer.

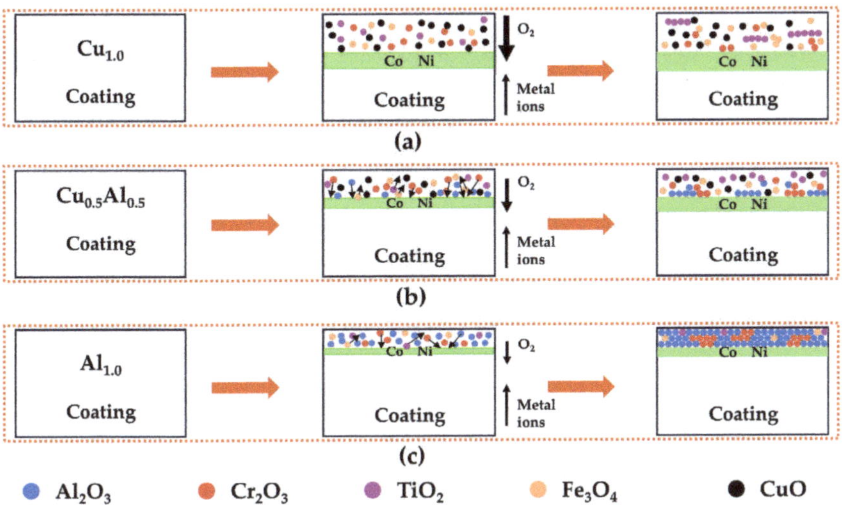

Figure 12. Schematic description of high-temperature oxidation of coatings [36]: (**a**) $Cu_{1.0}$; (**b**) $Cu_{0.5}Al_{0.5}$; (**c**) $Al_{1.0}$.

4. Conclusions

$Al_xCoCrCu_yFeNi$ HEA coatings were fabricated on the surface of a Ti6Al4V alloy using laser cladding technology. The influence of the Al/Cu ratio on the microstructure and high-temperature oxidation resistance of the $Al_xCoCrCu_yFeNi$ HEA coatings was thoroughly analyzed. The main conclusions drawn from this study are as follows.

- The coatings are primarily composed of FCC, BCC, and Ti-rich Laves-reinforced phases. The $Cu_{1.0}$ coating exhibits Co-Ni-Ti-rich FCC phases in the dendrites, Fe-Cr-rich BCC phases in the inter-dendrite regions, and Cu is concentrated at the grain boundaries to form Cu-rich phases (FCC2). The amount of Cu-rich phases decreases as the Cu content is reduced, and the FCC2 phases completely disappear when the Al content reaches 1.0. With the increase in the Al content, the negative mixing enthalpy between Ti and Al causes the FCC1 phases to transition from Co-Ni-Ti-rich to $AlNi_2Ti$- or $AlCo_2Ti$-rich phases.
- Compared to the Ti6Al4V alloy substrate, the solid solution strengthening induced by the doping of elements with larger atomic radii in the high-entropy alloy coatings, as well as the precipitation strengthening effect resulting from the diffusion of Ti from the

substrate into the coatings, significantly increases the microhardness of the coatings. The microhardness of the $Al_xCoCrCu_yFeNi$ HEA coatings is 2.01, 2.06, 2.08, 2.09, and 2.11 times that of the Ti6Al4V alloy substrate, respectively.

- The high-temperature oxidation resistance of the coating increases with the Al content, and the high-temperature oxidation resistance of the $Al_xCoCrCu_yFeNi$ HEA coating is the best when x = 1.0. The oxidation weight gain is 0.098 $mg{\cdot}mm^{-2}$, which is only 35% of that observed for the Ti6Al4V alloy substrate. The Al element easily reacts with oxygen to generate a continuous and dense Al_2O_3 film under high-temperature conditions, effectively preventing the high-temperature oxidation behavior of the substrate. Additionally, Cr atoms diffuse outward through the pores at the interface between the Al_2O_3 layer and the substrate, as well as through the oxide layer, promoting the formation of Cr_2O_3. As oxidation progresses, the formation of a Cr_2O_3-rich oxide layer in the steady-state stage impedes the further oxidation of the coating.

Author Contributions: L.Z.: Conceptualization, methodology, and writing—original draft. H.L.: Supervision, funding acquisition, and project administration. Q.Z.: Data curation. J.L.: Visualization. Y.P.: Investigation. X.H.: Formal analysis and investigation. C.Y.: Software. Y.L.: Data curation. Y.W.: Visualization. All authors have read and agreed to the published version of the manuscript.

Funding: This research was jointly supported financially by the National Nature Science Foundation of China (Grant No. 52475192), the Yunnan Fundamental Research Projects (Grant No. 202201AS070035), and the Scientific and Technologies Project of Yunnan Precious Metals Laboratory (YPML-20240502004, YPML-20240502080).

Data Availability Statement: Data will be made available on request.

Acknowledgments: The authors extend their appreciation to the China–Ukraine Institute of Welding of Guangdong Academy of Sciences, for providing equipment support, and Ben Niu for his valuable support throughout this project.

Conflicts of Interest: The authors declare that they have no known competing financial interests or personal relationships that could have appeared to influence the work reported in this paper.

References

1. Kumar, A.; Singh, G. Surface modification of Ti6Al4V alloy via advanced coatings: Mechanical, tribological, corrosion, wetting, and biocompatibility studies. *J. Alloys Compd.* **2024**, *989*, 174418. [CrossRef]
2. Li, S.; Yamaguchi, T. High-temperature oxidation performance of laser-cladded amorphous TiNiSiCrCoAl high-entropy alloy coating on Ti-6Al-4V surface. *Surf. Coat. Technol.* **2022**, *433*, 128123. [CrossRef]
3. Zhang, K.; Li, B.; Liu, W.; Liu, W.; Wang, W.; Wang, H.; Bian, H. Influences of friction stir processing on the microstructure and properties of TC4 titanium alloy by laser metal deposition. *J. Alloys Compd.* **2024**, *1008*, 176769. [CrossRef]
4. Han, X.; Ma, J.; Tian, A.; Wang, Y.; Li, Y.; Dong, B.; Tong, X.; Ma, X. Surface modification techniques of titanium and titanium alloys for biomedical orthopaedics applications: A review. *Colloids Surf. B Biointerfaces* **2023**, *227*, 113339. [CrossRef]
5. Ren, Y.; Wu, H.; Liu, B.; Liu, Y.; Guo, S.; Jiao, Z.B.; Baker, I. A comparative study on microstructure, nanomechanical and corrosion behaviors of AlCoCuFeNi high entropy alloys fabricated by selective laser melting and laser metal deposition. *J. Mater. Sci. Technol.* **2022**, *131*, 221–230. [CrossRef]
6. Xie, Y.; Wen, X.; Yan, J.; Huang, B.; Zhuang, J. Microstructure and wear resistance of AlCoCrFeNiCuSnX high-entropy alloy coatings by plasma cladding. *Vacuum* **2023**, *214*, 112176. [CrossRef]
7. Korobov, Y.; Antonov, M.; Astafiev, V.; Brodova, I.; Kutaev, V.; Estemirova, S.; Devyatyarov, M.; Okulov, A. Erosion Wear Behavior of HVAF-Sprayed WC/Cr_3C_2-Based Cermet and Martensitic Stainless Steel Coatings on AlSi7Mg0.3 Alloy: A Comparative Study. *J. Manuf. Mater. Process* **2024**, *8*, 231. [CrossRef]
8. Qadir, D.; Sharif, R.; Nasir, R.; Awad, A.; Mannan, H.A. A review on coatings through thermal spraying. *Chem. Pap.* **2023**, *78*, 71–91. [CrossRef]
9. Ahmed, A.; Mhaede, M.; Wollmann, M.; Wagner, L. Characteristics of sintered HA coating deposited by chemical method on AISI 316L substrate. *Mater. Des.* **2015**, *76*, 9–17. [CrossRef]

10. He, X.; Zhang, H.; Wu, Y.; Yang, T.; Li, R.; Nong, Z.; Cui, X. Research progress on effects of additives and pre- and post-treatment processes on corrosion resistance of micro-arc oxidation coatings on titanium and its alloys. *Int. J. Electrochem. Sci.* **2024**, *19*, 100836. [CrossRef]
11. Wu, H.; Wu, Y.; Yan, M.; Tu, B.; Li, Y. Microstructure and mechanical properties of surface coating prepared on grade 2 titanium by Ni-B composite electroplating and laser cladding. *Opt. Laser Technol.* **2023**, *164*, 109498. [CrossRef]
12. Xiang, Y.; Zou, C. Effect of Arc Currents on the Mechanical, High Temperature Oxidation and Corrosion Properties of CrSiN Nanocomposite Coatings. *Coatings* **2021**, *12*, 40. [CrossRef]
13. Ma, Q.; Zhang, H.; Lv, Y.; Song, C.; Guo, N.; Xiao, G.; Zhao, W. Microstructure and properties of $W_{0.5}Ta_{0.3}MoNbVAlTi_{1-x}Zr_x$ high entropy alloy coatings by laser cladding on the surface of 45# steel. *Ceram. Int.* **2023**, *49*, 36416–36428. [CrossRef]
14. Zhu, L.; Xue, P.; Lan, Q.; Meng, G.; Ren, Y.; Yang, Z.; Xu, P.; Liu, Z. Recent research and development status of laser cladding: A review. *Opt. Laser Technol.* **2021**, *138*, 106915. [CrossRef]
15. Zhao, H.; Zhao, C.; Xie, W.; Wu, D.; Du, B.; Zhang, X.; Wen, M.; Ma, R.; Li, R.; Jiao, J.; et al. Research Progress of Laser Cladding on the Surface of Titanium and Its Alloys. *Materials* **2023**, *16*, 3250. [CrossRef]
16. Zhang, Y.; Liu, Z.; Lv, Z.; Cao, J.; Tong, Y.; Sun, M.; Cui, C.; Wang, X. Effect of SiC and TiC content on microstructure and wear behavior of Ni-based composite coating manufactured by laser cladding on Ti–6Al–4V. *Wear* **2024**, *552–553*, 205431. [CrossRef]
17. Meng, Q.; Geng, L.; Ni, D. Laser cladding NiCoCrAlY coating on Ti-6Al-4V. *Mater. Lett.* **2005**, *59*, 2774–2777. [CrossRef]
18. Liu, X.-B.; Yu, R.-L. Microstructure and high-temperature wear and oxidation resistance of laser clad $\gamma/W_2C/TiC$ composite coatings on γ-TiAl intermetallic alloy. *J. Alloys Compd.* **2007**, *439*, 279–286. [CrossRef]
19. Kooi, B.J.; Pei, Y.T.; De Hosson, J.T.M. The evolution of microstructure in a laser clad TiB–Ti composite coating. *Acta Mater.* **2003**, *51*, 831–845. [CrossRef]
20. Barbatti, C.; Garcia, J.; Brito, P.; Pyzalla, A.R. Influence of WC replacement by TiC and (Ta,Nb)C on the oxidation resistance of Co-based cemented carbides. *Int. J. Refract. Met. Hard Mat.* **2009**, *27*, 768–776. [CrossRef]
21. Yang, X.; Wang, L.; Gao, Z.; Wang, Q.; Du, M.; Zhan, X. WC distribution, microstructure evolution mechanism and microhardness of a developed Ti-6Al-4 V/WC MMC coating fabricated by laser cladding. *Opt. Laser Technol.* **2022**, *153*, 108232. [CrossRef]
22. Li, J.; Cui, X.; Guan, Y.; Wan, S.; Jin, G.; Zheng, W.; Su, W. Enhancing coating performance: Design and characterization of a Ti-Al-Si-Cr-SiC system coating using ultrasonic and rare earth assisted laser cladding. *Mater. Charact.* **2024**, *210*, 113823. [CrossRef]
23. Das, A.K. Effect of rare earth oxide additive in coating deposited by laser cladding: A review. *Mater. Today Proc.* **2022**, *52*, 1558–1564. [CrossRef]
24. Yeh, J.-W.; Chen, S.-K.; Lin, S.-J.; Gan, J.-Y.; Chin, T.-S.; Shun, T.-T.; Tsau, C.-H.; Chang, S.-Y. Nanostructured High-Entropy Alloys with Multiple Principal Elements: Novel Alloy Design Concepts and Outcomes. *Adv. Eng. Mater.* **2004**, *6*, 299–303. [CrossRef]
25. Diao, H.Y.; Feng, R.; Dahmen, K.A.; Liaw, P.K. Fundamental deformation behavior in high-entropy alloys: An overview. *Curr. Opin. Solid State Mat. Sci.* **2017**, *21*, 252–266. [CrossRef]
26. Tokarewicz, M.; Grądzka-Dahlke, M. Review of Recent Research on AlCoCrFeNi High-Entropy Alloy. *Metals* **2021**, *11*, 1302. [CrossRef]
27. Dąbrowa, J.; Cieślak, G.; Stygar, M.; Mroczka, K.; Berent, K.; Kulik, T.; Danielewski, M. Influence of Cu content on high temperature oxidation behavior of AlCoCrCuxFeNi high entropy alloys (x = 0; 0.5; 1). *Intermetallics* **2017**, *84*, 52–61. [CrossRef]
28. Guo, S.; Liu, C.T. Phase stability in high entropy alloys: Formation of solid-solution phase or amorphous phase. *Prog. Nat. Sci. Mater. Int.* **2011**, *21*, 433–446. [CrossRef]
29. Miracle, D.B.; Senkov, O.N. A critical review of high entropy alloys and related concepts. *Acta Mater.* **2017**, *122*, 448–511. [CrossRef]
30. Du, X.; Du, Z.; Cao, J.; Zhang, S.; Gong, T.; Sun, B.; Zhang, C.; Huo, R.; Liu, J.; Cheng, J. Enhance the mechanical properties of CoCrFeNi high entropy alloy: Rare-earth element Gd was employed as a "cladding structure" to refine grains. *Mater. Today Commun.* **2024**, *41*, 110732. [CrossRef]
31. Zhang, S.; Sun, Y.; Cheng, W.; Chen, Y.; Gu, J. High temperature oxidation behavior of CoCrFeNiMo0.2 high-entropy alloy coatings produced by laser cladding. *Mater. Today Commun.* **2024**, *39*, 108639. [CrossRef]
32. Zhu, Z.-X.; Liu, X.-B.; Liu, Y.-F.; Zhang, S.-Y.; Meng, Y.; Zhou, H.-B.; Zhang, S.-H. Effects of Cu/Si on the microstructure and tribological properties of FeCoCrNi high entropy alloy coating by laser cladding. *Wear* **2023**, *512–513*, 204533. [CrossRef]
33. Ye, Y.F.; Wang, Q.; Lu, J.; Liu, C.T.; Yang, Y. High-entropy alloy: Challenges and prospects. *Mater. Today* **2016**, *19*, 349–362. [CrossRef]
34. Butler, T.M.; Weaver, M.L. Oxidation behavior of arc melted AlCoCrFeNi multi-component high-entropy alloys. *J. Alloys Compd.* **2016**, *674*, 229–244. [CrossRef]
35. Qin, L.; Ren, P.; Yi, Y.; Xie, T.; Hu, Y.; Chen, D.; Zhou, S. Effect of Al_2O_3 content on the high-temperature oxidation behaviour of CoCrAlYTa coatings produced by laser-induction hybrid cladding. *Corros. Sci.* **2022**, *209*, 110739. [CrossRef]

36. Wang, S.; Chen, Z.; Zhang, P.; Zhang, K.; Chen, C.L.; Shen, B.L. Influence of Al content on high temperature oxidation behavior of $Al_xCoCrFeNiTi_{0.5}$ high entropy alloys. *Vacuum* **2019**, *163*, 263–268. [CrossRef]
37. Huang, C.; Zhang, Y.; Shen, J.; Vilar, R. Thermal stability and oxidation resistance of laser clad TiVCrAlSi high entropy alloy coatings on Ti–6Al–4V alloy. *Surf. Coat. Technol.* **2011**, *206*, 1389–1395. [CrossRef]
38. Zhang, H.X.; Dai, J.J.; Sun, C.X.; Li, S.Y. Microstructure and wear resistance of TiAlNiSiV high-entropy laser cladding coating on Ti-6Al-4V. *J. Mater. Process. Technol.* **2020**, *282*, 116671. [CrossRef]
39. Geng, H.; Shen, J.; Hu, S.; Zhang, F.; Geng, K. Microstructure, corrosion and wear behavior of $(AlCu)_{3.5}CoCrNiFe$ and $(AlCu)_{3.5}CoCrNiTi$ high entropy alloy coatings prepared by laser cladding on AZ91 magnesium alloy. *J. Mater. Process Technol.* **2024**, *30*, 3383–3393. [CrossRef]
40. Yeh, J.-W.; Chang, S.-Y.; Hong, Y.-D.; Chen, S.-K.; Lin, S.-J. Anomalous decrease in X-ray diffraction intensities of Cu–Ni–Al–Co–Cr–Fe–Si alloy systems with multi-principal elements. *Mater. Chem. Phys.* **2007**, *103*, 41–46. [CrossRef]
41. Li, Y.; Shi, Y. Phase assemblage and properties of laser cladded TixCrFeCoNiCu high-entropy alloy coating on aluminum. *Mater. Res. Express* **2020**, *7*, 036519. [CrossRef]
42. Hao, X.; Liu, H.; Zhang, X.; Tao, J.; Wang, Y.; Yang, C.; Liu, Y. Microstructure and wear resistance of in-situ $TiN/(Nb, Ti)_5Si_3$ reinforced MoNbTaWTi-based refractory high entropy alloy composite coatings by laser cladding. *Appl. Surf. Sci.* **2023**, *626*, 157240. [CrossRef]
43. Jiang, Y.Q.; Li, J.; Juan, Y.F.; Lu, Z.J.; Jia, W.L. Evolution in microstructure and corrosion behavior of $AlCoCr_xFeNi$ high-entropy alloy coatings fabricated by laser cladding. *J. Alloys Compd.* **2019**, *775*, 1–14. [CrossRef]
44. Huang, Y. Characterization of dilution action in laser-induction hybrid cladding. *Opt. Laser Technol.* **2011**, *43*, 965–973. [CrossRef]
45. Li, K.; Chen, W. Recent progress in high-entropy alloys for catalysts: Synthesis, applications, and prospects. *Mater. Today Energy* **2021**, *20*, 100638. [CrossRef]
46. Yi, H.; Yang, X.; Yang, Y.; Yin, B.; Hu, J.; Liang, M.; Yue, J.; Yin, F. New insights in the oxidation behavior of $(FeCoCrNi)_{94}Al_4Ti_2Si$ high entropy alloys at 1100 °C. *Corros. Sci.* **2024**, *227*, 111673. [CrossRef]
47. Yu, K.; Zhao, W.; Li, Z.; Guo, N.; Xiao, G.; Zhang, H. High-temperature oxidation behavior and corrosion resistance of in-situ TiC and Mo reinforced AlCoCrFeNi-based high entropy alloy coatings by laser cladding. *Ceram. Int.* **2023**, *49*, 10151–10164. [CrossRef]
48. Chen, G.; Yang, H.; Sun, H.; Wang, F.; Wang, H.; Kong, Q.; An, X.; Zhang, Y.; Wang, J. Exploring the high-temperature steam oxidation behaviors of the lean-Cr (7–10 wt%) FeCrAl alloys. *Corros. Sci.* **2022**, *194*, 109927. [CrossRef]
49. Tsai, K.Y.; Tsai, M.H.; Yeh, J.W. Sluggish diffusion in Co–Cr–Fe–Mn–Ni high-entropy alloys. *Acta Mater.* **2013**, *61*, 4887–4897. [CrossRef]
50. Dąbrowa, J.; Kucza, W.; Cieślak, G.; Kulik, T.; Danielewski, M.; Yeh, J.-W. Interdiffusion in the FCC-structured Al-Co-Cr-Fe-Ni high entropy alloys: Experimental studies and numerical simulations. *J. Alloys Compd.* **2016**, *674*, 455–462. [CrossRef]
51. Sun, D.; Cai, Y.; Zhu, L.; Gao, F.; Shan, M.; Manladan, S.M.; Geng, K.; Han, J.; Jiang, Z. High-temperature oxidation and wear properties of TiC-reinforced CrMnFeCoNi high entropy alloy composite coatings produced by laser cladding. *Surf. Coat. Technol.* **2022**, *438*, 128407. [CrossRef]

materials

MDPI

Article

Antioxidant Behavior of Carbon/Carbon Composites with Hot Dip Plating and Electroplating for Single-Crystal Furnaces

Zuxing Qi [1], Chaofan Du [1], Guoying Bao [1], Shan Wang [1,*], Dedong Gao [1], Haixing Lin [1] and Yan An [2]

[1] Department of Mechanical Engineering, Qinghai University, Xining 810016, China; ys220805030205@qhu.edu.cn (Z.Q.); ys210805010157@ghu.edu.cn (C.D.); ys230805020201@qhu.edu.cn (G.B.); gaodd@qhu.edu.cn (D.G.); ys220802030171@qhu.edu.cn (H.L.)

[2] Sichuan Gokin Solar Technology Co., Ltd., Yibin 644600, China; yan.an@gokinsolar.com

* Correspondence: 2005990025@qhu.edu.cn

Abstract: In the Czochralski single-crystal silicon manufacturing industry, single-crystal furnaces often experience corrosion from silicon vapor, which reduces their operational lifespan. However, the preparation of metal coatings on the surface of C/C composites is challenging due to their low coefficient of thermal expansion and the intricate structure of carbon fibers. To address this issue and achieve high-quality alloy coatings, Ni-Al and Ni-Al/Si composite coatings are successfully prepared on the surface of C/C composites through a combination of electroplating and hot-dip plating, and their oxidation behavior at elevated temperatures is thoroughly investigated. The experimental results indicate that the Ni-Al composite coatings exhibit superior antioxidant properties compared to Ni coatings following thermal shock experiments, thereby significantly enhancing the antioxidant performance of C/C composites. This improvement is attributed to the preferential oxidation of surface aluminum, which forms a dense Al_2O_3 layer in aerobic and high-temperature environments, effectively preventing oxygen from reaching the underlying matrix. During the oxidation process, coating elements migrate outward along the concentration gradient, while oxygen molecules diffuse inward. Simultaneously, aluminum atoms diffuse inward, and Ni atoms diffuse outward, where they partially dissolve with oxygen. The inner coating's Ni enhances the bonding of the coating by connecting the substrate to the outer layer. Meanwhile, the added Si in the Ni-Al/Si composite coating further improves the antioxidant properties of the coating.

Keywords: carbon/carbon composites; composite coating; hot-dip plating; antioxidation performance

1. Introduction

Silicon solar cells are moving towards higher efficiency and lower cost. Cast-type single-crystal silicon is a promising material for achieving this goal [1]. In recent years, due to the growing demand for thermal field stability and improved growth quality in single-crystal furnaces, Carbon/Carbon (C/C) composites have been extensively utilized in the production of single-crystal furnaces. As a kind of high-temperature-resistant and high-strength material, C/C composites have attracted much attention in the field of aerospace and crystal manufacturing. In particular, the application of C/C composite materials has emerged as a predominant trend in the high-temperature regions of single-crystal furnaces, including components such as the crucible, flow guide tube, and upper and lower insulation tubes. As integral components of the single-crystal furnace, these elements directly influence the stability of the thermal field, as well as the quality and growth rate of monocrystalline silicon. Fu et al. consolidate recent advancements in micro/nano multiscale strategies—such as nanoparticles, nanowires, and graphene—to enhance high-temperature oxidation and ablation resistance in C/C composites. Their paper concludes with challenges and future directions for achieving superior mechanical and thermal performance, aiming to inspire scientific and industrial advancements in robust C/C materials [2–4]. However, the broader application of C/C composites is restricted by their limitations in temperature oxidation resistance. It has been found that C/C composites are oxidized at 450 °C and burned at 550 °C in an aerobic environment. Han et al. investigate the impact of oxidation on residual mechanical properties for safe in-service use. They propose a microscale degradation model based on mass conservation and diffusion to predict oxidation behavior and use simulations to show a 24–32% strength reduction after 30 min at 850 °C. Their findings reveal that oxidation redistributes stress, weakens load capacity, and increases oxygen diffusion in stressed areas of the pyrocarbon matrix [5–7]. C/C composites are primarily composed of carbon fiber mesh and pyrolytic carbon powder. The carbon mesh planes are typically characterized by unsaturated chemical bonds and free-electron fringe carbon atoms, which are regarded as active adsorption sites. These sites are prone to adsorption of surrounding oxidizing gasses, and oxidation reactions occur at temperatures of about 400 ° C. During oxidation, the initial erosion occurs at the material's surface, including pores and the fiber/matrix interface. This is followed by the erosion of the anisotropic and isotropic matrix carbon, then the fiber side surfaces extending to their ends, and ultimately the core of the fibers. Oxidative corrosion initially targets the matrix carbon with a low degree of graphitization, which is progressively removed as corrosion causes pore expansion. Concurrently, as the temperature rises, the carbon fibers experience gradual radial erosion [8,9].

To enhance the oxidation resistance of C/C composites, two primary methods are employed: substrate modification and coating preparation. Coating preparation involves applying a protective layer that prevents oxygen infiltration without altering the substrate's inherent properties. The antioxidant design idea of the coating is shown in Figure 1 below, which has the advantages of being simple, convenient, and cheap [10,11]. Composite coatings have been extensively studied by researchers worldwide for their superior antioxidant properties. Zhu et al. successfully prepared SiC-$ZrSi_2$ gradient composite coatings on the surface of C/C composites using the embedding and slurry methods. The results demonstrated that the coatings exhibited optimal antioxidant properties at 1200 °C, 1300 °C, and 1500 °C, and showed improved oxidation resistance at medium and high temperatures [12]. Li et al. investigated the protective effect of a three-layer gradient self-healing coating on C/C composites, consisting of an inner layer of B_4C and β-SiC, a middle layer of SiC, and an outer sealing layer of SiO_2. The results indicated that the coating exhibited effective self-healing at 1500 °C, with an oxidative weight change rate of 1.3% after 50 h at 1500 °C, along with excellent thermal shock resistance [13]. Huang et al. synthesized Al-doped SiC coatings on C/C composites using the filler cementation technique to study the microstructure, composition, and antioxidant properties of the coatings at different preparation temperatures (1500~2100 °C). The results showed that coatings prepared at 1500~1900 °C were discontinuous 3C-SiC/2H-SiC/C/Si/Al_2O_3 coatings with some pores and cracks. In contrast, coatings prepared at 2100 °C consisted of 4H-SiC with dissolved Al atoms and exhibited fewer cracks. The coatings obtained at 2100 °C demonstrated the best antioxidant properties due to the in situ formation of Al_2O_3 and SiO_2 oxide layers on the surface [14]. Chen et al. investigate Cr-coated Zr alloys as promising ATF cladding materials. In their study, they apply Cr coatings to 1400 mm N36 tubes using an industrial arc system. Orthogonal analyses identify key process parameters affecting coating characteristics, including surface roughness, defects, and crystal orientation [15]. In summary, compared to single coatings, composite coatings are more effective at mitigating or preventing high-temperature oxidation reactions, thereby reducing performance degradation and extending the lifespan of C/C composites. These coatings are crucial for enhancing the antioxidant properties of C/C composites [16].

Hot-dip plating, a straightforward method for preparing dense coatings, is frequently employed in coating metal substrates [17]. Hot-dip plating not only significantly enhances the corrosion resistance and high-temperature oxidation resistance of the substrate material but also improves the material's surface appearance and imparts a certain level of comprehensive mechanical properties [18,19]. The hot-dip plating process, currently encompassing two primary systems, involves the first method known as the protective gas method [20–22]. This technique employs a redox process to remove oil and oxides from the substrate surface, after which the substrate is introduced into a molten aluminum or aluminum alloy bath within a reducing atmosphere. The second system is the flux-assisted hot-dip plating method. This process entails removing contaminants and oxides from the substrate's surface, immersing the substrate in a flux solution to form a coating aid film, and finally submerging the treated part into molten aluminum to complete the aluminum plating process.

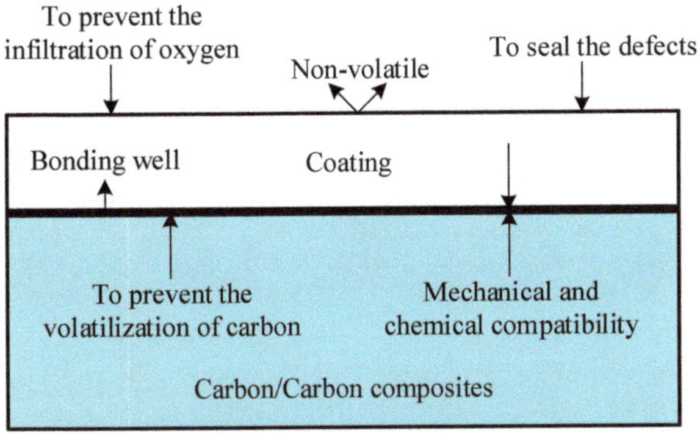

Figure 1. Antioxidant coating design conditions.

The use of auxiliary plating agents can activate the metal surface, improve the infiltration of aluminum to the steel matrix, and play an important role in protecting the surface of the plating from oxidation or contamination. In the molten aluminum solution, the film of the plating aid dissolves, allowing the exposed steel surface to come into direct contact with the aluminum. This promotes rapid wetting of the surface, leading to the formation of a continuous and uniform aluminum coating [23]. Cheng studied the microstructure evolution of intermetallic layers of hot-dip-aluminized low-carbon steel with silicon added [24]. The thickness of the coating decreased with the increase in Si content. The metal alloy layer consists of a $FeAl_3$ layer with an outer layer thickness of 5 μm and the main Fe_2Al_5 layer in the inner layer, which is mainly because silicon occupies the C-axis vacancy on Fe_2Al_5. The diffusion rate of iron and aluminum in the Fe_2Al_5 layer was slowed down. Zhang et al. employed hot-dip plating to deposit a WSi_2 coating on a tungsten substrate [25]. The results indicated that the tungsten–silicide coating consists of a WSi_2 layer and an interfacial W_5Si_3 layer, with the silicon concentration gradually decreasing and the tungsten concentration correspondingly increasing along the diffusion direction. The hot-dip plating temperature has a significant effect on the coating thickness and grain size, and the prepared coating has strong oxidation resistance. The combination of electroplating and melt hot-dip plating is applied to prepare coatings on C/C composites. This advances hot-dip plating techniques in C/C composite coatings and provides a novel approach for future development in the field.

Currently, C/C composite coating preparation methods include magnetic sputtering, chemical vapor deposition (CVD), the sol–gel method, and plasma spraying [10,26,27]. These techniques are characterized by complex processes, high costs, and limited applicability to small-scale components [28]. Meanwhile, rare-earth metals are often added to the coating materials to improve the antioxidant properties of the coatings, which further increases the cost [29,30]. In recent years, researchers have dedicated efforts to exploring novel coating preparation techniques, such as plasma-enhanced chemical vapor deposition (PCVD) and molten salt deposition, with the aim of addressing current challenges in the coating process, reducing production costs, and improving coating performance [31]. In addition, it is also crucial to continuously innovate the traditional coating preparation technology and explore new raw materials and processes, such as the development of new glass coatings, precious metal coatings, ceramic coatings, and so on [32].

Considering the challenges associated with depositing metal coatings on C/C composites, which arise from their low thermal expansion coefficient and intricate carbon fiber structure, a novel method is proposed for the preparation of Ni-Al and Ni-Al/Si composite coatings on C/C composite substrates, involving Ni electroplating followed by Al hot-dip plating. This method reduces corrosion of the boiler by silicon vapor in the Czochralski single-crystal silicon manufacturing industry.

2. Experimental Materials and Methods

2.1. Experimental Methods

The detailed experimental procedure is illustrated in Figure 2, which comprehensively represents the steps involved. The C/C composite material is processed as 15 mm × 15 mm × 5 mm. A Ni (Shanghai Macklin Biochemical Technology Co., Ltd., Shanghai, China) coating with a thickness of about 13 μm is deposited on the outer surface of the C/C composites by electroplating. Al (Shanghai Macklin Biochemical Technology Co., Ltd.) and Al-Si coatings are then applied on the Ni layer using hot-dip plating. Finally, Ni-Al and Ni-Al/Si composite coatings are formed.

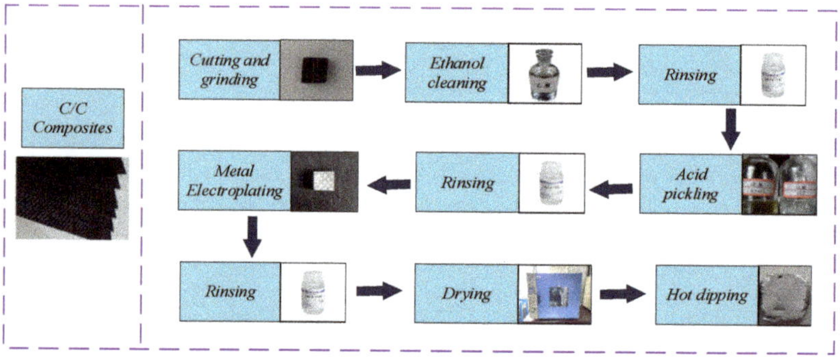

Figure 2. Experimental flowchart.

The C/C composites are polished with 400#, 600#, and 800# grit sandpaper prior to plating, resulting in a clean and smooth cut. Subsequently, ultrasonic cleaning is conducted with deionized water and anhydrous ethanol in succession, followed by drying in an oven at 130 °C for 3 h and storage in a desiccator. The plating solution requires temperature control during the plating process, maintaining the solution temperature at 55 °C, the pH between 5.8 and 6.2, the current density at 0.4 A/dm^2, the magnetic stirring speed at 500 rpm, and the plating time at 40 min.

The components of the plating solution for Ni deposition on the surface of C/C composites are listed in Table 1.

Table 1. Electroplating solution formulation.

$NiCO_3$	$NiCl_2$	H_3BO_3	pH	Stirring
406 gL^{-1}	45 gL^{-1}	175.5 gL^{-1}	5.8~6.2	500 rpm

$NiCO_3$ (Shanghai Macklin Biochemical Technology Co., Ltd.) and $NiCl_2$ (Shanghai Macklin Biochemical Technology Co., Ltd.) serve as sources of Ni ions for the solution, while boric acid acts as a buffer to regulate the pH of the electrolyte during the plating process. In the electrochemical reaction, variations in the pH of the solution influence the reaction rate at the electrode surface, the formation and precipitation of reaction products, and the protection of the electrode.

The Ni-coated samples are cleaned using ultrasonic bubbles in acetone and anhydrous alcohol, followed by low-temperature drying to remove surface oil and chemical residues.

The sample intended for hot dipping requires pre-treatment with an auxiliary plating process. The composition of the plating aid is shown in Table 2.

Table 2. Composition of hot-dip plating aids.

Reagent	NaCl	KCl	NaF	KF
Mass Fraction	3.6%	30%	2.4%	2.4%

The auxiliary plating is performed at 87 °C for 10 min, after which Al and Al/Si hot-dip coatings are applied under inert gas protection. Finally, Ni-Al and Ni-Al/Si composite coatings are successfully fabricated. The addition of silicon improves both the wear resistance and the corrosion resistance of the material by reducing the coating thickness [22]. Due to the significantly higher melting point of silicon compared to aluminum, Al/Si ingots are utilized to introduce silicon into the aluminum for easier melting. The effects of temperature, holding time, and silicon content on the antioxidant properties of the coating are investigated through three-factor, three-level orthogonal experiments. The corresponding parameters are provided in Table 3.

Table 3. Factor level table for orthogonal experiments.

Factor / Level	A/min	B/°C	C/Silicon Content (wt %)
1	60	720	0
2	120	740	2
3	180	760	5

2.2. *Coating Characterization*

The surface and cross-sectional morphologies of the coatings before and after oxidation are observed using a JEOL (JSM-7900F, JEOL Ltd., Tokyo, Japan) field emission electron microscope. The field FESEM offers high resolution, with secondary electron images reaching 0.7 nm at 15 kV. It has a magnification range from 25× to 1,000,000× and an acceleration voltage range of 0.01 kV to 30 kV. The microscope features a Schottky field emission electron gun with an electron beam current range of 1 pA to 500 nA. The working distance is automatically adjustable from 2 mm to 40 mm, with manual adjustment within 2 mm. The optimal working distance for EDS is 10 mm. An EDS is used for the quantitative analysis of the elemental composition in the microregions of the coating. Additionally, a D/max2500PC X-ray diffractometer (Rigaku, Tokyo, Japan) is used to characterize the phases of the prepared coatings and the oxide phases formed after high-temperature oxidation. The XRD (Rigaku, Tokyo, Japan) uses a copper (Cu) target as the anode, with a scintillation counter. Test mode settings include a step size of 0.02°and a speed of 8° per minute, using a 2-Θ/Θ linkage. Data analysis is conducted using JADE 9 (MDI-R98218) software.

2.3. *High-Temperature Oxidation Test*

A thermal shock test is conducted to evaluate the oxidation resistance of the coating [33]. Thermal shock tests are conducted in a box-type resistance furnace at temperatures of 450 °C, 500 °C, 550 °C, 600 °C, 650 °C, and 700 °C. When the furnace temperature reaches the specified level, the sample is placed in the furnace and held at that temperature for 30 min, followed by cooling at room temperature for 5 min. This constitutes 1 cycle, with a total of 20 cycles performed. The weight change rate of the thermal shock experimental coating at intermediate temperatures is used as the criterion for evaluating the results of the orthogonal experiment. Subsequently, the thermal shock experiment is conducted at temperatures ranging from 800 °C to 1400 °C for the C/C composite materials, Ni electroplated samples, and the optimal hot-dip plating solution, following the aforementioned experimental procedures. The macro-morphology is observed in the cycle and the weight change rate is calculated. The formula for calculating its oxidized weight rate is as follows:

$$\Delta W = (M_2 - M_1)/M_1$$

where ΔW is oxidized weight change in the specimen (%), M_2 is mass after oxidation (g), and M_1 is mass before oxidation (g).

2.4. Silicon Vapor Corrosion Test

Silicon vapor corrosion tests are conducted on the prepared Ni coating samples, composite coating samples, and C/C composite materials. The experimental process is as follows:

1. Sample pretreatment: The coated samples and C/C composites materials are subjected to degreasing and decontamination procedures to eliminate any impurity elements that could potentially influence the corrosion results during the experimental process.
2. Corrosion experiment: The sample is positioned in a crucible with a layer of monocrystalline silicon raw material added to the bottom. The crucible is then subjected to multiple vacuum cycles, followed by the introduction of argon to purge the furnace of air. The temperature is subsequently raised to 1450 °C and maintained for 5 h. After cooling, the sample is extracted and analyzed using microscopic morphology observation and energy-dispersive spectroscopy.

3. Results and Analysis

3.1. Analysis of Orthogonal Experiment Results

The effects of temperature, time, and silicon content on the oxidation resistance of the coating are evaluated through orthogonal experiments. The range R calculation method can be employed to ascertain the influence ranking of each factor on the coating quality, while mean value calculations can be used to identify the optimal parameter combination. The antioxidant performance of the coatings is used as a criterion, and the results are presented in Table 4 below. The order of influence on antioxidant performance, determined through the calculation of the extreme deviation R, is as follows: B > A > C. The optimal parameter combination is presented in Table 4 below. Based on the mean value calculation, the optimal parameters are $A_1B_2C_2$, which correspond to a time of 60 min, a temperature of 740 °C, and a silicon content of 2%. Temperature is the most critical factor affecting antioxidant properties, followed by time, while silicon content has a relatively minor impact. This indicates that, in improving the antioxidant properties of coatings, priority should be given to controlling the temperature, with appropriate extension of the treatment time. The limited effect of silicon content on antioxidant performance suggests that variations in silicon content have a minimal impact on the antioxidant properties of the coatings within the studied range. Details are in Appendix A.

Table 4. Orthogonal experimental data and computational analysis.

Experiment Number	A/min	B/°C	C/Silicon Content (%)	y/Oxidative Weight Change (%)
1	1	1	1	13.5108
2	1	2	2	17.2460
3	1	3	3	10.0913
4	2	1	2	11.2999
5	2	2	3	15.2080
6	2	3	1	9.6719
7	3	1	3	7.8758
8	3	2	1	10.2643
9	3	3	2	8.3390

Table 4. *Cont.*

Experiment Number	A/min	B/°C	C/Silicon Content (%)	y/Oxidative Weight Change (%)
I_{1j}	40.8481	32.6865	33.4470	
II_{2j}	36.1798	42.7183	36.8849	
III_{3j}	26.4791	28.1022	33.1751	
K_{1j}	13.6160	10.8955	11.1490	
K_{2j}	12.0599	14.2394	12.2950	
K_{3j}	8.8264	9.3674	11.0584	
R	4.7896	4.8720	1.2366	

The nine groups of samples undergo thermal shock experiments to investigate the optimal hot-dip plating parameters and the influence of each parameter. The results are presented in Figure 3. The figure shows that the weight change rate of the specimens increases with temperature during the thermal shock experiment. This increase is attributed to the accelerated oxidation reaction at higher temperatures, which facilitates greater oxygen diffusion into the coating. However, the weight change rate decreases significantly when the temperature exceeds 600 °C. This decrease is explained by the oxidation of the Al element on the surface and the formation of a protective Al_2O_3 layer, which impedes further oxygen penetration. The weight change observed in the specimens is primarily influenced by oxidizing reactions occurring at high temperatures. As the temperature rises, thermal energy accelerates chemical reactions, enhancing the rate of oxidation. Consequently, more oxygen atoms react with the coating material, leading to increased weight change. Additionally, the diffusion rate of oxygen through the coating is also affected by the oxidation temperature; higher temperatures facilitate faster oxygen diffusion, further contributing to the increased weight change. The weight change rate decreases above 600 °C, which can be attributed to the formation of a dense Al_2O_3 layer on the coating's surface. This Al_2O_3 layer acts as a barrier, preventing further oxygen penetration. It is formed through the oxidation of the Al element present on the coating's surface. This layer effectively restricts the interaction between oxygen and the underlying coating material, thereby reducing the rate of weight change.

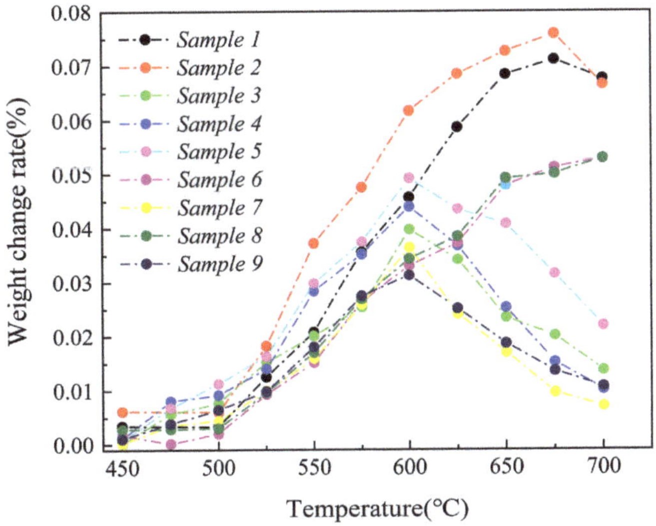

Figure 3. Comparison of weight change rate of the composite coating orthogonal experiment samples in the thermal shock experiment.

In conclusion, the thermal shock experiments demonstrated that the weight change rate of the samples increased with rising temperature due to the enhanced oxidation reactions. Beyond the critical temperature of 600 °C, the weight change rate decreased, primarily attributed to the formation of a protective Al_2O_3 layer that impedes further oxygen entry.

3.2. XRD Scanning Results After Coating and Coating Oxidation

As shown in Figure 4, the X-ray diffraction patterns of the oxidized composite coating, the hot-dip aluminum composite coating, and the electroplated Ni coating are presented in Figure 4a–c. After comparison with the PDF (98-000-0062) standard card, in Figure 4a, the peaks of Al are mainly 38.474° (111), 44.723° (200), 65.1° (220), 78.234° (311), and 82.441° (222). After comparison with the PDF (04-002-3621) standard card, in Figure 4a, the peak of Al_2O_3 is 35.133° (104). After comparison with the PDF (98-000-0062) standard card, in Figure 4b, the peaks of Al are mainly 38.474° (111), 44.723° (200), 65.1° (220), 78.234° (311), and 82.441° (222). After comparison with the PDF (04-002-8298) standard card, in Figure 4c, the peaks of Ni are 41.785° (002) and 58.532° (102). The absence of Ni and C diffraction peaks in the spectrum of the hot-dip aluminum plating indicates that an Al coating has been successfully formed on the surface of the substrate. The diffraction pattern after oxidation reveals a decrease in Al content and an increase in Al_2O_3 content following the oxidation process. Additionally, based on the XRD angles, the alumina formed is identified as α-Al_2O_3. This type of alumina possesses advantages such as high hardness, a high melting point, excellent thermal conductivity, and strong corrosion resistance, providing effective protection for the substrate [34]. Moreover, Al remains present on the surface of the coating, and no diffraction peak for the C element is detected. In summary, the prepared coating effectively protects the substrate and enhances the oxidation resistance of the coating.

The X-ray diffraction patterns lead to the following conclusions: The absence of C-element diffraction peaks in the hot-dip Al composite coating confirms the successful coverage of the C/C composite substrate by the coatings. Overall, the prepared coatings effectively protect the substrate and enhance its antioxidant properties.

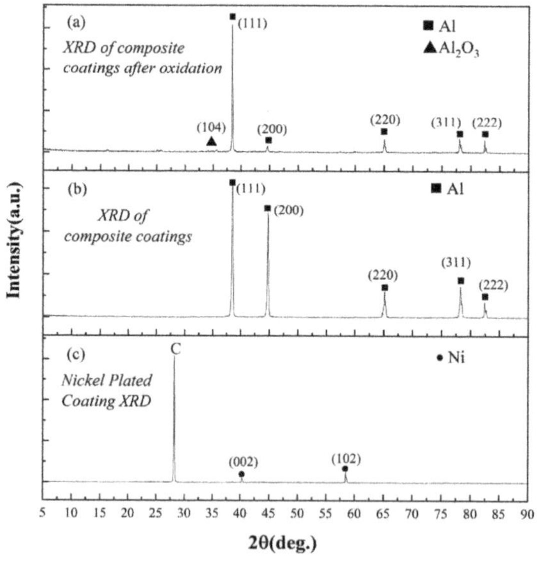

Figure 4. X-ray diffraction (XRD) pattern after coating and high-temperature oxidation. (**a**) XRD of composite coatings after oxidation, (**b**) XRD of composite coatings, and (**c**) XRD of nickel-plated coating.

3.3. Coating SEM Scanning Results and Microstructure

Figure 5 shows the microscopic morphology of the coating surface after the experiment. The Ni and Al coatings, formed through electroplating and hot-dip coating treatments, are successfully deposited on the surface of the C/C composites and effectively cover both the carbon fibers and the pyrolytic carbon substrate. The C/C composites mainly consist of carbon fibers and pyrolytic carbon as indicated in Figure 5a, and the surface of the C/C composites is not homogeneous with the simultaneous existence of pyrolytic carbon particles, so it is necessary to polish the substrate smooth before preparing for the coating process. In Figure 5b,c, the Ni coating exhibits a few surface holes, but the overall coverage is adequate. The C/C composites can produce hole defects during oxidation due to the thermal exposure area and oxidation time [35]. Adsorption occurs on the surface of the substrate despite the presence of defects. In contrast, the surface of the hot-dip aluminum coating is uniform and smooth, with no carbon fibers or other substrate materials visible. These coatings can effectively protect the substrate materials and improve their antioxidant properties. In contrast, the hot-dip Al/Si coating shown in Figure 5d has a rough surface. While pores and cracks are present, they are confined to the surface and do not cause immediate failure of the coating. The thermal shock experiments further demonstrate that the coating containing silicon exhibits better high-temperature oxidation resistance.

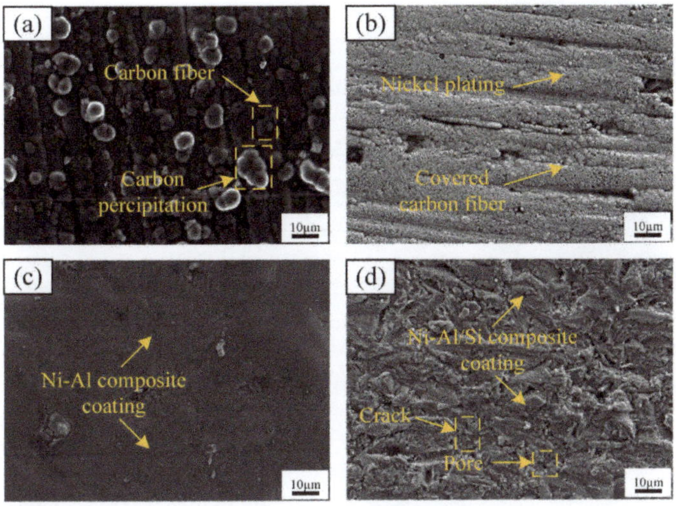

Figure 5. SEM image of coating surface: (**a**) C/C composite substrate, (**b**) Ni coating, (**c**) Ni-Al composite coating, and (**d**) Ni-Al/Si composite coating.

3.4. Coating EDS Analysis

The experimental results of the C/C composite surface coating preparation are thoroughly observed and analyzed, and the EDS energy spectra of the three coating surfaces are obtained through scanning electron microscopy. In the first set of experiments, the Ni coating is successfully prepared, as shown in Figure 6a, exhibiting a stable elemental composition primarily consisting of Ni, C, and O. The Ni coating is primarily derived from the C/C composite matrix. The carbon element originates mainly from the C/C composite substrate, while the oxygen element may be introduced from multiple sources, such as dissolved oxygen in the plating solution, oxygen exposure and adsorption from the air after electroplating, or oxygen introduced during sample preparation. The appearance of other element peaks mainly comes from impurities brought by sample preparation and instrument residues. Combined with the XRD analysis results, it can be seen that O only appears on the surface, and it has a relatively small content. It can be concluded that the Ni coating prepared in this experiment can be successfully deposited on the surface of C/C composites and has a stable element composition.

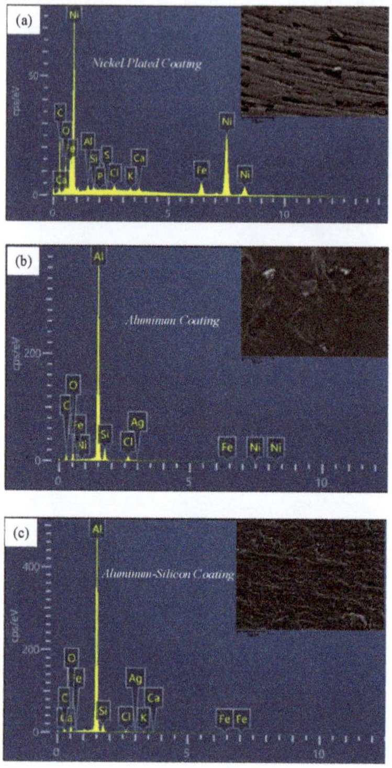

Spectrogram of the total number of distribution maps					
elemental	line type	apparent concentration	k specific value	Wt%	Wt% Sigma
C	K line system	11.77	0.11768	42	0.24
O	K line system	13.78	0.04636	10.31	0.08
Al	K line system	0.94	0.00678	0.36	0.01
Si	K line system	0.48	0.00382	0.15	0.01
P	K line system	0.23	0.00127	0.04	0.01
S	K line system	0.16	0.00134	0.05	0.01
Cl	K line system	1.24	0.01085	0.36	0.01
K	K line system	0.93	0.00792	0.24	0.01
Ca	K line system	1.52	0.01361	0.38	0.01
Fe	K line system	16.53	0.16527	3.88	0.04
Ni	K line system	155.73	1.55727	42.22	0.18
total:				100	

Spectrogram of the total number of distribution maps					
elemental	line type	apparent concentration	k specific value	Wt%	Wt% Sigma
C	K line system	4.45	0.04452	42.31	0.21
O	K line system	8.87	0.02985	12.45	0.09
Al	K line system	112.92	0.81102	39.7	0.15
Si	K line system	6.1	0.04832	3.05	0.02
Cl	K line system	3.31	0.02894	1.54	0.01
Fe	K line system	0.6	0.00602	0.26	0.02
Ni	K line system	0.9	0.00903	0.38	0.03
Ag	K line system	0.57	0.00568	0.31	0.02
total:				100	

Spectrogram of the total number of distribution maps					
elemental	line type	apparent concentration	k specific value	Wt%	Wt% Sigma
C	K line system	2.87	0.02871	36.55	0.5
O	K line system	6.83	0.02299	11.03	0.13
Al	K line system	118.91	0.85405	48.77	0.39
Si	K line system	3.98	0.03158	2.54	0.03
Cl	K line system	0.96	0.00836	0.54	0.01
K	K line system	0.44	0.00373	0.21	0.01
Ca	K line system	0.21	0.00188	0.1	0.01
Fe	K line system	0.14	0.00138	0.07	0.02
Ag	L line system	0.28	0.00276	0.18	0.02
total:				100	

Figure 6. Surface energy spectrum of coatings: (**a**) Ni coating, (**b**) hot-dip aluminum coating, and (**c**) hot-dip Al/Si coating.

In the second set of experiments, the Al coating is successfully prepared, as shown in Figure 6b, with an elemental composition primarily consisting of C, O, and Al. The mass fraction of Al is 39.7%, which is lower than that of Ni in the Ni coating, making the Al coating relatively more fragile. However, the Al coating still has good encapsulation and protection for the C/C composite substrate. It is important to note that the contents of C and O elements are 42.2% and 12.5%, respectively, primarily originating from the C/C composite substrate and various introduced factors. Other elements detected in the energy spectra are mostly impurities introduced during sample preparation or instrument residues, but their relative contents are minimal.

In the third set of experiments, Al/Si coatings are successfully prepared, as shown in Figure 6c. Their elemental composition primarily consists of C, O, Al, and Si, along with trace amounts of Ca, K, Ag, and other elements. The mass fraction of C is 36.6%, which is lower than the previous two experiments, while the mass fraction of Al is 48.8%, which is higher than the previous two experiments. The Si content in the Al/Si coatings meets the established requirement of 2%. The presence of Si contributes to increased hardness and wear resistance of the Al/Si coatings compared to the other two coatings, although it also makes the preparation process more challenging [36]. The energy spectra indicate the elements of Al/Si contained in the prepared coatings, and the Al/Si coatings are successfully prepared on the surface of the C/C composites. It should be noted that in all three coatings, the elements C and O have high contents, mainly from the C/C composite substrate as well as a variety of introduced factors. The energy spectra reveal other elements that are primarily impurities introduced during sample preparation and residues from the instrumentation, with their relative concentrations being minimal.

In summary, three coatings were successfully prepared in this study, and information on their elemental compositions and inclusions was obtained by SEM observation and EDS spectrogram analysis.

3.5. High-Temperature Oxidation Performance of Coatings

This experiment investigates the thermal shock properties of the C/C composite, electroplated Ni, and hot-dip Ni/Al composite coatings. The experimental results show that, as shown in Figure 7, the oxidation weight change in C/C composites gradually increases with the increase in temperature. Above 650 °C, the oxidation weight change accelerates, peaking at 800 °C. Subsequently, the rate of weight change exhibits undulating fluctuations within the temperature range of 800~1200 °C. Beyond 1200 °C, the rate of oxidation weight change decreases due to extensive oxidation of the C/C composites and the very limited number of remaining specimens.

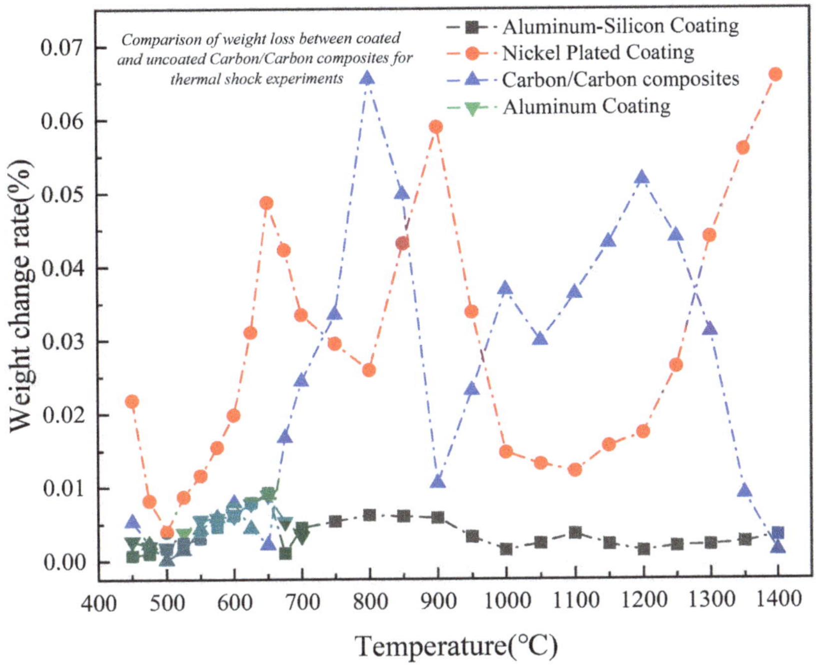

Figure 7. High-temperature oxidation weight change curve of coating and substrate.

For the electroplated Ni-coated specimens, the weight change rate increases sharply once the temperature exceeds 500 °C. This increase is attributed to the oxidizing reaction between the Ni in the coating and the air, with some of the oxidized material escaping from the specimen's surface, thereby increasing the weight change rate. During the oxidation process, the weight change rate decreases due to the presence of Ni oxide. After exceeding 900 °C, the coating starts to fail and the oxidation efficiency rises sharply. The oxidation weight change rate of the hot-dip aluminum and Al/Si alloy composite coatings remained relatively stable. However, as the temperature increased, the weight change rate of these coatings began to decrease. This decrease is attributed to the formation of Al_2O_3 on the surface of the coatings, which inhibits further oxidation. This indicates that the composite coatings effectively enhance the antioxidant performance of C/C composites and exhibit better antioxidant properties compared to the individual Ni coatings. They are still able to maintain a low weight change at 1400 °C.

3.6. Micro-Morphology of Coatings After Oxidation

Figure 8 shows the SEM microscopic morphology of the experimental specimens after the thermal shock resistance experiment. As shown in Figure 8, the coating adheres to the substrate materials. The SEM observations indicate that the unoxidized electroplated Ni coatings in Figure 8a exhibit a laminated structure. This structure provides effective protection to the substrate materials. In Figure 8b, after the oxidation experiment, an oxide layer forms on the surface of the Ni coating, which enhances its antioxidant performance. However, the Ni oxide exhibits poor adhesion to the substrate, resulting in coating delamination and a loss of protective effect on the substrate. For the samples treated with hot-dip aluminum plating (Figure 8c), the oxidation experiment results show that external oxygen preferentially reacts with the coating metal to form the α-Al_2O_3 oxide. This α-Al_2O_3 film acts as a barrier, preventing further oxygen erosion. Meanwhile, the internal aluminum remains, continuing to capture free oxygen and thereby maintaining the integrity of the substrate material. In Figure 8d, the layer thickness after the electroplating and hot-dipping methods is 36.62 μm. In summary, the preparation of this composite coating successfully enhances the antioxidant properties of C/C composites.

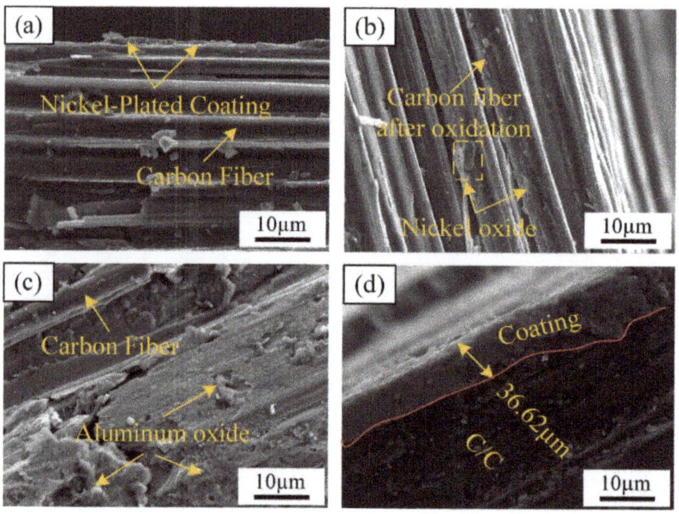

Figure 8. SEM morphology of oxidized coating cross-section: (**a**) electroplated Ni without oxidation, (**b**) electroplated Ni after oxidation, (**c**) hot-dip aluminum after oxidation, and (**d**) thickness of aluminum coating after hot-dip plating.

3.7. Silicon Vapor Corrosion Test Results for Coatings

The images obtained under the scanning electron microscope, shown below, reveal the results of silicon vapor etching conducted on C/C composites, electroplated Ni specimens, and hot-dip-plated aluminum specimens. The specific procedure for the silicon vapor etching experiment involved first cleaning the specimen's surface with anhydrous ethanol to remove oil and impurities. The specimen, along with monocrystalline silicon, was then placed in a tube furnace and held at 1450 °C for 5 h. Figure 9a,d demonstrate that, after silicon vapor etching, a significant amount of Si is deposited on the surface of the carbon fibers in the C/C composites, altering the mechanical properties of the materials. Ultimate tensile strength increases with increasing Si content in Al/Si alloys [36]. In contrast, Figure 9b,e show that the C/C composites with Ni coatings exhibit minimal Si deposition on the surface after corrosion. Due to the protection provided by the Ni coatings, Si deposition does not come into contact with the carbon fibers, thereby preserving the mechanical properties of the materials. Meanwhile, Figure 9c,f reveal that the aluminum coatings do not adsorb Si deposition, and the composite coatings formed at high temperatures effectively protect the matrix. In the EDS analysis, Si is not tested in the mapping pattern due to its low content. Overall, the composite coating treatment of C/C composites enhances their antioxidant properties and improves their resistance to silicon vapor corrosion, thus increasing their potential for application in single-crystal furnaces.

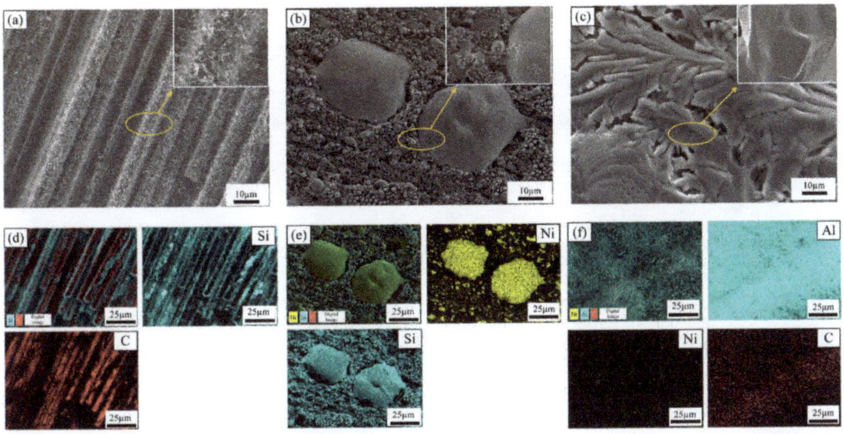

Figure 9. Scanning electron micrographs and elemental distribution of specimens after silicon vapor corrosion: (**a**) C/C composites after silicon corrosion. (**b**) Electroplated Ni specimens after silicon corrosion. (**c**) Hot-dip aluminum specimen after corrosion. (**d**) Elemental distribution of C/C composites after corrosion. (**e**) Elemental distribution of electroplated Ni specimens after corrosion. (**f**) Elemental distribution of hot-dip-plated specimens after silicon corrosion.

4. Conclusions

This study aims to develop composite coatings with superior resistance to high-temperature oxidation on the surface of C/C composites using a combination of electroplating and hot-dip plating methods. Ni coatings are successfully applied to the surface of C/C composites using the electroplating method and are shown to effectively enhance the antioxidant properties of the substrate. Subsequently, Ni-Al and Ni-Al/Si composite coatings are successfully prepared through hot-dip plating. Microscopic examination reveals good bonding between these coatings and the substrate materials. It is also observed that the Ni coatings are depleted during the hot-dip plating process. Orthogonal experiments are conducted to investigate the effects of temperature, holding time, and silicon content on the oxidation resistance of the coating. The results reveal that the influence on the oxidation resistance of the coating follows the order temperature > holding time > silicon content. Additionally, the optimal parameters for coating preparation are identified as $A_1B_2C_2$, which corresponds to a time of 60 min, a temperature of 740 °C, and a silicon content of 2%. The thermal shock test results demonstrate that all three coatings significantly enhance the oxidation resistance of the substrate, with the Ni-Al/Si composite coating exhibiting the best performance. This improvement is attributed to the formation of a dense α-Al_2O_3 film on the coating's surface during oxidation, which effectively blocks oxygen penetration. Additionally, the inclusion of Si elements fills oxidation gaps and further prevents oxygen immersion. In contrast, the oxidation performance of the Ni-Al composite coating is reduced due to the presence of a small number of voids that develop during the oxidation process. The hot-dip plating technology has been successfully applied to the surface coating preparation of C/C composite materials, offering a novel approach in this field. Future research should explore the effects of different hot-dip plating aids, materials, and other factors on coating properties to further optimize their performance and application.

The findings have both academic and practical significance, contributing to the enhancement of C/C composites' quality and the acceleration of their industrial applications.

Author Contributions: Validation, H.L.; Investigation, G.B.; Resources, Y.A.; Data curation, C.D.; Writing—original draft, Z.Q.; Writing—review & editing, visualization and project management, S.W.; Supervision, D.G. All authors have read and agreed to the published version of the manuscript.

Funding: This research received no external funding.

Institutional Review Board Statement: Not applicable.

Informed Consent Statement: Informed consent was obtained from all subjects involved in the study.

Data Availability Statement: The original contributions presented in the study are included in the article, further inquiries can be directed to the corresponding author.

Acknowledgments: This work has acknowledges the Xining City Bureau of Science and Technology major scientific and technological program [grant number 2022-Z-03] and the Yibin Introducing High-level Talents Program [grant number 2023YG02].

Conflicts of Interest: Author Zuxing Qi was employed by the Qinghai University. The remaining authors declare that the research was conducted in the absence of any commercial or financial relationships that could be construed as a potential conflict of interest.

Appendix A. Orthogonal Analysis Method

Table A1 shows the schema of range calculations in the $L_9(3^3)$ orthogonal experiments. In this table, the letters "A–C" represent the influencing factors investigated and the corresponding numbers "1–3" are the levels set for these factors. I_{1j}, II_{2j}, and III_{3j} are the sums of the results corresponding to the m-th level of the factor in column j. k_{mj} is the average value of I_{1j}, II_{2j}, and III_{3j}. R is the range, i.e., the difference between the maximum and the minimum values of I, II, and III.

Table A1. Schema of range calculations in the $L_9(3^3)$ orthogonal experiments.

No.	A	B	C	Index
1	1	1	1	X_1
2	1	2	2	X_2
3	1	3	3	X_3
4	2	1	2	X_4
5	2	2	3	X_5
6	2	3	1	X_6
7	3	1	3	X_7
8	3	2	1	X_8
9	3	3	2	X_9
I_{1j}	$I_{11} = X_1 + X_2 + X_3$	$I_{12} = X_1 + X_4 + X_7$	$I_{13} = X_1 + X_6 + X_8$	
II_{2j}	$II_{21} = X_4 + X_5 + X_6$	$II_{22} = X_2 + X_5 + X_8$	$II_{23} = X_2 + X_4 + X_9$	
III_{3j}	$III_{31} = X_7 + X_8 + X_9$	$III_{32} = X_3 + X_6 + X_9$	$III_{33} = X_3 + X_5 + X_7$	
K_{1j}	$I_{11}/3$	$I_{12}/3$	$I_{13}/3$	
K_{2j}	$II_{21}/3$	$II_{22}/3$	$II_{23}/3$	
K_{3j}	$III_{31}/3$	$III_{32}/3$	$III_{33}/3$	
R	(Max–Min) (I_{11}, II_{21}, III_{31})	(Max–Min) (I_{12}, II_{22}, III_{32})	(Max–Min) (I_{13}, II_{23}, III_{33})	

References

1. Song, L.; Yu, X. Defect Engineering in Cast Mono-like Silicon: A Review. *Prog. Photovolt. Res. Appl.* **2021**, *29*, 294–314. [CrossRef]
2. Fu, Q.; Zhang, P.; Zhuang, L.; Zhou, L.; Zhang, J.; Wang, J.; Hou, X.; Riedel, R.; Li, H. Micro/Nano Multiscale Reinforcing Strategies toward Extreme High-Temperature Applications: Take Carbon/Carbon Composites and Their Coatings as the Examples. *J. Mater. Sci. Technol.* **2022**, *96*, 31–68. [CrossRef]
3. Sanoj, P.; Kandasubramanian, B. Hybrid Carbon-Carbon Ablative Composites for Thermal Protection in Aerospace. *J. Compos.* **2014**, *2014*, 825607. [CrossRef]
4. Tzeng, S.-S.; Chr, Y.-G. Evolution of Microstructure and Properties of Phenolic Resin-Based Carbon/Carbon Composites during Pyrolysis. *Mater. Chem. Phys.* **2002**, *73*, 162–169. [CrossRef]
5. Han, M.; Zhou, C.; Silberschmidt, V.V.; Bi, Q. Oxidation Behaviour and Residual Mechanical Properties of Carbon/Carbon Composites. *Carbon Lett.* **2023**, *33*, 1241–1252. [CrossRef]
6. McKee, D.W. Oxidation Behavior and Protection of Carbon/Carbon Composites. *Carbon* **1987**, *25*, 551–557. [CrossRef]
7. Sheehan, J.E.; Buesking, K.W.; Sullivan, B.J. Carbon-Carbon Composites. *Annu. Rev. Mater. Res.* **1994**, *24*, 19–44. [CrossRef]
8. Jin, X.; Fan, X.; Lu, C.; Wang, T. Advances in Oxidation and Ablation Resistance of High and Ultra-High Temperature Ceramics Modified or Coated Carbon/Carbon Composites. *J. Eur. Ceram. Soc.* **2018**, *38*, 1–28. [CrossRef]
9. Shojaie-bahaabad, M.; Bozorg, M.; Najafizadeh, M.; Cavaliere, P. Ultra High Temperature Ceramic Coatings in Thermal Protection Systems (TPS). *Ceram. Int.* **2024**, *50*, 9937–9951. [CrossRef]
10. Chen, S.; Qiu, X.; Zhang, B.; Xu, J.; Zhong, F.; Zhu, B.; Zhang, Y.; Ou-Yang, J.; Yang, X. Advances in Antioxidation Coating Materials for Carbon/Carbon Composites. *J. Alloys Compd.* **2021**, *886*, 161143. [CrossRef]
11. Xu, L.; Yang, W.; Fan, Z.; Li, X.; Li, W.; Song, Y.; Feng, Z.; Zhang, D. Fabrication of Low-Density Carbon-Bonded Carbon Fiber Composites with an Hf-Based Coating for High Temperature Applications. *RSC Adv.* **2018**, *8*, 19171–19180. [CrossRef] [PubMed]
12. Zhu, X.; Zhang, Y.; Zhang, J.; Li, T.; Li, J.; Chen, R. A Gradient Composite Coating to Protect SiC-Coated C/C Composites against Oxidation at Mid and High Temperature for Long-Life Service. *J. Eur. Ceram. Soc.* **2021**, *41*, 123–131. [CrossRef]
13. Li, J.; Luo, R.; Lin, C.; Bi, Y.; Xiang, Q. Oxidation Resistance of a Gradient Self-Healing Coating for Carbon/Carbon Composites. *Carbon* **2007**, *45*, 2471–2478. [CrossRef]
14. Huang, J.; Guo, L.; Xu, M.; Zhang, P. Effect of Pack Cementation Temperatures on Component, Microstructure and Anti-Oxidation Performance of Al-Modified SiC Coatings on C/C Composites. *Ceram. Int.* **2020**, *46*, 8293–8298. [CrossRef]
15. Chen, H.; Ma, Z.; Wang, Y.; Wei, T.; Yang, H.; Du, P.; Wang, X.; Zhang, R. An Orthogonal Experimental Study on the Preparation of Cr Coatings on Long-Size Zr Alloy Tubes by Arc Ion Plating. *Materials* **2022**, *15*, 7177. [CrossRef]
16. Feng, G.; Chen, L.; Yao, X.; Yu, Y.; Li, H. Design and Characterization of Zirconium-Based Multilayer Coating for Carbon/Carbon Composites against Oxyacetylene Ablation. *Corros. Sci.* **2021**, *192*, 109785. [CrossRef]
17. Goodwin, F.E. Overview of Recent Advancements in Hot Dip Metallic Coatings for Steel. *Trans. Indian Inst. Met.* **2023**, *76*, 2295–2305. [CrossRef]
18. Dey, P.P.; Sahu, S.; Banerjee, P.S.; Ghosh, M. A Review on Metallurgical Features of Hot-Dip Aluminized Steel. *Eng. Res. Express* **2023**, *5*, 012002. [CrossRef]
19. Yu, Z.; Hu, J.; Meng, H. A Review of Recent Developments in Coating Systems for Hot-Dip Galvanized Steel. *Front. Mater.* **2020**, *7*, 74. [CrossRef]
20. Chakraborty, A.; Ghosh, R.; Sudan, M.; Mondal, A. Improvement in Hot Dip Galvanized Coating Microstructure and Properties by Pre-Metallic Deposition on Steel Surface: A Comprehensive Review. *Surf. Coat. Technol.* **2022**, *449*, 128972. [CrossRef]
21. Li, F.; Ni, H.; Yang, L.; Jiang, Y.; Wang, D.; Sun, B. Investigation of Fused Alumina Based-Mold Facecoats for DZ22B Directionally Solidified Blades. *Materials* **2019**, *12*, 606. [CrossRef] [PubMed]
22. Tu, H.; Zhang, Z.; Liu, Y.; Wu, C.; Peng, H.; Wang, J.; Su, X. Phase Equilibria of the Al-Ni-Sn System at 600 and 800 °C. *J. Phase Equilibria Diffus.* **2020**, *41*, 56–65. [CrossRef]
23. Mensinger, M.; Gaigl, C. Hot Dip Galvanized Steel Constructions under Fire Exposure. In Proceedings of the IFire SS 2nd International Fire Safety Symposium, Naples, Italy, 7–9 June 2017.
24. Cheng, W.-J.; Wang, C.-J. Microstructural Evolution of Intermetallic Layer in Hot-Dipp ed Aluminide Mild Steel with Silicon Addition. *Surf. Coat. Technol.* **2011**, *205*, 4726–4731. [CrossRef]
25. Zhang, Y.; Qie, J.; Cui, K.; Fu, T.; Fan, X.; Wang, J.; Zhang, X. Effect of Hot Dip Silicon-Plating Temperature on Microstructure Characteristics of Silicide Coating on Tungsten Substrate. *Ceram. Int.* **2020**, *46*, 5223–5228. [CrossRef]
26. Li, X.; Feng, J.; Jiang, Y.; Li, L.; Feng, J. Anti-Oxidation Performance of Carbon Aerogel Composites with SiCO Ceramic Inner Coating. *Ceram. Int.* **2019**, *45*, 9704–9711. [CrossRef]
27. Zou, X.; Fu, Q.; Liu, L.; Li, H.; Wang, Y.; Yao, X.; He, Z. ZrB2–SiC Coating to Protect Carbon/Carbon Composites against Ablation. *Surf. Coat. Technol.* **2013**, *226*, 17–21. [CrossRef]
28. Raphael, N.; Namratha, K.; Chandrashekar, B.N.; Sadasivuni, K.K.; Ponnamma, D.; Smitha, A.S.; Krishnaveni, S.; Cheng, C.; Byrappa, K. Surface Modification and Grafting of Carbon Fibers: A Route to Better Interface. *Prog. Cryst. Growth Charact. Mater.* **2018**, *64*, 75–101. [CrossRef]
29. Zhang, J.; Zhang, Y.; Fu, Y.; Hu, D.; Meng, J.; Li, T. Ablation Behavior of HfC Coating with Different Thickness for Carbon/Carbon Composites at Ultra-High Temperature. *J. Eur. Ceram. Soc.* **2021**, *41*, 1769–1778. [CrossRef]

30. Zuo, H.; Ruan, F.; Wang, H.; Wang, H.; Wang, X.; Huang, Y.; Wang, R.; Zou, L.; Xu, Z.; Li, D. Advances in Ablation or Oxidation Mechanisms and Behaviors of Carbon Fiber-Reinforced Si-Based Composites. *Molecules* **2023**, *28*, 6022. [CrossRef]
31. Nasonova, A.; Pham, H.C.; Kim, D.-J.; Kim, K.-S. NO and SO2 Removal in Non-Thermal Plasma Reactor Packed with Glass Beads-TiO2 Thin Film Coated by PCVD Process. *Chem. Eng. J.* **2010**, *156*, 557–561. [CrossRef]
32. Ren, X.; Li, H.; Chu, Y.; Li, K.; Fu, Q. ZrB2–SiC Gradient Oxidation Protective Coating for Carbon/Carbon Composites. *Ceram. Int.* **2014**, *40*, 7171–7176. [CrossRef]
33. Su, X.; Bao, Y.; Wan, D.; Zhang, H.; Xu, L.; Grasso, S.; Hu, C. Thermal Shock Resistance of Ti3SiC2 Ceramic under Extremely Rapid Thermal Cycling. *J. Alloys Compd.* **2021**, *866*, 158985. [CrossRef]
34. Nordahl, C.S.; Messing, G.L. Sintering of α-Al2O3-Seeded Nanocrystalline γ-Al_2O_3 Powders. *J. Eur. Ceram. Soc.* **2002**, *22*, 415–422. [CrossRef]
35. Mei, H.; Zhang, D.; Xia, J.; Yu, C.; Cheng, L. The Effect of Hole Defects on the Oxidation Behaviour of Two-Dimensional C/SiC Composites. *Ceram. Int.* **2016**, *42*, 15479–15484. [CrossRef]
36. Nikanorov, S.P.; Osipov, V.N.; Regel, L.I. Structural and Mechanical Properties of Directionally Solidified Al-Si Alloys. *J. Mater. Eng. Perform.* **2019**, *28*, 7302–7323. [CrossRef]

Journal of
Manufacturing and Materials Processing

MDPI

Article

Development of Polymer Hydrophobic Surfaces Through Combined Laser Ablation and Hot Embossing Processes

Esmaeil Ghadiri Zahrani [1,2], Amirmohmmad Fakharzadeh Jahromi [1] and Bahman Azarhoushang [1,*]

1 Institute for Advanced Manufacturing (KSF), Furtwangen University, 78532 Tuttlingen, Germany; esmaeil.ghadiri.zahrani@hs-furtwangen.de (E.G.Z.); amirmohammad.fakharzadehjahromi@hs-furtwangen.de (A.F.J.)
2 Department of Microsystems Engineering (IMTEK), University of Freiburg, 79110 Freiburg, Germany
* Correspondence: aza@hs-furtwangen.de

Abstract: The development of hydrophobicity on polymer surfaces in mass production is one of the most critical challenges in the plastic industry. This paper deals with a novel combined hot embossing process in which femtosecond laser ablation is utilized to texture the embossing stamps. By controlling the process temperature and axial forces, the laser textures were transferred to polymer surfaces, successfully resulting in hydrophobicity. Four different polymers, including ABS, PP, PA, and PC, along with two different laser textures, namely ball and pyramid, were tested. The laser and hot embossing parameters under which the textures were transferred to the polymers are introduced. The critical micro- and nano-features of the transferred textures that resulted in high hydrophobic contact angles are also discussed. The results indicate that PP and ABS have higher contact angles, respectively, while under the given parameters, PA and PC did not exhibit hydrophobic surfaces.

Keywords: polymer; hydrophobicity; femtosecond laser ablation; hot-embossing; micro- and nano-features

1. Introduction

Hydrophobicity, the property of repelling water, plays a crucial role in various applications. Surfaces with high hydrophobicity exhibit low water adhesion and are often characterized by water droplets forming spherical shapes or sliding off effortlessly with high values of contact angles [1].

Surface energy and roughness are the major contributing factors that determine whether a surface is wetting or non-wetting [2]. The exploration of wettability holds promise in designing self-cleaning surfaces [3] and surfaces with enhanced adhesive resistance [4]. Various techniques, such as photolithography [5], laser interference lithography [6], chemical etching [7], and layer-by-layer assembly [8], can be employed to create hydrophobic surfaces. However, these methods come with drawbacks, including multiple steps, the use of additional materials, and time-consuming processes, all contributing to increased costs.

To overcome these challenges, laser processing technology proves to be an effective technique for producing surfaces with nano- and micro-structures [9,10]. During the laser ablation process, the target material absorbs laser energy, melts, and evaporates, which causes morphological changes inside and around the ablation area [11]. Through the laser ablation process, different structures can result in hydrophobicity [12]. In one study, hydrophobicity on AISI 430 stainless steel was achieved in three laser patterns: octagonal donut patterns, ellipses at 90° angles, and a dimple array. The structure of ellipses at 90° angles showed a contact angle of 98° [13]. Both the structure type and the laser parameters are extremely important in producing hydrophobicity. The surface transitioned from hydrophilic to hydrophobic and eventually became superhydrophobic when laser processing was conducted at the highest repetition rate [14]. The stability of hydrophobicity on laser-textured surfaces is another criterion that is influenced by the laser fluence. It is shown that structured surfaces achieve stable hydrophobic conditions when the peak fluence exceeded a threshold, resulting in final contact angles above 140°. Conversely, when the fluence was below this threshold, the surface initially became hydrophobic, reaching a maximum contact angle, but its hydrophobicity diminished over time [15]. Although direct laser processing is a promising technique compared to other methods to produce hydrophobicity, the process represents some surface damages and it is still not an economical method for some industrial application. In the direct ablation process of polymers, a significant amount of energy is absorbed by the material, leading to heating, boiling, vaporization, and the formation of pores. These pores observed that the following laser ablation can be attributed to the rapid cooling of the superheated melt after intense boiling [16]. The presence of pores can negatively impact the performance of a final part, particularly if it is intended for use in harsh and abrasive environments. Therefore, the selection of laser parameters significantly influences the final performance. For instance, in terms of hydrophobicity, polymers exhibit different responses to various laser parameters. For example, PEEK displays varying contact angles depending on the laser wavelength: untreated PEEK has a contact angle of 92.4°, and laser treatment yield angles of 59.5° at 1064 nm, 131.0° at 532 nm, and 52.5° at 355 nm [17]. Therefore, a much more robust process is desirable. Combining laser processing with other methods, for example, hot embossing could be an alternative for producing hydrophobicity. Hot embossing is an intriguing process due to its ability to replicate fine micro- and nano-scale details with high precision and cost-effectiveness compared to other microfabrication techniques. It accommodates a variety of thermoplastic materials and can be scaled from small to large production volumes, making it suitable for industrial manufacturing. The process enhances material properties, such as lubricity. Advances in hot embossing technology have reduced cycle times and improved efficiency, enabling faster quicker production and lower costs [18].

Y. Gong et al. [19] focused on creating a superhydrophobic surface using PTFE/graphite composites through a hot embossing process. The surface morphology was analyzed by measuring the structural parameters of the replicated protrusions. The study further investigated the effects of embossing temperature and time on the polymer filling mechanism. Due to graphite's excellent lubricity, nanoscale mold cavities were filled efficiently, reducing the embossing time and enhancing mass production viability. Lowering the embossing temperature mitigates the impact of submicron fibers on the mold service life. Optimal process conditions were recommended to achieve hydrophobicity.

J. Sun et al. [20] explored the use of stainless steel stamps to create polymeric superhydrophobic surfaces with biomimetic hierarchical roughness through isothermal hot embossing. The resulting surfaces featured micro-platforms and microfibers formed during molding. This resulted in a water contact angle of 154° and an oil contact angle of 97°, which increased to 140° after fluorosilane modification. Water droplets rolled off the surface at a tilt of approximately 1°, indicating a very low sliding angle.

J. J. Victor et al. [21] introduced a template-based method for the creation of superhydrophobic polymer surfaces. The process structured PE, PP, and PTFE samples, creating nano-scale roughness on micro-scale protrusions, significantly increasing the contact angles and reducing the tilt angles. The effects of temperature, droplet size, and surfactant concentration on the wetting properties were also examined. Despite slight decreases in contact angles with increased droplet size and temperature, the surfaces remained hydrophobic up to 90 °C. High surfactant concentrations reduced hydrophobicity, but the pressed PTFE remained hydrophobic.

H.J. Lee et al. [22] introduced a repetitive ultrasonic embossing technique to create composite micro-patterns using a simply patterned mold. The process utilized ultrasonic imprinting to locally soften the polymer surface. This method was also used to fabricate composite micro-patterns combining prism and pyramid shapes on a single-polymer film.

In almost all the methods mentioned above, the efficiency of hydrophobicity is partial and depends on many factors such as stamp geometry, material properties, parameter variation, fillers, and other variables, making the optimization of parameters a complex problem. Therefore, it would be very useful to have a more tangible and applicable hot embossing method to avoid any uncertainty. In order to achieve this particular aim, in this study, a hybrid method of laser structuring and hot embossing is presented. Based on this explanation, as far as the authors know, there has been no prior publication on a hybrid process at an industrial level similar to our work. In this study, laser processing is employed to structure the stamp used in the embossing process. Various patterns were meticulously applied to the stamp to transfer texturing, which is explained in the next sections.

2. Experimental Set-Up

Figure 1 illustrates the sequential steps involved in the process of generating hydrophobic surfaces and assessing the hydrophobic properties of a polymer using a textured stamp. The process begins with the preparation stage, where a stamp without structure is prepared, and its surface roughness is measured to ensure the specific roughness quality. This is followed by the laser processing stage, which involves setting up the ultra-short-pulse laser machine, selecting appropriate laser parameters, and choosing a laser pattern to create the desired texture on the stamp. Once the laser processing is complete, the next stage is analyzing the texture of the stamp to determine its geometrical characteristics. Afterwards, the hot embossing process transfers the texture from the stamp to the polymer by applying force and temperature for a specific holding time. The detailed evaluation follows the analysis of the transferred texture on the polymer and measuring the contact angle.

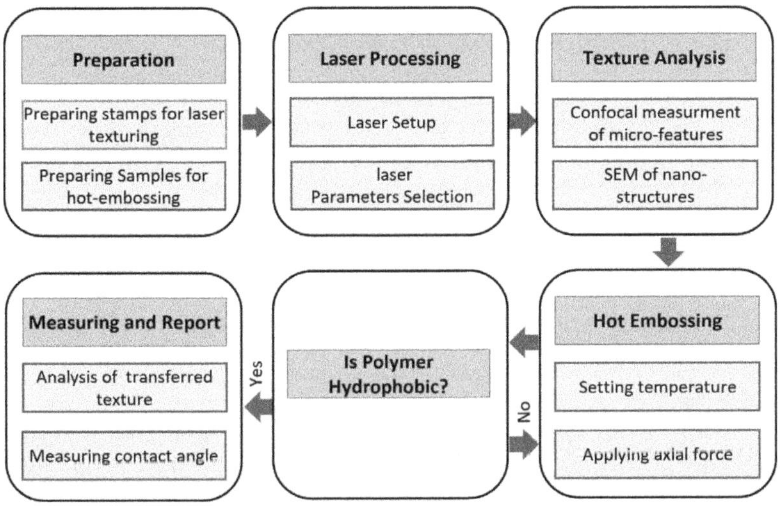

Figure 1. Overview of utilized steps for the combined processes of laser ablation and hot embossing.

The laser source employed was a solid-medium type known as Carbide-Model (CB3-40-0200-10-HB) from Light Conversion Company, Lithuania, with a wavelength of 1064 nm. The intensity profile of the laser beam was near-Gaussian ($M^2 < 1.2$), and the beam spot size ($2w_0$) was 22 μm. The laser system generates different infrared pulse widths, including tp = 750 fs, with a maximum average power output of 40 W. The laser source has been integrated into a 5-axis machine (+GF + LP400U), as shown in Figure 2a. The laser spot is positioned on the workpiece in nominal focus distance (F.D). The focus distance was kept in the nominal value of the F-theta lens equal to 100 mm. The used material is a disc (Ø28 × 12 mm) from stainless steel DIN1.4301 (Cr:17.5 to 19%, Ni: 8 to 10.5%, C < 0.07%_ Si < 1.0%_ Mn < 2.0%_ Pb < 0.045%_ S < 0.015%), which was machined before the laser machining. The initial surface roughness of the stamp and polymer workpieces was assessed using a tactile measurement instrument (Hommel-Etamic T 8000 by Jenoptik), with values of Ra = 1.38 μm (Rz = 7.79 μm) and Ra = 1.62 μm (Rz = 6.86 μm), respectively. A confocal microscope (NanoFocus Mobile μsurf) was used to measure the geometrical micro-features of the laser textures.

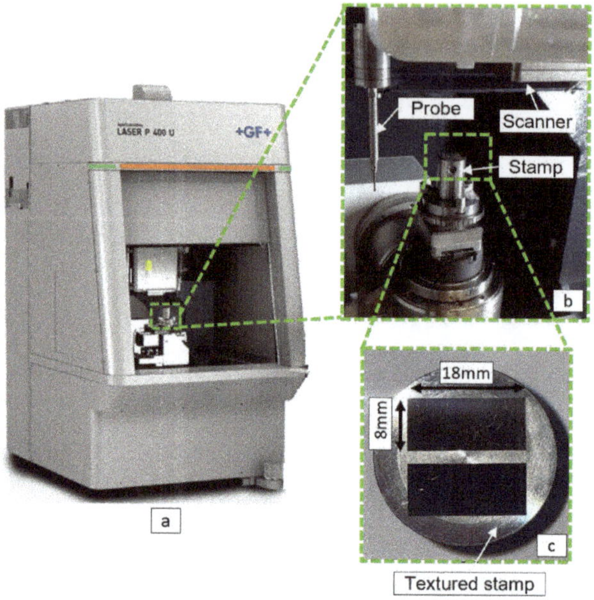

Figure 2. (**a**) Five-axis laser machine; (**b**) Positioning of stamp; and (**c**) Dimensions of laser-textured stamp.

The laser parameters used in this study are given in Table 1. The experiments were conducted in ambient air, and all the experiments were conducted three times. Screening tests for the final parameters had already been completed successfully. These tests encompassed a thorough examination of each parameter to ensure low thermal damage and the generation of nanostructures on micro-patterns.

Table 1. Laser parameter.

Average Power (W)	**3.2**
Frequency (kHz)	600
Scanning speed (mm/s)	2000
Laser pulse width (fs)	750
Jump speed (mm/s)	3500
Number of repetitions	80
Hatch distance (μm)	20

Laser experimentation was conducted with two surface texture patterns, as represented in Figure 3. These patterns were used for laser texturing due to their symmetrical properties. All these patterns create air pockets that can repel water, making the surface less prone to wetting. Figure 2c shows a laser-structured stamp in which two rectangles with dimensions of 18 × 8 mm were lasered. These dimensions ensure a sufficient sliding path for hydrophobicity tests.

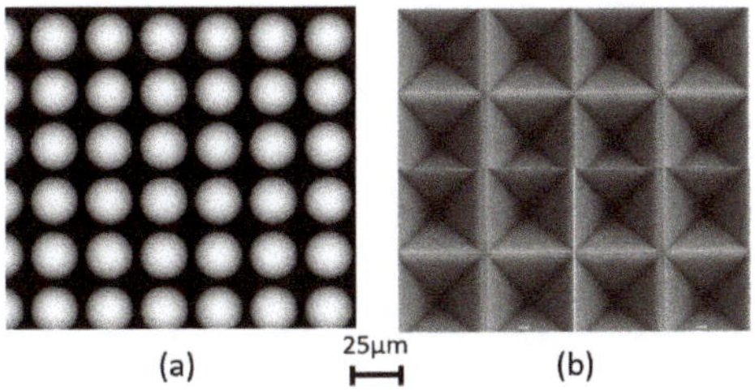

Figure 3. Patterns used for laser texturing: (**a**) Ball and (**b**) Pyramid.

Four types of polymers of ABS (acrylonitrile butadiene styrene), PP H (polypropylene), PC (polycarbonate), and PA (polyamide), were used as workpieces with dimensions of Ø28 × 10 mm. Table 2 summarizes the properties of the used polymers. The polymer workpieces were positioned between the laser-textured stamp and the holder ring, as illustrated in Figure 4.

Table 2. Polymers types and specification [23].

Polymer Type	ABS	PP	PC	PA
Glass transition temperature (°C)	85/104	−18	150	47–70
Melting point temperature (°C)	up to 230	165	260	220
Hardness (N/mm^2)	74	120	120	80

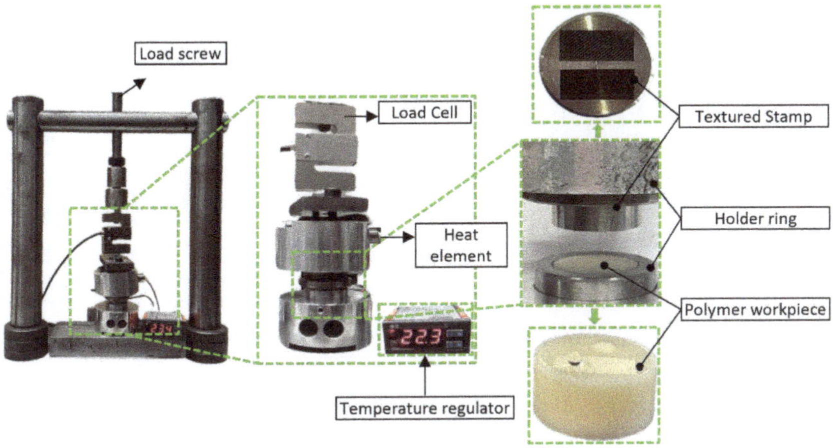

Figure 4. Hot embossing set-up.

In the hot-embossing setup (Figure 4), the stamp is positioned inside a holding ring which is connected to a load cell for measuring the axial force applied on the stamp by a load screw. A heating element was also inserted in the holding ring to provide the required temperature for the stamp. The temperature was also controlled via a regulator to ensure that the working temperature does not exceed the defined threshold. The temperature range was selected below the melting point of the polymer (Table 2). A holding time of 60 s was applied for all the workpieces during the hot embossing process. To determine the range of the hot-embossing parameters, some screening tests were conducted, and based on the initial results, the parameters in Table 3 were adopted. To measure the contact angle on the polymer workpieces, an optical analysis system (DataPhysics—OCA 200) with distilled water was utilized. The water volume used to measure the contact angle was between 7 and 8 μL.

Table 3. Hot-embossing parameter.

Parameter	Laser Textures & Polymers							
	Ball				Pyramid			
	ABS	PP	PC	PA	ABS	PP	PC	PA
Stamping temperature (°C)	120	160	120	190	120	160	120	190
Axial load (kg)	300	600	600	600	300	600	600	600
Holding time (s)	60	60	60	60	60	60	60	60

3. Results

Figure 5a shows the confocal images of the laser textures on the stamps as well as the micro-geometries that are transferred to the polymer surfaces. Comparing these images indicates that, for both ABS and PP, the micro-features are transferred almost one by one, while for PC and PA, the transformation is not complete. To compare and quantify these textures together, it is necessary to characterize the geometrical features of the structures. Figure 5b represents these features, and they are summarized in Table 4. The micro-features measured are height (A), landing length (B), width (C), peak-to-peak distance (D), and valley-to-valley distance (E). Overall, for the stamps, the pyramid texture exhibits the largest values in height, width, and valley-to-valley distance, while the ball texture tends to have lower measurements across the tip and peak-to-peak distance.

Table 4. Comparison between the micro-features of the laser-textured stamp and the hot-embossed polymer workpieces.

Micro-Feature	Laser Textures and Polymers									
	Ball					Pyramid				
	Stamp	ABS	PP	PC	PA	Stamp	ABS	PP	PC	PA
A (Height) [μm]	15.4	19	10	6.6	3	21.6	18.3	22.7	5.4	1.4
B (Landing length) [μm]	12	11	10.6	5.2	4.1	13.6	14.3	14.1	3.1	1.9
C (Width) [μm]	19	23.1	25.7	8.3	5	27	28.8	28.2	4.5	2.8
D (Peak to peak) [μm]	31.3	30.7	37.6	34.3	33.8	36	40.1	37.9	36.3	35.6
E (Valley to valley) [μm]	30.4	32.6	34.1	33.2	31.5	38.1	37.2	39.2	37.2	36.2

Based on the values in Table 4, all the micro-features of the stamp in the ball and pyramid textures were transferred for both ABS and PP polymers, while for PC and PA, these features were not transferred completely, despite a higher axial load and temperature being applied compared to ABS and PP. The differences in the transfer of micro-features in the hot embossing process could be attributed to several factors related to the thermal and rheological properties of these polymers. PC and PA have higher glass transition temperatures compared to ABS and PP, meaning that, at the applied temperature, these polymers might not soften enough to flow and conform to the micro-features of the stamp. Compared to ABS and PP, PC and PA generally have lower thermal conductivities; therefore, the temperature at the interface between the polymer workpiece and the stamp was not conducted uniformly. In addition, PC and PA have higher viscosities due to strong intermolecular forces and hydrogen bonding, respectively. ABS, as an amorphous polymer, does not have a crystalline structure that needs to be broken down [24]. This contributes to better flow characteristics at lower pressures.

By conducting hydrophobicity tests as shown in Figure 6, the contact angles for all the polymer surfaces as well as the stamp are summarized in Table 5. For the ABS material, the contact angle significantly increased from 43.9° without texture to 105.4° with the ball texture and slightly decreased to 94.0° with the pyramid texture. Similarly, for the PP material, the contact angle increased from 80.95° without texture to 118.2° with the ball texture and then slightly decreased to 112.0° with the pyramid texture. These results indicate that the ball texture generally leads to the highest contact angles for both ABS and PP, suggesting a greater hydrophobic effect with this texture compared to the pyramid and untextured surfaces.

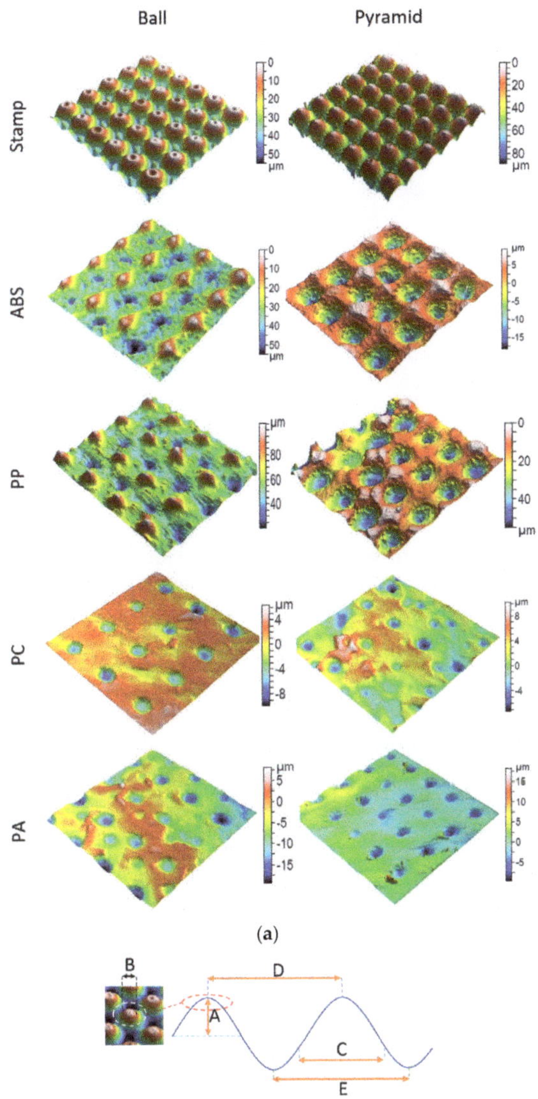

Figure 5. (**a**) Confocal images of the laser-textured stamps and (**b**) the 2D illustration of micro-features.

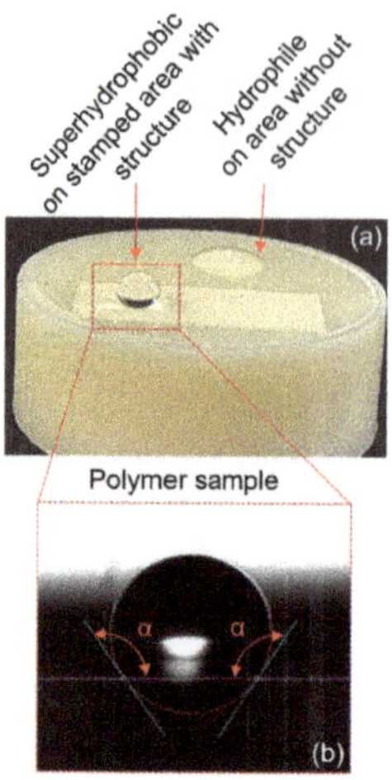

Figure 6. (**a**) Comparison between hydrophobic and hydrophile areas and (**b**) Measurement of contact angle.

Table 5. Contact angle measurements.

Texture	Material				
	Stamp	**ABS**	**PP**	**PC**	**PA**
Without texture	58°	43.9°	80.95°	68.2°	53.8°
Ball texture	90.8°	105.4°	118.2°	74.8°	65.5°
Pyramid texture	100.2°	94.0°	112.0°	85.7°	83.8°

Considering the contact angles (Table 5), the results in Table 4 indicate that achieving hydrophobicity requires surpassing a critical value for the geometrical micro-features, specifically for the height (A), landing length (B), and width (C). These geometrical features for both the laser textures of ball and pyramid can be defined as determining features. ABS with the pyramid texture achieves the highest height (22.7 μm), while PA with the ball texture shows the smallest height (1.4 μm). The pyramid texture on ABS yields the widest features (28.8 μm), while the PA with the ball results in the narrowest (2.8 μm). The pyramid consistently leads to larger distances compared to ball features, with slight material-dependent variation. Features for both textures tend to shrink (in terms of height, width, and spacing) as the material changes from ABS → PP → PC → PA. PA consistently has the smallest features across all measurements, indicating that it may be less receptive to deep or wide laser-textured features, while the ABS is more conducive to achieving pronounced textures. If the value of these features is met or exceeded after hot embossing, the surface becomes hydrophobic. If the geometrical features are smaller than the threshold values, the final transferred laser texture on the polymer surfaces does not result in hydrophobicity. The shape, depth, and spacing of the micro-features are crucial. For instance, deeper and narrower pockets can trap air more effectively, enhancing the stability of the Cassie–Baxter state, while shallow and wide pockets might not trap air as efficiently [25].

Increased roughness typically enhances hydrophobicity by trapping more air and reducing the contact area between the water and the solid surface. In fact, the morphology of the surface plays a significant role in this situation. To assess the effect of the surface morphology of the ablated area, a series of images was captured from the ablated texture using a scanning electron microscope (SEM-Zeiss, Crossbeam 550 L) (Figure 7). For both the ball and pyramid textures, the same nanostructures (Figure 7c,d) were observed. This is due to the fact that the same laser parameters were used for both structures; hence, the conditions for melt formation were identical. Vertically aligned ridges, as shown in Figure 7d, can effectively trap air, maintaining high contact angles and reducing the contact area between the liquid and the solid surface. These structures have a high aspect ratio, such that the height of the melt particles formed after laser ablation is larger than their average diameter; however, extremely high aspect ratio structures might be mechanically fragile. After the hot embossing process, the status of the nanostructures was again investigated. The results indicate that, due to the applied load on the nanostructures during the hot embossing process, the nanostructures tended to be flattened, as shown in Figure 7e. Evaluating the nanostructures of the stamp surface is very important from the perspective of whether the nanostructures or the micro-features of the laser textures are more decisive in the hydrophobicity of the polymer surfaces. Since the nanostructures were flattened after the first hot embossing experiment, and the stamps were utilized several times for the hot embossing tests, and the surfaces of the polymers remained hydrophobic despite the fact that it indicates that the micro-features are more important than the nanostructures.

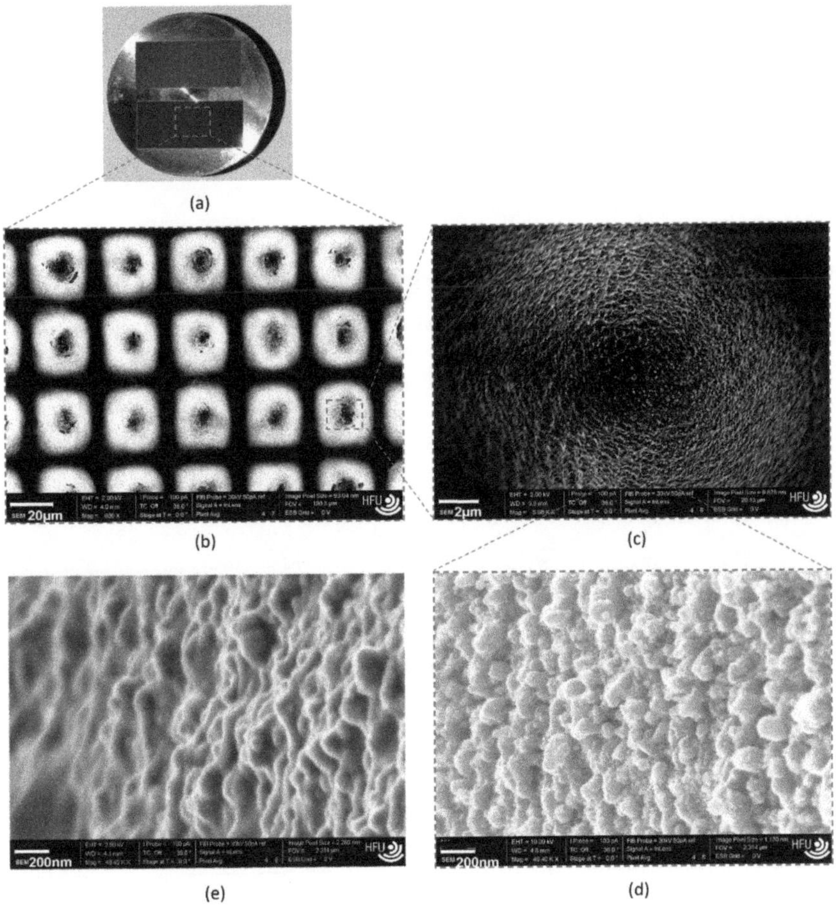

Figure 7. Investigation into nano-structures on the stamp—(**a**) laser-textured stamp, (**b**) pyramid texture array, (**c**) micro-feature, (**d**) ridge nano-structure before hot-embossing, and (**e**) ridge nano-structure after hot-embossing.

4. Conclusions

A novel, successful combined laser ablation and hot embossing process is presented in this study to produce hydrophobic surfaces on polymer workpieces. A femtosecond laser was utilized to texture the stamps with two different laser patterns. Four different types of polymers (ABS, PP, PA, and PC) were chosen as workpiece materials. The experimental results indicate that, among the five micro-features of the laser textures investigated herein, the height, landing length, and width are more decisive in achieving hydrophobicity. These features were completely transferred to the ABS and PP workpieces. The results showed that, for the PP samples compared to the ABS samples, the contact angles are considerably higher. Laser texturing for the ABS samples caused the contact angle to double.

Author Contributions: Conceptualization, B.A. and E.G.Z.; methodology, E.G.Z.; software, E.G.Z. and A.F.J.; validation, E.G.Z., A.F.J. and B.A.; formal analysis, E.G.Z.; investigation, E.G.Z. and A.F.J.; resources, B.A.; data curation, E.G.Z. and A.F.J.; writing—original draft preparation, E.G.Z.; writing—review and editing, E.G.Z. and B.A.; visualization E.G.Z. and B.A.; supervision B.A.; project administration, E.G.Z.; funding acquisition, B.A. All authors have read and agreed to the published version of the manuscript.

Funding: The results achieved in this study were part of the 'SuperSurf' project, funded by the Alfred Kärcher Foundation, for which the authors express their sincere gratitude.

Data Availability Statement: All the data are mentioned in the manuscript.

Acknowledgments: The results achieved in this study were part of the 'SuperSurf' project, funded by the Alfred Kärcher Foundation, for which the authors express their sincere gratitude. Also, many thanks to +GF+ Machining Solutions GmbH for their invaluable support in providing the laser machine. In addition, thanks are extended for the use of FIB-SEM, funded by the Deutsche Forschungsgemeinschaft (DFG, German Research Foundation), within the program Major Research Instrumentation (Project number: 505067193) under GG91b, and by the state of Baden-Württemberg, as well as by Furtwangen University. Thanks also to our colleague R. Bösinger for his help in preparing the SEM images. Many thanks also to Seyedehmina Mojabi for her help in experimentation.

Conflicts of Interest: The authors declare no conflict of interest.

References

1. Carré, A. (Ed.) *Superhydrophobic Surfaces*; VSP: Leiden, The Netherlands, 2009.
2. Chieng, B.W.; Ibrahim, N.A.; Ahmad Daud, N.; Talib, Z.A. Functionalization of Graphene Oxide via Gamma-Ray Irradiation for Hydrophobic Materials. In *Synthesis, Technology and Applications of Carbon Nanomaterials*; Elsevier: Amsterdam, The Netherlands, 2019; pp. 177–203.
3. Khan, S.A.; Boltaev, G.S.; Iqbal, M.; Kim, V.; Ganeev, R.A.; Alnaser, A.S. Ultrafast fiber laser-induced fabrication of superhydrophobic and self-cleaning metal surfaces. *Appl. Surf. Sci.* **2021**, *542*, 148560. [CrossRef]
4. Su, F.; Yao, K. Facile fabrication of superhydrophobic surface with excellent mechanical abrasion and corrosion resistance on copper substrate by a novel method. *ACS Appl. Mater. Interfaces* **2014**, *6*, 8762–8770. [CrossRef] [PubMed]
5. Limongi, T.; Schipani, R.; Di Vito, A.; Giugni, A.; Francardi, M.; Torre, B.; Allione, M.; Miele, E.; Malara, N.; Alrasheed, S.; et al. Photolithography and micromolding techniques for the realization of 3D polycaprolactone scaffolds for tissue engineering applications. *Microelectron. Eng.* **2015**, *141*, 135–139. [CrossRef]
6. Rodriguez, A.; Echeverría, M.; Ellman, M.; Perez, N.; Verevkin, Y.K.; Peng, C.S.; Berthou, T.; Wang, Z.; Ayerdi, I.; Savall, J.; et al. Laser interference lithography for nanoscale structuring of materials: From laboratory to industry. *Microelectron. Eng.* **2009**, *86*, 937–940. [CrossRef]
7. Liu, Y.; Yin, X.; Zhang, J.; Wang, Y.; Han, Z.; Ren, L. Biomimetic hydrophobic surface fabricated by chemical etching method from hierarchically structured magnesium alloy substrate. *Appl. Surf. Sci.* **2013**, *280*, 845–849. [CrossRef]
8. Zhao, Y.; Li, M.; Lu, Q.; Shi, Z. Superhydrophobic polyimide films with a hierarchical topography: Combined replica molding and layer-by-layer assembly. *Langmuir ACS J. Surf. Colloids* **2008**, *24*, 12651–12657. [CrossRef] [PubMed]
9. Sugioka, K.; Cheng, Y. Ultrafast lasers—Reliable tools for advanced materials processing. *Light Sci. Appl.* **2014**, *3*, e149. [CrossRef]
10. Malinauskas, M.; Žukauskas, A.; Hasegawa, S.; Hayasaki, Y.; Mizeikis, V.; Buividas, R.; Juodkazis, S. Ultrafast laser processing of materials: From science to industry. *Light Sci. Appl.* **2016**, *5*, e16133. [CrossRef] [PubMed]
11. Lei, S.; Zhao, X.; Yu, X.; Hu, A.; Vukelic, S.; Jun, M.B.G.; Joe, H.-E.; Yao, Y.L.; Shin, Y.C. Ultrafast Laser Applications in Manufacturing Processes: A State-of-the-Art Review. *J. Manuf. Sci. Eng.* **2020**, *142*, 031005. [CrossRef]
12. Alsaigh, R.A. Enhancement of Surface Properties Using Ultrashort-Pulsed-Laser Texturing: A Review. *Crystals* **2024**, *14*, 353. [CrossRef]

13. Moldovan, E.R.; Concheso Doria, C.; Ocaña, J.L.; Baltes, L.S.; Stanciu, E.M.; Croitoru, C.; Pascu, A.; Roata, I.C.; Tierean, M.H. Wettability and Surface Roughness Analysis of Laser Surface Texturing of AISI 430 Stainless Steel. *Material* **2022**, *15*, 2955. [CrossRef] [PubMed]
14. Gemini, L.; Faucon, M.; Romoli, L.; Kling, R. High throughput laser texturing of super-hydrophobic surfaces on steel. In Proceedings of the Laser-Based Micro- and Nanoprocessing XI, SPIE LASE, San Francisco, CA, USA, 28 January 2017; Klotzbach, U., Washio, K., Kling, R., Eds.; SPIE: Bellingham, WA, USA, 2017. 100921G. [CrossRef]
15. Gregorčič, P.; Conradi, M.; Hribar, L.; Hočevar, M. Long-Term Influence of Laser-Processing Parameters on (Super)hydrophobicity Development and Stability of Stainless-Steel Surfaces. *Materials* **2018**, *11*, 2240. [CrossRef] [PubMed]
16. Obilor, A.F.; Pacella, M.; Wilson, A.; Silberschmidt, V.V. Micro-texturing of polymer surfaces using lasers: A review. *Int. J. Adv. Manuf. Technol.* **2022**, *120*, 103–135. [CrossRef]
17. Riveiro, A.; Maçon, A.L.B.; Del Val, J.; Comesaña, R.; Pou, J. Laser Surface Texturing of Polymers for Biomedical Applications. *Front. Phys.* **2018**, *6*, 16. [CrossRef]
18. Worgull, M. *Hot Embossing. Theory and Technology of Microreplication*, 1st ed.; Micro & Nano Technologies; William Andrew: Oxford, UK, 2009.
19. Gong, Y.; Li, B.; Chen, L.; Lv, Q.; Wang, Q.; Liu, W.; Zou, L. Study on superhydrophobicity of hot embossed polytetrafluoroethylene/graphite composites. *J. Polym. Res.* **2023**, *30*, 168. [CrossRef]
20. Sun, J.; Li, H.; Huang, Y.; Zheng, X.; Liu, Y.; Zhuang, J.; Wu, D. Simple and Affordable Way To Achieve Polymeric Superhydrophobic Surfaces with Biomimetic Hierarchical Roughness. *ACS Omega* **2019**, *4*, 2750–2757. [CrossRef] [PubMed]
21. Victor, J.J.; Facchini, D.; Erb, U. A low-cost method to produce superhydrophobic polymer surfaces. *J. Mater. Sci.* **2012**, *47*, 3690–3697. [CrossRef]
22. Lee, H.-J.; Park, K. Development of composite micro-patterns on polymer film using repetitive ultrasonic imprinting. *Int. J. Precis. Eng. Manuf. Green Technol.* **2014**, *1*, 341–345. [CrossRef]
23. Wypych, G. *Handbook of Polymers*; ChemTec Publishing: Toronto, ON, Canada, 2012.
24. ASM International. *Engineered Materials Handbook. Desk Edition*, 1st ed.; ASM International: Materials Park, OH, USA, 1995.
25. Srinivasan, A.; Bandyopadhyay, S. (Eds.) *Advances in Polymer Materials and Technology*; Taylor & Francis CRC Press: Boca Raton, FL, USA, 2017.

materials

MDPI

Article

Intercomparison of Indexable Cutting Inserts' Wear Progress and Chip Formation During Machining Hardened Steel AISI 4337 and Austenitic Stainless Steel AISI 316 L

Karel Šramhauser [1,*], Pavel Kraus [2], František Špalek [1], Pavel Černý [1,3], Jean de Dieu Marcel Ufitikirezi [1], Tomáš Zoubek [1], Miroslav Strob [1], Yevhen Kononets [1], Pavel Kříž [1,3] and Vladimír Vochozka [3]

1 Department of Technology and Cybernetics, Faculty of Agriculture and Technology, University of South Bohemia in České Budějovice, 370 05 České Budějovice, Czech Republic; fspalek@fzt.jcu.cz (F.Š.); pcerny@pf.jcu.cz (P.Č.); ufitikirezi@fzt.jcu.cz (J.d.D.M.U.); zoubek@fzt.jcu.cz (T.Z.); miross00@fzt.jcu.cz (M.S.); kononets@fzt.jcu.cz (Y.K.); kriz@pf.jcu.cz (P.K.)

2 Institute of Technology and Materials, Faculty of Mechanical Engineering, Jan Evangelista Purkyně University in Ústí nad Labem, 400 96 Ústí nad Labem, Czech Republic; pavel.kraus@ujep.cz

3 Department of Applied Physics and Technology, Faculty of Education, University of South Bohemia in České Budějovice, 371 15 České Budějovice, Czech Republic; vvochozka@pf.jcu.cz

* Correspondence: sramhauser@fzt.jcu.cz; Tel.: +420-389-032-928

Abstract: This article deals with a mutual comparison of indexable cutting inserts of the CNMG 120408 type from two different manufacturers during the machining of hardened steel AISI 4337 and austenitic stainless steel AISI 316 L. The main goal is to analyse the different wear processes depending on the difference in the manufacturer's design and also depending on the properties of the different machined materials. The progress of the wear of the main spine of the tool, the types of wear and the service life of the cutting edge were monitored, with the achievement of the critical value VB_{max} = 300 μm being the standard. In addition to the wear of the inserts, the production of chips was monitored in terms of their shape, average size and number of chips per 100 g of chips produced. In order to understand the relationships arising from the obtained data, an SEM equipped with an elemental analyser was used to analyse the coating layers and the substrate of the unworn inserts and the types of wear and the intensity of the surface damage of the worn inserts. A several-fold difference in the lifetime of the cutting edge was found, both in terms of design and in terms of the selected machined material, while in both cases the cutting edge with Al_2O_3 and TiCN layers of half thickness achieved a better result in liveness. From the point of view of chip formation, very similar results in shape and average length were observed despite the different designs of chip breakers. Cutting inserts with half the thickness of the coating layers achieved longer cutting edge life in the non-primary material application compared to the target workpiece material. At the same time, it was observed that a thinner coating layer has a positive effect on chip formation in terms of its length and shape.

Keywords: indexable cutting insert; wear; hardened steel; stainless steel; chip production

1. Introduction

In the contemporary manufacturing sector, businesses are looking to implement a uniform carbide cutting element system that facilitates the easy exchanging of cutting components [1,2]. The geometric, physical and chemical properties of cutting elements directly influence their capabilities and performance [3–5]. These characteristics of each individual cutting element play a crucial role in influencing its performance when engaging with the workpiece and the surrounding conditions. Moreover, every cutting material exhibits an inherent inclination to alter cutting parameters, as well as to experience abrasive and adhesive wear as time progresses [3,6,7]. An inappropriate cutting tool for a specific manufacturing scenario possesses the potential to diminish the quality of the surface finish and escalate the rates of production defects. Moreover, it can considerably curtail the lifespan of the tool and augment the consumption of process fluid. Additionally, it leads to an upsurge in the expenses associated with the disposal of production waste [8–10].

The advancement of cutting materials has significantly improved since the introduction of the electric arc furnace. A key benefit of the arc furnace is its capability to melt tough materials like carbides. Carbides that are sintered exhibit superior bending strength, Young's modulus and fracture toughness compared to other cutting materials, making them ideal for high-speed machining and intermittent cutting in challenging environments [11,12]. The disadvantage is lower thermochemical and thermomechanical stability; for example, dry turning iron-based superalloy GH2132 limits the use at higher cutting speeds at >30 $m \cdot min^{-1}$ and higher than 250 $m \cdot min^{-1}$ when turning AISI 4340 steel (220 HV) [13,14]. Sintered carbide is a heterogeneous material consisting of solid carbide particles in a metal bond. For practical purposes, mainly hexagonal tungsten carbide (WC) and cubic carbides of titanium (TiC), tantalum (TaC) and niobium (NbC) are used in the manufacture of cutting tools [15–17]. These carbide particles improve strength, hardness and thermomechanical stability. Their geometrical form is of spherical shape and the size typically ranges from 0.1 to 19 μm [12]. The cobalt (Co) bond is the most prevalent method of anchoring carbide particles, typically comprising up to 13%, with the volume of the bond ranging from 5 to 40% of the total volume [18–20]. In order to enhance cutting capabilities, coatings are applied either across the entire surface or on specific functional areas. These coatings are typically administered in multiple layers, each possessing unique properties aimed at bolstering the resistance to mechanical wear of cutting inserts. Consequently, this process can significantly prolong the lifespan of these inserts by a factor of 2 to 10 times [21,22].

Sintered carbide cutting tools offer a distinct benefit through their well-established disposal and recycling process, which involves transforming them into atomic particles, purifying and separating them, and ultimately crystallising them into primary powders [23–25]. In machining, along with dulling of the cutting tool, the primary outcome is the removal of material from the workpiece in the form of chips. These metal chips are characterised by their dimensions and configuration, with particular emphasis on their length [26,27]. The assessment of metal chips in terms of both quantity and characteristics is frequently discussed within the realm of waste management. When dealing with long, uninterrupted shavings, the operational area of the machinery becomes filled rapidly, necessitating frequent emptying. The presence of substantial amounts of waste in the shape of long, continuous chips results in the occupation of excessive storage space and complicates subsequent machining processes. When elementary or fine mesh shavings are generated, storage space efficiency is increased, waste removal frequency in the production area is decreased and the recycling process operates at a higher level of performance [28–30]. The characteristics of a metal swarf, including its shape and length, are influenced by various factors. These factors include the properties of the material being machined, the specific cutting conditions employed, the application of process fluids at the cutting points, the geometry of the cutting tool, the properties of the cutting material and the type of coating used on the cutting tool [31]. The description and classification of the shape, size and direction of the chip emerging from the cutting point are covered in ISO 3685:1993 [32]. Numerous research studies have explored the enhancement of the cutting process through the utilisation of the Taguchi method and ANOVA (analysis of variance) [33–35]. When assessing cutting tool wear, the chip formation factor plays a crucial role in monitoring both the wear process and the overall machining process [36].

Machining is inherently a process that is influenced by many factors, each of which influences the outcome of the job. This result again consists of a set of goals that we want to achieve when creating the workpiece in terms of its surface properties, level of residual stresses, ideal handling of waste material, etc. The data analysed as part of this research demonstrate the influence of subtle manufacturing nuances of interchangeable inserts of the same type according to ISO 1832:2017 on their durability, course and rate of wear depending on the properties of the test materials [37]. At the same time, this article compares the chips produced from individual cutting processes and thus also adds to the factor of wear of the cutting tool the aspect of the influence of the mentioned production nuances, which are not subject to standardised marking, on the formation of chips in terms of their length and shape, which are important parameters when planning the cutting process and subsequent handling with waste material.

2. Materials and Methods

The aim of the experiment was to compare the CNMG 120408 type of indexable inserts in terms of their service life (specifically the progression of their wear) and their suitability for chip formation. For this purpose, a procedure was designed to ensure the comparability of the results obtained from the data. This experiment aimed to demonstrate the wear mechanisms of cutting tools and chip formation in relation to the selected cutting parameters as well as differences in the design of the specific tools. Additionally, the experiment incorporated the factor of the material properties of the workpieces, as demonstrated by the different types of steel used. The experiment was conducted in collaboration between the Faculty of Agriculture and Technology at the University of South Bohemia in České Budějovice and the Faculty of Mechanical Engineering at the Jan Evangelista Purkyně University, within the framework of research into tool wear mechanisms. The following sections describe the test materials, the selected indexable cutting inserts and the methodological approach, including the measuring equipment used.

2.1. Evaluated Cutting Inserts

The selected indexable cutting inserts were CNMG 120408 type inserts from two different manufacturers. These inserts were purposefully chosen with the same designation according to ISO 1832:2017 standards but with noticeable differences in design, depending on the manufacturer. Specifically, the selected indexable cutting inserts were:

- CNMG 120408-PM 4325: The cutting material designated by the manufacturer as GC4325 (HC) is presented as suitable for finishing to roughing applications when machining steels and cast steels, both in continuous and interrupted cuts, at high metal removal rates. The WC substrate, classified by the manufacturer as P25, is functionally graded using TiCN- and Al_2O_3-based coating layers. This was confirmed through elemental mapping of the VBD surface layer in a longitudinal section, as shown in Figure 1a. The chip breaker, labelled by the manufacturer as PM, is designed to allow a wide range of depth of cut (a_p) and feed rate (f) settings, as seen in Figure 1b. This chip breaker is considered one of the manufacturer's optimised designs with broad applicability, suitable for both continuous and interrupted cuts. It is recommended not to overload the cutting edge. An EDX analysis confirmed the presence of coating layers with a thickness of 11.5 ± 1.2 μm for the TiCN layer and 5.3 ± 0.3 μm for the Al_2O_3 layer.
- CNMG 120408E-MP3 WPP20S: The substrate is specified as a coated cemented carbide with CVD coating layers of TiCN and Al_2O_3. According to the manufacturer, the coating layer is optimised with a microstructure of Al_2O_3 designed to increase tool life and potentially reduce machining time. The main and secondary flanks are finished to improve wear identification. The EDX analysis of the coating layers confirmed the presence of the declared layers, with a TiCN layer thickness of 17.8 ± 0.4 μm and an Al_2O_3 layer thickness of 15.1 ± 0.3 μm, as shown in Figure 2a. The chip breaker labelled as MP3 is primarily intended for medium-duty machining of steel materials that produce long chips. The MP3 geometry is typically used for machining near-net-shape forgings and thinner-walled extruded parts. In the macro image of the VBD in Figure 2b, the so-called "Bullet Design" chip breaker in the shape of a hemisphere is visible, which, according to the manufacturer, should provide additional rigidity for optimal chip breaking.

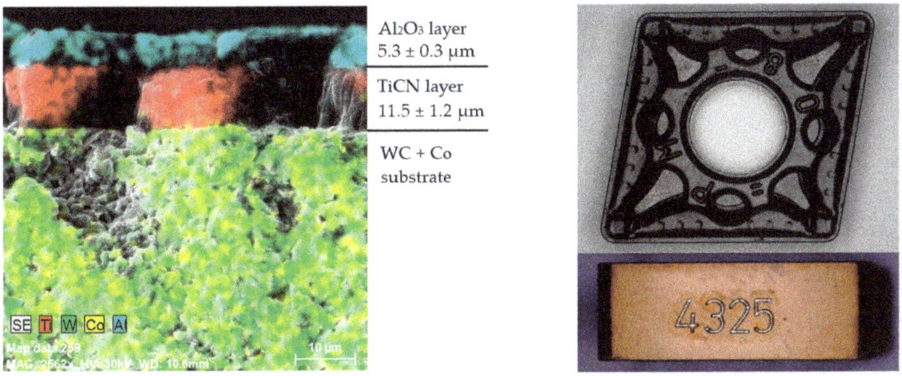

(**a**) (**b**)

Figure 1. Investigated CNMG 120408-PM 4325 cutting insert: (**a**) composition of deposited coating layers; (**b**) macro shot of functional surfaces.

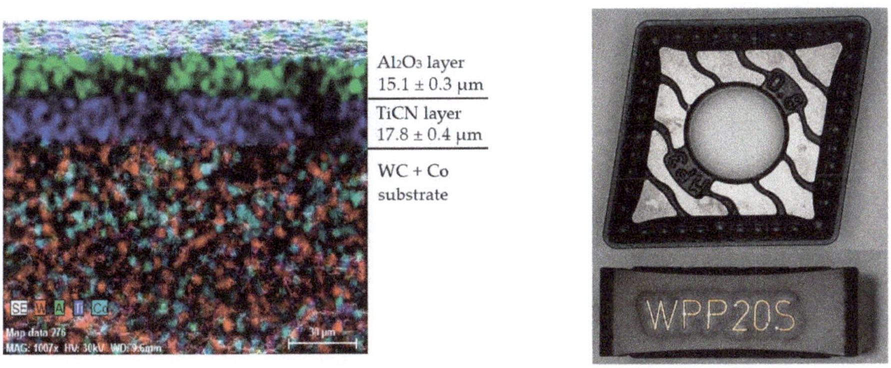

(**a**) (**b**)

Figure 2. Investigated CNMG 120408E-MP3 WPP20S cutting insert: (**a**) composition of deposited coating layers; (**b**) macro shot of functional surfaces.

Cutting edge parameters that are not identifiable from the marking according to ISO 1832:2017 are declared by the manufacturer and shown in Figure 3. These are chamfering structure parameters, where the rake angle is expressed by value a [°], the bevel of chamfering land is expressed by value b [°] and the land width is expressed by value c [mm].

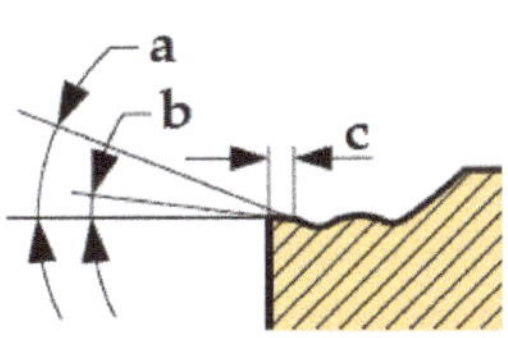

(**a**)

Indexable cutting insert	a [°]	b [°]	c [mm]
CNMG 120408-PM 4325	20	5	0.25
CNMG 120408E-MP3 WPP20S	22.5	5	0.25

(**b**)

Figure 3. Investigated chamfering structure parameters of evaluated cutting inserts: (**a**) schematic visualisation; (**b**) values for each cutting insert.

2.2. Experimental Materials

The first experimental material was AISI 4337 steel, a medium-alloyed chromium–molybdenum steel known for its high hardenability, making it suitable for highly stressed machine components. In its quenched and tempered condition, it offers a very favourable strength-to-yield ratio and high toughness. The high toughness of this steel helps to inhibit the propagation of fatigue cracks, resulting in excellent fatigue strength under alternating and combined loading conditions. This material is not susceptible to temper embrittlement. It is typically quenched in oil or synthetic polymer solutions. The chemical composition, as per ASTM A29:2004, is provided in Table 1 [38].

Table 1. Chemical composition of AISI 4337 according to ASTM A29:2004 (in weight %).

C	Mn	P	Cr	Mo	S	Ni	Si
0.3–0.38	0.5–0.8	< 0.4	1.3–1.7	1.3–1.7	0.15–0.3	<0.035	<0.035

The test material was supplied in the form of 50 mm diameter bars. For the experiment, the material was heat-treated to achieve a hardness of 40–44 HRC to a depth of 15 mm from the surface. The heat treatment consisted of a normalisation at 850 °C, followed by a quenching in oil and a tempering at 450 °C. Hardness measurements were taken from the centre of the workpiece to the surface using a CISAM-ERNST AT250 hardness tester (Cisam–Ernst, Induno Olona, Italy), yielding an average hardness value of 43 ± 1 HRC. The chemical composition of the supplied AISI 4337 material was verified using a Bruker Q4 TASMAN spark spectrometer (Bruker, Billerica, MA, USA). The results of the measurements (see Table 2) confirmed that the supplied material meets the standardised chemical composition requirements. According to the manufacturers of the tested indexable inserts, this material is ideal for machining with these tools.

Table 2. Chemical composition of AISI 4337 according to Bruker Q4 Tasman (in weight %).

C	Mn	P	Cr	Mo	S	Ni	Si
0.342	0.57	0.329	1.519	1.47	0.193	<0.005	<0.001

The second experimental material was AISI 316 L steel, an austenitic stainless steel known for its resistance to pitting corrosion in the presence of chlorides, as well as its good resistance to sulphuric and phosphoric acids. This material is highly weldable without the risk of intergranular corrosion in the heat-affected zone. It is suitable for cold forming and has good machinability, although achieving a mirror finish during polishing can be challenging. AISI 316 L is commonly used in welded structures in aggressive industrial environments, for the manufacture of pressure vessels and for components in the chemical, pharmaceutical and textile industries. Machinability is hindered by the material's thermal conductivity, higher ductility and tendency to harden during machining. The material's use in the food industry is influenced by its nickel content. The chemical composition according to ASTM F899:95 is provided in Table 3 [39].

Table 3. Chemical composition of AISI 316 L according to ASTM F899:95 (in weight %).

C	Mn	P	Cr	Mo	S	Ni	Si
<0.03	<2	<0.045	16.5–18.5	2–2.5	<0.015	10–13	<1

The material of the supplied AISI 316 L semi-finished product, in the form of a 50 mm diameter bar, was tested using a CISAM-ERNST AT250 hardness tester, and an average hardness of 18 ± 0.9 HRC was measured. The chemical composition of the supplied AISI 316 L material was analysed using a Bruker Q4 TASMAN spark spectrometer, and it was found to meet the specified tolerances, as shown in Table 4. According to the manufacturers of the tested cutting inserts, this material does not fall into the primary or secondary usage group for machining purposes.

Table 4. Chemical composition of AISI 316 L according to Bruker Q4 Tasman (in weight %).

C	Mn	P	Cr	Mo	S	Ni	Si
0.022	1.407	0.041	17.24	2.007	0.007	10.6	0.578

2.3. Experimental Setup

In the framework of the DoE (Design of Experiments), cutting parameters were established for both tested cutting inserts during machining on both test materials. The a_p was set to 1.5 mm and the f to 0.3 mm·ot^{-1}, which fall within the range of cutting parameters recommended by the manufacturers of both cutting inserts under examination. The cutting speed (v_c) was set at 100 m·min^{-1}, which corresponds to the cutting speeds typically used in conventional machining. The combination of selected cutting parameters was suitable for the given cutting tools and workpiece materials, meeting the requirements for the stability of the cutting process, as demonstrated by initial testing.

The criteria for concluding the experimental phase for each cutting insert was the achievement of a maximum flank wear (VB_{max}) of 300 µm. The VB_{max} value was selected for the experiment based on experience with the wear of carbide cutting tools, which typically ranges between 300 and 500 µm [32]. The experimental procedure was as follows: after each tool pass with a_p of 1.5 mm, the tested cutting insert was removed and the wear level on the main flank (VB_{max}) was measured using an Olympus SZX10 stereomicroscope (Olympus Corporation, Tokyo, Japan) equipped with a Bresser MicroCam-II digital camera (Rhede, Germany). Simultaneously, a sample of the produced chips, corresponding to the achieved flank wear, was collected. To evaluate the shape and size of the chips according to ISO 3685, the Olympus SZX10 stereomicroscope equipped with a Bresser MicroCam-II digital camera and measuring software (ver.: x64, 3.7.13814.20190120) was used again. The chip sample was also evaluated in terms of the number of chips per 100 g of produced chips, providing specific information about the chips set. While the average chip length gives insight into the cutting process itself, the count of chips per 100 g offers information about the quantity and volume of produced chips. This metric, rather than the median chip length, allows for better planning in handling and managing machining waste. Specifically, it may involve sizing waste containers collecting chips, which affects the frequency of waste disposal, movement on the shop floor, storage capacity before transportation for recycling, or even the scheduling of recycling operations like crushing and pressing. A higher chip count indicates the presence of shorter chips overall. After each measurement, the cutting insert was reinstalled in the PCLNR 2020 K12 holder (TaeguTec, Daegu, Republic of Korea), ensuring it returned to its original position. This process was repeated until the specified maximum flank wear (VB_{max} = 300µm) was reached. Each evaluated indexable cutting insert was tested three times to ensure data validity. After reaching the maximum wear limit, the experimental phase was concluded, and the collected data was subsequently analysed, as detailed below.

After reaching the limit wear of the inserts, the roughness of the workpiece was measured. The parameter R_z was monitored, which was the average maximum peak to valley of five consecutive sampling lengths within the measuring length. The reason for choosing this roughness parameter was its relevance in the evaluation of surface quality (especially in the automotive industry), and for this reason its numerous uses in practice [40]. A Mitutoyo Surftest SJ-210 portable roughness meter (Mitutoyo, Takatsu Ward, Japan) was used for measurement. The aim of measuring the roughness of the machined surface R_z was an informative comparison of the roughness in connection with the achieved limit flank wear of the selected cutting inserts for both test materials.

3. Results

The obtained data on the tool's flank wear were plotted against a time axis, where each measured wear value corresponded to a specific tool time in the run T [min]. An example of the continuous measurement of flank wear for a cutting insert is shown in Figure 4, which illustrates the change in the measurement and distribution of flank wear over time.

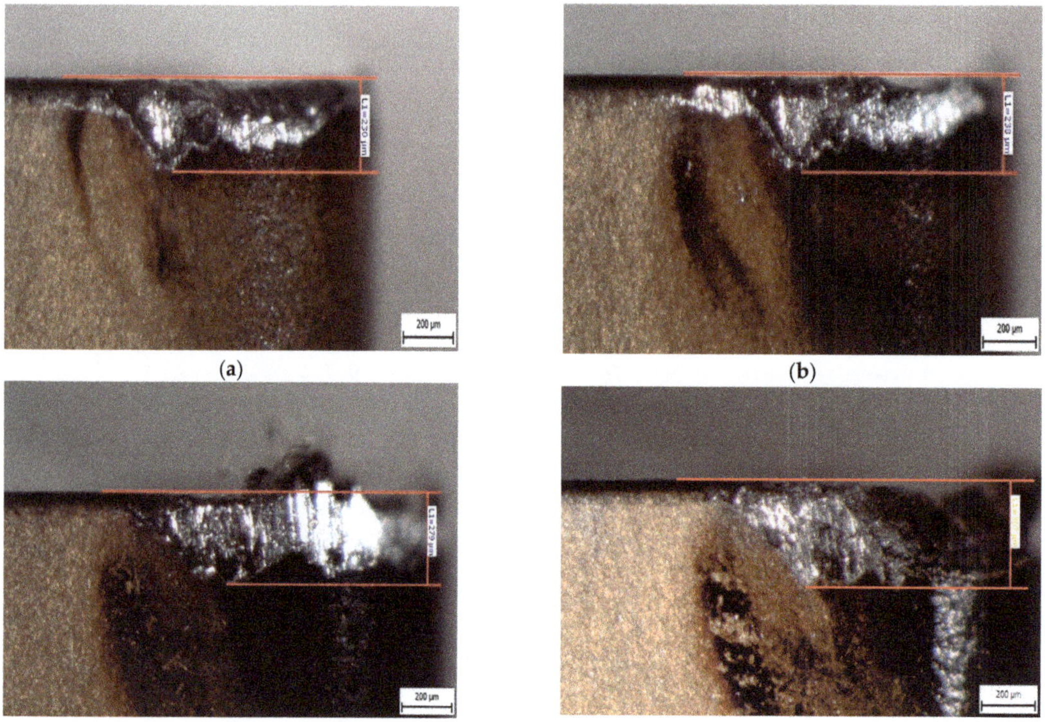

(a) (b) (c) (d)

Figure 4. Selected VB_{max} images of CNMG 120408-PM 4325 turning AISI 316 L steel, v_c = 100 m·min^{-1}: (**a**) T = 2.5 min; (**b**) T = 7.5 min; (**c**) T = 13 min; (**d**) T = 17.5 min.

Figure 5 shows the wear progression of the two cutting inserts examined during the machining of AISI 4337 steel. The CNMG 120408-PM 4325 insert reached the tool life of the cutting edge, i.e., the specified critical wear at T = 16 [min]. The CNMG 120408-MP3 WPP20S insert reached approximately one-third of the tool life of the cutting edge at T = 5.5 [min]. For both inserts, an initial slowing trend in the wear progression is visible.

The same analysis of the wear progression of the evaluated cutting inserts was conducted using the test material AISI 316 L. Figure 6 shows the wear progression of the CNMG 120408-PM 4325 insert, which reached the specified critical wear at time T = 17.3 [min], which was 1.3 minutes longer than with AISI 4337 material. The CNMG 120408-MP3 WPP20S insert reached critical wear at T = 1.5 [min], practically right after the first cut.

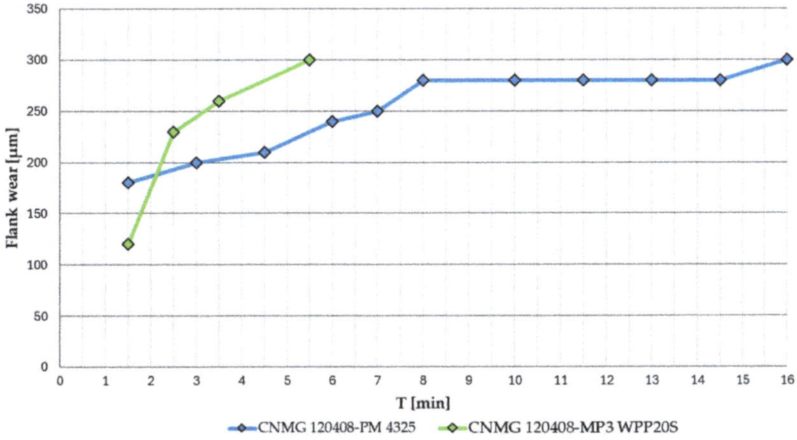

Figure 5. Flank wear progression of evaluated cutting inserts for AISI 4337 steel, v_c = 100 m·min^{-1}.

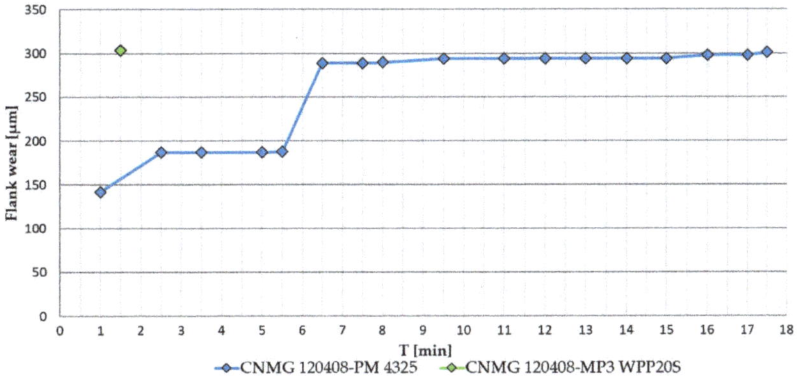

Figure 6. Flank wear progression of evaluated cutting inserts for AISI 316 L steel, v_c = 100 m·min^{-1}.

Using a Tescan Vega 3 SEM (Tescan Orsay Holding a.s., Brno, Czech Republic) equipped with a Bruker X-Flash analyser (Billerica, MA, USA), the cutting inserts under examination that had reached the specified critical flank wear were subjected to a visual assessment of the wear rate of the functional surfaces. Figure 7 shows the flank wear on the CNMG 120408-PM 4325 cutting insert. Figure 7a identifies the forms of flank face wear after machining AISI 4337 steel. The flank face surface was abrasively worn by chips in the area of chip exit from the cutting point. In Figure 7b, this worn condition is subjected to elemental (EDX) analysis, which confirms the presence of an Al_2O_3 layer in the flank face wear area down to the level of the support substrate. A portion of the workpiece material adhered in the chip exit area, revealing the Al_2O_3 layer to a small extent. Similarly, a visual assessment of the wear of the CNMG 120408-PM 4325 insert after machining AISI 316 L is shown in Figure 7c, where several types of wear were identified, namely flank face wear, abrasive wear by chips, chipping at the cutting edge and notch wear at the interface between the cutting tool edge and the workpiece surface. Figure 7d shows the wear at the notch wear point down to the level of the substrate, highlighting the wear limits of the TiCN and Al_2O_3 coatings. Additionally, a built-up edge (BUE) formed by the machined AISI 316 L material is visible. Overall, the wear rate was higher after machining AISI 316 L. Although similar flank face wear was achieved, the wear boundaries of each coating layer were more obvious after machining AISI 316 L stainless steel and there was more intense wear of the cutting edge in the form of chipping compared to machining AISI 4337.

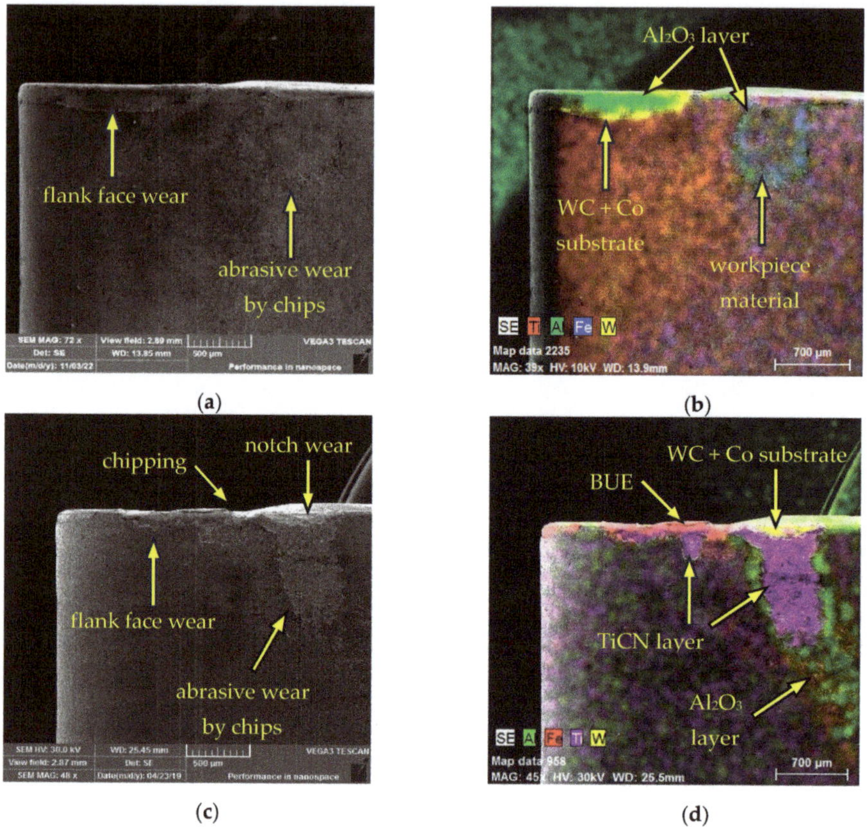

Figure 7. SEM analysis of worn CNMG 120408-PM 4325: (**a**) main flank wear types after turning AISI 4337; (**b**) main flank worn layers after turning AISI 4337; (**c**) main flank wear types after turning AISI 316 L; (**d**) main flank worn layers after turning AISI 316 L.

The identification of the types of wear and the degree of exposure of the individual coating layers, and possibly the substrate, on the face of the CNMG 120408-PM 4325 insert after reaching the critical wear limit are shown in Figure 7 for AISI 4337 steel. Beyond the rake face wear line that follows the edge of the cutting insert, wear in the form of crater wear was visible at the interface between the cutting edge of the insert and the workpiece edge, as shown in Figure 8a. As part of the experiment, a crater wear width (KB) of 785.7 ± 33.7 µm was measured on CNMG 120408-PM 4325 after reaching the wear limit VB_{max} when machining AISI 4337. In Figure 8b, wear is noticeable at the level of the TiCN coating on the rake face wear area and wear down to the level of the support substrate below the crater wear area. Figure 8c shows a picture of the worn sheet of rake face after turning AISI 316 L steel. A rake face wear line can be seen following the radius of the tip of the insert. Crater wear is scattered across the rake face. On CNMG 120408-PM 4325, KB = 709.2 ± 28.3 µm was measured after reaching the wear limit VB_{max} when machining AISI 316L. Within the rake face wear and crater wear, both the TiCN layer and the supporting substrate were exposed, which is visible in Figure 8d. Also, the workpiece material was identified in the form of BUE and notch wear on the cutting edge, which was also evident on the rake face. Similar to the main flank face, the rake face in the case of AISI

316 L machining experienced more intense surface wear and also more of the machined material was adhered to the face and the cutting edge.

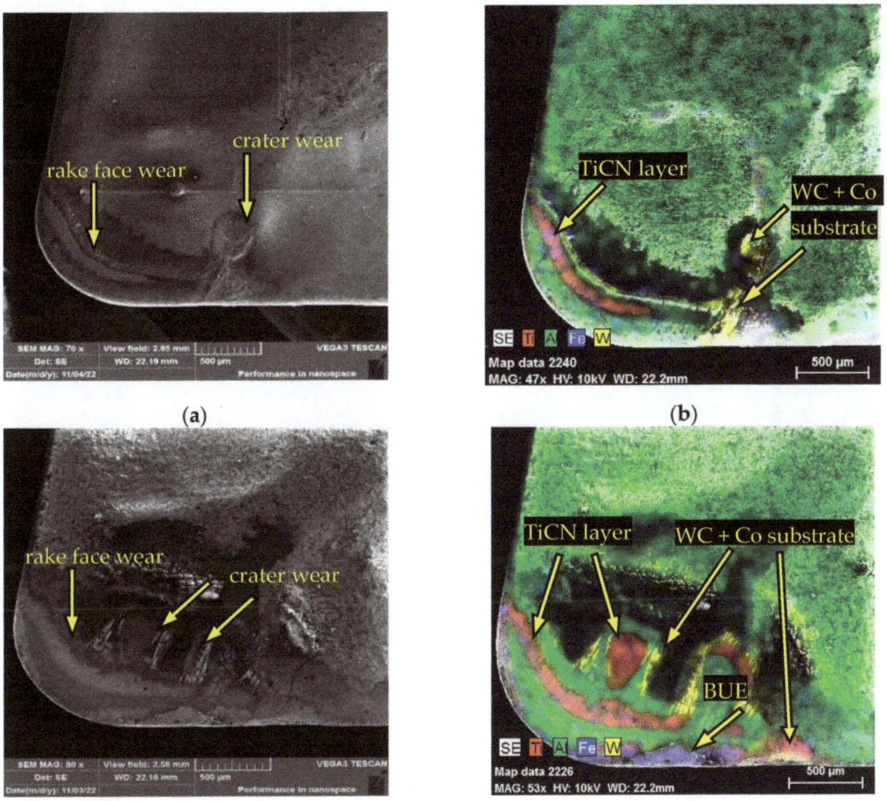

(a) (b)

(c) (d)

Figure 8. SEM analysis of worn CNMG 120408-PM 4325: (**a**) main rake wear types after turning AISI 4337; (**b**) main rake face worn layers after turning AISI 4337; (**c**) main rake wear types after turning AISI 316 L; (**d**) main rake face worn layers after turning AISI 316 L.

The wear patterns of the main flank face of the CNMG 120408-MP3 WPP20S insert after machining of AISI 4337 steel and the elemental analysis of the worn surfaces are shown in Figure 9a,b. The flank face wear and abrasive wear by chips are clearly visible and are defined with an exposed TiCN layer, locally adhered workpiece material (chips) and even partially exposed support substrate. The wear rate is therefore higher than in the case of flank face wear. In Figure 9c, after the machining of ASIS 316 L steel, chipping, flank face wear and notch wear can be seen at the interface between the cutting edge and the workpiece surface. In Figure 9d, EDX analysis confirms a more intense level of wear at the level of the TiCN layer in the area of the tip of the tool, and the substrate level in the notch wear area, which was partially filled by the workpiece material in the form of BUE. BUE is mainly observed at the areas where the underlying coating layers or substrate material have been exposed. Again, after machining AISI 316 L, the BUE was more extensive due to more intense wear at the level of the substrate.

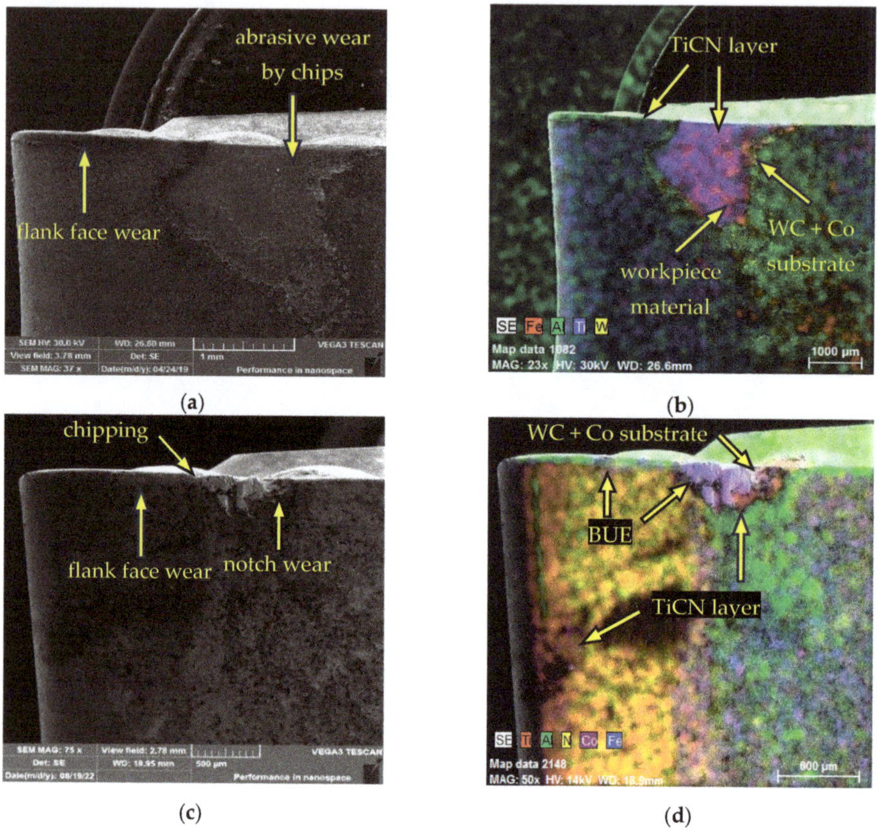

Figure 9. SEM analysis of worn CNMG 120408-MP3 WPP20S: (**a**) main flank wear types after turning AISI 4337; (**b**) main flank worn layers after turning AISI 4337; (**c**) main flank wear types after turning AISI 316 L; (**d**) main flank worn layers after turning AISI 316 L.

The wear patterns on the rake face of CNMG 120408-MP3 WPP20S after turning AISI 4337 steel are identified in Figure 10a. Notable observations include rake face wear and crater wear on the main surface near the chip formers in the form of hemispheres. As part of the experiment, KB = 726 ± 37.3 μm was measured on CNMG 120408-PM3 WPP20S after reaching the wear limit VB_{max} when machining AISI 4337. Notch wear was visible at the interface between the insert cutting edge and the workpiece. The EDX analysis in Figure 10b shows the extent of wear at the local level of the TiCN layer along the line of the cutting edge and the WC + Co substrate at the place where the crater wear was formed. In Figure 10c, the wear in the form of rake face wear, crater wear and notch wear after turning AISI 316 L is shown. On CNMG 120408-PM3 WPP20S, KB = 648.1 ± 36.1 μm was measured after reaching the wear limit VB_{max} when machining AISI 316L. EDX analysis depicted in Figure 10d identifies the exposure of the TiCN layer along the boundaries of the crater wear area. Additionally, the wear along the crater wear boundaries has reached the substrate level, which is visible even in the notch wear area. The adhered material could be observed both at the notch wear area and along the crater wear boundaries. Identical types of wear were identified on the rake face, but with less intensity or distribution in the case of AISI 4337 machining compared to AISI 316 L machining.

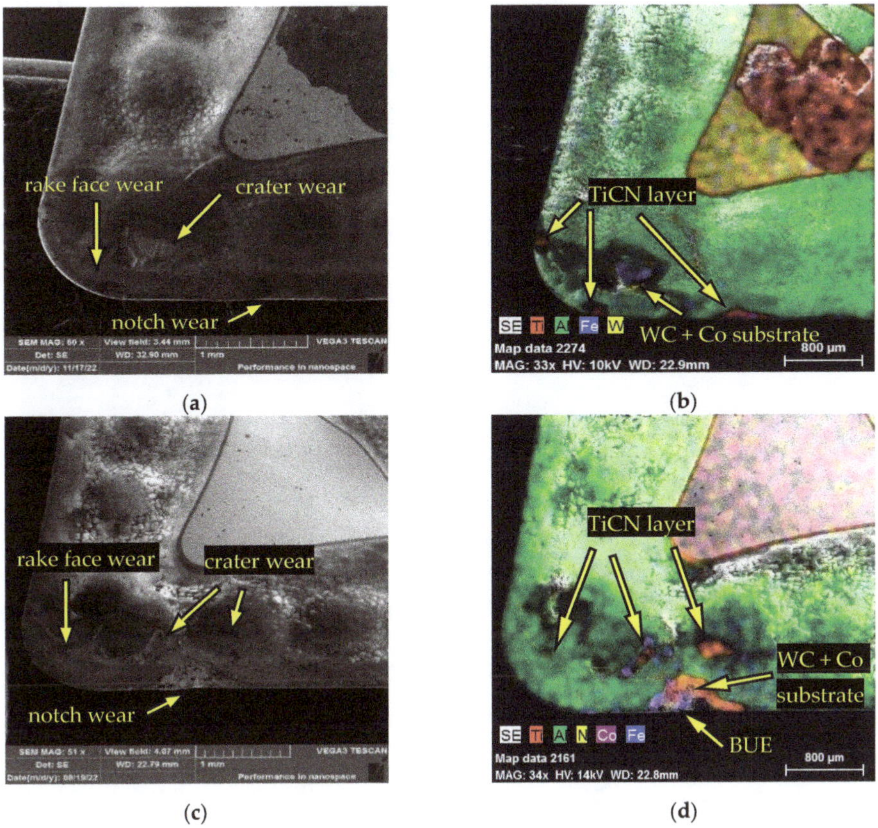

Figure 10. SEM analysis of worn CNMG 120408-MP3 WPP20S: (**a**) main rake wear types after turning AISI 4337; (**b**) main rake face worn layers after turning AISI 4337; (**c**) main rake wear types after turning AISI 316 L; (**d**) main rake face worn layers after turning AISI 316 L.

The experiment also included an analysis of the produced chips in terms of their length and shape. The resulting chronologically arranged photographic documentation of the chips provided the basis for the assessment of the shape of the chips (see Figure 11a,b), and also for the measurement of the length of the chips, which was carried out using the micrometric method (see Figure 11c). As part of the shape assessment, 5 to 15 chips were assessed per chip collection for each workpiece material removal. Length measurements were made on 3 to 5 pieces per each chip collection.

The analysis results of the produced chips are presented in Table 5. For both tested inserts, the data are divided according to the test material. The actual evaluation consisted of an assessment of chip shape/shapes, average length and average number of chips per 100 g of chips produced. From the measured data, it can be seen that despite different designs of the chip formers on the evaluated inserts, the same chip shapes (arc loose and arc connected) were identified, and their average length did not differ significantly within the overall production. A noticeable difference was found in the average number of chips per 100 g of chips. For both cutting inserts tested, this parameter was more favourable when machining AISI 4337 steel. In the case of the CNMG 120408-MP3 WPP20S insert, the difference in chip production between the tested materials was greater, and it achieved the best results in terms of chip production. When comparing the tested inserts after machining AISI 316 L, the results in terms of chip production were practically identical.

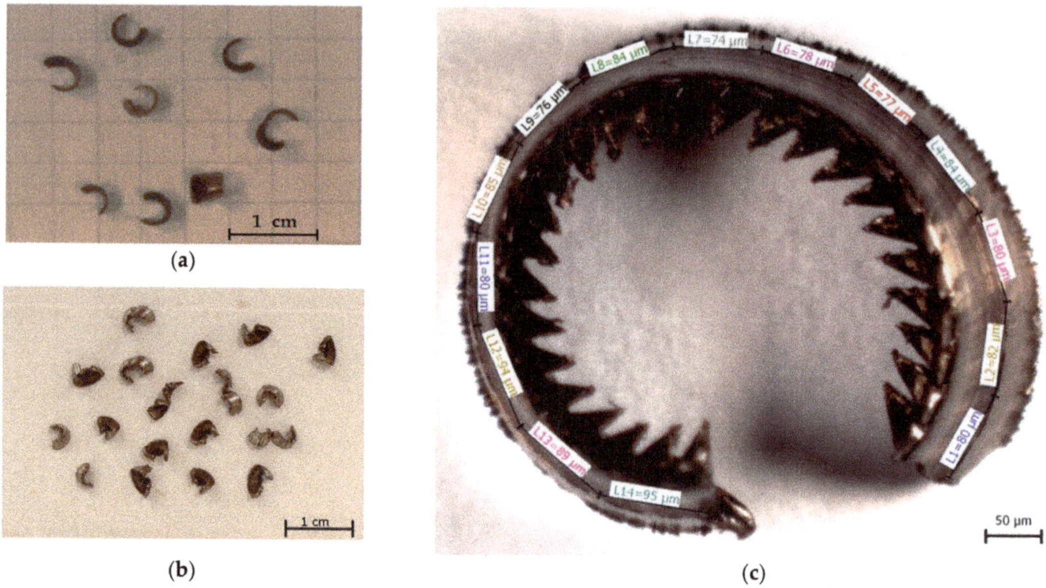

Figure 11. Demonstration of macroscopic chip evaluation: (**a**) CNMG 120408-PM 4325 chips after machining AISI 4337, T = 13 min; (**b**) CNMG 120408-PM 4325 chips after machining AISI 316 L, T = 17.5 min; (**c**) example of micrometric chip length measurement.

Table 5. Comparison of produced chips for individual cutting inserts and materials.

	CNMG 120408-PM 4325		**CNMG 120408E-MP3 WPP20S**	
	AISI 4337	**AISI 316 L**	**AISI 4337**	**AISI 316 L**
Chip Shape	Arc loose Arc connected	Arc loose Arc connected	Arc loose Arc connected	Arc loose Arc connected
Average Length [mm]	7.5 ± 0.6	6.2 ± 0.9	6.5 ± 1.3	6.3 ± 0.6
Avg. Number of Chips per 100 g of Chips [pcs]	2782 ± 350	2569 ± 243	3831 ± 691	2503 ± 408

The measured values of the roughness parameter R_z on the surface of the machined test blank after reaching the flank wear limit are shown in Table 6, where a better result is evident for the insert CNMG 120408-PM 4325 compared to CNMG 120408E-MP3 WPP20S.

Table 6. Comparison of R_z parameter for individual cutting inserts and materials after reaching the flank wear limit.

Roughness Parameter	**CNMG 120408-PM 4325**		**CNMG 120408E-MP3 WPP20S**	
	AISI 4337	**AISI 316 L**	**AISI 4337**	**AISI 316 L**
R_z [µm]	12.45 ± 0.229	10.017 ± 0.765	14.975 ± 0.259	15.722 ± 0.346

4. Discussion

Based on the analysed data, the wear process of the examined cutting inserts can be summarised as follows: When machining AISI 4337 steel, the CNMG 120408-PM 4325 insert achieved a longer tool life of 16 minutes, nearly three times that of the CNMG 120408E-MP3 WPP20S insert, which reached its critical wear after just 5.5 minutes. This difference may be attributed to the specific design of the CNMG 120408E-MP3 WPP20S insert, despite its classification under the same ISO 1832:2017 standard as the CNMG 120408-PM 4325. The main differences include the longitudinally curved cutting edge and the hemispherical chip breaker on the CNMG 120408E-MP3 WPP20S. The types of wear observed on the flank and rake faces of the tools are generally similar, with the exception of the notch wear observed on the rake face of the CNMG 120408E-MP3 WPP20S. A noticeable difference is in the wear of the coating layers, which are more clearly defined with distinct boundaries on the flank face of the CNMG 120408E-MP3 WPP20S. Conversely, the rake face of the CNMG 120408-PM 4325 insert exhibited more intense wear. However, in terms of dimensional accuracy of the workpiece, surface roughness and the potential introduction of stress into the surface layer, the monitoring of the main flank wear of the cutting tool is preferred.

The reason for the greater wear of the coating layer on the CNMG 120408E-MP3 WPP20S insert when machining AISI 4337 might paradoxically be due to the fact that this insert has coating layers of Al_2O_3 and TiCN that are up to twice as thick as those on the CNMG 120408-PM 4325 insert. This increased thickness could result in larger sections of the coating being detached, particularly in scenarios where the chip exits the cutting zone (causing abrasion against the tool's flank) and where there is a point contact between the workpiece and the cutting tool at the interface between the workpiece surface and the tool's cutting edge. It is possible that with thicker coatings, the layer composition may not adhere as well to the substrate, or the thicker layer might be subjected to greater stress from the cutting force during machining. This could mean that the coating used on the CNMG 120408E-MP3 WPP20S insert is less suitable for the specific combination of cutting conditions and workpiece material [41]. To verify this hypothesis, it is advisable to conduct a visual analysis of the cutting process using a high-speed camera focused on the main flank area of the tool.

When machining AISI 316 L steel, the CNMG 120408-PM 4325 insert again demonstrated superior tool life, lasting 17.5 minutes compared to the CNMG 120408E-MP3 WPP20S insert, which reached its critical wear limit on the main flank after just 1.5 minutes, immediately after removing the first layer of AISI 316 L material. This clearly shows that the CNMG 120408E-MP3 WPP20S insert is not suitable for machining austenitic stainless steel. Austenitic stainless steels are tough materials with poor thermal conductivity, requiring more energy for cutting. Consequently, a smaller proportion of the heat generated during machining is carried away by the chips, and more of this thermal energy is transferred through the tool itself. This can accelerate the wear process. The rapid wear observed when using the CNMG 120408E-MP3 WPP20S insert, despite the expectation that its thicker coating layers would provide greater durability when machining AISI 316 L, showed the exact opposite trend. The cause of this wear pattern could be attributed to the tool's geometry or the spherical shape of the chip breaker. On the other hand, the CNMG 120408-PM 4325 insert achieved a longer service life by 1.5 minutes when machining AISI 316 L, a material that is neither the primary nor an alternative recommendation for machining by the manufacturer. This outcome exceeded both theoretical and practical expectations, as it performed better in AISI 316 L machining than in the machining of hardened AISI 4337 steel.

The wear rate of the examined cutting inserts after machining AISI 316 L steel was more intense, exhibiting phenomena typically associated with the machining of austenitic steels, such as a higher degree of material adhesion in the form of built-up edge (BUE). On the CNMG 120408-PM 4325 insert, a greater degree of coating layer wear was observed on the flank, particularly at the interface between the machined material and the cutting edge, including more pronounced notch wear that exposed the substrate. Chipping of the cutting edge was also noted, along with more severe crater wear and the presence of adhered workpiece material in the form of BUE on the rake face. Through a combination of high cutting temperature and abrasion, the outgoing chips cause wear of the coating layers on the rake face and subsequently release the tungsten carbide grains from the basic structure of the substrate. As a result of diffusion, carbon leeches into the chips and thus the crater on the rake face increases. This mechanism is more intensive in the case of machining hard materials [42]. This explains the approximately 80 μm higher KB values when machining hardened steel AISI 4337. On the contrary, a larger layer of Al2O3 coating, which protects against the effect of the cutting temperature, in this case ensured a smaller crater wear on CNMG 120408-PM3 WPP20S [43]. However, this positive effect was observed only in the case of the formation of crater wear. The observed wear forms and their distribution correspond to the process of machining austenitic stainless steels and the previously described mechanisms of heat distribution that occur during the machining of this type of material. The rapid wear of the CNMG 120408E-MP3 WPP20S insert further underscores its unsuitability for applications involving austenitic stainless steels. A significant notch wear developed at the interface between the cutting edge and the workpiece surface, where material buildup in the form of BUE was evident. This occurrence is expected, as the freshly exposed layers of the cutting tool and the machined material surface in the form of chips are chemically clean. The formation of BUE on these surfaces is promoted by a combination of cutting temperature, pressure and the excellent adhesion properties of these chemically clean surfaces.

The process of continuous chip evaluation involved analysing their length and calculating the number of chips per 100 grams of produced chips. The outcome of processing these data was the average chip length and the average number of chips per 100 grams of produced chips for each cutting insert and machined material. In terms of average chip length, no significant differences were observed under the specified cutting conditions, either between the examined cutting inserts or between the machined materials, AISI 4337 and AISI 316 L. This fact contributes positively to the evaluation of chip formation by the examined cutting inserts within the application of the selected cutting conditions and machined materials, which differ significantly in terms of temperature distribution [44]. One would expect the average chip length to vary more as longer chips are formed during the machining of austenitic steels due to the usual higher ductility. This is also related to the austenitic structure of the machined material and the distribution of heat to the cutting tool, which creates chips of irregular shape. In contrast, shorter-length chips were expected to form during the machining of hardened steel [44,45]. Due to the cutting speed $v_c = 100\ m \cdot min^{-1}$, it is possible that during the machining of austenitic steel, adhesive wear was applied to a greater extent, which enabled the transfer of a larger amount of heat to the cutting insert, which could change the morphology of the formed chips [46]. From a production planning perspective, it is beneficial that the evaluated average lengths of the formed chips show relatively small deviations.

For the purposes of technological planning, the chips produced were also evaluated based on the number of chips generated per 100 grams of chips. This evaluation method provides a more detailed specification of the chips produced, taking into account their physical shape beyond just their average length, which is crucial for planning waste management. Similar to the average chip length, no significant differences in the number of chips were observed, except in the case of the CNMG 120408-E MP3 WPP20S cutting insert when machining AISI 4337 steel. In this scenario, the number of chips was 3831 ± 691 per 100 grams of chips, which is approximately 1000 more chips compared to the other combinations of cutting inserts and workpiece materials. This represents the best result concerning this parameter. The other combinations were comparably consistent. It can thus be stated that the designed spherical 'Bullet' chip breaker of the CNMG 120408-E MP3 WPP20S insert effectively shaped the chips in terms of their suitability in form and length under the given cutting conditions, particularly when machining hardened AISI 4337 steel. Given the primary purpose of the CNMG 120408-E MP3 WPP20S insert, as described in Chapter 2.1, this outcome was anticipated. In contrast, the CNMG 120408-PM 4325 insert, which achieved excellent durability results in the experiment on both tested materials, showed similar results regarding the number of chips per 100 grams across both test materials. This suggests that the CNMG 120408-PM 4325 insert is a more versatile cutting tool design in terms of its application.

As an informative addition, the roughness parameter R_z of the machined surface was measured after reaching the critical flank wear. The measured values showed a certain correlation with the achieved service life of the inserts when applied to specific test materials. Better results were achieved by the CNMG 120408-PM 4325 insert, with even better R_z values achieved by this insert when machining AISI 316 L. It is likely that similar mechanisms were involved in the formation of the machined surface profile as in chip formation in the results observed in this experiment [45]. There is an obvious connection between the roughness of the machined surface, the morphology of the formed chip and the wear mechanism of the cutting tool based on the cutting temperature, influenced by factors such as the properties and composition of the machined material and cutting conditions, especially the cutting speed [45,47]. For more complex results, it is recommended to analyse other components of surface integrity in connection with the wear of cutting tools.

5. Conclusions

CNMG 120408-coated indexable cutting inserts from two different manufacturers under identical cutting conditions were compared. As test materials, hardened steel AISI 4337, primarily intended for the examined cutting inserts, and stainless steel AISI 316L, which is not intended for use with the examined cutting inserts, were used. The course and character of the main flank wear were monitored until the set criterion VB_{max} = 300 μm was reached. At the same time, the produced chips were continuously assessed in terms of their shape, length and number of chips per 100g of chips produced according to ISO 3685. The results of the mutual comparison of indexable cutting inserts can be summarised in the following points:

- During wear testing during the machining of hardened AISI 4337 steel, which fell into the primary application group of the tested inserts, the CNMG 120408-PM 4325 insert achieved a life of T = 16 min, which was almost three times longer than the CNMG 120408E-MP3 WPP20S insert, which reached a service life of T = 5.5 min during the machining of hardened steel.
- During wear testing during the machining of AISI 316 L stainless steel, which according to the manufacturers of the examined inserts is not suitable for the application of the tested tools, the CNMG 120408-PM 4325 insert achieved a life of T = 17.3 min. The CNMG 120408E-MP3 WPP20S insert reached critical wear during machining of AISI 316L already at T = 1.5 min.
- The analysis of coating layers showed that CNMG 120408E-MP3 WPP20S has thicker layers of Al2O3 and TiCN coatings applied compared to CNMG 120408-PM 4325. Despite this, it achieved significantly worse durability results in the experiment and proved to be completely unsuitable during the machining of AISI 316L.
- The results of the evaluation of the morphology of the worn surface of the examined inserts indicate that a greater applied thickness of the coatings can reduce the occurrence of some types of wear (e.g., chipping), but it is not a guarantee that it will increase the overall tool life of the functional parts of the cutting tool within the application. Paradoxically, a lower thickness of the coating protection could be more advantageous from the point of view of reducing the risk of its separation even in the case of machining a material that does not dissipate heat well; see AISI 316 L described above.
- The shape and average length of the produced chips did not differ practically within the machining of both test materials. Despite the different designs of the chip formers, the evaluated inserts from both manufacturers showed similar results.
- As part of the evaluation of the number of chips per 100g of chips produced, CNMG 120408E-MP3 WPP20S achieved a better result during the machining of AISI 4337, when the number of chips produced per 100g of chips was set at 3831 ± 691 pcs, compared to CNMG 120408-PM 4325, when this figure was equal to 2782 ± 350. During the machining of AISI 316 L, the results of this parameter were practically identical for both evaluated inserts.

- It is possible that the so-called 'Bullet design' is more advantageous from the point of view of chip formation when applied to noble low to medium alloy steels; however, when applied to austenitic stainless steels, the importance decreases.
- Despite the identical marking of interchangeable inserts according to ISO 1832:2017 and the available parameters provided by the manufacturer, such as the composition of the coating layers or the basic geometry of the edge with the shape of the chip breaker, the differences in the functional properties of the cutting tools under the given conditions are striking, which increases the importance of choosing a suitable specific cutting tool.
- A connection was observed between the measured roughness parameter Rz of the machined surface, the morphology of the formed chip and the wear mechanism of the cutting tool with respect to the measured values after reaching the flank wear limit.

Author Contributions: Conceptualisation, K.Š. and P.K. (Pavel Kraus); methodology, K.Š., M.S. and P.K. (Pavel Kraus); validation, F.Š., T.Z. and P.Č.; formal analysis, K.Š., P.Č., Y.K. and P.K. (Pavel Kraus); investigation, K.Š., P.K. (Pavel Kříž), Y.K. and J.d.D.M.U.; resources, F.Š., Y.K. and K.Š.; data curation, K.Š., P.K. (Pavel Kraus) and F.Š.; writing—original draft preparation, K.Š. and T.Z.; writing—review and editing, P.K. (Pavel Kříž), P.Č. and V.V.; visualisation, K.Š., J.d.D.M.U. and P.Č.; supervision, P.K. (Pavel Kříž), M.S. and F.Š.; project administration, K.Š. and V.V.; funding acquisition, K.Š. and P.K. (Pavel Kraus). All authors have read and agreed to the published version of the manuscript.

Funding: This research was funded by the OP VVV Project Development of new nano and micro coatings on the surface of selected metallic materials—NANOTECH ITI II., Reg. No CZ.02.1.01/0.0/0.0/18_069/0010045.

Institutional Review Board Statement: Not applicable.

Informed Consent Statement: Not applicable.

Data Availability Statement: The original contributions presented in the study are included in the article, further inquiries can be directed to the corresponding author.

Conflicts of Interest: The authors declare no conflicts of interest.

References

1. García, J.; Ciprés, V.C.; Blomqvist, A.; Kaplan, B. Cemented carbide microstructures: A review. *Int. J. Refract. Met. Hard Mater.* **2019**, *80*, 40–68. [CrossRef]
2. Vasilko, K.; Murčinková, Z. Reduction in Total Production Cycle Time by the Tool Holder for the Automated Cutting Insert Quick Exchange and by the Double Cutting Tool Holder. *J. Manuf. Mater. Process.* **2023**, *7*, 99. [CrossRef]
3. Denkena, B.; Michaelis, A.; Herrmann, M.; Pötschke, J.; Krödel, A.; Vornberger, A.; Picker, T. Influence of tool material properties on the wear behavior of cemented carbide tools with rounded cutting edges. *Wear* **2020**, *456–457*, 203395. [CrossRef]
4. Ribeiro, C.A.C.; Ferreira, J.R.; e Silva, S.M.M.L. Thermal influence analysis of coatings and contact resistance in turning cutting tool using COMSOL. *Int. J. Adv. Manuf. Technol.* **2021**, *118*, 275–289. [CrossRef]
5. Shi, Z.; Li, X.; Duan, N.; Yang, Q. Evaluation of tool wear and cutting performance considering effects of dynamic nodes movement based on FEM simulation. *Chin. J. Aeronaut.* **2021**, *34*, 140–152. [CrossRef]
6. Boing, D.; de Oliveira, A.J.; Schroeter, R.B. Limiting conditions for application of PVD (TiAlN) and CVD ($TiCN/Al_2O_3/TiN$) coated cemented carbide grades in the turning of hardened steels. *Wear* **2018**, *416*, 54–61. [CrossRef]
7. Daicu, R.; Oancea, G. Methodology for Measuring the Cutting Inserts Wear. *Symmetry* **2022**, *14*, 469. [CrossRef]
8. Jahan, M.P.; Ma, J.; Hanson, C.; Arbuckle, G.K. Tool wear and resulting surface finish during micro slot milling of polycarbonates using uncoated and coated carbide tools. *Proc. Inst. Mech. Eng. Part B J. Eng. Manuf.* **2019**, *234*, 52–65. [CrossRef]
9. Mazid, A.M.; Hasan, S.; Ahsan, K.B. An Investigation on Optimum Process Parameters in Terms of Surface Roughness for Turning Titanium Alloy Ti-6Al-4V Using Coated Carbide. *Int. J. Eng. Mater. Manuf.* **2019**, *4*, 137–145. [CrossRef]
10. Zhuang, K.; Fu, C.; Weng, J.; Hu, C. Cutting edge microgeometries in metal cutting: A review. *Int. J. Adv. Manuf. Technol.* **2021**, *116*, 2045–2092. [CrossRef]
11. Amaro, P.; Ferreira, P.; Simões, F. Tool wear analysis during duplex stainless steel trochoidal milling. *AIP Conf. Proc.* **2018**, *1960*, 070001. [CrossRef]
12. Zeng, K.; Wu, X.; Jiang, F.; Shen, J.; Zhu, L.; Li, L. A comprehensive review on the cutting and abrasive machining of cemented carbide materials. *J. Manuf. Process.* **2023**, *108*, 335–358. [CrossRef]
13. Tian, X.; Yan, K.; Wang, Z.; Xie, F.; Liu, Y.; Wang, L.; Chen, X.; Yuan, J.; Liu, H. Performance of carbide tools in high-speed dry turning iron-based superalloys. *Proc. Inst. Mech. Eng. Part B J. Eng. Manuf.* **2021**, *236*, 427–439. [CrossRef]

14. Reis, B.C.M.; dos Santos, A.J.; dos Santos, N.F.P.; Câmara, M.A.; de Faria, P.E.; Abrão, A.M. Cutting performance and wear behavior of coated cermet and coated carbide tools when turning AISI 4340 steel. *Int. J. Adv. Manuf. Technol.* **2019**, *105*, 1655–1663. [CrossRef]
15. Fernandes, L.J.; Stoeterau, R.L.; Batalha, G.F.; Rodrigues, D.; Borille, A.V. Influence of sintering condition on the tool wear of NbC-based Ni binder-cemented carbide cutting tools. *Int. J. Adv. Manuf. Technol.* **2020**, *106*, 3575–3585. [CrossRef]
16. Rizzo, A.; Goel, S.; Grilli, M.L.; Iglesias, R.; Jaworska, L.; Lapkovskis, V.; Novak, P.; Postolnyi, B.O.; Valerini, D. The Critical Raw Materials in Cutting Tools for Machining Applications: A Review. *Materials* **2020**, *13*, 1377. [CrossRef] [PubMed]
17. Basha, M.M.; Basha, S.M.; Sankar, M.R. State of Art on Development of Diamond Reinforced Tungsten Carbide based Cutting Tool by Powder Metallurgy Routes. *Adv. Mater. Process. Technol.* **2022**, *8*, 37–48. [CrossRef]
18. Yang, X.; Wang, K.; Zhang, G.; Chou, K. Fabrication and performances of WC-Co cemented carbide with a low cobalt content. *Int. J. Appl. Ceram. Technol.* **2021**, *19*, 1341–1353. [CrossRef]
19. Gren, M.; Fransson, E.; Ångqvist, M.; Erhart, P.; Wahnström, G. Modeling of vibrational and configurational degrees of freedom in hexagonal and cubic tungsten carbide at high temperatures. *Phys. Rev. Mater.* **2021**, *5*, 033804. [CrossRef]
20. Lavigne, O.; Cinca, N.; Ther, O.; Tarrés, E. Effect of binder nature and content on the cavitation erosion resistance of cemented carbides. *Int. J. Refract. Met. Hard Mater.* **2022**, *109*, 105978. [CrossRef]
21. Derakhshandeh, M.; Eshraghi, M.; Razavi, M. Recent developments in the new generation of hard coatings applied on cemented carbide cutting tools. *Int. J. Refract. Met. Hard Mater.* **2023**, *111*, 106077. [CrossRef]
22. Chen, Y.; Zhang, L.; Zhu, J.-F.; Zhong, Z.-Q.; Gu, J.-H. Cohesive failure and film adhesion of PVD coating: Cemented carbide substrate phase effect and its micro-mechanism. *Int. J. Refract. Met. Hard Mater.* **2023**, *111*, 106066. [CrossRef]
23. Katiyar, P.K.; Randhawa, N.S. A comprehensive review on recycling methods for cemented tungsten carbide scraps highlighting the electrochemical techniques. *Int. J. Refract. Met. Hard Mater.* **2020**, *90*, 105251. [CrossRef]
24. Shemi, A.; Magumise, A.; Ndlovu, S.; Sacks, N. Recycling of tungsten carbide scrap metal: A review of recycling methods and future prospects. *Miner. Eng.* **2018**, *122*, 195–205. [CrossRef]
25. Muneer, L.; Al-Ethari, H. Tungsten Carbide Recovery from Hard Material Scrape. *Des. Eng.* **2021**, *7*, 14787–14797.
26. Niu, Z.; Jiao, F.; Cheng, K. An innovative investigation on chip formation mechanisms in micro-milling using natural diamond and tungsten carbide tools. *J. Manuf. Process.* **2018**, *31*, 382–394. [CrossRef]
27. Ahmed, Y.S.; Paiva, J.M.; Veldhuis, S.C. Characterization and prediction of chip formation dynamics in machining austenitic stainless steel through supply of a high-pressure coolant. *Int. J. Adv. Manuf. Technol.* **2019**, *102*, 1671–1688. [CrossRef]
28. Maksarov, V.V.; Khalimonenko, A.D.; Olt, J. Improvement of efficiency of metal recycling in metallurgical production. *Chernye Met.* **2021**, *1*, 45–51. [CrossRef]
29. Kuntoğlu, M.; Sağlam, H. Investigation of progressive tool wear for determining of optimized machining parameters in turning. *Measurement* **2019**, *140*, 427–436. [CrossRef]
30. Yılmaz, B.; Karabulut, Ş.; Güllü, A. A review of the chip breaking methods for continuous chips in turning. *J. Manuf. Process.* **2020**, *49*, 50–69. [CrossRef]
31. Molnar, T.G.; Berezvai, S.; Kiss, A.K.; Bachrathy, D.; Stepan, G. Experimental investigation of dynamic chip formation in orthogonal cutting. *Int. J. Mach. Tools Manuf.* **2019**, *145*, 103429. [CrossRef]
32. *ISO 3685:1993 (E)*; Tool-Life Testing with Single-Point Turning Tools. International Organization for Standardization (ISO Standard): London, UK, 1993.
33. Singh, K.; Karakoti, A. Optimization of Parameter for Surface Roughness by Using Taguchi Method. *Int. J. Adv. Res. Ideas Innov. Technol.* **2017**, *3*, 351–358.
34. Patel, N.; Parihar, P.L.; Makwana, J.S. Parametric optimization to improve the machining process by using Taguchi method: A review. *Mater. Today Proc.* **2021**, *47*, 2709–2714. [CrossRef]
35. Setyono, G.; Riyadi, S.; Muharom; Riyanto, O.A.W.; Pratama, S. Effect of Cutting Parameter Toward The Surface Roughness Applied In Turning Tool Steel Material. *Infotekmesin* **2022**, *13*, 233–238. [CrossRef]
36. Leksycki, K.; Feldshtein, E.; Królczyk, G.M.; Legutko, S. On the Chip Shaping and Surface Topography When Finish Cutting 17-4 PH Precipitation-Hardening Stainless Steel under Near-Dry Cutting Conditions. *Materials* **2020**, *13*, 2188. [CrossRef] [PubMed]
37. *ISO 1832:2017(E)*; Indexable inserts for cutting tools—Designation. International Organization for Standardization (ISO Standard): Geneva, Switzerland, 2017.
38. *ASTM A29:2004*; Standard Specification for Steel Bars, Carbon and Alloy, Hot-Wrought, General Requirements for. ASTM International: West Conshohocken, PA, USA, 2004.
39. *ASTM F899:95*; Standard Specification for Stainless Steel for Surgical Instruments. ASTM International: West Conshohocken, PA, USA, 1995.
40. Jędrzejczyk, D.; Mikołajczyk, J.R. *Defining the Correlation Between the Cutting Speed and Roughness Parameter Rz. Mik 21*; Spatium: Lublin, Poland, 2022; pp. 39–46; ISBN 978-83-67033-43-5.
41. Vereschaka, A.; Volosova, M.; Sitnikov, N.; Milovich, F.; Andreev, N.; Bublikov, J.; Sotova, C. Influence of the thickness of nanolayers in wear-resistant layer of Ti-TiN-(Ti, Cr, Al)N coating on the tool life and wear pattern of the carbide cutting tools in steel turning. *Procedia CIRP* **2021**, *101*, 262–265. [CrossRef]
42. Cappellini, C.; Abeni, A. Development and implementation of crater and flank tool wear model for hard turning simulations. *Int. J. Adv. Manuf. Technol.* **2022**, *120*, 2055–2073. [CrossRef]

43. Tooptong, S.; Nguyen, D.; Park, K.-H.; Kwon, P. Crater wear on multi-layered coated carbide inserts when turning three distinct cast irons. *Wear* **2021**, *484*, 203982. [CrossRef]
44. Uçak, N.; Aslantas, K.; Çiçek, A. The effects of Al_2O_3 coating on serrated chip geometry and adiabatic shear banding in orthogonal cutting of AISI 316L stainless steel. *J. Mater. Res. Technol.* **2020**, *9*, 10758–10767. [CrossRef]
45. Abidi, Y. Relationship between surface roughness and chip morphology when turning hardened steel. *Prod. Eng. Arch.* **2020**, *26*, 92–98. [CrossRef]
46. Moreno, M.; Andersson, J.M.; M'Saoubi, R.; Kryzhanivskyy, V.; Johansson-Jöesaar, M.P.; Johnson, L.J.; Odén, M.; Rogström, L. Adhesive wear of TiAlN coatings during low speed turning of stainless steel 316L. *Wear* **2023**, *524*, 204838. [CrossRef]
47. Kuntoğlu, M.; Gupta, M.K.; Aslan, A.; Salur, E.; Garcia-Collado, A. Influence of tool hardness on tool wear, surface roughness and acoustic emissions during turning of AISI 1050. *Surf. Topogr. Metrol. Prop.* **2022**, *10*, 015016. [CrossRef]

materials

MDPI

Article

Using the Ethaline Electropolishing Method on the Internal Surface of Additive Manufactured Tubes

Dongyi Zou [1], Chaojiang Li [1], Yuxin Yang [1,2], Xin Jin [1,*], Shenggui Liu [1], Hongyi Zhang [1] and Na Zhang [1]

[1] School of Mechanical Engineering, Beijing Institute of Technology, Beijing 100081, China; zoudy0610@163.com (D.Z.); mecjli@bit.edu.cn (C.L.); yyxxhzlj@gmail.com (Y.Y.); liusg_email@163.com (S.L.)
[2] Department of Electronic and Computer Engineering, University of Alberta, 9120 116 St NW, Edmonton, AB T6G 1C9, Canada
* Correspondence: goldking@bit.edu.cn

Abstract: Electropolishing is a widely used technique for polishing additive manufactured (AM) components, while complex internal surface polishing remains a challenge. In this study, we explore the use of ethaline as an electrolyte and investigate the effects of temperature, time, stirring speed, and voltage on the electropolishing effectiveness for AM tubes without pre-treatment through orthogonal experiments. The optimal combination of these factors is then applied in further electropolishing experiments on straight tubes with large length-to-diameter ratios and an angled tube. Our results indicate that temperature has the most significant impact on internal surface electropolishing performance, and other factors' effects are also analyzed. Ethaline can be a promising electrolyte for internal surface electropolishing of AM components because of its high viscosity, which is validated by flow field simulation of the hydrodynamic conditions inside the tubes.

Keywords: non-traditional machining; electropolishing; additive manufactured metal parts; deep eutectic solvents

1. Introduction

Additive manufacturing (AM, also known as 3D printing or rapid prototyping) is recognized as a key strategy for manufacturing parts with complex structures. Selective laser melting (SLM) is a state-of-the-art AM process that can produce higher specific mechanical properties by sintering metallic powder with negligible void content [1]. However, metal parts produced by SLM often have high surface roughness. Defects such as the step effect, the balling effect, and powder adhesion [2–4] limit further application of SLM in aviation, aerospace, and other fields. The balling effect may lead to turbulence in fluid flow or stress concentration points, while powder adhesion may result in contamination of the internal surface, affecting the tubes' properties such as corrosion resistance. In critical applications, surface defects can reduce performance over time, causing pressure losses and increased vulnerability to environmental factors [5]. For engine parts with inadequately polished internal walls, particles may adhere to the surface. Under harsh conditions, these particles can detach and enter the inner chamber, posing a risk to the engine.

Various surface finishing techniques are used to reduce surface roughness, including composite and non-traditional polishing methods. Reza et al. [6] proposed and experimentally validated effective process parameters in sequential processes to control surface topography and roughness, attaining surface quality with an arithmetic roughness value of 0.448 μm. Bai Yuchao et al. [7] used dry mechanical–electrochemical polishing to finish selective laser melted surfaces, reducing surface roughness from 8.72 μm to 0.75 μm. Wang Qinghua et al. [8] used pulsed laser micro-polishing to reduce the surface roughness (Sa) of the micro-end milled Ti-6Al-4V surface from 84 nm to 37 nm, an improvement of approximately 56%. Nevertheless, conventional mechanical machining is no longer suitable in terms of cost and time for additive manufactured structures which may present inner channels [9]. Among non-traditional surface finishing processes such as laser polishing which often generate heat-affected zones, reducing the surface consistency, electropolishing is a promising alternative because of its ability to treat inaccessible areas [10]. Zhao et al. [11] developed a novel electropolishing process to eliminate the adhesive powders and band protrusions generated during the laser melting process, and the surface roughness (Ra) of straight holes was reduced from 14.151 to 6.287 μm. Z. Chaghazardi et al. [12] investigated the electropolishing process of the internal walls of stainless steel 316 tubing considering potentials and times, and achieved an improvement of 60–80% for tubes with various lengths. The authors suggested that the interelectrode gap had a rather low impact on the final roughness but strongly impacted the final brightness.

The effectiveness of electrochemical polishing is influenced by many parameters, and the extent to which each parameter affects the final result varies. Therefore, many researchers have used the orthogonal experimental method to optimize the combination of these parameters. Prihandana GS et al. [13] optimized parameters through the Taguchi L9 orthogonal array to increase the material removal rate. Among the parameters, the electrolytes play a crucial role in controlling the material removal rate and preventing undesirable defects, which are typical directions in optimizing the electropolishing process [14]. A number of studies have investigated different kinds of electrolytes' performance on the electropolishing process, including acidic and neutral electrolytes [15–18]. The most frequently used water-based electrolytes contain highly concentrated acids and other additives which may be highly toxic and dangerous to people. In addition, for specific materials, the application of aqueous electrolytes may increase the risk of hydrogen embrittlement. Hence, non-aqueous electrolytes such as deep eutectic solvents (DESs) are applied to the electropolishing process [9]. DESs are a type of solvent formed from a mixture of two or more compounds that, when combined, have a melting point much lower than that of either of the individual components. In this study, ethaline, one kind of DES, which is a mixture of choline chloride and ethylene glycol, is introduced as the electrolyte because of its negligible gas evolution at the anode/solution interface and high current efficiency [19]. However, many studies which use DESs as electrolytes focus on the external surface of the metal components.

Therefore, the main roadblock in internal surface electropolishing using ethaline is the lack of optimization method of the processing parameters combination. The factors affecting electropolishing of the internal surface by ethaline remain unclear. In this study, an electrochemical simulation of rough surfaces is conducted to show the changing trend of the surface topography using ethaline. Then, electrochemical experiments based on the orthogonal experimental design are conducted. The range analysis reveals the main parameter which has the greatest impact on electropolishing quality (i.e., Ra in this study). The surface roughness is measured, and Tafel curves are investigated to figure out the corrosion resistance change of the internal surface. Moreover, flow field simulations of ethaline and NaCl solution are conducted, and the streamline charts are compared to validate the impact of hydrodynamic conditions on electropolishing. The present work aims to investigate the key factor in ethaline electropolishing of internal surfaces and validate the viscosity's impact on the electropolishing process in internal spaces.

2. Experiment and Simulation

2.1. Material Preparation

Samples with various lengths and angled tubes are prepared by SLM with an FS271M printer (Farsoon, Hunan, China). The corresponding manufacturing parameters are listed in Table 1. The diameter of the platinum electrodes is 1 mm.

Table 1. SLM fabrication parameters.

Manufacturing Parameters	Value
Powder layer thickness	0.03 mm
Scanning speed	1250 mm/s
Laser power	500 W
Ambient temperature	150 °C

The 3D models of the tubes are shown in Figure 1. The detailed size parameters are shown in Table 2. The angled tube's diameter is 8 mm and its bending radius is 13 mm. There are two kinds of straight tubes which are polished. The short one has a length of 10 mm, while the other has a length of 25 mm.

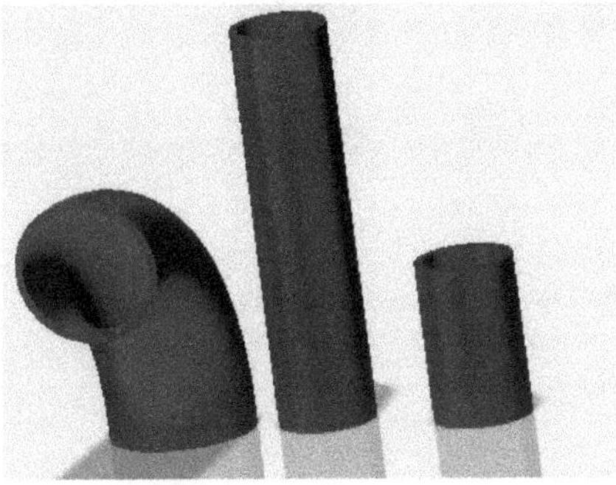

Figure 1. Three-dimensional models of tubes.

Table 2. Size of tubes.

Tube Type	Value
Short tube	Φ5 mm × 10 mm
Long tube	Φ5 mm × 25 mm
Angled tube diameter	8 mm
Angle	90°
Bending radius	13 mm

The ethaline is prepared by using glycol and choline chloride with a ratio of 2:1. The solutes are both from Sinopharm Chemical Reagent Co., Ltd. (Shanghai, China).

2.2. Simulation

In this section, we emphasize the qualitative changes in the electrochemical process rather than the quantitative changes. Therefore, a simplified two-dimensional illustration in which the rough surfaces of both the cathode and the component are represented by simplified linear outlines is introduced to simulate the side surface of the three-dimensional printed components. As shown in Figure 2, lines AC and BD are the boundaries of the electrolyte, and line AB represents the cathode, while line CD represents the rough surface of the anode which includes protrusions and craters, alongside distinctive defects like adhered spherical particles specific to additive manufactured components. The gray region surrounded by these lines represents the electrolyte.

(a)

(b)

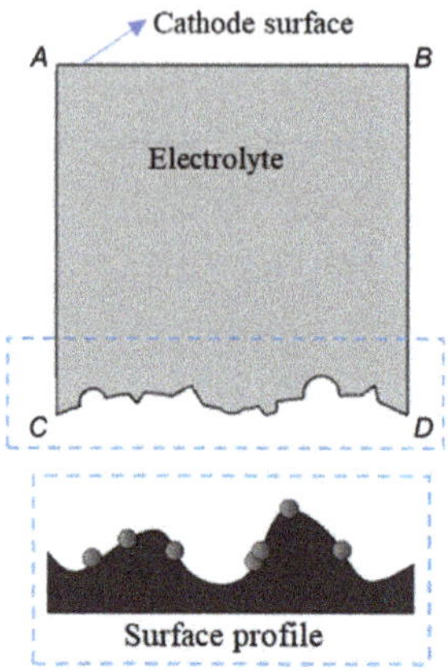

Figure 2. (**a**) Simplified simulation model; (**b**) partially melted metal particles sticking to as-printed side surface.

In addition, the boundaries AC and BD are set to be insulated, while the boundaries AB and CD are treated as equipotential surfaces due to their good conductivity. The boundary potentials are expressed as follows:

$$\phi|\Gamma_{AB} = 0, \tag{1}$$

$$\phi|\Gamma_{CD} = \phi, \tag{2}$$

$$\left.\frac{\partial\phi}{\partial n}\right|\Gamma_{AC,BD} = 0 \tag{3}$$

Equations (1) and (2) define the potentials of the cathode and anode, respectively. In the simulation, the cathode potential is set to 0, while the anode potential is set to the polishing voltage. Equation (3) indicates that the potential at the electrolyte boundary is 0.

The length of AB, AC, and BD is 9 mm. The material of the anode is 316 L stainless steel, and the material of the cathode is platinum. The voltage applied in the simulation is 7 V (i.e., Φ in Equation (2) is 7 V), and the conductivity of the electrolyte is 14 S/m with a density of 1.085 kg/m^3.

The electrochemical model in COMSOL is a pre-defined multi-physics model which consists of various current diffusion interfaces and deformed geometry interfaces. One corrosion interface which combines the Nernst–Planck interface, tertiary current distribution interface electroneutrality, and the deformed geometry formulation is selected in this work. The tertiary current distribution solves the full Nernst–Planck equation, describing the mass transport of each species through diffusion, migration, and convection, while adhering to the assumption of electroneutrality.

$$N_i = -D_i \nabla c_i - z_i u_{m,i} F c_i \nabla \phi_l + c_i u \tag{4}$$

where:

D_i is the diffusion coefficient and its value in the simulation is $1 \times 10^{-9} m^2/s$;

F is the Faraday constant;

R is the gas constant;

T is the temperature and is set to the polishing temperature;

z_i is the valence state of the compounds in the reaction.

According to the reaction, the main elements involved in 316 L dissolving are $Fe \rightarrow Fe^{2+} + 2e^-$, and the number of charges migrating from the electric field is 2.

c_i is the concentration of species i;
N_i is the flux of species i;
u is the velocity of the fluid (for convection).
The current density expression is shown as follows:

$$i = -\frac{F^2}{RT}\nabla\phi\sum D_i C_i z_i^2 - F\Sigma z_i D\sqsupset\nabla_{c_i} + vF\Sigma z_i c_i, \tag{5}$$

Since any solution should maintain electroneutrality, the expression can be simplified.

$$i = -\frac{F^2}{RT}\nabla\phi\sum D_i C_i z_i^2 - F\Sigma z_i D\sqsupset\nabla_{c_i} \tag{6}$$

In this work, a flow field simulation is conducted by using the mixture model in COMSOL, and a frozen rotor is introduced in this work. The model is shown in Figure 3, and the main parameters for the flow field simulation are listed in Table 3.

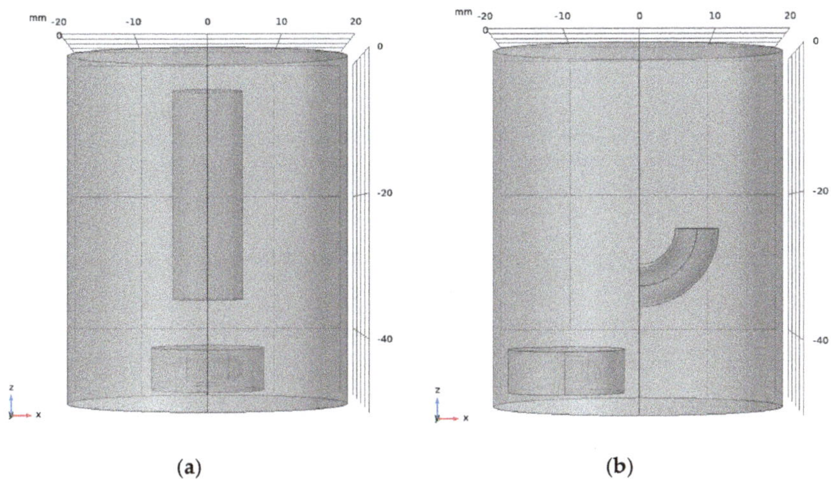

Figure 3. (**a**) Straight tube; (**b**) angled tube.

Table 3. Simulation parameters.

Simulation Parameter	Value
Stirring rate	800 rpm
Vessel height	50 mm
Vessel diameter	40 mm
Magnetic stirrer diameter	4 mm
Magnetic stirrer length	10 mm
Straight tube diameter	10 mm
Straight tube length	30 mm
Angled tube diameter	5 mm
Angled tube bending radius	8 mm
Ethaline viscosity	15 mPa·s
Ethaline density	1.085 kg/m^3
NaCl solution viscosity	1.409 mPa·s
NaCl solution density	1.212 kg/m^3

2.3. Experiment Setup

Figure 4 shows the experimental setup for the ECP process. An electrochemical workstation (CH Instruments Inc. (Austin, TX, USA)) was used to apply current between the electrodes. A beaker served as the reaction vessel which was filled with electrolytes and placed on the thermostatic magnetic heating stirrer. A surface roughness tester (Mitutoyo Measuring Instruments (Shanghai) Co., Ltd. (Shanghai, China)) was used to measure the surface roughness of the samples before and after polishing. Each time, the surface roughness was measured with a sampling length 0.8 mm and the evaluation length was 4 mm. The measurements were taken in both the horizontal and vertical directions of the sample, and the average value was calculated.

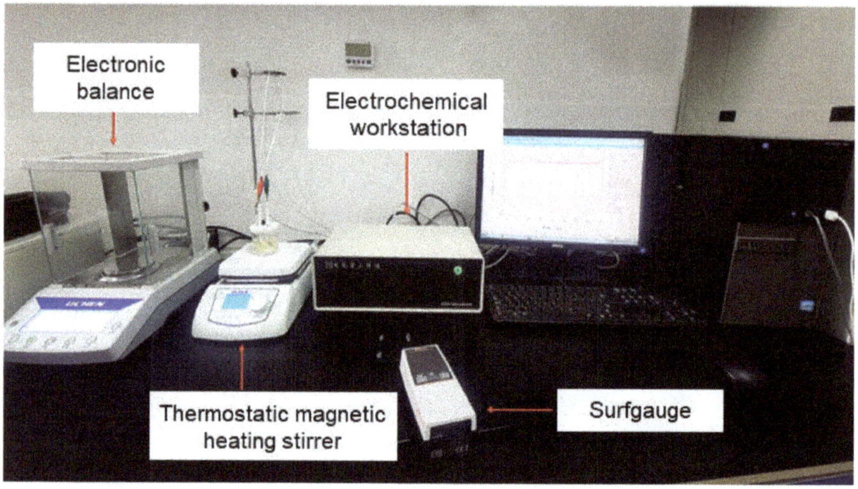

Figure 4. Experiment setup.

The experimental steps are detailed as follows: First, the electrolyte was prepared by using choline chloride and ethylene glycol with a molar ratio of 1:2.50 mL; ethaline was used in the experiments. Then, the samples were cleaned by ultrasonic cleaning in ethanol for 5 min and then in deionized water for 5 min. After ultrasonic cleaning, the samples were dried. Since there are many parameters that may have a potential impact on the polishing quality, orthogonal experiments were arranged to achieve the best combination of the parameters with the fewest tests. The optimal parameters were applied in the long tube and angled tube polishing experiments.

Temperature, polishing time, stirring speed, and voltage are the four main factors in the orthogonal experiments. The orthogonal array in this study is $L_9(3^4)$. The levels are shown in Table 4.

Table 4. Orthogonal experiment parameters of Φ5 mm tubes.

Level	Temperature (K)	Voltage (V)	Stirring Speed (rpm)	Time (s)
1	343	5	400	1600
2	348	6	600	1800
3	353	7	800	2000

Electrochemical testing was conducted by using an electrochemical workstation and based on the three-electrode system. The working electrode is the 316 L stainless steel samples, while the counter electrode is a platinum electrode. The reference electrode is composed of an Ag/AgCl electrode and KCl. Tafel curves are used to evaluate the corrosion resistance of the samples. The scanning range is from −1 V to 1 V, and the scanning rate is 0.005 V/s.

In tube polishing, range analysis is used to find out the influential sequence of parameters that have an impact on polishing quality. Tafel curves are also used in this part to identify the corrosion resistance of the tubes.

3. Results and Discussion

3.1. Electrochemical Simulation Results

The surface profile changes over time in the electrochemical simulation are shown in Figure 5. In the first 1000 s, most of the sharp characteristics are removed due to electrochemical corrosion and transform into smoother bulges, but the adhered particles still retain their distinctive morphology. The particle removal thumbnail shows the morphological changes of the adhered particles in the initial 800 s. After another 2000 s, the bulges are almost removed but the concavity still exists, which indicates that the edge effect has the main impact on the removal of the adhered particles. Moreover, another two interfaces using different current density distributions are used in the simulation. According to the polarization curve thumbnail, a linear polarization curve for the primary current density distribution interface is found because the kinetics of the electrode process and the effect of concentration dependence are neglected. The polarization curves of another two interfaces follow a similar trend but the current in the tertiary current density distribution interface is slightly smaller than the current in the other one due to concentration polarization. A passivation region is found and it follows a similar trend as the polarization curves for the ALM (additive layer manufactured) 316 L samples in choline chloride–ethylene glycol. However, the current density plateau is not shown in the simulation result, and the key factors affecting the electropolishing effect on the internal surface of the tubes remain unclear. Therefore, an orthogonal experiment is adopted in this work.

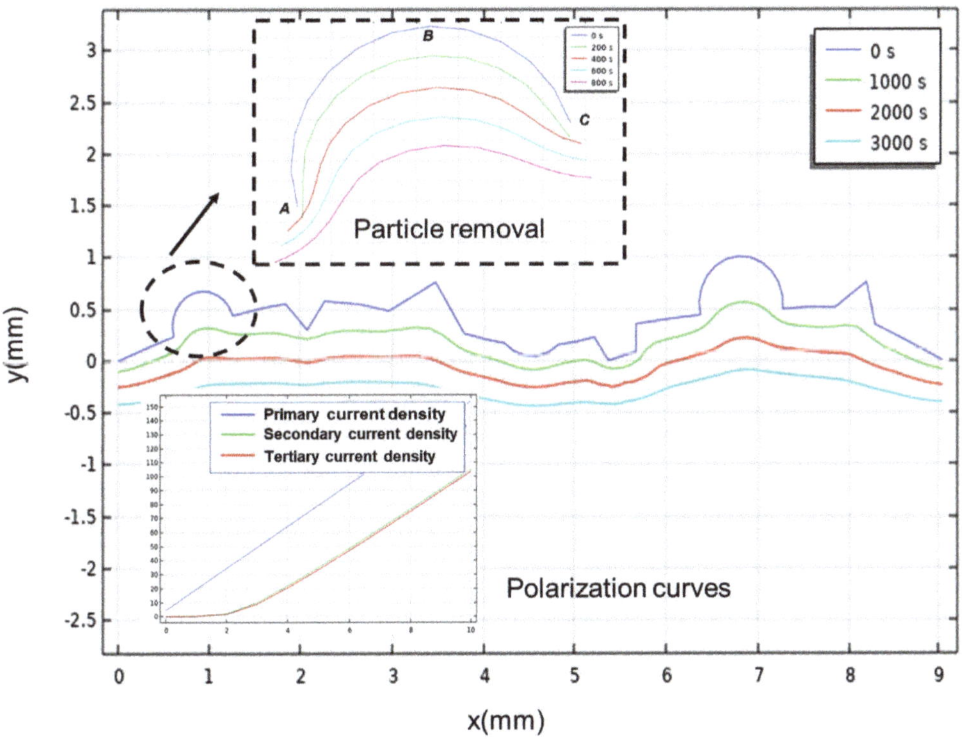

Figure 5. Plot of anode surface over time in tertiary current density distribution interface.

3.2. Orthogonal Experiment Results

The orthogonal experiment using four factors and three levels is adopted to plan the experimental scheme. Voltage, temperature, stirring speed, and time are selected as representative parameters. The orthogonal experimental design of the four factors and three levels, as shown in the first five columns in Table 5, is carried out to obtain the orthogonal calculation, while the last column lists the corresponding experimental results. The lowest surface roughness (Ra) is utilized as the index to investigate the electropolishing result by modifying the process parameters. Analyzing the results of the orthogonal test reveals that the minimum surface roughness in group 6 is 1.541 μm, which is the lowest value among all of the results. Table 6 displays the range analysis of the orthogonal experiment where $\overline{k_1}$, $\overline{k_2}$, and $\overline{k_3}$ refer to the mean values of level 1, 2, and 3, respectively.

Table 5. The orthogonal experimental results of the Φ5 mm tubes.

Group	Voltage (V)	Temperature (K)	Stirring Speed (rpm)	Time (s)	Ra (μm)
1	5	343	400	1600	2.486
2	5	348	600	1800	1.915
3	5	353	800	2000	1.742
4	6	343	600	2000	2.367
5	6	348	800	1600	1.974
6	6	353	400	1800	1.841
7	7	343	800	1800	2.149
8	7	348	400	2000	1.843
9	7	353	600	1600	2.069

Table 6. Range analysis.

Item	Parameters			
Surface Roughness Ra (μm)	Affected by Voltage	Affected by Temperature	Affected by Stirring Speed	Affected by Time
$\overline{k_1}$	2.048	2.334	2.057	2.176
$\overline{k_2}$	2.061	1.911	2.117	1.968
$\overline{k_3}$	2.02	1.884	1.955	1.984
Range	0.041	0.45	0.162	0.208
Rank	4	1	3	2

The analysis reveals that the influence of the three factors on the minimum surface roughness of the tubes' internal surface follows the order of temperature, time, stirring speed, and voltage, with the temperature having the greatest impact and voltage having the smallest impact. Figure 6 illustrates the influence of the factors on the internal surface roughness. The temperature leads to a decrease in the internal surface roughness. At factor level 1 (343 K), the reaction products cannot be discharged because of the low flow speed. When the temperature increases, the ionic mobility in the inner flow field also increases. As a result, the conductivity increases and the polishing efficiency increases. With the higher temperature, the conductivity and the ion migration speed keep increasing, thus leading to a lower surface roughness. The time scheme initially shows a decrease followed by an increase, suggesting that excessively raising the polishing time may cause an increase in internal surface roughness due to over-polishing.

In the early stage of electropolishing, surface materials are continuously removed, and the surface roughness decreases as the electrochemical reaction progresses. When the ideal polishing time is reached, the surface roughness achieves its optimal value. However, as the reaction continues, over-polishing occurs, leading to an increase in the surface roughness. Additionally, due to the large polished area, different positions inside may reach their optimal polishing state at different times; so, the ideal polishing time may fall within a range.

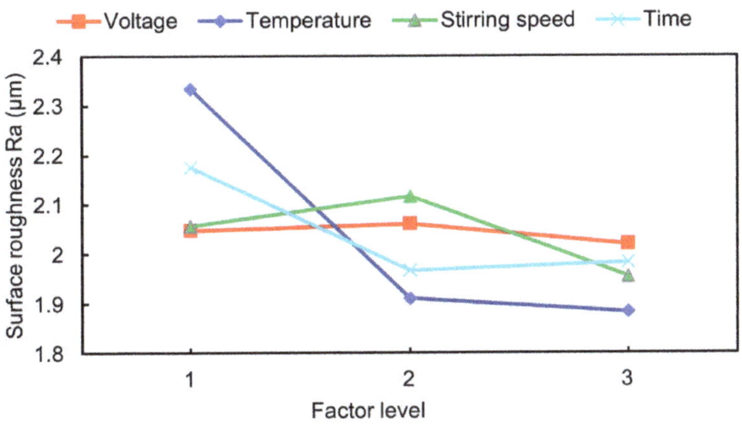

Figure 6. The influence trend of the factor levels on the internal surface roughness.

In addition, the voltage and stirring speed have an increase followed by a decrease. Unlike other studies, the effects of voltage and stirring speed on electropolishing performance are relatively smaller compared to other factors. This is mainly due to the different electrolytes used and the significantly larger polishing area (i.e., the whole internal surface in this work). Therefore, the increased polishing area results in a smaller improvement in the material removal rate per unit area with voltage change, thus reducing the impact of voltage on the surface roughness. On the other hand, the polishing rate is primarily influenced by the concentration of free ions in the electrolyte (governed by viscosity), rather than voltage. The stirring speed has little effect on altering the flow field, mainly due to the high viscosity of ethaline and the relatively small diameter of the tubes. Given the high viscosity of the electrolyte, increasing the stirring speed is likely less effective in improving the confined flow field limited by the tube structure. Hence, investigating the hydrodynamic conditions (i.e., considering viscosity or the electrolyte's other fluid properties) inside the tube in the electropolishing process should be a potential avenue to optimize the internal surface electropolishing parameter combination.

In summary, the optimal parameter combination to reduce the internal surface roughness is as follows: 353 K, 1800 s, 800 rpm, and 6 V. Then, Φ10, 8, and 5 mm tubes are polished with the optimal combination and the results are shown in Figure 7. In Figure 7a,d,g, the left section shows the initial surface, while the right section shows the polished surface. It can be observed in Figure 7b,e,h that the initial rough surface is covered with black particles formed by the spheroidization effect, along with adhered powder and noticeable unevenness. After electropolishing, according to Figure 7c,f,i, the surface flatness is significantly improved, with spherical particles and powder removed, and the surface becomes brighter, as shown in Figure 7a,d,g. However, compared to external/flat surface polishing, internal surface polishing is more prone to excessive local current density and elevated temperatures due to the significantly reduced gap between the electrodes and the restricted flow field area. Moreover, in the confined space filled with active Cl^- ions, pitting corrosion is observed. The internal surface roughness (Ra) of the Φ10, 8, and 5 mm tubes reduces from 8.419 μm, 7.684 μm, and 8.593 μm to 1.576 μm, 1.648 μm, and 1.535 μm (reduced by 81%, 78%, and 82%), respectively. For the Φ5 mm tube, the value of Ra reduces by 16.6% compared with the minimum Ra achieved in group 6 of the orthogonal experiment. The polished surfaces in Figure 7 exhibit a similar brightness and surface roughness in tubes with different diameters, indicating that ethaline performs uniformly across varying tube dimensions, effectively maintaining a consistent level of surface finish. Despite the pitting, ethaline is still a good electrolyte for tubes with various diameters.

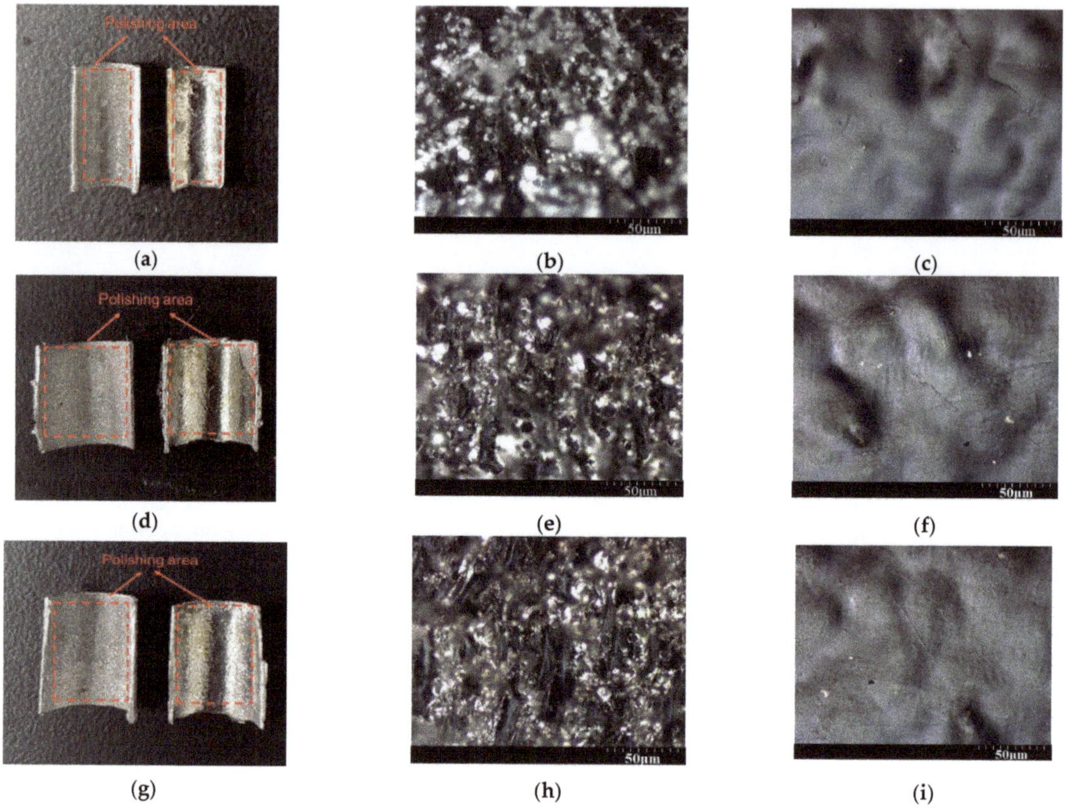

Figure 7. The polishing area of the Φ5 (**a**), Φ10 (**b**), and Φ8 mm (**c**) tubes and the internal surface image before (**d**,**e**,**h**) and after electropolishing (**g**,**f**,**i**) taken with a metallurgical microscope.

Furthermore, the corrosion resistance of the internal surfaces of the tubes is quite important. It is essential to investigate whether the corrosion resistance is improved after polishing. The corrosion potentials and corrosion current values of the tubes before and after polishing are reported in Table 7. A significant increase in the polarization resistance of the tubes before and after polishing is found. The polarization resistance is almost six times the resistance before polishing. The Tafel curves shown in Figure 8 illustrate that the corrosion resistance of the tubes is improved after the ECP process.

Table 7. Tafel curve data.

Group	β_a/mV	β_c/mV	E_{corr}/mV	i_{corr}/μA·cm^2	Rp/Ω
Φ10 mm initial	469.0	540.8	−0.583	12.18	3536.5
Φ10 mm polished	490.3	548.8	−0.378	5.023	8329.8
Φ8 mm initial	334.2	932.8	−0.585	7.664	4477.7
Φ8 mm polished	507.2	577.2	−0.384	5.706	7023.6
Φ5 mm initial	256.6	106.50	−0.654	22.45	1465.7
Φ5 mm polished	490.3	548.8	−0.378	5.023	8329.8

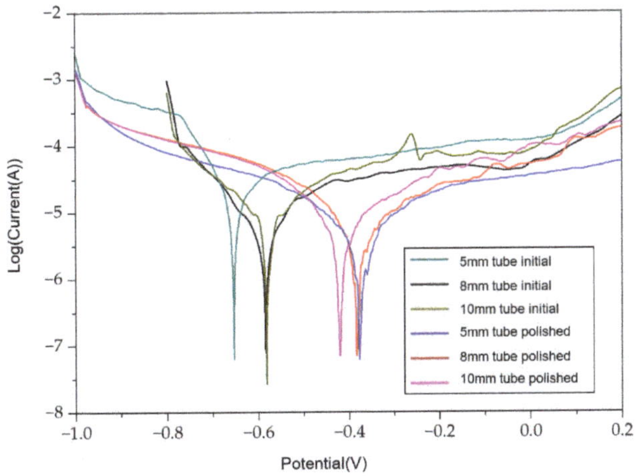

Figure 8. Tafel curves for short tubes.

3.3. *Optimal Solution Application Experiment*

The experiments described above focus on tubes with relatively small length-to-diameter ratios. In this section, the electropolishing results of using ethaline as an electrolyte for longer tubes (Φ5 mm × 25 mm tubes with a length–diameter ratio of 5) are presented. The optimal parameter combination is applied but the polishing time is increased because of the longer tubes and larger polishing area. Figure 9 shows the internal surface roughness change with polishing time.

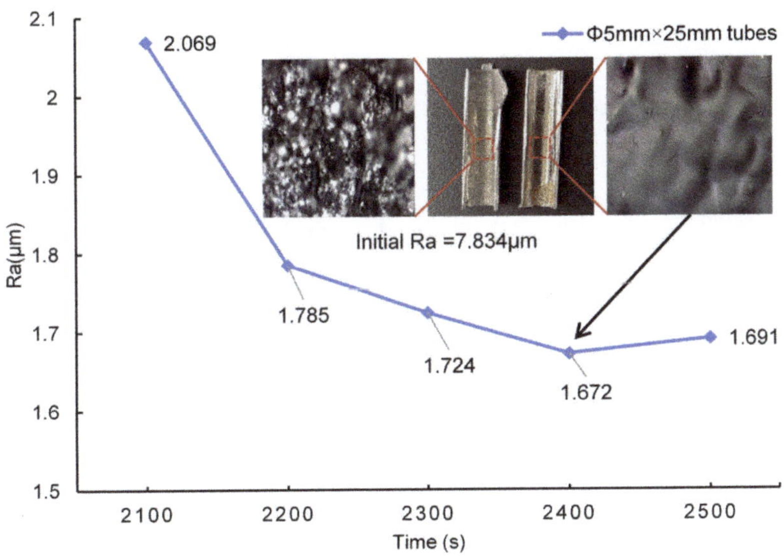

Figure 9. Ra vs. polishing time for electropolishing of long tubes.

The optimal parameter group is summarized in the previous part, showing that the surface roughness (Ra) is reduced from 7.834 μm to 1.672 μm (reduced by 78%). The internal surface has become as bright as that of short tube after polishing, as shown in Figure 7. In this section, it is shown that the polished surface shares a similar morphology to the short tube. The surface brightness in the middle section of the polished area of the long tube shows little difference compared to that of the short pipe. However, the surface brightness at the outlet of the long tube (i.e., the upper end of the long pipe in Figure 9) decreases due to the more turbulent flow field at the outlet, leading to reduced polishing consistency. The surface roughness (Ra) is approximately 9% higher than that of a shorter pipe with the same diameter. A larger length-to-diameter ratio may lead to a worse impact on internal surface quality.

In industry applications, angled tubes are also widely used. However, internal surface polishing for angled tubes is much more difficult compared to the internal surface of straight tubes. First, the complex geometry of angled tubes may lead to polishing inconsistency. Second, due to the bending, the time for electrolytes to flow through the tubes is varied, which can also have a negative impact on polishing quality because of the concentration difference. In summary, the complex geometric structure manufactured by additive manufacturing leads to complicated physical fields inside the components. In the following section, the optimal parameter combination is applied to the angled tube polishing process to investigate whether ethaline still shows a good performance as it does in short tubes. As shown in Figure 10, unlike long tubes, the surface roughness keeps decreasing with time.

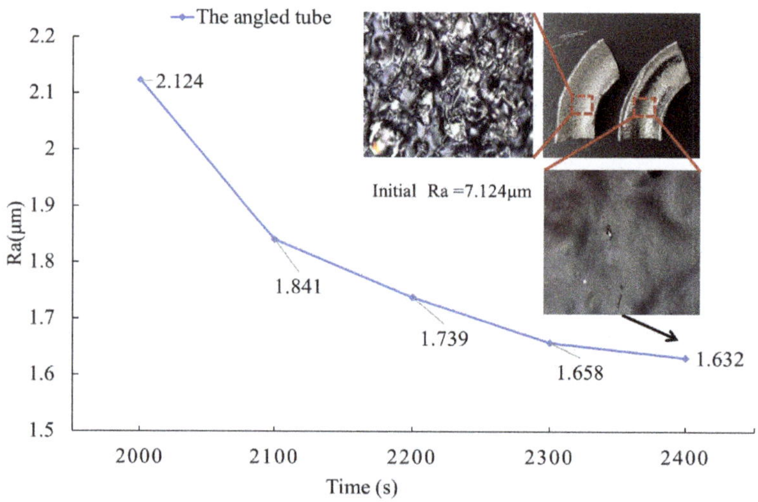

Figure 10. Ra vs. polishing time of angled tube polishing.

According to the figure which displays the comparison between the as-printed surface and the polished surface of the angled tube, the internal surface roughness (Ra) is reduced from 7.124 μm to 1.632 μm (reduced by 77%) and it shares a similar performance with the straight tubes. The brightness at the inner and outer diameters of the angled tube is lower compared to the middle section, but the brightness at the tube end (the lower end of the tube in Figure 10) is similar to that of the middle section, with no significant decrease. We believe this is due to the minimal effect of stirring on the flow field inside the bent tube since the orthogonal experimental results show that the stirring rate has a less noticeable impact on the polishing of the internal tube surface. It is noteworthy that some bulges which are not removed in the process still exist, as shown in Figure 10, while pitting is also found. It is considered that the cathode cannot match the curvature of the tubes perfectly so the current density does not keep consistent with other area. Furthermore, the magnetic stirrer generates a vortex which disturbs the flow field in the tubes. This indicates that the complex geometry makes it challenging to finish the internal surface of the angle tubes. As reported in the literature [20], the flow rate is an important factor for surface quality because the dissolution of the material in electropolishing is controlled by an adhesive layer which maintains a stable processing current. Therefore, with an inconsistent flow rate inside the angled tube, the bulges which are not removed completely appear more frequently than on the straight tubes. As the viscosity decreases, the flow rate will increase, and this leads to an unstable dissolution rate. Thus, selecting an electrolyte with relatively higher viscosity is a key factor in internal surface electropolishing.

3.4. Flow Field Simulation

To validate the advantages due to the high viscosity of electrolytes, in this section, a comparison is made between a NaCl electrolyte and ethaline by using COMSOL. The frozen rotor method is used to save time and calculation resources.

Figure 11 displays the simulation results of the velocity distribution inside the reaction beaker. The two types of electrolyte flow fields share similarities. On one hand, at the bottom of the flow field, the stirrer drives the fluid outward radially, generating two latitudinal vortices, resulting in higher flow velocity around the bottom stirrer. On the other hand, the fluid flows rapidly along the bottom of the reaction vessel to the vessel wall and then forms a transient high-speed upward zone. This is due to the upward movement caused by the electrolyte moving radially, encountering the structural obstruction of the wall, and thus appearing as a short segment of the high-speed zone on the velocity map. In the case of DESs, the wall velocity increases briefly before dropping sharply, whereas for the NaCl solution, due to its lower viscosity, the wall velocity remains high and can be maintained up to the top region. Moreover, as observed from the flow velocity distribution in Figure 11, although the electrolyte inside the beaker undergoes circulation due to the action of the stirrer, the flow velocity of the electrolyte inside the angled tube is lower compared to outside the tube. Therefore, the effect of stirring on improving the electrolyte flow field within the angled tube is not so significant.

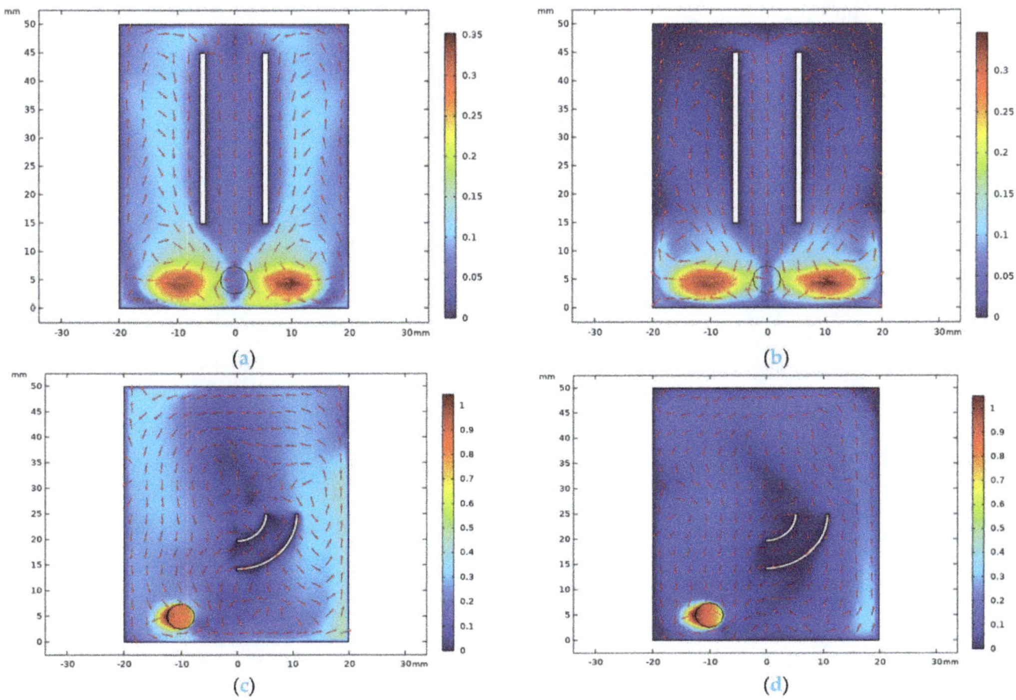

Figure 11. Velocity fields inside beaker: (**a**,**c**) NaCl solution; (**b**,**d**) ethaline.

Figure 12 displays the detailed streamline inside the tubes in the NaCl solution and ethaline. The flow velocity is higher at the center of the parts and lower near the wall. For the NaCl solution, a vortex is found at the entrance of the components, whereas when ethaline is used as the electrolyte, the streamline is uniform. As discussed in the previous section, a vortex refers to an unstable dissolution rate, leading to deteriorating results. As discussed in the previous section, temperature has the greatest impact on polishing performance. The viscosity value is directly related to the temperature, suggesting that viscosity should also be taken into consideration in electropolishing for internal surfaces. To quantitatively describe the flow rate, five points are set with the same interval, as shown in Figure 13. The velocity at each point exported directly from COMSOL is listed in Table 8.

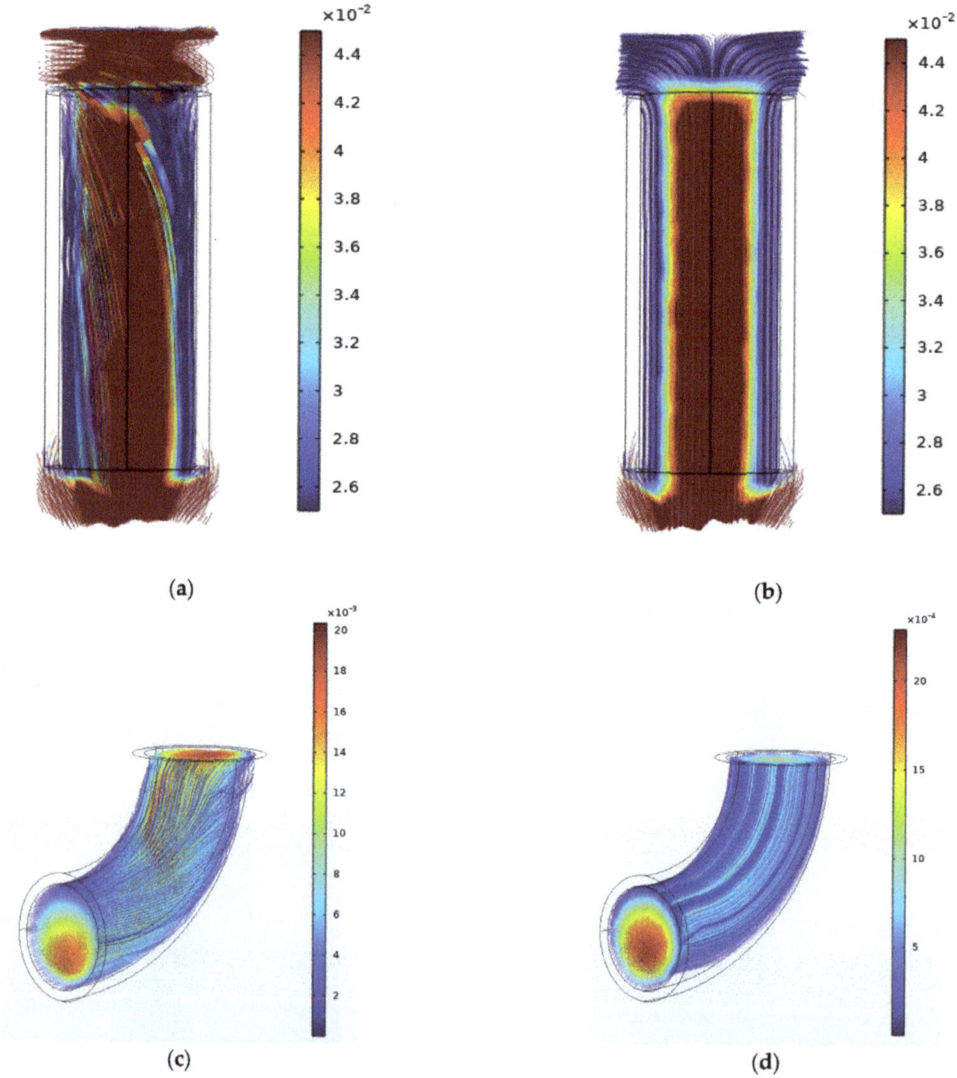

Figure 12. The streamline chart: (**a**,**c**) NaCl solution; (**b**,**d**) ethaline.

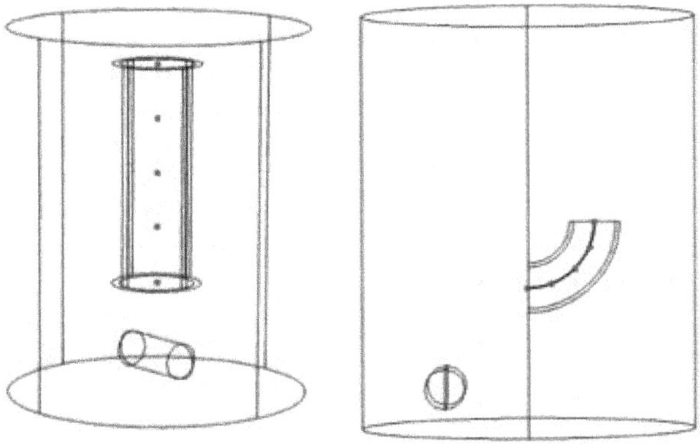

Figure 13. The point sketch map (red dots refer to the sampling points).

Table 8. Velocity distribution.

Sample Type	Solution	Point 1	Point 2	Point 3	Point 4	Point 5	Standard Deviation
Straight tube	Ethaline	31.8	18.6	19.2	18.8	15.8	5.62
	NaCl	102.8	36.7	51.1	47.7	31.0	25.5
Angled tube	Ethaline	2.17	0.863	0.174	0.754	0.109	0.6565
	NaCl	14.2	7.7	2.4	14.7	19.0	5.86

Table 8 shows that the velocities of both electrolytes at the first three points follow a pattern of initially decreasing and then increasing. This indicates that regardless of whether it is the high-viscosity ethaline or the low-viscosity NaCl solution, vortices are present when entering the pipe. However, due to the higher viscosity of ethaline, the change in velocity is not as noticeable. The standard deviation indicates that the flow velocity of ethaline is significantly less dispersed compared to that of the NaCl solution. On the other hand, in the angled tube, it is observed that the flow velocity is lowest at the bend (i.e., at the third point). Taking ethaline as an example, the flow velocity of ethaline in the bend decreases from 2.17 mm/s to 0.174 mm/s, with a reduction of approximately 92%. In actual processing, a low flow rate results in a low current density, which prevents further electropolishing. This explains the local polishing unevenness observed in the angled tube.

The flow field simulation of hydrodynamic conditions provides a new optimization avenue of research for the electropolishing process [12]. In this section, we compared the flow field distribution in straight tubes and angled tubes. According to the simulation results, we validated that ethaline had a higher electropolishing consistency due to its high viscosity. Also, the complex streamline at the outlet of the tube in the simulation explains the reason why the brightness at the outlet is not as good as that of the middle section. This simulation indicates that hydrodynamic conditions should be taken into consideration when it comes to the electropolishing process with a small interelectrode gap or a restricted flow field area.

4. Conclusions

In this work, ethaline is adopted as the electrolyte for electropolishing the internal surfaces of straight and angled tubes composed of additive manufactured 316 L stainless steel. The orthogonal experiments are conducted to rank four factors' impacts on the final surface roughness. The optimal factor combination is applied in further electropolishing experiments, and the flow field simulation is conducted. The following conclusions can be drawn:

In this study, we proposed an electropolishing method using ethaline for the internal surface of the tubes to reduce the surface roughness without any pre-treatment. The orthogonal experiments illustrate that temperature has the greatest impact on the surface roughness of the internal surfaces of 316 L tubes. For straight tubes with a diameter of 5 mm, 8 mm, and 10 mm, an optimal factor combination of 353 K, 1800 s, 800 rpm, and 6 V is applied and the surface roughness is reduced approximately by 82%, 78%, and 81%, respectively. For the Φ5 mm tube, the polarization resistance increases from 1465.7 Ω to 8329.8 Ω, nearly a sixfold improvement compared to the as-printed state.

The deep eutectic solvent, especially ethaline used in this work, indicates good processing repeatability when polishing the internal surface of 316 L tubes with a diameter of 5 mm and length-to-diameter ratios of 2 and 5; the surface roughness is 1.535 μm and 1.672 μm, respectively, using the same electrical parameters.

The flow field simulation illustrates that the viscosity of the electrolytes can also be a vital factor affecting polishing performance. However, the mechanism by which electrolyte viscosity affects polishing performance remains unclear and needs further research. Therefore, the simulation and modification of hydrodynamic conditions for a confined flow field of electropolishing can be a potential avenue of further research for the optimization of internal surface electropolishing processes.

Author Contributions: Conceptualization, D.Z.; methodology, D.Z.; validation, N.Z.; formal analysis, D.Z., Y.Y., S.L. and H.Z.; investigation, D.Z.; resources, C.L.; data curation, D.Z., Y.Y., S.L., H.Z. and N.Z.; writing—original draft preparation, D.Z.; writing—review and editing, X.J.; visualization, S.L.; supervision, C.L. and X.J.; project administration, C.L. and X.J.; funding acquisition, C.L. All authors have read and agreed to the published version of the manuscript.

Funding: This study was funded by the National Natural Science Foundation of China (Grant No. 52375402).

Institutional Review Board Statement: Not applicable.

Informed Consent Statement: Not applicable.

Data Availability Statement: The raw data supporting the conclusions of this article will be made available by the authors on request.

Conflicts of Interest: The authors declare no conflicts of interest.

References

1. Zoe, F.-S.; Tim, C.; Desi, B.; Chul, K.B. Effect of the surface morphology of SLM printed aluminium on the interfacial fracture toughness of metal-composite hybrid joints. *Int. J. Adhes. Adhes.* **2021**, *105*, 102779. [CrossRef]
2. Bhaduri, D.; Penchev, P.; Batal, A.; Dimov, S.; Soo, S.L.; Sten, S.; Harrysson, U.; Zhang, Z.; Dong, H. Laser polishing of 3D printed mesoscale components. *Appl. Surf. Sci.* **2017**, *405*, 29–46. [CrossRef]
3. Mumtaz, K.; Hopkinson, N. Top surface and side roughness of Inconel 625 parts processed using selective laser melting. *Rapid Prototyp. J.* **2009**, *15*, 96–103. [CrossRef]
4. Pyka, G.; Burakowski, A.; Kerckhofs, G.; Moesen, M.; Van Bael, S.; Schrooten, J.; Wevers, M. Surface Modification of Ti6Al4V Open Porous Structures Produced by Additive Manufacturing. *Adv. Eng. Mater.* **2012**, *14*, 363–370. [CrossRef]
5. Stimpson, C.K.; Snyder, J.C.; Thole, K.A.; Mongillo, D. Roughness effects on flow and heat transfer for additively manufactured channels. *J. Turbomach.* **2016**, *138*, 051008. [CrossRef]
6. Teimouri, R.; Sohrabpoor, H.; Grabowski, M.; Wyszyński, D.; Skoczypiec, S.; Raghavendra, R. Simulation of surface roughness evolution of additively manufactured material fabricated by laser powder bed fusion and post-processed by burnishing. *J. Manuf. Process.* **2022**, *84*, 10–27. [CrossRef]
7. Bai, Y.; Zhao, C.; Yang, J.; Fuh, J.Y.H.; Lu, W.F.; Weng, C.; Wang, H. Dry mechanical-electrochemical polishing of selective laser melted 316L stainless steel. *Mater. Des.* **2020**, *193*, 108840. [CrossRef]
8. Wang, Q.; Morrow, J.D.; Ma, C.; Duffie, N.A.; Pfefferkorn, F.E. Surface prediction model for thermocapillary regime pulsed laser micro polishing of metals. *J. Manuf. Process.* **2015**, *20*, 340–348. [CrossRef]
9. Rotty, C.; Doche, M.-L.; Mandroyan, A.; Hihn, J.-Y. Electropolishing Behavior of Additive Layer Manufacturing 316L Stainless Steel in Deep Eutectic Solvents. *ECS Trans.* **2017**, *77*, 1199–1207. [CrossRef]
10. Kim, U.S.; Park, J.W. High-Quality Surface Finishing of Industrial Three-Dimensional Metal Additive Manufacturing Using Electrochemical Polishing. *Int. J. Precis. Eng. Manuf.-Green Technol.* **2019**, *6*, 11–21. [CrossRef]

11. Zhao, C.; Qu, N.; Tang, X. Removal of adhesive powders from additive-manufactured internal surface via electrochemical machining with flexible cathode. *Precis. Eng.* **2021**, *67*, 438–452. [CrossRef]
12. Chaghazardi, Z.; Hof, L.; Wuthrich, R. Electropolishing of Inside Surfaces of Stainless Steel Tubing. *ECS Trans.* **2020**, *97*, 523. [CrossRef]
13. Prihandana, G.S.; Sriani, T.; Jamaludin, M.F.; Yusof, F.; Arifvianto, B.; Mahardika, M. Parameters Optimization for Electropolishing Titanium by Using Taguchi-Based Pareto ANOVA. *Metals* **2023**, *13*, 392. [CrossRef]
14. Kityk, A.; Pavlik, V.; Hnatko, M. Green electropolishing using choline chloride-based deep eutectic solvents: A review. *J. Mol. Liq.* **2023**, *392*, 123519. [CrossRef]
15. Han, W.; Fang, F. Eco-friendly NaCl-based electrolyte for electropolishing 316L stainless steel. *J. Manuf. Process.* **2020**, *58*, 1257–1269. [CrossRef]
16. Candela, V.; Pozzi, M.; Chyhyrynets, E.; Diaz, V.G.; Candela, S.; Dima, R.; Favero, G.; Pira, C.; Pepato, A.; Sonato, P. Smoothening of the down-skin regions of copper components produced via Laser Powder Bed Fusion technology. *Int. J. Adv. Manuf. Technol.* **2022**, *123*, 3205–3221. [CrossRef]
17. An, L.; Wang, D.; Zhu, D. Improvement on surface quality of 316L stainless steel fabricated by laser powder bed fusion via electrochemical polishing in $NaNO_3$ solution. *J. Manuf. Process.* **2022**, *83*, 325–338. [CrossRef]
18. Kim, S.-H.; Choi, S.-G.; Choi, W.-K.; Lee, E.-S. A study of the improvement surface roughness and optimum machining characteristic of L-shaped tube STS 316L by electropolishing. *Int. J. Adv. Manuf. Technol.* **2016**, *85*, 2313–2324. [CrossRef]
19. Kityk, A.; Hnatko, M.; Pavlik, V.; Boča, M. Electrochemical surface treatment of manganese stainless steel using several types of deep eutectic solvents. *Mater. Res. Bull.* **2021**, *141*, 111348. [CrossRef]
20. Wei, Y.; Fang, X.; Qu, N.; Zhu, D. Effects of electrolyte flushing and surface state on electropolishing TB2 titanium alloy. *Int. J. Adv. Manuf. Technol.* **2022**, *119*, 7865–7874. [CrossRef]

materials

MDPI

Article

Control and Modelling of Laser Shock Peening without Coating (LSPwC) Texture of AISI 9310 Steel

Ping Liu [1,2], Zhandiao Yang [1], Cenchao Xie [3], Fei Yang [3] and Liucheng Zhou [3,*]

1 School of Aeronautics, Chongqing Jiaotong University, Chongqing 400074, China; liuping@cqjtu.edu.cn (P.L.)
2 Chongqing Key Laboratory of Green Aviation Energy and Power, Chongqing Jiaotong University, Chongqing 401130, China
3 National Key Lab of Aerospace Power System and Plasma Technology, Air Force Engineering University, Xi'an 710038, China
* Correspondence: happyzlch@163.com; Tel.: +86-133-2454-3606

Abstract: LSPwC is an important development of Laser shock peening (LSP) technology, and surface texturing is an effective method to improve tribological properties. The combination of these is expected to innovate a new surface texturing technology with a strengthing effect, but no one has attempted it. In this paper, a new LSPTwC technology combining them is innovatively proposed and validated on AISI 9310 steel, which is commonly used in helicopter transmission components for surface texturing. The LSPTwC surface was studied using an optical microscope, electron microscope, energy dispersive spectrometer, and so on. The results proved that LSPTwC is an effective texturing method of AISI 9310 steel, which modulates the texture and improves the properties, such as the microhardness increased by more than 10%. A model for calculating the texture and process parameters is also constructed on a statistical basis, and a modeling method for textured surfaces is proposed. It is verified that the calculation results and the constructed model are highly consistent with the test, with a diameter deviation <3% and depth deviation <4%. The above results can provide the experimental basis, process design method, and calculation model for single-point LSPwC texturing of AISI 9310 steel parts for helicopters, which have application value.

Keywords: LSPTwC; LSPwC; LST; AISI 9310 steel; surface texturing; loss of lubrication

1. Introduction

Loss of lubrication (LOL) in helicopter gearboxes is a potentially catastrophic failure in which gearbox components such as gears and bearings rapidly lose lubrication, leading to a deterioration in frictional performance, resulting in an increase in surface temperatures, a reduction in material properties and a high risk of structural damage and gearbox failure [1–3], which is a global engineering challenge that has long plagued helicopter safety [4,5]. Surface texture is one of the most effective methods to improve tribological properties under different lubrication conditions [6–8]. By optimizing contact stress, cleanliness, and oil film distribution [9–11], tribological properties under lubrication-free conditions can be significantly improved, providing a potential solution to increase the LOL capacity [12].

LSPwC is a new type of LSP technology that does not require protective coatings, in which the laser is applied directly to the surface [13], and not only induces physical and chemical effects such as ablation, impact, remelting, and heat treatment but also creates a special strengthening surface texture that can improve tribological properties [14,15]. For example, the authors of this paper [16] carried out LSPwC treatment on AISI 9310 steel, which resulted in an increase in microhardness of about 50%, a reduction in the coefficient of friction of 12%, and an increase in wear resistance of 50 to 70%. In a further study, a work-hardening layer with a depth of more than 530 μm and a residual compressive stress layer of 510 μm were obtained, and the maximum residual compressive stress introduced was more than 748.9 MPa, and a regular weave of rhombus or rectangle was detected [17]. On the other hand, Song et al. [18] investigated the residual compressive stress introduced by LSP and found that the residual stress could significantly improve the rolling contact fatigue, and Yan et al. [19] reached similar conclusions in their study of conventional shot peening and further confirmed the beneficial effects of microstructural refinement and microhardness enhancement. In addition, Hereñú, Praveen, Zhao et al. [20–22] improved the high peripheral fatigue life, strength, and ductility of the material by LSPwC, Yu et al. [23] also found that it had an improved effect on micromotion wear at high temperatures, and Tong et al. [24,25] in their studies on aluminum alloys found that LSPwC improved the cavitation resistance and was further found to have the effect of improving surface wettability and pitting resistance. In conclusion, LSPwC is a surface-strengthening technique with obvious advantages and comprehensive effects, which can improve surface properties and prepare surface textures.

Although existing studies have identified the beneficial effects of LSPwC in improving tribological properties, high-temperature performance, and achieving surface texturization [16–18], there is no research precedent for controllable and designable texturing using LSPwC. Since material tribological properties, high-temperature performance, and fatigue strength are important factors affecting the loss of lubrication capability [11,26], it is highly expected that the application of LSPwC to surface texturing by combining the beneficial findings of existing studies will result in a novel surface texturization solution that significantly enhances the loss of lubrication capability of helicopters. Therefore, in this paper, a designable and controllable laser surface texturing (LST) based on LSPwC (LSPTwC for short) is innovatively proposed for the first time, and AISI 9310 steel, which is commonly used for helicopter transmission components, is used as the research object for the LSPTwC study, in order to validate the feasibility of the LSPTwC for helicopter transmission components and to provide experimental bases, design methodologies, and computational models for the regulation and design of the texturing structure.

2. Materials and Methods

2.1. Materials

AISI 9310 steel is a carburized bearing steel with excellent properties, which is one of the most commonly used materials for the drive train of aerospace equipment [27] and is widely used in the manufacture of helicopter transmission gears, bearings, and shafts [28]. In this paper, the selected raw material is a 300 mm diameter bar; its main constituents (mass fractions) are shown in Table 1 by measured, which is cut into 5 mm slices by AQ750L wire electric discharge machine (Sodick, Suzhou, China), set electrode wire diameter 0.2 mm, voltage 15 N, feed rate 200 mm/s, and then vacuum carburizing treatment using a RX3-20-12 high-temperature box resistance furnace (HengLi, Danyang, China). The carburizing process is as follows: preheating at 680 °C, carburizing at 920 °C, carburizing period of carbon potential of 1.3%, and then slow cooling in the furnace to 350 °C air cooling; and then heating to 870 °C water cooled quenching, and then slow cooling in the furnace for tempering at 680 °C, 800 °C salt bath oil cooling, and finally 200 °C air cooling tempering. The detailed process is shown in Figure 1. After carburisation, remeasured under the same test conditions, the main components were as shown in Table 2, the C content on the surface of the material increased to 5.2%.

Table 1. Main chemical elements and contents of AISI 9310 steel.

C	Mn	Si	Cr	Ni	Mo	Al	O	Fe
0.1	0.6	0.2	1.3	2.7	0.3	0.4	0	94.88

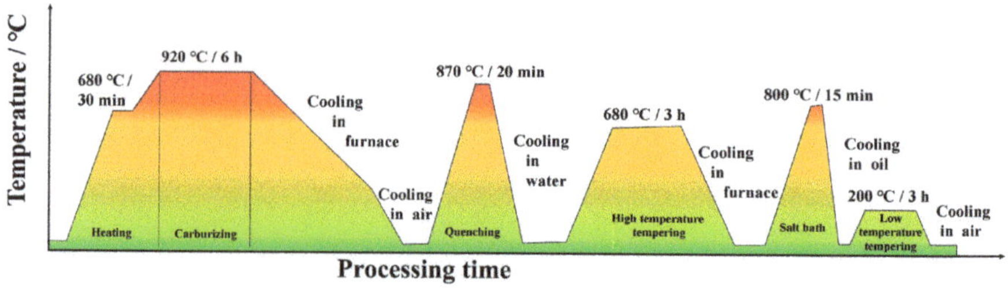

Figure 1. AISI 9310 steel carburizing process.

Table 2. Main chemical elements and contents of AISI 9310 steel after carburization.

C	Mn	Si	Cr	Ni	Mo	Al	O	Fe
5.2	0.6	0.3	1.3	3.1	0.3	0.4	0	89.7

2.2. LSPwC Scheme

The test was carried out using a YSM2000-C30A water-optical coaxial LSP device (Tyrida, Xi'an, China), as shown in Figure 2. The laser is a TABOR-200 Nd:YAG laser (Joram, Beijing, China). with a pulse width of 10 ns, a circular spot shape, a maximum pulse energy of 200 mJ, a wavelength of 532 nm, a divergence angle of <2 mrad, and a repetition frequency of 50 Hz. The amplification test program employed single-point amplification, as shown in Figure 3, and the control parameters included laser energy, spot diameter, and array spacing. The detailed process parameters are shown in Table 3, and the control parameters include laser energy, spot diameter, array spacing, etc. The detailed process parameters are shown in Table 3.

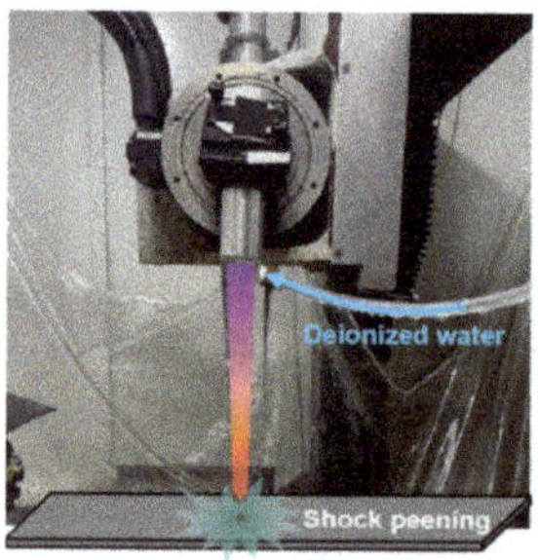

Figure 2. Water-optical coaxial LSP equipment.

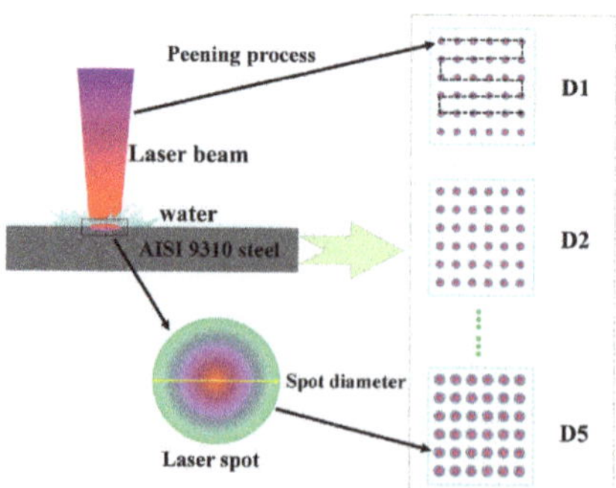

Figure 3. LSPTwC schemes with different spot diameters.

Table 3. Different spot diameters LSPTwC parameter list.

Parameter ID	Spot Diameter (μm)	Cols Spacing (μm)	Line Spacing (μm)	Laser Energy (mJ)
D1	350	1000	1000	90
D2	450	1000	1000	90
D3	550	1000	1000	90
D4	650	1000	1000	90
D5	750	1000	1000	90

2.3. *Surface Observation and Measurement Scheme*

Different from the traditional laser texturing technology, LSPwC is the direct action of the laser on the material surface, which not only induces plasma shock waves and forms the shock effect [29], but also directly occurs the energy transfer, resulting in ablation, remelting, and thermal impact on the material surface [30,31], resulting in the formation of multi-dimensional and multi-scale complex texture structures on the material surface. Therefore, through experimental research, obtaining the influence law of different spot diameters on the surface texture of AISI 9310 rigid under the condition of single-point LSPTwC and establishing the mapping relationship and calculation model between the laser diameter and the parameters of LSPTwC texture structure is a prerequisite for realizing the precise control of the LSPTwC surface texture structure design and peening process. Therefore, the research program of this paper is as follows:

Firstly, specimen preparation: the LSPTwC specimen and the untreated specimen were cut into 10 mm × 10 mm specimens for observation using a AP250Ls wire electric discharge machine (Sodick, Suzhou, China), set electrode wire diameter 0.08 mm, voltage 10 N, feed rate 50 mm/s, as shown in Figure 4, There were 3 specimens in each group LSPTwC parameter, to ensure that each set of data for all tests is not less than 3.

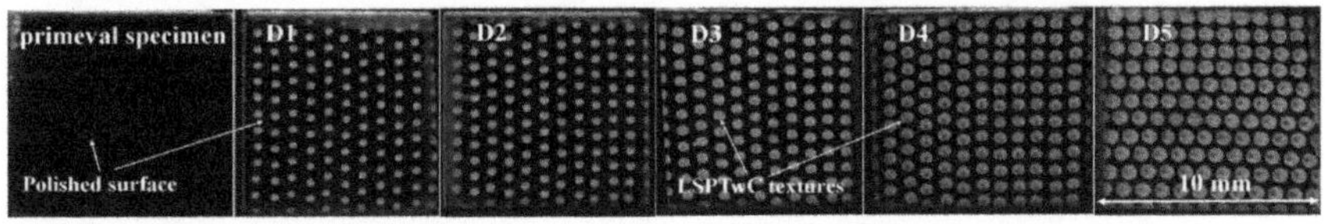

Figure 4. Sliced specimen of LSPTwC with different spot diameters.

Secondly, surface 2D characterization: the surface of the specimen LSPTwC and untreated were observed using a Axiocam 208 optical microscope (Zeiss, Suzhou, China) and an G4 UX electron microscope (Helios, Davisburg, MI, USA). The parameters of SEM were set as an HV of 5.00kV, det ETD, mode SE, and WD 4.3 mm. The surface chemical elemental analysis was carried out using an Ultim Max65 EDS (Ultim, Oxford, UK) with the accelerating voltage set to 20KV and a working distance of 4 mm. The metallographic organization was observed using a ZEISS-Axiocam 208 optical microscope, with the cut surfaces of the specimens were polished using 400 mesh, 800 mesh, 1000 mesh, 1500 mesh, and 2000 mesh sandpaper in that order, and then the cut surfaces were further polished using SiO2 polish, and then corroded using 4% ethanol nitric acid solution for 10 s. The microhardness in the depth direction of the cut surface was measured by the EM500-3A microhardness tester (HengYi, Shanghai, China) set to a load of 300 g, time 10 s, and one group was collected at 10 μm intervals from the surface to 100 μm, one group was collected at 50 μm intervals from 100 μm to 500 μm, one group was collected at 100 μm intervals from 500 μm to 1000 μm.

Finally, Surface 3D characterization: thirdly, a GT-K optical profiler (Bruker, Billerica, MA, USA) and a Bruker Dimension ICON atomic force microscope (Bruker, Billerica, MA, USA) were used to scan the surface profiles untreated and LSPTwC specimen to characterize the 3D profiles, sampling intervals are 0.1 μm and 1 nm. Through multi-dimensional and multi-scale observation and characterization, we provide the experimental basis and data support for analyzing the influence of spot diameter on the LSPTwC texture structure of AISI 9310 steel material and establishing the relationship model between spot diameter and texture structure.

3. Result and Discussion

3.1. 2D Texture Observation and Analysis

3.1.1. Optical Microscope Observations and Analysis

Different spot diameters of LSPTwC specimens were observed under a 10× light microscope, as shown in Figure 5. Regardless of the diameter, LSPTwC formed a well-bounded crater texture on the surface of AISI 9310 steel, and there was a clear difference in topography from the center to the edge. The center of the crater showed a bright metallic color, while the main body of the crater showed a black ablation feature, and the edge area showed an outwardly jetted extended texture. The highlighted area is not at the geometric center of the crater but is eccentric within the scan line, and the distance to the rim increases only slightly with increasing diameter. For the same crater, the further away from the highlighted area, the clearer the edge and the less jet-like the texture. The overall surface morphology of the crater indicates that ablation, melting, jetting, and remelting deposition occurred on the surface of the AISI 9310 steel during the LSPTwC process. The size of the highlighted area and the eccentricity of the crater increase with increasing spot diameter, suggesting that the change in spot diameter affects the laser energy distribution and its effect on the material surface.

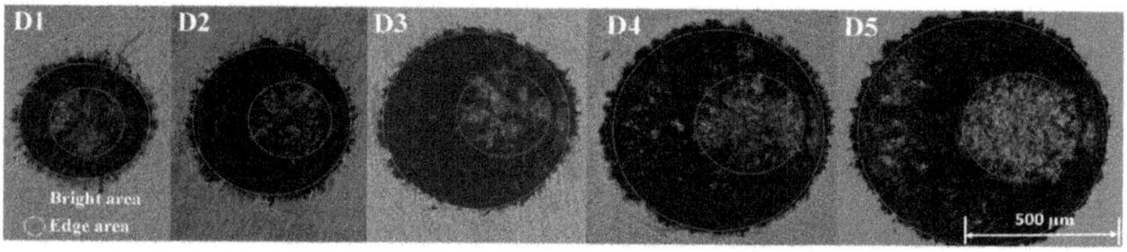

Figure 5. Overall view of LSPTwC surface with different spot diameters (OM 10×).

To further clarify the effect of spot diameter on the crater's microtexture structure, the microscope's resolution was adjusted to 50× and 100×, and the center and edge regions of the crater were compared, as shown in Figure 6. Jet-like textures are commonly found in the interior of the crater, and the smaller the diameter, the longer and thinner the jet-like textures are, and the more pronounced the outward jet feature is. All jet textures originate from the highlighted regions and show outward extension. The jet-like texture at the edge becomes shorter and shorter as the spot diameter increases, and remelting accumulation is found at the edge of the 650 um diameter crater. The distal edge of the 750 um diameter crater does not even have a jet-like texture, which is only characterized by ablation and remelting. In contrast, a wave-like remelting accumulation is found in the inner region near the center. The above texture and structural features are obviously the result of deposition and stacking of ablative and molten materials formed by splashing and flowing under the shock wave's action, and the shock's source is located in the highlight area. Therefore, the central area has less ablative and re-melted material deposited, giving it a metallic sheen, while the surrounding area has a large amount of ablative and re-melted material deposited, giving it a black ablative feature. In addition to the jet and wave textures, the surface of LSPTwC shows a large number of micro-nanometre-sized circular and reticulated pits under a high magnification microscope, with intact edges of the micro-pits and black bottoms and depositional structures in the centers of some of the micro-pits. Crater edges can be observed as heat-affected zones, and a large number of remelted depositional structures can be seen under a high-power magnification microscope. The remelted depositional structures decrease in diameter but increase in size. Large remelted deposits will show a slag structure. Combining the results of the optical characterization with different resolutions, it is easy to see that the impact effect is weakened as the spot diameter increases, the ablation and remelt residues increase, the ablation and remelt depositional structures increase, and the microtexture structure of the crater rim deteriorates.

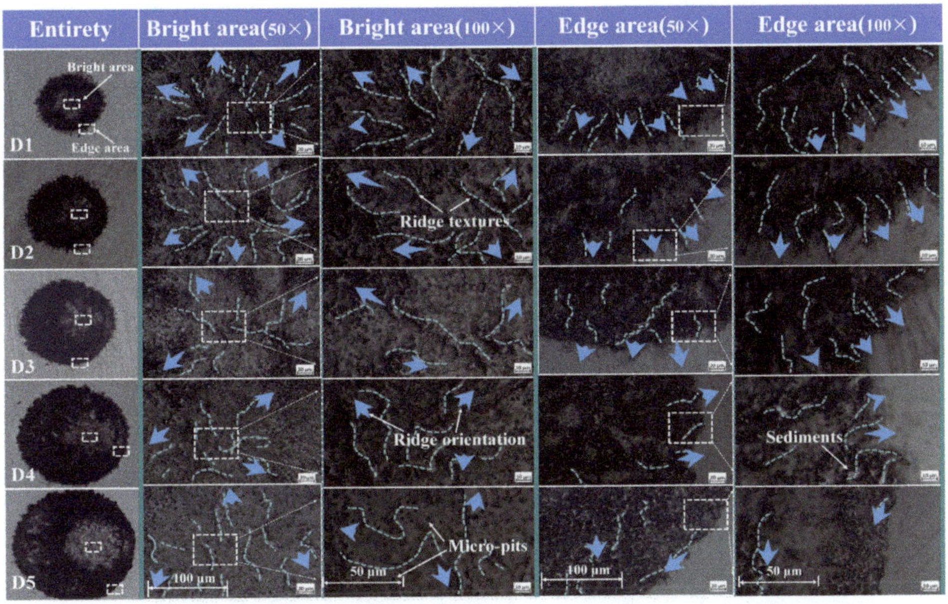

Figure 6. Comparison of LSPTwC surfaces with different spot diameter (multi-scale OM).

3.1.2. Electron Microscope Observations and Analysis

The observation and characterization of LSPTwC crater by scanning electron microscope (SEM) can avoid unfavorable interferences such as focal length and color of an optical microscope. Firstly, the SEM mag was adjusted to 300× for observation and characterization, and the results are shown in Figure 7. The center region of all craters is flatter than the edge region, and the texture structure is more delicate. A large amount of remelted deposition texture exists at the edge of craters, and even large aggregated remelted deposition material is found. The aggregated remelted deposits and their surrounding areas show distinct brightness variations, with a significant increase in the brightness of the deposits themselves and a significant decrease in the brightness of the surrounding areas. The brightness variation reflects the different conductivity of different regions of the crater, indicating that LSPTwC has caused complex physical and chemical changes on the surface of the AISI 9310 steel, resulting in surface modification of the material. The distribution of the re-melted deposits was significantly correlated with the crater diameter—the larger the diameter, the more deposits, the larger and closer to the center. However, when the diameter reached 650 μm, the phenomenon showed an improving trend.

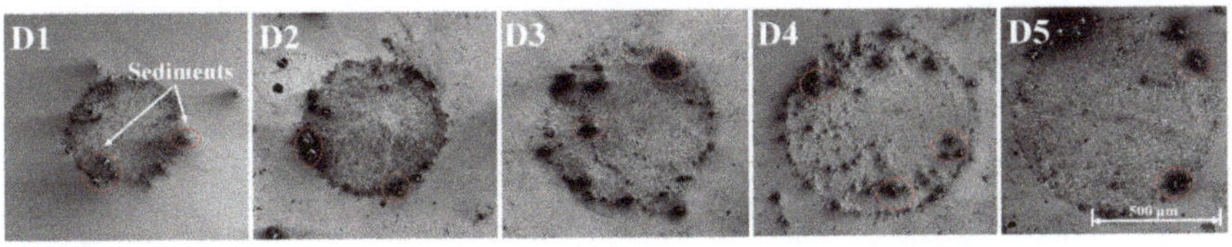

Figure 7. Overall of LSPTwC surface with different spot diameter (SEM mag300×).

The mag was then adjusted to 2500×, and the center and edge regions of the crater were locally enlarged and the characterization results are shown in Figure 8. It can be seen that the overall morphology of the crater center tends to become more straightforward with the increase of the spot diameter, and the texture and micropores decrease and become smaller. In contrast, the remelted deposit structures at the crater edge increase with the increase of the spot diameter, the micropores increase and become larger, the edges of the micropores gradually become indistinct, and even the cross-over phenomenon occurs. The loose deposit structures inside the micropores gradually increase and become larger. The large re-melted deposits showed welding and adhesion under 2500×, which were the adherent materials produced by splash dropping.

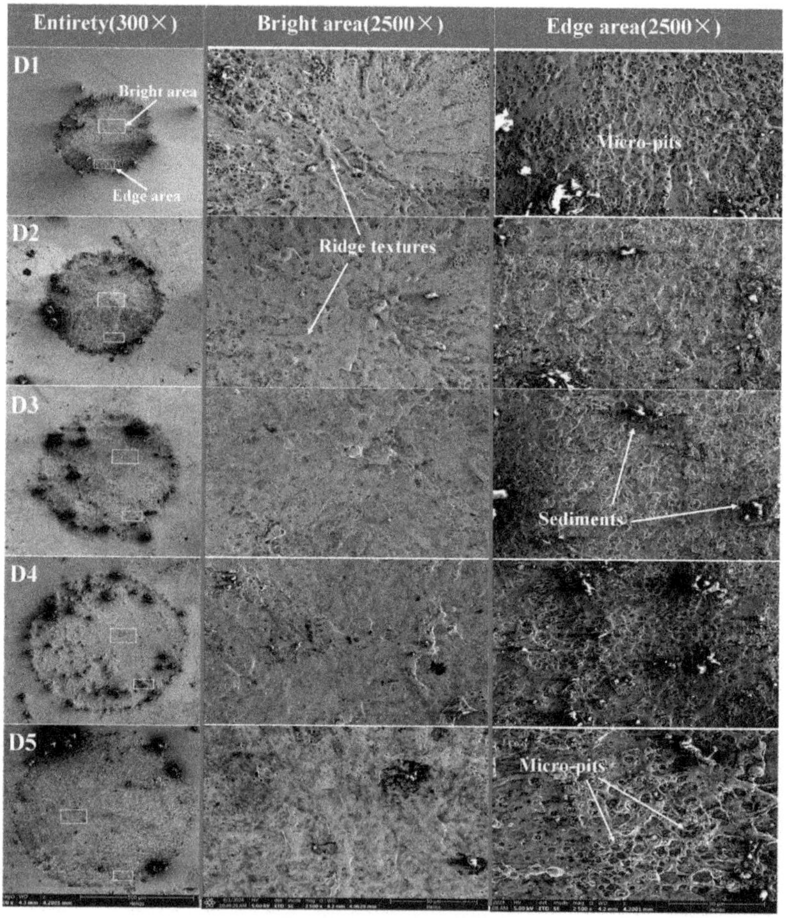

Figure 8. Comparison of LSPTwC surfaces with different spot diameter (multi-scale SEM).

Comparing the LSPTwC and untreated specimens, carbon and oxygen elements are the two components with the most significant changes, and the discovery of oxygen indicates the presence of oxidation phenomena such as ablation during the LSPTwC, and the increase in carbon has been similarly found in the existing studies as a change in favor of dry friction self-lubrication [16,17]. Comparing the same LSPTwC texture, the bright area and the edge area are typically different areas, and the carbon and oxygen content of the edge area is significantly higher than that of the bright area, indicating that remelted deposition enhances the change in the chemical element content, which is also evidenced by the enrichment of other elements in the remelted deposition texture. The edge area is relatively protruding, and its higher carbon and alloying elements content will be favorable for improving tribological properties [32,33]. The variation of the chemical element contents of the bright area and edge area with spot diameter is shown in Figure 9a,b; with the increase of spot diameter, the carbon content of edge area increases, the carbon content of bright area decreases, the oxygen content of edge area increases and then decreases, and the oxygen content of bright area decreases continuously, as shown in Figure 9c,d. The above results show that a smaller spot diameter can obtain a more uniform chemical composition of the texture, while the carbon and oxygen content of the bright area above 650 μm diameter is basically the same as that of the untreated samples, which indicates that LSPTwC has almost no modulation effect on the chemical composition under this condition.

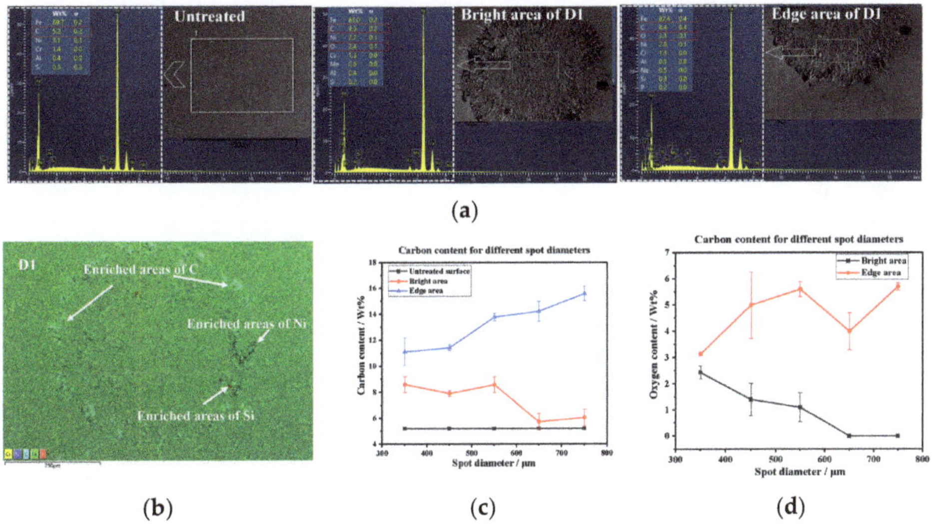

Figure 9. Analysis of chemical elements with different parameters and locations: (**a**) EDS testing; (**b**) EDS scanning; (**c**) carbon content analysis; (**d**) oxygen content analysis.

The metallographic corrosion characterization results comparing the LSPTwC of D1 and untreated specimens are shown in Figure 10. The metallographic phase of untreated specimens is a layered structure consistent with the heat treatment process. After LSPTwC, the original layered structure was changed, forming a continuous gradient structure from the matrix to the surface grains, which were gradually refined, as shown in Figure 10a, a typical LSP strengthening phenomenon [34]. The high-temperature tempering organization of the subsurface layer was transformed into martensite, reflecting the quenching effect of the fast heating and cooling of the laser and confined water layer. It was also found that many reticulated carburite structures precipitated at the martensitic grain boundaries, extending from the matrix to the surface, which is consistent with the elevated surface EDS carbon content. Grain refinement, martensite, and carburite all have a strengthening effect [35–37] in order to confirm the strengthening effect of LSPTwC on the material, the microhardness was measured as in Figure 10b, the microhardness of all specimens was elevated by more than 10%, and almost the surface is the hardest layer, which is consistent with the results of the metallurgical characterization, and the hardening depth of D3 was found to be the largest, which reached 700 μm as shown in Figure 10c, which is supposed to be a result of the combined effect of force and heat.

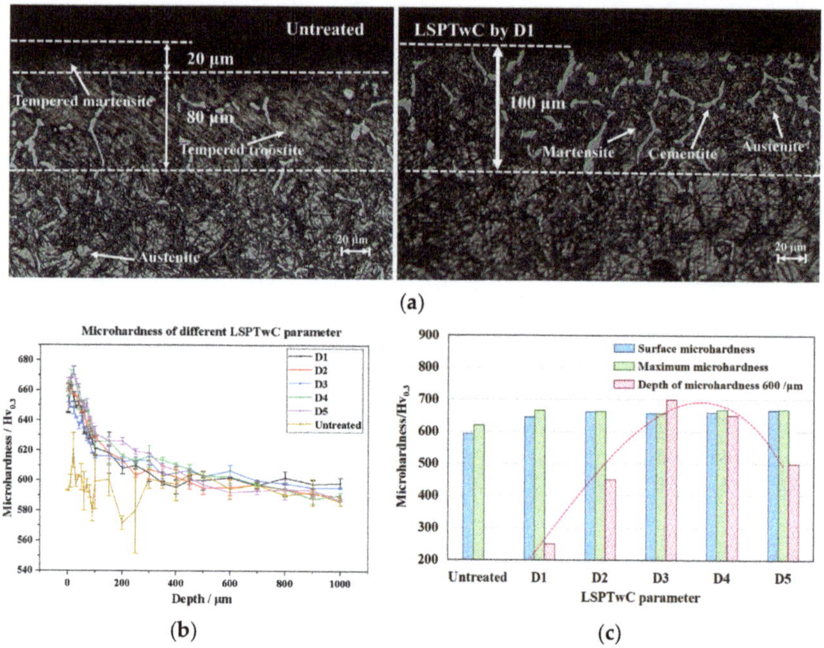

Figure 10. Metallographic and microhardness analyses. (**a**) Metallographic analysis of untreated and D1, (**b**) microhardness at different depths, and (**c**) maximum microhardness and hardening depth.

3.2. 3D Texture Observation Analysis and Modeling

3.2.1. 3D Texture Observation and Analysis

The optical profilometer characterization results provide the experimental basis and data support for us to quantitatively analyze the LSPTwC crater 3D texture structure of AISI 9310 steel. Figure 11 shows the overall 3D morphology of the original specimen and LSPTwC specimens with different spot diameters. The results show that LSPTwC produces a typical impact crater on the surface of AISI 9310 steel, and the 3D morphology of the crater textures formed by the same diameter spot has a good consistency, which almost does not affect the surface outside the impact area, except for some spatter and accumulated material at the edge.

Atomic force microscopy (AFM) was further used to observe and analyze the highlighted regions observed by OM at the nanoscale, as shown in Figure 12a,b. Before LSPTwC, the surface of the pristine specimen showed a capillary columnar structure characteristic of machining, with surface undulations up to about 40 nm. After LSPTwC, the surface of the specimen formed a complete and s mooth microcrystalline surface, and the rough peaks became coarse and rounded. This indicates that the polished surface of the sample, although bright and intact at the general microscopic scale, is not intact at the nanoscale. On the other hand, the remelting effect of LSPTwC can produce very intact crystalline surfaces at the micrometer and nanometre scales.

Since all craters have a symmetrical shape in the laser scanning direction, this paper extracts the crater surface profile data in the cross-section in the scanning direction and perpendicular to the cross-section in the scanning direction, using the laser scanning line as a reference and further quantitatively analyses the effect of spot diameter on the characteristic parameters of the crater surface profiles, as shown in Figure 13. The crater profile within the symmetric plane in the scanning direction is non-symmetric in shape and has a steeper distribution on one side, which is consistent with the eccentricity results observed by OM and SEM. However, the crater profile within the vertical section in the scanning direction is symmetrical. In addition, the LSPTwC texture has noisy micrometer-scale roughness peaks and valleys in addition to the crater, leading to a significant increase in surface roughness.

Obtaining crater texture structure characteristic parameters and their changing law is the key to realizing texture structure design and accurate control, but the raw surface profile data contains rough peaks and valleys, which poses challenges to crater structure characterization. Therefore, this paper adopts the Gaussian filtering method to separate the long- and short-period surface profile data, and the cutoff wavelength is set to 0.1 mm. The results of the long-wave cutoff filtering are shown in Figure 14a,b, and the results of the short-wave cutoff filtering are shown in Figure 15. The results show that with the increase of the spot diameter, the stability of the crater texture profile decreases, and the internal non-periodic fluctuating texture gradually increases. The rough peaks and valleys are basically consistent with the general rough surface shape and, thus, can be used to analyze the roughness.

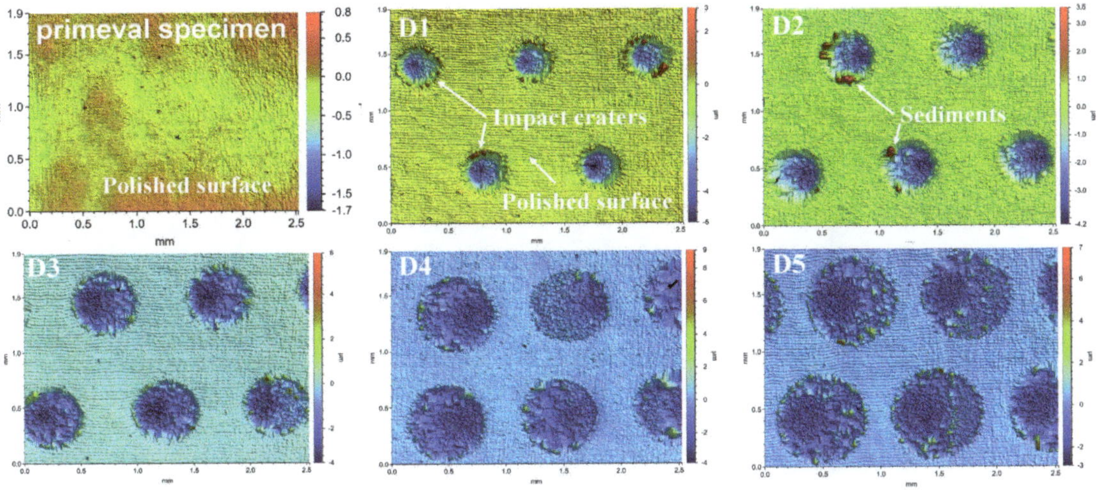

Figure 11. Overall LSPTwC surface with different spot diameters (OSP).

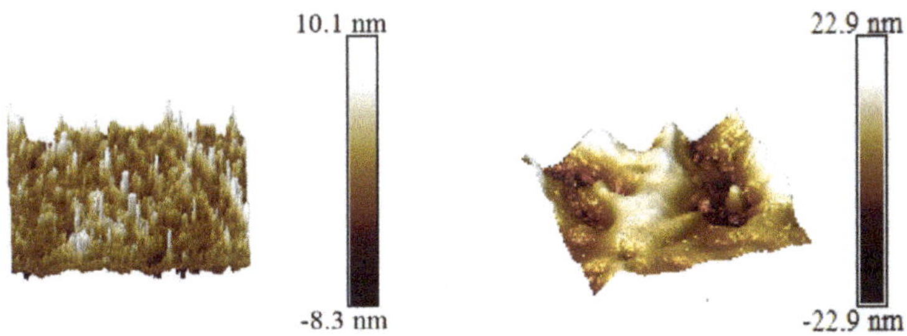

(**a**) (**b**)

Figure 12. Atomic force microscopy characterization of (**a**) original polished surface and (**b**) LSPTwC crater center.

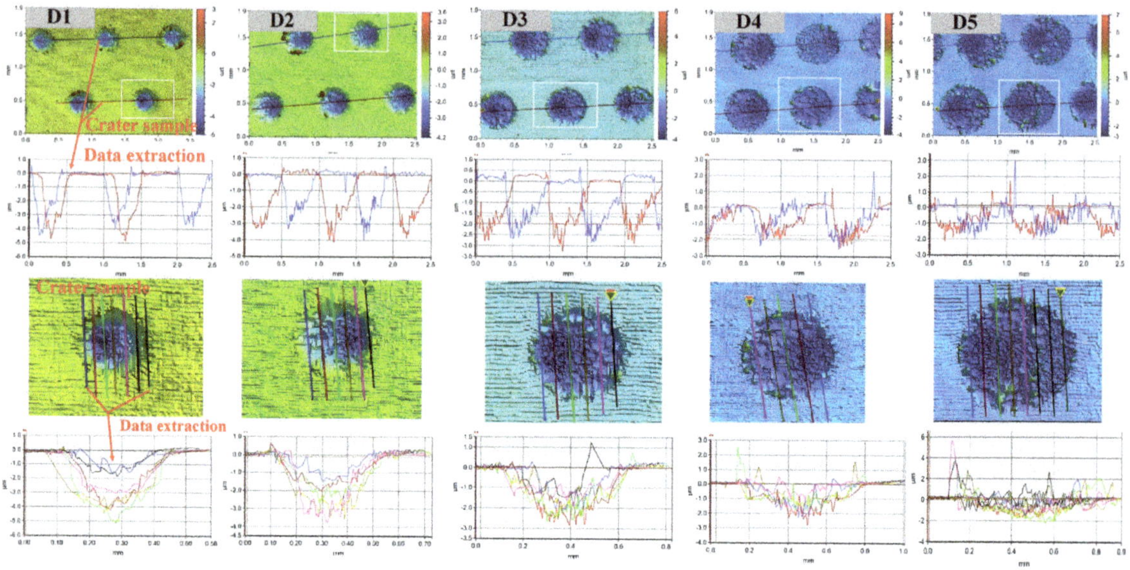

Figure 13. Extraction of surface profile data.

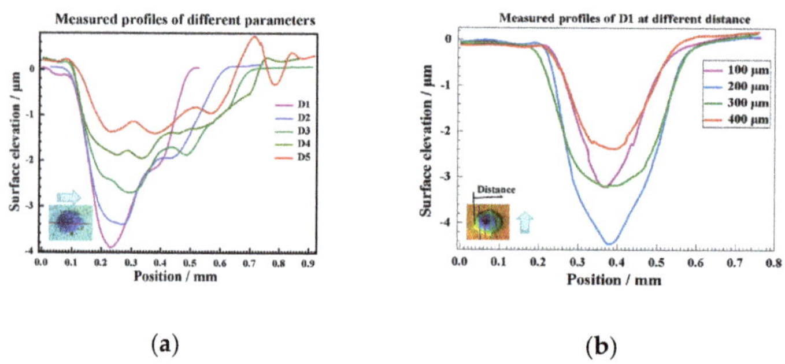

Figure 14. Long wavelength pass of (**a**) symmetry section and (**b**) vertical section.

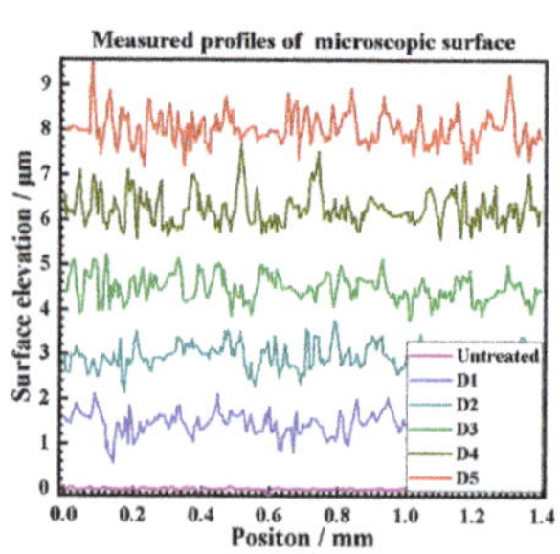

Figure 15. Short wavelength pass.

According to the characteristics of the laser energy distribution and the characteristics of the measured profile curve, this paper first tries to use normal and skewed functions to fit and analyze the crater profile curve, and the selected models are the origin basic functions Gauss Equation(1) and the LogNormal Equation (2), in which y is the surface elevation, x is the calculate position, y_0 reflects the elevation of the original surface, and

x_c is related to the polar transverse coordinates, and ω and A reflect the curve width and depth, respectively.

$$y = y_0 + \frac{A}{\omega\sqrt{\pi/2}} exp^{-2\frac{(x-x_c)^2}{\omega^2}} \quad (1)$$

$$y = y_0 + \frac{A}{\sqrt{2\pi}\omega x} exp^{\frac{-(ln\frac{x}{x_c})^2}{2\omega^2}} \quad (2)$$

The long wavelength pass data in Figure 13 are fitted by Origin2021, and the results are shown in Figure 16a,b. The LogNormal model can fit the symmetry section profiles with an average residual of less than 0.096 μm. Still, the fitting effect worsens when it is far away from the bottom or greatly increases the spot diameter. A large number of irregular textures are found in the area of optical observation, which is the result of the decrease in energy density. Gauss fits the profiles of the vertical section of D1 better; the average residual is less than 0.089 μm, as shown in Figure 16c,d. The results indicate that the morphological characteristics of the LSPTwC craters indeed match the laser energy distribution, and the eccentricity of the craters is likely to be caused by the biased distribution of the laser energy. Therefore, increasing the laser energy density can better control the crater structure; changing the spot energy distribution is an effective means to regulate the internal texture structure of the LSPTwC crater.

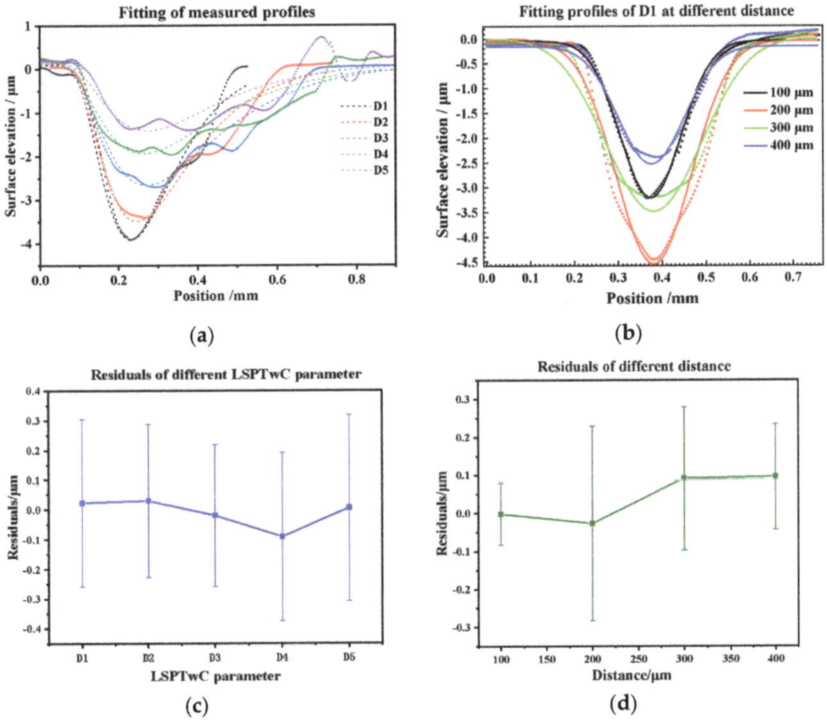

Figure 16. Fitting the long wavelength pass of (**a**) the symmetry section, (**b**) the vertical section profiles, (**c**) the symmetry section profiles fitting Residuals, and (**d**) the vertical section profiles fitting residuals.

According to the shape and change rule of crater profiles with different diameters, this paper further collected several sets of data. Statistically, it analyzed the filtered feature parameters to obtain the statistical curves of the diameter deviation and maximum depth of crater with the change of spot diameter, respectively. The results are shown in Figure 15.

The diameter deviation and maximum depth of the crater decrease with the increase of the spot diameter, as shown in Figure 17a,b. By using Equation (3) to calculate the roughness and peak-valley variance, it was found that the surface roughness increased more than 10 times relative to the original surface, as shown in Figure 17c. The surface roughness of the craters with different diameters had a good consistency, which increased slightly with the increase of the spot diameter, and the peak-valley variance also increased, and the maximum roughness was obtained at a spot diameter of 650 μm, indicating that the variation of the spot diameter parameter can also regulate the roughness of the LSPTwC surface.

$$Ra = \frac{1}{\updownarrow}\int_0^{\updownarrow} |Z(x)|dx \qquad (3)$$

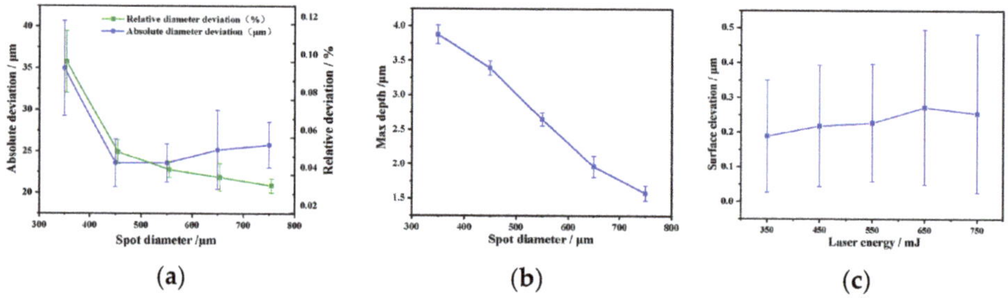

Figure 17. Statistical analysis of different spot diameters (**a**) diameter deviation, (**b**) max. depth, and (**c**) surface roughness.

3.2.2. 3D Texture Modeling and Verification

Since the spot diameter is only a limitation of the effective area of action of LSPTwC, it is difficult to reflect the intrinsic factors that affect the crater texture characteristics. Existing studies have shown that laser energy density is a key process parameter for LSP-induced shock waves [38]. Therefore, in this paper, we consider the energy density as the core parameter to replace the spot diameter as a reference for analysis, and its interconversion relationship is shown in Equation (4). Where E_d is the laser energy density, E is the laser energy, and d is the spot diameter.

$$E_d = \frac{E}{\pi d^2/4} \qquad (4)$$

The energy densities of LSPTwC of different diameters were calculated using Equation (2), as shown in Table 4:

Table 4. Calculated laser energy density (laser pulse energy 90 mJ).

Spot Diameter (μm)	350	450	550	650	750
Energy Density (mJ/mm^2)	936	566	379	271	204

Taking the laser energy density as the reference value, the relative deviation of crater diameter, maximum depth, and surface roughness is analyzed again, and the results are shown in Figure 18, and the statistical curves of the relative deviation of crater diameter, maximum depth and surface roughness with respect to the laser energy density are obtained. The maximum depth of the crater increases with the increase of the laser energy density, but the trend of increase is significantly slowed down and gradually approaches the limit value after 550 (mJ/mm^2), as shown in Figure 18a. After 550 (mJ/mm^2), the increase slows down significantly and gradually approaches the limit. The relative deviation of the spot diameter increases with increasing laser energy density and shows an accelerating trend, as shown in Figure 18b. The surface roughness and the variance of the roughness front valley have extreme values, and when the energy density deviates from the extreme value, the surface roughness will be improved, as shown in Figure 18c.

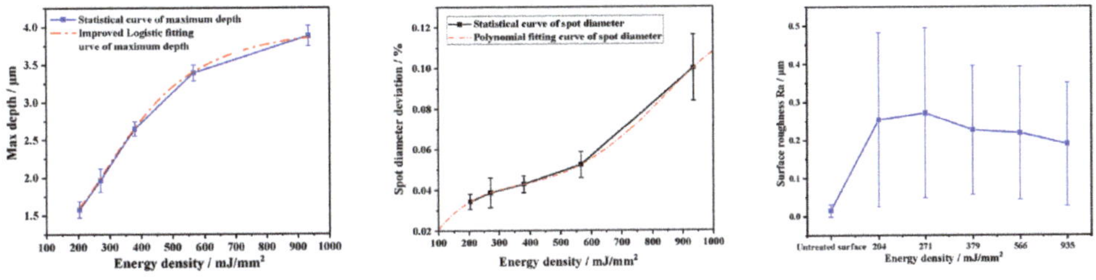

(a) (b) (c)

Figure 18. Statistical analysis of different spot diameters with laser energy density: (**a**) diameter deviation, (**b**) max. depth, and (**c**) surface roughness.

Further observation of the change rule of crater profile characteristic parameter with laser energy density shows that there is a significant correlation between the two, which creates the conditions for establishing the relationship model between crater profile characteristic parameter and laser energy density. For this purpose, according to the basic law that the shape of the curve and the LSP affect the amount of plastic deformation of the material [39], the improved logistic Equation (7) and the multinomial Equation (8) were used in this paper to fit the maximum depth and the spot diameter deviation. Among them, the improved logistic model was optimized based on the standard logistic Equation (5).

$$y = \frac{A}{1 + exp^{-k(x-x_0)}} \tag{5}$$

where A is the maximum saturation, k is represents the growth rate of the curve, x_0 is the centre position of the curve, and exp is the base of the natural logarithm.

Considering that the capacity of the equipment limits the laser energy in practical applications, a maximum and minimum value limit always exists, so the model is adjusted accordingly. Specifically, the standard logistic model is improved to Equation (6) by introducing non-zero maximum and minimum parameters A_1 and A_2.

$$y = A_2 + \frac{A_1 - A_2}{1 + e^{-k(x-x_0)}} \tag{6}$$

The relative exponential term was improved to $(x/x_0)^p$ to obtain more flexibility to get the final improved logistic Equation (7).

$$y = A_2 + \frac{A_1 - A_2}{1 + (x/x_0)^p} \tag{7}$$

The Multinomial Equation (8) is a standard polynomial model, where y is the relative deviation of the diameter, y_0 is the static relative deviation of the diameter, x is the laser energy density, and A_i is the weight of the different exponential phases of the laser energy.

$$y = y_0 + \sum_{1}^{n} A_i \times x^i \tag{8}$$

The fitting results are shown in Figure 16a,b, and both models have good fitting results. Plug in Equation (4), the spot diameter–maximum depth Equation (9) and spot diameter-diameter relative deviation Equation (10) were obtained from the fitting.

$$y = \frac{1.2 - 3.9}{1 + \left(\left(\frac{P}{\pi d^2/4}\right)/340\right)^{3.5}} + 3.9 \tag{9}$$

$$\begin{aligned} y = 3.495 \times 10^{-4} \times \left(\frac{P}{\pi d^2/4}\right) - 1.007 \times 10^{-6} \times \left(\frac{P}{\pi d^2/4}\right)^2 + 1.313 \times 10^{-9} \times \left(\frac{P}{\pi d^2/4}\right)^3 \\ -5.424 \times 10^{-13} \times \left(\frac{P}{\pi d^2/4}\right)^4 - 0.052 \end{aligned} \tag{10}$$

From the laser energy density calculation Equation (4), it can be seen that the energy density change is inversely proportional to the square of the diameter. Therefore, in this paper, the highest spot energy density position is taken as the benchmark, and the energy distribution is corrected according to the square of the diameter change to construct the crater profile coordinate calculation Equation(11). The profile depth of the crater is then scaled equally to the maximum profile depth to construct the profile depth calculation Equation (12).

$$\begin{cases} x = x_c - \left(\frac{x_c}{d_{cn}}\right)^2 \times \Delta d_{cn} \; x_c < 0 \\ x = x_c + \left(\frac{x_c}{d_{cp}}\right)^2 \times \Delta d_{cp} \; x_c > 0 \end{cases} \tag{11}$$

$$y = \frac{d_{max}}{d_{cmax}} \times d_c \tag{12}$$

The parameters are defined as shown in Figure 19. x is the computed profile coordinate, x_c is the calibrated profile coordinate of the computed point, d_{Cn} is the total length of the calibrated profile for $x_c < 0$, Δd_c is the increment of the profile computation in the direction of $x_c < 0$, d_{Cp} is the total length of the calibrated profile for $x_c > 0$, Δd_p is the increment of the profile computation in the direction of $x_c > 0$, which is defined in Figure 19a, and the increment of profile computation is corrected by Equation (8) in model (12), where y is the profile calculation depth, d_{max} is the maximum depth of the profile obtained by calculation, d_{cmax} is the maximum depth of the profile obtained by calibration, and d_c is the profile depth of the calibration point at the calculation point, which is defined in Figure 19b.

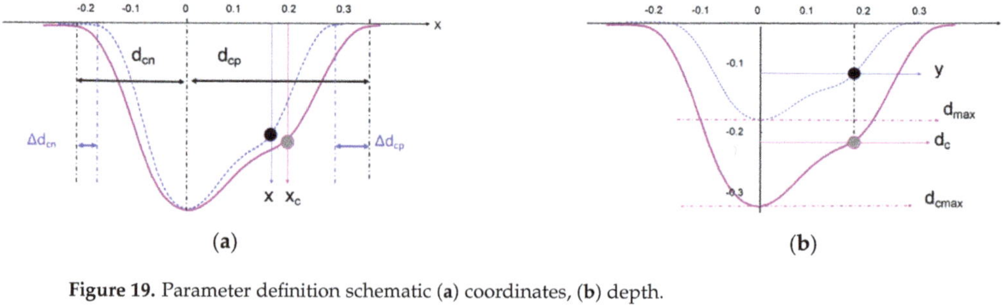

Figure 19. Parameter definition schematic (**a**) coordinates, (**b**) depth.

In order to verify the accuracy of the above models, the spot diameter–maximum depth Equation (9) was used to calculate the maximum depths of craters with different diameters and compared with the measured statistical results; the results are shown in Figure 20a. The comparison shows that the average inaccuracies with the statistical mean is 3.08%, and the maximum inaccuracies is 5.79%, which is within the range of statistical variance, indicating that the model can accurately calculate the maximum depths of craters with different spot diameters. The spot diameter–diameter relative deviation Equation (10) was used to calculate the crater diameter for different spot diameters and compare it with the measured statistical results shown in Figure 20b. The average inaccuracies are 0.03%, the maximum inaccuracies are 0.07% from the statistical mean, and the corrected crater diameter has a deviation of less than 1 μm from the statistical mean, which is much smaller than the statistical variance, indicating that the model can be used to accurately calculate the crater diameter.

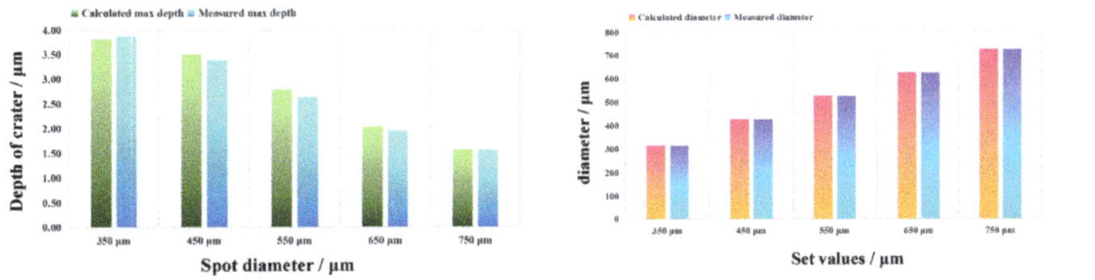

(**a**) (**b**)

Figure 20. Calculated LSPTwC texture feature parameters with different spot diameters: (**a**) max. depth and (**b**) spot diameter.

To make the calculation more accurate, the 350 μm peening crater with the best surface integrity was selected as the calibration base in this paper. Several groups of 350 μm crater profile data were collected for statistics, and the mean curve was obtained as calibration data, as shown in Figure 19. Based on the profile coordinate calculation Equation (11) and the profile depth calculation Equation (12), the profiles of 450 μm, 550 μm, 650 μm, and 750 μm were jointly calculated on the basis of the calibration data, as shown in Figure 21. By comparing the average value of the measured profile, the calculated profile has good consistency, as shown in Figure 22. In addition, under the condition of low energy density, the measured profiles also show irregular fluctuations of the internal profiles of the crater, which is consistent with the OM and SEM characterization results, as shown in Figure 23, the higher the energy, the better the consistency, that indicates that maintaining a higher energy density is conducive to obtaining more regular and controllable crater textures.

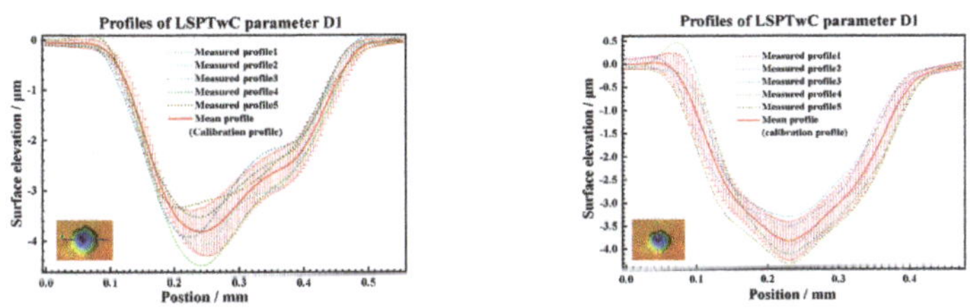

(**a**) (**b**)

Figure 21. Calibration profiles data of (**a**) symmetrical section and (**b**) vertical section.

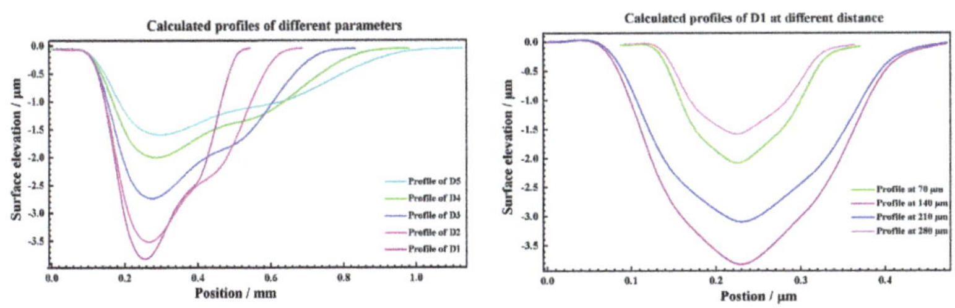

(**a**) (**b**)

Figure 22. Calculated profile data of (**a**) symmetrical section and (**b**) vertical section.

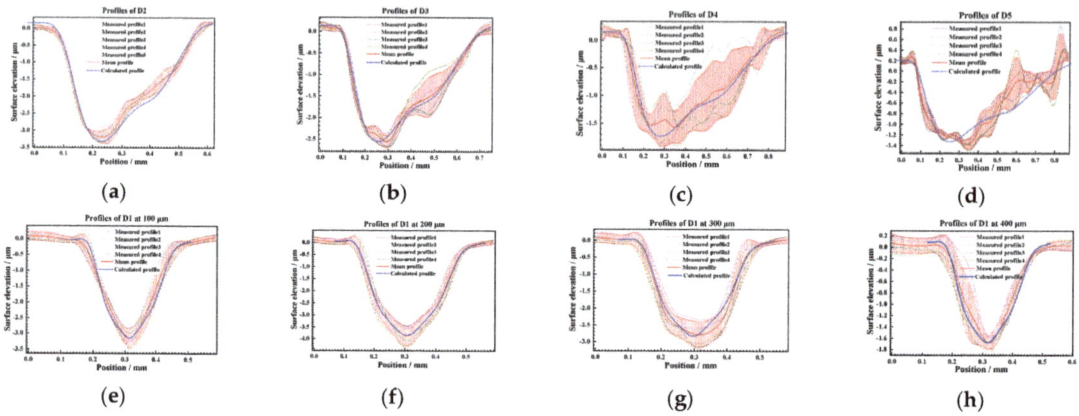

Figure 23. Comparison of measured and calculated profiles: (**a**–**d**) symmetrical section and (**e**–**h**) vertical section.

In conclusion, the calculation method of arbitrary laser spot diameter LSPTwC texture is as follows:

(1) The maximum depth and diameter are calibrated according to the test data.

(2) According to the laser energy, the maximum depth and coordinate correction are calculated using Equations (9) and (10).

(3) Calculate 2D profile data according to Equations (11) and (12).

(4) Use the same method to calibrate the reference profile in the vertical plane, then calculate the profile's maximum depth in the vertical plane, coordinate correction, and other characteristic parameters according to the profile data in the symmetrical plane.

(5) Finally, Equations (9) and (10) were used to calculate all the profile data in the vertical plane, and an LSPTwC textured 3D surface was constructed.

Through the above methods, using the 350 μm crater as the calibration data, LSPTwC crater textures of 450 μm, 550 μm, 650 μm, and 750 μm were successfully constructed, as shown in Figure 24. Compared with the experimental results, the calculated texture structures not only have the same main features but also can eliminate the texture defects of remelting accumulation. They can also be used for the design of LSPTwC texture structures and the calculation of process parameters.

Further attempts were made to construct texture models of 90 mJ, 75 mJ, 60 mJ, 45 mJ, and 30 mJ peening craters with a 500 μm spot. The maximum depth and diameter are shown in Figure 25. The average deviation of the maximum depth of the crater is 3.94%, and the maximum deviation is 9.67%, which is within the range of statistical variance. The average diameter deviation of the crater was 9 μm, which is less than the diameter calibration accuracy of the LSP instrument. The maximum deviation was 32 μm, which was less than the statistical variance. Then, the profiles of 90 mJ, 75 mJ, 60 mJ, 45 mJ, and 30 mJ under the condition of a 500 μm spot were calculated using the above profile coordinate calculation Equation (8) and the profile depth calculation Equation (9) as Figure 26. The symmetric section profiles of the craters with different energies are shown in Figure 27. Compared with the measured statistical data, the calculated profiles are in good agreement with the measured profiles. The vertical profile of the craters at different energies is shown in Figure 28. Compared with the measured statistical data, the calculated profiles also agree with the measured profile.

As shown in Figure 29, the LSPTwC crater textures of 90 mJ, 75 mJ, 60 mJ, 45 mJ, and 30 mJ were successfully constructed, as shown in Figure 29a. The shape and elevation distribution of LSPTwC textures with different energies calculated and constructed by this method still agree with the measured results, as shown in Figure 29b,c.

In summary, the above method can satisfy the calculation of spot diameter and energy regulation at the same time and can accurately design the surface texture of LSPTwC with different spot diameters and different laser energy, which can provide a reliable calculation method and model for the surface texture design, simulation, and process parameter calculation of a single-point LSPTwC scheme.

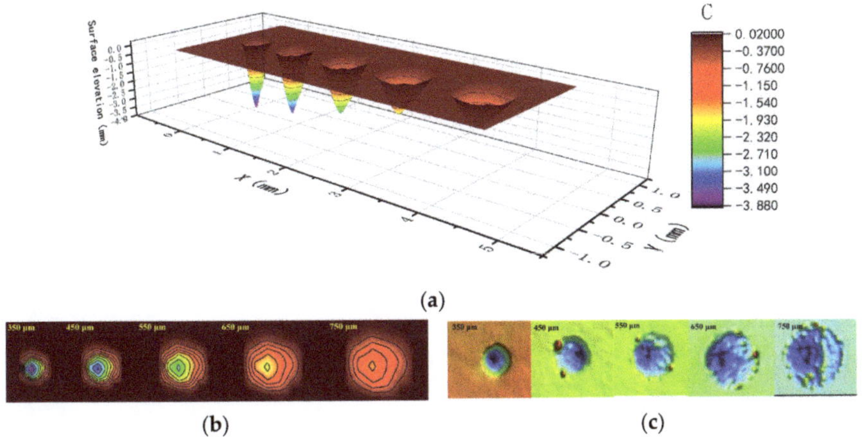

Figure 24. Comparison of calculated and measured textures of different diameters: (**a**) calculated 3D textures, (**b**) calculated 2D textures, and (**c**) measured 3D textures of overlooking.

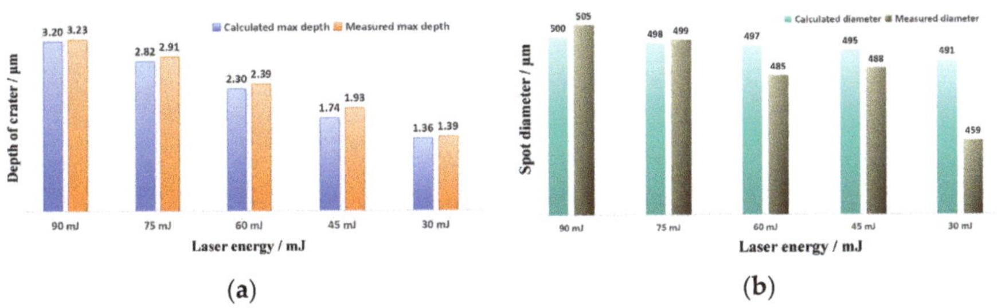

Figure 25. Calculation of the LSPTwC texture feature parameters with different laser energies: (**a**) max. depth and (**b**) spot diameter.

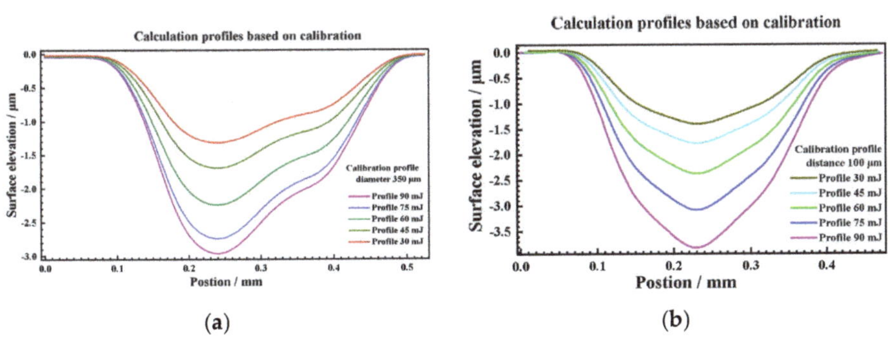

Figure 26. Calculated profiles data of (**a**) symmetrical section, (**b**) vertical section.

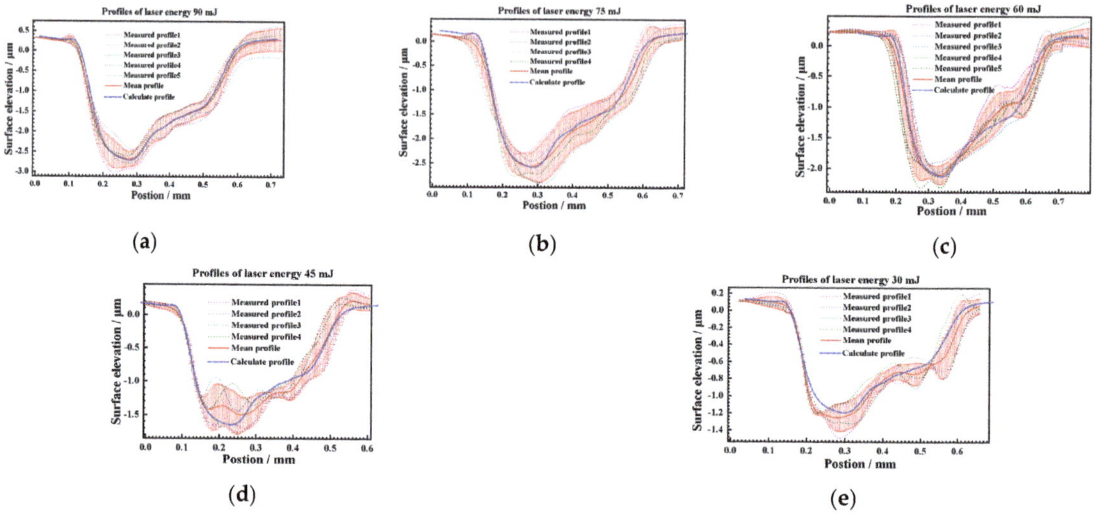

Figure 27. Comparison of calculated and measured profiles of different energies as. (**a**) 90 mJ, (**b**) 75 mJ, (**c**) 60 mJ, (**d**) 45 mJ, and (**e**) 30 mJ at symmetrical section.

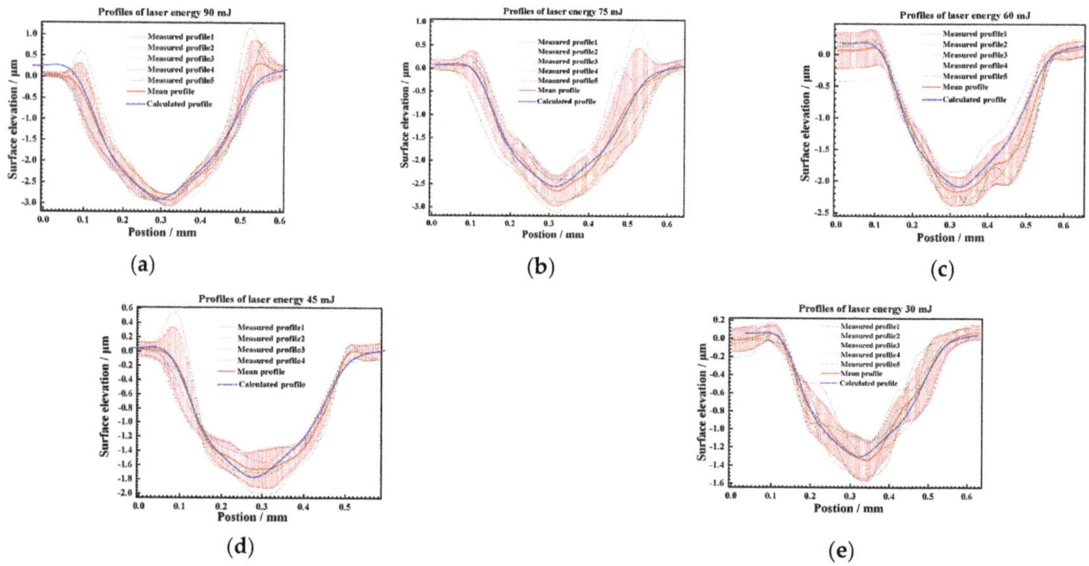

Figure 28. Comparison of calculated and measured profiles of different energies as. (**a**) 90 mJ, (**b**) 75 mJ, (**c**) 60 mJ, (**d**) 45 mJ, and (**e**) 30 mJ at vertical section.

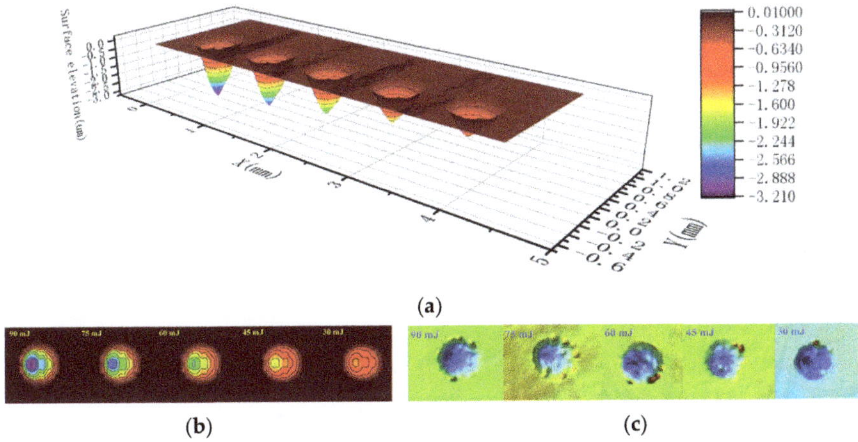

Figure 29. Comparison of calculated and measured textures of different energies: (**a**) calculated 3D textures, (**b**) calculated 2D textures, and (**c**) measured 3D textures of overlooking.

4. Conclusions

In this paper, the first LSPTwC scheme using LSPwC to achieve controllable and designable LST is proposed, and an LSPTwC study is carried out on AISI 9310 steel, which is commonly used in helicopter transmission components in response to the need for LOL-resistant texture on the surface of helicopter transmission components. The study not only verified the feasibility of LSPTwC and proved the combined force and thermal strengthening effect of LSPTwC but also proposed the LSPTwC formation modeling method, constructed the calculation model, and carried out experimental verification; this has never been done before [16,17]. The main conclusions of the whole study are as follows:

(1) The spot diameter, energy density, and distribution can precisely control the macro-structural parameters of the LSPTwC crater texture and the micro-texture structure inside the crater. Therefore, LSPwC can be fully applied to controllable and designable laser surface texturing, i.e., the innovative LSPTwC technology proposed in this paper is feasible.

(2) LSPTwC also has both physical and chemical effects, such as ablation, impact, remelting, and heat treatment, which can change the chemical composition, microstructure, metallographic structure, etc., of the surface material, thus forming a strengthening effect. For AISI 9310 steel, LSPTwC increases the microhardness by more than 10% with only 9.6% coverage and produces a gradient structure and a reticulated carburic organization that extends to the surface, demonstrating the technological advantages of texturizing and strengthening.

(3) Based on the fitting of the experimental data, the characteristic parameters of the LSPTwC crater texture can be identified more accurately, and a higher-accuracy computational model can be constructed. Based on the calibrated profile curves with higher energy density, the textured surface models with different LSPTwC parameters can be constructed more accurately with a diameter deviation of <3% and a depth deviation of <4%. The above results can provide a design method and calculation model for single-point LSPwC crater texturing and its process parameter design of AISI 9310 steel parts for helicopters, which has engineering value.

Author Contributions: Conceptualization, P.L. and L.Z.; methodology, P.L.; software, Z.Y.; validation, L.Z., Z.Y. and P.L.; formal analysis, F.Y., C.X. and P.L.; investigation, C.X., Z.Y., F.Y. and P.L.; resources, L.Z.; data curation, P.L.; writing—original draft preparation, P.L., F.Y. and Z.Y.; writing—review and editing, P.L., C.X.; visualization, P.L.; supervision, P.L. and L.Z.; project administration, L.Z.;

funding acquisition, L.Z. and P.L. All authors have read and agreed to the published version of the manuscript.

Funding: This research was funded by the "National Key R&D Plan of China, Grant number 2022YFB4601703" and the "Natural Science Foundation of Chongqing Municipality, grant number cstc2020jcyj-msxmX1040" and "the Science and Technology Research Program of Chongqing Municipal Education Commission, Grant number KJQN202000721".

Institutional Review Board Statement: Not applicable.

Informed Consent Statement: Not applicable.

Data Availability Statement: The data used to support the findings of this study are available from the corresponding author upon request.

Acknowledgments: Thanks to the Xi'an Tianruida Optoelectronic Technology Co. for their co-operation, providing LSP equipment and technical guidance for our research, and also thanks to the co-operation of the Chongqing College of Arts and Sciences for providing experimental equipment and technical guidance for our research.

Conflicts of Interest: The authors declare no conflicts of interest.

Nomenclature

ω	coefficient of maximum width
A	coefficient of maximum depth
A_1, A_2	maximum depth regulation interval
y	calculated elevation of profile (μm)
y_0	original elevation of surface (μm)
x	calculated position (μm)
x_c	polar transverse coordinates (μm)
Ra	surface roughness (μm)
$\updownarrow$	calculated length (mm)
Z	elevation of rough peaks and valleys (μm)
E_d	laser energy density (mJ.mm^{-2})
E	laser energy (mJ)
d	spot diameter (μm)
d_{cn}, d_{cp}	width relative to the vertex of the calibrated profile (μm)
Δd_{cn}, Δd_{cp}	width increment relative to the vertex of the calibrated profile (μm)
d_{max}, d_{cmax}	maximum depth of calculated and calibrated profile (μm)
d_c	depth of calibrated profile (μm)
D1~D5	LSPTwC parameter of different spot diameters

Abbreviations

LST	laser surface texturing
LSP	laser shock peening
LSPwC	laser shock peening without coating
LSPTwC	laser surface texturing by laser shock peening without coating
LOL	loss of lubrication
OM	optical microscope
SEM	scanning electron microscope
AFM	atomic force microscope

References

1. Liu, H.; Liu, H.; Zhu, C.; Parker, R.G. Effects of Lubrication on Gear Performance: A Review. *Mech. Mach. Theory* **2020**, *145*, 103701. [CrossRef]
2. Hu, B.; Zhou, C.; Wang, H.; Chen, S. Nonlinear Tribo-Dynamic Model and Experimental Verification of a Spur Gear Drive under Loss-of-Lubrication Condition. *Mech. Syst. Signal Process.* **2021**, *153*, 107509. [CrossRef]
3. Lu, F.; Mao, S.; Cao, X.; Bao, H.; Zhu, R. Multi-Time Scale Thermal Tribological Dynamics Coupling Model of the Helicopter Intermediate Reducer under Severe Loss of Lubrication. *Simul. Model. Pract. Theory* **2024**, *132*, 102884. [CrossRef]

4. Rashid, H.S.J.; Place, C.S.; Mba, D.; Keong, R.L.C.; Healey, A.; Kleine-Beek, W.; Romano, M. Reliability Model for Helicopter Main Gearbox Lubrication System Using Influence Diagrams. *Reliab. Eng. Syst. Saf.* **2015**, *139*, 50–57. [CrossRef]
5. Rashid, H.S.J.; Place, C.S.; Mba, D.; Lim, R.; Healey, A.; Kleine-Beek, W.; Romano, M. Helicopter MGB Oil System Failure Analysis Using Influence Diagrams and Random Failure Probabilities. *Eng. Fail. Anal.* **2015**, *50*, 7–19. [CrossRef]
6. He, C.; Yang, S.; Zheng, M. Analysis of Synergistic Friction Reduction Effect on Micro-Textured Cemented Carbide Surface by Laser Processing. *Opt. Laser Technol.* **2022**, *155*, 108343. [CrossRef]
7. Podgornik, B.; Vilhena, L.M.; Sedlaček, M.; Rek, Z.; Žun, I. Effectiveness and Design of Surface Texturing for Different Lubrication Regimes. *Meccanica* **2012**, *47*, 1613–1622. [CrossRef]
8. Morhard, B.; Lohner, T.; Stahl, K. Influence of Surface and Material Technologies on the Loss of Lubrication Performance of Gears. *J. Mater. Sci. Mater. Eng.* **2024**, *19*, 1. [CrossRef]
9. Hsu, S.M.; Jing, Y.; Zhao, F. Self-Adaptive Surface Texture Design for Friction Reduction across the Lubrication Regimes. *Surf. Topogr. Metrol. Prop.* **2015**, *4*, 014004. [CrossRef]
10. Zhang, S.; Zeng, X.; Igartua, A.; Rodriguez-Vidal, E.; Van Der Heide, E. Texture Design for Reducing Tactile Friction Independent of Sliding Orientation on Stainless Steel Sheet. *Tribol. Lett.* **2017**, *65*, 89. [CrossRef]
11. Li, Z.; Li, J.; An, B.; Li, R. The Design Method for Surface Texture of Sliding Friction Pairs Based on Machine Learning under Mixed Lubrication. *Tribol. Int.* **2024**, *194*, 109563. [CrossRef]
12. Vishnoi, M.; Kumar, P.; Murtaza, Q. Surface Texturing Techniques to Enhance Tribological Performance: A Review. *Surf. Interfaces* **2021**, *27*, 101463. [CrossRef]
13. Sano, Y. Laser Peening without Coating as a Surface Enhancement Technology. *JLMN* **2006**, *1*, 161–166. [CrossRef]
14. Trdan, U.; Porro, J.A.; Ocaña, J.L.; Grum, J. Laser Shock Peening without Absorbent Coating (LSPwC) Effect on 3D Surface Topography and Mechanical Properties of 6082-T651 Al Alloy. *Surf. Coat. Technol.* **2012**, *208*, 109–116. [CrossRef]
15. Wang, H.; Keller, S.; Chang, Y.; Kashaev, N.; Yan, K.; Gurevich, E.L.; Ostendorf, A. Effect of Laser Shock Peening without Protective Coating on the Surface Mechanical Properties of NiTi Alloy. *J. Alloys Compd.* **2022**, *896*, 163011. [CrossRef]
16. Li, X.; Zhou, L.; Zhao, T.; Pan, X.; Liu, P. Research on Wear Resistance of AISI 9310 Steel with Micro-Laser Shock Peening. *Metals* **2022**, *12*, 2157. [CrossRef]
17. Yang, F.; Zhao, T.; He, P.; Zhou, L.; Pan, X.; Liang, X.; Jia, W.; An, Z.; Liu, P.; Zhang, H. Effect of Pulsed Femtosecond Laser Shock Peening Surface Modification on Anti-Wear Failure Properties of AISI 9310 Gear Steel. *Eng. Fail. Anal.* **2024**, *159*, 108146. [CrossRef]
18. Song, J.; Luo, S.; Liang, X.; Cao, Z.; Zhao, W.; Pu, C.; He, W. Rolling Contact Fatigue and Damage Characteristic of AISI 9310 Steel with Pre-Laser Shock Peening Treatment. *Int. J. Fatigue* **2022**, *155*, 106588. [CrossRef]
19. Yan, H.; Wei, P.; Su, L.; Liu, H.; Wei, D.; Zhang, X.; Deng, G. Rolling-Sliding Contact Fatigue Failure and Associated Evolutions of Microstructure, Crystallographic Orientation and Residual Stress of AISI 9310 Gear Steel. *Int. J. Fatigue* **2023**, *170*, 107511. [CrossRef]
20. Hereñú, S.; Strubbia, R.; Rubio-González, C.; Spadaro, L.; Bolmaro, R.; Gomez-Rosas, G. High Cycle Fatigue Life Improvement of Superferritic Stainless Steel by Laser Shock Peening without Coating. *Opt. Laser Technol.* **2022**, *152*, 108083. [CrossRef]
21. Praveen, T.R.; Shivananda Nayaka, H.; Swaroop, S.; Gopi, K.R. Strength Enhancement of Magnesium Alloy through Equal Channel Angular Pressing and Laser Shock Peening. *Appl. Surf. Sci.* **2020**, *512*, 145755. [CrossRef]
22. Zhao, W.; He, W.; Zhu, D.; Liang, X.; Pang, Z.; Song, J.; Luo, S. Comparative Investigation on Microstructure and Mechanical Properties of GH4169 Superalloy after Laser Shock Peening with and without Coating. *J. Mater. Res. Technol.* **2024**, *29*, 276–285. [CrossRef]
23. Yu, Y.; Ma, T.; Gong, J.; Chen, T.; Zhou, L.; Cai, Z. Effect of Laser Shock Peening without Coating on Fretting Wear Behavior of GH4169 Superalloy at High-Temperature. *Wear* **2024**, *546*, 205349. [CrossRef]
24. Tong, Z.; Jiao, J.; Zhou, W.; Yang, Y.; Chen, L.; Liu, H.; Sun, Y.; Ren, X. Improvement in Cavitation Erosion Resistance of AA5083 Aluminium Alloy by Laser Shock Processing. *Surf. Coat. Technol.* **2019**, *377*, 124799. [CrossRef]
25. Tong, Z.; Zhang, Y.; Chen, J.; Xin, Z.; Zhou, W.; Ye, Y.; Ren, X. Dual Protection against Electrochemical Corrosion for AA5083 Al Alloys: Synergistic Action of Superhydrophobicity and Strain-Strengthening. *Surf. Coat. Technol.* **2023**, *472*, 129945. [CrossRef]
26. Radil, K.; Berkebile, S. Failure Progression of Spur Gears during a Simulated Loss-of-Lubrication Event in a Rotorcraft Drive System. *Tribol. Trans.* **2020**, *63*, 718–725. [CrossRef]
27. Wells, M.G.H. Advances in Steels for Aerospace Applications. *Key Eng. Mater.* **1992**, *77*, 71–80. [CrossRef]
28. Sebastian, J.; Grabowski, J.; Snyder, D. New High Performance Gear Steels for Rotorcraft Transmission Applications (Ferrium® C61™ and Ferrium C64™). In *Proceedings of the Volume 5: 25th International Conference on Design Theory and Methodology; ASME 2013 Power Transmission and Gearing Conference*; American Society of Mechanical Engineers: Portland, OR, USA, 2013; p. V005T11A009. [CrossRef]
29. Meng, S.; Yu, Y.; Wang, L.; Zhou, L.; Yu, Z.; Pan, X.; Wang, R.; Xie, C. Improving Tension Fatigue Performance of Gray Cast Iron by LSPwC-Induced Gradient Structure and Carbon Diffusion. *J. Mater. Res. Technol.* **2024**, *31*, 3580–3590. [CrossRef]
30. Jiang, J.; Zhao, W.; Sun, T.; Xie, L. Enhancing the Wear Resistance of PCD Tools in Cutting Cf/SiC Materials through Low-Energy Laser Shock Peening. *Surf. Coat. Technol.* **2024**, *486*, 130951. [CrossRef]
31. Dang, X.; Liang, X.; Luo, S.; Li, Y.; Jiao, Y.; Tian, Z.; He, W. Surface Strengthening and Fatigue Life Improvement of Single Crystal Ni-Based Superalloys via Laser Shock Peening without Coating. *Mater. Des.* **2023**, *232*, 112097. [CrossRef]

32. Xin, B.; Zhang, A.; Han, J.; Meng, J. The Tribological Properties of Carbon Doped Al0.2Co1.5CrFeNi1.5Ti0.5 High Entropy Alloys. *Wear* **2021**, *484–485*, 204045. [CrossRef]
33. Gangopadhyay, A. Mechanical and tribological properties of amorphous carbon films. *Tribol. Lett.* **1998**, *5*, 25–39. [CrossRef]
34. Deng, W.; Wang, C.; Lu, H.; Meng, X.; Wang, Z.; Lv, J.; Luo, K.; Lu, J. Progressive Developments, Challenges and Future Trends in Laser Shock Peening of Metallic Materials and Alloys: A Comprehensive Review. *Int. J. Mach. Tools Manuf.* **2023**, *191*, 104061. [CrossRef]
35. Ke, R.; Hu, C.; Zhong, M.; Wan, X.; Wu, K. Grain Refinement Strengthening Mechanism of an Austenitic Stainless Steel: Critically Analyze the Impacts of Grain Interior and Grain Boundary. *J. Mater. Res. Technol.* **2022**, *17*, 2999–3012. [CrossRef]
36. Zhang, K.; Liu, P.; Li, W.; Ma, F.; Guo, Z.; Rong, Y. Enhancement of the Strength and Ductility of Martensitic Steels by Carbon. *Mater. Sci. Eng. A* **2018**, *716*, 87–91. [CrossRef]
37. Kučerová, L.; Jirková, H.; Mašek, B. The Effect of Alloying on Mechanical Properties of Advanced High Strength Steels. *Arch. Metall. Mater.* **2014**, *59*, 1189–1192. [CrossRef]
38. Zhang, Z.; Qiu, W.; Zhang, G.; Liu, D.; Wang, P. Progress in Applications of Shockwave Induced by Short Pulsed Laser on Surface Processing. *Opt. Laser Technol.* **2023**, *157*, 108760. [CrossRef]
39. Zhou, W.; Ren, X.; Yang, Y.; Tong, Z.; Asuako Larson, E. Finite Element Analysis of Laser Shock Peening Induced Near-Surface Deformation in Engineering Metals. *Opt. Laser Technol.* **2019**, *119*, 105608. [CrossRef]

Journal of
Manufacturing and Materials Processing

MDPI

Article

Creation of Tool Coatings Based on Titanium Diboride for Highly Efficient Milling of Chromium–Nickel Alloys

Sergey N. Grigoriev, Marina A. Volosova *, Sergey V. Fedorov, Artem P. Mitrofanov, Vladimir D. Gurin and Anna A. Okunkova

Department of High-Efficiency Processing Technologies, Moscow State University of Technology STANKIN, Vadkovskiy per. 1, 127994 Moscow, Russia; s.grigoriev@stankin.ru (S.N.G.); sv.fedorov@icloud.com (S.V.F.); a.mitrofanov@stankin.ru (A.P.M.); v.gurin@stankin.ru (V.D.G.); a.okunkova@stankin.ru (A.A.O.)
* Correspondence: m.volosova@stankin.ru; Tel.: +7-499-972-94-29

Abstract: This paper describes the principles of obtaining wear-resistant coatings based on titanium diboride that are deposited on the cutting tool for use in the machining of chromium–nickel alloys. The spark plasma sintering of samples from the TiB_2/Ti powder composition was studied, and the influence of sintering modes on the characteristics of the ceramic targets was analyzed. The regularities of the magnetron sputtering of sintered targets were revealed. The dependences of the physical and mechanical properties of coatings formed on hard alloy substrates on deposition conditions were established. The wear resistance of carbide samples with TiB_2-based coatings under friction-sliding conditions and coated carbide ball-end mills in milling Inconel 718 chromium–nickel alloy that is widely used in the industry was assessed.

Keywords: chromium–nickel alloy; end mills; milling; PVD; spark plasma sintering; TiB_2 coatings; tribological tests; wear resistance

1. Introduction

The volume of structural heat-resistant chromium–nickel alloys in aircraft engineering is constantly increasing because of their excellent performance properties. Moreover, these alloys are characterized by poor cutting machinability, associated with low thermal conductivity and increased chemical adhesion with hard alloys for tool purposes, which causes intense wear of cutting tools [1–3]. The problem of ensuring the high wear resistance of cutting tools when machining chromium–nickel alloys is especially relevant in the manufacture of critical aircraft parts with complex-shaped surfaces, such as monowheels, turbine blades, combustor shells, turbine disks and shafts machined on multi-axis CNC machines. When machining such parts, it is necessary to ensure high cutting productivity (cutting speed) on the one hand and to preserve the cutting properties of the cutter throughout the entire machining cycle on the other hand. Today, wear-resistant coatings based on refractory metal nitrides see widespread industrial use, increasing the wear resistance of a wide range of carbide tools in cutting structural materials, particularly chromium–nickel alloys. Their use is aimed at increasing the microhardness of the contact areas of tools, ensuring their physical and chemical passivity, and reducing adhesive binding with chromium–nickel alloys under machining, which together can increase the wear resistance of tools when cutting at high speeds [4–7].

Over the past decade, one of the main directions in developing nitride coatings for tool purposes has been the improvement of their chemical composition and structure. A typical example is multicomponent coatings, such as TiAlCrSiYN and AlCrN-bSiTiN [8], and multilayer coatings, including various oxide, diamond-like and other layers in addition to nitride compounds such as TiN/AlTiN/AlCrN/AlCrO/AlTiCrN, CrN/TiCrN/DLC/Cr-DLC, etc. [9,10]. This approach involves increasing the complexity of the coating compositions and structure and has largely exhausted itself. In addition, the use of such coatings is accompanied by a significant increase in technological costs. Thus, the developed coatings often remain at the laboratory level. One of the leading global trends in the development of wear-resistant coatings for tool production is using transition metal borides of Ti and Zr (their most thermally stable modification is MeB_2) [11,12]. Diborides have higher hardness and less intense adhesive bonding with the counterbody during high-temperature friction than metal nitrides [13,14]. TiB_2 coatings can potentially minimize the frictional interaction between the tool and the materials to be machined in cutting due to protective compounds and tribofilms forming during high-temperature exposure. For example, the authors of [15] studied the wear characteristics of carbide inserts with TiB_2 coatings in turning Ti6Al4V titanium alloy. They found that these coatings form B_2O_3 tribofilms in cutting, which increase the lubricity of the coatings and reduce friction at the "tool–chip" interface, minimizing adhesion with the workpiece material. However, diboride coatings have their drawbacks, such as their quite high fragility and the relatively low strength of their adhesive bonds with tool substrates, which can be compensated by the formation of intermediate coatings, such as TiN, (CrAlSi)N, etc., before TiB_2 coating deposition [16–18].

Despite a relatively long period of laboratory research on TiB_2 coatings, the potential capabilities of which were investigated in 2000s [19,20], examples of real industrial applications of cutting tools with TiB_2 coatings have been sporadic. For example, the company Ceratizit S.A. (Mamer, Luxembourg) produces some standard sizes of milling inserts made of the CTCS245 hard alloy with TiB_2 coatings ~2.5 μm thick, having a TiN subcoating, deposited by chemical vapor deposition (CVD). Another example of the industrial application of TiB_2 coating technology is the production of vacuum plasma units of the PLATIT 411 series by Platit AG (Selzach, Switzerland). The units are equipped with a central rotating ceramic target for non-reactive magnetron sputtering (MS-PVD), allowing the deposition of TiB_2 coatings with a thickness of ~1.8 μm in an argon environment on cutting tools with a pre-formed intermediate (CrAlSi)N coating.

Magnetron sputtering is carried out at relatively low temperatures and does not involve the use of toxic reagents. Thus, this is the most suitable method for the formation of TiB_2 coatings. Therefore, the share of this method currently accounts for more than a third of all published research in the field of diboride coating deposition for various functional purposes. When using magnetron sputtering as a tool for the deposition of coatings based on refractory metal diborides, the most important role is played by the ceramic targets that are used in processing, the quality and composition of which directly determine the properties of the deposited coatings. Therefore, the issues of obtaining high-quality sputtering targets from powder and the deposition of coatings with improved properties are inextricably linked. The most promising methods of producing ceramic workpieces today include the processes of self-propagating high-temperature synthesis of composites [21], spark plasma sintering [22] and electro-thermal explosion under pressure [23]. These processes of obtaining targets from diboride powders are currently insufficiently studied, which hinders their practical application in the industry.

Spark plasma sintering is a relatively recently developed but already well-proven method of vacuum sintering powder placed in graphite dies under mechanical force and current pulses. Special electrodes supply electric current to the dies and create spark discharges between the sintered particles, ensuring heating to high temperatures and their binding. This process is attractive for manufacturing workpieces from difficult-to-sinter ceramics with high melting temperatures since they require high activation energy [24,25].

Within the frames of the current study, we aimed to solve the problem of studying the possibility of using spark plasma sintering to obtain high-density targets from TiB_2 powders and the subsequent deposition of wear-resistant coatings on carbide substrates by magnetron sputtering of sintered targets, and also investigate the effectiveness of using deposited TiB_2-based coatings in milling chromium–nickel alloys (using a case study of the alloy Inconel 718, which is common in the industry) with carbide end mills with a ball end, designed for machining complex shaped surfaces.

2. Materials and Methods

2.1. Technological Principles and Equipment for Spark Plasma Sintering of Targets from a Powder Composition Based on TiB_2

The spark plasma sintering method was used to produce a sputtered target. The technology consisted of sintering the initial powder under mechanical force and current pulses. TiB_2 powder (Plasmotherm, Moscow, Russia) with a purity of 99.8% and a density of 4.35 g/cm^3 was used to obtain targets. Figure 1 shows an SEM image of the TiB_2 powder and the results of its granulometric analysis using an Eyetech particle analyzer (AmbiValue B.V., Dussen, The Netherlands). The proportion of particles was 50% for a size of 1–10 μm and 30% for 10–16 μm. A total of 113.6 g of TiB_2 powder was used to sinter a disk-shaped target with a diameter of 80 mm and a height of 5 mm. Figure 2 shows an X-ray diffraction pattern of the original TiB_2 powder obtained on an Empyrean diffractometer (PANalytical, Almelo, The Netherlands). The resulting diffractogram of titanium diboride corresponds to pure TiB_2, as indicated in the electronic file. Ultrafine titanium powder with particle sizes of 1–5 μm in a quantity of 4 vol.% was added to the TiB_2 powder to reduce the sintered sample's porosity. It acted as a binding element during sintering, filling the gaps between the particles of the main TiB_2 component.

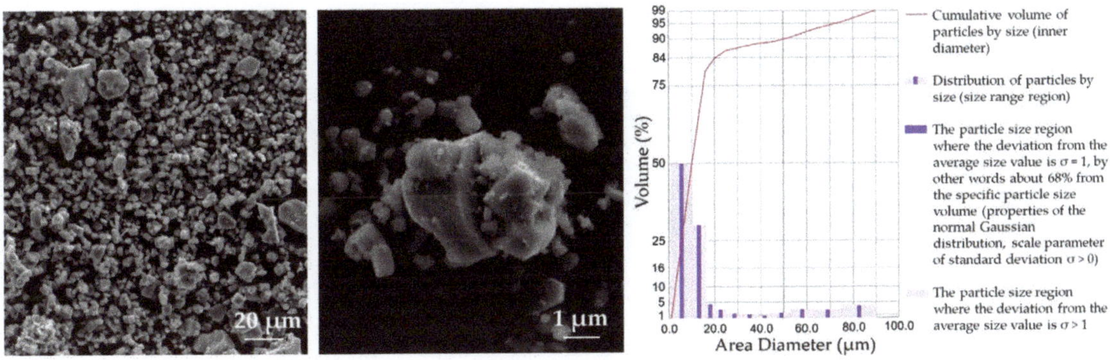

Figure 1. SEM image of TiB_2 powder (**a**) and volume fraction of TiB_2 particles of different sizes (**b**).

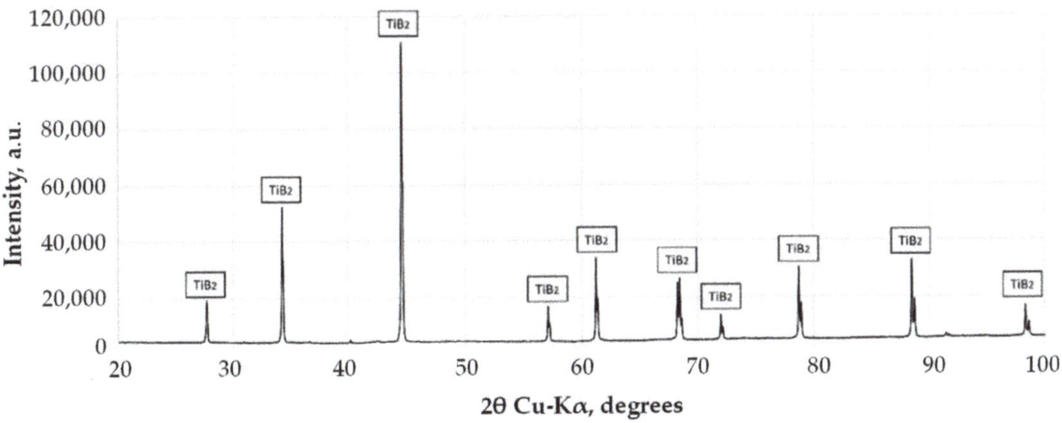

Figure 2. X-ray diffraction pattern of the initial TiB_2 powder used to produce targets by spark plasma sintering.

During sintering, the initial TiB_2/Ti powder was placed in a graphite mold under a press in the technological chamber of the unit under vacuum conditions. An H-HP D 25 SD Spark unit (FCT, Rauenstein, Germany) was used in the experiments. Electrodes are integrated into the mechanical part of the press. They supply electric current to the mold and create spark discharges between the TiB_2 particles to be sintered. The powder material in the impact zone is heated to high temperatures, and pulsed currents pass directly through the conductive TiB_2 + Ti powder and the mold, bypassing the entire volume of the chamber during the heating stage. This ensures the concentration of the thermal effect directly in the zone of the sintered target workpiece, which significantly reduces the required isothermal holding time and possible coagulation of powder particles using the approach compared with the widespread method of hot isostatic pressing [26].

The sequence of the main stages of sintering target blanks made of TiB_2 powder is shown in Figure 3. The process includes the preliminary assembly of a graphite mold (I), pouring sifted powder into the mold (II), the final assembly of the mold (III), the preliminary pressing of ceramic powder on a manual press at a pressure of 5 MPa (IV), setting up the mold in the sintering chamber and lifting it to the upper current supply (V) and the subsequent sintering of the powder and production of TiB_2 targets (VI). Information on the range of parameters varied during the spark plasma sintering of TiB_2 + Ti powder is provided in Table 1.

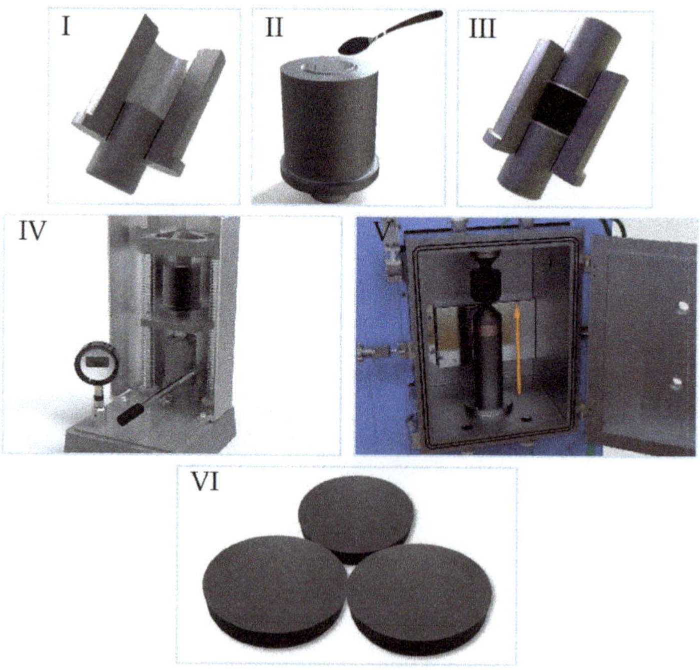

Figure 3. The sequence of the main stages of spark plasma sintering of targets of TiB_2 powder, where **(I)** is a graphite mold pre-assembly; **(II)** is sifted powder poured into the mold; **(III)** is the final mold assembly; **(IV)** is powder pre-pressing at 5 MPa; **(V)** is setting up the mold in the chamber and lifting it to the upper current supply; and **(VI)** is powder sintering for the production of TiB_2 targets.

The sintering was implemented as follows: a minimum pressure on the sample was maintained (43 MPa) until the sintering temperature of 300 °C was reached, and then the pressure and temperature were continuously increased to the required values with a heating rate of 100 °C/min. Before reaching the required maximum temperature, the last 100 °C was achieved at a heating rate of 25 °C/min. After sintering, the target blanks were subjected to finishing treatments, such as cleaning graphite foil residues on a grinding and polishing machine and the diamond grinding of horizontal surfaces on a grinding sharpening machine.

Table 1. Range of variable parameters in spark plasma sintering of targets made of TiB_2 powder.

Parameter	Value	Measuring Unit
Pre-pressing pressure	5	MPa
Sintering temperature	1500, 1600, 1700, 1800, 1900	°C
Sintering pressure	$5 \times 10^{-5}, 6.5 \times 10^{-5}, 8 \times 10^{-5}$	MPa
Isothermal holding time	3, 5, 10	min
Chamber operating pressure	5×10^{-6}	MPa

A set of studies was carried out to research the influence of process modes, such as temperature, pressure on the sintered material and isothermal holding time, on the characteristics of sintered materials (density and crack resistance) to identify rational conditions for sintering targets made of TiB_2 powder. The density of the samples was determined by hydrostatic weighing as a percentage of the theoretical density. The crack resistance was determined by microindentation with a Vickers pyramid for 10 s with a load of 98 N.

2.2. *Technological Principles and Equipment for Deposition of Coatings Based on TiB_2*

An experimental hybrid PVD setup STANKIN-APP3 (MSTU Stankin, Moscow, Russia) was used to study the deposition process of TiB_2 coatings. This setup allows subcoating deposition by vacuum-arc evaporation and TiB_2 coating by the magnetron sputtering of sintered targets in a single technological cycle. Figure 4 shows the general view and internal structure of the vacuum chamber of the experimental PVD setup.

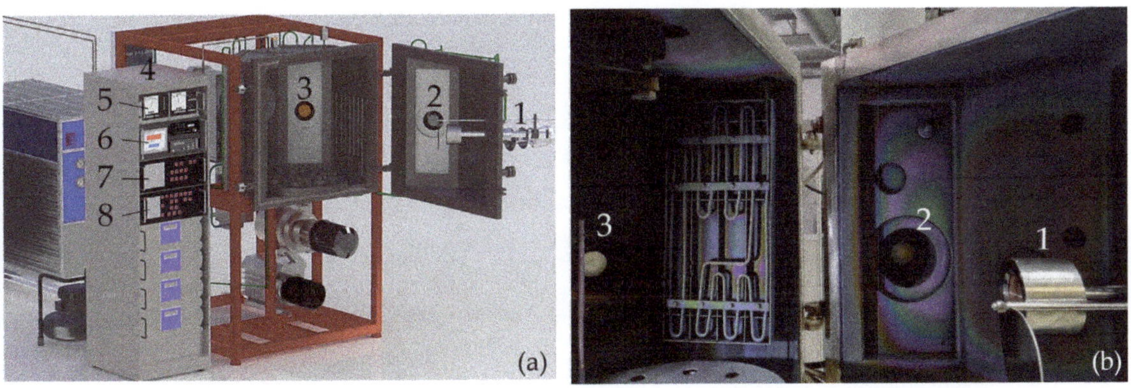

Figure 4. General view (**a**) and the vacuum chamber (**b**) of the experimental hybrid setup, equipped with a magnetron sputtering system with a sintered TiB_2 target (1) and vacuum-arc evaporators with AlSi (2) and Ti (3) cathodes, power cabinet (4), arc generators (TRUMPF Hüttinger GmbH + Co. KG, Freiburg im Breisgau, Germany), up to 4 pcs. (5), high-voltage unit arc cathode power supply ARM 25/6 (TRUMPF Hüttinger GmbH + Co. KG) (6), magnetron pulsed bias power supply PBP-26 (Applied Electronics Ltd., Tomsk, Russia) (7), pulsed power supply Apel-M-5PDC-650-2 (Applied Electronics Ltd.) (8).

A modern coating of three-component nitride (TiAlSi)N was used as a subcoating to improve the adhesion of the TiB_2 coating to carbide substrates. Today, this coating is recommended by cutting tool manufacturers for cutting difficult-to-cut alloys [27,28]. When choosing a coating design, we proceeded from the fact that by combining a nitride subcoating with a promising wear-resistant TiB_2 coating, it is possible to obtain an optimal system aimed at machining chromium–nickel alloys with an increased level of thermal and mechanical loads.

Before coating deposition, the surface layer of the carbide substrates that were mounted on a work table rotating at a frequency of 6 rpm was degassed for 30 min. The substrates were heated by tubular electric heaters to a temperature of 480 °C at a pressure of 0.03 Pa in the vacuum chamber, and argon was supplied through a multi-channel gas injection system. Subsequently, cleaning was carried out with gas and metal ions: cleaning for 20 min in a non-self-sustaining gas discharge at a pressure of 0.8 Pa in the vacuum chamber and a bias on the substrates of 650 V; cleaning with Ti ions for 20 min at an arc current of 90 A on a titanium cathode, with a pressure of 2.2 Pa and a bias of 750 V.

The (TiAlSi)N subcoating was deposited at a pressure of 1.2 Pa in a vacuum chamber in an Ar/N_2 gas mixture in a ratio of 5/95 vol% and a bias of 120 V on a rotating table with substrates. During the deposition, the temperature in the working chamber was 500 °C, and the arc current was 90 A on the titanium cathode and 80 A on the silumin cathode. The deposition time was 40 min, which ensured a subcoating thickness of ~2.8 μm. Information on the range of parameters that varied during the magnetron deposition of TiB_2 coatings is provided in Table 2.

Table 2. Range of variable parameters in magnetron sputtering of TiB_2-based targets and coating deposition.

Parameter	Value	Measuring Unit
Discharge bias	200, 300, 400, 500, 600	V
Argon pressure	0.2, 0.4, 0.6, 0.8, 1.0	Pa
Deposition time	30, 40, 50, 60, 70	min

The Berkovich pyramid nanoindentation method was used on a Nano Hardness Tester (CSM Instruments S.A., Peseux, Switzerland), with an indenter penetration depth of 380 nm and an applied load of 4.0 mN, to establish the effect of deposition modes on the hardness and elastic modulus of TiB_2 coatings. A Dektak 150 stylus profilometer (Bruker AXS, Billerica, MA, USA) was used to study the roughness of the coatings. The strength of the adhesion bond between the TiB_2 coating and hard alloy substrates was assessed using the Revetest RST3 scratch tester (Anton Paar GmbH, Graz, Austria) by the sclerometry method with recordings of the spectra of acoustic emission signals [29]. Tests were carried out at a linearly increasing load on the indenter (Rockwell diamond cone) from 0.1 to 40 N and a loading speed of 5 N/min. The characteristic levels of critical loads were determined based on the signal spectra: crack formation in the coating (L_{c1}) and delamination of the coating sections of the hard alloy substrate (L_{c2}). Tribological tests were carried out under heating conditions of up to 600 °C according to the "pin-on-disk" scheme on a Ducom POD-4.0 friction machine (Ducom Instruments Inc., Bohemia, NY, USA) with a load of 15 N, a rotation speed of 15 cm/s and a friction distance of 500 m. Inconel 718 chromium–nickel alloy was used as the material of the counterbody, in other words, as a pin, when a hard alloy substrate with various types of developed coatings was tested. The principal scheme of testing is provided in [30].

2.3. *Experimental Samples of End Mills and a Method for Wear Resistance Assessing in Cutting*

Carbide ball-end mills made of KFM 39 tungsten–cobalt hard alloy (WC—91 wt.%, Co—9 wt.%) were used as cutting tool samples, coated according to the technological principles described above. Figure 5 shows the design of the end mill samples used. Table 3 shows information about the design and geometric parameters. Four variants of cutting tools were used for testing: without coatings, with (TiAlSi)N and TiB_2 mono-coatings, and with a double-layer coating of "subcoating/external coating" structure of (TiAlSi)N/TiB_2 composition.

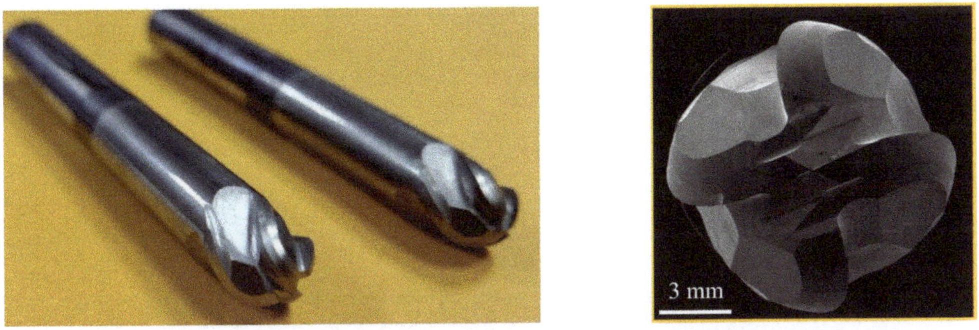

(a) (b)

Figure 5. General view (**a**) and end nose design (**b**) of carbide ball-end mills used for coating and subsequent testing in cutting Inconel 718 alloy.

Table 3. Design and geometric parameters of carbide ball-end mills.

Parameter	Value	Measuring Unit
Outer diameter	12	mm
Number of teeth	4	pcs.
Total length	75	mm
Cutting part length	30	mm
Tool material	KFM 39	-

The wear resistance of the end mills was assessed by cutting a workpiece made of the heat-resistant nickel alloy Inconel 718 (the chemical composition is provided in Table 4). Figure 6 shows the relative position of the end mill and the workpiece and the cutting movements during the experiments and the flank wear measurement technique, which is a quantitative assessment of the wear chamfer of the cutting tool's flank face over certain intervals until the critical wear criterion is reached (h = 300 μm). Data on the test conditions are provided in Table 5. The process was stopped after cutting for 5 min, and the flank wear on each end mill tooth was measured on an Axio Lab.A1 ZEISS binocular microscope (Carl Zeiss MicroImaging GmbH, Jena, Germany).

Table 4. Content of elements (%) in the Inconel 718 alloy for testing carbide end mills.

Element	Ni	Cr	Nb	Mo	Fe	Ti	Al	Co	Mn	Si	Rest
%	58.0	20.8	5.5	3.1	3.0	1.4	0.8	0.6	0.32	0.3	0.73

Table 5. Wear test conditions for carbide end mills.

Parameter	Value	Measuring Unit
Metal-working machine	CNC mill and turn center, CTX beta 1250 TC (DMG MORI Co., Ltd., Tokio, Japan)	-
Material and shape of the workpiece	Inconel 718, round rod	-
Workpiece diameter	60	mm
End mill rotation frequency	7000	min
End mill rotation speed	263.7	m/min
Workpiece rotation speed	1.75	m/min
Cutting feed	0.04	mm/tooth
Cutting depth	0.25	mm
Flank wear limit value (failure criterion)	300	μm

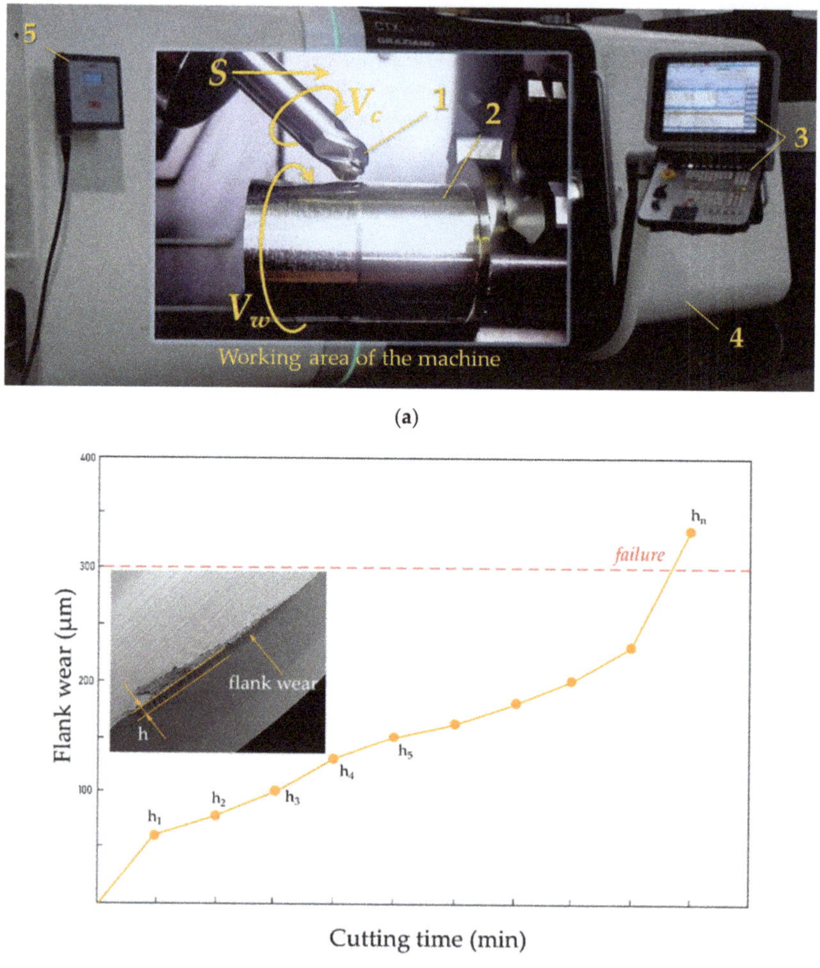

Figure 6. Methodology for testing carbide end mills on a CTX beta 1250 TC integrated mill and turn center when cutting a workpiece made of chromium–nickel alloy: (**a**): the scheme of the workpiece and cutting tool placement, where 1 is an end mill made of tungsten–cobalt hard alloy, and 2 is a workpiece made of chromium–nickel alloy, 3 is a computer numerical control system, 4 is a CNC machining center, 5 is a remote controller; (**b**) the flank wear measurement technique.

3. Results

3.1. Spark Plasma Sintering of Targets from a Powder Composition Based on TiB_2

The results of the experiments carried out with the spark plasma sintering modes showed that it is not possible to achieve the formation of pore-free high-density samples in the case of using pure TiB_2 powder: the sintered target blanks had numerous pores (Figure 7), and the density of the samples was no more than 84% (Table 6). It was possible to significantly reduce the porosity of the sintered samples and increase their relative density by adding Ti micropowder with particle sizes of up to 10 μm at 2, 4 and 6 vol.% to the TiB_2 powder (Table 6). In this case, the maximum density of the samples (~99%) after sintering was recorded with a Ti powder content of 4 vol.% in a powder composition based on TiB_2 and a sintering temperature of 1700 °C. A further increase in the content of Ti powder increased the porosity and contributed to a decrease in the density of the sintered samples. The significant compaction and reduction in porosity of the sintered samples established during the study of the SPS process due to the addition of spherical Ti powder to the TiB_2 powder can be explained by the fact that titanium particles act as a binding component that improves the interparticle binding of the solid phase the main component of TiB_2 and filling the pores during high-temperature sintering. A rather sharp decrease in the density of the target samples made of a powder composition based on TiB_2 is obviously associated with grain growth of the sintered material, as revealed during sintering in the 1700–1900 °C temperature range. Considering the obtained data, a powder composition of 96 vol.% TiB_2—4 vol.% Ti was used when sintering target blanks in subsequent studies. Figure 8 shows the results of the EDX analysis of a sintered target sample. The dosage of powder components to obtain one sample with a diameter of 80 mm and a thickness of 5 mm was established by calculation and amounted to 109.056 g of TiB_2 and 4.53 g of Ti.

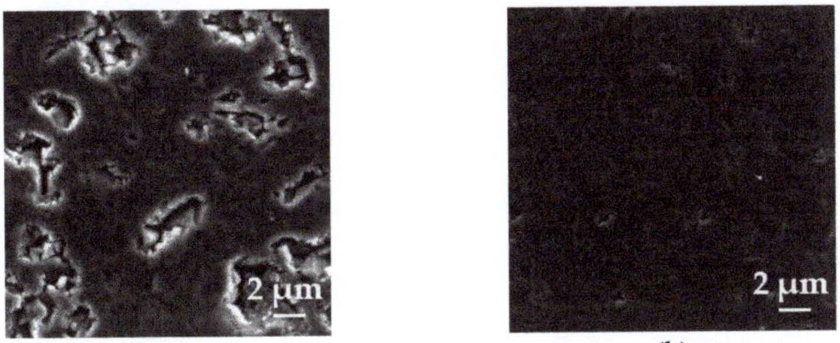

(a) (b)

Figure 7. SEM images of the microstructure of sintered samples produced from a powder composition based on TiB_2 without the addition of Ti (**a**), and with the addition of 4 vol.% Ti (**b**).

Table 6. The density of samples sintered with a powder composition based on TiB_2 with Ti powder additives at different temperatures.

Sintering Temperature, °C	Density (%) of Sintered TiB_2 Targets			
	0 vol.% Ti	2 vol.% Ti	4 vol.% Ti	6 vol.% Ti
1500	77	83	94	87
1600	81	85	97	89
1700	84	88	99	92
1800	82	87	98	90
1900	81	87	95	85

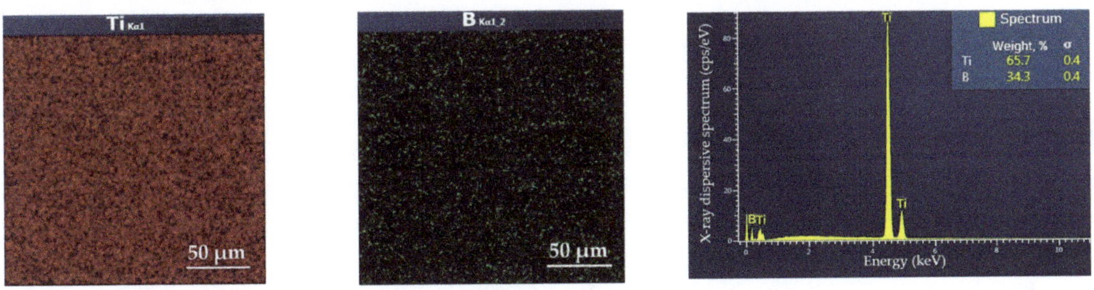

(a) (b) (c)

Figure 8. Elemental mapping (**a**,**b**) and elemental composition (**c**) obtained by EDX fracture analysis of a target sample sintered from a powder composition of 96 vol.% TiB_2—4 vol.% Ti.

Figure 9 shows the experimentally obtained dependences of the relative density and crack resistance of sintered TiB_2 targets on sintering temperature, sintering pressure and isothermal holding time. An increase in the sintering temperature in the range of 1500–1700 °C leads to a significant increase in the characteristics under study. In the range of 1700–1900 °C, a relatively sharp decrease is observed (especially for the density of the samples). The noted decrease is apparently associated with the phenomenon of some grain growth with increasing temperature. An increase in the pressure from 50 to 65 MPa significantly increases the density and crack resistance of the sintered material. It is definitely caused by a decrease in the volume and number of pores (shrinkage) in the sintered sample, which occurs with increasing pressure. With a further increase in pressure (from 65 to 80 MPa), the increase in sample density slows down somewhat, and the crack resistance remains virtually unchanged. With isothermal exposure for 3 min, the density of the sintered samples is about 95%, and then begins to increase sharply, reaching a maximum value (more than 99%) when held for 5 min. A subsequent increase in exposure to 10 min does not affect the change in the relative density of the samples. The performed studies made it possible to identify a rational combination of conditions for the spark plasma sintering of target blanks made of a powder composition of 96 vol.% TiB_2 + 4 vol.% Ti: the sintering temperature was 1700 °C, the pressure on the material was 80 MPa and the isothermal holding time was 5 min. Notably, the maximum density during sintering of a powder composed of 96 vol.% TiB_2—4 vol.% Ti was achieved in a short isothermal holding time (5 min), making the spark plasma process promising for industrial use.

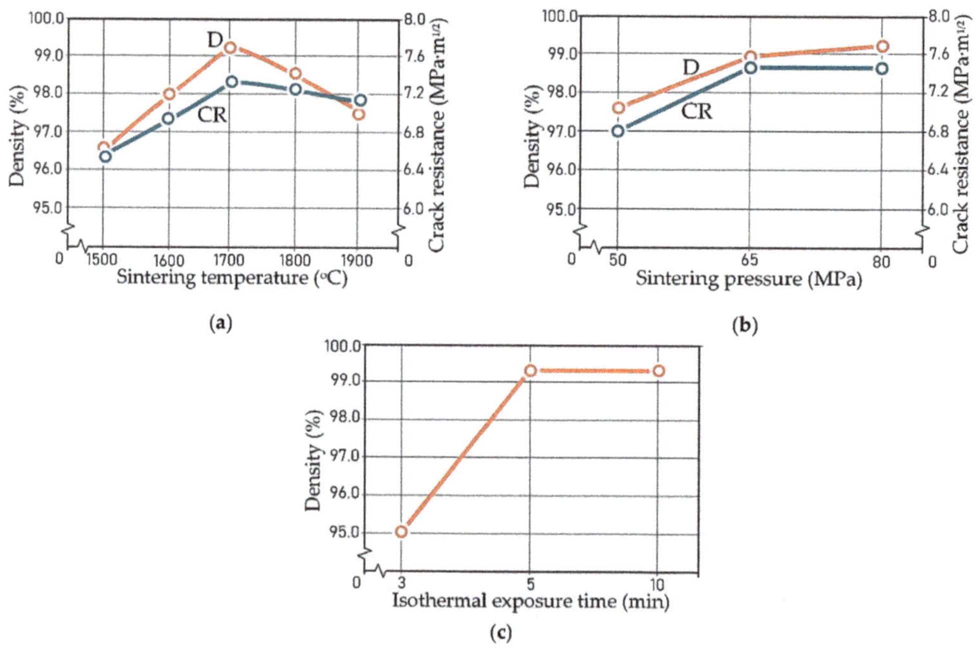

Figure 9. Dependences of the density (D) and crack resistance (CR) of TiB_2 targets on the sintering temperature (**a**), sintering pressure (**b**) and isothermal exposure time (**c**).

3.2. Deposition of TiB_2-Based Coatings on Carbide Substrates

Figure 10 shows an SEM image of a cross-section of a TiB_2 coating obtained by magnetron deposition on a hard alloy substrate and an X-ray diffraction pattern of the sample's surface layer. The deposited coating is characterized by a relatively dense columnar microstructure. The peaks of the TiB_2 and hard alloy substrate phases (WC and Co) are detected in the surface layer.

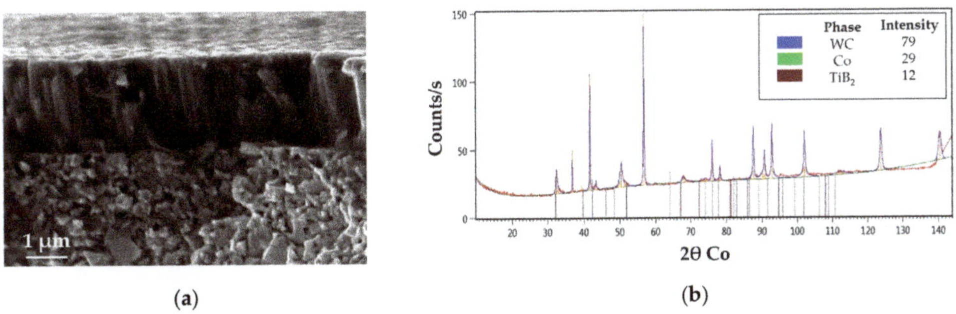

Figure 10. SEM image of the fracture cross-section of a TiB_2-coated hard alloy sample (**a**), and the X-ray diffraction pattern of the surface layer (**b**).

Figure 11 shows the experimentally obtained dependences of nanohardness, elastic modulus, roughness and thickness of TiB_2 coatings on the discharge bias, Ar pressure and deposition time during the magnetron sputtering of the target. It can be seen that the nanohardness and elastic modulus of TiB_2 coatings change extremely ambiguously depending on the discharge bias, varying within the ranges of 19–40 GPa and 380–560 GPa, respectively. The maximum ductility of coatings must be ensured, since coatings with increased hardness are simultaneously characterized by excessive rigidity, which sharply increases the tendency of brittle fracture during microdeformations under the mechanical loads in cutting. The ratio of their hardness to the elastic modulus (H/E parameter) is often used to measure the plasticity of coatings [31–33]. It was found that the nanohardness of TiB_2 coatings increases noticeably with an increase in Ar pressure from 0.2 to 1.0 Pa. At the same time, the roughness of the deposited coatings also increases.

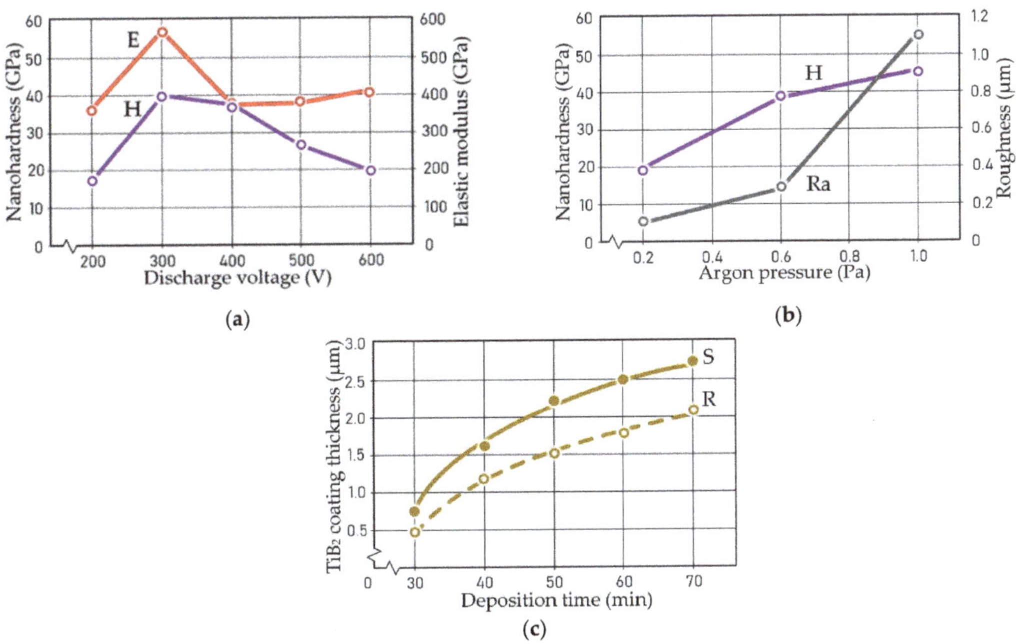

Figure 11. Dependences of nanohardness (H), elastic modulus (E) and roughness (Ra) and thickness of TiB_2 coatings when deposited on stationary (S) and rotating (R) hard alloy substrates on discharge voltage (**a**), argon pressure (**b**) and deposition time (**c**).

The deposition rate of TiB_2 coatings has a traditional character depending on the target sputtering time (Figure 11). In this case, the mobility of the hard alloy substrate noticeably affects the thickness of the formed coating: the deposition speed was ~1.7 μm/h for rotating samples and ~2.5 μm/h for stationary samples. Also, the influence of deposition time (coating thickness) on the morphology of the TiB_2 coating was revealed during this research. It has been established that with the increasing thickness of TiB_2 coatings, their morphology changes significantly, a significant number of dislocations appear and the misorientation of the grain structure is observed (Figure 12). The formation of dislocations during the growth of PVD coatings can be considered a natural process [34,35].

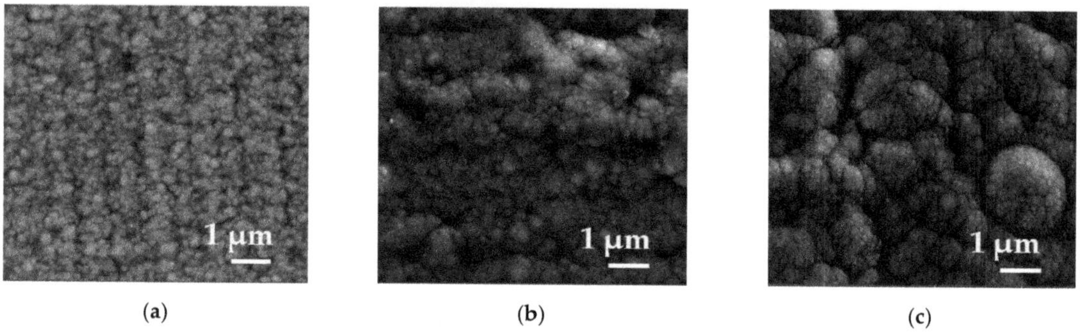

Figure 12. SEM images of the surface morphology of TiB_2 coatings with thicknesses of 1.0 μm (**a**), 2.0 μm (**b**) and 2.5 μm (**c**).

For tool purposes, the thickness of coatings is one of the critical parameters affecting the adhesion of coatings to substrates and wear resistance when exposed to thermal and mechanical loads. On the one hand, an increase in the coating thickness has a positive effect on the wear resistance of a surface, but, on the other hand, it traditionally leads to a decrease in the adhesion strength of the coating and the tool material and worsens the ability of the coating to resist brittle fracture during microdeformations [36]. Therefore, when conducting research, special attention was paid to choosing the rational thickness of TiB_2 coatings deposited on carbide substrates. Table 7 shows the results of assessing the adhesion strength of TiB_2 coatings with thicknesses of 1.0, 2.0 and 2.5 μm during scratch testing. Adhesion was assessed for both coatings deposited directly on hard alloy substrates and on pre-formed (TiAlSi)N subcoatings. It can be seen that an increase in the thickness of the coatings significantly reduces the critical loads at which cracking (L_{c1}) and peeling of sections of the coating (L_{c2}) occur. It was found that the deposition of TiB_2 coatings on substrates with a pre-formed (TiAlSi)N subcoating with a thickness of 2.8 μm ensures a shift in load to the region of higher values, i.e., it increases the adhesion of coatings to substrates. This effect is especially pronounced for L_{c2} loads, at which the peeling of the coatings occurs when L_{c1} does not increase significantly. For example, the average load L_{c2} was 23 N for a TiB_2 monocoating with a thickness of 2.0 μm, and the ultimate load increased by more than 30% and amounted to 31 N for the (TiAlSi)N/TiB_2 coating. Table 7 also presents the experimental data on the elastic moduli of TiB_2 coatings of different thicknesses formed on substrates without a subcoating and with a (TiAlSi)N subcoating. The presented results demonstrate a certain correlation between the adhesion strength of the TiB_2 coating to the hard alloy substrate and its elastic modulus. It is known that higher values of the elastic modulus usually correspond to higher values of residual stresses in the coating, which can lead to the development of cracking, and a lower elastic modulus ensures high resistance to plastic deformation [31,35]. This relation can be applied to the samples with TiB_2 coatings under study. An increase in its elastic modulus and a decrease in loads leading to cracking in the coating (L_{c1}) and delamination of coating areas (L_{c2}) are observed when the coating thickness increases (Table 7). However, it is possible to reduce this effect by introducing the (TiAlSi)N intermediate layer, providing lower stresses in the "hard alloy substrate—TiB_2 coating" system and the satisfactory adhesion strength of the TiB_2 coating. Thus, when the TiB_2 coating thickness was 2.0 µm, the elastic modulus of the coating deposited on a hard alloy substrate was ~390 GPa, and, with (TiAlSi)N subcoating, it was ~366 GPa.

Table 7. Results of assessing the adhesion strength of TiB_2 coatings to carbide substrates during scratch testing and the elastic moduli of those coatings.

Type of Coating	TiB_2 Coating Thickness (µm)	Critical Load Value (N)		Elastic Modulus, GPa
		L_{c1}	L_{c2}	
without subcoating	1.0	22 ± 1	29 ± 2	365 ± 7
	2.0	15 ± 3	23 ± 3	390 ± 8
	2.5	12 ± 2	18 ± 4	472 ± 8
with subcoating (TiAlSi)N	1.0	25 ± 1	34 ± 1	332 ± 6
	2.0	18 ± 2	31 ± 2	366 ± 7
	2.5	14 ± 3	23 ± 2	446 ± 8

Thus, the results of studies of the deposition of TiB_2 coatings on hard alloy substrates (WC-Co) using magnetron sputtering of sintered targets produced from a composition of 96 vol.% TiB_2—4 vol.% Ti show that it is possible to obtain coatings with a wide range of critical physical and mechanical properties by varying the deposition modes. For TiB_2 coating tool purposes, a combination of deposition modes was chosen to provide the maximum ratio of nanohardness and elastic modulus (H/E), i.e., making it possible to obtain coatings that simultaneously combine sufficiently high hardness (H = 38 GPa) and acceptable ductility (E = 390 GPa). The sufficient adhesion of TiB_2 coatings can be achieved with a thickness of 2.0 µm and deposition of coatings on carbide substrates with a (TiAlSi)N subcoating, which additionally ensures a reduction in the elastic modulus of the outer TiB_2 coating. The specified design of TiB_2-based coatings is potentially capable of resisting brittle fracture and peeling from the surface of a carbide tool under increased mechanical loads in milling. An analysis of the obtained experimental data made it possible to identify the preferred modes of TiB_2 coating deposition, which were used to deposit coatings on samples of carbide end mills: the discharge voltage was 400 V, argon pressure was 0.6 Pa and deposition time was 70 min.

3.3. *Wear Resistance of TiB_2-Based Coatings Deposited on Carbide Substrates during Tribological Tests*

Figure 13 shows the results of studies of the changes over time of the friction coefficient of WC-Co hard alloy substrates with various coatings under conditions of high-temperature sliding friction with counterbodies made of Inconel 718 alloy on distance. The large amplitude of friction coefficient fluctuations for a sample without a coating indicates that frictional interaction is accompanied by intense adhesive bonding with the material of the counterbody. This is illustrated by the SEM image of a section of a wear track (Figure 14), which shows destructive processes in the surface layer of the WC-Co sample, the formation of scuffing and the adhesion of the counterbody material. The deposition of a (TiAlSi)N monocoating reduces the amplitude of fluctuations in the friction coefficient but does not reduce its average value compared with the original sample (Figure 13). At the same time, destructive processes are significantly less pronounced on the friction path of a sample with a nitride coating. The TiB_2 monocoating reduces the average friction coefficient at the initial testing stage, but its value increases noticeably with increasing friction distance. It can be assumed that the TiB_2 monocoating is gradually destroyed under the influence of thermal and mechanical loads, and the friction conditions approach those characteristics equally compared to the uncoated samples. In this case, the type of friction path for the diboride monocoating and the observed destructive process are similar to the sample with nitride monocoating (Figure 14). For the (TiAlSi)N/TiB_2 coating, an increase in the friction coefficient is observed at the running-in stage. Then, it has stable values at the level of ~0.6 m with small amplitudes of its fluctuation throughout the entire friction distance (Figure 13). The SEM image of a section of a wear track of a sample with (TiAlSi)N/TiB_2 coating (Figure 14) allows us to conclude that destructive processes on the surface of the carbide substrate and the adhesion of the counterbody material are minimized.

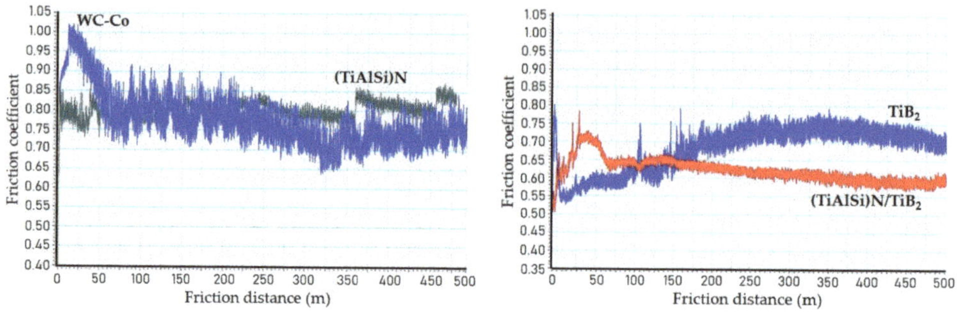

(**a**) (**b**)

Figure 13. The changes over time of the friction coefficient between substrates made of WC-Co hard alloy and counterbodies made of Inconel 718 alloy on friction distance: uncoated and with (TiAlSi)N coating (**a**), and with TiB_2 and (TiAlSi)N/TiB_2 coatings (**b**).

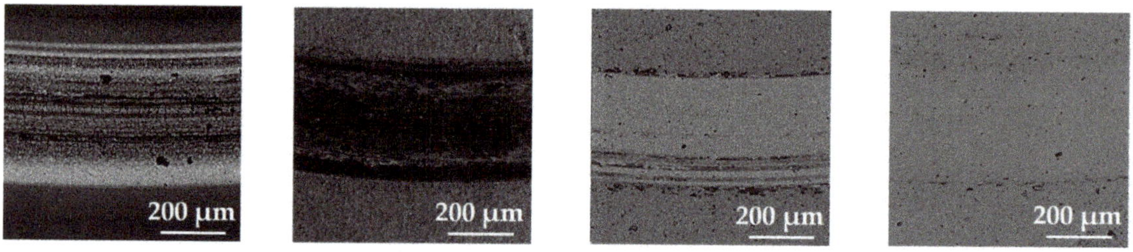

(**a**) (**b**) (**c**) (**d**)

Figure 14. SEM images of areas of wear tracks formed during tribological tests of carbide substrates without coating (**a**), and with a (TiAlSi)N coating (**b**), TiB_2 coating (**c**) and (TiAlSi)N/TiB_2 coating (**d**).

A comparative quantitative analysis of the wear resistance of the WC-Co hard alloy substrates with various coatings was performed based on a comparison of the 3D profiles and profilograms of sections of wear tracks (Figure 15). It can be seen that the formed wear areas for samples without a coating and with (TiAlSi)N, TiB_2 and (TiAlSi)N/TiB_2 coatings differ significantly as a result of contact interaction with chromium–nickel alloy during friction sliding (especially the maximum depth of wear tracks) and amount to 5.9, 3.0, 2.0 and 0.8 µm, respectively. The TiB_2 monocoating has some effect in comparison with samples without a coating and with the (TiAlSi)N monocoating, but is noticeably inferior in wear resistance to the (TiAlSi)N/TiB_2 coating. High thermomechanical loads during tribological tests lead to destructive processes to which the TiB_2 monocoating is poorly resistant. The results of scratch testing (Table 7) show that the forces expended on cracking and tearing off sections of the TiB_2 coating are noticeably less than for the (TiAlSi)N/TiB_2 coating. Throughout the entire high-temperature tribotests, the (TiAlSi)N/TiB_2 coating retains its functions, which allows us to conclude that there is a high potential for its use to increase the wear resistance of carbide tools during the cutting when the cutting edges experience an increased level of thermal and mechanical loads. It should be noted that since the tribological tests were carried out by the "pin-on-disk" scheme, the profile of the wear tracks was equal along the entire track and specific areas were absent.

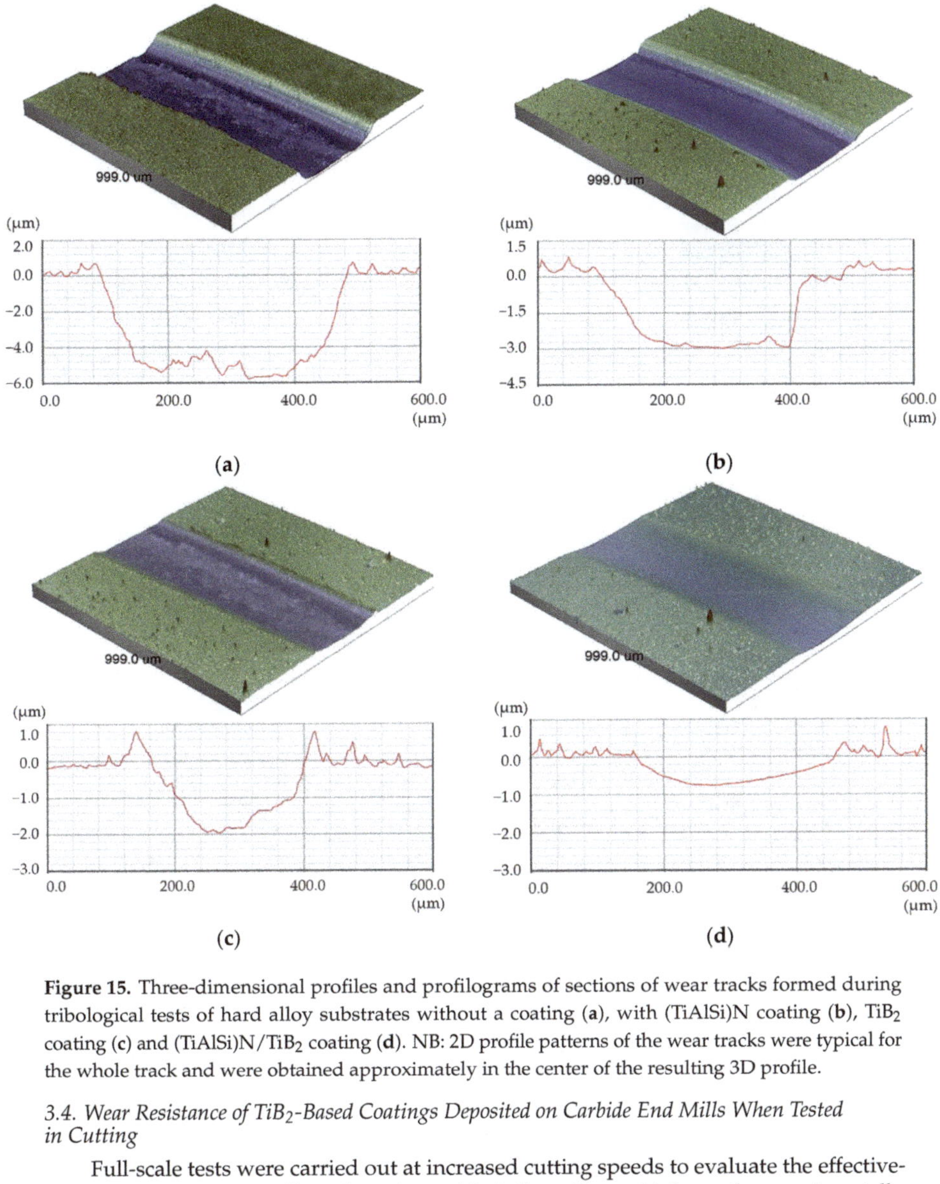

Figure 15. Three-dimensional profiles and profilograms of sections of wear tracks formed during tribological tests of hard alloy substrates without a coating (**a**), with (TiAlSi)N coating (**b**), TiB_2 coating (**c**) and (TiAlSi)N/TiB_2 coating (**d**). NB: 2D profile patterns of the wear tracks were typical for the whole track and were obtained approximately in the center of the resulting 3D profile.

3.4. Wear Resistance of TiB_2-Based Coatings Deposited on Carbide End Mills When Tested in Cutting

Full-scale tests were carried out at increased cutting speeds to evaluate the effectiveness of TiB_2 coatings in milling chromium–nickel alloys. Figure 16 shows the experimentally obtained curves of wear on the flank faces of carbide ball-end mills with various coatings versus cutting time in machining workpieces made of Inconel 718 alloy with a cutting speed of 263.7 m/min, a feed of 0.04 mm/tooth and a depth of 0.25 mm. The wear curves for all four samples under study had a rather classic character. At the short stage of running-in, the wear of the cutting part of the end mills occurs due to the microdeformation of the surface layer. Then, the stage of steady wear follows and is characterized by constant friction conditions and a monotonically increasing wear rate. It is known that the stage of catastrophic wear occurs when the optimal friction conditions typical for the previous stage are violated, which leads to a sharp increase in the wear rate [36,37]. The curves presented in Figure 16 vary significantly in wear rate on cutting time and correlate to some extent with the high-temperature tribological test results presented above.

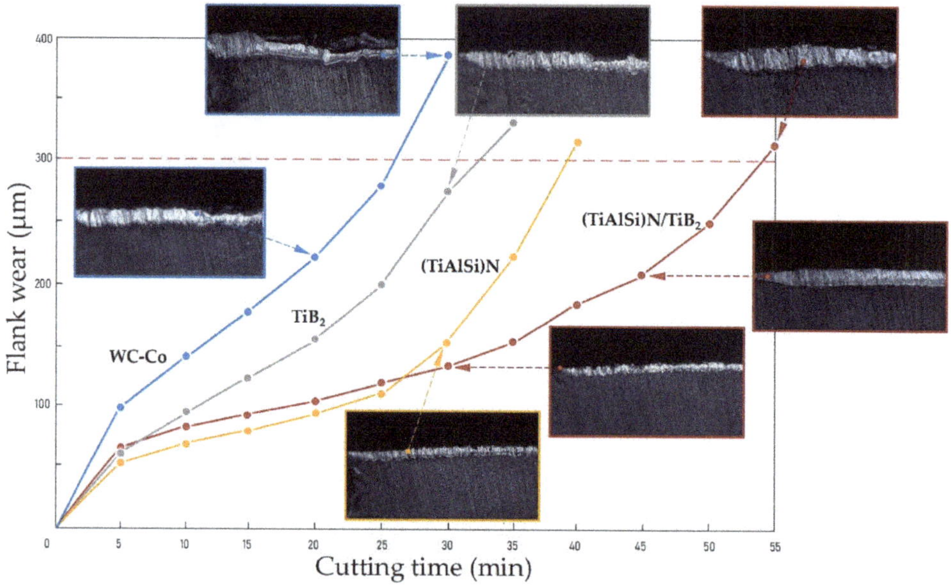

Figure 16. Dependence of flank wear of carbide end mills with various coatings on cutting time in machining workpieces of Inconel 718 alloy with a cutting speed of 263.7 m/min, feed of 0.04 mm/tooth and depth of 0.25 mm.

The wear resistance until critical wear (300 µm) was reached at ~26 min for uncoated end mills. The (TiAlSi)N monocoatings that are widely used in the industry significantly slowed down the wear rate and increased the wear resistance of cutters up to ~38 min. This effect of nitride coatings is explained by their ability to maintain high hardness under increased thermal loads [38,39]. The deposition of TiB_2 monocoatings on the cutting part of end mills slightly slows the wear rate compared to end mills without coatings (~33 min) but is inferior in wear resistance to the nitride coating. This is apparently due to the insufficient adhesive bond of TiB_2 coatings with carbide tools and destructive processes occurring under high thermal and mechanical loads, which have already been discussed when assessing the results of tribological tests. End mills with (TiAlSi)N/TiB_2 coatings are characterized by the longest stage of steady-state wear and the minimum wear rate, and their wear resistance before reaching critical wear is ~54 min, which is 2.0, 1.6 and 1.4 times more than for tools without coatings and with TiB_2 and (TiAlSi)N monocoatings, respectively (Figure 16).

The different behavior of the coating in the cutting zone correlates with the tribological behavior of the coating and surface processes in the contact zone between materials under corresponding thermal and mechanical loads. Not only does the coating material itself affect the wear behavior of the cutting tool, but also the substrate material, the quality of the surface before coating deposition, the adhesion of the coating to the underlying material, contact conditions, etc. During wear, the stability of the coating behavior and the friction coefficient plays an important role, which is also responsible for the release of heat during cutting, which in turn depends not only on the mechanical characteristics of the formed wear pad but also on the chemical composition of the coating and its physical properties, which determine the nature of the interaction of the coating with the workpiece material and the environment.

Diborides definitely have higher hardness and less intense adhesive bonding with the counterbody during high-temperature friction than nitrides of metal and form tribofilms during high-temperature exposure. During the development of our innovative coating structure, the three-nitride coating was chosen as a sublayer to improve the adhesion of the TiB_2 coating to the carbide cutting tool. Combining a three-nitride sublayer with a promising wear-resistant TiB_2 coating provides a solution for machining chromium–nickel alloys, where the TiB_2 outer layer is addressed to improve the resistance of the cutting tool and the three-nitride underlayer allows better adhesion to the hard alloy substrate.

4. Conclusions

The results of our experimental studies allow us to conclude that spark plasma sintering can be used as a high-performance process for producing high-density targets from TiB_2-based powder compositions. It has been established that it is possible to significantly reduce the porosity of sintered samples and increase their relative density by adding Ti micropowder of 4 vol.% to the TiB_2 powder. The revealed dependencies of the sintering process made it possible to select rational conditions (temperature of 700 °C, pressure of 80 MPa, exposure time of 5 min), which ensured the production of targets with a diameter of 80 mm and a thickness of 5 mm from a powder composition of 96 vol.% TiB_2—4 vol.% Ti with a relative density of at least 99% and a crack resistance of at least 7.3 $MPa{\cdot}m^{1/2}$.

The use of sintered targets for the TiB_2 coating deposition by magnetron sputtering makes it possible to deposit coatings on rotating WC-Co hard alloy substrates at a rate of 1.7 μm/h and to obtain coatings with a wide range of physical and mechanical properties by changing the discharge voltage and argon pressure. Taking into account the specific operation of cutting tools and the need for TiB_2 coatings to function under conditions of increased thermal and mechanical loads, it is recommended to use coatings that simultaneously combine a sufficiently high hardness (~38 GPa) and an acceptable elastic modulus (~390 GPa). The indicated properties can be obtained at a discharge voltage of 400 V and an argon pressure of 0.6 Pa. A satisfactory adhesion strength of TiB_2 coatings to carbide substrates can be achieved with a thickness of 2.0 μm (target sputtering time 70 min) and deposition of coatings onto a pre-formed (TiAlSi)N subcoating.

Tribological tests under high-temperature friction-sliding conditions of samples made of WC-Co with(TiAlSi)N, TiB_2 and (TiAlSi)N/TiB_2 coatings have established that the presence of a two-layer coating significantly stabilizes the frictional interaction with a counterbody made of chromium–nickel alloy and reduces the wear rate. Among all the studied samples, the (TiAlSi)N/TiB_2 coating had the most stable friction coefficient and a significantly smaller friction track depth throughout the entire testing period.

Full-scale tests carried out on milling workpieces made of Inconel 718 alloy with carbide ball-end mills at increased cutting speeds (263.7 m/min) showed the high efficiency of (TiAlSi)N/TiB_2 coatings. For end mills with (TiAlSi)N/TiB_2 coatings, the longest stage of steady wear and the minimum wear rate were established, and their wear resistance before reaching critical wear was ~54 min, which is 2.0, 1.6 and 1.4 times more than for the tools without coatings and with TiB_2 and (TiAlSi)N monocoatings, respectively. This allows us to consider TiB_2-based coatings as promising for solving various problems of machining parts made of difficult-to-cut materials using carbide tools.

Author Contributions: Conceptualization, S.N.G.; methodology, M.A.V. and V.D.G.; software, A.P.M. and A.A.O.; validation, S.V.F. and V.D.G.; formal analysis, S.V.F. and V.D.G.; investigation, M.A.V.; resources, A.P.M. and S.V.F.; data curation, A.P.M. and A.A.O.; writing—original draft preparation, M.A.V.; writing—review and editing, M.A.V.; visualization, M.A.V.; supervision, M.A.V. and S.N.G.; project administration, S.N.G.; funding acquisition, S.N.G. All authors have read and agreed to the published version of the manuscript.

Funding: The research was financially supported by the Ministry of Science and Higher Education of the Russian Federation (project No. FSFS-2023-0003).

Data Availability Statement: The original contributions presented in this study are included in the article; further inquiries can be directed to the corresponding author.

Conflicts of Interest: The authors declare no conflicts of interest.

References

1. Yin, Q.; Liu, Z.; Wang, B.; Song, Q.; Cai, Y. Recent progress of machinability and surface integrity for mechanical machining Inconel 718: A review. *Int. J. Adv. Manuf. Technol.* **2020**, *109*, 215–245. [CrossRef]
2. Bartolomeis, A.; Newman, S.T.; Jawahir, I.S.; Biermann, D.; Shokrani, A. Future research directions in the machining of Inconel 718. *J. Mater. Process. Technol.* **2021**, *297*, 117260. [CrossRef]
3. Grguras, D.; Kern, M.; Pusavec, F. Suitability of the full body ceramic end milling tools for high speed machining of nickel based alloy Inconel 718. *Procedia CIRP* **2018**, *77*, 630–633. [CrossRef]
4. Gusarov, A.V.; Migranov, M.S.; Mitrofanov, A.P.; Gusev, A.S.; Migranov, A.M.; Khmyrov, R.S. Effect of Surface Modification via Laser Irradiation on the Operability of Carbide End Mills When Cutting Aircraft Alloys. *Coatings* **2023**, *13*, 1823. [CrossRef]
5. Sortino, M.; Belfio, S.; Totis, G.; Kuljanic, E.; Fadelli, G. Innovative Tool Coatings for Increasing Tool Life in Milling Nickel-coated Nickel-Silver Alloy. *Procedia Eng.* **2015**, *100*, 946–952. [CrossRef]
6. Yu, J.; Wang, Y.; Zhao, X.; Li, Q.; Qiao, Q.; Zhao, J.; Zhai, S. Wear Resistance of Ni-Based Alloy Coatings. *Adv. Mater. Sci. Eng.* **2019**, *2019*, 1–7. [CrossRef]
7. Liu, Y.; Li, Z.; Li, G.; Tang, L. Friction and wear behavior of Ni-based alloy coatings with different amount of WC–TiC ceramic particles. *J. Mater. Sci.* **2023**, *58*, 1116–1126. [CrossRef]
8. Grigoriev, S.; Vereschaka, A.; Zelenkov, V.; Sitnikov, N.; Bublikov, J.; Milovich, F.; Andreev, N.; Mustafaev, E. Specific features of the structure and properties of arc-PVD coatings depending on the spatial arrangement of the sample in the chamber. *Vacuum* **2022**, *200*, 111047. [CrossRef]
9. Yuan, J.; Yamamoto, K.; Covelli, D.; Tauhiduzzaman, M.; Arif, T.; Gershman, I.; Veldhuis, S.; Fox-Rabinovich, G. Tribo-films control in adaptive TiAlCrSiYN/TiAlCrN multilayer PVD coating by accelerating the initial machining conditions. *Surf. Coat. Technol.* **2016**, *294*, 54–61. [CrossRef]
10. Alamgir, A.; Yashin, M.; Bogatov, A.; Viljus, M.; Traksmaa, R.; Sondor, J.; Lümkemann, A.; Sergejev, F.; Podgursky, V. High-Temperature Tribological Performance of Hard Multilayer TiN-AlTiN/nACo-CrN/AlCrN-AlCrO-AlTiCrN Coating Deposited on WC-Co Substrate. *Coatings* **2020**, *10*, 909. [CrossRef]
11. Lim, C.S.; Sofer, Z.; Mazanek, V.; Pumera, M. Layered titanium diboride: Towards exfoliation and electrochemical applications. *Nanoscale* **2015**, *7*, 12527–12534. [CrossRef] [PubMed]
12. Gild, J.; Zhang, Y.; Harrington, T.; Jiang, S.; Hu, T.; Quinn, M.C.; Mellor, W.M.; Zhou, N.; Vecchio, K.; Luo, J. High-Entropy Metal Diborides: A New Class of High-Entropy Materials and a New Type of Ultrahigh Temperature Ceramics. *Sci. Rep.* **2016**, *6*, 37946. [CrossRef] [PubMed]
13. Casais, R.; Baptista, A.M.; Silva, F.J.; Andrade, F.; Sousa, V.; Marques, M.J. Experimental study on the wear behavior of B4C and TiB_2 monolayered PVD coatings under high contact loads. *Int. J. Adv. Manuf. Technol.* **2022**, *120*, 6585–6604. [CrossRef]
14. Grishin, A.M.; Putrolaynen, V.V. Hard and Highly Adhesive $AlMgB_{14}$ Coatings RF Sputtered on Tungsten Carbide and High-Speed Steel. *Materials* **2023**, *16*, 6930. [CrossRef] [PubMed]
15. Chowdhury, M.S.I.; Bose, B.; Fox-Rabinovich, G.; Veldhuis, S.C. Investigation of the Wear Performance of TiB_2 Coated Cutting Tools during the Machining of Ti6Al4V Alloy. *Materials* **2021**, *14*, 2799. [CrossRef]
16. Schalk, N.; Tkadletz, M.; Mitterer, C. Hard coatings for cutting applications: Physical vs. chemical vapor deposition and future challenges for the coatings community. *Surf. Coat. Technol.* **2022**, *429*, 127949. [CrossRef]
17. Sroba, V.; Fiantok, T.; Truchly, M.; Roch, T.; Zahoran, M.; Grancic, B.; Svec, P.; Nagy, S.; Izai, V.; Kus, P.; et al. Structure evolution and mechanical properties of hard tantalum diboride films. *J. Vac. Sci. Technol. A* **2020**, *38*, 033408. [CrossRef]
18. Grigoriev, S.N.; Volosova, M.A.; Vereschaka, A.A.; Sitnikov, N.N.; Milovich, F. Properties of (Cr,Al,Si)N-(DLC-Si) composite coatings deposited on a cutting ceramic substrate. *Ceram. Int.* **2020**, *46*, 18241–18255. [CrossRef]
19. Rao, J.; Cruz, R.; Lawson, K.J.; Nicholls, J.R. Carbon and titanium diboride multilayer coatings. *Diam. Relat. Mater.* **2004**, *13*, 2221–2225. [CrossRef]
20. Lee, K.W.; Chung, Y.W.; Korach, C.; Keer, L.M. Tribological and dry machining evaluation of superhard TiB_2/TiC multilayer coatings deposited on Si(001), M2 steel, and C3 WC cutting tool inserts using magnetron sputtering. *Surf. Coat. Technol.* **2005**, *194*, 184–189. [CrossRef]
21. Levashov, E.A.; Mukasyan, A.S.; Rogachev, A.S.; Shtansky, D.V. Self-propagating high-temperature synthesis of advanced materials and coatings. *Int. Mater. Rev.* **2017**, *62*, 203–239. [CrossRef]
22. Twardowska, A.; Podsiadło, M.; Sulima, I.; Bryła, K.; Hyjek, P. Microstructure and Properties of TiB_2 Composites Produced by Spark Plasma Sintering with the Addition of Ti_5Si_3. *Materials* **2021**, *14*, 3812. [CrossRef] [PubMed]
23. Shcherbakov, V.A.; Gryadunov, A.N.; Semenchuk, I.E.; Alymov, M.I. Ta4HfC5 Ceramic by Electro-Thermal Explosion under Pressure: Thermal and Electrical Parameters of the Process. *Int. J. Self-Propag. High-Temp. Synth.* **2022**, *31*, 1–5. [CrossRef]

24. Oguntuyi, S.D.; Johnson, O.T.; Shongwe, M.B. Spark Plasma Sintering of Ceramic Matrix Composite of ZrB_2 and TiB_2: Microstructure, Densification, and Mechanical Properties—A Review. *Met. Mater. Int.* **2021**, *27*, 2146–2159. [CrossRef]
25. Grigoriev, S.N.; Volosova, M.A.; Peretyagin, P.Y.; Seleznev, A.E.; Okunkova, A.A.; Smirnov, A. The Effect of TiC Additive on Mechanical and Electrical Properties of Al_2O_3 Ceramic. *Appl. Sci.* **2018**, *8*, 2385. [CrossRef]
26. Cohen, S.; Ratzker, B.; Kalabukhov, S.; Frage, N. Diffusion bonding of transparent ceramics by spark plasma sintering (SPS) complemented by hot isostatic pressing (HIP). *J. Eur. Ceram. Soc.* **2023**, *43*, 6628–6633. [CrossRef]
27. Rajguru, R.R.; Vasudevan, H. A study of micro hardness in the machining of Inconel 625 using TiAlSiN coated tools under dry cutting conditions. *Adv. Mater. Process. Tec.* **2022**, *10*, 120–130. [CrossRef]
28. Yue, Q.B.; He, H.B.; Li, H.Y.; Zhang, J.; Li, Y.M.; Ma, L. Research on Friction Characteristics of AlCrN and TiAlSiN Coatings and Properties of Coated Tools. *Int. J. Precis. Eng. Manuf.* **2019**, *20*, 1581–1589. [CrossRef]
29. Drobný, P.; Mercier, D.; Koula, V.; Škrobáková, S.I.; Čaplovič, Ľ.; Sahul, M. Evaluation of Adhesion Properties of Hard Coatings by Means of Indentation and Acoustic Emission. *Coatings* **2021**, *11*, 919. [CrossRef]
30. Vithal, N.D.; Krishna, B.B.; Krishna, M.G. Impact of dry sliding wear parameters on the wear rate of A7075 based composites reinforced with ZrB2 particulates. *J. Mater. Res. Technol.* **2021**, *14*, 174–185. [CrossRef]
31. Beake, B.D. The influence of the H/E ratio on wear resistance of coating systems–Insights from small-scale testing. *Surf. Coat. Technol.* **2022**, *442*, 128272. [CrossRef]
32. Bouzakis, K.D.; Michailidis, N.; Skordaris, G.; Bouzakis, E.; Biermann, D.; M'Saoubi, R. Cutting with coated tools: Coating technologies, characterization methods and performance optimization. *CIRP Annals* **2012**, *61*, 703–723. [CrossRef]
33. Storchak, M.; Zakiev, I.; Zakiev, V.; Manokhin, A. Coatings strength evaluation of cutting inserts using advanced multi-pass scratch method. *Measurement* **2022**, *191*, 110745. [CrossRef]
34. Panjan, P.; Gselman, P.; Kek-Merl, D.; Čekada, M.; Panjan, M.; Dražić, G.; Bončina, T.; Zupanič, F. Growth defect density in PVD hard coatings prepared by different deposition techniques. *Surf. Coat. Technol.* **2013**, *237*, 349–356. [CrossRef]
35. Panjan, P.; Drnovšek, A.; Kovač, J. Tribological aspects related to the morphology of PVD hard coatings. *Surf. Coat. Technol.* **2018**, *343*, 138–147. [CrossRef]
36. Volosova, M.; Grigoriev, S.; Metel, A.; Shein, A. The Role of Thin-Film Vacuum-Plasma Coatings and Their Influence on the Efficiency of Ceramic Cutting Inserts. *Coatings* **2018**, *8*, 287. [CrossRef]
37. Shuster, L.S.; Fox-Rabinovich, G.S.; Chertovskikh, S.V. Influence of Cutting Conditions on the Wear Resistance of Tools with a TiB_2 Coating during Titanium Alloy Machining. *J. Frict. Wear* **2021**, *42*, 466–472. [CrossRef]
38. Su, K.; Liu, D.; Pang, H.; Shao, T. Improvement on thermal stability of TiAlSiN coatings deposited by IBAD. *Surf. Eng.* **2017**, *34*, 504–510. [CrossRef]
39. Kovalev, A.; Wainstein, D.; Konovalov, E.; Vakhrushev, V.; Fox-Rabinovich, G.; Fox-Rabinovich, M.; Dmitrievskii, S.; Tomchuk, A. Features of Tribooxidation of the High-Entropy Coating (AlCrZrTiTa)N during Dry High-Speed Cutting. *Coatings* **2023**, *13*, 1508. [CrossRef]

materials

MDPI

Article

Synthesis of $MnFe_2O_4$ Spinel on Rusted Q235 Steel Surface and Its Corrosion Resistance Properties

Bai-Ao Feng, Xu Miao and Ting-An Zhang *

Key Laboratory of Ecological Metallurgy of Multi-Metal Intergrown Ores of Ministry of Education, Special Metallurgy and Process Engineering Institute, School of Metallurgy, Northeastern University, Shenyang 110819, China; fba19960520@163.com (B.-A.F.); miaoxu0916@163.com (X.M.)
* Correspondence: zta2000@163.net

Abstract: Corrosion of steel is an issue that cannot be ignored in contemporary society. Due to large-scale corrosion, it is urgent to develop a surface treatment process that enhances the corrosion resistance of steel, allowing for application in various scenarios as needed. This study aims to investigate a novel surface treatment process to extend the service life of corroded Q235 steel, reduce its sensitivity to corrosion, and enable its use in multiple environments. This study employs the sol-gel method, using manganese nitrate solutions of varying concentrations to treat the surface of Q235 steel after different electrolysis times. The optimal conditions for precursor preparation were found to be a Mn^{2+} concentration of 0.1 mol/L and an electrolysis time of 2 h. Electrochemical tests using NaCl solutions of different concentrations revealed a significant reduction in the corrosion current for the composite coating based on Q235 steel treated with this method in NaCl solutions with wt.% = 1, 2, 3, 4, 5. Furthermore, the resistance to corrosion was strongest in the NaCl solution with a concentration of 1 wt.% where the corrosion current decreased from 24.8 $\mu A/cm^2$ to 6.79 $\mu A/cm^2$. Additionally, the coating was found to be diffusion-controlled in the early stages of the corrosion process and charge transfer-controlled in the later stages. The $MnFe_2O_4$ spinel coating demonstrated the greatest enhancement in corrosion resistance in the wt.% = 1 NaCl solution.

Keywords: Q235 steel; composite material; corrosion; spinel; passive films; SEM-EDS; electrochemical impedance spectroscopy

1. Introduction

In recent years, with the development of social productivity, the consumption of metal resources by humans has increased dramatically. A large number of metal products are utilized in human production and daily life, consequently leading to the issue of metal corrosion, which demands significant attention. According to an independent assessment report by the National Association of Corrosion Engineers (NACE), the direct economic loss caused by corrosion worldwide amounts to as much as $250 billion annually, which is equivalent to 3.4% of the global Gross Domestic Product (GDP) [1]. Additionally, corrosion leads to numerous indirect economic losses, including impacts on production safety and the environment [2–5]. It is estimated that more than 10 million tons of metal rusts in various industries worldwide each year [6].

Carbon steel, due to its excellent processing performance, good thermal conductivity, and outstanding plasticity and toughness, has comprehensive properties and is widely used in various fields. It finds extensive applications in aerospace, shipbuilding, construction steel, and petroleum and natural gas pipelines [7]. However, during usage, corrosion is inevitable [8,9].

To mitigate the corrosion of steel, numerous methods have been studied. Research has shown that carboxymethyl cellulose (CMC) can act as a corrosion inhibitor for Q235 steel in H_2SO_4 environments, with its inhibition efficiency increasing with the CMC concentration but decreasing with rising temperatures [10]. This is because CMC exists on the Q235 surface through physical adsorption, and its physical properties change with temperature, affecting its corrosion resistance. Similarly, an imidazoline-based inhibitor has been found to provide high protection efficiency for Q235 steel in NaCl solutions saturated with CO_2 [11], but its cost and environmental constraints limit its suitability for large-scale applications to Q235 steel. To reduce costs and the environmental impact on corrosion resistance, combining non-metal oxides with Q235 steel has been proposed. Scientists have used the sol-gel method to prepare SiO_2/ZrO_2-SiO_2 composites on the surface of Q235 steel [12], which exhibit excellent corrosion resistance and stability at room temperature. However, these composites tend to undergo brittle fracture at temperatures $\geq$ 1200 °C. Consequently, metal materials have become a mainstream research direction. It has been found that alloying steel with aluminum results in weathering steel with enhanced corrosion resistance [13]. This steel initially shows good protective performance, but the instability of the passivation film formed on its surface fails to provide long-term protection [14]. Hot-dip coating, an in situ synthesis method, shows promising potential. Studies have found that hot-dip aluminum coating on the inner surface of API X56 and API X70 steel samples forms an adhesive protective layer of aluminum–silicon alloy. This coating surface, through the formation of an Al_2O_3 passivation layer, provides strong corrosion resistance to Q235 steel, effectively protecting pipeline steel from marine or seawater corrosion [15].

A novel Al-Mg-Y alloy coating can be synthesized on the surface of Q235 steel through hot-dip coating. In corrosive solutions, the Al-Mg-Y alloy surface can form yttrium-rich structures and new alloy phases containing yttrium, which enhance the alloy's corrosion resistance [16,17]. However, an excess of yttrium can accelerate the corrosion reaction. Most metal oxides or metal–nonmetal composites exhibit stable properties and strong corrosion resistance. Scientists have developed a new type of Al_2O_3-anchored carbon quantum dot modified TiO_2 nanorod (Al/C/TNRs) photoelectrode material for efficient photo-cathodic protection of Q235 carbon steel under sunlight [18]. However, its high cost limits its application. Research has employed thermal flame spraying to deposit pure aluminum and aluminum/10% Ti_3SiC_2 coatings on carbon steel, resulting in excellent corrosion and wear resistance [19]. Nevertheless, given the widespread use of Q235 steel and the significant portion that has already suffered corrosion, this method has limitations. Therefore, it is crucial to develop a low-cost method to treat already corroded Q235 steel, stabilizing its properties and enhancing its corrosion resistance. Research on the treatment of corroded steel has become particularly important as it allows for a gradual reduction in the consumption of human and material resources, as well as the waste of resources caused by replacing long-used steel components. Moreover, this new study not only benefits resource conservation and recycling in steel-dependent industries such as marine, military, construction, and aerospace but also aids in the restoration and long-term preservation of steel artifacts, souvenirs, and historical items in people's daily lives.

Due to its poor corrosion resistance, Q235 steel, despite being widely used in many industries such as construction, bridges, ships, vehicles, and machinery manufacturing, has a relatively short lifespan and often requires the replacement of products that fail to meet usage requirements due to corrosion. Its corrosion products are mainly loose and porous Fe_2O_3, which allows H_2O and CO_2 from the air to penetrate and cause more severe internal corrosion. Therefore, surface treatment of already corroded Q235 steel can enhance its corrosion resistance and extend its service life. Currently, most research focuses on substrate modification and the preparation of multilayer composite materials. Scientists have not extensively explored the research area presented in this paper, leaving significant gaps. Previously, some scientists attempted to prepare self-healing fluid materials on steel surfaces to achieve sustained corrosion resistance [20]. A self-healing coating on steel substrate has been obtained by incorporating carbon nanotube (CNT) into the fluid matrix of epoxy resin (EP) or silicone oil (OIL). To further achieve the active corrosion resistance, 1H, 1H, 2H, $2H^-$ perfluorooctyltriethoxysilane (PTES), which can react with the water inside the coating, is added. The fluid self-healing anti-corrosion layer prepared by this method exhibits sustained anti-corrosion performance, but it has poor wear resistance and is highly expensive, with a high dependency on equipment. Therefore, it is necessary to research a low-cost, low-equipment-dependent method that can continuously treat the rust layer on the surface of Q235 steel, thereby enhancing its corrosion resistance.

The purpose of this study is to explore a method for treating the corrosion layer on the surface of Q235 steel that meets industrial requirements for low cost, reduces dependence on equipment, and has broad applicability. Thus, this study focuses on the corrosion susceptibility of Q235 steel in atmospheric environments, specifically addressing the reality of already corroded Q235 steel. An electrolytic acceleration method is used to simulate the actual application environment of the Q235 steel substrate [21]. To achieve a stable and uniform surface, the sol-gel method is employed [22,23], utilizing different concentrations of $Mn(NO_3)_2$ [24], citric acid, and ammonia as raw materials. Citric acid's -COOH functional group acts as a reducing agent, while the oxidizing property of NO^{3-} serves as the oxidizing agent. Upon heating, a redox reaction occurs, releasing a significant amount of heat, which facilitates the reaction between the rust layer of the substrate and the Fe^{3+} and Mn^{2+} in the gel layer, ultimately forming a $MnFe_2O_4$ spinel corrosion-resistant coating [25–27] (as shown in Figure 1). In different mass fractions of NaCl systems, the open circuit potential, polarization curves, and electrochemical impedance spectroscopy are used to investigate the corrosion susceptibility, electrode reaction parameters during corrosion, corrosion rate, and corrosion mechanisms of untreated and treated Q235 steel. XRD and FT-IR techniques are employed for compositional analysis of the corrosion products. The precursor prepared by the sol-gel method is subjected to high-temperature sintering to obtain the $MnFe_2O_4$ spinel coating, which is then characterized using XRD, FT-IR, and SEM.

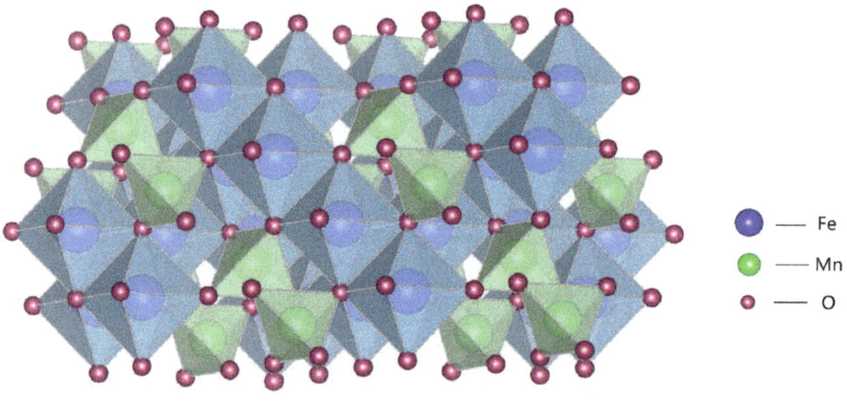

Figure 1. Structure diagram of $MnFe_2O_4$.

This study examines the effects of various process conditions, such as different electrolysis times and Mn^{2+} concentrations, on the $MnFe_2O_4$ spinel coating to determine the optimal process conditions. Different concentrations of NaCl solutions are used to simulate non-uniform application conditions, and the corrosion resistance of the $MnFe_2O_4$ spinel coating is investigated using polarization curves and electrochemical impedance spectroscopy. This research provides crucial scientific evidence for the development of new methods for remediating and enhancing the corrosion resistance of metal surface coatings.

2. Methods and Materials

2.1. Materials

Q235 steel (10 × 10 × 5 mm) was used as the anode and stainless steel as the cathode. Electrolysis was conducted at room temperature using a DC power (L305SPL, TongMen Electro Tech Co., Dongguan, China) supply with a current density of 1500 A/m^2 for 60, 90, 120, 150, and 180 min to obtain oxidation layers with varying degrees of reaction. By comparison, the optimal electrolysis time was determined. Prepared $Mn(NO_3)_2$ solutions with concentrations of 0.02 mol/L, 0.04 mol/L, 0.06 mol/L, 0.08 mol/L, and 0.1 mol/L, along with a small amount of citric acid, were ultrasonically mixed evenly. In order to closely approximate the actual industrial production environment and to avoid secondary corrosion caused by the reaction between citric acid and the substrate, ammonia was added to adjust the pH to 6. The electrolyzed Q235 steel was immersed in the prepared solution, followed by water bathing (HH-6, RongHua Instrument Manufacturing Co., Nanjing, China) at 60 °C for 2.0 h and then at 95 °C for 2.0 h. The samples were dried in a vacuum drying oven (DZF-6020, XinMiao Medical Equipment Manufacturing Co., Shanghai, China) at 60 °C for 12 h. The dried samples were then placed in a tube furnace (SK-G08143, Central Electric Furnace Co., Dongguan, China) under an argon atmosphere to maintain vacuum conditions. The calcination was performed at 650 °C with a heating rate of 10 °C/min, and the samples were held at this temperature for 1.0 h, resulting in Q235 steel samples with $MnFe_2O_4$ spinel corrosion-resistant coatings. The reaction is as follows:

$$1.8Mn(NO_3)_2 + 3.6FeOOH + C_6H_8O_7 \rightarrow 1.8MnFe_2O_4 + 1.8N_2(g) + 5.8H_2O(g) + 6CO_2 \quad (1)$$

2.2. *Electrochemical Measurements*

2.2.1. Electrode Preparation

A copper wire was soldered to the non-working surface of the Q235 steel using a welding torch and sealed with epoxy resin, leaving an exposed area of 1 cm^2 for testing. The exposed working surface was sequentially polished using 400, 600, 1000, 1500, 2000, and 3000 grit sandpapers. The surface was then polished to a mirror finish using chromium oxide powder on a metallographic polishing machine (P-2, Huaxing Testing Instruments Co., Jinan, China) and wiped with anhydrous ethanol. The surface morphological patterns of the samples were observed using a scanning electron microscope (SEM, ZEISS Gemini 300, Carl Zeiss AG, Oberkochen, Germany). The samples were characterized by X-ray diffraction (XRD) using a D8 Advance diffractometer (Karlsruhe, Germany) under conditions of 40 kV and 40 mA, 2θ = 10–90°, scan rate of 8°/min, and step length of 0.02°. Each group of samples was tested three times. FT-IR testing was conducted using a Fourier-transform infrared spectrometer (Nicolet IS 50, Thermo Fisher Scientific, Waltham, MA, USA).

2.2.2. Measuring Condition

In this experiment, all electrochemical tests were performed in triplicate using the same set of samples, utilizing an electrochemical workstation (Zennium Pro, Zahner, Kronach, Germany). A three-electrode system was utilized, with Q235 steel serving as the working electrode (dimensions: 10 × 10 × 5 mm), a platinum sheet as the counter electrode (dimensions: 10 × 10 × 1 mm), and a saturated calomel electrode (SCE) filled with saturated KCl solution as the reference electrode. The electrochemical testing conditions were as follows:

Open Circuit Potential (OCP): Testing duration: 7200 s, Frequency: 0.1 Hz. Polarization Curve: Conducted after OCP stabilization for 600 s, potential range: −0.6 V to 1.0 V, scan rate: 0.1 mV/s. Electrochemical Impedance Spectroscopy (EIS): Conducted after OCP stabilization for 600 s, sine wave potential amplitude: 10 mV, frequency range: 100 kHz to 5 mHz, every ten points was sampled five times.

3. Results and Discussion

3.1. *The Impact of Electrolysis Time on the Corrosion Resistance of $MnFe_2O_4$ Spinel Coatings*

3.1.1. XRD Analysis of Samples with Different Electrolysis Times

Figure 2 shows the XRD patterns for electrolysis times of 60, 90, 120, 150, and 180 min. Characteristic diffraction peaks corresponding to (111), (220), (311), (400), (511), and (440) are observed, which match the standard $MnFe_2O_4$ pattern (JCPDS 73-1964). As the electrolysis time increases, the intensity of the main peak initially increases and then decreases, indicating that the crystallinity first increases and then decreases. This suggests that the content of spinel-phase $MnFe_2O_4$ increases with electrolysis time initially, reaching a peak at 120 min, and then decreases. The intensity of the impurity peak $(FeO)_{0.099}(MnO)_{0.901}$ (JCPDS 77-2362) is higher at electrolysis times of 60, 90, 150, and 180 min compared to 120 min, indicating that the formation of $(FeO)_{0.099}(MnO)_{0.901}$ is related to electrolysis time. At electrolysis times of 60, 90, 150, and 180 min, the amount and growth of the non-target product $(FeO)_{0.099}(MnO)_{0.901}$ affect the formation of the target product $MnFe_2O_4$. This also explains why the $MnFe_2O_4$ crystals formed at 120 min of electrolysis show the best growth. The XRD results demonstrate that at an electrolysis time of 120 min, the crystallinity of $MnFe_2O_4$ on the surface is the highest, making it the optimal electrolysis time for the formation and accumulation of the spinel.

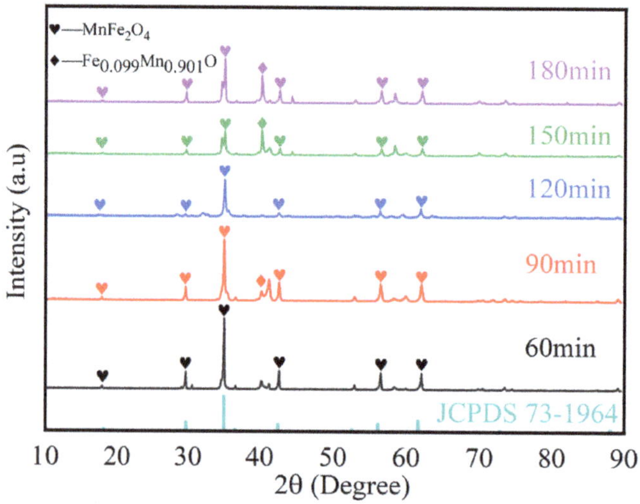

Figure 2. XRD results of samples after 60 min, 90 min, 120 min, 150 min, and 180 min of electrolysis with 0.1 mol/L Mn^{2+} concentrations.

3.1.2. FT-IR Analysis of Samples with Different Electrolysis Times

Figure 3a,b shows the FT-IR spectra of the precursors (the precursor is Q235 steel with an oxide layer, electrolyzed for 60, 90, 120, 150, and 180 min under a current of 1500 A/m^2). This study employs the electrolysis of Q235 steel to simulate the conditions of Q235 steel with a rust layer after prolonged use and calcined samples at different electrolysis times. In Figure 3a, the trends are consistent, indicating that the current density has no effect on the precursor. An absorption peak around 3240 cm^{-1} is observed, corresponding to the stretching vibration of -OH, indicating the presence of a small amount of bound water in the dried samples. The peak near 1562 cm^{-1} is attributed to the stretching vibration of N-O, possibly indicating the presence of NO^{3-} in the precursor. The absorption peak around 1406 cm^{-1} is due to the asymmetric stretching vibration of COO^-. The peaks near 1290 cm^{-1} and 1040 cm^{-1} correspond to the C-O stretching vibrations in -COOH. The absorption peaks below 1000 cm^{-1} are assigned to the γ_2 peaks of Fe-OH and Mn-OH.

In Figure 3b, an impurity peak is observed around 1130 cm^{-1} because calcination removes most organic impurities; however, the C-O bond is relatively stable, and calcination does not completely eliminate carbon compounds. Nevertheless, their presence does not affect the subsequent test results, while the characteristic absorption peaks of Mn-O and Fe-O are seen at 594 cm^{-1} and 472 cm^{-1}, respectively.

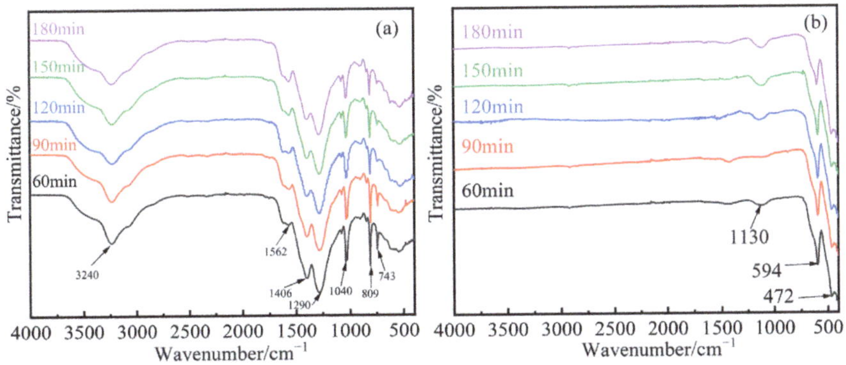

Figure 3. FT-IR results of samples after 60 min, 90 min, 120 min, 150 min, and 180 min of electrolysis with 0.1 mol/L Mn^{2+} concentrations. (**a**) Uncalcined precursor, (**b**) Calcined precursor.

3.1.3. Surface Morphology Analysis of Samples with Different Electrolysis Times

Figure 4a–d presents the morphological patterns of the samples at electrolysis times of 60, 90, 150, and 180 min, respectively. The morphological patterns of the samples at an electrolysis time of 120 min are shown after. In Figure 4a–d, the number of fine $MnFe_2O_4$ particles on the coating surface initially decreases and then increases [28,29]. In Figure 4a, the few fine $MnFe_2O_4$ particles and the presence of large continuous crystals suggest that the short electrolysis time results in insufficient Fe_3O_4 in the rust layer, leading to incomplete reactions. In Figure 4b, a mixed structure of large continuous crystals and $MnFe_2O_4$ particles is observed [30].

Figure 4. Scanning electron microscope photos of samples after (**a**) 60 min, (**b**) 90 min, (**c**) 150 min, and (**d**) 180 min of electrolysis with 0.1 mol/L Mn^{2+} concentrations. The circles show the characteristic structures of $MnFe_2O_4$ and Fe_3O_4.

Figure 5a–e presents the EDS maps for electrolysis times of 60, 90, 120, 150, and 180 min, respectively. From Figure 5a, it can be seen that the distribution of Mn atoms is not very uniform, with some areas lacking Mn atoms. The fine spherical aggregates and large crystalline blocks are loosely combined, with a large crack indicating high porosity. In Figure 5b, two different structures are observed: a large crystalline block covered with numerous fine spherical aggregates. This suggests different compositions. The EDS spectrum shows that Mn is barely present in the large crystalline regions, whereas in the spherical aggregate regions, the Mn content matches the overall average, indicating that the different morphologies arise from Mn content variations [31,32]. In Figure 5c, a relatively uniform $MnFe_2O_4$ layer formed on the coating surface. Figure 5d shows many spherical aggregates dispersed separately, resulting in high porosity between them. The morphological patterns of the samples in Figure 4c reveal numerous defects, possibly due to the long electrolysis time causing pits on the Q235 steel surface. Figure 5d demonstrates a uniform distribution of all elements. Combined with the morphological patterns of the sample images, this indicates successful rust layer conversion. The SEM spectra in Figure 5a–e show the appearance and reduction of large crystalline blocks, which are gradually replaced by spherical aggregates. This suggests that during the coating synthesis process, some base materials did not fully convert, and the conversion process likely involves the dissolution and recrystallization of crystals. This process leads to the different morphologies observed, indicating that with increased electrolysis time, the original rust layer more readily converts to $MnFe_2O_4$ in subsequent synthesis steps, resulting in an increased $MnFe_2O_4$ content in the final coating.

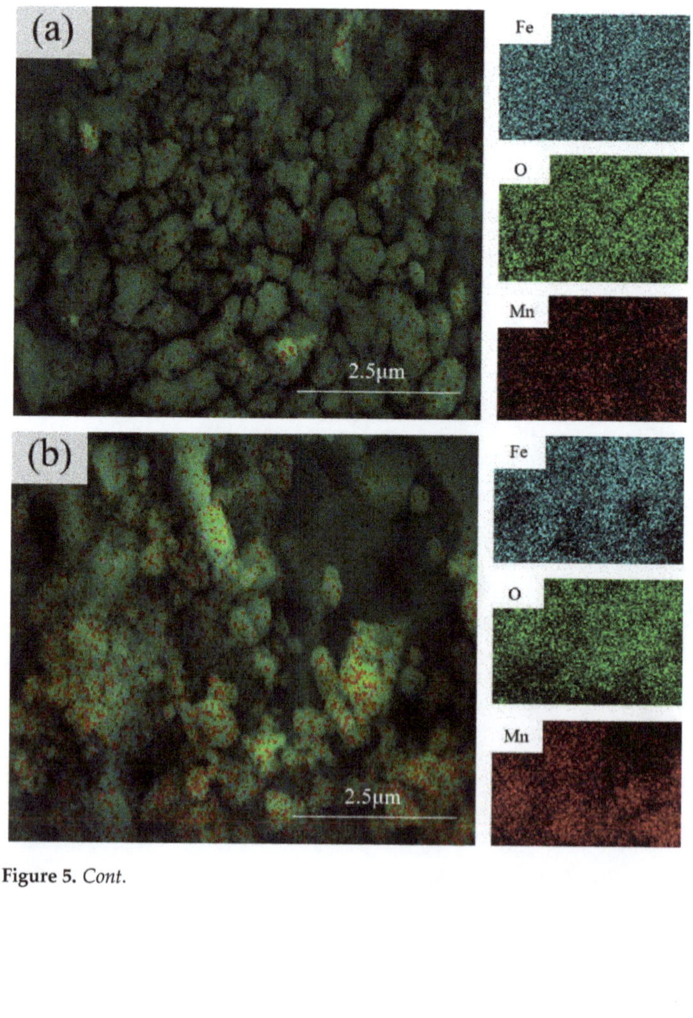

Figure 5. *Cont.*

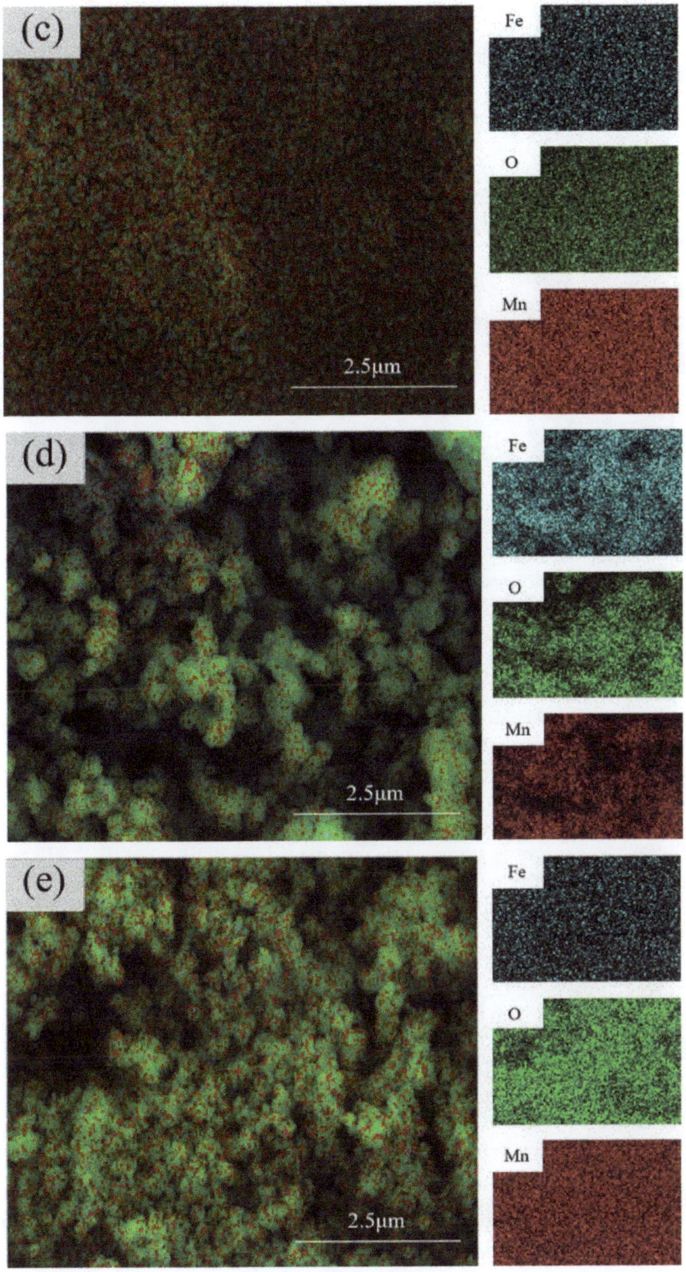

Figure 5. EDS diagram of samples after (**a**) 60 min, (**b**) 90 min, (**c**) 120 min, (**d**) 150 min, and (**e**) 180 min of electrolysis with 0.1 mol/L Mn^{2+} concentrations.

3.1.4. Cross-Sectional Morphology Analysis of Samples with Different Electrolysis Times

Figure 6 shows the cross-sectional morphologies at different electrolysis times, where Figure 6a, Figure 6b, Figure 6c, and Figure 6d represent electrolysis times of 60, 90, 150, and 180 min, respectively.

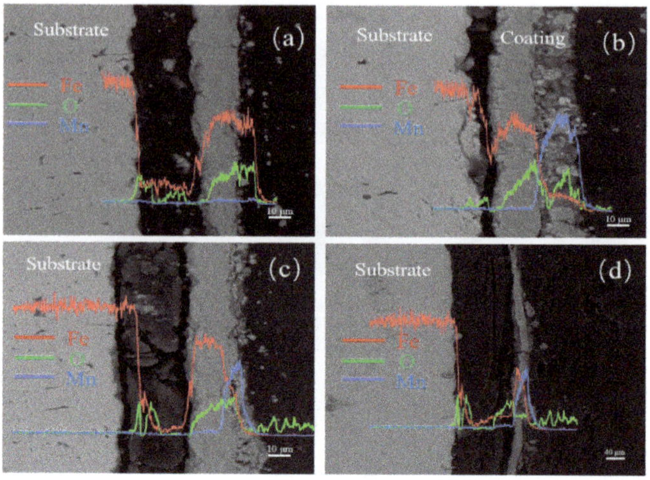

Figure 6. Cross-sectional morphology of samples after (**a**) 60 min, (**b**) 90 min, (**c**) 150 min, and (**d**) 180 min of electrolysis with 0.1 mol/L Mn^{2+} concentrations.

From Figure 6a, it can be observed that the area is divided into three layers. The leftmost layer is the substrate. According to the line scan results, the middle layer shows a decrease in Fe atoms and an increase in O atoms, possibly due to partial oxidation of the substrate. The prepared spinel coating is not observed, likely because the electrolysis time was too short, resulting in a thin rust layer and thus the $MnFe_2O_4$ being synthesized only on the surface of the substrate. In Figure 6b, a crack separates the middle layer from the substrate, possibly indicating delamination between the rust layer and the substrate during electrolysis. The O content in the middle layer is increased, and combined with the microstructural characterizations, this region is identified as the rust layer formed during electrolysis. Just to the right of the rust layer, small particles are seen filling in to form a new region, where the Mn content is significantly increased. According to the XRD and FT-IR characterizations, this region is identified as the $MnFe_2O_4$ spinel coating. Figure 6c shows that the spinel coating, located to the far right near the resin, has a low thickness and an uneven distribution, possibly due to defects in the rust layer caused by the prolonged electrolysis time. In Figure 6d, the bonding between the rust layer and the spinel coating is poor. The middle layer appears to be infiltrated by resin, which has penetrated into the substrate and rust layer. Combining the observations from Figure 6a–d, it can be concluded that with increasing electrolysis time, the rust layer initially increases and then decreases. This may be due to poor adhesion between the substrate and the rust layer, leading to delamination when the electrolysis time is too long.

3.2. *The Effect of Mn^{2+} Concentration on the Corrosion Resistance of $MnFe_2O_4$ Spinel Coatings*

3.2.1. XRD Analysis of Samples with Different Mn^{2+} Concentrations

Figure 7 illustrates the effect of different Mn^{2+} concentrations on the preparation of $MnFe_2O_4$ spinel coatings. As shown in the figure, under the condition of 0.02 mol/L Mn^{2+}, $MnFe_2O_4$ is not formed, likely due to the Mn^{2+} concentration being too low to initiate the reaction, resulting in Fe_3O_4 as the primary phase. As the Mn^{2+} concentration gradually increases, irregular iron-manganese oxides are initially formed, followed by the appearance of characteristic peaks of $MnFe_2O_4$. At 0.1 mol/L Mn^{2+}, the characteristic diffraction peaks of (111), (220), (311), (400), (511), and (440) appear, matching the standard $MnFe_2O_4$ pattern (JCPDS: 73-1964). Impurity peaks are almost eliminated, and the diffraction peaks exhibit high intensity and good crystallinity. Therefore, the optimal concentration of Mn^{2+} is determined to be 0.1 mol/L.

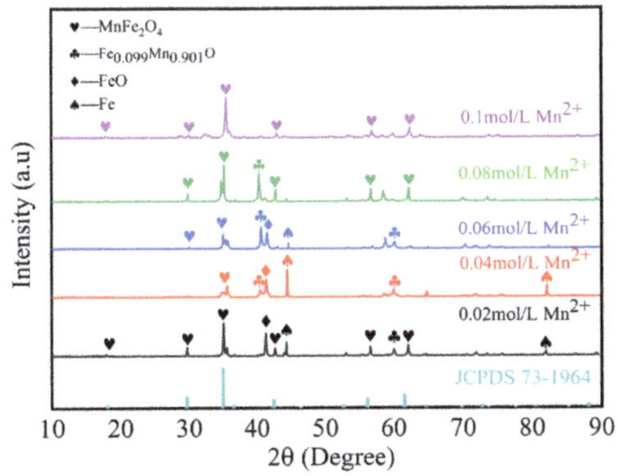

Figure 7. XRD of samples after 120 min of electrolysis with different Mn^{2+} concentrations: 0.02 mol/L, 0.04 mol/L, 0.06 mol/L, 0.08 mol/L, and 0.1 mol/L.

3.2.2. FT-IR Analysis of Samples with Different Mn^{2+} Concentrations

Figure 8a shows the FT-IR spectra of the precursors with different Mn^{2+} concentrations before calcination, and Figure 8b shows the FT-IR spectra after calcination. From Figure 8a, a broad absorption peak near 3230 cm^{-1} is observed, corresponding to the stretching vibration of -OH. The peak around 1570 cm^{-1} corresponds to the stretching vibration of N-O, possibly indicating the presence of NO^{3-} in the precursor. The peaks near 1290 cm^{-1} and 1040 cm^{-1} correspond to the C-O stretching vibrations in -COOH. The absorption peaks below 1000 cm^{-1} are attributed to the γ_2 peaks of Fe-OH and Mn-OH. As the Mn^{2+} concentration increases, the peak positions remain unchanged, but the widths increase. From Figure 8b, it is evident that $MnFe_2O_4$ exhibits two characteristic peaks in the 400–600 cm^{-1} range, corresponding to the Mn-O and Fe-O absorption peaks. At low concentrations, these characteristic peaks are either absent or have very low transmittance, and there is an impurity peak at 1360 cm^{-1}. When the Mn^{2+} concentration is 0.08 mol/L, two characteristic peaks appear around 594 cm^{-1} and 471 cm^{-1}, with very few impurity peaks. When the Mn^{2+} concentration is 0.1 mol/L, two characteristic peaks appear near 596 cm^{-1} and 476 cm^{-1}, with very few impurity peaks, indicating the formation of $MnFe_2O_4$ spinel at both 0.08 mol/L and 0.1 mol/L, which is consistent with the XRD analysis.

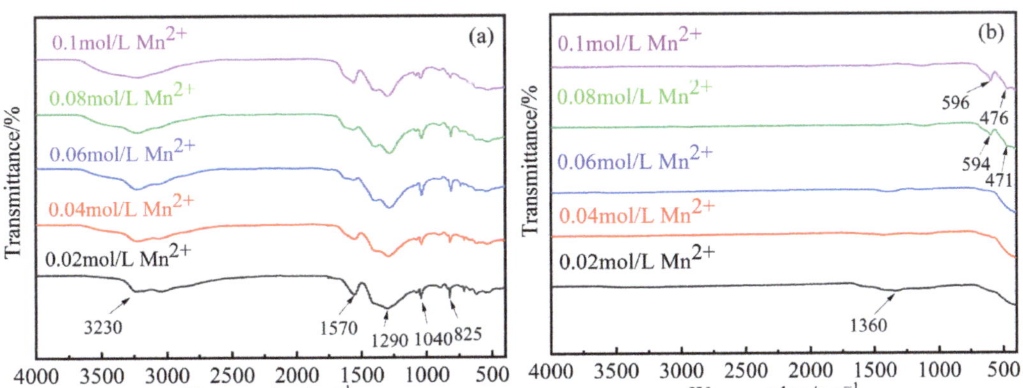

Figure 8. FT-IR of samples after 120 min of electrolysis with different Mn^{2+} concentrations: 0.02 mol/L, 0.04 mol/L, 0.06 mol/L, 0.08 mol/L, and 0.1 mol/L. (**a**) Uncalcined precursor, (**b**) Calcined precursor.

3.2.3. Surface Morphology Analysis of Samples with Different Mn^{2+} Concentrations

Figure 9a–e shows the morphological patterns of the samples of $MnFe_2O_4$ spinel coatings prepared with Mn^{2+} concentrations of 0.02, 0.04, 0.06, 0.08, and 0.1 mol/L, respectively. In Figure 9a, two main types of shapes can be observed: the majority are irregular protrusions identified as Fe_3O_4, while the less common shapes are three-dimensional convex polygons identified as Fe, which are scattered and disordered. Figure 9b shows, in addition to the previously mentioned shapes, a new layer-like structure identified as FeO. This structure is very smooth and closely bonded [33]. In Figure 9c, the synthesis of iron-manganese oxides begins, but due to the insufficient Mn^{2+} concentration, irregular iron-manganese compounds are formed. The morphological patterns of the samples appear spherical, with uneven surface morphology and noticeable agglomeration [34,35]. Figure 9d shows the formation of $MnFe_2O_4$ spinel with a polygonal surface morphology, but Fe_3O_4 is still present, indicating that the reaction is incomplete and that the Mn^{2+} concentration is still insufficient [36]. In Figure 9e, with a Mn^{2+} concentration of 0.1 mol/L, the reaction is complete. The sol-gel prepared $MnFe_2O_4$ shows well-developed polyhedral crystals with sizes around 125 ± 15 nm, exhibiting stacking and slight agglomeration. This is consistent with the microstructure of $MnFe_2O_4$ previously observed by scientists [37,38].

Figure 9. Morphological patterns of samples after 120 min of electrolysis with different Mn^{2+} concentrations: (**a**) 0.02 mol/L, (**b**) 0.04 mol/L, (**c**) 0.06 mol/L, (**d**) 0.08 mol/L, (**e**) 0.1 mol/L.

3.2.4. Cross-Sectional Morphology Analysis of Samples with Different Mn^{2+} Concentrations

After a current density of 1500 A/m^2 for 120 min of electrolysis, Figure 10a,b shows the cross-sectional morphologies of the coatings formed with 0.08 mol/L Mn^{2+} and 0.1 mol/L Mn^{2+}, respectively, with line scans included. From Figure 10a, the $MnFe_2O_4$ spinel coating thickness is around 15 μm. The coating exhibits noticeable voids and layering, with the layering located in the middle of the coating, showing distinct differences in morphology above and below this layer. This suggests a separation between Fe_3O_4 and $MnFe_2O_4$. The area near the substrate has a finer structure and shows good transition. From Figure 10b, the $MnFe_2O_4$ spinel coating thickness is around 20 μm. The coating contains fewer voids, and near the substrate, there is a visible transition between the substrate and the coating, with interlocking structures. The voids appear as fine particles in the morphological patterns of the samples. This indicates that the coating has a tight bond with the substrate, with a dense and strong exposed surface. Furthermore, the line scan results indicate that the Mn atomic ratio at the 0.08 mol/L Mn^{2+} concentration increases slightly compared to the substrate. However, at the 0.1 mol/L Mn^{2+} concentration, the Mn atomic ratio significantly increases compared to the substrate, indicating the successful preparation of the spinel coating.

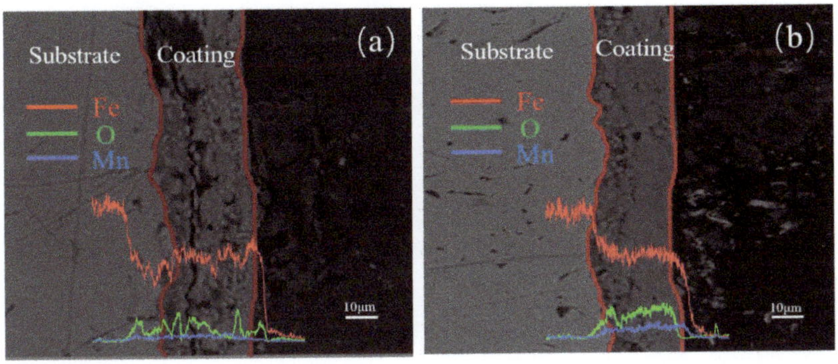

Figure 10. Cross-sectional morphology of samples after 120 min of electrolysis with different Mn^{2+} concentrations: (**a**) 0.08 mol/L, (**b**) 0.1 mol/L.

3.3. *Corrosion Resistance Testing of $MnFe_2O_4$ Spinel Coatings*

3.3.1. Potentiodynamic Polarization Measurement

Figure 11a,b shows the polarization curves of the prepared $MnFe_2O_4$ spinel coating and Q235 steel in NaCl solutions with mass fractions of 1 wt.%, 2 wt.%, 3 wt.%, 3.5 wt.%, 4 wt.%, and 5 wt.%, respectively. The scan rate was 0.1 mV/s. The polarization curves were fitted using Zahner Analysis software (https://zahner.de/products-details/software/zahner-analysis, accessed on 20 August 2024) to calculate the corrosion potential (E_{corr}), corrosion current density (I_{corr}), and polarization resistance (R_p) values, as shown in Tables 1 and 2. From Table 2, it can be seen that the self-corrosion potential of Q235 steel is around −400 mV, while the self-corrosion potential of the $MnFe_2O_4$ spinel coated Q235 steel is around −250 mV, an increase of 150 mV, indicating that the $MnFe_2O_4$ spinel coating reduces the corrosion tendency of Q235 steel. Comparing Tables 1 and 2, it is evident that the corrosion current of the $MnFe_2O_4$ spinel coated Q235 steel is consistently several times lower than that of Q235 steel in NaCl solutions of different mass fractions. Specifically, the corrosion current density of the $MnFe_2O_4$ spinel coating in 1 wt.% NaCl solution is 6.79 μA/cm^2, which is significantly lower than the 24.8 μA/cm^2 of Q235 steel in the same solution. This indicates that the prepared $MnFe_2O_4$ spinel coating exhibits significantly better corrosion resistance than uncoated Q235 steel. By comparing the data in Tables 1 and 2, we found that the corrosion currents of uncoated Q235 steel in NaCl solutions with concentrations of wt.% = 1%, 2%, 3%, 3.5%, 4%, and 5% are 24.8 μA/cm^2, 29.4 μA/cm^2, 53.2 μA/cm^2, 72.3 μA/cm^2, 62.5 μA/cm^2, and 55.7 μA/cm^2, respectively. In contrast, the corrosion currents of Q235 steel with $MnFe_2O_4$ spinel coating are 6.79 μA/cm^2, 10.6 μA/cm^2, 12.4 μA/cm^2, 15.6 μA/cm^2, 41.0 μA/cm^2, and 40.0 μA/cm^2, respectively, showing reductions of 72.6%, 63.9%, 76.6%, 78.4%, 34.4%, and 28.2%. This indicates that corrosion resistance is improved at each concentration of the corrosive solution. As the concentration of Cl^- in the corrosive solution increases, the protective effect of the spinel coating on Q235 steel is somewhat limited. This limitation is likely due to the Cl^- ions causing pitting corrosion on the coating, which then progresses to more extensive corrosion. However, compared to uncoated Q235 steel, the $MnFe_2O_4$ spinel coating still provides significant protection to the substrate. The concentration gradient of NaCl corrosive solutions chosen for this study is wt.% = 1, 2, 3, 3.5, 4, and 5, as the Cl^- concentration in seawater is usually less than wt.% = 5 NaCl, and in practical industrial production, the Cl^- concentration is typically less than wt.% = 3.5 NaCl. Therefore, we believe that the

$MnFe_2O_4$ spinel coating can meet the corrosion resistance requirements for steel in coastal industrial and craft applications.

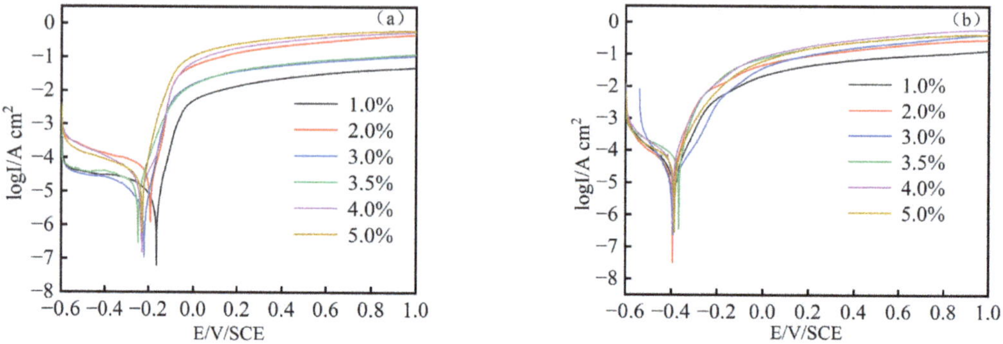

Figure 11. Polarization curve of samples after 120 min of electrolysis with or without coating in NaCl solution with different mass fractions: 1%, 2%, 3%, 3.5%, 4%, and 5%. (**a**) $MnFe_2O_4$ spinel coating, (**b**) Q235 steel.

Table 1. Electrochemical parameters of polarization curves of $MnFe_2O_4$ spinel coating under different mass fractions of NaCl: 1%, 2%, 3%, 3.5%, 4%, and 5%.

NaCl wt.%	$MnFe_2O_4$ Spinel Coating		
	E_{corr} (mV)	I_{corr} (μA/cm^2)	R_p (kΩ·cm^2)
1.0	−169	6.79	2.330
2.0	−231	10.6	1.860
3.0	−249	12.4	1.830
3.5	−253	15.6.	1.390
4.0	−247	41.0	0.561
5.0	−235	40.0	0.473

Table 2. Electrochemical parameters of polarization curves of Q235 steel under different mass fractions of NaCl: 1%, 2%, 3%, 3.5%, 4%, and 5%.

NaCl wt.%	Q235		
	E_{corr} (mV)	I_{corr} (μA/cm^2)	R_p (kΩ·cm^2)
1.0	−385	24.8	0.857
2.0	−396	29.4	0.586
3.0	−395	53.2	0.518
3.5	−379	72.3	0.292
4.0	−400	62.5	0.351
5.0	−397	55.7	0.497

3.3.2. Electrochemical Impedance Spectroscopy Analysis

Figure 12a shows the Nyquist plots for the $MnFe_2O_4$ spinel coating and Q235 steel in NaCl solutions with mass fractions of 1 wt.%, 2 wt.%, 3 wt.%, 3.5 wt.%, 4 wt.%, and 5 wt.%. The fitting of these plots results in the equivalent circuit diagram [39–41] shown in Figure 13a and the electrochemical impedance spectroscopy (EIS) fitting data for the $MnFe_2O_4$ spinel coating listed in Table 3. In this circuit, R_0 represents the solution resistance, R_1 represents the coating resistance, R_2 represents the charge transfer resistance, CPE represents the coating capacitance, and C_0 is the double-layer capacitance. Similarly, Figure 12b shows the Nyquist plots for Q235 steel in NaCl solutions with the same mass fractions, with the corresponding equivalent circuit diagram shown in Figure 13b and the EIS fitting data for Q235 steel listed in Table 4. In this circuit, R_0 represents the solution resistance, indicating the resistance between the saturated calomel electrode and Q235 steel, R_1 represents the rust layer resistance, R_2 represents the charge transfer resistance, and CPE represents the capacitance. The impedance response of the Constant Phase Element (CPE) is defined as:

$$Z_{CPE} = (j\omega)^{-n} / Y_0 \tag{2}$$

where j is an imaginary unit, ω is the angular frequency of the sinusoidal signal, Y_0 is the admittance at $\omega = 1$ rad/s $(1/|Z|)$, and n $(0 \leq n \leq 1)$ is the exponent. At $n = 1$, CPE behaves as a capacitive component, and at $n = 0.5$, it exhibits Warburg element characteristics, with a pure resistor at 0 [42]. Due to the non-ideal capacitive characteristics of non-uniform electrodes, C_0 substitutes the ideal double-layer capacitance (C_0). The capacitance of CPE, using $\omega_{\max}$ (the frequency at which the imaginary part of the impedance reaches its maximum value), is calculated as follows [43]:

$$C_{CPE} = Y_0(\omega_{max})^{n-1} \tag{3}$$

$$C_0 = Y_0^{1/n}\left(\frac{1}{R_1} + \frac{1}{R_2}\right)^{(n-1)/n} \tag{4}$$

Table 3. $MnFe_2O_4$ spinel coating fitting data of electrochemical impedance spectroscopy under different mass fractions of NaCl: 1%, 2%, 3%, 3.5%, 4%, and 5%.

NaCl wt.%	MnFe₂O₄ Spinel Coating					
	$R_0/\Omega\cdot cm^2$	$R_1/\Omega\cdot cm^2$	$R_2/\Omega\cdot cm^2$	$CPE/mF\cdot cm^{-2}$	n	$C_0/\mu F\cdot cm^{-2}$
1.0	7.76	77.7	536	762	0.780	67.4
2.0	4.26	3.08	523	679	0.712	52.4
3.0	2.51	6.37	476	638	0.637	285
3.5	2.25	5.75	466	722	0.721	670
4.0	2.35	17.7	553	709	0.754	44.2
5.0	1.35	8.61	501	712	0.712	127

Table 4. Q235 fitting data of electrochemical impedance spectroscopy under different mass fractions of NaCl: 1%, 2%, 3%, 3.5%, 4%, and 5%.

NaCl wt.%	Q235						
	$R_0/\Omega\cdot cm^2$	$R_1/\Omega\cdot cm^2$	$R_2/\Omega\cdot cm^2$	$CPE/mF\cdot cm^{-2}$	n_1	$CPE/mF\cdot cm^{-2}$	n_2
1.0	8.10	301	6.17	804	0.5	500	1
2.0	4.36	260	2.19	778	0.5	500	1
3.0	2.48	216	1.33	831	0.5	500	1
3.5	1.88	90.6	2.41	738	0.5	521	0.982
4.0	1.67	93.6	0.49	819	0.5	500	0.934
5.0	1.61	156	0.86	821	1	500	0.5

The film is considered a planar capacitor. From Figure 12a, it can be seen that at low NaCl mass fractions, the Nyquist plot initially appears as a straight line, transitioning to a semicircular shape at low frequencies. The $MnFe_2O_4$ spinel coating exhibits similar capacitive arcs in NaCl solutions of varying mass fractions, indicating that changes in the NaCl concentration do not affect the corrosion behavior of the $MnFe_2O_4$ spinel coating. The $MnFe_2O_4$ spinel coating shows a capacitive arc at low frequencies, with the arc radius initially decreasing and then increasing with the NaCl mass fraction, reflecting an inverse trend in corrosion resistance. The radius of the capacitive arc is largest when wt.(NaCl)% = 1. Compared to Figure 12b, the Q235 steel also exhibits capacitive arcs in low NaCl mass fraction solutions. From Table 3, it is evident that the charge transfer resistance R_2 initially decreases and then increases, with the minimum R_2 observed at 3.5 wt.% NaCl. However, compared to Table 4, R_2 for the $MnFe_2O_4$ spinel coating is several hundred times higher, indicating that the $MnFe_2O_4$ spinel coating exhibits varying corrosion resistance in NaCl solutions of different concentrations. This is because the increasing Cl^- ions in the corrosive solution challenge the pitting resistance of the $MnFe_2O_4$ spinel coating. Under low Cl^- conditions, the corrosion reaction of the spinel coating is less efficient, whereas at higher concentrations, the reaction rate increases. When pitting occurs on the coating surface, more Cl^- ions begin to accumulate at the pitting sites, leading to local concentration polarization and accelerating the corrosion reaction from within the pits. However, the $MnFe_2O_4$ spinel coating enhances the corrosion resistance of the substrate to varying degrees at all concentrations.

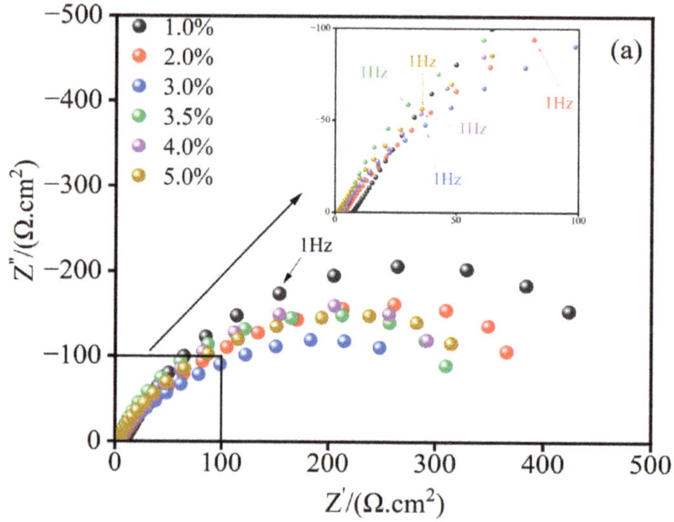

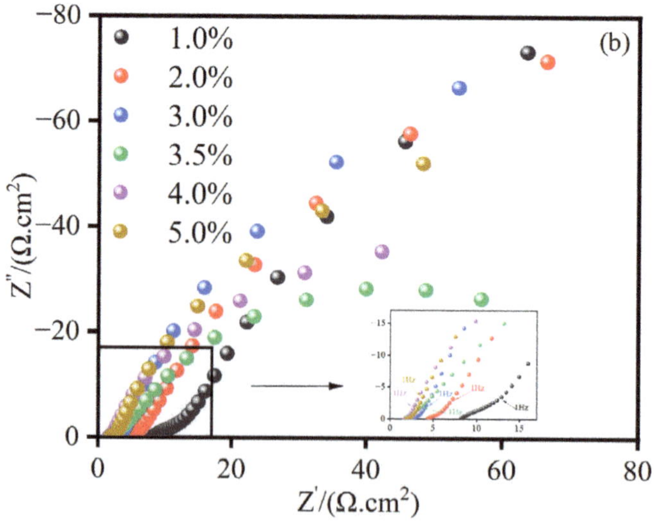

Figure 12. Nyquist curve of samples after 120 min of electrolysis with or without coating in NaCl solution with different mass fractions: 1%, 2%, 3%, 3.5%, 4%, and 5%. (**a**) $MnFe_2O_4$ spinel coatings, (**b**) Q235 steel.

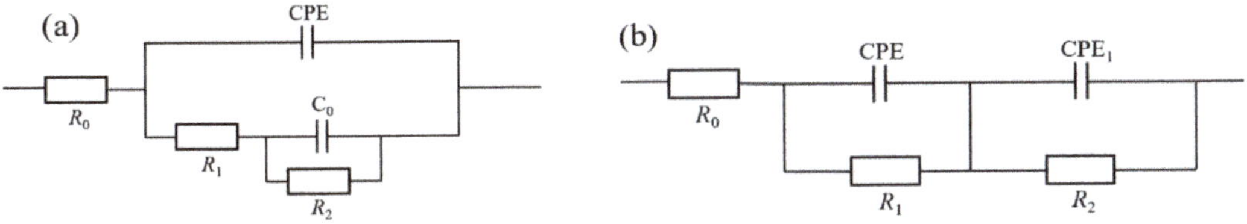

Figure 13. Equivalent circuit diagram of fitting impedance spectrum of different substances. (**a**) $MnFe_2O_4$ spinel coatings, (**b**) Q235 steel.

Figure 14 shows the Bode plots of coated Q235 steel in NaCl solutions with mass fractions of 1 wt.%, 2 wt.%, 3 wt.%, 3.5 wt.%, 4 wt.%, and 5 wt.%. In these plots, Figure 14a represents the variation of impedance with frequency, and Figure 14b represents the variation of the phase angle with frequency. From Figure 14a,b, it can be observed that a capacitive arc appears in the low-frequency region. As the NaCl mass fraction increases, the peak value of the phase angle generally shifts towards the high-frequency region, which explains the improvement in the barrier properties of the coating, suggesting that the coating is more effective in blocking corrosive agents. Furthermore, $|Z|$ decreases with the increase in NaCl concentration. Under the condition of wt. NaCl% = 1, the diameter of the capacitive loop and the absolute value of the impedance $|Z|$ are the highest, indicating that the $MnFe_2O_4$ spinel coating exhibits the strongest corrosion resistance under this condition. In Figure 14a, all curves show little change in $|Z|$ in the high-frequency region, and from Table 3, it can be seen that the change in R_2 is smaller than the change in R_1 with varying concentrations of the corrosive solution. This indicates that the corrosion reaction on the surface of the $MnFe_2O_4$ spinel coating is mainly hindered by the accumulation of corrosion products rather than being inhibited by the reaction itself. In Table 4, both R_1 and R_2 show significant changes, indicating that uncoated Q235 steel undergoes both diffusion and charge transfer during corrosion. The $MnFe_2O_4$ spinel coating can inhibit the charge transfer process.

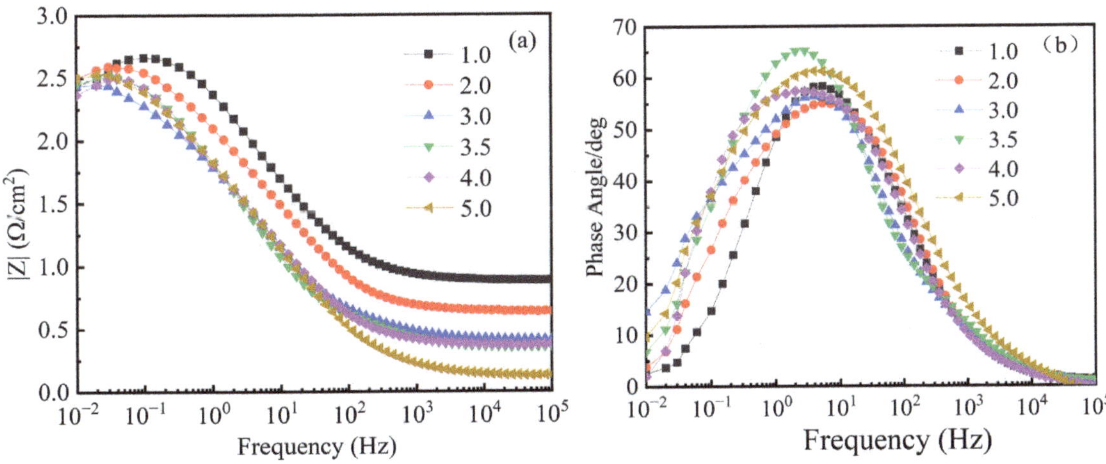

Figure 14. Bode diagram of $MnFe_2O_4$ spinel coating under different mass fractions of NaCl: 1%, 2%, 3%, 3.5%, 4%, and 5%. (**a**) Variation of impedance with frequency, (**b**) Variation of phase angle with frequency.

The corrosion resistance of $MnFe_2O_4$ spinel coatings in NaCl solutions with different mass fractions was investigated. The results of the polarization curves and electrochemical impedance tests indicate that $MnFe_2O_4$ spinel coatings exhibit higher corrosion resistance in low mass fraction NaCl solutions. Notably, in a 1 wt.% NaCl solution, the corrosion current density of the $MnFe_2O_4$ spinel coating is 6.79 μA/cm^2, which is significantly lower than the corrosion current density of Q235 steel under the same conditions at 24.8 μA/cm^2. This demonstrates that the $MnFe_2O_4$ spinel coating enhances the corrosion resistance of Q235 steel in NaCl solutions of varying concentrations. A schematic illustration of the corrosion process is presented in Figure 15.

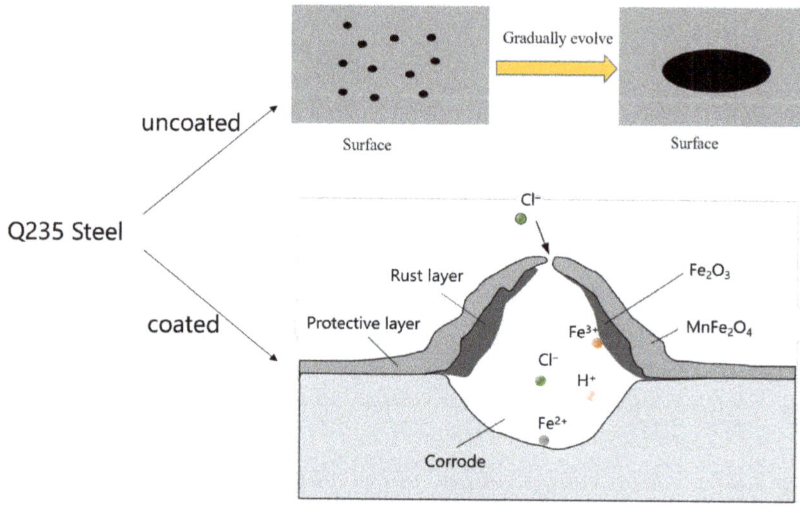

Figure 15. The schematic diagram of the corrosion mechanism.

4. Conclusions

This thesis primarily addresses the corrosion susceptibility of Q235 steel in atmospheric environments. It proposes a new approach involving the enhancement of rust formation to create a stable rust layer, followed by the in situ preparation of a precursor using the sol-gel method and high-temperature in situ processing to fabricate $MnFe_2O_4$ spinel corrosion-resistant coatings. The preparation process of $MnFe_2O_4$ spinel corrosion-resistant coatings is investigated, and their corrosion resistance is explored using polarization curves, electrochemical impedance spectroscopy, and corrosion weight loss methods. This method has low dependency on equipment, utilizes inexpensive and readily available materials, and can complete the reinforcement treatment of the rust layer on the substrate surface at ambient pressure and relatively low temperatures. Consequently, it significantly enhances the corrosion resistance of corroded Q235 steel. This makes the method applicable to practical industrial production, providing a strategy for the recycling of metals discarded due to corrosion. It can effectively reduce resource waste and environmental pollution. The main conclusions are as follows:

(1) Q235 steel undergoes electrolytic rust enhancement to form a stable rust layer. The precursor can then be prepared in situ using the sol-gel method, and $MnFe_2O_4$ spinel corrosion-resistant coatings can be fabricated through high-temperature in situ processing.
(2) Based on the XRD, FT-IR, morphological patterns of the samples, and cross-sectional morphology characterizations, the optimal conditions for precursor preparation are determined to be a Mn^{2+} concentration of 0.1 mol/L and an electrolysis time of 2 h.
(3) The results of the polarization curves indicate that $MnFe_2O_4$ spinel corrosion-resistant coatings exhibit higher corrosion resistance in low mass fraction NaCl solutions. Notably, in a 1 wt.% NaCl solution, the corrosion current density of the $MnFe_2O_4$

spinel coating is 6.79 μA/cm^2, which is significantly lower than the corrosion current density of Q235 steel under the same conditions at 24.8 μA/cm^2. This demonstrates that the $MnFe_2O_4$ spinel coating enhances the corrosion resistance of Q235 steel.

(4) The electrochemical impedance spectroscopy (EIS) results indicate that the $MnFe_2O_4$ spinel corrosion-resistant coating enhances corrosion resistance by inhibiting the charge transfer process. Compared to uncoated Q235 steel, the $MnFe_2O_4$ spinel coating significantly protects Q235 steel in corrosion solutions with wt.% NaCl = 1, 2, 3, 3.5, 4, and 5, exhibiting the strongest corrosion resistance at wt.% NaCl = 1.

Author Contributions: B.-A.F.: Conceptualization, Methodology, Data curation, Writing original draft, Software. X.M.: Software. T.-A.Z.: Resources, Project administration, Validation, Supervision, Writing—review and editing, Funding acquisition. All authors have read and agreed to the published version of the manuscript.

Funding: This research was funded by the National Natural Science Foundation of China, grant number 52174333.

Institutional Review Board Statement: Not applicable.

Informed Consent Statement: Not applicable.

Data Availability Statement: The raw data supporting the conclusions of this article will be made available by the authors on request.

Conflicts of Interest: The authors declare no conflict of interest.

References

1. IMPACT. *The International Measures of Prevention, Application, and Economics of Corrosion Technologies Study*; NACE International: Houston, TX, USA, 2016.
2. Kirchgeorg, T.; Weinberg, I.; Hörnig, M.; Baier, R.; Schmid, M.J.; Brockmeyer, B. Emissions from Corrosion Protection Systems of Offshore Wind Farms: Evaluation of the Potential Impact on the Marine Environment. *Mar. Pollut. Bull.* **2018**, *136*, 257–268. [CrossRef]
3. Oldfield, J.W.; Todd, B. Environmental Aspects of Corrosion in MSF and RO Desalination Plants. *Desalination* **1997**, *108*, 27–36. [CrossRef]
4. Sadeghi Meresht, E.; Shahrabi Farahani, T.; Neshati, J. Failure Analysis of Stress Corrosion Cracking Occurred in a Gas Transmission Steel Pipeline. *Eng. Fail. Anal.* **2011**, *18*, 963–970. [CrossRef]
5. Mandke, J.S. Corrosion Causes Most Pipeline Failures in Gulf of Mexico. *Oil Gas J.* **1990**, *88*, 44.
6. Panteleeva, M. Effective Modern Methods of Protecting Metal Road Structures from Corrosion. *IOP Conf. Ser. Earth Environ. Sci.* **2017**, *90*, 012119. [CrossRef]
7. Aslam, J.; Aslam, R.; Lone, I.H.; Radwan, N.R.E.; Mobin, M.; Aslam, A.; Parveen, M.; Al-Freedi, A.A.; Alzulaibani, A.A. Inhibitory Effect of 2-Nitroacridone on Corrosion of Low Carbon Steel in 1 M HCl Solution: An Experimental and Theoretical Approach. *J. Mater. Res. Technol.* **2020**, *9*, 4061–4075. [CrossRef]
8. Azam, M.A.; Sukarti, S.; Zaimi, M. Corrosion Behavior of API-5L-X42 Petroleum/Natural Gas Pipeline Steel in South China Sea and Strait of Melaka Seawaters. *Eng. Fail. Anal.* **2020**, *115*, 104654. [CrossRef]
9. Gao, F.; Li, J.; Zhou, N.; Luo, X.; Yang, H.; Chai, F.; Yang, C. Investigating the Corrosion Performance of Hull Steel with Different Microstructure in a Tropical Marine Atmosphere. *J. Mater. Res. Technol.* **2023**, *27*, 2600–2614. [CrossRef]
10. Solomon, M.M.; Umoren, S.A.; Udosoro, I.I.; Udoh, A.P. Inhibitive and Adsorption Behaviour of Carboxymethyl Cellulose on Mild Steel Corrosion in Sulphuric Acid Solution. *Corros. Sci.* **2010**, *52*, 1317–1325. [CrossRef]
11. Zhang, H.; Gao, K.; Yan, L.; Pang, X. Inhibition of the Corrosion of X70 and Q235 Steel in CO_2 -Saturated Brine by Imidazoline-Based Inhibitor. *J. Electroanal. Chem.* **2017**, *791*, 83–94. [CrossRef]
12. Yang, X.; Qing, W.; Feng, C.; Yin-Wei, M.; Dong-liang, Q. Sol-Gel Process and High-Temperature Property of SiO_2/ZrO_2-SiO_2 Composites. *Ceram. Int.* **2017**, *43*, 854–859. [CrossRef]
13. Chen, X.H.; Dong, J.H.; Han, E.H.; Ke, W. Effect of Al Alloying on Corrosion Performance of Steel. *Corros. Eng. Sci. Technol.* **2007**, *42*, 224–231. [CrossRef]
14. Guo, X.; Zhu, J.; Kang, J.; Duan, M.; Wang, Y. Rust Layer Adhesion Capability and Corrosion Behavior of Weathering Steel under Tension during Initial Stages of Simulated Marine Atmospheric Corrosion. *Constr. Build. Mater.* **2020**, *234*, 117393. [CrossRef]
15. Karim, M.R.A.; Daniel, S.; Ul Haq, E.; Ahmad, A.; Khan, K.I.; Raza, S.A. Improving the Corrosion Resistance of API X56 and API X70 Pipeline Steels by Hot Dip Aluminization. *KEM* **2021**, *875*, 315–321. [CrossRef]
16. Nouri, M.; Sun, X.; Li, D.Y. Beneficial Effects of Yttrium on the Performance of Mg–3%Al Alloy during Wear, Corrosion and Corrosive Wear. *Tribol. Int.* **2013**, *67*, 154–163. [CrossRef]

17. Liu, X.; Shan, D.; Song, Y.; Han, E.H. Influence of Yttrium Element on the Corrosion Behaviors of Mg–Y Binary Magnesium Alloy. *J. Magnes. Alloys* **2017**, *5*, 26–34. [CrossRef]
18. Feng, M.; Liu, Y.; Zhang, S.; Liu, Y.; Luo, N.; Wang, D. Carbon Quantum Dots (CQDs) Modified TiO_2 Nanorods Photoelectrode for Enhanced Photocathodic Protection of Q235 Carbon Steel. *Corros. Sci.* **2020**, *176*, 108919. [CrossRef]
19. Jamshidi, R.; Bayat, O.; Heidarpour, A. Tribological and Corrosion Behavior of Flame Sprayed Al–10 Wt.% Ti3SiC2 Composite Coating on Carbon Steel. *Surf. Coat. Technol.* **2019**, *358*, 1–10. [CrossRef]
20. Wu, Y.; Zhao, W.; Ou, J. Stable, Superfast and Self-Healing Fluid Coating with Active Corrosion Resistance. *Adv. Colloid Interface Sci.* **2021**, *295*, 102494. [CrossRef]
21. Huang, C.; Su, X.; Song, Q.; Wang, X. Effects of Temperature on Acceleration and Simulation of Indoor Corrosion Test of Q235 Carbon Steel. *Anti-Corros. Mater. Methods* **2021**, *68*, 564–570. [CrossRef]
22. Tiwari, S.K.; Sahu, R.K.; Pramanick, A.K.; Singh, R. Development of Conversion Coating on Mild Steel Prior to Sol Gel Nanostructured Al2O3 Coating for Enhancement of Corrosion Resistance. *Surf. Coat. Technol.* **2011**, *205*, 4960–4967. [CrossRef]
23. Kirtay, S. Preparation of Hybrid Silica Sol–Gel Coatings on Mild Steel Surfaces and Evaluation of Their Corrosion Resistance. *Prog. Org. Coat.* **2014**, *77*, 1861–1866. [CrossRef]
24. Wang, K.; Zhu, B.; Wang, Z.; Liu, Y.; Wang, L.; Zhang, Y.; Li, S. Successive Phase and Morphology Evolution of Galvannealed Coating in Hot Stamping and Diffusion Modeling of α-Fe(Zn)/Steel System Considering the Effect of Zn Concentration. *Surf. Coat. Technol.* **2019**, *380*, 125036. [CrossRef]
25. Oladijo, O.P.; Mathabatha, M.H.; Popoola, A.P.I.; Ntsoane, T.P. Characterization and Corrosion Behaviour of Plasma Sprayed Zn-Sn Alloy Coating on Mild Steel. *Surf. Coat. Technol.* **2018**, *352*, 654–661. [CrossRef]
26. Kang, J.-H.; Kim, D.; Kim, D.H.; Kim, S.-J. Fe-Zn Reaction and Its Influence on Microcracks during Hot Tensile Deformation of Galvanized 22MnB5 Steel. *Surf. Coat. Technol.* **2019**, *357*, 1069–1075. [CrossRef]
27. Imanieh, I.; Yousefi, E.; Dolati, A.; Mohammadi, M.R. Experiments Design for Hardness Optimization of the Ni-Cr Alloy Electrodeposited by Pulse Plating. *Acta Metall. Sin.* **2013**, *26*, 558–564. [CrossRef]
28. Li, X.; Kong, G.; Zhang, J.; Lai, D.; Che, C.; Yang, B.; Zhang, S.; Gu, J. Corrosion Behavior of Q235 Steel in Graphene Oxide Aqueous Dispersion. *Surf. Topogr. Metrol. Prop.* **2022**, *10*, 015039. [CrossRef]
29. Quan, B.; Li, J.; Chen, C. Effect of Sulfur on Corrosion Behavior of Q235 and 16Mn Steel in Sodium Aluminate Solutions. *Mater. Res. Express* **2020**, *7*, 035602. [CrossRef]
30. Liu, P.; Li, G.; Liu, B.; Liao, C.; Baker, I.; Wu, H. Tribocorrosion Behavior of Additively-Manufactured Fe-Mn Alloys. *Mater. Lett.* **2023**, *337*, 133949. [CrossRef]
31. Park, J.S.; Kim, S.O.; Jeong, Y.J.; Lee, S.G.; Choi, J.K.; Kim, S.J. Long-Term Corrosion Behavior of Strong and Ductile High Mn-Low Cr Steel in Acidic Aqueous Environments. *Materials* **2022**, *15*, 1746. [CrossRef]
32. Gu, D.-D.; Peng, J.; Wang, J.-W.; Liu, Z.-T.; Pan, F.-S. Effect of Mn Modification on the Corrosion Susceptibility of Mg–Mn Alloys by Magnesium Scrap. *Acta Metall. Sin.* **2021**, *34*, 1–11. [CrossRef]
33. McAdam, G.; Young, D.J. Morphological Development of Oxide-Sulfide Scales on Iron and Iron-Manganese Alloys. *Oxid. Met.* **1987**, *28*, 165–181. [CrossRef]
34. Santos Martinez, J.D.; Ladino, D.H.; Calderón Hernández, J.W.; Falleiros, N.A.; de Melo, H.G. Corrosion Characterization of New Low Manganese Microalloyed X65 Steels in Sour and NON-SOUR Synthetic Seawater. *J. Mater. Res. Technol.* **2022**, *18*, 3198–3214. [CrossRef]
35. Jeong, Y.J.; Kim, S.O.; Park, J.S.; Lee, J.W.; Hwang, J.-K.; Lee, S.G.; Choi, J.K.; Kim, S.J. Strong and Ductile Fe-24Mn-3Cr Alloy Resistant against Erosion-Corrosion. *NPJ Mater. Degrad.* **2021**, *5*, 1–10. [CrossRef]
36. Dehestani, M.; Trumble, K.; Wang, H.; Wang, H.; Stanciu, L.A. Effects of Microstructure and Heat Treatment on Mechanical Properties and Corrosion Behavior of Powder Metallurgy Derived Fe–30Mn Alloy. *Mater. Sci. Eng. A* **2017**, *703*, 214–226. [CrossRef]
37. Yan, X.; Kang, S.; Xu, M.; Li, P. Corrosion Product Film of a Medium-Mn Steel Exposed to Simulated Marine Splash Zone Environment. *Materials* **2021**, *14*, 5652. [CrossRef]
38. Aliyu, A.; Srivastava, C. Morphology and Corrosion Properties of FeMn-Graphene Oxide Composite Coatings. *J. Alloys Compd.* **2020**, *821*, 153560. [CrossRef]
39. Mobin, M.; Aslam, R.; Salim, R.; Kaya, S. An Investigation on the Synthesis, Characterization and Anti-Corrosion Properties of Choline Based Ionic Liquids as Novel and Environmentally Friendly Inhibitors for Mild Steel Corrosion in 5% HCl. *J. Colloid Interface Sci.* **2022**, *620*, 293–312. [CrossRef]
40. Quan, B.; Li, J.; Chen, C. Effect of Corrosion Time on the Synergistic Corrosion of Q235 Steel in Sodium Aluminate Solutions. *Metals* **2021**, *11*, 753. [CrossRef]
41. Quan, B.; Li, J.; Chen, C. Effect of Corrosion Temperature on the Corrosion of Q235 Steel and 16Mn Steel in Sodium Aluminate Solutions. *ACS Omega* **2021**, *6*, 25904–25915. [CrossRef]

42. Chang, B.-Y. Conversion of a Constant Phase Element to an Equivalent Capacitor. *J. Electrochem. Sci. Technol.* **2020**, *11*, 318–321. [CrossRef]
43. Hirschorn, B.; Orazem, M.E.; Tribollet, B.; Vivier, V.; Frateur, I.; Musiani, M. Determination of Effective Capacitance and Film Thickness from Constant-Phase-Element Parameters. *Electrochim. Acta* **2010**, *55*, 6218–6227. [CrossRef]

materials

MDPI

Article

Investigation of Electropolishing for High-Gradient 1.3 GHz and 3.9 GHz Niobium Cavities

Yue Zong [1,2], Jinfang Chen [1,2,3,4,*], Dong Wang [1,2,3,4], Runzhi Xia [4], Jiani Wu [3], Zheng Wang [1,2], Shuai Xing [3], Xiaowei Wu [5], Xuhao He [4] and Xiaohu Wang [4]

1 Shanghai Institute of Applied Physics, Chinese Academy of Sciences, Shanghai 201800, China; zongyue@sinap.ac.cn (Y.Z.); wangdong@sinap.ac.cn (D.W.); wangzheng@sinap.ac.cn (Z.W.)
2 University of Chinese Academy of Sciences, Beijing 100049, China
3 Shanghai Advanced Research Institute, Chinese Academy of Sciences, Shanghai 201210, China; wujn@sari.ac.cn (J.W.); xings@sari.ac.cn (S.X.)
4 School of Physical Science and Technology, ShanghaiTech University, Shanghai 201210, China; xiarzh2023@shanghaitech.edu.cn (R.X.); hexh2022@shanghaitech.edu.cn (X.H.); wangxh4@shanghaitech.edu.cn (X.W.)
5 Zhangjiang Laboratory, Shanghai 201210, China; wuxw@zjlab.ac.cn
* Correspondence: chenjinfang@sari.ac.cn

Abstract: Electropolishing (EP) has become a standard procedure for treating the inner surfaces of superconducting radio-frequency (SRF) cavities composed of pure niobium. In this study, a new EP facility was employed for the surface treatment of both 1.3 GHz and 3.9 GHz single-cell cavities at the Wuxi Platform. The stable "cold EP" mode was successfully implemented on this newly designed EP facility. By integrating the cold EP process with a two-step baking approach, a maximum accelerating gradient exceeding 40 MV/m was achieved in 1.3 GHz single-cell cavities. Additionally, an update to this EP facility involved the design of a special cathode system for small-aperture structures, facilitating the cold EP process for 3.9 GHz single-cell cavities. Ultimately, a maximum accelerating gradient exceeding 25 MV/m was attained in the 3.9 GHz single-cell cavities after undergoing the cold EP treatment. The design and commissioning of the EP device, as well as the electropolishing and vertical test results of the single-cell cavities, will be detailed herein. These methods and experiences are also transferable to multi-cell cavities and elliptical cavities of other frequencies.

Keywords: electropolishing; cold EP; SRF cavity

1. Introduction

Superconducting materials, which are characterized by their low surface resistance and hysteresis loss, enable SRF cavities to operate with minimal energy consumption and withstand higher power operation, thus demonstrating attractive features in continuous-wave (cw) particle accelerators. Currently, SRF cavities are primarily fabricated from high-purity niobium, and the surface conditions directly affect their radio-frequency performance.

Following the fabrication of bare cavities, the inner surfaces usually exhibit insufficient smoothness and even defects, thereby constraining the RF performance of cavities. To optimize RF performance, the inner surface of SRF cavities should closely approximate the theoretical ideal. Generally, the initial 100–200 μm of material can be removed via processes such as buffered chemical polishing (BCP) or electropolishing (EP) [1]. The EP on a bulk Nb cavity is characterized by the elimination of microroughness (leveling) and the absence of crystallographic and grain boundary attack (brightening), resulting in the production of smooth, bright surfaces [2–5]. Thus, in contrast to BCP, EP offers a smoother surface and potentially improved RF performance [6–11]. The EP process was first introduced for cavity surface treatment by Diepers [12]. The EP technique with rotating niobium cavities was pioneered by researchers at KEK [13]. Afterward, the EP process was successfully applied to polish SRF cavities worldwide by several institutes, such as JLab [14], FNAL [15–17], ANL [18], DESY [10], and IHEP [19]. In 2017, a novel approach termed "cold EP" was introduced by researchers at Fermilab, demonstrating the ability to achieve a smoother surface and uniform material removal throughout the cell [16,17]. Particularly for cavities treated with a nitrogen-doping recipe, the attainment of a smoother surface is of paramount importance [20–22].

The SHINE project requires 600 1.3 GHz nine-cell TESLA cavities and 16 3.9 GHz nine-cell cavities with exceptional performance [23]. For the 1.3 GHz nine-cell cavities, the target is to achieve an intrinsic quality factor (Q_0) of 2.7×10^{10} at 16 MV/m, along with a maximum accelerating gradient (E_{acc}) exceeding 19 MV/m. Similarly, for the 3.9 GHz nine-cell cavities, the objective is to attain a Q_0 value of 2.0×10^9 at 13.1 MV/m, with the maximum E_{acc} surpassing 16.5 MV/m. Hence, the implementation of the cold EP process is imperative in realizing high accelerating gradients, especially for 1.3 GHz nine-cell cavities. In 2019, we undertook an upgrade of a simple EP device located in Ningxia, China, incorporating an outer surface cooling system. This device was subsequently employed in nitrogen-doping experiments conducted on single-cell cavities [24,25]. Starting in 2021, a novel platform for the surface treatment of SRF cavities known as the Wuxi platform was constructed in Wuxi, China. The newly built EP devices at the Wuxi platform have been used in high-Q recipe studies, including those of nitrogen doping and medium temperature baking [26,27]. This paper primarily presents the newly established EP apparatus at the Wuxi platform, which was employed for the surface treatment of 1.3 GHz and 3.9 GHz single-cell cavities. Utilizing the cold EP mode and a two-step baking process [28–31], we achieved an elevated accelerating gradient for 1.3 GHz single-cell cavities.

2. EP devices at SHINE

2.1. EP Device for a Single-Cell Cavity

In 2021, two novel EP devices were constructed for the SHINE project on the Wuxi platform. One is the single-cell EP device (referred to as a "small EP"), as depicted in Figure 1; it possesses the capacity to accommodate the treatment processes for 1.3 GHz single-cell cavities and for 3.9 GHz single-cell and nine-cell cavities. The other one is specifically tailored for the treatment of 1.3 GHz nine-cell cavities [26]. Both EP devices share the same acid-mixing system.

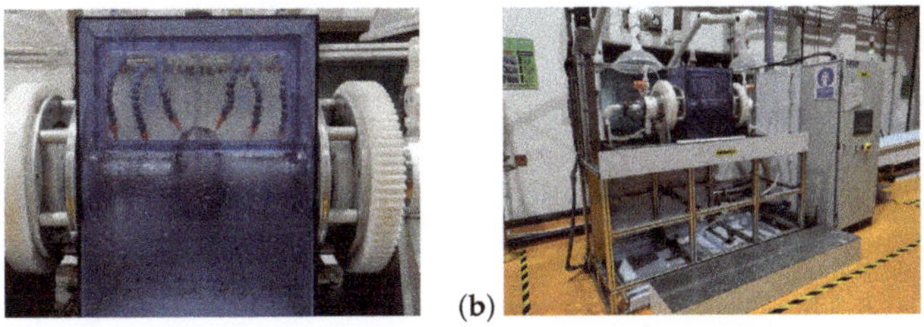

(a) (b)

Figure 1. (**a**,**b**) EP apparatus for 1.3 GHz single-cell cavities at the Wuxi Platform.

One of the most important objectives of the new EP system is to achieve steady temperature control. This is achieved by utilizing separate chiller units to independently regulate the temperatures of the external cooling water and the acid. The schematic of the single-cell EP device at the Wuxi platform is depicted in Figure 2a. Chiller 1 controls the acid temperature, while Chiller 2 regulates the cavity's external surface temperature. During the cold EP mode, the temperature of the cavity's external wall can be maintained below 8 °C, and the acid outlet temperature at both ends of the cavity can be maintained around 10 °C. Figure 1b illustrates the spray piping for external surface cooling water.

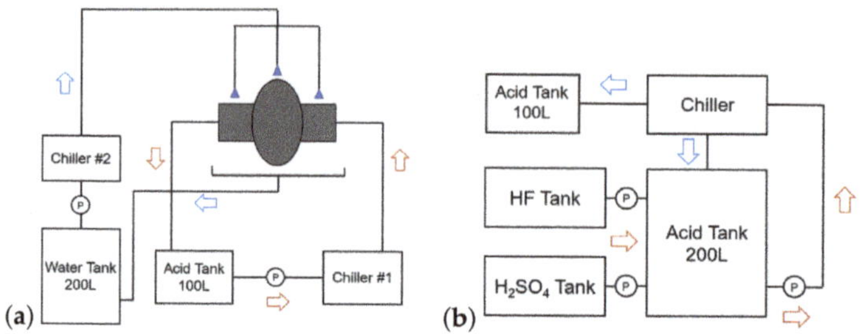

Figure 2. Schematics of single-cell EP device; (**a**) the polishing system and (**b**) the acid-mixing system.

The acid mixture ratio is based on the established classical proportions. The ratio of the electrolyte is H_2SO_4(98%):HF(48%) = 1:9 by volume. However, the newly designed EP system could automatically mix acid based on volume ratios, and a schematic diagram of the acid-mixing system is shown in Figure 2b. During the acid-mixing process, a separate refrigeration unit is employed for acid temperature control, thereby mitigating HF evaporation. The mixed acid is pumped into acid tanks in preparation for the EP treatment. To polish a 1.3 GHz single-cell cavity, around 60 L of electrolyte is needed every time. When the niobium concentration of the electrolyte is about 10 g/L, the used acid is dumped, and fresh acid is pumped into the acid tank. For light EP, fresh acid is used every time.

Both ends of the cathode rod are enveloped in polytetrafluoroethylene (PTFE) tape, exposing only the central region where the acid outlet is located, as shown in Figure 3. The diameter of this cathode rod is 24 mm. The area ratio between the cathode and anode is about 1:5 without PTFE tape and 1:10 during the EP process with PTFE tape. Overflow baffles are also installed at both ends of the device, and the cathode rod passes through the middle of the overflow baffle. When the acid level inside the cavity exceeds the height of the overflow baffle, the acid overflows and enters the EP return pipe, thereby maintaining the acid level inside the cavity at around 60%.

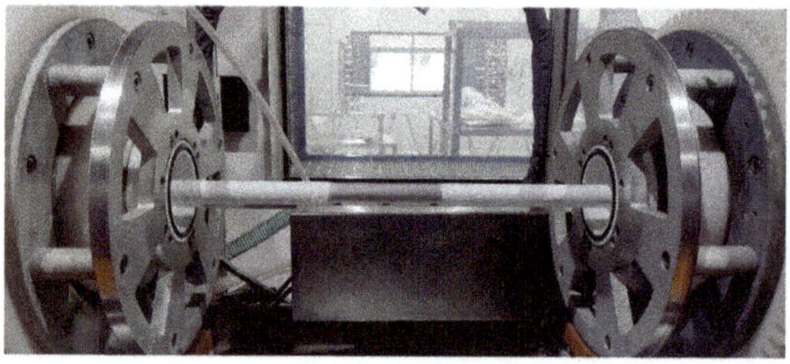

Figure 3. Cathode rod for the EP process of a 1.3 GHz single-cell cavity.

2.2. Parameters of the EP Device

In order to find the proper working point to polish a 1.3 GHz single-cell cavity, we measured the current–voltage (I–V) curves at different acid temperatures. Figure 4 shows the I–V curves for acid temperature ranging from 5 to 15 °C. To monitor the inlet and outlet acid temperatures, three acid temperature probes were strategically placed at the center of the cathode and at both ends of the cavity beam pipes. As shown in Figure 4a, with an increase in acid temperature, the reaction became more vigorous, leading to a slight uptick in the current. It can also be observed that there was no distinct plateau region in the curve, indicating an increase in the current even within the polishing region, which was consistent with the observations reported by Bertucci et al. [32]. Figure 4b, which presents the I–V curve at an acid temperature of 10 °C, shows that the voltage range from 0 to 8.5 V corresponded to a clear etching region, while the range from 8.5 to 13 V represented an oscillation region. Based on these observations, an operating voltage of 15 V was selected to ensure optimal polishing performance.

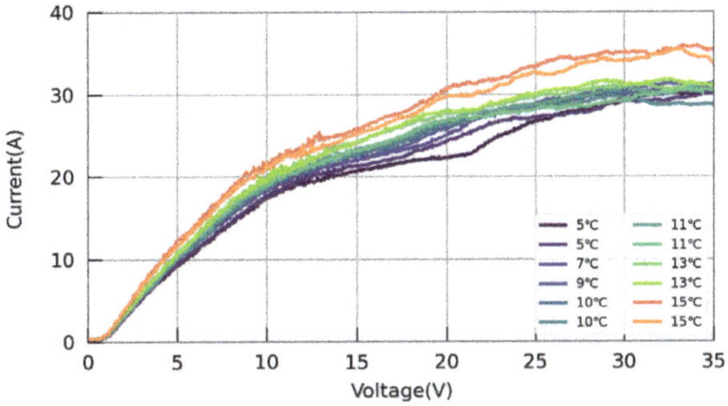

(**a**)

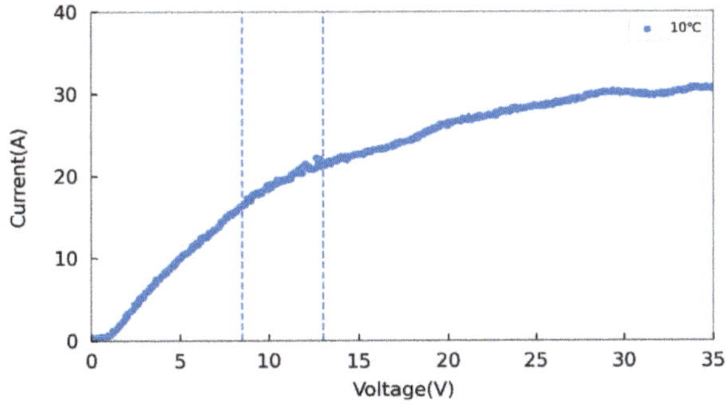

(**b**)

Figure 4. I–V curves of the EP device for a 1.3 GHz single-cell cavity. (**a**) I–V curves across acid temperatures ranging from 5 to 15 °C. (**b**) I–V curve at an acid temperature of 10 °C.

Previous studies [17] indicated that a proper temperature for cold EP polishing should be below 12 °C. Therefore, we chose 10 °C as the controlled temperature during the polishing process. The temperature control system adjusted the temperature of the inlet acid based on the feedback from the temperature probe inside the cavity, thereby maintaining the acid temperature inside the cavity at around 10 °C during the polishing process. Our experimental results indicated that an acid flow rate of 3 L/min was sufficient to control the acid temperature around 10 °C throughout the EP process. The parameters employed for the EP of 1.3 GHz single-cell cavities are detailed in Table 1 below.

During the EP process, the data on the current, temperature, and acid flow rate can be monitored and recorded in real time. With Equation (1), the theoretical removal thickness can be calculated with the current, where I is the current, S_0 is the inner surface area, M is the molecular weight of niobium at M = 92.9 g/mol, ρ is the density of niobium at $\rho = 8.57$ g/cm^3, and F is the Faraday constant. Thus, the average removal rate during the cold EP process for a 1.3 GHz single-cell cavity is around 7 μm/h or 0.12 μm/min. The typical curves for 150 μm bulk EP are summarized in Figure 5. From an examination of the complete curves, it is evident that the polishing process was stable and controllable. Furthermore, the final polished thickness aligned closely with the measurements obtained from the ultrasonic thickness gauge. Figure 6 shows a thickness removal measurement of targeted 50 μm along a 1.3 GHz single-cell cavity with an ultrasonic thickness gauge, demonstrating a removal non-uniformity of less than 20% at the cell part.

$$d\left[\frac{\mu m}{s}\right] = \frac{I}{S_0} \cdot \frac{M}{5\rho F} \quad (1)$$

Table 1. EP parameters chosen to polish 1.3 GHz and 3.9 GHz single-cell cavities.

<table>
<tr><th>Parameters</th><th>1.3 GHz Single-Cell Cavity</th><th>3.9 GHz Single-Cell Cavity</th></tr>
<tr><td>Voltage</td><td>15 V</td><td>20 V</td></tr>
<tr><td>Current</td><td>~15 A</td><td>~4 A</td></tr>
<tr><td>Electrolyte temp.</td><td>10 °C</td><td>10 °C</td></tr>
<tr><td>Average removal rate</td><td>~6 μm/h</td><td>~10 μm/h</td></tr>
<tr><td>Electrolyte flow rate</td><td>3 L/min</td><td>2 L/min</td></tr>
<tr><td>Cavity rotation speed</td><td colspan="2">1 rpm</td></tr>
<tr><td>Cooling water temp.</td><td colspan="2">7 °C</td></tr>
</table>

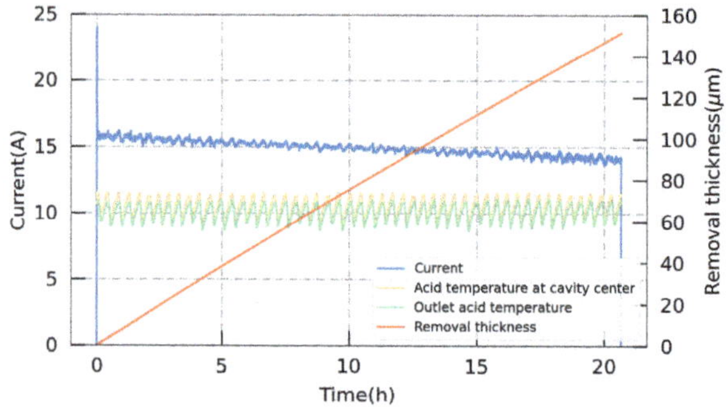

Figure 5. Recorded data on the current and temperature during the bulk EP process, along with the cumulative polishing amount calculated in real time.

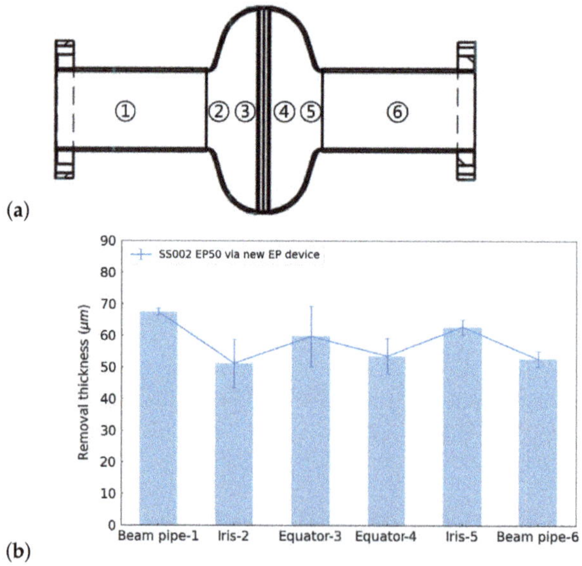

Figure 6. (**a**) Scheme of a 1.3 GHz single-cell cavity with the locations of the thickness measurement points. Points 1 and 6 correspond to measurement points on the beampipe, points 2 and 5 near the iris area, and points 3 and 4 on the equator area. (**b**) Measurement of the removal thickness with an ultrasonic probe at various positions of a 1.3 GHz single-cell cavity before and after 50 um polishing; the removal data at the six positions depicted in the figure represent the averages of measurements taken at four points of axial symmetry.

3. Surface Treatment of 1.3 GHz Single-Cell Cavities

3.1. The 1.3 GHz Single-Cell Cavity Treatment

Two 1.3 GHz single-cell cavities, namely, S02 and S04, were subjected to an EP baseline experiment. These two cavities were previously used for other studies. Initially, 50 μm removal was performed to reset the surfaces of both cavities. Subsequently, after undergoing a 900 °C baking process to release hydrogen and enhance the magnetic flux capabilities, a final 20 μm light EP process was applied to achieve the desired surface. Throughout the entire EP procedure, the cold EP mode was employed, and fresh acid was used for the final light EP process. These cavities were then baked in situ with a two-step baking method, consisting of baking at 75 °C for 4 h, followed by ramping up the temperature to 120 °C and baking for an additional 48 h. The equipment used in the two-step baking process is shown in Figure 7. The cavity was connected to the vacuum pump to maintain a high vacuum state during the baking process, and the heater enveloped the central region to bake the cavity. To further assess the recipe's stability, two newly manufactured cavities, SS002 and SS003, underwent treatment with the same procedure, but the bulk EP removal thickness was 150 μm. The main processing flow is illustrated in Figure 8.

Figure 7. A 1.3 GHz single-cell cavity undergoing the two-step baking process.

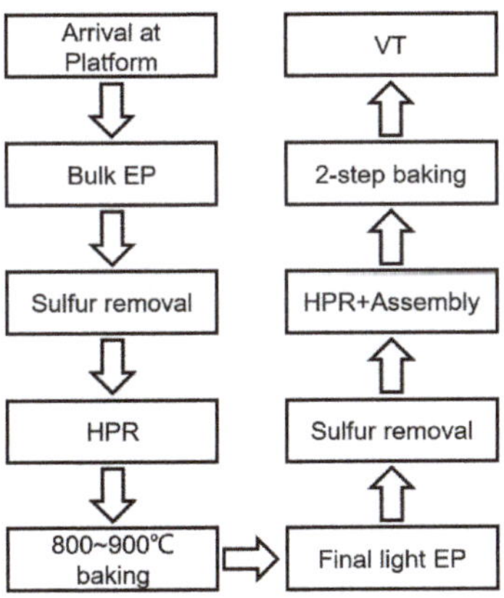

Figure 8. The main process flow for 1.3 GHz single-cell cavities.

3.2. Optical Inspection of the Inner Surface

After the bulk EP and high-temperature baking, the inner surface was examined with an inspection camera that is widely used in many SRF labs, such as KEK, DESY, and IHEP [19,33,34]. By comparing the inner surface before and after treatment, we were able to efficiently examine the quality of the EP process. Typically, the camera was set to take a picture for every four degrees, and it took 90 pictures in total for one circle.

The pictures used for the inspection of SS002 are summarized in Figure 9 and show that the inner surface was mirror-like after bulk EP.

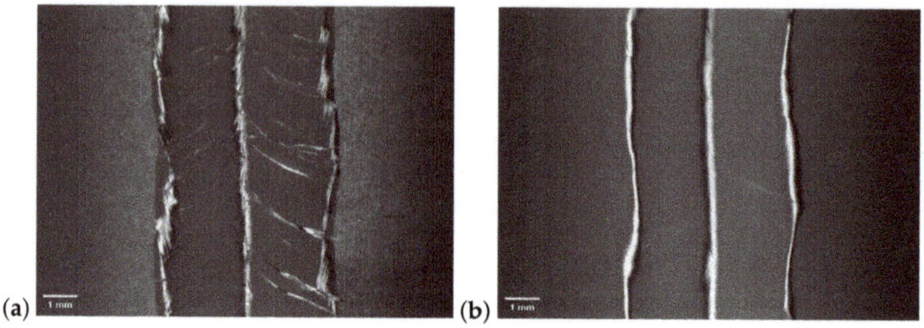

Figure 9. Inspection results of SS002 (**a**) before EP and (**b**) after 150 μm bulk EP.

3.3. *Vertical Test Results for the 1.3 GHz Single-Cell Cavities*

Following the surface treatment, the cavities underwent a vertical test at 2 K to evaluate their RF performance, which was primarily assessed via Q-E curves. Typically, such testing is carried out using a low-temperature vertical test stand (VTS).

After clean assembly in a clean room, the cavities underwent the two-step baking process and a vertical test. The vertical test results for the 1.3 GHz single-cell cavities are shown in Figure 10. The cavity SS003 was tested once before the two-step baking process, and the maximum E_{acc} value was 34.5 MV/m, with a strong high-field Q-slope that was limited by quenching. After the two-step baking process, the maximum E_{acc} value of SS003 was increased to 42 MV/m. The maximum E_{acc} values of all four cavities were over 40 MV/m and were limited by quenching; S04 and SS002 reached 46 MV/m, which verified the effectiveness of the treatment—especially the EP process. The Q-E curves of SS002 and SS003 exhibited differences from those of S02 and S04, with slightly lower Q_0 values being observed in the cavities of the SS series. A similar bifurcation of Q-E curves was also noted in a study by Bafia et al. [31], occurring within the same cavity during different tests. They primarily attributed this phenomenon to the suppression of niobium nano-hydride formation. However, in our instance, the niobium materials used in the cavities of the SS series and S series were from different suppliers, which could account for this discrepancy.

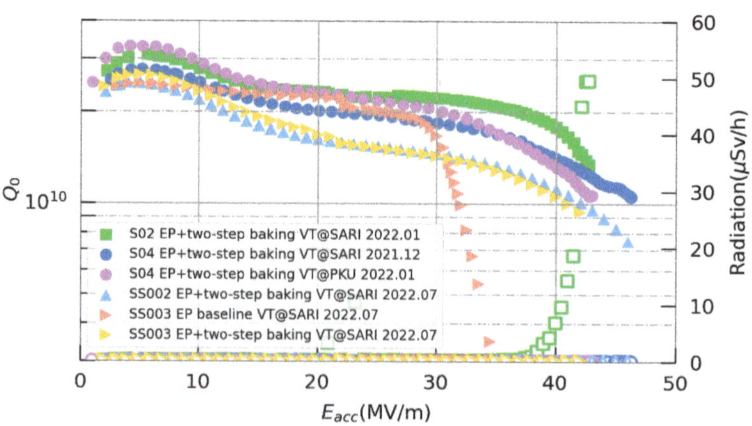

Figure 10. Vertical test results for the 1.3 GHz single-cell cavities at 2 K.

In order to verify the reliability of the vertical test results, the cavity S04 was sent to PKU for a comparison. This cavity was transferred to PKU directly after completing the vertical test at SARI while maintaining a vacuum inside. According to the results, S04 achieved similar results at both SARI and PKU, with a difference of less than 10%. It can be observed that there were still some differences between the two Q-E curves from PKU and SARI. SARI recorded a higher maximum E_{acc} value of 46 MV/m, whereas the maximum E_{acc} value recorded at PKU was slightly smaller at 43 MV/m. However, at PKU, higher Q_0 values at medium and low fields were observed. We mainly attribute this to the differences in the fast cooling processes employed by the two facilities during testing.

4. Surface Treatment of 3.9 GHz Single-Cell Cavities

4.1. Updates of the EP Device for 3.9 GHz Single-Cell Cavities

Due to their smaller dimensions, modifications to the EP equipment were necessary to meet the requirements for electrochemical polishing of the 3.9 GHz superconducting cavities. To address this, two adapter pieces made of PVDF material were initially designed and fabricated. These adapters facilitated the connection of the 3.9 GHz single-cell cavity bundle with the EP equipment, and this was supplemented by the addition of a cable to maintain electrical connectivity between the superconducting cavity and the equipment, as shown in Figure 11a. Furthermore, the overflow plates at both ends of the equipment underwent redesigning. Considering the compatibility of the narrower cathode rod with the 3.9 GHz superconducting cavity, the new design aimed to enhance the stability during the insertion of the cathode rod while also ensuring overflow functionality, thereby meeting the requirements for controlling the acid level within the cavity during the EP process. Figure 11b shows a sketch of this new design, where position 1 represents the overflow baffle, and position 2 denotes the limiting block. Simultaneously, adjustments were made to the cathode rod, primarily reducing its diameter to 16 mm, and the number of acid outlet holes on the cathode rod was reduced to 2 because of the shorter cell size of the 3.9 GHz cavity. However, due to space constraints, PTFE tape was not wrapped around the cathode rod. The actual installed configuration can be observed in Figure 11c.

Figure 11. (**a**) Electropolishing of a 3.9 GHz single-cell cavity, (**b**) the schematic design of the cathode system to adapt to the electropolishing of the 3.9 GHz cavity, including adapters and overflow plates, and (**c**) the cathode for 3.9 GHz single-cell cavities.

In the design of the cathode rod for the 3.9 GHz cavity, the initial objective was to maintain a comparable surface area ratio to that of the 1.3 GHz cavity by proportionally reducing the diameter of the cathode rod to approximately 10 mm. However, it was observed that a cathode rod with a diameter of 10 mm and an inner diameter of only 4 mm posed challenges in adjusting and stabilizing the acid flow rate. Consequently, the diameter of the cathode rod was increased to 16 mm while maintaining an inner diameter of approximately 10 mm, thereby allowing for a broader range of acid flow rate adjustments.

Furthermore, concerning the selection of the temperature and acid flow rate, efforts were made to uphold consistent polishing conditions akin to those of the 1.3 GHz cavity, namely, maintaining a temperature control of 10 °C and a proportional acid flow rate. However, experimentation revealed that different acid flow rates significantly affected temperature control. Figure 12 illustrates real-time temperature measurement curves across varying acid flow rates. The term 'acid temperature' in this figure refers to the temperature measured with the temperature control probe, which was positioned at the outlet and continuously provided real-time feedback to the refrigeration unit, enabling adjustments to its cooling power. Conversely, the 'refrigeration temperature' denotes the temperature of the refrigerant, as measured with an internal temperature probe within the refrigeration unit. This parameter reflects the operational temperature of the refrigeration unit at a given time and directly influenced the temperature of the acid as it entered the cavity. Notably, setting the temperature control to 10 °C resulted in a discernible delay in temperature regulation at a flow rate of 1 L/min, leading to an increase in the acid temperature to 15 °C before promptly decreasing. This fluctuation posed challenges for temperature control during the polishing process. Conversely, at a flow rate of 2 L/min, the temperature remained more stable around 10 °C.

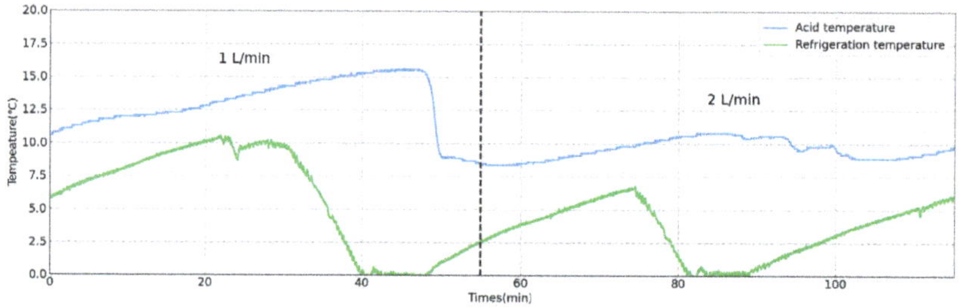

Figure 12. Temperature measurement curves across varying acid flow rates.

The reasons for this phenomenon can be attributed to two main factors. On the one hand, a decrease in the acid flow rate resulted in more heat absorption per unit volume of acid, leading to an increase in the outlet acid temperature. On the other hand, cooler acid from the acid tank also required a longer time to reach the inner cavity, thereby prolonging the adjustment time of the temperature control system. Our experimental results indicate that an acid flow rate of 2 L/min is more suitable for the 3.9 GHz single-cell cavity, as it maintains a more stable acid temperature.

The I–V curves of the 3.9 GHz cavity at acid temperatures of 8 and 10 °C are shown in Figure 13. The I–V curves exhibit a distinct plateau region, where the range from 0 to 4 V corresponds to the etching region, and that from 4 to 17 V represents the oscillation region. To avoid approaching the oscillation region too closely, a working voltage of 20 V was selected. The EP working parameters of 3.9 GHz single-cell cavities are shown in Table 1.

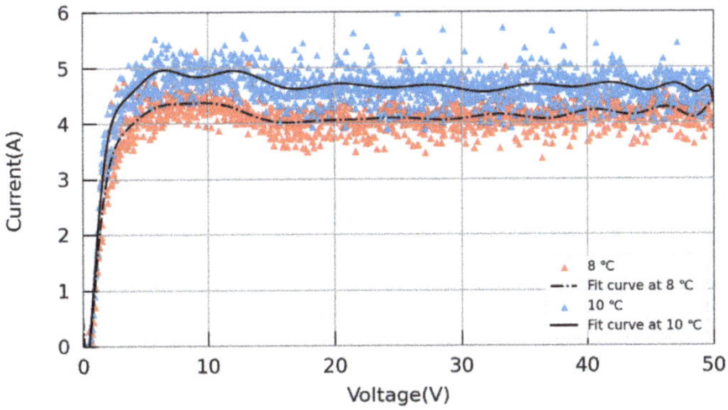

Figure 13. I–V curves of the EP device for the 3.9 GHz single-cell cavity.

4.2. Optical Inspection of the Inner Surface

To examine the internal surface conditions of the superconducting cavities before and after polishing, we devised a novel endoscopic optical inspection system tailored for the 3.9 GHz superconducting cavities. Utilizing a macro lens, the internal surfaces of the cavities could be clearly observed. Figure 14 illustrates comparative photographs of a 3.9 GHz cavity made of large-grain niobium before and after the EP process. It is evident from these images that, after polishing, the internal surfaces exhibited enhanced smoothness. Especially for large-grain cavities, the polished surfaces exhibited a noticeable mirror effect. It should be noted that fine-grain cavities appeared to have slightly poorer surface smoothness, which was most likely due to the previous BCP polishing treatment of these cavities, resulting in more prominent grain boundaries near the equator. After EP polishing, these grain boundaries still stood out, but the microroughness was significantly improved.

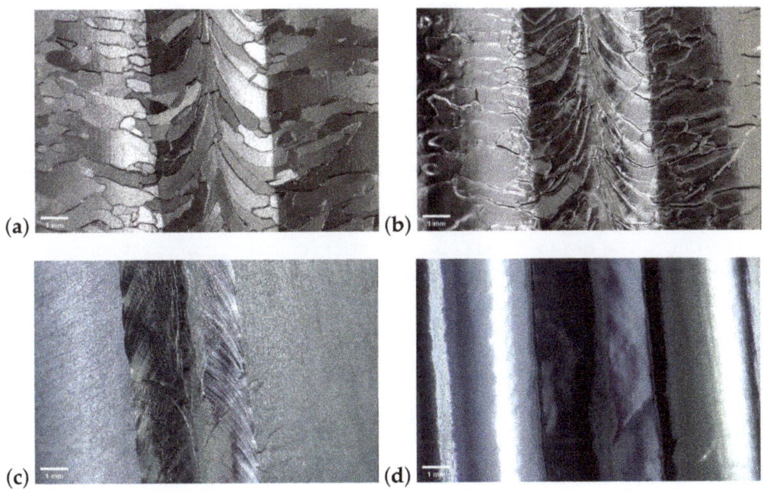

Figure 14. Optical inspection of 3.9 GHz single-cell cavities. (**a**,**b**) are from the fine-grain cavity H-S01, and they were shot before and after 150 um bulk EP, respectively; (**c**,**d**) are from the large-grain cavity H-L08, and they were shot before and after 150 um bulk EP, respectively.

4.3. Vertical Test Results for the 3.9 GHz Single-Cell Cavities

After determining the cold EP parameters for the 3.9 GHz single-cell cavity through experimentation, an EP baseline treatment was conducted on one fine-grained cavity, H-S01, and two large-grained cavities, H-L07 and H-L08. The procedure outlined in Figure 7 was followed, but the two-step baking process was omitted. Subsequently, vertical tests were conducted at SARI, and the results are summarized in Figure 15. All three cavities exhibited maximum accelerating gradients exceeding 23 MV/m, with the Q_0 values at 13 MV/m surpassing 2.7×10^9, thus meeting the SHINE specifications of a Q_0 value greater than 2.0×10^9 at 13 MV/m and a maximum accelerating gradient exceeding 16.5 MV/m. The Q-E curves of the two large-grained cavities showed a similar trend, which was characterized by a slight increase in Q_0 in the mid-field region, followed by a sudden decrease from 23 MV/m. Conversely, for the fine-grained cavity H-S01, Q_0 gradually declined starting from 15 MV/m. The underlying reasons for these differences warrant further investigation.

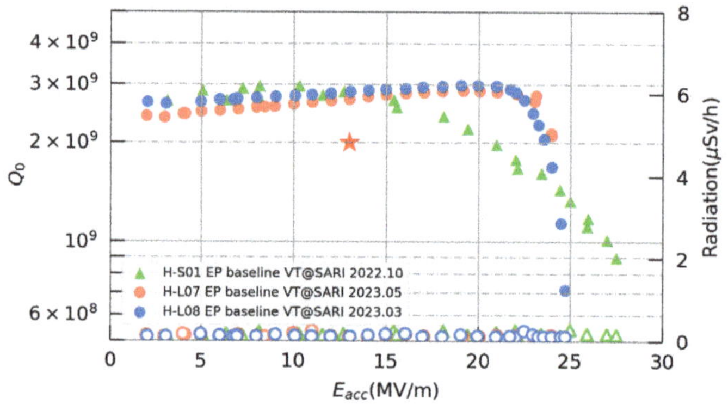

Figure 15. Vertical test results for the 3.9 GHz single-cell cavities.The red star represents the SHINE specification for the Q_0 value at 13 MV/m.

5. Discussion

We took notice of the significant influence of cathode rods on electrochemical polishing and cavity performance [35], which is particularly noteworthy for nine-cell cavities. In the current selection of EP parameters, we initially determined the diameter of the cathode rod and the position of the acid outlet on it. In the EP process for the 3.9 GHz single-cell cavity, the current polishing uniformity was not as good as that for the 1.3 GHz single-cell cavity. The main difference was that the absence of PTFE tape wrapped around the cathode rod in the EP process for the 3.9 GHz cavity contributed to a relatively large exposed area. Additionally, although the working voltage was chosen based on the I–V curves, it was comparatively higher than that in the 1.3 GHz EP process, exacerbating heat generation during the reaction. Drawing from our current experimental experience, the cold EP treatment of the 3.9 GHz cavity, given its smaller size, poses challenges in acid circulation and temperature control in comparison with the 1.3 GHz cavity. Therefore, in the future, we also hope to further optimize the working voltage and cathode rod size through sample experiments and simulations to enhance the polishing uniformity and cavity performance. Relevant experiments are currently in progress.

During the process of testing the I–V curves, we observed that, aside from the area ratio of the electrodes affecting the measured curve, the rate at which voltage was applied during testing could also cause a shift in the current plateau. As illustrated in Figure 16, applying a voltage at a rate of 0.1 V/s resulted in an onset voltage—at which the EP plateau appeared—of 17 V, while applying at 0.5 V/s led to an onset voltage of 25 V. We speculate that this may be due to the larger voltage step causing instability in the testing process; hence, a shorter voltage step is still required when measuring the I–V curve. Similar conclusions were also mentioned in Chouhan's study [35], where they maintained 12 s at each voltage step to obtain more accurate I–V curves. In our testing procedure, due to the inability to adjust the interval time for each step, we gradually increased the voltage with smaller steps. On average, the voltage was increased by 1 V every 10 s.

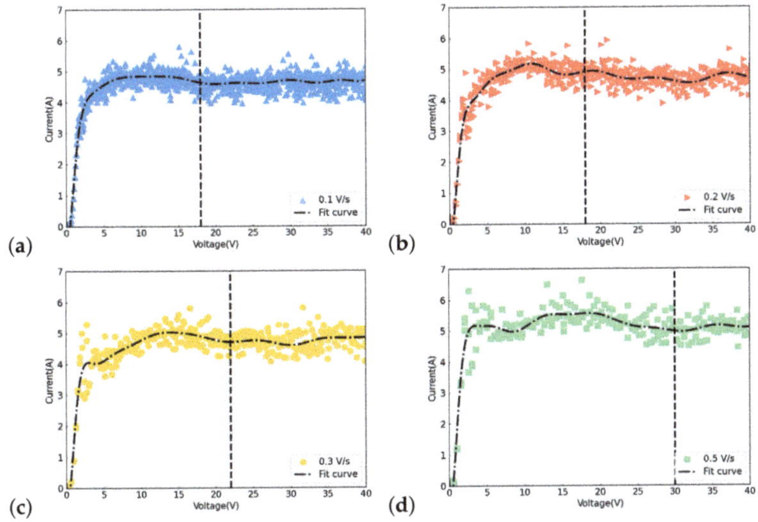

Figure 16. I–V curves with different voltage steps: (**a**) 0.1 V/s, (**b**) 0.2 V/s, (**c**) 0.3 V/s, (**d**) 0.5 V/s.

6. Conclusions

A new small EP device was successfully installed and operationalized on the Wuxi platform. Based on its experimental outcomes, the system demonstrated its remarkable proficiency in controlling polishing process temperatures and ensuring the uniformity of inner cavity surface polishing. Additionally, the facility is capable of accommodating EP polishing for both 1.3 GHz single-cell cavities and 3.9 GHz single-cell cavities, making it more versatile in practical applications. Optical inspection of the inner surfaces showed obvious improvements in smoothness after EP. When utilizing this device for cold EP treatment followed by a two-step baking process, all four 1.3 GHz single-cell cavities achieved a maximum E_{acc} that was higher than 42 MV/m, with half of them surpassing 46 MV/m. Furthermore, 3.9 GHz single-cell cavities subjected to cold EP treatment without the two-step baking process also exhibited elevated E_{acc} values exceeding 25 MV/m.

7. Patents

CN202311431082 was published on 12 January 2024 and is currently undergoing substantive examination.

Author Contributions: Conceptualization, J.C.; data curation, X.H. and X.W. (Xiaohu Wang); formal analysis, Y.Z.; funding acquisition, D.W.; investigation, Y.Z., R.X., J.W. and Z.W.; methodology, Y.Z. and J.C.; project administration, S.X. and J.W.; resources, S.X., J.W. and X.W. (Xiaowei Wu); software, Y.Z. and X.W. (Xiaohu Wang); supervision, J.C. and D.W.; visualization, Y.Z.; writing—original draft preparation, Y.Z.; writing—review and editing, J.C. All authors have read and agreed to the published version of the manuscript.

Funding: This work was supported by the Shanghai Municipal Science and Technology Major Project (No. 2017SHZDZX02), CAS Program (No. Y82G011252), and Double First-Class Initiative Fund of ShanghaiTech University.

Institutional Review Board Statement: Not applicable.

Informed Consent Statement: Not applicable.

Data Availability Statement: The datasets generated and analyzed in the current study are available from the corresponding author upon reasonable request.

Acknowledgments: The authors would like to thank the RF test team at SHINE for the vertical tests.

Conflicts of Interest: The authors declare no conflicts of interest.

Abbreviations

The following abbreviations are used in this manuscript:

EP	Electropolishing
SRF	Superconducting radio frequency
BCP	Buffered chemical polishing

References

1. Padamsee, H. The science and technology of superconducting cavities for accelerators. *Supercond. Sci. Technol.* **2001**, *14*, R28. [CrossRef]
2. Tian, H.; Corcoran, S.G.; Reece, C.E.; Kelley, M.J. The mechanism of electropolishing of niobium in hydrofluoric–sulfuric acid electrolyte. *J. Electrochem. Soc.* **2008**, *155*, D563. [CrossRef]
3. Sautebin, R.; Froidevaux, H.; Landolt, D. Theoretical and experimental modeling of surface leveling in ECM under primary current distribution conditions. *J. Electrochem. Soc.* **1980**, *127*, 1096. [CrossRef]
4. Sautebin, R.; Landolt, D. Anodic leveling under secondary and tertiary current distribution conditions. *J. Electrochem. Soc.* **1982**, *129*, 946. [CrossRef]
5. Landolt, D. Fundamental aspects of electropolishing. *Electrochim. Acta* **1987**, *32*, 1–11. [CrossRef]
6. Saito, K.; Kojima, Y.; Furuya, T.; Mitsunobu, S.; Noguchi, S.; Hosoyama, K.; Nakazato, T.; Tajima, T.; Asano, K. R and D of superconducting cavities at KEK. In Proceedings of the 4th Workshop on RF Superconductivity, Tsukuba, Japan, 14–18 August 1989; p. 60.
7. Saito, K.; Noguchi, S.; Ono, M.; Inoue, H.; Fujino, T.; Kako, E.; Shishido, T. Superiority of electropolishing over chemical polishing on high gradients. *Part. Accel.* **1998**, *60*, 193–217.
8. Kako, E.; Noguchi, S.; Ono, M.; Saito, K.; Shishido, T.; Aune, B.; Charrier, J.; Juillard, M.; Safa, H.; Knobloch, J. Improvement of cavity performance in the Saclay/Cornell/DESY's SC cavities. In Proceedings of the 1999 Workshop on RF Superconductivity, Santa Fe, NM, USA, 1999; pp. 179–186.
9. Saito, K. Techniques of SC cavity preparation for high gradient. In Proceedings of the 21st International Linear Accelerator Conference (LINAC2002), Gyeongju, Republic of Korea, 19–23 August 2002; Volume 2, p. 45.
10. Lilje, L.; Kako, E.; Kostin, D.; Matheisen, A.; Möller, W.D.; Proch, D.; Reschke, D.; Saito, K.; Schmüser, P.; Simrock, S.; et al. Achievement of 35MV/m in the superconducting nine-cell cavities for TESLA. *Nucl. Instrum. Methods Phys. Res. Sect. A Accel. Spectrometers Detect. Assoc. Equip.* **2004**, *524*, 1–12. [CrossRef]
11. Geng, R.L.; Castagnola, S.; Crawford, A.C.; Forehand, D.; Golden, B.A.; Reece, C.E.; Williams, S. Improving gradient and Q performance of BCP etched multi-cell cavities by applying a light EP. In Proceedings of the 14th International Conference on RF Superconductivity (SRF2009), Berlin, Germany, 20–25 September 2009.
12. Diepers, H.; Schmidt, O.; Martens, H.; Sun, F. A new method of electropolishing niobium. *Phys. Lett. A* **1971**, *37*, 139–140. [CrossRef]
13. Saito, K. Development of electropolishing technology for superconducting cavities. In Proceedings of the 2003 Particle Accelerator Conference, Portland, OR, USA, 12–16 May 2003; Volume 1, pp. 462–466.
14. Reece, C.E.; Crawford, A.C.; Geng, R.L. Improved performance of JLAB 7-cell cavities by electropolishing. In Proceedings of the 23rd Particle Accelerator Conference (PAC09), Vancouver, BC, Canada, 4–8 May 2009.
15. Cooper, C.; Brandt, J.; Cooley, L.; Ge, M.; Harms, E.; Khabiboulline, T.; Ozelis, J.; Boffo, C. Recent developments in electropolishing and tumbling R&D at fermilab. In Proceedings of the 14th International Conference on RF Superconductivity (SRF2009), Berlin, Germany, 20–25 September 2009; p. 806.
16. Crawford, A.C. Extreme diffusion limited electropolishing of niobium radiofrequency cavities. *Nucl. Instruments Methods Phys. Res. Sect. A: Accel. Spectrometers, Detect. Assoc. Equip.* **2017**, *849*, 5–10. [CrossRef]
17. Furuta, F.; Bice, D.; Crawford, A.C.; Ring, T. Fermilab EP Facility Implovements. In Proceedings of the 19th International Conference on RF Superconductivity (SRF'19), Dresden, Germany, 30 June–5 July 2019, p. 3.
18. Kelly, M.P.; Gerbick, S.M.; Bice, D.; Wu, G. Electropolishing at ANL/FNAL. In Proceedings of the 14th International Conference on RF Superconductivity (SRF2009), Berlin, Germany, 20–25 September 2009.
19. Jin, S.; Sha, P.; Pan, W.M.; Zhai, J.Y.; Mi, Z.H.; He, F.S.; Dong, C.; Ye, L.X.; He, X.C. Development and Vertical Tests of CEPC 650-MHz Single-Cell Cavities with High Gradient. *Materials* **2021**, *14*, 7654. [CrossRef] [PubMed]
20. Palczewski, A.; Geng, R.L.; Eremeev, G.; Reece, C. Quench studies of six high temperature nitrogen doped 9 cell cavities for use in the LCLS-II baseline prototype cryomodule at Jefferson Laboratory. In Proceedings of the 6th International Particle Accelerator Conference, Newport News, VA, USA, 3–8 May 2015. [CrossRef]
21. Bafia, D.; Checchin, M.; Grassellino, A.; Melnychuk, O.; Romanenko, A.; Sergatskov, D.; Zasadzinski, J. New insights on nitrogen doping. In Proceedings of the 19th International Conference on RF Superconductivity (SRF'19), Dresden, Germany, 30 June–5 July 2019. [CrossRef]

22. Ito, H.; Araki, H.; Takahashi, K.; Umemori, K. Influence of furnace baking on Q–E behavior of superconducting accelerating cavities. *Prog. Theor. Exp. Phys.* **2021**, *2021*, 071G01. [CrossRef]
23. Zhu, Z.; Zhao, Z.; Wang, D.; Liu, Z.; Li, R.; Yin, L.; Yang, Z. SCLF: An 8 GeV CW SCRF Linac-Based X-Ray FEL Facility in Shanghai. In Proceedings of the 38th International Free Electron Laser Conference, Santa Fe, NM, USA, 20–25 August 2017 [CrossRef]
24. Chen, J.; Hou, H.T.; Huang, Y.; Liu, Y.; Wang, D.; Wang, Y.; Wang, Z. N-Doping Studies With Single-Cell Cavities for the SHINE Project. In Proceedings of the 19th International Conference on RF Superconductivity (SRF'19), Dresden, Germany, 30 June–5 July 2019; pp. 102–105. [CrossRef]
25. Zong, Y.; Huang, X.; Wang, Z.; Chen, J.; Dong, P.; Hou, H.; Rong, J.; Wu, J.; Zhang, Y.; Wang, D.; et al. Development of Nitrogen-Doping Technology for SHINE. In Proceedings of the 12th International Particle Accelerator Conference (IPAC'21), Campinas, Brazil, 24–28 May 2021; pp. 1182–1185. [CrossRef]
26. Zong, Y.; Chen, J.F.; Wang, D.; Chen, Q.X.; Chen, Z.X.; Cheng, C.H.; Dong, P.C.; Hou, H.T.; Huang, X.; Huang, Y.W.; et al. Accelerating gradient improvement in nitrogen-doped superconducting radio-frequency cavities for SHINE. *Nucl. Instrum. Methods Phys. Res. Sect. A Accel. Spectrometers Detect. Assoc. Equip.* **2023**, *1057*, 168724. [CrossRef]
27. Zong, Y.; Chen, J.; Chen, Q.; Dong, P.; Hou, H.T.; Huang, X.; Huang, Y.; Pu, X.; Shi, J.; Sun, S.; et al. Realization of accelerating gradient larger than 25 MV/m on High-Q 1.3 GHz 9-cell cavities for SHINE. In Proceedings of the 21th International Conference on RF Superconductivity (SRF'23), Grand Rapids, MI, USA, 25–30 June 2023; pp. 658–661. [CrossRef]
28. Grassellino, A.; Romanenko, A.; Bice, D.; Melnychuk, O.; Crawford, A.C.; Chandrasekaran, S.; Sung, Z.; Sergatskov, D.A.; Checchin, M.; Posen, S.; et al. Accelerating fields up to 49 MV/m in TESLA-shape superconducting RF niobium cavities via 75C vacuum bake. *arXiv* **2018**, arXiv:1806.09824.
29. Ciovati, G. Effect of low-temperature baking on the radio-frequency properties of niobium superconducting cavities for particle accelerators. *J. Appl. Phys.* **2004**, *96*, 1591–1600. [CrossRef]
30. Ge, M.; Gruber, T.; Kaufman, J.; Koufalis, P.; Kulina, G.; Liepe, M.; Maniscalco, J. High-gradient SRF Cavity R&D at Cornell University. In Proceedings of the 10th International Particle Accelerator Conference (IPAC'19), Melbourne, Australia, 19–24 May 2019; pp. 3017–3019. [CrossRef]
31. Bafia, D.; Grassellino, A.; Melnychuk, O.; Romanenko, A.; Sung, Z.H.; Zasadzinski, J. Gradients of 50 MV/m in TESLA shaped cavities via modified low temperature bake. In Proceedings of the 19th International Conference on RF Superconductivity (SRF'19), Dresden, Germany, 30 June–5 July 2019; pp. 586–591. [CrossRef]
32. Bertucci, M.; Bosotti, A.; D'Ambros, A.; Michelato, P.; Monaco, L.; Pagani, C.; Paparella, R.; Sertore, D.; Gresele, A.; Visentin, A.; et al. Electropolishing of PIP-II Low Beta Cavity Prototypes. In Proceedings of the 19th International Conference on RF Superconductivity (SRF'19), Dresden, Germany, 30 June–5 July 2019, p. 5. [CrossRef]
33. Iwashita, Y.; Tajima, Y.; Hayano, H. Development of high resolution camera for observations of superconducting cavities. *Phys. Rev. Spec. Top.-Accel. Beams* **2008**, *11*, 093501. [CrossRef]
34. Wenskat, M. Optical surface properties and their RF limitations of European XFEL cavities. *Supercond. Sci. Technol.* **2017**, *30*, 105007. [CrossRef]
35. Chouhan, V.; Wu, G.; Guilfoyle, B.; Reid, T.; Ring, T.; Ng, M.; Furuta, F.; Bice, D.; Burk, D.; Melnychuk, O.; et al. Electropolishing parameters study for surface smoothening of low-β 650 MHz five-cell niobium superconducting radio frequency cavity. *Nucl. Instrum. Methods Phys. Res. Sect. A Accel. Spectrometers Detect. Assoc. Equip.* **2023**, *1051*, 168234. [CrossRef]

machines

MDPI

Article

Experiment and Simulation of Liquid Film Flow Driving Abrasive Particle Dispersion on the Surface of a Rotating Disk

Qiong Fu, Weibin Shi *, Nian Duan, Hui Huang and Yong Zhang

College of Mechanical Engineering and Automation, Huaqiao University, Xiamen 361021, China
* Correspondence: weibin.shi@hqu.edu.cn

Abstract: Controlling the distribution of the abrasive grains on the surface of the grinding tools in an appropriate way is important for improving the quality of grinding processing and meeting the workpiece precision requirements. In the present study, a novel method for the orderly arrangement of abrasive particles is proposed by using the liquid film flow on the surface of the rotating disk as the driving and controlling means for the uniform dispersion and position arrangement of abrasive particles. Computational fluid dynamics (CFD) simulations have been performed to clearly illustrate the trajectories of abrasive particles under the strong influence of liquid film flow on the rotating disk and reveal the effects of fluid flow, disk rotational motion, and the mixture viscosity on the particle distribution. A new abrasive grain arrangement device is designed and fabricated using this novel method. The operating parameters such as liquid volume flow rate, disk rotational speed, and liquid viscosity are adjusted to control the placement of abrasive grains on the surface of the grinding tool. An image processing tool is used to examine and analyze the arrangement results. The experimental results indicated that the application of the liquid film flow on a rotating disk to the abrasive grain arrangement can improve the arrangement of abrasive grains and get rid of the dependence on the template.

Keywords: particle distribution; abrasive particles; rotating disk; liquid film; CFD simulation

1. Introduction

As one of the key technologies in the field of precision machining, grinding has been considered as an indispensable surface processing method [1–3], which has been widely applied to difficult-to-machine materials, such as titanium alloys [4], superalloys [5], and 300M steel [6]. During the grinding process, the processing quality of the workpiece is determined by the combined action of the abrasive grains in different positions. The traditional consolidated abrasive tools exhibit randomness in the distribution of abrasive particle positions, leading to segregation and aggregation phenomena, and only a portion of the abrasive particles on the working surface of the tool contributes to the grinding process, while the rest generate heat through friction, which can result in severe adverse effects, such as abrasive wear, residual stress, and grinding burns [7]. The suitable control of the position of abrasive particles on the grinding tool surface can make the abrasive particles bear uniform force and wear, increase the chip tolerance space on the abrasive surface, and improve the grinding performance and service life of the abrasive tool comprehensively [8–10]. Therefore, it is very important to control the placement of the abrasive grains on the surface of the grinding tool.

It has been revealed that the abrasives prepared according to the specific pattern can produce workpieces of better quality. Denkena et al. investigated that patterned grinding wheels can result in a reduction of up to 30% in process force [11]. Fang et al. compared a radial structure polishing pad with a conventional polishing pad and found that a radial structure polishing pad could remove materials more evenly through trajectory calculation and corresponding experimental study [12]. Wen et al. reported that the grinding tool manufactured in a spiral pattern could process better surface quality of the workpiece, while the surface quality of the workpiece would be worse when processed by the grinding tool manufactured in a concentric circle pattern, by simulating the trajectory of the abrasive grains on the surface of the grinding tool [13]. Chen et al. experimentally measured that a bionic vein-like structured grinding wheel can significantly decrease normal and tangential force than an unstructured grinding wheel [14].

Currently, a number of methods have been applied to the manufacturing of grinding wheels with orderly grain arrangement. Sung used the template method to achieve a uniform and orderly arrangement of diamond abrasive grains by pressing diamond grains with a particle size of 238–358 μm into a thin layer of brazing powder mixed with an appropriate organic binder using a perforated metal template, layer by layer [15]. Zhu et al. positioned abrasive grains linearly by constructing a topological matrix, in which the abrasive grains are arranged in rows at a certain distance and angle. By changing the properties of the patterned grinding wheel, the surface roughness and subsurface damage were effectively changed [16]. Wu et al. arranged diamond grains on the surface of a grinding wheel, followed by brazing, and then utilized laser ablation to achieve a novel brazed grinding wheel with diamond grains uniformly distributed and featuring a microstructure on the surface [17]. Yin et al. prepared the grinding wheel by controlling the orientation of diamond grains by manipulating the magnetic field. The achieved surface roughness by grinding non-spherical parts made of YG8 hard alloy was 13 nm [18]. Considering that additive manufacturing (AM) technology makes it attainable to facilitate the production of parts with complex geometries [19], Li et al. employed 3D printing technology to fabricate porous metal-bonded grinding wheels with three structures, experimentally evaluating the grinding force and specific grinding energy of the fabricated grinding wheels, which were found to be significantly lower than those of electroplated grinding wheel [20]. Based on the principle of AM technology, Qiu designed and fabricated a three-dimensional controllable arrangement of abrasive grains, which combines abrasive grains with light-curing resin, and the arrangement of abrasive grains in the two-dimensional plane is controlled by a pre-designed template [21]. A kind of compliance grinding tool with a crystal structure using multi-jet fusion was proposed by Li et al., which can achieve effective machining with less material. The designed star-structured tool can achieve 89.27–94.72% of the material removal rate of a solid-structured tool [22].

It seems that either the external force field and the pre-designed template or the combination of both is inevitable to achieve the orderly arrangement of the abrasive grains. However, the above-mentioned methods are usually for cases with very few abrasive grains, which leads to a low grinding efficiency that is unfavorable to industrial applications. In this paper, the authors proposed a new method for the orderly arrangement of abrasive particles without using a template. Inspired by the flow field created by the classic spinning disk reactor [23–25], the proposed new method uses the liquid film flow on a rotating disk as the driving motion of the abrasive particles. Firstly, a computational method based on computational fluid dynamics (CFD) is used to carry out the simulation research. Then, a particle orderly arrangement device is designed and fabricated based on the results of simulations. In addition, a disk-shaped grinding tool has been manufactured with multiple layers of abrasive particles by using the new device. Finally, the parameters that affect the placement of abrasive particles in the grinding tools have been comprehensively studied in the present work, providing a deep insight into the mechanism of formation of arrangement patterns under the strong influence of liquid film flow.

2. Numerical Simulation

2.1. Computational Domain and Meshing

Figure 1a shows a three-dimensional computational domain. Since the liquid-phase resin is injected into the disk surface from the nozzle, the formation of the liquid film is affected by the flow conditions of the jet flow. Therefore, the nozzle is considered as the inlet, and the space between the nozzle and the disk is included in the computational domain for capturing the jet flow characteristics. A nozzle with a diameter of 4 mm is located symmetrically directly above the center of the disk. The solution enters the computational domain from this nozzle at a certain velocity and then flows over the surface of the stainless steel disk with a diameter of 100 mm. The computational domain with a height of 70 mm comprises only a limited portion of the air above the rotating disk to acquire a lower computational cost.

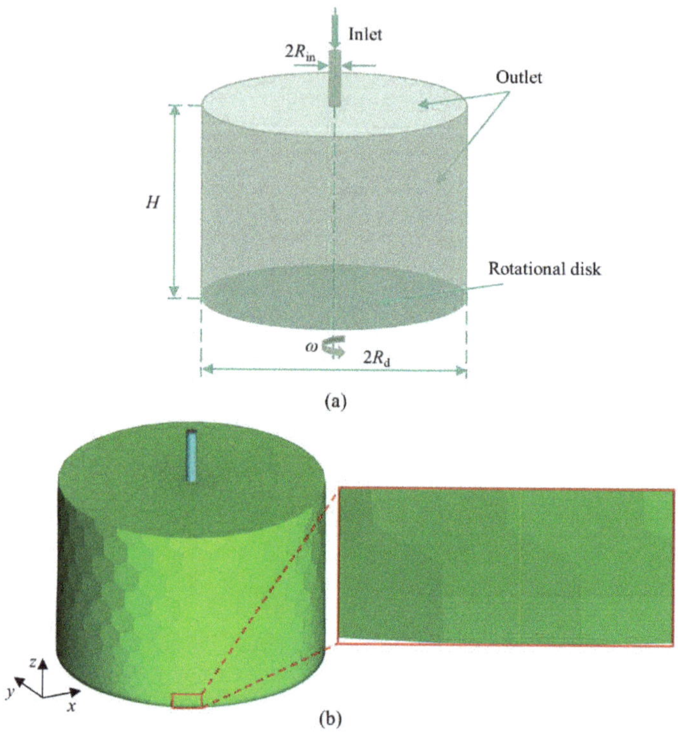

Figure 1. Schematic diagrams of (**a**) the computational domain and (**b**) the computational mesh.

Considering the high computational accuracy and better convergence of polyhedral meshing, a polyhedral grid meshed by the meshing module in ANSYS Fluent 2020 R2 was used for the present simulations. Figure 1b presents a global view of the three-dimensional computational grid and a partial view of the boundary layer grid near the wall surface. Since the surrounding air is not the main concern of the jet and the liquid film flow, the mesh cells applied to this region can be relatively coarser. In contrast, in order to capture more details of the liquid-film flow on the disk, 10 layers of mesh refinement are used right above the disk with a height of 0.1 mm for the first layer and a growth rate of 1.1. By using empirical correlation ($h = 0.886Q^{0.348}v^{0.328}\omega^{-0.676}r^{-0.70}$) [26], the thickness of the liquid film on the rotating disk is calculated at about 100 μm. In this case, the thickness of 0.1 mm for the first layer of the mesh is roughly the same as the thickness of the liquid film.

A grid-dependency test wherein the number of grid cells varied from 26,965 to 6898 was performed to measure the impact of grid density on the accuracy of simulation results. Under identical conditions, calculations were conducted to compare and analyze the velocity fields at specific locations (x = 0 mm, −50 mm ≤ y ≤ 50 mm, z = 0.1 mm) with different grid densities. Figure 2 shows that with the increase in grid density, the distribution of the velocity field gradually changes, and the theoretically calculated tangential velocity distribution for the rotating disk at different radial positions has been provided as a comparison. The disk rotational velocity refers to the disk surface (z = 0 mm), while the liquid-phase tangential velocities are taken at z = 0.1 mm. It can be seen that there is a small difference between the theoretical calculation and the simulation results. Although increasing grid density helps enhance the accuracy of simulations, it also leads to an increase in computational workload. Taking both the accuracy of simulation results and computational cost into account, 12,754 grids were selected for subsequent simulation work.

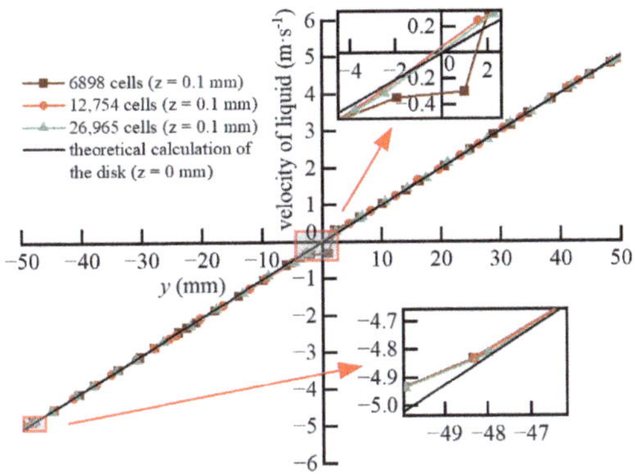

Figure 2. Velocity distribution of liquid under different grid numbers at monitoring position.

2.2. Mathematical Model

There are three phases present in the computational domain: gas, liquid, and solid. It is difficult to get rid of the gas phase in real operating conditions, even though it plays a less significant role than the liquid phase and solid particles. More importantly, the gas–liquid interaction especially the interphase drag force may affect the flow behaviors of the jet flow that comes from the nozzle. This cannot be ignored as the diameter/range of the jet decides the initial contact area on the rotating disk and hence the diameter of the hydraulic jump if any. These belong to the so-called non-machining area discussed in Section 4.1, which is also an important parameter of the disk when it is used as a grinding tool. Therefore, the simulation is carried out by coupling the multiphase flow model with the discrete phase model. It is assumed that the fluid under study is adiabatic and the energy equation is not included in the simulation. The following section provides a detailed introduction to the basic principles of these models and their corresponding equations.

2.2.1. Multiphase Flow Model

In the present work, an Euler–Euler–Lagrange approach [27] is employed to model the three-dimensional, transient, multiphase flow. The liquid film flow on the rotating disk is considered as the primary Eulerian phase and the surrounding air in the computation domain is considered as the secondary phase. The motion of the abrasive particles is tracked under a one-way coupled Lagrange framework. The mass and momentum conservations are given by Equations (1) and (2), respectively:

$$\frac{\partial(\alpha_k \rho_k)}{\partial t} + \nabla \cdot (\alpha_k \rho_k \boldsymbol{u}) = 0 \tag{1}$$

$$\frac{\partial(\alpha_k \rho_k \boldsymbol{u}_k)}{\partial t} + \nabla \cdot (\alpha_k \rho_k \boldsymbol{u}_k \boldsymbol{u}_k) = -\alpha_k \nabla p + \nabla \cdot \boldsymbol{\tau}_k + \alpha_k \rho_k \boldsymbol{g} + \boldsymbol{F} \tag{2}$$

where $\boldsymbol{u}$ is the velocity vector of the fluid phase, ρ is the fluid density, $\boldsymbol{g}$ is the gravitational acceleration, and $\boldsymbol{\tau}$ is the stress tensor. The volume fraction α_k represents the kth phase (k = 1, 2), and the sum of the phase must be satisfied:

$$\sum \alpha_k = 1 \tag{3}$$

The interphase forces $\boldsymbol{F}$ encompass various effects of forces, including but not limited to drag and lift forces [28]. The liquid-phase flow is considered as the continuous phase, which forms a jet flow when it comes out from the nozzle. At this stage, the gas–liquid interaction mainly occurs at the boundary layer of the jet as not many gas bubbles are blended into the liquid phase. This may lead to changes in the boundaries of the jet while the core is mainly affected by the drag force. As the liquid jet contacts with the rotating disk, the liquid film is quickly formed and the flow behaviors are much affected by the motion of the disk. Therefore, only the drag force was considered by making simplifications. The drag force can be described by the following Schiller–Naumann:

$$F_{\mathrm{D}} = \frac{C_{\mathrm{D}} Re}{24} \tag{4}$$

where C_{D} is the drag force coefficient, which can be calculated as follows:

$$\begin{cases} C_{\mathrm{D}} = 24\left(1 + 0.15{Re_{\mathrm{r}}}^{0.687}\right)/Re_{\mathrm{r}},\ 1 \le Re_{\mathrm{r}} \le 1000 \\ C_{\mathrm{D}} = 0.44,\ Re_{\mathrm{r}} > 1000 \end{cases} \tag{5}$$

and Re_{r} is the relative Reynolds number for the primary phase and secondary phase, which is obtained as follows:

$$Re_{\mathrm{r}} = \frac{|\boldsymbol{u}_{\mathrm{p}} - \boldsymbol{u}_{\mathrm{s}}| d_{\mathrm{b}} \rho_{\mathrm{s}}}{\mu_{\mathrm{s}}} \tag{6}$$

where $\boldsymbol{u}_{\mathrm{p}}$ is the velocity of the primary phase, $\boldsymbol{u}_{\mathrm{s}}$ is the velocity of the secondary phase, d_{b} is the droplet diameter, ρ_{s} is the density of the secondary phase, and μ_{s} is the viscosity of the secondary phase.

2.2.2. Discrete Phase Model

The motion of abrasive particles is predicted by the discrete phase model (DPM) method [29], which is calculated in a Lagrange reference frame. According to Newton's second law, the motion of particles is controlled as follows:

$$m\frac{\mathrm{d}\boldsymbol{u}_{\mathrm{p}}}{\mathrm{d}t} = \boldsymbol{F}_{\mathrm{d}} + \boldsymbol{F}_{\mathrm{G}} + \boldsymbol{F}_{\mathrm{F}} + \boldsymbol{F}_{\mathrm{af}} \tag{7}$$

$$\boldsymbol{F}_{\mathrm{G}} = m\boldsymbol{g} = \frac{1}{6}\pi d_{\mathrm{p}}^{3} \rho_{\mathrm{p}} \boldsymbol{g} \tag{8}$$

$$\boldsymbol{F}_{\mathrm{F}} = -\rho_{\mathrm{L}} \boldsymbol{g} V_{\mathrm{p}} = -\frac{1}{6}\pi d_{\mathrm{p}}^{3} \rho_{\mathrm{L}} \boldsymbol{g} \tag{9}$$

where $\boldsymbol{u}_{\mathbf{p}}$ is the velocity vector of the particle, m is the quality of the particle, ρ_{p} is the density of the particle, d_{p} is the diameter of the particle, and V_{p} is the volume of the particle. $\boldsymbol{F}_{\mathrm{G}}$, $\boldsymbol{F}_{\mathrm{F}}$, and $\boldsymbol{F}_{\mathrm{d}}$ are the gravity, buoyancy, and drag force acting on the abrasive particle, respectively. $\boldsymbol{F}_{\mathrm{af}}$ refers to the additional forces, with typical forces including the pressure gradient force, the virtual mass force, the Basset force, the Saffman force, and others. Throughout the entire motion process, the abrasive particles are treated as spherical rigid bodies. In this study, emphasis was placed on analyzing the influences of gravity, buoyancy, drag force, and the centrifugal force generated by rotational motion on the abrasive particles.

$\boldsymbol{F}_{\mathbf{d}}$ is the drag force acting on the abrasive particles, which can be calculated using the following equation:

$$\boldsymbol{F}_{\mathrm{d}} = \frac{C_{\mathrm{d}}}{8}\pi d_{\mathrm{p}}^{2} \rho_{\mathrm{p}} |\boldsymbol{u}_{\mathrm{L}} - \boldsymbol{u}_{\mathrm{p}}| (\boldsymbol{u}_{\mathrm{L}} - \boldsymbol{u}_{\mathrm{p}}) \tag{10}$$

where $\boldsymbol{u}_{\mathrm{L}}$ is the velocity of the liquid, and C_{d} is the drag force coefficient, which can be calculated as follows:

$$C_{\mathrm{d}} = \delta_1 + \frac{\delta_2}{Re} + \frac{\delta_3}{Re^2} \tag{11}$$

where δ_1, δ_2, and δ_3 are constants. Morsi and Alexander [30] provide values for these constants within different Reynolds number ranges.

2.2.3. Viscous Model

When numerically simulating the distribution of diamonds driven by the liquid film on the surface of a rotating disk, the flow state of the liquid film on the rotating disk surface often depends on the rotational Reynolds number. Under normal circumstances, the rotational Reynolds number for laminar flow typically does not exceed 150,000 [31], which was calculated as follows:

$$Re_R = \frac{\rho_L \omega R^2}{\mu} \tag{12}$$

where Re_R represents the rotational Reynolds number, ω is the angular velocity of the rotating disk, R is the radial distance from the measured point to the center of the disk, and ρ_L and μ are the density and viscosity of the liquid.

By substituting relevant physical parameters into Equation (12), the calculated Reynolds number ranges from 50 to 49,074, and the flow was assumed to be laminar throughout the simulation. However, this determination is not absolute because variations in the height of the liquid film may affect the flow characteristics. To accurately determine the flow state, the best approach is through experimental validation. Assuming that the liquid film flow remains in an approximately steady state, there is no significant fluctuation or only minor fluctuations at the edge of the disk. Therefore, the laminar model is used. Also considering the incompressible nature of the liquid-phase fluid, the viscous term in the Navier–Stokes equation can be simplified as follows:

$$\nabla \cdot \tau = \mu \begin{bmatrix} \left(\frac{\partial^2 u}{\partial x^2} + \frac{\partial^2 u}{\partial y^2} + \frac{\partial^2 u}{\partial z^2}\right) + \frac{\partial}{\partial x}\left(\frac{\partial u}{\partial x} + \frac{\partial v}{\partial y} + \frac{\partial w}{\partial z}\right) \\ \left(\frac{\partial^2 v}{\partial x^2} + \frac{\partial^2 v}{\partial y^2} + \frac{\partial^2 v}{\partial z^2}\right) + \frac{\partial}{\partial y}\left(\frac{\partial u}{\partial x} + \frac{\partial v}{\partial y} + \frac{\partial w}{\partial z}\right) \\ \left(\frac{\partial^2 w}{\partial x^2} + \frac{\partial^2 w}{\partial y^2} + \frac{\partial^2 w}{\partial z^2}\right) + \frac{\partial}{\partial z}\left(\frac{\partial u}{\partial x} + \frac{\partial v}{\partial y} + \frac{\partial w}{\partial z}\right) \end{bmatrix} = \mu \nabla^2 \boldsymbol{u} \tag{13}$$

2.3. Boundary Conditions and Simulation Details

The liquid inlet is positioned 90 mm above the disk. Since the liquid is incompressible, velocity inlet boundary conditions are applied. The abrasive phase is solved using DPM. The running time for the particles is set to 500 s. This duration is sufficiently long compared to the entire simulation process, making it continuous to replenish the particle quantity, thereby ensuring that the volume fraction of particles carried by the liquid remains constant upon entry. In this study, a surface injection source is set at the inlet, with the abrasive type being inert particles. Based on the properties of diamond, the particle density is set to 3520 kg/m^3, and the particle diameter is 28 μm.

The disk is set to have a no-slip boundary condition and is assigned an absolute rotational velocity to simulate the actual rotation effect. The boundary in contact with the atmosphere is defined as an open boundary, employing a pressure outlet condition to allow the free outflow of liquid and particles. Due to the presence of a pressure outlet, the operating pressure is set to atmospheric pressure, i.e., 101,325 Pa. Table 1 lists the specific simulated conditions.

For all simulated cases, the SIMPLE scheme was used for pressure–velocity coupling, the Least Squares Cell-Based method was selected for the discretization of the gradient term, the pressure was discretized using a second-order scheme, and the first-order upwind method was used for the convective transport of volume fraction and velocity. The solution of all equations employed a first-order upwind implicit method. The convergence criterion was set as 10^{-5}, with a time step of 0.001 s chosen, and 1000 iterations were performed. The calculations were performed on a workstation equipped with 48 physical cores, 31.7 GB of memory, and a clock speed of 2.30 GHz.

Table 1. Specifications of the simulated conditions.

Operating Conditions	Values
Liquid flow rate, Q (mL/min)	100, 200, 300, 400
Rotational speed, n (rpm)	760, 860, 960, 1060
Liquid viscosity, μ (mPa·s)	50, 150, 250, 350

3. Experimental Setup and Procedure

Figure 3 shows a schematic of the experimental apparatus. The experimental media were a diamond-resin solution with a 10% mass fraction, and the physical parameters were the same as the simulation as listed in Table 1. The selected light-curing resin is transparent in color and is compatible with a wavelength of 405 nm. Single crystal diamond micropowder with a particle size of W28 was selected as the abrasive. The solution as the working fluid was mixed in the storage tank by the stirrer and then transported to the nozzle through hoses via a peristaltic pump. The solution entered the disk surface vertically from the central position of the disk through the nozzle. The nozzle with an inner diameter of 4 mm was 70 mm away from the surface of the disk. The disk was made of 304 stainless steel with a diameter of 200 mm and a thickness of 1 mm. The disk was secured by a bracket to the rotating shaft and was driven by a motor through a gear mechanism. As the disk rotated, the solution spread as a film over the disk surface and then reached the edge of the disk. With the continuous rotation of the disk, the solution then left the edge of the disk at a certain velocity, traveled until it impacted the inner wall of the collecting housing, and finally fell to the bottom of the collecting housing. It flowed along the inclined bottom of the collecting housing and was discharged at the outlet into recovering vessels. The UV curing lamp was positioned directly above the rotating disk. After the solution ceased to enter the disk surface, the UV curing lamp started operating to cure the liquid film on the surface of the rotating disk, forming a single layer of cured diamond resin. By repeating the above operation, layer by layer, grinding tools of different thicknesses can be prepared. This study focuses on the research and analysis of grinding tools with only one layer of diamond resin. A high-speed camera (Phantom V2511, AMETEK, Berwyn, PA, USA) with a shooting speed of 1000 pps and a resolution of 1280 × 800 was placed above the rotating disk to observe the flow state of the liquid film on the disk surface. Figure 4 shows the photograph of the experiment setup. As shown in Figure 4, the experimental platform mainly consists of supporting aluminum materials, a rotating disk, a peristaltic pump, a nozzle, an elevating screw mechanism, a guide rail mechanism, and a high-speed camera. The dimensions of the aluminum frame are 540 × 540 × 800 mm.

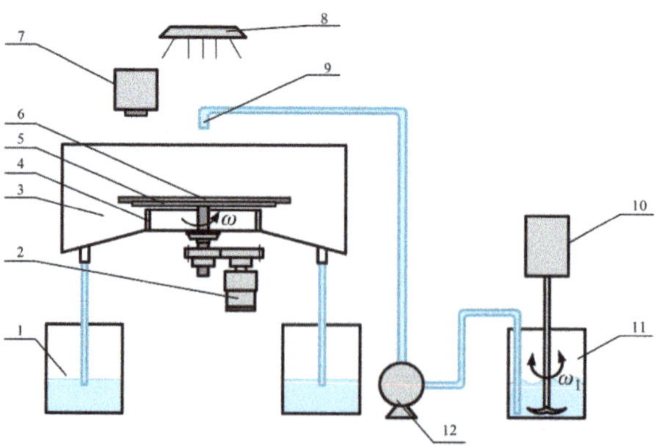

Figure 3. Schematic diagram of the experimental setup: (1) recovering vessel, (2) motor, (3) collecting housing, (4) shaft case, (5) mounting bracket, (6) rotating disk, (7) high-speed camera, (8) curing light, (9) nozzle, (10) stirrer, (11) storage tank, and (12) peristaltic pump.

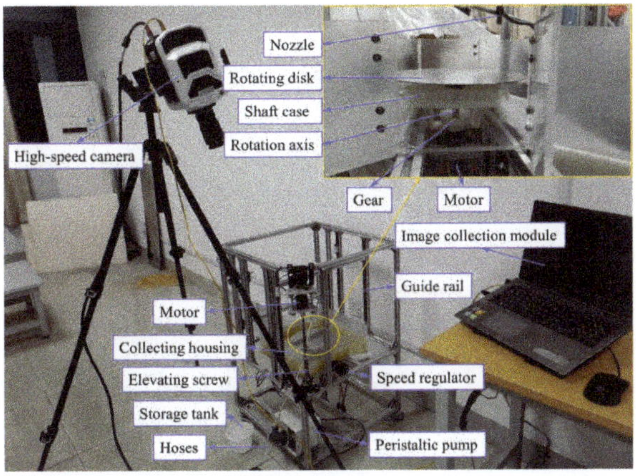

Figure 4. Photograph of the experimental setup.

4. Analysis Method and Simulation Validation

4.1. Uniformity Evaluation

A grid counting method initially proposed by Christiansen [32] was quoted to assess the distribution characteristics of abrasive particles on the disk surface. This method was used to assess the distribution characteristics of sprinkler water. It is also being applied to describe the distribution characteristics of diamonds, as Chen et al. [33] also used the same formula to evaluate the pattern of distribution. In contrast with the liquid droplet distribution due to the sprinkling spray, the present study focuses on the particle distribution, which comes together with the liquid resin from the nozzle and is driven by the liquid film. In terms of the pattern of distribution, both the droplet distribution due to irrigation by sprinkling and the particle distribution due to jet flow impinging on a disk are considered as dispersed fluid particles, even though the flow types are indeed different. In addition, the purpose of the above two flows is similar, which is to make the pattern of distribution as uniform as possible.

This method quantifies the uniformity of particle distribution by calculating the deviation between the number of particles in each grid region and the average number of particles in the entire observation area, comparing the sum of these deviations to the total number of particles, thus characterizing the distribution characteristics of particles across the entire observation area. The calculation formula is as follows:

$$\begin{cases} U = 100 \times \left[1 - \frac{\sum\limits_{i=1}^{N} |p_i - \overline{p}|}{N\overline{p}}\right], U > 0 \\ U = 0,\ U \leq 0 \end{cases} \tag{14}$$

In this equation, N is the number of grids, p_i is the number of particles in the i-th grid, and $\overline{p}$ is the average number of particles. The expression $\sum\limits_{i=1}^{N} |p_i - \overline{p}|/N$ in Equation (14) is similar to the mathematical formula of the standard deviation $\sigma = \sqrt{\sum\limits_{i=1}^{N} (x - \overline{x})^2/N}$. In this case, $1 - \sum\limits_{i=1}^{N} |p_i - \overline{p}|/N$ can be equivalently understood as 1-σ, thus being defined as uniformity. However, rather than the number of particles that deviated from the averaged value, Equation (14) considered a percentage difference by dividing the average particle number, which yields $\sum\limits_{i=1}^{N} |p_i - \overline{p}|/N\overline{p}$.

A systematic grid partitioning strategy was adopted in this paper. Firstly, along the circumference of the disk, its surface is divided into eight equally spaced concentric circular regions. The width of each concentric circular region is equal, ensuring uniformity in the circumferential partitioning. Next, along the radial direction of the disk, eight line segments are drawn outward from the center point, with the length of each line segment equal to the radius of the disk. By employing both circumferential and radial partitioning, a total of 64 grid cells were obtained, with each grid cell representing a specific region on the surface of the disk, as illustrated in Figure 5a,b.

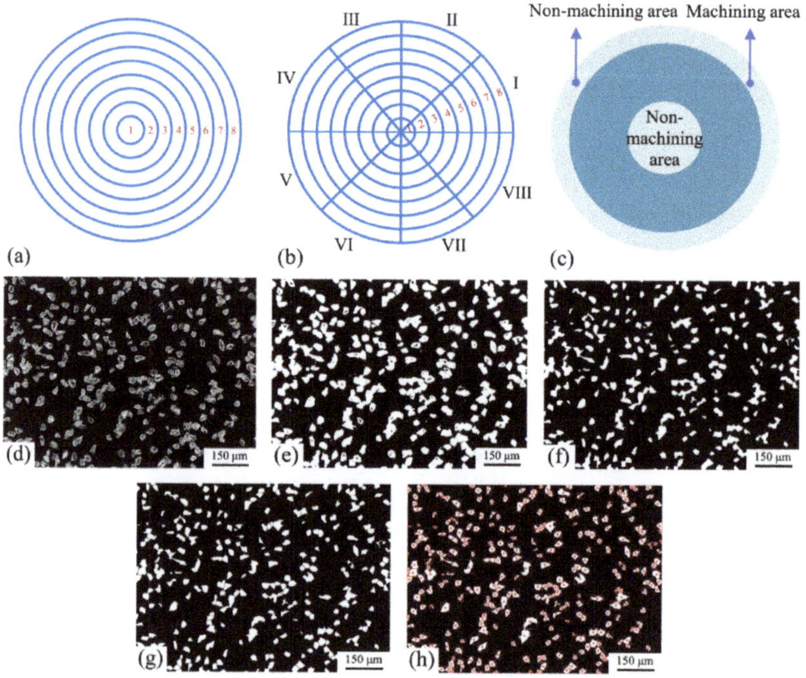

Figure 5. (**a**) Circumferential grid division; (**b**) radial grid division; (**c**) the division of machining area and non-machining area; (**d**–**h**) image processing including original image, binary image, image noise removal, image segmentation, and particle counting.

Considering the actual machining conditions, the surface of the disk was divided into machining and non-machining areas, as shown in Figure 5c. The diameter of the regions measured is within 100 mm, corresponding to the diameters of the eight concentric circular regions divided along the circumference: 12.5 mm, 25 mm, 37.5 mm, 50 mm, 62.5 mm, 75 mm, 87.5 mm, and 100 mm. Combining with Figure 5a,b, it can be observed that the diameter range of the machining area's concentric circles is from 37.5 mm to 75 mm, corresponding to regions 3 to 6 in the figures.

In the simulation, the coordinates of the abrasive particles are exported to determine whether they are on the disk. Subsequently, the number of particles on the disk or within each region of the disk is further counted. To ensure comparability with simulation data, the selected measurement area during experiments remains consistent with the corresponding area in the simulation model. During the experimental process, three points randomly selected within a single grid cell in the machining area were observed using a three-dimensional video microscope, calculating the total number of abrasive particles at the selected three points. Since the single measurement range of the three-dimensional video microscope is limited, the total number of abrasive particles at three points within a grid cell cannot represent the total number of abrasive particles within the entire grid cell. Therefore, an appropriate adjustment was made to Equation (14) when calculating uniformity, as follows:

$$\begin{cases} U = 100 \times \left[1 - \dfrac{\sum\limits_{i=1}^{N} \left| \dfrac{p_i}{S_i} - \dfrac{\overline{p}}{\overline{S}} \right|}{\dfrac{N\overline{p}}{\overline{S}}} \right], U > 0 \\ U = 0, \ U \leq 0 \end{cases} \tag{15}$$

where S_i represents the area of the i-th grid cell, and $\overline{S}$ denotes the average grid cell area. The formula for calculating the area of a single grid cell is as follows:

$$S_i = \frac{1}{8}\pi T^2(2i - 1) \tag{16}$$

where T represents the distance between two concentric circles. Considering the particles fall into different zones and sections with different areas, only taking into account the percentage difference of particles may still not be adequate, e.g., a larger area might have a higher possibility for the particles to be located. Therefore, it is reasonable that we introduced a concept of a number of particles per unit area in Equation (15), to eliminate the influence of different zone areas.

Image processing technologies, which have been widely applied in feature detection [34], were used to count the number of abrasive particles in this case. Figure 5d–h illustrate the image process of binarization, noise removal, image segmentation, and particle counting. Figure 5d shows the original image measured by the KH-8700 three-dimensional video microscope (HIROX, Tokyo, Japan). The original image was globally adaptively binarized using the Otsu algorithm, as shown in Figure 5e. The noise in the image is removed using morphological opening operations to further enhance the image quality, as shown in Figure 5f. After image preprocessing and noise removal, the watershed transformation algorithm was utilized to segment each diamond particle in the image. The segmented image is shown in Figure 5g. Finally, the number of diamond particles in the image is counted, as shown in Figure 5h, with diamond particles marked with red dots.

4.2. Simulation Validation

Figure 6a shows the axial view of the overall model after convergence of the computation under the operating parameters of liquid volume flow rate of 100 mL/min, disk speed of 960 rpm, and liquid viscosity of 250 mPa·s. It displays the 3D trajectories of abrasive particles throughout the entire computational domain. It can be observed that the abrasive particles enter vertically from the nozzle onto the disk surface, with the highest concentration in the central region of the disk, then dispersing in a spiral shape from the center towards the edge of the disk. The direction of rotation of the disk is counterclockwise. Visibly, the spiral direction of the abrasive particles' movement on the disk surface is opposite to the direction of disk rotation.

An abrasive tool with a single cured layer was prepared under the experimental conditions of a liquid volume flow rate of 100 mL/min, disk rotational speed of 960 rpm, and resin viscosity of 250 mPa·s, as shown in Figure 6b. It can be observed that the liquid film on the disk surface spreads in a spiral pattern, opposite to the direction of disk rotation, which is consistent with the simulation results under the same operating conditions.

Comparative analysis of the uniformity of abrasive particle distribution on the disk surface between the simulated model is shown in Figure 6a, and the experimental data obtained from the detection of the abrasive tool with a single cured layer are shown in Figure 6b. The uniform distribution of abrasive particles calculated by the simulation model is 63.66%, and the uniform distribution of abrasive grains measured experimentally is 70.9%. The ratios of the number of abrasives within circumferential sections I~VIII to the total number of abrasives within radial section 3 were calculated for both the simulation and the experiment. The same was achieved for sections 4~6, resulting in the ratio chart shown in Figure 6c. Although the ratio in each zone still varies under the current evaluation frame, the excessive fluctuations in the simulation result are due to the insufficient number of released particles, resulting in an inadequate sample size. If enough particles are released, which requires much more computational resources, it is believed that these fluctuations can be gradually eliminated. However, a coarser comparison can still be made to assess the sensitivity of the simulation result, which is to compare the ratio of particle number in each circumferential section to the total number of particles, as shown in Figure 6d.

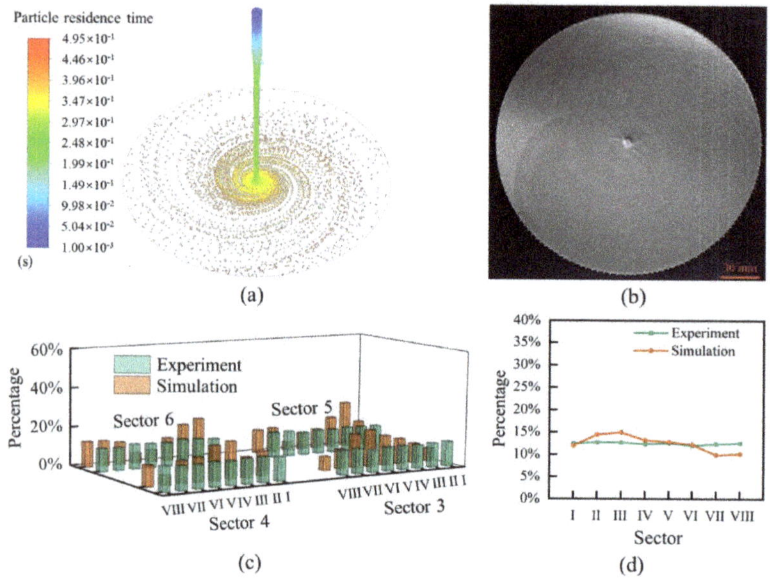

Figure 6. (**a**) Simulation analysis of abrasive particle distribution on the disk surface; (**b**) the abrasive tool with a single cured layer; (**c**) comparison of simulation and experiment in the radial direction; (**d**) comparison of simulation and experiment in the circumferential direction.

5. Results and Discussion

5.1. *Influence of Flow Rate*

Figure 7a–d depict the distribution of abrasive particles colored by particle residence time under the conditions of a disk rotation speed of 960 rpm and a liquid kinematic viscosity of 250 mPa·s, with liquid volumetric flow rates of 100 mL/min, 200 mL/min, 300 mL/min, and 400 mL/min, respectively. It can be observed from Figure 7a–c that the particle tracks exhibit a spiral distribution on the surface of the disk, with a rotation direction opposite to the rotation direction of the disk. As the flow rate increases, the diffusion speed of abrasive particles on the disk surface accelerates. Notably, in Figure 7d, when the liquid propels the abrasive particles to impact the disk surface, a distinct ejection occurs from the center towards the periphery. The abrasive particles are densely distributed at the center of the disk, forming a unique "fireworks-like" pattern. The reason for the above-mentioned distribution of abrasive particles is that the liquid jet impacts the disk surface, and a liquid film is formed on the surface of the disk. Under the driving force of the liquid film and the centrifugal force and Coriolis force generated by the rotation of the disk, the abrasive particles form a spiral distribution on the disk surface. At a certain rotational speed, the centrifugal force remains constant, while the inertial force acting on the liquid film increases with the increase in flow rate, and the viscous resistance does as well. Under the influence of these forces, especially at higher liquid flow rates, a "fireworks-like" pattern distribution of abrasive particles is formed. Figure 7e shows the variation in the number of abrasive particles on the disk surface over time at different liquid volumetric flow rates. It can be observed that with increasing time, the number of abrasive particles on the disk surface continuously increases. At flow rates of 200 mL/min, 300 mL/min, and 400 mL/min, the number of abrasive particles stabilizes after 0.55 s, 0.45 s, and 0.35 s, respectively. The larger the flow rate, the smoother the increase in the number of abrasive particles on the disk surface, and the fewer abrasive particles in the stable state. At a flow rate of 100 mL/min, the number of abrasive particles shows an increasing trend before 0.5 s, followed by about a 5% decrease (from 7600 to 7200). The decrease can be attributed to the mismatch of the liquid flow rate and the disk rotating speed (e.g., the flow rate is too small while the disk rotates too fast), which leads to fluctuations in the number of diamond particles. As time proceeds, the number of particles may not reach a stable state, but the amplitude of fluctuations can be quite small.

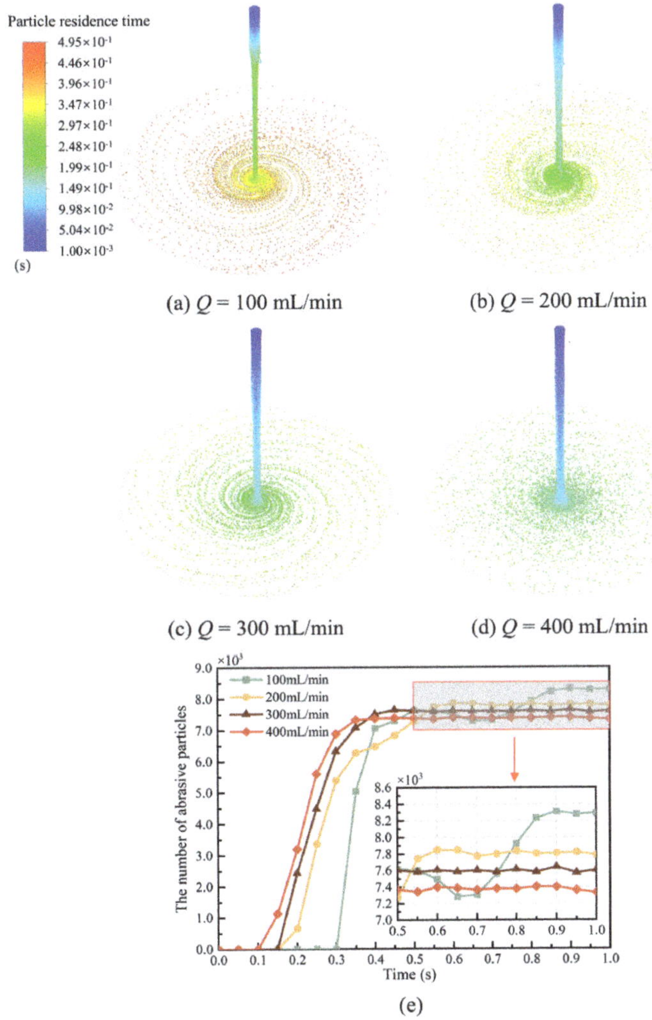

Figure 7. The distribution (**a**–**d**) and the number (**e**) of abrasive particles on the disk surface under different volume flow rates.

5.2. Influence of Rotational Speed

In Figure 8a–d, a comparison of the distribution of abrasive particles on the disk surface can be made by maintaining a liquid volumetric flow rate of 300 mL/min and a liquid viscosity of 250 mPa·s, with disk rotation speeds of 760 rpm, 860 rpm, 960 rpm, and 1060 rpm, respectively. The observed trajectories of the abrasive particles exhibited a spiral pattern, distributing outward from the center of the disk. As the rotational speed increases, the diffusion speed of abrasive particles on the disk surface accelerates. Near the center of the disk, the dominant influence on the liquid film is the inertial force, leading to a relatively high concentration of abrasive particles in the central region of the disk. With the same operating time, as the rotational speed of the disk increases, the degree of aggregation of abrasive particles in the central region of the disk decreases, showing a decreasing trend in the overall distribution of abrasive particles on the disk surface. As the rotational speed of the disk increases, the velocity of the liquid film enhances correspondingly. Consequently, the number of abrasive particles carried out of the disk by the liquid film increases within the same time frame. Thus, the quantity of abrasive particle distribution on the disk surface decreases. Figure 8e provides a clearer observation that with the increase in rotational speed, the quantity of abrasive particles distributed on the disk surface decreases. Under different disk rotational speeds, the growth trend of abrasive particle quantity on the disk surface remains consistent. A higher disk rotational speed results in a shorter time for the abrasive particle quantity to stabilize. It can be observed that at rotational speeds of 760 rpm, 860 rpm, 960 rpm, and 1060 rpm, the number of abrasive particles stabilizes after 0.5 s, 0.45 s, 0.45 s, and 0.4 s, respectively.

5.3. Influence of Viscosity

As shown in Figure 9a–d, the distribution of abrasive particles on the disk surface was analyzed with a constant liquid volume flow rate of 300 mL/min and a rotational speed of 960 rpm, while the liquid viscosities varied from 50 mPa·s to 350 mPa·s. It can be observed that when the liquid viscosity is low, the abrasive particles exhibit significant ejection in the central region of the disk, forming a "fireworks-like" distribution from the center to the edge of the disk surface, as shown in Figure 9a,b. Increasing the liquid viscosity leads to a transition of abrasive particle distribution on the disk surface from a "fireworks-like" pattern to a spiral shape, as shown in Figure 9c,d. When the viscosity is low, the diffusion rate of abrasive particles on the disc surface is slow. As the viscosity increases, the diffusion rate also increases. Figure 9e shows the number of abrasive particles on the disk surface under different liquid viscosities. It can be observed that viscosity has a minor impact on the distribution of the number of abrasive particles on the disk surface. The number of abrasive particles stabilizes after 0.4 s for all four viscosity conditions, followed by minor fluctuations. The fluctuations are greater with lower viscosities.

5.4. Uniformity under Different Operating Parameters

Comparison and analysis of the uniformity of diamond abrasive particle distribution in the processing area under different liquid volume flow rates, different disk rotational speeds, and different liquid viscosities are shown in Figure 10. It can be observed that as the liquid flow rate changes from 100 mL/min to 400 mL/min, the total number of diamond particles in the statistical area remains relatively constant, while the uniformity of particle distribution gradually increases, as shown in Figure 10a. During the process of the disk speed changing from 760 rpm to 1060 rpm, the total number of diamond particles in the statistical area gradually decreases, while the uniformity of particle distribution remains relatively consistent, as shown in Figure 10b. During the process of liquid viscosity changing from 50 mPa·s to 350 mPa·s, the total number of diamond particles in the statistical area initially increases and then decreases, while the uniformity of particle distribution gradually decreases, as shown in Figure 10c. The analysis results indicate that under conditions of higher flow rates, the distribution of abrasive particles on the disk surface can achieve a relatively uniform state. The rotational speed of the disk has a minor impact on the uniformity of abrasive particle distribution on the disk surface, but it does affect the density of abrasive particle distribution. Reducing the viscosity of the liquid facilitates the formation of a more uniform distribution of abrasive particles on the disk. Conversely, increasing the viscosity appropriately helps to enhance the density of abrasive particle distribution on the disk.

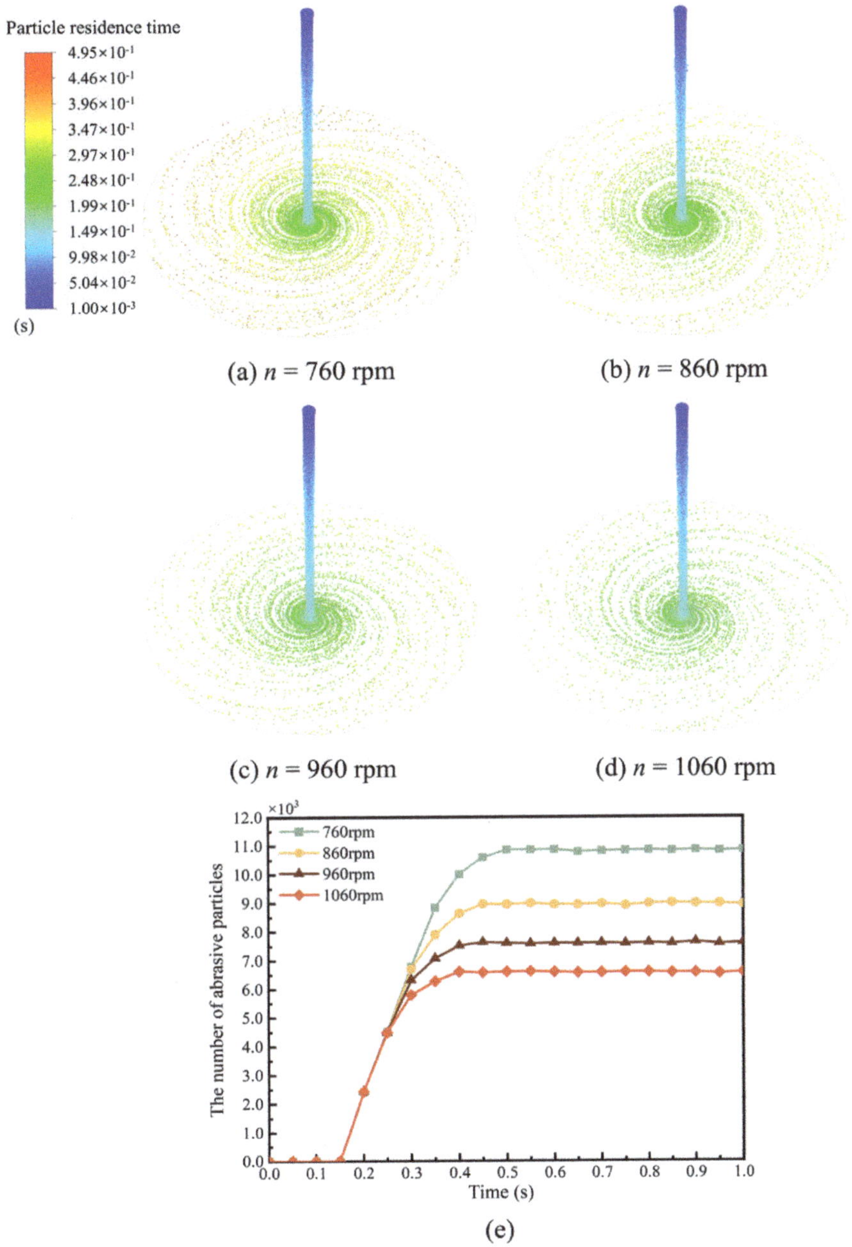

Figure 8. The distribution (**a**–**d**) and the number (**e**) of abrasive particles on the disk surface under different rotational speeds.

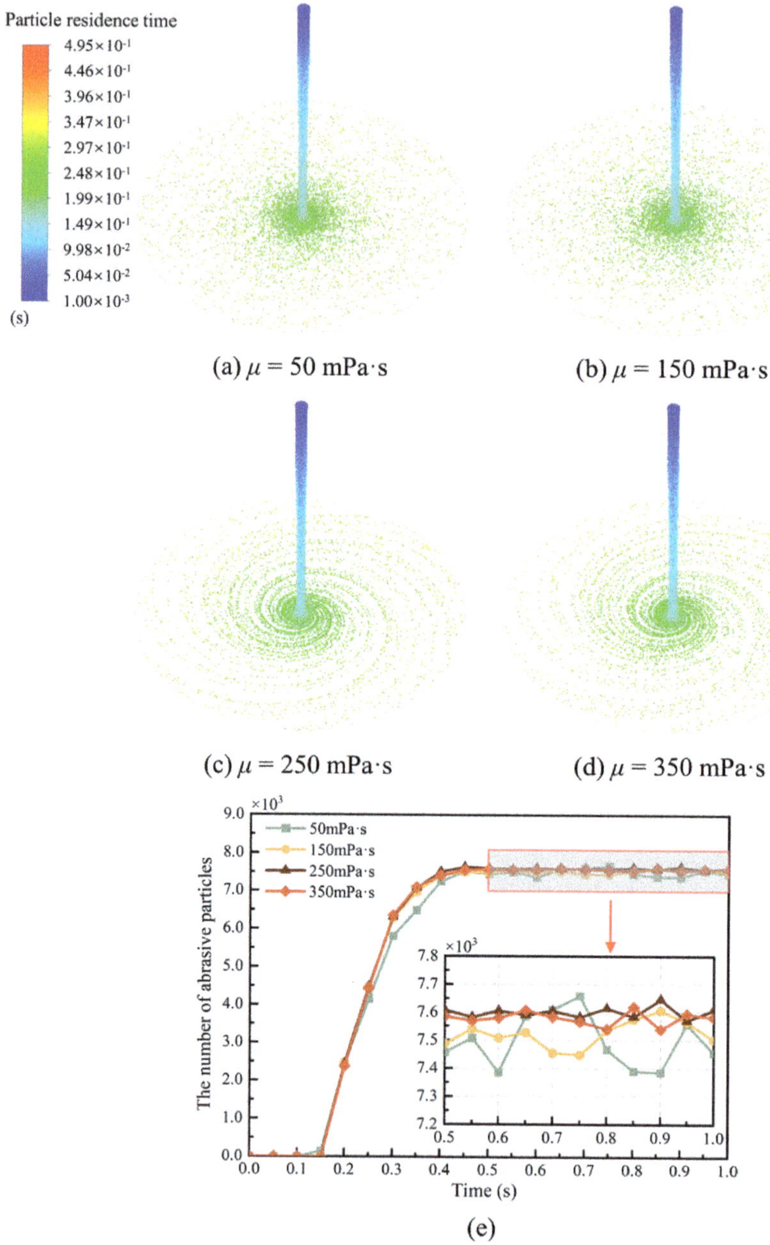

Figure 9. The distribution (**a–d**) and the number (**e**) of abrasive particles on the disk surface under different liquid viscosities.

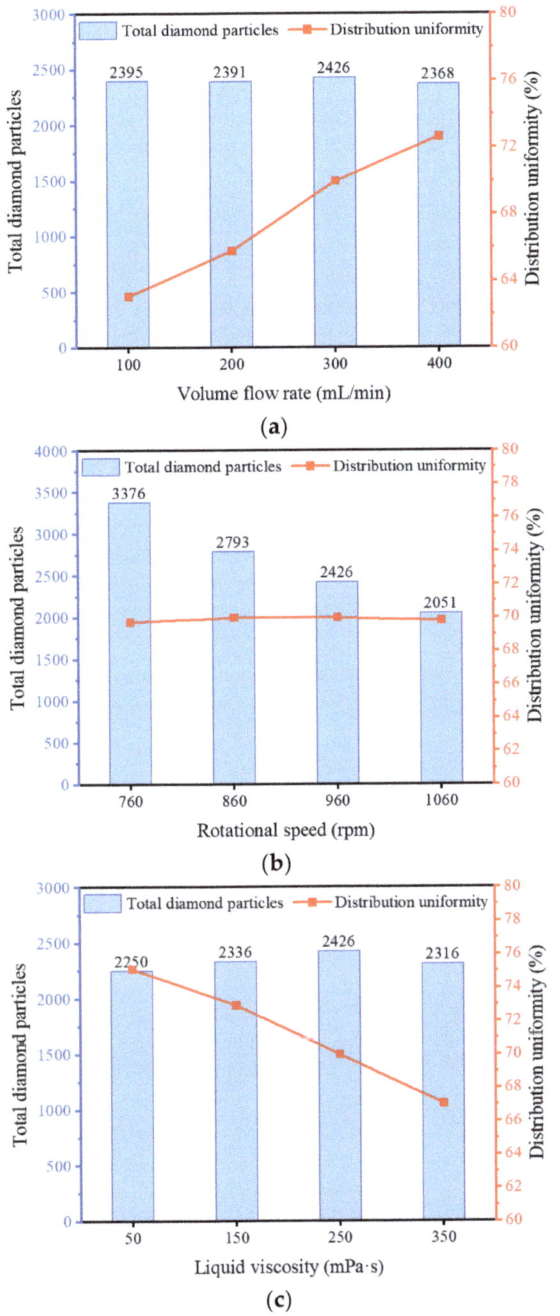

Figure 10. Distribution uniformity and total number of diamond particles under different (**a**) liquid volume flow rates, (**b**) different disk rotational speeds, and (**c**) different liquid viscosities.

5.5. Orthogonal Simulation to Optimize Operating Parameters

From the above analysis, it is evident that the liquid volume flow rate, the speed of the rotating disk, and the liquid viscosity all have varying degrees of influence on the distribution of abrasive particles on the disk surface. The effects produced by these three factors were examined by means of an orthogonal test, where three different values were taken for each factor, with the factor levels shown in Table 2. The three test factors, liquid volume flow rate, disk speed, and liquid viscosity, are denoted as A, B, and C, respectively.

In the orthogonal test, the factors 100 mL/min, 760 rpm, and 50 mPa·s are excluded. For the liquid volume flow rate 100 mL/min, there are two reasons. Firstly, as shown in Figure 7e, the number of particles remained on the disk surface for 100 mL/min is quite fluctuating. It is believed that this can be attributed to the mismatch of the liquid flow rate and the disk rotating speed. Additionally, as shown in Figure 10a, the distribution uniformity calculated from 100 mL/min is the lowest with the same group of tests, which suggests that it may not be necessary to include this set of tests in the orthogonal test and significantly increase the test scale. For the disk rotational speed of 760 rpm, it seems that the result of the number of total diamond particles is the highest within the same group of tests, as shown in Figure 10b. However, once taking a close look at Figure 8a, the distribution of these particles is much more concentrated at the central region of the disk. Once the diamond particles are light-cured and the disk is used as a grinding tool, the grinding efficiency and effective machining area will also be concentrated in the central region. Therefore, ensuring the surface roughness of the workpiece after grinding is more difficult. For dynamic viscosity of 50 mPa·s, it should be noticed that the viscosity values are actually the viscosities of the liquid–solid mixture phase, which suggests that the amount of solid particles affects the mixture's viscosity. Compared with the other parameters within this group of tests, the mass fraction that corresponds to 50 mPa·s is the lowest, and the total number of particles that remained on the disk is the lowest, as shown in Figure 10c. Considering the disk with arranged particles is to be used as a grinding or polishing tool, the number of abrasive particles cannot be too little, or the machining efficiency and service life could be significantly affected. Therefore, this set of data has been removed from the orthogonal test.

Table 2. Orthogonal factor levels.

Level	Test Factors		
	A/mL·min^{-1}	B/rpm	C/mPa·s
1	200	860	150
2	300	960	250
3	400	1060	350

Parameter settings and simulation analysis were conducted based on the orthogonal test table, and the two-dimensional distribution of abrasives on the disk surface was extracted as shown in Figure 11. It can be observed that, as mentioned above, there is a certain regularity in the distribution of abrasives. Abrasives are more concentrated in the central area of the disk and then spread out in a spiral or "fireworks-like" pattern on the disk surface. The uniformity of distribution under each parameter is statistically recorded, and the results are listed in Table 3.

The analysis started with factor A, which represents the effect of liquid volume flow rate on the test indicators. Here, A_1 denotes the 1 level of factor A and so forth for the subsequent levels. As can be seen from Table 3, tests numbered 1, 2, and 3 reflect the influence of A_1, tests numbered 4, 5, and 6 reflect the influence of A_2, and tests numbered 7, 8, and 9 reflect the influence of A_3, where y denotes the uniformity of abrasive distribution in the machining area on the disk surface. Columns without factors or interactions are referred to as blank columns. The fluctuations in the test results of blank columns can be regarded as test errors. If the range of the blank column is greater than the ranges of all other factors, it suggests that there may be significant interactions between the factors or that other important factors affecting the test results have been omitted. It should be noted that blank columns do not affect the test design.

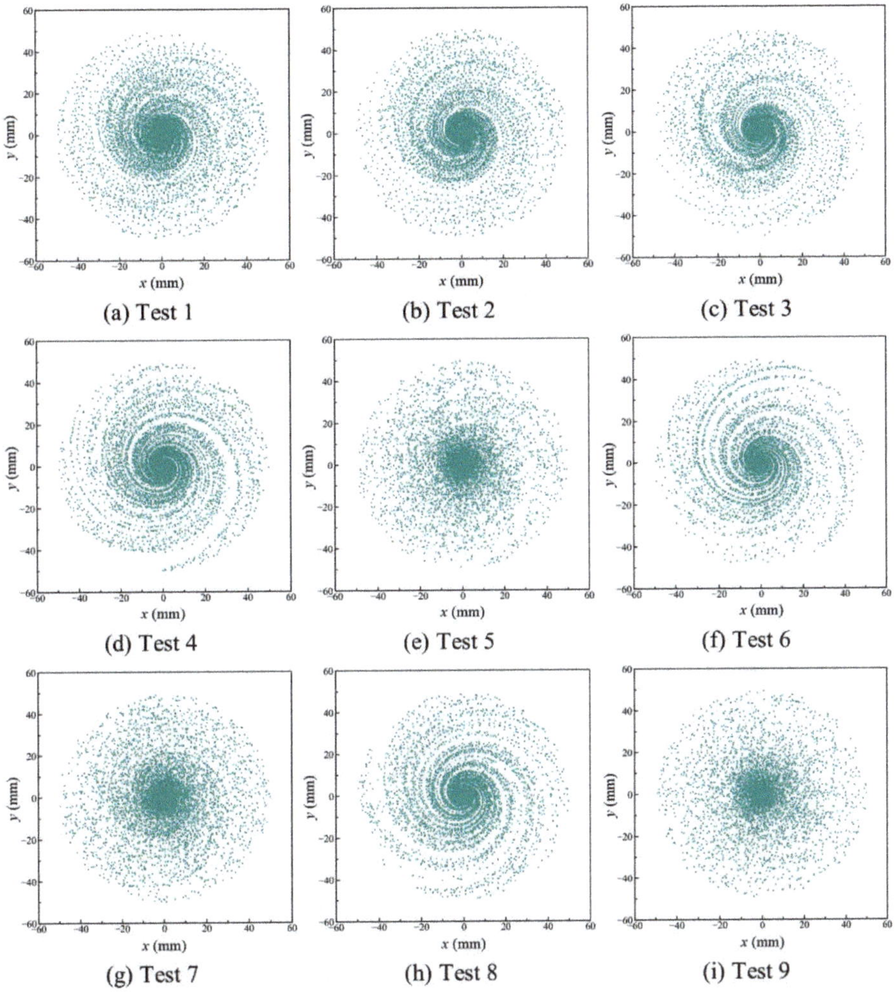

Figure 11. Orthogonal simulation of abrasive particles' distribution on the disk surface (top view extracted from 3D simulations).

Table 3. Orthogonal test results.

Test Number	Test Factors				y/%
	A	B	Blank Column	C	
1	1	1	1	1	69.69
2	1	2	2	2	65.81
3	1	3	3	3	63.89
4	2	1	2	3	70.53
5	2	2	3	1	72.95
6	2	3	1	2	69.82
7	3	1	3	2	71.06
8	3	2	1	3	70.20
9	3	3	2	1	73.37

The analysis of the effects of A, B, and C on the test indicators is presented in Table 4, where K_j (j = 1, 2, or 3) is the sum of the corresponding test results at a certain level. k_j represents the arithmetic mean of K_j. R represents the range, which can be calculated in any column by Equation (17):

$$R = \max\{k_1, k_2, k_3\} - \min\{k_1, k_2, k_3\} \quad (17)$$

The sum of the experimental results corresponding to A_1 is denoted by K_1 and is calculated as 69.69 + 65.81 + 63.89 = 199.39, then k_1 is 66.46. The sum corresponding to A_2, denoted as K_2, is calculated as 70.53 + 72.95 + 69.82 = 213.30, then k_2 is 71.10. The sum corresponding to A_3, denoted as K_3, is calculated as 71.06 + 70.20 + 73.37 = 214.63, then k_3 is 71.54. Range R_A is calculated as 71.54 − 66.46 = 5.08. Similarly, corresponding calculations and analyses are conducted for factors B and C.

Table 4. Influence analysis of indicators.

Parameters	A	B	Blank Column	C
K_1	199.39	211.28	209.71	216.01
K_2	213.30	208.96	209.71	206.69
K_3	214.63	207.08	207.90	204.62
k_1	66.46	70.43	69.90	72.00
k_2	71.10	69.65	69.90	68.90
k_3	71.54	69.03	69.30	68.21
R	5.08	1.40	0.60	3.79

It can be observed from Table 4 that the ranges are unequal, with $R_A > R_C > R_B$, indicating that the variations in the levels of each factor have different degrees of impact on the test results. The larger the range, the greater the variation in the test indicators in the numerical values, which is caused by the change in the values of the factors in that column within the test scope. So that the column with the largest range is also the factor with the greatest impact on the test results. Therefore, the liquid volume flow rate corresponding to factor A is the primary influencing factor, and the range of the blank column is smaller than the ranges of all other factors, indicating that there are no significant interactions between the factors and that the three factors considered are relatively comprehensive.

According to the characteristics of orthogonal tests, for factors A, B, and C, the influence of each factor on the test results at different level settings can be evaluated independently. If factor A has no effect on the results of the test, then the values of K_1, K_2, and K_3 obtained at different levels, namely conditions A_1, A_2, and A_3, should be equal. However, the actual calculation results show that the values of K_1, K_2, and K_3 are not equal, indicating that the variation in the levels of factor A has an impact on the test results. By comparing the specific values of K_1, K_2, and K_3, the degree of influence of factor A on the test indicators can be assessed. This test indicator refers to the uniformity of abrasive particles in the machining area on the surface of the disk. The greater the uniformity, the more favorable the result of the test. Thus, based on the order $K_3 > K_2 > K_1$, it can be determined that A_3 is the optimal level of factor A.

Similarly, the optimal levels for factors B and C are B_1 and C_1, respectively. The optimal combination of the three factors in this test is $A_3B_1C_1$, indicating that the liquid volume flow rate is 400 mL/min, the disk rotational speed is 860 rpm, and the liquid viscosity is 150 mPa·s, resulting in the best uniformity of abrasive distribution on the disk surface in the machining area. A simulation was conducted using the optimal level combination, as shown in Figure 12a. The calculated uniformity of abrasive distribution in the machining area was 75.36%. This compared to a uniformity of 63.89% for Test 3 as shown in Figure 12b, indicating an increase in uniformity of 11.47%. This may not seem to be a significant number for the uniformity improvement. However, the pattern of particle distribution has obviously changed. The abrasive particle distribution pattern at the optimal level resembles a "firework" shape, while the distribution pattern with poorer uniformity appears spiral. These phenomena suggest that the statistical method of using only one parameter to evaluate the uniformity may not be able to reflect the detailed changes of the particle distributions, especially under the driving media of the liquid film flow. In this case, a more comprehensive method to evaluate the abrasive particle distribution is required in future studies.

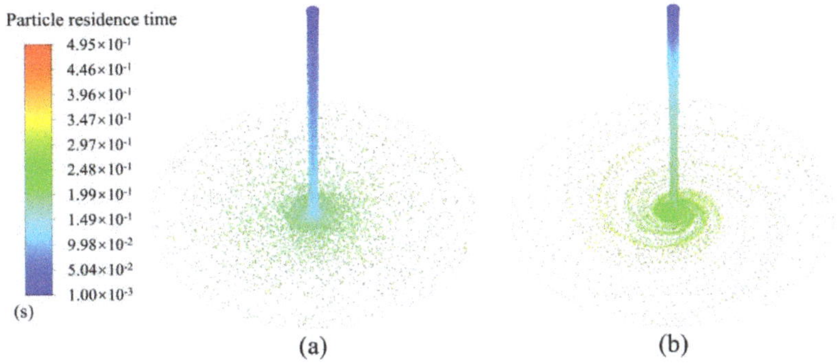

Figure 12. Simulation analysis of (**a**) the optimal level combination of A3B3C1 and (**b**) Test 3 combination of A1B3C3.

6. Conclusions

An Euler–Euler–Lagrange model was applied to simulate the distribution of abrasive particles driven by the liquid film on the rotating disk surface. A device based on the rotating disk system was designed and constructed, and comparative validation of simulation and experimental results was performed using image processing techniques. Based on the current study, the main conclusions are as follows:

(1) The grinding tools with uniformly distributed abrasives can be obtained based on the liquid film generated by the rotating disk system. The distribution patterns of abrasive particles on the disk surface are mainly spiral and fireworks-shaped.

(2) In general, the abrasive particles are more concentrated in the vicinity of the disk center and move against the direction of disk rotation. As time proceeds, the abrasive particles gradually spread to the entire disk surface, and the number of abrasive grains on the disk surface initially increases and then stabilizes.

(3) Establishing a three-factor, three-level orthogonal test table, the changes in distribution uniformity under nine test conditions were compared. It was found that the liquid volume flow rate had the greatest influence on the uniformity of abrasive particle distribution, followed by liquid viscosity, while the rotational speed of the disk had the least impact. At the same time, the optimal combination of the three factors was found to be a liquid volume flow rate of 400 mL/min, disk rotational speed of 860 rpm, and liquid viscosity of 150 mPa·s. Compared with the statically calculated result at the liquid volume flow rate of 200 mL/min, disk rotational speed of 1060 rpm, and liquid viscosity of 350 mPa·s, the optimized uniformity has increased by 11.47%. Although the improvement of uniformity is not a significant number, it is worth noticing that the pattern of particle distribution has been altered obviously.

Author Contributions: Conceptualization, Q.F., W.S. and N.D.; methodology, Q.F. and W.S.; software, Q.F.; validation, Q.F.; formal analysis, Q.F. and W.S.; investigation, Q.F.; resources, W.S., N.D. and H.H.; data curation, Q.F.; writing—original draft preparation, Q.F.; writing—review and editing, W.S., N.D., H.H. and Y.Z.; visualization, N.D.; supervision, W.S.; project administration, W.S.; funding acquisition, W.S. All authors have read and agreed to the published version of the manuscript.

Funding: This research was funded by the Natural Science Foundation of Fujian Province (2021J01296, 2022J01111551) and the Promotion Program for Yong and Middle-aged Teacher in Science and Technology Research of Huaqiao University (ZQN-1105).

Data Availability Statement: The data that support the findings of this study are available from the corresponding author, W.B.S. (Weibin Shi), upon reasonable request.

Conflicts of Interest: The authors declare no conflicts of interest.

References

1. Lin, B.; Zhou, K.; Guo, J.; Liu, Q.Y.; Wang, W.J. Influence of Grinding Parameters on Surface Temperature and Burn Behaviors of Grinding Rail. *Tribol. Int.* **2018**, *122*, 151–162. [CrossRef]
2. Zhao, B.; Wang, X.; Ding, W.; Wang, Y.; Fu, Y.; Zhao, Y.; Zhu, J. Grain Erosion Wear Properties and Grinding Performance of Porous Aggregated Cubic Boron Nitride Abrasive Wheels. *Chin. J. Aeronaut.* **2023**, *36*, 446–459. [CrossRef]
3. He, Y.; Xiao, G.; Zhu, S.; Liu, G.; Liu, Z.; Deng, Z. Surface Formation in Laser-Assisted Grinding High-Strength Alloys. *Int. J. Mach. Tools Manuf.* **2023**, *186*, 104002. [CrossRef]
4. Xiao, G.; Zhang, Y.; Zhu, B.; Gao, H.; Huang, Y.; Zhou, K. Wear Behavior of Alumina Abrasive Belt and Its Effect on Surface Integrity of Titanium Alloy during Conventional and Creep-Feed Grinding. *Wear* **2023**, *514–515*, 204581. [CrossRef]
5. Zhao, Z.; Qian, N.; Ding, W.; Wang, Y.; Fu, Y. Profile Grinding of DZ125 Nickel-Based Superalloy: Grinding Heat, Temperature Field, and Surface Quality. *J. Manuf. Process.* **2020**, *57*, 10–22. [CrossRef]
6. Dang, J.; Zang, H.; An, Q.; Ming, W.; Chen, M. Feasibility Study of Creep Feed Grinding of 300M Steel with Zirconium Corundum Wheel. *Chin. J. Aeronaut.* **2022**, *35*, 565–578. [CrossRef]
7. Yin, Y.; Chen, M. Analysis of Grindability and Surface Integrity in Creep−Feed Grinding of High−Strength Steels. *Materials* **2024**, *17*, 1784. [CrossRef] [PubMed]
8. Peng, R.; Liu, K.; Tong, J.; Tang, X. Performance of the Internal-Cooling Grooved Grinding Wheel with Patterned Abrasives. *Int. J. Adv. Manuf. Technol.* **2020**, *106*, 1633–1644. [CrossRef]
9. Herzenstiel, P.; Aurich, J.C. CBN-Grinding Wheel with a Defined Grain Pattern—Extensive Numerical and Experimental Studies. *Mach. Sci. Technol.* **2010**, *14*, 301–322. [CrossRef]
10. Ding, W.; Dai, C.; Yu, T.; Xu, J.; Fu, Y. Grinding Performance of Textured Monolayer CBN Wheels: Undeformed Chip Thickness Nonuniformity Modeling and Ground Surface Topography Prediction. *Int. J. Mach. Tools Manuf.* **2017**, *122*, 66–80. [CrossRef]
11. Denkena, B.; Grove, T.; Göttsching, T. Grinding with Patterned Grinding Wheels. *CIRP J. Manuf. Sci. Technol.* **2015**, *8*, 12–21. [CrossRef]
12. Fang, C.; Zhao, Z.; Lu, L.; Lin, Y. Influence of Fixed Abrasive Configuration on the Polishing Process of Silicon Wafers. *Int. J. Adv. Manuf. Technol.* **2017**, *88*, 575–584. [CrossRef]
13. Wen, D.; Qi, H.; Ma, L.; Lu, C.; Li, G. Kinematics and Trajectory Analysis of the Fixed Abrasive Lapping Process in Machining of Interdigitated Micro-Channels on Bipolar Plates. *Precis. Eng.* **2016**, *44*, 192–202. [CrossRef]
14. Chen, Z.; Zhang, X.; Wen, D.; Li, S.; Wang, X.; Gan, L.; Rong, X. Improved Grinding Performance of SiC Using an Innovative Bionic Vein-like Structured Grinding Wheel Optimized by Hydrodynamics. *J. Manuf. Process.* **2023**, *101*, 195–207. [CrossRef]
15. Sung, C.M. Brazed Diamond Tools by Infiltration 2000. U.S. Patent US006039641A, 21 March 2000.
16. Zhu, Y.; Hou, Z.; Huang, Z.; Li, B.; Zhang, Z.; Xu, J.; Ding, W. Study on Morphology Reconstruction of SiC Ceramics Ground by Monolayer-Patterned Grinding Wheel Considering Strain Rate Effect. *Int. J. Adv. Manuf. Technol.* **2024**, *130*, 2675–2686. [CrossRef]
17. Wu, S.; Zhang, F.; Ni, Y.; Chen, F.; Yan, Z. Grinding of Alumina Ceramic with Microtextured Brazed Diamond End Grinding Wheels. *Ceram. Int.* **2020**, *46*, 19767–19784. [CrossRef] [PubMed]
18. Yin, S.H.; Chen, F.J.; Yu, J.W.; Wang, M. A Novel Orderly Arrangement Method Controlled by Magnetic Field for Diamond Abrasives of Grinding Wheel. *Adv. Mater. Res.* **2012**, *497*, 6–9. [CrossRef]
19. González, M.; Rodríguez, A.; Pereira, O.; de Lacalle, L.N.L. Surface Roughness Evaluation When Brushing Heat-Resistant Alloy Components. *Int. J. Adv. Manuf. Technol.* **2024**. [CrossRef]
20. Li, X.; Wang, C.; Tian, C.; Fu, S.; Rong, Y.; Wang, L. Digital Design and Performance Evaluation of Porous Metal-Bonded Grinding Wheels Based on Minimal Surface and 3D Printing. *Mater. Des.* **2021**, *203*, 109556. [CrossRef]
21. Qiu, Y.; Huang, H. Research on the Fabrication and Grinding Performance of 3-Dimensional Controllable Abrasive Arrangement Wheels. *Int. J. Adv. Manuf. Technol.* **2019**, *104*, 1839–1853. [CrossRef]
22. Li, M.; Huang, Y.; Wang, W.; Yan, S.; Liu, Y.; Zou, L. A Novel 3D Printed Compliant Ball-End Grinding Tool with Crystal Structure: Feasibility and Performance Analysis. *Mater. Des.* **2024**, *237*, 112591. [CrossRef]
23. Aoune, A.; Ramshaw, C. Process Intensification: Heat and Mass Transfer Characteristics of Liquid Films on Rotating Discs. *Int. J. Heat Mass Transf.* **1999**, *42*, 2543–2556. [CrossRef]
24. Wang, D.; Jin, H.; Ling, X.; Peng, H.; Yu, J.; Cui, Z. Regulation of Velocity Zoning Behaviour and Hydraulic Jump of Impinging Jet Flow on a Spinning Disk Reactor. *Chem. Eng. J.* **2020**, *390*, 124392. [CrossRef]
25. Sharma, K.; Vijay, N.; Mabood, F.; Badruddin, I.A. Numerical Simulation of Heat and Mass Transfer in Magnetic Nanofluid Flow by a Rotating Disk with Variable Fluid Properties. *Int. Commun. Heat Mass Transf.* **2022**, *133*, 105977. [CrossRef]
26. Leshev, I.; Peev, G. Film Flow on a Horizontal Rotating Disk. *Chem. Eng. Process. Process Intensif.* **2003**, *42*, 925–929. [CrossRef]

27. Fluent, A.N. Eulerian Model Theory. In *ANSYS Fluent Theory Guide*; ANSYS Inc.: Canonsburg, PA, USA, 2020.
28. Naeeni, S.K.; Pakzad, L. Droplet Size Distribution and Mixing Hydrodynamics in a Liquid–Liquid Stirred Tank by CFD Modeling. *Int. J. Multiph. Flow* **2019**, *120*, 103100. [CrossRef]
29. Cundall, P.A.; Strack, O.D.L. A Discrete Numerical Model for Granular Assemblies. *Geotechnique* **1979**, *29*, 47–65. [CrossRef]
30. Morsi, S.A.; Alexander, A.J. An Investigation of Particle Trajectories in Two-Phase Flow Systems. *J. Fluid Mech.* **1972**, *55*, 193. [CrossRef]
31. Soma, T.; Katayama, T.; Tanimoto, J.; Saito, Y.; Matsushita, Y.; Aoki, H.; Nakai, D.; Kitamura, G.; Miura, M.; Asakawa, T.; et al. Liquid Film Flow on a High Speed Rotary Bell-Cup Atomizer. *Int. J. Multiph. Flow* **2015**, *70*, 96–103. [CrossRef]
32. Christiansen, J.E. *Irrigation by Sprinkling*; University of California, Agricultural Experiment Station: Berkeley, CA, USA, 1942.
33. Chen, F.; Zhang, L.; Yin, S.; Huang, S.; Tang, Q. Study on Distribution Characteristics of Diamond Particles under High-Voltage Electrostatic Field. *Int. J. Adv. Manuf. Technol.* **2018**, *96*, 1393–1401. [CrossRef]
34. González, M.; Rodríguez, A.; López-Saratxaga, U.; Pereira, O.; López De Lacalle, L.N. Adaptive Edge Finishing Process on Distorted Features through Robot-Assisted Computer Vision. *J. Manuf. Syst.* **2024**, *74*, 41–54. [CrossRef]

lubricants

MDPI

Article

Effect of CeO_2 Content on Microstructure and Wear Resistance of Laser-Cladded Ni-Based Composite Coating

Bingqing Zhang [1], Wenqing Shi [1,2,3,4,*], Yiming Lin [1], Longwei Jiang [1], Lijun Wang [2] and Kuanfang He [3]

[1] College of Electronic and Information Engineering, Guangdong Ocean University, Zhanjiang 524088, China; zbqingxj@163.com (B.Z.); lymgdou0125@163.com (Y.L.); jlwgdou@163.com (L.J.)
[2] School of Ship and Shipping, Guangdong Ocean University, Zhanjiang 524088, China; wanglj20010803@163.com
[3] College of Mechanical and Electrical Engineering and Automation, Foshan University, Foshan 528000, China; hkf791113@163.com
[4] School of Materials Science and Engineering, Guangdong Ocean University, Yangjiang 529500, China
* Correspondence: swqafj@126.com

Abstract: In order to improve the wear resistance of 45 steel, in this study, WC/Ni60 composite coatings with different CeO_2 additions (0, 1, 2, and 3 wt%) were prepared on 45 steel by the laser cladding technique; the experimental analysis was carried out by means of scanning electron microscopy (SEM), energy-dispersive spectroscopy (EDS), X-ray diffraction (XRD), a Vickers hardness tester, and a friction and wear tester. The results show that CeO_2 had little effect on the phase composition of the coatings; however, with the increase in CeO_2 content, the CeO_2 played a key role in refining the grains of the coating, thus reducing the generation of cracks. In addition, CeO_2 could effectively strengthen the internal structure of the coating and improve its microhardness and wear resistance. Particularly noteworthy is the observed reduction in both the friction coefficient and mass loss of the coating when the CeO_2 addition reached 2%. This suggests an enhancement in the tribological performance of the coating at this concentration.

Keywords: laser cladding; rare-earth oxides; microstructure; grain refinement; wear resistance

1. Introduction

To date, 45 steel has been widely used in machinery manufacturing, automobiles, construction, mining equipment, and other fields because of its commendable mechanical properties, processing performance, and low cost [1–4]. However, the wear resistance of 45 steel is insufficient, meaning it may not meet the criteria in some application scenarios requiring high wear resistance and results in the premature wear of parts. In addition, its fatigue performance is also limited. Therefore, the surface modification of 45 steel is of great significance.

With its excellent high hardness and strength advantages, ceramic materials have achieved a prominent position in the field of reinforcement and repair of key mechanical equipment components [5]. Through the application of laser cladding technology, the constraints faced by ceramic materials in the processing process have been successfully overcome, thus achieving the preparation of high-quality ceramic coatings on metal surfaces [6]. For example, Hu et al. [7] prepared nickel-based fusion-coated coatings containing different WC additions on stainless steel substrates and observed inhomogeneity in the distribution of WC particles. Further analysis shows that with the increase in WC mass fraction, the degree of grain refinement and microhardness of the coating show an increasing trend. Xia et al. [8] investigated the role of adding WC particles with different particle sizes in nickel-based composite coatings. The results show that the average microhardness of the composites increases by 59.70% (coarse WC) and 74.66% (fine WC), and the wear rate decreases by 84.87% (coarse WC) and 89.17% (fine WC), respectively. The study also elucidated that WC particles have physical shielding and solid solution strengthening effects. The WC reinforced Fe-based amorphous composite coating prepared using rectangular spot laser cladding technology by Zhou et al. [9] can significantly improve the surface quality and improve the microhardness after laser remelting.

Laser cladding technology is of great significance in improving the wear resistance of metal parts. It not only prolongs the service life of key components but also optimizes resource utilization. Hence, in recent years, researchers have employed rare -earth elements in ceramic coatings through laser cladding technology to further improve their performance. These results show that the introduction of rare-earth elements can have a positive impact on coating performance in many aspects. Firstly, rare-earth elements can refine the microstructure of the coating, enhance the fluidity of the molten pool, and promote the discharge of gas from the molten pool, thereby effectively reducing the porosity of the coating and improving the overall quality [10]. Secondly, by adding rare earthearths or rare -earth oxides (such as Y_2O_3, CeO_2, and La_2O_3) to the cladding powder, the cracking tendency of the coating can be further reduced, the formation of defects such as pores can be inhibited, the grains can be refined, and the microstructure uniformity of the coating can be improved [11]. For example, Quazi et al. showed that the moderate addition of rare-earth oxides (e.g., Y_2O_3, CeO_2, and La_2O_3) improves the appearance and microstructure of coatings, reduces the crack susceptibility and porosity, and effectively prevents defects, such as cracks, pores, and inclusions, from occurring during laser melting and cladding [12], which is of great significance for studying the mechanism of the role of rare earths in laser melting and cladding composite coatings. Ye et al. [13] found that the addition of CeO_2 has a significant effect on the microstructure, hardness, and corrosion resistance of a laser cladding coating on the surface of titanium alloy, and proper addition can improve the performance. Zhang et al. [14] studied the effect of Y_2O_3 addition on the quality, microstructure, and microhardness of a multi-pass laser cladding Ti-6Al-4V coating by the coaxial powder feeding method. The results show that the addition of Y_2O_3 can completely eliminate the formation of pores in the coating and improve the quality and hardness of the coating. At the same time, the introduced Y_2O_3 has a great influence on the grain boundary and grain, resulting in grain boundary strengthening and grain refinement.

Building upon prior domestic and international research, the primary objective of this study was to fabricate WC/Ni60/CeO_2 composite coatings using laser cladding technology to enhance the microhardness and wear resistance of 45 steel surfaces. This research holds significant implications for the investigation of surface reinforcement techniques for critical components. Furthermore, the study investigated the impact of varying weight percentages of CeO_2 on the geometric morphology, microstructure, wear resistance, and hardness of the coatings. To achieve this, distinct CeO_2 content additives were formulated, followed by the execution of laser cladding experiments and comprehensive evaluation of the resultant coatings.

2. Test Materials and Methods

2.1. Test Materials

In this experiment, 45 steel was chosen as the base material, with a size of 50 (length) × 50 (width) × 5 (height) (unit: mm). Ni60 and WC powders of 300~500 mesh and CeO_2 powders of 500 mesh were mechanically mixed by a FP400 planetary ball mill (180 r/min, 1 h). The mixed powder was dried in a vacuum oven (200 °C, 2 h). In the comparative experimental design, the percentage of rare-earth oxide CeO_2 was set to 0%, 1%, 2%, and 3%, respectively. The CeO_2 powder is shown in Figure 1a, the WC powder is shown in Figure 1b, the Ni60 powder is shown in Figure 1c, and the chemical composition of 45 steel is shown in Table 1. The exact volume fraction of the powder is shown in Table 2.

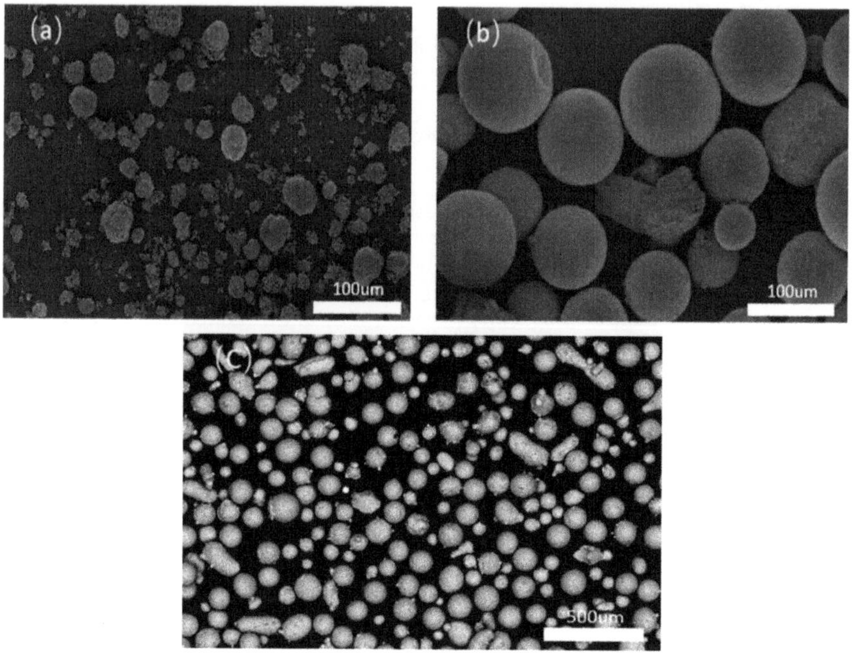

Figure 1. Morphology of cladding material: (**a**) CeO_2 powder; (**b**) WC powder; (**c**) Ni60 powder.

Table 1. Chemical composition of 45 steel (wt%).

Materials	**C**	**Si**	**Mn**	**Ni**	**Cu**	**Fe**
Value	0.42–0.05	0.17–0.37	0.5–0.8	$\leq$0.30	$\leq$0.25	Bal

Table 2. Ni60, WC, and CeO_2 component ratio (mass fraction, %).

Sample	**A**	**B**	**C**	**D**
Ni60	7	6.3	5.6	4.9
WC	3	2.7	2.4	2.1
CeO_2	0	1	2	3

2.2. Test Method

The XL-F2000 W fiber laser cladding system was selected for the laser cladding experiments, and the maximum output laser power of the equipment is 2000 W. For the powder feeding method, we employed the prefabricated technique. Prior to experimentation, the substrate surface underwent sanding using sandpapers of varying specifications (240, 600, 800, 1500, 3000, and 5000 mesh) to eliminate surface oxides and impurities, followed by thorough cleaning with anhydrous ethanol to remove any residual debris and oil stains, and was subsequently air-dried in a ventilated environment. The mixed powder was uniformly applied onto the substrate surface at a thickness of 1 $\pm$ 0.1 mm. Through preliminary experiments, the optimum process parameters were determined as follows: a laser power of 1500 W, scanning speed of 10 mm/s, and defocusing amount of +5 mm, all of which induce a better cladding effect. Subsequently, a multi-channel laser cladding test was conducted using these established parameters. The schematic diagram depicting the laser cladding process is presented in Figure 2. Post-cladding samples underwent further processing including wire cutting (utilizing a Deep Yang QC350 K, Sichuan, China), inlaying, grinding, and polishing, after which the cross-sectional morphology of the coating was examined using an XJL-302 optical metallographic microscope (Guangzhou, China). Additionally, the microstructure of the coating was analyzed using a JSM6460 scanning electron microscope (SEM) (Brno, Czech Republic). The elements and compounds of the composite coating were analyzed by an X-ray diffractometer (XRD-6100) produced by Shimadzu, Shimane-ken, Japan, and the radiation source was CuKα. Furthermore, the microhardness of the cladding layer was tested by a digital microhardness tester (MHVD-1000AT) (Shanghai, China). The test was conducted with a 150 μm spacing from the top of the cladding layer to the substrate. Measurements are taken at three points, each spaced 50 μm apart on either side of the vertical test direction. The applied load was 200 g, and the duration of loading was 10 s. The wear resistance of the composite coating was tested at room temperature using the SFT-2M pin-on-disc friction and wear tester produced by Lanzhou Zhongke Kaihua Technology Co., Ltd. (Lanzhou, China). Dry sliding friction was tested. The grinding material was a Si_3N_4 steel ball with a diameter of 4 mm. The applied load was 40 N, the rotation speed was 200 r/min, the rotation radius was 2 mm, and the frequency was 1 Hz. The test time of each sample was 30 min.

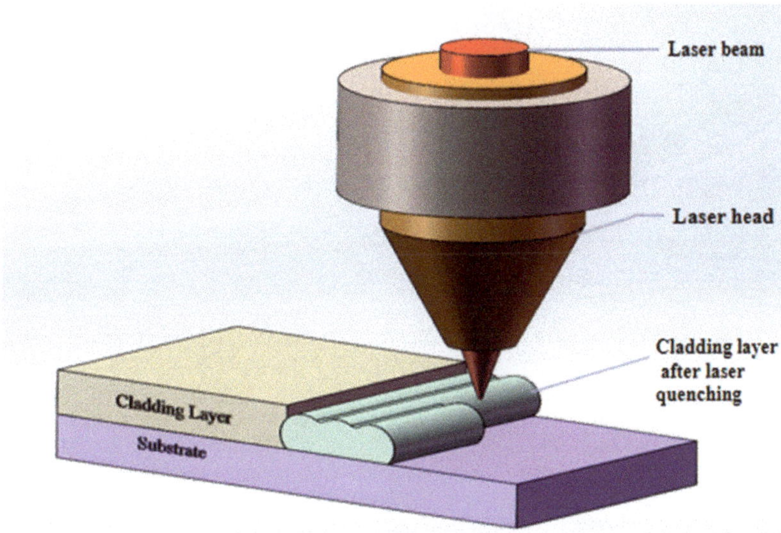

Figure 2. Schematic diagram of laser cladding processing.

3. Results and Discussion

3.1. Coating Macromorphology Analysis

Figure 3 presents the cross-sectional SEM images (a1–a4) and surface macrographs (b1–b4) of the laser-cladded Ni60/WC composite coatings with varying CeO_2 contents. From top to bottom are the obtained cladding coatings with CeO_2 mass fractions of 0%, 1%, 2%, and 3%. The results show that no defects such as cracks or pores were observed in the cross-sections of all samples, as shown in Figure 3(a1–a4). Figure 3(b1) shows that for the coating without CeO_2, the surface is smooth, but there are long cracks. These cracks extend along the transverse direction and form an angle of nearly 90° with the cladding direction. Observing Figure 3(b2–b4), it is evident that the addition of CeO_2 resulted in a smoother surface for the coating, with no noticeable cracks. Given the susceptibility of nickel-based alloy coatings to crack, this outcome can be primarily attributed to the significant residual tensile stress concentrated within the coating and at the interface between the coating and the substrate [15], which commonly facilitates crack propagation. Therefore, the addition of cerium oxide (CeO_2) can optimize the forming quality of the coating.

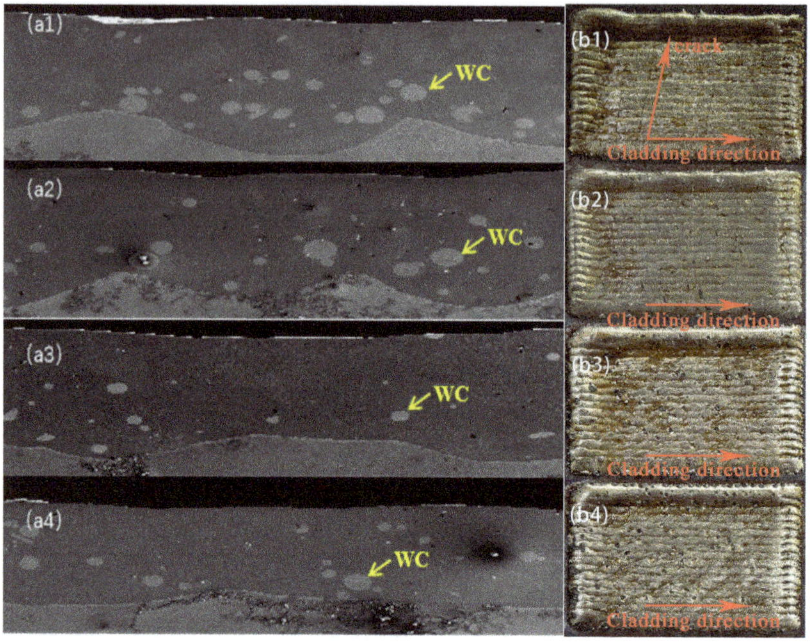

Figure 3. Macroscopic morphology of the coating: typical cross-sectional morphology (**a1–a4**) and surface macrograph (**b1–b4**).

3.2. Microscopic Morphology

Figure 4 shows the X-ray diffraction (XRD) patterns of Ni/WC composite coatings with different cerium oxide contents after laser cladding. The strong peaks of C, WC, and W2C were observed for four different contents of cerium oxide, which verified the fact that WC can be synthesized in situ [16,17]. The existence of WC proves the existence of unmelted WC particles in the coating. The existence of W2C phase indicates that during the solidification process, W reacts with C, resulting in non-equilibrium solidification during the cladding process [18]. Remarkably, with a cerium oxide content surpassing 2%, there is a notable increase in the quantity of W2C hard phase. In addition, under intense laser irradiation, iron is considered to be diluted from the substrate due to high-temperature melting and liquid-phase flow, while the remaining tungsten is dissolved in the Ni matrix after the in situ reaction. Therefore, there is also a certain proportion of Fe0.64Ni0.32 phase in the coating, which verifies the existence of the non-equilibrium solidification phenomenon in the cladding process again.

Based on the X-ray diffraction (XRD) analysis results of the aforementioned coatings, it is evident that all samples exhibit a similar crystal phase composition. Thus, to investigate the impact of CeO_2 on the coating, samples 0, 1, 2, and 3 were chosen for microstructural comparison. Moreover, the multi-channel coating was segmented into two regions (bottom region and middle region) to probe the microstructural evolution of the coating, allowing for a detailed analysis of the structural changes and hardness values within the coating.

Figure 5 shows the morphology of the bottom region of the coating with a different cerium oxide content. The grains at the bottom of the cladding layer show an obvious orientation, and the columnar crystals mainly extend along the direction perpendicular to the coating fusion line. This phenomenon is due to the slower heat dissipation rate at the bottom of the molten pool near the substrate, resulting in the growth of particles at the bottom of the coating along the top direction [19]. Before the addition of CeO_2, the grains at the bottom of the cladding layer are coarse, including large planar grains and columnar crystals. The addition of CeO_2 results in significant grain refinement [20,21]. With the increase in CeO_2 content, the grains at the bottom of the cladding layer are refined first and then coarsened. When the CeO_2 content is 2%, the grain reaches the finest state, and when the CeO_2 content is 3 wt%, the grain becomes coarser.

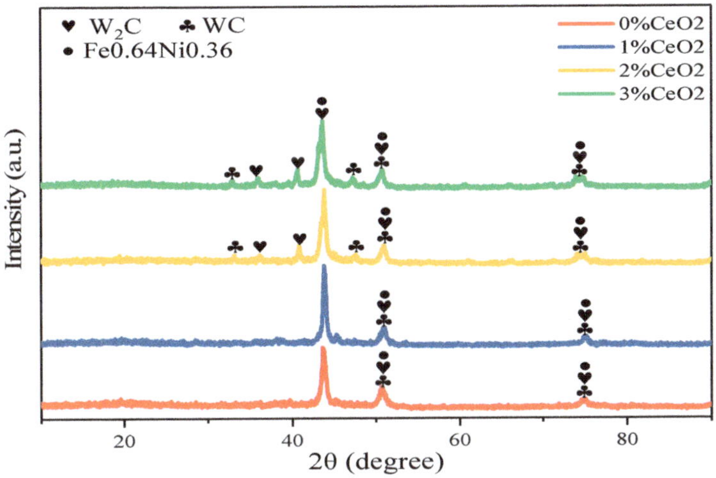

Figure 4. Phase composition of Ni/WC composite coating.

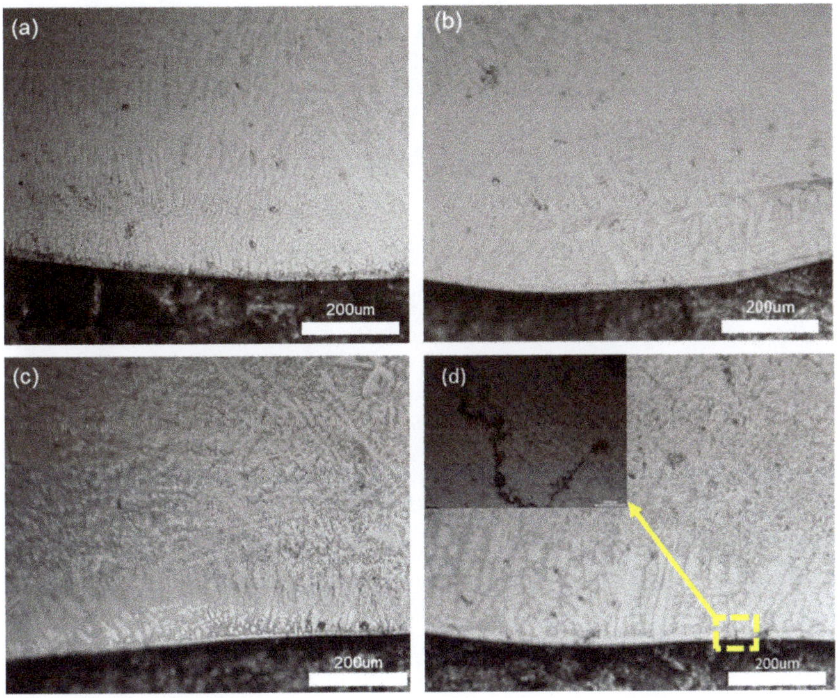

Figure 5. Microstructure of Ni/WC composite coating at the bottom: (**a**) S0; (**b**) S1; (**c**) S2; (**d**) S3.

Figure 6a–d shows the morphology of the middle region of the coating with different cerium oxide contents. Before the addition of CeO_2, the grains in the middle of the cladding layer are mainly composed of larger dendritic and rod-like crystals. After the addition of CeO_2, the WC particles in the middle of the cladding layer grow radially around [17], which leads to the refinement of the grains around the WC particles, thus significantly enhancing the fine grain effect [22]. With the increase in CeO_2 content, the grains of WC particles in the middle of the cladding layer are refined first and then gradually coarsened. When the CeO_2 content is 2%, the most dendrites and secondary dendrites are observed around the WC particles.

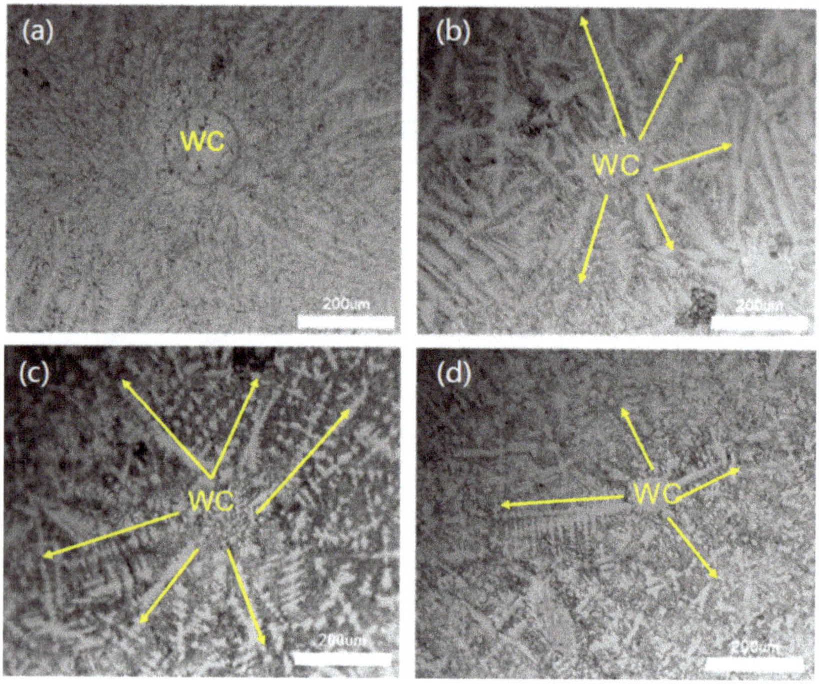

Figure 6. Microstructure of the middle part of Ni/WC composite coatings: (**a**) S0; (**b**) S1; (**c**) S2; (**d**) S3.

Cracks are evident in the magnified area of Figure 5d. The underlying reasons for this phenomenon are multifaceted: Firstly, the uniformity of the microstructure within the cladding layer emerges as a pivotal factor influencing cracking [20]. The excessive addition of CeO_2 to form refractory compounds diminishes the fluidity of the molten pool, exacerbating organizational stresses [21], thus precipitating crack formation. Secondly, during the solidification phase, the excess of rare-earth elements accelerates the absorption rate of the cladding material, thereby enhancing energy within the molten pool and facilitating the release of more WC particles [23]. The WC content is proportional to the hardness range; that is, the higher the WC content, the higher the hardness of the coating, but the worse the ductility. The precipitated carbides will increase the tensile stress, thus increasing the crack sensitivity [24,25]. Therefore, the higher WC content deposited at the bottom will also lead to cracks.

3.3. Hardness Analysis

Microhardness serves as a crucial parameter for assessing coating performance. Figure 7 illustrates the microhardness variation along the cross-section of the samples, spanning from the coating's apex to the substrate. As depicted in Figure 7, all sample coatings exhibited a notable increase in microhardness. Figure 8 illustrates a progressive increase in the average microhardness of the samples. Specifically, the microhardness of sample S2's coating reached 610.83 HV, approximately threefold that of the substrate (202.86 HV). Similarly, sample S3's coating achieved a microhardness of 634.63 HV, approximately 3.1 times greater than that of the substrate. The results demonstrate that the addition of CeO_2 to the coating led to a more pronounced increase in hardness compared to the S0 sample. The results show that the addition of CeO_2 can improve the hardness of the coating. The reason for the higher hardness of the S1 and S2 samples is that the addition of a low content of CeO_2 makes the microstructure of the cladding layer denser and the grains gradually smaller [21]. The grain refinement and solid solution strengthening effect improve the strength and microhardness of the cladding layer [26]. Furthermore, the heightened surface hardness observed in sample S3 can be attributed to two primary factors: Firstly, Figure 3(a4) reveals a higher presence of unmelted WC particles in the middle and lower sections of the coating, contributing to enhanced coating hardness [16,17]. Secondly, the increase in WC content leads to the decomposition of WC into W and C atoms, subsequently dissolved within the matrix, thereby enhancing the solid solution strengthening effect [27,28].

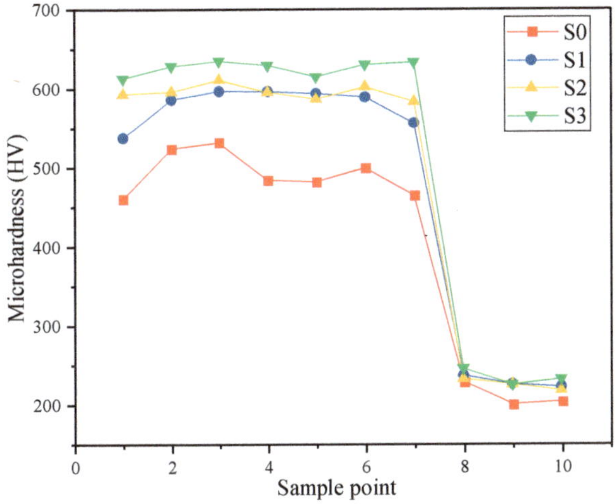

Figure 7. Microhardness distribution of Ni/WC composite coating.

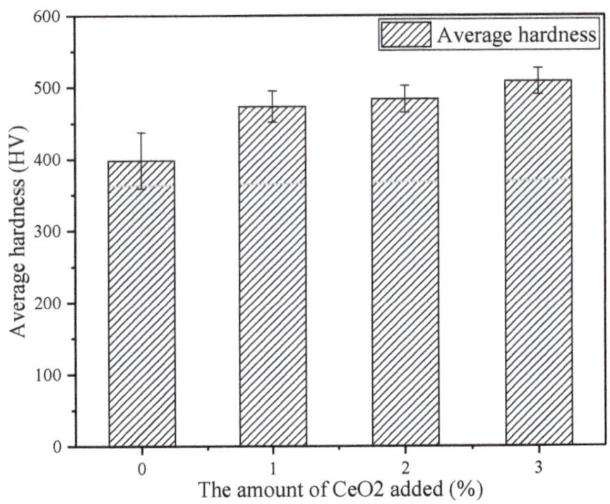

Figure 8. Average microhardness distribution of Ni/WC composite coating.

3.4. Wear Resistance Analysis

Figure 9 shows the variation curve of the friction coefficient of coating samples with different ratios. Under the load condition of 40 N, the sliding friction enters the stable wear stage after a run-in stage. Comparative analysis reveals a significant reduction in the average friction coefficient of samples S0, S1, S2, and S3 relative to the substrate, indicative of superior wear resistance exhibited by the coatings. Notably, in comparison to sample S0 lacking CeO_2, samples S1, S2, and S3 also demonstrate a noteworthy decrease in the average friction coefficient. Furthermore, the wear loss of samples S0, S1, S2, and S3 is also significantly reduced compared to that of the substrate, as depicted in Figure 10. The friction coefficient of the S0 sample exhibits considerable fluctuation within 20 min, indicating unstable wear resistance. This instability may stem from the presence of incompletely dissolved tungsten carbide particles within the coating. The interaction of these substances with the friction balls could induce violent friction phenomena [18]. Conversely, the friction coefficient curves of samples S1, S2, and S3 display a stable trend during the stable wear stage.

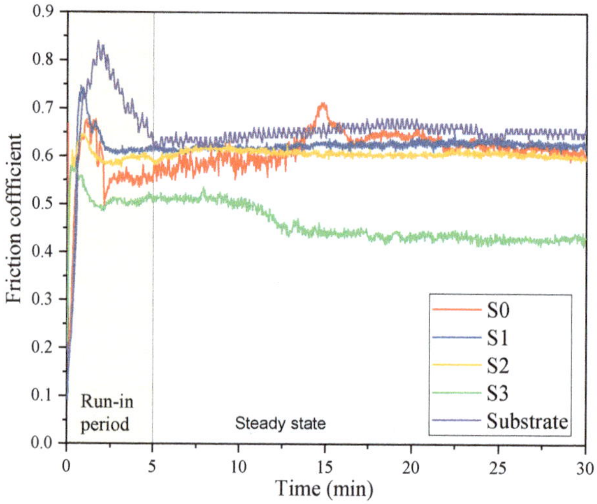

Figure 9. The friction coefficient curves of Ni/WC composite coatings with different samples.

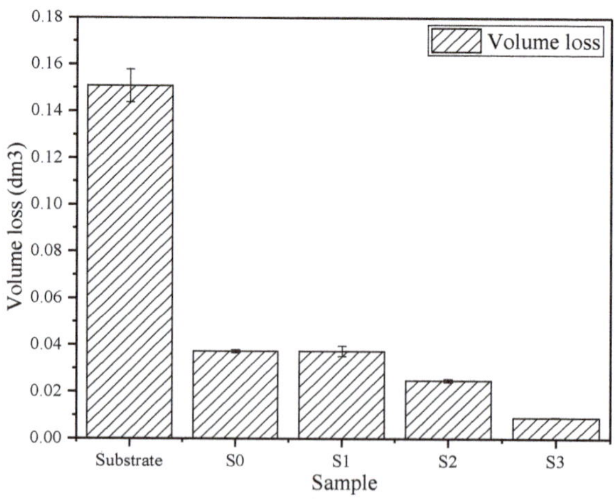

Figure 10. Wear loss of Ni/WC composite coatings with different samples.

In order to further study the wear mechanism of CeO_2/Ni60 coatings, the worn surface morphology of each sample in Figure 11 was observed by scanning electron microscopy (SEM). As illustrated in Figure 11a, the wear surface of sample S0 exhibits flake or substantial material shedding marks, along with the presence of plough-shaped grooves, indicative of abrasive wear and severe adhesive wear. Since the CeO_2-free coating is softer than the matched parts, the hard micro-convex bodies on the surface of the matched parts can easily penetrate into the sliding surface during the sliding process, forming a large wear contact area between the two. This leads to the formation of severe adhesive wear. Following the incorporation of CeO_2, the wear surface of the coating manifests increased smoothness relative to the non-CeO_2-incorporated counterpart, concomitant with significant reductions in both the friction coefficient and wear volume. Inspection reveals comparable wear degrees among samples S1, S2, and S3; nonetheless, furrows persist on the coating's wear surface, as depicted in Figure 11b–d. The heightened hardness of the coating impedes plastic deformation during interaction with mating components throughout the wear process, consequently diminishing wear contact areas. Moreover, the low plastic deformation contributes to the suppression of adhesive wear, aligning with Archard's well-known law of adhesive wear [29]. Therefore, the wear mechanism of the CeO_2-added coating is mainly mild adhesive wear and abrasive wear [30]. This shows that the wear resistance of the coating with CeO_2 is significantly improved. The hardness of the coating material is proportional to the wear resistance of the coating [31]. The higher the hardness of the coating material, the better the wear resistance of the coating. The reason for sample S3 having the best wear resistance is as follows: First, the submicron CeO_2 particles are exposed to the surface of the friction pair during the wear process, providing lubrication [30]. Secondly, the WC particles on the coating surface are preferentially worn to protect the softer 45-steel substrate from direct contact with the Si_3N_4 steel ball. In addition, at high temperatures generated by dry sliding friction, broken WC decomposes into W and C elements. The new carbides formed by C, W, and other elements in the matrix are hard phases, which can significantly improve the hardness and wear resistance of the matrix [32]. The results show that the addition of CeO_2 can improve the wear resistance of the composite coating [27]. Combined with the above microstructure observation and hardness test, micro-cracks are formed inside the sample S3 coating. Therefore, when the added amount of CeO_2 is 2%, the tribological properties of the coating, such as the friction coefficient and wear rate, are superior.

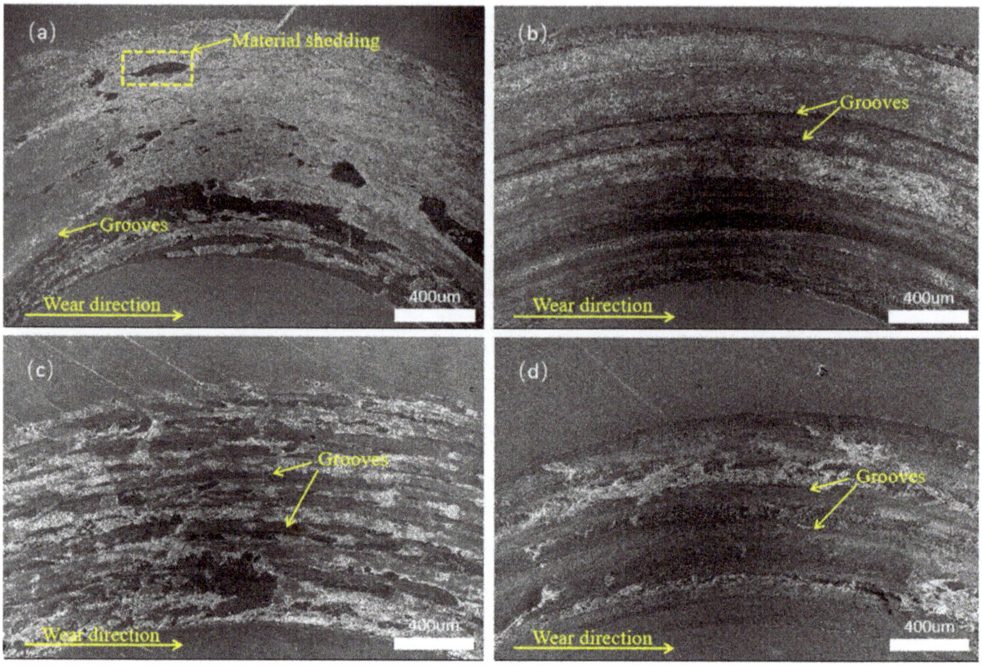

Figure 11. OM wear morphology of Ni/WC composite coatings for different specimens: (**a**) S0; (**b**) S1; (**c**) S2; (**d**) S3.

4. Conclusions

Ni60/WC/CeO_2 composite coatings were fabricated on the surface of a 45-steel substrate utilizing laser cladding technology. The surface modification effect of trace CeO_2 on the composite coatings was investigated and substantiated through an examination of macroscopic and microscopic features, hardness, and wear resistance. The principal findings are summarized as follows:

1. The incorporation of CeO_2 into the cladding powder demonstrated a capacity to mitigate surface cracks in nickel-based composite coatings, thereby enhancing the overall quality of the formed composite coating.
2. An optimal quantity of CeO_2 was shown to refine grain structures, promote uniformity in the microstructure of the cladding layer, and enhance solid solution strengthening within the cladding layer, thereby augmenting both the hardness and wear resistance properties of the coating.
3. When the CeO_2 addition reached 2%, notable enhancements in both the hardness and wear resistance of the composite coating were observed, resulting in optimized mechanical properties.

Author Contributions: Conceptualization, B.Z. and W.S.; project administration, W.S.; formal analysis, Y.L., L.J., L.W. and K.H.; validation, Y.L., L.J. and L.W.; investigation, Y.L., L.J. and K.H.; writing—original draft preparation, B.Z. and W.S. All authors have read and agreed to the published version of the manuscript.

Funding: This research was funded by the National Natural Science Foundation of China, grant No. 62073089; Zhanjiang science and technology plan project, topic number 2023A216; and Yangjiang science and technology plan project, project no. SDZX2023004. The authors are also grateful for the support from Guangdong Provincial.

Institutional Review Board Statement: Not applicable.

Informed Consent Statement: Not applicable.

Data Availability Statement: No new data were created or analyzed in this study. Data sharing is not applicable to this article.

Acknowledgments: The authors sincerely thank Wenqing Shi of Guangdong Ocean University for his supervisory work on this thesis.

Conflicts of Interest: The authors declare no conflicts of interest.

References

1. Liu, B.X.; Fan, K.Y.; Yin, F.X.; Feng, J.H.; Ji, P.G. Effect of Caliber Rolling Reduction Ratios on the Microstructure and Mechanical Properties of 45 Medium Carbon Steel. *Mater. Sci. Eng. A* **2020**, *774*, 138954. [CrossRef]
2. Liu, H.; Wei, J.; Dong, J.; Zhou, Y.; Chen, Y.; Wu, Y.; Babu, S.D.; Umoh, A.J.; Ke, W. The Synergy between Cementite Spheroidization and Cu Alloying on the Corrosion Resistance of Ferrite-Pearlite Steel in Acidic Chloride Solution. *J. Mater. Sci. Technol.* **2021**, *84*, 65–75. [CrossRef]
3. Sun, L.; Yang, Q.; Zou, J.; Ni, L.; Song, X.; Zhang, R.; Liang, S. Effect of Hot-Dip Tin Coating Treatment on the Cusn10pb10/45 Steel Joint Fabricated by Diffusion Bonding. *J. Mater. Res. Technol.* **2022**, *18*, 1147–1154. [CrossRef]
4. Sun, Y.-G.; Qi, Y.-S.; Li, J.; Du, Z.-M.; Chen, L.-L.; Chen, L.-H. Fabrication and Performance Analyses of 45# Steel Supports Using Liquid Forging. *J. Zhejiang Univ.-Sci. A* **2020**, *21*, 908–922. [CrossRef]
5. Su, Z.; Li, J.; Shi, Y.; Ren, S.; Zhang, Z.; Wang, X. Effect of Process Parameters on Microstructure and Tribological Properties of Ni60a/Cr_3C_2 Laser Cladding on 60Si2Mn Steel. *Surf. Coat. Technol.* **2023**, *473*, 130005. [CrossRef]
6. Deng, C.; Yi, Y.; Jiang, M.; Hu, L.; Zhou, S. Microstructure and High-Temperature Resistance of Al_2O_3/Conicraly Coatings by Laser Cladding. *Ceram. Int.* **2023**, *49*, 32885–32895. [CrossRef]
7. Hu, Y.; Wang, Z.; Pang, M. Effect of Wc Content on Laser Cladding Ni-Based Coating on the Surface of Stainless Steel. *Mater. Today Commun.* **2022**, *31*, 103357. [CrossRef]
8. Xia, Y.; Chen, H.; Liang, X.; Lei, J. Circular Oscillating Laser Melting Deposition of Nickel-Based Superalloy Reinforced by WC: Microstructure, Wear Resistance and Electrochemical Properties. *J. Manuf. Process.* **2021**, *68*, 1694–1704. [CrossRef]
9. Zhou, S.; Xu, Y.; Liao, B.; Sun, Y.; Dai, X.; Yang, J.; Li, Z. Effect of Laser Remelting on Microstructure and Properties of WC Reinforced Fe-Based Amorphous Composite Coatings by Laser Cladding. *Opt. Laser Technol.* **2018**, *103*, 8–16. [CrossRef]
10. Li, M.; Huang, K.; Yi, X. Crack Formation Mechanisms and Control Methods of Laser Cladding Coatings: A Review. *Coatings* **2023**, *13*, 1117. [CrossRef]
11. Weng, F.; Chen, C.; Yu, H. Research Status of Laser Cladding on Titanium and Its Alloys: A Review. *Mater. Des.* **2014**, *58*, 412–425. [CrossRef]
12. Quazi, M.M.; Fazal, M.A.; Haseeb, A.S.M.A.; Yusof, F.; Masjuki, H.H.; Arslan, A. Effect of Rare Earth Elements and Their Oxides on Tribo-Mechanical Performance of Laser Claddings: A Review. *J. Rare Earths* **2016**, *34*, 549–564. [CrossRef]
13. Ye, F.; Shao, W.; Ye, X.; Liu, M.; Xie, Y.; Bian, P.; Wang, X.; Liu, L.; Wu, H. Microstructure and Corrosion Behavior of Laser-Cladding CeO_2-Doped Ni-Based Composite Coatings on TC4. *J. Chem.* **2020**, *2020*, 8690428. [CrossRef]
14. Zhang, T.; Xiao, H.; Zhang, Z.; Yao, B.; Yang, F. Effect of Y_2O_3 Addition on Microstructural Characteristics and Microhardness of Laser-Cladded Ti-6Al-4V Alloy Coating. *J. Mater. Eng. Perform.* **2020**, *29*, 8221–8235. [CrossRef]
15. Sun, S.; Fu, H.; Chen, S.; Ping, X.; Wang, K.; Guo, X.; Lin, J.; Lei, Y. A Numerical-Experimental Investigation of Heat Distribution, Stress Field and Crack Susceptibility in Ni60A Coatings. *Opt. Laser Technol.* **2019**, *117*, 175–185. [CrossRef]
16. Li, W.; Yang, X.; Xiao, J.; Hou, Q. Effect of WC Mass Fraction on the Microstructure and Friction Properties of WC/Ni60 Laser Cladding Layer of Brake Discs. *Ceram. Int.* **2021**, *47*, 28754–28763. [CrossRef]
17. Shu, D.; Dai, S.; Wang, G.; Si, W.; Xiao, P.; Cui, X.; Chen, X. Influence of CeO_2 Content on Wc Morphology and Mechanical Properties of WC/Ni Matrix Composites Coating Prepared by Laser In-Situ Synthesis Method. *J. Mater. Res. Technol.* **2020**, *9*, 11111–11120. [CrossRef]
18. Zhao, J.; Li, R.; Feng, A.; Feng, H. Effect of Rare Earth La_2O_3 Particles on Structure and Properties of Laser Cladding WC-Ni60 Composite Coatings. *Surf. Coat. Technol.* **2024**, *479*, 130569. [CrossRef]
19. Chang, W. Study on the Fabrication of Laser Cladding Cermet Cutting Tools and Its Cutting Performance. Master's Thesis, Qilu University of Technology, Jinan, China, 2023.
20. Wang, C.; Gao, Y.; Wang, R.; Wei, D.; Cai, M.; Fu, Y. Microstructure of Laser-Clad Ni60 Cladding Layers Added with Different Amounts of Rare-Earth Oxides on 6063 Al Alloys. *J. Alloys Compd.* **2018**, *740*, 1099–1107. [CrossRef]
21. Zheng, C.; Huang, K.; Mi, T.; Li, M.; Li, S.; Yi, X. Impact of Ceo2 Modification on the Quality and Wear Performance of Al_2O_3/SiC Reinforced Metal-Based Coatings. *Mater. Charact.* **2024**, *208*, 113641. [CrossRef]
22. Lu, J.Z.; Cao, J.; Lu, H.F.; Zhang, L.Y.; Luo, K.Y. Wear Properties and Microstructural Analyses of Fe-Based Coatings with Various Wc Contents on H13 Die Steel by Laser Cladding. *Surf. Coat. Technol.* **2019**, *369*, 228–237. [CrossRef]
23. Chen, Y.; Huang, X.; Jiang, J.; Lian, G.; Chen, C. Study on the Effect of Rare Earth Oxide Addition on the Microstructure and Properties of Ni60/WC-Ni Coatings Prepared by Laser Cladding. *Materials* **2023**, *16*, 7263. [CrossRef]

24. Shen, X.; Peng, H.; Xue, Y.; Wang, B.; Su, G.; Zhu, J.; Li, A. Microstructure and Properties of WC/Ni-Based Laser-Clad Coatings with Different WC Content Values. *Materials* **2022**, *15*, 6309. [CrossRef]
25. Wang, Q.; Li, Q.; Zhang, L.; Chen, D.X.; Jin, H.; Li, J.D.; Zhang, J.W.; Ban, C.Y. Microstructure and Properties of Ni-WC Gradient Composite Coating Prepared by Laser Cladding. *Ceram. Int.* **2022**, *48*, 7905–7917. [CrossRef]
26. Zhang, Z.; Yang, Q.; Yang, F.; Zhang, H.; Zhang, T.; Wang, H.; Ma, Q. Comparative Investigation on Wear Properties of Composite Coatings with Varying CeO_2 Contents. *Coatings* **2022**, *12*, 906. [CrossRef]
27. Liang, F.; Li, K.; Shi, W.; Zhu, Z. Effect of Y_2O_3 Content on Microstructure and Corrosion Properties of Laser Cladding Ni-Based/WC Composite Coated on 316L Substrate. *Coatings* **2023**, *13*, 1532. [CrossRef]
28. Ren, M.; Li, R.; Zhang, X.; Gu, J.; Jiao, C. Effect of Wc Particles Preparation Method on Microstructure and Properties of Laser Cladded Ni60-WC Coatings. *J. Mater. Res. Technol.* **2023**, *22*, 605–616. [CrossRef]
29. Rupert, T.J.; Schuh, C.A. Sliding Wear of Nanocrystalline Ni–W: Structural Evolution and the Apparent Breakdown of Archard Scaling. *Acta Mater.* **2010**, *58*, 4137–4148. [CrossRef]
30. Li, S.; Liu, D.; Liu, G.; Xin, S.; Deng, Z.; Li, C.; Chen, T. Study on the Preparation of CeO_2-Doped Alcocrfeniti High Entropy Alloy Bioinert Coatings by Laser Cladding: The Effect of CeO_2 Particle Size on the Tribological Properties of the Coatings. *Surf. Coat. Technol.* **2023**, *475*, 130155. [CrossRef]
31. Cui, C.; Wu, M.; He, R.; Gong, Y.; Miao, X. Effect of CeO_2 Addition on Grain Refinement and Mechanical Properties of Stellite-6 Coating Fabricated by Laser Cladding. *J. Therm. Spray Technol.* **2022**, *31*, 2621–2634. [CrossRef]
32. Li, W.; Di, R.; Yuan, R.; Song, H.; Lei, J. Microstructure, Wear Resistance and Electrochemical Properties of Spherical/Non-Spherical WC Reinforced Inconel 625 Superalloy by Laser Melting Deposition. *J. Manuf. Process.* **2022**, *74*, 413–422. [CrossRef]

machines

MDPI

Review

Effects of Cryogenic- and Cool-Assisted Burnishing on the Surface Integrity and Operating Behavior of Metal Components: A Review and Perspectives

Jordan Maximov * and Galya Duncheva

Department of Material Science and Mechanics of Materials, Technical University of Gabrovo, 5300 Gabrovo, Bulgaria; duncheva@tugab.bg
* Correspondence: jordanmaximov@gmail.com

Abstract: When placed under cryogenic temperatures (below −180 °C), metallic materials undergo structural changes that can improve their service life. This process, known as cryogenic treatment (CrT), has received extensive research attention over the past five decades. CrT can be applied as either an autonomous process (for steels and non-ferrous alloys, tool materials, and finished products) or as an assisting process for conventional metalworking. Cryogenic impacts and conventional machining or static surface cold working (SCW) can also be performed simultaneously in hybrid processes. The static SCW, known as burnishing, is a widely used environmentally friendly finishing process that achieves high-quality surfaces of metal components. The present review is dedicated to the portion of the hybrid processes in which burnishing under cryogenic conditions is carried out from the viewpoint of surface engineering, namely, finishing–surface integrity (SI)–operational behavior. Analyzes and summaries of the effects of cryogenic-assisted (CrA) burnishing on SI and the operational behavior of the investigated materials are made, and perspectives for future research are proposed.

Keywords: cryogenic treatment; cryogenic-assisted burnishing; surface integrity; roughness; microhardness; residual stresses; fatigue behaviour; wear resistance; corrosion resistance

1. Introduction

High-temperature treatments of metals have been performed since ancient times [1]. Such treatments are based on three fundamental properties of matter (adaptability, memory, and reflection) and consist of changing the temperature of metal components according to a corresponding law in order to obtain a desired structure. However, due to the aforementioned properties of matter, the structure of the metal components changes even at negative temperatures (in Celsius). The benefits of low-temperature effects on the structural modification of metallic components have only been realized and exploited during the last 100 years and are a direct consequence of advances in science and technology that enable the creation of ultra-low-temperature environments [2]. Previous studies [3] have shown that such approaches were already attempted in the first half of the last century, whereby negative temperatures were used to improve the operational behavior of watch parts, engine blocks, military aircraft components, and chainsaw blade links. Liquefied carbon dioxide was used as a coolant in machine operations as early as 1919 [2]. The term cryogenic treatment (CrT) was introduced in 1966, but systematic guidelines for the application of cryogenic engineering date from only 1976 [2]. Cryogenic is derived from the Greek word "κρύος" meaning cold. Cryogenic conditions have a very wide range of applications in modern engineering: to improve the properties and operational behavior of steels (martensitic [3–7] and austenitic [8–12]) and non-ferrous alloys [13,14], workpieces [15], finished components [3] and tools [16–18], and as a cooling medium for machining [19–22] and plastic deformation [2]. It is, perhaps, for this reason of broad applicability that there is no consensus among researchers on temperature-based terminology.

Jawahir et al. [2], following the recommendations of research and standards organizations, indicated three limits (−150 °C, −153 °C, and −180 °C) below which cryogenic temperatures can be defined. The authors noted that −180 °C is a logical limit since the boiling points of the stable gases helium, nitrogen, neon, hydrogen, and oxygen are below −180 °C. Analyzing the processing of steels under negative temperature conditions, Diekman [4] applied the concepts of cold treating and cryogenic treatment, stating that the optimal temperature for cold treating is −84 °C, whereas typical CrT processes consist of a slow cool-down from ambient temperature to approximately −193 °C. In a comprehensive review paper dedicated to martensitic steels [3], the following concepts were summarized: cold treatment (>−80 °C); shallow CrT (from −80 °C to −160 °C); and deep CrT (<−160 °C); however, some authors using different cooling methods during machining have used the term cryogenic to denote any measured temperature from 0 °C to −100 °C [23–26]. It should be noted that the CrT is sometimes termed cryogenic processing, deep cryogenic processing, deep CrT, cryogenic tempering, and deep cryogenic tempering [4].

For cryogenic conditions (below −180 °C) [2], it is difficult to trace to which group of engineering objects (steels, tools, details, or processes) CrT was first applied. Figure 1 illustrates the application areas of CrT.

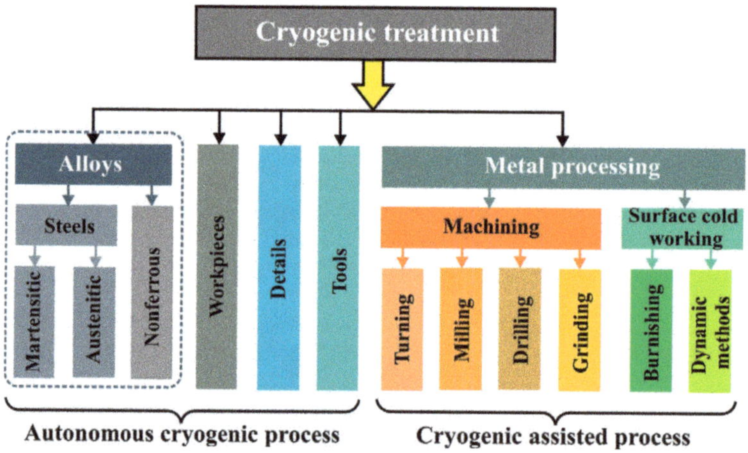

Figure 1. Fields of application of cryogenic treatment.

In summary, the benefits of applying CrT are as follows [2,4]: (1) relaxation of unwanted residual stresses; (2) strong reduction in residual austenite in martensitic steels; (3) precipitation of fine carbides in ferrous alloys; (4) improving the properties of steels and non-ferrous alloys by increasing their hardness and strength, dimensional stability, and thermal and electrical conductivity; (5) improvement of operational behavior in terms of increase in wear and corrosion resistances and fatigue life; (6) eliminations of damage induced by the heat, which is generated during machining and surface cold working processes.

As shown in Figure 1, CrT is applied as either an autonomous or an assisting process in conventional metalworking. Cryogenic-assisted (CrA) processes are based on promising directions of technology development involving different types of impacts. According to the way in which impacts are coordinated in time, two types of processes are possible: (1) hybrid processes, whereby the cryogenic impact and conventional machining are performed simultaneously; and (2) combined processes, in which impacts are applied sequentially. In practice, hybrid processes are implemented in two ways, according to the type of conventional machining used: (1) CrA machining (turning [2,25,27]; milling [2,23,28–

30]; drilling [2,24]; grinding [2,26]); and (2) CrA burnishing [31–66]. Combined processes (CrA hard turning and subsequent burnishing) have been demonstrated in previous studies [15,67–70]. Chronologically, CrA machining precedes CrA burnishing. According to Jawahir et al. [2], the term cryogenic machining was first used by K. Uehara and S. Kumagai in 1968.

In summary, there are three types of cryogenic processes, illustrated in Figure 2: (1) autonomous CrT (see also Figure 1); (2) hybrid processes; and (3) combined processes.

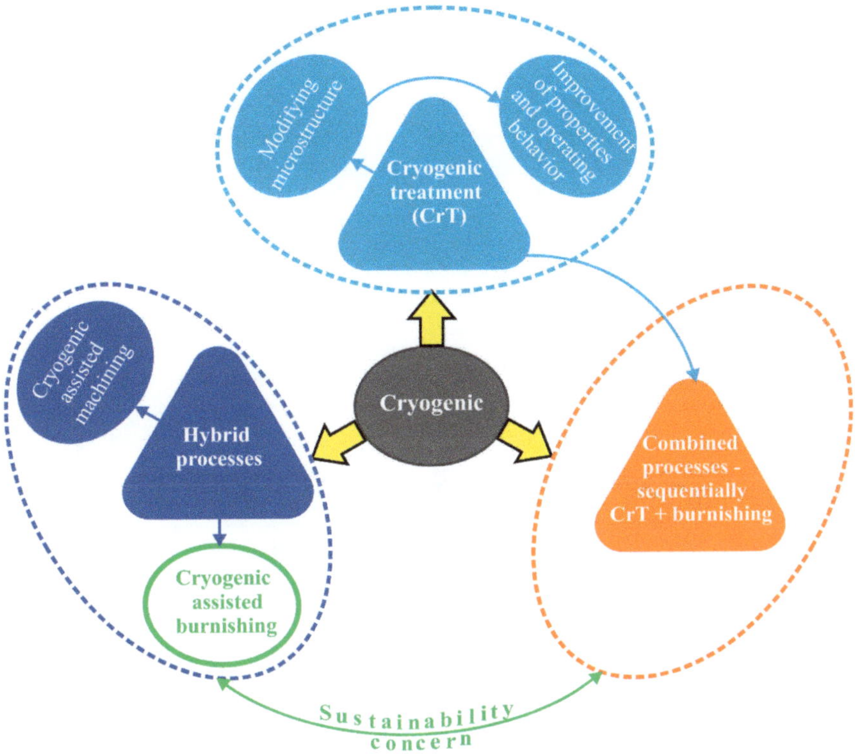

Figure 2. Types of cryogenic processes.

During the last two decades, numerous review papers have been published, primarily dedicated to (1) autonomous CrT of steels [3,71–73], tool materials [74], cutting tools [75], steels and non-ferrous alloys [76], and (2) CrA machining [2,77–80]. However, excluding Jawahir et al. [2], in which some attention is paid to burnishing, there is no comprehensive review paper dedicated to CrA burnishing.

Burnishing of metal components is a static surface coldworking method, which creates three effects [81]: smoothing, cold work, and residual compressive stresses at the surface and in adjacent subsurface layers. Following [82], the present work distinguishes between the concepts of the burnishing method and the burnishing process. A burnishing method is a coherent time–space mechanical arrangement of two bodies (the deforming element and the surface being treated) with a defined geometry and known physical and chemical properties. In contrast, the burnishing process is an energy–force exchange resulting from coherent interactions between the two bodies with clearly defined quantitative characteristics. Therefore, using the same method but with different magnitudes of the governing factors, many deforming processes can be undertaken, and, as a result, the processed components will have different surface integrity (SI) values.

Classifications of conventional burnishing methods have been made according to different characteristics [82]: (1) according to the shape of the deforming element (roller or ball), termed roller/ball burnishing (RB/BB); and (2) according to the type of contact (sliding or rolling) between the deforming element and the treated surface, termed slide burnishing (SB) and RB/BB. Depending on the desired SI characteristics of the processed component, Ecorol [81] defined two processes: roller burnishing (RB); and deep rolling (DR). The primary goal of the first process is to produce mirror-like surfaces, while the accompanying beneficial effects of cold work (e.g., increased surface microhardness and compressive residual stress) are less pronounced. Conversely, the second process introduces significant compressive residual stresses and substantially increases the microhardness. Both processes are implemented either by the single-roller RB method or the BB method (including hydrostatic sphere). According to previous work [83,84], SB and diamond burnishing (DB) methods can implement three processes: smoothing; hardening; and mixed burnishing. Using the ball burnishing with hydrostatic sphere (BBHS) method, Lambda Research invented the low-plasticity burnishing (LPB) process [85], which was intended to create large compressive residual stresses at significant depths, whereby the equivalent plastic strain (degree of cold work) is automatically controlled in order to not exceed the set limit. All conventional processes can be converted into hybrid processes by means of cryogenic assistance. Figure 3 shows the schematics of CrA burnishing methods.

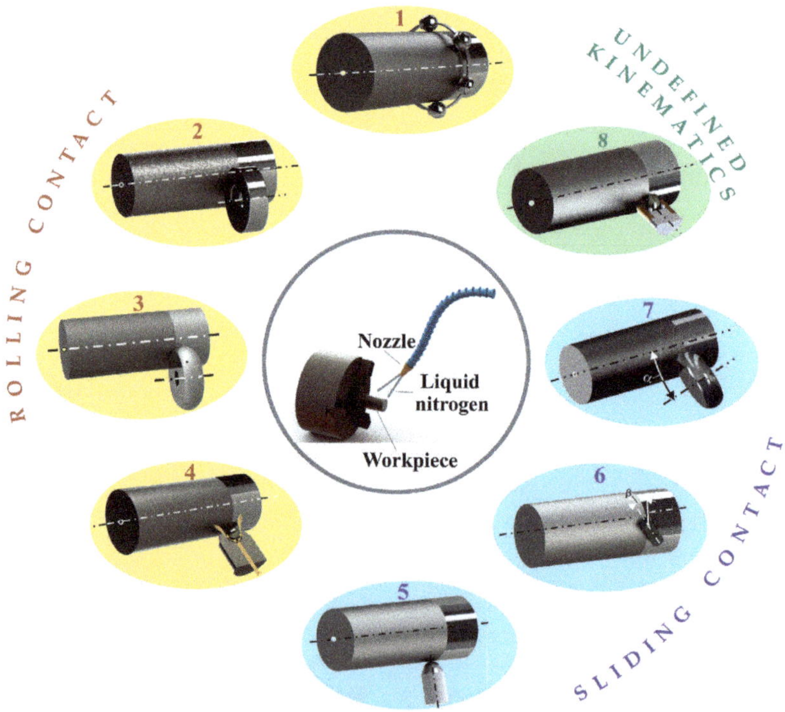

Figure 3. Schemes of cryogenic-assisted burnishing methods: 1—multiple ball burnishing; 2—roller burnishing with cylindrical roller; 3—roller burnishing with toroidal roller; 4—ball burnishing with hydrostatic sphere; 5—slide burnishing with spherical-ended insert; 6—slide burnishing with cylindrical-ended insert; 7—slide-roller burnishing; 8—ball burnishing with undefined ball motion.

This review paper is dedicated to the portion of the hybrid processes in which static surface cold working under cryogenic or cool conditions occurs, i.e., CrA or cool-assisted (CoA) burnishing. This review considers the viewpoint of surface engineering [86], i.e., finishing–SI–operational behavior, with the aim of summarizing the effects of CrA and CoA burnishing on the SI characteristics and operational behavior of the investigated materials and proposing future research directions.

2. Investigations of CrA and CoA Burnishing—General Overview

In the present review, we follow the previous [2] in adopting the term cryogenic-assisted burnishing (CrAB) when the temperature in the surface plastic deformation zone is below −180 °C. At higher temperatures (for example, when using dry ice), the term cool-assisted burnishing (CoAB) is used. Although cryogenic machining [2] has been used since 1968, research on CrAB and CoAB has been performed more recently (2011–2024) and is significantly lower in volume than that dedicated to cryogenic machining. Much research on CrAB has been performed at the University of Kentucky, USA [32–36,42–45,60–66], although studies dedicated to CrAB and CoAB by other authors have also been performed worldwide [31,37–41,46–59]. Studies concerning CrAB and CoAB are systematized and classified in Tables 1 and 2.

Table 1. Investigations on hybrid processes—general overview.

	Material	Burnish. Method/ Process	Agent	Hybrid Process	Other Cooling/Heating/Lubricating Conditions					Reference
					Flood	Dry	MQL	Hybrid (Cooling+ MQL)	Heating	
Magnesium alloys	AZ31B-O	SB	LN2	CrA						[42,43]
	AZ31B-O	SB	LN2	CrA		√				[44,45]
Titanium alloys	Ti-6Al-4V	RB/DR	LN2	CrA	√	√				[32]
	Ti-6Al-4V	RB/DR	LN2	CrA	√	√	√	√		[33]
	Ti-6Al-4V	RB/DR	LN2	CrA	√		√	√		[34]
	Ti-6Al-4V	RB/DR	LN2	CrA		√	√			[48,49]
	Ti-6Al-4V	SB	LN2	CrA		√				[57]
	Ti-6Al-4V	SB	LN2	CrA						[58]
Aluminum alloy	7050-T7451	RB/DR	LN2	CrA		√				[35,36]
Additively manufactured materials	Inconel 718/ Laser Powder Bed Fusion	RB/RB	LN2	CrA		√			√	[37]
	laser-clad Stellite 6	BBHS/LPB	LN2	CoA						[59]
Carbon steel	SS 400	SB/ Friction stir	CO2	CoA		√	√			[38,47]
	SS 400	SB/ Friction stir	CO2	CoA		√	√	√		[46]
Tool steel	AISI D3	BBHS/DR	CO2	CoA	√					[39]
	AISI D3	BBHS/DR	CO2	CoA	√					[40]
Martensitic stainless steel	17-4 PH	DB	LN2	CrA		√	√			[50,52]
	17-4 PH	DB	LN2	CrA						[51,53–55]

Table 1. *Cont.*

	Material	Burnish. Method/ Process	Agent	Hybrid Process	Other Cooling/Heating/Lubricating Conditions					Reference
					Flood	Dry	MQL	Hybrid (Cooling+ MQL)	Heating	
Austenitic stainless steel	AISI 304	BBHS/DR	LN2	CrA	√				√	[41]
Biomaterials	Cast homogenized Mg-4Zn-2Sr	BB/RB	LN2	CrA						[31]
	Co-Cr-Mo AZ31	SB	LN2	CrA		√				[60]
	Co-Cr-Mo	SB	LN2	CrA		√				[61–63,66]
	Co-Cr-Mo	SB	LN2	CrA		√				[64]
	Co-Cr-Mo	SB	LN2	CrA						[65]
Thermal spray coating		SB	LN2	CrA		√				[56]

Abbreviation: BB—ball burnishing; BBHS—ball burnishing with hydrostatic sphere; DR—deep-rolling process; RB—roller-burnishing method (roller burnishing process); LPB—low-plasticity burnishing; LN2—liquid nitrogen; SB—slide burnishing; DB—diamond burnishing; CrA—cryogenic-assisted; CoA—cool-assisted. √ is an indicator of the presence of studies in the literature.

Table 2. Effect of hybrid processes on SI and operating behavior of materials.

	Material	Burnish. Method/ Process	Hybrid Process	SI Characteristics				Operating Behavior			Reference
				R	MH	RS	M	F	WR	CR	
Magnesium alloys	AZ31B-O	SB	CrA	√	√		√			√	[42,43]
	AZ31B-O	SB	CrA		√		√			√	[44,45]
Titanium alloys	Ti-6Al-4V	RB/DR	CrA	√	√						[32]
	Ti-6Al-4V	RB/DR	CrA		√		√				[33]
	Ti-6Al-4V	RB/DR	CrA	√	√						[34]
	Ti-6Al-4V	RB/DR	CrA	√	√				√		[48,49]
	Ti-6Al-4V	SB	CrA	√			√			√	[57]
	Ti-6Al-4V	SB	CrA				√			√	[58]
Aluminum alloy	7050-T7451	RB/DR	CrA		√		√				[35,36]
Additively manufactured materials	Inconel 718/ Laser Powder Bed Fusion	RB/DR	CrA	√	√						[37]
	laser-clad Stellite 6	BBHS/LPB	CoA	√	√	√					[59]

Table 2. *Cont.*

Material		Burnish. Method/ Process	Hybrid Process	SI Characteristics				Operating Behavior			Reference
				R	MH	RS	M	F	WR	CR	
Carbon steel	SS 400	SB/Friction stir	CoA	√							[38,46]
Carbon steel	SS 400	SB/Friction stir	CoA		√						[47]
Tool steel	AISI D3	BBHS/DR	CoA	√	√		√				[39]
Tool steel	AISI D3	BBHS/DR	CoA	√	√		√				[40]
Martensitic stainless steel	17-4 PH	DB	CrA		√	√					[50]
Martensitic stainless steel	17-4 PH	DB	CrA	√	√	√	√				[52]
Martensitic stainless steel	17-4 PH	DB	CrA	√	√						[51,53–55]
Austenitic stainless steel	AISI 304	BBHS/DR	CrA			√	√	√			[41]
Biomaterials	Cast homogenized Mg-4Zn-2Sr	BB/RB	CrA	√	√						[31]
Biomaterials	Co-Cr-Mo AZ31	SB	CrA		√		√				[60]
Biomaterials	Co-Cr-Mo	SB	CrA		√		√				[61,63]
Biomaterials	Co-Cr-Mo	SB	CrA	√	√		√				[62,66]
Biomaterials	Co-Cr-Mo	SB	CrA		√				√		[64]
Biomaterials	Co-Cr-Mo	SB	CrA				√				[65]
Thermal spray coating		SB	CrA	√	√	√	√				[56]

Abbreviation: BB—ball burnishing; BBHS—ball burnishing with hydrostatic sphere; DR—deep-rolling process; RB—roller-burnishing method (roller burnishing process); LPB—low-plasticity burnishing; SB—slide burnishing; DB—diamond burnishing; CrA—cryogenic-assisted; CoA—cool-assisted; R—roughness; MH—microhardness; RS—residual stresses; M—microstructure; F—fatigue behavior; WR—wear resistance; CR—corrosion resistance. √ is an indicator of the presence of studies in the literature. √ is an indicator of improvement after CrA (CoA) hybrid processes.

Figure 4 illustrates the percentage share of the investigated materials. CrAB and CoAB are most commonly applied to four types of steel, followed by three types of biomaterials. The effect of hybrid processes on coatings has received the least investigation. It is important to note that there is a general lack (with one exception [35,36]) of research on the application of high-strength aluminum alloys in the aerospace industry. The SB method (Figure 5), using either a diamond or non-diamond deforming element, is most often used, followed by the RB and BBHS methods. The latter methods are used to implement DR processes. In 86.11% of cases, the hybrid CrA process is implemented using liquid nitrogen. In the remaining cases, the hybrid CoA process is carried out using dry ice as the cooling medium. Figure 6a shows the percentage share of SI characteristics investigated in previous research. The most frequently studied feature is microhardness, followed by microstructure; in contrast, the least attention has been paid to the effects of the hybrid process on residual stresses. Limited research has been devoted to the influence of CrA and CoA processes on operational behavior (Figure 6b), whereas fatigue was only investigated in [41].

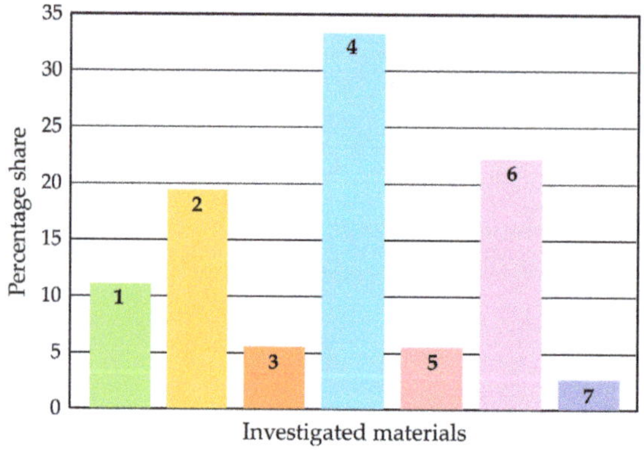

Figure 4. Percentage share of the studied materials: 1—magnesium alloys; 2—titanium alloys; 3—aluminum alloys; 4—steels; 5—additively manufactured materials; 6—biomaterials; 7—thermal spray coatings.

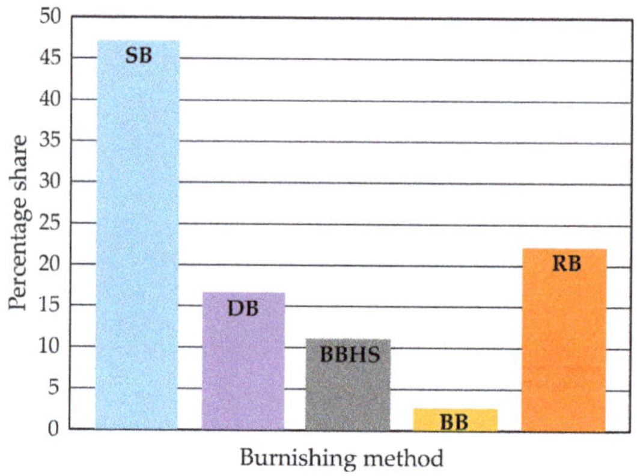

Figure 5. Percentage share of the used burnishing methods: SB—slide burnishing; DB—diamond burnishing; BBHS—ball burnishing with hydrostatic sphere; BB—ball burnishing; RB—roller burnishing.

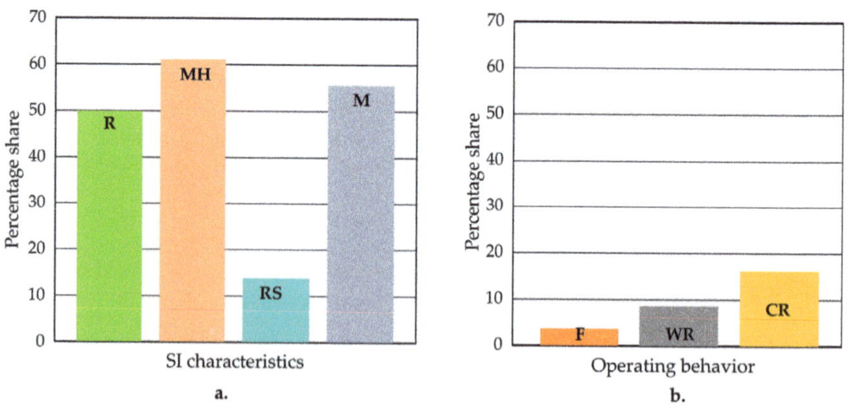

Figure 6. Percentage share of SI characteristics (**a**) and operating behavior (**b**): R—roughness; MH—microhardness; RS—residual stresses; M—microstructure; F—fatigue; WR—wear resistance; CR—corrosion resistance.

3. Effects of CrA and CoA Burnishing on SI

The effects of hybrid processes on SI characteristics are analyzed for each of the studied groups of materials.

3.1. *Magnesium Alloy AZ31B-O*

The effect of CrA SB implemented by high-speed steel cylindrical-shaped deforming elements on the SI characteristics of cylindrical specimens was studied by Pu et al. [42–45], who established that grinding at the room temperature provides a lower value of the Ra roughness parameter (Ra $= 0.18\ \mu$m) than CrA SB (Ra $= 0.22\ \mu$m). Due to CrA SB, the surface microhardness (as well as the microhardness in the layers adjacent to the surface) increases to 1.35 GPa, and the bulk material microhardness reaches 0.9 GPa. In terms of residual stresses, Pu et al. [42–45] established the following: (1) CrA SB introduces positive hoop stresses; and (2) the axial stresses introduced are compressive but are smaller in absolute value than those introduced via dry SB. The explanation for the positive circumferential stresses lies in the relatively large tangential force applied (500 N). This force is known to be minimal in the RB process. The CrA SB homogenizes and strongly grinds grains in the surface and near-subsurface layers, and cryogenic temperatures suppress grain growth due to dynamic recrystallization. Grain refinement is carried out by a mechanism of dynamic recrystallization.

The SI characteristics of the same magnesium alloy were also investigated by Yang et al. [60], who implemented CrA SB using a carbide fixed roller as the deforming element and employed an outer cylindrical treated surface. Comparing with the results achieved by conventional dry SB, the authors have found that CrA SB provides (1) significantly higher microhardness up to a depth of approximately 230 μm and (2) remarkably refined microstructures on the burnished surface (grains smaller than 300 nm were observed).

3.2. *Titanium Alloy Ti-6Al-4V*

The effects of CrA hybrid processes on the SI of this alloy were studied by Caudill et al. [32–34], Rotella et al. [48,49], and Tang et al. [57,58].

3.2.1. Roughness

Starting from an initial roughness of Ra $=$ 1.31 μm (after turning), Caudill et al. [32] found that the CrA DR process achieved a smaller reduction (56.7%) of initial roughness than that achieved by dry (63.4%) and flood-cooled (58.2%) processes. All three processes are realized by means of the RB method using a 3.04 mm diameter tungsten carbide roller and four magnitudes of burnishing force. The influence of the burnishing force on roughness was found to be less than that of the burnishing conditions. Figure 7 summarizes the effects of CrA DR on the investigated characteristics of SI [32]. To compare the change trends of heterogeneous quantities, the Ra roughness parameter and the values of microhardness are normalized to unity, i.e., the maximum value of the corresponding quantity is taken as unity.

In a previous study [34], it was shown that CrA and flood-cooled DR processes, implemented by a burnishing force of 1500 N, achieved a reduction in roughness by 61% and 64.8% of the initial value, respectively. The increased surface tension and greater friction explain the higher values of the roughness parameter Ra obtained by the CrA DR process compared to conventional flood-cooled DR. It was found that when the number of tool passes is greater than seven, the trend reverses. Rotella et al. [48,49] reported higher values of the integral height parameter Ra achieved by the CrA DR process compared to DR under the minimum quantity of lubricant (MQL) condition; however, the values observed were lower than those with the dry DR process. Tang et al. [57,58] implemented the CrA SB process using a cemented carbide deforming cylindrical-shaped element, whose axis was orthogonally crossed with the axis of the rotating cylindrical workpiece. The resulting roughness was Ra $=$ 0.32 μm. Dry SB was found to significantly degrade the roughness.

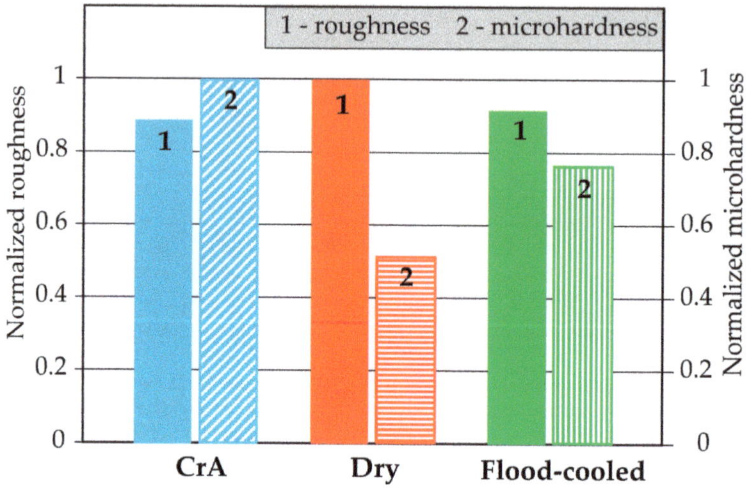

Figure 7. Comparison of the effects of DR on the roughness and microhardness of Ti-6Al-4V under different cooling conditions.

3.2.2. Surface Microhardness

Caudill et al. [32] identified a significant increase in surface microhardness due to CrA DR treatment relative to DR under dry and flood-cooled conditions for all investigated magnitudes of burnishing force (1000, 1500, 2000, and 2500 N). The maximum increase relative to turning in each case was as follows: CrA DR—64.2%; flood-cooled DR—48.9%; and dry DR—32.8% (see Figure 7).

CrA DR resulted in the greatest depth of the hardened layer, followed by DR under flood-cooled and dry conditions, respectively. The explanation for this phenomenon is as follows. The presence of a cooling mechanism in lubricated DR and CrA DR compensates for the thermal softening effect observed in dry DR. The coolant removes the heat generated during plastic deformation, allowing for the material to be compressed at higher burnishing forces. As the force increases, the density of dislocations in the surface and subsurface layers increases, leading to strain hardening. Therefore, two opposing mechanisms, namely, strain hardening and thermal softening, act on the material. Without coolant, thermal softening predominates over strain hardening because the heat accumulates near the surface, essentially annealing the dislocation mesh and reducing the hardness. Conventional flood cooling removes some of the generated heat, but liquid nitrogen provides a significantly more efficient coolant. Therefore, CrA surface plastic deformation is the most effective method to increase surface microhardness due to the strong cooling effect of liquid nitrogen.

A previous study [33] established that implementing the CrA DR process (using a pneumatically actuated tool equipped with a tungsten carbide roller with a 1.5 mm radius of curvature) in combination with MQL, provides maximum surface microhardness, i.e., higher than any of the following burnishing conditions: dry; MQL; flood-cooled; and cryogenic. It was found that increasing the number of passes (up to eight) increases the microhardness under all burnishing conditions. For example, eight-pass CrA DR with 1500 N burnishing force increases the microhardness by 45% relative to the initial state. Almost the same increase is obtained from single-pass CrA DR with a 2500 N burnishing force. On this basis, multipass CrA DR implemented with less force has been found to be a suitable finishing for parts with lower bending stiffness, such as turbine and compressor blades [33]. A comparison has also been made between the effects of the DR process, implemented with a burnishing force of 1500 and 2500 N, on the surface microhardness for four types of burnishing conditions: MQL; flood-cooled; cryogenic; and cryogenic + MQL, depending on the feed rate [34]. For all combinations of burnishing parameters, the cryogenic and cryogenic + MQL conditions provided the highest surface microhardness and the greatest depth of the affected layer. The increase in surface microhardness is approximately 41% more than the initial state. A similar result was reported by Rottela et al. [48,49] by implementing the DR process under three burnishing conditions (dry, MQL, and cryogenic), showing that the optimal surface microhardness and depth of the affected layer were provided by the CrA DR process.

3.2.3. Microstructure

Caudill et al. [33] established that the CrA DR process creates a nanostructured layer with a thickness of 1 μm and grain sizes ranging from 51 to 212 nm, causing the surface microhardness to increase. Below this layer, the grain refinement becomes increasingly less pronounced with depth until the microstructure transitions to its bulk state. Tang et al. [57] reported that the grain size in the nanostructured layer obtained by single-pass CrA SB is 24.3 nm, and the length ratio is 1.74. Using a two-pass process, the grain size can be reduced to 20.7 nm while the length ratio is reduced to 1.45.

3.3. Aluminium Alloy 7050-T7451

The effects of the CrA DR process on the SI characteristics of this alloy have been investigated by Huang et al. [35,36], who found that the CrA process increased surface microhardness by an average of 20–30% within a layer of 200 μm depth; this is more significant than the 5–10% increase when using dry conventional burnishing. The authors' explanation is that tangential burnishing forces occurring during the CrA process are higher than those of the dry conventional burnishing due to rapid cooling and material work hardening. The reason for the increased surface microhardness is the resulting nanostructured surface layer, which has an average grain size of 40 nm.

3.4. Additively Manufactured Materials

3.4.1. Inconel 718/Laser Powder Bed Fusion

Kaya et al. [37] studied the effects of the CrA RB process (using the RB method) on the roughness, microhardness, and microstructure of this alloy. A comparison was made with the effects of conventional DR under two other burnishing conditions: drying and preheating at 200 °C. CrA RB achieved the highest integral height 3D roughness parameters. The greatest increase in surface microhardness (compared to the as-built state), or 21%, was provided by CrA RB. Figure 8 illustrates the influence of the burnishing force and burnishing conditions on the obtained surface microhardness, presented in a form normalized to unity. As the force increases, the degree of plastic deformation, and hence the microhardness, increases, whereas the increased amount of heat generated (and imported from outside via preheating) causes the surface microhardness to decrease.

At a depth of approximately 25 μm, CrA RB provides the highest microhardness; at greater depths, the microhardness obtained from CrA RB is the least. The greatest depth of the affected layer is achieved by conventional RB after preheating at 200 °C due to the softening effect. Using X-ray diffraction (XRD) patterns, it was found that CrA RB has the weakest effect on the intensity in the (111) and (200) planes compared to dry RB and RB after preheating the treated surface. After CrA RB, a thin layer of white color was observed on the surface, which is a sign of the nanocrystalline structure of the material.

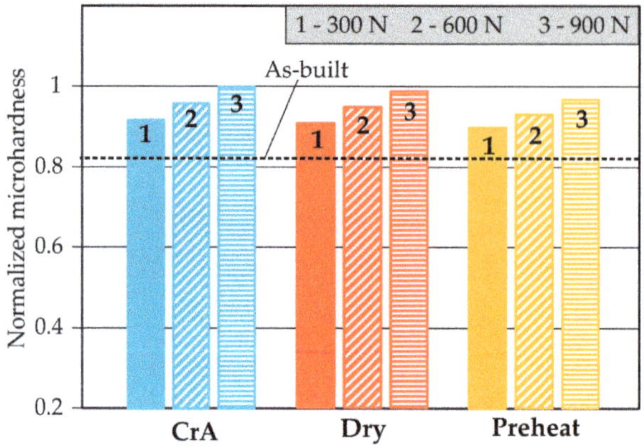

Figure 8. Influence of the burnishing force and the burnishing conditions on the obtained surface microhardness of Inconel 718.

3.4.2. Laser-Clad Stellite 6 and 420 Stainless Steel as Substrates

Anirudh et al. [59] studied the effects of the CoA low-plasticity burnishing (LPB) process (implemented by the BBHS method) on roughness, microhardness, and residual stresses while maintaining the temperature of the surface at approximately −40 °C using liquid nitrogen and a 6 mm ceramic deforming ball coated with silicon nitride. A comparison was made with the effects of conventional LPB and vertical face milling, finding that CoA LPB provides the following: (1) a minimum value of the roughness parameter Sa, with a negative skewness and a kurtosis above three; (2) a maximum surface microhardness; and (3) maximum compressive residual stresses.

3.5. Steel

3.5.1. Carbon Steel SS 400

CoA (supercritical carbon dioxide + MQL) single-pass SB with a tool rotating around its axis (termed friction stir burnishing) and with tungsten carbide grade K10 insert was performed in several previous studies [38,46,47]. A comparison was made with the effect of SB under dry and MQL conditions, showing that (1) the roughness parameter Ra was reduced by approximately 56.2% and 52.4% relative to dry and MQL conditions, respectively, and (2) the highest microhardness was obtained by cryogenic burnishing close to the surface and up to a depth of 0.27 mm.

3.5.2. Tool Steel AISI D3

The CoA DR process (implemented by the BBHS method) has been applied to this steel [39,40] with the aim of SI improvement. Dry ice cooling was applied. A comparison was made against the effects of conventional DR at room temperature. Higher values of the height integral parameters Ra and Rz were obtained relative to conventional DR, which was explained in terms of the degraded roughness resulting from the increased brittleness of the material at lower temperatures, leading to reduced deformability. However, CoA DR was also found to provide a significantly higher hardness (HV0.5) at depths up to approximately 1.4 mm and a significant reduction in residual austenite relative to DR at room temperature.

3.5.3. Martensitic Stainless Steel 17-4 PH

CrA DB was implemented by Sachin et al. [50–55] with the objective of improving the SI of this steel. Maximum surface microhardness values of 395 HV, 369 HV, and 357 HV corresponded to maximum surface residual stresses of −352 MPa, −282 Mpa, and −195 MPa, respectively, under cryogenic, MQL, and dry environment conditions [50].

Figure 9 shows the influence of DB conditions on surface axial residual stress and surface microhardness, normalized to unity. Both SI characteristics decrease with increasing heat generation, with the effect being more pronounced for residual stresses.

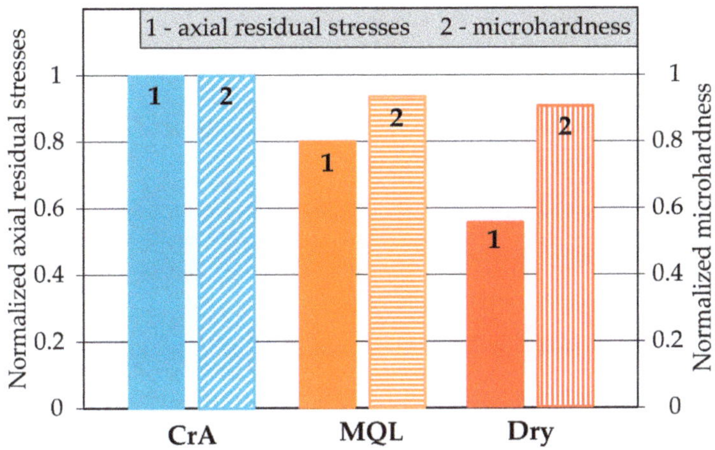

Figure 9. Influence of DB conditions on the surface axial residual stress and the surface microhardness of 17-4 PH martensitic stainless steel.

A previous study [51] reported that CrA DB provided a roughness parameter of Ra below 0.14 μm, under which conditions the surface microhardness was increased to 410 HV (340 HV of the base material). In another study [52], it was established that (1) for all investigated combinations of magnitudes of the governing factors, CrA DB exhibited the lowest roughness parameter Ra (below 0.05 μm) and the highest surface microhardness (up to a depth of 120 μm) relative to MQL and dry conditions; (2) the highest compressive surface residual stress of −356 MPa was observed after CrA DB (−298 MPa and −215 MPa under MQL and dry DB conditions); (3) uniform surfaces were formed after CrA DB, whereas MQL and dry conditions resulted in the formation of microdefects. It has also been reported that by means of CrA DB, a roughness Ra = 0.03 μm could be achieved at an optimal burnishing velocity of 85 m/min and a surface microhardness of 413 HV (radius of the deforming diamond insert r = 3 mm) [53]. Using the same radius, an optimal microhardness was achieved at a depth of up to 130 μm, as well as a maximum residual compressive stress of −335 MPa. Similar results were obtained in other studies [54,55].

3.5.4. Austenitic Stainless Steel AISI 304

Nikitin et al. [41] applied the CrA DR process, implemented by the BBHS method, to this austenitic steel, establishing that regardless of the large amount of dissipated generated heat, the cryogenic burnishing condition (−192 °C) leads to a minimum surface residual axial compressive stress of approximately −60 MPa and a positive surface residual hoop stress (close to zero). Increasing the temperature of the processed surface (using halogenic radiant heating) to +300 °C causes the residual axial surface stress to reach −834 MPa (maximum) so that the hoop is approximately −220 MPa. With a further gradual increase in temperature to +550 °C, the absolute value of the axial residual stress decreases to −300 MPa, and the circumferential stress reaches −100 MPa. Nikitin et al. [41] also found that CrA DR produced significantly more (68%) strain-induced α′-martensite to depths of approximately 60 μm relative to room temperature conditions (6%). Conventional DR at room temperature forms a nanocrystalline layer with a thickness of 2 μm and grain size of approximately 20 nm. It was reported that with the temperature increasing, grain size increased; however, no information is available to constrain the effect of CrA DR on grain refinement. For the martensitic stainless steel 17-4 PH, a positive effect of CrA DB on the residual stresses was reported [50–55]. The behavior of the austenitic stainless steel 304 under CrA DR is the inverse [41]. The reason for this difference in behavior is probably rooted in the inherent characteristics of the austenite phase and in the differing contacts between the deforming element and the processed surface, sliding friction [50–55], and rolling friction [41] during the application of the burnishing method.

3.6. Biomaterials

3.6.1. Co-Cr-Mo Alloy

The effects of CrA SB on the SI characteristics of Co-Cr-Mo bioalloy specimens were investigated by Yang et al. [60–66]. In one such study [64], the cylindrical specimen was fixed, and one of its flat surfaces was burnished. The deforming roller did not rotate around its axis but rather around the axis of the specimen. The two axes (of the roll and of the specimen) intersected orthogonally. In other studies [60–63,65,66], the high-speed tool steel deforming roller is stationary; the sample rotates around its axis, and the treated surface is cylindrical. In [62], it was shown that CrA SB provided an approximately 40% decrease in the Ra roughness parameter (for two depths of penetration used) relative to dry SB. The authors explain the lower value of Ra with the reduced temperature and the increased wear resistance of the deforming element due to the effective cooling with liquid nitrogen. In other studies [60–63], it was found that CrA SB provided significantly higher microhardness on the surface and at depths of up to approximately 230 μm compared to dry SB. The residual stresses were also investigated in [64], showing that CrA SB with a depth of penetration of 0.127 mm introduces greater axial compressive residual stresses, both at the surface and in depth, than dry SB. As a result, the compressive zone has a greater depth. When the depth of penetration increases to 0.254 mm, CrA SB introduces larger compressive stresses than dry SB at a depth of approximately 50 μm, after which the trend reverses. As a result, the compressive zone introduced by CrA SB is smaller than that of dry SB. The residual hoop stress distribution is the opposite of that described. CrA SB leads to remarkably refined microstructures on the burnished surface [60–63], and grains with sizes below 300 nm have been observed. Compared to dry SB, the depth of the affected layer is increased by up to 170%. An infrared camera was used to measure the temperature and temperature distribution in depth and at the surface, showing that the contact zone temperature was above 700 °C for dry SB but decreased below 300 °C for CrA SB [63]. Lower temperatures inhibit grain growth due to dynamic recrystallization. The latter was studied and modeled in [65]. CrA SB provides two times smaller grains beneath the processed surface than dry SB.

3.6.2. Mg-4Zn-2Sr Alloy

The effects of the CrA RB process (implemented by BB method) on roughness and microhardness were investigated by Akshay et al. [31], who reported a reduction in roughness (roughness parameter not specified) from 3.376 μm to 1.853 μm and an increase in microhardness from 54 HV to 78 HV.

3.7. Thermal Spray Coating

Singh et al. [56] applied a CrA SB onto a high-velocity oxy-fuel thermal spray coating (WC-10Co-4Cr) on one face of grit blasted substrate using a rotating carbide-based disc type tool and a workpiece performing translation. The treated surface was planar. These authors have found the following: (1) a reduction in the average roughness of the coating from 1.94 μm to 0.84 μm; (2) an increased surface microhardness; (3) an isotropic region of surface residual compressive stresses greater than those introduced by dry SB; and (4) reduced porosity within the surface layer.

4. Effects of CrA and CoA Burnishing on Operating Behavior

The effects of the hybrid processes on the operating behavior are analyzed for each of the studied groups of materials.

4.1. Magnesium Alloy AZ31B-O

The effect of CrA SB on the corrosion resistance of this magnesium alloy was studied by Pu et al. [42–45], who found a significant improvement in corrosion resistance due to smaller grain size. Near the uppermost surface, the grain size reduces to 523 nm, and the smallest size reaches up to 263 nm. The reason for this small size is the strain-induced dynamic recrystallization and effective cooling with liquid nitrogen, which prevents grain growth.

4.2. Titanium Alloy Ti-6Al-4V

Rotella et al. [48,49] studied the effects of the CrA DR process on the wear resistance of this alloy via a linearly reciprocating ball-on-flat sliding wear test. A comparison was made to determine the effects of DR under dry and MQL conditions, showing that the lowest wear rate of the burnished surface occurred after CrA DR. The favorable combination of cryogenic conditions and the use of a coated deforming tool lead to the best surface hardness (an increase of about 60% compared to dry and MQL conditions) and, thus, the highest wear resistance.

The effect of CrA SB on the corrosion resistance of this alloy was investigated by Tang et al. [57,58], who demonstrated its increased corrosion resistance due to the formation of a more rapid, stable, and less defective passive film over the surface nanocrystalline layer due to its high density of grain boundaries and dislocations.

4.3. Austenitic Stainless Steel AISI 304

Nikitin et al. [41] studied the effects of the CrA DR process (using the BBHS method) on the fatigue behavior of specimens made of this steel and made a comparison with the effects of DR at temperatures between −192 °C and +800 °C. The rotary bending fatigue test was conducted at room temperature (25 °C), and the cycle parameters include the cycle asymmetry factor $R = -1$ and a stress amplitude of 360 MPa. These authors have established that CrA DR led to the lowest number of cycles before fatigue failure. This was interpreted to be due to the large content of strain-induced α'-martensite in the surface and near-subsurface layers. The longest fatigue life (48,000 cycles) was shown by the samples heated under burnishing at 550 °C. The influence of the temperature during the DR process on the surface axial residual stresses and the corresponding number of cycles to failure in a normalized form to unity is shown in Figure 10, which clearly shows that the maximum surface compressive residual stress does not automatically lead to a maximum number of cycles to failure.

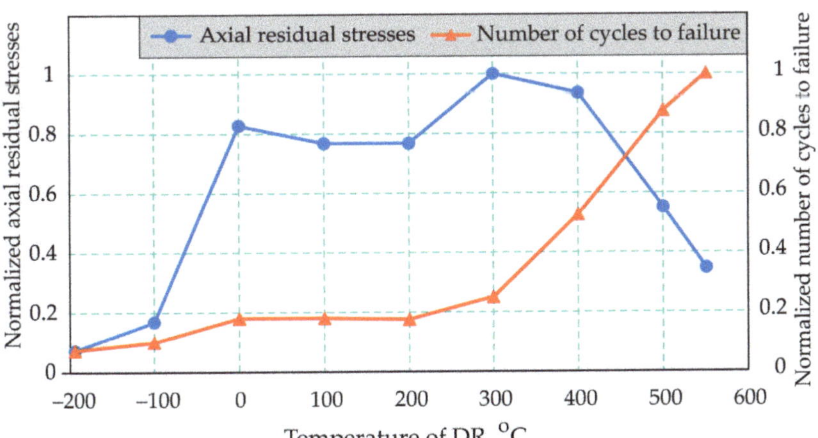

Figure 10. Influence of the temperature during the DR process on the surface axial residual stresses and the corresponding number of cycles to failure of 304 austenitic stainless steel.

The authors' explanation [41] is that at a temperature of 550 °C, the strain aging effect occurs, which multiplies the conventional coldworking effect. As a result, the most favorable combination of SI characteristics is obtained: an absence of martensite; formation of carbides; near-surface nanocrystallisation; high dislocation densities; surface quality; and (to a lesser extent) compressive residual stresses. The S-N curves constructed by these authors (Figure 11) show that the phenomenon described above is valid over the entire high-cycle fatigue field.

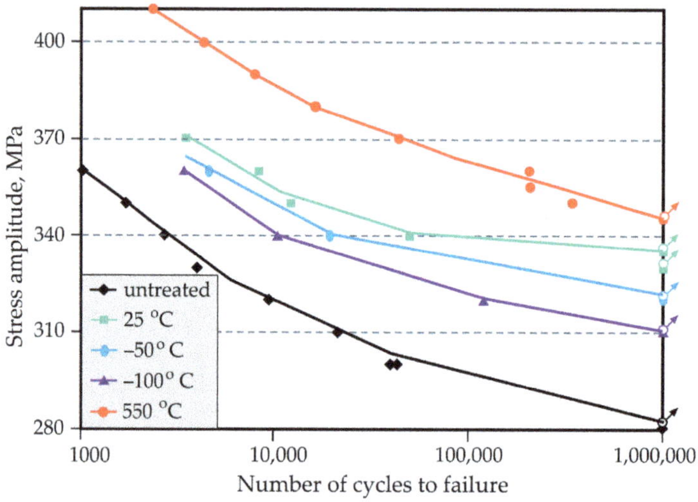

Figure 11. S-N curves of 304 austenitic stainless steel [41].

4.4. Biomaterial Co-Cr-Mo

Yang et al. [64] subjected this alloy to CrA SB to increase its wear resistance using a pin-on-disk tester, finding that the cryogenic condition provides minimal wear volume losses compared to SB under dry conditions. The authors explain this phenomenon by microstructure refinement, compressive residual stresses, and the occurrence of a preferred hexagonal close-packed phase; moreover, this phase has been revealed to be the property with the greatest degree of influence over wear resistance.

5. Conclusions and Future Research Perspectives

This review paper has summarized the state-of-the-art hybrid processes of CrA and CoA burnishing, allowing for the following conclusions to be drawn:

- For all investigated materials, the CrA and CoA burnishing processes significantly increase surface microhardness relative to other burnishing conditions. This is a direct consequence of the nanostructured surface layer produced. The dynamic recrystallization induced by surface plastic deformation takes place under intense cooling conditions, which inhibit grain growth;
- The benefit from the hybrid processes to improve the roughness compared to the effect of burnishing under other conditions (dry, flood, MQL, preheating), as well as compared to other finishings (for instance, grinding), is not unambiguously defined for the different materials and burnishing methods. For the magnesium alloy AZ31 B-O, grinding at room temperature achieves a lower value of the Ra roughness parameter than CrA SB. For the Ti-6Al-4V alloy, the CrA DR process realized by the RB method achieves a higher value of the Ra parameter than DR under other conditions, but the hybrid CrA SB process reduces Ra relative to dry SB. The roughness worsens when CrA DR is applied to Inconel 718 and tool steel AISI D3. In the remaining analyzed cases (Table 2), hybrid CrA and CoA processes improve the resulting roughness. It

is important to note that when the hybrid process is implemented using SB or DB methods, the roughness is always improved relative to SB (respectively DB) under dry, flood, and MQL conditions;

- With the exception of CrA DR using 304 austenitic stainless steel [41], hybrid processes increase the compressive residual stresses in the surface layer. Intensive cooling greatly reduces the thermal effect, leading to softening. As a result, the surface microhardness and surface residual compressive stresses increase. The resulting effect is analogous to the effect of burnishing hardened steel;
- CrA burnishing increases wear and corrosion resistance because the nanostructured surface layer exhibits increased microhardness;
- Only one study has considered the effects of CrA and CoA burnishing on fatigue behavior. CrA DR (implemented via the BBHS method) was applied to austenitic stainless steel 304 [41], showing that CrA DR leads to the lowest number of cycles to fatigue failure relative to cases in which DR is conducted on preheated surfaces. This occurs as a result of the large content of strain-induced α'-martensite in the surface and near-subsurface layers.

Based on the analyses conducted and the conclusions outlined above, the following directions for future investigations can be proposed:

- It is of interest to study the effects of CrA and CoA burnishing on the SI and operating behavior of important material groups, such as aluminum bronzes, which are widely used in the marine industry, shipbuilding, aviation, railway, offshore platform applications, and other fields. For example, single-phase aluminum bronzes do not undergo heat treatment, and their surface microhardness may only be increased by surface cold working. It is, therefore, of interest for engineering practices to determine the effects of burnishing under cryogenic or cool conditions with the objective of improving the microstructure and surface microhardness of bronzes, in addition to improving bronzes with β-transformation;
- The effects of CrA and CoA burnishing on SI and the operating behavior of high-strength aluminum alloys, which are widely used in the aerospace and automotive industries, have not been fully evaluated. The only exception is 7050-T7451 AA [35,36]; but for this alloy, no information has been reported regarding the effect of CrA and CoA burnishing on its operational behavior;
- Chromium–nickel austenitic stainless steels form 70% of the total share of stainless steels used for various industrial applications. However, only one study has considered the CrA DR of AISI 304 steel [41]. Other studies have shown a significant increase in the mechanical characteristics of such steels when subjected to cryogenic temperatures [8,9]. Therefore, the effect of conventional surface cold working on austenitic steels that have been previously subjected to autonomous cryogenic treatment remains of interest;
- There is a lack of research on the effect of the cryogenic conditions during the burnishing process on the dimensional accuracy of the treated surfaces. The results of such research are essential for engineering practice since burnishing is a finishing process;
- Increasing the fatigue life of metal components is essential to engineering practice. Therefore, investigating the effect of CrA and CoA burnishing on the fatigue behavior of metallic components would be of great interest.

Author Contributions: Conceptualization, J.M. and G.D.; methodology, J.M. and G.D.; software, J.M. and G.D.; validation, J.M. and G.D.; formal analysis, G.D. and J.M.; investigation, G.D. and J.M.; resources, J.M. and G.D.; data curation, J.M. and G.D.; writing—original draft preparation, J.M. and G.D.; writing—review and editing, J.M. and G.D.; visualization, J.M. and G.D.; supervision, J.M.; project administration, J.M. and G.D.; funding acquisition, J.M. and G.D. All authors have read and agreed to the published version of this manuscript.

Funding: This research was supported by the European Regional Development Fund within the OP. "Science and Education for Smart Growth 2014–2020", Project CoC "Smart Mechatronics, Eco- and Energy Saving Systems and Technologies", No. BG05M2OP001-1.002-0023.

Data Availability Statement: The original contributions presented in this study are included in the article; further inquiries can be directed to the corresponding authors.

Conflicts of Interest: The authors declare no conflicts of interest.

Abbreviations

BB	ball burnishing
BBHS	ball burnishing with hydrostatic sphere
CrT	cryogenic treatment
CoA	cool-assisted
CoAB	cool-assisted burnishing
CR	corrosion resistance
CrA	cryogenic-assisted
CrAB	cryogenic-assisted burnishing
DB	diamond burnishing
DR	deep rolling
F	fatigue
LPB	low plasticity burnishing
M	microstructure
MH	microhardness
MQL	minimum quantity of lubrication
R	roughness
RB	roller burnishing
RS	residual stresses
SB	slide burnishing
SI	surface integrity
WR	wear resistance
XRD	X-ray diffraction

References

1. Balevski, A.T. *Metal Science*; Technika: Sofia, Bulgaria, 1988. (In Bulgarian)
2. Jawahir, I.S.; Attia, H.; Biermann, D.; Duflou, J.; Klocke, F.; Meyer, D.; Newman, S.T.; Pusavec, F.; Putz, M.; Rech, J.; et al. Cryogenic manufacturing processes. *CIRP Ann.-Manuf. Technol.* **2016**, *65*, 713–736. [CrossRef]
3. Jurči, P.; Dlouhý, I. Cryogenic Treatment of Martensitic Steels: Microstructural Fundamentals and Implications for Mechanical Properties and Wear and Corrosion Performance. *Materials* **2024**, *17*, 548. [CrossRef] [PubMed]
4. Diekman, F. Cold and Cryogenic Treatment of Steel. In *ASM Handbook, Volume 4A, Steel Heat Treating Fundamentals and Processes*; ASM International: Almere, The Netherlands, 2013; pp. 382–386. Available online: https://dl.asminternational.org/handbooks/edited-volume/18/Steel-Heat-Treating-Fundamentals-and-Processes (accessed on 28 April 2024).
5. Pellizari, M. Influence of deep cryogenic treatment on the properties of conventional and PM high speed steels. *Metall. Ital.* **2008**, *9*, 17–22.
6. Jurči, P.; Bartkowska, A.; Hudáková, M.; Dománková, M.; Caplovičová, M.; Bartkowski, D. Effect of Sub-zero Treatments and Tempering on Corrosion Behaviour of Vanadis 6 tool steel. *Materials* **2021**, *14*, 3759. [CrossRef] [PubMed]
7. Ptačinová, J.; Durica, J.; Jurči, P.; Kusy, M. Influence of sub-zero treatment in liquid helium and tempering on the microstructure of tool steel Vanadis 6. In Proceedings of the 28th International Conference on Metallurgy and Materials METAL 2019, Brno, Czech Republic, 22–24 May 2019; Tanger Ltd.: Greensboro, NC, USA, 2019; pp. 569–574.
8. Baldissera, P. Deep cryogenic treatment of AISI 302 stainless steel: Part I—Hardness and tensile properties. *Mater. Des.* **2010**, *31*, 4725–4730. [CrossRef]
9. Baldissera, P. Deep cryogenic treatment of AISI 302 stainless steel: Part II—Fatigue and corrosion. *Mater. Des.* **2010**, *31*, 4731–4737. [CrossRef]
10. Zhang, H.; Gu, K.; Guo, J.; Xue, X.; Wang, J. Cryogenic Treatment of 06Cr19Ni10 Austenitic Stainless Steel. *Mater. Sci. Forum* **2013**, *749*, 187–191. [CrossRef]
11. Liu, Z.; Zhang, Q.; Meng, F.; Miao, Z. Research on modification treatment on austenitic stainless steel. *IOP Conf. Ser. Earth Environ. Sci.* **2020**, *512*, 012173. [CrossRef]
12. Fedorikova, A. Effect of plastic deformation on mechanical properties of austenitic stainless steel under cryogenic condition. In Proceedings of the Metallurgy Junior 2018, Herlany, Slovakia, 10–11 May 2018; pp. 39–42.

13. Wei, L.; Wang, D.; Li, H.; Xie, D.; Ye, F.; Song, R.; Zheng, G.; Wu, S. Effect of Cryogenic Treatment on the Microstructure and Residual Stress of 7075 Aliminum Alloy. *Metals* **2018**, *8*, 273. [CrossRef]
14. Kučerova, L.; Hajek, J.; Vitek, J. The effect of Cryogenic Treatment on Mechanical Properties, Wear and Corrosion Resistance of Aluminium Alloy AW7075. *Manuf. Technol.* **2020**, *20*, 60–65. [CrossRef]
15. Cinto, C.J. Hardened parts produced by hard turning and ball burnishing operations. *Int. Res. J. Eng. Technol.* **2022**, *9*, 1178–1185.
16. Ramji, B.R.; Narasimha, M.H.N.; Krishna, M.; Raghu, M.J. Performance study of cryogenically treated HSS drills in drilling gray cast iron using orthogonal array technique. *Res. J. Appl. Sci. Eng. Technol.* **2010**, *2*, 487–491.
17. Seah, K.H.W.; Rahman, M.; Yong, K.H. Performance evaluation of cryogenically treated tungsten carbide cutting tool inserts. *Proc. Inst. Mech. Eng. Part B* **2003**, *217*, 29–43. [CrossRef]
18. Gill, S.S.; Singh, J.; Singh, H.; Singh, R. Investigation on wear behavior of cryogenically treated TiAlN coated tungsten carbide inserts in turning. *Int. J. Mach. Tools Manuf.* **2011**, *51*, 25–33. [CrossRef]
19. Dhar, N.R.; Paul, S.; Chattopadhyay, A.B. Role of Cryogenic Cooling on Cutting Temperature in Turning Steel. *J. Manuf. Sci. Eng.* **2002**, *124*, 146–154. [CrossRef]
20. Nguyen, T.; Zarudi, I.; Zhang, L.C. Grinding-hardening with Liquid Nitrogen: Mechanisms and Technology. *Int. J. Mach. Tool Manuf.* **2007**, *47*, 97–106. [CrossRef]
21. Jerold, B.D.; Kumar, M.P. Experimental Comparison of Carbon-dioxide and Liquid Nitrogen Cryogenic Coolants in Turning of AISI 1045 Steel. *Cryogenics* **2012**, *52*, 569–574. [CrossRef]
22. Klocke, F.; Lung, D.; Kramer, A.; Cayli, T.; Sangermann, H. Potential of Modern Lubricoolant Strategies on Cutting Performance. *Key Eng. Mater.* **2013**, *554–557*, 2062–2071. [CrossRef]
23. Yalcin, B.; Ozgur, A.; Koru, M. The Effects of Various Cooling Strategies on Surface Roughness and Tool Wear during Soft Materials Milling. *Mater. Des.* **2009**, *30*, 896–899. [CrossRef]
24. Dix, M.; Wertheim, R.; Schmidt, G.; Hochmuth, C. Modeling of Drilling Assisted by Cryogenic Cooling for Higher Efficiency. *CIRP Ann.-Manuf. Technol.* **2014**, *63*, 73–76. [CrossRef]
25. Machai, C.; Biermann, D. Machining of Beta-Titanium-Alloy Ti-10V-2Fe-3Al Under Cryogenic Conditions—Cooling with Carbon Dioxide Snow. *J. Mater. Process. Technol.* **2011**, *211*, 1175–1183. [CrossRef]
26. Tsai, H.H.; Hocheng, H. Investigation of the Transient Thermal Deflection and Stresses of the Workpiece in Surface Grinding with the Application of a Cryogenic Magnetic Chuck. *J. Mater. Process. Technol.* **1998**, *79*, 177–184. [CrossRef]
27. Khanna, N.; Agrawal, C.; Dogra, M.; Pruncu, C.I. Evaluation of tool wear, energy consumption, and surface roughness during turning of inconel 718 using sustainable machining technique. *J. Mater. Res. Technol.* **2020**, *9*, 5794–5804. [CrossRef]
28. Aramcharoen, A.; Chuan, S.K. An experimental investigation on cryogenic milling of Inconel 718 and its sustainability assessment. *Procedia CIRP* **2014**, *14*, 529–534. [CrossRef]
29. Pereira, O.; Celaya, A.; Urbikain, G.; Rodriguez, A.; Fernandez-Valdivielso, A.; de Lacalle, L.N.L. CO_2 cryogenic milling of Inconel 718: Cutting forces and tool wear. *J. Mater. Res. Technol.* **2020**, *9*, 8459–8468. [CrossRef]
30. Shokrani, A.; Dhokia, V.; Newman, S. Investigation of the effects of cryogenic machining on surface integrity in CNC end milling of Ti-6Al-4V titanium alloy. *J. Manuf. Process.* **2016**, *21*, 172–179. [CrossRef]
31. Akshay; Shravan; Suhas; Sumanth. Effect of Cryogenic Ball Burnishing Parameters on Mechanical Properties of Mg-4Zn-2Sr—A Review. *Int. J. Sci. Res. Sci. Eng. Technol.* **2022**, *9*, 60–64.
32. Caudill, J.; Huang, B.; Arvin, C.; Schoop, J.; Meyer, K.; Jawahir, I.S. Enhancing the surface integrity of Ti-6Al-4V alloy through cryogenic burnishing. *Procedia CIRP* **2014**, *13*, 243–248. [CrossRef]
33. Caudill, J.; Schoop, J.; Jawahir, I.S. Producing sustainable nanostructures in Ti-6Al-4V alloys for improved surface integrity and increased functional life in aerospace applications by cryogenic burnishing. *Procedia CIRP* **2019**, *80*, 120–125. [CrossRef]
34. Caudill, J.; Schoop, J.; Jawahir, I.S. Correlation of surface integrity with processing parameters and advanced interface cooling/lubrication in burnishing of Ti-6Al-4V alloy. *Adv. Mater. Process. Technol.* **2019**, *5*, 53–66. [CrossRef]
35. Huang, B.; Kaynak, Y.U.; Sun, Y.; Jawahir, I.S. Surface Layer Modification by Cryogenic Burnishing of Al 7050-T7451 Alloy and Validation with FEM-Based Burnishing Model. *Procedia CIRP* **2015**, *31*, 1–6. [CrossRef]
36. Huang, B.; Kaynak, Y.; Sun, Y.; Khraisheh, M.K.; Jawahir, I.S. Surface Layer Modification by Cryogenic Burnishing of Al 7050-T7451 Alloy with Near Ultra-Fine Grained Structure. *J. Manuf. Sci. Eng.* **2022**, *144*, 031002. [CrossRef]
37. Kaya, M.; Yaman, N.; Tascioglu, E.; Kaynak, Y. Influence of burnishing process on surface integrity of inconel 718 fabricated by laser powder bed fusion additive manufacturing. *Sigma J. Eng. Nat. Sci.* **2024**, *42*, 335–343. [CrossRef]
38. Mahalil, K.; Rahim, E.A.; Mohid, Z. Performance Evaluation of Sustainable Coolant Techniques on Burnishing Process. *IOP Conf. Ser. Mater. Sci. Eng.* **2019**, *494*, 012001. [CrossRef]
39. Meyer, D. Cryogenic deep rolling—An energy based approach for enhanced cold surface hardening. *CIRP Ann.-Manuf. Technol.* **2012**, *61*, 543–546. [CrossRef]
40. Meyer, D.; Brinksmeier, E.; Hoffmann, F. Surface Hardening by Cryogenic Deep Rolling. *Procedia Eng.* **2011**, *19*, 258–263. [CrossRef]
41. Nikitin, I.; Altenberger, I.; Scholtes, B. Effects of deep rolling at elevated and low temperatures on the isothermal fatigue of AISI304. *Altern. Process.* **2005**, *70*, 185–190.

42. Pu, Z.; Song, G.L.; Yang, S.; Dillon, O.W., Jr.; Puleo, D.A.; Jawahir, I.S. Cryogenic Burnishing of AZ31B Mg Alloy for Enhanced Corrosion Resistance. In *Magnesium Technology*; TMS (The Minerals, Metals & Materials Society): Pittsburgh, PA, USA, 2011; pp. 513–518.
43. Pu, Z.; Yang, S.; Song, G.L.; Dillon, O.W., Jr.; Puleo, D.A.; Jawahir, I.S. Ultrafine-grained surface layer on Mg-Al-Zn alloy produced by cryogenic burnishing for enhanced corrosion resistance. *Scr. Mater.* **2011**, *65*, 520–523. [CrossRef]
44. Pu, Z.; Song, G.-L.; Yang, S.; Outeiro, J.C.; Dillon, O.W.; Puleo, D.A.; Jawahir, I.S. Grain refined and basal textured surface produced by burnishing for improved corrosion performance of AZ31B Mg alloy. *Corros. Sci.* **2012**, *57*, 192–201. [CrossRef]
45. Pu, Z. Cryogenic machining and burnishing of AZ31B magnesium alloy for enhanced surface integrity and functional performance. Theses and Dissertations—Mechanical Engineering. Doctoral Dissertation, University of Kentucky, Lexington, KY, USA, 2012.
46. Rachmat, H.; Rahim, E.A.; Mohid, Z.; Mahalil, K.; Kasah, A.A.; Nadzri, A. Effect of Burnishing Tool Diameter and Coolant Strategies on Burnishing Performance. *J. Phys. Conf. Ser.* **2019**, *1150*, 012070. [CrossRef]
47. Rachmat, H.; Mahalil, K.; Mohid, Z.; Rahim, E.A. Comparison between Dry, MQL, and Cryogenic Cooling Technique on Surface Integrity of Burnished Surface. *Int. J. Integr. Eng.* **2019**, *11*, 35–41. [CrossRef]
48. Rotella, G.; Caruso, S.; Del Prete, A.; Filice, L. Prediction of Surface Integrity Parameters in Roller Burnishing of Ti6Al4V. *Metals* **2020**, *10*, 1671. [CrossRef]
49. Rotella, G.; Rinaldi, S.; Filice, L. Roller burnishing of Ti6Al4V under different cooling/lubrication conditions and tool design: Effects on surface integrity. *Int. J. Adv. Manuf. Technol.* **2020**, *106*, 431–440. [CrossRef]
50. Sachin, B.; Narendranath, S.; Chakradhar, D. Effect of cryogenic diamond burnishing on residual stress and microhardness of 17-4 PH stainless steel. *Mater. Today Proc.* **2018**, *5*, 18393–18399. [CrossRef]
51. Sachin, B.; Narendranath, S.; Chakradhar, D. Experimental evaluation of diamond burnishing for sustainable manufacturing. *Mater. Res. Express* **2018**, *5*, 106514. [CrossRef]
52. Sachin, B.; Narendranath, S.; Chakradhar, D. Effect of working parameters on the surface integrity in cryogenic diamond burnishing of 17-4 PH stainless steel with a novel diamond burnishing tool. *J. Manuf. Process.* **2019**, *38*, 564–571.
53. Sachin, B.; Narendranath, S.; Chakradhar, D. Enhancement of Surface Integrity by Cryogenic Diamond Burnishing Toward the Improved Functional Performance of the Components. *J. Braz. Soc. Mech. Sci. Eng.* **2019**, *41*, 396. [CrossRef]
54. Sachin, B.; Narendranath, S.; Chakradhar, D. Application of desirability approach to optimize the control factors in cryogenic diamond burnishing. *Arab. J. Sci. Eng.* **2020**, *45*, 1305–1317. [CrossRef]
55. Sachin, B.; Charitha, M.R.; Naik, G.; Puneet, N.P. Influence of slide burnishing process on the surface characteristics of precipitation hardenable steel. *SN Appl. Sci.* **2021**, *3*, 223. [CrossRef]
56. Singh, P.; Kumar, P.; Virdi, R.L. Effect of In-Process Cryogenic Cooling in the Burnishing Process on the Solid Particle Erosion Behavior of HVOF Cermet Coating. *J. Therm. Spray Technol.* **2023**, *32*, 2068–2080. [CrossRef]
57. Tang, J.; Luo, H.Y.; Zhang, Y.B. Enhancing the surface integrity and corrosion resistance of Ti–6Al–4V titanium alloy through cryogenic burnishing. *Int. J. Adv. Manuf. Technol.* **2017**, *88*, 2785–2793. [CrossRef]
58. Tang, J.; Luo, H.; Qi, Y.; Xu, P.; Ma, S.; Zhang, Z.; Ma, Y. The effect of cryogenic burnishing on the formation mechanism of corrosion product film of Ti–6Al–4V titanium alloy in 0.9% NaCl solution. *Surf. Coat. Technol.* **2018**, *345*, 123–131. [CrossRef]
59. Anirudh, P.V.; Kumar, B.; Girish, G.; Shailesh, S.; Oyyaravelu, R.; Kannan, C.; Balan, A.S.S. Effect of Cryogenics-Assisted Low-Plasticity Burnishing on Laser-Clad Stellite 6 over SS420 Substrate. *J. Mater. Eng. Perform.* **2020**, *29*, 6861–6869. [CrossRef]
60. Yang, S.; Pu, Z.; Puleo, D.; Dillon, O.W.; Jawahir, I.S. Cryogenic Processing of Biomaterials for Improved Surface Integrity and Product Sustainability. In *Advances in Sustainable Manufacturing*; Springer: Berlin/Heidelberg, Germany, 2011; pp. 177–182.
61. Yang, S.; Puleo, D.; Dillon, O.W., Jr.; Jawahir, I.S. Surface Layer Modifications in Co-Cr-Mo Biomedical Alloy from Cryogenic Burnishing. *Procedia Eng.* **2011**, *19*, 383–388. [CrossRef]
62. Yang, S.; Dillon, O.W., Jr.; Puleo, D.A.; Jawahir, I.S. Effect of cryogenic burnishing on surface integrity modifications of Co–Cr–Mo biomedical alloy. *J. Biomed. Mater. Res. B Appl. Biomater.* **2013**, *101*, 139–152. [CrossRef]
63. Yang, S.; Umbrello, D.; Dillon, O.W., Jr.; Puleo, D.; Jawahir, I.S. Cryogenic cooling effect on surface and subsurface microstructural modifications in burnishing of Co–Cr–Mo biomaterial. *J. Mater. Process. Technol.* **2015**, *217*, 211–221. [CrossRef]
64. Yang, S.; Dillon, O.W., Jr.; Puleo, D.A.; Jawahir, I.S. Enhancement of wear resistance for improved functional performance of Co-Cr-Mo hip implants through cryogenic surface treatment: A case study. *Mach. Sci. Technol.* **2021**, *25*, 455–476. [CrossRef]
65. Yang, S.; Umbrello, D.; Dillon, O.W., Jr.; Jawahir, I.S. Numerical Investigation of Dynamic Recrystallization Induced Microstructural Evolution in Cryogenic Burnishing of Co-Cr-Mo Biomaterial. *J. Mater. Eng. Perform.* **2022**, *31*, 6904–6921. [CrossRef]
66. Yang, S. Cryogenic burnishing of Co-Cr-Mo biomedical alloy for enhanced surface integrity and improved wear performance. Theses and Dissertations—Mechanical Engineering. Doctoral Dissertation, University of Kentucky, Lexington, KY, USA, 2012.
67. Grzesik, W.; Zak, K. Producing high quality hardened parts using sequential hard turning and ball burnishing operations. *Precis. Eng.* **2013**, *37*, 849–855. [CrossRef]
68. Zak, K.; Grzesik, W.; Prazmowski, M. Investigation og Sequential Cryogenic Hard Turning and ball Burnishing. *Metalurgija* **2014**, *53*, 521–525.
69. Zak, K.; Grzesik, W. Investigation of technological effects of ball burnishing after cryogenic turning. *Adv. Manuf. Sci. Technol.* **2014**, *38*, 37–52.
70. Grzesik, W.; Zak, K.; Prazmowski, M.; Storch, B.; Pałka, T. Effects of cryogenic cooling on surface layer characteristics produced by hard turning. *Arch. Mater. Sci. Eng.* **2012**, *54*, 5–12.

71. Savyasachi, N.; Reji, R.; Sajan, J.A.; Rafi, A.M. A Review on the Cryogenic Treatment of Stainless Steels, Tool Steels and Carburized Steels. *Int. J. Innov. Sci. Res. Technol.* **2020**, *5*, 31–38.
72. Ekaputra, R.C.; Mochtar, M.A. The Effect of Parameters in Cryogenic Treatment on Mechanical Properties of Tool Steel: A Review. *J. Mater. Explor. Find.* **2023**, *2*, 101–113.
73. Kumar, T.V.; Thirumurugan, R.; Viswanath, B. Influence of cryogenic treatment on the metallurgy of ferrous alloys: A review. *Mater. Manuf. Process.* **2017**, *32*, 1789–1805. [CrossRef]
74. Kalsi, N.S.; Sehgal, R.; Sharma, V.S. Cryogenic treatment of tool materials: A review. *Mater. Manuf. Process.* **2010**, *25*, 1077–1100. [CrossRef]
75. Akincioğlu, S.; Gökkaya, H.; Uygur, I. A review of cryogenic treatment on cutting tools. *Int. J. Adv. Manuf. Technol.* **2015**, *78*, 1609–1627. [CrossRef]
76. Baldissera, P.; Delprete, C. Deep Cryogenic Treatment: A bibliographic Review. *Open Mech. Eng. J.* **2008**, *2*, 1–11. [CrossRef]
77. Khanna, N.; Agrawal, C.; Pimenov, D.Y.; Singla, A.K.; Machado, A.R.; da Silva, L.R.R.; Gupta, M.K.; Sarikaya, M.; Krolczyk, G.M. Review on design and development of cryogenic machining setups for heat resistant alloys and composites. *J. Manuf. Process.* **2021**, *68*, 398–422. [CrossRef]
78. Zindani, D.; Kumar, K. A brief review on cryogenics in machining process. *SN Appl. Sci.* **2020**, *2*, 1107. [CrossRef]
79. Yildiz, Y.; Nalbant, M. A review of cryogenic cooling in machining processes. *Int. J. Mach. Tool Manuf.* **2008**, *48*, 947–964. [CrossRef]
80. Shokrani, A.; Dhokia, V.; Munoz-Escalona, P.; Newman, S.T. State-of-the-art cryogenic machining and processing. *Int. J. Comp. Integr. Manuf.* **2013**, *26*, 616–648. [CrossRef]
81. Ecoroll Catalogue. *Tools and Solutions for Metal Surface Improvement*; Ecoroll Corporation Tool Technology: Milford, OH, USA, 2006.
82. Maximov, J.T.; Duncheva, G.V.; Anchev, A.P.; Ichkova, M.D. Slide burnishing—Review and prospects. *Int. J. Adv. Manuf. Technol.* **2019**, *104*, 785–801. [CrossRef]
83. Maximov, J.T.; Duncheva, G.V.; Anchev, A.P.; Dunchev, V.P. Smoothing, deep, or mixed diamond burnishing of low-alloy steel components—Optimization procedures. *Int. J. Adv. Manuf. Technol.* **2020**, *106*, 1917–1929. [CrossRef]
84. Maximov, J.T.; Duncheva, G.V.; Anchev, A.P.; Dunchev, V.P.; Anastasov, K.; Daskalova, P.H. Effect of Roller Burnishing and Slide Roller Burnishing on Surface Integrity of AISI 316 Steel: Theoretical and Experimental Comparative Analysis. *Machines* **2024**, *12*, 51. [CrossRef]
85. Prevey, P.S. Burnishing Method and Apparatus for Providing a Layer of Compressive Residual Stress in the Surface of a Workpiece. US Patent 5826453, 27 October 1998.
86. Burakowski, T.; Wierzchon, T. *Surface Engineering of Metals: Principles, Equipment, Technologies*; Tyrkiel, E.A., Dearnley, P., Eds.; CRC Press: London, UK, 1998.

Journal of
*Functional
Biomaterials*

MDPI

Article

A New Hyaluronic Emulgel of Hesperetin for Topical Application—An In Vitro Evaluation

Raquel Taléns-Visconti [1,*,†], Yousra Belarbi [1,2,†], Octavio Díez-Sales [1,2], Jesus Vicente de Julián-Ortiz [3], Ofelia Vila-Busó [4] and Amparo Nácher [1,2]

1 Department of Pharmacy and Pharmaceutical Technology and Parasitology, Faculty of Pharmacy and Food Sciences, University of Valencia, Av. Vicent Andrés Estellés s/n, 46100 Valencia, Spain; belarbiyousra31@gmail.com (Y.B.); octavio.diez@uv.es (O.D.-S.); amparo.nacher@uv.es (A.N.)
2 Instituto Interuniversitario de Investigación de Reconocimiento Molecular y Desarrollo Tecnológico (IDM), Universitat Politècnica de València, Av. Vicent Andrés Estellés s/n, 46100 Valencia, Spain
3 Molecular Topology and Drug Design Research Unit, Department of Physical Chemistry, Faculty of Pharmacy and Food Sciences, University of Valencia, Av. Vicent Andrés Estellés s/n, 46100 Valencia, Spain; jesus.julian@uv.es
4 Colloids Research Unit, Department of Physical Chemistry, Faculty of Pharmacy and Food Sciences, University of Valencia, Av. Vicent Andrés Estellés s/n, 46100 Valencia, Spain; ofelia.vila@uv.es
* Correspondence: raquel.talens@uv.es
† These authors contributed equally to this work.

Abstract: The present study aimed to formulate and characterize a hesperetin formulation to achieve adequate deposition and retention of hesperetin in the epidermis as a target for some cosmetic/dermatological actions. To derive the final emulgel, various formulations incorporating different proportions of Polysorbate 80 and hyaluronic acid underwent testing through a Box–Behnken experimental design. Nine formulations were created until the targeted emulgel properties were achieved. This systematic approach, following the principles of a design of experiment (DoE) methodology, adheres to a quality-by-design (QbD) paradigm, ensuring a robust and purposeful formulation and highlighting the commitment to a quality-driven design approach. The emulsions were developed using the phase inversion method, optimizing the emulgel with the incorporation of hyaluronic acid. Physically stable optimized emulgels were evaluated for their globule size, surface charge, viscosity, pH, electrical conductivity, and hesperetin content. These assays, along with the temperature swing test, were used to select the optimal formulation. It was characterized by a droplet size, d[4,3], of 4.02 μm, a Z-potential of −27.8 mV, an O/W sign, a pH of 5.2, and a creamy texture and proved to be stable for at least 2 months at room temperature. Additionally, in vitro release kinetics from the selected emulgel exhibited a sustained release profile of hesperetin. Skin assays revealed adequate retention of hesperetin in the human epidermis with minimum permeation. Altogether, these results corroborate the promising future of the proposed emulgel in cosmetic or dermatological use on healthy or diseased skin.

Keywords: flavonoid; hesperetin; emulgel; hyaluronic acid; stability; topical application; sustained release kinetics

1. Introduction

The topical route of administration has a long history, but contemporary advancements in methods and technologies have driven its resurgence [1,2]. This mode of administration, whether for cosmetic [3] or dermatologic [4,5] purposes, enjoys widespread utilization across both healthy and diseased skin.

Diverse forms—solid, semi-solid, or liquid—exist for topical applications, spanning powders, creams, ointments, gels, lotions, solutions, and others. In this context, emulsions are one of the most popular physicochemical forms used in cosmetics and pharmaceutical products, e.g., creams, balms, masks, etc. can be presented in the form of an emulsion. Additionally, the emergence of combined systems comprising multiple classical formulations represents a burgeoning field, albeit they are less prevalent in the market despite their high acceptability. Notably, emulgels stand out as captivating and challenging products to explore [6–9]. These systems arise from the fusion of classic gels with emulsions, resulting in highly functional yet complex dosage forms. Consequently, the gel component extends the formulation's residence time on the skin by augmenting the viscosity [10]. Simultaneously, the emulsion part shields the active ingredients from degradation and hydrolysis while enhancing skin permeation [11]. Thus, achieving optimal outcomes relies on both retaining higher concentrations of the active ingredient for a sufficient duration and facilitating its effective penetration through the skin [8,12–14].

In addition, the exploration of novel formulations, such as emulgels, coincides with the search for effective delivery systems. These intricate systems, which bring together the characteristics of gels and emulsions, underline the persistent drive to optimize topical drug delivery. Moreover, the exploration of natural compounds, particularly polyphenols derived from various food sources, resonates with the ongoing quest for effective dermal preparations. These compounds, renowned for their diverse pharmacological activities, have emerged as promising alternatives to synthetic chemicals in both medical and cosmetic applications. Their intrinsic properties, such as antioxidant, anti-inflammatory, and antibacterial attributes, hold significant potential in mitigating conditions linked to free radical accumulation and skin-related pathologies. However, despite their remarkable bioactivities, the widespread application of polyphenols in medicine and cosmetics faces challenges due to their limited solubility in water. Addressing these formulation limitations becomes pivotal, especially considering the necessity for innovative delivery strategies to harness the full potential of polyphenols in dermal applications [15].

Notably, hesperetin, a flavanone identified as 3,5,7-trihydroxy-4-methoxyflavanone ($C_{16}H_{14}O_6$) (Figure 1), is the aglycone of hesperidin. It is obtained from the mesocarp (middle layer of the pericarp of the fruit) of the peel of *Citrus sinensis* (sweet orange) and has different types of pharmacological activities, including some that are very favorable for topical administration [16,17].

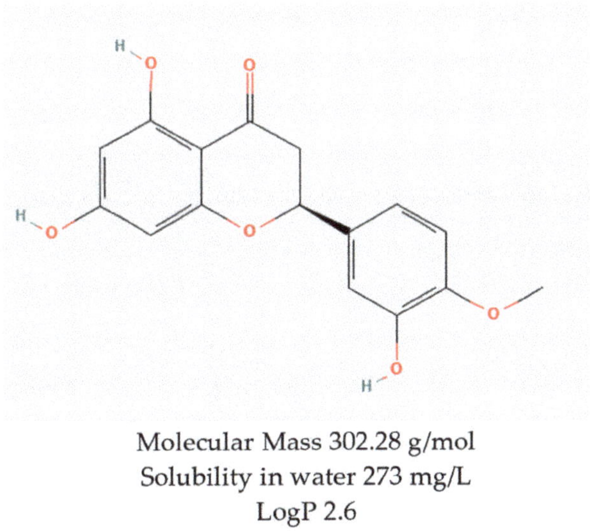

Figure 1. Chemical structure and some properties of hesperetin. Reprinted from Ref. [18].

Hesperetin is primarily acknowledged for its antioxidant, anti-inflammatory, antibacterial, and immunomodulatory properties [19,20], presenting the potential to address conditions associated with free radical accumulation. Furthermore, it demonstrates anti-carcinogenic and neuroprotective effects [20–23]. Additionally, hesperetin has showcased the ability to absorb ultraviolet radiation, offering skin protection and influencing collagen biosynthesis in dermal fibroblasts [24,25]. However, despite these impressive attributes, hesperetin's utility in medicine and cosmetics has been hindered by its insolubility in water.

Within this context, this study aims to devise an effective topical formulation that ensures adequate deposition and retention of hesperetin within the epidermis.

2. Materials and Methods

2.1. Materials

Hesperetin (>95% purity) obtained from Sigma-Aldrich (Steinheim production, Steinheim, Germany) was used as the active ingredient. The semi-solid vehicle was prepared using the following compounds: Phospholipid S75 (Lipoid®, Abaran Materias Primas S.L., Madrid, Spain), Almond oil (Acofarma®, Barcelona, Spain), and Transcutol CG (Gattefossé®, Neuilly-Sur-Seine, France). Polysorbate 80, tocopherol, cetostearyl alcohol, propylene glycol, ethanol, methylparaben, propylparaben, and hyaluronic acid of low molecular weight were obtained from Guinama® (Valencia, Spain), and MilliQ water was also used.

2.2. Hesperetin Quantification

The hesperetin content was quantified using a reverse phase HPLC method employing a Perkin Elmer® Series 200 HPLC equipped with a UV detector (Perkinelmer España S L, Madrid, Spain), detecting the molecule's absorbance at 288 nm. The analysis utilized a Teknokroma® Brisa "LC2" C18 column (150 × 4.6 mm, 5.0 µm) (Teknokroma, Barcelona, Spain) and a mobile phase consisting of a methanol: water MilliQ mixture in a 70:30 proportion, respectively. The flow rate was maintained at 1 mL/min and the sample injection volume was 20 µL. Hesperetin was eluted after 3.5 min and the method was linear in the concentration range of 0.24–62.7 µg/mL with a correlation coefficient of 0.9999. The limits of detection (*LOD*, Equation (1)) and quantification (*LOQ*, Equation (2)) were estimated using the calibration curve procedure, as follows:

$$LOD = \frac{3.3 \cdot SD}{S} \tag{1}$$

$$LOQ = \frac{10 \cdot SD}{S} \tag{2}$$

where SD and S are the standard deviation of the y-intercept and the slope of the linear calibration curve, respectively.

Under these conditions, the limit of quantification was 0.918 µg/mL and the limit of detection was 0.303 µg/mL.

2.3. Preformulation Studies

2.3.1. Experimental Design

To obtain the final emulgel to be used, several formulations incorporating different proportions of Polysorbate 80 and hyaluronic acid were designed using a Box–Behnken experimental design [26] using the STATISTICA program (v 8.0) [27].

2.3.2. Preparation of Formulations

To obtain the final emulgel to be used during this study, nine formulations with three distinct proportions of Polysorbate 80 (2, 3.5 and 5% *w*/*w*) and hyaluronic acid (0.5, 1, and 1.5% *w*/*w*) were tested (Table 1).

Table 1. Varying composition of emulgel formulations (% *w*/*w*).

Ingredient (% *w*/*w*)	F1	F2	F3	F4	F5	F6	F7	F8	F9
Hesperetin	0.5	0.5	0.5	0.5	0.5	0.5	0.5	0.5	0.5
Phospholipid	5	5	5	5	5	5	5	5	5
Polysorbate 80	2	3.5	2	2	3.5	5	3.5	5	5
Cetostearyl alcohol	3	3	3	3	3	3	3	3	3
Almond oil	10	10	10	10	10	10	10	10	10
Transcutol GC	4	4	4	4	4	4	4	4	4
Tocopherol	0.1	0.1	0.1	0.1	0.1	0.1	0.1	0.1	0.1
Propylene glycol	5	5	5	5	5	5	5	5	5
Methylparaben	0.1	0.1	0.1	0.1	0.1	0.1	0.1	0.1	0.1
Propylparaben	0.05	0.05	0.05	0.05	0.05	0.05	0.05	0.05	0.05
Hyaluronic acid	1.5	0.5	0.5	1	1	1.5	1.5	1	0.5
Ethanol	q.s	q.s	q.s	q.s	q.s	q.s	q.s	q.s	q.s
Water	68.75	68.25	69.75	69.25	67.75	65.75	67.25	66.25	66.75

The procedure followed for the preparation of the formulations consisted of phase inversion and subsequent homogenization. In a thermostatic bath, the ingredients of the oil phase and the aqueous phase were heated separately to 75 °C. The aqueous phase was then slowly poured over the oil phase under continuous and vigorous stirring to form emulgels, which were cooled to room temperature. Cetostearyl alcohol added to the oil phase and hyaluronic acid added to the aqueous phase were included to stabilize the formulations. In order to ensure the smallest possible uniform droplet size, a homogenization process using an Ultraturrax® (Janke & Kunkel, IKA-Labortechnik, Staufen, Germany) was completed for 5 min at 24,000 rpm. A hesperetin dispersion of 0.5% *w*/*v* was then added in a sufficient amount of ethanol. The prepared formulations were stored in tightly closed containers and kept at room temperature until testing.

2.3.3. Characterization of Emulgels and Stability Assessment

- Physical Evaluation

The developed formulations were visually and physically inspected for appearance, color, odor, phase separation, pH, electrical conductivity, and viscosity. For pH measurement, a 10% *w*/*v* dispersion of the emulgels was prepared in distilled water and measured using a calibrated Hanna® (Hanna Instruments Ltd., Leighton Buzzard, UK) pH meter. A conductivity meter (Hanna®, Hanna Instruments Ltd., Leighton Buzzard, UK) was employed to validate the presence of a continuous aqueous phase, confirming that the emulgels were O/W type. Viscosity was measured using a Brookfield viscometer (ViscoStar plus R.V5002, Fungilab, I.C.T, S.L, Lardero, Spain) at room temperature and with a fixed rotation speed (60 rpm) using an R6 spindle. Centrifugation at 3500 rpm for 30 min determined the emulgels' stability regarding phase separation, cracking, or creaming (See Table 2). All the measurements were performed in triplicate.

Table 2. Conditions for forced stability studies.

Forced sedimentation test	30 min at 3500 rpm
Temperature swing test	Cold–heat cycle 4 °C to 40 °C
	Freeze–thaw cycle −20 °C to 25 °C

- Optical Microscopy

Microscopic observations of each formulation were carried out using a Carl Zeiss (ScienceServices GmbH company, München, Germany) model GSZ stereoscopic microscope in order to observe the homogeneity of the formulation and to qualitatively verify whether coalescence or an increase in drop size occurred over time.

- Evaluation of Droplet Size and Zeta Potential

The droplet size was measured using a Mastersizer 2000® (Malvern Instruments, Worcestershire, UK). The value used was d[4,3], which corresponds to the weighted average of the volume of the emulgel droplets [28,29]. These data were counterbalanced with size scatter plots. Additionally, the Z-potential was assessed using a Zetasizer nano-S (Malvern Instruments, Worcestershire, UK), with three measurements per sample.

- Temperature Swing Tests

The formulations were subjected to swing tests following two temperature cycles. One of these cycles corresponded to the temperature changes observed in extreme hot–cold conditions such as summer–winter changes, and the other cycle referred to the temperature changes that would occur if the product was frozen and thawed. For this, the samples were stored successively at a temperature of 4 °C and 40 °C for 48 h at each temperature, which was considered one cycle. On the other hand, the respective formulations were stored for 48 h at −20 °C in the deep freezer and at room temperature (25 ± 2 °C), which was considered one freeze–thaw cycle. These procedures were repeated for three cycles each (Table 2). Once the three stages were completed, the samples were subjected to visual, physical, and microscopic evaluation, as described above. In addition, the droplet size, zeta potential, and hesperetin concentration were also measured.

These tests served as selection criteria to identify the stable optimized formulation for subsequent experiments.

- Long-Term Stability Study

The chosen optimized formulation underwent a 2-month long-term stability study at room temperature. After this period, the samples were subjected to physical, visual, and microscopic evaluation, as described above. In addition, the droplet size, zeta potential, and hesperetin concentration were also measured.

2.4. In Vitro Hesperetin Release and Skin Retention Studies

The release and skin permeability assessments of hesperetin were conducted on the selected emulgel using a Franz vertical cell (effective diffusion area: 0.785 cm^2) with a sample size of $n = 3$. The assembled cell was placed in a thermostatic bath at 37 ± 1 °C and maintained under constant agitation. For this purpose, the 'infinite dose' condition was used to determine the release and skin permeation coefficients. Briefly, the experiment consisted of adding 500 mg of the test formulation to the donor chamber and a 50% (v/v) hydroalcohol mixture to the receptor chamber. In the release tests, a cellulose acetate membrane separated the chambers, while the human epidermis was used for the skin permeability tests. The human epidermis was abdominal female skin, which was obtained from patients after they signed informed consent for inclusion before participating in the study. The study was conducted in accordance with the Declaration of Helsinki, and the protocol was approved by the Ethical Committee of the University of Valencia (protocol number: H 1540295606992, approved date: 8 November 2018). The epidermal membranes were obtained via heat separation and immersed in water at 60 °C for 60 s [30] or frozen and stored after fatty tissue removal.

Samples of 200 µL were collected from the receptor chamber, and an equal volume was replenished at different time intervals for up to 24 h. Simultaneously, the accumulated hesperetin in the skin was quantified after methanol extraction. To achieve this, after the 24-h period, the epidermis was cut, placed into a glass vial containing methanol, and sonicated for 2 min to ensure the complete extraction of hesperetin. The resulting dispersion underwent centrifugation at 3000 rpm for 10 min. Subsequently, the supernatant was filtered, and the hesperetin content was measured using HPLC.

The in vitro release kinetics of hesperetin from the emulgel were modeled mathematically. For this purpose, the data were fitted to zero-order (Equation (3)), first-order (Equation (4)) and Higuchi (Equation (5)) equations.

$$Q_t = K_0 t \tag{3}$$

$$Q_t = Q_0 \cdot e^{-k1t} \tag{4}$$

$$\frac{Q_t}{Q_\infty} = K_h \cdot t^{0.5} \tag{5}$$

where t is time, Q_t is the amount of hesperetin released at time t, Q_0 is the initial amount of hesperetin, Q_∞ is the amount of hesperetin released at time ∞, and K_0, K_1, and K_h are the hesperetin release rate constants of each of the kinetics.

To determine the mechanism of hesperetin release, Korsmeyer–Peppas (Equation (6)) and Peppas–Sahlin (Equation (7)) mathematical models were employed and fitted to the experimental data using Sigmaplot 10.0® (Systat Software, Inc., San Jose, CA, USA).

$$\text{Korsmeyer} - \text{Peppas}: \frac{Q_t}{Q_\infty} = K \cdot t^n \tag{6}$$

$$\text{Peppas} - \text{Sahlin}: \frac{Q_t}{Q_\infty} = K_1 t^n + K_2 t^{2n} \tag{7}$$

where Q_t/Q_∞ represents the fraction released of the active ingredient, K, K_1, and K_2 are the diffusion constants, t is the time in which the active substance is released, and n is the exponent that characterizes the diffusion process. So for $n < 0.5$, the release is Fickian, between $0.5 < n < 1.0$, it indicates that an anomalous process has occurred, while for $n = 1$, the release obeys zero-order kinetics [31].

Additionally, for the skin permeation studies, as a representative of the diffusion process in the experimental conditions, the Scheuplein equation was used, which relates the quantities (Q, mg/cm^2) permeated with time (t, hours) (Equation (8)):

$$Q_{(t)} = A \cdot P \cdot L \cdot C \cdot \left[D \cdot \frac{t}{L^2} - \frac{1}{6} - \frac{2}{\pi^2} \cdot \sum_{n=1}^{\infty} \frac{(-1)^n}{n^2} \cdot Exp\left(\frac{-D \cdot n^2 \cdot \pi^2 \cdot t}{L^2} \right) \right] \tag{8}$$

where A is the useful diffusion area (cm^2), P is the distribution coefficient of the drug between the skin and the donor vehicle, L is the thickness of the membrane (cm), C is the drug concentration in the donor solution (mg/mL), D is the diffusion coefficient in the membrane (cm^2/h), and n is a value integer between one and infinity.

Equation (8) has been simplified to Equation (9) and represents the linear section of the percutaneous absorption process:

$$Q_{(t)} = A \cdot P \cdot L \cdot C \cdot \left[D \cdot \frac{t}{L^2} - \frac{1}{6} \right] \tag{9}$$

The accumulated hesperetin quantities against time were fitted to Equation (9). The values of P and D were obtained as the primary parameters of the fit. From these values, the latency time (t_L, h), the permeability coefficient (K_p, cm/h), and the flow (J, μg/cm^2·h) were calculated:

$$t_L = \frac{1}{6 \cdot D} \tag{10}$$

$$K_p = \frac{P \cdot D}{L} \tag{11}$$

$$J = K_p \cdot C \tag{12}$$

2.5. Statistical Analysis

To observe whether there were significant differences between the formulas tested, statistical tests were performed using the IBM SPSS 22.2 Statistics program (Systat Software, Inc., San Jose, CA, USA) with a fixed level of significance of $p = 0.05$. Initially, the Levene statistic test was employed to ascertain the homogeneity of variances in the results. Subsequently, for the cases meeting this condition, the parametric Scheffe's multiple comparison test was applied, while the non-parametric Dunnett T3 test was utilized for the cases not meeting the homogeneity condition. The results are presented as mean values alongside their standard deviations.

3. Results and Discussion

The aim of this work was to provide an effective formulation strategy for enhanced skin retention of a highly lipophilic molecule, such as hesperetin, to have the potential to provide topical formulations for the skin. In that context, a hyaluronic emulgel could be the ideal vehicle, as an emulsion allows for the incorporation of hydrophilic/lipophilic substances due to its biphasic nature, and hyaluronic acid permits the retention of active substances longer on the skin, prolonging their action [32]. Additionally, semi-solid formulations prepared using this polymer are characterized by excellent sensory and rheological properties and have shown excellent qualities as wound-healing formulations [33,34].

3.1. Experimental Design

The Box–Behnken experimental design encompassed three distinct proportions of Polysorbate 80 and hyaluronic acid. The design, as detailed in Table 3, comprised nine trials (three different variable levels applied to each component).

Table 3. Coded representations of the actual values for the variables.

Independent Variable	F1	F2	F3	F4	F5	F6	F7	F8	F9
V1: Concentration of Polysorbate 80	−1	0	−1	−1	0	+1	0	+1	+1
V2: Concentration of hyaluronic acid	+1	−1	−1	0	0	+1	+1	0	−1

Wherein:

V1 levels of Polysorbate 80 concentrations are represented as follows:

- (−1) = 2.0% *w*/*w*
- (0) = 3.5% *w*/*w*
- (+1) = 5.0% *w*/*w*

V2 levels of hyaluronic acid concentrations are depicted as:

- (−1) = 0.5% *w*/*w*
- (0) = 1.0% *w*/*w*
- (+1) = 1.5% *w*/*w*

The specific values for each level were determined based on prior formulation experiences, contributing to the assignment of these variable levels.

3.2. Physical Evaluation

All the developed emulgels (F1–9) were smooth, and most of them were creamy in texture without clumps, with a white color in appearance and a pleasant odor. The pH values were observed to be 5.2–6.3. Moreover, all pH values were around that of the skin pH (4.5–6), suggesting an acceptable and non-irritating value [34]. Moreover, the creamy formulations that did not show phase separation were those with a higher hyaluronic acid concentration (1.5% *w*/*v*) with a viscosity of around 3020 ± 7.4 mPa·s (Figure 2).

Finally, the electrical conductivity showed values of 430–585 μS thus confirming that all the formulations were of an external aqueous phase.

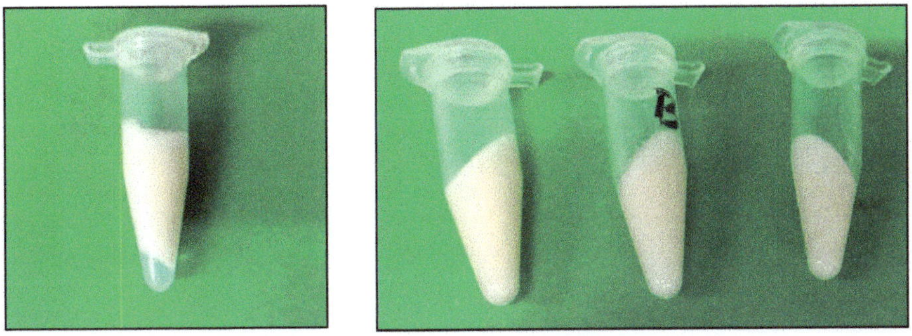

Figure 2. Representative images of results of centrifugation assays (3500 rpm/30 min) in different formulations (F4, showing phase separation; F1, F6, and F7, showing stability).

3.3. *Optical Microscopy*

The different formulations were observed under the visual microscope at zero time as well as after the cycle tests. In addition, the formulation that showed the best results (F7) was also observed at one and two months (long term). Homogeneity and the absence of aggregates were verified. A representative image is shown in Figure 3.

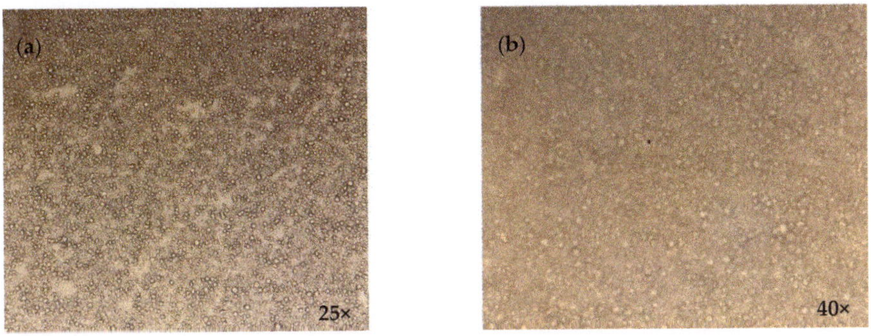

Figure 3. Representative image of emulgel formula F7 obtained using Carl Zeiss stereo microscope model GSZ at 25× (**a**) and 40× (**b**).

3.4. *Temperature Swing Test*

The formulations that did not show flocculation, creaming, or cracking during physical evaluation were subsequently subjected to the temperature swing test. After the three cycles of temperature changes, the formulations were characterized following the analyses mentioned in the methodology section, thus allowing for observation of the changes produced in the emulgels. The results before and after the temperature tests showed that the values of the surface charge, droplet size, pH, conductivity, and hesperetin content (%) were more stable and consistent for formulation 7. A representative image of emulgel formula F7 obtained using a Carl Zeiss stereo microscope model GSZ is shown in Figure 3. Additionally, a control emulgel was used with the same composition but without hesperetin to verify that the introduction of this molecule did not modify any characteristics (Table 4).

Table 4. Results of characterization and stability of the emulgel.

Emulgel	State	Z-Potential	Droplet Size d[4,3] μm	pH	Conductivity (μS)	% Hesperetin
Control		−28.3 ± 1.39	2.07	5.89	487	-
F7 (selected formula with 0.5% hesperetin)	Initial	−27.8 ± 0.68	4.02	5.16	507	0.58
	Cycle 4/40 °C	−25.6 ± 2.89	6.52	5.13	564	0.56
	Cycle −20/25 °C	−25.8 ± 0.70	3.28	5.36	587	0.56

Moreover, the size distribution of the droplets was evaluated under these temperature changes with respect to the same emulgel at room temperature (initial state). The aim was to investigate whether the droplets underwent a change in size with respect to temperature. The plots obtained using the MasterSizer showed the particle size distribution. Figure 4 shows the particle size distribution of the F7 emulgel and the distribution after temperature cycling.

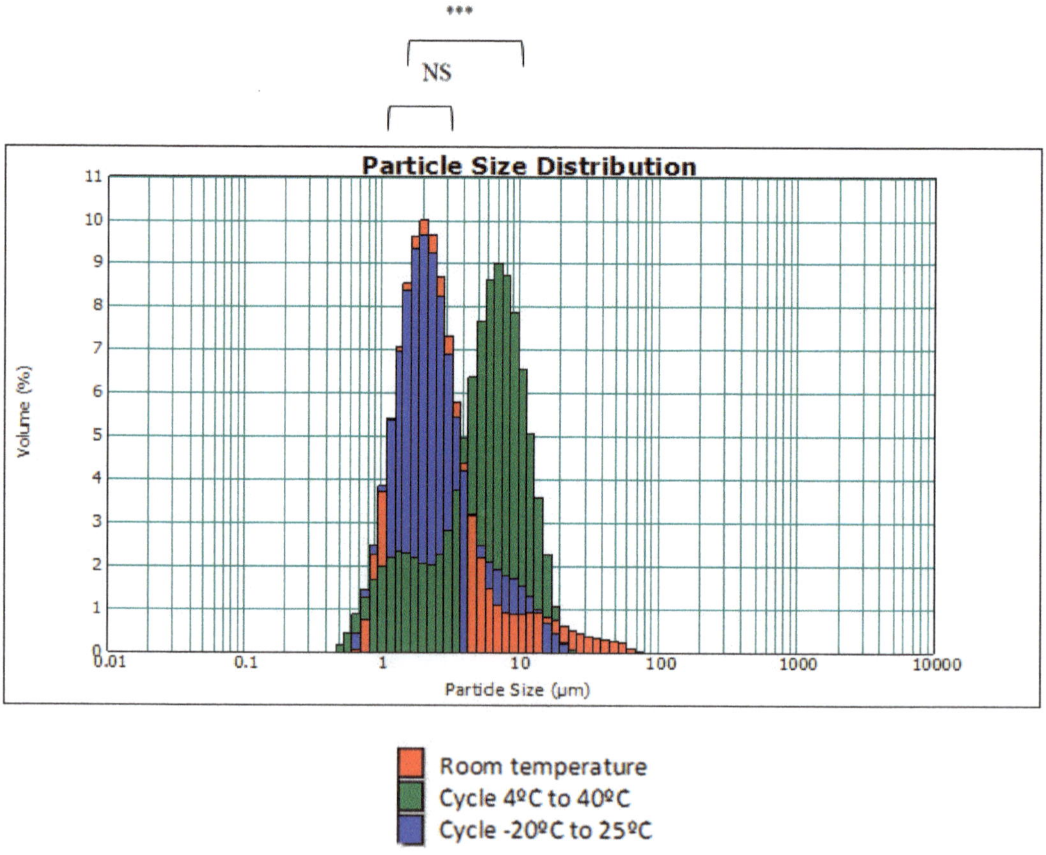

Figure 4. Drop size distribution (d[4,3]) in accelerated temperature tests (NS: no significant difference; ***: $p < 0.001$).

As seen in Figure 4, with increasing temperature, the droplets undergo a slight increase in size as the size dispersion curve of the 4 °C to 40 °C cycle has shifted slightly to the right. This may be due to the fact that the affinity of surfactants varies with temperature, resulting in some degree of aggregation of smaller droplets. Although there is statistical significance between the 4 °C to 40 °C cycle curve and the rest of the results, these differences have no practical repercussion and will not influence the quality of the cosmetic cream to be developed, since the range of measurements is still between 0.2–12 µm, as in the case of room temperature and the −20–25 °C cycle.

On the other hand, the formulations are shown to be stable after temperature cycling, as the Z-potential (Figure 5), sign, and pH are not affected, with no statistical significance. Furthermore, preliminary chemical stability tests with the formulations incorporating hesperetin have shown that temperature cycling does not affect their initial concentration. Altogether, this affirms that the emulgel is stable regarding temperature changes; therefore, it may be possible to store it without refrigeration at room temperature without seasonal changes altering the amount of the active ingredient.

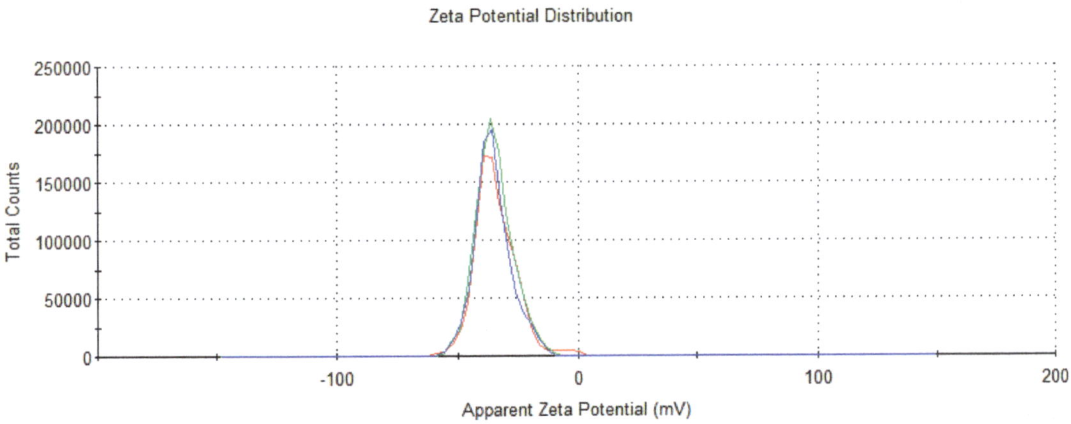

Figure 5. Representative image of Z-potential determination Zetasizer nano-S (Malvern Instruments, Worcestershire, UK).

3.5. Long-Term Stability Study

Due to the satisfactory results of the preliminary studies described in the previous section, the F7 emulgel was subjected to long-term studies. These consisted of measuring the Z-potential, conductivity, pH, droplet size, and quantification of hesperetin for two months, once per month, at room temperature.

Figure 6 shows the droplet size distribution as a function of time. It can be seen that the majority of droplets are between 0.6–6 µm, which is similar to the size distribution of the formula without hesperetin.

Figure 7 shows the values of pH, Z-potential, and conductivity and the percentage of hesperetin with respect to time.

Overall, these results allow us to conclude that the emulgel is stable after 2 months at room temperature since it does not undergo significant variations in terms of pH, conductivity, droplet size, or percentage of the active ingredient. In fact, although significant differences in the Z-potential data were observed, in practice, the emulgel was considered to be stable since it favored the repulsion between the droplets and reduced the tendency to aggregate. On the other hand, the pH remained around 5, which is within the favorable range for topical use. The conductivity remained stable, indicating that the emulgel did not undergo any change in its sign and continued to be O/W. Finally, the percent of hesperetin did not decrease after two months.

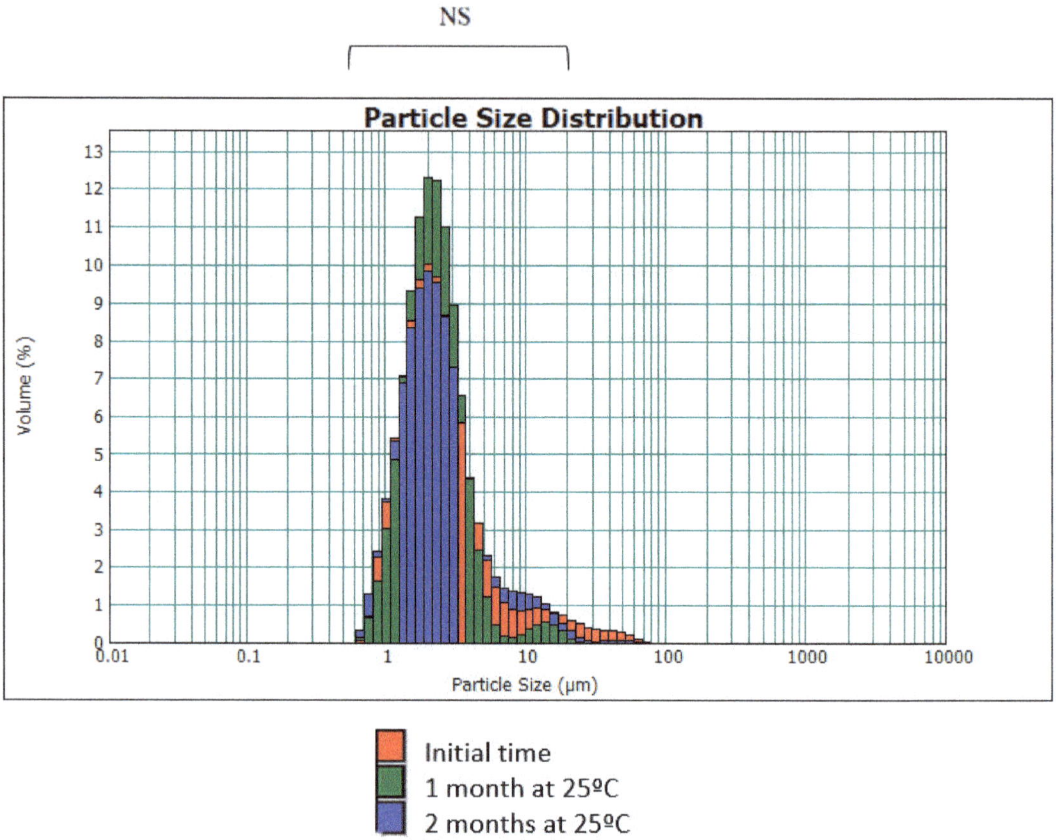

Figure 6. Distribution of droplet size (d[4,3]) as a function of time (NS: no significant differences).

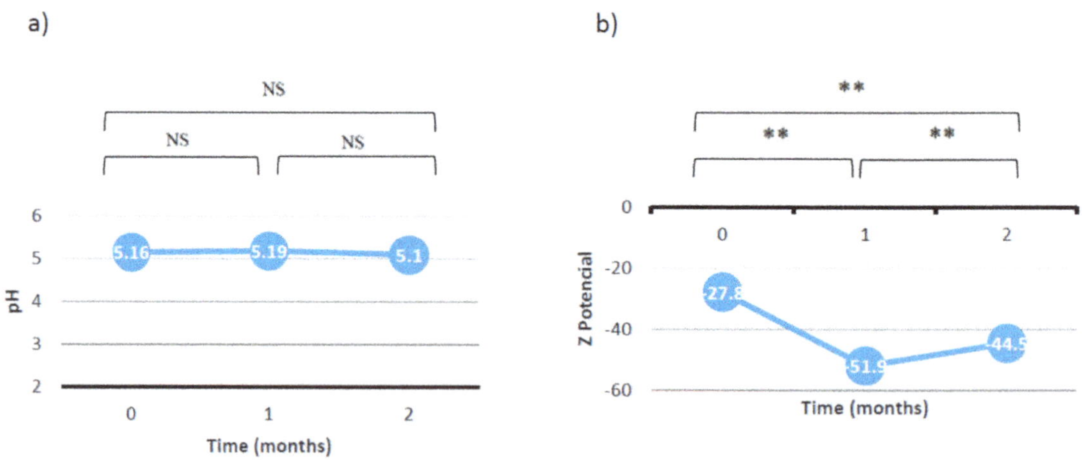

Figure 7. *Cont.*

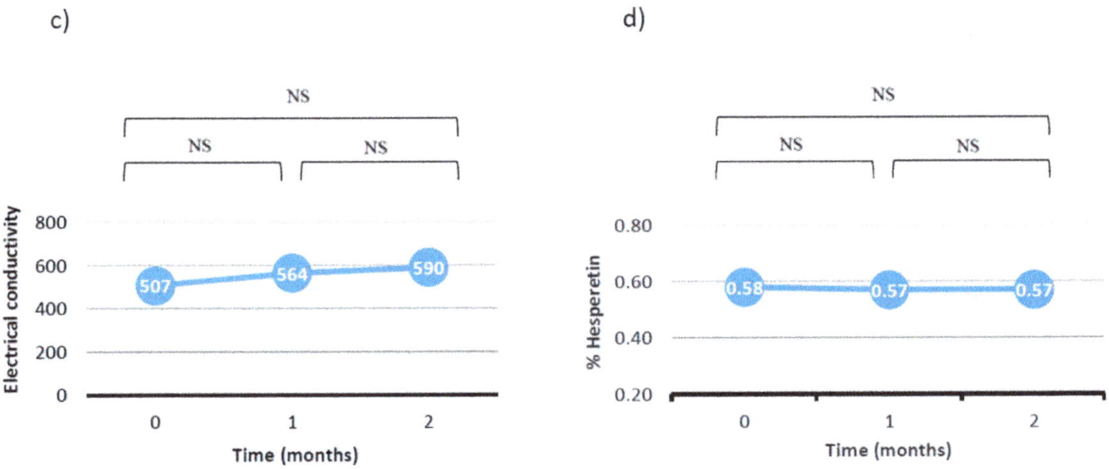

Figure 7. Results of (**a**) pH, (**b**) Z—potential, (**c**) conductivity, and (**d**) percent of hesperetin with respect to the time (NS: no significant differences; **: $p < 0.01$).

3.6. Release Studies

The release of a drug is a determining parameter in any pharmaceutical development as the therapeutic efficacy of any drug is dependent upon the release of the drug from the pharmaceutical preparation [35]. To establish the release kinetics of the active ingredient from the emulgel with hesperetin (F7), the diffusion technique in "Franz" type cells was used. In this technique, the vehicle located in the donor compartment is kept in contact with a porous support membrane. This contact allows the free diffusion of drug molecules into the receiving compartment.

The percentages of hesperetin released as a function of time from the formulation assayed are expressed in Figure 8. In total, 10% of the active ingredient was released after 24 h for a semi-solid vehicle and approximately 20% was released after 48 h.

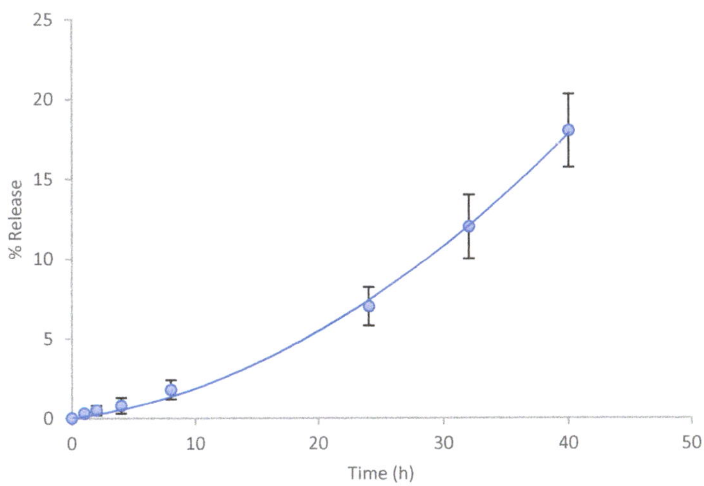

Figure 8. Percentage of hesperetin released as a function of time (h) for the emulgel ($n = 4$).

To describe the release kinetics of hesperetin, first-order, zero-order, and Higuchi models were tested and compared (Table 5).

Table 5. Release kinetics from emulgel, statistical summary of the fitted models (SS: sum of squares, AIC: Akaike Information Criterion).

	Zero-Order Kinetics	First-Order Kinetics	Square-Root Kinetics
SS	1.87	64.58	86.77
AIC	5.7S4	27.01	28.78

According to the results of the release kinetics, the data were best fitted to the zero-order model with the lower sum of squares and Akaike's criterion values (K_0 = 0.43%/h, r = 0.994). Moreover, the Kosmeyer–Peppas and Peppas–Shalin models were used to characterize and predict the mechanism of hesperetin release from the emulgel. The values of the parameters obtained (Table 6) indicate that both the Peppas–Sahlin and Korsmeyer–Peppas kinetic models can appropriately describe the release profile data. Considering that K_2 is practically equal to zero, it can be concluded that the release process is of zero order for the emulgel (n = 1.05). These results agree with those previously shown in Table 5, where the lowest sum of squares and AIC among the tested formulations was a zero-order model and were in accordance with the study of Zillich et al. [36] in which zero-order kinetic models were obtained for hydrophobic polyphenol release with a similar molecular weight to hesperetin from various cosmetic O/W emulgels. The release of actives from topical formulations depends upon several factors including gelling agents, emulsifying agents, and viscosity [37]. It is likely that the hydrophobic molecules, located apparently inside the dispersed lipid phase, had to overcome the interface O/W prior to the diffusion to the release media. Moreover, the controlled release rate in the emulgel could be explained by the fact that the hesperetin-loaded oil droplets were additionally coated by the polymeric surface. These results were in agreement with information found in the literature, which describes the release profile of combined systems [7,38].

Table 6. Parameters obtained from those adjusted using the equations indicated in the characterization of solid—vehicle formulation from release kinetics for the models used.

Diffusion Model	Parameters	Semi-Solid Vehicle
Korsmeyer–Peppas	K	0.10 ± 0.02
	n	1.06 ± 0.06
	r	0.9990
Peppas–Shalin	K_1	0.13 ± 0.08
	K_2	0.002 ± 0.004
	n	1.05 ± 0.33
	r	0.9993

3.7. Skin Permeation Studies

In order to understand the action of this new vehicle, a permeation study using the human epidermis was carried out by applying an emulgel of hesperetin or a hydroalcohol solution of hesperetin (0.5 mg/mL in both cases) on the skin.

The accumulative hesperetin amounts (%) obtained from the different formulations in the donor chamber, epidermis skin, and receptor chamber are shown in Figure 9. As can be seen, the solution provided the greatest penetration rate of hesperetin in the receptor compartment at 24 h compared to that of the hesperetin emulgel. This fact could be attributed to the formulation characteristics, like the higher oil content and viscosity [37], which are in accordance with the zero-order release kinetics previously observed with the emulgel (see Figure 8).

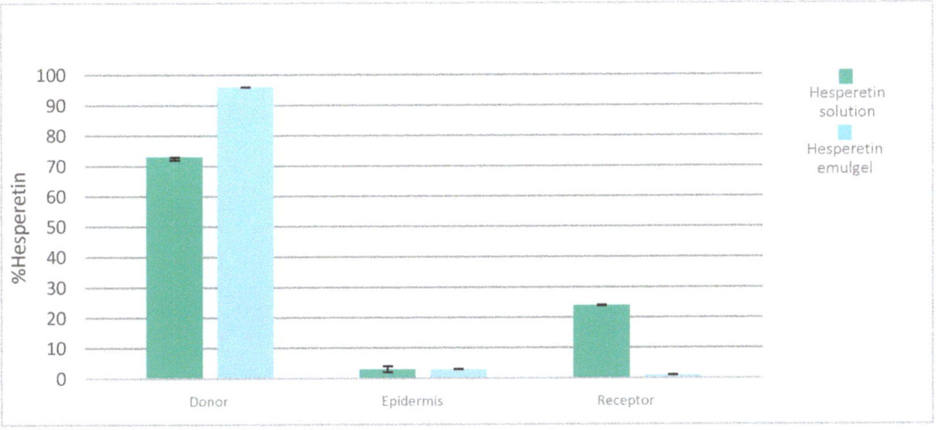

Figure 9. Distribution of hesperetin (%) after 24 h human epidermis permeation experiments after application of a solution and an emulgel, respectively.

Nevertheless, the accumulation of hesperetin in the epidermis from the solution or from the emulgel occurred in equal amounts, which would mean that the skin was saturated in hesperetin with both formulations (Figure 9).

Figure 10 shows the permeation kinetics of hesperetin (cumulative quantities as a function of epidermis surface (Q, $\mu g/cm^2$) versus time) through the human epidermis from the solution and the emulgel, respectively. In addition, the skin-vehicle partition coefficients, P, (Equation (9)), the diffusion coefficients within the epidermis, D (cm^2/h) (Equation (9)), the latency time, t_L (h) (Equation (10)), the permeability coefficient, K_p (cm/h) (Equation (11)), and the flow, J ($\mu g/cm^2 \cdot h$) (Equation (12)), were calculated for the solution and the emulgel. These values are presented in Table 7.

In accordance with Table 7, the lag time (t_L) was 2.2 h for the solution and 6.7 h for the emulgel. This indicates that the steady state (constant flow) was reached three times later for the emulgel, as expected.

Significant differences between the flux values and the amount of permeated hesperetin provided by the emulgel preparation and hesperetin solution were detected ($p < 0.01$).

The flux (=24 $\mu g/cm^2 \cdot h$) was much higher in the case of the hesperetin solution, which indicates greater penetration of the tested molecule through the epidermis and, therefore, a greater amount in the receptor compartment (Figure 9). A high flow rate is not necessarily better when the objective is for the active ingredient to reach its target layer (in our case, the epidermis). In this case, a slower flow is preferred, and the active ingredient should not reach the systemic circulation since a local action is of interest for the cosmetic effect. In this sense, the emulgel is a preparation that improves the topical application of hesperetin as a dermatological product and exerts greater control over the release and penetration of the active ingredient. (flux = 0.004 $\mu g/cm^2 \cdot h$). Moreover, the skin permeation coefficients, K_p, were approximately ten times lower than the respective release coefficients, K (Table 6). These observations indicate that the epidermis is a significant barrier against the permeation of hesperetin when the vehicle is the emulgel.

The diffusion coefficients (D, cm^2/h) did not differ between the control and the emulgel. Nevertheless, the human epidermis-vehicle (P) partition coefficient was much lower for the emulgel. This implies that the skin is saturated sooner (shorter lag time) with the solution. Therefore, the permeability coefficient of the hesperetin showed faster access to the receptor compartment, as would be expected under in vivo conditions.

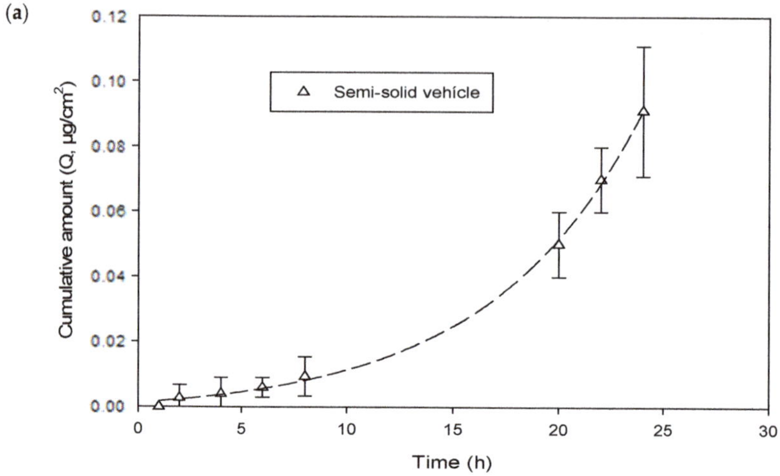

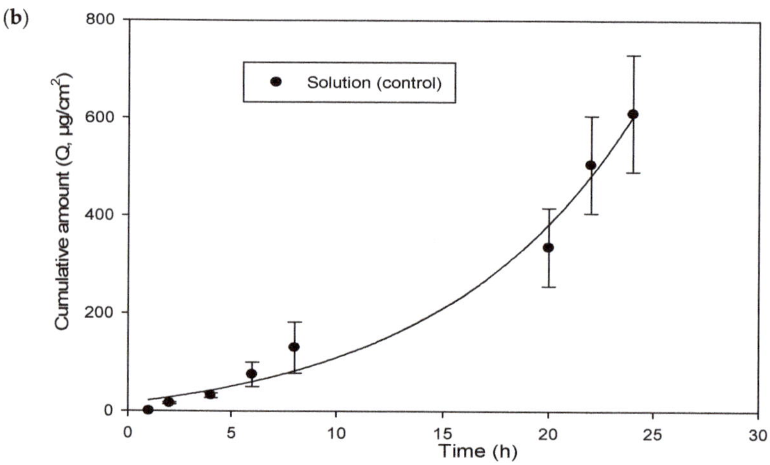

Figure 10. Cumulative amount of hesperetin across human skin. (**a**) Semi-Solid Vehicle. (**b**) Solution (Control).

Table 7. Parameters obtained for the percutaneous absorption for the solution and semi-solid vehicle (Equation (9)).

Parameter	Solution (Control)	Semi-Solid Vehicle
P	648.35 ± 367.67	0.374 ± 0.077
D (cm^2/h)	0.075 ± 0.038	0.025 ± 0.004
t_L (h)	2.2 ± 0.4	6.7 ± 0.7
$K_p \cdot 10^{-3}$ (cm/h)	48.7 ± 5.9	0.010 ± 0.002
J (µg/cm$^2 \cdot$h)	24.2 ± 2.8	0.004 ± 0.001
r>	0.98	0.96

4. Conclusions

Preformulation trials allowed the design and optimization of a formulation in the form of an aqueous external phase emulgel that was shown to have the appropriate characteristics to administer hesperetin topically. The presence of 3.5% w/w Polysorbate 80 and 1.5% w/w hyaluronic acid exhibits a stable formulation and a zero-order release of hesperetin, which contributes to achieving appropriate epidermal retention, which is desirable to develop a local action both for dermatological or cosmetical use, for example, in possible vitiligo therapy or solar photoprotection. Overall, the proposed formulation could be an alternative with great potential for the development of topical formulations based on hesperetin.

Author Contributions: Conceptualization, R.T.-V., A.N. and O.D.-S.; methodology and investigation, Y.B., R.T.-V., A.N., O.V.-B. and O.D.-S.; formal analysis, all authors; writing—original draft preparation, R.T.-V. and A.N.; writing—review and editing, R.T.-V., A.N., O.V.-B., J.V.d.J.-O. and O.D.-S.; supervision and project administration, R.T.-V. and A.N.; funding acquisition, R.T.-V., A.N, J.V.d.J.-O. and O.D.-S. All authors have read and agreed to the published version of the manuscript.

Funding: This research was funded by Bioplatino SL (OTR2021-21431INVES).

Institutional Review Board Statement: The study was conducted in accordance with the Declaration of Helsinki, and approved by the Ethical Committee of the University of Valencia (protocol code H 1540295606992 and date of approval 8 November 2018).

Informed Consent Statement: Informed consent was obtained from all subjects involved in the study.

Data Availability Statement: The data presented in this study are available on request from the corresponding author.

Conflicts of Interest: The authors declare no conflict of interest. The funder had no role in the design of the study; in the collection, analysis, or interpretation of the data; in the writing of the manuscript; or in the decision to publish the results.

References

1. Rathbone, M.J.; Hadgraft, J.; Roberts, M.S. (Eds.) *Modified-Release Drug Delivery Technology*; Boca Raton CRC Press: Boca Raton, FL, USA, 2002. [CrossRef]
2. Yu, Y.-Q.; Yang, X.; Wu, X.-F.; Fan, Y.-B. Enhancing permeation of drug molecules across the skin via delivery in nanocarriers: Novel strategies for effective transdermal applications. *Front. Bioeng. Biotechnol.* **2021**, *9*, 646554. [CrossRef] [PubMed]
3. de Araújo Andrade, T.; Heimfarth, L.; dos Santos, D.M.; dos Santos, M.R.; de Albuquerque-Júnior, R.L.; dos Santos-Neto, A.G.; de Araujo, G.R.; Lira, A.A.; Matos, S.S.; Frank, L.A.; et al. Hesperetin-based hydrogels protect the skin against UV radiation-induced damage. *AAPS PharmSciTech* **2022**, *23*, 170. [CrossRef] [PubMed]
4. Taléns-Visconti, R.; Perra, M.; Ruiz-Saurí, A.; Nácher, A. New vehiculation systems of mometasone furoate for the treatment of inflammatory skin diseases. *Pharmaceutics* **2022**, *14*, 2558. [CrossRef] [PubMed]
5. Lionberger, D.R.; Brennan, M.J. Topical nonsteroidal anti inflammatory drugs for the treatment of pain due to soft tissue injury: Diclofenac epolamine topical patch. *J. Pain Res.* **2010**, *3*, 223–233. [CrossRef] [PubMed]
6. Pednekar, A.; Dandagi, P.; Gadad, A.; Mastiholimath, V. Formulation and characterization of meloxicam loaded emulgel for topical application. *Int. J. Pharm. Pharm. Sci.* **2015**, *7*, 216–222.
7. Talat, M.; Zaman, M.; Khan, R.; Jamshaid, M.; Akhtar, M.; Mirza, A.Z. Emulgel: An Effective Drug Delivery System. *Drug Dev. Ind. Pharm.* **2021**, *47*, 1193–1199. [CrossRef] [PubMed]
8. Donthi, M.R.; Munnangi, S.R.; Krishna, K.V.; Saha, R.N.; Singhvi, G.; Dubey, S.K. Nanoemulgel: A novel nano carrier as a tool for topical drug delivery. *Pharmaceutics* **2023**, *15*, 164. [CrossRef] [PubMed]
9. Padaraju, A.; Dwivedi, F.; Kumar, G. Microemulsions, nanoemulsions and Emulgels as carriers for antifungal antibiotics. *Ther. Deliv.* **2023**, *14*, 721–740. [CrossRef] [PubMed]
10. Kolawole, O.M.; Cook, M.T. In situ gelling drug delivery systems for topical drug delivery. *Eur. J. Pharm. Biopharm.* **2023**, *184*, 36–49. [CrossRef]
11. N'Da, D. Prodrug strategies for enhancing the percutaneous absorption of drugs. *Molecules* **2014**, *19*, 20780–20807. [CrossRef]
12. Naga Sravan Kumar Varma, V.; Maheshwari, P.V.; Navya, M.; Reddy, S.C.; Shivakumar, H.G.; Gowda, D.V. Calcipotriol delivery into the skin as emulgel for effective permeation. *Saudi Pharm. J.* **2014**, *22*, 591–599. [CrossRef] [PubMed]
13. Kaur, J.; Kaur, J.; Jaiswal, S.; Gupta, G. A review on novel approach of antifungal emulgel for topical delivery in fungal infections. *Indo Am. J. Pharm. Res.* **2016**, *6*, 6321–6324.

14. Purushottam, S.S.; Bhaskarrao, G.S.; Ravindra, S. Gellified emulsion: A new born formulation for topical delivery of hydrophobic drugs. *Pharm. Tech. Med.* **2013**, *2*, 370–376.
15. Bae, J.-Y.; Seo, Y.-H.; Oh, S.-W. Antibacterial activities of polyphenols against foodborne pathogens and their application as antibacterial agents. *Food Sci. Biotechnol.* **2022**, *31*, 985–997. [CrossRef] [PubMed]
16. Ferreira de Oliveira, J.M.; Santos, C.; Fernandes, E. Therapeutic potential of Hesperidin and its Aglycone Hesperetin: Cell cycle regulation and apoptosis induction in cancer models. *Phytomedicine* **2020**, *73*, 152887. [CrossRef] [PubMed]
17. Ding, H.-W.; Huang, A.-L.; Zhang, Y.-L.; Li, B.; Huang, C.; Ma, T.; Meng, X.-M.; Li, J. Design, synthesis and biological evaluation of hesperetin derivatives as potent anti-inflammatory agent. *Fitoterapia* **2017**, *121*, 212–222. [CrossRef] [PubMed]
18. Hesperetin. Available online: https://pubchem.ncbi.nlm.nih.gov/compound/72281 (accessed on 16 February 2024).
19. Choi, S.-S.; Lee, S.-H.; Lee, K.-A. A comparative study of Hesperetin, Hesperidin and hesperidin glucoside: Antioxidant, anti-inflammatory, and antibacterial activities in vitro. *Antioxidants* **2022**, *11*, 1618. [CrossRef]
20. Jiang, S.; Wang, S.; Zhang, L.; Tian, L.; Li, L.; Liu, Z.; Dong, Q.; Lv, X.; Mu, H.; Zhang, Q.; et al. Hesperetin as an adjuvant augments protective anti-tumour immunity responses in b16f10 melanoma by stimulating cytotoxic CD8+ T cells. *Scand. J. Immunol.* **2020**, *91*, e12867. [CrossRef]
21. Khan, A.; Ikram, M.; Hahm, J.R.; Kim, M.O. Antioxidant and anti-inflammatory effects of citrus flavonoid hesperetin: Special focus on neurological disorders. *Antioxidants* **2020**, *9*, 609. [CrossRef]
22. Tan, Y.Q.; Chiu-Leung, L.C.; Lin, S.; Leung, L.K. The citrus flavonone Hesperetin attenuates the nuclear translocation of aryl hydrocarbon receptor. *Comp. Biochem. Physiol. Part C Toxicol. Pharmacol.* **2018**, *210*, 57–64. [CrossRef]
23. Wang, Y.; Liu, S.; Dong, W.; Qu, X.; Huang, C.; Yan, T.; Du, J. Combination of Hesperetin and platinum enhances anticancer effect on lung adenocarcinoma. *Biomed. Pharmacother.* **2019**, *113*, 108779. [CrossRef] [PubMed]
24. Usach, I.; Taléns-Visconti, R.; Magraner-Pardo, L.; Peris, J.-E. Hesperetin induces melanin production in adult human epidermal melanocytes. *Food Chem. Toxicol.* **2015**, *80*, 80–84. [CrossRef] [PubMed]
25. Sheen, Y.; Huang, H.; Liao, Y. The efficacy and safety of an antiaging topical serum containing hesperetin and sodium cyclic lysophosphatidic acid: A single-center clinical trial. *J. Cosmet. Dermatol.* **2021**, *20*, 3960–3967. [CrossRef] [PubMed]
26. Pereira, L.M.S.; Milan, T.M.; Tapia-Blácido, D.R. Using response surface methodology (RSM) to optimize 2G bioethanol production: A Review. *Biomass Bioenergy* **2021**, *151*, 106166. [CrossRef]
27. *STATISTICA, v 8.0*; Stat Soft Inc.: Tulsa, OK, USA, 2007.
28. Noboa, G.; Márquez, L.; López, J.C. Tamaño de gota: Factor determinante sobre la velocidad de clarificación de una emulsión o/w. *Cienc. E Ing.* **2017**, *38*, 259–264.
29. Available online: https://www.montana.edu/eal-lres/documents/Mastersizer-3000-user-manual-English-MAN0474-2-1.pdf (accessed on 16 February 2024).
30. Chilcott, R.P.; Jenner, J.; Hotchkiss, S.A.; Rice, P. In vitro skin absorption and decontamination of sulphur mustard: Comparison of human and Pig-Ear Skin. *J. Appl. Toxicol.* **2001**, *21*, 279–283. [CrossRef]
31. Ritger, P.L.; Peppas, N.A. A simple equation for description of solute release I. Fickian and non-Fickian release from nonswellable devices in the form of slabs, spheres, cylinders or discs. *J. Control. Release* **1987**, *5*, 23–36. [CrossRef]
32. Daneluz, J.; da Silva Favero, J.; dos Santos, V.; Weiss-Angeli, V.; Gomes, L.B.; Mexias, A.S. The influence of different concentrations of a natural clay material as active principle in cosmetic formulations. *Mater Res.* **2020**, *23*, e20190572. [CrossRef]
33. Zanini, M. Gel de ácido tricloroacético—Uma nova técnica para um antigo ácido. *Med. Cutánea Ibero Lat. Am.* **2007**, *35*, 14–17.
34. Alshangiti, D.M.; El-Damhougy, T.K.; Zaher, A.; Madani, M.; Mohamady Ghobashy, M. Revolutionizing biomedicine: Advancements, applications, and prospects of nanocomposite macromolecular carbohydrate-based hydrogel biomaterials: A review. *RSC Adv.* **2023**, *13*, 35251–35291. [CrossRef]
35. Khan, B.A.; Khan, A.; Khan, M.K.; Braga, V.A. Preparation and properties of high sheared poly(vinyl alcohol)/chitosan blended hydrogels films with Lawsonia inermis extract as wound dressing. *J. Drug Deliv. Sci. Technol.* **2021**, *61*, 102227. [CrossRef]
36. Zillich, O.V.; Schweiggert-Weisz, U.; Hasenkopf, K.; Eisner, P.; Kerscher, M. Release and in vitro skin permeation of polyphenols from cosmetic emulsions. *Int. J. Cosmet. Sci.* **2013**, *35*, 491–501. [CrossRef]
37. Islam, N.; Irfan, M.; Zahoor, A.F.; Iqbal, M.S.; Syed, H.; Khan, I.; Rasul, A.; Khan, S.; Alqahtani, A.M.; Ikram, M.; et al. Development of Transfersomal Oral Films. Improved bioavailability of ebastine through development of transfersomal oral films. *Pharmaceutics* **2021**, *13*, 1315. [CrossRef] [PubMed]
38. Yeo, E.; Yew Chieng, C.J.; Choudhury, H.; Pandey, M.; Gorain, B. Tocotrienols-rich naringenin nanoemulgel for the management of diabetic wound: Fabrication, characterization and comparative in vitro evaluations. *Curr. Res. Pharmacol. Drug Discov.* **2021**, *2*, 100019. [CrossRef] [PubMed]

materials

MDPI

Article

Modelling and Optimization of Machined Surface Topography in Ball-End Milling Process

Renwei Wang [1], Bin Zhao [2,*], Dingzhong Tan [1,*] and Wenjie Wan [1]

[1] College of Mechanical and Electrical Engineering, Harbin Engineering University, Harbin 150001, China; wan1771256485@163.com (W.W.)

[2] College of Power and Energy Engineering, Harbin Engineering University, Harbin 150001, China

* Correspondence: donglizhaobin@hrbeu.edu.cn (B.Z.); tandingzhong@hrbeu.edu.cn (D.T.)

Abstract: In order to optimize machined surface topography, this paper presents a novel algorithm for simulating the surface topography and predicting the surface roughness of a ball-end milling process. First, a discrete workpiece model was developed using the Z-map method, and the swept surface of a cutter edge was represented using triangular approximation. The workpiece surface was updated (i.e., material removal process) using the intersection between the vertical reference line and the triangular facet under a cutting judgement. Second, the proposed algorithm was verified by comparing the simulated 3D surface topography as well as 2D surface profile and average roughness (*Sa*) with experimental measurements. Then, numerical simulation examples planed by the Box–Behnken design methods were carried out to investigate the *Sa* in the ball-end milling operation. The correlations of S_a and cutting parameters were represented by a response surface reduced quadratic model based on the ANOVA results. Finally, the feed per tooth, radial depth of cut, and tilt and lead angles were optimized for improving the machining efficiency under the *Sa* constraints. This study presents an effective method for simulating surface topography and predicting the *Sa* to optimize the cutting parameters during ball-end milling process.

Keywords: surface topography and roughness; ball-end milling; triangular approximation; Z-map model; optimal cutting parameters

1. Introduction

Surface topography has a clear effect on the performance of components, such as their contact, friction, wear, and lubrication properties [1,2]. Surface topography is characterized by the surface roughness quantity for evaluating surface quality. Ball-end milling is one of the most widely used machining operations in various industries, such as aerospace, automobile, bioengineering and molding, as a semi-finish and finish method to replace grinding or polishing for improving the machining efficiency [3]. Therefore, the surface finish and dimensional accuracy of the final part are mainly determined by the ball-end milling operation in the finishing stage. According, the surface topography is easily influenced by many factors (cutting parameters) via ball-end milling. Chen et al. [4] investigated the effects of cutter inclination angle on the surface topography geometrical features using cutting experiments. Zhang et al. [5] investigated the ratio and product of the feed per tooth and radial depth of cut using an analytical model for reducing the surface roughness at a constant material removal rate. Peng et al. [6] investigated the effects of the cutter initial phase angle on the surface topography in micro-ball-end milling, and the surface texture was controlled by planning the noncutting tool path. Zhang et al. [7] used a modified cutting edge to develop a surface topography model considering the tool flank wear effect. Zhang et al. [8] considered stochastic tool wear for simulating the surface topography in micro-milling. Lotfi et al. [9] developed a modified cutter/workpiece engagement algorithm for simulating the surface model in five-axis ball-end milling considering the cutter deflection and cutter runout. Hao et al. [10] introduced a correction model of ball-end milling by wear and force deformation for simulating the surface topography. Omar et al. [11] proposed an integral model for simulating the surface topography and predicting the cutting force simultaneously, and investigated the effects of cutter axis tilt, cutter runout, tool deflection, systems dynamics, and face wear for the surface topography in the ball-end milling. Wang et al. [12] investigated the effects of cutting vibration on the surface topography in ball-end milling of thin-walled parts using a numerical method. Li et al. [13] investigate the geometrical errors imposed by the machine tool on the surface topography in five-axis ball-end milling processes, and evaluated quantity by means of modal coefficients. Biondani et al. [14] developed a surface topography model for investigating the effect of cutter edge micro-geometry on the surface topography and roughness based on an actual cutting-edge model and a judgment of minimum chip thickness. Therefore, numerical simulation has more applications compared with milling experiments for investigating the surface topography in ball-end milling operation. The surface topography simulation involves lower costs and is less time-consuming.

The surface topography simulation methods can be classed into three groups: solid modeling, iterative methods, and discretization. In the solid modeling methods [15], the material removal process is simulated by the Boolean operations between the cutter swept volume and workpiece. The iterative method was first proposed by Zhang et al. [16]. Mathematically, it is like solving the intersection of the swept surface of the cutter edge and the reference line at a selected workpiece point by solving the nonlinear equation using iterative methods. In the discretization methods, first, the cutter is dispersed into a series of units. The workpiece is constructed by means of the Z-map, N-buffer, Dexel and Voxel elements, cut plane, points set, etc. Then, machining time (or tool path) is dispersed, and at each discrete time point (or discrete cutter location point), the surface topography value (i.e., coordinate of workpiece surface) can be determined by the cutter edge unit intersecting with the workpiece. According to related research works, these are the main problems with the solid modeling as well as iterative methods and discretization. Indeed, in the solid modeling techniques, the computation of the intersection between geometrical objects is difficult and time-consuming. In iterative methods, a good initial value (also called starting point) must be determined first, while the determination of the initial value is not an easy task due to the nonlinearity of the equation of the swept surface and multiple intersections with the reference lines. In discretization methods, the computation efficiency and accuracy are low because of the plentiful discretization for the cutter and workpiece. However, the discretization method is still widely employed for modeling as a result of its extensibility and robustness with respect to solid modeling and iterative methods. Moreover, some study works have been performed to improve the accuracy and efficiency of the discretization method. Li et al. [17] introduced iterative methods to improve the discrete methods. In other words, they proposed a new initial value selection algorithm using discretization methods for the iterative methods. Xu et al. [18], Yang et al. [19], and Chen et al. [20] improved the discrete method using different interpolation methods. Therefore, the discretization method was adopted to develop the surface topography model. Meanwhile, an efficient algorithm for calculating the intersection point of the swept surface and workpiece was proposed, which is the key technology for simulating surface topography successfully.

The optimization of cutting parameters is simple and effective in machining process for obtaining the desired surface topography and roughness. Dikshit et al. [21,22] optimized the cutting parameters for minimizing the surface roughness using the response surface methodology in the ball-end milling of AL2014-T6. Zhou et al. [23] developed a modified grey relational analysis model for optimizing the surface roughness and residual stress in the ball-end milling of Inconel 718. Buj-Corral et al. [24] developed an artificial neural network for correlating surface roughness with cutting parameters to obtain the optimal machining strategy in the five-axis ball-end milling of W-Nr. 12344. Venkata-Rao et al. [25] considered the surface roughness as a constraint for reducing power consumption in the micro-ball-end milling of D2 steel. Sonawane et al. [26] used a response surface methodology for optimizing the cutting parameters and obtaining the minimum surface roughness in the high-speed ball-end milling of Inconel-718 thin cantilevers. Torres et al. [27] proposed a structured numerical simulation using several friction coefficient approximations, and adapted it to a pre-textured martensitic precipitation hardening stainless-steel. Cross et al. [28] investigated the surface topography of material measures using confocal microscope and atomic force microscope measurements, and it was found that the surface topography depends primarily on the spindle speed and much less on the feed per tooth, whereas differences between the milling tools are observed in the tendency to form substructures. Klauer et al. [29] found that the ploughing of the material instead of a cutting occurs when too-small tilt angles are applied, and investigated the tilt angle's influence on the resulting roughness and machined geometry using micro-milling sinusoidal freeform surfaces at different tilt angles. Arruda et al. [30] optimized the surface roughness using the Taguchi method in the ball-end milling of AISI P20. Despite all this, there is a lack of research on improving the machining efficiency when surface roughness is applied as a constraint.

The objective of this study is to present a surface topography simulation algorithm for replacing a milling experiment to study the surface roughness during ball-end milling processes. As mentioned earlier, in order to model the surface topography, the intersection point between the swept surface of the cutter edge and the workpiece must be solved first. Hence, in the current study, an efficient numerical algorithm was proposed for calculating the intersection point between the swept surface and workpiece using an improved discretization method. To this end, first, the swept surface of the cutter edge was formulated by means of the homogeneous coordinate transformation and the kinematics of the ball-end milling cutter. Then, a discrete workpiece model was developed using the Z-map method, and the swept surface was represented using a triangular approximation. Finally, the intersection is calculated from the vertical reference lines and triangular facet. Once the intersection is solved, the mesh point of the workpiece surface can be updated by the intersection point with a smaller z-coordinate. The main contribution of the current study pertains to the central part of the proposed model, i.e., an algorithm for calculating the intersection point of the swept surface and the workpiece for updating the workpiece surface.

Henceforth, the paper is organized as follows. In Section 2, the surface topography simulation algorithm is presented in detail. In Section 3, milling experiments are carried out to valid the proposed algorithm. A response surface model has been developed for improving the machining efficiency under the average roughness (*Sa*) constraint. Finally, the conclusions of the work are summarized in Section 4.

2. Modeling of Machined Surface Topography

In order to obtain the equation of the swept surface of a cutter edge, the coordinate systems involved in the ball-end milling of flat surface processes are established as shown in Figure 1.

Machine tool coordinate system, $O_m - X_m Y_m Z_m = \{O_m; e_1^M, e_2^M, e_3^M\}$: the global coordinate system, O_m, is placed at the origin of the machine tool, and e_1^M, e_2^M and e_3^M can be determined by the kinematic structure of the machine tool.

Cutter coordinate system, $O_c - X_c Y_c Z_c = \{O_c; e_1^C, e_2^C, e_3^C\}$: the local coordinate system is fixed on the ball-end milling cutter. O_c is placed at the ball center and defined as the cutter location (CL) point. e_3^C is along the cutter axis. e_1^C is along the tangent of the cutter edge at the cutter tip. e_2^C is the cross product of e_3^C and the e_1^C, $e_2^C = e_3^C \times e_1^C$. The cutter revolves around the spindle axis at an angular speed ω, and the cutter axis and the spindle axis show eccentricity, e.

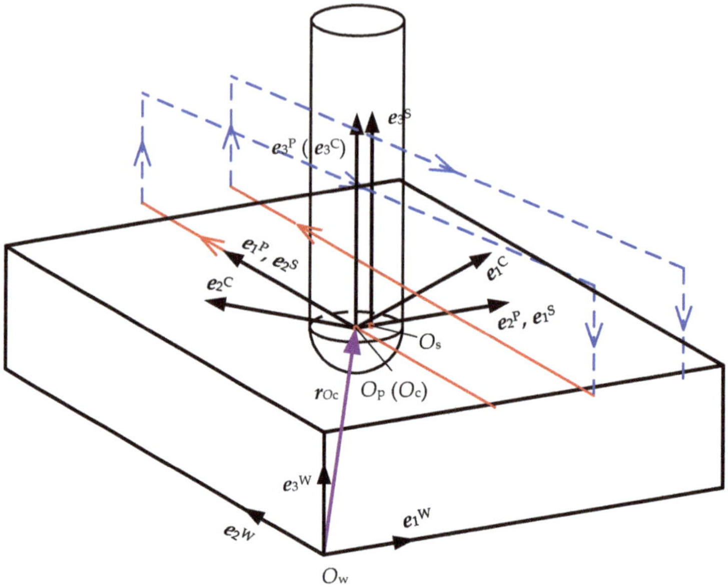

Figure 1. Definition of coordinate systems in ball-end milling of flat surface.

Spindle coordinate system, $O_s - X_sY_sZ_s = \{O_s; \boldsymbol{e}_1{}^S, \boldsymbol{e}_2{}^S, \boldsymbol{e}_3{}^S\}$: local coordinate system attached to the spindle. O_s is placed at the intersection between the axis of the spindle and the X_cY_c plane. $\boldsymbol{e}_3{}^S$ is placed along the axis of the spindle, which is parallel to the $\boldsymbol{e}_3{}^C$. The spindle translates along the cutter path at $\boldsymbol{r}_{Oc} = (x_{Oc}, y_{Oc}, z_{Oc})$. When the tilt and lead angles are equal to zero, $e_1{}^S$ is parallel to $\boldsymbol{e}_1{}^M$ and $\boldsymbol{e}_2{}^S$ is parallel to $\boldsymbol{e}_2{}^M$, while $\boldsymbol{e}_3{}^S$ is parallel to $e_3{}^M$.

Process coordinate system, $O_p - X_pY_pZ_p = \{O_P; \boldsymbol{e}_1{}^P, \boldsymbol{e}_2{}^P, \boldsymbol{e}_3{}^P\}$: the local coordinate system is placed at the cutter path. O_p is placed at the cutter path and consistent with O_s. $e_1{}^P$ is along the feed direction and assumed to be $\boldsymbol{e}_2{}^W$, $\boldsymbol{e}_3{}^P$ is along the normal direction of the cutter center location surface and is parallel to $\boldsymbol{e}_3{}^M$, $\boldsymbol{e}_2{}^P$ is along the cross-feed direction and is also the cross product of $\boldsymbol{e}_3{}^P$ and $\boldsymbol{e}_1{}^P$, $\boldsymbol{e}_2{}^P = \boldsymbol{e}_3{}^P \times \boldsymbol{e}_1{}^P$.

Workpiece coordinate system, $O_w - X_wY_wZ_w = \{O_w; \boldsymbol{e}_1{}^W, \boldsymbol{e}_2{}^W, \boldsymbol{e}_3{}^W\}$: the global coordinate system is attached to the workpiece, in which the cutter path and surface topography are described. O_w is placed at an arbitrarily selected point on the workpiece and the directions of $\boldsymbol{e}_1{}^W$, $\boldsymbol{e}_2{}^W$, and $\boldsymbol{e}_3{}^W$ are assigned to be consistent with the $O_m - X_mY_mZ_m\boldsymbol{e}_1{}^M$, $\boldsymbol{e}_2{}^M$ and $\boldsymbol{e}_3{}^M$, respectively.

The transformation of a selected point from $O_c - X_cY_cZ_c$ to $O_s - X_sY_sZ_s$ (see Figure 2) can be performed by a rotation around $\boldsymbol{e}_3{}^C$ and a translation along the O_sO_c. The transformation matrices $\boldsymbol{M}^{CS}$ can be expressed as

$$\boldsymbol{M}^{CS} = \begin{bmatrix} \cos(-\omega \cdot t + \mu_0) & -\sin(-\omega \cdot t + \mu_0) & 0 & e \cdot \cos(-\omega \cdot t + v_0) \\ \sin(-\omega \cdot t + \mu_0) & \cos(-\omega \cdot t + \mu_0) & 0 & e \cdot \sin(-\omega \cdot t + v_0) \\ 0 & 0 & 1 & 0 \\ 0 & 0 & 0 & 1 \end{bmatrix}, \quad (1)$$

where $\omega = 2\pi n/60$ (rad/s) is the angular speed of the cutter and n (r/min) is the rotational speed of the spindle, t is the machining time, eccentricity e is the distance from O_c and O_s, μ_0 is the initial angle of $\boldsymbol{e}_1{}^C$ and $\boldsymbol{e}_1{}^S$, and v_0 is the initial phase angle of the cutter runout.

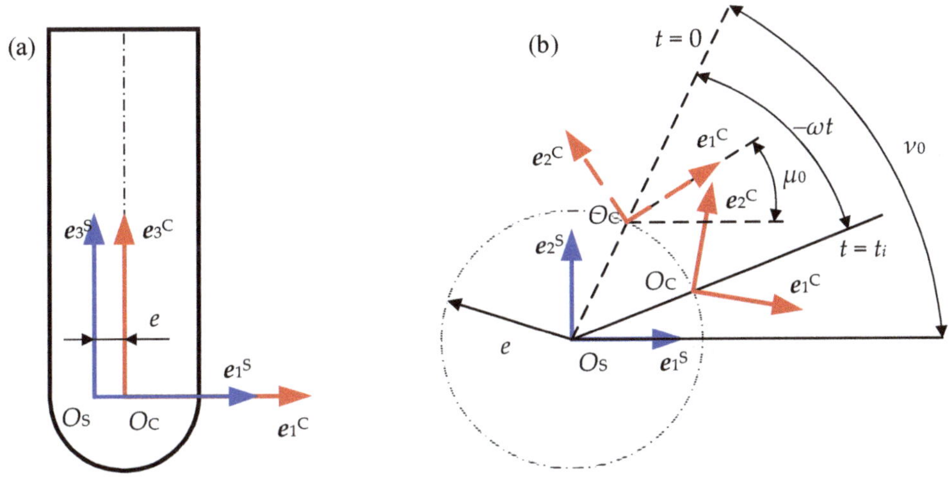

Figure 2. Definition of cutter runout: (**a**) front view; (**b**) top view.

For a type-AB five-axis machining center, we must keep the cutter location point r_{Oc} in the proper position and the cutter axis in proper orientation. The spindle translates along the cutter path, and rotates around $e_1{}^M$ at β_2, and rotates around $e_2{}^M$ at β_1 relative to the body of the machine tool (see Figure 3). Therefore, the transformation matrix M^{SW} from $O_s - X_sY_sZ_s$ to $O_w - X_wY_wZ_w$ is defined as follows:

$$M^{SW} = \begin{bmatrix} \cos\beta_1 & \sin\beta_1\cdot\sin\beta_2 & -\sin\beta_1\cdot\cos\beta_2 & x_{Oc} \\ 0 & \cos\beta_2 & \sin\beta_2 & y_{Oc} \\ \sin\beta_1 & -\cos\beta_1\cdot\sin\beta_2 & \cos\beta_1\cdot\cos\beta_2 & z_{Oc} \\ 0 & 0 & 0 & 1 \end{bmatrix}, \tag{2}$$

where (x_{Oc}, y_{Oc}, z_{Oc}) is the coordinate of in $O_w - X_wY_wZ_w$, β_1 is the tilt angle, and β_2 is the lead angle.

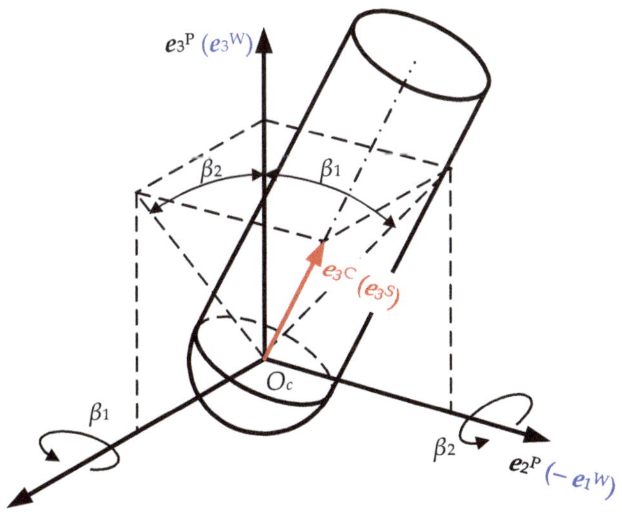

Figure 3. Definition of lead and tilt angles.

As shown in Figure 4, with the parameter of the axial position angle θ of a selected point p along the cutter edge, in a cutter coordinate system, $O_c - X_cY_cZ_c$, the spherical cutter edge can be expressed as:

$$\begin{bmatrix} x_p{}^C \\ y_p{}^C \\ z_p{}^C \end{bmatrix} = \begin{bmatrix} R_0 \cdot \cos\theta \cdot \cos\varphi \\ R_0 \cdot \cos\theta \cdot \sin\varphi \\ -R_0 \cdot \cos\theta \end{bmatrix}, \quad (0 \le \theta \le \tan\beta_0), \tag{3}$$

where $(x_p{}^C, y_p{}^C, z_p{}^C)$ is the coordinate of the point P in $O_c - X_cY_cZ_c$. R_0 is the radius of the cutter. For a spherical cutter edge with equal lead, the lag angle φ can be calculated as follows:

$$\varphi = \tan\beta_0(1 - \cos\theta) + (j-1)2\pi/N_c, \tag{4}$$

where β_0 is the nominal helix angle as measured in the connection of the spherical and cylindrical cutter edges, j is the index of the cutter edge, and N_c is the number of cutter edges.

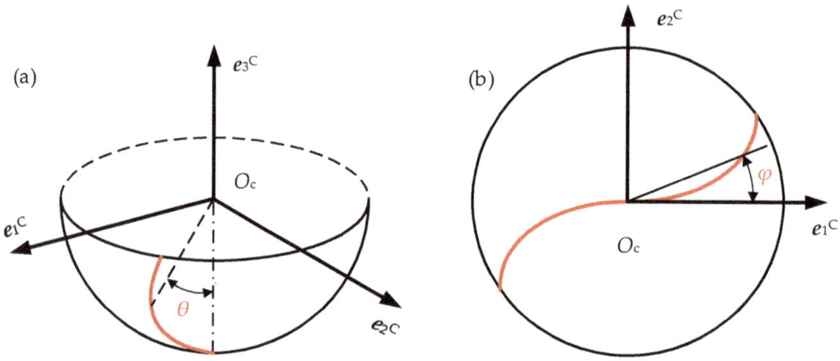

Figure 4. Definition of cutting edge model: (**a**) general view; (**b**) top view.

Given the cutter geometry as well as the cutter position and the cutter orientation, the swept surface of the cutter edge can be calculated in terms of homogeneous coordinate transformation. Mathematically, it is expressed as follows:

$$\begin{bmatrix} x_p{}^W \\ y_p{}^W \\ z_p{}^W \\ 1 \end{bmatrix} = \boldsymbol{M}^{SW}\boldsymbol{M}^{CS} \begin{bmatrix} x_p{}^C \\ y_p{}^C \\ z_p{}^C \\ 1 \end{bmatrix}, \tag{5}$$

where $(x_p{}^W, y_p{}^W, z_p{}^W)$ is the coordinate of a selected point P on the swept surface in O_W-$X_WY_WZ_W$.

The workpiece surface is updated (i.e., material removal process) based on a ray-triangle algorithm. First, the workpiece surface is dispersed into a series of discrete points using Z-map methods, as shown in Figure 5. Second, the swept surface of the cutter edge is dispersed into a series of triangle facets, as shown in Figure 6. Then, we update the discrete points of the workpiece surface using the intersection point between the vertical line of the workpiece surface and the triangular facet of the swept surface. Finally, we draw the surface topography as well as the surface profile, and calculate the surface roughness according to the coordinate of the discrete point of the workpiece surface. A sketched map of the surface topography simulation algorithm is shown in Figure 7.

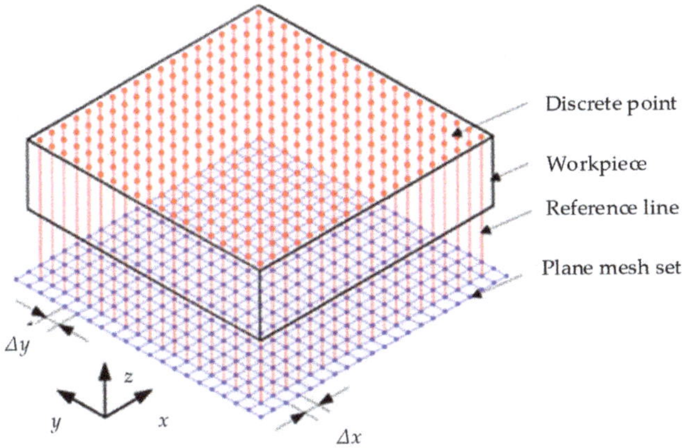

Figure 5. Z-map model of the workpiece.

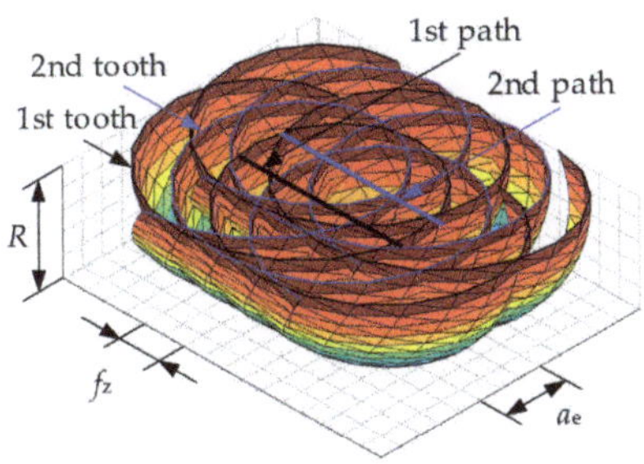

Figure 6. Triangular approximation of the swept surface.

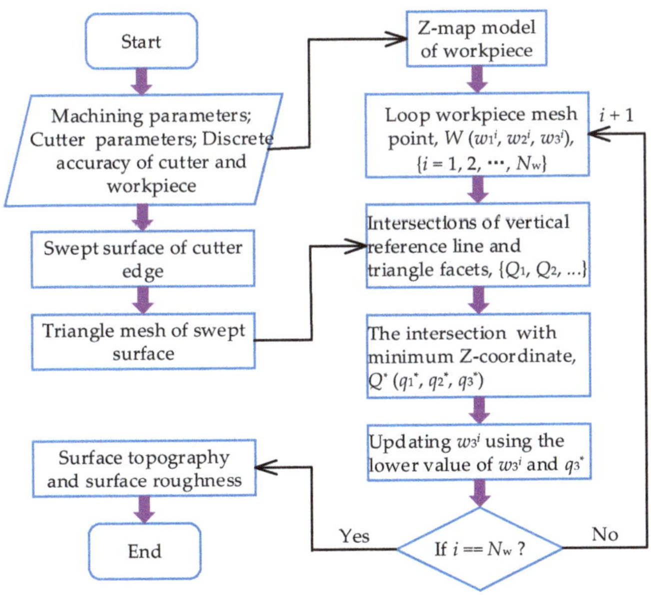

Figure 7. Sketched map of surface topography simulation algorithm.

3. Results and Discussion

3.1. Experimental Verification

In this study, the typical hot work die steel, AISI P20/3Cr2Mo (33 ± 3 HRC), combines a very good machinability with mirror polishing performance, and covers a wide variety of applications; it was chosen as the workpiece material here. The nominal chemical compositions and mechanical properties of the workpiece material are listed in Tables 1 and 2, respectively. The workpiece samples were prepared as rectangular blocks (250 mm × 150 mm × 40 mm). Four samples were prepared. The top surfaces of each workpiece with sizes of 250 mm × 150 mm were divided into eight areas. Before milling experiments, the blocks were face-milled on the top and bottom surfaces to remove the heat treatment-related surface defects and flatness to eliminate errors in the experimental results. The cutter utilized in milling experiments was an integral carbide ball-end mill with two teeth (JH970100-Tribon, Seco Tools Company®, Troy, MI, USA) with a diameter of 10 mm and a 30° helix angle.

Table 1. Nominal chemical composition of AISI P20 steel (wt.%).

C	Si	Mn	Cr	Mo	Ni	Fe
0.28~0.40	0.20~0.80	0.60~1.00	1.40~2.00	0.30~0.55	0.05~0.10	Bal

Table 2. Mechanical properties of AISI P20 steel at room temperature.

Density (kg/m^3)	Young's Modulus (GPa)	Hardness (HRC)	Yield Strength (MPa)	Thermal Conductivity (W/m·K)
7800	207	30~36	1140	29.0

The flat surface milling tests were carried out on a five-axis computer numerical controlled vertical machining center (DMU 60P duo BLOCK, DMG®, DMP, Frankfurt, Germany) in a dry cutting environment with air-cooling. The milled samples were degreased by ultrasonic cleaning for 15 min in acetone and then rinsed with deionized water. The surface topography, as well as surface profile and *Sa*, of the machined surface was measured using an interferometer (Wyko NT9300, Bruker Corporation®, Billerica, MA, USA), and the vertical resolution was 0.1 nm. A rectangular region (1 mm × 1 mm) was chosen in order to compared and analyze the characteristics of surface topography during the ball-end milling process. In order to reduce accidental errors in the measurement, three different regions of the machined surface topography were measured under the same machining parameters. According to the engineering and cutting parameters manual, the value range of feed per tooth is 0~1 mm, the value range of the radial depth of cut is 0~1 mm, the value range of the axial depth of cut is 0~1 mm, the value of the spindle speed is 3000~10,000 r/min, and the lead and tilt angles are 0~45°. In order to deeply understand the influences of changes in cutting parameters on the five-axis ball-end milling process, five values of each factor have been selected according to the actual engineering and cutting parameters manual, as shown in [31]. Three arbitrary experiments under different cutting parameters (see Table 3) were selected to validate the developed surface topography model.

Table 3. Machining parameters of ball-end milling experiments.

No.	Spindle Speed, n (r/min)	Feed per Tooth, f_z (mm/tooth)	Radial Depth of Cut, a_e (mm)	Axial Depth of Cut, a_p (mm)	Tilt Angle, β_1 (°)	Lead Angle, β_2 (°)
1#	5000	0.36	0.25	0.3	−8	4
2#	6500	0.36	0.3	0.15	−12	0
3#	6500	0.56	0.2	0.1	−16	4

The discrete precision of the workpiece was set to 0.01 mm, which is the less than the minimum value of $0.1f_z$ and $0.1a_e$. The discrete precision of the swept surface is that the angle interval $\Delta\theta = 0.2°$ and Δt is the time it takes the cutter to rotate 2°.

The predicted surface 3D topography and 2D surface profiles were compared with the experimentally measured results from the second trial, as shown in Figures 8 and 9. The results show that the experiments and simulations were very similar to each other, both in the numbers of peaks and valleys and their distribution. There is a certain difference between the simulated and experimental results, such as the fact that the surface topography and profile were smooth in the simulations, while they were rough in the experiments. The main difference between the simulated and experimental results is in the sliding; the abrasion and elastic strain of the material sliding were not considered in the simulation.

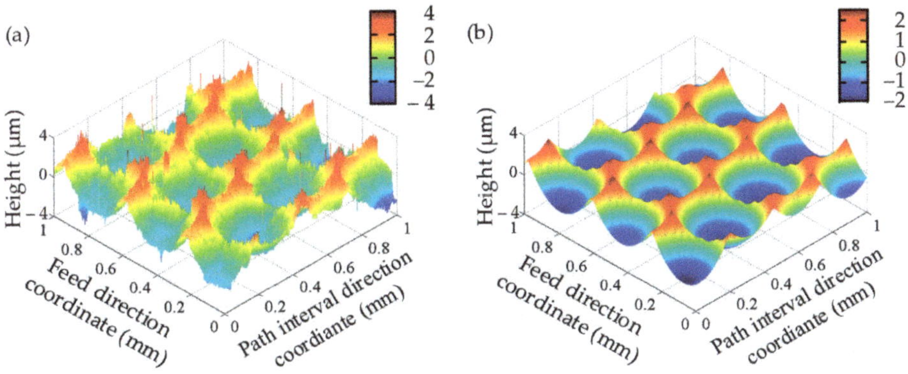

Figure 8. 3D surface topography: (**a**) experimental; (**b**) simulated.

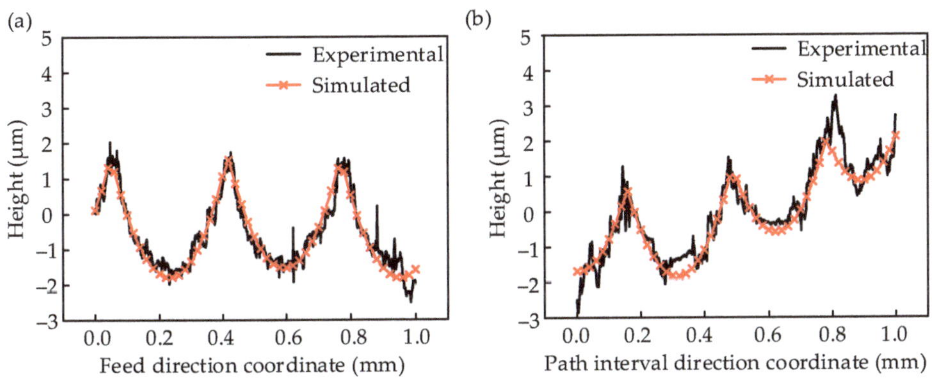

Figure 9. 2D surface profile: (**a**) experimental; (**b**) simulated.

According to the simulated and experimental surface topography results, the average roughness, *Sa*, has been calculated and used to evaluate the developed algorithm quantitatively. The *Sa* of trials 1#, 2# and 3# are listed in Table 4. The derivation between the simulated and experimental results with a maximum deviation is 13.48%, which shows that the developed model has a high prediction accuracy when predicting the surface roughness. The standard deviations are all less 0.1, which indicates that the measurement results are relatively concentrated, and the values of dispersion are small.

The verified results from Figures 8 and 9 and Table 4 indicate that the surface topography simulation algorithm developed for the ball-end milling of AISI P20 is reliable.

Therefore, this surface simulation algorithm can accurately simulate the surface topography and roughness, and thus replace milling experiments.

Table 4. Experimental and simulated roughness.

No.	Experimental *Sa* (μm)		Predicted *Sa* (μm)	Relative Deviation (%)
	Average Value	Standard Deviation		
1#	0.8158	0.0656	0.8227	0.85
2#	0.9289	0.0871	0.9087	−2.17
3#	1.8723	0.0985	2.1246	13.48

3.2. Cutting Parameters Optimization

The *Sa* is a basic requirement of the machining product's quality, and the surface quality is classified into several grades using the value of *Sa*. Therefore, the *Sa* has been taken as a response parameter of the milled surface, which is calculated with the help of simulated surface topography. The feed per tooth (f_z), radial depth of cut (a_e), tilt (β_1) and lead (β_2) angle are selected as the numeric factors because they all have clear effects on the surface topography and can be changed by adjusting the feed rate, tool path and tool orientation. The ranges of the level of each numeric factor were set to feed per tooth—0.1~0.5 mm/tooth, radial depth of cut—0.1~0.5 mm, tilt angle—−8~8° and lead angle—−8~8° for the response analysis using the Box–Behnken design. The designs of the surface topography simulation examples and the simulated *Sa* values are presented in Table 5.

Table 5. Box–Behnken design and simulated results of average roughness.

Std	Feed per Tooth, f_z (mm)	Radial Depth of Cut, a_e (mm)	Tilt Angle, β_1 (°)	Lead Angle, β_2 (°)	Simulated Roughness, *Sa* (μm)
1	0.1	0.1	4	4	0.0937
2	0.5	0.1	4	4	2.7981
3	0.1	0.5	4	4	1.6139
4	0.5	0.5	4	4	2.9981
5	0.3	0.3	0	0	1.644
6	0.3	0.3	8	0	0.8779
7	0.3	0.3	0	8	0.8779
8	0.3	0.3	8	8	0.8536
9	0.1	0.3	4	0	0.5844
10	0.5	0.3	4	0	3.2895
11	0.1	0.3	4	8	0.5847
12	0.5	0.3	4	8	2.2598
13	0.3	0.1	0	4	0.9095
14	0.3	0.5	0	4	1.6984
15	0.3	0.1	8	4	0.6976
16	0.3	0.5	8	4	1.6937
17	0.1	0.3	0	4	0.5844
18	0.5	0.3	0	4	3.2895
19	0.1	0.3	8	4	0.5847
20	0.5	0.3	8	4	2.2598
21	0.3	0.1	4	0	0.9095
22	0.3	0.5	4	0	1.6984
23	0.3	0.1	4	8	0.6976
24	0.3	0.5	4	8	1.6937
25	0.3	0.3	4	4	0.925
26	0.3	0.3	4	4	0.925
27	0.3	0.3	4	4	0.925
28	0.3	0.3	4	4	0.925
29	0.3	0.3	4	4	0.925

The results in Table 6 show that the model's *F*-value of 145.73 implies the model is significant. There is only a 0.01% chance that a model *F*-value this large could occur due noise. Values of "Prob>*F*" less than 0.05 indicate that the model terms are significant. In this case, f_z, a_e, β_1, β_2, $f_z \times a_e$, $f_z \times \beta_1$, $f_z \times \beta_2$, ${f_z}^2$ and ${a_e}^2$ are significant model terms. In contrast, values greater than 0.10 indicate the model terms are not significant; in this case, $a_e \times \beta_1$, $a_e \times \beta_2$, ${\beta_1}^2$ and ${\beta_2}^2$ are not significant model terms, and are reduced to simplify and improve the regression model, as shown in Equation (6). As shown in Figure 10, the comparison of predicted and simulated *Sa* indicates the high accuracy of the response surface model.

$$\begin{aligned} Sa = & -0.2235 + 0.3955 f_z + 1.118 a_e + 0.0078(\beta_1 + \beta_2) - 8.2513 f_z \cdot a_e \\ & -0.3219 f_z(\beta_1 + \beta_2) + 0.0116 \beta_1 \cdot \beta_2 + 16.681 f_z^2 + 5.936 a_e^2 \end{aligned} \quad (6)$$

Table 6. ANOVA for response surface quadratic model.

Source	Sum of Squares	df	Mean Square	*F* Value	*p*-Value (Prob>*F*)
Model	21.11	10	2.11	145.73	<0.0001
f_z	13.76	1	13.76	949.88	<0.0001
a_e	2.33	1	2.33	161.02	<0.0001
β_1	0.35	1	0.35	23.86	0.0001
β_2	0.35	1	0.35	23.86	0.0001
$f_z \times a_e$	0.44	1	0.44	30.08	<0.0001
$f_z \times \beta_1$	0.27	1	0.27	18.31	0.0005
$f_z \times \beta_2$	0.27	1	0.27	18.31	0.0005
$a_e \times \beta_1$	0.011	1	0.011	0.79	0.3883
$a_e \times \beta_2$	0.011	1	0.011	0.79	0.3883
$\beta_1 \times \beta_2$	0.14	1	0.14	9.50	0.0064
${f_z}^2$	3.07	1	3.07	212.15	<0.0001
${a_e}^2$	0.39	1	0.39	26.87	<0.0001
${\beta_1}^2$	0.03	1	0.03	2.18	0.1615
${\beta_2}^2$	0.03	1	0.03	2.18	0.1615
Residual	0.26	14	0.014		
Lack of fit	0.26	14	0.019		
Pure error	0.000	4	0.000		
Cor Total	21.37	28			

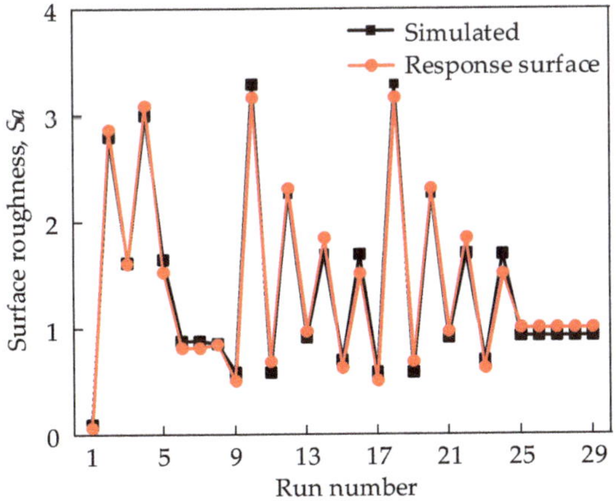

Figure 10. Comparison of average roughness between surface topography simulation and response surface models.

As shown in Figure 10, the predicted *Sa* using the response surface model was consistent with the simulated *Sa* using the proposed surface topography model, which indicates the high accuracy of the response surface model. The Studentized residuals are in the range of ±3, which is consistent with the results of the test. In addition, no abnormal data points were found in the residual plot (see Figure 11). The effects of machining parameters on the *Sa* are shown in Figure 12. The *Sa* increases with the increase in feed per tooth (f_z) and radial depth of cut (a_e) (see Figure 12a–c). The *Sa* decreases with the increase in tilt angle (β_1) and lead angle (β_2) (see Figure 12b–d). The feed per tooth (f_z) and radial depth of cut (a_e) have significant effects on material residual height, and the material residual height increases with the feed per tooth (f_z) and radial depth of cut (a_e). The tilt angle (β_1) and lead angle (β_2) can improve the interaction between the tool and workpiece, and reduce the *Sa*.

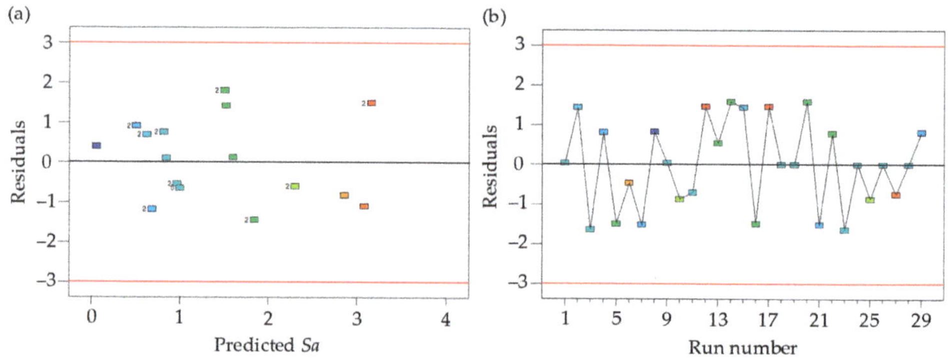

Figure 11. Residuals of response surface quadratic model: (**a**) residuals to predict surface roughness; (**b**) residuals to run number. The numbers represents the amount of similar results, and the colors make it easier to identify corresponding data points in (**a**,**b**).

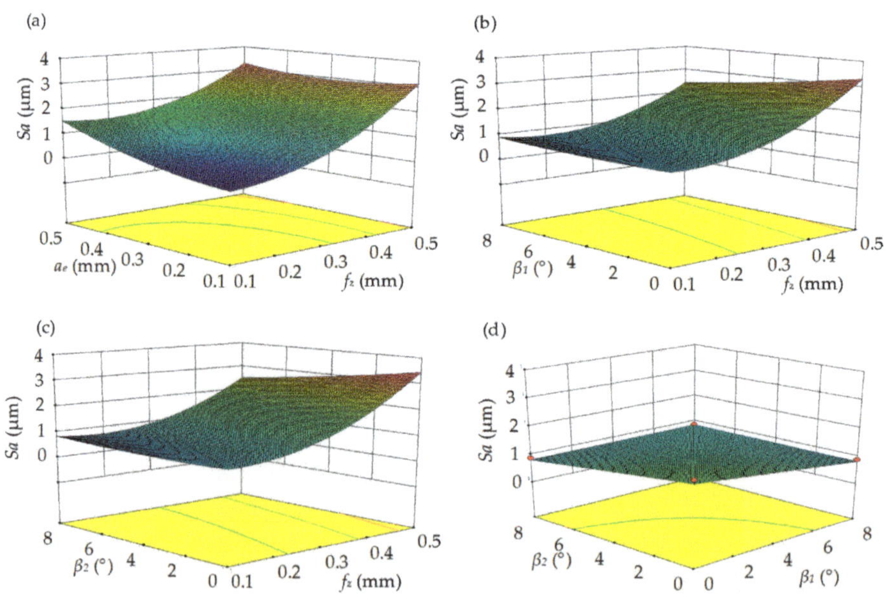

Figure 12. Interaction effect of machining parameters on average roughness: (**a**) feed per tooth and radial depth of cut; (**b**) feed per tooth and tilt angle; (**c**) feed per tooth and lead angle; (**d**) tilt and lead angles. Colors represent the design points above predicted value.

The response surface model's accuracy has here been investigated. Next, the ability of the model to improve the machining efficiency by selecting the cutting parameters is analyzed (see Equation (7)). For this purpose, an optimal cutting parameter with the maximum material removal rate is obtained with an *Sa* of less than 0.7 by solving the response surface model; the optimal cutting parameters are f_z = 0.2 mm/tooth, a_e = 0.34 mm, β_1 = 7.23° and β_2 = 2.54°. The predicted *Sa* is 0.6841 μm, in contrast to the simulated *Sa*, which is 0.7791 μm. The devotion is −12.19%, which denotes the regression model is effective. In contrast to 20#, 21# and 23#, for which the *Sa* is less than 0.7 μm, the material removal rate can be improved from 0.03 to 0.068 at a growth of 126.67%. The result of the cutting parameters' optimization is a substantial improvement in the machining efficiency under the surface roughness constraints.

$$\begin{cases} \max & f_z \times a_e \\ s.t. & Sa \le 0.7, 0 < f_z \le 0.5, 0 < a_e \le 0.5, 0 \le \beta_1 \le 8, 0 \le \beta_2 \le 8 \end{cases} \tag{7}$$

4. Conclusions

In this paper, a novel surface topography simulation algorithm was developed for investigating the surface roughness in the five-axis ball-end milling process. The main conclusions as follows:

(1) A novel surface topography model was developed using triangular approximation and Z-map methods. The consistency between the simulated and experimental results shows that the model can replace the milling experiment when studying the surface topography and roughness during ball-end milling processes;

(2) A response surface-reduced quadratic model was developed based on the proposed surface topography simulation algorithm. The model can effectively characterize the correlation of *Sa* and cutting parameters (i.e., feed per tooth, radial depth of cut, tilt, and lead angles) based on ANOVA results;

(3) An optimization model was developed for improving the machining efficiency by means of the response surface model. The material removal rate (i.e., product of feed per tooth and radial depth of cut) can be improved effectively under the surface roughness constraints;

(4) The complex interaction between cutting edge and workpiece is neglected in the proposed model, so the cutting edge trajectory error, cutting edge micro-geometry and workpiece material deformation should be considered in the next study to secure a reliable prediction of surface topography.

Author Contributions: Conceptualization, R.W.; methodology, R.W.; software, R.W.; validation, R.W.; investigation, W.W.; writing—original draft preparation, R.W.; writing—review and editing, B.Z. and D.T.; project administration, R.W. and B.Z. All authors have read and agreed to the published version of the manuscript.

Funding: This research was funded by the Fundamental Research Funds for the Central Universities (3072023CFJ0701), and the National Natural Science Foundation of China (52271312).

Institutional Review Board Statement: Not applicable.

Informed Consent Statement: Not applicable.

Data Availability Statement: Data are contained within the article.

Conflicts of Interest: The authors declare no conflicts of interest.

References

1. Zhao, B.; Zhang, S.; Man, J.; Zhang, Q.; Chen, Y. A modified normal contact stiffness model considering effect of surface topography. *Proc. Inst. Mech. Eng. J.-J. Eng.* **2015**, *229*, 677–688. [CrossRef]
2. Milan, N.; Sorgato, M.; Parenti, P.; Annoni, M.; Lucchetta, G. Effects of micromilled NiP mold surface topography on the optical characteristics of injection molded prismatic retroreflectors. *Precis. Eng.* **2020**, *61*, 126–135. [CrossRef]

3. Ghorbani, M.; Movahhedy, M.R. Extraction of surface curvatures from tool path data and prediction of cutting forces in the finish milling of sculptured surfaces. *J. Manuf. Process.* **2019**, *45*, 273–289. [CrossRef]
4. Chen, X.; Zhao, J.; Dong, Y.; Han, S.; Li, A.; Wang, D. Effects of inclination angles on geometrical features of machined surface in five-axis milling. *Int. J. Adv. Manuf. Technol.* **2012**, *65*, 1721–1733. [CrossRef]
5. Zhang, Q.; Zhang, S. Effects of feed per tooth and radial depth of cut on amplitude parameters and power spectral density of a machined surface. *Materials* **2020**, *13*, 1323. [CrossRef]
6. Peng, F.; Wu, J.; Fang, Z.; Yuan, S.; Yan, R.; Bai, Q. Modeling and controlling of surface micro-topography feature in micro-ball-end milling. *Int. J. Adv. Manuf. Technol.* **2013**, *67*, 2657–2670. [CrossRef]
7. Zhang, C.; Zhang, H.; Li, Y.; Zhou, L. Modeling and on-line simulation of surface topography considering tool wear in multi-axis milling process. *Int. J. Adv. Manuf. Technol.* **2015**, *77*, 735–749. [CrossRef]
8. Zhang, X.; Yu, T.B.; Zhao, J. Surface generation modeling of micro milling process with stochastic tool wear. *Precis. Eng.* **2020**, *61*, 170–181. [CrossRef]
9. Lotfi, S.; Wassila, B.; Gilles, D. Cutter workpiece engagement region and surface topography prediction in five-axis ball-end milling. *Mach. Sci. Technol.* **2017**, *22*, 181–202. [CrossRef]
10. Hao, Y.; Liu, Y. Analysis of milling surface roughness prediction for thin-walled parts with curved surface. *Int. J. Adv. Manuf. Technol.* **2017**, *93*, 2289–2297. [CrossRef]
11. Omar, O.; El-Wardany, T.; Ng, E.; Elbestawi, M.A. An improved cutting force and surface topography prediction model in end milling. *Int. J. Mach. Tools Manuf.* **2007**, *47*, 1263–1275. [CrossRef]
12. Wang, Z.; Wang, B.; Yuan, J. Modeling of surface topography based on cutting vibration in ball-end milling of thin-walled parts in ball-end milling of thin-walled parts. *Int. J. Adv. Manuf. Technol.* **2018**, *101*, 1837–1854. [CrossRef]
13. Li, B.; Cao, Y.; Chen, W.; Pan, J. Geometry simulation and evaluation of the surface topography in five-axis ball-end milling. *Int. J. Adv. Manuf. Technol.* **2017**, *93*, 1651–1667. [CrossRef]
14. Biondani, F.G.; Bissacco, G. Effect of cutting edge micro geometry on surface generation in ball end milling. *CIRP Ann.-Manuf. Technol.* **2019**, *68*, 571–574. [CrossRef]
15. Imani, B.M.; Elbestawi, M.A. Geometric simulation of ball-end milling operations. *J. Manuf. Sci. Eng.* **2001**, *123*, 177–184. [CrossRef]
16. Zhang, W.; Tan, G.; Wan, M.; Gao, T.; Bassir, D.H. A new algorithm for the numerical simulation of machined surface topography in multiaxis ball-end milling. *J. Manuf. Sci. Eng.* **2008**, *130*, 0110031. [CrossRef]
17. Li, S.J.; Dong, Y.H.; Li, Y.; Li, P.Y.; Yang, Z.C.; Landers, R.G. Geometrical simulation and analysis of ball-end milling surface topography. *Int. J. Adv. Manuf. Technol.* **2019**, *102*, 1885–1900. [CrossRef]
18. Xu, J.T.; Zhang, H.; Sun, Y.W. Swept surface-based approach to simulating surface topography in ball-end CNC milling. *Int. J. Adv. Manuf. Technol.* **2018**, *98*, 107–118. [CrossRef]
19. Yang, L.; Wu, S.; Liu, X.L.; Liu, Z.J.; Zhu, M.W.; Li, Z.H. The effect of characteristics of free-form surface on the machined surface topography in milling of panel mold. *Int. J. Adv. Manuf. Technol.* **2018**, *98*, 151–163. [CrossRef]
20. Chen, H.Q.; Wang, Q.H. Modeling and simulation of the surface topography in ball-end milling based on biharmonic spline interpolation. *Int. J. Adv. Manuf. Technol.* **2018**, *99*, 2451–2466. [CrossRef]
21. Dikshit, M.K.; Puri, A.B.; Maity, A. Optimization of surface roughness in ball-end milling using teaching-learning-based optimization and response surface methodology. *Proc. Inst. Mech. Eng. B-J. Eng.* **2017**, *231*, 2596–2607. [CrossRef]
22. Dikshit, M.K.; Puri, A.B.; Maity, A. Modelling and application of response surface optimization to optimize cutting parameters for minimizing cutting forces and surface roughness in high-speed, ball-end milling of Al2014-T6. *J. Braz. Soc. Mech. Sci.* **2017**, *39*, 5117–5133. [CrossRef]
23. Zhou, J.H.; Ren, J.X.; Yao, C.F. Multi-objective optimization of multi-axis ball-end milling Inconel 718 via grey relational analysis coupled with RBF neural network and PSO algorithm. *Measurement* **2017**, *102*, 271–285. [CrossRef]
24. Buj-Corral, I.; Ortiz-Marzo, J.A.; Costa-Herrero, L.; Vivancos-Calvet, J.; Luis-Pérez, C. Optimal machining strategy selection in ball-end milling of hardened steels for injection molds. *Materials* **2019**, *12*, 860. [CrossRef] [PubMed]
25. Venkata-Rao, K. Power consumption optimization strategy in micro ball-end milling of D2 steel via TLBO coupled with 3D FEM simulation. *Measurement* **2019**, *132*, 68–78. [CrossRef]
26. Sonawane, H.A.; Joshi, S.S. Modeling of machined surface quality in high-speed ball-end milling of Inconel-718 thin cantilevers. *Int. J. Adv. Manuf. Technol.* **2015**, *78*, 1751–1768. [CrossRef]
27. Torres, A.; Amini, C.; Cuadrado, N.; Travieso-Rodriguez, J.A.; Llumà, J.; Vilaseca, M. Experimental validation of ball burnishing numerical simulation on ball-end milled martensitic stainless-steel considering friction and the initial surface topography. *J. Mater. Res. Technol.* **2023**, *22*, 3352–3361. [CrossRef]
28. Cross, K.; Eifler, M.; Klauer, K.; Huttenlochner, K.; Kirsch, B.; Ziegler, C.; Aurich, J.C.; Seewig, J. Determination of the surface topography of ball end micro milled material measures. *Eng. Sci. Technol.* **2021**, *24*, 543–555. [CrossRef]
29. Klauer, K.; Eifler, M.; Kirsch, B.; Seewig, J.; Aurich, J.C. Ball end micro milling of areal material measures: Influence of the tilt angle on the resulting surface topography. *Prod. Eng.-Res. Dev.* **2020**, *14*, 239–252. [CrossRef]

30. Arruda, E.M.; Brandão, L.C.; Ribeiro Filho, S.L.M.; De Oliveira, J.A. Integrated optimization using mixture design to confirm the finishing of AISI P20 using different cutting strategies and ball nose end mills. *Measurement* **2014**, *47*, 54–63. [CrossRef]
31. Wang, P.; Zhang, S.; Yan, Z.G. Study on surface defects in five-axis ball-end milling of tool steel. *Int. J. Adv. Manuf. Technol.* **2016**, *89*, 599–609. [CrossRef]

materials

MDPI

Article

The Influence of Flame Exposure and Solid Particle Erosion on Tensile Strength of CFRP Substrate with Manufactured Protective Coating

Przemysław Golewski [1,*] and Michał Budka [2]

1 Department of Solid Mechanics, Faculty of Civil Engineering and Architecture, Lublin University of Technology, Nadbystrzycka 40 Str., 20-618 Lublin, Poland

2 Laboratory of Civil Engineering, Faculty of Civil Engineering and Architecture, Lublin University of Technology, Nadbystrzycka 40 Str., 20-618 Lublin, Poland; m.budka@pollub.pl

* Correspondence: p.golewski@pollub.pl

Abstract: This paper presents the results of laboratory tests for new materials made of a carbon fibre-reinforced polymer (CFRP) composite with a single-sided protective coating. The protective coatings were made of five different powders—Al_2O_3, aluminium, quartz sand, crystalline silica and copper—laminated in a single process during curing of the prepreg substrate with an epoxy matrix. The specimens were subjected to flame exposure and solid particle erosion tests, followed by uniaxial tensile tests. A digital image correlation (DIC) system was used to observe the damage location and deformation of the specimens. All coatings subjected to solid particle erosion allowed an increase in tensile failure force ranging from 5% to 31% compared to reference specimens made of purely CFRP. When exposed to flame, only three of the five materials tested, Al_2O_3, aluminium, quartz sand, could be used to protect the surface, which allowed an increase in tensile failure force of 5.6%.

Keywords: solid particle erosion; flame exposure; carbon fibre-reinforced polymer (CFRP); protective layer; tensile test

1. Introduction

PMC (polymer matrix composite) materials have been used since the mid-1930s, when Owens Corning was the first company in the world to start an industry in this field using glass fibres as reinforcement [1]. Since then, composites have become widespread in many technical fields, such as aerospace, the automotive industry, shipbuilding, wind turbines and civil engineering. Despite their many advantages, such as favourable weight-to-strength ratio, composites also have disadvantages, such as, among others, the ease with which they can be damaged, e.g., by erosion or high temperatures, and the difficulty of repair and recycling. Therefore, the entire design and manufacturing process must take place with great care to avoid failures early on in the life of the product. Despite this, some application ranges remain unattainable or very limited. It becomes necessary to apply suitable coatings on their surface to protect or provide functional properties. In terms of coatings on the surface of PMC composites, various solutions can be found in the literature, e.g., the use of:

- Flame retardant resin additives [2,3];
- Nanopaper with carbon nanofibres [4];
- Intumescent or ceramic mats [5–7];
- Flame spray or cold spray [8,9];
- Sol–gel technique [10];
- Powders mixed with resin [11].

Another example would be unwanted coatings that form naturally as oxides and need to be removed. Ref. [12] proposes the use of a numerical method to calculate the effects of particles through blasting on a rust-covered surface, and points out that the material parameters of the tested coatings are crucial in the model simulation process.

Coatings applied to the surface of PMC composites can serve different functions. Polyurea as an outer layer can be used as protection against low-velocity impact or point static loading [13–15]. Energy absorption during impact can be increased by as much as 94% compared to pure CFRP. Structures such as aircraft fuselages or wind turbine blades are exposed to lightning strikes [16]. In this case, the authors in [17] used polyaniline to produce the coating. It was found that with a 345 µm-thick PANI-based LSP system, 83% and 100% of the original strength of the sample can be achieved for composites such as CFRP and GFRP, respectively, loaded with a discharge of 40 kA to 100 kA. Electrical conductivity of the coating can also be achieved by chemical application of nickel [18] or by using a copper grid [19]. A difficult issue is the protection of the composite substrate against high temperatures arising, for example, as a result of a fire or when part of the structure has to operate at elevated temperatures. The authors in [20] used gypsum-based coatings. These are relatively thick (25 mm) and have an irregular surface, but allow the CFRP composite to be protected in a fire for 2 h, which meets the British standards for building materials. An order of magnitude less thick are the coatings proposed in [21] that also serve as a barrier against high temperatures. In this case, a poly(vinyl alcohol) (PVA)-based coating containing ammonium polyphosphate (APP) and sepiolite nanofillers (SP) was used. A glass fabric interlayer was still required between this material and the laminate. This allowed the temperature of the flame-retardant surface of the sample to be lowered. There is no information on the mechanical strength of the coatings in the two articles mentioned above. In terms of mechanical loading, both the gypsum-based and PVA-based material may not have sufficient strength for particle erosion or impact. Hence, other methods are sought, e.g., based on different types of powders mixed with resin and applied to the sample surface. In Ref. [22], a new thermal barrier was proposed using Al_2O_3 particles. The fabrication of the samples was carried out in two stages. First, a thermal barrier was produced by combining Al_2O_3 powder with a ceramic binder, while in the second stage, RTM (resin transfer moulding) technology was used to produce a composite with the previously prepared barrier. The resin partially fills the porous space of the thermal barrier and a mechanical bond between the two materials is created. Bending tests carried out for specimens loaded with a 700 °C flame prove that 50% of the substrate strength can be retained. The authors in [23] proposed an original technology for producing the protective layer using a special powder that is a mixture of diglycidyl ether of bisphenol A (DGEBA), polyester resins, mineral fillers and pigment. The whole procedure is carried out in several steps and involves firstly curing the layer, and only then is the prepreg layer laid down and re-cured. Hence, the authors in [24] proposed a new, less complex technology, which has the advantage of making the coating in one process while the prepreg is cured. This type of coating can be formed from different metal powders or oxides, while the formed coating has very good adhesion to the CFRP substrate. Table 1 summarizes the latest developments in coating manufacturing on PMC composites.

To date, there are no results in the literature on the effects of flame or solid particle erosion for a coating formed such as that presented in [24]. In the previous article, the authors used the same technology as now, but the specimens were not subjected to any thermal or erosive effects and were tested in three-point bending tests using acoustic emission, while the objective was to determine whether sudden and early delamination of the coatings would occur. The results of the tests showed that delamination does not occur, even with large deformations of the substrate; hence, in the present work, samples with manufactured coatings of powders—Al_2O_3, aluminium, quartz sand, crystalline silica and copper—were subjected to flame exposure and solid particle erosion tests followed by uniaxial tensile tests. In the case of specimens subjected to solid particle erosion, all of the proposed materials fulfilled their function and allowed an increase in tensile failure force in the range of 5% to 31%, while for flame exposure, only the first three of the above-mentioned can be used for which this increase was at the level of 5.6%.

Table 1. Literature reports of coating manufacturing on composite substrates.

Type and Thickness of the Substrate	Type and Thickness of Coating Material	Method of Manufacturing/Joining the Coating	Type of Load/Test	References
CFRP 4.1 mm	Ti/TiN, 1.5 to 11 μm	PVD (physical vapour deposition)	sand erosion, rain erosion	[25]
CFRP 2.4–3.85 mm	Al_2O_3 powder and ceramic binder, 1.45 mm	VARTM (vacuum-assisted resin transfer moulding)	500–700 °C flame exposure, flexural strength	[22]
GFRP 3.7 mm	(PVA)-based coating 2.8 mm	vacuum infusion process	cone calorimeter tests 50 kW/m^2	[21]
CFRP 2 mm and GFRP 2.4 mm	polyaniline coating, 249–427 μm	manual application, curing at 130 °C	lightning strike	[17]
CFRP 1.5 mm	copper, quartz sand, Al_2O_3, aluminium, crystalline silica, microballoon, 0.3–1 mm	curing process in the autoclave	3-point bending tests	[24]
CFRP 2 mm	polyurea, 0.5 mm–1 mm	spraying process, curing for 7 days	quasi-static indentation and low velocity impact	[13]
CFRP 1.5 mm	aluminium bond coat + top coat (four different ceramic materials coating, 0.8 mm)	plasma sprayed	mechanical properties—Young's modulus	[26]
CFRP 1.7 mm	graphene nano platelet-based coating, prepreg 62 μm	compression moulding process (3 bar), cured for 1 h at 120 °C	laser heating with power density 25–150 kW/m^2	[27]
CFRP 1.2–3 mm	non-intumescent fireproof coating, 15 mm, 25 mm, 35 mm	curing process for a duration of 20 days	large-space fire, standard fire, bending tests	[20]
GFRP 3.2 mm	CNF-based nanopaper, 0.3 mm	RTM (Resin Transfer Moulding)	cone calorimeter tests, post fire 3-point bending tests	[4]
GFRP 2.4 mm	flame-retarded epoxy resin and ceramic particles (Ce) 1.09 mm, (Zr) 0.89 mm, (Re) 0.94 mm	hand lay-up method	cone calorimeter tests: 20, 30, 40, 50 kW/m^2	[28]

2. Materials and Methods

A total of five chemically and physically different powders—Al_2O_3, aluminium, quartz sand, crystalline silica and copper (PolyCore, Świdnik, Poland)—were selected to produce the coatings. The properties of the materials are summarised in Table 2, while images of the powder grains taken using SEM are shown in Figure 1 for the same ×200 magnification. The first material (Al_2O_3) is widely used in technology and commercially available for different powder gradations. It is relatively inexpensive and is used for the manufacture of grinding wheels or blasting. It has a high compressive strength and Young's modulus, as well as a much lower thermal conductivity compared to metals, and hence it can be a candidate for both erosion and high-temperature coatings. The second material, aluminium, has the advantage that it can deform plastically, which will not result in brittle cracking under high mechanical loads. In addition, compared to low-carbon steel, it does not corrode as quickly atmospherically and is almost a third of the weight. Quartz sand is the cheapest and most accessible material. Like Al_2O_3, it has high strength and low thermal conductivity. Quartz sand has been used on a large scale in façade composite panels on a building of the Museum of Modern Art in San Francisco [29], serving both a decorative and protective function. The fourth material is a lightweight filler in the form of crystalline silica. The grains are spherical and the inside is hollow, as shown in Figure 1e. The wall thickness is approximately 3–4 µm. Hence, a coating made of this material will have the advantage that the weight of the final product will not increase significantly. The last material is copper. It is the most expensive considering all the materials presented, so its use must be well thought out and justified. The copper-coated samples shown in Appendix A are highly aesthetic and attractive. Hence, such a product can find use in the construction industry, e.g., replacing copper in sheets for roofing. Copper has another advantage, antimicrobial properties [30], which can be used as an advantage in the construction of facade panels, which are often covered by fungi and moulds [31].

Figure 1. SEM images: (**a**) Al_2O_3, (**b**) aluminium, (**c**) quartz sand, (**d**) crystalline silica (×200), (**e**) crystalline silica (×2000) (**f**) copper.

Table 2. Properties of materials used in the manufacture of coatings [32].

	Al_2O_3	Aluminium	Quartz Sand	Crystalline Silica	Copper
Density (kg/m^3)	3690	2700	2650	500	8940
Young's modulus (GPa)	370	70	74	-	110
Tensile strength (MPa)	300	240	155	-	365
Compressive strength (MPa)	3000	240	1600	-	365
Vickers hardness	1365	83	1100	-	90
Melting temperature (°C)	2050	630	1650	1200	1066
Thermal conductivity (25 °C) (W/mK)	46	200	1.5	-	260

All samples were made using Kordcarbon epoxy prepreg (Fiberpreg GmbH, Neu-Ulm, Germany) by Wit-Composites. The technology for making the samples comprised the following steps:

1. Preparation of the flat mould, cutting and layering of the prepreg;
2. Laying of a 1 mm thick tool to form the powder layer on the prepreg;
3. Manual moulding of the powder within the limits of the tool;

4. Removing the tool and preparing the vacuum pack;
5. Forming of samples in the autoclave according to the prepreg manufacturer's instructions.

The photographs of the workflow are analogous to those presented in [24]. The whole process is also explained in Figure 2. In each case, the thickness of the stationary layer-forming tool was the same. However, densification occurred during the process—less for coarse grains and more for fine powder. This resulted in different thicknesses (Table 3). The microstructure of the manufactured coatings can be seen in the Appendix A in Figure A1.

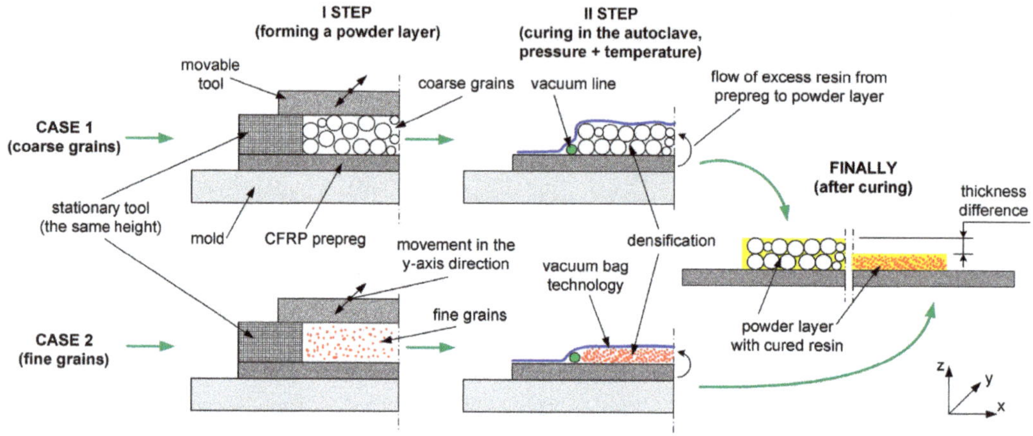

Figure 2. Process of making the protective coating.

Table 3. Sample thicknesses [mm].

	Al_2O_3	Aluminium	Quartz Sand	Crystalline Silica	Copper	References
Sample 1	2.41	2.40	3.46	2.48	2.24	2.05
Sample 2	2.41	2.54	3.53	2.75	2.25	2.13
Sample 3	2.54	2.43	3.40	2.60	2.33	2.05
Sample 4	2.46	2.54	3.05	2.72	2.16	
Sample 5	2.45	2.49	2.91	2.56	2.18	
Mean	2.45	2.48	3.27	2.62	2.23	2.07

After curing, the specimens were cut with a CNC plotter to a size of 30 mm × 250 mm. Then, the samples were subjected to thickness measurements using a digital micrometre with a range of 0–25 mm with output to a computer made by Mitutoyo, whose readings in millimetres were to three decimal places.

The samples were divided into two main groups:

T—specimens exposed to flame.

E—samples exposed to solid particle erosion.

There were 5 batches of samples with protective coatings in each group. There were 5 samples in each batch. In addition, there were reference samples (3 each) in both groups that did not have a protective coating. Hence, the total number of samples was 56.

Group T specimens were exposed to the flame using the test stand shown in Figure 3a. This type of stand is often used for comparative tests, e.g., [33,34], where a gas torch and digital temperature measurement are used. The distance of the burner from the surface of the sample was set using a K-type thermocouple so that the temperature in front of the sample was in the range 750–800 °C. This is a temperature that can occur, for example, during a developed fire in a building or in hot sections of turbine engines. For this purpose, an ORCA 3 burner (Metalurgica Orca Ltd., Sao Paulo, Brazil) and isobutane gas (Alpen Camping, Schilpario, Austria) were used. A FLIR SC5000 thermal imaging camera was used to record the temperature distribution on the opposite surface of the sample. The flame exposure time in each case was 30 s. The results from the thermal imaging camera were processed in the Altair software (version 5.91.010.).

(a) **(b)**

Figure 3. Schematics of laboratory stands: (**a**) for flame exposure (1—thermal imaging camera, 2—sample, 3—gas burner); (**b**) for solid particle erosion test.

Group E specimens were subjected to a point solid particle erosion, as shown in Figure 3b, using an abrasive blast gun (Tagred TA 1358 (Nowe, Poland)). The angle between the eroding jet and the sample surface was 90° analogous to [35]. The erosion was carried out at one point of the specimen in its axis. The exposure time of the blast was 3 s, the installation pressure was 0.6 MPa, the inner diameter of the nozzle was 7.5 mm and the distance of the nozzle from the sample surface was 38 mm. Electrocorundum with a gradation of F60 was used as the abrasive.

After flame and particle erosion tests, the specimens were prepared for uniaxial tensile tests by bonding 30 mm × 50 mm tabs to the ends. The tabs were made from 1 mm thick PFCC 201 laminate (Izo-Erg, Gliwice, Poland) and bonded with Epidian 5 epoxy resin with PAC hardener (CIECH Sarzyna S.A., Nowa Sarzyna, Poland).

Microscopic observations were made using a Keyence VHX-7000 (Osaka, Japan) digital microscope. Uniaxial tensile tests were performed using an MTS 100 kN testing machine (Eden Prairie, MO, USA) and with an Aramis digital image correlation system (Lenso Ltd., Poznań, Poland). The speed at which the specimens were loaded was 1 mm/min. The results from the testing machine were processed in Diadem 2019.

3. Results

In the presented work, the results are divided into two subsections on the effects of flame and erosion. In the first case, the results are more extensive, as a thermal imaging camera was used to record the temperature distribution and further process these results. In the case of particle erosion, the results relate only to mechanical tensile tests.

The results of the work carried out are so extensive that photographs of the flame exposure tests and the solid particle erosion tests (the stage before the strength tests) are included in Appendix A.

3.1. Flame Exposure Test Results

Flame exposure tests for each material were carried out for 30 s. The time was chosen so that damage to the coating was visible, but there was no complete degradation of the substrate. Figure 4 shows the flame effects for each material. In almost every batch, the surface of the sample was ignited after several seconds. An exception is the coating made of quartz sand. This is the material with the largest grains of those tested. Crystalline silica, which is a lightweight filler, behaves very unfavourably; it is in the form of spheres that are hollow inside. When these are damaged, a porous structure with a large heat transfer surface area is formed (Figure 1e), which is easily heated.

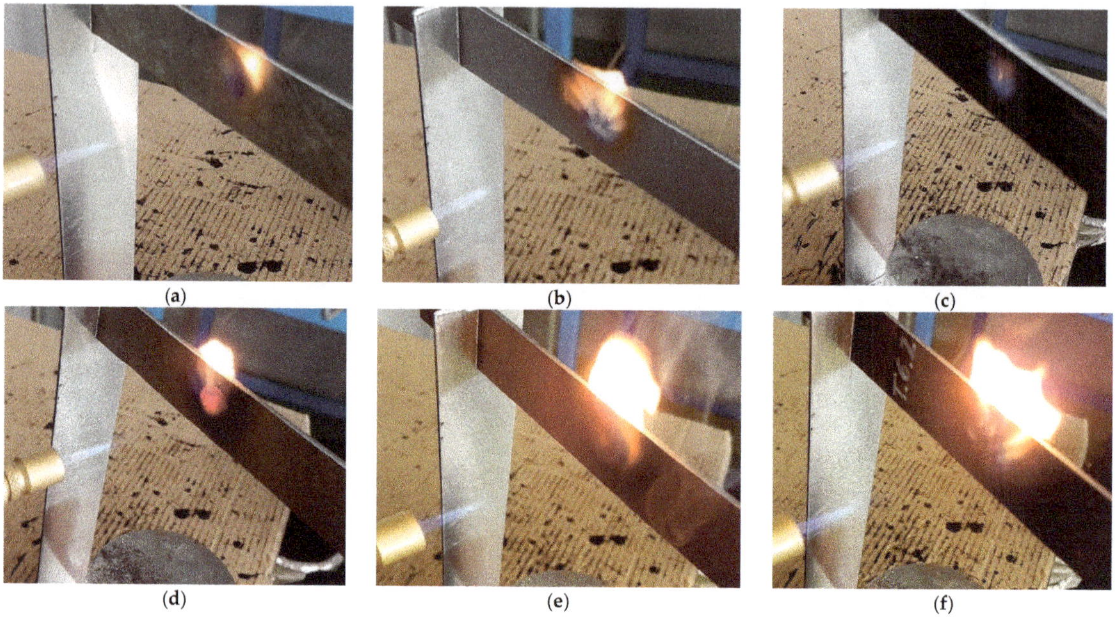

Figure 4. Performing flame exposure tests: (**a**) Al_2O_3, (**b**) aluminium, (**c**) quartz sand, (**d**) crystalline silica, (**e**) copper, (**f**) reference.

Throughout the flame exposure, the temperature on the back of the sample was recorded. Due to the extensive nature of the results, Figure 5 shows images of the temperature distribution after 30 s for each of the materials tested. For each sample, a 30 mm × 30 mm area was defined in Altair software, for which the minimum, maximum and average temperatures could be determined. Both temperature distribution maps and values are extremely important from a practical point of view as well as for numerical modelling.

A summary of the results for 30 s of heating for a 30 mm × 30 mm field is presented in Figure 6. These are the averaged results for each batch. The standard deviation bars are included. To simplify the analysis, horizontal lines were also drawn in Figure 5 to refer to the batch of reference samples. The most favourable material for building a protective coating is quartz sand. It has a thermal conductivity coefficient of 1.5 W/m·K. The decrease in the average temperature value compared to the reference samples was approximately 27.5%. The worst-performing material is copper. Copper is a good thermal conductor with a heat conduction coefficient of 260 W/m·K, and thus there was an increase in the mean surface temperature of approximately 14.5% compared to the reference samples. Furthermore, it should be noted that, as shown in Table 3, uniform coating thicknesses were not achieved. A low thermal conductivity of 46 W/m·K is also possessed by Al_2O_3, but with this grain size and coating thickness, the advantages of this material did not become apparent. Similar results to Al_2O_3 were also obtained for aluminium and crystalline silica-coated samples.

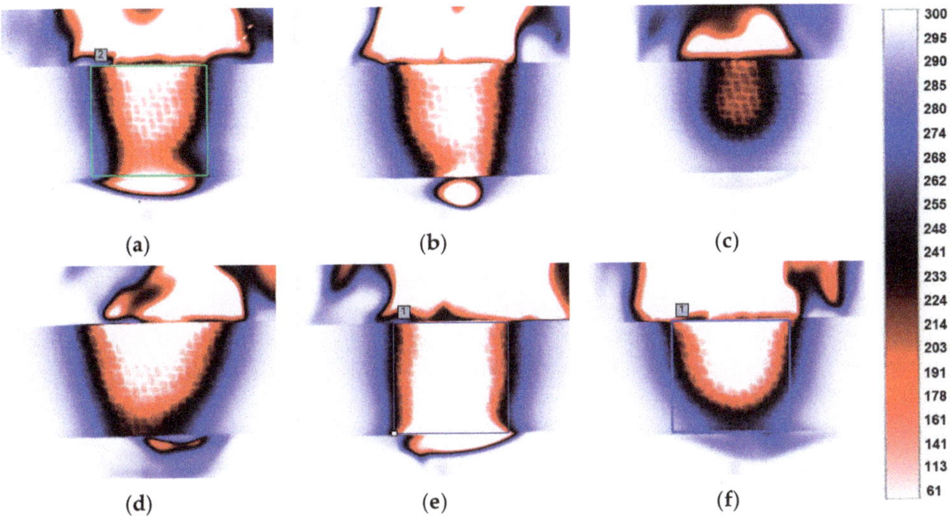

Figure 5. Thermal camera images for 30 s flame exposure: (**a**) Al_2O_3, (**b**) aluminium, (**c**) quartz sand, (**d**) crystalline silica, (**e**) copper, (**f**) reference.

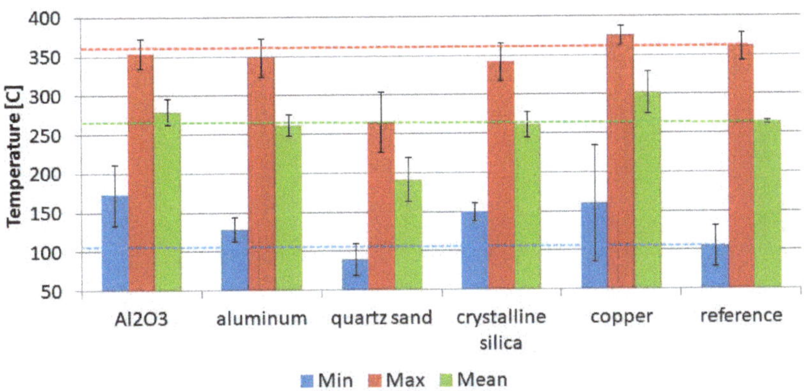

Figure 6. Batch-averaged temperature values on a 30 mm × 30 mm field for 30 s flame exposure.

Figure 7 presents the results from uniaxial tensile tests on specimens subjected to flame exposure. Throughout the tensile process, the displacements of the specimen surface were recorded using the DIC system. After processing, these were presented in the form of principal strain maps. For each specimen, three images were selected for different loading levels: at an early stage, at half load and at the final stage before the specimen failed. In each case, the strain maps refer to the first sample in the batch.

The force–displacement diagrams are all linear, with the failure of the specimen occurring in a sudden manner. However, the strain maps revealed that coating damage can occur much earlier, even before the maximum force is reached. This is particularly evident for two materials, Al_2O_3 and quartz sand. For these materials, the fields are not homogeneous and horizontal lines appear at an early stage of tensile strain, indicating that cracks are occurring in the coating.

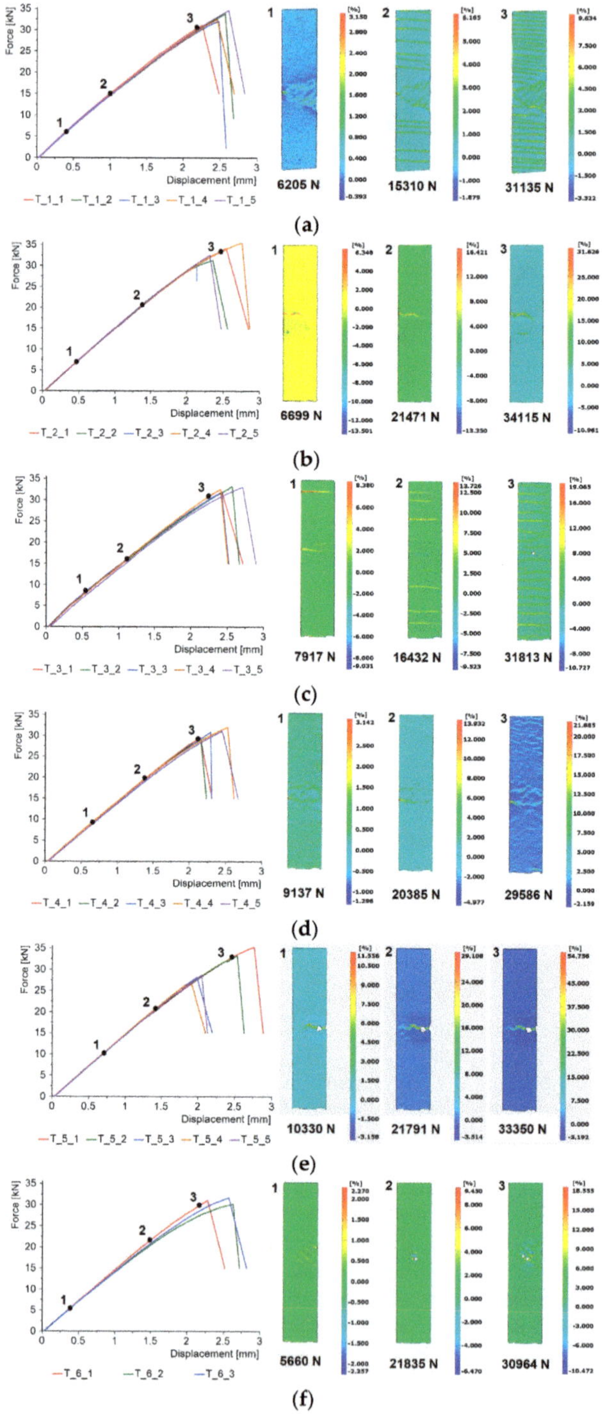

Figure 7. Force–displacement diagrams and principal strain maps for specimens subjected to flame exposure: (**a**) Al_2O_3, (**b**) aluminium, (**c**) quartz sand, (**d**) crystalline silica, (**e**) copper, (**f**) reference.

However, these cracks do not grow and further ones appear in the sample. The reason for the formation of cracks is that both oxide materials do not deform plastically. In the case of copper and aluminium, the strain fields are uniform outside the flame area. Similarly for the reference sample, homogeneous strain fields are also obtained. The visible white blank fields in the copper sample and the reference sample in the flame-affected zone are due to the fact that the camera system has lost its reference points, probably as a result of paint spalling due to significant deformation and poor adhesion.

3.2. Solid Particle Erosion Test Results

The results for the specimens subjected to solid particle erosion tests are shown in Figure 8.

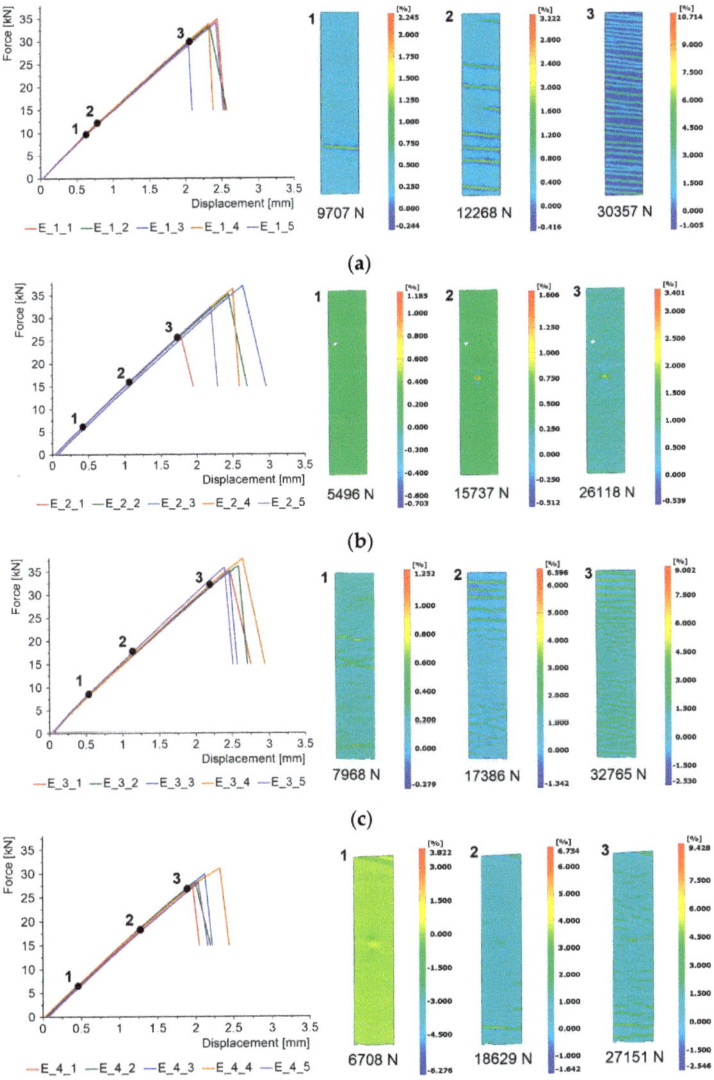

Figure 8. *Cont.*

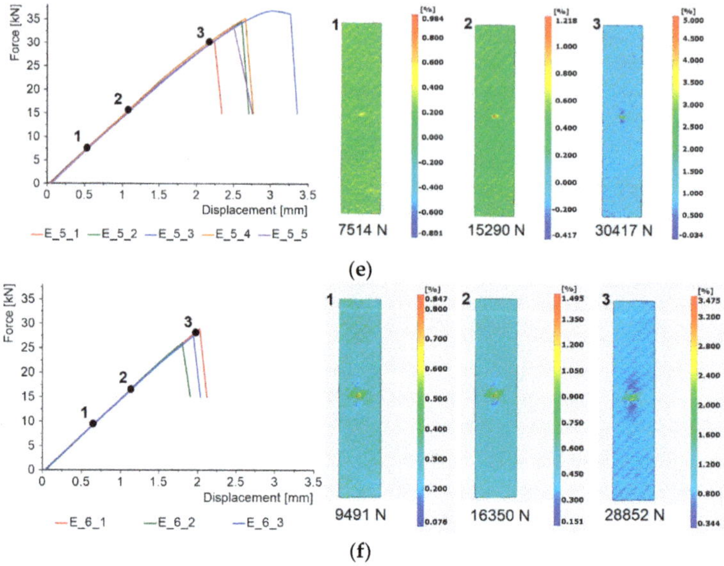

Figure 8. Force–displacement diagrams and principal strain maps for specimens subjected to erosion exposure: (**a**) Al_2O_3, (**b**) aluminium, (**c**) quartz sand, (**d**) crystalline silica, (**e**) copper, (**f**) reference.

As with the specimens subjected to flame exposure tests, the results for the principal strain fields for the three load levels are also presented. The graphs are also linear and there is a sudden failure of the specimen. For specimens with a coating made of oxides (Al_2O_3 and quartz sand), horizontal lines representing brittle cracking in the coating also appear. However, the intensity of these cracks is more pronounced than for the flame-exposed samples. The reason for this may be that the effect of the high temperature was not limited to the flame area, but the sample was heated along its entire length. Hence, the mechanical properties of the epoxy resin matrix may be reduced, its stiffness decreases, and thus the intensity of the strain maps also decreases.

Considering the strain maps for the coatings made of aluminium and copper, these are homogeneous. However, throughout the tensile stage, the point in the specimen axis where the intensity of erosion was greatest is visible.

4. Discussion

The thickness of the achieved coatings should be considered the first issue in the discussion. The average thickness of the substrate was 2.075 mm and the bulk thickness of the powder layers was 1 mm. This means that theoretically, the thickness of the whole sample should be around 3 mm. However, as Table 3 shows, this dimension is not achievable. This is due to the fact that the samples were autoclaved under pressure and the grains fill the voids between each other and further indent into the substrate, which becomes elastic during curing. Such a phenomenon can be considered positive for two reasons. Firstly, as a result of the reduction of the voids between the grains, there is no significant outflow of resin from the prepreg. Secondly, such a layer becomes more compact and stronger than if there are loose grains with a large amount of resin, which could occur during moulding with, for example, the non-pressurised manual mixing of powder and resin.

However, the disadvantage of such a technology is the difficulty in designing a suitable layer thickness, because each time the powder material, grain size and planned thickness are changed, a technological test must be carried out in advance. This is thus a topic for future research.

In the following analysis, it is important to take a closer look at applied technology. On the one hand, it is innovative and has advantages such as:

- The production of the coating in one process, during the curing of the prepreg;
- Very good adhesion to the CFRP substrate, as presented in [24];
- The possibility of using different powders and mixtures of powders;
- High aesthetic qualities of the outer surface.

On the other hand, it should be noted that this technology is imperfect due to the manual formation of the layer. This may even become impossible with small thicknesses of the order of 0.1 mm. Therefore, the use of mechanical feeders and numerically controlled powder distribution on the surface should be considered.

Another disadvantage of the presented solution is that as a result of the difference in thermal expansion of the substrate material and the forming coating, the sample will be slightly bent when cooled and removed from the autoclave (Figure 9).

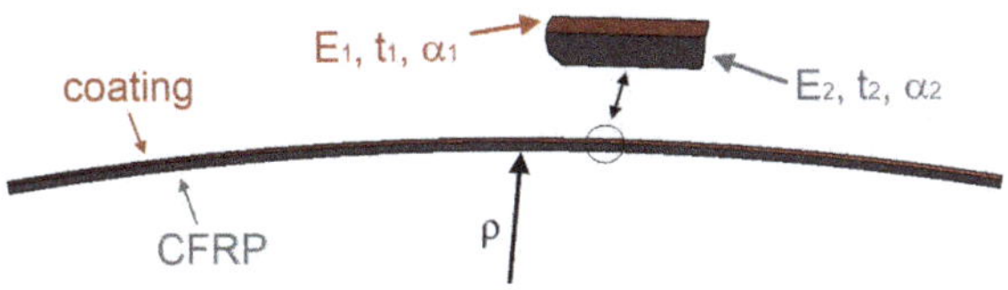

Figure 9. Deformation of the sample after removal from the autoclave.

The deformation will depend on both the coating material used and the thickness ratio of the substrate to the coating and can be determined analytically using Equation (1) [36].

$$\rho = \frac{t\left[3 \cdot (1+m)^2 + (1+m \cdot n)\left(m^2 \cdot \frac{1}{m \cdot n}\right)\right]}{6 \cdot (\alpha_1 - \alpha_2)(T_h - T_c)(1+m)^2} \tag{1}$$

where:

$m = \frac{t_1}{t_2}$—ratio of coating thickness to composite thickness,
$n = \frac{E_1}{E_2}$—ratio of Young's modulus of the coating to the composite,
$t = t_1 + t_2$—sum of coating and substrate thickness,
α_1, α_2—coefficients of thermal expansion for the coating and the composite,
T_h, T_c—sample curing temperature and ambient temperature.

Equation (1) applies to isotropic materials. In this case, it would be necessary, using a representative volume element, to determine the Young's modulus and coefficient of expansion for CFRP and the protective coating. On the other hand, literature data [37,38] and Table 2 can be used to make calculations for the quartz sand-coated sample for which the smallest radius of curvature was obtained. If the following data are provided—t_1 = 1.2 mm, t_2 = 2.07 mm, E_1 = 74 GPa, E_2 = 40 GPa, $\alpha_1 = 16.41 \times 10^{-6}$ 1/°C, $\alpha_2 = 10 \times 10^{-7}$ 1/°C, T_h = 125 °C, T_c = 22 °C—then the radius of curvature will be 1.128 m.

In the considered specimens, the radii of curvature and the curvature for each specimen are summarised in Table 3. These are results based on sample measurements using a digital sensor. If we compare the result from Table 4 for the quartz sand coating with the calculation from Equation (1), the difference is 18.7%.

Table 4. Results of curvature measurements of the samples.

	Al_2O_3	Aluminium	Quartz Sand	Crystalline Silica	Copper
radius "r" of curvature [m]	2.538	2.811	1.388	2.443	9.527
curvature 1/r [1/m]	0.39	0.36	0.72	0.41	0.1

Specimens with a quartz sand coating have the highest curvature, while those with a copper coating have the lowest. Therefore, when designing products, especially flat products, using this technology, numerical simulations should be carried out in advance to verify the deformation of the composite product. The occurrence of curvature also leads to the appearance of additional residual stresses in the structure. However, in order to prevent this, it is possible, for example, to form the coating on both sides.

At present, almost all research papers that can be found in the literature focus on forming the coating on a flat surface. An exception is [39], in which both the original technology and a curved moulder were used. The use of flat specimens, on the one hand, greatly simplifies the whole technological and research process, but on the other hand, it should be kept in mind that composite parts such as wind turbine blades or boat hulls have a curved shape. Hence, the authors of this article are currently working on a technology that will allow coatings to be formed easily and efficiently also on curved surfaces.

Turning to the results of the flame exposure tests, it should be noted that extremely different results were obtained with the five coatings. After the flame exposure tests, the samples were cleaned of soot and loose particles with a steel brush. A view of the surface from the flame exposure side and the side of the specimen is presented in Figure 10. A view of the whole specimens is also available in Appendix A in Figure A2.

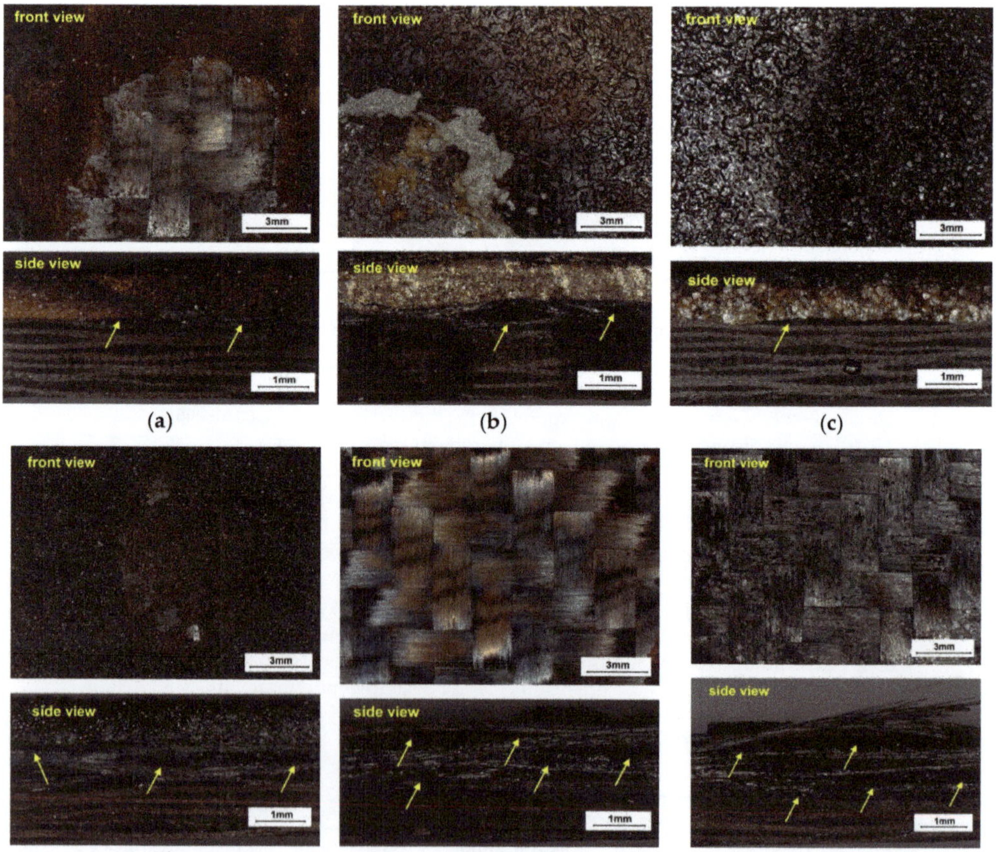

(a) (b) (c) (d) (e) (f)

Figure 10. Surface view of specimens for flame exposure tests: (**a**) Al_2O_3, (**b**) aluminium, (**c**) quartz sand, (**d**) crystalline silica, (**e**) copper, (**f**) reference.

Flame effects can be divided into two main groups: when there is no uncovered substrate and when the substrate is uncovered. The first group will include only two materials: aluminium and quartz sand. The second group, in order of the most uncovered substrate, will include copper, Al_2O_3 and crystalline silica. However, no matter which layer is used, a barrier effect still arises if we compare the effects with Figure 10f for a pure CFRP substrate. The absence of any barrier results in significant degradation and deformation in the PMC composite. Important information is shown by the images in Figure 10 on the side of the sample. For the coating made of quartz sand, no negative changes in the substrate are visible. Only a slight delamination of the coating occurs. This type of delamination without damage to the substrate itself is also visible for the Al_2O_3 coating, but in this case, as mentioned above, there is an uncovering of a significant area of the substrate. In the aluminium coating, evidence of thermal damage to the substrate are already visible, but these do not extend as deeply as for the copper or crystalline silica coating.

The samples subjected to particle erosion can be seen in Appendix A in Figure A3. In this case, two effects can also be distinguished: uncovering of the composite substrate and no uncovering. In fact, in each batch, there was a minimum of one sample in which the composite surface was uncovered to a greater or lesser extent. Such an effect could be due to a locally smaller coating thickness or a different intensity of stream exposure, e.g., a pressure surge in the compressed air system caused by switching on the compressor. However, there are undoubtedly two materials whose application is not favourable: crystalline silica and Al_2O_3. For the first material, the substrate was uncovered in every sample, and for the second in three samples. The most surprising effect is for the copper coating, as it was the thinnest. For copper, aluminium and quartz sand, substrate uncovering was observed for only one sample in the series.

Figure 11 summarises the results of the microscopic observations for the samples in which substrate uncovering occurred. The images show the boundary zone between the coating and the substrate. The erosion is so intense that not even a single grain remains on the CFRP surface.

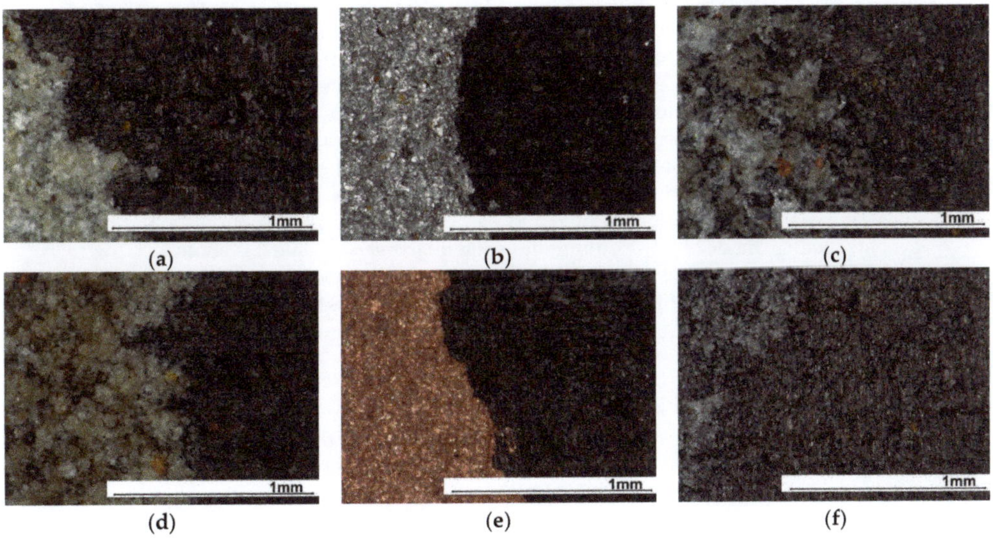

Figure 11. Surface view of specimens for solid particle erosion: (**a**) Al_2O_3, (**b**) aluminium, (**c**) quartz sand, (**d**) crystalline silica, (**e**) copper, (**f**) reference.

For the designer, who could possibly apply one of the selected coatings, it is important how the aggressive environment will affect the strength. The effectiveness of the respective coatings was verified in a uniaxial tensile test, and the results are collected in Figure 12a for maximum forces and Figure 12b for absorbed energy. In the present tests, for the substrate subjected to the exposure to flame and particle erosion, values of 30.96 kN and 27.46 kN of failure forces were obtained, respectively. These are the levels to which the results for the other samples will be related, and hence the horizontal lines in Figure 12a for easier analysis. The reason that the reference specimens obtained a higher force for the flame test than for the erosion test can be explained by the fact that the high temperature does not damage the fibres themselves, but the matrix, while the fibres are degraded by erosion.

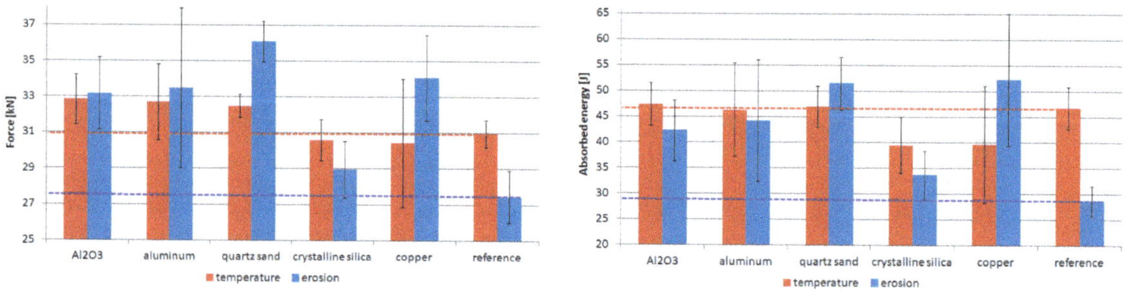

(a) (b)

Figure 12. Results of uniaxial tensile tests: (**a**) summary of results averaged over a batch of maximum forces, (**b**) summary of results averaged over a batch of absorbed energy.

Considering the effect of flame exposure, these results confirm previous microscopic observations. Materials such as crystalline silica and copper resulted in a reduction in maximum force of approximately 1.7%. For other materials, an increase of about 5.6% was obtained, which is a favourable phenomenon.

When considering the results for specimens subjected to solid particle erosion, in this case, each of the proposed coatings performs well and an increase in failure force ranging from 5% to 31% was obtained for crystalline silica and quartz sand respectively. This increase is due to the fact that part of the load is transferred through the coating, which is a great advantage in contrast to paints or intumescent mats, which are characterised by low stiffness and thus cannot take part in the load transfer. It is also worth noting that the smallest standard deviation was obtained for the quartz sand samples.

Figure 12b displays the results for the energy absorbed during the tensile tests, which corresponds to the area under the force–displacement diagram. In this case, similar conclusions can be drawn. Considering the flame exposure, the absorbed energy is at the same level for coatings made of Al_2O_3, aluminium and quartz sand, while it is about 15% lower for the coating made of crystalline silica and copper. In the case of erosion, an increase in absorbed energy ranging from 17% (crystalline silica) to 82% (copper) was obtained for all coatings.

5. Conclusions

This paper presents the application of a new technology to produce five different coatings on CFRP substrates and discusses its advantages and disadvantages in relation to current methods and future expectations. A total of 56 samples were made, which were divided into two groups and subjected to flame and solid particle erosion exposure. In the final stage, the specimens were subjected to uniaxial tensile tests. The conducted work allows the following conclusions to be drawn.

- The use of coating formation technology in a single process during the curing of the prepreg results in residual stresses that, depending on the ratio of substrate thickness to layer thickness, can lead to deformation of the product.

- With the quartz sand coating, a 27.5% reduction in average substrate temperature was achieved compared to the reference samples. For the copper-coated samples, the situation is reversed and the substrate temperature increases by 14.5%.
- The application of the DIC method allowed the surface strain of the samples to be observed. For coating materials such as Al_2O_3 and quartz sand, the appearance of horizontal cracks was observed, the number of which intensifies when the load is increased. Such cracks occur over the entire surface irrespective of the location of flame or erosion damage. Samples with coatings made of metal powders (aluminium and copper) are characterised by homogeneous strain fields outside the damage area.
- Considering flame exposure, an increase in failure force compared to the reference value was obtained for samples with coatings made of Al_2O_3, aluminium and quartz sand. The increments in failure force were at the level of 5.6%. This shows that the fabricated coatings can be considered barriers against the effects of high temperature.
- All of the proposed coatings can be used for erosion protection. Increases in tensile failure force ranging from 5% to 31% were obtained for crystalline silica and quartz sand, respectively.
- Of all the proposed coatings, considering both flame and erosion exposure, the quartz sand coating shows the best results. This is evidenced not only by the force increases compared to the reference samples and microscopic images but also by the smallest standard deviation, which was 0.65 kN and 1.13 kN for flame and erosion exposure, respectively.

Author Contributions: Conceptualization, P.G.; methodology, P.G.; software, P.G. and M.B.; validation, P.G.; formal analysis, P.G.; investigation, P.G. and M.B.; resources, P.G.; data curation, P.G. and M.B.; writing—original draft preparation, P.G.; writing—review and editing, P.G.; visualization, P.G.; supervision, P.G.; project administration, P.G.; funding acquisition, P.G. All authors have read and agreed to the published version of the manuscript.

Funding: This research was funded by the National Centre for Research and Development of Poland (grant LIDER XIII 0135/L-13/2022).

Institutional Review Board Statement: Not applicable.

Informed Consent Statement: Not applicable.

Data Availability Statement: The data presented in this study are available on reasonable request from the corresponding author.

Acknowledgments: The authors would like to acknowledge the Wit-Composites company for sample preparation and Marcin Kneć for the help with sample testing.

Conflicts of Interest: The authors declare no conflicts of interest.

Appendix A

Figure A1. Microstructure of coatings: (**a**) cross-sectional view of the sample, (**b**) Al_2O_3, (**c**) aluminium, (**d**) quartz sand, (**e**) crystalline silica, (**f**) copper.

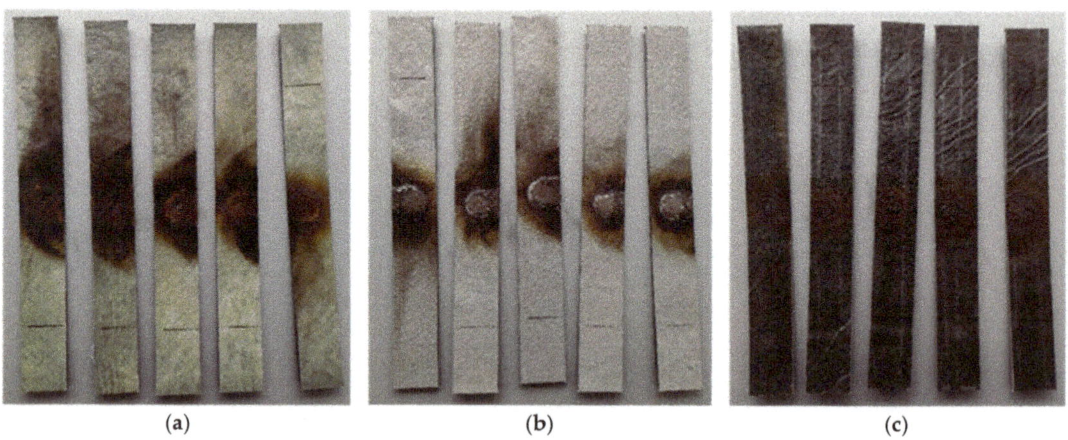

Figure A2. *Cont.*

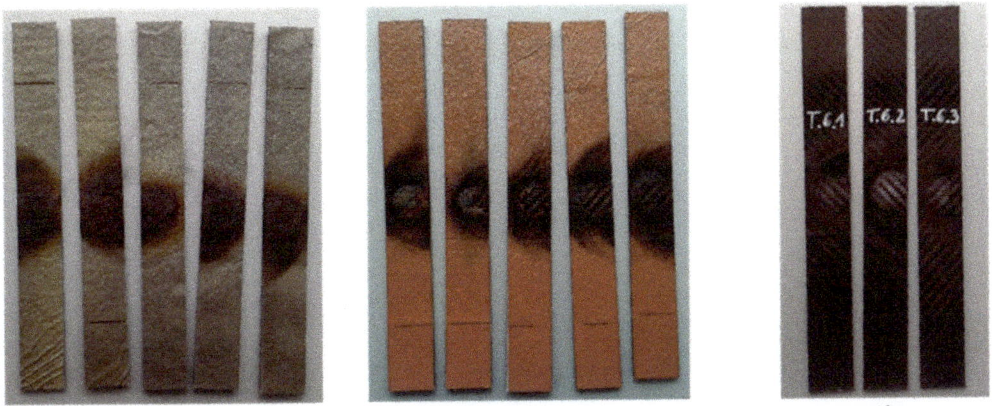

(**d**) (**e**) (**f**)

Figure A2. View of samples after flame exposure tests: (**a**) Al_2O_3, (**b**) aluminium, (**c**) quartz sand, (**d**) crystalline silica, (**e**) copper, (**f**) reference.

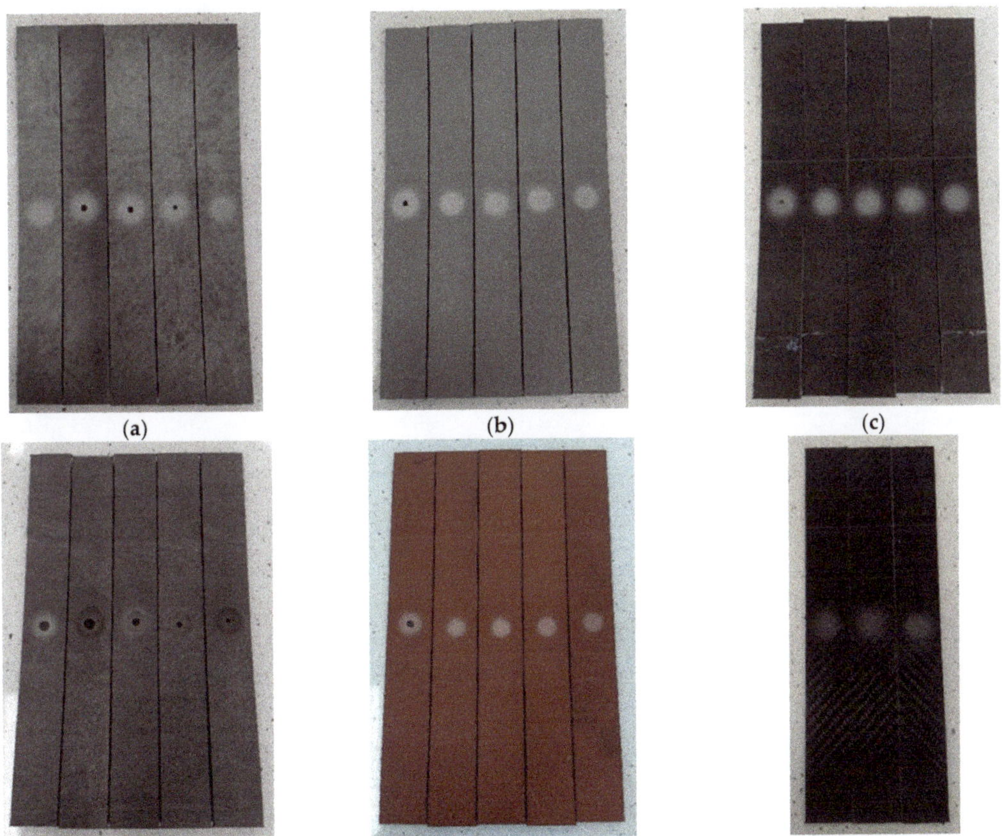

(**d**) (**e**) (**f**)

Figure A3. View of samples after erosion tests: (**a**) Al_2O_3, (**b**) aluminium, (**c**) quartz sand, (**d**) crystalline silica, (**e**) copper, (**f**) reference.

References

1. Sivasubramanian, P.; Mayandi, K.; Arumugaprabu, V.; Rajini, N.; Rajesh, S. History of Composites and Polymers. In *Polymer-Based Composites*; CRC Press: Boca Raton, FL, USA, 2021; pp. 1–21. [CrossRef]
2. Petersen, M.R.; Chen, A.; Roll, M.; Jung, S.J.; Yossef, M. Mechanical properties of fire-retardant glass fiber-reinforced polymer materials with alumina tri-hydrate filler. *Compos. Part B Eng.* **2015**, *78*, 109–121. [CrossRef]
3. Kamalbabu, P.; Kumar, G.C.M. Effects of Particle Size on Tensile Properties of Marine Coral Reinforced Polymer Composites. *Procedia Mater. Sci.* **2014**, *5*, 802–808. [CrossRef]
4. Zhuge, J.; Gou, J.; Chen, R.H.; Kapat, J. Finite difference analysis of thermal response and post-fire flexural degradation of glass fiber reinforced composites coated with carbon nanofiber based nanopapers. *Compos. Part A Appl. Sci. Manuf.* **2012**, *43*, 2278–2288. [CrossRef]
5. Kandare, E.; Griffin, G.J.; Feih, S.; Gibson, A.G.; Lattimer, B.Y.; Mouritz, A.P. Fire structural modelling of fibre-polymer laminates protected with an intumescent coating. *Compos. Part A Appl. Sci. Manuf.* **2012**, *43*, 793–802. [CrossRef]
6. Golewski, P.; Sadowski, T. Description of thermal protection against heat transfer of carbon fiber reinforced plastics (CFRP) coated by stiffened ceramic mat (TBC). *Compos. Struct.* **2019**, *229*, 111489. [CrossRef]
7. Golewski, P.; Rusinek, A.; Sadowski, T. Material characterization of PMC/TBC composite under high strain rates and elevated temperatures. *Materials* **2020**, *13*, 167. [CrossRef] [PubMed]
8. Lopera-Valle, A.; McDonald, A. Flame-sprayed coatings as de-icing elements for fiber-reinforced polymer composite structures: Modeling and experimentation. *Int. J. Heat. Mass. Transf.* **2016**, *97*, 56–65. [CrossRef]
9. Lupoi, R.; O'Neill, W. Deposition of metallic coatings on polymer surfaces using cold spray. *Surf. Coat. Technol.* **2010**, *205*, 2167–2173. [CrossRef]
10. Huang, W.; Wang, Z.; Xu, J.; Fan, X.; Wang, Y.; Zou, B.; Cao, X. Novel thermal protection coating based on Zr0.75Ce0.25O2/phosphate duplex system for polyimide matrix composites fabricated via a combined sol-gel/sealing treatment process. *Corros. Sci.* **2013**, *74*, 22–34. [CrossRef]
11. Luangtriratana, P.; Kandola, B.K.; Myler, P. Ceramic particulate thermal barrier surface coatings for glass fibre-reinforced epoxy composites. *Mater. Des.* **2015**, *68*, 232–244. [CrossRef]
12. Zhang, J.; Dong, X.; Zhou, Y.; Liu, J. Surface derusting of rust-covered metallic surface via shot blasting: A smoothed particle hydrodynamics simulation. *Wear* **2023**, *514–515*, 204566. [CrossRef]
13. Ao, W.; Zhou, Q.; Xia, Y. Impact resistance and damage mechanism of polyurea coated CFRP laminates under quasi-static indentation and low velocity impact. *Thin-Walled Struct.* **2023**, *193*, 111258. [CrossRef]
14. Liu, Q.; Guo, B.; Chen, P.; Su, J.; Arab, A.; Ding, G.; Yan, G.; Jiang, H.; Guo, F. Investigating ballistic resistance of CFRP/polyurea composite plates subjected to ballistic impact. *Thin-Walled Struct.* **2021**, *166*, 108111. [CrossRef]
15. Liu, Q.; Guo, B.; Chen, P.; Zhai, H.; Guo, Y.; Tang, S. Experimental investigation blast resistance of CFRP/polyurea composite plates under blast loading. *Thin-Walled Struct.* **2022**, *181*, 110149. [CrossRef]
16. Zhao, Y.; Zhang, Y.; Zhou, Q.; Bian, X.; Lyu, W. Lightning damage on GFRP materials of wind turbines under positive first return stroke. *Electr. Power Syst. Res.* **2023**, *215*, 108978. [CrossRef]
17. Das, S.; Kumar, V.; Lee, J.; Yokozeki, T.; Okada, T. Thickness threshold study of polyaniline-based lightning strike protection coating for carbon/glass fiber reinforced polymer composites. *Compos. Struct.* **2022**, *280*, 114954. [CrossRef]
18. Pang, J.; Qin, W.; You, W.; Zhong, M.; Meng, Q. Induction Heating Characteristics of Electroless Ni-Coated CFRP for Deicing and Anti-Icing. *Coatings* **2023**, *13*, 1531. [CrossRef]
19. Chen, H.; Wang, K.; Chen, Y.; Le, H. Mechanical and Thermal Properties of Multilayer-Coated 3D-Printed Carbon Fiber Reinforced Nylon Composites. *J. Compos. Sci.* **2023**, *7*, 297. [CrossRef]
20. Qiang, X.; Wang, K.; Jiang, X.; Xiao, Y.; Yibo, E. Effect of Fireproof Coatings on the Post-Fire Behavior of CFRP Composite Sheets. *Appl. Sci.* **2023**, *13*, 10369. [CrossRef]
21. Floch, L.; Da Cruz Chiochetta, B.; Ferry, L.; Perrin, D.; Ienny, P. Fire protective surface coating containing nanoparticles for marine composite laminates. *J. Compos. Sci.* **2021**, *5*, 6. [CrossRef]
22. Kim, H.; Kim, J.; Lee, J.; Lee, M.W. Thermal barrier coating for carbon fiber-reinforced composite materials. *Compos. Part B Eng.* **2021**, *225*, 109308. [CrossRef]
23. Lafabrier, A.; Fahs, A.; Louarn, G.; Aragon, E.; Chailan, J.F. Experimental evidence of the interface/interphase formation between powder coating and composite material. *Prog. Org. Coat.* **2014**, *77*, 1137–1144. [CrossRef]
24. Golewski, P.; Sadowski, T. Technological and Strength Aspects of Layers Made of Different Powders Laminated on a Polymer Matrix Composite Substrate. *Molecules* **2022**, *27*, 1168. [CrossRef]
25. Coto, B.; Hallander, P.; Mendizabal, L.; Pagano, F.; Kling, H.; Ortiz, R.; Barriga, J.; Selegård, L. Particle and rain erosion mechanisms on Ti/TiN multilayer PVD coatings for carbon fibre reinforced polymer substrates protection. *Wear* **2021**, *466–467*, 203575. [CrossRef]
26. Semmler, C.; Gyoktepeliler-Akin, E.; Killinger, A. Plasma sprayed ceramic coatings for the thermal protection of carbon fiber reinforced plastics (CFRP): Thermal and mechanical properties of YSZ, aluminum titanate, cordierite and mullite coatings. *Surf. Coat. Technol.* **2023**, *462*, 129509. [CrossRef]
27. Cilento, F.; Leone, C.; Genna, S.; Giordano, M.; Martone, A. Graphene NanoPlatelet-based coating as thermal protection from high-power radiative fluxes. *Compos. Struct.* **2023**, *319*, 117157. [CrossRef]

28. Kandola, B.K.; Luangtriratana, P. Evaluation of thermal barrier effect of ceramic microparticulate surface coatings on glass fibre-reinforced epoxy composites. *Compos. Part B Eng.* **2014**, *66*, 381–387. [CrossRef]
29. Bedon, C. Review on the use of FRP composites for façades and building skins. *Am. J. Eng. Appl. Sci.* **2016**, *9*, 713–723. [CrossRef]
30. Montero, D.A.; Arellano, C.; Pardo, M.; Vera, R.; Gálvez, R.; Cifuentes, M.; Berasain, M.A.; Gómez, M.; Ramírez, C.; Vidal, R.M. Antimicrobial properties of a novel copper-based composite coating with potential for use in healthcare facilities 06 Biological Sciences 0605 Microbiology. *Antimicrob. Resist. Infect. Control* **2019**, *8*, 3. [CrossRef]
31. Asmone, A.S.; Chew, M.Y.L. A study on the effectiveness of biological growth resistant coatings on external building façade systems in the tropics. *J. Build. Eng.* **2020**, *31*, 101377. [CrossRef]
32. Bauccio, M. *ASM Engineered Materials Reference Book*, 2nd ed.; ASM International: Detroit, MI, USA, 1994.
33. Kalkan, M.; Erenson, C. Direct flame test performance of boards containing waste undersized pumice materials. *J. Eng. Res.* **2023**. [CrossRef]
34. Golewski, P.; Sadowski, T. A novel application of alumina fiber mats as TBC protection for CFRP/epoxy laminates –Laboratory tests and numerical modeling. *J. Eur. Ceram. Soc.* **2018**, *38*, 2920–2927. [CrossRef]
35. Hassan, M.K.; Redhwi, A.M.N.; Mohamed, A.F.; Backar, A.H.; Abdellah, M.Y. Investigation of Erosion/Corrosion Behavior of GRP under Harsh Operating Conditions. *Polymers* **2022**, *14*, 5388. [CrossRef] [PubMed]
36. Khatkhate, A.M.; Singh, M.P.; Mirchandani, P.T. A Parametric Approximation for the Radius of Curvature of a Bimetallic Strip. *Int. J. Eng. Res.* **2017**, *V6*, 68–78. [CrossRef]
37. Bobrowski, A.; Drożyński, D.; Jakubski, J.; Szumera, M.; Kaczmarska, K.; Grabowska, B. Thermal deformation of moulding and core sands with an inorganic binder containing a relaxation additive. *Arch. Foundry Eng.* **2018**, *18*, 93–98. [CrossRef]
38. Yates, B.; McCalla, B.A.; Phillips, L.N.; Kingston-Lee, D.M.; Rogers, K.F. The thermal expansion of carbon fibre-reinforced plastics—Part 5 The influence of matrix curing characteristics. *J. Mater. Sci.* **1979**, *14*, 1207–1217. [CrossRef]
39. Zhou, X.L.; Chen, A.F.; Liu, J.C.; Wu, X.K.; Zhang, J.S. Preparation of metallic coatings on polymer matrix composites by cold spray. *Surf. Coat. Technol.* **2011**, *206*, 132–136. [CrossRef]

materials

MDPI

Article

Design and Study of Machine Tools for the Fly-Cutting of Ceramic-Copper Substrates

Chupeng Zhang, Jiazheng Sun, Jia Zhou and Xiao Chen *

School of Mechanical Engineering, Hubei University of Technology, Wuhan 430068, China;
20181048@hbut.edu.cn (C.Z.) 102110079@hbut.edu.cn (J.S.); 15272757302@139.com (J.Z.)
* Correspondence:20221010@hbut.edu.cn

Abstract: Ceramic-copper substrates, as high-power, load-bearing components, are widely used in new energy vehicles, electric locomotives, high-energy lasers, integrated circuits, and other fields. The service length will depend on the substrate's copper-coated surface quality, which frequently achieved by utilising an abrasive strip polishing procedure on the substrate's copper-coated surface. Precision diamond fly-cutting processing machine tools were made because of the low processing accuracy and inability to match the production line's efficiency. An analysis of the fly-cutting machining principle and the structural makeup of the ceramic-copper substrate is the first step in creating a roughness prediction model based on a tool tip trajectory. This model demonstrates that a shift in the tool tip trajectory due to spindle runout error directly impacts the machined surface's roughness. The device's structural optimisation design is derived from the above analyses and implemented using finite element software. Modal and harmonic response analysis validated the machine's gantry symmetrical structural layout, a parametric variable optimisation design optimised the machine tool's overall dimensions, and simulation validated the fly-cutterring's constituent parts. Enhancing the machine tool's stability and motion accuracy requires using the LK-G5000 laser sensor to measure the guideway's straightness. The result verified the machine tool's design index, with the Z- and Y-axes' straightness being better than 2.42 μm/800 mm and 2.32 μm/200 mm, respectively. Ultimately, the device's machining accuracy was confirmed. Experiments with flying-cut machining on a 190 × 140 mm ceramic-copper substrate yielded a roughness of Sa9.058 nm. According to the experimental results, the developed machine tool can fulfil the design specifications.

Keywords: ceramic-copper substrate; fly-cutting; roughness

1. Introduction

Electric locomotives, green cars, integrated circuits, and high-energy lasers are just a few of the quickly developing industries that extensively use heat-resistant,+ ceramic-copper substrates. These substrates boast exceptional thermal conductivity and power efficiency, which means their surface quality directly impacts their longevity [1–4]. Currently, surface of copper cladding is usually processed by abrasive belt grinding, which is time-consuming and less accurate. Moreover, its requires higher levels of accuracy and efficiency to meet production line demands. It might be beneficial to develop a type of diamond precision fly-cutting machine tool that can be applied to the ceramic-copper substrate because this type of ultraprecision and precision machining technology can realise the rapid removal of surface materials on metal and ceramics and effectively improve the machining surface finish and machining accuracy [5–8].

The design of precision machine tools has been thoroughly researched by academics from around the world. A typical precision machining technology is the diamond precision fly-cutting machining process. In the design and optimisation of precision machines, to reduce the impact of dynamic errors within the machine on the stability of the device and the accuracy of the machining, the static and dynamic characteristics and the machine's thermal deformation must be analysed [9–13]. The use of a gantry symmetrical structure in the design of precision machine tools can effectively reduce the impact of thermal deformation on the device; in order to ensure its performance, the machine tool should be considered, first of all, with reference to the effect of gravity on its static deformation [14,15]In research to improve the machining accuracy of precision machine tools, Wang et al. [16] promulgated a machine tool error model based on helix theory, identified the machine tool's weak points, and then optimised its design with harmonic response analysis to assess the device's layout form. The dynamic properties of machine tools greatly influence the precision of a device's machining [17]. Lei et al. [18] exploited measurements to determine weak subsystems while enhancing the machine's settings for operation to increase machine effectiveness. By implementing a dynamic model based on the hypothesis of multibody systems, Ding et al. [19,20] investigated the impact of the spindle and tool's dynamic properties on the manufactured surface's midfrequency ripple magnitude. Chen et al. [21] hold that the spindle is just like the main component of the precision device, and the spindle motion accuracy will directly affect the face shape accuracy of the machined surface. In the study of surface quality in fly-cutting machining, the diamond precision fly-cutting process is intermittent, and tool wear and deflection affect the quality of the machined surface and thus the evaluation of surface roughness factors [22–26]. Li et al. [27] conducted tool tip position response experiments by modelling the tip position dynamics to predict the roughness of the machined surface. Bai et al. [28] adopted a neural network based on the ResNet-50 architecture for deep learning to forecast machined surface roughness. Tan et al. [29] used experimental research to determine the influence law of various process parameters on the machined surface roughness and confirmed the ideal process parameters. Cao et al. [30] constructed a theoretical surface morphology model by combining the cutting path and working condition parameters. They confirmed it through examinations to discover the quantitative link between the condition parameters on the surface roughness of the workpiece. The present presents the development of a diamond precision fly-cutting machine for ceramic-copper substrates to increase the productivity and machining accuracy of copper cladding surfaces on ceramic-copper substrates.

The overall layout scheme was determined by performing a harmonic response analysis of the machine's layout form using finite element analysis, following the study of the structural model of the ceramic-copper substrate and the principle of the diamond fly-cutting machining process. The residual height model of the workpiece surface was established based on the tool tip trajectory equation to investigate the impact of spindle runout error on the machined surface roughness. A multistation design was employed to increases the efficiency of the manufacturing line; the device's stability was improved through selection of the best design parameters, and parametric variables were employed for determining the critical dimensions of the beam-column structure. The fly-cutter disc assembly was subjected to static and transient analyses to confirm its stability. Lastly, the machine assembly was concluded through a checking of the precision of the Y and Z guides, and the tool's operation was ensured via fly-cutting machining trials.

2. Design Principle of the Diamond Precision Fly-Cutting Device

2.1. Ceramic-Copper Substrate Composition

The ceramic-copper substrate structure consists of a ceramic substrate, copper cladding, and photographic film, as shown in Figure 1. One of the copper claddings, through the electroplating process, is coated with copper on the surface of the ceramic substrate due to the effect of the electroplating process causing uneven thickness of the copper cladding surface phenomenon. However, the quality of the copper cladding surface directly impacts the service life of the substrate and, therefore, needs to be in the electroplating after the copper cladding surface processing. In order to ensure the performance of ceramic-copper substrates, it is usually necessary to control the surface accuracy of the copper cladding processing to 15 nm.

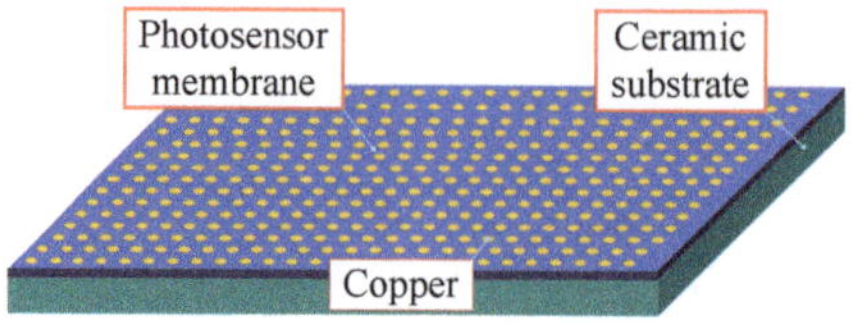

Figure 1. Ceramic-Copper Substrate Composition Mechanism.

2.2. Fly-Cutting Processing Method

There are two typical precision diamond fly-cutting machining (or fly-cutting machining) methods related to the installation arrangement of the diamond tool and the machining requirements. Figure 2 displays the two modes of operation: Figure 2a shows the flat surface machining, where the diamond tools are positioned axially along the spindle at the lower end of the fly cutter disc, and Figure 2b shows the free-form surface machining, where the diamond tools are positioned radially along the spindle at the side of the fly-cutter disc.

For different types of workpiece clamping, there are two types of spindle layout: vertical and horizontal. To increase the machining efficiency of the fly-cutting method, the device laid out in this research only needs to remove material from the copper-clad surface of the ceramic-copper substrate, with consideration to the workpiece's processing requirements and the clamping method Figure 2a.

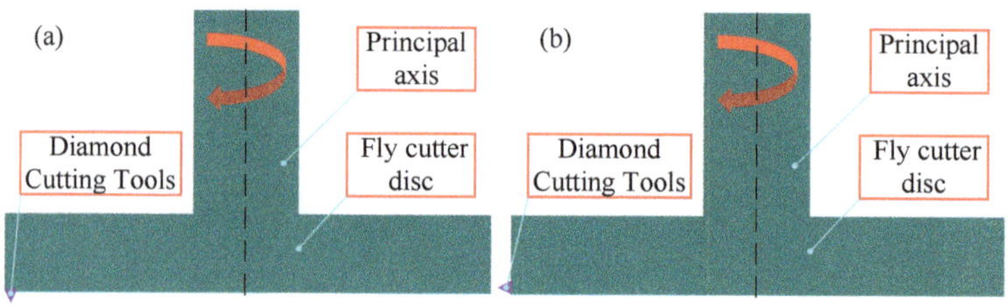

Figure 2. Diamond fly-cutting processing method. (**a**) End fly-cutting. (**b**) Radial fly-cutting.

The machining schematic depicted in Figure 3 shows the fly-cutter disc connected to the spindle along the z-axis vertically feeding to the cutting height; the fly-cutter disc is at the lower end of the installation of counterweights to adjust the dynamic balance in the fly-cutter disc, and the workpiece is placed by the vacuum fixture adsorption, along the horizontal direction of the y-axis for uniform speed feeding.

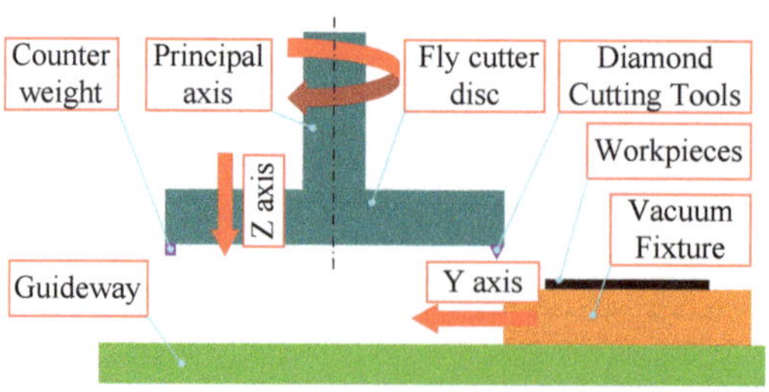

Figure 3. Schematic diagram of the fly-cutting process.

2.3. Fly-Cutting Principle and Accuracy Analysis

Fly-cutting principle and accuracy analysis Figure 4 indicates a type of noncontinuous machining called "flying-cut machining". During machining, the workpiece is fed uniformly along the horizontal plane while the spindle swings the diamond cutter on the fly-cutter disc at a high speed. The blade only comes in contact with the workpiece once.

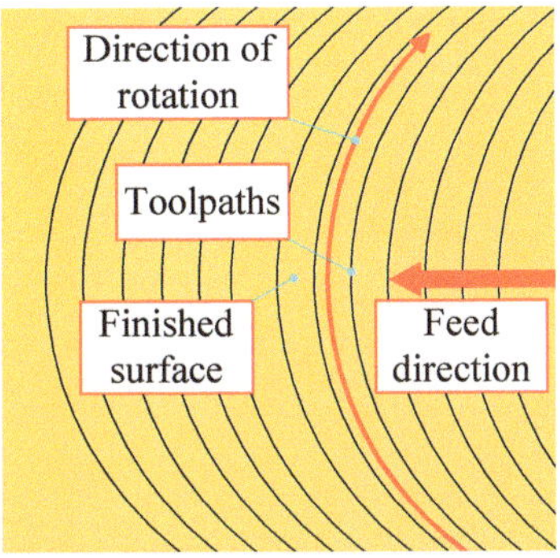

Figure 4. Fly cutting trajectory diagram.

By examining the motion trajectory of the tooltip on the flying tool disc and using the previously discussed analysis of the motion of the machining surface and the flying tool disc, we can determine the three-dimensional trajectory equation of the tip motion concerning the tooltip and the coordinate system of the workpiece surface as follows:

$$\begin{aligned} x(t) &= R\cos(2\pi nt) \\ y(t) &= R\sin(2\pi nt) + ft \\ z(t) &= A[1 + \varepsilon\sin(2\pi nt)] \end{aligned} \quad (1)$$

where R represents the radius of the fly-cutting disc (mm), f represents the feed speed (mm/s), n represents the spindle speed of spindle rotation (r/min), t is the fly-cutting machining time (s), A is the removal volume (mm), and ε is the runout error of the spindle.

As shown in Figure 5, the axial runout error of the spindle in the form of sinusoidal waves during fluctuating causes the fly-cutting disc to deflect, affecting the tooltip's relative position on the workpiece's surface.

$$E_\varepsilon = \varepsilon\sin(\omega t) \quad (2)$$

where: E_ε is the runout error of the flying tool disc, and ω implies the angular frequency of spindle rotation (rad/s). This can be expressed as follows:

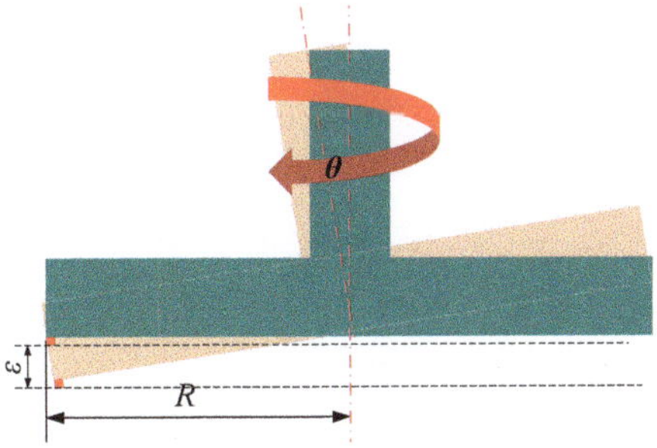

Figure 5. Schematic diagram of fly-cutting process.

During fly-cutting, the shape of the material's surface removed is related to the tool's geometry through the proximity between the tool and the workpiece, as shown in Figure 4. As a noncontinuous machining process, fly-cutting can be viewed as a motion with two degrees of freedom during the surface removal process: the workpiece's horizontal feed motion and the spindle's rotary motion. The residual height is the amount of uncut material on the workpiece's surface between every two cuts. The theoretical maximum material residual height appears in Figure 6 below [31].

$$R_{max} = r - \sqrt{r^2 - \left(\frac{f}{2}\right)^2} \tag{3}$$

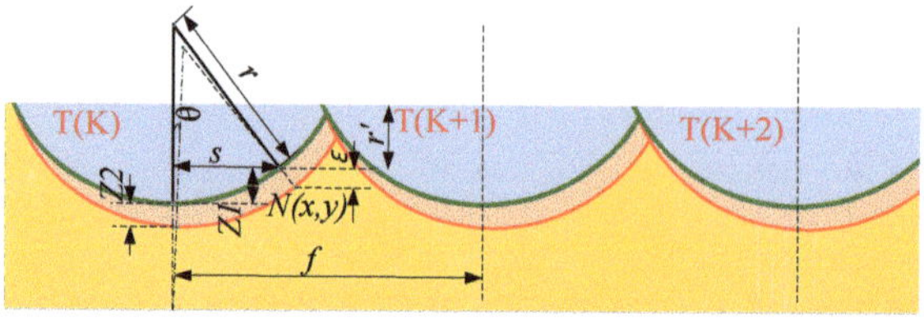

Figure 6. Residual height model after the fly-cutting process.

The current theory states that the maximum material residual height can respond to the level and trend of the theoretical value of surface roughness to some extent. As a result, the material residual height that results can be considered the workpiece's theoretical roughness value after the fly-cutting process. The theoretical surface roughness Raof the workpiece can be known as $\approx R_{max}/4$ [32], so the theoretical surface roughness is related to the academic maximum material residual height.

Considering the spindle runout error, the machining surface was initially discretised as an X–Y plane comprised of nodes. The N(x,y) coordinates for any node may count as appointed, as illustrated in the figure: position any node N(x,y) at the machining surface of the workpiece.

$$T(K) = k \cdot dt - \frac{1}{2}dt \cdot \arcsin\left(\frac{y}{r}\right)/\pi \tag{4}$$

Its actual maximum residual height at any node N can be expressed as follows:

$$R'_{max} = r - \sqrt{r^2 - s^2} + E_\varepsilon \tag{5}$$

$$s = x - \sqrt{r^2 - y^2} + r - f \cdot T(K) \tag{6}$$

$$R'_{max} = \frac{r - \sqrt{r^2 - \left(x - \sqrt{r^2 - y^2} + r - f \cdot \left(k \cdot dt - \frac{1}{2}dt \cdot \arcsin\left(\frac{y}{r}\right)/\pi\right)\right)^2} + \varepsilon \sin(\omega t)}{4} \tag{7}$$

where T (K) is the machining time at node N relative to the Kth tool tip motion trajectory (s), r is the radius of the tool's arc (mm), f is the workpiece's horizontal feed rate (mm/s), s is the horizontal distance between node N(x,y) and the tool tip trajectory at any place (mm), and R'_{max} is the roughness inclusive of the runout error of the spindle.

Equation (7), as can be observed, shows a relationship between the spindle's runout error and the roughness of the workpiece's actual machined surface. In the machining process, the solid surface roughness results from the superposition of various factors, in which the spindle will produce sinusoidal periodic runout error, affecting the surface accuracy of the machined surface. Therefore, it will cause the tool to generate an offset caused by the inconsistency between the theoretical machining surface and the actual machining surface. Consequently, it is necessary to consider analysing the influence of spindle runout error on fly-cutting.

3. Design of the Ceramic-Copper Substrate Fly-Cutting Device

3.1. Form of the Installation's General Layout

In the design of precision devices, the layout form of the device will affect the static and dynamic characteristics of the device itself, as well as thermal deformation, which directly affects the performance of the device, so the layout form of the precision device plays a crucial role in the machining quality of the machined surface. The common precision device layout form has a gantry layout and cantilever layout of these two forms.

A gantry layout frequently occurs in the form of a symmetrical beam-column structure, which has good static and dynamic characteristics and can effectively reduce the impact of thermal deformation of the machine tool. A cantilever layout typically includes an open-column structure, which has the advantage being compact and easy to assemble, but its resistance to thermal deformation is poor, and there is considerable flexural deformation. The two layout schemes' gantry layout and the cantilever layout were comparedto guarantee the performance of the fly-cutting machine designed in this study. The harmonic response results of the modal and harmonic response analysis of these two layout schemes are displayed in Figure 7, where the first-order intrinsic frequencies of the cantilever layout form and the gantry layout form are 245 Hz and 346 Hz, respectively. In the range of 0–1000 Hz, there are three obvious resonances of the cantilever layout form and two apparent resonances of the gantry layout form. The gantry layout form has good harmonic response properties. Given that the gantry layout form exhibited favourable dynamic properties, the gantry layout was selected as the machine tool's layout scheme in this study to enhance the designed device's performance.

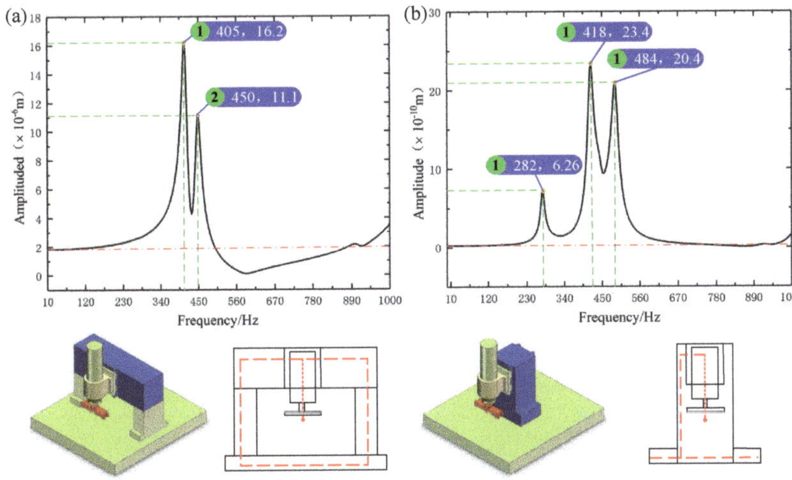

Figure 7. Comparison of the harmonic response analysis of the two schemes. (**a**) Gantry structure. (**b**) Cantilever layout.

Based on the above finite element analysis results, the overall layout was then designed after the symmetrical gantry layout was determined to be the layout form of the device developed in this study. A marble base, bed, column, beam, electro spindle, flying cutter disc, pneumatic fixture, x/y-axis skateboard, and others were the main components of the device designed in this study, depicted in Figure 8. The main body of the device described in this paper was made of concrete and vibration-resistant marble to improve the stability of the device.

3.2. Design Indicators for Machine Tools

Requirements for the designed device in the fly-cutting processing were a machining surface roughness Sa better than 15 nm and a parallelism of the guideway better than

4 μm/800 mm. This was assumed to ensure the machining performance of the machine tool as well as to improve the machining efficiency of the ceramic-copper substrate machining surface according to the machining requirements for developing the device design indicators shown in Table 1.

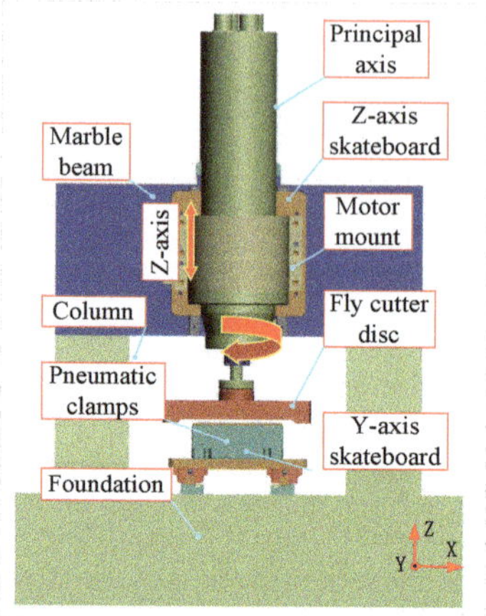

Figure 8. The general structure of the device.

Table 1. Design index of precision fly-cutting device.

Typology	Fly-Cutting
Distance ravelled	Y-axis: 800 mm Z-axis: 5 mm
Guideway straightness error	Y-axis: 4 μm/800 mm Z-axis: 3 μm/200 mm
Minimum feed speed	Y-axis: 10 μm/s Z-axis: 1 μm/s
Spindle runout error	Spindle: 0.5 μm
Workpiece material	Copper
Spindle speed	0–1000 rpm
Size of workpiec	190 mm × 140 mm
Workpiece roughness	Sa $\leq$ 15 nm

3.3. Structural Design of the Beam-Column Arrangement

To achieve rapid thinning of the ceramic-copper substrate copper cladding surface and improve production efficiency, the fly-cutting device designed in this study employed a double-station design, which includes two stations symmetrically arranged on the same marble beam, which serves to semifinish and finish the ceramic-copper substrate machining surface, respectively. Within the crossbeam guide mounting surface, a remote force of 150 N was applied. The single-station and dual-station schemes were verified by employing Workbench. Figure 9 displays the results of the analysis. The fly-cutting device with a dual-station design had a maximum deformation at the crossbeam of 0.16488 μm, which was less than the maximum deformation of the single-station design. Consequently, the maximum deformation may be made by using the dual-station design. Thus, while adhering to the design index, the double-station design can increase production efficiency. Consequently, the two-station design scheme appears within this paper.

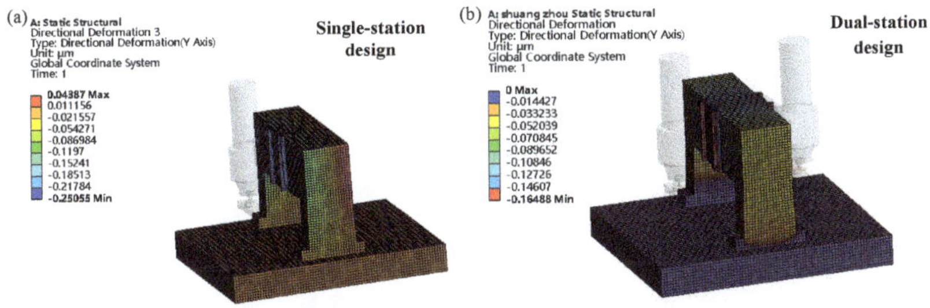

Figure 9. The general structure of the device. (**a**) Single station. (**b**) Dual station.

After the overall structure of the device was determined, the details of the device needed to be optimised and designed. The main body of the device included a symmetrical gantry beam-column structure. To further improve the static and dynamic performance of the device, the positioning and setting dimensions in the beam-column structure were parametrically varied in the structure design. As shown in Figure 10, the distance between the setting-up column and the two ends of the beam is a, the width of the column is b, the span is c, and the total length of the beam is d. The machining object of the device is a ceramic-copper substrate with a size of 190 × 140 mm, and to ensurethe Y and Z sliding plate, the vacuum jig, and the spindle's mounting and workspace, the dimensions of the device's column structure were determined to be 300 × 220 mm (c × g).

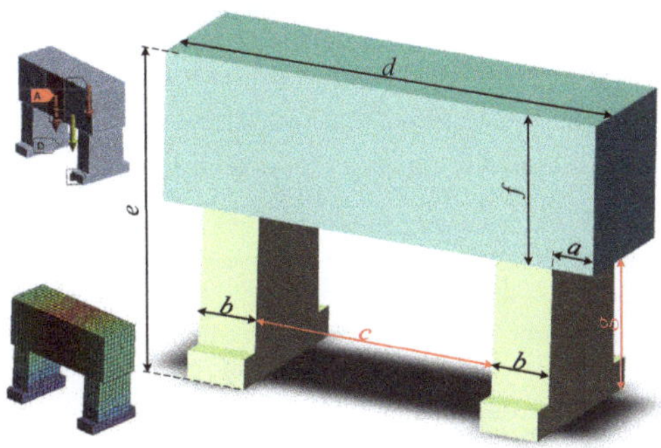

Figure 10. Parametric variable in the design of the beam-column structure.

The equations for the other parameters in the beam-column structure are as follows:

$$
\begin{aligned}
d - 2b - 2a &= 300 \\
e - f &= 220 \\
0 \leq a &\leq 40 \\
60 \leq b &\leq 80 \\
420 \leq d &\leq 540 \\
420 \leq e &\leq 440 \\
200 \leq f &\leq 220
\end{aligned}
\tag{8}
$$

The beam-column structure's parametric structural design was determined according to the methodology described above, and with the displacement of the column from the end face of the beam as the variable, finite element analysis was carried out, with the ideal design parameters derived based on the optimal simulation results. Table 2 displays the design parameters both before and after optimisation.

Table 2. Parameters before and after optimisation of the crossbeam-column structure.

Parametric	a	b	c	d	e	f	g	Frequency
Initial size (mm)	0	75	300	450	440	220	220	405 Hz
Optimised size (mm)	20	80	300	500	420	200	220	420 Hz

To consider the effect gravity, a remote force of 150 N was exerted at the Z-rail mounting surface of the crossbeam, and finite element analysis was carried out on the crossbeam-column structure under the two materials of marble and structural steel by using Workbench software. The simulation results appear in Figure 11, which clearly shows that selecting marble as the cross-beam column structure can effectively reduce the beams' static deformation. Hence, a marble crossbeam-column structure was chosen.

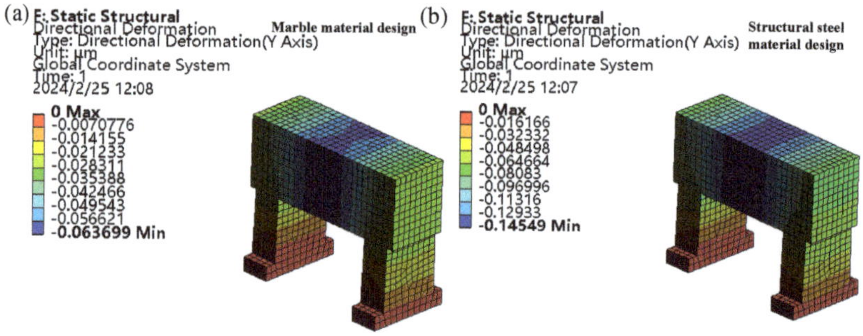

Figure 11. Static deformation of the beam column. (**a**) Marble material design. (**b**) Structural steel material design.

Marble was chosen as the material for the crossbeam-column structure to lessen the crossbeam's distortion. In Figure 12, the machine's harmonic response curve before the optimisation is represented by the blue dashed line, while the red solid line represents the machine's harmonic response curve after optimisation. The machine's first-order intrinsic frequency increased from 405 to 420 Hz when compared to the preoptimisation parameters, and the maximum deformation of the Z-directional orientation of the marble crossbeam of the device slumped from 0.55433 μm to 0.063699 μm. Thus, optimising the beam-column structure utilising parametrised variables can enhance the device's static and dynamic performance.

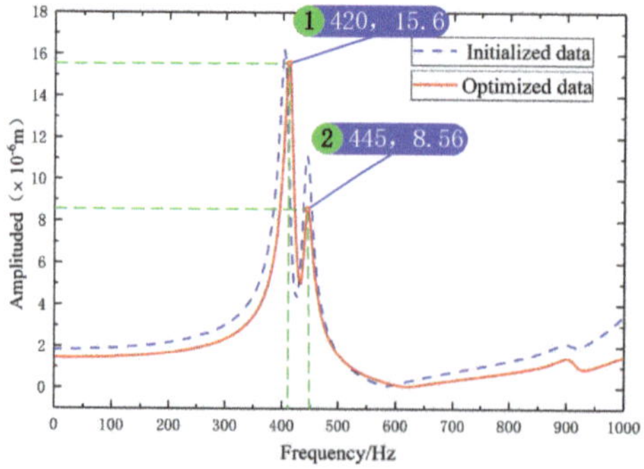

Figure 12. Optimised machine harmonic response.

3.4. Structural Design of the Flying-Cutting Disc Components

The fly-cutting approach requires that the flying-disc components mounted on the diamond tool rotate the workpiece surface continuously at high speed. As a result, the flying-disc components must have good dynamic and static properties. A fly-cutting device used at the end of the fly-cutting processing mode was designed in this study,with the diamond tool migrating to the fly-cutter disc side.The influence of spindle runout on the tool path was reduced by using a tapered adapter to connect the spindle to the flying tool disc. The fly-cutter disc components included a symmetrical structure design to facilitate dynamic balance debugging of the fly-cutter disc.The results above indicate that the fly-cutting disc's overall layout performs as intended. Figure 13 depicts the overall configuration of the fly-cutter disc described in this paper, which includes the counterweight block, diamond tool, electric spindle, and fly-cutter disc.

As the fly-cutting disc has a high linear speed, which directly affects the machining quality, to ensure the flying cutter's processing stability, two schemes were implemented to analyse the fly-cutter disc components via Workbench software. To account for gravity, the rotary support constraints on the fly cutter disc were set to a rotational speed of 800 r/min, and the overall structure of the fly-cutter disc of structural steel material was subjected to finite element analysis.As shown in Figure 14, the b scheme had a static self-weight maximum deformation of 0.16736 μm and a transient maximum deformation of 0.16772 μm; the use b scheme's stability was better than that of the a scheme Based on the above analysis, to guarantee the device's design index, the b scheme was chosen as the design scheme of the flying knife disc.

Figure 13. Fly-cutter disc component structure.

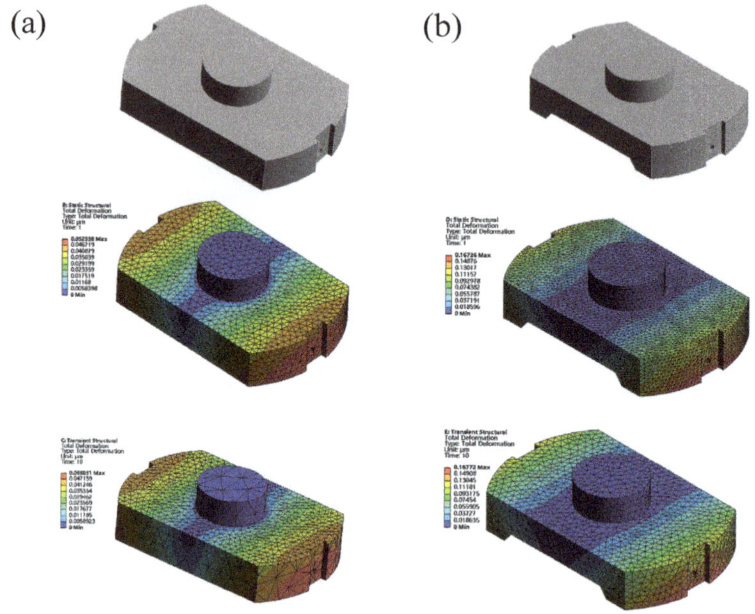

Figure 14. Comparison of the fly- cutter disc options. (**a**) Original programme. (**b**) Lightweight programme.

4. Measurement and Verification Experiments

4.1. Complete Machine Assembly

After the design and processing were concluded, the machine tool created in this study needed to be constructed and debugged. The machine tool made for this paper, as illustrated in Figure 15, has two machining stations that allow for semifinishing and finishing. The holding motor drives the screw-nut sub of the Z-axis to produce the high-precision linear motion of the Z-direction skid; the axes S1 and S2 originate from the electric spindle to obtain a higher rotational speed to remove surface materials rapidly. The fly-cutter disc spindles S1 and S2 are positioned within the Z-direction skids corresponding to the two stations. A pair of nuts serve for anchoring the Y-skate at the bottom, and a servomotor rotates the screw to provide the skate's linear motion. Spindles drive axes S1 and S2 to achieve high speeds and quick removal of surface materials. To ensure the precision of the Z-direction movement,the spindles go the axes S1 and S2 to obtain high speeds and achieve fast removal of surface materials. The servomotor works to rotate the screw, enabling the Y-skate to move linearly. The nut pair sits at the bottom of the skate. To ensure the positioning accuracy of the Z-direction movement of the Z-direction skateboard during fly-cutting, the Z-axis adopts a linear scale with a resolution of 0.05 μm, which realises the high-precision positioning of the Z-direction skateboard.

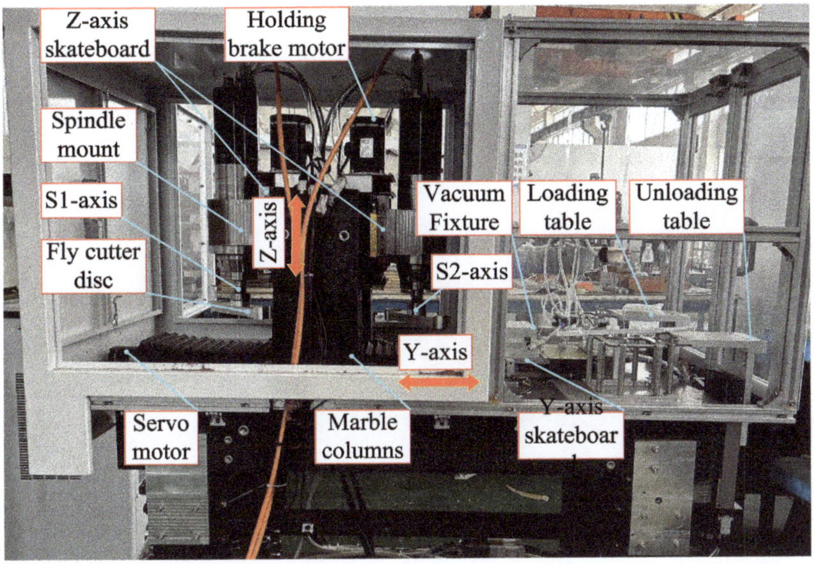

Figure 15. Complete machine assembly.

4.2. Measurement of Guideway Straightness

In the fly-cutting process, the Z- and Y-direction guides must move in a straight line, which has an essential influence on the accuracy of the machined surface. As a result, measuring and verifying the precision of the Y and Z guide rails is required to increase the accuracy of fly-cutting machining. The data from the Matlab fitting curve were collected,with a micrometre measurement through the Y-axis guide being employed, along with a rail measurement range of 800 mm. The skateboard is moved at a constant speed of 20 mm/min along the Y-axis from the start to the end of the movement. Figure 16 depicts the tilt-removal principle model. The tilt slope of the fitted picture curve, k_1, can be written as follows:

$$z_1(i) = k_1 \times x(i) \tag{9}$$

$$x(i) = L + v_x \cdot t \tag{10}$$

$$z'(i) = z(i) - z_1(i) \tag{11}$$

$$z'(i) = z(i) - \frac{\sum\limits_{i=1}^{n} x(i)z(i) - \sum\limits_{i=1}^{n} x(i)z(i)/n}{\sum\limits_{i=1}^{n} x^2(i) - \sum\limits_{i=1}^{n} x^2(i)/n} x(i) \tag{12}$$

where $z(i)$ is the measurement of the guideway vice parallelism data, $x(i)$ is the measurement of real-time X directions, n is the measurement of the number of acquisition points, $z_1(i)$ is the guideway vice-tilt measurement of the quantity of error, $z'(i)$ is the principle of removing the tilt error in the processing of the data, v_x is the data measurement and acquisition of the speed of the movement, and t is the detailed measurement and acquisition of time.

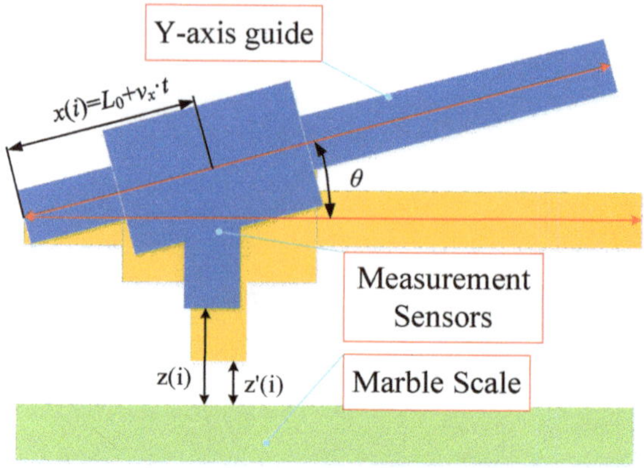

Figure 16. Corrected model for guideway straightness skew.

As illustrated in Figure 17, the marbling size was utilised as the standard for guide rail straightness measurement. The surface face shape accuracy on the scale is less than $\lambda/10$. The least squares method was used to eliminate the degree of tilt in the data to accurately obtain the straightness data of the rail sub.

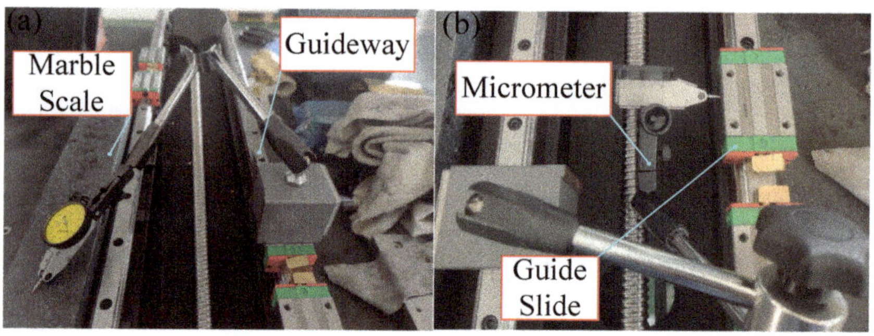

Figure 17. Guideway accuracy Measurement. (**a**) Straightness measurement. (**b**) Flatness measurement.

The formula utilises the collected $z(i)$ data to obtain the $z'(i)$ data through Matlab fitting; in Figure 18, the red line represents the conclusion of fitting the measurement data via removal of the tilt error amount of the guideway vice tilt measurement, and the blue line indicates the initial data fitting image after measurement; the fitted image demonstrates that the X-axis motion straightness error is 2.42 μm/800 mm, which satisfies the design indexes.

The accuracy measurement results of the same approach to the Z-way guide are shown in Figure 18; the Z-axis motion straightness error is 2.32 μm/200 mm, thus meeting the design indicators.

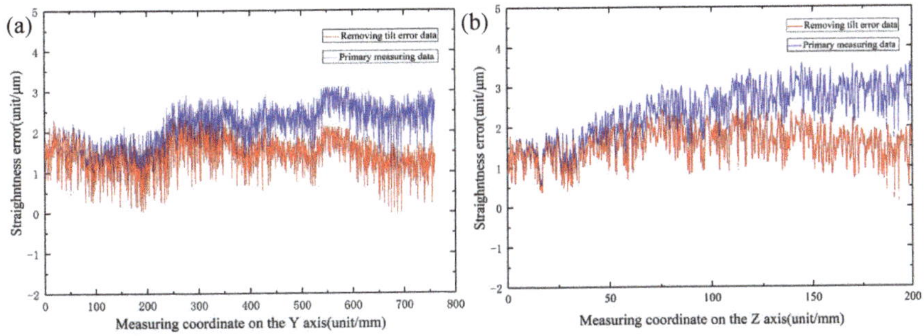

Figure 18. Guideway straightness measurement results. (**a**) Y-axis guide measurement results. (**b**) Z-axis guide measurement results.

4.3. *Fly Cutting Processing Experiment*

After the assembly of the device designed in this study and completion of the accuracy measurements, the processing accuracy of the device was verified by carrying out experiments and measurements on the fly-cutting machining of a 190 mm × 140 mm ceramic-copper substrate. Two vacuum cleaners were installed at each station to prevent the loss of waste debris on the machined surface during the removal process, which could affect the machined surface's machining quality.

After being sucked through a vacuum fixture, the machined components were used in fly-cutting experiments. The fly-cutting machining experimental parameters included a 0.1 mm cutting amount, an 800 rpm cutting spindle speed, a 100 mm/min workpiece feed speed, an arc radius of 4791.27 μm, a front angle of 0°, a back angle of 10°, and a diamond tool height of 6.189 mm, as shown in Figure 19.

Figure 19. Diamond cutting tools.

The roughness of the ceramic-copper substrate's copper-coated surface, which is ignored by fly-cutting, was measured with the ZYGO white light interferometer to be Sa133.67 nm, as shown in Figure 20. After the sample was fly cut via the device proposed in this paper, in order to meet the device's design parameters, the surface roughness was measured as Sa9.058 nm.

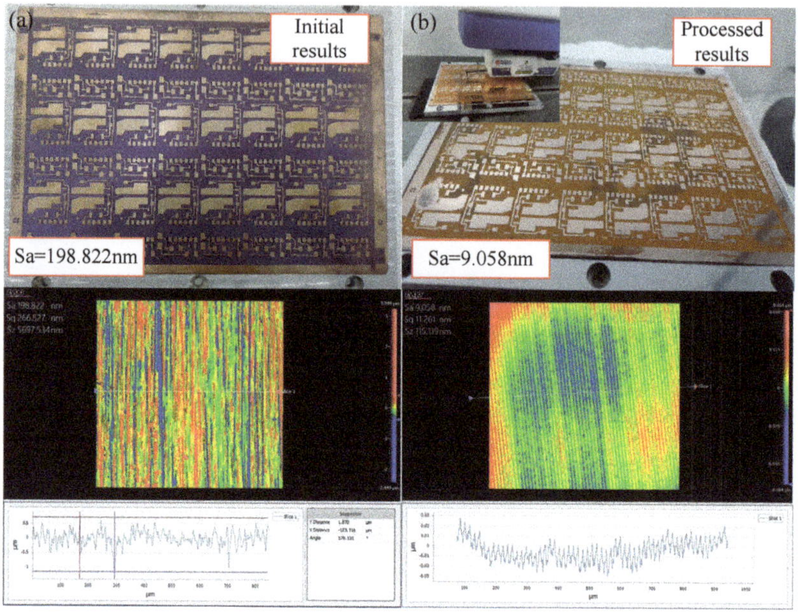

Figure 20. Roughness measurement results. (**a**) Initial results. (**b**) Processed results.

5. Conclusions

We created a precision diamond fly-cutting tool that was applied specifically to the ceramic-copper substrate to increase the machining accuracy and productivity of the substrate's machining surface.

(1) By analysing the working principle of fly-cutting processing and considering the spindle movement accuracy, we constructed a roughness prediction model based on the tool tip trajectory, and it appeared that the runout error of the spindle led to the change of the tool tip trajectory, which directly affected the machining surface roughness.

(2) Through an harmonic response analysis and comparison results, the gantry symmetric structure was chosen as the layout form of the device to improve the stability of the machine. The use of a double-station design was employed to enhance productivity, and the overall layout of the critical dimensions underwent parametric variable design. Static, modal, and harmonic response analyses were used to optimize the general layout of the device and the fly-cutting disc components. The optimised simulation results showed that after the optimisation of the device's first-order, the first-order intrinsic frequency of the device was increased to 420 Hz, and the maximum deformation of the crossbeam's Z-direction was reduced to 0.063699 μm; the maximum deformation of the flying-cutter disc component's static deadweight was 0.16736 μm, and the maximum deformation of the transient state was 0.16772 μm, which could effectively improve the machine tool's static and dynamic characteristics.

(3) The LK-G5000 laser sensor was used for measuring the precision of the diamond fly-cutting device. The measurement results showed that the Y-axis guideway straightness was better than 2.42 μm/800 mm and that the Y-axis guideway straightness was better than 2.32 μm/200 mm, which met the design indicators.

(4) Fly-cutting processing experiments and measurements were conducted on the developed precision diamond fly-cutting device with 190 mm × 140 mm ceramic-copper substrate.The measurement of the surface roughness of the processed surface Sa was 9.058 nm, which met the device's design specifications.

This paper mainly discusses the complete design of a precision fly-cutting machine tool and the use of spindle runout error to examine its impact on machining surface roughness. As the design of the fly-cutting machine also needs to account for manufacturing costs, the online detection of tool wear, compensation, and other issues, a more in-depth understanding of the fly-cutting process is needed. To this end, further experimental follow-up should be conducted.

Author Contributions: Methodology, C.Z.; writing—original draft, J.S.; software, J.Z.; funding acquisition, X.C. All authors have read and agreed to the published version of the manuscript.

Funding: This work has been supported by the following funds: Hubei Provincial Key Research and Development Programme Project Tasks, 2023BAB067. 2023 Open Fund for Hubei Provincial Key Laboratory of Modern Manufacturing Quality Engineering, KFJJ-2023017.

Institutional Review Board Statement: Not applicable.

Informed Consent Statement: Not applicable.

Data Availability Statement: Data are contained within the article.

Conflicts of Interest: The authors declare no conflicts of interest.

References

1. Guzowski, B.; Rosowski, A.; Lisik, Z.; Jakubowska, M.; Sharp, M. The integration of optical interconnections on ceramic substrates. *Microelectron. Eng.* **2016**, *157*, 19–23. [CrossRef]
2. Goto, T.; Matsubayashi, Y.; Akedo, J. Ceramic coating on rubber by aerosol deposition with cryogenic substrate cooling. *Ceram. Int.* **2024**, *50*, 892–896. [CrossRef]
3. Zhu, J.; An, R.; Wang, C.; Wen, G. Fabrication of Al_2O_3-Mullite-AlN Multiphase Ceramic Layer on W-Cu Substrates for Power Semiconductor Packaging. *IEEE Trans. Components Packag. Manuf. Technol.* **2015**, *5*, 182–187. [CrossRef]
4. Kim, Y.; Ahn, C.W.; Choi, J.J.; Ryu, J.; Kim, J.W.; Yoon, W.H.; Park, D.S.; Yoon, S.Y.; Ma, B.; Hahn, B.D. Next Generation Ceramic Substrate Fabricated at Room Temperature. *Sci. Rep.* **2017**, *7*, 6637 . [CrossRef] [PubMed]
5. Schubert, M.; Hahn, M.; Exner, J.; Kita, J.; Moos, R. Effect of substrate hardness and surface roughness on the film formation of aerosol-deposited ceramic films. *Funct. Mater. Lett.* **2017**, *10*, 1750045. [CrossRef]
6. Ganesan, G.; Malayath, G.; Mote, R.G. A review of cutting tools for ultra-precision machining. *Mach. Sci. Technol.* **2022**, *26*, 923–976. [CrossRef]
7. Lukin, E.S.; Anufrieva, E.V.; Popova, N.A.; Morozov, B.A.; Preobrazhenskii, V.S.; Bezlepkin, V.A.; Ukhvatova, L.S.; Brzhezinskii, G.V. Analysis of Surface Microstructure and Quality and Properties of Aluminum Oxide Substrates. *Glass Ceram.* **2011**, *67*, 271–276. [CrossRef]
8. Andre, J.; Botrel, R.; Schunck, J.; Pinay, A.; Chicanne, C.; Theobald, M. Specific Adaptations of Mechanical Machining Processes for Laser Target Manufacturing. *Fusion Sci. Technol.* **2016**, *70*, 237–243. In Proceedings of the 21st Target Fabrication Specialists Meeting (TFM), Las Vegas, NV, USA, 21–25 June 2015, [CrossRef]
9. Lyu, D.; Liu, Q.; Liu, H.; Zhao, W.H. Dynamic error of CNC machine tools: A state-of-the-art review. *Int. J. Adv. Manuf. Technol.* **2020**, *106*, 1869–1891. [CrossRef]
10. Bougharriou, A.; Bouzid, W.; Sai, K. Analytical modeling of surface profile in turning and burnishing. *Int. J. Adv. Manuf. Technol.* **2014**, *75*, 547–558. [CrossRef]
11. Liu, P.F.; Chen, W.Q.; Su, H.; Chen, G.D. Dynamic design and thermal analysis of an ultra-precision flycutting machine tool. *Proc. Inst. Mech. Eng. Part B-J. Eng. Manuf.* **2018**, *232*, 404–411. [CrossRef]
12. Yuan, J.C.; Li, J.S.; Wei, W.; Liu, P.K. Operational modal identification of ultra-precision fly-cutting machine tools based on least-squares complex frequency-domain method. *Int. J. Adv. Manuf. Technol.* **2022**, *119*, 4385–4394. [CrossRef]
13. Lu, H.J.; Ding, Y.Y.; Chang, Y.; Chen, G.L.; Rui, X.T. Dynamics Modelling and Simulating of Ultra-precision Fly-Cutting Machine Tool. *Int. J. Precis. Eng. Manuf.* **2020**, *21*, 189–202. [CrossRef]
14. Li, W.; Li, B.Z.; Yang, J.G. Design and dynamic optimization of an ultra-precision micro grinding machine tool for flexible joint blade machining. *Int. J. Adv. Manuf. Technol.* **2017**, *93*, 3135–3147. [CrossRef]
15. Wang, H.; Li, T.J.; Sun, X.Z.; Mynors, D.; Wu, T. Optimal Design Method for Static Precision of Heavy-Duty Vertical Machining Center Based on Gravity Deformation Error Modelling. *Processes* **2022**, *10*, 1930. [CrossRef]
16. Wang, T.Z.; Cheng, J.; Liu, H.N.; Chen, M.J.; Wu, C.Y.; Su, D.N. Ultra-precision grinding machine design and application in grinding the thin-walled complex component with small ball-end diamond wheel. *Int. J. Adv. Manuf. Technol.* **2019**, *101*, 2097–2110. [CrossRef]
17. Yang, X.; An, C.H.; Wang, Z.Z.; Wang, Q.J.; Peng, Y.F.; Wang, J. Research on surface topography in ultra-precision flycutting based on the dynamic performance of machine tool spindle. *Int. J. Adv. Manuf. Technol.* **2016**, *87*, 1957–1965. [CrossRef]
18. Lei, P.L.; Wang, Z.Z.; Shi, C.C.; Peng, Y.F.; Lu, F.; Liu, Z.H. Experimental study on the reliability and the precision maintenance of the ultra-precision grinding machine based on the key subsystem platforms. *Int. J. Adv. Manuf. Technol.* **2023**, *124*, 3923–3934. [CrossRef]

19. Ding, Y.Y.; Rui, X.T.; Chang, Y.; Lu, H.J.; Chen, Y.H.; Ding, J.G.; Shehzad, A.; Chen, G.L.; Gu, J.J. Optimal design of the dynamic performance of the ultra-precision fly cutting machine tool. *Int. J. Adv. Manuf. Technol.* **2023**, *124*, 567–585. [CrossRef]
20. Ding, Y.Y.; Rui, X.T.; Lu, H.J.; Chang, Y.; Chen, Y.H. Research on the dynamic characteristics of the ultra-precision fly cutting machine tool and its influence on the mid-frequency waviness of the surface. *Int. J. Adv. Manuf. Technol.* **2020**, *106*, 441–454. [CrossRef]
21. Chen, G.D.; Sun, Y.Z.; Zhang, F.H.; An, C.H.; Chen, W.Q.; Su, H. Influence of ultra-precision flycutting spindle error on surface frequency domain error formation. *Int. J. Adv. Manuf. Technol.* **2017**, *88*, 3233–3241. [CrossRef]
22. Asilturk, I.; Cunkas, M. Modeling and prediction of surface roughness in turning operations using artificial neural network and multiple regression method. *Expert Syst. Appl.* **2011**, *38*, 5826–5832. [CrossRef]
23. Wang, J.H.; Yazawa, T.; Han, J.G.; Otsubo, T.; Kato, T. Study on improving machined surface accuracy using diamond fly cutting on an ordinary milling machine. *Int. J. Adv. Manuf. Technol.* **2022**, *121*, 8409–8421. [CrossRef]
24. Azami, S.; Kudo, H.; Mizumoto, Y.; Tanabe, T.; Yan, J.; Kakinuma, Y. Experimental study of crystal anisotropy based on ultra-precision cylindrical turning of single-crystal calcium fluoride. *Precis.-Eng.-J. Int. Soc. Precis. Eng. Nanotechnol.* **2015**, *40*, 172–181. [CrossRef]
25. Sung, A.N.; Ratnam, M.M.; Loh, W.P. Effect of wedge angle on surface roughness in finish turning: Analytical and experimental study. *Int. J. Adv. Manuf. Technol.* **2014**, *74*, 139–150. [CrossRef]
26. Kull Neto, H.; Diniz, A.E.; Pederiva, R. The influence of cutting forces on surface roughness in the milling of curved hardened steel surfaces. *Int. J. Adv. Manuf. Technol.* **2016**, *84*, 1209–1218. [CrossRef]
27. Li, J.S.; Wei, W.; Huang, X.J.; Liu, P.K. Study on dynamic characteristics of ultraprecision machining and its effect on medium-frequency waviness error. *Int. J. Adv. Manuf. Technol.* **2020**, *108*, 2895–2906. [CrossRef]
28. Bai, L.; Yang, Q.Z.; Cheng, X.; Ding, Y.; Xu, J.F. A hybrid physics-data-driven surface roughness prediction model for ultra-precision machining. *Sci. China-Technol. Sci.* **2023**, *66*, 1289–1303. [CrossRef]
29. Tan, R.K.; Zhao, X.S.; Lin, F.T.; Jin, S.J.; Guo, X.M.; Chen, X.; Sun, T. Analytical modelling and experimental study of surface roughness in ultrasonic elliptical vibration assisted ultra-precision cutting of Ti-6Al-4 V alloy. *Int. J. Adv. Manuf. Technol.* **2023**, *126*, 1863–1875. [CrossRef]
30. Cao, Y.; Zhao, X.S.; Li, G.; Zong, W.J.; Sun, T. Study regarding the influence of process conditions on the surface topography during ultra-precision turning. *J. Manuf. Process.* **2023**, *102*, 23–36. [CrossRef]
31. Chen, D.; Li, S.; Fan, J. Effect of KDP-Crystal Material Properties on Surface Morphology in Ultra-Precision Fly Cutting. *Micromachines* **2020**, *11*, 802. [CrossRef] [PubMed]
32. Peterka, J. A New Approach to Calculating the Arithmetical Mean Deviation of a Profile during Copy Milling. *Stroj. Vestn.* **2004**, *50*, 594–597.

Journal of
Manufacturing and Materials Processing

MDPI

Article

Predicting the Dynamic Parameters for Milling Thin-Walled Blades with a Neural Network

Yu Li [1,2], Feng Ding [1,*], Dazhen Wang [3], Weijun Tian [4] and Jinhua Zhou [5,*]

1 Mechatronic Engineering, Xi'an Technological University, Xi'an 710021, China; liyuu@mdit.edu.cn
2 School of Intelligent Manufacturing and Control Technology, Xi'an Mingde Institute of Technology, Xi'an 710124, China
3 Xi'an Aerospace Propulsion Testing Technology Research Institute, Xi'an 710025, China; wangdazhen@mail.nwpu.edu.cn
4 Engineering Practice Training Center, Northwestern Polytechnical University, Xi'an 710129, China; tianwj@nwpu.edu.cn
5 School of Mechanical Engineering, Northwestern Polytechnical University, Xi'an 710072, China
* Correspondence: dfeng@xatu.edu.cn (F.D.); zhoujinhua@mail.nwpu.edu.cn (J.Z.)

Abstract: Accurately predicting the time-varying dynamic parameters of a workpiece during the milling of thin-walled parts is the foundation of adaptively selecting chatter-free machining parameters. Hence, a method for accurately and quickly predicting the time-varying dynamic parameters for milling thin-walled parts is proposed, which is based on the shell FEM and a three-layer neural network. The time-dependent dynamics of the workpiece can be calculated using the FEM by obtaining the geometrical parameters of the arc-faced junctions within the discrete cells of the initial and machined workpiece. It is unnecessary to re-divide the mesh cells of the thin-walled parts at each cutting position, which enhances the computational efficiency of the workpiece dynamics. Meanwhile, in comparison with the three-dimensional cube elements, the shell elements can reduce the number of degrees of freedom of the FEM model by 74%, which leads to the computation of the characteristic equation that is about nine times faster. The results of the modal test show that the maximum error of the shell FEM in predicting the natural frequency of the workpiece is about 4%. Furthermore, a three-layer neural network is constructed, and the results of the shell FEM are used as samples to train the model. The neural network model has a maximum prediction error of 0.409% when benchmarked against the results of the FEM. Furthermore, the three-layer neural network effectively enhances computational efficiency while guaranteeing accuracy.

Keywords: thin-walled parts; milling chatter; dynamic parameters; shell element; neural network

1. Introduction

To reduce the weight of aircraft and spacecraft, thin-walled parts have been widely used in the aerospace industry, such as in aircraft engine casings and blades. The machining of thin-walled parts is mainly carried out through machining and non-traditional energy-assisted machining methods [1], but machining is still the main material removal process for the vast majority of aerospace parts as non-traditional methods are affected by efficiency and quality control. The structural characteristics of thin-walled parts make them prone to chatter during CNC machining, thereby leading to machining defects on the surface of parts. This results in difficulties in achieving the required machining precision and surface integrity for the machined parts. To select reasonable machining parameters, the dynamic modeling of thin-walled parts is required to calculate the stable machining parameter domain. However, due to material removal, the machining system of thin-walled parts exhibits strong time-varying behavior; therefore, it is necessary to obtain the dynamic parameters during the machining process of thin-walled parts. It would be very time-consuming to use the finite element method to calculate the dynamic parameters of a thin-walled part at any tool position. In this paper, a finite element model based on shell elements is developed to obtain the workpiece thickness and dynamic parameter datasets, which are then used to train a neural network model to highly improve computational efficiency while ensuring prediction accuracy.

There are already some results from studies on the prediction of the time-varying kinetic parameters for cutting thin-walled parts. After using modal tests to obtain the kinetic parameters of thin-walled parts at different cutting steps, Bravo et al. [2] point out that the variation in workpiece kinetic parameters affects the milling stability limit of thin-walled parts. However, since it is somewhat time-consuming to frequently repeat tests on the workpiece during the cutting process, the authors were subsequently prompted to choose numerical methods instead of investigating the effect of the workpiece dynamic parameter variation on the milling stability. Seguy et al. [3] use the finite element method (FEM) to study the evolution of workpiece dynamic parameters during the milling process. Arnaud et al. [4] analyze the variation in workpiece kinetic parameters during the cutting process through finite element analysis and investigate the impact of material removal and cutting position on the chatter-free boundary. Thevenot et al. [5] use both experimental tests and finite element modeling to investigate the variation in workpiece kinetic parameters during machining. Song et al. [6] first correct the finite element calculation values using experiment data, and then, with the corrected finite element model, they manage to establish the cutting dynamic equations that take into account the time-varying nature of the workpiece kinetic parameters. This method guarantees prediction accuracy as long as the boundary conditions of the model are correctly defined. However, since reconstructing the finite element model and re-computing the modal analysis are to a certain degree quite time-consuming, several algorithms have been proposed by scholars to avoid the frequent modeling and meshing of the finite element model of the workpiece as well as to improve the computational efficiency of the workpiece kinetic parameters. Meshreki et al. [7] propose an analytical model for calculating the kinetic parameters of thin-walled parts during milling, taking into account the variation in the workpiece thickness. Song et al. [8] opt for the Sherman–Morrison–Woodbury equation to predict the Frequency Response Functions (FRFs) of the workpiece during the milling process of thin-walled parts. Moreover, Song et al. [9] also propose a spatio-temporal discrete method for milling the stability prediction of thin-walled parts based on thin-plate theory and the principle of modal superposition, which take into account the influence of the tool–workpiece engagement position and the system's multimodal state. Ahmadi [10] applies FSMs (Finite Strip Models) to obtain the kinetic parameters of the workpiece during the milling process of frame structures. Tuysuz et al. [11] propose a process model for updating the kinetic parameters of the workpiece in the frequency domain, which is based on complete- and reduced-order dynamic substructures.

All the aforementioned studies are conducted on flat-plate-like structures. In terms of the machining of surfaced parts, Yang et al. [12] propose a method to calculate the workpiece modal parameters when the tool cuts at an arbitrary position that is based on structural modal parameter modification, which only calculates the modal vibration patterns and natural frequencies of the workpiece during machining through the modal analysis of the initial workpiece's FEM model without re-modeling or re-meshing the workpiece used for each cutting state. Subsequently, they investigate the effect of time-varying modal parameters on the milling stability of thin-walled parts. Yang et al. [13] put forth a decomposition–condensation method to predict the modal parameters of a workpiece for the finishing/semi-finishing of a large thin-walled workpiece. The method decomposes the workpiece into the machined workpiece and the material to be removed using structural dynamic modification techniques to update the time-varying model of the material to be removed, through which the reconstruction of the finite element model of the workpiece is avoided. Dang et al. [14] advance a method combining the degree of freedom reduction and structural modification techniques to calculate the FRFs of a workpiece that is used in machining, which aims to reduce the degree of freedom of the workpiece's mass matrix and stiffness matrix during machining. After that, they look into the impact of tool position and material removal on modal parameters such as the workpiece natural frequency and modal vibration pattern, and they establish three-dimensional stability lobe diagrams for the milling of flat thin-walled and surfaced thin-walled parts.

For the machining of complex curved-blade parts, Budak et al. [15] utilize commercial finite element software to build a finite element model of the blade structure based on 3D cubic cells to obtain the initial FRFs of the workpiece. They update the workpiece modal parameters with the cells removed during the machining process. Zhang et al. [16] choose the matrix perturbation method to calculate the modal parameters of the blade structure. Luo et al. [17] apply numerical simulations to study the evolution of workpiece modal parameters and cutting stability during the milling of thin-walled blades. Tian et al. [18] calculate the natural frequency of the workpiece during the process of thin-walled blade milling with the matrix perturbation method. Ju [19] delve into the multi-axis milling of thin-walled parts with variable levels of thickness on complex surfaces. The entire machining process is discretized into a finite number of elements along the tool path, and the cutting process is viewed as a process of removing each discrete element one by one. The Sherman–Morrison–Woodbury formula is later applied to predict the variation in the workpiece modal parameters during the machining process. Tuysuz et al. [20] introduce a model for predicting instantaneous modal parameters of blade parts during the cutting process based on reduced structural matrix decoupling and modal mode perturbation, verifying the proposed model through cutting experiments and FRFs testing. With an analytical method, Liu [21] seeks to solve the variation in workpiece modal parameters during impeller blade machining based on theoretical mechanics and the shallow shell theory.

To sum up, the literature [2–11] focuses on flat-plate structures. In terms of the curved blades, the literature [17,18] attempts to simplify the curved-blade structure to a flat-plate one. However, there is a significant discrepancy in the mechanical properties between flat-plate structures and curved-blade parts, which can lead to significant calculation errors. The literature [15] employs three-dimensional cubic elements for the finite element modeling of thin-walled blades, which results in the large model of degrees of freedom. Additionally, it is time-consuming to simulate and analyze the cutting process using finite element software. The method proposed in the literature [21] inherently presumes a vibration equation to the workpiece based on boundary conditions. There will be a significant computational mistake if the assumed vibration that is equation is unreasonable. The study yields a maximum error of 8% for the first two modal natural frequencies. Based on the matrix perturbation method, the literature [15] predicts the natural frequency of blade structures with a maximum error of 7.3%, which is relatively high. To implement the curved-blade machining requirements, it is necessary to establish a workpiece modal parameter prediction model that factors in the blade's surface characteristics and ensures both calculation accuracy and efficiency. Currently, machine learning is widely used in various fields such as system identification, pattern recognition, and intelligent control [22]. Its highly robust and fault-tolerant ability allows for the adequate approximation of complex nonlinear relationships.

At present, neural networks have been widely used for predicting cutting force, tool wear, and workpiece surface roughness [23–27]. Yeganefar et al. [23] predicted and optimized the surface roughness and cutting forces in the slot milling of a 7075-T6 aluminum alloy using regression analysis, support vector regression (SVR), artificial neural network (ANN), and multivariate analysis methods. The effect of the process parameters, including cutting speed, feed per tooth, depth of cut, and tool type, on the response was investigated by analysis of variance (ANOVA). Pavlenko et al. [24] evaluated the effect of the process parameters on the power characteristics of the cutting process in cylindrical feed grinding based on artificial neural network (ANN) and multiparameter proposed linear regression analysis methods as well as numerical experiments in order to investigate the cutting force and evaluate the key indicators of the cutting force. SK et al. [25] used an artificial neural network approach to accurately predict the back face wear of a tool based on the response of the cutting force and surface roughness. Vasanth et al. [26] predicted the surface roughness by using a regression model and an artificial neural network model to fuse the cutting force, cutting temperature, tool wear, and vibrational displacements of the tool, which showed that the predicted surface roughness was better than that predicted by a multi-parameter proposed linear regression analysis and multi-parameter proposed linear regression analysis. The results show that the predicted surface roughness is more accurate than that of the regression model. Wang et al. [27] predicted the cutting force by using a neural network approach and built a "transmission network" based on the data obtained from the simulation. Compared with the "normal network", the transmission network has obvious performance advantages. Heitz1 et al. [28] used model training as a new feature of the mechanical force model in the time and frequency domains, and the cutting force was predicted efficiently by the extreme gradient boosting optimization algorithm, which improved the optimization accuracy, efficiency, user-friendliness, and efficiency. Currently, machine learning neural network models are rarely studied for modal parameter predictions.

Hence, a method that is based on both the shell FEM and a three-layer neural network and that also considers workpiece geometry is proposed for predicting the time-varying dynamics during the milling of the thin-walled curved-blade structure. The built FEM based on shell elements only needs to mesh the initial workpiece and the machined workpiece once, and it only needs to calculate the geometric parameters of the initial workpiece and the machined workpiece. It does not need to re-mesh the thin-walled parts at each cutting position nor to repeatedly calculate the geometric parameters of the workpiece during the cutting process. Meanwhile, in comparison with the three-dimensional cube elements used in reference [15], the shell elements can reduce the number of degrees of freedom of the FEM and improve computing efficiency. Meanwhile, a three-layer neural network model is constructed to predict the time-varying dynamic parameters for milling the thin-walled parts. The three-layer neural network model can highly improve calculation efficiency while ensuring calculation accuracy.

A dynamic model is established in Section 1 for the milling of the thin-walled blades. In Section 2, a shell unit-based time-varying dynamic parameter prediction model is constructed for the milling of the thin-walled parts. In Section 3, a three-layer neural network model is constructed to achieve the prediction of the time-varying kinetic parameters for milling the thin-walled parts. In Section 4, the validity of the shell unit finite element model is verified through modal test experiments. Meanwhile, the neural network model is trained using the calculation results of the shell unit finite element model as training samples, and the calculation results of the neural network model and the shell unit finite element model are compared and analyzed.

2. Modeling of Thin-Walled Blade Dynamics Based on the Removal Process

2.1. Dynamic Model of the Thin-Walled Blade Structure

Aviation blades are structurally characteristic of their thin-walled parts, and the cross-section thickness varies significantly along the chord direction, which makes it challenging to characterize their structural characteristics through specific geometric parameters. Analyzing this type of continuous medium dynamic problem requires the discretization of the spatial region using the finite element method to obtain a discretized multi-degree-of-freedom dynamic model, as shown in Figure 1.

The structural dynamic equations of the discretized model of the blade in the physical coordinate system can be put as follows:

$$M\ddot{u} + C\dot{u} + Ku = 0 \quad (1)$$

where u, $\dot{u}$, and $\ddot{u}$ denote the displacement vector, velocity vector, and acceleration vector, respectively. M, C, and K, respectively, stand for the mass matrix, damping matrix, and stiffness matrix of the workpiece. If the workpiece's damping is disregarded, the aforementioned equation can be simplified as follows:

$$M\ddot{u} + Ku = 0 \tag{2}$$

The characteristic equation of Equation (2) can be written as follows:

$$(K - \lambda M)\Phi = 0 \tag{3}$$

where λ and Φ denote the eigenvalues and eigenvectors of the eigenequations. The blade structure's nature frequency ω, and its corresponding mode shapes are $\sqrt{\lambda}$ and Φ, respectively.

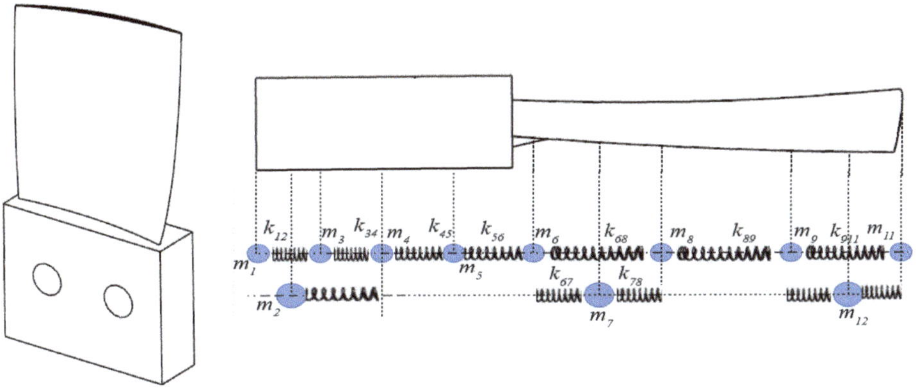

Figure 1. Simplification and discretization of the blade process model.

The complete process constitutes the continuous removal of the blade material from semi-finishing to the end of finishing. When meshing the blade body, the structural properties of the original workpiece state can be expressed in terms of the mass matrix M_0, stiffness matrix K_0, eigenvalues, and eigenvectors Φ_0. Hence, the workpiece nature frequency and modal formation can be expressed as:

$$K_0\Phi_{i0} = \lambda_{i0}M_0\Phi_{i0} \ \ (i = 1 \sim n) \tag{4}$$

The material removal process can be considered as a continuous removal process of a number of grid cells on the original conditions of the workpiece, and the amount of change in the mass and stiffness induced by a number of grid cells can be represented as ΔK, ΔM. After resection is complete, the mass and stiffness matrices of the new process system can be put as follows:

$$K = K_0 - \Delta K; \ M = M_0 - \Delta M \tag{5}$$

The structural properties of the modified system can be stated as follows:

$$(K_0 - \Delta K)x_i = \lambda(M_0 - \Delta M)x_i \tag{6}$$

2.2. Milling Process Grid Cell Definitions

Due to its unique structural properties, the wall thickness of the blade is significantly less than 1/10 of the total structural size of the blade. Using the shell unit can simplify the three-dimensional geometric structure by extracting the mid-plane, provided that the stress along the thickness direction is neglected. It can highly reduce the solution scale and improve the calculation efficiency while maintaining geometrical accuracy. Therefore, shell units are used herein to simulate the structural characteristics of thin-walled blades, and the original 3D eight-node hexahedron is converted into a 2D first-order planar body unit by extracting the mid-plane, as illustrated in Figure 2.

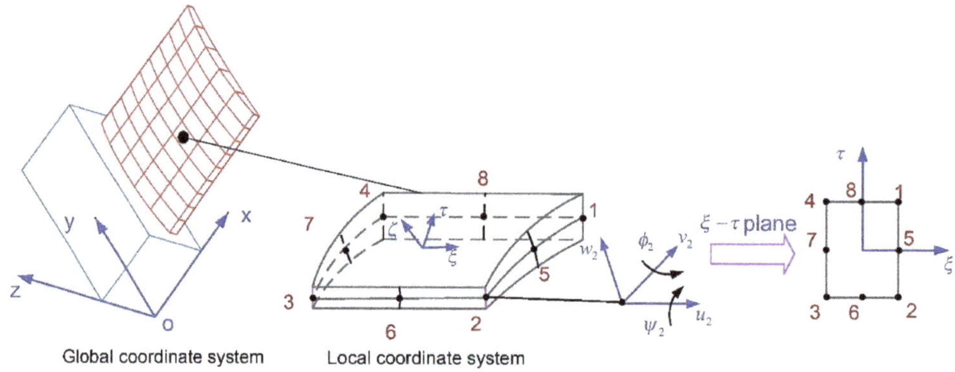

Figure 2. The eight-node thick shell cell.

To further describe the change process of the structural parameters of the material removal process when milling thin-walled blades, a novel approach is proposed herein to divide the workpiece into three categories including initial workpiece, workpiece during the cutting process, and machined workpiece. Additionally, the cutting process workpiece is divided into two parts as follows: the uncut part and the machined part, as shown in Figure 3.

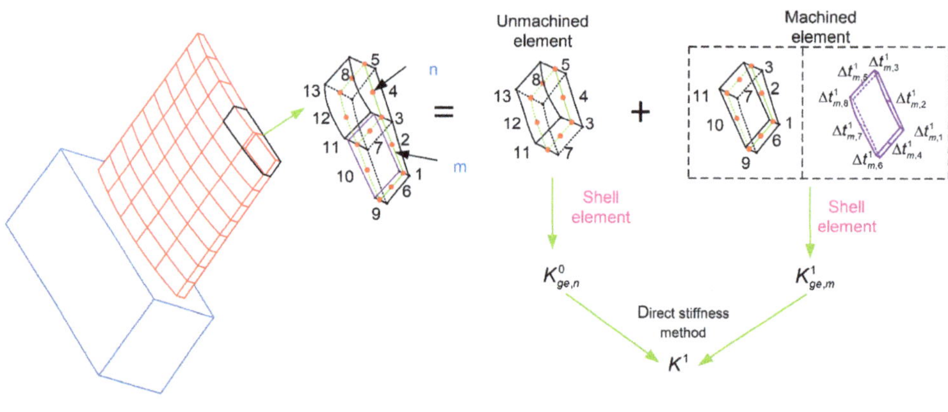

Figure 3. Flow of the workpiece stiffness matrix calculation during the thin-walled blade cutting process.

The same discrete cell form is then used to mesh the three states of the workpiece. Subsequently, the geometrical parameters of the arc-face nodes in the discrete cells of the initial and machined workpiece are calculated. By combining the corresponding discrete cells of the initial and machined workpiece, it is possible to obtain the geometrical parameters of the discrete cell nodes of the workpiece during the cutting process. Then, the stiffness and mass matrices of the discrete units of the workpiece during the cutting process are constructed by the shell unit, and, lastly, the overall stiffness and mass matrices of the workpiece during the cutting process are obtained by synthesis, and the workpiece's time-varying dynamic parameters are calculated. To improve the efficiency of the calculation of the kinetic parameters of the workpiece, this method only needs to mesh the initial workpiece and the processed workpiece once and calculate the geometric parameters of the initial workpiece and the processed workpiece. It also dispenses with the need to re-divide the mesh cells of the thin-walled parts and repeat the calculation of the workpiece geometric parameters during the cutting process at each cutting position. The specific synthesis process is as follows.

Based on the above finite element idea, the discretized unit stiffness matrix and mass matrix are represented as follows using the dynamic equations [29]:

$$\begin{cases} K_{le}((x_i, y_i, z_i), t_i, m_i) = \int_{-1}^{1}\int_{-1}^{1}\int_{-1}^{1} B^{\mathrm{T}} D B |J| d\xi d\tau d\zeta \\ M_{le}((x_i, y_i, z_i), t_i, m_i) = \int_{-1}^{1}\int_{-1}^{1}\int_{-1}^{1} \rho N^{\mathrm{T}} N |J| d\xi d\tau d\zeta \end{cases}, (i = 1, 2, \cdots, 8) \tag{7}$$

where ρ signifies the density of the workpiece and the elasticity matrix D is expressed as follows:

$$D = \frac{E}{1-\mu^2}\begin{bmatrix} 1 & \mu & 0 & 0 & 0 \\ \mu & 1 & 0 & 0 & 0 \\ 0 & 0 & \frac{1-\mu}{2} & 0 & 0 \\ 0 & 0 & 0 & \frac{1-\mu}{2k} & 0 \\ 0 & 0 & 0 & 0 & \frac{1-\mu}{2k} \end{bmatrix} \tag{8}$$

where E is the modulus of elasticity, μ is Poisson's ratio, and the coefficient k is introduced to account for the impact of the inhomogeneity of the shear stresses and takes the value of 1.2. From Equation (7), it is evident that to compute the mass and stiffness matrices of the discrete shell cell, it is necessary to define the matrices of J, N, and B. The matrices J, N, and B are defined as follows:

Define the matrix J: The global coordinates of any point within the shell cell in the global coordinate system can be represented as follows:

$$\begin{Bmatrix} x \\ y \\ z \end{Bmatrix} = \sum_{i=1}^{8} N_i \begin{Bmatrix} x_i \\ y_i \\ z_i \end{Bmatrix}_{MID} + \sum_{i=1}^{8} N_i \frac{\zeta}{2} \begin{Bmatrix} H_{3i}(1) \cdot t_i \\ H_{3i}(2) \cdot t_i \\ H_{3i}(3) \cdot t_i \end{Bmatrix} \tag{9}$$

where N_i signifies the cell shape function, MID the shell cell midplane, (x_i, y_i, z_i) the coordinates of the nodes on the midplane of the shell cell, H_{3i} the midplane normal vector, and t_i the shell cell thickness. For the cell node numbering shown in Figure 1, the cell shape function can be put as follows:

$$\begin{cases} N_1 = \frac{1}{4}(1+\xi)(1+\tau)(\xi+\tau-1), N_2 = \frac{1}{4}(1+\xi)(1-\tau)(\xi-\tau-1) \\ N_3 = \frac{1}{4}(1-\xi)(1-\tau)(-\xi-\tau-1), N_4 = \frac{1}{4}(1-\xi)(1+\tau)(-\xi+\tau-1) \\ N_5 = \frac{1}{2}(1+\xi)(1-\tau^2), N_6 = \frac{1}{2}(1-\xi^2)(1-\tau) \\ N_7 = \frac{1}{2}(1-\xi)(1-\tau^2), N_8 = \frac{1}{2}(1-\xi^2)(1+\tau) \end{cases} \tag{10}$$

The Jacobi matrix J is defined as:

$$J = \begin{bmatrix} \frac{\partial x}{\partial \xi} & \frac{\partial y}{\partial \xi} & \frac{\partial z}{\partial \xi} \\ \frac{\partial x}{\partial \tau} & \frac{\partial y}{\partial \tau} & \frac{\partial z}{\partial \tau} \\ \frac{\partial x}{\partial \zeta} & \frac{\partial y}{\partial \zeta} & \frac{\partial z}{\partial \zeta} \end{bmatrix} \tag{11}$$

Define the matrix N: The shell cell node i has three angular displacements and two linear displacements. Construct the three orthogonal vectors H_{1i}, H_{2i}, and H_{3i} to define the angular displacements of node i. H_{3i} denotes the vector normal to the midplane of the cell at node i, and H_{2i} and H_{3i} are two vectors that are tangent to the midplane and perpendicular to the vector and can be constructed in the following form:

$$H_{2i} = H_{3i} \times i(x - axis), H_{1i} = H_{2i} \times H_{3i} \tag{12}$$

Then, we can obtain the directional cosines l_{1i}, m_{1i}, n_{1i} of vector H_{1i}, the directional cosines l_{2i}, m_{2i}, n_{2i} of vector H_{2i}, and the directional cosines $l_{3i} = H_{3i}(1)$, $m_{3i} = H_{3i}(2)$ and $n_{3i} = H_{3i}(3)$ of vectors H_{3i}. Subsequently, the node displacements $\delta_i = [u_i,\ v_i,\ w_i,\ \varphi_i,\ \psi_i]^T$ and the cell shape function $N_i(\xi,\ \tau)$ can be used to compute the displacements of any point within the cell, defining the matrix N as follows:

$$N = \begin{bmatrix} N_1 & 0 & 0 & \frac{\zeta N_1 t_1}{2} l_{11} & \frac{\zeta N_1 t_1}{2} l_{21} & \cdots N_8 & 0 & 0 & \frac{\zeta N_8 t_8}{2} l_{18} & \frac{\zeta N_8 t_8}{2} l_{28} \\ 0 & N_1 & 0 & \frac{\zeta N_1 t_1}{2} m_{11} & \frac{\zeta N_1 t_1}{2} m_{21} & \cdots 0 & N_8 & 0 & \frac{\zeta N_8 t_8}{2} m_{18} & \frac{\zeta N_8 t_8}{2} m_{28} \\ 0 & 0 & N_1 & \frac{\zeta N_1 t_1}{2} n_{11} & \frac{\zeta N_1 t_1}{2} n_{21} & \cdots 0 & 0 & N_8 & \frac{\zeta N_8 t_8}{2} n_{18} & \frac{\zeta N_8 t_8}{2} n_{28} \end{bmatrix} \tag{13}$$

Defining the Geometric Matrix B: To calculate the strain of the shell unit in the local coordinate system, it is necessary to construct the local coordinate system. The local coordinate system axes can be constructed in the following form:

$$z\prime = N_1 H_{31} + N_2 H_{32} + \cdots + N_8 H_{38} \tag{14}$$

where the y' and x' axes are expressed as follows:

$$y' = z' \times x = \begin{vmatrix} i & j & k \\ l_3 & m_3 & n_3 \\ 1 & 0 & 0 \end{vmatrix}, x' = z' \times y' \tag{15}$$

Then, the direction cosines l_3, m_3, n_3 of the axis z'; the direction cosines l_2, m_2, n_2 of the axis y'; and the direction cosines l_1, m_1, n_1 of the axis x' can be calculated.

The components of the geometric matrix are defined as follows:

$$B_i = \begin{bmatrix} l_1\alpha_1 & m_1\alpha_1 & n_1\alpha_1 & \beta_1\gamma_1 & \beta_1\lambda_1 \\ l_2\alpha_2 & m_2\alpha_2 & n_2\alpha_2 & \beta_2\gamma_2 & \beta_2\lambda_2 \\ l_1\alpha_2 + l_2\alpha_1 & m_1\alpha_2 + m_2\alpha_1 & n_1\alpha_2 + n_2\alpha_1 & \beta_1\gamma_2 + \beta_2\gamma_1 & \beta_1\lambda_2 + \beta_2\lambda_1 \\ l_2\alpha_3 + l_3\alpha_2 & m_2\alpha_3 + m_3\alpha_2 & n_2\alpha_3 + n_3\alpha_2 & \beta_2\gamma_3 + \beta_3\gamma_2 & \beta_2\lambda_3 + \beta_3\lambda_2 \\ l_3\alpha_1 + l_1\alpha_3 & m_3\alpha_1 + m_1\alpha_3 & n_3\alpha_1 + n_1\alpha_3 & \beta_3\gamma_1 + \beta_1\gamma_3 & \beta_3\lambda_1 + \beta_1\lambda_3 \end{bmatrix} \tag{16}$$

The parameters in Equation (16) are as follows:

$$\begin{cases} \alpha_s = l_s \frac{\partial N_i}{\partial x} + m_s \frac{\partial N_i}{\partial y} + n_s \frac{\partial N_i}{\partial z} \\ \beta_s = (l_s \frac{\partial P_i}{\partial x} + m_s \frac{\partial P_i}{\partial y} + n_s \frac{\partial P_i}{\partial z}) \frac{t_i}{2} \\ \gamma_s = l_s l_{1i} + m_s m_{1i} + n_s n_{1i} \\ \lambda_s = l_s l_{2i} + m_s m_{2i} + n_s n_{2i} \end{cases} \tag{17}$$

where $P_i = N_i\zeta$. The geometric matrix is denoted as $B = [B_1\ B_2\ \cdots\ B_8]$.

Matrix combination of cells: To derive the mass and stiffness matrices of the blade structure, it is important to transform the matrices obtained from Equation (17) in the local coordinate system to the global coordinate system. The transformation matrix is expressed as follows:

$$L = \begin{bmatrix} \gamma_1 & 0 & 0 & 0 & 0 & 0 & 0 & 0 \\ 0 & \gamma_2 & 0 & 0 & 0 & 0 & 0 & 0 \\ 0 & 0 & \gamma_3 & 0 & 0 & 0 & 0 & 0 \\ 0 & 0 & 0 & \gamma_4 & 0 & 0 & 0 & 0 \\ 0 & 0 & 0 & 0 & \gamma_5 & 0 & 0 & 0 \\ 0 & 0 & 0 & 0 & 0 & \gamma_6 & 0 & 0 \\ 0 & 0 & 0 & 0 & 0 & 0 & \gamma_7 & 0 \\ 0 & 0 & 0 & 0 & 0 & 0 & 0 & \gamma_8 \end{bmatrix} \gamma_i = \begin{bmatrix} l_{1i} & m_{1i} & n_{1i} & 0 & 0 \\ l_{2i} & m_{2i} & n_{2i} & 0 & 0 \\ l_{3i} & m_{3i} & n_{3i} & 0 & 0 \\ 0 & 0 & 0 & l_{1i} & m_{1i} \\ 0 & 0 & 0 & l_{2i} & m_{2i} \end{bmatrix} \tag{18}$$

The stiffness matrix and mass matrix of the shell cell in the global coordinate system are represented as follows:

$$K_{ge} = L^T K_{le} L;\ M_{ge} = L^T M_{le} L \tag{19}$$

The overall stiffness matrix and mass matrix of the blade structure are obtained by processing the cell stiffness matrix and mass matrix with the direct stiffness method [30].

$$K = \sum_{n=1}^{Ne} K_{ge,n};\ M = \sum_{n=1}^{Ne} M_{ge,n} \tag{20}$$

where Ne denotes the number of cells in the finite element model of the blade.

2.3. Solving for the Kinetic Parameters

To obtain a finite model of the blade structure in the shell cell, it is first necessary to obtain the coordinates (x_i, y_i, z_i) at the node i on the mid-arc surface of the blade, the corresponding blade thickness at the node, and the normal vector $m_i = (H_{3i}(1), H_{3i}(2), H_{3i}(3))$ of the mid-arc surface. The steps for calculating the geometrical parameters of the node on the mid-arc surface of the blade structure are illustrated in Figure 4. The specific calculation steps are as follows:

(1) Establish a three-dimensional model of the thin-walled blade in NX;
(2) Insert the point sets in the U and V directions of the basin and the back of the thin-walled blade, and set the number of points in the U direction to 11 and the number of points in the V direction to 19;
(3) Compile the UG secondary development plug-in in C++, and export the point sets on the same cross-section to the same file through the function UC4524;
(4) Take the exported point set from step (3) through the curve fitting to generate the two curves above and below the blade's cross-section;
(5) Calculate the coordinates of the discrete points on the mid-arc and the thickness of the blade through the mid-arc solution algorithm, and generate the mid-arc curves through the curve fitting;
(6) Generate the mid-arc face through the mid-arc curve fitting;
(7) Use C++ to compile the UG secondary development plug-ins, obtain the coordinates of the discrete points of the mid-arc surface through the UG/Open function UF_MODL_ask_face_parm, and obtain the normal vector of the mid-arc surface through the function UF_MODL_ask_face_props.

Figure 5 shows the steps for calculating the workpiece dynamic parameters in the cutting process, the steps are described as follows:

(1) Use the same number of units to discretize the uncut workpiece and the machined workpiece, and use the method shown in Figure 3 to calculate the coordinates of the nodes on the arc surface of the discrete unit, the thickness of the blade corresponding to the nodes, and the normal vector of the mid-arc surface, and store them in the file.
(2) Adopt the discrete method in step (1) to discretize the workpiece in the cutting process into the machined units and uncut units, and then number them.
(3) According to the discrete unit number of the workpiece in the cutting process in step (2), read the corresponding geometric parameters in step (1), establish a finite element model of the workpiece in the cutting process based on the shell unit, and calculate the stiffness matrix and mass matrix of the workpiece.
(4) Using the calculated stiffness matrix and mass matrix of the workpiece, the nature frequency and mode shapes of the workpiece in the cutting process are calculated by solving the characteristic equations.

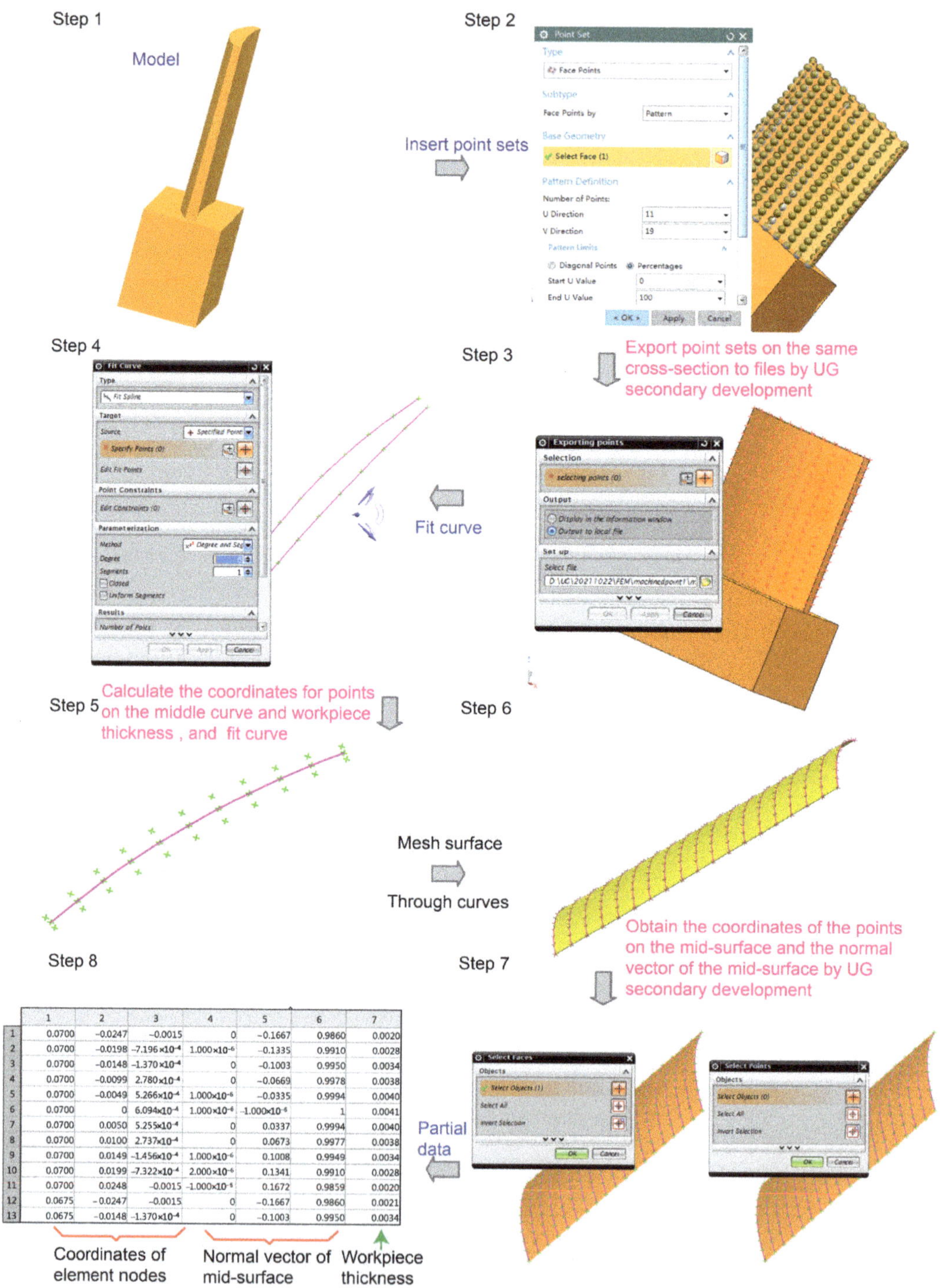

	1	2	3	4	5	6	7
1	0.0700	−0.0247	−0.0015	0	−0.1667	0.9860	0.0020
2	0.0700	−0.0198	-7.196×10^{-4}	1.000×10^{-6}	−0.1335	0.9910	0.0028
3	0.0700	−0.0148	-1.370×10^{-4}	0	−0.1003	0.9950	0.0034
4	0.0700	−0.0099	2.780×10^{-4}	0	−0.0669	0.9978	0.0038
5	0.0700	−0.0049	5.266×10^{-4}	1.000×10^{-6}	−0.0335	0.9994	0.0040
6	0.0700	0	6.094×10^{-4}	1.000×10^{-6}	-1.000×10^{-6}	1	0.0041
7	0.0700	0.0050	5.255×10^{-4}	0	0.0337	0.9994	0.0040
8	0.0700	0.0100	2.737×10^{-4}	0	0.0673	0.9977	0.0038
9	0.0700	0.0149	-1.456×10^{-4}	1.000×10^{-6}	0.1008	0.9949	0.0034
10	0.0700	0.0199	-7.322×10^{-4}	2.000×10^{-6}	0.1341	0.9910	0.0028
11	0.0700	0.0248	−0.0015	-1.000×10^{-5}	0.1672	0.9859	0.0020
12	0.0675	−0.0247	−0.0015	0	−0.1667	0.9860	0.0021
13	0.0675	−0.0148	-1.370×10^{-4}	0	−0.1003	0.9950	0.0034

Figure 4. Flow of calculating the geometrical parameters of the discrete unit nodes on the curved surfaces of thin-walled blades.

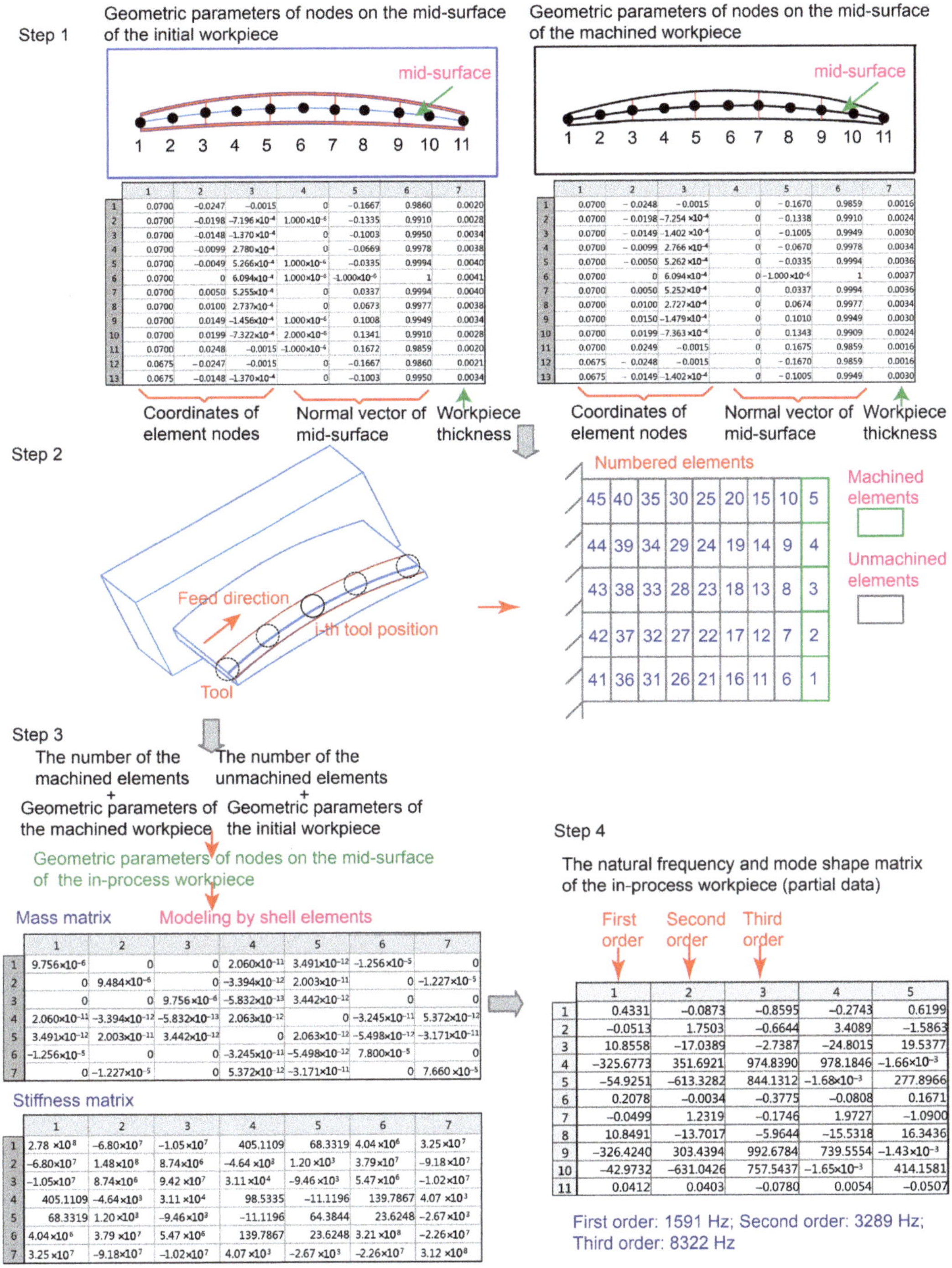

Figure 5. The process of calculating the parameters of the workpiece dynamics during the cutting process.

3. Neural Network-Based Prediction of Kinetic Parameters

Section 2 herein proposes a method for calculating the time-varying kinetic parameters for milling thin-walled parts based on shell cells, which only requires the geometrical parameters at the discrete cell nodes of the uncut and machined workpiece to calculate the workpiece kinetic parameters during the cutting process through finite element modeling. However, the abovementioned method consumes too much time to solve the characteristic equations for the finite element model of each cutting state. With the development of deep learning technology, neural networks have become widely used for cutting force predictions, tool life predictions, and predictions made in other fields. As demonstrated in Figure 6, neural networks offer a strong parallel computing capacity, which can predict the dynamic parameters of the workpiece in multiple cutting states at the same time, thereby significantly reducing the calculation time.

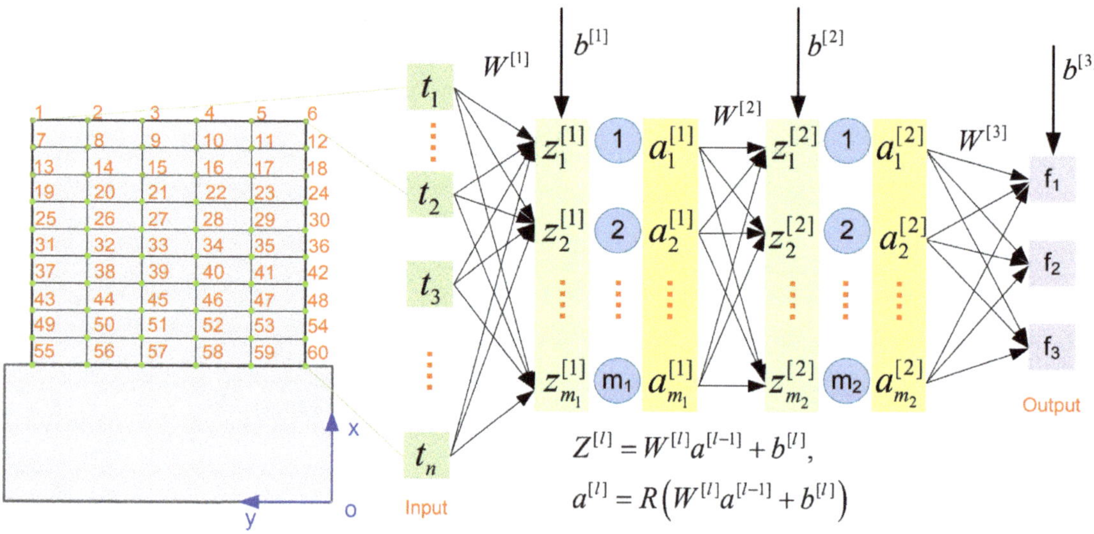

Figure 6. Neural network structure diagram.

Herein, a three-layer neural network, as shown in Figure 6, is used for the thin-walled blade milling dynamic parameter prediction. Taking one single sample point as an example to describe the calculation process, the input quantity t marks the thickness of the unit node and the output variable f marks the first three orders of the nature frequency of the workpiece, which are denoted as follows:

$$\begin{array}{l} t = \begin{bmatrix} t_1 & t_2 & \cdots & t_n \end{bmatrix}^T \\ f = \begin{bmatrix} f_1 & f_2 & f_3 \end{bmatrix}^T \end{array} \tag{21}$$

The ReLU function is used for the activation function, which is expressed as follows [31]:

$$R(x) = \max(0, x) \tag{22}$$

From the input t, the weight matrix $W^{[1]}$, the offset value $b^{[1]}$ of the first layer of the network, and the activation function calculation, we can obtain [32] as follows:

$$Z^{[1]} = \begin{bmatrix} w_{1,1}^{[1]} & w_{1,2}^{[1]} & \cdots & w_{1,n}^{[1]} \\ w_{2,1}^{[1]} & w_{2,2}^{[1]} & \cdots & w_{2,n}^{[1]} \\ \vdots & \vdots & \ddots & \vdots \\ w_{m_1,1}^{[1]} & w_{m_1,2}^{[1]} & \cdots & w_{m_1,n}^{[1]} \end{bmatrix} \begin{bmatrix} t_1 \\ t_2 \\ \vdots \\ t_n \end{bmatrix} + b^{[1]} = W^{[1]}t + b^{[1]}, \quad a^{[1]} = R(Z^{[1]}) \tag{23}$$

From the output $a^{[1]}$ of the first layer of the network, the weight matrix $W^{[2]}$, the offset value $b^{[2]}$ of the second layer of the network, and the activation function calculation, one can obtain the following:

$$Z^{[2]} = \begin{bmatrix} w_{1,1}^{[2]} & w_{1,2}^{[2]} & \cdots & w_{1,m_1}^{[2]} \\ w_{2,1}^{[2]} & w_{2,2}^{[2]} & \cdots & w_{2,m_1}^{[2]} \\ \vdots & \vdots & \ddots & \vdots \\ w_{m_2,1}^{[2]} & w_{m_2,2}^{[2]} & \cdots & w_{m_2,m_1}^{[2]} \end{bmatrix} \begin{bmatrix} a_1^{[1]} \\ a_2^{[1]} \\ \vdots \\ a_{m_1}^{[1]} \end{bmatrix} + b^{[2]} = W^{[2]}a^{[1]} + b^{[2]}, \quad a^{[2]} = R(Z^{[2]}) \tag{24}$$

From the output $a^{[2]}$ of the second layer of the network, the weight matrix $W^{[3]}$, and the offset value $b^{[3]}$ calculation of the third layer of the network, one can obtain the following:

$$f = \begin{bmatrix} f_1 \\ f_2 \\ f_3 \end{bmatrix} = \begin{bmatrix} w_{1,1}^{[3]} & w_{1,2}^{[3]} & \cdots & w_{1,m_2}^{[3]} \\ w_{2,1}^{[1]} & w_{2,2}^{[1]} & \cdots & w_{2,m_2}^{[1]} \\ w_{3,1}^{[1]} & w_{3,2}^{[1]} & \cdots & w_{3_1,m_2}^{[1]} \end{bmatrix} \begin{bmatrix} a_1^{[2]} \\ a_2^{[2]} \\ \vdots \\ a_{m_2}^{[2]} \end{bmatrix} + b^{[3]} = W^{[3]}a^{[2]} + b^{[3]} \tag{25}$$

The mean square error loss function between the predicted and nominal values of the nature frequency of the workpiece is established as follows:

$$L = \frac{1}{2}\left\| f - \hat{f} \right\|^2 \tag{26}$$

Thus, the prediction of the milling dynamic parameters for thin-walled parts is transformed into an optimization problem as follows:

$$\theta^* = \underset{\theta}{\operatorname{argmin}} L(\theta) \tag{27}$$

where parameter θ signifies the neural network weight matrix W and the offset vector b, and the parameters θ are solved iteratively using the most rapid descent method.

The partial derivative of the loss function L with respect to the weight matrix W and the offset vector b is expressed as follows:

$$\frac{\partial L}{\partial W^{[3]}} = (f - \hat{f})(a^{[2]})^T \tag{28}$$

$$\frac{\partial L}{\partial W^{[2]}} = (W^{[3]})^T (f - \hat{f})(a^{[1]})^T \tag{29}$$

$$\frac{\partial L}{\partial W^{[1]}} = (W^{[2]})^T (W^{[3]})^T (f - \hat{f})(t)^T \tag{30}$$

$$\frac{\partial L}{\partial b^{[3]}} = f - \hat{f} \tag{31}$$

$$\frac{\partial L}{\partial b^{[2]}} = (f - \hat{f})^T W^{[3]} \tag{32}$$

$$\frac{\partial L}{\partial b^{[1]}} = (f - \hat{f})^T W^{[3]} W^{[2]} \tag{33}$$

The parameters are iteratively updated using the most rapid descent method.

$$\begin{array}{l} W = W - \eta \frac{\partial L}{\partial W}, \\ b = b - \eta \frac{\partial L}{\partial b}, \end{array} \tag{34}$$

After the weight matrix W and the offset vector b are calculated, they are substituted into Equations (21)–(23), to calculate the nature frequency of the thin-walled part.

The whole frequency calculation steps are as follows.

(1) Take the thickness of the unit node as the input quantity and the first three orders of the workpiece's nature frequency as the output variable, and define the forward computation model with a fixed frequency, as shown in Equations (23)–(25).
(2) Using the mean square deviation between the predicted values f and nominal values $\hat{f}$ of the workpiece's nature frequency as the loss function, the prediction of the milling dynamic parameters of thin-walled parts is transformed into an optimization problem, as shown in Equation (27).
(3) Iterative solution parameters, i.e., the neural network weight matrix W and offset vectors b, are used through the most rapid descent method to calculate the nature frequency of the thin-walled parts.

4. Prediction and Validation of Milling Dynamic Parameters for Thin-Walled Parts

4.1. Finite Element Model Calculation and Test Verification

The titanium alloy TC4 test piece, with a density of 4500 kg/m^3, a modulus of elasticity of 104 Gpa, and a Poisson's ratio of 0.305, has its design dimensions shown in Figure 7.

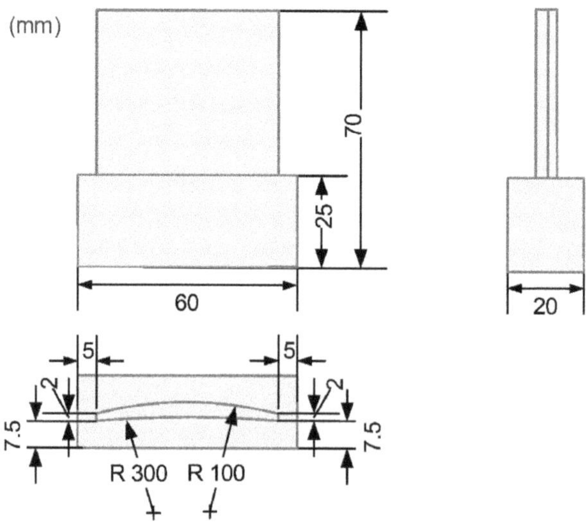

Figure 7. Structural design dimensions of thin-walled parts.

After the mass and stiffness matrices of the FEM model have been obtained, the natural frequencies and the mode shapes of the workpiece can be calculated. However, degrees of freedom (DOF) of the FEM model affect the calculation speed of the characteristic equation. For reducing the number of DOF of the model on the premise of guaranteeing the calculation accuracy, a convergence analysis is performed. Meanwhile, the number of DOF of the model that is built by shell element and the model that is built by 3D cube element, which are used in the reference [15], are compared when the calculated natural frequencies achieve convergence. As the base of the blade is clamped by a fixture during machining, only the blade profile is modeled in the FEM modeling process in order to reduce the number of DOF of the model.

Matlab (R2014a) is used to build the FEM model of the blade based on the shell elements. To facilitate the following experimental verification, one element is set in the thickness direction, and five elements are set in the y-direction. The convergence of the natural frequencies is studied by increasing the number of elements (NOEs) in the x-direction from 6 to 9. Abaqus (6.14) is used to calculate the natural frequencies when the 3D cube elements with the type of C3D20R are used to model the blade. When the blade is meshed, two and five elements are set in the thickness direction and y-direction, respectively. The convergence of the natural frequencies is studied by increasing the NOEs in the x-direction from 4 to 8. As the high-order natural frequencies of the blade are difficult to be excited during machining, only the first three frequencies are studied in this paper. The convergence processes of the frequencies are shown in Figure 8. The values in parentheses show the variations the frequencies obtained by the present model and the previous model. It is supposed that when the variations are less than 0.1%, the natural frequencies reach convergence.

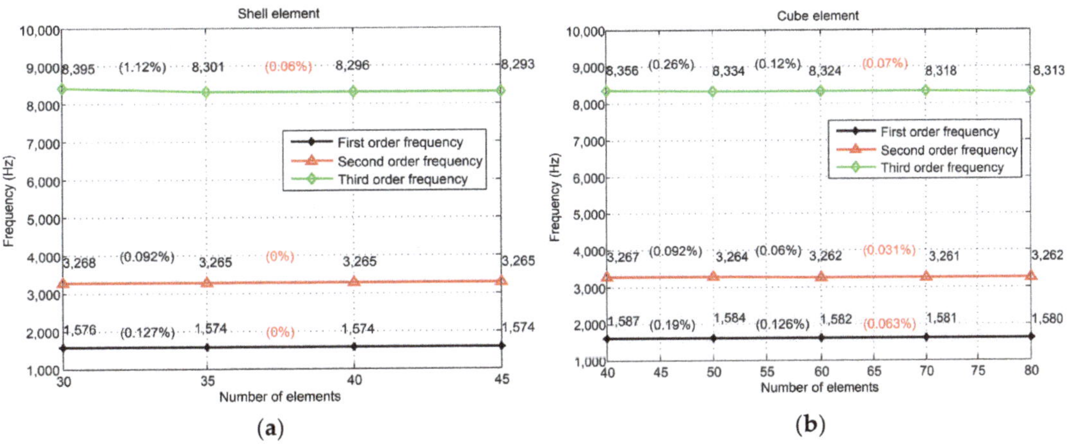

Figure 8. Convergence process of the natural frequencies (Hz) (**a**) obtained by shell element and (**b**) obtained by cube element.

As shown in Figure 8a, when the NOES increases from 35 to 40, the maximum variation in the first three natural frequencies is 0.06%. Therefore, when the number of shell elements in the y-direction and x-direction are five and eight, respectively, the calculated frequencies reach convergence. At that point, the number of DOF of the shell model is 735, and when the clamped-free boundary conditions are imposed on the model, the number of DOF is 680. Meanwhile, Figure 8b shows that when the NOEs increases from 60 to 70, the maximum variation is 0.07%. At that point, the number of DOF of the cube model is 2916, and when the clamped-free boundary conditions are applied, the number of DOF is 2646. One can see the number of DOF of the FEM model established by the 3D cube elements is 3.89 times the size of the model established by the shell elements. This leads to about the nine times faster computation of the eigenvalues when the shell elements are used. When the value obtained by the 3D cube elements is regarded as a standard, the maximum error of the first three frequencies calculated by the shell elements is 0.38% and the average error is 0.237%. Therefore, under the premise of guaranteeing the calculation accuracy, the shell elements can reduce the number of DOF of the FEM model and thus reduce the time for calculating the dynamics of the blade. In the following point, the shell model is used to calculate the workpiece dynamics. Furthermore, one element is set in the thickness direction, five elements are set in the y-direction, and nine elements are set in the x-direction.

To verify the accuracy of the method proposed herein, the milling and modal testing of the thin-walled parts are performed on a five-axis machining center JD GR200-A10SH with a maximum spindle speed of 28,000 rpm, as shown in Figure 9. In the experiment, double-sided milling is used to simulate the actual machining condition of the blade. A four-tooth ball-end milling cutter with a diameter of 8 mm, a spindle speed of 3000 rpm, a feed rate of 300 mm/min, an axial depth of cut of 5 mm, and a radial depth of cut of 0.2 mm are used. The parameters of machining are shown in Table 1.

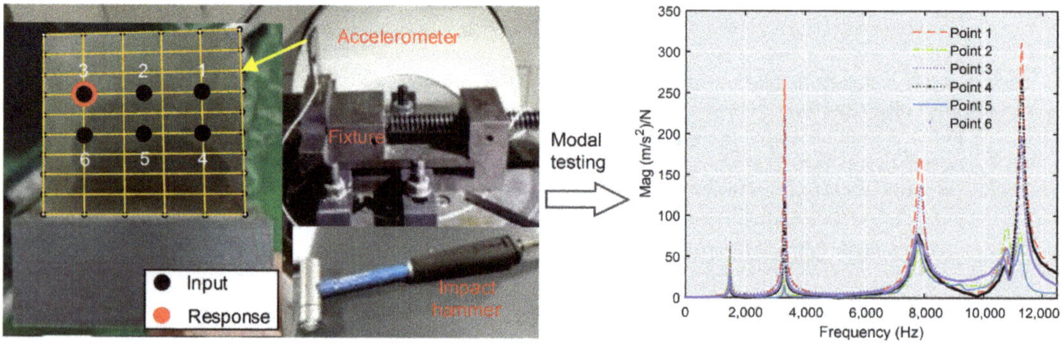

Figure 9. Milling modal test experiment for the thin-walled parts of a blade.

Table 1. The parameters used in the test.

Spindle Speed (rpm)	Feed Rate (mm/min)	Axial Depth of Cut (mm)	Radial Depth of Cut (mm)
3000	300	5	0.2

The force hammer model is the Dytran Model 5800B (Marilla, CA, USA), the acceleration sensor model is Dytran 3224A1 (Marilla, CA, USA), and the data acquisition system is ECON-AVANT-8008 (Plymouth, MN, USA). This modal analysis software developed by Econ Company (North York, ON, USA) is used to extract the kinetic parameters of the workpiece. To ensure the validity of the test and shorten the test time, the modal test was conducted on the workpiece in six cutting stages. Step 0 represents the initial workpiece state when no cutting is performed and steps 1 to 5 each represents the material with a depth of 5 mm, 10 mm, 15 mm, 20 mm, and 25 mm along the direction of the removal. The moving force hammer test method with a fixed sensor is adopted herein, and the sticking position of the sensor and the hammering test points are shown in Figure 9, where point 3 is the response point of all the tests, namely the reference point. According to the structural characteristics of the blade, hammering is carried out at each test point along the direction of the blade.

The mode shapes extracted from the RFRs obtained through the tests and the FEM are compared, as shown in Figure 10, and the figures on the left show the results of the tests. It can be seen that the FEM results are consistent with the test results, and the mode shapes of modes 1 to 3 are the first bending mode, first torsional mode, and second bending mode, respectively.

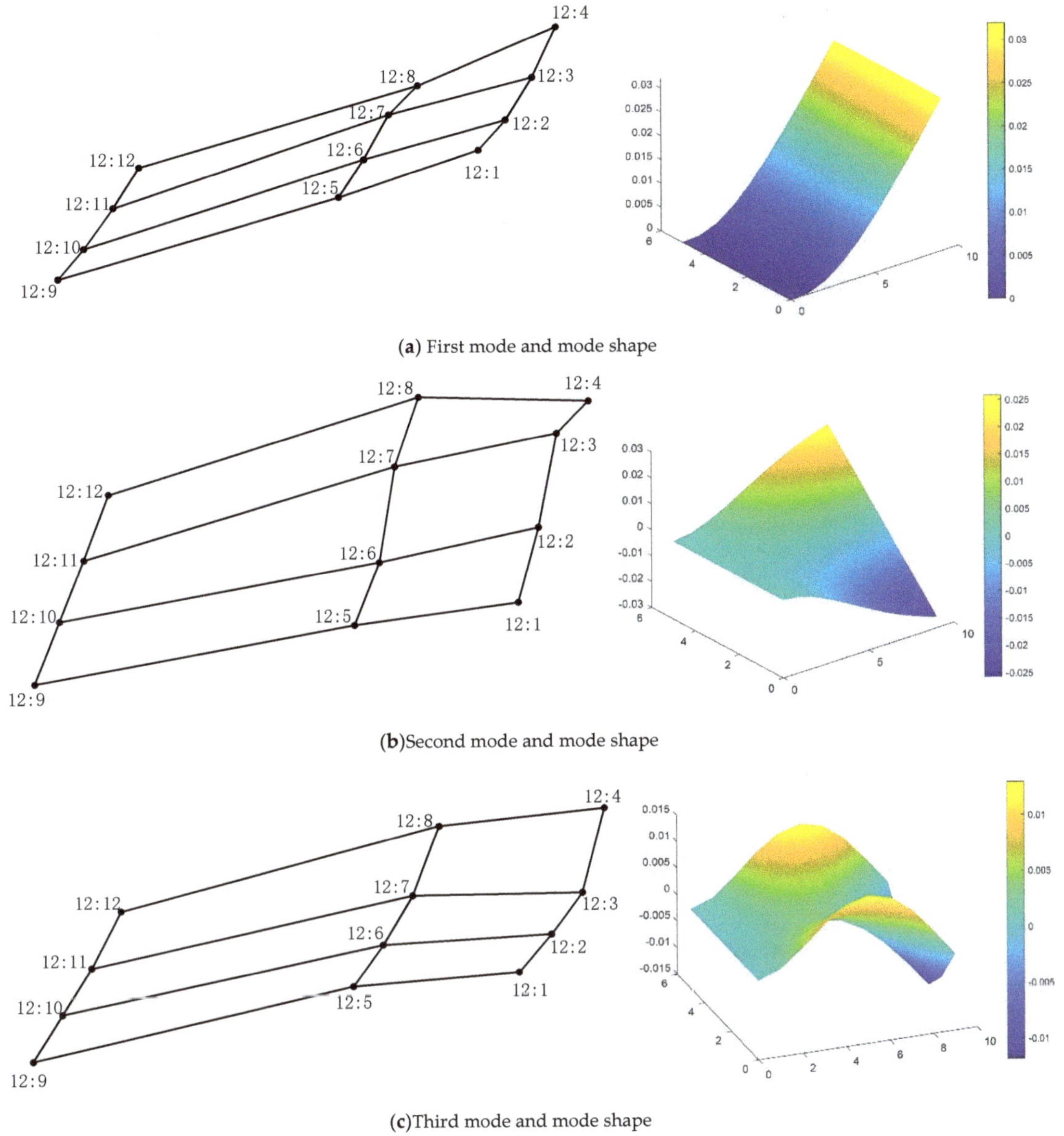

Figure 10. Mode shapes obtained through tests and the FEM as follows: (**a**) mode 1, (**b**) mode 2, and (**c**) mode 3.

After comparing and analyzing the nature frequency of the workpiece calculated using the finite element model with that of the workpiece obtained from the test during the milling process, the relative error is expressed as follows:

$$\eta_1 = \frac{f_{FEM} - f_{Test}}{f_{Test}} \times 100\% \tag{35}$$

where f_{FEM} and f_{Test} denote the frequencies of the workpiece obtained from the shell cell finite element model and the modal test, respectively. The calculated results are shown in Table 2, and the maximum errors of the results obtained from the shell cell finite element model are 2.51%, 2.53%, 3.42%, 4.04%, 4.01%, and 3.81%, respectively, compared with the test results. The experiment analysis shows that the shell finite element modeling method can effectively predict the time-varying dynamic parameters of the blade during the milling process.

Table 2. Comparison of blade intrinsic frequencies (Hz) for the different cutting steps obtained from the shell cell finite element model and hammering test.

Steps	Modes	1	2	3
0	Tests	1535	3290	8152
	FEM	1574	3265	8293
	η_1 (%)	(2.51)	(−0.76)	(1.72)
1	Tests	1552	3313	8094
	FEM	1591	3289	8233
	η_1 (%)	(2.53)	(−0.73)	(1.71)
2	Tests	1551	3312	8101
	FEM	1604	3302	8299
	η_1 (%)	(3.42)	(−0.32)	(2.45)
3	Tests	1548	3268	7972
	FEM	1611	3301	8244
	η_1 (%)	(4.04)	(1.02)	(3.41)
4	Tests	1551	3239	7965
	FEM	1613	3286	8174
	η_1 (%)	(4.01)	(1.44)	(2.62)
5	Tests	1550	3210	7867
	FEM	1609	3259	8112
	η_1 (%)	(3.81)	(1.52)	(3.11)

4.2. Neural Network Model Predictive Analysis

For the thin-walled parts shown in Figure 7, the unit structure is discretized, and when the thickness **t** of the nodes is taken as a fixed value, the first three orders of the nature frequency of the workpiece are calculated through the shell cell finite element model established herein, which is treated as a sample point. A total of 154 samples are obtained from the calculations herein of which 123 samples are taken as training samples. The model training learning rate is set to 0.01, the training data is disrupted, and the batch size is set to 30. The model is trained before the nature frequency of the workpiece is normalized to the first-order frequency because the nature frequency of the workpiece is higher relative to the thickness of the workpiece's discrete unit node, and all the samples in the same batch will have the same predicted value when the model is being trained as per the example as follows:

$$f_N^1 = \frac{f^1 - f_{\min}^1}{f_{\max}^1 - f_{\min}^1} \tag{36}$$

where f^1 represents the first-order frequency, $f_{\min}^1$ and $f_{\max}^1$ represent the minimum and maximum values of the first-order frequency of all the samples, respectively, and f_N^1 represents the normalized first-order frequency.

With the increase in the number of model training rounds, the convergence of the workpiece's first three orders of nature frequency is displayed in Figure 11, where one can see that the predicted value reaches convergence when the epoch = 300, and the epoch = 300 is adopted herein for calculation. The comparison of the nature frequency of the workpiece at various cutting stages, which is calculated using the neural network model and the shell cell finite element model, is shown in Table 3. When the results of the shell cell finite element model are treated as the benchmark, the maximum error of the first three orders of nature frequency calculated using the three-layer neural network model reaches

0.409%. Therefore, the kinetic parameters of the workpiece during the cutting process can be efficiently predicted by training the three-layer neural network model.

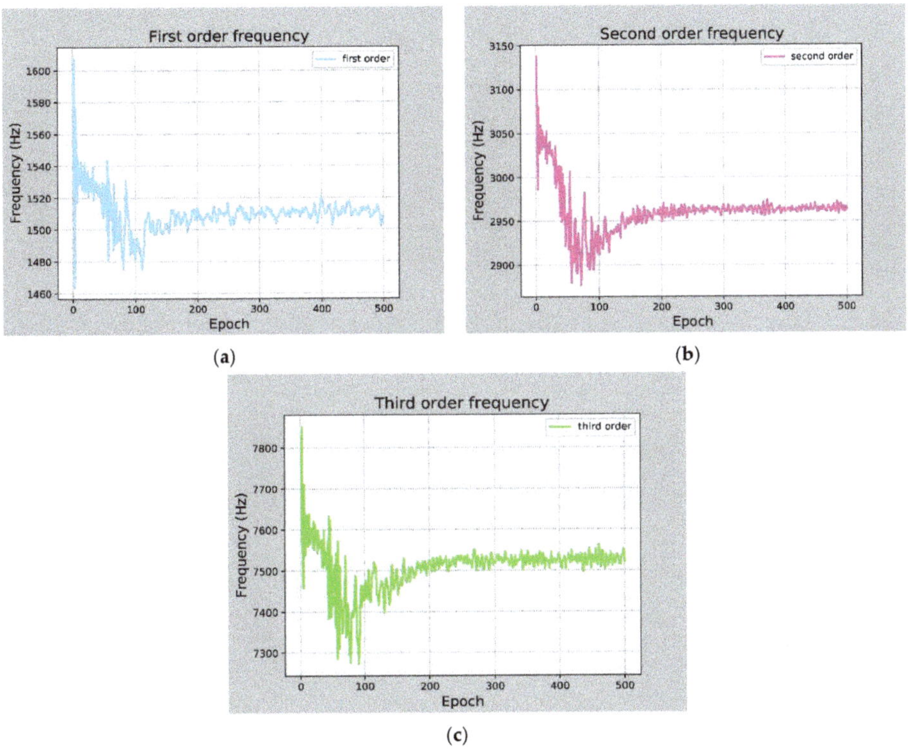

Figure 11. Convergence analysis of the neural network model's predictive values. (**a**) First-order nature frequency, (**b**) second-order nature frequency, (**c**) third-order nature frequency.

Table 3. Comparison of blade intrinsic frequencies (Hz) during the cutting calculated using the neural network model (NN) and shell cell finite element model (FEM).

State	Modes	1	2	3	State	Modes	1	2	3
1	FEM	1536	3002	7538	7	FEM	1589	3032	7370
	NN	1534	3004.5	7545.1		NN	1584	3037.1	7368.6
	η (%)	(−0.12)	(0.082)	(0.095)		η (%)	(−0.31)	(0.168)	(−0.019)
2	FEM	1512	2965	7523	8	FEM	1557	2957	7322
	NN	1510	2967	7526.4		NN	1556	2964.6	7333.7
	η (%)	(−0.11)	(0.067)	(0.046)		η (%)	(−0.03)	(0.255)	(0.16)
3	FEM	1553	3099	7806	9	FEM	1505	2881	7289
	NN	1549	3098	7821.4		NN	1509	2889.6	7296.3
	η (%)	(−0.29)	(−0.001)	(0.197)		η (%)	(0.25)	(0.298)	(0.099)
4	FEM	1581	3131	7748	10	FEM	1496	2954	7487
	NN	1576	3131.9	7766.3		NN	1500	2960.9	7517.1
	η (%)	(−0.30)	(0.0027)	(0.236)		η (%)	(0.28)	(0.233)	(0.401)
5	FEM	1598	3131	7624	11	FEM	1511	2970	7459
	NN	1593	3132.5	7628.7		NN	1514	2977.3	7489.5
	η (%)	(−0.31)	(0.047)	(0.062)		η (%)	(0.20)	(0.247)	(0.409)
6	FEM	1602	3095	7479	12	FEM	1519	2969	7400
	NN	1598	3094.3	7486.3		NN	1522	2977.6	7420.7
	η (%)	(−0.23)	(−0.023)	(0.098)		η (%)	(0.23)	(0.291)	(0.28)

Next, the computational efficiency of the shell cell finite element model is compared with the three-layer neural network model. The stiffness matrix and mass matrix of the shell unit finite element model of the thin-walled parts herein have 765 degrees of freedom, and the characteristic equation is solved by using Matlab software to calculate the nature frequency of the workpiece. For one sample, the solution time reaches 11.771 s. When epoch = 300 and batch size = 30, the CPU is used for computation, and the training time of the three-layer neural network model is shown in Figure 12a, which is about 9 s. After obtaining the neural network computational model from the training data, the kinetic parameters of the workpiece at different cutting stages can be batch-computed using the model. The time spent predicting the kinetic parameters of the workpiece through the neural network model is shown in Figure 12b. The loading time of the neural network model and its input data loading time amount to 1.6611 s. The prediction time is only 0.00315 s for the 150-numbered samples. The model training time is only 4 s when the GPU is used for the computation.

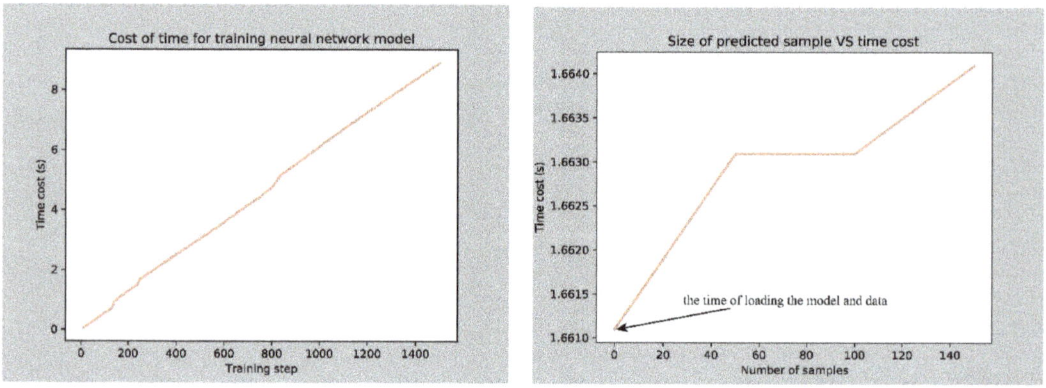

(a) (b)

Figure 12. Design neural network model training and prediction time. (**a**) Neural network model training time, (**b**) neural network batch prediction time for the milling dynamic parameters of the thin-walled parts.

The modal testing results showed that the maximum prediction error of the FEM was 4.04%. In comparison with the calculations using the FEM, the neural network model exhibited a maximum prediction error of 0.409%, thereby suggesting that the neural network model's maximum prediction error is approximately 4% when compared with the modal testing results. The abovementioned analysis shows that compared with the shell cell finite element model, the computation efficiency can be significantly improved by utilizing the batch computation function of the neural network while guaranteeing prediction accuracy.

5. Conclusions

In this paper, the neural network model is introduced into the prediction of the dynamic parameters of thin-walled parts to achieve the real-time and rapid prediction of workpiece dynamics during the milling of thin-walled parts. Firstly, the FEM based on shell elements was built. After the node coordinates, workpiece thickness, and normal vector of the intermediate surface of the initial workpiece and the machined workpiece were obtained, the established shell FEM was used to calculate the time-varying dynamic parameters of the workpiece during the milling process. To improve the computational efficiency, a three-layer neural network model was constructed to predict the time-varying dynamic parameters for milling the thin-walled parts. Then, the dataset obtained using the shell FEM was used as the training samples to train the neural network model. Finally, the effectiveness of the model was verified through modal testing experiments. The following conclusions can be drawn.

(1) The proposed method only needs to mesh the initial workpiece and the machined workpiece once, and it only needs to calculate the geometric parameters of the initial workpiece and the machined workpiece. It does not need to re-mesh the thin-walled parts at each cutting position and to repeatedly calculate the geometric parameters of the workpiece during the cutting process. Meanwhile, in comparison with the three-dimensional cube elements, the shell elements can reduce the number of degrees of freedom of the FEM model by 74%, which leads to about nine times faster computation of the characteristic equation.
(2) Modal test results show that the maximum prediction error of the workpiece nature frequency of the built shell FEM is about 4%.
(3) Compared with the calculation results of the FEM, the maximum prediction error of the neural network model is 0.409%. It can be inferred that compared with the modal test results, the maximum prediction error of the neural network model is about 4%.
(4) When the training samples are 132, the epoch is 300, the batch size is 30, and the CPU is used for calculation, the training time is about 9 s. When using the model to batch-calculate the dynamic parameters of a thin-wall part at different cutting stages, the loading time of the neural network model and the model input data is about 1.6611 s. When the number of predicted cutting states is 150, the prediction time is only 0.00315 s. Therefore, the three-layer neural network model can highly improve calculation efficiency while ensuring calculation accuracy.

Meanwhile, the present study has certain limitations. As compared with modal testing, the prediction error of the FEM is about 4%, and this results in an approximate 4% prediction error in the neural network model. In the future, a dataset of workpiece thickness and natural frequencies will be constructed based on modal testing experiments. Then, the prediction accuracy of the neural network model will be enhanced.

Author Contributions: Y.L.: methodology, investigation, formal analysis and draft writing. D.W.: formal analysis, review, and editing. W.T.: project administration, formal analysis, review, and editing. J.Z.: project administration, formal analysis, review, and editing. F.D.: supervision, project administration, and review. All authors have read and agreed to the published version of the manuscript.

Funding: This work is supported by the National Natural Science Foundation of China (Grant No. 52075451), the Natural Science Basic Research Program of Shaanxi (Program No. 2023-JC-YB-347), and the Aeronautical Science Foundation of China (Grant No. 2019ZE053008).

Data Availability Statement: All data and materials used or analyzed during the current study are included in this manuscript. Written informed consent has been obtained from the patient(s) to publish this paper.

Conflicts of Interest: The authors declare no conflicts of interest.

References

1. Zhao, G.L.; Zhao, B.; Ding, W.F.; Xin, L.J.; Nian, Z.W.; Peng, J.H.; He, N.; Xu, J.H. Nontraditional energy-assisted mechanical machining of difficult-to-cut materials and components in aerospace community: A comparative analysis. *Int. J. Extreme Manufacturing* **2023**, *6*, 022007. [CrossRef]
2. Bravo, U.; Altuzarra, O.; De Lacalle, L.N.L.; Sánchez, J.A.; Campa, F.J. Stability limits of milling considering the flexibility of the workpiece and the machine. *Int. J. Mach. Tools Manuf.* **2005**, *45*, 1669–1680.
3. Seguy, S.; Campa, F.J.; De Lacalle, L.N.L.; Arnaud, L.; Dessein, G.; Aramendi, G. Toolpath dependent stability lobes for the milling of thin-walled parts. *Int. J. Mach. Mach. Mater.* **2008**, *4*, 377. [CrossRef]
4. Arnaud, L.; Gonzalo, O.; Seguy, S.; Jauregi, H.; Peigné, G. Simulation of low rigidity part machining applied to thin-walled structures. *Int. J. Adv. Manuf. Technol.* **2011**, *54*, 479–488. [CrossRef]
5. Thevenot, V.; Arnaud, L.; Dessein, G.; Cazenave-Larrocheet, G. Influence of material removal on the dynamic behavior of thin-walled structures in peripheral milling. *Mach. Sci. Technol.* **2006**, *10*, 275–287. [CrossRef]
6. Song, Q.H.; Ai, X.; Tang, W.X. Prediction of simultaneous dynamic stability limit of time–variable parameters system in thin-walled workpiece high-speed milling processes. *Int. J. Adv. Manuf. Technol.* **2011**, *55*, 883–889. [CrossRef]

7. Meshreki, M.; Attia, H.; Kövecses, J. Development of a new model for the varying dynamics of flexible pocket-structures during machining. *J. Manuf. Sci. Eng. Am. Soc. Mech. Eng.* **2011**, *133*, 41002. [CrossRef]
8. Song, Q.H.; Liu, Z.Q.; Wan, Y.; Ju, G.G.; Shi, J.H. Application of Sherman-Morrison-Woodbury formulas in instantaneous dynamic of peripheral milling for thin-walled component. *Int. J. Mech. Sci.* **2015**, *96*, 79–90. [CrossRef]
9. Song, Q.H.; Shi, J.H.; Liu, Z.Q.; Wan, Y. A time-space discretization method in milling stability prediction of thin-walled component. *Int. J. Adv. Manuf. Technol.* **2017**, *89*, 2675–2689. [CrossRef]
10. Ahmadi, K. Finite strip modeling of the varying dynamics of thin-walled pocket structures during machining. *Int. J. Adv. Manuf. Technol.* **2017**, *89*, 2691–2699. [CrossRef]
11. Tuysuz, O.; Altintas, Y. Frequency domain updating of thin-walled workpiece dynamics using reduced order substructuring method in machining. *J. Manuf. Sci. Eng. Am. Soc. Mech. Eng.* **2017**, *139*, 71013. [CrossRef]
12. Yang, Y.; Zhang, W.H.; Ma, Y.C.; Wan, M. Chatter prediction for the peripheral milling of thin-walled workpiece with curved surfaces. *Int. J. Mach. Tools Manuf.* **2016**, *109*, 36–48. [CrossRef]
13. Yang, Y.; Zhang, W.H.; Ma, Y.C.; Wan, M.; Dang, X.B. An efficient decomposition-condensation method for chatter prediction in milling large-scale thin-walled structures. *Mech. Syst. Signal Process.* **2019**, *121*, 58–76. [CrossRef]
14. Dang, X.B.; Wan, M.; Yang, Y.; Zhang, W.H. Efficient prediction of varying dynamic characteristics in thin-wall milling using freedom and mode reduction methods. *Int. J. Mech. Sci.* **2019**, *150*, 202–216. [CrossRef]
15. Budak, E.; Tun, L.T.; Alan, S.; Nevzat-Özgüvenet, H. Prediction of workpiece dynamics and its effects on chatter stability in milling. *CIRP Ann.* **2012**, *61*, 339–342. [CrossRef]
16. Zhang, X.; Zhu, L.; Ding, H. Matrix perturbation method for predicting dynamic modal shapes of the workpiece in high-speed machining. *Proc. Inst. Mech. Eng. Part B J. Eng. Manuf.* **2010**, *224*, 177–183. [CrossRef]
17. Luo, M.; Zhang, D.H.; Wu, B.H.; Tang, M. Modeling and analysis effects of material removal on machining dynamics in milling of thin-walled workpiece. *Adv. Mater. Res.* **2011**, *223*, 671–678. [CrossRef]
18. Tian, W.J.; Ren, J.X.; Zhou, J.H.; Wang, D.Z. Dynamic modal prediction and experimental study of thin-walled workpiece removal based on perturbation method. *Int. J. Adv. Manuf. Technol.* **2018**, *94*, 2099–2113. [CrossRef]
19. Ju, G.G. Research on Instability Characteristics of Thin-walled Parts with Complex Curved Surface and Variable Thickness during Multi-Axis Milling. Master's Thesis, Shandong University, Jinan, China, 2016.
20. Tuysuz, O.; Altintas, Y. Time-domain modeling of varying dynamic characteristics in thin-wall machining using perturbation and reduced-order substructuring methods. *J. Manuf. Sci. Eng.* **2018**, *140*, 011015. [CrossRef]
21. Liu, Y.L. Analysis of Milling Dynamics and Chatter Detection and Control Methods for Thin-Walled Workpiece. Ph.D. Thesis, Northwestern Polytechnical University, Xi'an, China, 2017.
22. Kim, D.H.; Kim, T.Y.; Wang, X.L.; Kim, M.; Quan, Y.J.; Oh, J.W.; Min, S.H.; Kim, H.J.; Bhandari, B.; Yang, I.; et al. Smart Machining Process Using Machine Learning: A Review and Perspective on Machining Industry. *Int. J. Precis. Eng. Manuf.-Green Technol.* **2018**, *5*, 555–568. [CrossRef]
23. Yeganefar, A.; Niknam, S.A.; Asadi, R. The use of support vector machine, neural network, and regression analysis to predict and optimize surface roughness and cutting forces in milling. *Int. J. Adv. Manuf. Technol.* **2019**, *105*, 951–965. [CrossRef]
24. Pavlenko, I.; Saga, M.; Kuric, I.; Kotliar, A.; Basova, Y.; Trojanowska, J.; Ivanov, V. Parameter identification of cutting forces in crankshaft grinding using artificial neural networks. *Materials* **2020**, *13*, 5357. [CrossRef]
25. SK, T.; Shankar, S.; T, M.; K, D. Tool wear prediction in hard turning of EN8 steel using cutting force and surface roughness with artificial neural network. *Proc. Inst. Mech. Eng. Part C J. Mech. Eng. Sci.* **2020**, *234*, 329–342. [CrossRef]
26. Vasanth, X.A.; Paul, P.S.; Varadarajan, A.S. A neural network model to predict surface roughness during turning of hardened SS410 steel. *Int. J. Syst. Assur. Eng. Manag.* **2020**, *11*, 704–715. [CrossRef]
27. Wang, J.; Zou, B.; Liu, M.; Li, Y.; Ding, H.; Xue, K. Milling force prediction model based on transfer learning and neural network. *J. Intell. Manuf.* **2021**, *32*, 947–956. [CrossRef]
28. Heitz, T.; He, N.; Ait-Mlouk, A.; Bachrathy, D.; Chen, N.; Zhao, G.; Li, L. Investigation on eXtreme Gradient Boosting for cutting force prediction in milling. *J. Intell. Manuf.* **2023**. [CrossRef]
29. Zhu, B.F. ; *The Finite Element Method Theory and Applications*; China Institute of Water Resources and Hydropower Research: Beijing, China, 2018.
30. Logan, D.L. *A First Course in the Finite Element Method*; Cengage Learning: Boston, MA, USA, 2011.
31. Agarap, A.F. Deep learning using rectified linear units (relu). *arXiv* **2018**, arXiv:1803.08375.
32. Goodfellow, I.; Bengio, Y.; Courville, A. *Deep Learning*; MIT Press: Cambridge, MA, USA, 2016.

machines

MDPI

Article

Compensation Method for Correcting the Topography Convolution of the 3D AFM Profile Image of a Diffraction Grating

Kai Zhang *, Yang Bai and Zhimin Zhang

Mechanics and Acoustics Department, National Institute of Metrology, Beijing 102200, China
* Correspondence: zhang-kai@nim.ac.cn

Abstract: Any 3D AFM image is a convolution of the geometry of the AFM tip and the profile of the scanned sample, especially when the dimensions of the scanned sample are comparable to those of the AFM tip shape. The precise profile of the scanned sample can be extracted from the 3D AFM image if the geometry of the AFM tip is known. Therefore, in order to separate the geometry of the AFM probe tip from the 3D AFM image of a diffraction grating with a rectangular profile and to correct for the topographic convolutions induced by the AFM probe tip, a method is used to quantitatively evaluate the geometry of the AFM probe tip, including the tip radius and the included angle. A model for reconstructing the measured AFM image is proposed to correct topography convolutions caused by the AFM tip shape when scanning a diffraction grating with rectangular profiles. A series of experiments were performed to verify the effectiveness of the proposed AFM tip geometry evaluation method, and comparison experiments were conducted to demonstrate the feasibility and reliability of the proposed reconstruction model.

Keywords: atomic force microscopy; AFM probe tip; image convolution; edge reversal method; diffraction grating

1. Introduction

Reliable topographic profile characterization of diffraction gratings is essential in many fields, such as semiconductors [1], materials and metrology [2–4]. There are several methods for topographic characterization of diffraction gratings. Optical sensors can provide non-contact detection of the grating topography [5,6]. However, the lower resolution in the horizontal direction limits the measurement accuracy of the topographic characterization of diffraction gratings, especially for the grating with vertical sidewalls. Scanning electron microscopes (SEMs), which have the advantages of nanometer spatial resolution and large depth of field, represent an effective approach to the straightforward determination of the in-plane dimensions of gratings from the top view [7]. However, in order to use the SEM to determine the other dimensions of the gratings, such as heights and sidewall angles, the sample should be coated, and the focused ion beam method should be applied [8], which is time-consuming and complicated. In comparison with optical sensors and SEMs, atomic force microscopes (AFMs) have nanoscale in-plane resolution and sub-angstrom vertical resolution [9,10]. An AFM, which has a 3D metrology capability, can be used to carry out a non-destructive topographic characterization of the diffraction gratings. However, 3D AFM images are recognized as the convolutions of the geometry of the AFM probe tip and the actual profile of the grating [11]. The geometry of the AFM probe tip, including the tip radius and included angle, significantly affects the measurement accuracy of the 3D AFM image, especially when the dimensions of the scanned grating are comparable to the size of the AFM probe tip [12].

Many methods have been proposed to reduce the geometry of the topography convolutions induced by the AFM tip and to improve the measurement accuracy of the AFM.

One method, blind reconstruction algorithms [13], has been successfully developed to deconvolve the 3D topographic images of the sample and the AFM tip. This method allows the prior measurement of the actual structure of the sample, and the shape of the AFM tip is not necessary in compensating the effect of the geometry of the AFM tip on the measured 3D image [14]. However, it is difficult to guarantee the effectiveness and accuracy of this method because the convergence of the algorithms is highly dependent on the determination of the input values. For example, the Goldilocks threshold value can have a significant impact on the final reconstructed profile in the practical application of the blind reconstruction algorithms, but there are no clear guidelines for the selection of the optimal value [14]. A critical dimension AFM (CD-AFM) probe tip [15] and a carbon nanotube (CNT-AFM) probe tip [16] have been developed as alternative techniques to reduce the topography convolutions in the horizontal direction [17,18]. The CD-AFM and CNT-AFM probe tips have a vertical shape, unlike the conventional AFM probe, which has a triangular pyramid structure. These tips were designed to overcome the limitations of the conventional AFM when measuring narrow and deep trenches. They are commonly used to determine the angle, height and line edge roughness of micro/nanostructures [19]. However, both types of AFM probes only have the capability to mitigate the drawbacks caused by the included angle of the traditional AFM probe tip. Despite this, the topography convolution induced by the AFM tip radius is still evident in the recorded 3D profile of gratings [20]. In addition, the high cost of both types of AFM probes makes them unsuitable for academic laboratories in mainstream universities [15,16]. Another technique to reduce distortion in AFM measurements is to apply arithmetic operations based on the geometric relationship between the sample and the AFM probe tip [21,22]. These equations can be derived effortlessly when working with gratings that have a clear structure. The actual surface profile of the grating can be reconstructed by combining the measured profile data of the grating with the equations. The precision of the reconstructed 3D profile image of the grating is contingent on the accuracy of the method used to assess the geometry of the AFM probe tip that is employed during the AFM measurement [23]. Therefore, choosing an appropriate approach to evaluate the geometry of the AFM probe tip, encompassing the tip radius and included angle, can significantly improve the accuracy of reconstructing the 3D image of the specimen.

There are several techniques that can assess the geometry of the AFM probe tip, including measuring the included angle and tip radius of the AFM probe. A common approach is to visually inspect the AFM tip geometry using a high-resolution SEM or TEM [24–26]. However, contamination of the AFM probe tip surface by deposited plasma in the case of the SEM and the limited range for measuring thickness when using the TEM restrict the usefulness of these techniques in the measurement of an AFM probe tip with a pyramidal structure. In addition, without proper alignment, SEMs and TEMs cannot measure the same 2D projection of the AFM tip that was utilized in the AFM measurement [26]. Accurately characterizing the topography of the AFM probe tip can be achieved by scanning a reference artifact with a well-known nanostructure [12,27]. Typically, a reference artifact with a line space structure and steep sidewalls in the cross-section is fabricated for this purpose. The discrepancy between the profiles of the reference artifact and the AFM image being measured is used to assess the geometry of the AFM probe tip. This technique performs admirably when assessing the included angle of the AFM probe but is inadequate for determining tip radii under 30 nm [27]. Therefore, the practical application of this method is limited by the high precision and complex procedure required to fabricate a reference artefact. An edge reversal method has been proposed to improve the accuracy of AFM tip radius evaluation to overcome the limitation of measuring tip radii larger than 30 nm [28]. The edge reversal method has effectively compensated for the influence of the AFM tip radius on the 3D tool edge profile of a single-point diamond tool [29]. Compared to other tip radius measurement methods, the advantage of this method is that the tip radius of the AFM probe can be accurately evaluated without considering the influence of the convolution. However, the combination of the tip radius and the included angle of an AFM probe can significantly affect the 3D AFM image of a diffraction grating with vertical sidewalls [17,20]. Therefore, the geometry of the AFM probe tip can be accurately determined by combining the advantages of two techniques, scanning a reference artefact to measure the included angle and using the edge reversal method to measure the tip radius.

This paper proposes a correction method aimed at eliminating geometric distortions resulting from the geometry of the AFM probe tip when measuring a diffraction grating with vertical sidewalls. The topography distortion of the diffraction grating is corrected by accurately evaluating the geometry of the AFM probe tip using an edge reversal method and scanning a line space structure. Furthermore, a reconstruction model is proposed to correct the influence of the tip radius and included angle of the AFM probe in accurately measuring a diffraction grating with a rectangular profile. The proposed method and model can decrease the errors in the reconstruction of the profile image of the diffraction grating compared with other methods. A series of experiments were conducted to demonstrate the feasibility and reliability of the proposed method and model.

2. Measurement Principle

2.1. Principle for AFM Measurement of a Diffraction Grating with a Rectangular Profile

Figure 1 displays an illustration of a diffraction grating which was scanned using an AFM in the contact mode. A diffraction grating with a rectangular profile could be produced by combining the exposure and etching processes [7]. The resulting sidewalls are nearly perpendicular to the substrate's surface, and the corner rounding in the intersection between the top surface and sidewalls is typically evaluated to be around several to ten nanometers [20]. Schematic diagrams of AFM measurements of a diffraction grating using different types of AFM probe tips can be seen in Figure 1a,b. The accuracy of diffraction grating measurements is significantly influenced by the tip shape of the AFM probe, including the tip radius and included angle, since the measured 3D AFM image is treated as a convolution of the diffraction grating surface and the AFM tip shape, according to [14]. Hence, it is crucial to reconstruct the 3D profile of a diffraction grating without considering the tip shape of the AFM probe to improve the accuracy of AFM measurements.

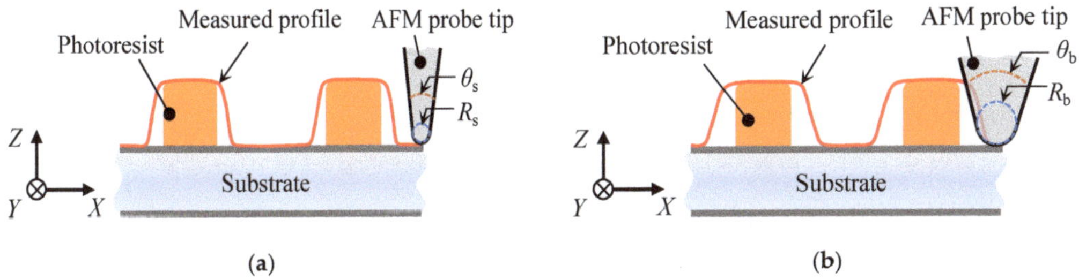

Figure 1. AFM measurement of a diffraction grating with rectangular profiles. (**a**) The trajectory of the tip apex of the AFM probe with high resolution (utilization of an AFM probe with a tip radius of R_s and an included angle of θ_s); (**b**) the trajectory of the tip apex of the AFM probe with low resolution (utilization of an AFM probe with a tip radius of R_b and an included angle of θ_b, where $R_b > R_s$ and $\theta_b > \theta_s$).

A grid with a corner-rounding radius of R_{corner} and tilt angle α of sidewalls was extracted from the entire diffraction grating for further research. An AFM probe with a tip radius of R_{tip} and an included angle of θ ($\theta_1 + \theta_2$) was employed to scan the surface of the extracted grid. It was assumed that the AFM probe tip has a spherical shape, which can fit most AFM cantilevers. Figure 2 illustrates the geometric relationship between the measured and actual profiles of the grid in the cross-section. The measured profile of the grid in the cross-section is considered to be the convolution of the AFM tip shape and the actual profile of the measured surface due to the finite tip shape of the AFM probe. The tilt angles ε_1 and ε_2 of the measured profile on the sidewalls matched the corresponding tilt angles θ_1 and θ_2 of the AFM tip shape. This convolution of the geometries between the AFM tip shape and the actual profile of the grid results in the AFM image measurements of the right and left side of the grid in regions C_1C_3' and C_4C_6' corresponding to the actual profile in regions C_1C_3 and C_4C_6. The coordinates of point A have been adjusted from (x_1, z_1) to (x_1', z_1) due to the convolution effect of R_{tip}. Consequently, the actual curve C_1C_2 in region C_1C_3 of the grid has been replaced by the measured curve C_1C_2'. Similarly, the coordinates of point B and the right sidewall (Line C_2C_3) of the grating in region C_1C_3 have been altered to (x_2', z_2) and Line $C_2'C_3'$ in the AFM image, respectively, owing to the convolution effect of the tip radius (R_{tip}) and the included angle (θ) of the AFM probe tip. Based on the geometric relationship in the convolution area on the right side, it is feasible to formulate a reconstruction model that counteracts the impact of the AFM tip shape by using the following equation:

$$x_i = \begin{cases} \sqrt{\left(\sqrt{x_i'^2 + (z_i + R_{tip})^2} - R_{tip}\right)^2 - z_i^2} & x_i' \in C_1C_2 \\ x_i' - \left(\tan\frac{\pi - 2\theta_1}{4} \cdot R_{tip} + z_i \cdot (\tan\theta_1 - \tan\alpha_1)\right) & x_i' \in C_2C_3 \end{cases} \quad (1)$$

where (x_i, z_i) represents the coordinates of points in the region C_1C_3 and (x_i', z_i) represents the corresponding coordinates of points in the region C_1C_3'. θ_1 and α_1 represent the inclined angle of the AFM tip shape on the left side and the tilt angle of the actual profile image on the right sidewall, respectively.

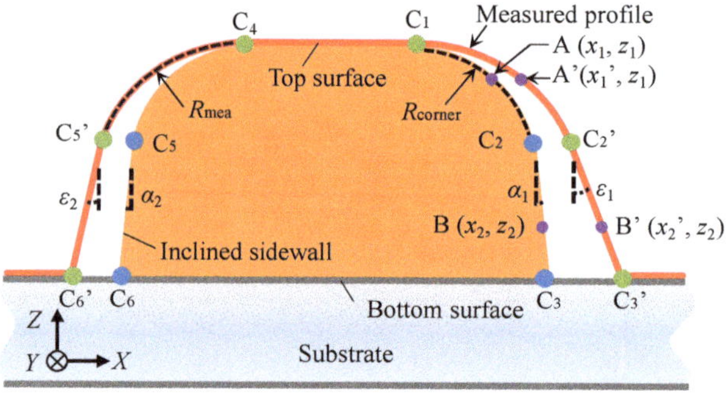

Figure 2. The schematic of the geometrical relationship between the measured profile image and the real surface of a grid.

Similarly, the reconstruction model on the left side can be established using the same methodology and can be articulated as follows:

$$x_j = \begin{cases} \sqrt{\left(\sqrt{x_j'^2 + (z_j + R_{tip})^2} - R_{tip}\right)^2 - z_j^2} & x_j' \in C_4C_5 \\ x_j' + \left(\tan\frac{\pi - 2\theta_2}{4} \cdot R_{tip} + z_j(\tan\theta_2 - \tan\alpha_2)\right) & x_j' \in C_5C_6 \end{cases} \quad (2)$$

where (x_j, z_j) represents the coordinates of points in the region C_4C_5 and (x_j', z_j) represents the corresponding coordinates of points in the region C_4C_5'. θ_2 and α_2 represent the inclined angle of the AFM tip shape on the right side and the tilt angle of the actual profile image on the left sidewall, respectively. The derivations of the reconstruction model can be found in Appendix A.

As explained above, the accuracy of the AFM measurement is significantly affected by the geometry of the AFM probe tip, including its tip radius and included angle. As a result, it is crucial to quantitatively estimate the tip radius and included angle of the AFM probe to eliminate the convolution effect in the AFM image. It is possible to evaluate the included angle of the AFM probe by scanning a line pattern structure [27]. However, accurately evaluating the tip radius of the AFM probe, which typically ranges in the nanometer scale, poses a challenge using conventional methods. Therefore, an edge reversal method is proposed to accurately assure the tip radius of the AFM probe in this paper [28].

2.2. Method for Evaluating the Geometry of an AFM Probe Tip

As explained in Section 2.1, it is not possible to accurately evaluate the geometry of an AFM probe tip by only scanning a reference artefact with steep sidewalls, as the radius of the AFM tip is too small for accurate measurement using this method. Hence, the assessment of an AFM probe tip's geometry is divided into two parts. Firstly, the included angle θ of the AFM probe is assessed through the measurement of an artifact with a known topography. Secondly, the tip radius R_{tip} of the AFM probe is assessed using the edge reversal method.

Figure 2 illustrates that the convolution effect of the AFM probe causes the radius R_{mea} of the corner rounding to exceed the radius R_{corner} of the artifact. Moreover, the edge slopes ε_1 and ε_2 of the measured profiles deviate from the actual edge slopes α_1 and α_2 of the standard artifact. Denoting the inclined angles of the left and right surfaces of the AFM probe as θ_1 and θ_2, respectively, the following equation is satisfied:

$$\theta = \theta_1 + \theta_2 = \varepsilon_1 + \varepsilon_2 \tag{3}$$

since $\varepsilon_1 > \alpha_1$ and $\varepsilon_2 > \alpha_2$, the included angle θ of the AFM probe can be determined by extracting ε_1 and ε_2 from the measured profile.

Quantitatively evaluating the AFM tip radius R_{tip}, which typically has a nominal value of several nanometers, poses a challenge compared to the measurement method for θ. The proposed solution to this issue is the edge reversal method [28,30]. The proposed method is designed in such a way that the AFM image of an object has an edge whose apex radius is comparable to the AFM tip radius and that of the indentation mark generated by using the edge of the object. Through the arithmetic operation with the obtained AFM images, the AFM tip radius can be evaluated quantitatively while separating the influence of the apex radius of the edge of the object.

Figure 3a,b show the schematic of the principle of the edge reversal method in the measurement of the indentation mark and the diamond tool (a single-point diamond tool was selected as the object of this study), respectively. The AFM tip radius can be evaluated with Equation (4) if the effect of elastic recovery on the indentation mark can be ignored [31,32]:

$$R_{tip} = \frac{R_{tool\_m} - R_{mark\_m}}{2} \tag{4}$$

where R_{tool_m} and R_{mark_m} represent the measured apex radius of the indentation mark and measured cutting edge radius of the diamond tool, respectively. A description of the edge reversal method in detail can be found in the relevant papers [28,33]. As described above, the procedure of evaluating the tip shape of the AFM probe is complex and takes a lot of time in a single measurement. Simplification of this method will be researched in the future.

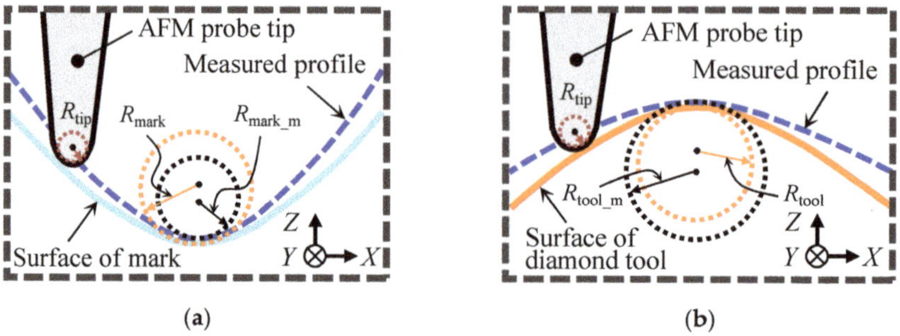

Figure 3. Principle of the edge reversal method for evaluating the tip radius of an AFM probe. (**a**) Trajectory of the AFM tip apex in the measurement of the indentation mark in the cross-section; (**b**) trajectory of the AFM tip apex in the measurement of the diamond tool in the cross-section.

3. Experimental Results and Discussion

A series of experiments was carried out to verify the effectiveness of the proposed methods. A commercial two-dimensional (2D) diffraction grating (7110-00, Vecco, New York, NY, USA), which was fabricated using photolithography technology and has a nominal step height of 200 nm and pitch of 10 μm, was selected as the sample in this study. The SEM image of the diffraction grating is shown in Figure 4. The AFM image of the diffraction grating in the X-direction was reconstructed after obtaining the θ and R_{tip} of the AFM probe.

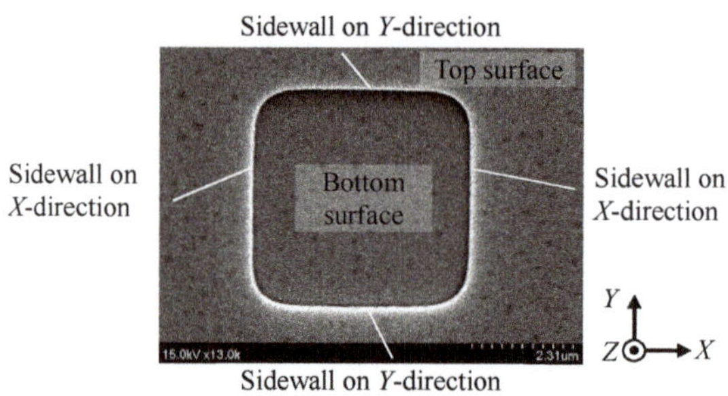

Figure 4. Top view of the SEM image of a commercial diffraction grating.

3.1. Evaluation of the Included Angle of the AFM Probe

The diffraction grating was also regarded as an artifact to evaluate the included angle θ of the AFM probe tip, since its sidewalls are almost perpendicular to the substrate. An AFM cantilever (OMCL-AC240TS, Olympus, Tokyo, Japan) with a nominal included angle of 36° and a nominal tip radius of 7 nm was employed to measure the topography of the diffraction grating on a commercial AFM instrument (Innova, Bruker, Billerica, MA, USA). The alignment between the diffraction grating and the AFM probe was meticulously carried out using an optical microscope to guarantee that the scan trace of the AFM probe was perpendicular to the sidewall of the diffraction grating in the X-direction, making it possible to accurately evaluate the included angle of the AFM probe. The scanning area was set to 10 μm × 10 μm with a scanning rate of 4 μm/s. Both the X- and Y-directions had 1024 scanning lines.

The measured AFM image of the diffraction grating is shown in Figure 5a. The enlarged 3D AFM images with an area of 1 μm × 1 μm on the left and the right sides are shown in Figure 5b,c, respectively. The sectional profiles on A-A′ and B-B′ extracted from the left and right sides of the 3D AFM image to estimate the included angle are shown in Figure 6a,b, respectively. The sidewalls on the left and right side show a slight slope, which is due to the geometry of the AFM probe. Tilt angles (ε_1 and ε_2) were obtained by fitting points in the sidewalls of the sectional profile using the least squares method. The values obtained for ε_1 and ε_2 of the diffraction grating were 19.02° and 18.45°, respectively. Substituting these values into Equation (3), the included angle θ of the AFM probe tip used during the measurement was calculated to be 37.47°. The estimated value was very close to the nominal value of 36° given by the manufacturer, which effectively demonstrated the estimation of θ for the AFM tip. Moreover, the measured corner radii on the left (R_{mea_L}) and right (R_{mea_R}) sides were found to be 54.24 nm and 78.37 nm, respectively.

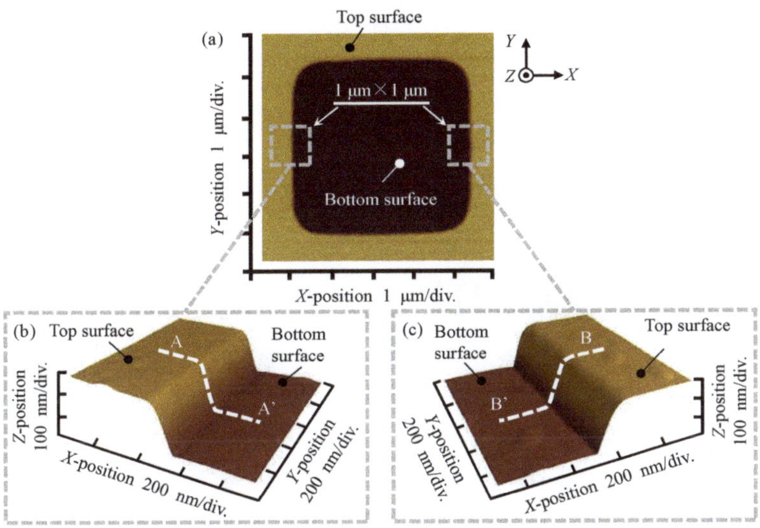

Figure 5. A 3D AFM image of the diffraction grating. (**a**) Top view; (**b**) enlarged 3D AFM image on the left side and (**c**) enlarged 3D AFM image on the right side.

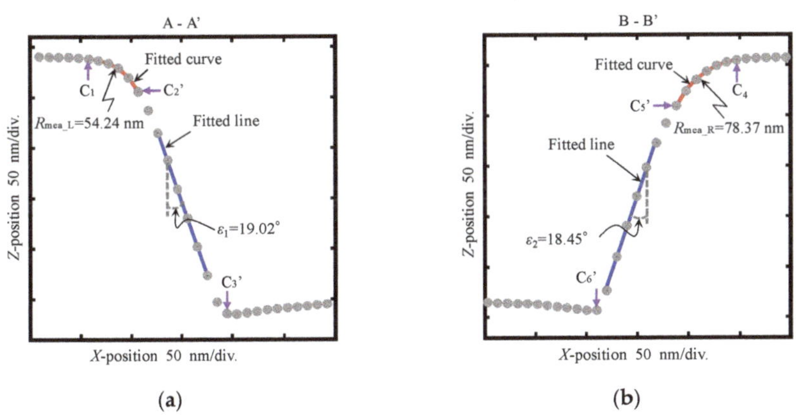

Figure 6. Sectional profiles of the 3D AFM image: (**a**) A-A′ on the left side and (**b**) B-B′ on the right side.

3.2. Evaluation of the Tip Radius of the AFM Probe

Prior to using the AFM to measure the topography of the indentation mark and the diamond tool, an ultra-precision nanoindentation instrument was designed to accurately replicate the geometry of the cutting edge of a diamond tool with a nominal nose radius of 1 mm onto a soft copper substrate by indenting the copper substrate with the diamond tool [30]. The quality of the replicated indentation mark can be affected by the surface roughness of the workpiece. To minimize this influence, a workpiece with an RMS roughness of 2.43 nm was used in the experiments. Figure 7 shows the SEM image of the indentation mark. The results obtained by molecular dynamics (MD) simulations indicated that elastic recovery of the indentation mark can be ignored when the indentation depth is greater than 200 nm. The principle of the ultra-precision nanoindentation instrument and the procedure for creating the indentation mark can be found in references [30,33] and will not be repeated in this paper for the sake of brevity.

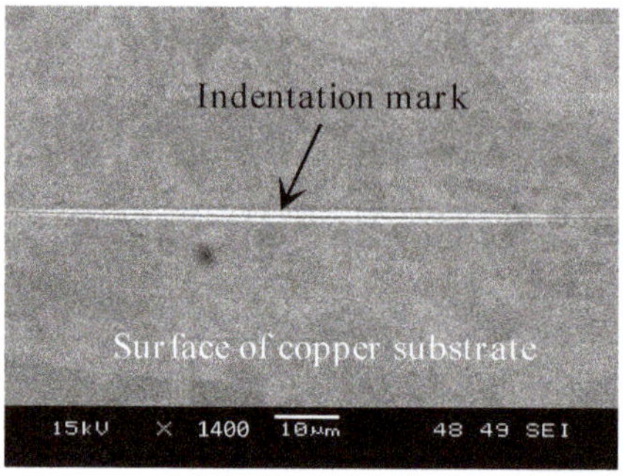

Figure 7. SEM image of the indentation mark replicated using the designed nanoindentation instrument.

The AFM cantilever previously utilized for measuring the diffraction grating as described in Section 3.1 was also employed to measure the 3D profiles of the indentation mark and diamond tool. The indentation mark was initially scanned at a rate of 0.8 μm/s with a scanning area of 2 μm (*X*) × 2 μm (*Y*). The number of scanning lines was set to 1024 in both the *X*- and *Y*-directions. Figure 8 shows the 3D AFM image and the cross-sectional profile extracted from the AFM image of the replicated artifact with an area of 0.4 μm × 0.4 μm. The apex radius of the indentation mark was obtained by fitting the points in the apex of the cross-sectional profile of the indentation mark using the least squares method. The radius of the fitted arc R_{mark_m}, which is represented by a black solid line with circle markers in Figure 8, is approximately 17.08 nm.

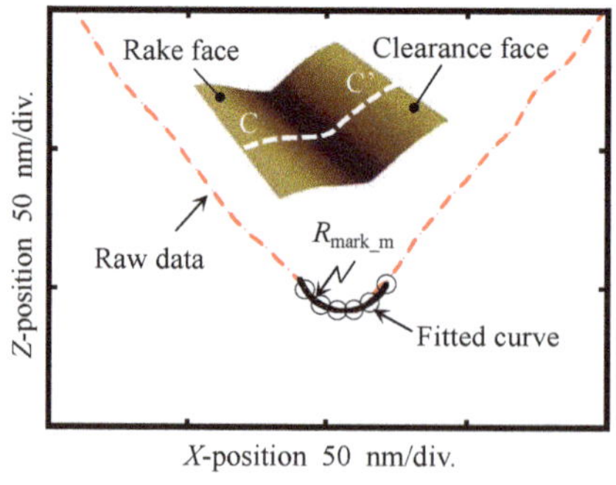

Figure 8. The apex radius (R_{mark_m}) achieved from section C-C′ of the AFM topography of the indentation mark.

The diamond tool was mounted on a tool holder and positioned on the *Z*-scanner of the AFM instrument. The bisector of the included angle of the diamond tool, between its rake face and clearance face, was aligned parallel to the *Z*-direction of the AFM. Subsequently, the diamond tool was scanned with the same AFM tip, using the same parameters as in the measurement of the indentation mark, including the scan rate, the scan area and the number of data points in the *X*- and *Y*-directions.

Figure 9 shows the 3D AFM image and cross-sectional profile extracted from the AFM image of the diamond tool, covering an area of 0.4 μm × 0.4 μm. A discernible edge separates the rake face and the clearance face, which can be useful for determining R_{tool_m}. The radius of the fitted arc, R_{tool_m}, indicating the apex radius of the cutting edge, was calculated to be approximately 32.19 nm using the least squares method. Therefore, R_{tip} was evaluated to be approximately 7.55 nm by substituting the evaluated R_{mark_m} and R_{tool_m} into Equation (4). The evaluated R_{tip} corresponded well to the nominal R_{tip} of 7 nm.

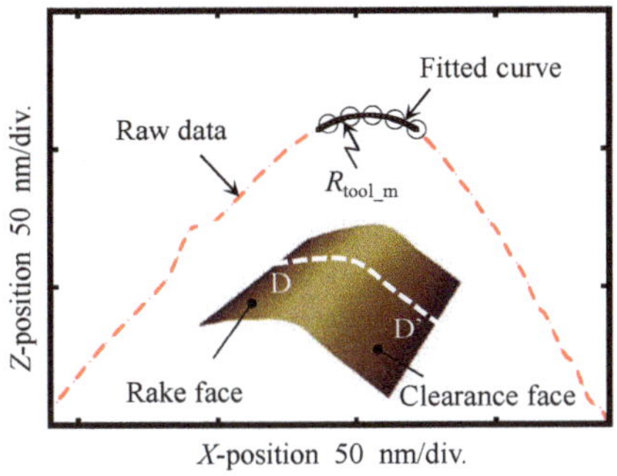

Figure 9. The apex radius (R_{tool_m}) achieved from section D-D′ of the AFM topography of the diamond tool.

The experimental results above demonstrate that accurate characterization of the geometry of the tip radius and included angle of the AFM probe tip on-machine can be achieved through the edge reversal method and scanning of a line pattern structure, respectively. Please note that the measured parameters of the selected AFM cantilever with a triangular pyramid structure can only reflect the outline of the AFM tip shape in contact with the sample surface during the measurement. There will be a slight difference in the AFM profile images of the scanned sample if the alignment between the AFM cantilever and sample is altered prior to the AFM measurement.

3.3. *Reconstruction of 3D AFM Image of the Diffraction Grating*

As mentioned above, the geometry of the AFM probe tip, including R_{tip} and θ, can significantly influence the measurement accuracy of the grating of an AFM. After R_{tip} and θ were obtained, the topography convolutions induced by the AFM probe tip in the 3D topographic characterization of the diffraction grating were corrected using the proposed reconstruction model.

Figure 10a,b depict the reconstructed 3D AFM images of the diffraction grating on the right and left sides, respectively, corresponding to the 3D AFM images exhibited in Figure 5b,c. Both the top and bottom surfaces of the reconstructed 3D image can be clearly seen from Figure 10a,b. A significant decrease in the inclined angle of the sidewall can be observed when compared to Figure 5b,c. This is because the lateral resolution of the white light interferometer used to measure the actual surface profile of the diffraction is too large, reaching several hundred nanometers and exceeding the requirement of several nanometers in lateral resolution to measure the sidewall and corner rounding in the X-axis direction. Currently, there is no effective method in our laboratory that can measure the actual profile image of the diffraction grating. A comparison between the reconstructed and actual profile images of the grating based on a designed standard artefact will be carried out in the future.

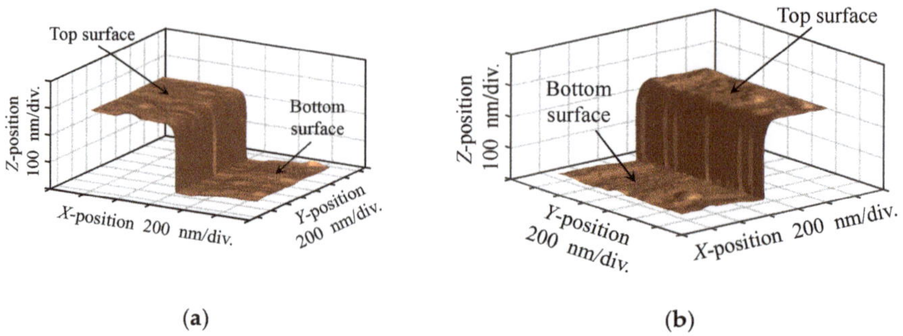

Figure 10. Reconstructed 3D AFM image of the diffraction grating on (**a**) the left side and (**b**) the right side.

The cross-sectional profiles extracted from Figure 10 are shown in Figure 11. It is apparent from Figure 11 that both sidewalls became vertical, and the corner radii decreased. The reconstructed corner radius (R_{rec_L}) on the left side was found to be 46.70 nm, while that on the right side (R_{rec_R}) was 70.86 nm. The differences of the reconstructed and the measured corner radii on both left and right sides were calculated to be 7.54 nm and 7.51 nm, respectively. These values were almost the same as those of the measured tip radius of the AFM probe. The tilt angles of the sidewalls on the left and right sides were both reconstructed to be zero, which corresponded well with the assumed tilt angle of the sidewalls in this experiment. The sectional profiles without reconstruction were additionally plotted in Figure 11 for comparison. It is evident that the influence of the tip radius (R_{tip}) and included angle (θ) of the AFM cantilever was successfully removed, thus demonstrating the feasibility of the proposed reconstruction model. Please note that the actual tilt angle of the sidewall can be evaluated using a CD-AFM or CNT-AFM. However, the purpose of this research is to verify the feasibility of the proposed reconstruction method for compensating the profile convolutions induced by the tip shape of the AFM probe in the measurement of a diffraction grating. Therefore, we assume that the sidewall is perpendicular to the substrate before reconstruction.

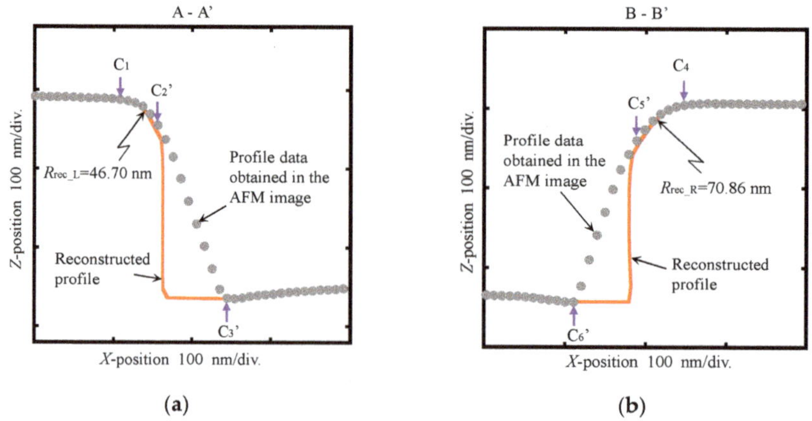

Figure 11. Reconstructed sectional profiles of the diffraction grating on (**a**) the left side and (**b**) the right side.

Figure 12 presents the corner radius values of the reconstructed AFM image of the diffraction grating in ten distinct sections. It can be seen that the corrected corner radii were smaller than the measured corner radii. The differences between the measured and the corrected corner radii on the left and right sides were evaluated to be 7.544 nm with a standard deviation of 0.018 nm and 7.534 nm with a standard deviation of 0.013 nm, respectively. The obtained values were close to that of the AFM tip radius of 7.55 nm, which was determined through the edge reversal method. The proposed reconstruction method compensated for twenty cross-sections in the 3D image. It was able to accurately reconstruct the tilt angle and corner radius in each cross-section. Therefore, the repeatability of the proposed reconstruction method can be confirmed with the above results.

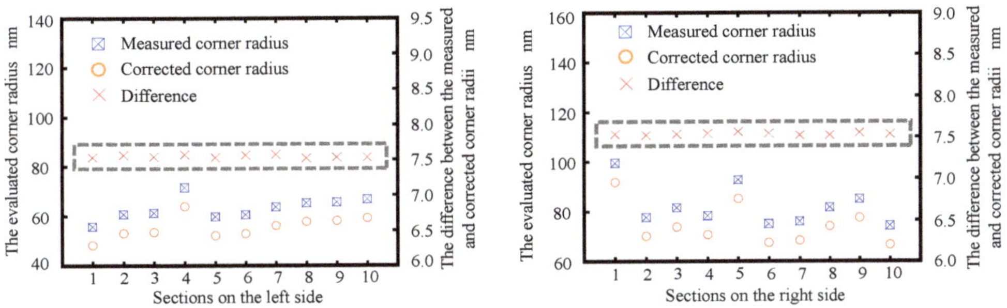

(a) (b)

Figure 12. Summarization of the corner radii of (**a**) the left side and (**b**) the right side of the reconstructed AFM image of the diffraction grating in ten different sections.

4. Conclusions

A method of reconstruction has been proposed with the objective of removing the impact of the geometry convolution of the AFM probe tip on the scanned AFM image of a rectangular diffraction grating. The geometry of the AFM probe tip has been assessed through the edge reversal method and a commercial diffraction grating, resulting in a tip radius of 7.55 nm and an included angle of 37.47 degrees. The measured results corresponded well with the nominal geometry of the AFM probe tip provided by the manufacturer. The diffraction grating was also used as a scanned sample, and its 3D AFM image was reconstructed using the proposed model. The corner radii at the left and right sides (R_{mea_L} and R_{mea_R}, respectively) of the diffraction grating were assessed to be 54.24 nm and 78.37 nm before reconstruction. A tilt angle was observed in the measured sidewall of the diffraction grating. Following the reconstruction process, the values for R_{rec_L} and R_{rec_R} were adjusted to 46.70 nm and 70.86 nm, respectively, while the tilt angles were determined to be zero. This provides evidence for the validity of the proposed reconstruction model. Experimental results have confirmed that the influence of the geometry of the AFM probe tip on the 3D AFM image of the diffraction grating can be successfully eliminated using the proposed reconstruction method.

Author Contributions: Conceptualization, Methodology, Validation, Visualization and Writing: K.Z.; Supervision and Review: Y.B.; Supervision and Review: Z.Z. All authors have read and agreed to the published version of the manuscript.

Funding: This work is supported by the Japanese Government Scholarship (MEXT), the China Scholarship Council (CSC), the Fundamental Research Fund of National Institute of Metrology, China (Grant No. AKYCX2302), China National Key Research and Development Plan (Grant No. 2021YFF0603803) and the National Natural Science Foundation of China (Grant No. 52175527).

Data Availability Statement: Data and optimization codes can be shared upon request.

Acknowledgments: We thank Wei Gao (Tohoku University) for kindly supporting the related experiments, Yuki Shimizu (Hokkaido University) for general discussions related to visualization and Yindi Cai (Dalian University of Technology) for careful advice on the manuscript.

Conflicts of Interest: The authors declare no conflicts of interest.

Appendix A

An AFM probe with a tip radius of R_{tip} and an included angle of θ ($\theta_1 + \theta_2$) is utilized to scan the surface of the diffraction grating. As shown in Figure A1, due to the finite tip shape of the AFM probe, the measured profile image of the diffraction grating in the cross-section is considered as the convolution of the tip shape of the AFM probe and the actual profile of the measured surface. The measured profile image in the convolution region from C_1 to C_3 can be classified into two sections based on the geometrical relationship. In the curve region between C_1 and C_2, the measured profile image is only affected by the tip radius of the AFM probe. The coordinates of point A_1 on the actual surface of the diffraction grating are denoted as (x_1, z) while the coordinates of point A_2 in the measured profile image are denoted as ($x_1{}'$, z_1). According to the geometrical relationship between the actual and the measured profiles, the following equations can be derived:

$$x_1 = \sqrt{R_{corner}^2 - z^2} \tag{A1}$$

$$R_{mea} = R_{corner} + R_{tip} = \sqrt{x_1'^2 + (z + R_{tip})^2} \tag{A2}$$

The X-coordinates of point A_1 can be expressed as follows by combining the above equations:

$$x_i = \sqrt{\left(\sqrt{x_i'^2 + (z_i + R_{tip})^2} - R_{tip}\right)^2 - z_i^2} \qquad x_i' \in C_1C_2 \tag{A3}$$

In the straight-line region from C_2 to C_3, the measured profile is affected by the tip radius and the included angle of the AFM probe. The coordinates of point B_1 on the actual surface of the diffraction grating are denoted as (x_2, z) while the coordinates of point B_3 in the measured profile image are denoted as ($x_2{}'$, z_1). According to the geometrical relationship between the actual and the measured profiles, the following equations can be derived:

$$x_2' = x_2'' - \tan\frac{\pi - 2\theta_1}{4} \cdot R_{tip} \tag{A4}$$

$$x_2 = x_2' - z \cdot (\tan\theta_1 - \tan\alpha_1) \tag{A5}$$

The X-coordinates of point B_1 can be expressed as follows by combining the above equations:

$$x_i = x_i'' - \left(\tan\frac{\pi - 2\theta_1}{4} \cdot R_{tip} + z_i(\tan\theta_1 - \tan\alpha_1)\right) \qquad x_i'' \in C_2C_3 \tag{A6}$$

where θ_1 and α_1 represent the inclined angle of the AFM tip shape on the left side and the tilt angle of the actual profile image on the right sidewall, respectively.

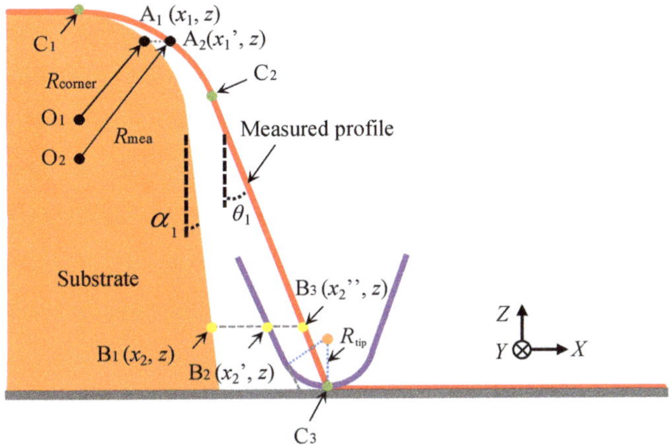

Figure A1. The geometrical relationship between the measured and the actual profiles of the diffraction grating scanned by an AFM probe tip.

References

1. Yan, J.-N.; Chang, C.-W.; Lee, Y.-C. A metal-embedded photo-mask for contact photolithography with application on patterned sapphire substrate. *Microelectron. Eng.* **2014**, *122*, 20–24. [CrossRef]
2. Gao, W. *Precision Nanometrology–Sensors and Measuring Systems for Nanomanufacturing*; Springer: London, UK, 2010.
3. Li, X.; Gao, W.; Muto, H.; Shimizu, Y.; Ito, S.; Dian, S. A six-degree-of-freedom surface encoder for precision positioning of a planar motion stage. *Precis. Eng.* **2013**, *37*, 771–781. [CrossRef]
4. Gao, W.; Haitjema, H.; Fang, F.; Leach, R.; Cheung, C.; Savio, E.; Linares, J. On-machine and in-process surface metrology for precision manufacturing. *CIRP Ann.* **2019**, *68*, 843–866. [CrossRef]
5. Quan, L.; Shimizu, Y.; Xiong, X.; Matsukuma, H.; Gao, W. A new method for evaluation of the pitch deviation of a linear scale grating by an optical angle sensor. *Precis. Eng.* **2021**, *67*, 1–13. [CrossRef]
6. Shimizu, Y.; Uehara, K.; Matsukuma, H.; Gao, W. Evaluation of the grating period based on laser diffraction by using a mode-locked femtosecond laser beam. *J. Adv. Mech. Des. Syst. Manuf.* **2018**, *12*, JAMDSM0097. [CrossRef]
7. Yu, Z.; Chen, L.; Wu, W.; Ge, H.; Chou, S.Y. Fabrication of nanoscale gratings with reduced line edge roughness using nanoimprint lithography. *J. Vac. Sci. Technol. B Microelectron. Nanometer Struct. Process. Meas. Phenom.* **2003**, *21*, 2089–2092. [CrossRef]
8. Shin, D.W.; Quan, L.; Shimizu, Y.; Matsukuma, H.; Cai, Y.; Manske, E.; Gao, W. In-Situ Evaluation of the Pitch of a Reflective-Type Scale Grating by Using a Mode-Locked Femtosecond Laser. *Appl. Sci.* **2021**, *11*, 8028. [CrossRef]
9. Dai, G.; Koenders, L.; Pohlenz, F.; Dziomba, T.; Danzebrink, H.-U. Accurate and traceable calibration of one-dimensional gratings. *Meas. Sci. Technol.* **2005**, *16*, 1241–1249. [CrossRef]
10. Dai, G.; Pohlenz, F.; Dziomba, T.; Xu, M.; Diener, A.; Koenders, L.; Danzebrink, H.-U. Accurate and traceable calibration of two-dimensional gratings. *Meas. Sci. Technol.* **2007**, *18*, 415–421. [CrossRef]
11. Gao, W.; Asai, T.; Arai, Y. Precision and fast measurement of 3D cutting edge profiles of single point diamond micro-tools. *CIRP Ann.-Manuf. Technol.* **2009**, *58*, 451–454. [CrossRef]
12. Itoh, H.; Fujimoto, T.; Ichimura, S. Tip characterizer for atomic force microscopy. *Rev. Sci. Instrum.* **2006**, *77*, 103704. [CrossRef]
13. Villarrubia, J. Algorithms for scanned probe microscope image simulation, surface reconstruction, and tip estimation. *J. Res. Natl. Inst. Stand. Technol.* **1997**, *102*, 425–454. [CrossRef]
14. Dongmo, L.; Villarrubia, J.; Jones, S.; Renegar, T.; Postek, M.; Song, J. Experimental test of blind tip reconstruction for scanning probe microscopy. *Ultramicroscopy* **2000**, *85*, 141–153. [CrossRef]
15. Orji, N.G.; Dixson, R.G. Higher order tip effects in traceable CD-AFM-based linewidth measurements. *Meas. Sci. Technol.* **2007**, *18*, 448–455. [CrossRef]
16. Nguyen, C.V.; Ye, Q.; Meyyappan, M. Carbon nanotube tips for scanning probe microscopy: Fabrication and high aspect ratio nanometrology. *Meas. Sci. Technol.* **2005**, *16*, 2138–2146. [CrossRef]
17. Dai, H.; Hafner, J.H.; Rinzler, A.G.; Colbert, D.T.; Smalley, R.E. Nanotubes as nanoprobes in scanning probe microscopy. *Nature* **1996**, *384*, 147–150. [CrossRef]
18. Martin, Y.; Wickramasinghe, H.K. Method for imaging sidewalls by atomic force microscopy. *Appl. Phys. Lett.* **1994**, *64*, 2498–2500. [CrossRef]
19. Dahlen, G.; Osborn, M.; Liu, H.-C.; Jain, R.; Foreman, W.; Osborne, J.R. Critical dimension AFM tip characterization and image reconstruction applied to the 45-nm node. In Proceedings of the SPIE 31st International Symposium on Advanced Lithography, San Jose, CA, USA, 19–24 February 2006; Volume 6152, pp. 945–955.

20. Hussain, D.; Ahmad, K.; Song, J.; Xie, H. Advances in the atomic force microscopy for critical dimension metrology. *Meas. Sci. Technol.* **2016**, *28*, 012001. [CrossRef]
21. Bykov, V.; Novikov, Y.; Rakov, A.; Shikin, S. Defining the parameters of a cantilever tip AFM by reference structure. *Ultramicroscopy* **2003**, *96*, 175–180. [CrossRef]
22. Vesenka, J.; Miller, R.; Henderson, E. Three-dimensional probe reconstruction for atomic force microscopy. *Rev. Sci. Instrum.* **1994**, *65*, 2249–2251. [CrossRef]
23. Yue, X.; Lei, D.; Cui, H.; Zhang, X.; Xu, M.; Kong, L. An integrated method for measurement of diamond tools based on AFM. *Precis. Eng.* **2017**, *50*, 132–141. [CrossRef]
24. Liu, J.; Notbohm, J.K.; Carpick, R.W.; Turner, K.T. Method for Characterizing Nanoscale Wear of Atomic Force Microscope Tips. *ACS Nano* **2010**, *4*, 3763–3772. [CrossRef] [PubMed]
25. Chen, L.; Cheung, C.L.; Ashby, P.D.; Lieber, C.M. Single-Walled Carbon Nanotube AFM Probes: Optimal Imaging Resolution of Nanoclusters and Biomolecules in Ambient and Fluid Environments. *Nano Lett.* **2004**, *4*, 1725–1731. [CrossRef]
26. Chen, Z.; Luo, J.; Doudevski, I.; Erten, S.; Kim, S.H. Atomic Force Microscopy (AFM) Analysis of an Object Larger and Sharper than the AFM Tip. *Microsc. Microanal.* **2019**, *25*, 1106–1111. [CrossRef]
27. Morgenroth, W.; Meyer, H.; Sulzbach, T.; Brendel, B.; Hübner, U.; Mirandé, W. Downwards to metrology in nanoscale: Determination of the AFM tip shape with well-known sharp-edged calibration structures. *Appl. Phys. A Mater. Sci. Process.* **2003**, *76*, 913–917. [CrossRef]
28. Chen, Y.-L.; Cai, Y.; Xu, M.; Shimizu, Y.; Ito, S.; Gao, W. An edge reversal method for precision measurement of cutting edge radius of single point diamond tools. *Precis. Eng.* **2017**, *50*, 380–387. [CrossRef]
29. Zhang, K.; Shimizu, Y.; Matsukuma, H.; Cai, Y.; Gao, W. An application of the edge reversal method for accurate reconstruction of the three-dimensional profile of a single-point diamond tool obtained by an atomic force microscope. *Int. J. Adv. Manuf. Technol.* **2021**, *117*, 2883–2893. [CrossRef]
30. Cai, Y.; Chen, Y.-L.; Xu, M.; Shimizu, Y.; Ito, S.; Matsukuma, H.; Gao, W. An ultra-precision tool nanoindentation instrument for replication of single point diamond tool cutting edges. *Meas. Sci. Technol.* **2018**, *29*, 054004. [CrossRef]
31. Cai, Y.; Chen, Y.-L.; Shimizu, Y.; Ito, S.; Gao, W. Molecular dynamics simulation of elastic–plastic deformation associated with tool–workpiece contact in force sensor–integrated fast tool servo. *Proc. Inst. Mech. Eng. Part B J. Eng. Manuf.* **2018**, *232*, 1893–1902. [CrossRef]
32. Cai, Y.; Chen, Y.-L.; Shimizu, Y.; Ito, S.; Gao, W.; Zhang, L. Molecular dynamics simulation of subnanometric tool-workpiece contact on a force sensor-integrated fast tool servo for ultra-precision microcutting. *Appl. Surf. Sci.* **2016**, *369*, 354–365. [CrossRef]
33. Zhang, K.; Cai, Y.; Shimizu, Y.; Matsukuma, H.; Gao, W. High-Precision Cutting Edge Radius Measurement of Single Point Diamond Tools Using an Atomic Force Microscope and a Reverse Cutting Edge Artifact. *Appl. Sci.* **2020**, *10*, 4799. [CrossRef]

materials

MDPI

Article

Effect of Duty Cycle on Cutting Force for Ultrasonic Vibration-Assisted Milling Carbon Fiber-Reinforced Polymer Laminates

Yukun Zhang [1,2], Junxue Ren [1,2] and Jinhua Zhou [1,2,*]

[1] Key Laboratory of High Performance Manufacturing for Aero Engine, Ministry of Industry and Information Technology, School of Mechanical Engineering, Northwestern Polytechnical University, Xi'an 710072, China; ykzhang@mail.nwpu.edu.cn (Y.Z.); rjx1968@nwpu.edu.cn (J.R.)

[2] Engineering Research Center of Advanced Manufacturing Technology for Aero Engine, Ministry of Education, School of Mechanical Engineering, Northwestern Polytechnical University, Xi'an 710072, China

[*] Correspondence: zhoujinhua@mail.nwpu.edu.cn; Tel.: +86-15332469815 or +86-02988495232; Fax: +86-02988494060

Abstract: Cutting force is an important factor that affects the surface quality of machining carbon fiber-reinforced polymer (CFRP). High cutting force can lead to surface damage such as the burrs and the delamination in the machining process of CFRP. Ultrasonic vibration-assisted machining (UVAM) can reduce the cutting force in the machining process. This work is focused on the relationship between the duty cycle and the cutting force in UVAM of CFRP. Based on the kinematics of UVAM, the movement of the cutting tool edge and the tool–workpiece separation in UVAM were analyzed, and a calculation formula for the duty cycle was obtained. The milling experiment of CFRP was conducted to compare the cutting force between UVAM and conventional machining (CM), and the relationship between the reduction in the cutting force in UVAM and the duty cycle was determined. The experimental results showed that when the duty cycle was 0.2916, the cutting force of UVAM was reduced by 7.4% to 27% compared with that of CM. When the duty cycle was 1, the cutting force of UVAM was reduced by −4.5% to 7.5% compared with that of CM. Therefore, the effect of reducing the cutting force of UVAM can be enhanced by adjusting the process parameters to reduce the duty cycle of UVAM, and a lower cutting force can be obtained.

Keywords: CFRP; ultrasonic vibration-assisted machining; duty cycle; cutting force

1. Introduction

Ultrasonic vibration-assisted machining (UVAM) is a machining process that combines ultrasonic vibration with conventional machining (CM) by attaching high-frequency vibration to the tool or workpiece. Compared with CM, UVAM with appropriate process parameters has the advantages of reducing cutting forces [1–3], lowering cutting temperatures [4], decreasing surface roughness [5,6], and improving tool life [7,8]. Carbon fiber reinforced polymer (CFRP) is a heterogeneous anisotropic material widely used in the aerospace industry due to its high specific strength, high specific modulus, and corrosion resistance. Typically, CFRP requires machining such as milling or grinding after the forming process. However, when cutting CFRP, problems exist such as a short tool life and easy machining damage [9]. Based on the advantages of UVAM, some researchers have conducted research on machining CFRP by using UVAM and compared it with CM. The results demonstrate that using UVAM to process CFRP can reduce cutting force [10,11], lower cutting temperature [12,13], reduce surface roughness [14,15], and improve tool life [16].

During the machining process, ultrasonic vibration changes the motion trajectory of the tool or workpiece, which may cause periodic separation between the tool and workpiece. High spindle speed, low ultrasonic frequency, or low ultrasonic amplitude, may lead to the disappearance of tool–workpiece separation. Brehl et al. [17] used the duty cycle to explain tool–workpiece separation in UVAM. The duty cycle refers to the ratio of the time when the tool contacts the workpiece to the total machining time, and they stated that a decrease in the duty cycle will reduce the cutting force. Chen et al. [18] proposed that different vibration modes result in different separation situations, categorizing the tool–workpiece separation process in one-dimensional and two-dimensional UVAM into three types. Ni et al. [19] presented an analytical model to calculate the tool–workpiece contact ratio (TWCR) for UVAM. They stated that a decrease in the TWCR corresponds to a reduction in net cutting time, which leads to lower cutting force, reduced accumulation of cutting heat, improved surface quality, and extended tool life. Sui et al. [20] found that the tool–workpiece separation and the duty cycle calculation in high-speed ultrasonic vibration cutting (HUVC) are different from the other UVAM methods. The tool–workpiece separation of HUVC depends on the feed rate, the amplitude of ultrasonic vibration, and the fraction part of the cutting frequency ratio, and the duty cycle can be changed by adjusting the spindle speed to alter the phase difference. At present, many references mention the duty cycle of UVAM. However, no existing reference has been found that investigates the relationship between the duty cycle and cutting force in UVAM.

Fiber-reinforced polymers are widely used in various fields such as automotive [21], aerospace [22], and construction [23] due to their excellent material properties. However, this also brings challenges such as difficult processing and surface damage. In the 1980s, some researchers verified the feasibility of using UVAM to process fiber-reinforced composites [24]. They stated that UVAM can reduce the average cutting force and surface roughness and reduce burrs and subsurface damage. Since then, researchers have conducted theoretical and experimental studies on machining CFRP using UVAM. Xu et al. [25–27] found that fiber orientation can significantly affect the subsurface damage and cutting force. At the same time, elliptical vibration-assisted cutting can reduce the effect of fiber orientation, and the subsurface damage of elliptical vibration-assisted cutting of carbon fiber composites is less than that of conventional cutting. Yuan et al. [28] developed a cutting force model for rotary ultrasonic drilling of CFRP-T700 based on brittle fracture material mechanism and found that feed rate and spindle speed are the main parameters affecting cutting force. Abd Halim et al. [29] compared UVAM and CM machining CFRP based on experiments. They found that UVAM results in lower maximum cutting forces but faster tool wear, the surface roughness was slightly worse, and resin degradation occurred in both machining methods. At present, there are many studies on using UVAM to reduce the cutting force on CFRP, but the cutting force reduction rate by using UVAM is different in various references. How to adjust process parameters to prevent UVAM from being unable to achieve the effect of reducing cutting force, or to make UVAM achieve a better effect of reducing cutting force, is an issue that needs to be considered when applying UVAM in practical machining.

To address the challenges above, this study started from the tool–workpiece separation characteristic of UVAM, investigated the influence of process parameters on duty cycle and cutting force through theoretical calculations and experiments, and explored the relationship between the duty cycle and the cutting force. First, the relative motion trajectory of the tool and workpiece in UVAM milling was analyzed, the calculation method of the duty cycle was obtained, and the influence factors of duty cycle are discussed. Second, a CFRP cutting force experiment was carried out to compare the cutting force of UVAM and CM, and the impact of process parameters on the UVAM cutting force reduction effect was analyzed. Finally, based on the theoretical study of the duty cycle and the experimental results of cutting force, the influence of the duty cycle on the UVAM cutting force was obtained. This study provides theoretical support for the optimization of UVAM process parameters and determines the process parameter conditions suitable for UVAM, which ensures that UVAM can achieve an effective reduction in cutting force.

2. Theoretical Calculation of Duty Cycle for UVAM

2.1. Kinematic Analysis of Tool Motion Trajectory

The motion of UVAM milling CFRP laminates is shown in Figure 1. The workpiece feed direction is parallel to the x-axis, the tool rotational direction is always in the XY plane, and the ultrasonic vibration direction is parallel to the z-axis. In this configuration, the motion trajectory of any point on the tool can be expressed as

$$\begin{cases} x = v_f \cdot t + \frac{D}{2} \cdot sin\left[\frac{\pi n}{30}t - \frac{2\pi(Z_i - 1)}{Z}\right] \\ y = \frac{D}{2} \cdot cos\left[\frac{\pi n}{30}t - \frac{2\pi(Z_i - 1)}{Z}\right] \\ z = a_p \cdot sin(2\pi f_z t) \end{cases} \quad (1)$$

while D is the tool diameter, n is the spindle speed, t is the time, v_f is the feed rate, Z is the number of tool teeth, i is the i-th tool teeth, a_p is the ultrasonic amplitude, and f_z is the ultrasonic frequency.

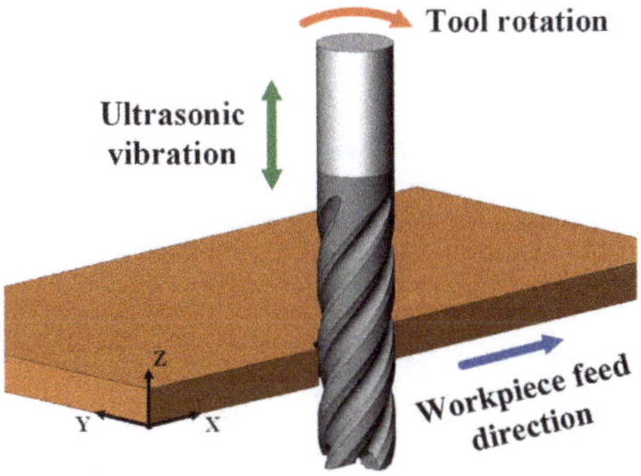

Figure 1. Movement of tool and workpiece in UVAM.

Figure 2 shows the machining trajectory at any point on the tool relative to the workpiece in UVAM and CM. It can be observed that axial ultrasonic vibration significantly changes the motion trajectory of the point, changing the machining trajectory from a planar curve in CM to a more complex three-dimensional curve in UVAM. In CM, the motion trajectories of points on a cutting edge are distributed in different planes, so the motion trajectories of each point will not overlap. However, in UVAM, the motion trajectories of points on the tool edge are all located in the same three-dimensional space, so the trajectories of different points may overlap. To facilitate the kinematic analysis, the cutting space is unfolded into a two-dimensional space, where the helical curve of the cutting edge becomes a straight line, and the rotational motion of the tool becomes linear motion. Figure 3 shows the motion of the tool edge relative to the workpiece in UVAM in two-dimensional space, and the sinusoidal curve in the figure is the motion trajectory of a point on the tool edge. The tool rotation speed v_{rot} is decomposed into v_a and v_b, where v_a is parallel to the direction of the tool edge, and v_b is parallel to the direction of the ultrasonic vibration velocity. The velocity v_{mix} can be calculated as follows:

$$v_{mix} = v_{vib} + v_b \quad (2)$$

From Figure 3, it can be seen that when the tool edge moves along the curve OA, the velocity v_{mix} is in the same direction as v_b. In this case, the tool performs normal-thickness cutting. When the cutting edge moves along the curve AB, v_{mix} is in the opposite direction to v_b. In this situation, the tool edge and the workpiece are separated, and no cutting is performed. When the tool edge moves along the curve BC, v_{mix} is in the same direction as v_b again. Since OA has already been processed, the tool edge moves along curve BC, resulting in a very small chip thickness. When the tool edge moves along the curve CD, the chip thickness returns to the normal state, and it can be observed that within one ultrasonic vibration cycle, the tool edge goes through three processes, including normal-thickness cutting, non-cutting, and low-thickness cutting. This cycle repeats continuously throughout the entire machining process.

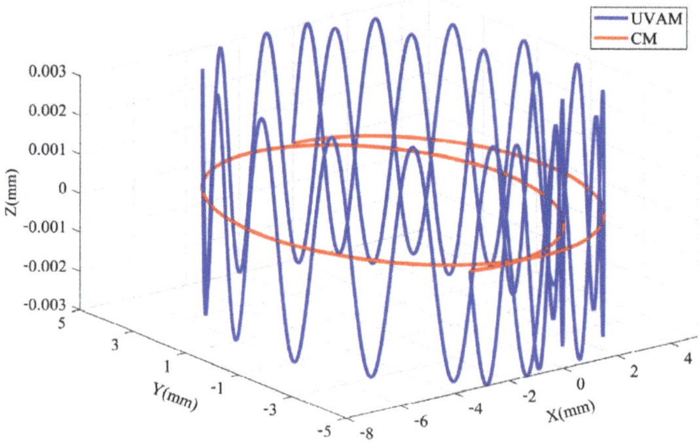

Figure 2. Comparison of the UVAM and the CM tool trajectories.

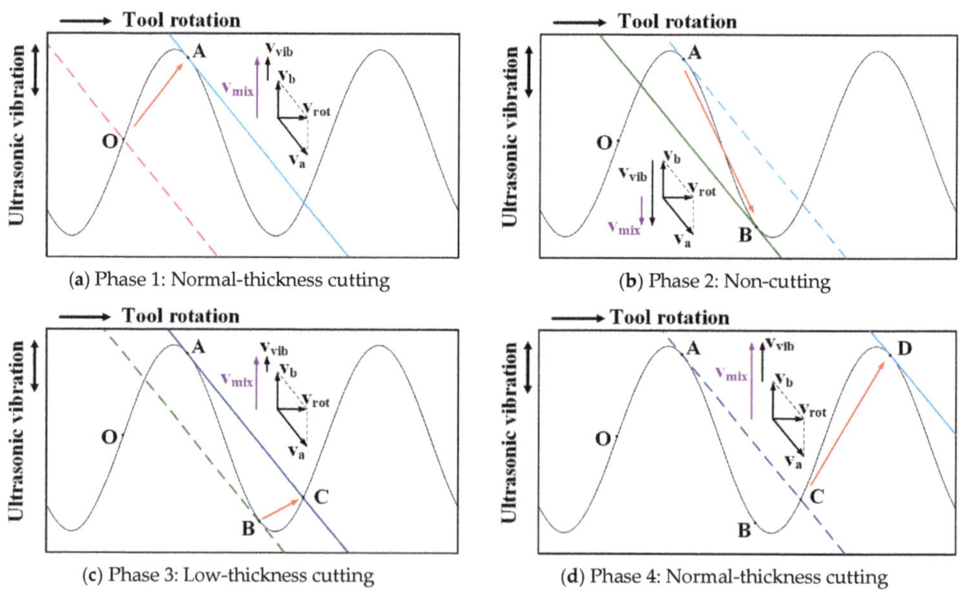

Figure 3. Motion of UVAM in two-dimensional space.

2.2. *Theoretical Investigation of Duty Cycle*

To study the contact separation situation between the tool and the workpiece in UVAM, it is necessary to obtain the times t_1, t_2, and t_3 for the tool to move to points A, B, and C, respectively. Since the feed rate is typically much smaller than the tool edge rotation speed and the speed of ultrasonic vibration, the influence of the feed rate can be neglected in the two-dimensional analysis. As shown in Figure 4, an xOy coordinate system is established, where the angle i is the helix angle of the tool. At this time, the equation for the motion trajectory of a point on the tool edge in the two-dimensional space is given by

$$\begin{cases} x = \frac{\pi D n}{60} t \\ y = a_p \cdot sin(2\pi f_z t) \end{cases} \tag{3}$$

The velocity of the tool edge along the x and y directions is

$$\begin{cases} v_{rot} = \frac{\pi D n}{60} \\ v_{vib} = 2\pi f_z a_p \cdot cos(2\pi f_z t) \end{cases} \tag{4}$$

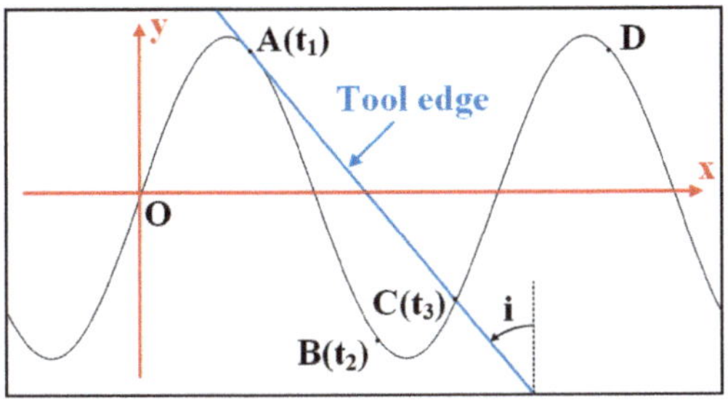

Figure 4. Coordinate system with O as origin.

From the previous analysis, it can be seen that the tool–workpiece separation is related to the tool rotation speed, ultrasonic vibration speed, and tool helix angle. Therefore, the critical condition for tool-workpiece separation and contact is given by

$$v_{vib} = -v_b = -\frac{v_{rot}}{tani} \tag{5}$$

The term $-v_b$ represents that v_b is in the opposite direction to v_{vib}. When $v_{vib} < -v_b$, the tool and the workpiece are in the separation state. When $v_{vib} > -v_b$, the tool and workpiece are in contact and the cutting state. Combining Equations (4) and (5), the calculation formulas for the times t_1, t_2, and t_3 for the tool to reach points A, B, and C are

$$t_1 = \frac{arccos\left(\frac{-Dn}{120 f_z a_p \cdot tani}\right)}{2\pi f_z} \tag{6}$$

$$t_2 = \frac{1}{f_z} - \frac{arccos\left(\frac{-Dn}{120 f_z a_p \cdot tani}\right)}{2\pi f_z} \tag{7}$$

$$\begin{aligned} a_p sin(2\pi f_z t_3) &= a_p sin\left(arccos\left(\frac{-Dn}{120 f_z a_p \cdot tani}\right)\right) - \frac{\pi D n}{60 tani} t_3 \\ &+ \frac{Dn \cdot arccos\left(\frac{-Dn}{120 f_z a_p \cdot tani}\right)}{120 f_z tani} \end{aligned} \tag{8}$$

Equation (8) needs to be solved numerically. After obtaining the values of t_1, t_2, and t_3, the times for normal-thickness cutting, non-cutting, and low-thickness cutting can be calculated, and the duty cycle of UVAM can be calculated. The duty cycle refers to the proportion of the time the tool is in contact with the workpiece to the total machining time. Since the low-thickness cutting phase occurs on the machined surface, and the chip thickness during this phase is much smaller than that during the normal-thickness cutting phase, the period from t_1 to t_3 is defined as the time when the tool does not cut. In this case, the UVAM duty cycle DC is calculated as

$$DC = \frac{T - (t_3 - t_1)}{T} \tag{9}$$

From the critical condition of tool–workpiece separation and contact, it can be seen that when the maximum speed of ultrasonic vibration $v_{vibmax} > v_{rot}/tani$, tool–workpiece separation occurs in machining, and when $v_{vibmax} < v_{rot}/tani$, there is no tool–workpiece separation. From Equation (9), it can be seen that the duty cycle $DC \in [0,1]$. When $DC = 0$, the tool and workpiece are always separated, and no cutting occurs. When $DC = 1$, the tool is always in contact with the workpiece, continuously in the cutting state. The duty cycle of CM is always $DC = 1$. Therefore, when $v_{vibmax} > v_{rot}/tani$, $0 \leq DC < 1$, and only when $v_{rot} = 0$ or $i = \pi/2$, $DC = 0$. When $v_{vibmax} < v_{rot}/tani$, $DC = 1$.

2.3. Effect of Process Parameters on the Duty Cycle

According to Equations (6), (8) and (9), the duty cycle DC is related to five parameters: tool helix angle i, ultrasonic amplitude a_p, ultrasonic frequency f_z, tool diameter D, and spindle speed n. The relationship between the process parameters and the duty cycle is shown in Figure 5. It can be observed that the duty cycle gradually decreases with the increase in the tool helix angle, ultrasonic amplitude, and ultrasonic frequency, while it increases with the increase in the tool diameter and spindle speed. From Figure 5a,b, it can be seen that when the ultrasonic amplitude and frequency are below the critical value, the duty cycle $DC = 1$. When they exceed the critical value, the duty cycle rapidly decreases with increasing parameters, and then the rate of decrease gradually slows down. In Figure 5c, it can be observed that the duty cycle decreases as the tool helix angle increases; when the helix angle $i = \pi/2$, the duty cycle $DC = 0$. Figure 5d,e show that as the tool diameter and spindle speed increase, the duty cycle gradually increases until the duty cycle $DC = 1$. Therefore, to achieve tool–workpiece separation in UVAM, the helix angle, ultrasonic amplitude, and frequency need to exceed critical values, while the tool diameter and spindle speed should be controlled below critical values. Moreover, it can be considered to increase the tool helix angle, ultrasonic amplitude, or frequency and decrease the tool diameter or spindle speed to increase the tool–workpiece separation time, i.e., to reduce the duty cycle.

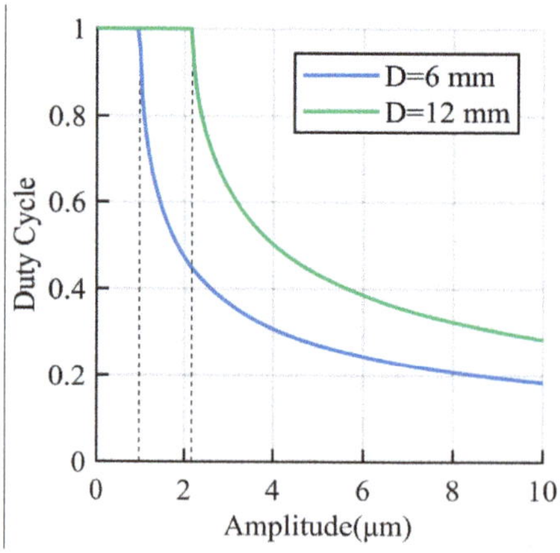

(**a**) Ultrasonic amplitude a_p

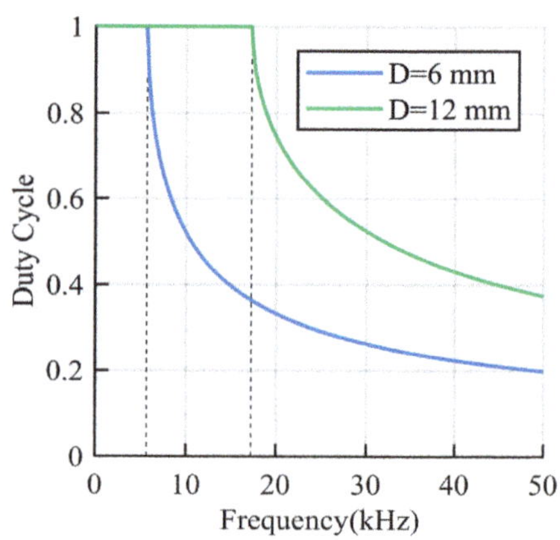

(**b**) Ultrasonic frequency f_z

Figure 5. *Cont.*

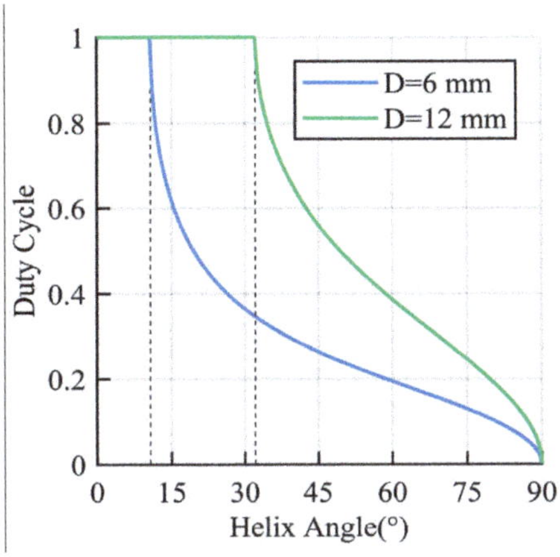

(**c**) Tool helix angle *i*

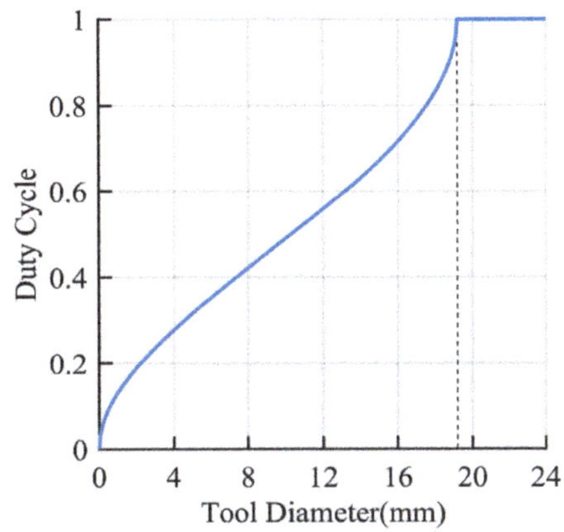

(**d**) Tool diameter *D*

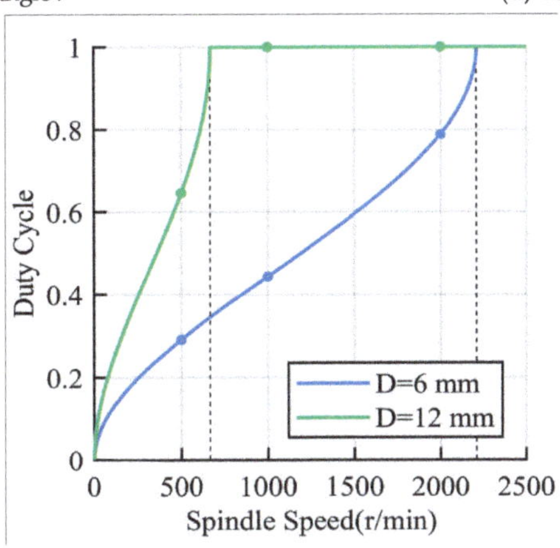

(**e**) Spindle speed *n*

Figure 5. Effect of process parameters on the duty cycle. (**a**–**c**) duty cycle decreases as amplitude, frequency, and helix angle increase, (**d**,**e**) duty cycle increases as tool diameter and spindle speed increase.

3. UVAM Cutting Force Experiment

In this experiment, the cutting forces of UVAM under different process parameters were measured and compared with the cutting force of CM. Thus, the effect of UVAM in reducing the cutting force in CFRP machining was verified, and the relationship between the duty cycle and the cutting force in UVAM was determined. The workpiece used in this experiment was a multidirectional CFRP laminate with a thickness of 4.3 mm, and the fiber used in the laminate was T800 carbon fiber. The tools were diamond-coated carbide end mills with tool diameters of 6 mm and 12 mm and a tool helix angle of 40°. Both tools were experimented with according to the process parameters given in Table 1, and each set of experiments was conducted to compare the cutting forces of UVAM and CM. In this experiment, the axial depth of cut a_p = 4.3 mm, the feed rate range was 50~200 mm/min, the radial depth of cut range was 0.3~1 mm, and the spindle speed range was 500~3000 n/min. The specific experimental design is shown in Table 1.

Table 1. Process parameters in the experiment.

Number	v_f (mm/mim)	a_e (mm)	n (r/min)
1	50, 100, 200	0.5	500, 1000
2	100	0.3, 0.7, 1	500, 1000
3	100	0.5	2000, 3000

The experiment was conducted on the Johnford VMC850 machine tool equipped with an axial ultrasonic vibration tool holder. Before the experiment, the displacement of the tool under ultrasonic vibration was measured using the Keyence LK-G5000 laser displacement sensor with the Keyence LK-H020 sensor head. The displacement sampling frequency was 200 kHz, and the displacement data were processed using the Fourier transform to obtain the ultrasonic frequency and amplitude of UVAM. The cutting force measurement system used during the machining process consisted of the Kistler 9119A multicomponent dynamometer, the Kistler 5080A charge amplifier, and the Kistler 5697A data acquisition system. The cutting force data were displayed and stored through the Dynoware (version 3.3.2.0). The sampling frequency for the cutting force was 6000 Hz, and low-pass filtering was used to process the cutting force data. Figure 6 shows the cutting force experimental equipment and measurement software.

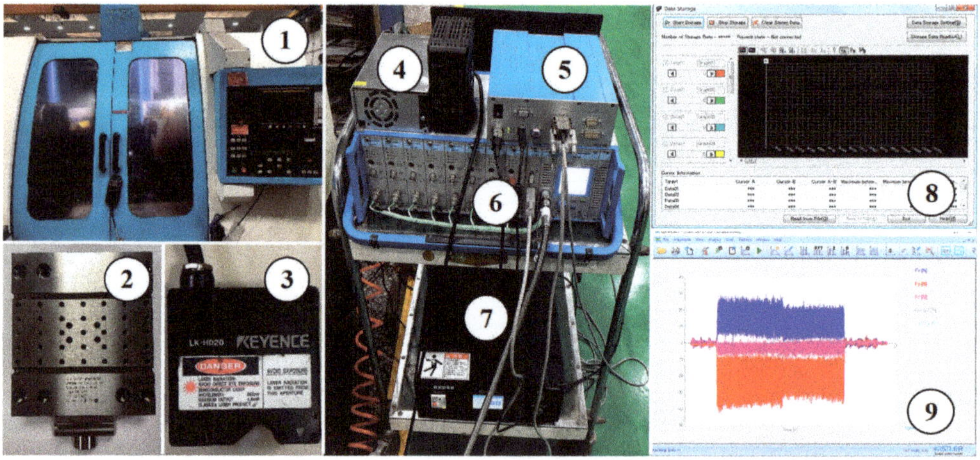

Figure 6. Cutting force experimental equipment and measurement software. (1) VMC850 machine tool, (2) Kistler 9119A multicomponent dynamometer, (3) Keyence LK-H020 sensor head, (4) Keyence LK-G5000 laser displacement sensor, (5) Kistler 5697A data acquisition system, (6) Kistler 5080A charge amplifier, (7) ultrasonic generator, (8) LK-Navigator2 (version 01.04.0.3), and (9) Dynoware software.

According to the measurements, the ultrasonic frequency of the 6 mm diameter milling tool was found to be 29,932 Hz, with an amplitude of 4.4 μm. The ultrasonic frequency of the 12 mm diameter milling tool was measured to be 27,467 Hz, with an amplitude of 2.9 μm. The cutting force measurement results are shown in Table 2. Compared with CM, the cutting force using UVAM for machining showed a different reduction rate.

Table 2. Cutting force measurement data.

No.	v_f (mm/min)	a_e (mm)	n (r/min)	$F(D = 6$ mm$)$ (N)				$F(D = 12$ mm$)$ (N)			
				Fx		Fy		Fx		Fy	
				CM	UVAM	CM	UVAM	CM	UVAM	CM	UVAM
1	50	0.5	500	18.64	13.81	25.73	22.99	6.08	5.91	9.46	9.75
2	100	0.5	500	23.52	19.25	25.27	23.28	7.90	7.39	10.23	10.03
3	200	0.5	500	32.67	27.30	30.35	26.70	12.22	11.53	12.51	12.20
4	50	0.5	1000	15.56	12.95	24.85	22.71	5.02	4.83	9.34	9.29
5	100	0.5	1000	18.14	15.26	23.89	21.61	5.83	5.66	8.18	8.16
6	200	0.5	1000	22.74	19.32	29.11	27.23	9.42	9.29	11.35	11.39
7	100	0.3	500	15.87	11.57	23.35	20.85	8.46	7.38	11.45	10.89
9	100	0.7	500	31.03	25.11	32.68	30.26	17.77	16.44	17.74	17.52
10	100	1	500	40.33	32.16	34.32	31.57	23.71	22.23	18.50	18.33
11	100	0.3	1000	13.39	10.92	22.29	19.03	5.90	5.59	9.45	9.50
12	100	0.7	1000	24.57	21.52	29.39	31.29	12.77	12.04	16.15	16.03
13	100	1	1000	32.31	28.30	32.34	33.88	16.78	15.81	16.60	16.45
14	100	0.5	2000	14.51	13.24	21.12	20.62	5.83	5.66	8.18	8.16
15	100	0.5	3000	12.05	11.66	17.12	17.90	5.09	5.20	7.77	8.11

4. Results and Discussion

The experimental results of the cutting force at different feed rates are shown in Figure 7. As the feed rate increased, both CM and UVAM exhibited a gradual increase in cutting forces. The UVAM cutting force was lower than the CM cutting force under the same process parameters, which proved that the application of ultrasonic vibration reduced the cutting force in milling CFRP. To study the reduction effect of ultrasonic vibration on the cutting force, the reduction rate in the cutting force of UVAM compared with that of CM was calculated as follows:

$$\delta F = \frac{F(CM) - F(UVAM)}{F(CM)} \times 100 \tag{10}$$

The calculation results are shown in Figure 8, where the y-axis represents δF, which indicates the reduction rate of cutting force of UVAM compared with that of CM under the same process parameters. In Figure 8, the average cutting force reduction rate for the $D = 6$ mm tool was as follows: at $n = 500$ r/min, $\delta F_x = 20.2\%$, and $\delta F_y = 10.2\%$; at $n = 1000$ r/min, $\delta F_x = 15.9\%$, and $\delta F_y = 8.2\%$. For the $D = 12$ mm tool, the average reduction rate was as follows: at $n = 500$ r/min, $\delta F_x = 10.3\%$, and $\delta F_y = 4.5\%$; at $n = 1000$ r/min, $\delta F_x = 5.0\%$, and $\delta F_y = 0.4\%$. It can be seen that the cutting force reduction rate for the $D = 6$ mm tool was higher than that of the $D = 12$ mm tool. Additionally, at the same tool diameter, the reduction rate at $n = 500$ r/min was higher than that at $n = 1000$ r/min. According to the calculated duty cycle shown in Figure 5, the duty cycle of the $D = 6$ mm tool was 0.2916 and 0.445, and that of the $D = 6$ mm tool was 0.6467 and 1. This indicated that the value of the duty cycle was inversely proportional to the cutting force reduction rate. Therefore, using a smaller tool diameter and lower spindle speed to achieve a lower duty cycle can enhance the effect of UVAM in reducing the cutting force.

According to Figures 7 and 8, when the other process parameters were the same, the cutting forces in UVAM did not converge to the cutting forces in CM as the feed rate changes. Instead, there was always a certain difference between the cutting forces in UVAM and CM, which indicated that the change in the feed rate had little effect on the ultrasonic vibration effect of UVAM. Using UVAM at different feed rates can still reduce cutting forces during the machining process.

The cutting force experimental results under different radial depths of cut are shown in Figure 9. It can be seen that as the radial cutting depth increased, the cutting forces of CM and UVAM gradually increased. When the process parameters were the same, the cutting force of UVAM was smaller than that of CM. Using Equation (10) to calculate, the cutting force reduction rate under different radial cutting depths was obtained, as shown in Figure 10. In Figure 10, the average cutting force reduction rate for the $D = 6$ mm tool was as follows: at $n = 500$ r/min, $\delta F_x = 21.1\%$, and $\delta F_y = 8.5\%$; at $n = 1000$ r/min, $\delta F_x = 14.8\%$, and $\delta F_y = 3.2\%$. For the $D = 12$ mm tool, the average reduction rate was as follows: at $n = 500$ r/min, $\delta F_x = 8.7\%$, and $\delta F_y = 2.2\%$; at $n = 1000$ r/min, $\delta F_x = 5.8\%$, and $\delta F_y = 0.8\%$. It can be observed that when the tool diameter and spindle speed were the same, which meant the duty cycle was the same, the cutting force reduction rates obtained from the radial cutting depth experiment were similar to those obtained from the feed rate experiment. This indicated that when the duty cycle remained constant, the effect of UVAM in reducing the cutting force did not show significant variation. The experimental results showed the same trend as in Figure 8; when the process parameters were the same, the cutting force reduction rate of the $D = 6$ mm tool was higher than that of the $D = 12$ mm tool, and under the same tool diameter, the cutting force reduction rate at $n = 500$ r/min was higher than that at $n = 1000$ r/min.

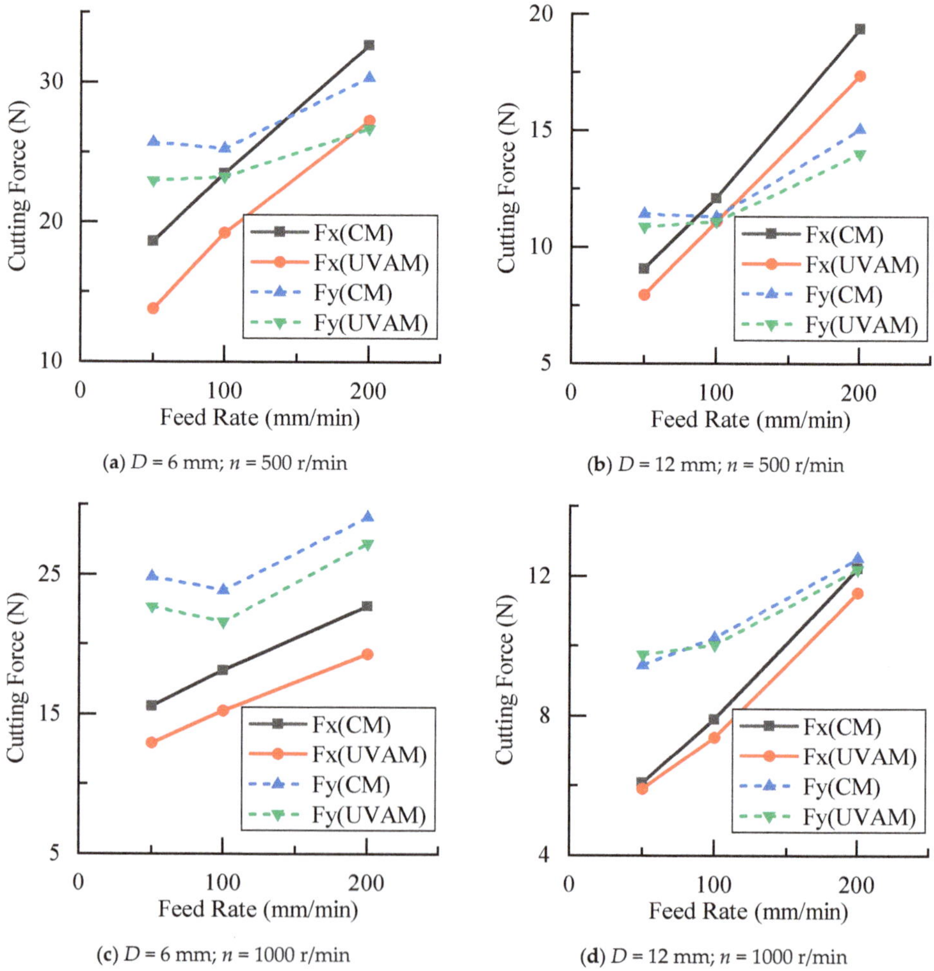

Figure 7. Comparison of cutting forces between CM and UVAM at different feed rates.

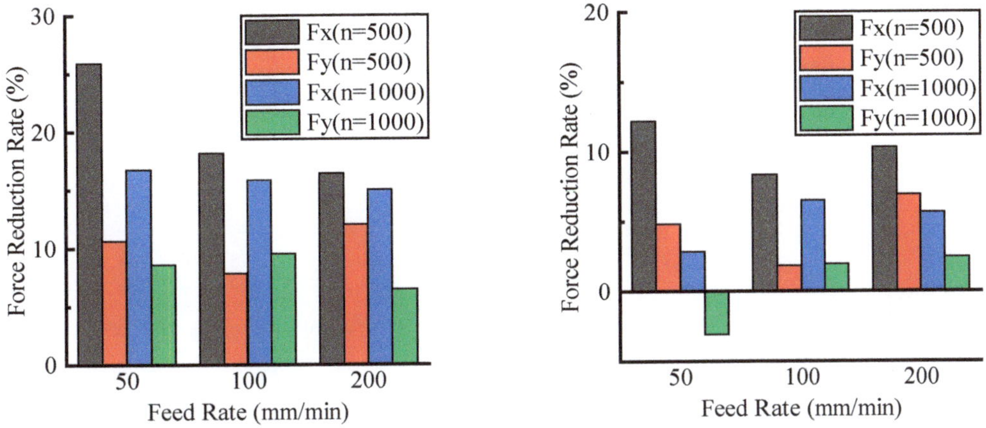

(a) $D = 6$ mm (b) $D = 12$ mm

Figure 8. Effect of feed rate on cutting force reduction.

According to Figures 9 and 10, it can be concluded that when other parameters were the same, the change in the radial cutting depth did not make the cutting force of UVAM converge to the cutting force of CM. Instead, there was always a certain difference between the cutting forces in UVAM and CM. This indicated that the variation in radial cutting depth had little effect on the cutting force reduction effect of UVAM.

The experimental results of cutting forces at different spindle speeds are shown in Figure 11. It can be seen that as the spindle speed increased, the cutting force gradually decreased. In most cases, the cutting force in UVAM was smaller than that in CM. However, as the spindle speed increased, the cutting force of UVAM gradually approached the cutting force of CM, until at $n = 3000$ r/min, the cutting force of UVAM was close to or greater than the cutting force of CM. Figure 12 is calculated through Equation (10), it shows the decrease in the cutting force reduction rate as the spindle speed increased. At $n = 3000$ r/min, both tools with different diameters struggled to reduce the cutting force of the machining process by using UVAM. At this time, the duty cycle was 1, and the cutting force reduction rate ranged from 3.2% to −4.6%. Abootorabi et al. [1] obtained similar results, that is, as the spindle speed increased, the cutting force of UVAM approached that of CM. This indicated that the spindle speed affected the reduction in the cutting force by UVAM, and the lower the spindle speed, the lower the duty cycle, and the better the effect of using UVAM to reduce the cutting force. If the spindle speed was too high, UVAM could not effectively reduce the cutting force.

The process parameters used for plotting Figure 5e were the same as those in the experiment of this work. Therefore, as shown in Figure 5e, in this experiment, the tool with a diameter of $D = 12$ mm only had the duty cycle $DC < 1$ at the spindle speed of $n = 500$ r/min, while the tool with a diameter of $D = 6$ mm had the duty cycle of $DC = 1$ only at the spindle speed of $n = 3000$ r/min. This indicated that compared with the $D = 12$ mm tool, when using the $D = 6$ mm tool for the experiment, there were more occurrences of tool–workpiece separation, and under the same process parameters, the duty cycle of the $D = 6$ mm tool was smaller, resulting in a longer tool–workpiece separation time.

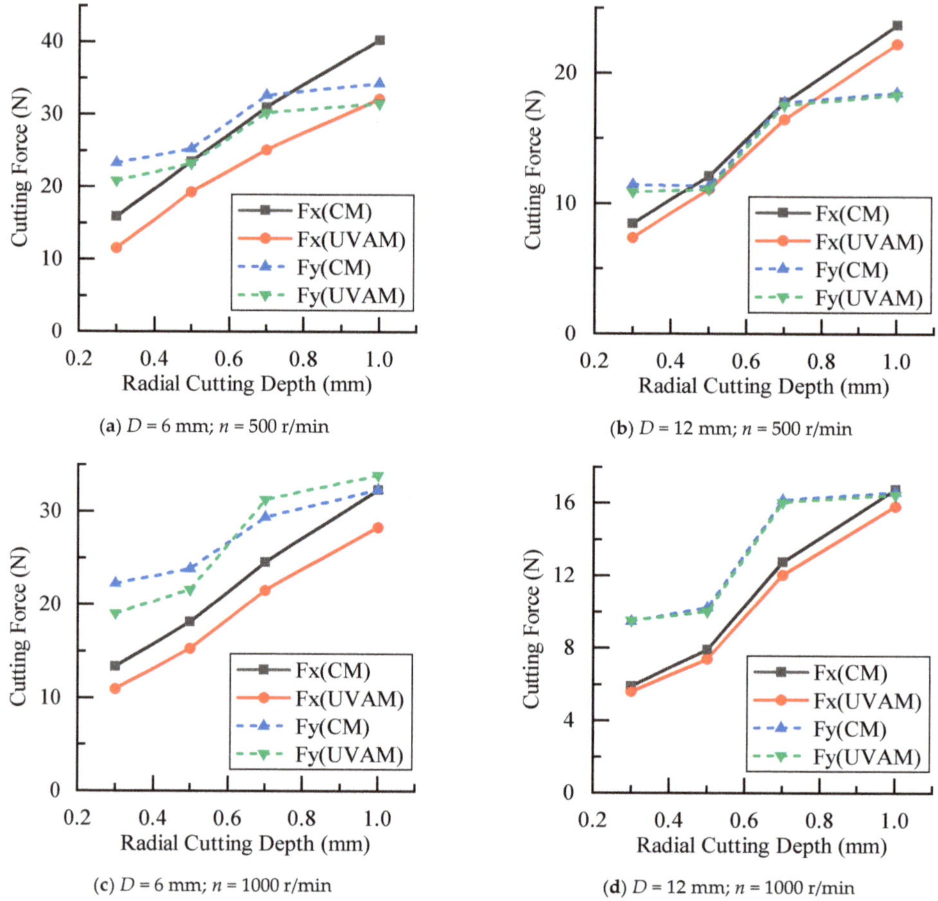

(**a**) D = 6 mm; n = 500 r/min

(**b**) D = 12 mm; n = 500 r/min

(**c**) D = 6 mm; n = 1000 r/min

(**d**) D = 12 mm; n = 1000 r/min

Figure 9. Comparison of cutting forces between CM and UVAM at different radial cutting depths.

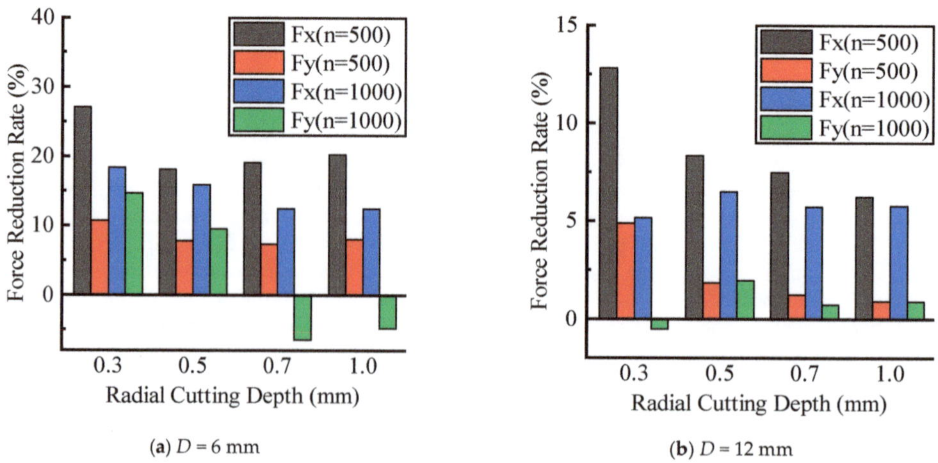

(**a**) D = 6 mm

(**b**) D = 12 mm

Figure 10. Effect of radial cutting depth on cutting force reduction.

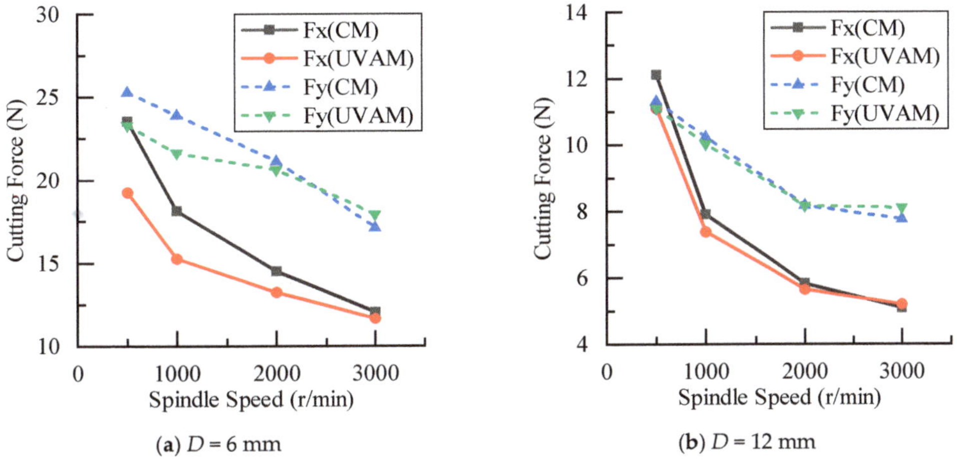

Figure 11. Comparison of cutting forces between CM and UVAM at different spindle speeds.

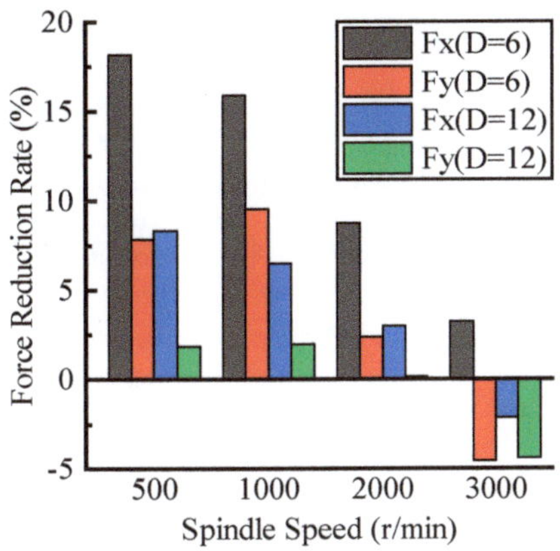

Figure 12. Effect of spindle speed on cutting force reduction.

All the experimental data in this work were plotted to show the relationship between the duty cycle and the magnitude of the cutting force decrease, as shown in Figure 13. It can be observed that as the duty cycle increased, the proportion of the tool–workpiece separation time to the total machining time decreased. The tool–workpiece contact time of UVAM gradually approached that of CM. The cutting forces of UVAM and CM also gradually approached each other, and the cutting force reduction rate in UVAM gradually decreased. Ni et al. [19] reached a similar conclusion in their study, they found that a shorter tool–workpiece contact time can reduce the cutting force. When $D = 6$ mm and $n = 500$ r/min, the duty cycle $DC = 0.2916$; at this time, the cutting force in UVAM was 7.4% to 27% lower than that in CM. When the duty cycle was $DC = 1$, the cutting force in UVAM was -4.5% to 7.5% lower than that in CM.

Based on the analysis above, we can draw the following conclusions: the duty cycle, which represents the proportion of the tool–workpiece contact time to the total machining time, has an influence on the effect of UVAM in reducing the cutting force. The smaller the duty cycle, the longer the tool–workpiece separation time, allowing UVAM to significantly reduce the cutting force. Conversely, a larger duty cycle, indicating a shorter tool–workpiece separation time, results in a diminished effect of UVAM in reducing the cutting force. Therefore, to achieve the effect of reducing the cutting force in UVAM and make it more obvious, it should be considered to have a smaller duty cycle during machining. This can be achieved by using a smaller spindle speed and tool diameter and a larger tool helix angle, ultrasonic amplitude, and ultrasonic frequency while meeting the machining requirements.

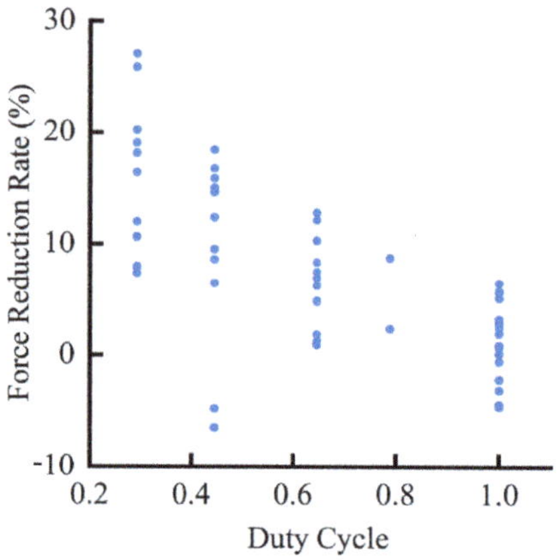

Figure 13. Relationship between duty cycle and cutting force reduction rate.

5. Conclusions

In this work, a method for calculating the duty cycle in UVAM was provided, which was used to analyze the tool–workpiece separation in UVAM. An experimental study on the cutting force of milling CFRP was conducted to compare the cutting forces between UVAM and CM, and the relationship between the duty cycle and the UVAM cutting force was analyzed. The following conclusions were drawn:

1. UVAM significantly reduces the cutting force during CFRP machining. In this study, UVAM achieved a maximum reduction of 27% in the cutting force.
2. The reduction rate of the cutting force in UVAM is related to the duty cycle. A larger duty cycle results in a shorter tool–workpiece separation time, which leads to UVAM and CM having closer cutting forces and a smaller reduction rate in the cutting force.
3. The duty cycle is influenced by the spindle speed, tool diameter, tool helix angle, ultrasonic amplitude, and ultrasonic frequency. Increasing the spindle speed and tool diameter will increase the duty cycle, while increasing the tool helix angle, ultrasonic amplitude, and ultrasonic frequency will decrease the duty cycle.
4. To maximize the effect of UVAM in reducing the cutting force, it is recommended to reduce the duty cycle while ensuring the requirements of the machining process parameters. This can be achieved by reducing the spindle speed or tool diameter and increasing the tool helix angle, ultrasonic amplitude, or ultrasonic frequency.

This study mainly focused on the duty cycle and cutting force of UVAM. In future research, further investigation will be conducted to explore the mechanism of reducing the cutting force in UVAM and establish a cutting force model for UVAM. In addition, for the application of UVAM in CFRP machining, it is also needed to study the impact of UVAM on the surface quality of CFRP and to explore the role of UVAM in reducing the surface roughness and minimizing machining damage.

Author Contributions: Y.Z.: methodology, writing, experiment, and formal analysis. J.Z.: project administration, formal analysis, review, and editing. J.R.: supervisor and review. All authors have read and agreed to the published version of the manuscript.

Funding: This work is supported by the National Key R&D Program of China (2021YFB3703900), the National Natural Science Foundation of China (Grant No. 52075451), the National Science and Technology Major Project (Grant No. J2019-VII-0001-0141), the Science Center for Gas Turbine Project (Grant No. P2022-B-IV-012-001), the Aeronautical Science Foundation of China (Grant No. 2019ZE053008), and the China Postdoctoral Science Foundation (Grant No. 2020M683551).

Institutional Review Board Statement: The manuscript is approved by all authors for publication. This manuscript is original, and the entire paper or any part of its content has never been published in other journals. The data and results are true and clear. We guarantee the transparency and objectivity of research and strictly abide by the ethical and professional standards of the industry. All the authors listed agree to participate in this manuscript.

Informed Consent Statement: All co-authors agree to publish the version of this work in Materials.

Data Availability Statement: All data and materials used or analyzed during the current study are included in this manuscript.

Conflicts of Interest: The authors declare no conflict of interest.

References

1. Zarchi, M.M.A.; Razfar, M.R.; Abdullah, A. Influence of ultrasonic vibrations on side milling of AISI 420 stainless steel. *Int. J. Adv. Manuf. Technol.* **2012**, *66*, 83–89. [CrossRef]
2. Feng, Y.; Zhang, M.; Zhu, Z.; Jia, B.; Wang, X. Axial cutting force prediction model of titanium matrix composites TiBw/TC4 in ultrasonic vibration–assisted drilling. *Int. J. Adv. Manuf. Technol.* **2019**, *105*, 121–135. [CrossRef]
3. Niu, Y.; Jiao, F.; Zhao, B.; Gao, G. Investigation of Cutting Force in Longitudinal-Torsional Ultrasonic-Assisted Milling of Ti-6Al-4V. *Materials* **2019**, *12*, 1955. [CrossRef] [PubMed]
4. Feng, Y.; Hsu, F.-C.; Lu, Y.-T.; Lin, Y.-F.; Lin, C.-T.; Lin, C.-F.; Lu, Y.-C.; Liang, S.Y. Temperature prediction of ultrasonic vibration-assisted milling. *Ultrasonics* **2020**, *108*, 106212. [CrossRef] [PubMed]
5. Ren, W.; Xu, J.; Lin, J.; Yu, Z.; Yu, P.; Lian, Z.; Yu, H. Research on homogenization and surface morphology of Ti-6Al-4V alloy by longitudinal-torsional coupled ultrasonic vibration ball-end milling. *Int. J. Adv. Manuf. Technol.* **2019**, *104*, 301–313. [CrossRef]
6. Liu, X.; Wang, W.; Jiang, R.; Xiong, Y.; Lin, K.; Li, J. Investigation on surface roughness in axial ultrasonic vibration–assisted milling of in situ TiB2/7050Al MMCs. *Int. J. Adv. Manuf. Technol.* **2020**, *111*, 63–75. [CrossRef]
7. Gong, H.; Fang, F.; Hu, X. Kinematic view of tool life in rotary ultrasonic side milling of hard and brittle materials. *Int. J. Mach. Tools Manuf.* **2010**, *50*, 303–307. [CrossRef]
8. Liang, W.; Xu, J.; Ren, W.; Liu, Q.; Wang, X.; Yu, H. Study on the influence of tool point angle on ultrasonic vibration–assisted drilling of titanium alloy. *Int. J. Adv. Manuf. Technol.* **2019**, *105*, 1069–1082. [CrossRef]
9. Karpat, Y.; Bahtiyar, O.; Değer, B. Mechanistic force modeling for milling of unidirectional carbon fiber reinforced polymer laminates. *Int. J. Mach. Tools Manuf.* **2012**, *56*, 79–93. [CrossRef]
10. Liu, J.; Zhang, D.; Qin, L.; Yan, L. Feasibility study of the rotary ultrasonic elliptical machining of carbon fiber reinforced plastics (CFRP). *Int. J. Mach. Tools Manuf.* **2012**, *53*, 141–150. [CrossRef]
11. Ning, F.; Cong, W.; Pei, Z.; Treadwell, C. Rotary ultrasonic machining of CFRP: A comparison with grinding. *Ultrasonics* **2016**, *66*, 125–132. [CrossRef] [PubMed]
12. Xu, W.; Zhang, L. Heat effect on the material removal in the machining of fibre-reinforced polymer composites. *Int. J. Mach. Tools Manuf.* **2019**, *140*, 1–11. [CrossRef]
13. Geng, D.; Lu, Z.; Yao, G.; Liu, J.; Li, Z.; Zhang, D. Cutting temperature and resulting influence on machining performance in rotary ultrasonic elliptical machining of thick CFRP. *Int. J. Mach. Tools Manuf.* **2017**, *123*, 160–170. [CrossRef]
14. Cong, W.; Pei, Z.; Treadwell, C. Preliminary study on rotary ultrasonic machining of CFRP/Ti stacks. *Ultrasonics* **2014**, *54*, 1594–1602. [CrossRef] [PubMed]
15. Feng, Q.; Cong, W.L.; Pei, Z.J.; Ren, C.Z. Rotary ultrasonic machining of carbon fiber-reinforced polymer: Feasibility study. *Mach. Sci. Technol.* **2012**, *16*, 380–398. [CrossRef]
16. Geng, D.; Zhang, D.; Xu, Y.; He, F.; Liu, D.; Duan, Z. Rotary ultrasonic elliptical machining for side milling of CFRP: Tool performance and surface integrity. *Ultrasonics* **2015**, *59*, 128–137. [CrossRef]
17. Brehl, D.E.; Dow, T.A. Review of vibration-assisted machining. *Precis. Eng.* **2008**, *32*, 153–172. [CrossRef]
18. Chen, W.; Huo, D.; Hale, J.; Ding, H. Kinematics and tool-workpiece separation analysis of vibration assisted milling. *Int. J. Mech. Sci.* **2018**, *136*, 169–178. [CrossRef]
19. Ni, C.; Zhu, L.; Liu, C.; Yang, Z. Analytical modeling of tool-workpiece contact rate and experimental study in ultrasonic vibration-assisted milling of Ti–6Al–4V. *Int. J. Mech. Sci.* **2018**, *142–143*, 97–111. [CrossRef]
20. Sui, H.; Zhang, X.; Zhang, D.; Jiang, X.; Wu, R. Feasibility study of high-speed ultrasonic vibration cutting titanium alloy. *J. Mater. Process. Technol.* **2017**, *247*, 111–120. [CrossRef]
21. Fonseca, J.H.; Han, G.; Quagliato, L.; Kim, Y.; Choi, J.; Keum, T.; Kim, S.; Han, D.S.; Kim, N.; Lee, H. Design and numerical evaluation of recycled-carbon-fiber-reinforced polymer/metal hybrid engine cradle concepts. *Int. J. Mech. Sci.* **2019**, *163*, 105115. [CrossRef]
22. Soutis, C. Fibre reinforced composites in aircraft construction. *Prog. Aerosp. Sci.* **2005**, *41*, 143–151. [CrossRef]

23. Chaiyasarn, K.; Ali, N.; Phuphasuwan, P.; Poovarodom, N.; Joyklad, P.; Mohamad, H.; Zhou, M.; Hussain, Q. Flexural Behavior of Natural Hybrid FRP-Strengthened RC Beams and Strain Measurements Using BOTDA. *Polymers* **2021**, *13*, 3604. [CrossRef] [PubMed]
24. Takeyama, H.; Iijima, N. Machinability of Glassfiber Reinforced Plastics and Application of Ultrasonic Machining. *CIRP Ann.* **1988**, *37*, 93–96. [CrossRef]
25. Xu, W.; Zhang, L. Mechanics of fibre deformation and fracture in vibration-assisted cutting of unidirectional fibre-reinforced polymer composites. *Int. J. Mach. Tools Manuf.* **2016**, *103*, 40–52. [CrossRef]
26. Xu, W.; Zhang, L. On the mechanics and material removal mechanisms of vibration-assisted cutting of unidirectional fibre-reinforced polymer composites. *Int. J. Mach. Tools Manuf.* **2014**, *80–81*, 1–10. [CrossRef]
27. Xu, W.X.; Zhang, L.; Wu, Y.B. Micromechanical Modelling of Elliptic Vibration-Assisted Cutting of Unidirectional FRP Composites. *Adv. Mater. Res.* **2012**, *591–593*, 531–534. [CrossRef]
28. Yuan, S.; Zhang, C.; Amin, M.; Fan, H.; Liu, M. Development of a cutting force prediction model based on brittle fracture for carbon fiber reinforced polymers for rotary ultrasonic drilling. *Int. J. Adv. Manuf. Technol.* **2015**, *81*, 1223–1231. [CrossRef]
29. Halim, N.F.H.A.; Ascroft, H.; Barnes, S. Analysis of Tool Wear, Cutting Force, Surface Roughness and Machining Temperature During Finishing Operation of Ultrasonic Assisted Milling (UAM) of Carbon Fibre Reinforced Plastic (CFRP). *Procedia Eng.* **2017**, *184*, 185–191. [CrossRef]

lubricants

MDPI

Article

Analysis, Modeling and Experimental Study of the Normal Contact Stiffness of Rough Surfaces in Grinding

Yuzhu Bai, Xiaohong Jia *, Fei Guo and Shuangfu Suo

State Key Laboratory of Tribology in Advanced Equipment, Tsinghua University, Beijing 100084, China; baiyz18@mails.tsinghua.edu.cn (Y.B.); guof2014@mail.tsinghua.edu.cn (F.G.); sfsuo@mail.tsinghua.edu.cn (S.S.)
* Correspondence: jiaxh@mail.tsinghua.edu.cn

Abstract: Grinding is the most important method in machining, which belongs to the category of precision machining processes. Many mechanical bonding surfaces are grinding surfaces. Therefore, the contact mechanism of grinding a joint surface is of great significance for predicting the loading process and dynamic characteristics of precision mechanical products. In this paper, based on the collected grinding surface roughness data, the profile parameters and topography characteristics of the asperity were analyzed, the rough surface data were fitted, the asperity profile was reconstructed, and the parabola $y = nx^2 + mx + l$ of the cylindrical asperity model was established. After analyzing the rough surface data of the grinding process, the asperity distribution height was fitted with a Gaussian distribution function, which proved that asperity follows the Gaussian distribution law. The validity of this model was confirmed by the non-dimensional processing of the assumed model and the fitting of six plasticity indices. When the pressure is the same, the normal stiffness increases with the decrease in the roughness value of the joint surface. The experimental stiffness values are basically consistent with the fitting stiffness values of the newly established model, which verifies the reliability and effectiveness of the new model established for the grinding surface. In this paper, a new model for grinding joint surface is established, and an experimental platform is set up to verify the validity of the model.

Keywords: grinding surface; contact stiffness; modeling; asperity; rough surface

1. Introduction

The precision of grinding processing is high, and the combined surface of many parts is obtained by grinding. Therefore, it is of great theoretical significance to study the characteristics of grinding joint surfaces for the design of precision machinery. Normal contact stiffness is an important characteristic of grinding joint surfaces, which determines the running stability of mechanical parts. It is of practical value to study normal contact stiffness in order to improve the stability of mechanical equipment. In this paper, the asperity contact model is first established, then the contact stiffness formula is obtained, and the reliability of the model is verified by experiments. The establishment of an accurate and effective contact stiffness model for joint surfaces is the basis for further precision mechanical design, dynamic modeling, and analysis of machine tools. The modeling of a rough surface contact mechanism can be divided into a statistical model and a fractal model, according to the distribution of micro-convex bodies.

Greenwood and Williamson [1] were the first to study the rough surface contact problem with statistical methods, namely the GW model. Based on Hertz's contact theory and combined with statistical principles, they systematically studied the contact mechanism of rough surfaces. In this model, the contact between the upper and lower rough bonded surfaces is equivalent to the contact between a spherical asperity and a smooth rigid plane. All the asperities are uniformly distributed on the matrix, and the asperities have the same curvature radius and are isotropic. Matrix deformation is ignored, only elastic deformation occurs in the slight asperity, and plastic deformation is ignored. The height of the asperities accords with Gaussian distribution, and the interaction between asperities is ignored. The asperities are assumed to be independent of each other. In view of the above limitations of the GW model, scholars have further improved on his research and put forward many revised models. Chang et al. [2] proposed a volume-invariant statistical model, namely the C.E.B. model, based on the volume conservation principle of asperities. The traditional model is compared with the numerical results of this model. The comparison results show that the previous analysis, which did not consider the volume conservation of asperities, is slightly improved. Kogut and Etsion [3] established an elastic–plastic finite element model (the KE model) considering the elastic–plastic deformation factors of asperities. The contact mechanism of rough surface asperity in the pure elastic, elastoplastic, and completely plastic deformation stages was analyzed. Wenzhen Xie et al. [4] studied the influence of roughness size on critical interference through an molecular dynamics (MD) simulation and proposed a critical interference fitting formula considering the effect of roughness size based on the results of an MD analysis and the traditional contact model. Hossein Jamshidi et al. [5] considered the influence of elastic substrate deformation on the friction coefficient, and established a statistical model of frictional contact between a hard rough surface and an elastic substrate by using the multi-rough surface contact theory. Hehe Kang et al. [6] proposed a statistical model to describe the normal and tangential contact behavior of multi-scale rough interfaces, considering the coupling effects of macro-profile deviation and mesoscopic roughness on contact behavior. Bo Yuan et al. [7] proposed an acoustic model combined with the microwave transmission line theory that considered contact interface and boundary effects. Li wei et al. [8] established the tangential contact model of the lap interface with a non-Gaussian surface.

Majumdar and Bhushan [9] used a fractal function (the WM function) to characterize the contours of a rough surface, and proposed, for the first time, a contact model based on fractal geometry theory, namely the MB model. When the power law spectrum of the surface profile falls within the fractal range, the multi-scale roughness structure can be characterized by fractal geometry. Xie yuanhong et al. [10] calculated the contact stiffness formula of a non-uniform interface of nodes with multi-rough surface contact characteristics at the microscopic level by using the fractal contact theory, and built a creep classification model of composite materials based on the elastic–viscoplastic theory. The elastic–viscoplastic contact mechanism of the fractal contact theory varying with time was further studied. Shen fei [11] used the three-dimensional Weierstrass–Mandelbrot (WM) function to characterize the fractal function parameters of rough surfaces. Yu xin et al. [12] equated the rough surface area to oil film thickness, derived the analytical formula for oil film thickness, and established a generalized closed fractal model of the friction coefficient between fractal surfaces in mixed lubrication. Chen hongxu et al. [13] analyzed the effects of fractal dimensions, fractal roughness parameters and Meyer index on contact stiffness, and constructed a modified fractal roughness model to calculate the contact stiffness of fixed joints. Chen Jian et al. [14] analyzed and studied the dynamic contact performance of engineering surfaces according to the scale dependence of key contact parameters of rough surfaces and the size distribution of contact points and proposed a fractal contact model considering multi-scale effects. Wang et al. [15] analyzed the loading and unloading process of a cylindrical friction surface and established a fractal contact model for a spherical micro-convex body. Yuan et al. [16] established an elastic–plastic contact model for the loading and unloading of three-dimensional rough fractal surfaces. Zhou et al. [17] used the maximum peak-to-peak ratio (VPR) to establish a three-dimensional convex and convex model of rough surfaces.

Most of the traditional models are based on the G.W. model, which is based on the spherical micro-convex hypothesis. In addition, there are elliptic micro-convex, two-dimensional asperity, normal curve asperity, residual curve asperity and other different shape asperity models. Fei shen et al. [18] proposed a multi-micro-convex type contact model based on the size distribution law of truncated region of rough surface. Qiuping

Yu et al. [19] established the shoulder–shoulder contact geometric model of the rough sealing surface. Linbo zhu et al. [20] proposed a shoulder–shoulder contact normal contact stiffness model between rough body pairs considering three different contact modes of three-dimensional topography and rough surface elasticity, elastic–plastic and full-plastic. Wenzhen xie et al. [21] simulated the microscopic contact behavior of rough surfaces through molecular dynamics and established a semi-analytical model based on this. Lan zhang et al. [22], based on statistical methods, proposed a modified rectangular micro-convex model for the continuous observation of the length of non-adherent rough contact. Guangbin yu [23] comprehensively analyzed the geometric structure and mechanical properties of elliptical concave and convex points on rough surfaces, and established a contact mechanical model of elliptical asperities on rough surfaces containing elliptical concave and convex points. In order to accurately predict the leakage rate of the sealed interface, Bao lv et al. [24] established a two-dimensional finite element model of proton-exchange membrane fuel cells by analyzing the microscopic surface morphology and the rough contact process of components. Chaodong Zhang et al. [25] established a new contact stiffness model with sawtooth, normal function, and trapezoid contact stiffness of a textured rough surface by analysing the three kinds of machining surface textures: turning, milling and grinding. An qi et al. [26] adopted novel cosine curve roughness and traditional Gaussian distribution to simulate rough surfaces, and established a cosine curve asperity contact stiffness model. Li ling et al. [27] introduced the triangular distribution function and established the triangular distribution function contact stiffness model. Jiang kai et al. [28], considering the different surface contact characteristics of rough matrix interactions, used mesh regeneration technology to reconstruct the real surface topography, estimated the fractal parameters using the fractal parameters estimation neural network method, and established a mesh regeneration finite element model.

Based on the above review and an analysis of existing models, this paper collects the rough surface data of grinding processing, analyzes the profile parameters and topography characteristics of asperity, fits the rough surface data, resimulates the profile of asperity, and establishes a new parabola $y = nx^2 + mx + l$ cylindrical asperity model. An experimental platform is set up to verify the effectiveness of the new model. In Figure 1, for ease of understanding, the technical route flow chart is presented.

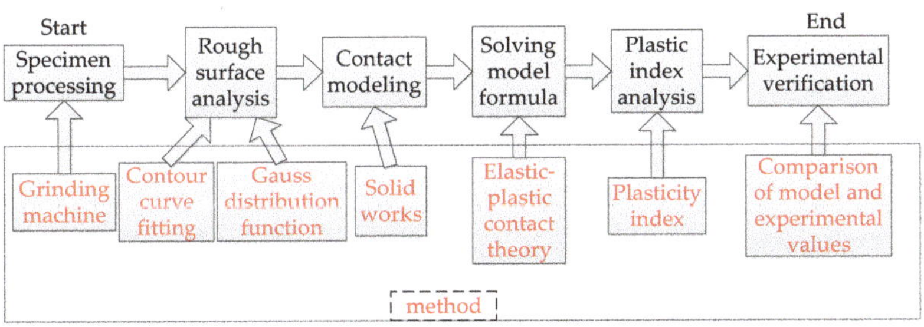

Figure 1. Flow chart.

2. Hypothetical Surface

The study of the contact mechanism of rough bonded surfaces starts with a microscopic morphology of rough surfaces. The surface microstructure features were revealed, and the contact law was analyzed based on mechanical principles. In order to reveal the micro-morphology of the rough surface, a 3D white light interferometer was used to measure the grinding surface and obtain the precise data of the surface micro-morphology. By fitting the real topography data obtained from the measurement, the grinding simulation surface was established.

As shown in Figure 2a, ZYGONexView (ZYGO Connecticut, Middlefield, CT, USA), a practical 3D white light interferometer tested in this paper, is a high-linear-capacitance sensor. The accuracy of the instrument is to ensure the full range of 0.1 nm in 0–150 μm to ensure real and effective measurement results. Figure 2b shows the specimen obtained by grinding.

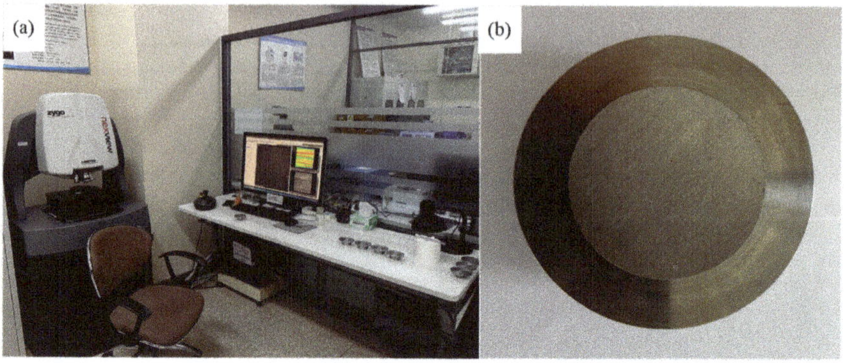

Figure 2. (**a**) White light interference surface topography instrument; (**b**) specimen.

Figure 3 shows the 3D micromorphologies of grinding specimens obtained by 3D white light interferometer test. By observing their appearance, many strips of asperity are shown to be evenly and neatly distributed on the surface of the specimen, and the surface asperity is undulating, concave and convex, and the groove texture is clear.

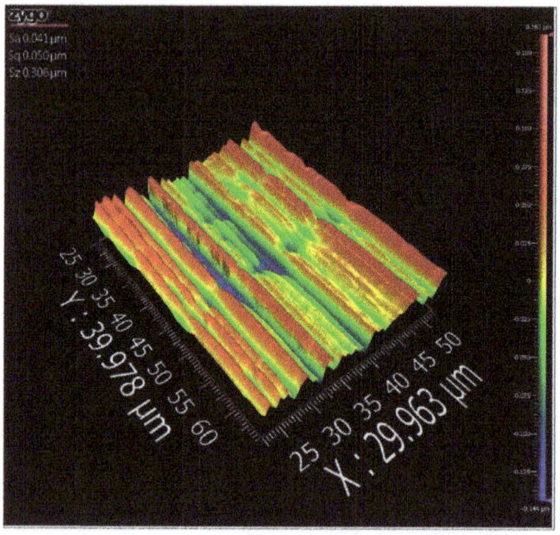

Figure 3. 3D micro-topography of grinding surface.

In order to reveal the real topography characteristics of the rough surface, the actual contour characteristics of a single asperity are tested. Then, the parabolic function $y = nx^2 + mx + l$ is used to fit the profile data points of the asperity cross-section. The fitting results are shown in Figure 4, and the fitting data points of the profile of the asperity cross-section are basically consistent with the curve of the parabola function. This proved that the distribution of data points of a single asperity cross-section on the rough surface obtained by grinding is close to the parabolic function curve.

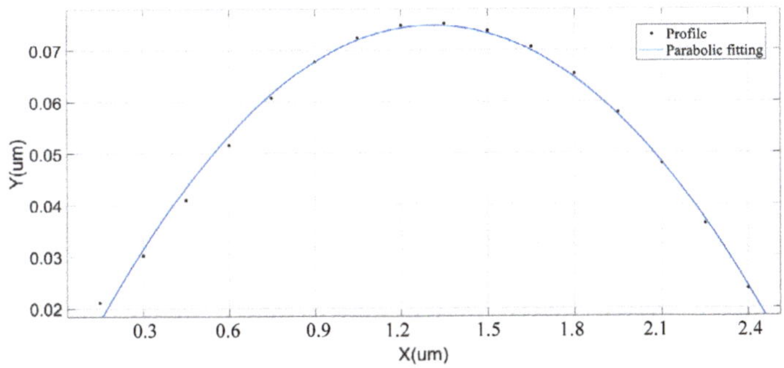

Figure 4. Results of parabolic function fitting of asperity contour.

In order to verify the superiority of the function $y = nx^2 + mx + l$ fitting, parabolic function $y = cx^3$ and trigonometric function gsin (x) were used to fit the profile data points of the asperity cross-section. A root mean square error (RMSE) analysis was performed on the fitting results, and the analysis results are shown in Table 1. It can be seen from the table that the root mean square error of fitting function $y = nx^2 + mx + l$ is the lowest, which proves that fitting function $y = nx^2 + mx + l$ is the most suitable for fitting data points of the asperity cross-section.

Table 1. Root mean square error.

Fitting function	$y = nx^2 + mx + l$	$y = cx^3$	$y = \text{gsin}(x)$
RMSE	0.016	0.018	0.117

In order to obtain the height distribution of the rough surface, Gaussian distribution function is fitted to the data obtained by white light interferometer. The fitting results are shown in Figure 5, and the height distribution of the asperity on the rough surface conforms to the Gaussian distribution law.

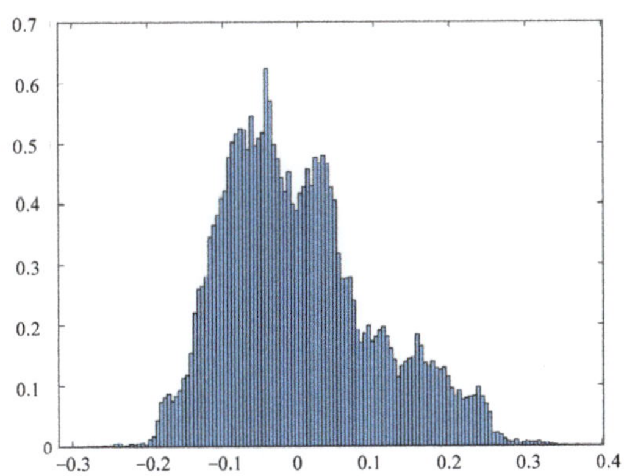

Figure 5. Height distribution of asperity.

According to the above fitting analysis, the distribution of the data points of the cross-section of a single asperity on the rough surface is close to the curve of the parabolic function, and the height distribution of the asperity conforms to the Gaussian distribution law. Based on the above research results, this paper establishes a hypothetical surface with parabolic cylindrical asperity cross-section profile, and the height of asperity follows the

Gaussian distribution law. As shown in Figure 6, it is assumed that the surface is more consistent with the real surface topography characteristics of grinding.

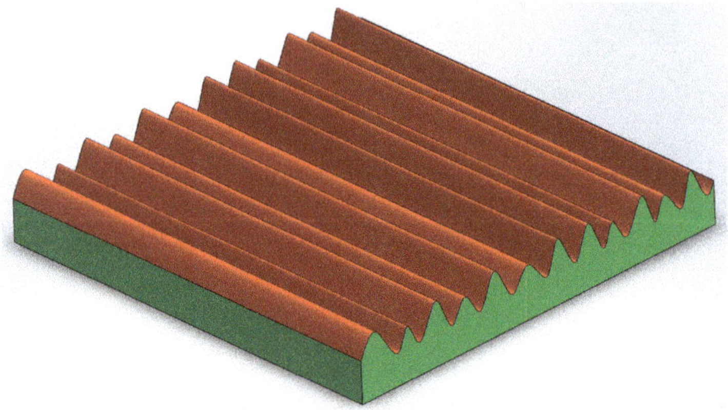

Figure 6. The hypothetical surface.

3. Normal Contact Stiffness Theoretical and Analytical Model

3.1. Contact Model Assumptions

In the previous section, a hypothesis surface with a parabolic cylindrical asperity cross-section profile was established by analyzing the micro-topography of the grinding surface. According to the surfaces assumed above, a theoretical analytical model of the normal contact stiffness of rough grinding surfaces is established in this section. The model made the following assumptions: (1) The contact of two rough surfaces is equivalent to the contact between two planes with evenly distributed asperity surfaces, and the cross-section outline of the asperity is a parabolic cylindrical asperity. (2) The angle between the cylindrical asperity of the upper and lower rough surfaces is 45° on average. (3) The height of the asperities follows the Gaussian distribution law. (4) The curvature radius of all asperity profiles is the same. (5) The asperities are independent of each other, ignoring the interaction between the asperities. (6) The texture direction of the asperity distribution on the rough surface is aligned. (7) Ignore the deformation of the matrix and only calculate the deformation of the asperity. Figure 7 shows the schematic diagram of the model hypothesis surface.

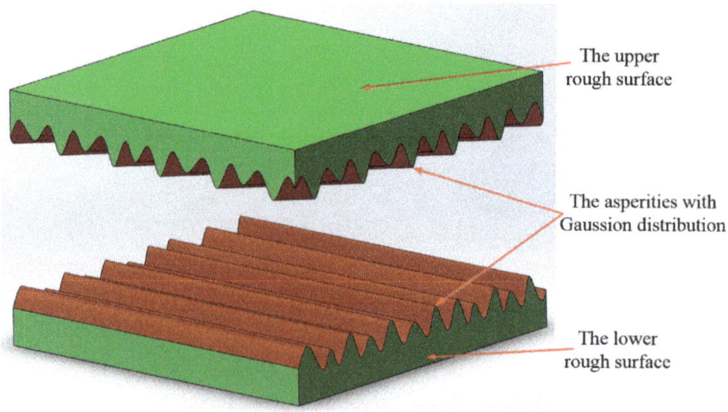

Figure 7. The surface representation of the 3D model hypothesis.

Based on the rough surface hypothesis model established above, the analytic formula of normal contact stiffness in the contact process of a single asperity is derived based on Hertzian contact theory. Then, according to the distribution law of the asperity and the statistical theory, the analytical formula of the contact stiffness of the whole joint surface is obtained.

3.2. Contact Stiffness of Mechanical Joint Surface

According to Reference [29], the stiffness of the whole rough surface of the asperity at different deformation stages is obtained below.

The relation between contact load and displacement variable is as follows:

$$\begin{aligned} F_t(d) &= F_{et}(d) + F_{ept}(d) + F_{pt}(d) \\ &= N\int_d^{d+\frac{\omega_1}{2}} F_{et}\Phi(z)dz + N\int_{d+\frac{\omega_1}{2}}^{d+\frac{\omega_2}{2}} F_{ept}\Phi(z)dz + N\int_{d+\frac{\omega_2}{2}}^{\infty} F_{Pt}\Phi(z)dz \\ &= \eta A_n E\int_d^{d+\frac{\omega_1}{2}} \frac{32\sqrt{R(z-d)^3}}{3\lambda^{\frac{3}{2}}} \frac{1}{\sqrt{2\pi}\sigma}e^{-\frac{(z-d)^2}{2\sigma^2}}dz + \frac{9\eta A_n R^2(\pi\alpha\beta)^3}{16E^2}\int_{d+\frac{\omega_1}{2}}^{d+\frac{\omega_2}{2}} \left(H-0.6H\frac{\ln\omega_2-\ln 2(z-d)}{\ln\omega_2-\ln\omega_1}\right)^3 \cdot \\ &\left[1-2\left(\frac{2z-2d-\omega_1}{\omega_2-\omega_1}\right)^3+3\left(\frac{2z-2d-\omega_1}{\omega_2-\omega_1}\right)^2\right]^3 \frac{1}{\sqrt{2\pi}\sigma}e^{-\frac{(z-d)^2}{2\sigma^2}}dz + 4\pi\eta A_n RH\int_{d+\frac{\omega_2}{2}}^{\infty}(z-d)\frac{1}{\sqrt{2\pi}\sigma}e^{-\frac{(z-d)^2}{2\sigma^2}}dz \end{aligned} \tag{1}$$

The relation between contact stiffness and displacement variable is as follows:

$$\begin{aligned} K_t(d) &= K_{et}(d) + K_{ept}(d) + K_{pt}(d) \\ &= N\int_d^{d+\frac{\omega_1}{2}} K_{et}\Phi(z)dz + N\int_{d+\frac{\omega_1}{2}}^{d+\frac{\omega_2}{2}} K_{ept}\Phi(z)dz + N\int_{d+\frac{\omega_2}{2}}^{\infty} K_{Pt}\Phi(z)dz \\ &= \eta A_n E\int_d^{d+\frac{\omega_1}{2}} \frac{8\sqrt{R(z-d)}}{\lambda^{\frac{3}{2}}} \frac{1}{\sqrt{2\pi}\sigma}e^{-\frac{(z-d)^2}{2\sigma^2}}dz + \frac{27\eta A_n(\pi\alpha\beta)^3 R^2}{16E^2}\int_{d+\frac{\omega_1}{2}}^{d+\frac{\omega_2}{2}} \left\{\frac{0.6H}{2(z-d)(\ln\omega_2-\ln\omega_1)}\cdot\right. \\ &\left[H-0.6H\frac{\ln\omega_2-\ln 2(z-d)}{\ln\omega_2-\ln\omega_1}\right]^2\left[1-2\left(\frac{2z-2d-\omega_1}{\omega_2-\omega_1}\right)^3+3\left(\frac{2z-2d-\omega_1}{\omega_2-\omega_1}\right)^2\right]^3 \\ &+\left[H-0.6H\frac{\ln\omega_2-\ln 2(z-d)}{\ln\omega_2-\ln\omega_1}\right]^3\left[1-2\left(\frac{2z-2d-\omega_1}{\omega_2-\omega_1}\right)^3+3\left(\frac{2z-2d-\omega_1}{\omega_2-\omega_1}\right)^2\right]^2 \\ &\left.\left[\frac{6(2z-2d-\omega_1)}{(\omega_2-\omega_1)^2}-\frac{6(2z-2d-\omega_1)^2}{(\omega_2-\omega_1)^3}\right]\right\}\frac{1}{\sqrt{2\pi}\sigma}e^{-\frac{(z-d)^2}{2\sigma^2}}dz + 2\pi\eta A_n RH\int_{d+\frac{\omega_2}{2}}^{\infty}\frac{1}{\sqrt{2\pi}\sigma}e^{-\frac{(z-d)^2}{2\sigma^2}}dz \end{aligned} \tag{2}$$

3.3. Dimensionless Processing of the Model

The characteristics of the contact stiffness model reveal the essential characteristics of the mechanical bonding surface, involving the displacement, loading size, loading time and other complex factors, involving more dimensions. Therefore, the features it embodies should avoid the influence of dimensions. For easier comparison between the models, it is necessary to carry out a dimensionless processing of the models.

Dimensionless load F^* and dimensionless stiffness K^* are obtained by dividing Equations (1) and (2) by A_nE and $(A_nE)/\sigma$, and all corresponding variables and length parameters in the formula are divided by σ for dimensionless processing.

$$\begin{aligned} F_t^*(d) &= \frac{F_t(d)}{A_n E} \\ &= \eta\int_{h^*-y^*}^{h^*-y^*+\frac{\omega_1^*}{2}} \frac{32\sqrt{R(z^*-h^*+y^*)^3}}{3\lambda^{\frac{3}{2}}}\frac{\sigma}{\sqrt{2\pi}\sigma_s}e^{-\frac{\sigma^2(z^*-h^*+y^*)^2}{2\sigma_s^2}}dz^* \\ &+\frac{9\eta R^2(\pi\alpha\beta)^3}{16E^3}\int_{h^*-y^*+\frac{\omega_1^*}{2}}^{h^*-y^*+\frac{\omega_2^*}{2}}\left(H-0.6H\frac{\ln\omega_2^*-\ln 2(z^*-h^*+y^*)}{\ln\omega_2^*-\ln\omega_1^*}\right)^3\cdot \\ &\left[1-2\left(\frac{2(z^*-h^*+y^*)-\omega_1^*}{\omega_2^*-\omega_1^*}\right)^3+3\left(\frac{2(z^*-h^*+y^*)-\omega_1^*}{\omega_2^*-\omega_1^*}\right)^2\right]^3\frac{\sigma}{\sqrt{2\pi}\sigma_s}e^{-\frac{\sigma^2(z^*-h^*+y^*)^2}{2\sigma_s^2}}dz^* \\ &+\frac{4\pi\eta RH}{E}\int_{h^*-y^*+\frac{\omega_2^*}{2}}^{\infty}(z^*-h^*+y^*)\frac{\sigma}{\sqrt{2\pi}\sigma_s}e^{-\frac{\sigma^2(z^*-h^*+y^*)^2}{2\sigma_s^2}}dz^* \end{aligned} \tag{3}$$

By substituting $\beta = \eta R\sigma$ into the formula:

$$\begin{aligned} F_t^*(d) &= \frac{F_t(d)}{A_n E} \\ &= \beta\left(\frac{\sigma}{R}\right)^{\frac{1}{2}} \int_{h^*-y^*}^{h^*-y^*+\frac{\omega_1^*}{2}} \frac{32\sqrt{(z^*-h^*+y^*)^3}}{3\lambda^{\frac{3}{2}}} \frac{\sigma}{\sqrt{2\pi}\sigma_s} e^{-\frac{\sigma^2(z^*-h^*+y^*)^2}{2\sigma_s^2}} dz^* \\ &+ \frac{\beta R}{\sigma}\frac{9(\pi\alpha\beta)^3}{16E^3} \int_{h^*-y^*+\frac{\omega_1^*}{2}}^{h^*-y^*+\frac{\omega_2^*}{2}} \left(H - 0.6H\frac{\ln\omega_2^* - \ln 2(z^*-h^*+y^*)}{\ln\omega_2^* - \ln\omega_1^*}\right)^3 \cdot \\ &\left[1 - 2\left(\frac{2(z^*-h^*+y^*)-\omega_1^*}{\omega_2^*-\omega_1^*}\right)^3 + 3\left(\frac{2(z^*-h^*+y^*)-\omega_1^*}{\omega_2^*-\omega_1^*}\right)^2\right]^3 \frac{\sigma}{\sqrt{2\pi}\sigma_s} e^{-\frac{\sigma^2(z^*-h^*+y^*)^2}{2\sigma_s^2}} dz^* \\ &+ \frac{4\beta\pi H}{E} \int_{h^*-y^*+\frac{\omega_2^*}{2}}^{\infty} (z^*-h^*+y^*) \frac{\sigma}{\sqrt{2\pi}\sigma_s} e^{-\frac{\sigma^2(z^*-h^*+y^*)^2}{2\sigma_s^2}} dz^* \end{aligned} \quad (4)$$

$$\begin{aligned} F_t^*(d) &= \frac{K_t(d)\sigma}{A_n E} \\ &= \eta\sigma \int_{h^*-y^*}^{h^*-y^*+\frac{\omega_1^*}{2}} \frac{8\sqrt{R(z^*-h^*+y^*)}}{\lambda^{\frac{3}{2}}} \frac{\sigma}{\sqrt{2\pi}\sigma_s} e^{-\frac{\sigma^2(z^*-h^*+y^*)^2}{2\sigma_s^2}} dz^* \\ &+ \frac{27\eta\sigma(\pi\alpha\beta)^3 R^2}{16E^3} \int_{h^*-y^*+\frac{\omega_1^*}{2}}^{h^*-y^*+\frac{\omega_2^*}{2}} \left\{\frac{0.6H}{2(z^*-h^*+y^*)(\ln\omega_2^* - \ln\omega_1^*)} \cdot\right. \\ &\left[H - 0.6H\frac{\ln\omega_2^* - \ln 2(z^*-h^*+y^*)}{\ln\omega_2^* - \ln\omega_1^*}\right]^2 \left[1 - 2\left(\frac{2(z^*-h^*+y^*)-\omega_1^*}{\omega_2^*-\omega_1^*}\right)^3 + 3\left(\frac{2(z^*-h^*+y^*)-\omega_1^*}{\omega_2^*-\omega_1^*}\right)^2\right]^3 \\ &+ \left[H - 0.6H\frac{\ln\omega_2^* - \ln 2(z^*-h^*+y^*)}{\ln\omega_2^* - \ln\omega_1^*}\right]^3 \left[1 - 2\left(\frac{2(z^*-h^*+y^*)-\omega_1^*}{\omega_2^*-\omega_1^*}\right)^3 + 3\left(\frac{2(z^*-h^*+y^*)-\omega_1^*}{\omega_2^*-\omega_1^*}\right)^2\right]^2 \\ &\left.\left[\frac{6(2(z^*-h^*+y^*)-\omega_1^*)}{(\omega_2^*-\omega_1^*)^2} - \frac{6(2(z^*-h^*+y^*)-\omega_1^*)^2}{(\omega_2^*-\omega_1^*)^3}\right]\right\} \frac{\sigma}{\sqrt{2\pi}\sigma_s} e^{-\frac{\sigma^2(z^*-h^*+y^*)^2}{2\sigma_s^2}} dz^* \\ &+ \frac{2\pi\eta\sigma RH}{E} \int_{h^*-y^*+\frac{\omega_2^*}{2}}^{\infty} \frac{\sigma}{\sqrt{2\pi}\sigma_s} e^{-\frac{\sigma^2(z^*-h^*+y^*)^2}{2\sigma_s^2}} dz^* \end{aligned} \quad (5)$$

By substituting $\beta = \eta R\sigma$ into the formula:

$$\begin{aligned} F_t^*(d) &= \frac{K_t(d)\sigma}{A_n E} \\ &= \beta\left(\frac{\sigma}{R}\right)^{\frac{1}{2}} \int_{h^*-y^*}^{h^*-y^*+\frac{\omega_1^*}{2}} \frac{8\sqrt{(z^*-h^*+y^*)}}{\lambda^{\frac{3}{2}}} \frac{\sigma}{\sqrt{2\pi}\sigma_s} e^{-\frac{\sigma^2(z^*-h^*+y^*)^2}{2\sigma_s^2}} dz^* \\ &+ \beta\frac{R}{\sigma}\frac{27(\pi\alpha\beta)^3}{16E^3} \int_{h^*-y^*+\frac{\omega_1^*}{2}}^{h^*-y^*+\frac{\omega_2^*}{2}} \left\{\frac{0.6H}{2(z^*-h^*+y^*)(\ln\omega_2^* - \ln\omega_1^*)} \cdot\right. \\ &\left[H - 0.6H\frac{\ln\omega_2^* - \ln 2(z^*-h^*+y^*)}{\ln\omega_2^* - \ln\omega_1^*}\right]^2 \left[1 - 2\left(\frac{2(z^*-h^*+y^*)-\omega_1^*}{\omega_2^*-\omega_1^*}\right)^3 + 3\left(\frac{2(z^*-h^*+y^*)-\omega_1^*}{\omega_2^*-\omega_1^*}\right)^2\right]^3 \\ &+ \left[H - 0.6H\frac{\ln\omega_2^* - \ln 2(z^*-h^*+y^*)}{\ln\omega_2^* - \ln\omega_1^*}\right]^3 \left[1 - 2\left(\frac{2(z^*-h^*+y^*)-\omega_1^*}{\omega_2^*-\omega_1^*}\right)^3 + 3\left(\frac{2(z^*-h^*+y^*)-\omega_1^*}{\omega_2^*-\omega_1^*}\right)^2\right]^2 \\ &\left.\left[\frac{6(2(z^*-h^*+y^*)-\omega_1^*)}{(\omega_2^*-\omega_1^*)^2} - \frac{6(2(z^*-h^*+y^*)-\omega_1^*)^2}{(\omega_2^*-\omega_1^*)^3}\right]\right\} \frac{\sigma}{\sqrt{2\pi}\sigma_s} e^{-\frac{\sigma^2(z^*-h^*+y^*)^2}{2\sigma_s^2}} dz^* \\ &+ \frac{2\pi\beta H}{E} \int_{h^*-y^*+\frac{\omega_2^*}{2}}^{\infty} \frac{\sigma}{\sqrt{2\pi}\sigma_s} e^{-\frac{\sigma^2(z^*-h^*+y^*)^2}{2\sigma_s^2}} dz^* \end{aligned} \quad (6)$$

In the formula:

$$\omega_1^* = \omega_1/\sigma = 1.709\left[\frac{9}{128R}F_1^2\left(\frac{1-\nu_1^2}{E_1} + \frac{1-\nu_2^2}{E_2}\right)^2\right]^{\frac{1}{3}} = 0.000332811\frac{R}{\sigma} \quad (7)$$

$\omega^* = \omega/\sigma$ is the normal deformation of a single asperity after non-dimensionalization. $\omega_1^* = \omega_1/\sigma$ is the normal critical deformation of the asperity after dimensionless transition from the stage of complete elastic deformation to the stage of elastoplastic deformation. $\omega_2^* = \omega_2/\sigma$ is the normal critical deformation of the asperity from the elastoplastic deformation stage to the complete plastic deformation stage after dimensionless deformation.

$z^* = z/\sigma$ is the height of a single asperity without dimensionalization; $y^* = y/\sigma$ is the distance between the mean height plane of the asperity and the mean height plane of the surface after non-dimensionalization. $h^* = h/\sigma$ is the distance between the average height planes of the upper and lower bound surfaces after being dimensionless.

3.4. Plasticity Index

The plasticity index, which is considered to be the standard of elasticity or plasticity of a rough surface, is essentially the ratio of elastic hardness to actual hardness. It can be seen from the plastic index that the contact of rough surface is determined by two material properties of plane stress elastic modulus and hardness, and three shape properties of rough surface asperity density, standard deviation of height distribution, and average radius. The plasticity index is a generalized surface morphology parameter that combines the material and morphology characteristics. For most rough surfaces, the deformation mode is not affected by the change of load, and the low plasticity index is elastic, with high plasticity index is plastic. According to the literature [2], the calculation formula of plasticity index can be obtained as follows::

$$\psi = \frac{2E}{1.5\pi KH}\left(\frac{\sigma}{R}\right)^{\frac{1}{2}}\left(1 - \frac{3.717 \times 10^{-4}}{\beta^2}\right)^{\frac{1}{4}} \tag{8}$$

It can be seen from the plasticity index formula that the contact characteristics of rough surfaces are determined by the values of β and σ/R. The plasticity index parameters were obtained by testing typical engineering rough surfaces and referring to the experimental data of K. A. NURI and J. HALLING [30], as shown in Table 2.

Table 2. Plasticity index and rough surface engineering parameters.

Ψ	β	σ/R
0.5	0.0302	8.75×10^{-5}
0.7	0.0339	1.60×10^{-4}
1.0	0.0414	3.02×10^{-4}
1.5	0.0476	6.57×10^{-4}
2.0	0.0541	1.144×10^{-3}
2.5	0.0601	1.77×10^{-3}

3.5. Dimensionless Stiffness Is Fitted by Simulation

Figure 8 shows the numerical simulation curves of the dimensionless contact stiffness and dimensionless contact pressure of the assumed rough surface, with plasticity indices of 0.5, 0.7, 1.0, 1.5, 2.0 and 2.5, respectively. The material is 304 steel; elastic modulus is E = 195 GPa; hardness H = 1870 MPa. Figure 8 shows the influence of different plasticity indices on normal contact stiffness. It can be seen from the figure that the contact stiffness increases with the increase in contact pressure. When the contact pressure is the same, the normal contact stiffness increases with the increase in plastic index. This is because the physical meaning of the plasticity index is to determine the elastic–plastic contact characteristics between two surfaces. When the contact load increases, the contact stiffness increases, and the increase in the contact stiffness of the surface with a high plasticity index is higher than that of the surface with a low plasticity index, which verifies the validity of this model.

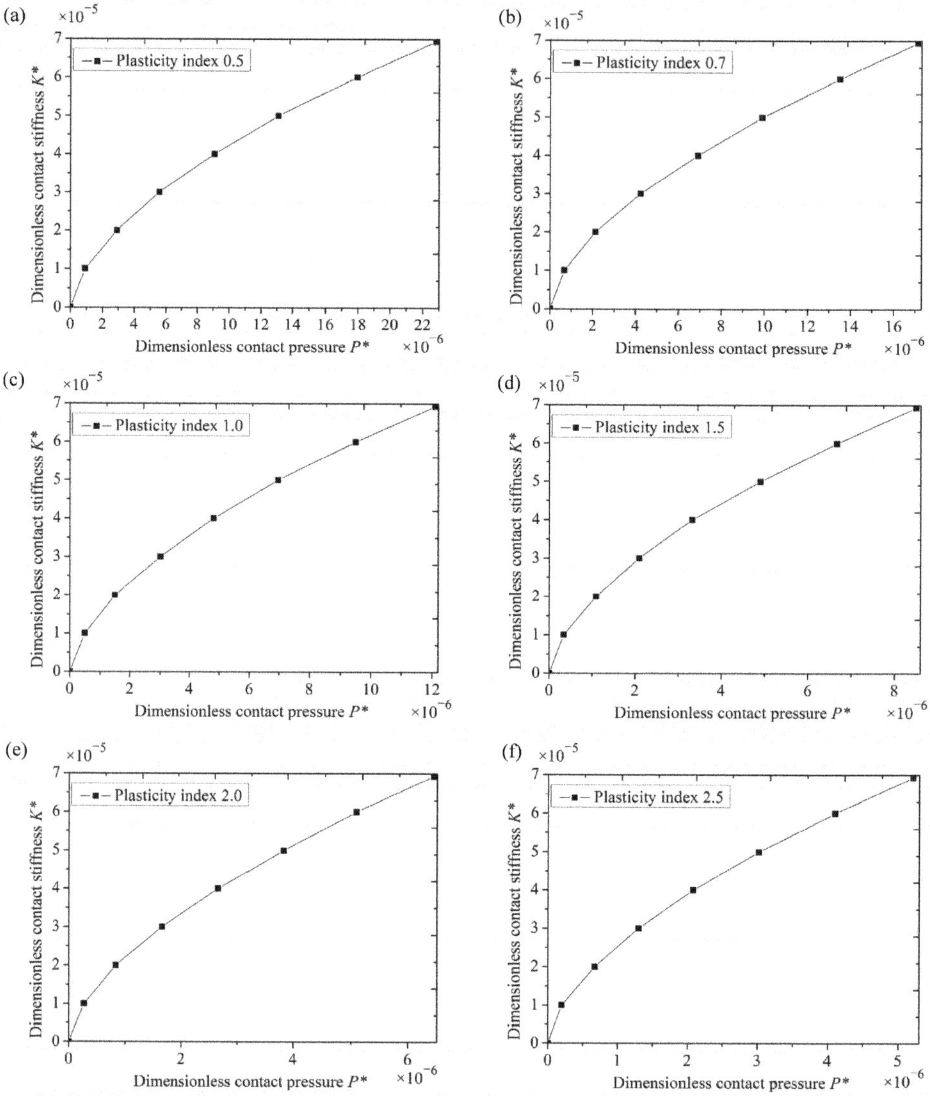

Figure 8. Stiffness at different plasticity indices (**a**) 0.5; (**b**) 0.7; (**c**) 1.0; (**d**) 1.5; (**e**) 2.0; (**f**) 2.5.

4. Experiment

4.1. Experimental Specimen

The rough surface of the specimen in this experiment was obtained by grinding, and the material was 304 steel. The parameters of No. 304 steel are obtained by mechanical experiments: Young's modulus $E_1 = E_2 = 195$ GPa; Poisson's ratio $\nu_1 = \nu_2 = 0.29$; hardness $H = 1870$ MPa. Figure 9a shows the processing and molding equipment of this experiment specimen. The model is an okamoto ultra-precision static press surface grinding machine (Okamoto, Tokyo, Honshu, Japan), upg series. Figure 9b shows the normal contact stiffness test specimen. The bonding surface is the convex surface obtained by grinding, and the height of the convex surface is 2 mm. The roughness of the joint surface was Sa 0.854 μm, Sa 1.173 μm, Sa 1.391 μm and Sa 1.524 μm, respectively.

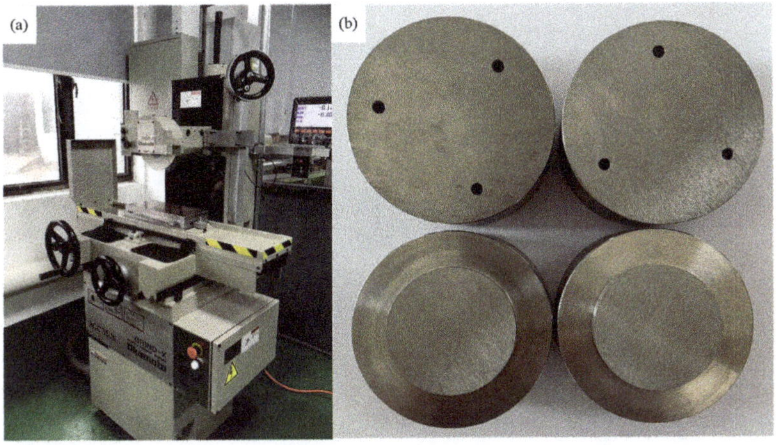

Figure 9. (**a**) Grinding machine; (**b**) contact stiffness test specimen.

4.2. Experimental Apparatus

As shown in Figure 10, in order to verify the accuracy of the model presented in this paper, a normal contact stiffness test platform was built. Figure 10a shows the execution devices of the test platform, including four parts: pressure testing machine, compression module, data acquisition module and control module. The pressure testing machine is a W.D.W. series microcomputer controlled electronic universal testing machine (5–20 KN); the equipment test error is less than $\pm 0.5\%$. Figure 10b shows the compression module, which consists of seven parts: compression rod, upper specimen, eddy current displacement sensor, lower specimen, link disk, thrust roller bearing and fixed column. The eddy current sensor is a KAMAN(KAMAN Bloomfield, CT, USA) high-precision eddy current displacement sensor, model KD2306-2U. The measuring accuracy is ± 0.1 μm, the measuring range is 0–1 mm, and the output voltage is 0–10 V. The data acquisition module uses an Altai data collector, model PCIe5653.

Figure 10c reveals the working principle of the contact stiffness test platform of the grinding joint surface. The compression rod is connected to the pressure sensing part of the pressure testing machine, and the lower end of the compression rod is connected to the upper specimen. Three eddy current displacement sensors are installed in three holes, uniformly distributed along the edge of the upper specimen. The lower specimen is connected with a connecting disc, and a thrust roller bearing is installed between the connecting disc and the fixed column. The thrust roller bearing can effectively adjust the contact between the upper and lower two specimens, offset the installation error and ensure complete normal contact between the upper and lower two specimens, so that the pressure is evenly distributed throughout the joint surface. At the beginning of the test, the universal testing machine applies a normal load on the upper specimen through the compression rod so that pressure is generated between the upper and lower specimens. At this time, the joint surface is subjected to normal static force, and relative normal displacement will be generated. The eddy current displacement sensor converts the displacement into voltage and transmits it to the data acquisition module for storage. Combined with the loading data of the universal testing machine, the static compression force and relative deformation data of the joint surface are finally collected. In order to ensure the measurement accuracy, the average displacement value of three eddy current displacement sensors is taken as the test data.

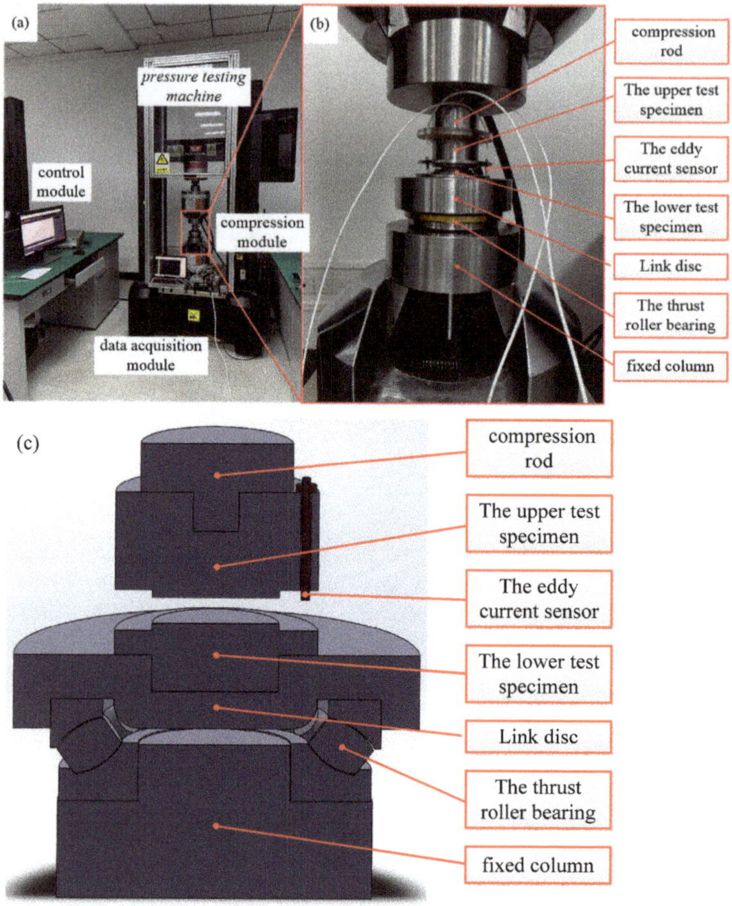

Figure 10. (**a**) The test platform; (**b**) compression module; (**c**) schematic diagram.

5. Results and Discussion

In order to verify the reliability of the proposed model, the simulation stiffness of the proposed model is compared with the experimental stiffness. Roughness is an important parameter that determines the precision of the machined rough surface. The roughness values of four rough surfaces, Sa 0.854 µm, Sa 1.173 µm, Sa 1.391 µm and Sa 1.524 µm, were obtained by grinding in this experiment to verify the validity of the model. Figure 11 shows the comparison result curve between model simulation stiffness and experimental stiffness.

It can be seen from the figure that the experimental value of normal stiffness increases with the increase in normal pressure. When the normal pressure is the same, the normal stiffness increases with the decrease in roughness value. The roughness is Sa 0.854 µm, Sa 1.173 µm, Sa 1.391 µm and Sa 1.524 µm at different scales. When the normal contact pressure reaches 5 MPa, the simulated normal contact stiffness of the model reached 10.57 MPa/µm, 8.519 MPa/µm, 6.81 MPa/µm and 5.814 MPa/µm, respectively. When the asperity is subjected to normal load, the force on the top of the asperity is small, and the deformation is within the pure elastic range. In this case, the deformation of the asperity can be regarded as pure elastic deformation, and the relationship between normal load and stiffness can be regarded as linear. When the pressure increases, the deformation of the asperity gradually enters the elastoplastic deformation stage, the deformation of the asperity will be mainly elastoplastic, and the normal stiffness will gradually increase. When the normal contact load increases to a certain extent, most of the asperity deformation enters the plastic deformation stage, and the contact deformation of asperity is mainly plastic. With the increase in load, the deformation of the asperity finally enters the stage of complete plastic deformation. Therefore, the normal stiffness gradually increases with the increase in normal pressure. From the perspective of surface topography, when the pressure is the same, the density of asperity on the surface with high roughness is lower than that on the surface with small roughness, so the amount of asperity involved in the contact is relatively small, and the relative displacement is larger than that on the surface with small roughness, making the contact stiffness lower than that on the surface with small roughness. Through the above research, it is confirmed that the grinding surface topography affects the contact stiffness between mechanical parts, and the roughness affects the running state between components. By improving the grinding precision and reducing the roughness of the joint surface, the stability of mechanical equipment can be effectively improved.

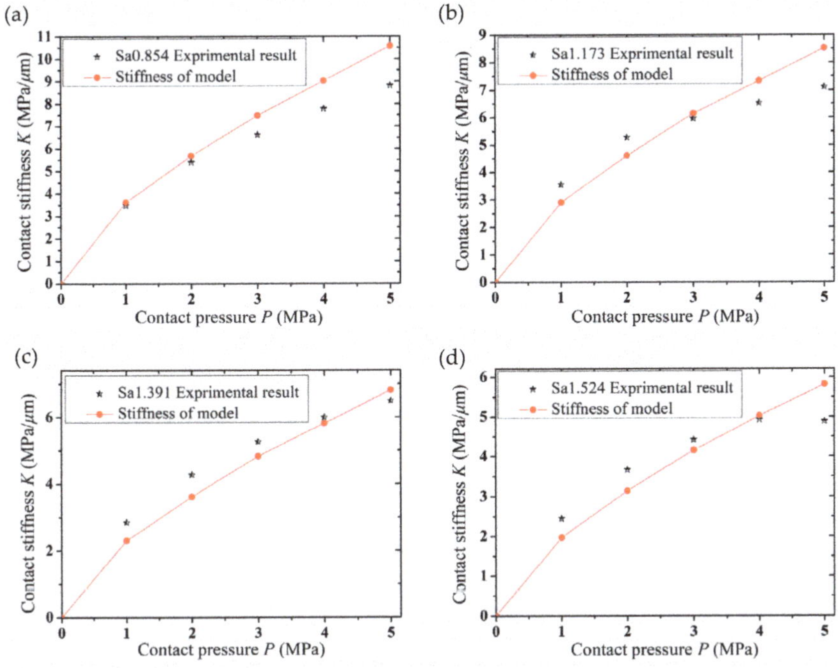

Figure 11. (**a**) Sa 0.854 μm; (**b**) Sa 1.173 μm; (**c**) Sa 1.391 μm; (**d**) Sa 1.524 μm.

It can be seen from the figure that the experimental stiffness curve value and the model stiffness curve value are basically consistent under four different roughness conditions, which proves that the model has certain application value. Firstly, by analyzing the micro-morphology of the grinding surface, this paper refits the profile curve of the asperity and establishes a parabolic cylindrical asperity hypothesis, which is more suitable for the grinding surface. The hypothesis of the model is more suitable for the actual contact state, which is conducive to improving the accuracy of the simulation stiffness value of the model. Secondly, according to the elastic–plastic theory, the contact deformation of the asperity is divided into three different stages, namely, elastic, elastic–plastic and complete plastic deformation stages, which reduces the fitting error of the model. Finally, this is used for the precision design of the measuring device in this experiment, and the testing mechanism is more reasonable and effective. In this experiment, a high-precision measuring instrument was used to improve the measuring accuracy.

In order to verify the accuracy of the new model, the stiffness values of the present model, GW model, CEB model and KE model were compared with those of the experimental test. Figure 12 shows the comparison between the experimental contact stiffness values of Sa 0.854 μm, Sa 1.173 μm, Sa 1.391 μm and Sa 1.524 μm of test samples with different roughness values and the simulated stiffness values of each model.

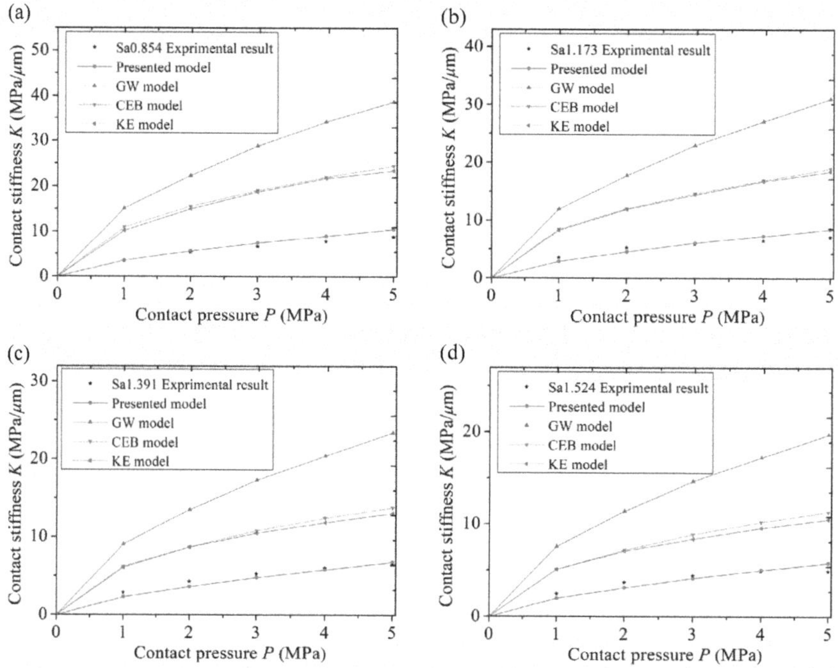

Figure 12. (**a**) Sa 0.854 μm; (**b**) Sa 1.173 μm; (**c**) Sa 1.391 μm; (**d**) Sa 1.524 μm.

As can be seen from Figure 12, the normal contact stiffness value keeps increasing with the increase in normal contact pressure. When the normal pressure reaches 5 MPa, the stiffness error of GW model is the largest, and the stiffness error of CEB model and KE model is relatively low compared with the GW model. The trend of the new model curve is basically the same as that of the experimental curve, and the simulated and experimental stiffness values are close to each other.

Figure 13 shows the relative error histogram of the experimental value of the normal contact stiffness of rough surfaces with different roughness values compared with the stiffness value of this model, GW model, CEB model and KE model, respectively. As can be seen from the figure, the relative error of the stiffness of the GW model is the largest, the relative error of stiffness of the CEB model and KE model is smaller than that of the GW model, and the error of the new model is the smallest.

The comparison and relative error analysis results of the above models show that the GW model has the largest relative error because the GW model simplifies the contact between two rough surfaces to the contact between a spherical asperity and a rigid plane, and assumes that only elastic deformation occurs in the asperity. Without considering the influence of plastic deformation, it does not conform to the real contact mechanism, resulting in a large error. The CEB model assumes that the volume of the asperity remains unchanged during the contact process, the plastic deformation factor is considered, and the error is smaller than that of the GW model. However, the CEB model assumes that the asperity is in contact with the rigid surface as a spherical asperity, and the model topography assumption is not consistent with the real contact situation, resulting in poor model accuracy. The KE model assumes that the rough surface asperity is spherical asperity, and the deformation of asperity is divided into the elastic stage, elastoplastic stage and complete plastic stage. Compared with the GW model, the error is lower. However, the KE model assumes that the contact of the bonding surface is the contact between a spherical asperity and a rigid plate. This deviates from the real contact condition, resulting in error in the model simulation value. The model proposed in this paper establishes a parabolic cylindrical asperity model by fitting the real rough surface data. This is consistent with the real contact of the bonding surface. In this model, the deformation of asperity is divided into the elastic deformation stage, elastoplastic deformation stage and complete plastic deformation stage. Therefore, the simulation results of this model are better than those of the GW, KE and CEB models.

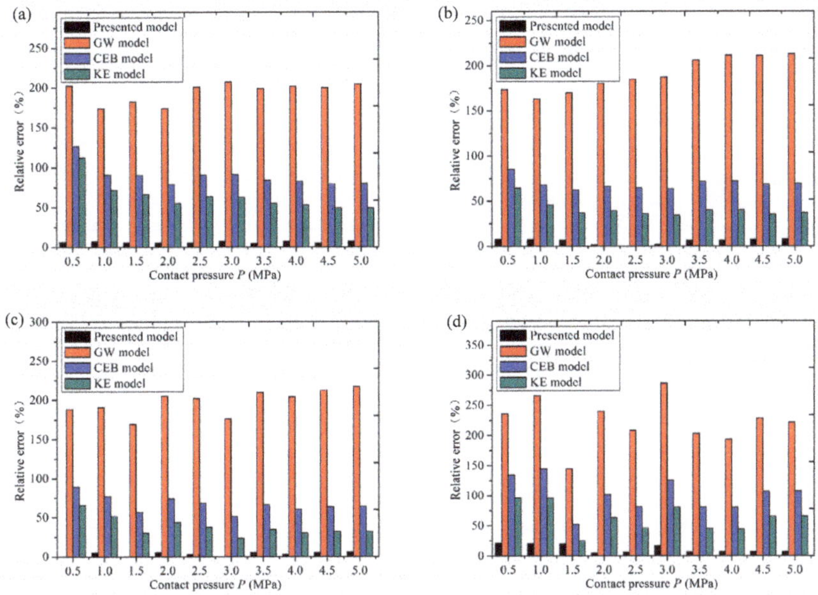

Figure 13. (**a**) Sa 0.854 μm; (**b**) Sa 1.173 μm; (**c**) Sa 1.391 μm; (**d**) Sa 1.524 μm.

6. Conclusions

In this paper, the rough surface grinding data were collected, the profile parameters and topography characteristics of the asperity were analyzed, the rough surface data were fit, the asperity profile was resimulated, and the parabola $y = nx^2 + mx + l$ cylindrical asperity model was established. After analysis and research, the following conclusions were reached:

- The rough surface data of grinding were analyzed, and the distribution height of asperity was fitted with the Gaussian distribution function to confirm that the height of asperity obeyed the Gaussian distribution law.
- The hypothesized model was treated without dimensions and fitted with six plasticity indices to confirm the validity of the model.
- The experimental stiffness values are basically consistent with the fitting stiffness values of the new model, which verifies the reliability and effectiveness of the new model established for the grinding surface.
- When the pressure is the same, the normal stiffness increases with the decrease in the roughness value of the joint surface. Roughness affects the contact stiffness between components. In practice, the stability of mechanical equipment can be effectively

improved by improving the grinding precision and reducing the roughness of the joint surface.

Author Contributions: Conceptualization and methodology, Y.B.; validation, X.J., F.G. and S.S.; writing—original draft preparation and visualization, Y.B. All authors have read and agreed to the published version of the manuscript.

Funding: This work is financially supported by the National Natural Science Foundation of China (No. 52174154), and the National Natural Science Foundation of China (No. 11802035).

Data Availability Statement: Data is contained within the article.

Conflicts of Interest: The authors declare no conflict of interest.

References

1. Greenwood, J.A.; Williamson, J.B.P. Contact of nominally flat surfaces. *Proc. R. Soc. Lond. Ser. A Math. Phys. Sci.* **1966**, *295*, 300–319. [CrossRef]
2. Chang, W.R.; Etsion, I.; Bogy, D.B. An Elastic-Plastic Model for the Contact of Rough Surfaces. *J. Tribol.* **1987**, *109*, 257–263. [CrossRef]
3. Kogut, L.; Etsion, I. Elastic-plastic contact analysis of a sphere and a rigid flat. *J. Appl. Mech.-T Asme* **2002**, *69*, 657–662. [CrossRef]
4. Xie, W.; Liu, C.; Huang, G.; Qin, Z.; Zong, K.; Jiang, D. Trans-scale rough surface contact model based on molecular dynamics method: Simulation, modeling and experimental verification. *Eur. J. Mech. A/Solids* **2023**, *100*, 105021. [CrossRef]
5. Jamshidi, H.; Tavakoli, E.; Ahmadian, H. Modeling polymer-metal frictional interface using multi-asperity contact theory. *Mech. Syst. Signal Process.* **2022**, *164*, 108227. [CrossRef]
6. Kang, H.; Li, Z.-M.; Liu, T.; Zhao, G.; Jing, J.; Yuan, W. A novel multiscale model for contact behavior analysis of rough surfaces with the statistical approach. *Int. J. Mech. Sci.* **2021**, *212*, 106808. [CrossRef]
7. Yuan, B.; Wang, Y.; Sun, W.; Mu, X.; Zhang, C.; Sun, Q. Theoretical and experimental study on interface stiffness measurement of rough surface using improved acoustic model. *Mech. Syst. Signal Process.* **2023**, *186*, 109839. [CrossRef]
8. Li, W.; Zhan, W.; Huang, P. A physics-based model of a dynamic tangential contact system of lap joints with non-Gaussian rough surfaces based on a new Iwan solution. *AIP Adv.* **2020**, *10*, 035207. [CrossRef]
9. Majumdar, A.; Bhushan, B. Role of Fractal Geometry in Roughness Characterization and Contact Mechanics of Surfaces. *J. Tribol.* **1990**, *112*, 205–216. [CrossRef]
10. Xie, Y.; Xiao, Y.; Lv, J.; Zhang, Z.; Zhou, Y.; Xue, Y. Influence of creep on preload relaxation of bolted composite joints: Modeling and numerical simulation. *Compos. Struct.* **2020**, *245*, 112332. [CrossRef]
11. Shen, F.; Li, Y.-H.; Ke, L.-L. On the size distribution of truncation areas for fractal surfaces. *Int. J. Mech. Sci.* **2023**, *237*, 107789. [CrossRef]
12. Yu, X.; Sun, Y.; Wu, S. Analytically decoupling of friction coefficient between mixed lubricated fractal surfaces. *Int. J. Mech. Sci.* **2023**, *255*, 108465. [CrossRef]
13. Chen, H.; Yin, Q.; Dong, G.; Xie, L.; Yin, G. Stiffness model of fixed joint considering self-affinity and elastoplasticity of asperities. *Ind. Lubr. Tribol.* **2019**, *72*, 128–135. [CrossRef]
14. Chen, J.; Liu, D.; Wang, C.; Zhang, W.; Zhu, L. A fractal contact model of rough surfaces considering detailed multi-scale effects. *Tribol. Int.* **2022**, *176*, 107920. [CrossRef]
15. Wang, H.; Jia, P.; Wang, L.; Yun, F.; Wang, G.; Liu, M.; Wang, X. Modeling of the Loading–Unloading Contact of Two Cylindrical Rough Surfaces with Friction. *Appl. Sci.* **2020**, *10*, 742. [CrossRef]
16. Yuan, Y.; Xu, K.; Zhao, K. The Loading–Unloading Model of Contact Between Fractal Rough Surfaces. *Int. J. Precis. Eng. Manuf.* **2020**, *21*, 1047–1063. [CrossRef]
17. Zhou, C.; Wang, H.; Wang, H.; Hu, B. Three-dimensional asperity model of rough surfaces based on valley–peak ratio of the maximum peak. *Meccanica* **2021**, *56*, 711–730. [CrossRef]
18. Shen, F.; Li, Y.-H.; Ke, L.-L. A novel fractal contact model based on size distribution law. *Int. J. Mech. Sci.* **2023**, *249*, 108255. [CrossRef]
19. Yu, Q.; Sun, J.; Ji, Z. Mechanics Analysis of Rough Surface Based on Shoulder-Shoulder Contact. *Appl. Sci.* **2021**, *11*, 8048. [CrossRef]
20. Zhu, L.; Chen, J.; Zhang, Z.; Hong, J. Normal contact stiffness model considering 3D surface topography and actual contact status. *Mech. Sci.* **2021**, *12*, 41–50. [CrossRef]
21. Xie, W.; Liu, C.; Huang, G.; Jiang, D. Numerical and Experimental Study on Rod-Fastened Rotor Dynamics Using Semi-Analytical Elastic-Plastic Model. *J. Eng. Gas Turbines Power* **2022**, *144*, 64501. [CrossRef]
22. Zhang, L.; Wen, J.; Liu, M.; Xing, G. A Revised Continuous Observation Length Model of Rough Contact without Adhesion. *Mathematics* **2022**, *10*, 3764. [CrossRef]
23. Yu, G.; Mao, H.; Jiang, L.; Liu, W.; Valerii, T. Fractal Contact Mechanics Model for the Rough Surface of a Beveloid Gear with Elliptical Asperities. *Appl. Sci.* **2022**, *12*, 4071. [CrossRef]

24. Lv, B.; Han, K.; Wang, Y.; Li, X. Analysis and Experimental Verification of the Sealing Performance of PEM Fuel Cell Based on Fractal Theory. *Fractal Fract.* **2023**, *7*, 401. [CrossRef]
25. Zhang, C.; Yu, W.; Yin, L.; Zeng, Q.; Chen, Z.; Shao, Y. Modeling of normal contact stiffness for surface with machining textures and analysis of its influencing factors. *Int. J. Solids Struct.* **2023**, *262–263*, 112042. [CrossRef]
26. An, Q.; Suo, S.; Lin, F.; Shi, J. A Novel Micro-Contact Stiffness Model for the Grinding Surfaces of Steel Materials Based on Cosine Curve-Shaped Asperities. *Materials* **2019**, *12*, 3561. [CrossRef]
27. Li, L.; Wang, J.; Pei, X.; Chu, W.; Cai, A. A modified elastic contact stiffness model considering the deformation of bulk substrate. *J. Mech. Sci. Technol.* **2020**, *34*, 777–790. [CrossRef]
28. Jiang, K.; Liu, Z.; Zhang, T. Research on contact mechanics of a 3D self-affine surface topography using mesh regeneration technology. In Proceedings of the 2022 2nd International Conference on Algorithms, High Performance Computing and Artificial Intelligence (AHPCAI), Guangzhou, China, 21–23 October 2022; pp. 312–318.
29. Bai, Y.; An, Q.; Suo, S.; Wang, W.; Jia, X. An Analytical Model for the Normal Contact Stiffness of Mechanical Joint Surfaces Based on Parabolic Cylindrical Asperities. *Materials* **2023**, *16*, 1883. [CrossRef]
30. Nuri, K.A.; Halling, J. The normal approach between rough flat surfaces contact. *Wear* **1975**, *32*, 81–93. [CrossRef]

materials MDPI

Article

Innovative Post-Processing for Complex Geometries and Inner Parts of 3D-Printed AlSi10Mg Devices

Martí Calvet *, Anna Domènech , Sergi Vilaró, Toni Meseguer and Lorenzo Bautista

Surface Chemistry Area, Applied Chemistry & Materials Department, Leitat Technological Center, 08225 Terrassa, Barcelona, Spain; adomenech@leitat.org (A.D.); svilaro@leitat.org (S.V.); ameseguer@leitat.org (T.M.); lbautista@leitat.org (L.B.)

* Correspondence: mcalvet@leitat.org

Abstract: A new technology consisting of new and sustainable chemical polishing treatment for aluminum components with complex shapes, such as heat exchangers, manifolds, busbars, aerospace devices, etc., manufactured by Additive Manufacturing (AM) technologies is described in this paper. This technology will contribute to the development of a more efficient manufacturing process driven by AM, reinforcing the main idea of AM, which is based on reducing the amount of material and achieving cost savings through smart and improved designs. The present study shows a significant reduction in the surface roughness of consolidated AlSi10Mg metal parts manufactured by the SLM technique after carrying out the new chemical polishing post-process investigated in this work. Roughness values have been measured by mechanical and optical profilometry. The results obtained demonstrate the effectiveness of the chemical polishing, decreasing the roughness by up to 40%, being a reproducible and repeatable post-process. The presence of smut as solid residues on such types of chemical treatments has been also analyzed with XRF and ICP-MS techniques. The results obtained show that Si and Mg precipitates are removed from the metal surface at the last step of the investigated post-process. The percentages of the elements decrease from 25.0% to 8.09% Si and from 0.86% to 0.42% Mg, achieving the alloy smut-free composition on the metal surface. Tensile strength measurements have shown that the post-process described not only maintains the mechanical properties of the bulk material but, in comparison with non-post-processed parts, a slight improvement is observed with respect to the initial values, Young modulus (61.1 GPa to final 62.2 GPa), yield strength (from 236.8 to 246.7 MPa), and tensile strength (from 371.9 to 382.5 MPa) is observed, suggesting that the post-process has positive impact on the printed metal part.

Keywords: post-process; chemical polishing; roughness; additive manufacturing; 3D printing; selective laser melting

1. Introduction

After iron and steel, aluminum alloys are the most widely used structural materials for automotive, aerospace, and aeronautic applications, among other industries, thanks to their high strength-to-weight ratio, good corrosion resistance, excellent electrical and thermal conductivity, and recyclability. As a result, the additive manufacturing industry has been focused on these areas for many years. The possibility to design and manufacture new optimized geometries using less material has a huge impact on reducing the weight of structural metal parts, which play a big role in fuel consumption in transport applications and, therefore, on cost savings [1,2].

Engineering parts made of aluminum alloys using traditional manufacturing processes includes casting, forging, extrusion, and powder metallurgy. The selected process depends on the final application and the nature of the alloy itself, with 2XXX (Al-Cu or Al-Cu-Mg), 6XXX (Al-Mg-Si), and 7XXX (Al-Zn-Mg) being the commercial Al alloy series that can be strengthened by various heat treatments, while the non-heat-treatable Al series include

1XXX, 3XXX, and 4XXX containing only Si, and 5XXX alloys being those whose strength is increased by mechanical deformation processes [3,4]. These traditional processes do not allow the production of 3D complex structures. Therefore, new manufacturing routes have recently been developed to consolidate complex metal parts, such as Powder Bed Fusion (PBF) [5,6], Directed Energy Deposition (DED) [7], Sheet Lamination (SL) [8,9], Binder Jetting (BJ) [10], and Cold Spray (CS) [11]. Of the powder bed fusion techniques, Selective Laser Melting/Sintering (SLM/SLS) has become the most promising one. This manufacturing technique consists of a powder bed that is melted layer by layer in a protective atmosphere using a high-energy laser beam, and then this molten metal solidifies rapidly [12]. It allows parts to be produced without the need for tooling and appears to be revolutionary, significantly reducing design time and cost. However, conventional aluminum alloys used in traditional manufacturing are neither suitable for laser melting nor for sintering; there are some difficulties overcome such as: a low absorptivity and a high reflectivity to the laser beam, a high cooling rate, an aluminum oxide layer easily formed on aluminum surfaces, the fluctuation of low-melting-point alloying elements as Mg, a high thermal expansion, a high moisture absorption, and poor fluidity due to the formation of powder agglomerates. A detailed study has therefore been undertaken to provide an ideal aluminum powder for SLM technology that meets all these requirements. At present, casting-grade Al-Si alloys with good weldability due to a near eutectic composition, especially AlSi10Mg, have been a key finding to fulfil the conditions required for laser beam manufacturing processes [13–15]. Even so, SLS/SLM technique has a major drawback that has already been mentioned: a poor surface quality that does not meet the requirements for industrial applications.

While the focus for Additive Manufacturing (AM) has been put on designing, post-processing steps of great importance have also resulted. The implementation of AM parts has been more challenging than expected due to the need to remove supports and reduce the high roughness typically obtained after AM processing. After support removal, surface finishing has been identified as the top trend in the AM technological sector for 2022. In fact, nowadays, post-printing processes represent almost 30% of the total project budget related to manufacturing of AM parts [16]. One of the current more challenging demands is not only to improve the AM surface finishing of external surfaces but also to reduce surface roughness of AM inner parts, pipes, and complex shapes. Surface roughness can have a critical effect on properties such as fatigue, strength, friction, and heat transfer [14,17]. It is also of great importance for the joining of metal parts, where excessive roughness can lead to the failure of a compact part or cause corrosion issues [13,18]. In addition, the final finish of a part has a major impact on its cost from the customer's point of view, representing up to 40% of its market value [19].

Therefore, it can be concluded that it is necessary to implement post-processes which provide a substantial improvement of the final finishing for AM metal parts, not only to improve the appearance of the metal parts but also to guarantee a long service life that better fulfils its function.

External parts of AM end products have been successfully surface-treated by various technologies, with abrasive processes being the most widely used. Vibratory finishing has been successfully implemented in the industry because of its simplicity. It does not require a fixed geometry tool. It is clean and demands little manual labor intervention. AM parts with highly complex shapes are placed in a vessel where fine particles of abrasives are used to achieve a smoothing of the surface by the impacts caused by vibration [20–24]. Compressive residual stresses can occur during the mechanical process and may eventually form the so-called Beilby layer, which contains a mixture of oxides of the base metal and the compounds used for polishing; therefore, the physicochemical properties of the surface layer obtained by mechanical polishing can be different depending on the nature of the underlying alloy, causing mechanical stress which, under certain conditions, can give rise to processes of corrosion [25]. Mechanical post-processes such as vibratory finishing present a potential clogging of abrasive particles particularly for AM inner parts and complex geometries. Surface post-processing of this type of AM parts with complex geometries is still a challenge.

In addition to mechanical finishing, chemical treatments such as alkaline etching, acid pickling, and electropolishing can be used to tackle unevenness of the metallic surface, without altering the bulk properties of the workpiece. Aluminum pitting can be used for surface processing where localized corrosion of the metal is used as an advantage to create uniform surface roughness [26]. Electropolishing is an alternative surface treatment in which the metal to be polished acts as an anode in an electrolytic cell and dissolves. The micro- and macroprojections, or high points of the rough surface, just as regions with burrs, are areas of higher current density than the rest of the surface and dissolve at a faster rate, giving rise to smoother, more level, and/or burred surface. This is why electropolishing is mainly used to achieve mirror finishing quality. The difficulty lies in polishing conducts, cavities, pipes, and those regions shielded by a complex geometry, which are common for AM parts [27].

An alternative post-processing method to improve the surface quality of 3D-printed samples is chemical polishing. In contrast to electropolishing, the lack of current during this process results in a uniform attack all over the surface, avoiding the intrinsic shortcomings of electropolishing on treating cavities and areas with higher current density [28]. Nevertheless, the chemical polishing described to date is very aggressive and uses strong acids to reduce the surface roughness of the object to be polished. Polishing in acid medium is carried out at high temperatures and is not suitable because too much material is removed from the part, and surface defects (pitting) are then created [29].

On the other hand, all the chemical methods described above must face the formation of a smut layer as a by-product of the process. This smut layer for aluminum alloys is primarily composed of an amorphous aluminum hydroxide that can have a significant impact on the resulting morphology of the final surface [26]. In addition, in magnesium- and silicon-containing aluminum alloys (AlSi10Mg being the ones used for 3D printing), coarse constituent Mg_2Si intermetallic particles and fine Mg_2Si precipitates that are commonly formed during casting and processing of aluminum alloys can further interfere in the polishing step by forming $Mg(OH)_2$ and $SiO_2 \cdot xH_2O$ sublayers that are particularly difficult to eliminate [30]. As such, removing smut is a key step in order to obtain a functional polished aluminum piece.

According to the needs outlined above, Leitat has recently developed an innovative post-processing method that is particularly useful for complex geometries and inner parts of aluminum 3D-printed metal parts obtained by means of additive manufacturing [19]. This method combines a series of steps which end up with a smut-free polished metal part. First, the metal surface is activated and sensitized to be prepared for the following steps. Secondly, the roughness is faced and reduced in a controlled manner by selective etching of the peaks. Next, the smut formed during the overall process is removed, and, finally, the methodology is repeated n times to achieve the roughness required for the end case. All the steps are carried out by immersion in aqueous solutions to guarantee a homogeneous attack to the metal part. With less processing time, less residues, and lower temperatures, this process also contributes to reducing the environmental footprint.

In this study, a novel post-processing method to face complex shapes and inner parts of AlSi10Mg devices consolidated by SLM technique is presented. It is an environmentally friendly post-process involving a low risk for operators and easy automation, that is presented as an alternative to mechanical, electropolishing, and other chemical polishing processes [31–33]. Promising results have been obtained in this work, which allow us to envisage a successful future implementation at the industrial level.

2. Materials and Methods

Aluminum samples were obtained from selective laser melting technique (SLM) using a Renishaw AM 500 M metal additive manufacturing machine (Renishaw Iberica S.A.U., Barcelona, Spain), spot size 80 μm, layer thickness 60 μm, scanning rotation 67° and argon atmosphere. Al powder was obtained from Renishaw AlSi10Mg commercial powder (Renishaw Ibérica S.A.U., Barcelona, Spain)with the following composition (in wt %) 9.00 to 11.00 Si, 0.25 to 0.45 Mg, <0.25 Fe, <0.20 N, <0.2 O, <0.15 Ti, <0.10 Zn, <0.10 Mn, <0.05 Ni, <0.05 Cu, <0.02 Pb, <0.02 Sn, Al balance, with a layer thickness of 25 μm and 400 W of laser power [34].

2.1. *Surface Treatment Experimental Part*

For this study, several replicates of AlSi10Mg samples were consolidated by SLM to ensure the reproducibility and repeatability of the test. In order to test the efficiency of the post-treatment, a preliminary complex geometry was designed with two pipes and edges and corners to check that they are not rounded during the process (Figure 1).

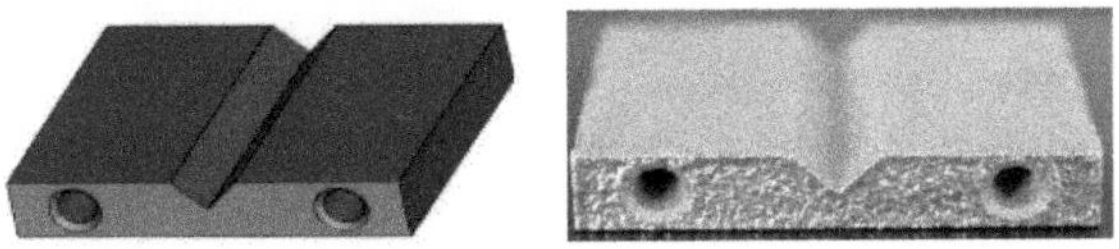

Figure 1. AlSi10Mg metal parts design for post-processing tests; on the left is a schematic design, on the right is a picture of the consolidated metal part.

After checking the initial roughness of the samples described previously, the samples were post-processed following the chemical polishing method described in WO 2021/094641 [19]. Activation of the surface is performed under alkaline conditions using sodium hydroxide; then, two polishing steps are performed also by means of an alkaline solution combining sodium hydroxide (1.5%) and potassium carbonate (25%), and finally, the desmutting step is performed using acetic acid (5%). The overall process is performed at room temperature and can be repeated by *n* cycles depending on the initial roughness of the metal parts and the final requirements. The schematic process is represented in Figure 2.

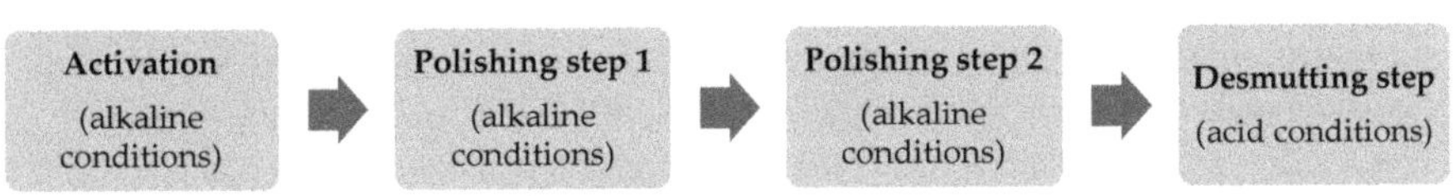

Figure 2. Schematic representation of steps followed on polishing AlSi10Mg using patent WO 2021/094641 [19].

The procedure illustrated in Figure 2 was repeated 8 times according to the initial values from the aluminum samples and the knowledge acquired working with this alloy in previous projects.

A KLA-TENCOR ALPHA STEP D 600 3D mechanical profilometer (KLA, Milpitas, CA, USA) was used to determinate the superficial roughness of planar samples in terms of arithmetical mean height of the surface (Sa), maximum height of the surface (Sz), maximum height of valleys (Sv), maximum height of peaks (Sp), kurtosis of height distribution (Sku), skewness of height distribution (Ssk), and root mean square height of the surface (Sq) following ISO 25178. The probe used is 5 μm in diameter, and the scan parameters were: 0.10 mm/s speed and stylus force of 10.0 mg. 3D mapping was performed with 0.025 mm of spacing between profiles. In order to evaluate superficial roughness of nonplanar samples, a JENOPTIK Waveline W10 optic profilometer (JENOPTIK Industrial Metrology Germany GmbH, Villingen-Schwenningen, Germany) following ISO 4287:1999/AC/AC1:2010 [35,36] standard was used. Surface roughness arithmetical mean height (Ra), maximum height of profile (Rz), maximum profile peak height (Rp), and maximum profile valley depth (Rv) values [37] were obtained at the evaluation conditions: λc 0.80 mm, lp 8.000 mm, lr 0.8 mm, lw 0.8 mm, traverse length 8.00 mm, speed 0.50 mm/s, measuring range 400 μm, probe type T3.

The elemental surface composition of the samples in different polishing stages was studied using the X-ray spectrophotometer NITON XL3t (Thermo Fisher Scientific, Waltham, MA, USA). This handheld equipment allows nondestructive testing of a sample by processing the spectrum obtained from X-ray photons and converting the information into elemental analysis. This technique quantifies the amount percent of alloying elements by analyzing a specific area, in this case a 3 mm diameter spot. Three different samples were evaluated. The first one consisted of a raw sample without any treatment. The second one was a smut-containing surface sample (Activation + Step 1), and the last one was a sample treated using a desmutting step (Activation + Step 1 + Step 2 + Desmutting step).

Elemental composition of the smut powder deposited on the aluminum surface formed in the alkaline polishing step was analyzed by Induced Coupled Plasma analysis using Agilent 8900 triple quadrupole ICP-MS (ICP-QQQ) (Agilent, Santa Clara, CA, USA).

2.2. Mechanical Properties Experimental Part

Samples were manufactured with a circular cross section and a central part standardized with a constant section to submit them to tensile testing. The shape of the ending sections of the specimen facilitates fastening into the universal machine jaws (Figure 3).

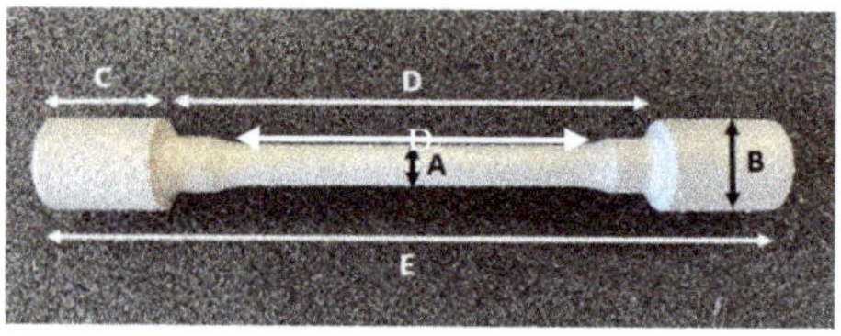

Figure 3. Specimen measurements. A—5.08 mm; B—10.00 mm; C—12.25 mm; D—51.47 mm; D_2—36.04 mm; E—76.02 mm.

As these consolidated parts prepared for tensile strength test are not planar, to characterize the roughness of such devices it was necessary to measure it by JENOPTIK Waveline W10 optic profilometer (JENOPTIK Industrial Metrology Germany GmbH, Villingen-Schwenningen, Germany), and 4 sections were established to check the effect before and after the post-treatment in the same area (Figure 4).

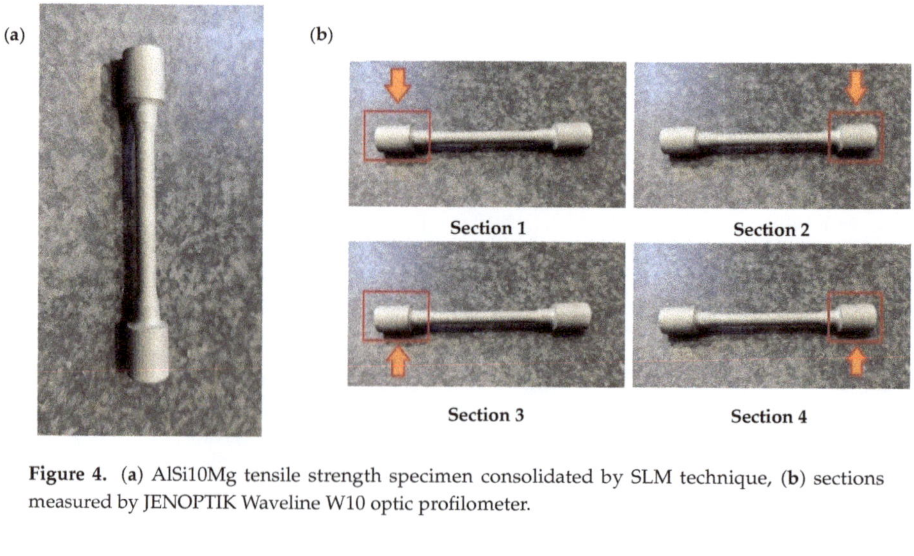

Figure 4. (**a**) AlSi10Mg tensile strength specimen consolidated by SLM technique, (**b**) sections measured by JENOPTIK Waveline W10 optic profilometer.

The samples were post-treated following the patented process described previously. The roughness was checked again using the same equipment.

Tensile strength of consolidated metal parts was evaluated following UNE EN-ISO 6892-1:2020 [38]. Prior to conducting the test of the specimen, it was ensured that the testing machine was aligned according to the standard ASTM E-1012-19 [39]. The test speed in the elastic zone was 0.0042 mm/mm/min in strain control. The test speed in the plastic zone was 4.32 mm/min in displacement control until fracture. The test was conducted at a room temperature of 23 ± 3 °C and relative humidity of 50 ± 10%. Yield strength was obtained using the offset method according to section 13 of UNE-EN ISO 6892-1:2020 [38] at an offset of 0.2%. The information about the equipment is summarized in Table 1.

Table 1. Equipment used in tensile strength test.

Equipment	Model
Universal Testing Machine "100 kN"	MTS 370.25
Extensometer	MTS 632.11F-90
Digital Calliper	MITUTOYO CD-15APX

3. Results and Discussion

The obtained-as-designed SLM consolidated metal parts with pipes and corners, the appearance of an aluminum sample before/after the post-treatment, and a sample which has not been treated with the desmutting step are shown in Figure 5.

Figure 5. From left to right, a consolidated aluminum metal part as-printed, smut-treated, and a desmutted sample.

In this study, an innovative chemical post-treatment process is described which respects complex geometries, sharps, and performs in inner parts and cavities. Figure 5 shows how edges and corners are maintained, and the diameter of the pipe is slightly increased from 2.85 mm to 3.01 mm after the post-treatment. When the desmutting step is not performed, the sample becomes almost black, and the chemical post-treatment is inefficient. The desmutting step provides brightness thanks to the reduction in the surface roughness. The results of initial roughness measured with a KLA-TENCOR ALPHA STEP D 600 3D mechanical profilometer (KLA, Milpitas, CA, USA) are represented in Table 2.

As can be observed, in order to acquire accurate information on the metal surface roughness, a 3 × 3 mm square area was analyzed. The initial roughness of the consolidated aluminum parts was 20.2 µm Sa, and 162 µm Sz, these two parameters being the ones which provide the most valuable information.

Table 2. Roughness results in terms of quadratic area of AlSi10Mg consolidated metal parts, and 2D/3D surface profiles.

Parameters	Sq	Ssk	Sku	Sp	Sv	Sz	Sa
Values (μm)	25.5	−0.13	3.00	79.4	82.2	162	20.2

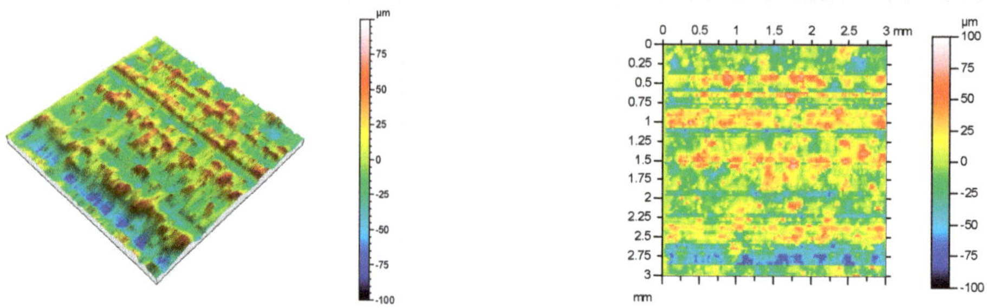

After this initial analysis, the obtained roughness after performing the post-treatment described in Section 2, and repeated 8 times, is shown in Table 3.

Table 3. Roughness results in terms of quadratic area of AlSi10Mg post-processed metal parts, and 2D/3D surface profiles.

Parameters	Sq	Ssk	Sku	Sp	Sv	Sz	Sa
Values (μm)	13.9	−0.287	3.46	43.8	66.1	110	10.9

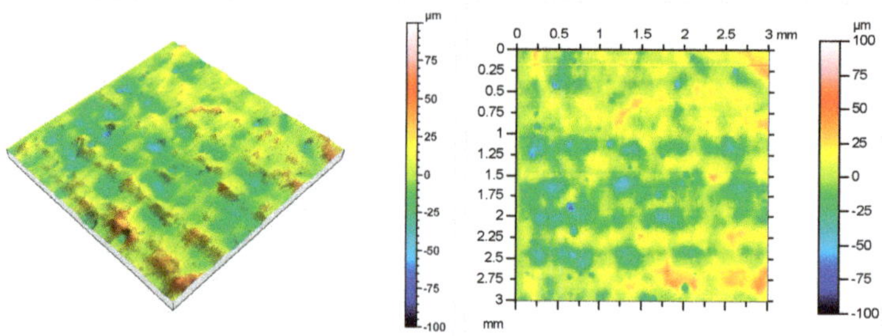

After the post-treatment, a significant improvement on AlSi10Mg superficial roughness is demonstrated compared to the initial values. A decrease in Sa, from 20.2 μm (Table 2) to 10.9 μm (Table 3), and from 162 μm (Table 2) to 110 μm (Table 3) for Sz has been observed, meaning a reduction of 46% and 32%, respectively. A matte and smut-free finishing has been obtained after carrying out our novel chemical finishing post-processing on AlSi10Mg metal parts. Common Al/Si/Mg oxide precipitates generated typically during chemical polishing of Al/Si alloys and deposited usually onto their surfaces have not been observed.

Nevertheless, to confirm that our novel chemical polishing post-process removes the smut upper layer, an exhaustive study has been performed on the chemical composition of an aluminum sample on different stages of the polishing method. AlSi10Mg metal parts derived from different stages of chemical polishing method were analyzed using XRF technique. The results obtained are presented in Table 4.

Table 4. Element sorting (%) by XRF technique on different aluminum samples.

Element (%)	Raw Sample	Sample with Smut	Desmutted Sample
Si	8.684 ± 0.115	25.042 ± 0.179	8.086 ± 0.108
Mg	0.587 ± 0.174	0.858 ± 0.173	0.421 ± 0.201
Fe	0.244 ± 0.007	0.250 ± 0.007	0.222 ± 0.007
Zn	0.005 ± 0.001	0.005 ± 0.001	0.005 ± 0.001
Ni	0.012 ± 0.003	0.014 ± 0.003	0.012 ± 0.003
Cu	0.004 ± 0.001	0.005 ± 0.001	0.004 ± 0.001
Al	90.095 ± 0.144	73.512 ± 0.127	90.940 ± 0.139

The findings revealed that the, as expected, major constituents in AlSi10Mg alloy are aluminum, silicon, and magnesium. Once the piece was treated with alkaline conditions and smut appeared (giving a black appearance on the metal surface), XRF analysis showed an increase of Si (from 8.684 to 25.042%) and Mg (from 0.587 to 0.858%). The black powder formed, then, was a result of a deposition of $Mg(OH)_2$ and $SiO_2{\cdot}xH_2O$ on the aluminum surface [26]. Al detection decreases from 90.095% of the raw sample, which corresponds to mainly aluminum oxide, to 73.512% measured on the sample with smut. This is because smut layer is deposited on top of the metal surface and interferes with Al identification. Once smut is removed, Al is again revealed (90.940%).

Finally, smut formed in the alkaline treatment step was removed from the AlSi10Mg samples using the acid desmutting step liquid and in order to submit it to analysis. The presence of elements was measured by ICP-MS using a semiquantitative method ("Semi-Quant") (Agilent 8900 triple quadrupole ICP-MS (ICP-QQQ),Agilent, Santa Clara, CA, USA) in which both the composition and approximate (typically better than ±30%) concentration of a sample can be measured in a single analysis without any previous knowledge of the sample [40]. The results are shown in Table 5.

Table 5. Approximate elemental composition of smut obtained in the alkaline treatment step.

Element	Si	Mg	Fe	Al
Composition (%)	83	1	2	14

This table shows that the main component of the black powder is Si (83%), so this explains that most of the product formed on the alkaline treatment is a component formed by silicon ($SiO_2{\cdot}xH_2O$). These results are correlated with the XRF results shown in Table 4 that also explain that the main element present in smut is Si (25.042%). The subproducts formed may be other oxides from magnesium and iron (elements present in the alloy). Aluminum which comes from the metal piece is also detected because the smut powder extraction is performed by etching the piece to undergo ICP-MS analysis.

Illustrated also in Table 4, Si values were reduced to initial numbers when the sample was treated with the desmutting step (from 25.042 to 8.086%), which was related with the elimination of the oxides formed during alkaline treatment. Furthermore, Mg percentage is reduced (from 0.858% to 0.421%), meaning that desmutting treatment not only eliminates silicon oxides formed but also magnesium ones.

3.1. Sample Preparation for Tensile Strength Test

As the investigated post-process consists of a dip treatment, the liquid could seep through the porosity given in metal parts when manufactured by SLM manufacturing routes. To prove that the novel chemical polishing method does not have any influence on bulk properties, and so that the chemical components from different formulations do not further attack the interior of the metal, the samples manufactured with a circular cross section were post-treated and submitted to tensile testing.

The samples were post-treated following the patented process described previously, obtaining a modification of 15% from the initial measurements of the samples (Table 6); Section A goes from 5.08 to 4.25 cm.

Table 6. Tensile strength samples section before and after polishing. Sections corresponding to Figure 3.

(cm)	Section A	Section B	Section C	Section D	Section D2	Section E
Before polishing	5.08	10.00	12.25	51.47	36.04	76.02
After polishing	4.25	9.80	12.18	51.41	35.96	75.96

Also, samples were weighed before and after the treatment. The initial average weight corresponding to the samples is 7.55 g and the average final weight is 6.96 g, thus obtaining an average mass decrease of 7.77%. The obtained roughness results for consolidated samples before (Table 7) and after the post-processing (Table 8) are represented below.

Table 7. Optic profilometer average roughness measurement values obtained from 3 replicates before post-processing. (**a**) Average results per section, (**b**) average roughness results of the samples.

(a)	Section 1	Section 2	Section 3	Section 4
	Average	Average	Average	Average
Ra (μm)	15.52 ± 1.69	14.33 ± 1.23	13.48 ± 0.82	12.60 ± 0.81
Rz (μm)	72.14 ± 5.80	66.48 ± 4.12	67.68 ± 1.96	63.34 ± 4.66
Rp (μm)	36.00 ± 4.46	32.69 ± 1.95	35.90 ± 1.58	30.99 ± 3.63
Rv (μm)	36.13 ± 1.61	33.79 ± 3.09	31.78 ± 0.87	32.35 ± 1.06
(b)	**Global Average**			
Ra (μm)	13.98 ± 0.65			
Rz (μm)	67.41 ± 2.26			
Rp (μm)	33.89 ± 1.61			
Rv (μm)	33.51 ± 1.02			

Table 8. Optic profilometer average roughness measurement values obtained from 3 replicates after post-processing. (**a**) Average results per section, (**b**) average roughness results of the samples.

(a)	Section 1	Section 2	Section 3	Section 4
	Average	Average	Average	Average
Ra (μm)	8.07 ± 0.18	9.08 ± 0.37	7.84 ± 0.24	8.52 ± 0.80
Rz (μm)	41.86 ± 2.62	43.26 ± 3.54	41.33 ± 3.19	41.58 ± 2.54
Rp (μm)	21.08 ± 2.17	22.56 ± 2.89	22.05 ± 1.72	22.61 ± 1.35
Rv (μm)	20.79 ± 0.44	20.70 ± 0.65	19.28 ± 1.47	18.97 ± 1.35
(b)	**Global Average**			
Ra (μm)	8.38 ± 0.31			
Rz (μm)	42.01 ± 1.28			
Rp (μm)	22.07 ± 0.81			
Rv (μm)	19.93 ± 0.65			

The roughness results obtained after the samples were post-processed, as can be observed in Table 8, show an improvement in roughness values (Ra goes from average 13.98 to 9.38 μm and Rz from 67.41 to 42.01 μm), also represented in terms of percent reduction in Table 9 (reduction of 40% in Ra and 38% in Rz). Repeatability is demonstrated by means of standard deviation, analyzing Ra value in average reduction is only 0.31 μm and Rz standard deviation is higher (1.28 μm) due to the fact that the height of peaks and valleys is dissimilar from the beginning. Additionally, peaks and valleys heights, represented by Rp and Rv, respectively (total average reduction of 35 and 41%), also decrease compared to the initial metal part.

Table 9. Total average reduction of roughness values after AlSi10Mg parts post-treatment.

	Roughness Reduction (μm)	Reduction (%)
Ra	5.6 ± 0.31	40
Rz	35.4 ± 1.28	38
Rp	11.8 ± 0.81	35
Rv	13.6 ± 0.65	41

3.2. *Tensile Strength Test According to UNE EN-ISO 6892-1:2020 [38]*

Once the samples were polished following the novel chemical polishing process, tensile strength tests were evaluated according to UNE EN-ISO 6892-1:2020 [38] and compared with nontreated samples. The results are shown in Table 10, and the elongation curve of each sample tested is represented in Appendix A.

Table 10. Numeric tensile strength results.

Treatment	Specimen	Young Modulus (Gpa)	Yield Strength Re (Mpa)	Tensile Strength Rm (Mpa)
Nontreated	Al 1	62.7	238	373
	Al 2	60.0	237	364
	Al 3	60.4	236	378
	Average	**61.1 ± 1.5**	**236.8 ± 0.8**	**371.9 ± 7.1**
Treated	Al 1	59.3	242	388
	Al 2	62.9	249	378
	Al 3	64.5	249	382
	Average	**62.2 ± 2.6**	**246.7 ± 4.0**	**382.5 ± 5.0**

The results obtained indicate that final mechanical properties of chemically polished AlSi10Mg parts are maintained from initial values corresponding to nontreated AlSi10Mg parts (Table 10). Young modulus, yield strength, and tensile strength provide data on mechanical behavior directly. The first one, also known as elastic modulus, provides information about the stiffness of the material. There are no differences between nontreated and treated specimens' values, which go from 61.1 ± 1.5 to 62.2 ± 2.6 GPa.

However, despite the differences not being very significant, the comparison between yield strength and tensile strength shows a trend which provides information about other potential advantages of polishing apart from finishing (i.e., reduction in surface roughness) and aesthetics (i.e., optical aspect of AM part surfaces). Treated samples have major values of both properties, which are shown in Table 10: yield strength is increased by 4% (average values go from 236.8 to 246.7 MPa), while tensile strength by about 3% on average (from 371.9 to 382.5 MPa). During the manufacturing process, parameters such as printing direction, laser parameters, and initial powder properties can cause cracks. These cracks increase the risk of mechanical failures and premature fractures during the operation of parts or during a mechanical test [41]. An inefficient post-process could have increased the depth of these microcracks leading to a negative impact in yield strength and tensile strength. On the contrary, the obtained results demonstrate that the present polishing process is responsible for mitigating this issue by removing the upper layer of the metal part, and thus these inclusions and microcracks present on the metal surface serve as failure initiation points. Parts produced from SLS technique are built up in a layer-by-layer manner and their properties depend on the direction in which they are deposited. Specimens of this report are produced on Z-direction (vertical) and cause samples to be more sensitive to crack initiation due to the presence of more borderline porosity [42].

4. Conclusions

The surface roughness of AlSi10Mg metal parts manufactured by SLM technique can be notably decreased by utilizing a novel environmentally friendly chemical polishing post-process. The final roughness requirements are achieved through a loop treatment that blends alkaline and acidic baths. The roughness parameters can be decreased up to 40% thanks to the described chemical polishing. The surface toughness values have been confirmed through an extensive study of the surface roughness by means of two different techniques: mechanical and optical profilometry. The process has demonstrated repeatability and reproducibility, with multiple samples tested, as is described in the report. Moreover, the new chemical polishing method's desmutting efficiency has been evaluated. XRF analysis demonstrates that AlSi10Mg alloy maintains its alloy elements after the post-treatment, and so, the typical precipitates from smut, which are Si and Mg, are removed thanks to the desmutting step. This methodology has been further supported by the ICP-MS technique that analyzed the removed outer layer. Finally, the tensile strength of consolidated pre- and post-treated AlSi10Mg metal components revealed that the new chemical surface treatment introduced in this study does not have detrimental effects on the bulk material. In fact, there has been a slight improvement due to a reduction in defects found on the metal surface, which act as failure initiation points. It is evident that this new chemical treatment can be beneficial for enhancing the overall quality of AlSi10Mg metal parts.

A technique has been developed to polish the surface of AlSi10Mg, which improves superficial roughness without compromising its mechanical properties. This will help to consolidate the implementation of additive manufacturing technologies in the industry, by providing not only an aesthetic finish but also the surface functionalities and overall performance of the printed device will be enhanced. The novel methodology outlined in this paper is ideal for use on printed parts with complex geometries, cavities, and pipes. This feature sets it apart from conventional polishing processes (both mechanical and electrochemical). The development of post-processing solutions unlocks fresh applications that were previously unattainable, spanning from healthcare to industrial machinery.

Author Contributions: Conceptualization, M.C. and T.M.; methodology, M.C. and A.D.; validation, M.C.; investigation, M.C., A.D. and S.V.; data curation, T.M.; writing—original draft preparation, M.C. and A.D.; writing—review and editing, L.B., M.C, T.M., A.D. and S.V.; visualization, L.B.; supervision, M.C., T.M. and L.B.; project administration, T.M. and L.B. All authors have read and agreed to the published version of the manuscript.

Funding: This research has been carried out thanks to the funding from the Ministry of Science and Innovation of Spain within the framework of the "3D Printing HUB project".

GOBIERNO DE ESPAÑA
MINISTERIO DE CIENCIA E INNOVACIÓN

Institutional Review Board Statement: Not applicable.

Informed Consent Statement: Not applicable.

Data Availability Statement: Not applicable.

Acknowledgments: The authors acknowledge Applus Laboratories for their support in the mechanical analyses.

Conflicts of Interest: The authors declare no conflict of interest.

Appendix A

(a)

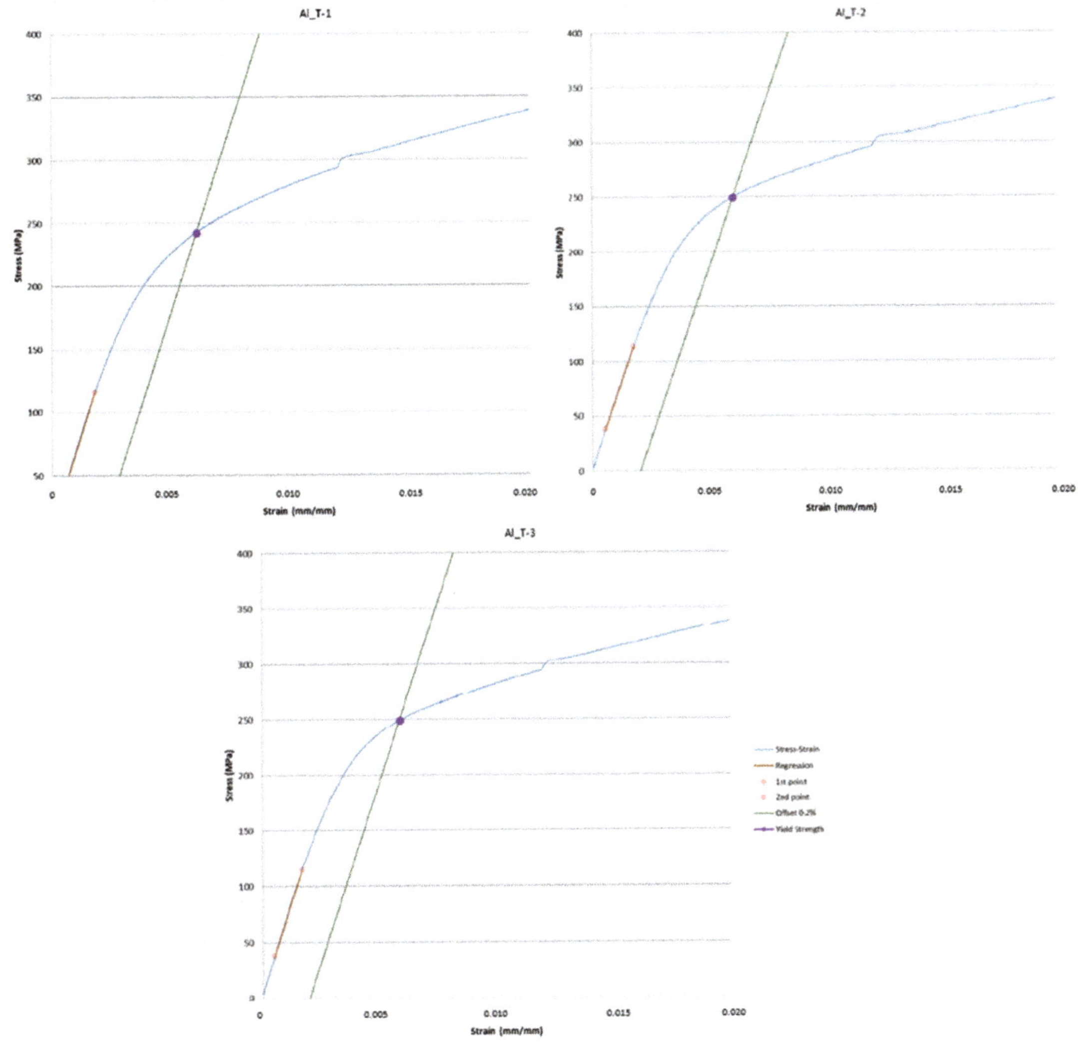

Figure A1. *Cont.*

(b)

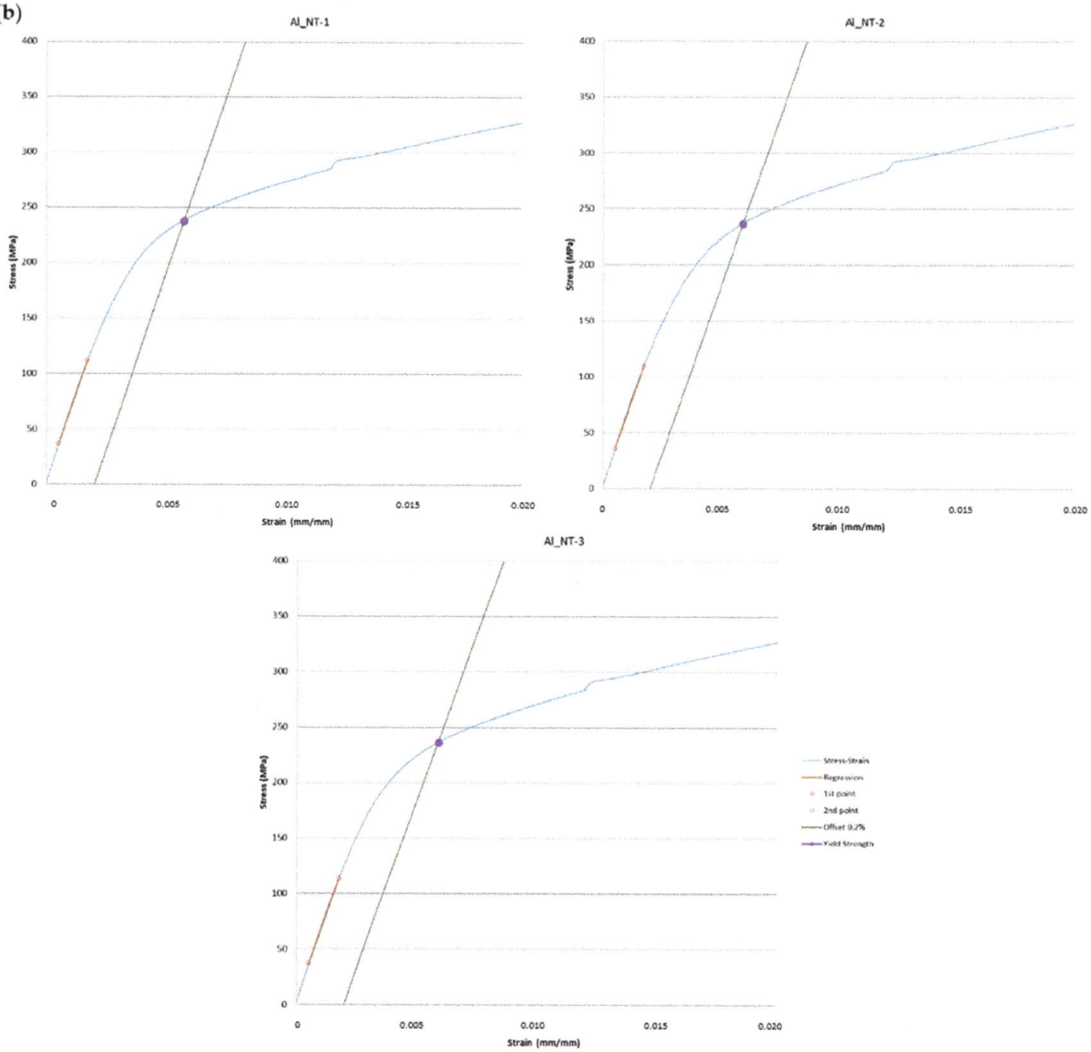

Figure A1. Elongation curves from the stress–strain tests corresponding to the treated aluminum samples (**a**) compared with the nontreated ones (**b**).

References

1. Zhang, J.; Song, B.; Qinsong, W.; Bourell, D.; Yusheng, S. A Review of Selective Laser Melting of Aluminum Alloys: Processing, Microstructure, Property and Developing Trends. *J. Mater. Sci. Technol.* **2019**, *35*, 270–284. [CrossRef]
2. Olakanmi, E.O.; Cochrane, R.F.; Dalgarno, K.W. A Review on Selective Laser Sintering/Melting (SLS/SLM) of Aluminium Alloy Powders: Processing, Microstructure, and Properties. *Prog. Mater. Sci.* **2015**, *74*, 401–477. [CrossRef]
3. Mondolfo, L.F. *Aluminum Alloys: Structure and Properties*; Elsevier: Amsterdam, The Netherlands, 2013; ISBN 978-1-4831-4482-5.
4. Polmear, I.; StJohn, D.; Nie, J.-F.; Qian, M. *Light Alloys: Metallurgy of the Light Metals*; Butterworth-Heinemann: Oxford, UK, 2017; ISBN 978-0-08-099430-7.
5. Murr, L.E. Metallurgy Principles Applied to Powder Bed Fusion 3D Printing/Additive Manufacturing of Personalized and Optimized Metal and Alloy Biomedical Implants: An Overview. *J. Mater. Res. Technol.* **2020**, *9*, 1087–1103. [CrossRef]
6. Mostafaei, A.; Zhao, C.; He, Y.; Reza Ghiaasiaan, S.; Shi, B.; Shao, S.; Shamsaei, N.; Wu, Z.; Kouraytem, N.; Sun, T.; et al. Defects and Anomalies in Powder Bed Fusion Metal Additive Manufacturing. *Curr. Opin. Solid State Mater. Sci.* **2022**, *26*, 100974. [CrossRef]

7. Liu, W.; Wei, H.; Liu, A.; Zhang, Y. Multi-Index Co-Evaluation of Metal Laser Direct Deposition: An Investigation of Energy Input Effect on Energy Efficiency and Mechanical Properties of 316l Parts. *J. Manuf. Process.* **2022**, *76*, 277–290. [CrossRef]
8. Lee, J.-Y.; An, J.; Chua, C.K. Fundamentals and Applications of 3D Printing for Novel Materials. *Appl. Mater. Today* **2017**, *7*, 120–133. [CrossRef]
9. Gao, W.; Zhang, Y.; Ramanujan, D.; Ramani, K.; Chen, Y.; Williams, C.B.; Wang, C.C.L.; Shin, Y.C.; Zhang, S.; Zavattieri, P.D. The Status, Challenges, and Future of Additive Manufacturing in Engineering. *Comput.-Aided Des.* **2015**, *69*, 65–89. [CrossRef]
10. Shad, A.; Stache, R.; Rütjes, A. Effects of Fumed Silica Flow Aids on Flowability and Packing of Metal Powders Used in Binder-Jetting Additive Manufacturing Process. *Mater. Des.* **2021**, *212*, 110253. [CrossRef]
11. Prashar, G.; Vasudev, H. A Comprehensive Review on Sustainable Cold Spray Additive Manufacturing: State of the Art, Challenges and Future Challenges. *J. Clean. Prod.* **2021**, *310*, 127606. [CrossRef]
12. Liu, Y.; Yang, Y.; Mai, S.; Wang, D.; Song, C. Investigation into Spatter Behavior during Selective Laser Melting of AISI 316L Stainless Steel Powder. *Mater. Des.* **2015**, *87*, 797–806. [CrossRef]
13. Leon, A.; Aghion, E. Effect of Surface Roughness on Corrosion Fatigue Performance of AlSi10Mg Alloy Produced by Selective Laser Melting (SLM). *Mater. Charact.* **2017**, *131*, 188–194. [CrossRef]
14. Brandl, E.; Heckenberger, U.; Holzinger, V.; Buchbinder, D. Additive Manufactured AlSi10Mg Samples Using Selective Laser Melting (SLM): Microstructure, High Cycle Fatigue, and Fracture Behavior. *Mater. Des.* **2012**, *34*, 159–169. [CrossRef]
15. Altıparmak, S.C.; Yardley, V.A.; Shi, Z.; Lin, J. Challenges in Additive Manufacturing of High-Strength Aluminium Alloys and Current Developments in Hybrid Additive Manufacturing. *Int. J. Light. Mater. Manuf.* **2020**, *4*, 246–261. [CrossRef]
16. Clancy, K. Annual Additive Manufacturing 3D Post-Printing Survey Results. *Javelin 3d Solut.* 2022. Available online: https://www.javelin-tech.com/3d/annual-additive-post-printing-survey-trends-report/ (accessed on 23 March 2023).
17. Malakizadi, A.; Mallipeddi, D.; Dadbakhsh, S.; M'Saoubi, R.; Ktajnik, P. Post-Processing of Additively Manufactured Metallic Alloys—A Review. *Int. J. Mach. Tools Manuf.* **2022**, *179*, 103908. [CrossRef]
18. Hu, Z.; Wan, L.; Lü, S.; Zhu, P.; Wu, S. Research on the Microstructure, Fatigue and Corrosion Behavior of Permanent Mold and Die Cast Aluminum Alloy. *Mater. Des.* **2014**, *55*, 353–360. [CrossRef]
19. Meseguer, T.; Soldi, L.; Calvet, M.; Domenech, A.; Dominguez, E. Method for Polishing Parts Made of Aluminum. WO 2021/094641 A1, 15/11/2019.
20. Sood, A.; Mullany, B. Advanced Surface Analysis to Identify Media-Workpiece Contact Modes in a Vibratory Finishing Processes. *Procedia Manuf.* **2021**, *53*, 155–161. [CrossRef]
21. Metal Parts Vibrating and Dry Vibration Polishing. Available online: https://www.coniex.com/en/vibration/ (accessed on 23 March 2023).
22. Rotary Vibrators—Rösler Oberflächentechnik GmbH. Available online: https://uk.rosler.com/uk-en/products/mass-finishing/rotary-vibrators/ (accessed on 23 March 2023).
23. Uhlmann, E.; Eulitz, A. Influence of Ceramic Media Composition on Material Removal in Vibratory Finishing. *Procedia CIRP* **2018**, *72*, 1445–1450. [CrossRef]
24. Bhaduri, D.; Penchev, P.; Dimov, S.; Essa, K.; Carter, L.N.; Pruncu, C.I.; Jiang, J.; Pullini, D. On the Surface Integrity of Additive Manufactured and Post-Processed AlSi10Mg Parts. *Procedia CIRP* **2020**, *87*, 339–344. [CrossRef]
25. Mahmood, M.A.; Chioibasu, D.; Ur Rehman, A.; Mihai, S.; Popescu, A.C. Post-Processing Techniques to Enhance the Quality of Metallic Parts Produced by Additive Manufacturing. *Metals* **2022**, *12*, 77. [CrossRef]
26. Wilson, B.P.; Dotremont, A.; Biesemans, M.; Willem, R.; Campestrini, P.; Terryn, H. Effect of Additives on Smut-Layer Formation and Pitting during Aluminum Etching in Hydrochloric Acid. *J. Electrochem. Soc.* **2008**, *155*, C22. [CrossRef]
27. Basha, M.M.; Basha, S.M.; Jain, V.K.; Sankar, M.R. State of the Art on Chemical and Electrochemical Based Finishing Processes for Additive Manufactured Features. *Addit. Manuf.* **2022**, *58*, 103028. [CrossRef]
28. Tyagi, P.; Goulet, T.; Riso, C.; Stephenson, R.; Chuenprateep, N.; Schlitzer, J.; Benton, C.; Garcia-Moreno, F. Reducing the Roughness of Internal Surface of an Additive Manufacturing Produced 316 Steel Component by Chempolishing and Electropolishing. *Addit. Manuf.* **2019**, *25*, 32–38. [CrossRef]
29. Chen, G.S.; Wan, K.-C.; Gao, M.; Wei, R.P.; Flournoy, T.H. Transition from Pitting to Fatigue Crack Growth—Modeling of Corrosion Fatigue Crack Nucleation in a 2024-T3 Aluminum Alloy. *Mater. Sci. Eng. A* **1996**, *219*, 126–132. [CrossRef]
30. Jin, Z.; Cai, C.; Hashimoto, T.; Yuan, Y.; Kang, D.; Hunter, J.; Zhou, X. Alkaline Etching and Desmutting of Aluminium Alloy: The Behaviour of Mg2Si Particles. *J. Alloys Compd.* **2020**, *842*, 155834. [CrossRef]
31. Mesicek, J.; Ma, Q.-P.; Hajnys, J.; Zelinka, J.; Pagac, M.; Petru, J.; Mizera, O. Abrasive Surface Finishing on SLM 316L Parts Fabricated with Recycled Powder. *Appl. Sci.* **2021**, *11*, 2869. [CrossRef]
32. Maleki, E.; Bagherifard, S.; Unal, O.; Sabouri, F.; Bandini, M.; Guagliano, M. Effects of Different Mechanical and Chemical Surface Post-Treatments on Mechanical and Surface Properties of as-Built Laser Powder Bed Fusion AlSi10Mg. *Surf. Coat. Technol.* **2022**, *439*, 128391. [CrossRef]
33. Limbasiya, N.; Jain, A.; Soni, H.; Wankhede, V.; Krolczyk, G.; Sahlot, P. A Comprehensive Review on the Effect of Process Parameters and Post-Process Treatments on Microstructure and Mechanical Properties of Selective Laser Melting of AlSi10Mg. *J. Mater. Res. Technol.* **2022**, *21*, 1141–1176. [CrossRef]

34. Data Sheet: AlSi10Mg-0403 (400 W) Powder for Additive Manufacturing. Available online: https://www.renishaw.com/resourcecentre/en/details/data-sheet-alsi10mg-0403-400-w-powder-for-additive-manufacturing{-}{-}73122 (accessed on 24 March 2023).
35. Geometrical Product Specifications (GPS)—Surface Texture: Profile Method—Terms, Definitions and Surface Texture Parameters—Amendment 1: Peak Count Number (ISO 4287:1997/Amd 1:2009). Available online: https://www.une.org/encuentra-tu-norma/busca-tu-norma/norma?c=norma-une-en-iso-4287-1999-a1-2010-n0046045 (accessed on 23 March 2023).
36. Geometrical Product Specifications (GPS)—Surface Texture: Profile Method—Terms, Definitions and Surface Texture Parameters (ISO 4287:1997/Cor 1:1998/Cor 2:2005). Available online: https://www.une.org/encuentra-tu-norma/busca-tu-norma/norma/?c=N0045298 (accessed on 23 March 2023).
37. *Surface Texture: Surface Roughness, Waviness, and Lay*; American Society of Mechanical Engineers: New York, NY, USA, 2020; ISBN 978-0-7918-7325-0.
38. Metallic Materials—Tensile Testing—Part 1: Method of Test at Room Temperature (ISO 6892-1:2019). Available online: https://www.une.org/encuentra-tu-norma/busca-tu-norma/norma/?c=N0064441 (accessed on 23 March 2023).
39. Standard Practice for Verification of Testing Frame and Specimen Alignment Under Tensile and Compressive Axial Force Application. Available online: https://www.astm.org/e1012-19.html (accessed on 23 March 2023).
40. Wilbur, S. *Applications of ICP-MS in Homeland Security*; American Laboratory: Bellevue, WA, USA, 2004; Volume 36, pp. 20–26.
41. Galy, C.; Le Guen, E.; Lacoste, E.; Arvieu, C. Main Defects Observed in Aluminum Alloy Parts Produced by SLM: From Causes to Consequences. *Addit. Manuf.* **2018**, *22*, 165–175. [CrossRef]
42. Kempen, K.; Thijs, L.; Van Humbeeck, J.; Kruth, J.-P. Mechanical Properties of AlSi10Mg Produced by Selective Laser Melting. *Phys. Procedia* **2012**, *39*, 439–446. [CrossRef]

Journal of
Manufacturing and Materials Processing

MDPI

Article

Wire Electrical Discharge Machining of AISI304 and AISI316 Alloys: A Comparative Assessment of Machining Responses, Empirical Modeling and Multi-Objective Optimization

Mona A. Aboueleaz [1], Noha Naeim [2], Islam H. Abdelgaliel [3,4], Mohamed F. Aly [3,*] and Ahmed Elkaseer [2,5,6]

1 Production Engineering and Mechanical Design Department, Faulty of Engineering, Mansoura University, Mansoura 35516, Egypt; maboueleaz@mans.edu.eg

2 Department of Production Engineering and Mechanical Design, Faculty of Engineering, Port Said University, Port Fuad 42526, Egypt; noha.fouaad@eng.psu.edu.eg (N.N.); ahmed.elkaseer@kit.edu (A.E.)

3 Department of Mechanical Engineering, School of Sciences and Engineering, The American University in Cairo, AUC Avenue, New Cairo 11835, Egypt; ihamdys@aucegypt.edu

4 Department of Mechanical Engineering, Faculty of Engineering, Fayoum University, Fayoum 63514, Egypt

5 Department of Mechanical Engineering, Faculty of Engineering, The British University in Egypt (BUE), El-Sherouk City 11837, Egypt

6 Institute for Automation and Applied Informatics, Karlsruhe Institute of Technology, 76344 Eggenstein-Leopoldshafen, Germany

* Correspondence: mfawzyaly@aucegypt.edu

Abstract: This research investigates the multi-response of both material removal rate (MRR) and surface roughness (Ra) for the wire electrical discharge machining (WEDM) of two stainless steel alloys: AISI 304 and AISI 316. Experimental results are utilized to compare the machining responses obtained for AISI 316 with those obtained for AISI 304, as previously reported in the literature. The experimental work is conducted through a full factorial experimental design of five running parameters with different levels: applied voltage, transverse feed, pulse-on/pulse-off times and current intensity. The machined workpieces are analyzed using an image processing technique in order to evaluate the size of cut slots to allow the calculation of the MRR. Followed by the characterization of the surface roughness along the side walls of the slots. Different mathematical regression techniques were developed to represent the multi-response of both materials using the MATLAB regression toolbox. It was found that WEDM process parameters have a fuzzy influence on the responses of both material models. This allowed for multi-objective optimization of the regression models using four different techniques: multi-objective genetic algorithm (MOGA), multi-objective pareto search algorithm (MOPSA), weighted value grey wolf optimizer (WVGWO) and osprey optimization algorithm (OOA). The optimization results reveal that the optimal WEDM parameters of each response are inconsistent to the others. Hence, the optimal results are considered a compromise between the best results of different responses. Noteworthily, the multi-objective pareto search algorithm outperformed the other candidates. Eventually, the optimal results of both materials share the high voltage, high transverse feed rate and low pulse-off time parameters; however, AISI 304 requires low pulse-on time and current intensity levels while AISI 316 optimal results entail higher pulse-on time and current levels.

Keywords: WEDM; stainless steel 304; stainless steel 316; surface integrity; MRR; productivity; multi-objective optimization

1. Introduction

The challenging issue of manufacturing functional products with high accuracy and superior mechanical and thermal properties encourage several researchers to develop new material processing techniques or improve the existing ones [1,2]. In return, this leads to the necessity of the optimization of process parameters in order to obtain the highest quality and productivity [3]. Each material cutting process has its own effect on the material structure and properties. In addition, the capability of the cutting methods does not have to include the processing of all material types as some materials require high cutting force or power [4]. Wire electrical discharge machining (WEDM) is one of the key thermo-electric processes of cutting materials, irrespective of their high hardness and strength properties [5,6], such as high-carbon steels [7,8], stainless steel 316 (AISI 316) [9], etc. In WEDM, the cutting process is carried out by means of eroding the workpiece (as a cathode) with a travelling thin wire tensioned between two guide rollers (as an anode) by initiating a series of sparks between the electrodes in the presence of a dielectric medium [10,11]. One of the advantages of WEDM is to process difficult-to-cut materials, irrespective of their hardness, with acceptable surface properties and accurate dimensions, which promotes WEDM as a prime cutting method to produce punch dies and molds of high-strength materials [1,8,10].

There are a wide range of WEDM controlling parameters, such as servo voltage, pulse-on/pulse-off times, transverse feed rate, peak current, machined material thermo-physical properties, wire tension, wire material, wire diameter, and dielectric fluid type and pressure [12–18]. The most challenging problem in WEDM is that most of the mentioned controlling parameters have a fuzzy influence on the desired output, such as surface roughness (Ra) and material removal rate (*MRR*). In return, the process is considered a stochastic problem that requires sufficient mathematical representation and optimization [19–22]. However, using a mathematical model and MATLAB simulated annealing (SA) algorithm, it is claimed that feed rate, peak current and pulse-one time are the most influential parameters of the WEDM of EN31 high-strength steel alloy [23].

Sharma et al. [14] used Taguchi L_9 and ANOVA to analyze the *MRR*, gap current and machining time responses of D2 tool steel. It was found that pulse-off time was the most influential and significant parameter for all responses, followed by pulse-on time, while the peak current and wire tension influence were negligible. Using the amalgamation of response surface modeling (RSM), genetic algorithm (GA) and ANOVA, Sharma et al. [24] obtained the optimal WEDM parameters of the minimum overcut (9.9922 μm) of machined high-strength low-alloy (HSLA) steel using brass wire as an electrode at pulse-on/pulse-off times of 117 and 50 μs, respectively; gap voltage of 49 V; current of 180 A; and wire tension of 6 g. Meanwhile, using Taguchi L_9, analysis of variance (ANOVA) and signal-to-noise (S/N) ratio, the optimal WEDM process parameters of machining AISI 1045 alloy were obtained. The most influential parameter is current, with a p-value of 0.026. Moreover, the best *MRR* is 0.7112 mm^3/min, which was obtained at current = 16 A, voltage = 50 V and pulse-on time = 100 μs. The authors recommended further experimental-based multi-objective optimization that includes more responses and variable parameters [25]. Involving eight parameters and using a larger Taguchi array (L_{18} $2^1 \times 3^7$) followed by the ANOVA and S/N ratio, the optimal cutting conditions of Skd 61 alloy steel were obtained separately for the *MRR* and Ra—64.79 mm^2/min and 1.279 μm, respectively—as two single objectives, with a maximum relative error to the experimental data that did not exceed 9.8%. In addition, the study stated that pulse-on time is the most influential factor on both *MRR* and Ra models [26]. Additionally, Huang and Liao [27] confirmed that pulse-on time has the main influence on both *MRR* and Ra using grey relational analysis (GRA) and S/N ratio. In a recent study using similar methods, Kumar et al. [28] reduced the number of process parameters to four. In addition, the optimal parameters of the best *MRR* of D2 steel were a wire speed of 21 m/min, flushing pressure of 159 kgs/m^2, voltage of 81 V and current of 61 A, while the best Ra was obtained at 31 m/min, 121 kgs/m^2, 82 V and 80 A, with the same aforementioned order.

Furthermore, an optimization study using GRA on the WEDM machining of tungsten carbide material found that the optimal surface roughness (0.3435 μm) can be obtained at current = 2 A, voltage = 5 V and pulse-off time = 8 μs [29]. Chen et al. [30] used a combined multi-objective algorithm based on support vector machine and particle swarm optimization (SVR-PSO) to obtain the minimum Ra of 3.6 μm and the maximum *MRR* of 0.261 mm^2/s of 65 vol.% SiCp/Al composite. The optimal WEDM parameters were pulse-on time = 258.2 ns, pulse-off time = 125.3 ns, voltage = 41 V, wire feed = 9.452 mm/s and wire tension = 12.7 N. To wrap up, a review study on the optimization techniques of WEDM used in the literature showed that most researchers are mainly interested in multi-responses such as *MRR*, Ra and tool wear rate (TWR). The authors also stated that the most used optimization technique is Taguchi's methodology followed by GRA and fuzzy logic. Eventually, it was confirmed that current is a dominant factor requiring optimal control in order to obtain optimal multi-response answers [31].

The aim of this study is to widely investigate and optimize WEDM process parameters in order to achieve the trade-off between maximizing the *MRR* and keeping the Ra at the minimum level of two different steel alloys: stainless steels 304 and 316 (AISI 304 and AISI 316). A previously reported experimental investigation of WEDM of AISI 304 is experimentally extended for AISI 316, for which five parameters are investigated and modeled using the MATLAB regression toolbox. A combination of four multi-objective algorithms is used and compared. The algorithms are: (1) multi-objective genetic algorithm (MOGA), (2) multi-objective pareto search algorithm (MOPSA), (3) weighted value grey wolf optimizer (WVGWO) and (4) osprey optimization algorithm.

2. Materials and Methods

As formerly mentioned, this is a significant extension of the work reported by Naeim et al. in [10], in which the MRR and obtainable surface roughness in the WEDM of AISI 304 were analyzed. Herein, AISI 316 samples are machined under similar cutting conditions to those reported in [10] with the aim of comparatively assessing the machining responses of both materials under similar conditions. In addition, statistical analyses of the experimental results for both materials are carried out and discussed. The materials' compositions and properties are presented in depth. In addition, the equipment setup of the WEDM process is illustrated. The details of the design of the experiment are also discussed. Finally, the measurement instruments and conditions are provided.

2.1. Materials

Stainless steel is the most commonly utilized material for contact surfaces in dairy processing equipment. This metal possesses corrosion resistance, mechanical strength, hardness and ease of manufacture (weldability). AISI types 304 and 316 are the most suitable grades for general process fluid heating, storage and distribution. Because of the presence of molybdenum, type 316 is more expensive but offers superior corrosion resistance. Given that it is primarily required to protect the machine from the atmosphere, water and any spilled liquids, AISI 304 is almost always utilized externally or for the outside vessel jacket. Not only are they known for their resistance to corrosion but also for their clean appearance and overall cleanliness [32].

The experiments were conducted on AISI 304 and AISI 316 specimens with geometry of 120 mm × 30 mm × 3.1 mm. The chemical compositions of both materials are listed in Table 1. The physical and mechanical properties of both materials are listed in Table 2.

Table 1. Chemical compositions of used materials (%).

Grade	Mn	C	S	P	Si	Ni	Cr	Mo	N	V	Fe
AISI 304 [10]	2.00	0.08	0.03	0.045	0.75	8	18–20	-	0.10	-	Balance
AISI 316 [33]	1.97	0.077	0.005	0.0004	0.49	10.18	17.13	1.853	-	0.0615	Balance

Table 2. Physical and mechanical properties of used materials [34].

Property	AISI 304	AISI 316
Density (g/cm^3)	8.00	8.00
Melting Point (°C)	1450	1400
Modulus of Elasticity (GPa)	193	193
Electrical Resistivity (Ω·m)	0.72×10^{-6}	0.74×10^{-6}
Thermal Conductivity (W/m·K at 100 °C)	16.2	16.3
Thermal Expansion (10^{-6}/K at 100 °C)	17.2	15.9

2.2. Equipment Setup

The WEDM machine used is an ONA NX3, which has a positioning resolution of 1 μm. The cutting wire is molybdenum with diameter 0.18 mm. The wire tension is 8 (index). Water and gel were used as dielectrics. For preprocessing, the specimen surface is machined by grinding using a silicon carbide abrasive wheel in order to ensure the specimen's top and bottom surfaces are flat and parallel in all experiment trials (see Figure 1).

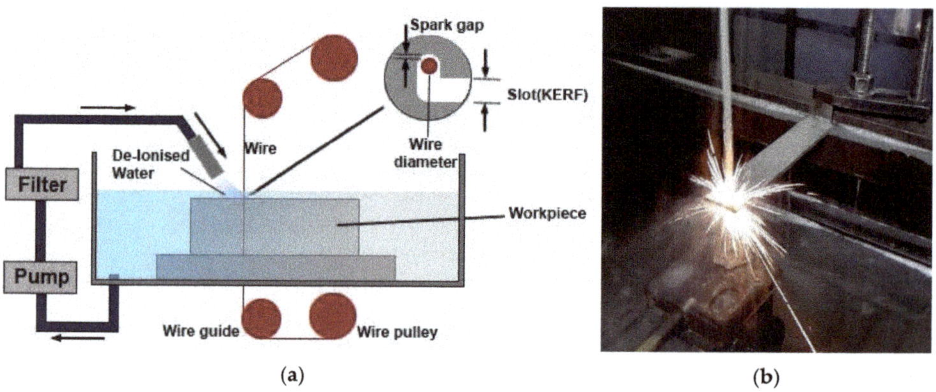

Figure 1. Experiment setup of WEDM: (**a**) schematic illustration of WEDM process and (**b**) real experiment on ONA NX3 EDM machine.

2.3. Design of Experiment

The selected running conditions for the WEDM, as reported in [10], are voltage (V) in V, transverse feed (f) in mm/min, pulse-off time (P_{off}) in μs, pulse-on time (P_{on}) in μs and current intensity (C) in A. Each parameter has certain levels, from 2 to 3 levels, as shown in Table 3.

Table 3. WEDM selected parameters and their levels.

Parameter	Levels		
Voltage (V), V	Low		High
Transverse feed (f), mm/min	80		120
Pulse-off time (P_{off}), μs	6		7
Pulse-on time (P_{on}), μs	25	30	40
Current intensity (C), A	1	2	4

The design of experiment (DOE) used in this research is a full factorial design. This led to conducting 72 experimental runs for both 304 and 316 stainless steels. Each run was carried out 3 times and the average of trial results was considered. The order of the 72 experiment's running conditions is illustrated in Figure 2.

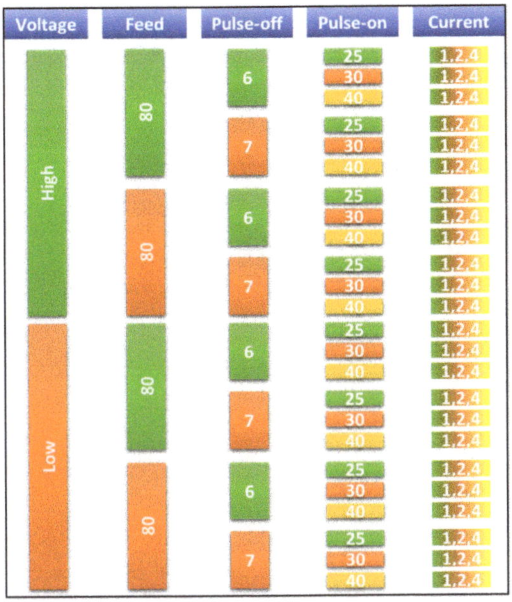

Figure 2. Full factorial DOE of experiment.

2.4. Measurement and Characterization

Each experimental trial of a certain set of cutting parameters had two wire cuts. The first cut was 15 mm in length and was machined in order to evaluate the cut width. The machined specimens were photographed by a high-resolution camera under appropriate lighting conditions. Following that, the collected images were processed and analyzed using the image processing techniques presented in [10] to detect the edge and precise width of the cut slot along its entire length. In addition, the machining time was calculated alongside the cut width. This allowed the material removal rate (*MRR*) to be calculated. The second cut was a full-width cut of 30 mm that allowed the characterization of the surface quality of the side walls of the cut. A Mitutoyo SJ-210 surface profilometer (Mitutoyo Corp., Kawasaki, Japan) was used to measure the surface roughness (Ra) along the full length of the side walls. To eliminate measurement uncertainty, seven readings were measured for each specimen and the average measurement was considered; further details of the characterization procedures can be found in [10].

3. Results and Discussion

3.1. Experimental Results

Figures 3 and 4 illustrate the experimental results of the *MRR* and Ra of the AISI 304 alloy. Remarkably, the *MRR* increases proportionally with the increase in the feed rate of 120 mm/min and high voltage. Also, the *MRR* is higher at P_{off} = 6 µs than at 7 µs, as shown in Figure 3c,d. Figure 3c–h show the *MRR* behavior to the change of the P_{on} and *C*. The *MRR* tends to achieve the best results at high P_{on} and *C* on the top-right corner of the contours. The *MRR* can reach 7 mm^3/min at high voltage, f = 120 mm/min, P_{off} = 6 µs, P_{on} = 40 µs and *C* = 2.4 A, as shown in Figure 3c. On the contrary, Figure 4a–h show that the Ra is tailored with the best surface quality at low P_{on} and *C* on the bottom-left corner of the contours. This leads to the best surface quality of Ra = 3.39 µm at high voltage and at the lowest level of the remaining parameters, as shown in Figure 4a.

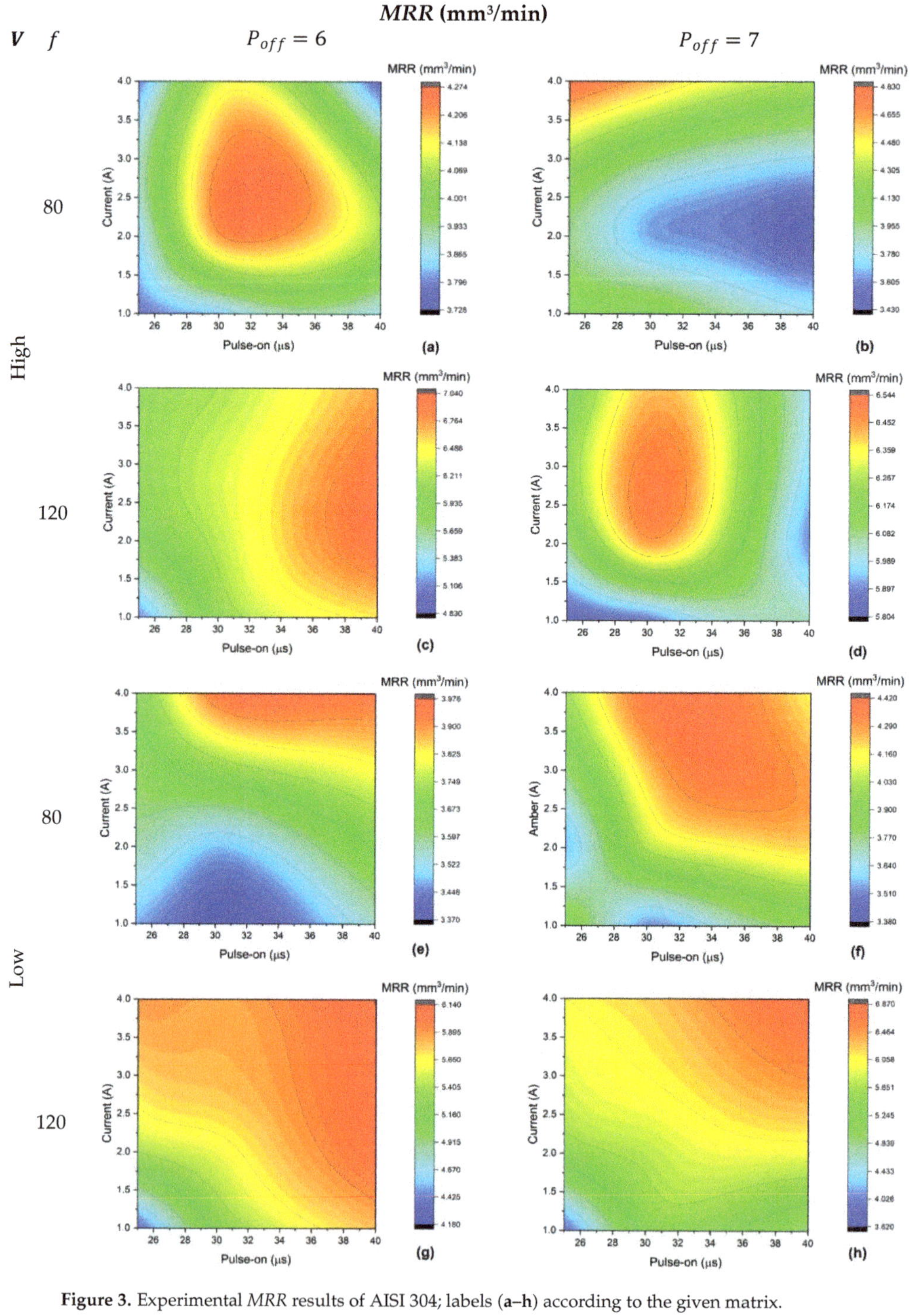

Figure 3. Experimental *MRR* results of AISI 304; labels (**a–h**) according to the given matrix.

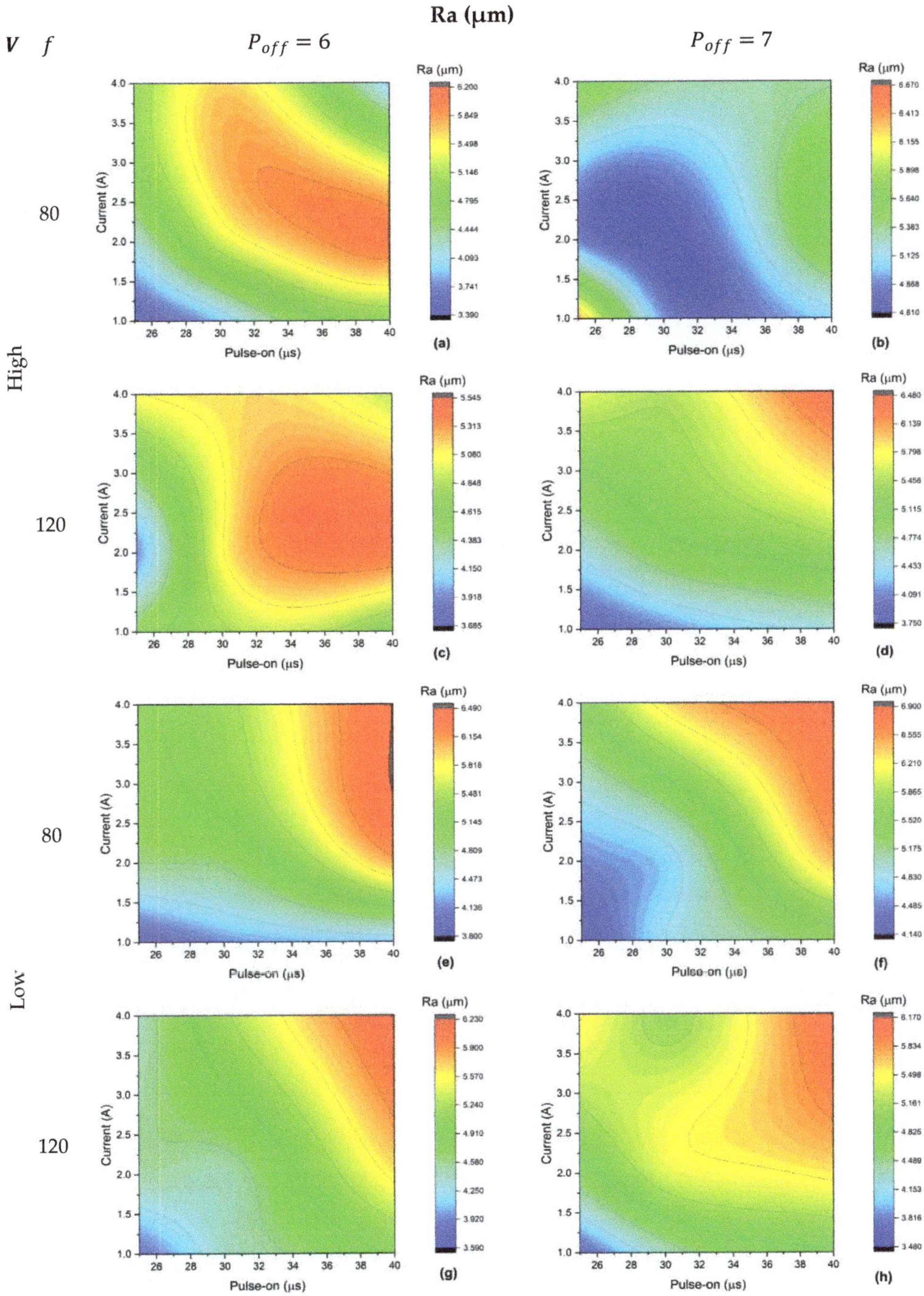

Figure 4. Experimental Ra results of AISI 304; labels (**a–h**) according to the given matrix.

Figures 5 and 6 explain the parameters' effect on the *MRR* and Ra of the AISI 316 alloy. Again, the *MRR* is high at the upper bounds of voltage, feed rate, pulse-on time and current; conversely, low pulse-off time is recommended for higher productivity of AISI 316, as shown in Figure 5a–d. Hence, the *MRR* entails the best productivity at 6.59 mm^3/min, with the corresponding parameters as shown in Figure 5c. Regarding the surface quality of wired AISI 316, the most desired surface quality (3.64 μm) can be obtained at high voltage, low feed rate, low pulse-on and pulse-off times and low current, as shown in Figure 6a.

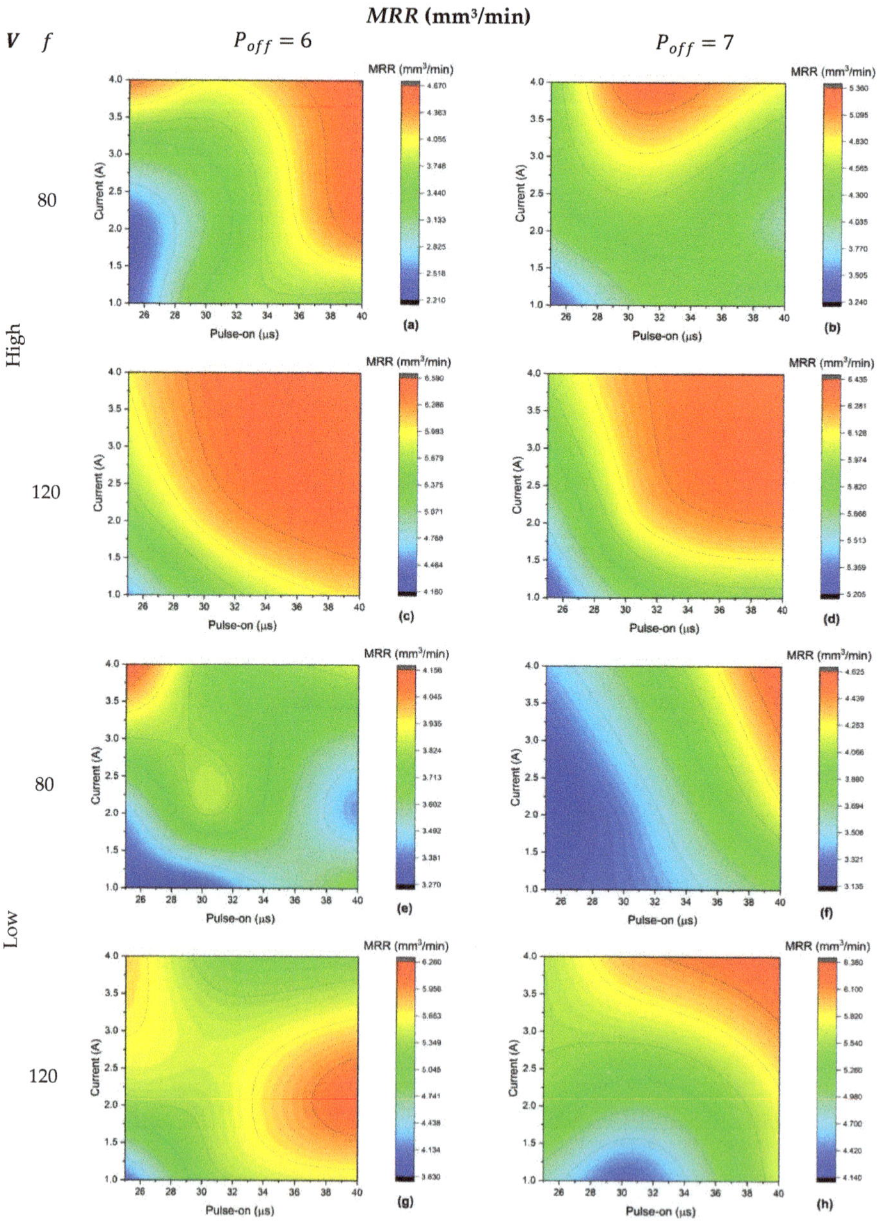

Figure 5. Experimental *MRR* results of AISI 316; labels (**a–h**) according to the given matrix.

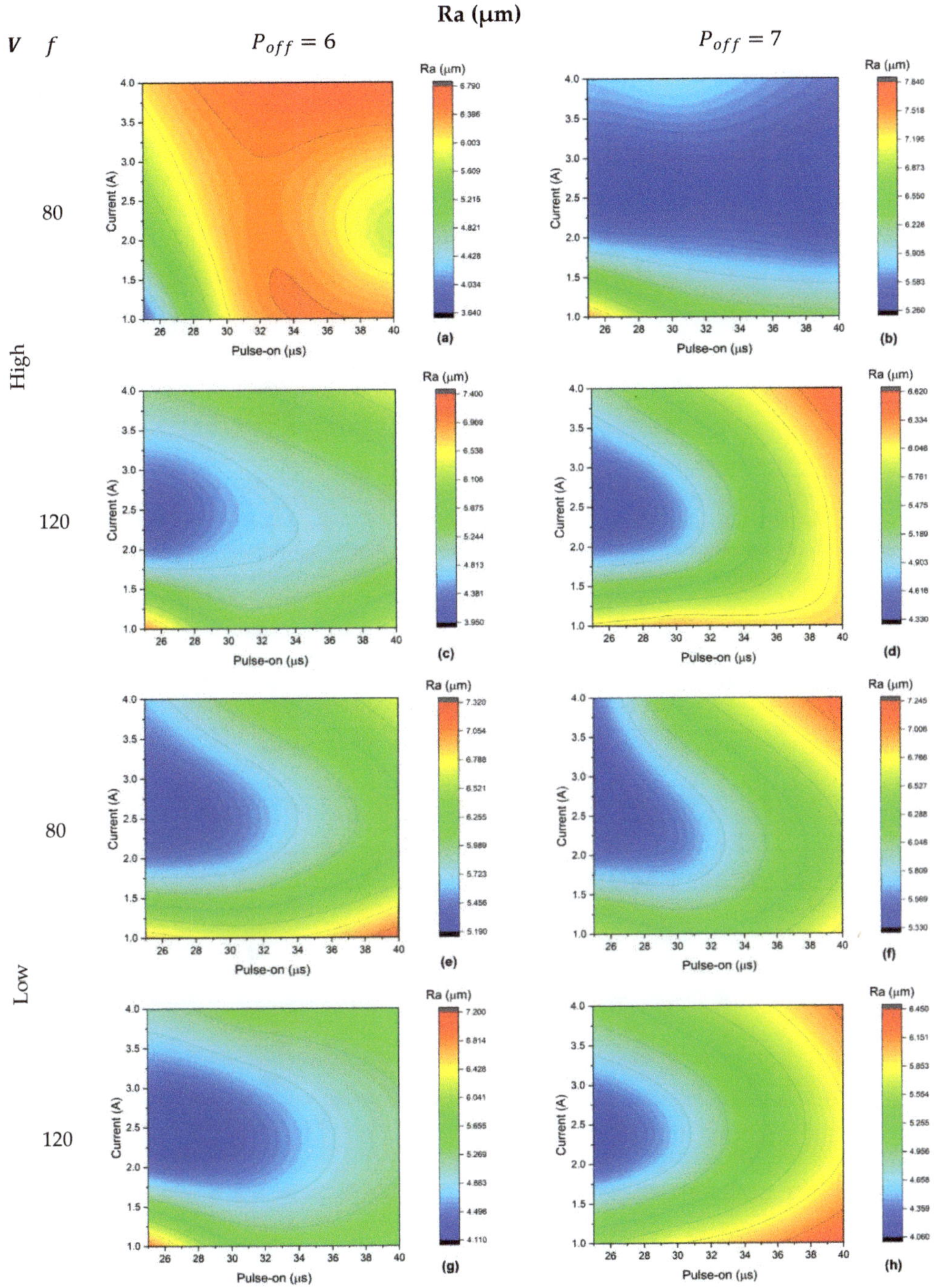

Figure 6. Experimental Ra results of AISI 316; labels (**a**–**h**) according to the given matrix.

3.2. *Mathematical Model Regression*

The mathematical models of both used materials are developed using the MATLAB regression learner toolbox. In order to obtain close accurate fitting, the parameters are normalized to the interval [−1,1] using Equation (1). Hence, the subscript *n* stands for the normalized parameter (see Table 4).

$$x_n = 2\left(\frac{x - x_{min}}{x_{max} - x_{min}}\right) - 1 \quad (1)$$

Table 4. WEDM normalized parameters and their new levels.

Parameter	Levels		
Voltage (V), V	−1		1
Transverse feed (f), mm/min	−1		1
Pulse-off time (P_{off}), µs	−1		1
Pulse-on time (P_{on}), µs	−1	−0.3333	−1
Current intensity (C), A	−1	−0.3333	−1

3.2.1. AISI 304 Model

MATLAB quadratic regression is used to develop the material removal rate (*MRR*) model of AISI 304, as shown in Equation (2). The values of the R-squared = 90.1% and R-adjusted = 88.3%. Meanwhile, the model p-value = 5.92×10^{-26}. Also, a comparison between the developed model and the experimental trials is conducted, as shown in Figure 7. It was found that the maximum error between the developed model and the experiment results is 36.47% and the average error is 5.74%. The maximum error appeared in one trial that the model was incapable of fitting with.

$$\begin{gathered} MRR_{304} = 4.9286 + 0.17772\ V_n + 1.0069\ f_n + 0.037764\ P_{off_n} + \\ 0.14666\ P_{on n} + 0.28073\ C_n + 0.085168\ V_n f_n - 0.11894\ V_n C_n - \\ 0.0766\ f_n P_{off_n} + 0.14842\ f_n P_{on n} + 0.093122\ f_n C_n - 0.099665\ P_{off_n} P_{on n} \end{gathered} \quad (2)$$

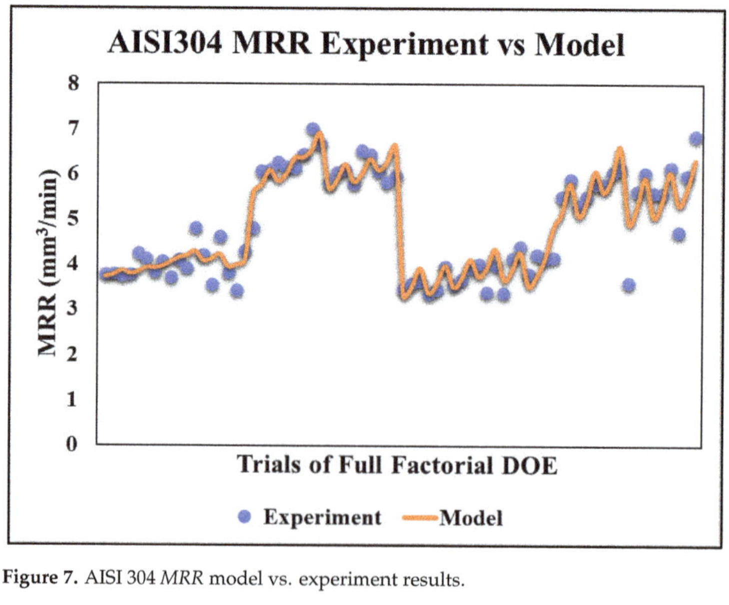

Figure 7. AISI 304 *MRR* model vs. experiment results.

For the roughness (Ra) model, the linear stepwise regression method on MATLAB is used (see Equation (3)). The fitting is quite inappropriate, as the R-squared = 60% and R-adjusted = 50%; however, the model p-value is 5.64×10^{-6}, which is a good indicator of significance. The comparison between the regression model and the actual experiment results, shown in Figure 8, revealed that the maximum error is 34.04% while the average error increased to 7.88% due to the low R-squared and R-adjusted values. To wrap up, the model can be incapable of fitting with the exact values of the experimental results; however, the model maintained the proportionality of results in a desired shape.

$$\begin{gathered} Ra_{304} = 5.3725 - 0.090425\ V_n - 0.091921\ f_n + 0.17402\ P_{off_n} + \\ 0.40703\ P_{on_n} + 0.51585\ C_n + 0.044755\ V_n f_n + 0.028745\ V_n P_{off_n} - \\ 0.1892\ V_n P_{on_n} - 0.13684\ V_n C_n - 0.091468\ f_n P_{off_n} + 0.076865\ P_{off_n} C_n - \\ 0.055949\ P_{on_n} C_n - 0.33842\ {C_n}^2 \end{gathered} \quad (3)$$

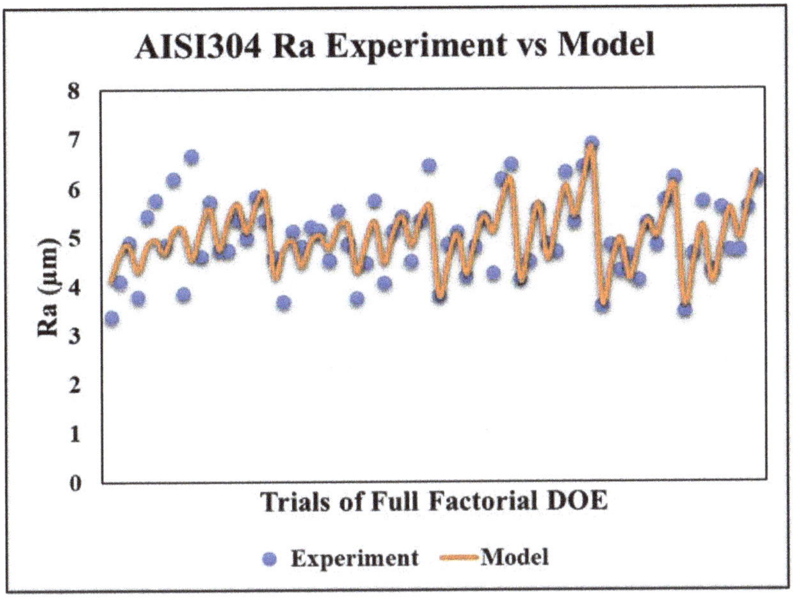

Figure 8. AISI 304 Ra model vs. experiment results.

3.2.2. AISI 316 Model

The *MRR* model of the second material AISI 316 is developed using the MATLAB linear interaction method, which is considered a first-order linear regression, as in Equation (4). Figure 9 depicts a comparison between the model and the experiment with R-squared = 87.8%, R-adjusted = 84.5% and p-value = 3.53×10^{-20}. The calculated errors are maximum error = 21% and average error = 7.77%. Fruitfully, this promotes the model to be reliable for further calculations.

$$\begin{gathered} MRR_{316} = 4.7202 + 0.20988\ V_n + 0.91948\ f_n + 0.10544\ P_{off_n} + \\ 0.34751\ P_{on_n} + 0.39651\ C_n + 0.049668\ V_n f_n + 0.073788\ V_n P_{off_n} + \\ 0.072056\ V_n P_{on_n} + 0.072247\ V_n C_n - 0.061812\ f_n P_{off_n} - 0.18745\ P_{on_n} C_n \end{gathered} \quad (4)$$

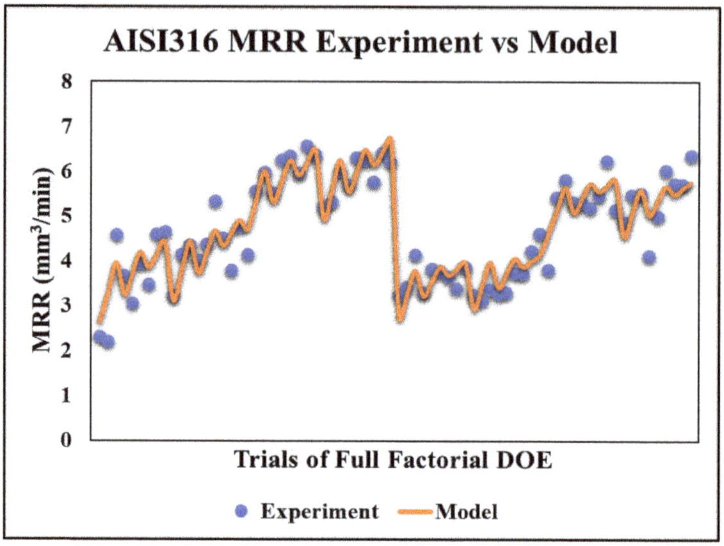

Figure 9. AISI 316 *MRR* model vs. experiment results.

Equation (5) is obtained by using the MATLAB linear stepwise regression method. The previously calculated values—R-squared = 61%, R-adjusted = 60%, p-value = 2.24×10^{-10}, maximum error = 21% and average error = 6.44%—are obtained. Again, the Ra model slightly lacks a fit with the experiment; however, the fitting is in good proportion, as shown in Figure 10.

$$\begin{gathered} \mathrm{Ra}_{316} = 5.0791 - 0.031417\ V_n - 0.28337\ f_n + 0.05213\ P_{off_n} + \\ 0.32202\ P_{on n} - 0.21772\ C_n + 0.13506\ V_n f_n + 0.25158\ P_{on n} C_n + 0.99187\ {C_n}^2 \end{gathered} \quad (5)$$

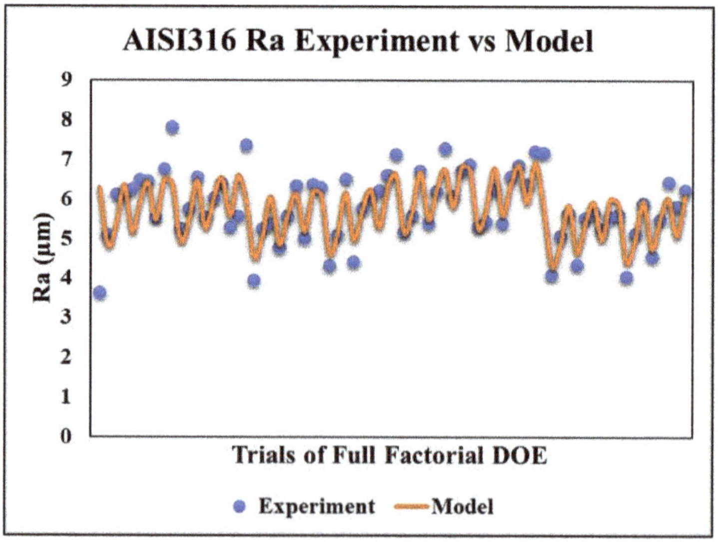

Figure 10. AISI 316 Ra model vs. experiment results.

3.3. Multi-Objective Optimization Results

In this stage, the developed models in the previous Section 3.2 are promoted to be used as objective functions in the multi-objective optimization model. In order to equalize the optimization results for a better solution, the model results of *MRR* and Ra are normalized from 0 to 1. Hence, the new symbols MRR_n and Ra_n are used. Then, the objective functions minimize Ra_n and maximize MRR_n. Almost all optimization algorithms tend to minimize the objective function; although, the minimization of $(1-MRR_n)$ is considered. The lower and upper bounds of the parameters are set to be -1 and 1, respectively, as the inputs parameters are normalized. However, the voltage parameter is forced to be equal to -1 (if $V_n \leq 0$) for a low value and 1 (if $V_n > 0$) for a high value. In this work, four multi-objective algorithms are used; two of them are built-in MATLAB functions—the multi-objective genetic algorithm (MOGA) [35,36] and multi-objective pareto search algorithm (MOPSA) [37–39]—while the other two are brought from the literature—the weighted value gray wolf optimizer (WVGWO) [40,41] and osprey optimization algorithm (OOA) [42]. The MOGA, MOPSA and WVGWO algorithms showed impressive solutions in recent work by the authors of [43,44]. Meanwhile, the OOA was introduced in 2023, and the author claimed that OOA outperformed many recent and late, well-known algorithms [42]. The optimization models of both materials are quite similar for all algorithms, as shown in Table 5.

Table 5. The multi-objective optimization model.

Model Item	Values
Number of Variables	5
Lower Bounds	$\begin{pmatrix} V_n & f_n & P_{off_n} & P_{on_n} & C_n \\ -1 & -1 & -1 & -1 & -1 \end{pmatrix}$
Upper Bounds	$\begin{pmatrix} V_n & f_n & P_{off_n} & P_{on_n} & C_n \\ 1 & 1 & 1 & 1 & 1 \end{pmatrix}$
Linear Inequality	[ ]
Linear Equality	[ ]
Initial Starting Point	$\begin{pmatrix} V_n & f_n & P_{off_n} & P_{on_n} & C_n \\ 0 & 0 & 0 & 0 & 0 \end{pmatrix}$
Objective Function 1	Min $(1-MRR_n)$
Objective Function 2	Min (Ra_n)

3.3.1. MOGA Model

The two material models are inserted into the MOGA toolbox in MATLAB. The additional options are maximum stall generations, set to 1000, and the maximum generations are 2000. In fact, the MOGA's output is a 2D x–y plot of the two normalized objectives, see Table 5, on both axes considering all data on the plot as a solution. This graph shows the trade-off between the two objectives—that is, the so called "Pareto front". Figure 11 illustrates the Pareto front graph of both AISI 304, Figure 11a, and AISI 316, Figure 11b. The gold dash box surrounds the desirable solution region, which is called the "Feasible Solution Area". One can resize the feasible solution area as required, for example, the box is set to be [0.4, 0.4] in size. The optimal solution is selected depending on the furthest point on a parallel line to a trend line connecting the plot data. There is a remarkable difference between the two materials' optimal solutions. Both optimal solutions are triggered at high voltage. The AISI 304 model is optimal at f = 119.37 mm/min, P_{off} = 6.16 µs, P_{on} = 28.66 µs and C = 1.06 A; in addition, AISI 316 achieved the optimal solution at f = 119.97 mm/min, P_{off} = 6.06 µs, P_{on} = 39.53 µs and C = 3.36 A. At these running conditions, the *MRR*s of AISI 304 and AISI 316 are 5.707 mm^3/min and 6 mm^3/min, respectively. Also, the surface roughness (Ra) is 4.353 µm and 4.741 µm for the same mentioned order. The *MRR* of AISI 316 is 5% better than that of AISI 304; however, the surface quality is reduced by 9%.

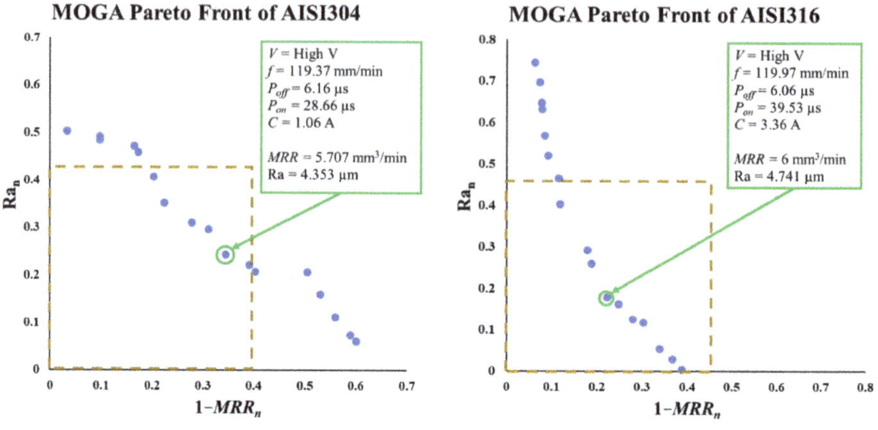

(a) (b)

Figure 11. Optimal solutions by MOGA Pareto front of both (**a**) AISI 304 and (**b**) AISI 316.

3.3.2. MOPSA Model

Similar to the MOGA, the input parameters in Table 5 are the same except for the options used. The maximum iterations are 1000. Also, the output is a Pareto front graph. The MOPSA works by defining a mesh grid around an initial point. This grid expands or shrinks depending on the previous search fitness. Figure 12 shows the Pareto front of both materials plus the selected optimal solutions. As mentioned before, the graph is a trade-off between the two objective functions; hence, one cannot say whether the MOPSA outperformed the MOGA as the MOPSA's solution is another feasible solution in the feasible area. Both materials have their optimal solution at high voltage and f = 120 mm/min. However, the other three parameters are different; AISI 304 is optimal at P_{off} = 6.44 µs, P_{on} = 25 µs and C = 1 A, while AISI 316 is optimal at P_{off} = 6. µs, P_{on} = 39.53 µs and C = 3.34 A. For AISI 304, the *MRR* is 5.933 mm^3/min (4% higher than the MOGA's solution) and the surface roughness Ra is 4.448 µm (2% difference to the MOGA). In addition, the optimal *MRR* of AISI 316 is 5.96 mm^3/min and the optimal Ra is 4.676 µm (better surface quality and lower productivity than the MOGA's solution).

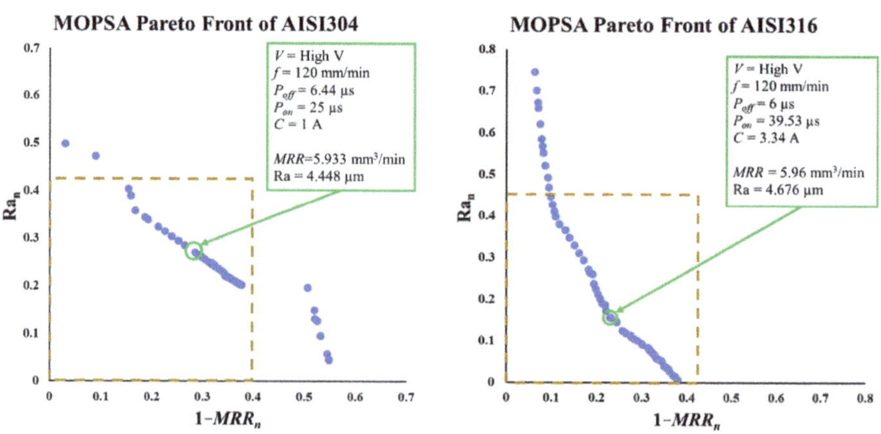

(a) (b)

Figure 12. Optimal solutions by MOPSA Pareto front of both (**a**) AISI 304 and (**b**) AISI 316.

3.3.3. WVGWO Model

This algorithm is a bio-inspired simulation of the hunting journey of a wolf pack. Usually, the pack has an alpha wolf (leader); lower rank wolves, beta and delta; and the rest of the pack are called omega wolves. The WVGWO begins with the initial wolf pack (population) that is searching for prey (best solution) and moving together in a certain direction. If the pack find prey in a certain location (parameter set in the current population), this promotes the winner wolf to a higher rank in the next search. These promotions are stored in an archive during the whole searching process. This archive is named the non-dominated wolves, which are the red plotted data on Figure 13. The final population is called the grey wolves, which includes the non-dominated wolves. Similar to the previously illustrated algorithms, the non-dominated wolves are considered as the Pareto front date; however, the graph keeps the other solutions of the final population as grey wolves who cannot outperform the non-dominated wolves. Clearly, it is found that the optimal solution of the WVGWO is very similar to the MOPSA's solution (see Figures 12 and 13). Another remark, on achieving the same *MRR* for both stainless steel materials, AISI 304 attains better surface quality (around 4 to 5%) than AISI 316. Also, this final remark is found in the MOPSA model.

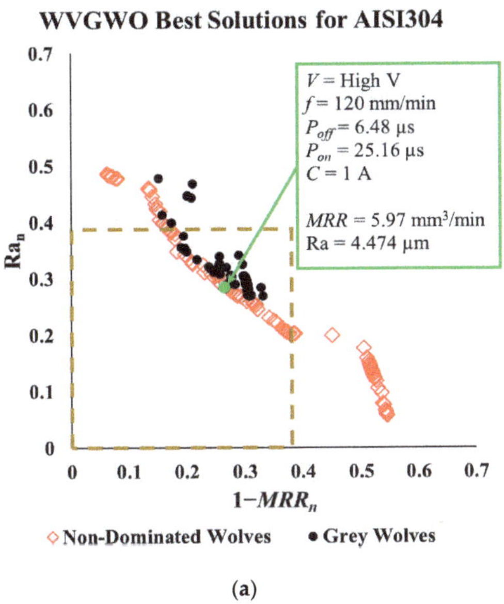

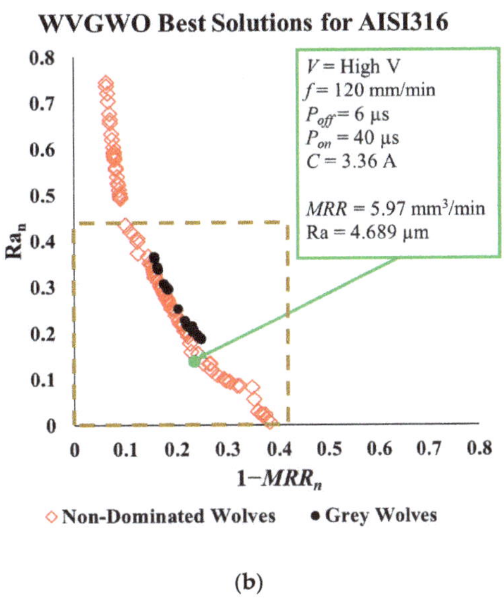

Figure 13. Optimal solutions by WVGWO of both (**a**) AISI 304 and (**b**) AISI 316.

3.3.4. OOA Model

Another bio-inspired algorithm is used called the osprey optimization algorithm (OOA). This algorithm is quite similar to the WVGWO; however, the hunting population moves separately and randomly, unlike the grey wolves that move together in one random direction. Ospreys start hunting fishes from underwater. The winner osprey indicates a hunt location that updates the current population by the evaluation of the objective function. The proposed MATLAB code by [42], obtained from the MathWorks website, creates one generation and finds the optimal solution after a certain number of iterations. In this work, a loop of generations is added while the optimal solution of each generation is stored in an archive. The optimal solutions are compared with the previous algorithms in a way that considers Ra as the first criterion. If a better surface quality solution is selected, the corresponding MRR solution is chosen—that is, the surface roughness objective of the OOA is the best amongst the algorithms studied; however, the MRR solution is lower than that of the other algorithms. Table 6 provides the optimal solutions obtained by the OOA for both materials. Once more, the ratio between the surface quality of both materials is 4 to 5% (AISI 304 better than AISI 316) on achieving the same productivity (MRR).

Table 6. Optimal solutions by OOA of both stainless steels 304 and 316.

Parameter	AISI 304	AISI 316
Voltage (V), V	High	High
Transverse feed (f), mm/min	120	120
Pulse-off time (P_{off}), µs	6.38	6
Pulse-on time (P_{on}), µs	26.85	39.97
Current intensity (C), A	1	3.16
Material removal rate (MRR), mm^3/min	5.886	5.89
Surface roughness (Ra), µm	4.427	4.61

Table 6 entails the optimal solutions obtained by the OOA for both materials. Once more, the ratio between the surface quality of both materials is 4 to 5% (AISI 304 better than AISI 316) for achieving the same productivity (*MRR*).

To summarize, Scott et al. [45] concluded that no single combination can be an optimal solution for multi-responses of WEDM. However, the trade-off optimal solutions of both addressed responses (*MRR* and Ra) of all algorithms for both stainless steel materials are presented in Table 7.

Table 7. Optimal solutions comparison by all used algorithms.

Model	AISI 304							AISI 316						
	V	f	P_{off}	P_{on}	C	MRR	Ra	V	f	P_{off}	P_{on}	C	MRR	Ra
MOGA	High	119.37	6.16	28.66	1.06	5.707	4.353	High	119.97	6.06	39.53	3.36	6	4.741
MOPSA	High	120	6.44	25	1	5.933	4.448	High	120	6	39.53	3.34	5.96	4.677
WVGWO	High	120	6.48	25.16	1	5.97	4.474	High	120	6	40	3.36	5.97	4.689
OOA	High	120	6.38	26.85	1	5.886	4.427	High	120	6	39.97	3.16	5.89	4.61

Obviously, both materials' optimal running voltage is high and feed rate is 120 mm/min. For AISI 304, mid-range pulse-off, low pulse-on and low current are optimal running parameters. On the other hand, the running conditions of AISI 316 are optimal at low pulse-off, high pulse-on and high current.

4. Conclusions

The findings of this optimization experimental-based study on the surface roughness and material removal rate of stainless steels 304 and 316 by WEDM are presented in this paper. The wire used in this research is molybdenum wire with a diameter of 0.18 mm. The investigated parameters are voltage, transverse feed rate, pulse-on/pulse-off times and current intensity. The surface roughness and material removal rate were evaluated and analyzed. Mathematical regression models were developed using MATLAB. Multi-objective optimization is carried out on these regression models. The main outputs of this research are as follows:

- Despite the fact that the WEDM process has a fuzzy proportion with the running parameters, the developed mathematical regression models represented the experimental results with small negligible errors that promote the models for optimization.
- The most influential parameters on both *MRR* and Ra are pulse-on time (P_{on} and current (C).

- For the optimization model of AISI 304, the MOGA algorithm attained the best surface roughness at 4.353 μm, while the optimal *MRR* was obtained by the WVGWO at a value of 5.97 mm^3/min. However, the MOPSA provided a trade-off multi-response solution as Ra = 4.448 μm (−2.18% from the optimal solution by the MOGA) and *MRR* = 5.933 mm^3/min (−0.62% from the best solution by the WVGWO). The optimal parameters obtained by the MOPSA are high voltage, f = 120 mm/min, P_{off} = 6.44 μs, P_{on} = 25 μs and C= 1 A.
- Similarly, for the AISI 316 model, the optimal Ra of 4.61 μm is obtained by the OOA and the optimal *MRR* = 5.97 mm^3/min by the WVGWO. Again, the MOPSA outperformed the other algorithms and resulted optimal *MRR* and Ra values of 5.96 mm^3/min and 4.677 μm, respectively. In this case, the obtained optimal parameters by the MOPSA are high voltage, f = 120 mm/min, P_{off} = 6 μs, P_{on} = 39.53 μs and C = 3.43 A.
- The optimal solution by the WVGWO of both materials in Table 7 show that the machining of AISI 304 and AISI 316 have the same productivity of *MRR* = 5.97 mm^3/min; however, AISI 304 has better surface roughness (Ra = 4.474 μm) than AISI 316 (Ra = 4.689 μm), making AISI 304 better by 4.58%.
- Obviously, the workpiece material's thermo-physical properties play a great role in the influence of WEDM parameters on the responses in terms of *MRR* and Ra.

Author Contributions: Conceptualization, A.E; methodology, M.A.A. and N.N.; formal analysis and investigation, A.E., M.A.A. and N.N.; writing—original draft preparation, I.H.A. and M.F.A.; writing—review and editing, I.H.A., M.F.A. and A.E.; optimization and modeling analysis, I.H.A. and M.F.A.; resources, A.E., M.A.A. and N.N.; supervision, A.E. and M.F.A. All authors have read and agreed to the published version of the manuscript.

Funding: This research received no external funding.

Data Availability Statement: The data presented in this study are available on request from the corresponding author.

Conflicts of Interest: The authors declare no conflict of interest.

References

1. Somashekhar, K.P.; Mathew, J.; Ramachandran, N. A feasibility approach by simulated annealing on optimization of micro-wire electric discharge machining parameters. *Int. J. Adv. Manuf. Technol.* **2012**, *61*, 1209–1213. [CrossRef]
2. Gupta, K.; Jain, N.K. On surface integrity of miniature spur gears manufactured by wire electrical discharge machining. *Int. J. Adv. Manuf. Technol.* **2014**, *72*, 1735–1745. [CrossRef]
3. Huang, J.T.; Liao, Y.S.; Hsue, W.J. Determination of finish-cutting operation number and machining-parameters setting in wire electrical discharge machining. *J. Mater. Process. Technol.* **1999**, *87*, 69–81. [CrossRef]
4. Robert, F.; Prince, A.A.; Jac Fredo, A.R. Influence of wire electrical discharge machine cutting parameters on the magnetization characteristics of electrical steel laminations. *Mater. Today Proc.* **2022**, *52*, 746–750. [CrossRef]
5. Sen, R.; Paul, S.; Choudhuri, B. Investigation on wire electrical discharge machining of AISI 304 stainless steel. *Mater. Today Proc.* **2022**, *62*, 1210–1214. [CrossRef]
6. Chary, E.A.; Basha, S.U.A.; Dheeraj, Y.; Kumar, D.S. Taguchi optimization in machining dual-phase steel (DP980) with WEDM using brass wire. *Mater. Today Proc.* **2022**, *62*, 3416–3420. [CrossRef]
7. Rao, M.; Vijayan, V.; Anil, A.; Rai, P.K.; Jain, N.R. Effect of wire electrode discharge machining process parameters on the surface roughness, hardness, and microstructure of the high carbon steels. *Mater. Today Proc.* **2021**, *46*, 2625–2629. [CrossRef]
8. Pattnaik, S.; Sutar, M.K. Analysis of surface roughness of high carbon steel by WEDM. *Mater. Today Proc.* **2021**, *44*, 1765–1769. [CrossRef]
9. Ishfaq, K.; Mufti, N.A.; Ahmed, N.; Mughal, M.P.; Saleem, M.Q. An investigation of surface roughness and parametric optimization during wire electric discharge machining of cladded material. *Int. J. Adv. Manuf. Technol.* **2018**, *97*, 4065–4079. [CrossRef]
10. Naeim, N.; AbouEleaz, M.A.; Elkaseer, A. Experimental Investigation of Surface Roughness and Material Removal Rate in Wire EDM of Stainless Steel 304. *Materials* **2023**, *16*, 1022. [CrossRef]
11. Ukey, K.; Sahu, A.R.; Gajghate, S.S.; Behera, A.K.; Limbadri, C.; Majumder, H. Wire electrical discharge machining (WEDM) review on current optimization research trends. *Mater. Today Proc.* **2023**, *in press*. [CrossRef]
12. Ghuge, N.; Sagare, S.; Pawar, S.; Kolambe, C.; Palande, D. Parameter optimization on WEDM machine of oil hardened nickel (Non-Shrinking) steel. *Mater. Today Proc.* **2022**, *65*, 3150–3160. [CrossRef]

13. Ikram, A.; Mufti, N.A.; Saleem, M.Q.; Khan, A.R. Parametric optimization for surface roughness, kerf and MRR in wire electrical discharge machining (WEDM) using Taguchi design of experiment. *J. Mech. Sci. Technol.* **2013**, *27*, 2133–2141. [CrossRef]
14. Sharma, S.; Vates, U.K.; Bansal, A. Parametric optimization in wire EDM of D2 tool steel using Taguchi method. *Mater. Today Proc.* **2021**, *45*, 757–763. [CrossRef]
15. Rehman, M.; Khan, S.A.; Naveed, R. Parametric optimization in wire electric discharge machining of DC53 steel using gamma phase coated wire. *J. Mech. Sci. Technol.* **2020**, *34*, 2767–2773. [CrossRef]
16. Nawaz, Y.; Maqsood, S.; Naeem, K.; Nawaz, R.; Omair, M.; Habib, T. Parametric optimization of material removal rate, surface roughness, and kerf width in high-speed wire electric discharge machining (HS-WEDM) of DC53 die steel. *Int. J. Adv. Manuf. Technol.* **2020**, *107*, 3231–3245. [CrossRef]
17. Manjaiah, M.; Laubscher, R.F.; Kumar, A.; Basavarajappa, S. Parametric optimization of MRR and surface roughness in wire electro discharge machining (WEDM) of D2 steel using Taguchi-based utility approach. *Int. J. Mech. Mater. Eng.* **2016**, *11*, 7. [CrossRef]
18. Guha, S.; Das, P.P.; Routara, B.C. Parametric optimization of wire electric discharge machining on AISI4140 alloy steel using regression analysis and grey-fuzzy approach. *Mater. Today: Proc.* **2021**, *46*, 4734–4740. [CrossRef]
19. Asgar, M.E.; Singholi, A.K. A Review on Wire Electrical Discharge Machining of Advanced Conductive Materials. *Int. Trans. J. Eng.* **2021**, *12A10Q*, 110. [CrossRef]
20. Patel, J.D.; Maniya, K.D. A Review on: Wire cut electrical discharge machining process for metal matrix composite. *Procedia Manuf.* **2018**, *20*, 253–258. [CrossRef]
21. Slătineanu, L.; Dodun, O.; Coteaţă, M.; Nagîţ, G.; Băncescu, I.B.; Hriţuc, A. Wire Electrical Discharge Machining—A Review. *Machines* **2020**, *8*, 69. [CrossRef]
22. Sharma, V.; Patel, D.S.; Jain, V.K.; Ramkumar, J. Wire electrochemical micromachining: An overview. *Int. J. Mach. Tools Manuf.* **2020**, *155*, 103579. [CrossRef]
23. Sankar, V.A.; Suresh, P.; Sridharan, P.; Vignesh, R.; Gowtham, M. Mathematical and MATLAB based process optimization by simulated annealing algorithm for wire electrical discharge machining of EN31. *Mater. Today Proc.* **2021**, *47*, 6941–6946. [CrossRef]
24. Sharma, N.; Khanna, R.; Gupta, R.D. WEDM process variables investigation for HSLA by response surface methodology and genetic algorithm. *Eng. Sci. Technol. Int. J.* **2015**, *18*, 171–177. [CrossRef]
25. Zaman, U.K.U.; Khan, U.A.; Aziz, S.; Baqai, A.A.; Butt, S.U.; Hussain, D.; Siadat, A.; Jung, D.W. Optimization of Wire Electric Discharge Machining (WEDM) Process Parameters for AISI 1045 Medium Carbon Steel Using Taguchi Design of Experiments. *Materials* **2022**, *15*, 7846. [CrossRef]
26. Kumar, A. Performance Analysis of Wire Electric Discharge Machining (W-EDM). *Int. J. Eng. Res.* **2012**, *1*, 4.
27. Huang, J.T.; Liao, Y.S. Optimization of machining parameters of Wire-EDM based on Grey relational and statistical analyses. *Int. J. Prod. Res.* **2003**, *41*, 1707–1720. [CrossRef]
28. Kumar, A.; Jagota, V.; Shawl, R.Q.; Sharma, V.; Sargam, K.; Shabaz, M.; Tanveer Khan, M.; Rabani, B.; Gandhi, S. Wire EDM process parameter optimization for D2 steel. *Mater. Today Proc.* **2021**, *37 Pt 2*, 2478–2482. [CrossRef]
29. Sudharsan, S.; Arul, R.; Ajay, C.V.; Veerakumar, S. Multiple response optimization of WEDM parameters using grey relational method. *Mater. Today Proc.* **2022**, *59*, 305–308. [CrossRef]
30. Chen, Z.; Zhou, H.; Yan, Z.; Han, F.; Yan, H. Machining characteristics of 65 vol.% SiCp/Al composite in micro-WEDM. *Ceram. Int.* **2021**, *47*, 13533–13543. [CrossRef]
31. Kumar, N.; Kumari, S.; Abhishek, K.; Nandi, G.; Ghosh, N. Study on various parameters of WEDM using different optimization techniques: A review. *Mater. Today Proc.* **2022**, *62*, 4018–4024. [CrossRef]
32. Cronin, K.; Cocker, R. Plant and Equipment | Materials and Finishes for Plant and Equipment. In *Encyclopedia of Dairy Sciences*; Elsevier: Amsterdam, The Netherlands, 2011; pp. 134–138. [CrossRef]
33. Abbas, A.T.; Anwar, S.; Abdelnasser, E.; Luqman, M.; Qudeiri, J.E.A.; Elkaseer, A. Effect of Different Cooling Strategies on Surface Quality and Power Consumption in Finishing End Milling of Stainless Steel 316. *Materials* **2021**, *14*, 903. [CrossRef] [PubMed]
34. Aalco-Ferrous and Non-Ferrous Metals Stockist, "Grade 304 Stainless Steel: Properties, Fabrication and Applications." AZoM, Jul. 2023. Available online: https://www.azom.com/article.aspx?ArticleID=2867 (accessed on 20 July 2023).
35. Conn, A.R.; Gould, N.I.M.; Toint, P. A Globally Convergent Augmented Lagrangian Algorithm for Optimization with General Constraints and Simple Bounds. *SIAM J. Numer. Anal.* **1991**, *28*, 545–572. [CrossRef]
36. Goldberg, D.E.; Holland, J.H. Genetic Algorithms and Machine Learning. *Mach. Learn.* **1988**, *3*, 95–99. [CrossRef]
37. Bratley, P.; Fox, B.L. Algorithm 659: Implementing Sobol's quasirandom sequence generator. *ACM Trans. Math. Softw.* **1988**, *14*, 88–100. [CrossRef]
38. Custódio, A.L.; Madeira, J.F.A.; Vaz, A.I.F.; Vicente, L.N. Direct Multisearch for Multiobjective Optimization. *SIAM J. Optim.* **2011**, *21*, 1109–1140. [CrossRef]
39. Fleischer, M. The Measure of Pareto Optima Applications to Multi-objective Metaheuristics. In *Evolutionary Multi-Criterion Optimization*; Lecture Notes in Computer Science; Fonseca, C.M., Fleming, P.J., Zitzler, E., Thiele, L., Deb, K., Eds.; Springer: Berlin/Heidelberg, Germany, 2003; Volume 2632, pp. 519–533. [CrossRef]
40. Mirjalili, S.; Mirjalili, S.M.; Lewis, A. Grey Wolf Optimizer. *Adv. Eng. Softw.* **2014**, *69*, 46–61. [CrossRef]

41. Malik, M.R.S.; Mohideen, E.R.; Ali, L. Weighted distance Grey wolf optimizer for global optimization problems. In Proceedings of the 2015 IEEE International Conference on Computational Intelligence and Computing Research (ICCIC), Madurai, India, 10–12 December 2015; IEEE: Madurai, India, 2015; pp. 1–6. [CrossRef]
42. Dehghani, M.; Trojovský, P. Osprey optimization algorithm: A new bio-inspired metaheuristic algorithm for solving engineering optimization problems. *Front. Mech. Eng.* **2023**, *8*, 1126450. [CrossRef]
43. Abbas, A.T.; Al-Abduljabbar, A.A.; Alnaser, I.A.; Aly, M.F.; Abdelgaliel, I.H.; Elkaseer, A.A. Closer Look at Precision Hard Turning of AISI4340: Multi-Objective Optimization for Simultaneous Low Surface Roughness and High Productivity. *Materials* **2022**, *15*, 2106. [CrossRef]
44. Abbas, A.T.; Al-Abduljabbar, A.A.; El Rayes, M.M.; Benyahia, F.; Abdelgaliel, I.H.; Elkaseer, A. Multi-Objective Optimization of Performance Indicators in Turning of AISI 1045 under Dry Cutting Conditions. *Metals* **2023**, *13*, 96. [CrossRef]
45. Scott, D.; Boyina, S.; Rajurkar, K.P. Analysis and optimization of parameter combinations in wire electrical discharge machining. *Int. J. Prod. Res.* **1991**, *29*, 2189–2207. [CrossRef]

materials

MDPI

Article

Evolution of CrC_x Ceramic Induced by Laser Direct Energy Deposition Multilayered Gradient Ni204-dr60 Coating

Yu Zhao [1,2], Ruobing Wang [1], Jian Zhang [1], Muhammad Imran Farid [1], Wenzheng Wu [1,2,*] and Tianbiao Yu [3]

[1] School of Mechanical and Aerospace Engineering, Jilin University, Changchun 130025, China; yzhao0109@jlu.edu.cn (Y.Z.); wrb12182023@163.com (R.W.); 13065201199@163.com (J.Z.); farid9919@mails.jlu.edu.cn (M.I.F.)
[2] Chongqing Research Institute, Jilin University, Chongqing 401133, China
[3] School of Mechanical Engineering and Automation, Northeastern University, Shenyang 110819, China; tianbiaoyudyx@gmail.com
* Correspondence: wzwu@jlu.edu.cn

Abstract: The manufacturing process for many large components of machines leads to a difference in their properties and performances based on changes in location. Functionally graded materials can meet these requirements and address the issue of generation and expansion of interface cracks. Ni204–dr60 gradient coatings were successfully fabricated using laser direct energy deposition (LDED). Microstructure mechanism evolution and microhardness of the gradient coating were comprehensively investigated. The change in the precipitated phase at the grain boundary and the intergranular zones resulted in a change in microstructural characteristics and also affected the microhardness distribution. The reinforced phase of the Ni204 → dr60 gradient zone from Ni204 to dr60 gradually precipitated and was rich in Mo and Nb phase, lath-shaped CrC_x phase, network-shaped CrC_x phase, block shape (Ni, Si) (C, B) phase, block CrC_x phase, and block Cr (B, C) phase. The gradient coating thus acts as a potential candidate to effectively solve the problem of crack generation at the interface of dr60 and the substrate.

Keywords: laser direct energy deposition; Ni204–dr60 multilayered gradient coating; microstructure evolution; microhardness

1. Introduction

Widespread and progressive development of the remanufacturing process has become the most requisite part of the advancement. Laser direct energy deposition (LDED) technology offers some fascinating characteristics, such as die-less manufacturing, short production cycle, and excellent laser-cladded performance. Relying on the three-dimensional model of parts, the laser trajectory is controlled to manufacture parts, showing high flexibility and a tendency to remain unaffected by the complexity of the parts [1–7]. Therefore, LDED technology was found to be favorable for enhancing recycling benefits, resulting in waste parts by enabling component refurbishment, thus gradually increasing its share in the remanufacturing industry [8]. Yu et al. [9] used additive the metal-layer deposition technique and directly deposited two specific types of metal powders, AISI-P21 and AISI-H13, onto the FC300, respectively. Zhang et al. [10] investigated the effect of building strategies on remanufacturing cylindrical parts using laser metal deposition. Notably, the hardness and toughness of castings and forgings such as gears and spindles exhibited various distribution patterns from outside to inside. Thus, laser repair using a single material cannot meet the requirements. Moreover, the quality of the interface between the LDED zone and the pending-repair component affects the remanufacturing quality and service life. Therefore, improvements in metallurgical bonding strength in interface zones, gradient of microhardness distribution, and surface performance constitute the key elements of LDED application in the field of remanufacturing. To solve the issues, gradient coating, composite materials, and parameter optimization have frequently been taken into account. It is worth noting that waste components with different degrees of defects increase material and coating performance requirements. Nevertheless, physical differences between component property and LDED material, which meet the performance requirements, are responsible for laser repair quality. Therefore, the property gradient distribution of laser repair coatings has become a focus of research [11].

Li et al. [12] used a carbon dioxide (CO_2) laser to weld the dissimilar metals Inconel 625 and SUS 304. Element segregation in the inter-dendritic region was found to be responsible for the reduced tensile strength and toughness. Lee et al. [13] obtained similar results in the course of laser welding of the dissimilar metal alloys 690 and SUS 304L. They pointed out that the distribution of elements in the fusion zone near SUS 304L was not homogeneous. Nb was primarily distributed in the inter-dendritic zone; however, its distribution throughout the fusion zone was not homogeneous. Although fusion welding can join similar and dissimilar metals, the different coefficients of thermal expansion, element segregation, cracks, pores, and generation of brittle intermetallic phases are still the main factors affecting laser welding quality [14]. Defects and irregular compositional distribution in the fusion zone are still encountered, leading to the failure of the welded component. The use of functional gradient materials has proven to be an effective way of improving metallurgical bonding quality and performance gradient distribution [15–17].

Shah et al. [18] produced SS316L and Inconel 718 functionally graded material using laser direct metal deposition. The distribution of hardness in the sample exhibited an approximate parabolic curve that was not evident. Carroll et al. [19] fabricated a specimen from 304L stainless steel incrementally graded to Inconel 625, which had gradients of hardness and density. Secondary phases primarily emerged in the gradient zone and were found to be responsible for the undesirable properties. Wang et al. [20] characterized the gradient microstructure fusion interface of dissimilar metals between high Cr steel and alloy 617 (Ni-based alloy). The results revealed that the formation of the gradient microstructure was determined by the angle between the tangents of the fusion line and the interface of adjacent layers. Savitha et al. [21] prepared discrete and compositional graded samples of SS316 and Inconel 625 alloys using laser-wise deposition and appraised the properties of the interface region. The microhardness of discrete composite coatings containing Inconel 625 alloys ranging from 25 to 75% exhibited small variations in their values.

According to previous studies, functional gradient materials show good performance, can improve the strength of the joint region, and effectively meet the various requirements of the component. The graded materials can reduce the probability of defects induced by different physical parameters of dissimilar materials. Compared to ultrasonic welding and friction stir welding, the widely graded structure can solve the issues of large thermal and mechanical residual stresses resulting from the connection of dissimilar materials [16,22]. Direct laser fabrication is an ideal technology for preparing functional gradient materials. However, cracks still emerge easily in the interface between different materials. The combination of the quality between dissimilar materials still meets the diversity requirements, albeit with difficulties.

Nickel (Ni)-based alloys are one of the most widely used metal alloys with good self-embellishment [23]. Although numerous studies are available on Ni-based alloy coatings, only a few studies on Ni–Ni gradient coatings have been reported to date. According to our previous study, the Ni204 alloy has good wettability with 45 steel, good toughness, and corrosion resistance, but low hardness [24]. Dr60 alloys (Nickel-based alloys) have high microhardness that can meet the requirements of machine components; nonetheless, the poor toughness and large difference in physical property parameters lead to bad bonding quality and a high probability of crack generation [25,26]. Therefore, corrosion-resistant Ni204 alloy and dr60 alloy were selected herein to study the effect of Ni204-based alloy as a transition material between different types of high-hardness powders (Ni-based) and substrates on coating properties. These two alloys primarily contain Fe, Ni, and Cr elements, which have good solubility. The fabrication of graded materials from Ni-based alloy to stainless steel has been previously studied [18,27,28]. These selected materials can effectively meet the criteria for wrought components of high microhardness and wear resistance, such as gear or spindle, although reduced toughness and defects, including cracks and uneven hardness gradient distribution, still exist. Therefore, functional gradient materials can be considered effective alternatives for laser repair of large, damaged parts. Functional gradient Ni204–dr60 materials were reported in this study for the first time, while previous studies were mainly focused on process optimization [29].

In this study, the discrete layer $Ni204_Xdr60_Y$ ($0 \leq x \leq 80\%$, $20\% \leq y \leq 100\%$, and $x + y = 1$, x and y stand for weight percent) and Ni204–dr60 functional gradient materials were fabricated using LDED by varying the composition of the powder between layers to obtain a hardness gradient. The microstructure of CrC_X ceramic, the elemental composition, and the microhardness of discrete and compositional materials with graded Ni204–dr60 coatings were subjected to contrastive analyses. Additionally, the relationship between CrC_X ceramic content and the characteristic and hardness gradient distribution was systematically investigated.

2. Experimental Section

2.1. Materials and Equipment

The substrate material employed in this study is $45^{\#}$ steel. Ni204 (Zhongke Yuchen (Beijing, China), spherical, 53 μm < particle size < 150 μm) and dr60 (Stellite, spherical, 53 μm < particle size < 150 μm) powders were used as initial materials for the fabrication of Ni204–dr60 gradient samples. Table 1 gives the chemical compositions of the powders and substrate used.

Table 1. Chemical compositions of powders and substrate (wt.%).

	C	Si	B	Cr	Mn	Mo	Nb	Ni	Fe
Ni204	≤0.03	0.4		21	-	9	4	Bal.	1.5
dr60	0.71	3.93	2.72	14.93	-	-	-	Bal.	3.56
$45^{\#}$ steel	0.42–0.5	0.17–0.37		≤0.25	0.5–0.8	-	-	≤0.25	Bal.

The open-loop control laser direct energy deposition (LDED) system was employed to fabricate the gradient materials. The IPG fiber (YLR-500W), with a wavelength of 1020 nm, was employed in the LDED experiments. Pure argon (99.99%) was used as the shielding gas and powder-sending gas. A laser power of 450 W with 1.1 mm diameter spots, a scan speed of 5.5 mm s^{-1}, and a powder sending rate of 0.7 r min^{-1} were used as the main LDED parameters. In this study, discrete layers of $Ni204_Xdr60_Y$ ($20\% \leq Y \leq 90\%$, $X + Y = 1$) and gradient samples consisting of 10 different layers (as presented in Table 2) with compositions ranging from pure Ni204 to pure dr60 were prepared. Before the experiment was conducted, the cladding powder was homogeneously mixed using Al_2O_3 ball milling at 30 rpm for 2 h to reduce segregation and then dried in a drying oven at 120 °C for 6 h to remove moisture.

Table 2. Chemical compositions of the different layers of the fabricated gradient sample (in wt.%).

Layer No.	Composition	Layer No.	Composition
1–4	100% Ni204	21–24	40% Ni204 60% dr60
5–8	80% Ni204 20% dr60	25–28	30% Ni204 70% dr60
9–12	70% Ni204 30% dr60	29–32	20% Ni204 80% dr60
13–16	60% Ni204 40% dr60	33–36	10% Ni204 90% dr60
17–20	50% Ni204 50% dr60	37–40	100% dr60

Figure 1 shows a schematic illustration of the equipment, forming path, and forming mechanism. The program switch obtained using the off-line programming software-controlled KUKA robot drives the LDED head to move and realize the LDED part formation [30]. The distances between the two layers and the adjacent clad were 0.4 mm and 0.9 mm, respectively. To avoid oxidation and guarantee the bunching of the powder, the shielding gas rate and the sending powder gas rate used were 16 L min^{-1} and 8 L min^{-1}. All samples were prepared using the same LDED parameters, as indicated in Table 3.

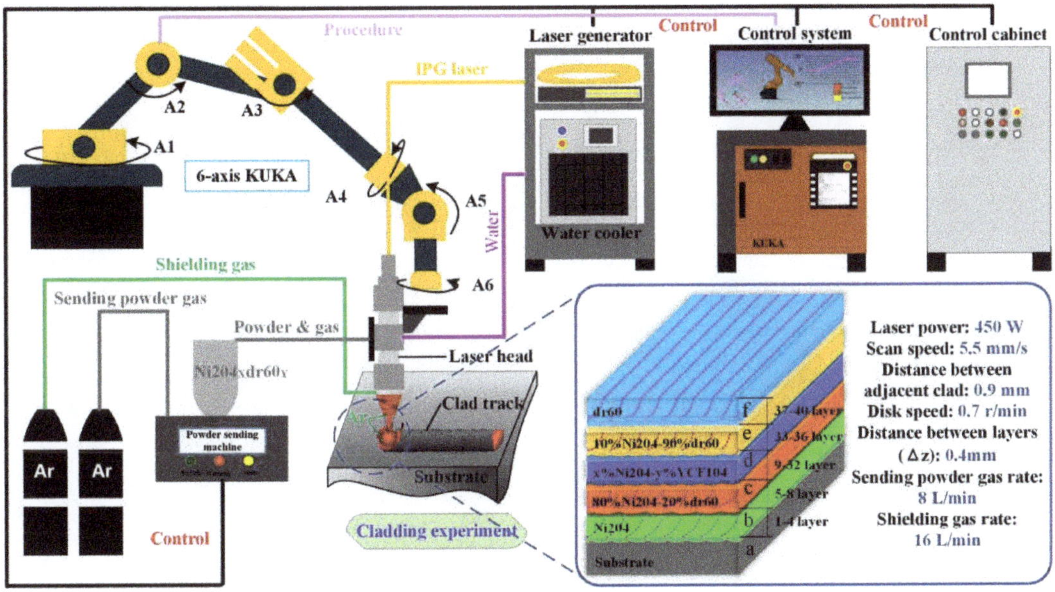

Figure 1. Schematic illustration of gradient fabrication.

Table 3. LDED parameters.

Parameters	Laser Power	Scan Speed	Sending Powder Rate	Distance between the Two Layers	Distance between the Adjacent Clad
Value	450 W	5.5 mm s^{-1}	0.7 r min^{-1}	0.9 mm	0.4 mm

2.2. Experimental Test

After LDED, the cross-section of the specimen was polished and etched 120 S–150 S using an etching solution ($HCl:HNO_3:H_2O$ = 2:1:1). The microstructures, cross-sectional morphologies, and elemental distributions of the discrete and compositional graded samples were observed using a field emission scanning electron microscope (FESEM, Zeiss Ultra Plus, Jena, Germany) equipped with EDS module and a laser confocal microscope (OLYMPUS LEXTOLS4100, Tokyo, Japan). Microhardness distribution from the substrate to the layer surface was measured using a Vickers microhardness tester (EM500-2A) (Shanghai, China) with a load of 500 gf applied for 12 s. Two adjacent test points were marked 100 μm apart.

3. Results and Discussion

3.1. Microstructure and Microhardness of Various Discrete Samples

The cross-sectional morphologies and microstructures of discrete composite samples are presented in Figure 2. As seen in Figure 2a–i, cracks appeared in the interface between the substrate and the $Ni204_Xdr60_Y$ composite coating when the dr60 content was more than 70%. In this case, the residual stress at the interface of the composite coating and the substrate is greater than the maximum allowable stress. Figure 2(a1–i1,j) displays an obvious difference in the microstructure. The microstructure of Ni204Xdr60Y resembles that of Ni204 when the Ni204 content is more than 70 wt.%, as shown in Figure 2(a1,b1,j). The composition ranges from 40 to 60 wt.%. Dr60 (Figure 2(c1–e1)) exhibits significant diversity in the microstructure. Clear dendrite structures can still be seen in these samples; however, the grain boundaries exhibit variations. Some novel phases precipitated in the grain boundary, and their contents and sizes increased with the increase in dr60 content. In the dr60 range of 70–90 wt.% (Figure 2(f1–h1)), microstructural characteristics transformed from dendritic structures to deposited dr60 (Figure 2(i1)).

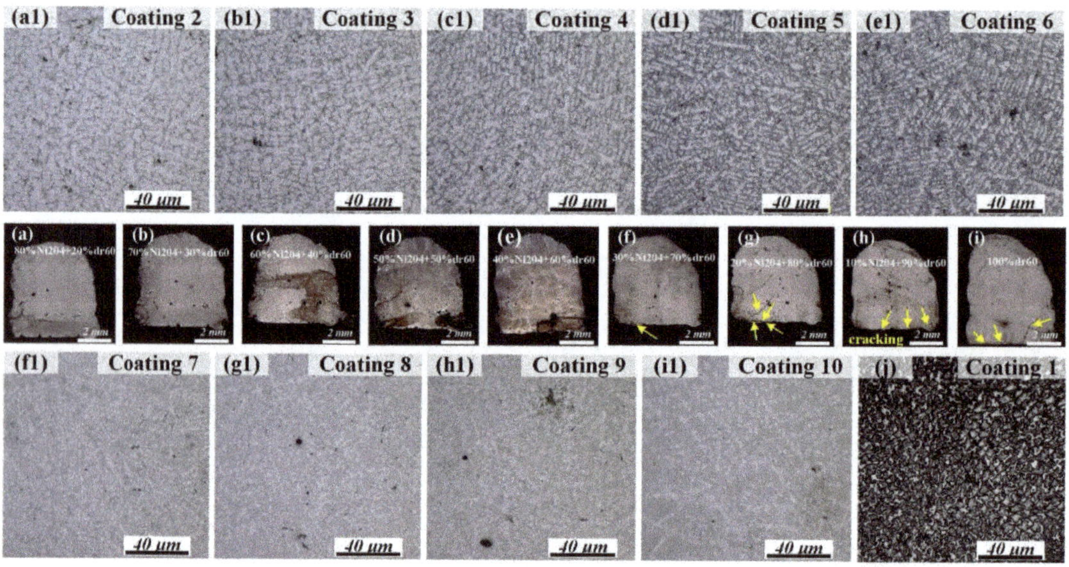

Figure 2. Cross-sectional morphologies of discrete composite samples: (**a–j**) samples $Ni204_Xdr60_Y$ ($0 \leq x \leq 80\%$, $20\% \leq y \leq 100\%$); (**a1–i1**) enlarged confocal microscope microstructure morphologies corresponding to samples (**a–i**).

The results of microhardness tests for $Ni204_Xdr60_Y$ composite coatings with various mass fraction ratios showed that the dr60 content significantly influenced microhardness distribution in composite coatings (Figure 3). The measurement results indicate that the microhardness of the coatings increased with the dr60 content. The increase in the microhardness of $Ni204_Xdr60_Y$ coatings varies approximately linearly with the $dr60_Y$ content at levels ranging from 274.6 $HV_{0.5}$ ($Ni204_{100\%}$) to 764.1 $HV_{0.5}$ ($dr60_{100\%}$). Surprisingly, hardness increased dramatically from 615.5 $HV_{0.5}$ ($Ni204_{10\%}dr60_{90\%}$) to 764.1 $HV_{0.5}$ ($dr60_{100\%}$), and $Ni204_Xdr60_Y$ coatings exhibited small microhardness fluctuations, with stable microhardness distributions in the coatings. It is worth noting that the differences in the microstructures shown in Figure 2 correspond to the microhardness presented in Figure 3, that is, the microhardness values are associated with the phases that precipitated in the grain boundaries. In order to further investigate the results, the microstructure evolutions of Ni204 $\rightarrow$ dr60 gradient coatings are analyzed in the subsequent studies.

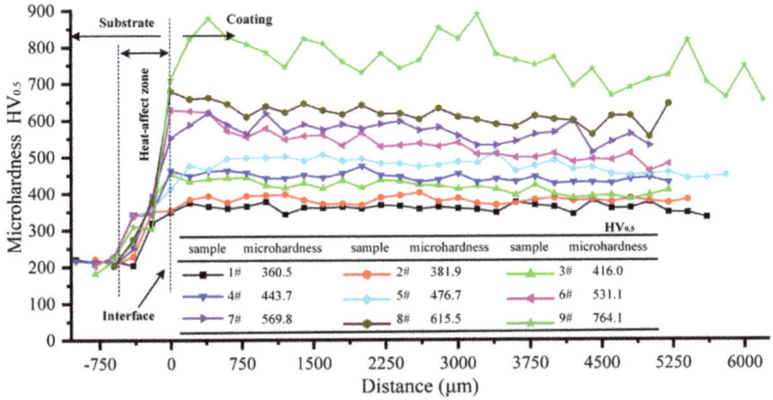

Figure 3. Microhardness corresponding to different Ni204/dr60 mass ratios.

3.2. Characterization of Microstructures and Phase Structures of Graded Samples

Figure 4 shows the distributions of Nb, Mo, B, C, Fe, Cr, and Ni elements in the Ni204 → dr60 gradient coatings. Marginal fluctuations in the Nb and Mo distribution curves from the substrate to the sample end surface become smaller when the dr60 content becomes greater than 50%. In contrast, the fluctuations in C, B, and Cr become larger. EDS line scan results show that many Nb- and Mo-rich phases precipitated in the layers when the content of the added Ni204 was more than 50%, while phases rich in Cr, B, and C precipitated in the layers when the dr60 content was more than 50%. The distribution of the elements in the layers corresponded to the graded materials (Figure 4a). Figure 4b shows that the interface between dissimilar materials has good metallurgical bonding. The concentrations of Nb, Mo, and Cr decreased, while those of B and C increased. Both changes in elements affected the precipitated phases, changing the grain structure and thus affecting the microhardness of the gradient layer. In order to further analyze the influence of these elements, the microstructures of the gradient coatings were anatomized in the following section.

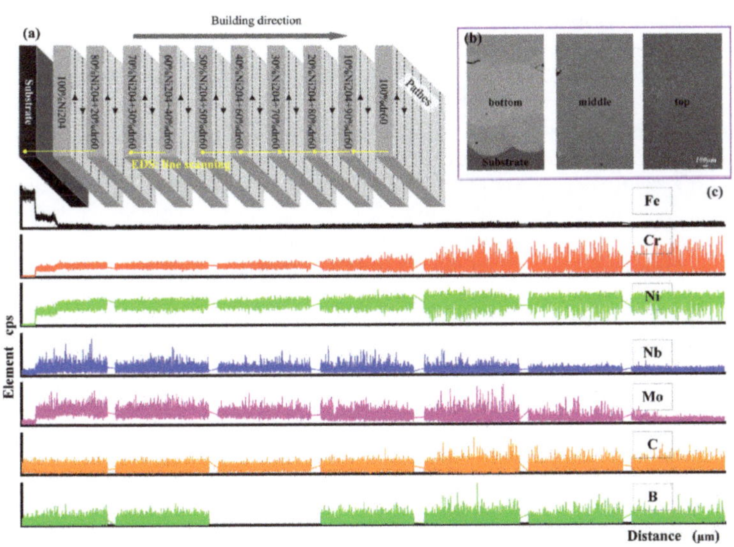

Figure 4. EDS line scan results of the distributions of Nb, Mo, C, B, Fe, Cr, and Ni in Ni204 → dr60 gradient coating. (**a**) EDS line scan test position; (**b**) SEM image of test position; (**c**) EDS results.

From the BSE images in Figure 5a–i, graded microstructure characteristics can be observed in coatings 1–10 of the produced gradient sample. The precipitation phases agglomerated along the grain boundaries, the content and size of precipitated phases increased with the addition of dr60, and the grain boundary subsequently disappeared. Moreover, the interface between the different discrete composite layers rarely showed cracks and pores, as shown in Figure 5j. For coating 1 (Ni204), the cell, columnar, and equiaxed crystals are the main microstructures, as shown in Figure 5a. The long-chain phase (white phase) emerges primarily along grain boundaries and many small bulges are observed in the interface between Ni204 and the substrate, indicating the occurrence of interface diffusion. When the dr60 content is less than 30 wt.%, the grain structure is similar to that of the Ni204 layer (Figure 5b). According to Figures 6 and 7, the white phase mainly contains Nb and Mo as the second phase precipitated at the grain boundaries. Ni, Cr, and Fe are detected in the intragranular zone [Ni, Cr] solid solution. Figure 5c–e shows the initiation of the dramatic change in microstructural characteristics, and long-chain white precipitated phases were gradually replaced with gray lath-shaped phases. With increasing dr60 content, the gray lath-shaped phase varied from a single lath-shaped structure to a network to blocky, and the milk-white phase began to precipitate (Figure 5e). Cr and C atoms were detected in the gray blocky structure phase, indicating that the gray blocky phase is CrC_x. This phenomenon is related not only to increasing concentrations of B and C but also to the drop in Nb and Mo concentrations. The increase in the content of B and C promoted the precipitation of carbides and borides in the layer, thus reinforcing microhardness. Cr, Nb, and Mo are found mainly at the grain boundary and in the intergranular zones in the course of solidification. Mo and Nb precipitated as the [Nb, Mo] binary phase, Cr combined with C to form a single lath-shaped gray phase CrC_x, and Cr and C dissolved in the intragranular region near the grain boundaries also precipitated, forming a network phase. With the continuous increase in dr60 content, the black blocky phase gradually superseded the gray network phase. A fresh phase, i.e., the black phase, appears and increases by reuniting to form black blocky phases. At the same time, small milk-white blocky phases appear and gradually increase. At this point, the honeycomb grain structure disappears completely and milk blocky, white, black blocky, and gray blocky phases become uniformly distributed in the matrix, as shown in Figure 5f–i. The results of energy spectrum analysis (Table 4) at points 1–4 in Figure 5i indicate that the black blocky phase at point 1 contains primarily 38.59 au.% Cr, 30.30 au.% B, and 26.00 au.% C atoms, indicating that the phase is a Cr (C, B) ternary phase. The gray phase at point 2 mainly contains 56.29 au.% Cr and 30.61 au.% C atoms (Cr: C = 1.83: 1), indicating that the gray phase is CrC_x. The 54.38 au.% Ni, 21.53 au.% B, 5.46 au.% Si, and 17.32 au.% C atoms were detected at point 4, indicating that the milk blocky phase is (Ni, Si) (C, B) ternary phase. The matrix areas primarily contain Ni, Cr, and C elements, and the matrix is rich in Ni. The C and B contents increase with the addition of dr60. The CrC_x phases were first to precipitate from the Ni–Cr–C–B–Si five-element liquid phase, causing a drop in the concentration of C in the liquid phase. The increased value of the B/C ratio promoted the precipitation of borides, B atoms combined with Cr to form (Cr, B) and dissolved into the CrC_x phase to form the Cr (C, B) ternary phase—the black phase. Residues B and C dissolved in the matrix precipitated in the form of multiple (Ni, Si) (B, C) phases. The matrix contained Ni and a small number of Fe, Cr, and other elements, and the solid Ni solution formed via matrix solidification. In the Ni204 $\rightarrow$ dr60 graded layers, the Mo, Nb, C, and B contents were responsible for the variations in microstructural characteristics and the gradient distribution of microhardness.

Figure 6 illustrates the distributions of Fe, Ni, Mo, Nb, Cr, C, B, and Si elements in the interface between the substrate and coating 1. Nb and Mo are largely distributed in the grain boundaries. Obvious thin bands formed in the interface between the substrate and the Ni204 coating, indicating good metallurgical bonding [31]. However, there were no obvious segregation or diffusion phenomena for Fe, Ni, Mo, Nb, Cr, C, B, and Si elements at the interface.

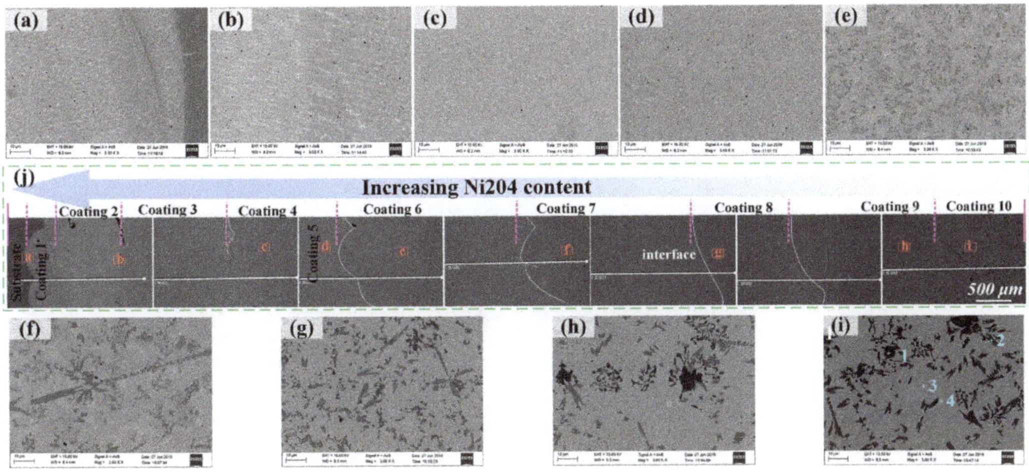

Figure 5. SEM images of cross-sectional microstructures of coatings 1–10. (**a**) Interface between coating 1 and substrate; (**b**) Interface between coating 2 and coating 3, (**c**) Coating 4, (**d**) Coating 5, (**e**) Coating 6, (**f**) Coating 7, (**g**) Coating 8, (**h**) Coating 9, and (**i**) Coating 10.

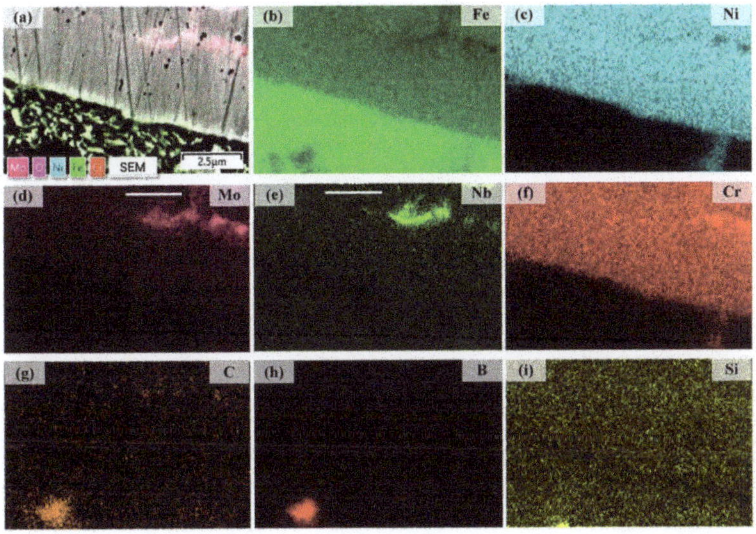

Figure 6. (**a**) SEM image of the interface between substrate and coating 1; (**b**–**i**) EDS maps showing the distributions of Fe, Ni, Mo, Nb, Cr, B, C, and Si.

The distribution of Fe, Ni, Mo, Nb, Cr, C, B, and Si elements in coating 3 is given in Figure 7. Mo, Nb, Cr, B, and C elements are detected in grain boundaries (Figures 5c and 7d–h). The concentrations of these elements at grain boundaries are significantly higher than in the matrix and are precipitated in the form of strengthened phases. Area 'L' mainly contains Mo, Cr, and B elements, and area 'M' mainly contains Nb, Mo, and B elements, while element C is detected in both areas 'L' and 'M'. With the increase in the concentrations of elements C and B, the grain boundary of the Ni204 coating with [Nb, Mo] as the capital precipitated phase changed to carbide and boride. At this time, the addition of elements C and B promotes the precipitation of Cr, Nb, and Mo in the form of multiple phases at the grain boundaries.

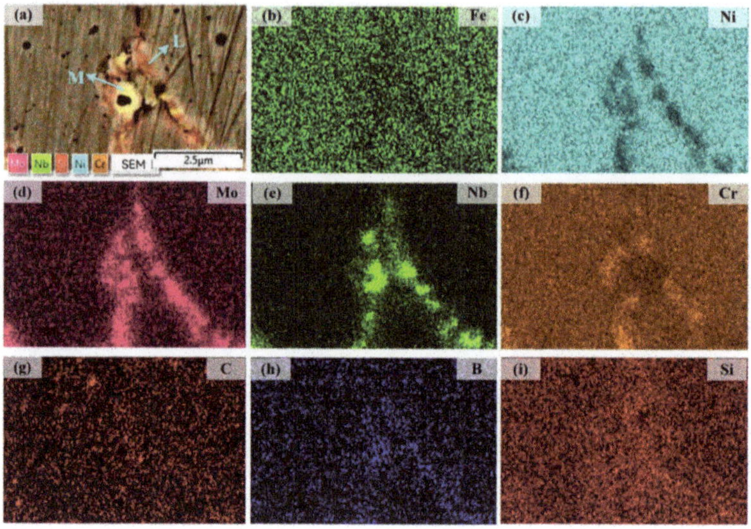

Figure 7. (**a**) SEM image of coating 4; (**b–i**) EDS maps showing the distributions of Fe, Ni, Mo, Nb, Cr, C, B, and Si.

Table 4. EDS analysis results of Ni204–dr60 gradient coatings (au. %).

Marked Locations	Fe	Cr	Ni	Nb	Mo	C	B	Si
Point 1	1.47	38.59	3.46	0	0	26.00	30.30	0.18
Point 2	2.79	56.29	9.97	0	0	30.61	0	0.34
Point 3	1.12	11.15	68.25	0	0	16.95	0	2.53
Point 4	1.29	3.72	54.38	0	0	17.32	21.53	5.46

Figure 8 shows the distributions of Si, Fe, Ni, B, C, Cr, Nb, and Mo at the interface between coating 3 and coating 4. There was good metallurgical bonding quality at the interface; the contents of Si, Fe, and Nb visibly changed, and the gradual content curve indicated the diffusion of elements at the interface. The gradual distribution of elements at the interface reduces the sensitivity of the material to the stress concentration resulting from the sudden change in properties. From the line scan results of elements B, C, Cr, Nb, and Mo at the grain boundaries, one can see that the main elements contained at the grain boundaries are elements Mo, Nb, Cr, B, and C, which are consistent with the results in Figures 5 and 7.

Figure 9 shows EDS mapping analysis results for elements Fe, Ni, Mo, Nb, Cr, C, B, and Si in coating 5. As seen in Figure 9b,c, the matrix mainly contains Fe and Ni. The milky white phase in Figure 9a largely contains Mo and Nb. EDS line scan results for elements Si, Fe, Ni, B, C, Cr, Nb, and Mo in the interface between coatings 4 and 5 (Figure 10), except elements Nb and Mo, B, C, and Si, are detected in the milky white phase. Elements B, C, and Si dissolved in the [Nb, Mo] phase formed by (Nb, Mo, Si) (C, B). From Figure 9d–h, elements Cr, C, and B are detected in network and blocky black phases, confirming the results in Figure 5.

Figure 11 shows EDS mapping analysis results for elements Fe, Ni, Mo, Nb, Cr, C, B, and Si in coating 9. The coating was fabricated using 10% Ni204 and 90% dr60. According to Figure 11b, Figure 11c, Figure 11f, and Figure 11i, elements Fe, Ni, Cr, and Si are detected in the matrix. Combined with Figure 5, the matrix is confirmed as [Ni, Fe, Si, Cr]. From Figure 11c,g–i, the gray phase mainly contains Ni and Si; combined with Figure 5, the gray phase is confirmed as (Ni, Si) (C, B). From Figure 11b–i, element Cr is detected in the black phase and there are no Fe, Ni, and Si. The black block phase in the left zone of Figure 11a mainly contains Cr and B, and few Mo and C elements are also detected; combined with Figure 5i, the Cr (C, B) phase is confirmed. The hindering effect of Mo on carbide contributes to the accumulation of Mo around carbides. The network black block phase in the right zone of Figure 11a mainly contains elements Cr, Mo, and C; combined with Figure 5i, the CrC_x phase is confirmed. EDS mapping analysis of coating 8 verified the correctness of the phase evolution analysis in Figure 5.

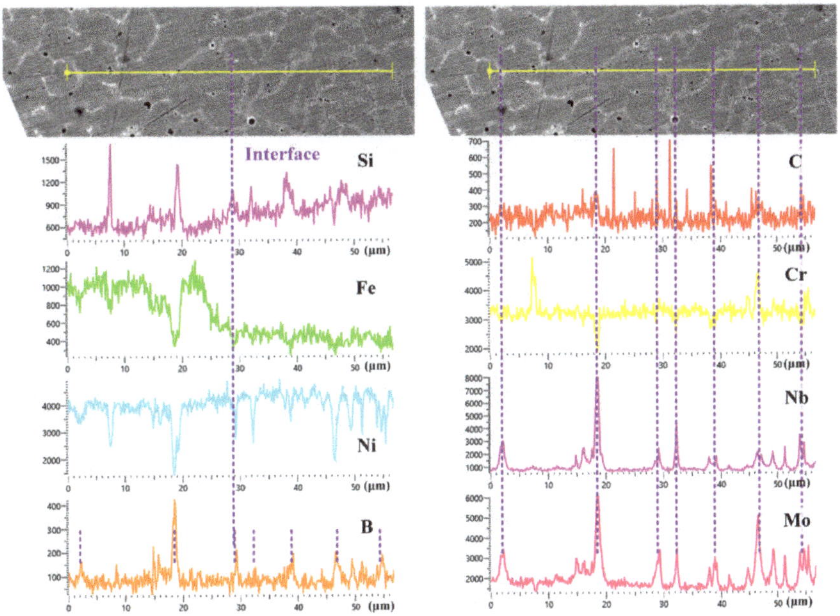

Figure 8. EDS line scan results of Si, Fe, Ni, B, C, Cr, Nb, and Mo at the interface between coating 3 and coating 4.

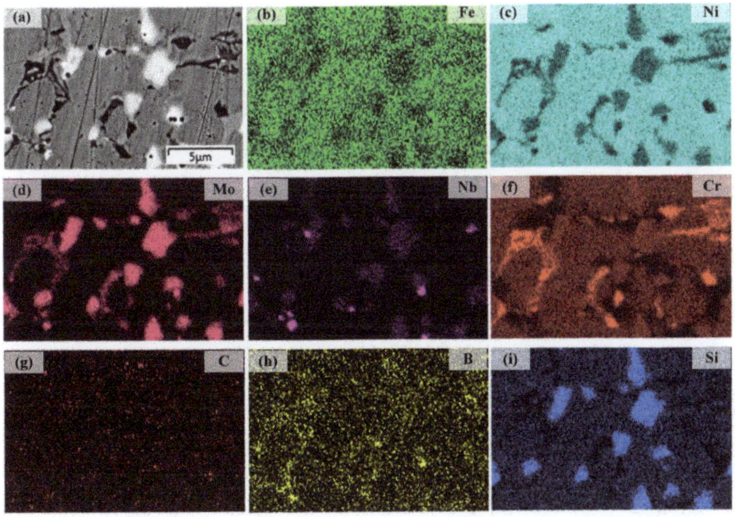

Figure 9. (**a**) SEM image of coating 5; (**b**–**i**) EDS maps showing the distributions of Fe, Ni, Mo, Nb, Cr, C, B, and Si.

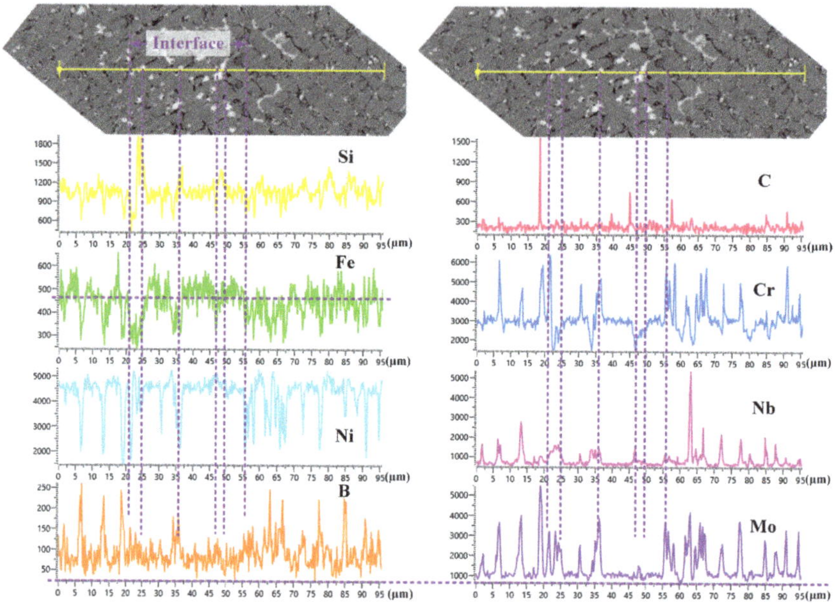

Figure 10. EDS line scan results for Si, Fe, Ni, B, C, Cr, Nb, and Mo at the interface between coating 4 and coating 5.

Through phase structure analysis of different coatings and interfaces of the gradient coatings, one can see that with the increase in the contents of elements C and B and the decrease in elements Nb and Mo, the precipitation of equal typical phase structures in grain boundaries and intergranular zones gradually evolves, consistent with the results of the analysis of phase structure evolution of gradient coatings in Figure 5. The change in the structure of the precipitated phase of Cr is the main reason for the change in coating microhardness.

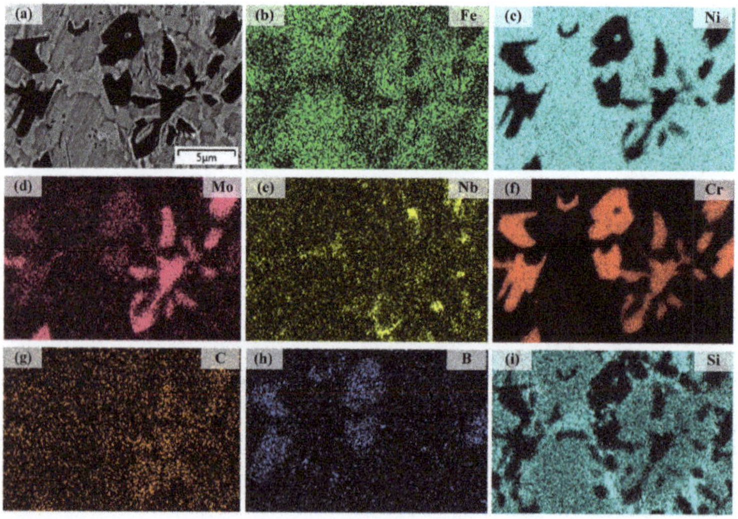

Figure 11. (**a**) SEM image of coating 9; (**b**–**i**) EDS maps showing the distributions of Fe, Ni, Mo, Nb, Cr, C, B, and Si.

3.3. *Microhardness of Ni204–dr60 Gradient Composite Materials*

Measurement of the microhardness of the Ni204 → dr60 graded coating revealed that the graded coating has a gradient distribution in the depth direction, as shown in Figure 12. Furthermore, and more remarkably, the microhardness values exhibited a sudden increase in the heat-affected zone [24]. The average microhardness of the substrate is about 200 HV0.5, which amounts to heat treatment of the substrate, reinforcing the substrate properties. The variations in the concentrations of elements B and C cause an increase in the precipitation phase at the grain boundary and in the intergranular zones, thus altering the microstructure characteristics and microhardness (Figure 5). Notable positive correlations were observed between microhardness and CrC_X, Cr (C, B), and (Ni, Si) (C, B) contents. The microhardness of gradient zones arranged in the range established based on the hardness of deposited Ni204 and dr60. Ni, Mo, Nb, B, and C diffused in the interfaces between dissimilar materials. Therefore, the microhardness in the dr60 layer is 704.2$HV_{0.5}$ and drops by 7.8% (60 HV) compared to the discrete dr60 coating. The microhardness showed gradient distribution; however, the value was lower than the value of the corresponding discrete coating ($Ni204_X dr60_Y$, 10% < x, y < 90%), and the diffusions of C, B, Mo, and Nb were primarily responsible for this.

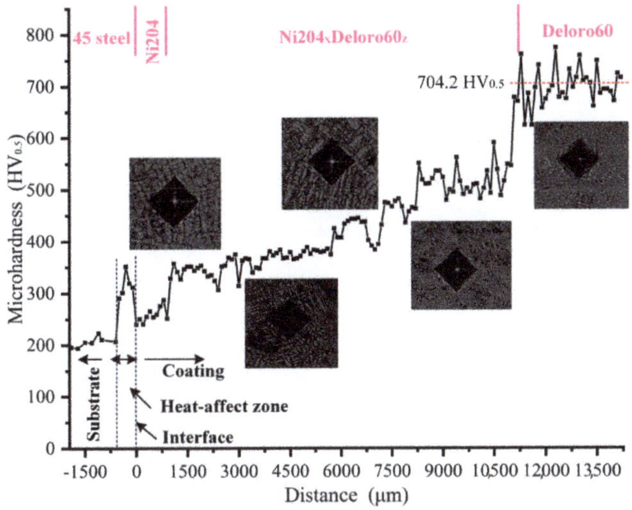

Figure 12. Microhardness of the Ni204–dr60 gradient coating.

4. Conclusions

Graded components of Ni204 → dr60 without visible interface defects and significant microstructure evolution were successfully fabricated. An obvious gradient distribution of hardness was measured in the Ni204 → dr60 gradient coating. Columnar and columnar dendrite crystals were observed near the Ni204 end of the gradient zone (more than 70% of Ni204 added), and an Nb- and Mo-rich precipitated phase existed at the grain boundary and in the intergranular zones. Low hardness was mainly attributed to this microstructure characteristic. However, large numbers of black phase Cr (B, C), gray phase CrC_X, and milk-white phase (Ni, Si) (C, B) were observed near the dr60 end of the gradient zone (more than 80% of dr60 added). The carbide and boride contents were responsible for the improvement in microhardness in the gradient zone. The reinforced phase of the gradient zone switched from Mo- and Nb-rich precipitated phase → lath-shaped CrC_X phase → network-shaped CrC_X → network-shaped CrC_X, block-shaped (Ni, Si) (C, B) phase, and block CrC_X and (Ni, Si) (C, B) phase → block CrC_X, (Ni, Si) (C, B) and Cr (B, C) phase. The changes in the composition of the precipitated phase are mainly responsible for the gradient distribution of microhardness in the graded zone. The graded coating effectively solved the issue associated with crack generation at the interface between dr60 and the substrate. The occurrence of (Fe, Ni) diffusion in various $Ni204_X dr60_Y$ dissimilar materials gives rise to its microhardness, which is slightly lower than that of the corresponding material.

Author Contributions: Y.Z.: experimentation, data curation, methodology, resources, and writing the original draft; R.W.: data collection and manuscript revision; J.Z.: experimentation and data curation; M.I.F.: data curation, writing—review & editing; W.W.: supervision, conceptualization, and methodology; T.Y.: conceptualization, methodology, and resources. All authors have read and agreed to the published version of the manuscript.

Funding: This work was supported by the Special Financial Aid to China Postdoctoral Science Foundation (No. 2022TQ0116), the China Postdoctoral Science Foundation (No. 2022M721315), the General Program of Chongqing Natural Science Foundation (No. CSTB2022NSCQ-MSX1623), the Science Foundation of Education Department of Jilin Province (JJKH20231150KJ), and the Ministry of Industry and Information Technology of China (No. 201675514).

Institutional Review Board Statement: Not applicable.

Informed Consent Statement: Not applicable.

Data Availability Statement: Not applicable.

Conflicts of Interest: The authors declare that they have no financial or personal relationships with other people or organizations that can inappropriately influence the work. There is no professional or other personal interest of any nature or kind in any product, service, and/or company that could be construed as influencing the position presented in, or the review of the work.

References

1. Martucci, A.; Aversa, A.; Lombardi, M. Ongoing challenges of laser-based powder bed fusion processing of Al alloys and potential solutions from the literature—A Review. *Materials* **2023**, *16*, 1084. [CrossRef] [PubMed]
2. Zhang, W.; Shang, X.; Chen, X.; Chen, S.; Liu, Z.; Zhang, L. Quenching and tempering-dependent evolution on the microstructure and mechanical performance based on a laser additively manufactured 12CrNi2 alloy steel. *Materials* **2023**, *16*, 3443. [CrossRef] [PubMed]
3. Zhao, Y.; Yu, T.; Sun, J.; Xi, W.; Bi, X. Effect of laser cladding on forming qualities of YCF101 alloy powder in the different lap joint modes. *Int. J. Adv. Manuf. Technol.* **2018**, *96*, 1991–2001. [CrossRef]
4. Yu, T.; Zhao, Y.; Sun, J.; Chen, Y.; Qu, W. Process parameters optimization and mechanical properties of forming parts by direct laser fabrication of YCF101 alloy. *J. Mater. Process. Technol.* **2018**, *262*, 75–84. [CrossRef]
5. Li, P.; Yang, T.P.; Li, S.; Liu, D.S.; Hu, Q.W.; Xiong, W.H.; Zeng, X.Y. Direct laser fabrication of nickel alloy samples. *Int. J. Mach. Tools Manuf.* **2005**, *45*, 1288–1294.
6. Abioye, T.E.; Farayibi, P.K.; Kinnel, P.; Clare, A.T. Functionally graded Ni-Ti microstructures synthesised in process by direct laser metal deposition. *Int. J. Adv. Manuf. Technol.* **2015**, *79*, 843–850. [CrossRef]
7. Tong, Z.; Ren, X.; Jiao, J.; Zhou, W.; Ren, Y.; Ye, Y.; Larson, E.A.; Gu, J. Laser additive manufacturing of FeCrCoMnNi high-entropy alloy: Effect of heat treatment on microstructure, residual stress and mechanical property. *J. Alloys Compd.* **2019**, *785*, 1144–1159. [CrossRef]
8. Ferreira, N.M.; Vila Pouca, M.; Fernandes, C.; Seabra, J.; Lesiuk, G.; Parente, M.; Jesus, A. Direct energy deposition parametric simulation investigation in gear repair applications. *Materials* **2023**, *16*, 3549. [CrossRef]
9. Yu, J.H.; Lee, K.Y.; Shim, D.S.; Park, S.H. Characterization of mechanical behavior in repaired FC300 using directly deposited AISI-P21 and AISI-H13 metal powders. *Proc. Inst. Mech. Eng. Part B J. Eng. Manuf.* **2020**, *234*, 157–169. [CrossRef]
10. Zhang, X.; Cui, W.; Li, W.; Liou, F. Effects of tool path in remanufacturing cylindrical components by laser metal deposition. *Int. J. Adv. Manuf. Technol.* **2019**, *100*, 1607–1617. [CrossRef]
11. Yoo, S.-W.; Lee, C.-M.; Kim, D.-H. Effect of functionally graded material (FGM) interlayer in metal additive manufacturing of inconel-stainless bimetallic structure by laser melting deposition (LMD) and wire arc additive manufacturing (WAAM). *Materials* **2023**, *16*, 535. [CrossRef] [PubMed]
12. Li, G.; Huang, J.; Wu, Y. An investigation on microstructure and properties of dissimilar welded Inconel 625 and SUS 304 using high-power CO_2 laser. *Int. J. Adv. Manuf. Technol.* **2014**, *76*, 1203–1214. [CrossRef]
13. Lee, H.T.; Jeng, S.L.; Kuo, T.Y. The microstructure and fracture behavior of the dissimilar alloy 690-SUS 304L joint with various Nb addition. *Metall. Mater. Trans. A Phys. Metall. Mater. Sci.* **2003**, *34*, 1097–1105. [CrossRef]
14. Sun, Z.; Karppi, R. The application of electron beam welding for the joining of dissimilar metals: An overview. *J. Mater. Process. Technol.* **1996**, *59*, 257–267. [CrossRef]
15. Frazier, W.E. Metal additive manufacturing: A review. *J. Mater. Eng. Perform.* **2014**, *23*, 1917–1928. [CrossRef]

16. Hofmann, D.C.; Kolodziejska, J.; Roberts, S.; Otis, R.; Dillon, R.P.; Suh, J.O.; Liu, Z.K.; Borgonia, J.P. Compositionally graded metals: A new frontier of additive manufacturing. *J. Mater. Res.* **2014**, *29*, 1899–1910. [CrossRef]
17. Birman, V.; Byrd, L.W. Modeling and analysis of functionally graded materials and structures. *Appl. Mech. Rev.* **2007**, *60*, 195–216. [CrossRef]
18. Shah, K.; Haq, I.; Khan, A.; Shah, S.A.; Khan, M.; Pinkerton, A.J. Parametric study of development of Inconel-steel functionally graded materials by laser direct metal deposition. *Mater. Des.* **2014**, *54*, 531–538. [CrossRef]
19. Carroll, B.E.; Otis, R.A.; Borgonia, J.P.; Suh, J.O.; Dillon, R.P.; Shapiro, A.A.; Hofmann, D.C.; Liu, Z.K.; Beese, A.M. Functionally graded material of 304L stainless steel and inconel 625 fabricated by directed energy deposition: Characterization and thermodynamic modeling. *Acta Mater.* **2016**, *108*, 46–54. [CrossRef]
20. Wang, Y.; Cui, H.; Fan, M.; Chen, Y.; Lu, F. Characterization on the gradient microstructure near the fusion interface of dissimilar metal between high Cr heat-resistant steel and Ni-based Alloy 617. *Mater. Charact.* **2019**, *151*, 227–236. [CrossRef]
21. Savitha, U.; Jagan Reddy, G.; Venkataramana, A.; Sambasiva Rao, A.; Gokhale, A.A.; Sundararaman, M. Chemical analysis, structure and mechanical properties of discrete and compositionally graded SS316-IN625 dual materials. *Mater. Sci. Eng. A.* **2015**, *647*, 344–352. [CrossRef]
22. Sobczak, J.J.; Drenchev, L. Metallic functionally graded materials: A specific class of advanced composites. *J. Mater. Sci. Technol.* **2013**, *29*, 297–316. [CrossRef]
23. Soffel, F.; Lin, Y.; Keller, D.; Egorov, S.; Wegener, K. Laser remelting process simulation and optimization for additive manufacturing of Nickel-based super alloys. *Materials* **2022**, *15*, 177. [CrossRef]
24. Zhao, Y.; Yu, T.; Guan, C.; Sun, J.; Tan, X. Microstructure and friction coefficient of ceramic (TiC, TiN and B_4C) reinforced Ni-based coating by laser cladding. *Ceram. Int.* **2019**, *45*, 20824–20836. [CrossRef]
25. Sun, S.; Fu, H.; Chen, S.; Ping, X.; Wang, K.; Guo, X.; Lin, J.; Lei, Y. A numerical-experimental investigation of heat distribution, stress field and crack susceptibility in Ni60A coatings. *Opt. Laser Technol.* **2019**, *117*, 175–185. [CrossRef]
26. Xia, M.; Gu, D.; Ma, C.; Zhang, H.; Dai, D.; Chen, H.; Li, C.; Zhou, Z.; Chen, G.; Kelbassa, I. Fragmentation and refinement behavior and underlying thermodynamic mechanism of WC reinforcement during selective laser melting of Ni-based composites. *J. Alloys Compd.* **2019**, *777*, 693–702. [CrossRef]
27. Lin, X.; Yue, T.M. Phase formation and microstructure evolution in laser rapid forming of graded SS316L/Rene88DT alloy. *Mater. Sci. Eng. A* **2005**, *402*, 294–306. [CrossRef]
28. Lin, X.; Yue, T.M.; Yang, H.O.; Huang, W.D. Laser rapid forming of SS316L/Rene88DT graded material. *Mater. Sci. Eng. A* **2005**, *391*, 325–336. [CrossRef]
29. Zhang, K.; Wang, S.; Liu, W.; Shang, X. Characterization of stainless steel parts by laser metal deposition shaping. *Mater. Des.* **2014**, *55*, 104–119. [CrossRef]
30. Zhao, Y.; Yu, T.; Sun, J.; Chen, L.; Chen, Y. Effect of laser cladding on forming microhardness and tensile strength of YCF101 alloy powder in the different full lap joint modes. *J. Alloys Compd.* **2020**, *820*, 150230. [CrossRef]
31. Zhao, Y.; Chen, L.Y.; Sun, J.; Wu, W.; Yu, T. Microstructure evolution and wear resistance of in-situ synthesized (Ti, Nb) C ceramic reinforced Ni204 composite coatings. *Ceram. Int.* **2022**, *48*, 17518–17528. [CrossRef]

Journal of
Functional
Biomaterials

MDPI

Article

Biological Performance of Duplex PEO + CNT/PCL Coating on AZ31B Mg Alloy for Orthopedic and Dental Applications

Morteza Daavari [1], Masoud Atapour [1,*], Marta Mohedano [2], Endzhe Matykina [2], Raul Arrabal [2] and Dobrila Nesic [3,*]

1 Department of Materials Engineering, Isfahan University of Technology, Isfahan 84156-83111, Iran; morteza.daavari@gmail.com
2 Departamento de Ingeniería Química y de Materiales, Facultad de Ciencias Químicas, Universidad Complutense, 28040 Madrid, Spain; mmohedan@ucm.es (M.M.); e.matykina@quim.ucm.es (E.M.); rarrabal@ucm.es (R.A.)
3 Division of Fixed Prosthodontics and Biomaterials, University Clinic of Dental Medicine, University of Geneva, Rue Michel-Servet 1, CH-1211 Geneva, Switzerland
* Correspondence: m.atapour@iut.ac.ir (M.A.); dobrila.nesic@unige.ch (D.N.)

Abstract: To regulate the degradation rate and improve the surface biocompatibility of the AZ31B magnesium alloy, three different coating systems were produced via plasma electrolytic oxidation (PEO): simple PEO, PEO incorporating multi-walled carbon nanotubes (PEO + CNT), and a duplex coating that included a polycaprolactone top layer (PEO + CNT/PCL). Surfaces were characterized by chemical content, roughness, topography, and wettability. Biological properties analysis included cell metabolism and adhesion. PEO ± CNT resulted in an augmented surface roughness compared with the base material (BM), while PCL deposition produced the smoothest surface. All surfaces had a contact angle below 90°. The exposure of gFib-TERT and bmMSC to culture media collected after 3 or 24 h did not affect their metabolism. A decrease in metabolic activity of 9% and 14% for bmMSC and of 14% and 29% for gFib-TERT was observed after 3 and 7 days, respectively. All cells died after 7 days of exposure to BM and after 15 days of exposure to coated surfaces. Saos-2 and gFib-TERT adhered poorly to BM, in contrast to bmMSC. All cells on PEO anchored into the pores with filopodia, exhibited tiny adhesion protrusions on PEO + CNT, and presented a web-like spreading with lamellipodia on PEO + CNT/PCL. The smooth and homogenous surface of the duplex PEO + CNT/PCL coating decreased magnesium corrosion and led to better biological functionality.

Keywords: biodegradable Mg-based implant; plasma electrolyte oxidation (PEO); multi-walled carbon nanotubes (MWCNT); polycaprolactone (PCL); cell adhesion; cell metabolism

1. Introduction

The demand for metallic biomaterials has grown significantly over the past decades due to a constant increase in the aging population, bone diseases, and accidents [1]. Biomaterials developed and used in orthopedics and dentistry for fracture fixation and joint replacement are non-degradable metals, such as titanium, titanium alloys, cobalt-chrome alloys, and stainless steel [2,3]. Despite their excellent mechanical properties and biocompatibility, non-degradable metals present several drawbacks, including stress-shielding effects; undesirable host responses, including inflammation and the release of wear particles; and a necessity for a second removal surgery [4,5]. Among degradable metallic materials, magnesium (Mg) and its alloys have found their place in orthopedic and dental surgeries due to a favorable combination of biocompatibility, strength and elastic modulus, and relatively low weight compared with other implantable metals [6–9]. However, the rapid degradation of Mg alloys in aqueous environments, which leads to the loss of mechanical strength and ultimately implant failure as well as an accumulation of excessive OH^- and H_2, limits their clinical application [10,11]. Therefore, improving the corrosion resistance of Mg by alloying design approaches [12], microstructural modification strategies [13], and surface coating methods [14] have been demonstrated as important techniques to prolong the Mg lifetime in physiological environments [15]. Moreover, surface coatings can not only reduce and control Mg alloys' degradation behavior but also enhance their surface biocompatibility [16,17] to ultimately accelerate tissue regeneration.

Different coatings on Mg alloys, including metal oxide [15], polymer coatings [16], inorganic nonmetallic coatings [18], and composite coatings [14], have been investigated to increase their corrosion resistance, biocompatibility, and osteointegration [19–21]. Among various methods, plasma electrolytic oxidation (PEO) is considered an ideal and simple technique for the surface modification of Mg alloys due to its high adhesion strength to the substrate and the adjustable composition [22,23]. PEO coatings are typically produced in modified alkaline electrolytes, such as phosphate, silicate, or sodium hydroxide solutions [24]. Depending on their thickness, microstructure, porosity, and composition, such coatings can provide an optimal combination of wear and corrosion resistance [25]. However, the formation of microspores and cracks due to PEO discharges and the limited range of composition associated with PEO coatings are the main hindrances to their applications in the biomaterial field [26]. These limitations account for the deterioration of the long-term protection and biocompatibility of PEO coatings. Therefore, the in-situ incorporation of particles into PEO coatings and the fabrication of multilayer structures are considered the best potential solutions to overcome these limitations [27].

Particle incorporation can improve various surface properties and add new functionality to a PEO coating toward the fabrication of a "smart" biomaterial [28,29]. ZnO [30], hydroxyapatite (HA) [31], and Ag particles [32] are among the most frequently studied particles that simultaneously improve the corrosion resistance and biocompatibility of the PEO surface. For instance, Liu et al. [31] investigated the effect of the incorporation of HA particles into a PEO coating and reported a significant enhancement in corrosion resistance due to the formation of a denser and more stable outer later. Because of their biocompatibility, excellent tensile strength, and high aspect ratio to inhibit crack propagation, multi-walled carbon nanotubes (MWCNTs) represent a promising potential material for improving PEO coatings for tissue engineering [33]. An important feature of carbon nanotubes is that they can incorporate drugs and corrosion inhibitors [34]. The use of MWCNT-containing electrolytes for the fabrication of PEO coatings has been previously studied, and they were shown to exhibit high corrosion and high wear resistance [35]. However, very limited information on MWCNT-containing PEO coatings on Mg alloys is available. Notably, their biofunctional features remain unknown.

An additional approach to enhance the corrosion resistance and biological properties of PEO coatings on Mg alloys is to apply polymeric and nanocomposite coatings to fill the pores resulting from PEO layer formation [36–38]. The degradation rate of magnesium substrates could thus be controlled [39,40]. Furthermore, these coatings can incorporate drugs and allow for their prolonged and sustained delivery [41]. Due to attributes such as biocompatibility and biodegradability, polycaprolactone (PCL) is considered a promising polymer in a wide range of biomedical applications, including bone plates, screws, and stents [42,43]. Furthermore, PCL has also been recently employed as a biomaterial for fabricating the bone scaffold for bone tissue regeneration [42,44,45]. The duplex coating with a PCL overlay on PEO with MWCNT coating can be an effective approach to improve the corrosion resistance and biological properties of Mg-based implants.

In continuation of our previous studies, in which the effects of MWCNTs and PCL on the biocorrosion and biotribology of the PEO coating were investigated [35,46], the current work is dedicated to the assessment of the potential effects of MWCNT incorporation and PCL-based duplex treatment on the biological performance of the PEO-coated AZ31B Mg alloy. Indeed, obtaining a significant improvement in biotribology by incorporating MWC-

NTs into the PEO coating [35] and enhancing corrosion protection by combining MWCNT incorporation and PCL duplex treatment [46] evoke the necessity for the assessment of the mentioned modification strategy on the biological response as a determining factor of the success of an implantation of a load-bearing Mg-based bone implant. The obtained results elucidate the pathway for choosing the most effective solutions and eventually reaching the desired life span of the Mg-based load-bearing implant.

2. Materials and Methods

2.1. Sample Preparation

Rectangular-shaped AZ31B Mg alloy samples with dimensions of 100 mm × 30 mm and a chemical composition (wt%) of 3.1% Al, 73% Zn, 0.25% Mn, <0.001% Cu, <0.02% Si, and 95.9% Mg were purchased from Magnesium Elektron Ltd. (Manchester, UK) and used as the base (substrate) material in this study. During preparation, the surfaces of each sample were ground using silicon carbide (SiC) abrasive papers (Struers, Copenhagen, Denmark) until achieving an average surface roughness (Sa) of 0.11 ± 0.01 µm. Subsequently, they were sequentially washed in deionized water and pure ethanol and dried with warm airflow.

2.2. Plasma Electrolytic Oxidation (PEO) Process and PCL Deposition

PEO treatment was conducted as previously described [46]. In brief, PEO coating was performed in a silicate-phosphate-based solution containing 10 g/L $Na_3PO_4{\cdot}12H_2O$, 10 g/L Na_2SiO_3, 3 g/L CaO, and 1 g/L KOH. This electrolyte was prepared using deionized water and analytical grade chemicals at pH = 13 and an electrical conductivity of 29.7 mS/cm. The coatings were fabricated in a 2 L double jacket glass beaker. The PEO process was conducted for 600 s with a square-shaped electrical signal, a duty cycle of 50%, an initial ramp of 60 s, a frequency of 400 Hz, and a current density limit of 100 mA/cm^2 using an EAC-S2000 power supply (ET system electronic GmbH, Altlußheim, Germany). A homemade cylindrical mesh made of 316 stainless steel was used as a counter electrode. The anodic and cathodic voltages were adjusted at +405 V and −25 V, respectively. The electrolyte was stirred during the PEO process using a magnetic stirrer (Bunsen, Madrid, Spain). The temperature of the PEO electrolyte was around 20 °C, and the electrolyte was cooled using a chiller WK1200 (Lauda, Lauda-Königshofen, Germany) during the PEO treatment.

To produce PEO coating containing MWCNT, 5 g/L MWCNTs (length: 10–30 µm, outer diameter: 20–30 nm, inner diameter: 5–10 nm, supplier US Research Nanomaterials Inc., Houston, TX, USA) were added to the electrolyte mentioned above, and the same PEO deposition conditions were applied. After the PEO process, the coated specimens were rinsed sequentially with deionized water and ethanol and dried with warm airflow.

To produce the duplex coating, a polymeric (PCL) overlay was applied on top of the MWCNT-incorporated PEO coating. For this purpose, 10 g of polycaprolactone (PCL) granules (polydispersity index (PDI): 1.63, NaturePlast, Normandy, France) were dissolved in 100 mL of chloroform solvent (Analytical grade, Sigma-Aldrich, Darmstadt, Germany) under magnetic stirring for 8 h. A ND-DC 11/1 150 automatic dip coater machine (NadeTech Innovations, Pamplona, Spain) was used to deposit two PCL layers by immersing and withdrawing the specimens at a 150 mm/min speed. Coated samples were dried at room temperature for 24 h. The production of different surfaces is illustrated in Figure 1.

The following acronyms were used in this study: BM for the uncoated base material, PEO for PEO coating, PEO + CNT for PEO with incorporated MWCNTs, and PEO + CNT/PCL for the MWCNT-incorporated PEO with a PCL overlay.

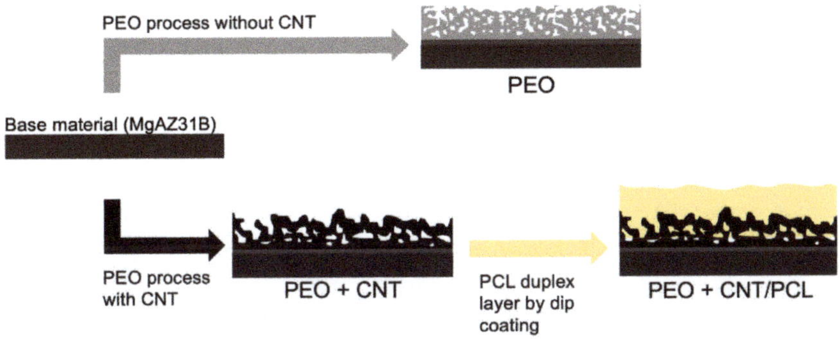

Figure 1. Schematic summary of the coating development procedures.

2.3. Characterization of Base Material and Coated Surfaces

The coating thickness was measured using an eddy current meter (Fischer ISOSCOPE-FMP10, Helmut Fischer GmbH, Sindelfingen, Germany) by applying the probe randomly over the entire surface of the specimen. The cited values are an average of 10 measurements with corresponding standard deviations. The surface and cross-sectional chemistry and morphology of different coatings were investigated using a scanning electron microscope equipped with an energy dispersive spectrometer (EDS) (Philips XL30, FEI, Eindhoven, The Netherland). The average thickness was verified, and the porosity level of the coatings was assessed on the cross-section and surface of the SEM images using ImageJ software (version 1.6, National Institutes of Health, Bethesda, MD, USA). To observe the cross-sections, the specimens were first polished with a series of SiC papers with successively finer grades up to 2400 grit, followed by polishing with alumina particles (0.05 μm). Finally, the specimens were washed ultrasonically in deionized water and ethanol and dried with cold airflow.

The topography of the surfaces was evaluated by an optical profilometer InfiniteFocus SL (Alicona, Gmbh, Graz, Austria) at different magnifications. High-resolution 3D measurements (20 nm vertical resolutions with $\times$ 50 objective) were obtained. Three measurements were performed for each surface, and data are expressed as Sa (arithmetic mean height of the selected surface area)/Sz (distance between the highest peak and the deepest valley in the selected area) values.

The wettability of the uncoated and coated surfaces was assessed by determining the contact angle of a sessile droplet (2 μL) 5 min after the dropping moment using a homemade contact angle measurement device (Sharif solar Ca-500A, Tehran, Iran). The measurements were performed at the ambient temperature and in a humid chamber to minimize the evaporation of the liquid. Simulated body fluid (SBF) (8.035 g/L NaCl, 0.355 g/L $NaHCO_3$, 0.225 g/L KCl, 0.231 g/L $K_2HPO_4{\cdot}3H_2O$, 0.311 g/L $MgCl_2{\cdot}6H_2O$, 0.292 g/L $CaCl_2$, and 0.072 Na_2SO_4) was used as the test liquid instead of deionized water to consider the effect of soluble salts on the surface's wettability.

2.4. Biological Assays: Effects on Cell Metabolism and Cell Adhesion

For cell culture experiments, full-size rectangular-shaped specimens had to be cut into 1 $\times$ 1 cm cuboids. To facilitate the cutting process, deionized water was utilized as a lubricant. The process was carried out at room temperature. The outer edges were covered with wax to prevent the corrosion of the base material from the sides. For sterilization, samples were left overnight in 70% ethanol, dried, and exposed to UV light for 45 min on each side. BM, PEO, PEO + CNT, and PEO + CNT/PCL specimens (n = 4 per group) were incubated in cell culture media consisting of DMEM/F12 (Gibco, Thermo Fisher Scientific, Zug, Switzerland) supplemented with 10% fetal bovine serum substitute FetalClone III (HyClone, Thermo Fisher Scientific, Zug, Switzerland) and 100 U/mL penicillin + 100 mg/mL streptomycin (Sigma, Buchs, Switzerland) for 21 days. Culture media was collected after 3 h, 1, 3, 7, 14, and 21 days and frozen and kept at −20 °C until further use. TERT-immortalized human gingival fibroblasts (gFib-TERT, abm, Richmond, BC, Canada), and human bone marrow mesenchymal stem cells, bmMSCs (own cell bank), were seeded at 10,000 cells/96 wells in triplicate per time point, corresponding to the exposed media collection and material/surface type. After 48 h, the cell culture medium was removed and replaced with the collected exposed medium. After 24 h, cell metabolism was evaluated with a resazurin colorimetric assay. Optical density was read at 570 nm and 630 nm with an LEDETECT 96 microplate reader (Labexim Products, Lengau, Austria) and expressed as the ratio 530 nm/630 nm. Dimethyl sulfoxide, DMSO, 5%, (Sigma, Buchs, Switzerland), was used as a positive control to induce cell death. Cell incubation in a standard medium was used as a negative control. After the experiment, the specimens were rinsed 3 times in sterile water and air-dried.

BM, PEO, PEO + CNT, and PEO + CNT/PCL specimens previously incubated for 3 h in a cell culture medium were used for the direct cell adhesion experiment. gFib-TERT, bmMSC, and Saos-2, a human osteosarcoma cell line, were seeded at 5000 cells/cm^2 in a cell culture medium and incubated for 24 h. After rinsing in PBS and 1 h of fixation in 10% PFA at room temperature, the specimens were rinsed in PBS and dehydrated in ascending ethanol concentrations (30, 50, 60, 70, 80, 90, 95, and 100%) for 10 min each. After drying in the ventilated hood, samples were sputtered with gold and examined under an SEM (Zeiss, Sigma 300 VP, Jena, Germany).

2.5. Statistical Analysis

The effect of different surfaces on cell metabolism over time was analyzed with two-way ANOVA, followed by a Bonferroni post hoc test, using Prism Version 9 for Macintosh (GraphPad Software, Inc., San Diego, CA, USA). All data are expressed as the mean ± SD. A p-value < 0.05 was considered statistically significant.

3. Results and Discussion

3.1. Characterisation of the Mg Alloy Surfaces

Surface characteristics, including the presence, distribution, and percentage of chemical elements; topography; roughness; and wettability, all influence cell responses to biomaterials [47,48]. The results of the chemical and structural analysis of the base material, namely AZ31B Mg alloy, and the three coatings are shown in Figure 2.

The presence of Al, as the main magnesium alloying element, and Mn and Fe led to the formation of intermetallic particles in the microstructure of the base material (Figure 2a). The cathodic nature of these particles can increase the degradation rate of the Mg matrix through the formation of the galvanic couples led by the high potential difference [49].

A coating could suppress the adverse effects of intermetallic particles and allow for controlled magnesium degradation rates. Three different PEO-based coatings were developed, and their surface microstructure, EDS-Map, and EDS-Line were analyzed. PEO coating without and with MWCNTs resulted in the typical creation of micro-pores, ceramic granules, and micro-cracks (Figure 2c,f). The formation of micro-pores is related to the generation of the discharge channels and gas entrapment. At the same time, micro-cracks originate from thermal stresses due to the rapid solidification of molten oxide. The porosity percentages of ~4.3% and ~15.9% were obtained for the PEO and PEO + CNT coatings, respectively, as reported previously [46], suggesting that the addition of nanoparticles leads to an increase in the irregular shape pore opening number and diameter. The EDS analysis results identified Mg, O, C, Ca, P, and Si as the major elements on the surface of all three coatings, albeit in different ratios, independent of areas or structural appearance (Figure 2d,g,j). The PEO and PEO + CNT coatings demonstrated no significant difference between their surface chemistry, with a predominance of magnesium and oxygen. The EDS-Line analyses across the deposited layers indicated that the distributions of the elements were similar for PEO and PEO + CNT. Moreover, the thickness of the PEO coating increased after the incorporation of MWCNTs (Figure 2e,h). The deposition of the PCL layer onto the PEO + CNT coating resulted in the formation of a uniform 10 μm thick polymeric layer, efficiently sealing pores and cracks (Figure 2i–k). The presence of pores was shown to improve the bonding strength between the base metal and the PCL layer [50]. The EDS-Map of the PCL coating indicated a decreased percentage of Mg and increased percentages of carbon and oxygen on the surface while detecting the other elements from the underlying PEO + CNT coating (Figure 2j), in line with previous work [50].

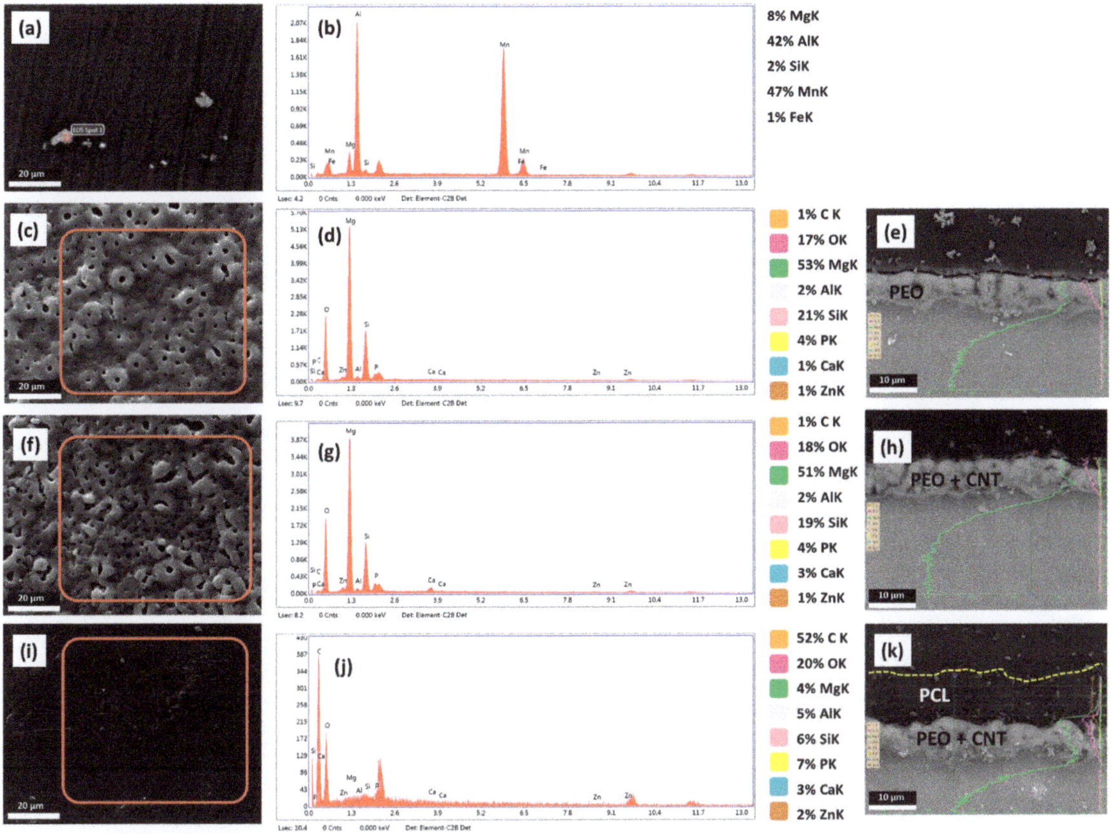

Figure 2. Characterization of Mg alloy surfaces. The SEM micrograph depicting the surface of BM (**a**), EDS point analysis of BM (**b**), SEM micrograph of PEO surface (**c**), EDS analysis of the selected PEO area (**d**), PEO cross-section EDS-Line (**e**), SEM micrograph PEO + CNT surface (**f**), EDS analysis of the selected PEO + CNT area (**g**), PEO + CNT cross-section EDS-Line (**h**), SEM micrograph of PEO + CNT/PCL surface (**i**), EDS analysis of the selected PEO + CNT/PCL area (**j**), PEO + CNT/PCL cross-section EDS-Line (**k**). The percentages of elements based on EDS analysis are depicted. The semi-quantitative EDS analysis results are given in %.

The uniformity and the thickness of the PCL layer are important factors in controlling magnesium degradation. Due to the dipping process constraints, the thickness of the deposited PCL layer depends on the specimen shape, particularly the length, and on the different immersion times of the specimen into the polymeric solution during each dipping cycle. Although the same conditions were employed, due to the longer specimen in the current study, the PCL layer was thinner (10 µm) than that in our previous study (60 µm) [46].

The surface roughness of the coatings is one of the critical factors controlling the biological reaction to the implants [51]. In comparison with the BM roughness of Sa/Sz 0.16 ± 0.04/3.20 ± 0.15 μm, the deposition of PEO resulted in an increase of Sa/Sz to (0.46 ± 0.04)/(6.69 ± 0.62) μm, and the deposition of PEO + CNT further increased the Sa/Sz to 0.66 ± 0.08/13.32 ± 1.43 μm, respectively. These results are in agreement with the formation of larger irregular pores (openings) in the PEO + CNT coating. Applying the PLC layer to the PEO + CNT decreased the surface roughness to Sa/Sz to 0.24 ± 0.02/ 2.53 ± 0.15 μm. The surface roughness of the PEO coating with the PCL polymer overlay strongly decreased due to the filling of the pores and cracks. The 2D micrographs showing the topography of different coatings revealed that PEO coating resulted in an uneven surface compared with BM (Figure 3a,b), and the incorporation of MWCNTs resulted in a more pronounced uneven surface (Figure 3c). By contrast, the duplex PEO + CNT/PCL polymer coating exhibited a very smooth surface (Figure 3d).

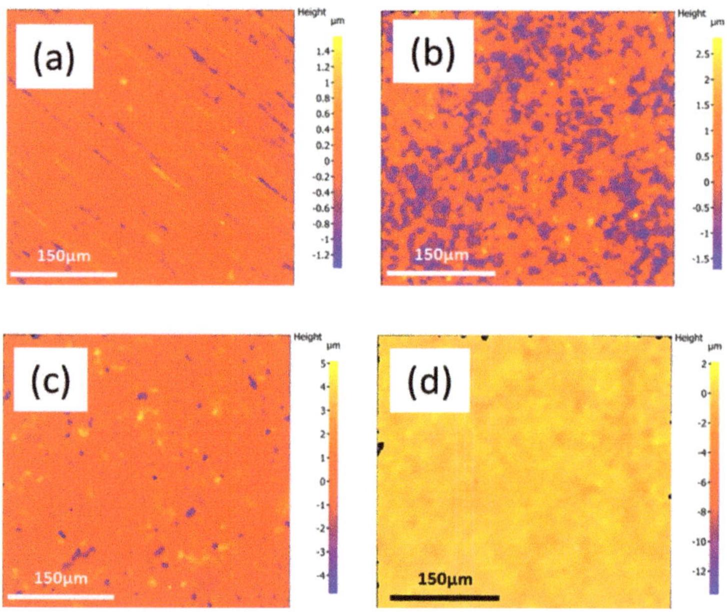

Figure 3. Evaluation of surface topography. Two-dimensional optical images showing the topography of the BM (**a**) PEO, (**b**) PEO + CNT (**c**), and PEO + CNT/PCL (**d**) coatings. Please note the differences in the sidebar reference scale.

3.2. Wettability

The wettability of the coated specimens was identified by measuring the contact angles of SBF, and the results are shown in Figure 4. Compared with the uncoated base metal, an increase in hydrophilicity, i.e., lower values of contact angles for the PEO-treated sample (~55.2°) and PEO + CNT (~60.3°) were observed, in line with previous work [50]. The hydrophobic nature of the MWCNT particles [52] could explain the slight increase in the contact angle compared with that in the PEO coating. The addition of CNT may slow down the penetration of SBF into the pores of the PEO coating. In comparison with BM, lower values for PEO + CNT/PCL specimens were observed.

Figure 4. Contact angles for different magnesium surfaces. Contact angles were measured on (**a**) BM, (**b**) PEO, (**c**) PEO + CNT, and (**d**) PEO + CNT/PCL surfaces.

3.3. Cell Metabolism and Cell Adhesion

Understanding the interactions between different surface properties and biological responses at the cellular level is a prerequisite for designing innovative, functional biomaterials and surface coatings. The dynamic interplay between cell materials influences cell adhesion, proliferation, and finally, differentiation [48]. Biomaterial biocompatibility relies on the cell response to the material degradation products and the surface properties the cells encounter, including wettability, roughness, surface charge, chemical functionalities, and stiffness [53–56]. Additionally, the cell type, the culturing conditions, as well as the medium composition influence cell behavior [55].

The effect of potentially toxic agents released from the different materials/surfaces on cell metabolism was examined with two human cell types: gingival fibroblasts, gFib-TERTs, and bone marrow MSCs. In clinics, both cell types would naturally be in contact with the tested materials/surfaces. The obtained results indicated similar responses of both cell types to the media collected after incubation times from 3 h to 21 days with the studied surfaces (Figure 5a,b).

Cells were exposed to collected media for 24 h. After 3 h of incubation with media collected from BM, PEO, PEO + CNT, and PEO + CNT/PCL specimens, the metabolism of both cell types remained unperturbed and similar to that of the cells incubated in fresh cell culture media. After one day, however, while gFib-TERTs displayed similar metabolic rates independently of the surface exposure medium, there was a significantly higher metabolism in bmMSCs exposed to media collected from PEO + CNT and PEO + CNT/PCL compared with PEO. At day 3, compared with the fresh medium control, a 9% and 14% decrease in cell metabolism was observed for bmMSCs and gFib-TERTs, respectively, independent of the specimen analyzed. Exposed media collected from the BM after 7 days of incubation induced the complete death of bmMSCs and gFib-TERTs. At the same time, metabolic levels decreased by 14% for bmMSCs and 29% for gFib-TERTs, independent of the surface.

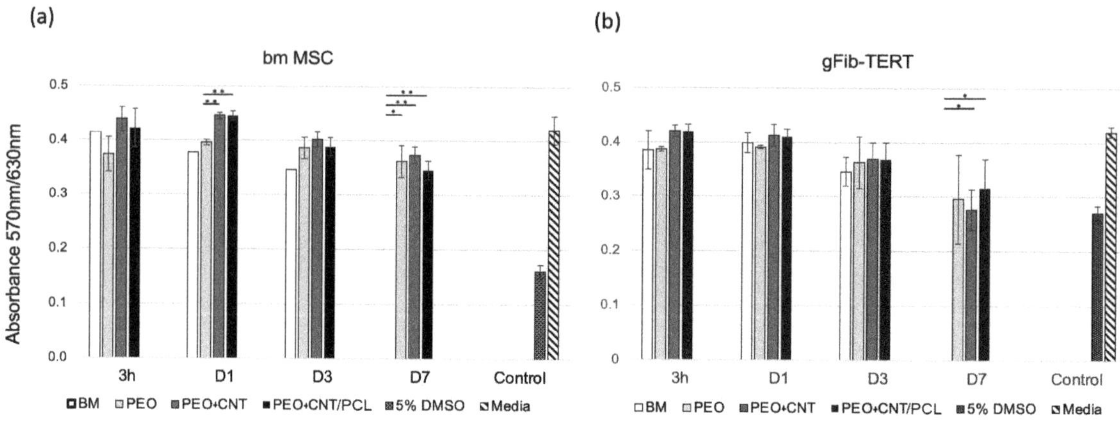

Figure 5. Cell metabolism on different surfaces. Bone marrow MSCs (**a**) and gFib-TERTs (**b**) metabolism expressed as an absorbance ratio measured at 570 nm and 630 nm. Cells were exposed for 24 h to the culture media collected upon incubation with BM, PEO, PEO + CNT, and PEO + CNT/PCL for 3 h, 1, 3, and 7 days. Controls comprised exposure to 5% DMSO (positive control) and fresh cell culture media (negative control). Significant differences ($p < 0.05$) are indicated: * for $p < 0.05$, and ** for $p < 0.01$.

In comparison, exposure to 5% DMSO resulted in metabolic decreases of 62% and 36% for bmMSCs and gFib-TERTs, respectively, compared with fresh media. After 15 days, surface degradation and ensuing corrosion lead to complete cell death in both cell types, regardless of the coating differences (data not shown). Previous studies have demonstrated that the corrosion process can significantly change the pH of the medium and affect cell metabolism [57,58]. The pH value of human tissue is in the range of 7 ~ 7.4. Hence, a higher pH value beyond this optimal range, resulting from the release of OH- ions [59], is expected to negatively influence cell metabolism [60], in line with the presented data. DMEM is a complex solution containing salts, glucose, amino acids, and various proteins indispensable for cell survival. The precise corrosion mechanisms induced by the medium, particularly by the rapid surface adhesion of proteins and carbon, necessitate a more comprehensive analysis. The difference in the PCL layer thickness could also influence the release rate of toxic components during surface and material degradation and affect cell survival. In comparison with our previous study, the PCL layer was thinner, which could lead to its faster degradation and, consequently, the earlier exposure of cells to the corrosive environment. A thicker PCL layer should be investigated in future studies.

Cell adhesion represents the first contact between a cell and a surface. The adhesion process comprises three stages: cell attachment to the substrate, flattening and spreading, and the formation of focal adhesion points [61]. The optimal surface for cell adhesion depends on cell type, culture conditions, surface chemistry, and topography [62,63]. The surface adhesion of three human cell types was investigated: the cell line derived from the primary osteosarcoma, Saos-2, representing osteogenic cells; bone marrow MSCs, implicated in bone regeneration; and hFib-TERTs, representing oral soft tissue cells that may encounter the investigated surfaces during the healing process. The different cell types showed different morphologies in the monolayer tissue culture: Saos-2 had a cobblestone-like appearance, bmMSCs were large and spread out, and hFib-TERTs presented a typical fibroblastic spindle shape (Figure 6a). Depending on different topographies and surface roughness, the cells displayed different adhesion characteristics. After 24 h on the base material, Saos-2 and gFib-TERTs showed signs of cell detachment and were small (Figure 6b). By contrast, bmMSCs were flat and spread on the surface, albeit with marked cell/surface boundaries and a lack of visible lamellipodia. On PEO, all three cell types were spread and produced thick cytoplasmic protrusions, filopodia, anchored into the pores (Figure 6c). Tiny pores on the order of a few microns were shown to facilitate cell attachment by providing anchoring sites [64]. A previous study demonstrated efficient adhesion of human sarcoma U2OS cells to the PEO surface [57]. A different cell response was observed with the PEO + CNT surface: the thick filopodia were not observed despite the presence of irregular openings. Instead, cells adhered to the surface but without completely spreading (Figure 6d). A previous study demonstrated a similar morphology of the Saos-2 cells on MWCNTs [65]. Such cell reaction can be due to the presence of the MWCNTs in the superficial layers of the coating. The presence of MWCNTs was shown to facilitate cell adhesion via integrin binding [66], which may be explained by the fact that their surface roughness and surface area are similar to those of collagen fibers, favoring cell attachment, proliferation, and osteogenic differentiation [67]. Additionally, Mg_2SiO_4, which was detected on the PEO + CNT surface, was reported to improve osteoblast adhesion [68]. Finally, the formation of very fine web-like lamellipodia with extensive and thin cell adhesion surface was observed on the PEO + CNT/PCL surface, suggesting ideal cell adhesion (Figure 6e). Similar spreading was demonstrated for the human osteoblastic cell line (MG63) on the PCL surface deposited directly onto Mg AZ31 [69].

The surface appearance of the PEO + CNT specimens used as a negative control for cell adhesion contained cracks, which contrasts our previous finding [46]. The difference could be due to the surface exposure to processes before the cell experiment, including water during specimen cutting and ethanol during sterilization. Moreover, the cutting process could have caused cracking of the PEO + CNT coating. Another potential explanation is the different composition of the solution to which the surfaces were exposed in the current study (DMEM with FCS) and a previous study (SBF).

All surfaces demonstrated hydrophilic properties, i.e., a contact angle below 90°, indicating reasonably good wettability and a potentially cell-friendly surface. For the polymer surface, optimal cell adhesion is expected in the range of 40°–70° [55]. Given the similar hydrophilicity of PEO + CNT/PCL (80°) and BM (88°), both higher than the 55° and 60° observed for PEO and PEO + CNT, respectively, a correlation between surface wettability and cell adhesion could not be established. Similarly, a comparative analysis of the effect of the hydrophilicity of different surfaces on gingival fibroblasts did not reveal a correlation [70]. Although a different solution was used for wettability assessment in this study from that in our previous study [46], leading to different contact angle values per surface, the order of the most hydrophobic to the most hydrophilic surface remained similar. Since the presence of soluble salt increases the surface tension of the SBF compared with deionized water, the contact angle values in the current study were higher.

Surface roughness is another known parameter influencing cell adhesion. Interestingly, different cell types respond differently: while osteoblasts preferred a moderately rough surface (Sa = 1.0 to 2.0 μm) [71], a smooth surface of Sa = 0.2 μm proved optimal for human gingival fibroblast attachment and proliferation [72,73]. In this study, PEO + CNT/PCL had a lower surface roughness compared with the other surfaces and demonstrated the most efficient cell adhesion.

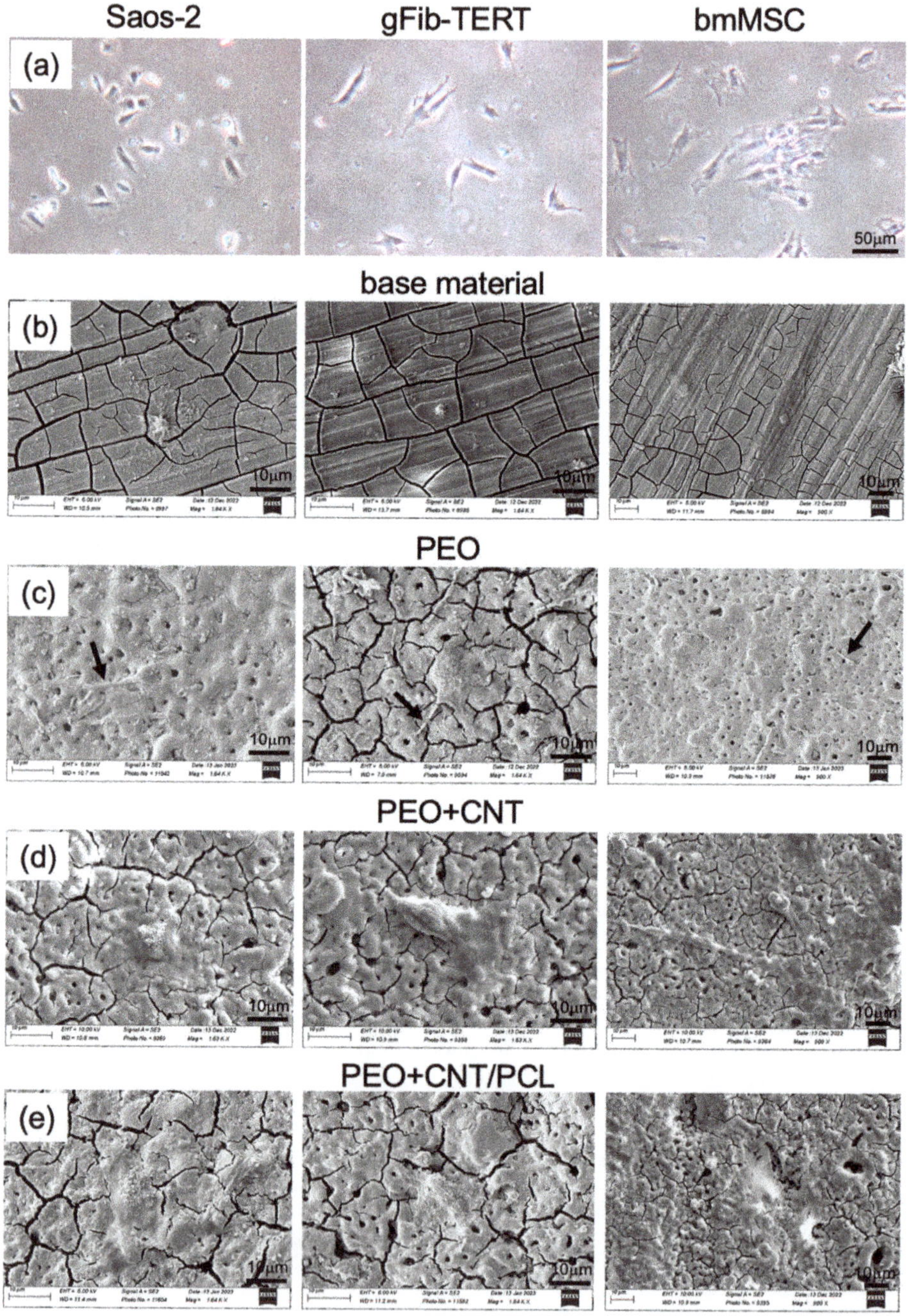

Figure 6. Cell adhesion and spreading on different surfaces. Phenotypic appearance of osteoblastic cell line Saos-2, oral gingival fibroblasts gFib-TERTs, and bone marrow MSCs on tissue culture plate (**a**), base material (**b**), PEO (**c**), PEO + CNT (**d**), and PEO + CNT/PCL (**e**) after 24 h. Filopodia are indicated with arrows.

4. Conclusions

In this study, we employed a step-by-step surface modification approach to improve the biological performance of the Mg-based alloy for dental and orthopedic applications. PEO coating alone or containing MWCNTs and PEO containing MWCNTs with a duplex PCL overlayer coating were produced on the magnesium alloy AZ31B and characterized. Based on the similar metabolic cell reactions to all coated surfaces yet favorable cell adhesion on the PEO + CNT/PCL surface, the duplex PEO + CNT/PCL coating may offer a better biological functionality of the magnesium alloy. The PEO and PEO + CNT surfaces provided a platform for successful cell anchoring. The lower wettability and surface roughness of the PCL coating system compared with PEO and PEO + CNT did not affect the positive biological response of the cells. Future work should investigate the in vivo effects of these Mg alloy coatings in an animal model.

Author Contributions: Conceptualization, M.D., M.A. and D.N.; methodology, M.D., M.A. and D.N.; formal analysis, M.D., M.A. and D.N.; investigation, M.D., M.A. and D.N.; writing original draft preparation, M.A. and D.N.; writing review and editing, M.D., M.A., M.M., E.M., R.A. and D.N.; visualization, M.D. and D.N.; supervision, M.A., M.M., E.M. and R.A.; project administration, M.A. and D.N.; funding acquisition, M.A., M.M., E.M. and R.A. All authors have read and agreed to the published version of the manuscript.

Funding: This work was supported by the Avicenna Center of Excellence (ACE) and Isfahan Research Network, as well as by MCIU/AEI/FEDER, UE, grant numbers PID2021-124341OB-C22 and ADITIMAT-CM, the Regional Government of Madrid and EU Structural Funds, grant number S2018/NMT-4411. M. Mohedano received support from the Ministerio de Ciencia e Innovación grant number RYC-2017-21843.

Institutional Review Board Statement: Not applicable.

Informed Consent Statement: Not applicable.

Data Availability Statement: The data are available upon a reasonable request to the corresponding authors.

Conflicts of Interest: The authors declare no conflict of interest.

References

1. Kaur, M.; Singh, K. Review on Titanium and Titanium Based Alloys as Biomaterials for Orthopaedic Applications. *Mater. Sci. Eng. C* **2019**, *102*, 844–862. [CrossRef] [PubMed]
2. Geetha, M.; Singh, A.K.; Asokamani, R.; Gogia, A.K. Ti Based Biomaterials, the Ultimate Choice for Orthopaedic Implants—A Review. *Prog. Mater. Sci.* **2009**, *54*, 397–425. [CrossRef]
3. Niinomi, M.; Nakai, M.; Hieda, J. Development of New Metallic Alloys for Biomedical Applications. *Acta Biomater.* **2012**, *8*, 3888–3903. [CrossRef]
4. Sumner, D.R. Long-Term Implant Fixation and Stress-Shielding in Total Hip Replacement. *J. Biomech.* **2015**, *48*, 797–800. [CrossRef] [PubMed]
5. Zhou, Z.; Shi, Q.; Wang, J.; Chen, X.; Hao, Y.; Zhang, Y.; Wang, X. The Unfavorable Role of Titanium Particles Released from Dental Implants. *Nanotheranostics* **2021**, *5*, 321. [CrossRef] [PubMed]
6. Kamrani, S.; Fleck, C. Biodegradable Magnesium Alloys as Temporary Orthopaedic Implants: A Review. *Biometals* **2019**, *32*, 185–193. [CrossRef] [PubMed]
7. Wang, J.; Xu, J.; Hopkins, C.; Chow, D.H.; Qin, L. Biodegradable Magnesium-based Implants in Orthopedics—A General Review and Perspectives. *Adv. Sci.* **2020**, *7*, 1902443. [CrossRef] [PubMed]
8. Wei, S.; Ma, J.-X.; Xu, L.; Gu, X.-S.; Ma, X.-L. Biodegradable Materials for Bone Defect Repair. *Mil. Med. Res.* **2020**, *7*, 54. [CrossRef]
9. Zhao, D.; Witte, F.; Lu, F.; Wang, J.; Li, J.; Qin, L. Current Status on Clinical Applications of Magnesium-Based Orthopaedic Implants: A Review from Clinical Translational Perspective. *Biomaterials* **2017**, *112*, 287–302. [CrossRef]
10. Lee, J.-W.; Han, H.-S.; Han, K.-J.; Park, J.; Jeon, H.; Ok, M.-R.; Seok, H.-K.; Ahn, J.-P.; Lee, K.E.; Lee, D.-H. Long-Term Clinical Study and Multiscale Analysis of in Vivo Biodegradation Mechanism of Mg Alloy. *Proc. Natl. Acad. Sci. USA* **2016**, *113*, 716–721. [CrossRef]
11. Yu, Y.; Lu, H.; Sun, J. Long-Term in Vivo Evolution of High-Purity Mg Screw Degradation—Local and Systemic Effects of Mg Degradation Products. *Acta Biomater.* **2018**, *71*, 215–224. [CrossRef]
12. Amukarimi, S.; Mozafari, M. Biodegradable Magnesium-based Biomaterials: An Overview of Challenges and Opportunities. *MedComm* **2021**, *2*, 123–144. [CrossRef] [PubMed]

13. Yan, K.; Liu, H.; Feng, N.; Bai, J.; Cheng, H.; Liu, J.; Huang, F. Preparation of a Single-Phase Mg–6Zn Alloy via ECAP-Stimulated Solution Treatment. *J. Magnes. Alloy.* **2019**, *7*, 305–314. [CrossRef]
14. Yin, Z.-Z.; Qi, W.-C.; Zeng, R.-C.; Chen, X.-B.; Gu, C.-D.; Guan, S.-K.; Zheng, Y.-F. Advances in Coatings on Biodegradable Magnesium Alloys. *J. Magnes. Alloy.* **2020**, *8*, 42–65. [CrossRef]
15. Ali, M.; Elsherif, M.; Salih, A.E.; Ul-Hamid, A.; Hussein, M.A.; Park, S.; Yetisen, A.K.; Butt, H. Surface Modification and Cytotoxicity of Mg-Based Bio-Alloys: An Overview of Recent Advances. *J. Alloys Compd.* **2020**, *825*, 154140. [CrossRef]
16. Liu, Y.; Zhang, Y.; Wang, Y.-L.; Tian, Y.-Q.; Chen, L.-S. Research Progress on Surface Protective Coatings of Biomedical Degradable Magnesium Alloys. *J. Alloys Compd.* **2021**, *885*, 161001. [CrossRef]
17. Peng, F.; Li, H.; Wang, D.; Tian, P.; Tian, Y.; Yuan, G.; Xu, D.; Liu, X. Enhanced Corrosion Resistance and Biocompatibility of Magnesium Alloy by Mg–Al-Layered Double Hydroxide. *ACS Appl. Mater. Interfaces* **2016**, *8*, 35033–35044. [CrossRef]
18. Zhengjie, L.; Ying, Z.; Zhixiong, Z.; Yanfei, X.; Kelvin, Y. Antibacterial Properties, Hemolysis and Biocompatibility of Biodegradable Medical Magnesium Alloys. *Rare Met. Mater. Eng.* **2018**, *47*, 403–408.
19. Tokunaga, T.; Ohno, M.; Matsuura, K. Coatings on Mg Alloys and Their Mechanical Properties: A Review. *J. Mater. Sci. Technol.* **2018**, *34*, 1119–1126. [CrossRef]
20. Agarwal, S.; Curtin, J.; Duffy, B.; Jaiswal, S. Biodegradable Magnesium Alloys for Orthopaedic Applications: A Review on Corrosion, Biocompatibility and Surface Modifications. *Mater. Sci. Eng. C* **2016**, *68*, 948–963. [CrossRef]
21. Ghanbari, A.; Bordbar-Khiabani, A.; Warchomicka, F.; Sommitsch, C.; Yarmand, B.; Zamanian, A. PEO/Polymer Hybrid Coatings on Magnesium Alloy to Improve Biodegradation and Biocompatibility Properties. *Surf. Interfaces* **2023**, *36*, 102495. [CrossRef]
22. Lin, Z.; Wang, T.; Yu, X.; Sun, X.; Yang, H. Functionalization Treatment of Micro-Arc Oxidation Coatings on Magnesium Alloys: A Review. *J. Alloys Compd.* **2021**, *879*, 160453. [CrossRef]
23. Esmaeili, M.; Tadayonsaidi, M.; Ghorbanian, B. The Effect of PEO Parameters on the Properties of Biodegradable Mg Alloys: A Review. *Surf. Innov.* **2021**, *9*, 184–198. [CrossRef]
24. Yao, Z.; Gao, H.; Jiang, Z.; Wang, F. Structure and Properties of ZrO_2 Ceramic Coatings on AZ91D Mg Alloy by Plasma Electrolytic Oxidation. *J. Am. Ceram. Soc.* **2008**, *91*, 555–558. [CrossRef]
25. Kyrylenko, S.; Warchoł, F.; Oleshko, O.; Husak, Y.; Kazek-Kęsik, A.; Korniienko, V.; Deineka, V.; Sowa, M.; Maciej, A.; Michalska, J. Effects of the Sources of Calcium and Phosphorus on the Structural and Functional Properties of Ceramic Coatings on Titanium Dental Implants Produced by Plasma Electrolytic Oxidation. *Mater. Sci. Eng. C* **2021**, *119*, 111607. [CrossRef] [PubMed]
26. Jiang, P.; Blawert, C.; Zheludkevich, M.L. The Corrosion Performance and Mechanical Properties of Mg-Zn Based Alloys—A Review. *Corros. Mater. Degrad.* **2020**, *1*, 7. [CrossRef]
27. Sikdar, S.; Menezes, P.V.; Maccione, R.; Jacob, T.; Menezes, P.L. Plasma Electrolytic Oxidation (PEO) Process—Processing, Properties, and Applications. *Nanomaterials* **2021**, *11*, 1375. [CrossRef]
28. O'Hara, M.; Troughton, S.C.; Francis, R.; Clyne, T.W. The Incorporation of Particles Suspended in the Electrolyte into Plasma Electrolytic Oxidation Coatings on Ti and Al Substrates. *Surf. Coatings Technol.* **2020**, *385*, 125354. [CrossRef]
29. Lu, X.; Mohedano, M.; Blawert, C.; Matykina, E.; Arrabal, R.; Kainer, K.U.; Zheludkevich, M.L. Plasma Electrolytic Oxidation Coatings with Particle Additions–A Review. *Surf. Coatings Technol.* **2016**, *307*, 1165–1182. [CrossRef]
30. Keyvani, A.; Zamani, M.; Bahamirian, M.; Nikoomanzari, E.; Fattah-Alhosseini, A.; Sina, H. Role of Incorporation of ZnO Nanoparticles on Corrosion Behavior of Ceramic Coatings Developed on AZ31 Magnesium Alloy by Plasma Electrolytic Oxidation Technique. *Surfaces and Interfaces* **2021**, *22*, 100728. [CrossRef]
31. Liu, C.; Zhang, W.; Xu, T.; Li, H.; Jiang, B.; Miao, X. Preparation and Corrosion Resistance of a Self-Sealing Hydroxyapatite-MgO Coating on Magnesium Alloy by Microarc Oxidation. *Ceram. Int.* **2022**, *48*, 13676–13683. [CrossRef]
32. Han, Q.; Li, Y.; Lu, X.; Mei, D.; Chen, Q.; Zhang, T.; Wang, F. Fabrication of Ag Containing Antibacterial PEO Coatings on Pure Mg. *Mater. Lett.* **2021**, *293*, 129731. [CrossRef]
33. Xiao, Y.; Gong, T.; Zhou, S. The Functionalization of Multi-Walled Carbon Nanotubes by in Situ Deposition of Hydroxyapatite. *Biomaterials* **2010**, *31*, 5182–5190. [CrossRef] [PubMed]
34. Luo, X.; Matranga, C.; Tan, S.; Alba, N.; Cui, X.T. Carbon Nanotube Nanoreservior for Controlled Release of Anti-Inflammatory Dexamethasone. *Biomaterials* **2011**, *32*, 6316–6323. [CrossRef] [PubMed]
35. Daavari, M.; Atapour, M.; Mohedano, M.; Arrabal, R.; Matykina, E.; Taherizadeh, A. Biotribology and Biocorrosion of MWCNTs-Reinforced PEO Coating on AZ31B Mg Alloy. *Surf. Interfaces* **2021**, *22*, 100850. [CrossRef]
36. Gnedenkov, A.S.; Sinebryukhov, S.L.; Filonina, V.S.; Ustinov, A.Y.; Sukhoverkhov, S.V.; Gnedenkov, S.V. New Polycaprolactone-Containing Self-Healing Coating Design for Enhance Corrosion Resistance of the Magnesium and Its Alloys. *Polymers* **2022**, *15*, 202. [CrossRef]
37. Moreno, L.; Wang, C.; Lamaka, S.V.; Zheludkevich, M.L.; Rodríguez-Hernández, J.; Arrabal, R.; Matykina, E. Ciprofloxacin Release and Corrosion Behaviour of a Hybrid PEO/PCL Coating on $Mg_3Zn_{0.4}Ca$ Alloy. *J. Funct. Biomater.* **2023**, *14*, 65. [CrossRef]
38. Gnedenkov, A.S.; Sinebryukhov, S.L.; Filonina, V.S.; Ustinov, A.Y.; Gnedenkov, S. V Hybrid Coatings for Active Protection against Corrosion of Mg and Its Alloys. *Polymers* **2023**, *15*, 3035. [CrossRef]
39. Rai, R.; Tallawi, M.; Grigore, A.; Boccaccini, A.R. Synthesis, Properties and Biomedical Applications of Poly (Glycerol Sebacate) (PGS): A Review. *Prog. Polym. Sci.* **2012**, *37*, 1051–1078. [CrossRef]
40. Kim, M.J.; Hwang, M.Y.; Kim, J.; Chung, D.J. Biodegradable and Elastomeric Poly (Glycerol Sebacate) as a Coating Material for Nitinol Bare Stent. *Biomed Res. Int.* **2014**, *2014*, 956952. [CrossRef]

41. Stewart, S.A.; Domínguez-Robles, J.; McIlorum, V.J.; Gonzalez, Z.; Utomo, E.; Mancuso, E.; Lamprou, D.A.; Donnelly, R.F.; Larrañeta, E. Poly (Caprolactone)-Based Coatings on 3D-Printed Biodegradable Implants: A Novel Strategy to Prolong Delivery of Hydrophilic Drugs. *Mol. Pharm.* **2020**, *17*, 3487–3500. [CrossRef] [PubMed]
42. Dwivedi, R.; Kumar, S.; Pandey, R.; Mahajan, A.; Nandana, D.; Katti, D.S.; Mehrotra, D. Polycaprolactone as Biomaterial for Bone Scaffolds: Review of Literature. *J. Oral Biol. Craniofacial. Res.* **2020**, *10*, 381–388. [CrossRef]
43. Zhang, W.; Ullah, I.; Shi, L.; Zhang, Y.; Ou, H.; Zhou, J.; Ullah, M.W.; Zhang, X.; Li, W. Fabrication and Characterization of Porous Polycaprolactone Scaffold via Extrusion-Based Cryogenic 3D Printing for Tissue Engineering. *Mater. Des.* **2019**, *180*, 107946. [CrossRef]
44. Liu, D.; Nie, W.; Li, D.; Wang, W.; Zheng, L.; Zhang, J.; Zhang, J.; Peng, C.; Mo, X.; He, C. 3D Printed PCL/SrHA Scaffold for Enhanced Bone Regeneration. *Chem. Eng. J.* **2019**, *362*, 269–279. [CrossRef]
45. Murugan, S.; Parcha, S.R. Fabrication Techniques Involved in Developing the Composite Scaffolds PCL/HA Nanoparticles for Bone Tissue Engineering Applications. *J. Mater. Sci. Mater. Med.* **2021**, *32*, 93. [CrossRef]
46. Daavari, M.; Atapour, M.; Mohedano, M.; Sánchez, H.M.; Rodríguez-Hernández, J.; Matykina, E.; Arrabal, R.; Taherizadeh, A. Quasi-in Vivo Corrosion Behavior of AZ31B Mg Alloy with Hybrid MWCNTs-PEO/PCL Based Coatings. *J. Magnes. Alloy.* **2022**, *10*, 3217–3233. [CrossRef]
47. Chaudhary, S.; Ghosal, D.; Tripathi, P.; Kumar, S. Cellular Metabolism: A Link Connecting with Physiochemical Properties of Biomaterials for Bone Tissue Engineering. *Biomater. Sci.* **2023**, *11*, 2277–2291. [CrossRef]
48. Jurak, M.; Wiącek, A.E.; Ładniak, A.; Przykaza, K.; Szafran, K. What Affects the Biocompatibility of Polymers? *Adv. Colloid Interface Sci.* **2021**, *294*, 102451. [CrossRef]
49. Zeng, R.-C.; Yin, Z.-Z.; Chen, X.-B.; Xu, D.-K. Corrosion Types of Magnesium Alloys. In *Magnesium Alloys: Selected Issue*; Books on Demand GmbH: Norderstedt, Germany, 2018; pp. 29–53.
50. Kim, Y.-K.; Lee, K.-B.; Kim, S.-Y.; Jang, Y.-S.; Kim, J.H.; Lee, M.-H. Improvement of Osteogenesis by a Uniform PCL Coating on a Magnesium Screw for Biodegradable Applications. *Sci. Rep.* **2018**, *8*, 13264. [CrossRef]
51. Zhang, Y.; Fan, Z.; Xing, Y.; Jia, S.; Mo, Z.; Gong, H. Effect of Microtopography on Osseointegration of Implantable Biomaterials and Its Modification Strategies. *Front. Bioeng. Biotechnol.* **2022**, *10*, 981062. [CrossRef]
52. Saji, V.S. Carbon Nanostructure-Based Superhydrophobic Surfaces and Coatings. *Nanotechnol. Rev.* **2021**, *10*, 518–571. [CrossRef]
53. Li, J.; Mou, X.; Qiu, J.; Wang, S.; Wang, D.; Sun, D.; Guo, W.; Li, D.; Kumar, A.; Yang, X. Surface Charge Regulation of Osteogenic Differentiation of Mesenchymal Stem Cell on Polarized Ferroelectric Crystal Substrate. *Adv. Healthc. Mater.* **2015**, *4*, 998–1003. [CrossRef]
54. Kaivosoja, E.; Barreto, G.; Levon, K.; Virtanen, S.; Ainola, M.; Konttinen, Y.T. Chemical and Physical Properties of Regenerative Medicine Materials Controlling Stem Cell Fate. *Ann. Med.* **2012**, *44*, 635–650. [CrossRef] [PubMed]
55. Oliveira, S.M.; Alves, N.M.; Mano, J.F. Cell Interactions with Superhydrophilic and Superhydrophobic Surfaces. *J. Adhes. Sci. Technol.* **2014**, *28*, 843–863. [CrossRef]
56. Sergeeva, Y.N.; Huang, T.; Felix, O.; Jung, L.; Tropel, P.; Viville, S.; Decher, G. What Is Really Driving Cell–Surface Interactions? Layer-by-Layer Assembled Films May Help to Answer Questions Concerning Cell Attachment and Response to Biomaterials. *Biointerphases* **2016**, *11*, 019009. [CrossRef] [PubMed]
57. Husak, Y.; Michalska, J.; Oleshko, O.; Korniienko, V.; Grundsteins, K.; Dryhval, B.; Altundal, S.; Mishchenko, O.; Viter, R.; Pogorielov, M. Bioactivity Performance of Pure Mg after Plasma Electrolytic Oxidation in Silicate-Based Solutions. *Molecules* **2021**, *26*, 2094. [CrossRef]
58. Wang, Y.; Ouyang, Y.; Peng, X.; Mao, L.; Yuan, G.; Jiang, Y.; He, Y. Effects and Degradable Mg-Nd-Zn-Zr Alloy on Osteoblastic Cell Function. *Int. J. Immunopathol. Pharmacol.* **2012**, *25*, 597–606. [CrossRef] [PubMed]
59. Farshid, S.; Kharaziha, M.; Atapour, M. A Self-Healing and Bioactive Coating Based on Duplex Plasma Electrolytic Oxidation/Polydopamine on AZ91 Alloy for Bone Implants. *J. Magnes. Alloy.* **2023**, *11*, 592–606. [CrossRef]
60. Tian, P.; Peng, F.; Wang, D.; Liu, X. Corrosion Behavior and Cytocompatibility of Fluoride-Incorporated Plasma Electrolytic Oxidation Coating on Biodegradable AZ31 Alloy. *Regen. Biomater.* **2017**, *4*, 1–10. [CrossRef]
61. Ahmad Khalili, A.; Ahmad, M.R. A Review of Cell Adhesion Studies for Biomedical and Biological Applications. *Int. J. Mol. Sci.* **2015**, *16*, 18149–18184. [CrossRef]
62. Ghibaudo, M.; Trichet, L.; Le Digabel, J.; Richert, A.; Hersen, P.; Ladoux, B. Substrate Topography Induces a Crossover from 2D to 3D Behavior in Fibroblast Migration. *Biophys. J.* **2009**, *97*, 357–368. [CrossRef] [PubMed]
63. Ranella, A.; Barberoglou, M.; Bakogianni, S.; Fotakis, C.; Stratakis, E. Tuning Cell Adhesion by Controlling the Roughness and Wettability of 3D Micro/Nano Silicon Structures. *Acta Biomater.* **2010**, *6*, 2711–2720. [CrossRef] [PubMed]
64. Gariboldi, M.I.; Best, S.M. Effect of Ceramic Scaffold Architectural Parameters on Biological Response. *Front. Bioeng. Biotechnol.* **2015**, *3*, 151. [CrossRef] [PubMed]
65. Akasaka, T.; Yokoyama, A.; Matsuoka, M.; Hashimoto, T.; Abe, S.; Uo, M.; Watari, F. Adhesion of Human Osteoblast-like Cells (Saos-2) to Carbon Nanotube Sheets. *Biomed. Mater. Eng.* **2009**, *19*, 147–153. [CrossRef]
66. Imaninezhad, M.; Schober, J.; Griggs, D.; Ruminski, P.; Kuljanishvili, I.; Zustiak, S.P. Cell Attachment and Spreading on Carbon Nanotubes Is Facilitated by Integrin Binding. *Front. Bioeng. Biotechnol.* **2018**, *6*, 129. [CrossRef]
67. Xu, J.; Hu, X.; Jiang, S.; Wang, Y.; Parungao, R.; Zheng, S.; Nie, Y.; Liu, T.; Song, K. The Application of Multi-Walled Carbon Nanotubes in Bone Tissue Repair Hybrid Scaffolds and the Effect on Cell Growth in Vitro. *Polymers* **2019**, *11*, 230. [CrossRef]

68. Ni, S.; Chou, L.; Chang, J. Preparation and Characterization of Forsterite (Mg_2SiO_4) Bioceramics. *Ceram. Int.* **2007**, *33*, 83–88. [CrossRef]
69. Zomorodian, A.; Garcia, M.P.; Moura e Silva, T.; Fernandes, J.C.S.; Fernandes, M.H.; Montemor, M.d.F. Biofunctional Composite Coating Architectures Based on Polycaprolactone and Nanohydroxyapatite for Controlled Corrosion Activity and Enhanced Biocompatibility of Magnesium AZ31 Alloy. *Mater. Sci. Eng. C* **2015**, *48*, 434–443. [CrossRef]
70. Paul, S.; Hanisch, O.; Nesic, D. Human Gingival Fibroblast Proliferation on Materials Used for Dental Implant Abutments: A Systematic Review. *Int. J. Prosthodont.* **2021**, *34*, 811–828. [CrossRef]
71. Albrektsson, T.; Wennerberg, A. Oral Implant Surfaces: Part 1--Review Focusing on Topographic and Chemical Properties of Different Surfaces and in Vivo Responses to Them. *Int. J. Prosthodont.* **2004**, *17*, 536–543.
72. Kim, Y.-S.; Shin, S.-Y.; Moon, S.-K.; Yang, S.-M. Surface Properties Correlated with the Human Gingival Fibroblasts Attachment on Various Materials for Implant Abutments: A Multiple Regression Analysis. *Acta Odontol. Scand.* **2015**, *73*, 38–47. [CrossRef] [PubMed]
73. Mustafa, K.; Odén, A.; Wennerberg, A.; Hultenby, K.; Arvidson, K. The Influence of Surface Topography of Ceramic Abutments on the Attachment and Proliferation of Human Oral Fibroblasts. *Biomaterials* **2005**, *26*, 373–381. [CrossRef] [PubMed]

Journal of
Manufacturing and Materials Processing

MDPI

Article

Dynamic Analysis of the Thermo-Deformation Treatment Process of Flat Surfaces of Machine Parts

Volodymyr Gurey [1], Pavlo Maruschak [2,*], Ihor Hurey [1,3], Volodymyr Dzyura [2], Tetyana Hurey [4] and Weronika Wojtowicz [5]

1 Department of Robotics and Integrated Mechanical Engineering Technologies, Lviv Polytechnic National University, 12, Bandera St., 79013 Lviv, Ukraine; vgurey@gmail.com (V.G.); ihor.hurey@gmail.com (I.H.)
2 Department of Wheel Vehicles, Ternopil Ivan Puluj National Technical University, 56, Ruska St., 46001 Ternopil, Ukraine; volodymyrdzyura@gmail.com
3 Faculty of Mechanics and Technology, Rzeszow University of Technology, 12, Powstancow Warszawy St., 35-959 Rzeszow, Poland
4 Department of Transport Technologies, Lviv Polytechnic National University, 12, Bandera St., 79013 Lviv, Ukraine; tetiana.a.hurei@lpnu.ua
5 Faculty of Mechanical Engineering and Aeronautics, Rzeszow University of Technology, 12, Powstancow Warszawy St., 35-959 Rzeszow, Poland; wwktmiop@prz.edu.pl
* Correspondence: maruschak.tu.edu@gmail.com

Abstract: Thermo-deformation treatment refers to methods of strengthening during which strengthened layers with a nanocrystalline structure are formed in the surface layers by modifying the metal surface layer, which changes its phase and structural and chemical compositions, reduces grain size, and improves performance. Grinding of the metal structure was achieved by combining two methods simultaneously during this treatment: the action of a highly concentrated energy source on the surface layer and intense plastic deformation. The source of highly concentrated energy was generated in the contact zone of the tool-disc, which rotates at high speed during friction on the treated surface. Intense deformation was achieved due to the grooves on the tool's working surface. Dynamic analysis of the thermo-deformation treatment process of flat surfaces of machine parts and a calculation scheme of the surface grinder machine's elastic system, which is the three-mass model, were developed. When the groove width increased from 4 mm to 8 mm, the force amplitude in the contact zone increased from 10 N to 75 N. Accordingly, the thickness of the nanocrystalline layer increased from 190–220 μm to 250–260 μm, and its hardness increased from 9.3 GPa to 11.1 GPa.

Keywords: friction hardening; strengthening; white layer; nanocrystalline structure; mathematical model

1. Introduction

The operational reliability of products is one of the most important parameters of the quality of machine parts and is defined by such indicators as wear resistance, corrosion resistance, fatigue and corrosion-fatigue strength, the strength of fits and mating parts, etc. [1,2]. The accuracy and quality of manufactured surfaces and the condition of the surface layers of parts directly affect the performance of machines. During any type of operation, destruction begins on the surface of the part [3,4]. Increasing the durability of machine parts is achieved by forming the appropriate stereometric parameters of the machined surfaces of the parts and the properties of the metal surface layer [5,6]. The set parameters of the treated surfaces can be obtained using technological methods that belong to a new direction—surface engineering [7,8]. It allows for the carrying out of surface strengthening of machine parts by a method of modification of the metal's surface layer [9,10]. The surface layer of the metal is strengthened by changing its structure and chemical and phase composition, reducing the grain size to the nanoscale, etc., during modification. It should be noted that the crystal lattice remains the same, as only its dimensions change [11,12].

Modification of the surface layers of metal in machine parts can be carried out by technological methods using highly concentrated energy sources, namely laser [13], plasma [14], electron beam [15,16], frictional strengthening [17,18], and other treatments [19–21]. In these methods of processing under the influence of concentrated energy flows, small volumes of the surface layer of the metal are heated at a high rate to temperatures above the point of phase transformations [22,23]. After removing the heat source, the surface layers of the metal are cooled at a high rate by transferring the heat to the depth of the part [24,25]. Strengthened nanocrystal layers with specific physical, mechanical, and operational properties are formed in the surface layers of the metal.

Forming, changing the structure, and reducing the grain size of the strengthened surface layer of metal are also obtained by methods of intense plastic deformation [26,27]. The surface layer formed during rapid heating and cooling, as well as after intense plastic deformation, has changed physical and mechanical characteristics, different phase states and chemical compositions of the metal, the presence of residual compressive stresses, and formed specific structures [28]. Technological methods that modify the surface layer of metal reduce the grain size when it is in at least one direction that is less than 100 nm (the process of forming the nanostructure), which is significantly smaller than the base material. Surface layers with nanocrystalline structures have unique physical, mechanical, and chemical properties that improve the performance of parts [9,11,27,29,30].

The developed method of thermo-deformation treatment (strengthening) of machine parts—which belongs to the methods of surface strengthening that use high-intensity heat sources—using a tool with transverse grooves on the working surface allows for the combination of two parallel methods of strengthening—strengthening that uses a high-intensity heat source and intensive plastic deformation [9,11,30]. The friction of the working surface of the tool at high speed on the processed surface forms a highly concentrated source of energy, and transverse grooves on its periphery provide high-intensity shock loads.

Thermo-deformation treatment refers to the methods of forming hardened surface layers with a nanocrystalline structure on massive parts. During this treatment, the action of a highly concentrated heat energy flow and intense plastic deformation are simultaneously combined, the source of which is the high-speed friction of a metal tool-disc on the treated surface in the zone of mutual contact. The intensity of deformation of the metal surface layer is increased by increasing the width of the grooves on the tool's working surface.

The method of thermo-deformation treatment belongs to finishing operations. During finishing operations, it is attempted to reduce the dynamic components of the processing procedure. In this case, the dynamic components of the process are increased to intensify the process of hardening (strengthening) and to improve the conditions for the formation of hardened (strengthened) layers. The aim of this work was the dynamic analysis of the thermo-deformation treatment process of flat surfaces of machine parts to determine the magnitude of impact loads in the zone of contact of the tool with the treated flat surface. Thermo-deformation treatment is one of the surface strengthening methods that uses highly concentrated energy flows. Highly concentrated energy flow is formed during high-speed (60–90 m/s) friction of a tool on the treated part (sample) in the zone of contact. The heating rate of the contact zone reaches $4 \cdot 10^5$–$5 \cdot 10^6$ K/s. The surface layers of the metal are heated to temperatures above the point of phase transformations (Ac_3). After removing the heat source, the heated surface layer is cooled at a high rate ($6 \cdot 10^4$–$9 \cdot 10^5$ K/s) by transferring heat to the depth of the metal part. There is simultaneous high-speed shear deformation of the treated surface in the contact zone. Nanocrystalline structures (white layer) are formed in the surface layer of parts due to high-speed heating and cooling and intensive shear deformation.

2. Materials and Methods

Thermo-deformation strengthening of the flat surfaces of machine parts in terms of the performance and kinematics of the process is similar to technological grinding methods (Figure 1a). For their performance, surface grinding machines are used with the modernization of the rotation drive unit to ensure a linear speed of 60–90 m/s at the tool periphery. A metal tool-disc is mounted instead of an abrasive wheel. The overall and mounting dimensions of the metal tool-disc correspond to the dimensions of the abrasive wheel that is used on this grinding machine.

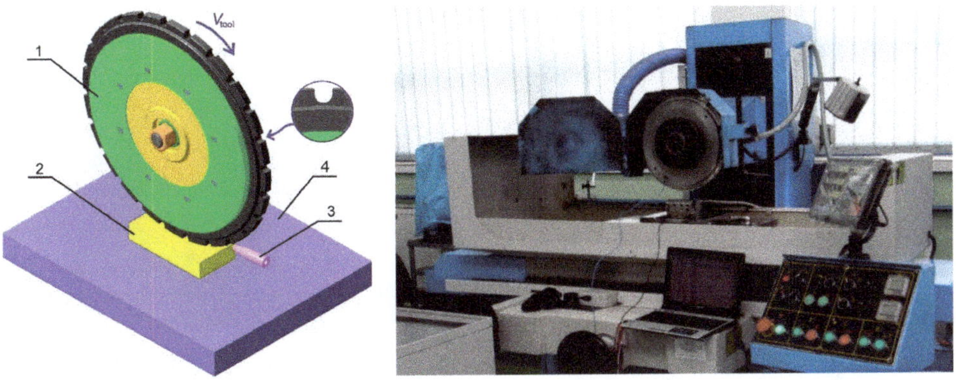

(a) (b)

Figure 1. Scheme of thermo-deformation treatment of flat surfaces of machine parts (**a**): 1—tool, 2—treated flat surface of a part, 3—nozzle for provision of the technological medium, 4—machine table, V_{tool}—linear speed; (**b**) HFS 3063 VS surface grinder from Knuth (Wasbek, Germany).

An upgraded HFS 3063 VS surface grinder from Knuth (Wasbek, Germany) was used to perform thermo-deformation strengthening of flat surfaces (Figure 1b). The machine's spindle drive was replaced with a different one with a higher frequency and the ability to control speed using a DC motor and a control panel from Siemens (Munich, Germany). The bearings of the spindle assembly were calculated for operation at higher speeds and with higher loads. The tool was made of stainless steel. The external diameter D_{tool} of the tool-disc was 360 mm, and the width of the tool's working part b_{tool} was 6–8 mm (Figure 2). The tool was set up into a mandrel holder that was included in the kit with the surface grinder machine. We performed static balancing before mounting the tool-disc assembly with the mandrel holder on the surface grinder machine [9,30].

To improve the processes of formation, a strengthened layer with a nanocrystal structure, namely increasing its thickness and microhardness and reducing the grain size of the structure, a tool with transverse grooves on its periphery was used during thermo-deformation strengthening. There is a patent in Ukraine (No. 123883) [31] and a patent in Poland (No. 240972) [32] for the design of this tool. When using a tool with transverse grooves on its periphery, a discontinuous, highly concentrated energy flow (heating and cooling cycles) is formed, and a source of high-frequency deformation is formed in the zone of contact of the tool with the treated surface. During thermo-deformation strengthening with the tool with transverse grooves on its periphery, in the surface layers of the metal of the treated surfaces of the parts, two processes take place simultaneously: the action of a highly concentrated energy source and intensive shear deformation.

On the working part, the tool has transverse grooves with the same circular pitch, alternating smooth parts, and a groove with the same frequency. The minimum value of the groove width was selected based on the condition that the tool's working surface was guaranteed to completely lose contact with the sample. The tool is pressed to the sample's treated surface during thermo-deformation strengthening with a force of 500–1200 N, and the effect of this force in the contact zone forms a radial component of the force between the tool and the sample. In addition, there is a tangential component of the force between the tool and the sample that occurs by rotation of the tool in the contact zone, which determines the value of the heat flow. When the groove is over the contact zone between the tool and the part at this time point, the contact zone is unloaded, and the normal and tangential components of the force of the tool on the part equal zero. The influence of the heat source is suspended. A sharp (impact) load is applied to the contact zone, followed by high-speed friction of the smooth part of the tool on the samples' surfaces when the tool's next working smooth surface comes into contact. The action of the source of intensive thermal energy is reset in the contact zone. Thermal energy, shear deformation, and impact loads are impulsively applied to the contact zone. The frequency of impact loads is determined by the number of grooves on the working part of the tool. These processes form strengthened layers with a nanocrystalline structure, which have greater thickness, increased hardness, and more fine-grained structure compared to treatment with a tool that has a smooth working surface [11].

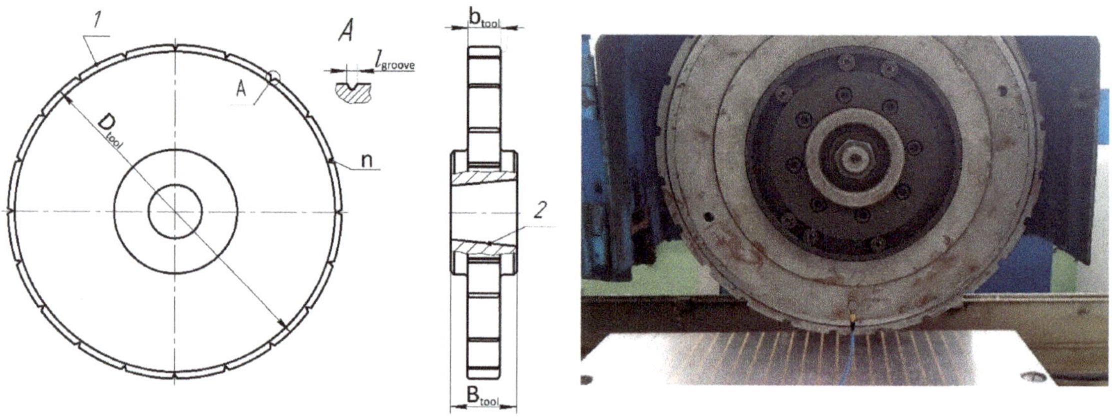

(a) (b)

Figure 2. The tool with transverse grooves on its periphery: sketch (**a**) and tool mounted on surface grinder (**b**): 1—working smooth surface of the tool, 2—Morse taper, D_{tool}—external diameter, B_{tool}—width of the tool, b_{tool}—width of the working part of the tool, l_{groove}—width of the groove, n—number of grooves.

Thermo-deformation strengthening of flat surfaces according to the kinematics of the process is similar to the grinding of flat surfaces (Figure 3), so the friction force in the contact area between the tool and the part can be divided into three components: normal component, which acts on the radius of the tool, perpendicular to the axis of its rotation; tangential component, which acts tangentially to the working surface of the tool parallel to the workpiece surface; and transverse component, which acts parallel to the axis of rotation of the tool in the direction opposite to the feed movement. Measurements of the components of friction forces in the contact area between the tool and the part were performed using a three-component type 9121 dynamometer from Kistler (Winterthur, Switzerland) (Figure 4). Experimental samples were made of steel 41Cr4 (quench-hardening and low-temperature tempering) with a size of 15mm × 30mm × 100 mm. The chemical composition of the steel is as follows: mass %: 0.40 C; 0.78 Mn; 0.26 Si; 1.12 Cr; 0.01 S; 0.01 P; Fe: balance. A three-component dynamometer with a fixed sample was mounted on the magnetic plate of the surface grinder machine.

The Type 9121 Kistler three-component dynamometer uses a piezoelectric force component measurement system. Due to its high natural frequencies, it can measure high-speed processes. The piezoelectric dynamometer is connected to a multi-channel high-quality amplifier–converter (National Instruments, Austin, TX, USA), which has an interface for connection to a computer. LabView software was used to process the results of the force component studies.

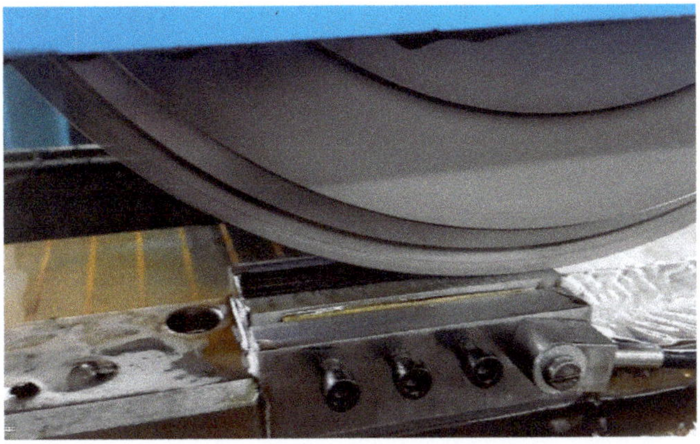

Figure 3. View of the sample's surface during processing (the black surface is the already-processed surface).

Figure 4. Three-component dynamometer type 9121 (Kistler, Switzerland): 1—tool, 2—dynamometer, 3—sample, 4—machine table.

Experimental studies were conducted to determine the normal and tangential components of the friction force in the contact zone between the tool and the treated surface during thermo-deformation strengthening of the samples' flat surfaces made of steel 41Cr4 (quench-hardening and low-temperature tempering). A metal tool-disc made of stainless steel X10CrNiTi18–10 (EU) was installed on a modernized machine (HFS 3063 VS by KNUTH). The tool's outer diameter was 360 mm, the width of the working surface was 6 mm, and 24 grooves were formed on the working surface of the tool. The linear speed on the tool's peripheral working part was 70 m/s, and the speed of movement of the grinder machine table was 4 m/min.

Mathematical Model

Notation and units: ω—angular velocity of the tool's rotation, s^{-1}; x_1, y_1—displacements of the spindle headstock in the horizontal and vertical planes, m; x_2, y_2—displacements of the tool in the horizontal and vertical planes, m; x_3, y_3—displacements of the machine table in the horizontal and vertical planes; D_{tool}—external diameter of the tool, m; n—the number of grooves or working smooth surfaces of the tool; F_{rc}—the force of friction between the treated surface of the tool's working smooth surface and the sample, N; m_1—the spindle head's mass, kg; m_2—the tool's mass, kg; m_3—the parts' (workpiece's) mass, kg; c_1 and c_2—stiffness between the column guideways and the machine headstock in the horizontal and vertical planes, N/m; c_3 and c_4—stiffness of the machine spindle supports in the horizontal and vertical planes, N/m; c_5 and c_6—stiffness of the machine table in the horizontal and vertical planes, N/m; c_7—contact stiffness between the part and the tool, N/m; μ_1 and μ_2—damping coefficient between the column guideways and the machine headstock in the horizontal and vertical planes, Ns/m; μ_3 and μ_4—damping coefficient of the machine spindle supports in the horizontal and vertical planes, Ns/m; μ_5 and μ_6—damping coefficient of the machine table in the horizontal and vertical planes, Ns/m; μ_7—damping coefficient between the tool and sample (internal damping), Ns/m.

The first stage of dynamic analysis of the thermo-deformation treatment process of the flat surfaces of machine parts is to develop a calculation scheme of the machine's elastic system and describe it as a three-mass model. To do this, the surface grinder machine was divided into a number of units, each one of which represents a separate mass (Figure 5). It is very important to define the part or unit that will be selected as a part with a conditionally infinite mass relative to which the movements of the other masses are determined. The bed of the surface grinder machine to which all the other masses (the weight of the spindle head, the weight of the tool, and the weight of the machine table with the part) are "attached" was accepted as a part with conditionally unlimited mass. Elastic and damping bonds describe the relationship between individual masses.

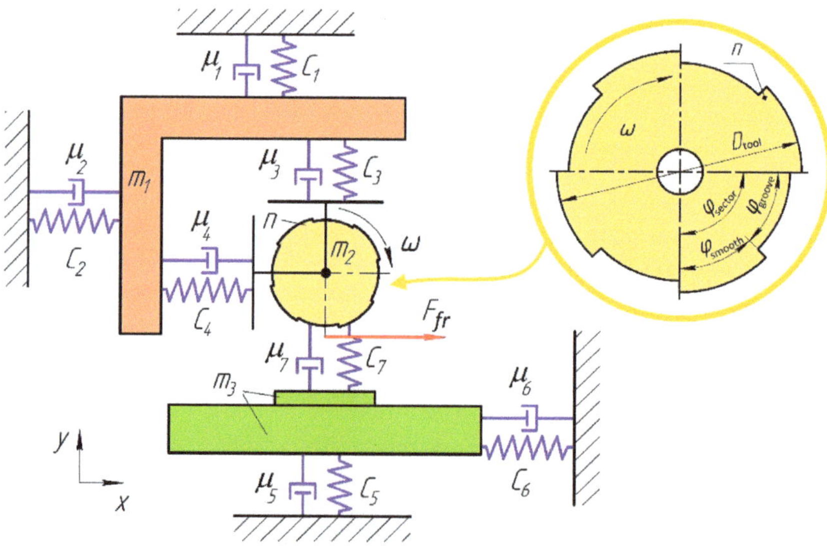

Figure 5. Calculation scheme of the grinder machine.

The pressing force of the tool to the sample was proposed to be defined in an implicit form, i.e., to set through the relative movement (Δy) of the spindle head in the direction of the surface grinder machine's table in this mathematical model. Having set a certain amount of displacement Δy, the pressing force is generated by the composition of the stiffness of all the elements in the machine's oscillatory circuit. Most machines cannot measure the components of the forces that occur during the treatment process. Nevertheless, they all have a limb or optical ruler that can be used to determine tool movement, which served as this choice's main idea. The normal component of the force that occurs during processing has been pre-determined by experimental studies [29].

The generalized coordinates $q_{i,j}$ (where the number of masses is i = 1, 2, 3 and the number of coordinates is j = 1, 2) in our case will be:

- for the spindle head with the mass m_1: $q_{11} = x_1$; $q_{12} = y_1$.

- for the tools with the mass m_2: $q_{21} = x_2$; $q_{22} = y_2$.
- for the parts (workpiece) with the mass m_3: $q_{31} = x_3$; $q_{32} = y_3$.

The motion of the system can be described by differential equations built based on Euler–Lagrange equations [33–36]:

$$\frac{d}{dt}\left(\frac{\partial T}{\partial \dot{q}_{i,j}}\right) - \frac{\partial T}{\partial q_{i,j}} + \frac{\partial V}{\partial q_{i,j}} + \frac{\partial D}{\partial q_{i,j}} = Q_{q_i}, \tag{1}$$

where T is the kinetic energy of the system; V is the potential energy of the system; D is the energy dissipation function in the system (Rayleigh dissipation function); and $Q_{q_{i,j}}$ are the generalized forces that refer to the corresponding generalized coordinates x_i and y_i.

We should accept the following assumptions when forming a mathematical model: we assume that the law of change of rigidity in the elastic elements of the system does not go beyond linearity, and it corresponds to Hooke's law. This is justified, provided that there are small deviations of the spring from the equilibrium position; we will consider the mechanical system of the machine as one consisting of absolutely rigid bodies connected by ideal holonomic ties and elastic elements of strictly defined rigidity. Into the dynamic model, we will introduce coefficients of viscous friction that are proportional to the speed of movement of the moving sliders along the corresponding guide axes and that reflect the energy dissipation in the corresponding elastic elements of the system in the form of dampers [33,34,36,37].

Thus, using the ratio of the forces' virtual work $\sum \delta A(F_k)$ that acts on each element to the increment of a certain generalized coordinate $\delta q_{i,j}$ for the corresponding masses, the generalized force can be determined from the next equation:

$$Q_{i,j} = \frac{\delta A_{i,j}}{\delta q_{i,j}}. \tag{2}$$

Therefore, for the spindle head, the generalized force will look as follows:

$$\begin{gathered} Q_{x_1} = 0, \\ Q_{y_1} = \frac{N \cdot \delta y_1}{\delta y_1} = N. \end{gathered} \tag{3}$$

For the tool:

$$\begin{gathered} Q_{x_2} = f\frac{(N - N_{23}) \cdot \delta x_2}{\delta x_2} = f(N - N_{23}), \\ Q_{y_2} = \frac{N_{23} \cdot \delta y_2}{\delta y_2} = N_{23}. \end{gathered} \tag{4}$$

For the part:

$$Q_{x_3} = f\frac{(N - N_{23}) \cdot \delta x_3}{\delta x_3} = f(N - N_{23}); \quad Q_{y_3} = \frac{N_{23} \cdot \delta y_3}{\delta y_3} = N_{23}. \tag{5}$$

where f is the friction coefficient between the tool and the sample and N is the tool's force of the normal pressure to the sample, which is formed due to the displacement y_0 of the spindle head in the direction of the sample (depth of cut during the technological operation of grinding), i.e.,:

$$N = c_y^* \cdot y_0, \tag{6}$$

where y_0 is the displacement of the spindle head in the direction of the sample when the tool's working smooth surface is in contact with the treated surface of the part, i.e., by means of this displacement, the mutual pressing force is formed; c_y^* is the reduced stiffness of the system in the vertical plane:

$$\frac{1}{c_y^*} = \frac{1}{c_1 + c_3} + \frac{1}{c_5 + c_7}. \tag{7}$$

The formation of the pressing force, as described above, happens by moving the spindle head (lowering) in the direction of the part through the transmission from the screw and the nut to the value y_0. It is assumed to be constant during processing (i.e., displacement, which is controlled by the machine limb and forms the pressing force between the tool's working smooth surface and the treated surface of the part) [9,11,30]. Due to the geometry of the tool (alternating the tool's working smooth surface with the groove), there is a decrease in the pressing force between the tool and the part by the amount of displacement y_{23}.

$$y_{23} = \begin{cases} 0,\ when\ \omega t = (0 \ldots \varphi_{smooth}) + \frac{2\pi}{n} \cdot m \\ (R - R \cdot \cos \omega t),\ when\ \omega t = \left(\varphi_{smooth} \cdots \frac{\varphi_{groove}}{2}\right) + \frac{2\pi}{n} \cdot m \\ (R - R \cdot \cos \omega t) - \left(R - R \cdot \cos\left(\frac{\varphi_{sector} - \varphi_{smooth}}{2}\right)\right),\ when\ \omega t = \left(\frac{\varphi_{groove}}{2} \ldots \varphi_{smooth}\right) + \frac{2\pi}{n} \cdot m \end{cases} \quad (8)$$
$$m = 0, 1, 2 \ldots (n-1)$$

where R is the diameter of the tool $R = D_{tool}/2$ and n is the number of smooth parts of the tool.

The normal force changes in accordance with the condition of contact of the tool's periphery with the sample's treated surface, i.e., the contact of the tool's working smooth part with the sample's treated surface, as well as the position of the tool's groove above the sample's treated surface, form the analytical dependences of the generalized forces Q_i, as follows:

$$\begin{array}{l} Q_{y_1} = N; \quad Q_{y_2} = N_{23}; \\ Q_{x_1} = 0; \quad Q_{x_2} = F_{fr} = f \cdot (N - N_{23}); \\ |Q_{y_2}| = |-Q_{y_3}|; \quad |Q_{x_2}| = |-Q_{x_3}|. \end{array} \quad (9)$$

where N_{23} is the force that reduces the force of the normal pressure of the tool to the part, which is formed due to the displacement y_{23}, caused by alternating changes of the smooth part and the groove during rotation of the tool:

$$N_{23} = c_y^* \cdot y_{23}, \quad (10)$$

Additionally, it is necessary to describe the case when the groove on the periphery of the tool is narrow and allows for the stopping of the heat flow but does not allow the tool to freely get out of contact with the part (Figure 6). Therefore, the minimum value of the pressing force is determined by the dependence:

$$\begin{array}{l} If\ y_0 < \Delta R, \\ then\ |N_{min}| = c_y^* \cdot (y_0 - \Delta R)\ when\ \omega t = (\varphi_{smooth} \cdots \varphi_{sector}) + \frac{2\pi}{n} \cdot m \\ \Delta R = R - R' = R - R \cdot cos\left(\frac{\varphi_{sector} - \varphi_{groove}}{2}\right). \end{array} \quad (11)$$

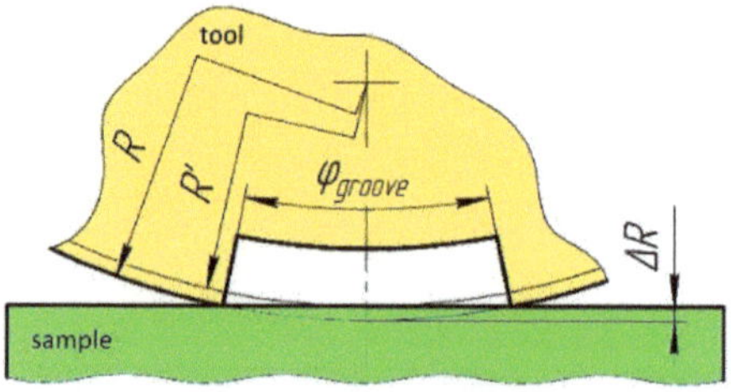

Figure 6. Contact of the tool in the contact area (groove above the sample's treated surface): R—radius of the tool, R′—value from the center of the tool normal to the processing sample when the groove is symmetrical on the processing zone, ΔR—difference between R and R′, and φ_{groove}—size of the groove.

To start writing the left part of the Euler–Lagrange equations, we need to find the derivatives of the expression of kinetic energy T:

For the spindle head:

$$\begin{aligned} &\frac{\partial T}{\partial \dot{q}_{11}} = \frac{\partial T}{\partial \dot{x}_1} = m_1 \dot{x}_1; \quad \frac{\partial T}{\partial \dot{q}_{12}} = \frac{\partial T}{\partial \dot{y}_1} = m_1 \dot{y}_1; \\ &\frac{d}{dt}\left(\frac{\partial T}{\partial \dot{x}_1}\right) = m_1 \ddot{x}_1; \quad \frac{d}{dt}\left(\frac{\partial T}{\partial \dot{y}_1}\right) = m_1 \ddot{y}_1. \end{aligned} \tag{12}$$

For the tool:

$$\begin{aligned} &\frac{\partial T}{\partial \dot{q}_{21}} = \frac{\partial T}{\partial \dot{x}_2} = m_2 \dot{x}_2; \quad \frac{\partial T}{\partial \dot{q}_{22}} = \frac{\partial T}{\partial \dot{y}_2} = m_2 \dot{y}_2; \\ &\frac{d}{dt}\left(\frac{\partial T}{\partial \dot{x}_2}\right) = m_2 \ddot{x}_2; \quad \frac{d}{dt}\left(\frac{\partial T}{\partial \dot{y}_2}\right) = m_2 \ddot{y}_2. \end{aligned} \tag{13}$$

For the part:

$$\begin{aligned} &\frac{\partial T}{\partial \dot{q}_{31}} = \frac{\partial T}{\partial \dot{x}_3} = m_3 \dot{x}_3; \quad \frac{\partial T}{\partial \dot{q}_{32}} = \frac{\partial T}{\partial \dot{y}_3} = m_3 \dot{y}_3; \\ &\frac{d}{dt}\left(\frac{\partial T}{\partial \dot{x}_3}\right) = m_3 \ddot{x}_3; \quad \frac{d}{dt}\left(\frac{\partial T}{\partial \dot{y}_3}\right) = m_3 \ddot{y}_3. \end{aligned} \tag{14}$$

And

$$\frac{\partial T}{\partial x_i} = 0; \quad \frac{\partial T}{\partial y_i} = 0. \tag{15}$$

The elastic elements accumulate the potential energy V in the system, and the following dependence is used to find it:

$$\mathrm{V} = \frac{c_1 \cdot y_1^2}{2} + \frac{c_3 (y_1 - y_2)^2}{2} + \frac{c_7 (y_2 - y_3)^2}{2} + \frac{c_5 \cdot y_3^2}{2} + \frac{c_2 \cdot x_1^2}{2} + \frac{c_4 (x_1 - x_2)^2}{2} + \frac{c_6 \cdot x_3^2}{2}. \tag{16}$$

The next elements on the left part of the Euler–Lagrange equations are the derivatives of the expression of potential energy:

For the spindle head:

$$\frac{\partial \mathrm{V}}{\partial x_1} = c_2 x_1 + c_4 x_1 - c_4 x_2; \quad \frac{\partial \mathrm{V}}{\partial y_1} = c_1 y_1 + c_3 y_1 - c_3 y_2. \tag{17}$$

For the tool:

$$\frac{\partial \mathrm{V}}{\partial x_2} = -c_4 x_1 + c_4 x_2; \quad \frac{\partial \mathrm{V}}{\partial y_2} = -c_3 y_1 + c_3 y_2 + c_7 y_2 - c_7 y_3. \tag{18}$$

For the part:

$$\frac{\partial \mathrm{V}}{\partial x_3} = c_6 x_3; \quad \frac{\partial \mathrm{V}}{\partial y_3} = -c_7 y_2 + c_7 y_3 + c_5 y_3. \tag{19}$$

We assume that the energy dissipation is directly proportional to the velocity, so the dissipative function D for the system is calculated using the following expression:

$$\mathrm{D} = \frac{\mu_1 \cdot \dot{y}_1^2}{2} + \frac{\mu_3 (\dot{y}_1 - \dot{y}_2)^2}{2} + \frac{\mu_7 (\dot{y}_2 - \dot{y}_3)^2}{2} + \frac{\mu_5 \cdot \dot{y}_3^2}{2} + \frac{\mu_2 \cdot \dot{x}_1^2}{2} + \frac{\mu_4 (\dot{x}_1 - \dot{x}_2)^2}{2} + \frac{\mu_6 \cdot \dot{x}_3^2}{2}. \tag{20}$$

The last elements on the left part of the Euler–Lagrange equations are the derivatives of the expression of the dissipative function:

For the spindle head:

$$\frac{\partial \mathrm{D}}{\partial x_1} = \mu_2 \dot{x}_1 + \mu_4 \dot{x}_1 - \mu_4 \dot{x}_2; \quad \frac{\partial \mathrm{D}}{\partial y_1} = \mu_1 \dot{y}_1 + \mu_3 \dot{y}_1 - \mu_3 \dot{y}_2. \tag{21}$$

For the tool:

$$\frac{\partial \mathrm{D}}{\partial x_2} = -\mu_4 \dot{x}_1 + \mu_4 \dot{x}_2; \quad \frac{\partial \mathrm{D}}{\partial y_2} = -\mu_3 \dot{y}_1 + \mu_3 \dot{y}_2 + \mu_7 \dot{y}_2 - \mu_7 \dot{y}_3. \tag{22}$$

For the part:

$$\frac{\partial D}{\partial y_3} = -\mu_7\dot{y}_2 + \mu_7\dot{y}_3 + \mu_5\dot{y}_3; \qquad \frac{\partial D}{\partial x_3} = \mu_6\dot{x}_3. \tag{23}$$

Now, the dynamics of a surface grinder machine's mechanical system can be described by the following mathematical model:

$$\begin{cases} m_1\ddot{x}_1 + c_2x_1 + c_4x_1 - c_4x_2 + \mu_2\dot{x}_1 + \mu_4\dot{x}_1 - \mu_4\dot{x}_2 = Q_{x_1} \\ m_2\ddot{x}_2 - c_4x_1 + c_4x_2 - \mu_4\dot{x}_1 + \mu_4\dot{x}_2 = Q_{x_2} \\ m_3\ddot{x}_3 + c_6x_3 + \mu_6\dot{x}_3 = -Q_{x_3} \\ m_1\ddot{y}_1 + c_1y_1 + c_3y_1 - c_3y_2 + \mu_1\dot{y}_1 + \mu_3\dot{y}_1 - \mu_3\dot{y}_2 = Q_{y_1} \\ m_2\ddot{y}_2 - c_3y_1 + c_3y_2 + c_7y_2 - c_7y_3 - \mu_3\dot{y}_1 + \mu_3\dot{y}_2 + \mu_7\dot{y}_2 - \mu_7\dot{y}_3 = Q_{y_2} \\ m_3\ddot{y}_3 - c_7y_2 + c_7y_3 + c_5y_3 - \mu_7\dot{y}_2 + \mu_7\dot{y}_3 + \mu_5\dot{y}_3 = -Q_{y_3} \end{cases} \tag{24}$$

We assume that at the initial moment of the simulation, the tool is in contact with the sample with a smooth working part; then, the pressure at the initial moment of the simulation will be equal:

$$F_0 = y_0 \cdot c_y^*. \tag{25}$$

Initial conditions:

$$\begin{aligned} &\dot{x}_i\big|_{t=0} = 0, \quad \dot{y}_i\big|_{t=0} = 0, \\ &x_1\big|_{t=0} = 0, \quad x_2\big|_{t=0} = 0, \quad x_3\big|_{t=0} = 0. \\ &y_1\big|_{t=0} = \tfrac{F_0}{c_1+c_3}, \quad y_2\big|_{t=0} = \tfrac{F_0}{c_1+c_3+c_7}, \quad y_3\big|_{t=0} = \tfrac{F_0}{c_1+c_3+c_5+c_7}. \end{aligned} \tag{26}$$

Modeling of the interaction of the tool's working smooth part with the sample is carried out using the contact stiffness and damping of the energy of local elastic–plastic deformation. For this reason, it is necessary to introduce an additional condition of checking the mutual contact between the tool's working smooth part and the sample's treated surface. In other words, if the tool moves in the opposite direction to the sample, contact stiffness and damping will be lost:

$$If\ y_3 - y_2 < 0, then\ c_7 = 0, \mu_7 = 0. \tag{27}$$

3. Results

We started the simulation from no-load running (I-stage); next was the gradual contact of the tool's working smooth surface with the sample's treated surface (from the starting point of the tool's working smooth surface) in two segments (II-stage) (by segments, we mean the tool's working smooth part and the one groove); and then, the full value of the force of the tool pressure to the part was formed (III-stage). The rotation of the tool happens clockwise with an angular velocity (the speed of the movement of the table with the sample is significantly lower, and in the simulation, we neglected it; the linear speed of rotation of the tool was 60–70 m/s, and the speed of the table of the grinder machine was 2–4 m/min).

Modeling of the dynamic processes of the technological operation of thermo-deformation strengthening of flat surfaces was performed using MATLAB-Simulink software, and the differential equations were solved using the Runge–Kutta method. The process simulation was performed using a tool with transverse grooves, which with the same circular pitch alternated with the smooth surface in turn.

The process of the technological operation of thermo-deformation strengthening of flat parts is similar to the technological operation of flat grinding by kinematics and is performed on a surface grinder. The tool alternately comes into contact with the processed part; after passing the strengthening surface, it comes out of contact with it. Therefore, in the simulation of the load on the part in this model when the tool comes into contact, the pressure should be given by a variable that increases from zero to the operating value. That is, for ease of testing the process, we set the following procedure for forming the pressure of the tool to the part: the first stage was no-load running (idling); with a lack of pressure when rotating the first segment, the tool was outside the workpiece (tool's smooth surface and groove, i.e., $\pi/4$); the second stage was a linear increase in the pressure to its specified value during the rotation of the next three segments, the contact of the tool with the part (from $\pi/4$ to 2π); in the third stage, the working pressure, i.e., all the other rotations of the tool, was taken as that which set a constant pressure equal to $F = 500$ N (Figure 7). Additionally, the initial conditions for the movement of all masses in the vertical direction will be zero, i.e., the right-hand side of Equation (24) will be zero.

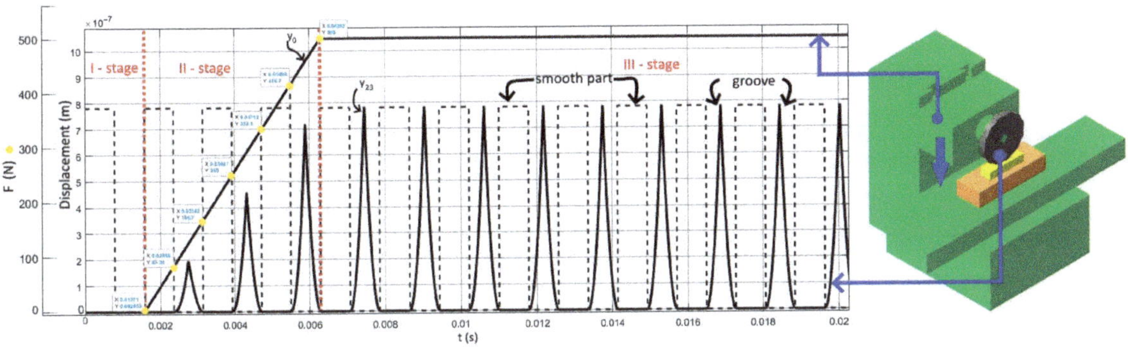

Figure 7. Simulation of pressure by moving the spindle head of the surface grinder machine and the geometry of the tool (The dashed line only shows the location of the smooth parts of the tool (the y-axis has a representative scale)).

Numerical solutions of displacements were made for the following parameters: number of grooves $n = 16$, groove width $l_{groove} = 4$ mm; the force of the pressure of the tool to the sample $F = 500$ N; the tool's external diameter $D = 260$ mm; and the angular spindle speed $\omega = 500$ s^{-1}. The displacements in the vertical direction y_1, y_2, y_3 of the corresponding mass will be stable and of the same frequency after the value y_0 is equal to the working displacement (Figure 8), i.e., the full working value of the pressure will be formed (full contact of the tool with the processed surface of the part); the velocity of displacement of the tool in the vertical direction is shown in Figure 9.

Displacement of the tool in the horizontal direction (Figure 10a) was caused by friction force, which depends on the pressure of the tool to the part and is variable during the strengthening process. The velocity of displacement of the tool in the horizontal direction is shown in Figure 10b. An important parameter in the study is the reaction that occurs in the contact zone during the processing, which is presented in Figure 11. Modeling of the process of thermo-deformation strengthening of flat parts was performed at several groove widths and pressures (Figures 12 and 13).

It is known that the formation of nanocrystalline structures in the surface layers is particularly influenced by high-frequency deformations of low-amplitude values. When analyzing the obtained numerical solution of system displacements using different values of pressure and groove width, it can be argued that the value of the groove width has a significant effect on the reaction amplitude taking place in the processing area between the tool and the workpiece. For example, when using a tool with a groove in which the groove width is 4 mm, the pressure is 750 N, and the number of grooves on the periphery of the tool is equal to 8 pcs, the reaction amplitude is equal to 10 N. With a groove width of 8 mm, the difference between the maximum and minimum values of the reaction amplitude is 7.5 times higher and is 75 N (Figure 14). However, as the number of grooves on the periphery of the tool increases, the difference between the maximum and minimum values of the amplitude of the reaction decreases. For example, with the same pressure of the tool to the part with the number of grooves equal to 24 pcs and a groove width of 4 mm, the amplitude difference will be equal to 5 N, and with a groove width of 8 mm, the amplitude difference will be 38 N. It should also be noted that the value of the difference between the maximum and minimum values of the amplitude of the reaction is the same when using the same width of tool at different pressures. That is, if the amount of displacement of the tool to the part that forms the clamping force is greater than the value ΔR (see Figure 6), the difference between the maximum and minimum values of the reaction amplitude will be the same at different clamping forces, but the groove width will be the same.

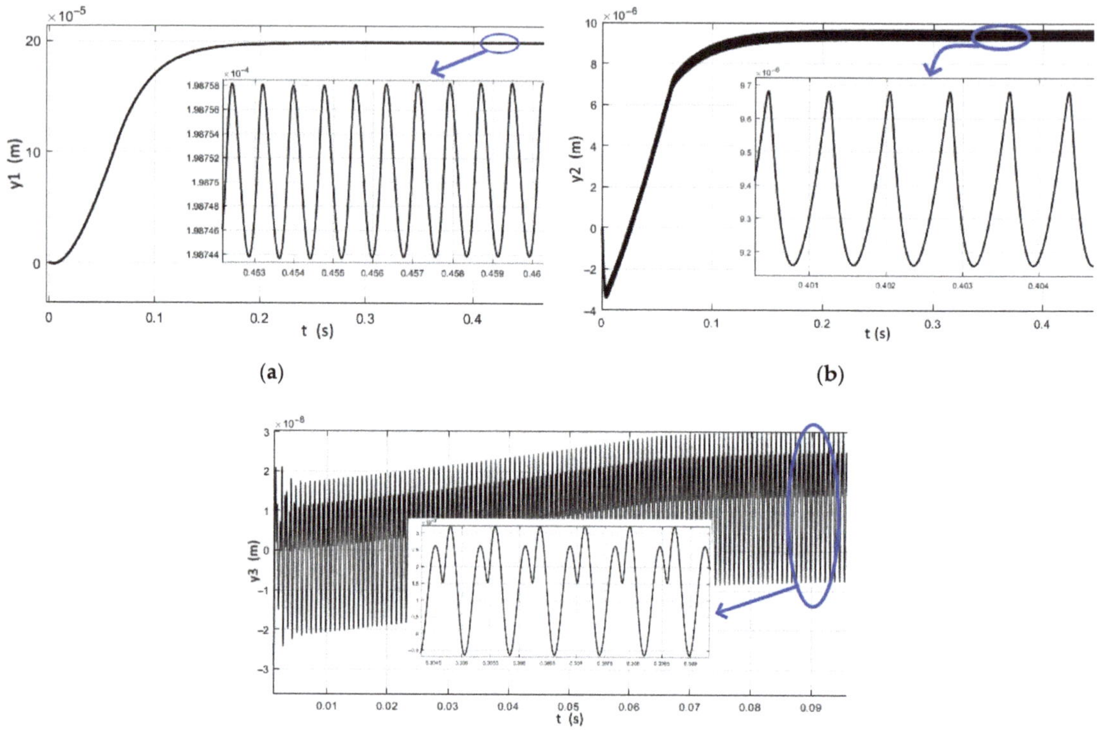

(a) (b)

(c)

Figure 8. Displacement in the vertical plane: (**a**) spindle headstock; (**b**) spindle with tool; (**c**) table with part.

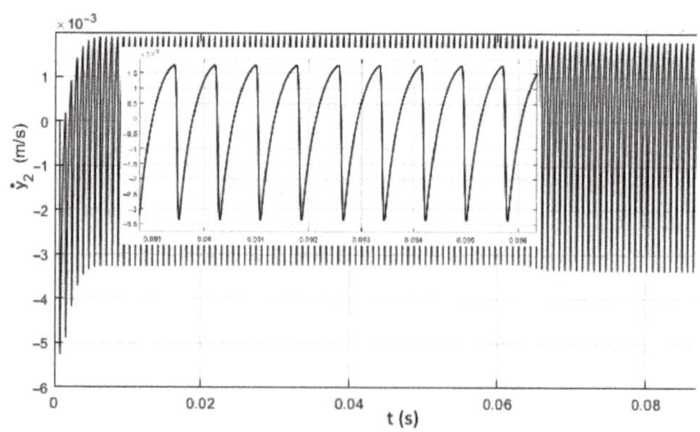

Figure 9. Velocity of displacement of the tool in the vertical plane.

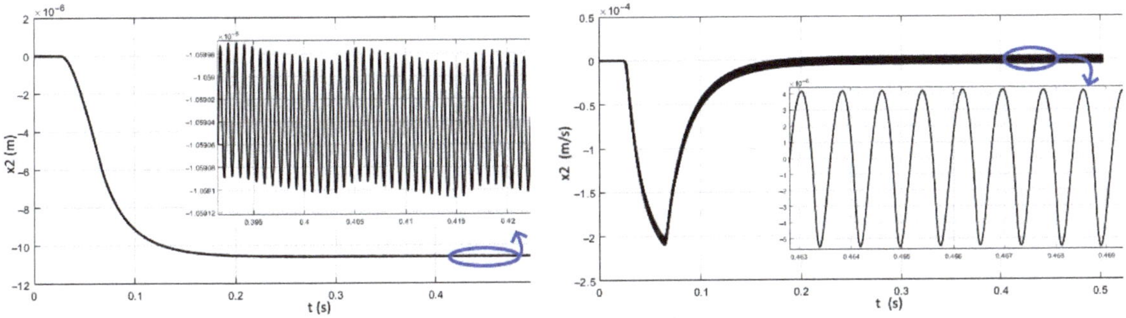

(a) (b)

Figure 10. Displacement (**a**) and velocity (**b**) of the tool in the horizontal plane.

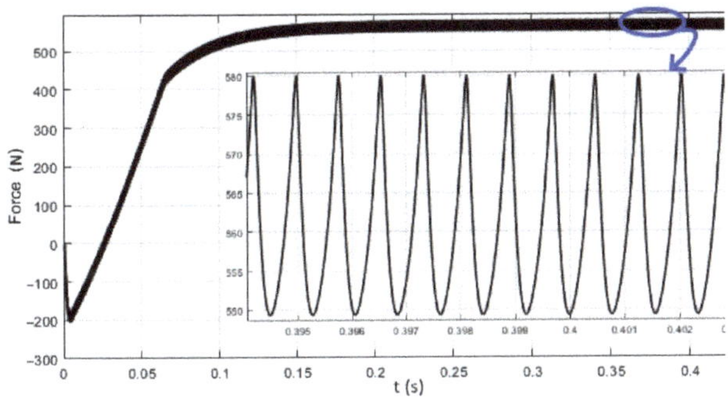

Figure 11. Amount of force between the tool and the sample.

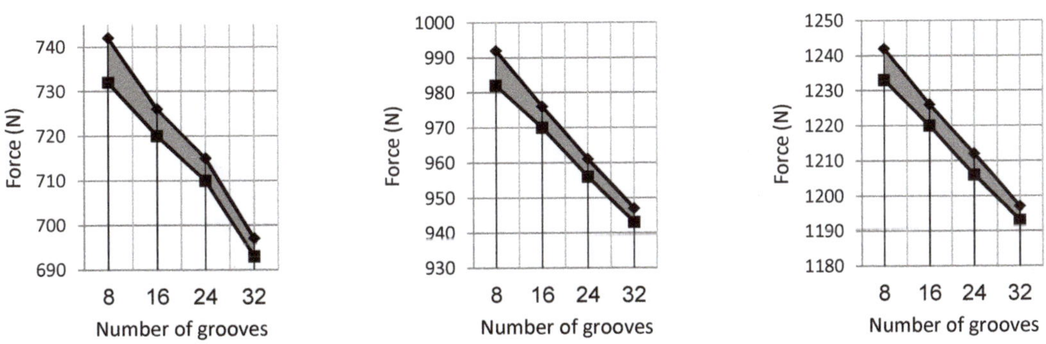

(a) (b) (c)

Figure 12. The amplitude of the force of the tool with a groove width of 4 mm on the part in the contact zone: (**a**) pressure of 750 N; (**b**) pressure of 1000 N; (**c**) pressure of 1250 N (■—minimum value of force; ♦—maximum value of force).

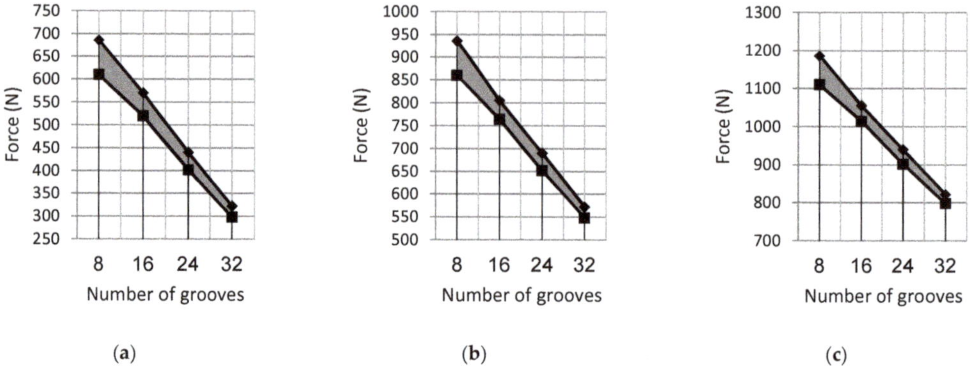

Figure 13. The amplitude of the force of the tool with a groove width of 8 mm on the part in the contact zone: (**a**) pressure of 750 N, (**b**) pressure of 1000 N, and (**c**) pressure of 1250 N (■—minimum value of force; ◆—maximum value of force).

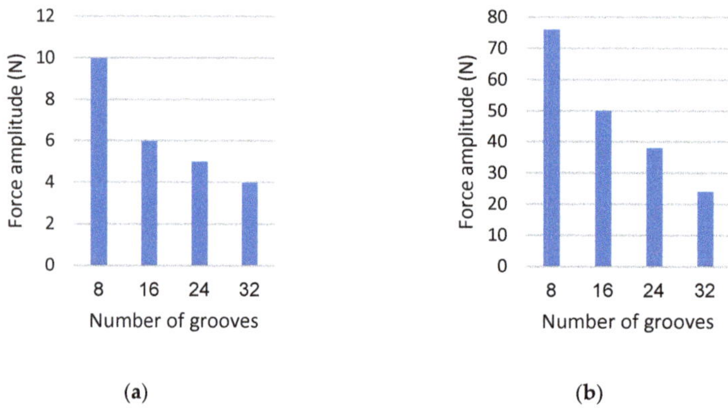

Figure 14. Force amplitude of the reaction of the tool to the part: (**a**) groove width: 4 mm; (**b**) groove width: 8 mm.

An important factor for the formation of nanocrystal layers is the frequency of impact of the tool with the part. It has been shown that as the number of grooves increases, the frequency of impacts in the treatment area increases significantly. Thus, when the number of grooves is equal to eight, the frequency of impact of the tool on the part is equal to 750 Hz, and when the quantity of grooves is equal to 32, the frequency of impact of the tool on the part is equal to 2.5 kHz (Figure 15).

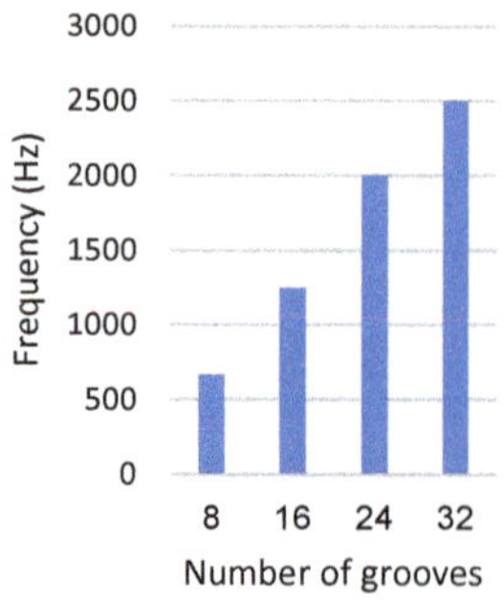

Figure 15. The frequency of the mutual displacement of the part and the tool in the area of their contact.

Experimental studies have shown that during thermo-deformation strengthening of flat surfaces of samples made of steel 41Cr4, the tangential and normal components of the friction force in the tool–sample contact zone have a wave-like character (Figure 16). The amplitude of the change in the components of the friction force is 0.15–0.25 of the magnitude of the force. The highest value is reached by the normal component of the friction force in the tool–part contact zone. The processing modes significantly affect the value, especially the amount of vertical tension. It determines the amount of pressing of the tool to the processed surface of the part. The nature of the change in the tangential component of the force, which occurs in the contact zone of the tool with the processed surface of the samples, is similar to that of the normal component and corresponds to the change in the geometry of the working surface of the tool. The amount of loading and unloading of a unit area of contact of the tool with the processed surface depends on the processing modes (linear speed on the periphery of the tool and speed of movement of the machine table) and the number of grooves. Accordingly, the movement of the tool along the processed surface is interrupted. Due to the specified processes, we obtained a wave-like character of changes in the components of the friction force during thermo-deformation strengthening.

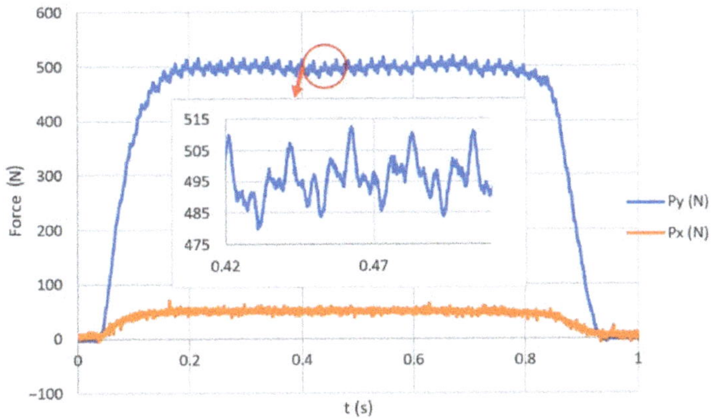

Figure 16. Normal (Py) and tangential (Px) components of the forces that occur during the strengthening of flat surfaces.

The formation of strengthened layers with a nanocrystalline structure is intensified by increasing the intensity of deformation of the contact zone on the treated surface by a tool with transverse grooves on its periphery. The metal of the surface layer is cyclically loaded and unloaded, as well as heated and cooled, during the alternate passage of the smooth surface and the groove along the surface of the tool's contact zone with the sample. The contact zone of the tool with the processed surface during thermo-deformation strengthening is heated to temperatures close to the melting point [24]. At such high temperatures, the metal softens, and its plasticity increases. After the groove passes over the contact zone, the next smooth surface comes into contact and creates a shock load on the contact zone. Such processes intensify the deformation of the surface layer of the metal to a greater depth, contribute to grain grinding, and form a strengthened layer to a greater depth.

The thickness of the strengthened layer and its hardness obtained during thermo-deformation strengthening of flat samples made from steel 41Cr4 (quench-hardening and low-temperature tempering) significantly affect the shape of the working surface of the tool, as shown by conducted metallographic studies. Thus, during thermo-deformation strengthening with a tool with a smooth working surface, the thickness of the hardened layer was 140–160 microns, and the hardness was 8.6 GPa (with a hardness of the base metal of 5.6 GPa) (Figure 17). During thermo-deformation strengthening with a tool with transverse grooves on its working part, the thickness of the hardened layer and its hardness increase. When processing with a tool with 24 transverse grooves with a width of 3–4 mm, the thickness of the layer increased to 190–220 μm, and the hardness increased to 9.3 Gpa. When the width of the transverse grooves increased to 8–9 mm, the thickness of the layer increased to 250–260 μm, and the hardness increased to 11.1 Gpa. A further increase in the width of the grooves led to a certain decrease in the thickness of the strengthened layer. The use of a tool with transverse grooves leads to an intensification of the plastic deformation of the surface layer of the metal during thermo-deformation strengthening, which also leads to a decrease in grain size, which increases by about 10 nm near the surface. The grain size of the strengthened layers obtained by strengthening with a tool with a smooth part is much larger and is 60–80 nm. The resulting strengthened layers are nanocrystalline. The metallographic structure of steel 41Cr4 is shown in Figure 18.

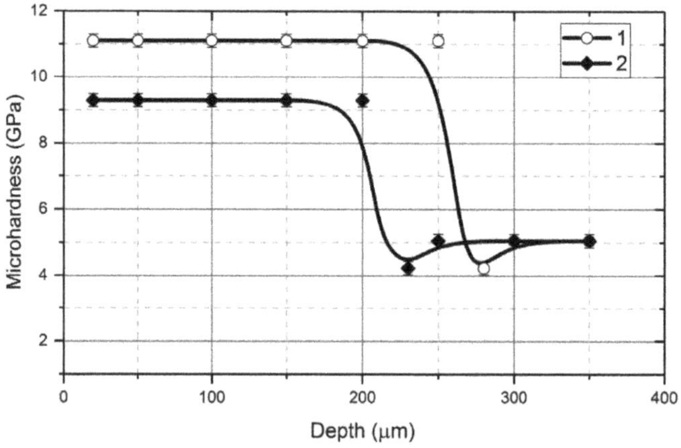

Figure 17. Microhardness of steel after thermo-deformation treatment 41Cr4: 1—length of grooves: 4 mm; 2—length of grooves: 8 mm.

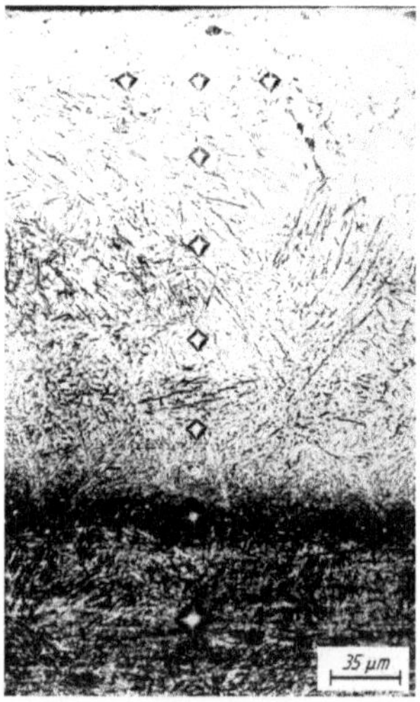

Figure 18. Structure of steel 41Cr4 after thermo-deformation treatment (length of grooves: 8 mm).

The conducted experimental studies confirm our theoretical calculations.

4. Conclusions

Thermo-deformation strengthening of the surface layers of parts using a tool with transverse grooves on its working peripheral surface combines two methods of forming a strengthened layer with a nanocrystalline structure—the action of a high-intensity heat source and intense plastic deformation.

Intensive plastic deformation of the metal of the surface layer of the sample is achieved by alternately passing the smooth tool surface and the groove through the contact zone of the tool and the sample, during which time the contact zone is cyclically loaded and unloaded.

The design scheme of the elastic system of a surface grinding machine was developed to study the dynamic processes that take place during thermo-deformation strengthening of flat surfaces of samples in the contact zone between the tool and the treated surface.

When the width of the groove increases, the difference between the maximum and minimum amount of the amplitude of the reaction force, which occurs in the contact zone between the tool and the processed surface, increases. Thus, when using a tool with eight grooves with a width of 4 mm on its working surface (pressing force 750 N), the difference between the amounts of the force amplitude was equal to 10 N. When the groove width was 8 mm, the difference in the force amplitude increased by 7.5 times and was 75 N. With an increase in the force of pressing the tool to the processed surface with the same width of the groove, it did not affect the magnitude of the change in the amplitude of the force in the contact zone.

Experimental studies have shown that during the thermal deformation treatment (strengthening) of samples made of steel 41Cr4, the thickness and microhardness of the strengthened layer increase with the increase in the groove width on the working part of the tool. For example, with a groove width of 3–4 mm, the thickness of the strengthened layer was 190–220 μm, and its microhardness was 9.3 GPa. With an increase in the groove width of 9 mm, the thickness and microhardness of the layer increased to 250–260 μm and 11.1 GPa, respectively.

Author Contributions: Conceptualization, V.G., I.H. and T.H.; methodology, I.H., V.G. and P.M.; validation, I.H., P.M. and T.H.; formal analysis, V.G., P.M., I.H., V.D., T.H. and W.W.; investigation, I.H., V.G. and W.W.; writing—original draft preparation, V.G., P.M., I.H., V.D., T.H. and W.W.; writing—review and editing, I.H., P.M. and V.G.; visualization, V.G. and V.D.; supervision, I.H. and P.M. All authors have read and agreed to the published version of the manuscript.

Funding: This research received no external funding.

Data Availability Statement: The data presented in this study are available on request from the corresponding author.

Conflicts of Interest: The authors declare no conflict of interest.

References

1. Davis, J.R. *Surface Engineering for Corrosion and Wear Resistance*; ASM International: Geauga County, OH, USA, 2001; ISBN 0-87170-700-4.
2. Dwivedi, D.K. *Surface Engineering: Enhancing Life of Tribological Components*; Springer: New Delhi, India, 2018; ISBN 978-81-322-3777-8.
3. Troshchenko, V.T. *Deformation and Destruction of Metals under Multicycle Loading*; Naukova Dumka: Kyiv, Ukraine, 1981.
4. Pineau, A.; McDowell, D.L.; Busso, E.P.; Antolovich, S.D. Failure of Metals II: Fatigue. *Acta Mater.* **2016**, *107*, 484–507. [CrossRef]
5. Santecchia, E.; Hamouda, A.M.S.; Musharavati, F.; Zalnezhad, E.; Cabibbo, M.; El Mehtedi, M.; Spigarelli, S. A Review on Fatigue Life Prediction Methods for Metals. *Adv. Mater. Sci. Eng.* **2016**, *2016*, 9573524. [CrossRef]
6. Melters, P.M. *Introduction to Metal Fatigue*; Aarhus University: Arhus, Denmark, 2018; ISBN 2245-4594.
7. Troshchenko, V.T.; Khamaza, L.A. Stages of Fatigue Failure of Metals and Alloys. *Rep. Natl. Acad. Sci. Ukr.* **2018**, 2, 56–63. [CrossRef]
8. Yushchenko, K.A.; Borysov, Y.S.; Kuznetsov, V.D.; Korzh, V.M. *Surface Engineering*; Naukova Dumka: Kyiv, Ukraine, 2007.
9. Gurey, V.; Hurey, I. The Effect of the Hardened Nanocrystalline Surface Layer on Durability of Guideways. In *Proceedings of the Lecture Notes in Mechanical Engineering*; Tonkonogyi, V., Ivanov, V., Trojanowska, J., Oborskyi, G., Edl, M., Kuric, I., Pavlenko, I., Dasic, P., Eds.; Springer International Publishing: Cham, Switzerland, 2020; pp. 63–72.

10. Gurey, V.; Hurey, I.; Hurey, T.; Wojtowicz, W. Fatigue Strength of Steel Samples After Friction Treatment. *Lect. Notes Mech. Eng.* **2023**, *1*, 274–283. [CrossRef]
11. Kyryliv, V.I.; Gurey, V.I.; Maksymiv, O.V.; Hurey, I.V.; Kulyk, Y.O. Influence of the Deformation Mode on the Force Conditions of Formation of the Surface Nanostructure of 40Kh Steel. *Mater. Sci.* **2021**, *57*, 422–427. [CrossRef]
12. Shaw, L.L. The Surface Deformation and Mechanical Behavior of Nanostructured Alloys. In *Nanostructured Metals and Alloys*; Elsevier: Cambridge, UK, 2011; pp. 481–506, ISBN 9781845696702.
13. Bartkowska, A. Production and Properties of FeB-Fe_2B-Fe_3(B,C) Surface Layers Formed on Tool Steel Using Combination of Diffusion and Laser Processing. *Coatings* **2020**, *10*, 1130. [CrossRef]
14. Korzhyk, V.; Tyurin, Y.; Kolisnichenko, O. *Theory and Practice of Plasma-Detonation Technology of Surface Hardening Metal Products*; Privat Company Technology Center: Kharkiv, Ukraine, 2021; ISBN 9786177319466.
15. Lukaszkowicz, K. Review of Nanocomposite Thin Films and Coatings Deposited by PVD and CVD Technology. In *Nanomaterials*; Rahman, M.M., Ed.; InTech: Rijeka, Croatia, 2011.
16. Quintino, L. Overview of Coating Technologies. In *Surface Modification by Solid State Processing*; Elsevier: Cambridge, UK, 2014; pp. 1–24, ISBN 9780857094681.
17. Hurey, I.; Hurey, T.; Lanets, O.; Dmyterko, P. The Durability of the Nanocrystalline Hardened Layer during the Fretting Wear. *Lect. Notes Mech. Eng.* **2021**, *2*, 23–32. [CrossRef]
18. Vilaça, P. Friction Surfacing. In *Surface Modification by Solid State Processing*; Elsevier: Cambridge, UK, 2014; pp. 25–72.
19. Zhang, F.; Duan, C.; Wang, M.; Sun, W. White and Dark Layer Formation Mechanism in Hard Cutting of AISI52100 Steel. *J. Manuf. Process.* **2018**, *32*, 878–887. [CrossRef]
20. Zhang, F.; Duan, C.; Sun, W.; Ju, K. Effects of Cutting Conditions on the Microstructure and Residual Stress of White and Dark Layers in Cutting Hardened Steel. *J. Mater. Process. Technol.* **2019**, *266*, 599–611. [CrossRef]
21. Javaheri, V.; Sadeghpour, S.; Karjalainen, P.; Lindroos, M.; Haiko, O.; Sarmadi, N.; Pallaspuro, S.; Valtonen, K.; Pahlevani, F.; Laukkanen, A.; et al. Formation of Nanostructured Surface Layer, the White Layer, through Solid Particles Impingement during Slurry Erosion in a Martensitic Medium-Carbon Steel. *Wear* **2022**, *496–497*, 204301. [CrossRef]
22. Inoue, A.; Hashimoto, K. (Eds.) *Amorphous and Nanocrystalline Materials*; Advances in Materials Research; Springer: Berlin/Heidelberg, Germany, 2001; Volume 3, ISBN 978-3-642-08664-9.
23. Hosseini, S.B.; Klement, U. A Descriptive Phenomenological Model for White Layer Formation in Hard Turning of AISI 52100 Bearing Steel. *CIRP J. Manuf. Sci. Technol.* **2021**, *32*, 299–310. [CrossRef]
24. Gurey, V.; Shynkarenko, H.; Kuzio, I. Mathematical Model of the Thermoelasticity of the Surface Layer of Parts during Discontinuous Friction Treatment. *Lect. Notes Mech. Eng.* **2021**, *2*, 12–22. [CrossRef]
25. Wu, S.; Liu, G.; Zhang, W.; Chen, W.; Wang, C. Formation Mechanism of White Layer in the High-Speed Cutting of Hardened Steel under Cryogenic Liquid Nitrogen Cooling. *J. Mater. Process. Technol.* **2022**, *302*, 117469. [CrossRef]
26. Lowe, T.C.; Zhu, Y.T.; Semiatin, S.L.; Berg, D.R. Overview and Outlook for Materials Processed by Severe Plastic Deformation. In *Investigations and Applications of Severe Plastic Deformation*; Lowe, T.C., Valiev, R.Z., Eds.; Springer: Dordrecht, The Netherlands, 2000; pp. 347–356, ISBN 978-94-011-4062-1.
27. Hahn, H. Unique Features and Properties of Nanostructured Materials. In *Nanomaterials by Severe Plastic Deformation*; Wiley-VCH Verlag GmbH & Co., KGaA: Weinheim, Germany, 2005; pp. 2–17, ISBN 9783527602469.
28. Guo, Y.B.; Warren, A.W.; Hashimoto, F. The Basic Relationships between Residual Stress, White Layer, and Fatigue Life of Hard Turned and Ground Surfaces in Rolling Contact. *CIRP J. Manuf. Sci. Technol.* **2010**, *2*, 129–134. [CrossRef]
29. Cicero, S.; Álvarez, J.A. Fracture, Fatigue, and Structural Integrity of Metallic Materials. *Metals* **2019**, *9*, 913. [CrossRef]
30. Kyryliv, V.; Maksymiv, O.; Gurey, V.; Hurey, I.; Kyryliv, Y.; Zvirko, O. The Mode Deformation Effect on Surface Nanocrystalline Structure Formation and Wear Resistance of Steel 41Cr4. *Coatings* **2023**, *13*, 249. [CrossRef]
31. Hurey, I.; Gurey, V.; Kyryliv, V. A Tool for Alloying the Surface Layers of Machine Parts. Patent No. 123883, 2021.
32. Burek, J.; Hurey, I. Tool for Forming a Nanocrystalline Hardened Surface Layer of an Object and a Method for Forming a Nanocrystalline Hardened Surface Layer of an Object. Patent No. 240972, 2022.
33. Gans, R.F. *Engineering Dynamics*; Springer: New York, NY, USA, 2013; ISBN 978-1-4614-3929-5.
34. Timoshenko, S.; Weaver, W. *Vibration Problems in Engineering*; Wolfenden Press: Singapore, 2007; ISBN 978-1406774658.
35. Torby, B.J. *Advanced Dynamics for Engineers*; Holt Rinehart & Winston: New York, NY, USA, 1984; ISBN 0-03-063366-4.
36. Goldstein, H.; Poole, C.; Safko, J. *Classical Mechanics*, 3rd ed.; Addison Wesley: Boston, MA, USA, 2001; ISBN 9780201657029.
37. Taylor, R.J. *Classical Mechanics*; University Science Books: Melville, NY, USA, 2005; ISBN 1-891389-22-X.

Journal of
Manufacturing and Materials Processing

MDPI

Article

Determination of the Cutting-Edge Microgeometry Based on Process Forces during Peripheral Milling of Ti-6Al-4V Using Machine Learning

Matthias Wimmer *, Roman Hartl and Michael F. Zaeh

TUM School of Engineering and Design, Technical University of Munich, 85748 Garching, Germany
* Correspondence: matthias.wimmer@iwb.tum.de

Abstract: The residual stress state of the machined sub-surface influences the service quality indicators of a component, such as fatigue life, tribological properties, and distortion. During machining, the radius of the cutting edge changes due to tool wear. The cutting-edge rounding significantly affects the residual stress state in the part and the occurring process forces. This paper presents a tool wear prediction model based on in-process measured cutting forces. The effects of the cutting-edge geometry on the force behavior and the machining-induced residual stresses were examined experimentally. The resulting database was used to realize a Machine Learning algorithm to calculate the hidden value of tool wear. The predictions were validated by milling experiments using rounded cutting edges for different process parameters. The microgeometry of the cutting edge could be determined with a Root Mean Square Error of 8.94 μm.

Keywords: milling; titanium alloy Ti-6Al-4V; residual stresses; process forces; cutting-edge radius; machine learning; tool wear prediction; supervised learning; multilayer perceptron

1. Introduction

The titanium alloy Ti-6Al-4V is widely used in the aerospace and medical industries due to its advantageous material properties. These include an excellent strength-to-weight ratio, corrosion resistance and low thermal conductivity [1]. These properties also present challenges in the manufacturing process, where machining is often used to produce the components. Titanium and its alloys are among the most difficult materials to machine [2]. The mechanical and chemical properties of Ti-6Al-4V result in high cutting forces and heat generation, causing excessive tool wear of the cutting edges [2]. In addition, the integrity of the surface of the machined components is altered by the machining process. In order to ensure the quality, functionality, and durability of the finished part, this is one of the most critical factors. In this respect, an important area of finishing machining is the improvement in the functional performance of high-value components [1,3,4].

In particular, the thermo-mechanical load applied by machining alters the residual stress state of the sub-surface [5]. Residual stress represents the remaining stress in the machined material after the external load has been removed, reflecting the total macro-stresses in the machined sub-surface zone. The residual stress state significantly affects the performance of the component such as corrosion resistance and fatigue life [6]. The change in residual stresses can therefore be used to assess the overall structural strength of the component and its propensity to fatigue failure in service [7]. Mechanical loads tend to cause compressive stresses, while thermal effects generally cause residual stresses in the tensile range [8,9]. Compressive residual stresses have been found to positively affect the fatigue life, creep behavior and crack propagation resistance, while tensile residual stresses are usually detrimental [1,10]. The amount of residual stress change is defined by the stress and heat sources that affect the sub-surface during the machining process. These can be characterized by examining the primary shear zone and the tool-workpiece interface, and are significantly dependent on the microgeometry of the cutting edge [8].

2. Influence of the Cutting-Edge Radius

2.1. Residual Stresses

In machining with a geometrically defined cutting edge, the main contact area between the tool and the workpiece is formed by cutting-edge rounding. Consequently, the cutting process is significantly influenced by the micro-geometry of the cutting edge [11,12]. The geometry of the cutting edge is a significant factor in influencing the tool–workpiece–chip contact, the occurring thermo-mechanical loads in the contact area, and the residual stresses induced by machining [5,13]. Therefore, the influence of rounded cutting edges on the resulting surface integrity has been evaluated for different materials and processes in the literature. To investigate the impact of the edge radius on the machining-induced residual stresses, Denkena et al. performed turning experiments on AISI 52100 [11]. They found that larger-edge radii caused higher compressive residual stresses and the affected surface layer increased. Changing the cutting-edge radius from 40 μm to 105 μm, the authors measured an increase in the maximum compressive stresses for both the circumferential and axial stress directions from 570 MPa to 1050 MPa. As a result of adopting a larger cutting-edge rounding, they witnessed higher passive forces, which led to an increase in the mechanical load and consequently to more plastic deformation in the sub-surface. The effect of the cutting-edge radius on the surface and near-surface residual stresses in up- and down-milling was examined by Wyen et al. using a one-fluted milling tool [14]. In this study, the titanium alloy Ti-6Al-4V was machined. Measurements of the residual stresses after machining showed an increase in maximum compressive stress on the surface due to the change in the cutting-edge radius from 310 MPa using a cutting-edge rounding of 6 μm to 600 MPa with a radius of 50 μm, whereas no uniform trend was observed for down-milling.

Furthermore, the surface compressive residual stress increased, but the depth of the maximum compressive stress d_{RS} (Figure 1d) and the range of influence of the machining-induced residual stress decreased as the edge radius changed from 0.01 mm to 0.03 mm. Tools with sharp and rounded cutting edges were used by Coelho et al. and Li et al. [12,15]. After turning Inconel 718 and milling AISI H13 steel, both identified the generation of compressive residual stresses independently of the process. Higher machining-induced compressive stresses were induced with rounded cutting edges rather than with sharp tools. Nespor compared the influence of the edge geometry on residual stresses after machining Ti-6Al-4V using ball end milling and orthogonal turning and planing [13]. The surface residual stresses in the cutting direction increased with a larger edge radius for all performed machining processes, and the absolute values were higher for turning than for planing. Ozel and Ulutan compared the effect of the cutting-edge radius on the resulting residual stress profiles for turning of Ti-6Al-4V [16]. In the radial direction, all experiments induced compressive residual stresses both on the surface and at the sub-surface of the material. However, in the circumferential direction, tensile residual stresses on the machined surface were measured, which gradually became compressive in the depth direction. Wyen and Wegener confirmed that the increased thermal and mechanical loads due to higher cutting-edge radii result in an increase in the process forces that act directly on the surface during machining [17]. Additional material deformation occurs at the front of the cutting edge due to its rounded shape, particularly where the cutting edge is in contact with the workpiece. The plastic deformation of the sub-surface (Figure 1d) led to compressive stresses in both up- and down-milling. The stresses are of different magnitudes depending on the direction. The compressive stresses in the cutting direction were smaller than those in the direction perpendicular to the cutting speed.

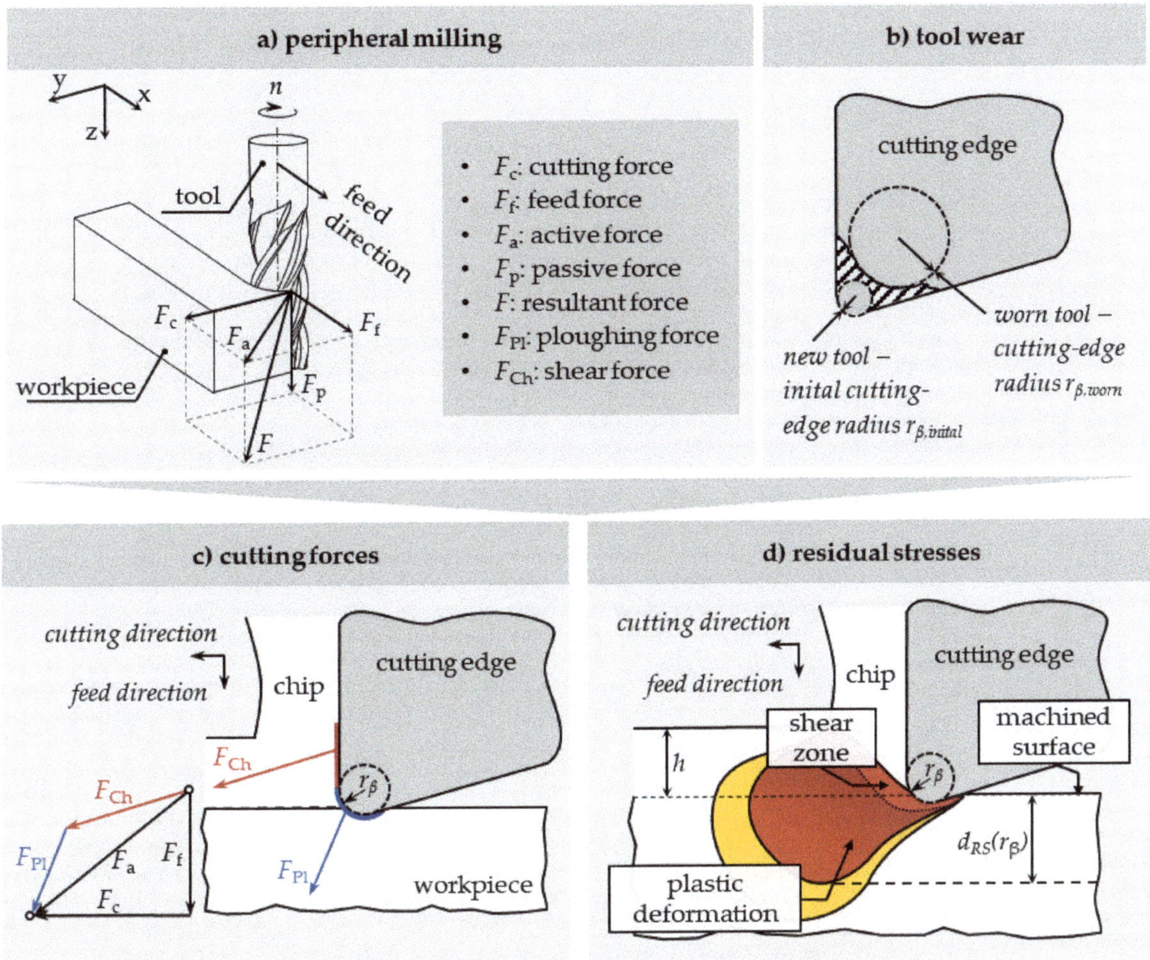

Figure 1. (**a**) Process forces in peripheral milling. (**b**) Change in the cutting-edge radius due to tool wear. (**c**) Ploughing forces due to the cutting-edge radius. (**d**) Influence of the cutting-edge rounding on plastic deformation according to Schoop et al., Albrecht, DIN 6580 and DIN 6584 [8,18–20].

2.2. Process Forces

Due to the tool-workpiece contact during the peripheral milling process, a resulting force F occurs. It comprises the cutting force F_c, the feed force F_f, and the passive force F_p (Figure 1a) [19]. In general, total forces recorded in a cutting process are the sum of individual forces acting on the tool flank face, its cutting edge, and the rake face [21]. The cutting edge is assumed to have an ideal sharp geometry in orthogonal cutting theory [22]. In practice, however, the cutting edge is rounded (Figure 1b), which causes a severe elastic and plastic deformation of the material around the cutting edge to occur simultaneously with the cutting process. This mechanism is referred to as the ploughing effect and can be described in terms of the ploughing force F_{Pl} (Figure 1c) acting directly on the cutting edge [18]. The percentage of this force in the total cutting force has often been considered insignificant in the previous literature and sometimes been neglected in force modeling for simplification reasons. A cutting force prediction model that considers the edge geometry was proposed in 1960 by Albrecht [18]. On the basis of this early model, Moufki et al.

pointed out that the importance of the ploughing force depends on the relation between the uncut chip thickness h and the cutting-edge rounding r_β [23]. The ploughing force can dominate when the uncut chip thickness is comparable to the cutting-edge radius. The rounded cutting edges influence the active force components including the ploughing forces and tool face friction, which are especially important in machining titanium alloys such as Ti-6Al-4V. Wegener and Wyen investigated the effect of different cutting-edge radii on the cutting and feed forces during the orthogonal turning of Ti-6Al-4V [14,17]. The force components increased with the increasing cutting-edge radius where the cutting force was less sensitive than the feed force. Experiments conducted during milling of SUS-316L steel by Lv et al. showed that a change in the cutting-edge radius from 4 to 15 μm led to an increase of 23% in the cutting force and 56% in the feed force for an uncut chip thickness h of 0.2 mm [24].

The radius of the cutting edge significantly influences the occurring cutting forces and therefore the mechanical load on the workpiece during milling. Particularly for titanium, which is used in applications requiring high mechanical integrity, information on the dependence of the surface integrity on the cutting-edge geometry is essential. The cutting-edge radius changes during milling as a result of tool wear (Figure 1b). In order to be able to determine the resulting residual stress state, a prediction of the microgeometry of the cutting edge is required. Therefore, the aim of this paper is to predict the cutting-edge radius. None of the currently available analytical, numerical, or empirical wear monitoring methods is sufficient to determine a nominal cutting-edge rounding value during milling.

For this reason, a new approach in artificial intelligence is presented, which uses the relationships between the process parameters, the cutting-edge radius and the measured cutting forces. In particular, machine learning (ML) algorithms can reveal the unknown underlying structure in the data and are therefore applied. To obtain the tool wear prediction model, experiments were first performed to generate a data set, including

- the measured process forces,
- the process parameters, and
- the measured cutting-edge radius.

The measured data and the process parameters were used as input and output values for the ML method. Characteristic values, such as the average forces, were extracted from the measured force signals through signal processing and data analysis. The cutting-edge radii were determined using a laser scanning microscope and extracted from the obtained profile. Because the objective is the determination of the cutting-edge rounding, the preprocessed force data and process parameters are input values. The cutting-edge radius for each data sample is assigned to the output value. This provides labeled data that define the problem at hand as a problem in the supervised learning setup. Furthermore, because the goal is to determine a nominal value for the cutting-edge rounding, the problem is one of regression [25]. The multilayer perceptron (MLP) algorithm was used as a universal function approximator [25,26].

3. Experiments and Procedures

In order to evaluate the influence of the microgeometry of the cutting edge on the cutting forces during peripheral milling of Ti-6Al-4V, down-milling experiments were conducted on a GROB G352-T 5-axis machining center. The geometry of the plates as samples was 100 mm × 100 mm × 10 mm. A constant cutting speed v_c of 30 m/min and a radial depth of cut a_e of 0.5 mm were used. The workpiece thickness of 10 mm was equal to the axial depth of cut. The machining parameters corresponded to those typically used for machining titanium. The end mills were OptiMill®-Titan-HPC SCM39 uncoated solid carbide milling tools with a diameter of 12 mm and four teeth with a helix angle β of 40° from MAPAL Dr. Kress KG. According to the manufacturer's specification regarding the cutting edges, the rake angle γ was $5.0 \pm 0.13°$ and the clearance angle α_c was $11.4 \pm 0.12°$. In addition, no cutting fluids were applied, thus eliminating them as a possible factor influencing the process forces.

3.1. Force Measurement

During the peripheral milling experiments, a tool holder with integrated sensors was used to measure the forces [27]. The occurring forces generated a bending moment in the tool holder, which can be converted into the cutting forces in the Cartesian coordinate system, F_x, F_y, and F_z. Using strain gauges arranged at an angle of 90° to each other, the force components in the x-, y- and z-directions were measured [28]. Data acquisition was performed with a sampling rate of 2 kHz and the recorded forces were transmitted to the data receiver via Wi-Fi. The experimental setup is illustrated in Figure 2.

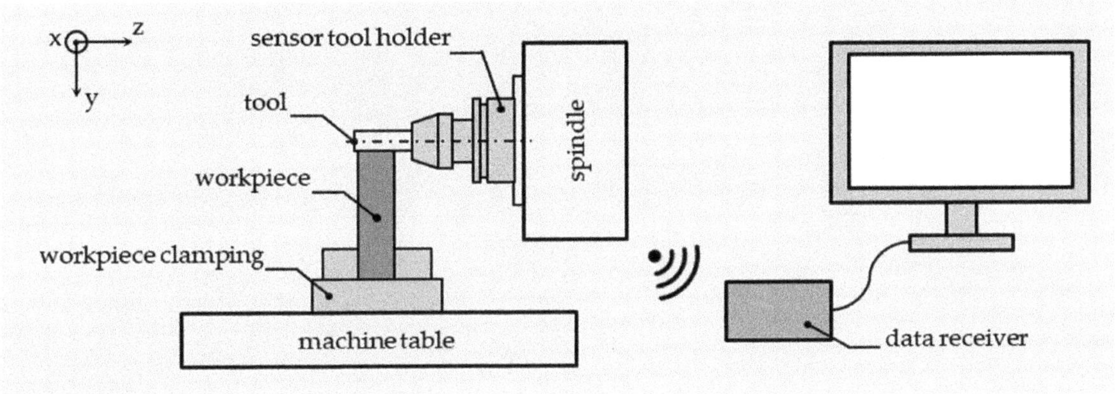

Figure 2. Experimental setup.

3.2. Process Parameters

The process parameters used for the experiments are given in Table 1. A constant depth of cut a_e of 0.5 mm was selected because the investigations focus on the finishing process (a_e between 0.1 and 0.5 mm). In order to reduce thermal effects leading to an increase in tool wear, a low cutting speed v_c of 30 m/min was used for the peripheral milling of Ti-6Al-4V. The first set of experiments involved nine tools (No. 1–9) with new cutting edges. To ensure repeatability, three tools were used, each with the same feed per tooth f_z. Tool wear tests were performed until each initially sharp tool had reached a milling length of 35.0 m. This limit was set because wear marks appeared on the tool as well as visible bumps on the component surface at this distance. Therefore, to obtain a reasonable data set size, the measurement intervals were chosen in steps of 0.5 m. A total of 630 samples was generated.

Table 1. Process parameters.

No.	Tool	a_e in mm	v_c in m/min	f_z in mm	r_β in µm
1	1/2/3	0.5	30	0.04	variable
2	4/5/6	0.5	30	0.06	variable
3	7/8/9	0.5	30	0.08	variable

3.3. Measurement of the Cutting-Edge Radius

The cutting edge is defined by the DIN 6581 standard [28]. It is the part of the tool that effectively produces the surface of the component. While the characterization of the macrogeometry of the cutting edge is internationally standardized [28,29], there is no standard method for describing the microgeometry of the cutting edge, which is the actual shape of the intersection of the flank and the rake face of a tool [30]. The cutting-edge rounding was formed by an approximating circle. Extensive research on the characterization and measurement of the cutting-edge profile has been performed by Wyen [30].

The method developed is shown in Figure 3, which was used for the measurement of the radius of the cutting edge. The procedure can be divided into five individual steps, with the aim of determining a reference circle (dotted red circle in Figure 3) that is adapted to the contour of the microgeometry of the cutting edge. The resulting radius r_β of the circle corresponds to the radius of the rounded cutting edge. In the first step, a straight-line fitting is performed according to the least squares method on the flank and rake face. Using the determined straight lines, the bisector of the wedge angle β can be calculated, which intersects the cutting-edge profile at the point p_{int} (step 2). Based on the intersection point p_{int}, a circle can be inserted that is tangent to the two fitting lines. The new upper bound for the least squares fitting range results from the points where the circle contacts the straight fitting lines (step 3). By repeating steps 2 and 3, the iterative geometric fitting can minimize the difference between the points where the straight lines are tangent to the circle and the upper fitting limit of the previous step to approximately zero (step 4). This allows the determination of the absolute limit of the transition between the macro- and micro-geometry, which defines the area for the circle fitting to determine the cutting-edge radius r_β (step 5).

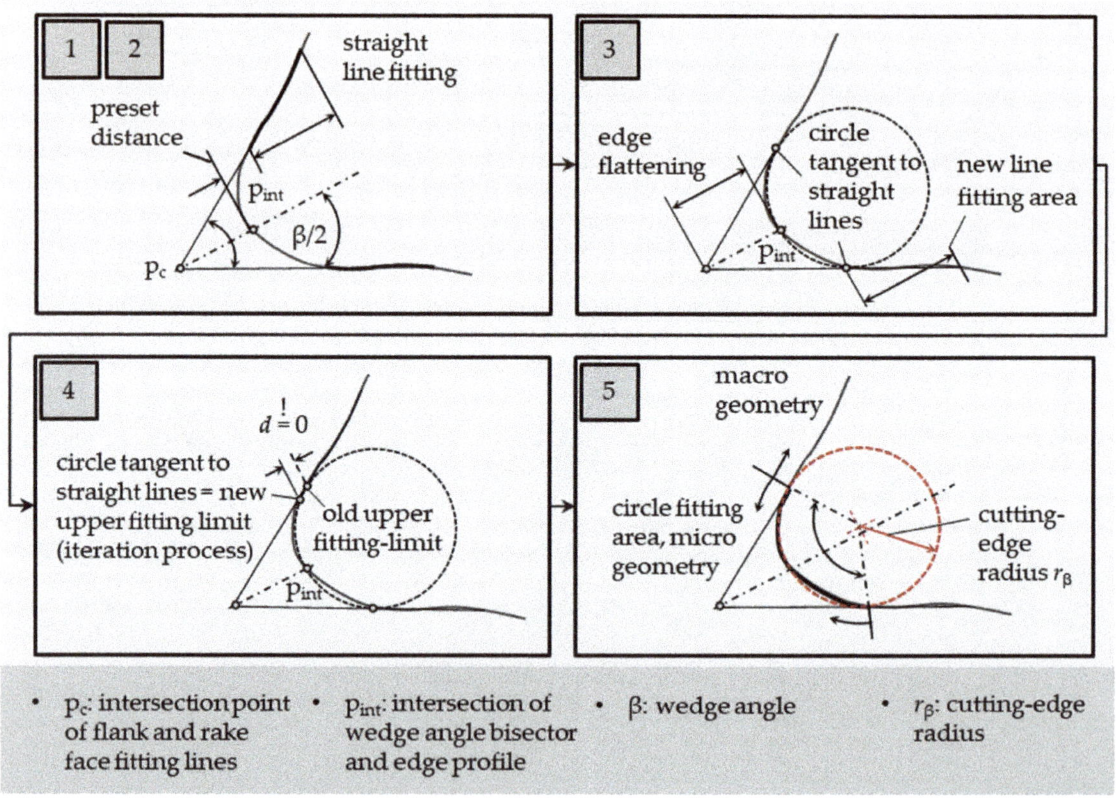

Figure 3. Method for measuring of the cutting-edge radius according to Wyen [30].

This approach drastically reduces the uncertainty of the point sensitivity and the deviations of a single point. After defined intervals of the milling length of 0.5 m, the cutting edges of the milling tool were measured. In order to determine the maximum cutting-edge rounding $r_{\beta,max}$ due to tool wear during the experiments described in Table 1, the tool was removed from the machine and placed in an ultrasonic bath to remove adhering chips. The microgeometry of the tools was measured using a KEYENCE VK-X1050 laser scanning microscope. The height profile obtained by the microscope was analyzed using the MultiFile Analyzer software included in the VK-X1050.

3.4. Residual Stress Measurement

The individual cutting edges of the tool were prepared to create defined cutting-edge radii. The edges were symmetrically rounded by tool grinding using a CNC tool grinder. The radii were determined using the characterization method described in Section 3.2. For the second series of experiments (Table 2), eight different tools (No. 4–11) were prepared, with cutting edge radii ranging from 5 ± 1 µm (new tool) to 60 ± 1 µm. The cutting speed v_c and the radial depth of cut a_e were kept constant and the feed per tooth was varied from 0.04 to 0.08 mm in steps of 0.2 mm. The objective was to determine the quantitative relationship between the cutting-edge rounding and the machining-induced residual stresses on the machined surface.

Table 2. Experimental plan for residual stress measurements.

No.	a_e in mm	v_c in m/min	f_z in mm	r_β in µm
4	0.5	30	0.04, 0.06, 0.08	5 (sharp)
5	0.5	30	0.04, 0.06, 0.08	10
6	0.5	30	0.04, 0.06, 0.08	20
7	0.5	30	0.04, 0.06, 0.08	25
8	0.5	30	0.04, 0.06, 0.08	30
9	0.5	30	0.04, 0.06, 0.08	40
10	0.5	30	0.04, 0.06, 0.08	50
11	0.5	30	0.04, 0.06, 0.08	60

The residual stresses on the machined surface along and perpendicular to the feed direction were measured using X-ray diffraction (XRD). Three repetitions were performed in each case and the scatter was calculated. The measurement was performed using a Stresstech Xstress3000 XRD device with a G2R goniometer. The stresses were determined using the sin-2ψ method. This method is based on determining the elastic lattice strains, which are converted to stresses by the 3D form of Hooke's law [27,31]. The macro stresses distort the crystal lattice, resulting in an angular shift of the diffraction peak selected to measure the residual stresses. These measurements provide information on the residual stresses on the surface because the X-ray beam penetrates only to a depth of about 5 µm [32].

4. Experimental Results

4.1. Force Behavior

During the peripheral milling tests, the force signals in the Cartesian directions (F_x, F_y, and F_z) were recorded. Afterward, a filter was applied to reduce the impact of the noise on the force signal. Therefore, a Kalman filter was applied, which corrects its prediction of the system's state using the available data [33]. After these pre-processing steps, the average forces $F_{i,ave}$ were calculated in each Cartesian direction. The average cutting forces obtained for the different feeds per tooth f_z over the milling length l are depicted in Figure 4. The average forces increase for higher values of f_z with an initial cutting-edge radius of 5 µm (sharp). The cutting force components in x- and z-directions slightly increase with the ascendant milling length. The most significant effect of a change in the microgeometry of the cutting edge over the milling path caused by tool wear can be observed in the cutting force component F_y.

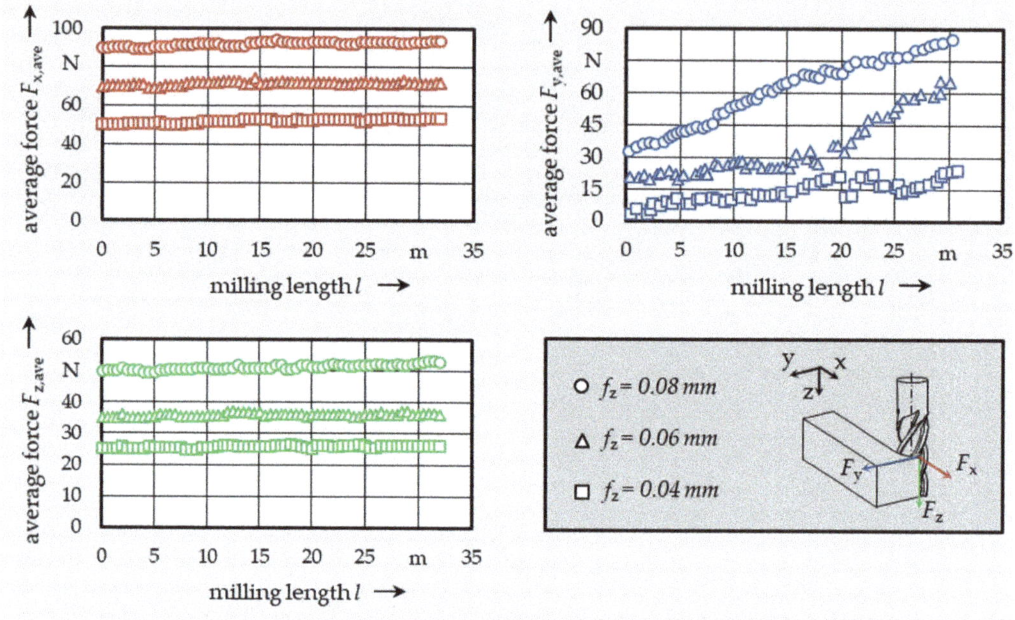

Figure 4. Measured cutting forces in x-, y- and z-directions over the milling length for different feeds per tooth.

4.2. Tool Wear Behavior

The cutting-edge radius is assumed to increase with progressive tool wear and the experiments conducted in this study confirm this trend. The individual cutting edges exhibit a uniform wear behavior (Figure 5). The results show an increase in the cutting-edge radius over the milling length for all three tested feeds per tooth f_z (0.04 mm, 0.06 mm, and 0.08 mm).

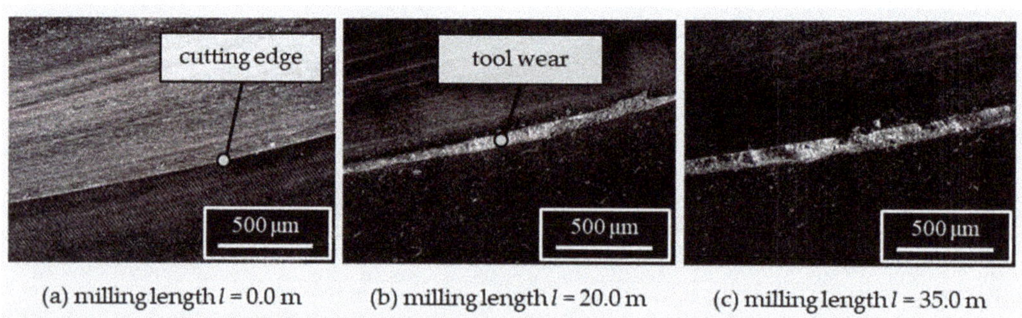

Figure 5. Progressive tool wear for a feed per tooth f_z of 0.06 mm at the cutting edge for different milling lengths l recorded with the KEYENCE VK-X1050 with a magnification of 10×.

Additionally, the experiments revealed that the tools with the same feed per tooth (i.e., Tools 1–3, Tools 4–6, and Tools 7–9) exhibited similar wear behavior. In this case, the change in the cutting-edge rounding was increased by selecting a higher feed per tooth. After a milling length of 35 m, a maximum cutting-edge radius of 31 µm for a feed per tooth of 0.04 mm, 43 µm for 0.06 mm and 54 µm for 0.08 mm was measured. The values for the measured cutting-edge radius r_β over the milling length l for each of the three feed per tooth values are shown in Figure 6.

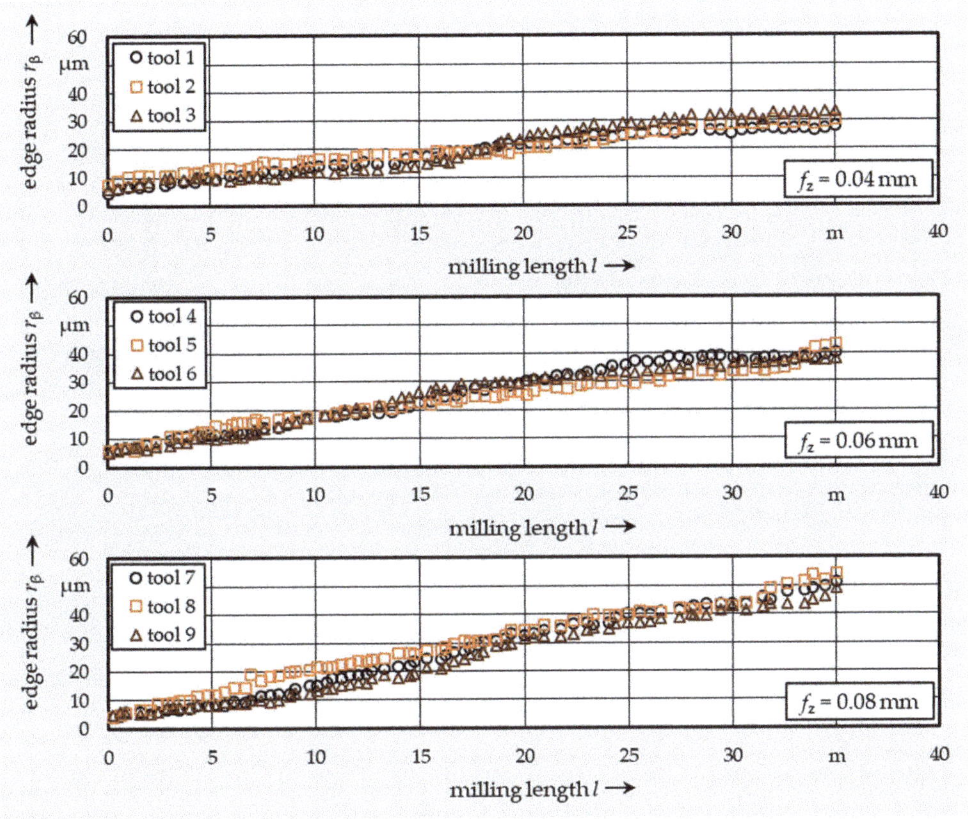

Figure 6. Change in the cutting-edge radius over the milling length.

4.3. *Influence of the Edge Radius on the Surface Residual Stress*

The results of the residual stress measurements after peripheral milling with defined cutting-edge radii always show compressive residual stresses on the surface of the machined components made of Ti-6Al-4V (Figure 7). The magnitude of the compressive residual stresses perpendicular to the feed direction (Figure 4, z-direction) and in the feed direction (Figure 4, x-direction) increases with higher cutting-edge radii. For a feed per tooth of 0.04 mm, the residual compressive stresses perpendicular to the feed direction showed a change from −114 MPa with a cutting-edge radius of 5 µm (sharp) to −289 MPa for a cutting edge rounding of 60 µm (minus sign indicates compressive stresses). Additionally, the machining-induced residual stresses in the feed direction showed a change from −52 MPa to −253 MPa. A higher feed per tooth reduces the maximum residual compressive stresses on the surface for the same cutting-edge microgeometry. With a feed per tooth of 0.06 mm, maximum residual compressive stresses of −246 MPa perpendicular to the feed direction and −244 MPa in the feed direction were induced in the machined surface with a cutting-edge rounding of 60 µm.

The cutting-edge radius thus significantly affects the resulting residual stresses after peripheral milling of Ti-6Al-4V. Consequently, knowing the cutting-edge rounding, the residual stress state on the surface and in the sub-surface could be derived. However, the microgeometry of the cutting edge is unknown during the milling process and must therefore be determined.

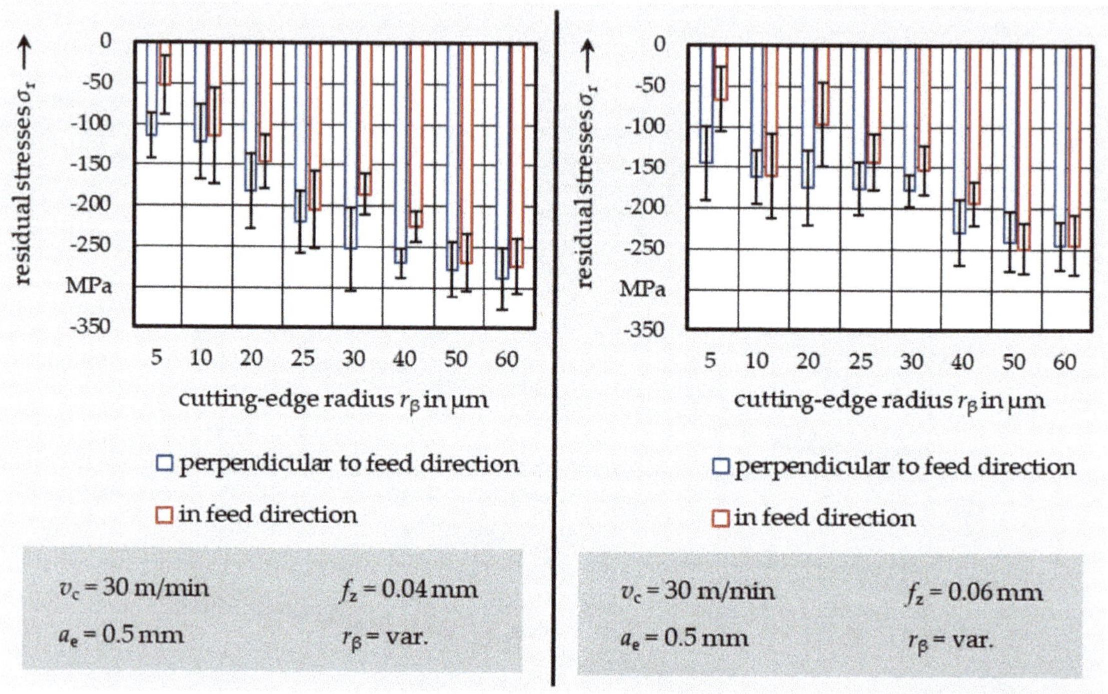

Figure 7. Effect of the cutting-edge radius on the surface residual stresses for two different feeds per tooth f_z.

5. Tool Wear Prediction Model

5.1. Data Acquisition

For determining the cutting-edge radius using ML, a database was first built up which contains

- the process forces in the Cartesian directions (F_x, F_y, and F_z),
- the process parameters (v_c, a_e, and f_z), and
- the current maximum cutting-edge radius (r_β).

For this purpose, the process forces were recorded during the peripheral milling process of Ti-6Al-4V using the sensor tool holder. In addition, the cutting-edge radius was measured offline in defined intervals of 0.5 m. A ML model was trained to accurately predict the cutting-edge radius at any given time during the milling process by building up a database of in-process data and offline cutting-edge radius measurements.

5.2. Signal Processing and Data Analysis

The relevant input data for the tool wear prediction model are the forces in x-, y-, and z-directions and the feed per tooth f_z. These input data must be pre-processed to use them in the tool wear prediction model (Figure 8). For the forces, the mean values were subtracted from the signal. Because the feed values were 0.04 mm, 0.06 mm and 0.08 mm while the measured forces were in a two- to three-digit range, min–max scaling was applied. Thus, the input data were scaled to the interval [0, 1].

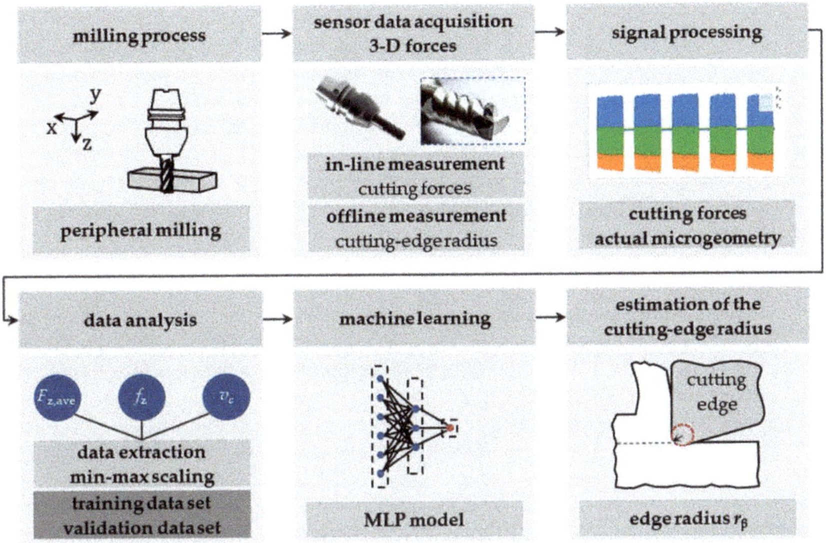

Figure 8. Methodology for determining the cutting-edge radius using an MLP regressor based on the measured cutting forces.

5.3. Machine Learning

After the generation and the pre-processing of the input data, ML algorithms were applied to discover the relationship between the cutting forces F_x, F_y, and F_z and the cutting-edge rounding r_β for the different feeds per tooth f_z. For the data analysis, a universal function approximator in the form of an MLP was applied and the performance of the data set was evaluated. The MLP is a type of artificial neural network (ANN) composed of multiple layers of interconnected neurons. The most common type of MLP consists of a single input layer, one or more hidden layers, and a single output layer. The neurons in each layer are connected to neurons in the previous and following layers with connections. The input layer receives the input data and has several neurons equal to the inputs in the data set. The hidden layers process the data using weights and biases. The output layer has a single neuron and produces the final result. An MLP has a fully connected architecture. This denotes that each of the neurons in one layer is connected to every neuron in the following layer [34].

The given problem is one of regression in the supervised learning case because the data points are labeled with an actual number, the current cutting-edge radius. MLPs are commonly used for supervised learning tasks. They are also known as feedforward neural networks, as the data flows through the network in only one direction, from input to output. The limited memory BFGS (L-BFGS) algorithm from the family of Quasi-Newton solvers was applied next. Solvers are one of the hyperparameters of an ANN. A solver is an algorithm used in the backpropagation process to calculate the weights of the neural network. For performance and flexibility concerning the solver implementations provided, a sklearn.neural_network.MLP regressor class was used. This class provides the L-BFGS algorithm [25].

The MLP regressor uses activation functions to introduce non-linearity into the prediction model. The activation function is a function that maps the input of the neuron onto its output [35]. Basic activation functions are commonly designed to keep the signal within a wanted range (tanh), or to eliminate unwanted elements (ReLU). To predict the cutting-edge radius based on the cutting forces and the process parameters, the tanh function and the ReLU function were compared. To optimize the model architecture, several hyperparameters were varied, such as the number of layers, the number of neurons in each layer, the learning rate, and the used activation function. Additionally, the maximum number of gradient steps was considered a hyperparameter, which was increased by steps of 50 within a range of [50, 300]. The learning rate α was chosen to be 0.001, which was defined as the default setting. To generate a training and validation data set, an 80:20 split was applied. To preselect all possible parameter combinations, test runs of the prediction were performed. Due to the prevalence of the maximum root mean square error (RMSE) of 15.97 µm, a threshold value of 16 µm was set. The methodology for determining the cutting-edge radius using an MLP regressor based on the measured cutting forces is illustrated in Figure 8.

6. Modeling Results

The predicted cutting-edge radii were compared with the measured radii of the data set to evaluate the performance of the tool wear prediction model. The RMSE was calculated to assess the prediction results. The number of gradient steps that led to minor errors was the minimum limit of the tested range, i.e., 50. An RMSE of 15.76 µm was obtained. The range was therefore decreased, and the limits were lowered to values ranging from 10 to 50 in increments of 5. The errors decreased for the lower number of allowed gradient steps. The produced errors for the different architectures using the L-BFGS solver with the ReLU function are shown in Table 3. The prediction model using the ReLU function as an activation function consisting of three hidden layers with four hidden neurons each achieved an error of 13.23 µm for a learning rate α of 0.00868 and 20 gradient steps.

Table 3. RMSE for different architectures using L-BFGS with ReLU activation.

Activ. Function	Hidden Layers	Hidden Neurons	Max. Gradient Steps	RMSE
ReLU	2	3	50	15.76 µm
ReLU	3	4	50	15.40 µm
ReLU	1	4	10	13.75 µm
ReLU	3	4	20	13.23 µm

Using the tanh activation function (Table 4), a non-linear dependency between r_β and F_y was found. As the dependency between r_β and F_y was the most remarkable (see Figure 4), an MLP model was built using $F_{y,norm}$ as the only input. The result was an RMSE of 15.56 µm for four hidden layers with four hidden neurons and 50 gradient steps. By reducing the maximum gradient steps, the RMSE could be lowered. In addition, the number of hidden layers and hidden neurons was changed. Using the normalized force $F_{y,norm}$ as input, the lowest error was 8.94 µm for four hidden layers including three hidden neurons each and a number of 25 maximum gradient steps.

Table 4. RMSE for different architectures using L-BFGS with tanh activation.

Activ. Function	Hidden Layers	Hidden Neurons	Max. Gradient Steps	RMSE
tanh	4	4	50	15.56 µm
tanh	1	4	25	14.16 µm
tanh	4	2	20	13.99 µm
tanh	4	3	35	9.00 µm
tanh	3	3	35	8.98 µm
tanh	3	3	30	8.97 µm
tanh	4	3	25	8.94 µm

The predicted values of the cutting-edge roundings $r_{\beta,pred}$ that were generated using the elaborated model were plotted against the measured cutting-edge radii $r_{\beta,true}$. The corresponding graph is shown in Figure 9. The results for evaluating this model on the complete data set are promising.

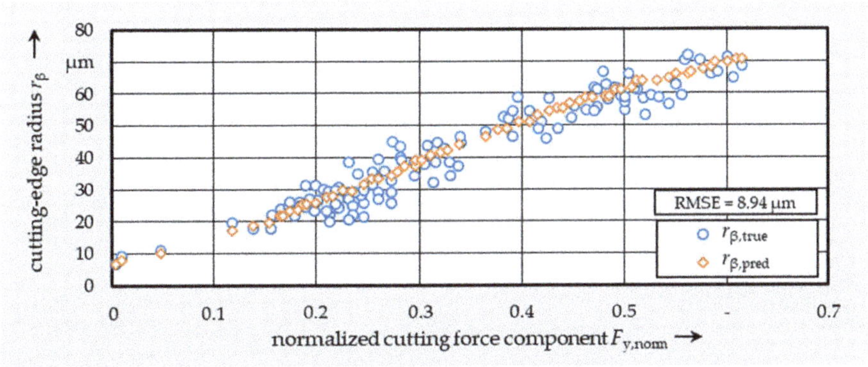

Figure 9. Predicted cutting-edge radii $r_{\beta,pred}$ vs. true response $r_{\beta,true}$ for an MLP with L-BFGS solver, a learning rate α of 0.001, tanh activation function, four hidden layers with three neurons and 25 gradient steps based on the normalized cutting force component $F_{y,norm}$.

7. Conclusions

The aim of this work was the determination of the current maximum cutting-edge radius based on the cutting forces during peripheral milling using Machine Learning. The investigations presented in this publication concern the change in the cutting-edge radius due to tool wear during peripheral milling of the titanium alloy Ti-6Al-4V. Experiments using different feeds per tooth were performed. The cutting forces were measured using a sensor tool holder during the milling tests. The cutting-edge radius was measured offline at regular intervals using a laser scanning microscope. In addition, tests were performed with prepared cutting-edge radii with subsequent measurements of the residual stresses on the machined surface. The main findings of the analysis of the experimental studies are as follows:

- The forces in x-, y- and z-directions increased over the milling length. In particular, an increase in the forces in the y-direction F_y was observed.
- An increase in the cutting-edge rounding over the milling length could be observed. Due to abrasive tool wear, material was removed from the cutting edges.
- Residual stresses were consistently obtained within the compression range. A change in the cutting-edge radii from 5 µm to 60 µm led to a significant increase in the residual compressive stresses.

The machining-induced residual stresses after the peripheral milling process strongly depend on the actual cutting-edge geometry. The radius of the cutting edges cannot be measured directly during machining and must therefore be determined based on a model.

To determine the microgeometry of the cutting edges during the milling process, the relationship between the size of the cutting-edge radius and the resulting cutting forces during the milling process was used to generate a database. A training and a validation data set was built through signal processing and data analysis. Finally, the MLP method was applied using different architectures. The performance of the algorithm was measured with the RMSE. The best results were achieved for an MLP with an L-BFGS solver with a tanh activation function. In conclusion, the developed tool wear prediction model performed well on the data set. The forces are suitable for predicting the cutting-edge radius, especially the force component in the y-direction.

Overall, the experiments and the developed prediction model provide insights into the effect of changing edge microgeometry due to tool wear on the cutting forces and the resulting residual stresses during the peripheral milling of Ti-6Al-4V. During the machining process, the state of residual stresses in the sub-surface of the workpiece is influenced by the choice of the process parameters and the micro-geometry of the cutting edge. The most important parameter for the residual stresses induced by machining is the cutting-edge radius. This significant factor makes a difference in the cutting behavior in advanced manufacturing, covering the occurring mechanical loads and the machining-induced surface integrity. A change in the cutting-edge radius due to tool wear induces higher compressive residual stresses, which have a positive effect on the service life of the component. These observations could be correlated with higher ploughing forces. When milling Ti-6Al-4V, this factor influences the residual stress state due to peripheral milling. The predicted cutting-edge radii can be used in future work as an input variable for the model-based determination of the residual stress state, depending on the cutting-edge geometry. In addition, further work should investigate the influence of tool wear on other surface integrity properties, such as surface roughness and hardness.

Author Contributions: Conceptualization, M.W.; methodology, M.W.; software, M.W.; validation, M.W.; formal analysis, M.W.; investigation, M.W.; resources, M.W.; data curation, M.W.; writing—original draft preparation, M.W.; writing—review and editing, M.W., M.F.Z. and R.H.; visualization, M.W.; supervision, M.F.Z. and M.W.; project administration, M.W.; funding acquisition, M.F.Z. All authors have read and agreed to the published version of the manuscript.

Funding: This research was funded by the German Research Foundation (DFG) in the priority program SPP 2086, grant numbers ZA 288/72-1, project number 359102403.

Data Availability Statement: The reference data presented in this study are available upon request from the corresponding author.

Conflicts of Interest: The authors declare that they have no conflict of interest.

References

1. Ulutan, D.; Ozel, T. Machining induced surface integrity in titanium and nickel alloys: A review. *Int. J. Mach. Tools Manuf.* **2011**, *51*, 250–280. [CrossRef]
2. Niknam, S.A.; Khettabi, R.; Songmene, V. Erratum to: Machinability and Machining of Titanium Alloys: A Review. In *Machining of Titanium Alloys*; Davim, J.P., Ed.; Springer: Berlin/Heidelberg, Germany, 2014; pp. 1–30. ISBN 978-3-662-43901-2.
3. Sridhar, B.; Devananda, G.; Ramachandra, K.; Bhat, R. Effect of machining parameters and heat treatment on the residual stress distribution in titanium alloy IMI-834. *J. Mater. Process. Technol.* **2003**, *139*, 628–634. [CrossRef]
4. Che-Haron, C. Tool life and surface integrity in turning titanium alloy. *J. Mater. Process. Technol.* **2001**, *118*, 231–237. [CrossRef]
5. Wimmer, M.; Woelfle, C.; Krempaszky, C.; Zaeh, M. The influences of process parameters on the thermo-mechanical workpiece load and the sub-surface residual stresses during peripheral milling of Ti-6Al-4V. *Procedia CIRP* **2021**, *102*, 471–476. [CrossRef]
6. Zhang, P.; Liu, Z. Effect of sequential turning and burnishing on the surface integrity of Cr–Ni-based stainless steel formed by laser cladding process. *Surf. Coatings Technol.* **2015**, *276*, 327–335. [CrossRef]
7. Yang, D.; Xiao, X.; Liu, Y.; Sun, J. Peripheral milling-induced residual stress and its effect on tensile–tensile fatigue life of aeronautic titanium alloy Ti–6Al–4V. *Aeronaut. J.* **2019**, *123*, 212–229. [CrossRef]
8. Schoop, J.; Hasan, M.; Zannoun, H. Physics-Informed and Data-Driven Prediction of Residual Stress in Three-Dimensional Machining. *Exp. Mech.* **2022**, *62*, 1461–1474. [CrossRef]
9. Schoop, J. In-Situ Calibrated Modeling of Residual Stresses Induced in Machining under Various Cooling and Lubricating Environments. *Lubricants* **2021**, *9*, 28. [CrossRef]
10. Li, X.; Wang, Z.; Yang, S.; Guo, Z.; Zhou, Y.; Han, K. Influence of turning tool wear on the surface integrity and anti-fatigue behavior of Ti1023. *Adv. Mech. Eng.* **2021**, *13*, 16878140211011278. [CrossRef]
11. Denkena, B.; Grove, T.; Maiss, O. Influence of the cutting-edge radius on surface integrity in hard turning of roller bearing inner rings. *Prod. Eng. Res. Devel.* **2015**, *9*, 299–305. [CrossRef]
12. Li, B.; Zhang, S.; Yan, Z.; Jiang, D. Influence of edge hone radius on cutting forces, surface integrity, and surface oxidation in hard milling of AISI H13 steel. *Int. J. Adv. Manuf. Technol.* **2018**, *95*, 1153–1164. [CrossRef]
13. Nespor, D. Randzonenbeeinflussung durch die Rekonturierung Komplexer Investitionsgüter aus Ti-6Al-4V. Ph.D. Thesis, Leibniz Universität Hannover, Hannover, Germany, 2015.
14. Wyen, C.-F.; Jaeger, D.; Wegener, K. Influence of cutting-edge radius on surface integrity and burr formation in milling titanium. *Int. J. Adv. Manuf. Technol.* **2013**, *67*, 589–599. [CrossRef]
15. Coelho, R.; Silva, L.; Braghini, A.; Bezerra, A. Some effects of cutting-edge preparation and geometric modifications when turning INCONEL 718™ at high cutting speeds. *J. Mater. Process. Technol.* **2004**, *148*, 147–153. [CrossRef]
16. Özel, T.; Ulutan, D. Prediction of machining induced residual stresses in turning of titanium and nickel-based alloys with experiments and finite element simulations. *CIRP Ann.* **2012**, *61*, 547–550. [CrossRef]
17. Wyen, C.-F.; Wegener, K. Influence of cutting-edge radius on cutting forces in machining titanium. *CIRP Ann.* **2010**, *59*, 93–96. [CrossRef]

18. Albrecht, P. New Developments in the Theory of the Metal-Cutting Process: Part I. The Ploughing Process in Metal Cutting. *J. Eng. Ind.* **1960**, *82*, 348–357. [CrossRef]
19. *DIN 6580-10*; Begriffe Der Zerspantechnik; Bewegungen und Geometrie des Zerspanvorganges. Beuth Verlag GmbH.: Berlin, Germany, 1985.
20. *DIN 6584*; Begriffe Der Zerspantechnik; Kräfte, Energie, Arbeit, Leistungen. Beuth Verlag GmbH.: Berlin, Germany, 1982.
21. Merchant, M.E. Mechanics of the Metal Cutting Process. I. Orthogonal Cutting and a Type 2 Chip. *J. Appl. Phys.* **1945**, *16*, 267–275. [CrossRef]
22. Altintas, Y. *Manufacturing Automation: Metal Cutting Mechanics, Machine Tool Vibrations, and Cnc Design*, 2nd ed.; Cambridge University Press: New York, NY, USA, 2012; ISBN 9781107001480.
23. Moufki, A.; Dudzinski, D.; Le Coz, G. Prediction of cutting forces from an analytical model of oblique cutting, application to peripheral milling of Ti-6Al-4V alloy. *Int. J. Adv. Manuf. Technol.* **2015**, *81*, 615–626. [CrossRef]
24. Lv, D.; Wang, Y.; Yu, X. Effects of cutting-edge radius on cutting force, tool wear, and life in milling of SUS-316L steel. *Int. J. Adv. Manuf. Technol.* **2020**, *111*, 2833–2844. [CrossRef]
25. Murphy, K.P. *Machine Learning: A Probabilistic Perspective*; MIT Press: Cambridge, MA, USA, 2012; ISBN 9780262305242.
26. Bishop, C.M. *Pattern Recognition and Machine Learning*; Springer: New York, NY, USA, 2006; ISBN 0387310738.
27. Wimmer, M.; Hameed, M.Z.S.; Wölfle, C.; Weisbrodt, V.; Zaeh, M.F.; Werner, E.; Krempaszky, C.; Semm, T. The influence of the process parameters on the surface integrity during peripheral milling of Ti-6Al-4V. *TM-Tech. Mess.* **2020**, *87*, 721–731. [CrossRef]
28. *DIN 6581*; Bezugssysteme und Winkel am Schneidteil des Werkzeuges. German Institute for Standardization: Berlin, Germany, 1985.
29. *ISO 3002-1*; Basic Quantities in Cutting and Grinding; General Terms, Reference Systems, Tool and Working Angles, Chip Breakers. 1st ed. International Organization for Standardization: Geneva, Switzerland, 1982.
30. Wyen, C.-F.; Knapp, W.; Wegener, K. A new method for the characterization of rounded cutting edges. *Int. J. Adv. Manuf. Technol.* **2012**, *59*, 899–914. [CrossRef]
31. Eigenmann, B.; Macherauch, E. Röntgenographische Untersuchung von Spannungszuständen in Werkstoffen. Teil III. Fortsetzung von Matwiss. und Werkstofftechn. 3/1995, 148–160 and 4/1995, 199–216. *Materialwiss. Werkst.* **1995**, *27*, 426–437. [CrossRef]
32. X-ray Diffraction Residual-Stress Techniques. In *X-ray Diffraction Residual Stress Techniques In: Metals handbook*; American Society for Metals (Ed.) ASM International: Metals Park, OH, USA, 1986; pp. 440–458. ISBN 978-1-62708-213-6.
33. Chui, C.K.; Chen, G. (Eds.) *Kalman Filtering*, 5th ed.; Springer International Publishing: Cham, Switzerland, 2017; ISBN 978-3-319-47610-0.
34. Hastie, T.; Tibshirani, R.; Friedman, J. *The Elements of Statistical Learning: Data Mining, Inference, and Prediction*, 2nd ed.; Springer Series in Statistics; Springer: New York, NY, USA, 2009; ISBN 978-0-387-84857-0.
35. Goodfellow, I.; Bengio, Y.; Courville, A. *Deep Learning*; MIT Press: Cambridge, MA, USA, 2016; ISBN 0-262-33737-1.

Journal of
*Functional
Biomaterials*

MDPI

Article

Engineered Cross-Linked Silane with Urea Polymer Thin Durable Coatings onto Polymeric Films for Controlled Antiviral Release of Activated Chlorine and Essential Oils

Elisheva Sasson [1,†], Omer Agazani [2,†], Eyal Malka [1], Meital Reches [2,*] and Shlomo Margel [1,*]

1 Bar-Ilan Institute of Nanotechnology and Advanced Materials (BINA) and Department of Chemistry, Bar-Ilan University, Ramat-Gan 5290002, Israel; elisheva.sa@gmail.com (E.S.)
2 Institute of Chemistry and the Center for Nanoscience and Nanotechnology, The Hebrew University of Jerusalem, Jerusalem 9190401, Israel
* Correspondence: meital.reches@mail.huji.ac.il (M.R.); shlomo.margel@biu.ac.il (S.M.)
† These authors contributed equally to this work.

Abstract: In March 2020, the World Health Organization announced a pandemic attributed to SARS-CoV-2, a novel beta-coronavirus, which spread widely from China. As a result, the need for antiviral surfaces has increased significantly. Here, the preparation and characterization of new antiviral coatings on polycarbonate (PC) for controlled release of activated chlorine (Cl^+) and thymol separately and combined are described. Thin coatings were prepared by polymerization of 1-[3-(trimethoxysilyl)propyl] urea (TMSPU) in ethanol/water basic solution by modified Stöber polymerization, followed by spreading the formed dispersion onto surface-oxidized PC film using a Mayer rod with appropriate thickness. Activated Cl-releasing coating was prepared by chlorination of the PC/SiO_2-urea film with NaOCl through the urea amide groups to form a Cl-amine derivatized coating. Thymol releasing coating was prepared by linking thymol to TMSPU or its polymer via hydrogen bonds between thymol hydroxyl and urea amide groups. The activity towards T4 bacteriophage and canine coronavirus (CCV) was measured. PC/SiO_2-urea-thymol enhanced bacteriophage persistence, while PC/SiO_2-urea-Cl reduced its amount by 84%. Temperature-dependent release is presented. Surprisingly, the combination of thymol and chlorine had an improved antiviral activity, reducing the amount of both viruses by four orders of magnitude, indicating synergistic activity. For CCV, coating with only thymol was inactive, while SiO_2-urea-Cl reduced it below a detectable level.

Keywords: COVID-19; antiviral coatings; N-halamine compounds; T4 bacteriophage; essential oils; canine coronavirus

1. Introduction

COVID-19, the pandemic attributed to a novel coronavirus, had a drastic effect worldwide, with over 4.3 million confirmed deaths at the initial stage in 2020 [1]. The persistence of SARS-CoV-2 on surfaces triggered a focus on strategies that decrease their viral contamination. The virus is infectious for up to 8 h on cardboard and copper, and 72 h on steel/plastic surfaces [2,3]. Other human coronaviruses are infectious for 9 days [4,5]. Antiviral surfaces and coatings have been explored to address this phenomenon. Between 2020 and 2022, 512 papers were published regarding polymers and COVID-19 [6]. Despite great efforts to develop effective antiviral drugs against SARS-CoV-2, approval by the Food and Drug Administration (FDA) has been quite limited [7]. However, over the years, most reports focus on anti-bacterial abilities, while there are fewer reports on antiviral functionality, coatings or surfaces. The reason for this is that in the absence of host cells, some viruses immediately become inactive [5,8]. Moreover, some viruses are not able to spread without a host [5,9]. SARS-CoV-2, however, persists on surfaces, and therefore has the potential to be transmitted by surfaces [4,10].

Reports on antiviral surfaces include a comparison between copper and stainless steel surfaces. Noyce et al. showed that copper surfaces can reduce the amount of Influenza A by three orders of magnitude compared to stainless steel [11]. Nanoparticles and ions of copper also show antiviral activity [12,13]. Other inorganic materials such as silver and zinc are also active [14–18]. Moreover, organic compounds such as polycations are used in antiviral coatings [5,19–21].

Interest in compounds with covalent bonds between nitrogen and halogen is growing, owing to their excellent antimicrobial ability. Moreover, such N-halamines exhibit prolonged stability in aqueous solution, weak toxicity and low cost, and can be readily regenerated [22–24]. Such compounds are attractive for disinfection of water and food [25]. Their aqueous dissociation constant is quite low, and increases from imide to amide to amine; however, oxidative transfer of chlorine to microorganisms rather than hydrolysis is significantly favored (Figure 1). Although the hydrolysis constant is around 10^{-8}, the transfer of oxidative chlorine from amide-halamine compounds to the bacteria is much more favorable [26–28]. For industrial applications, such compounds are considered the most practical due to their moderate chlorine transfer rate.

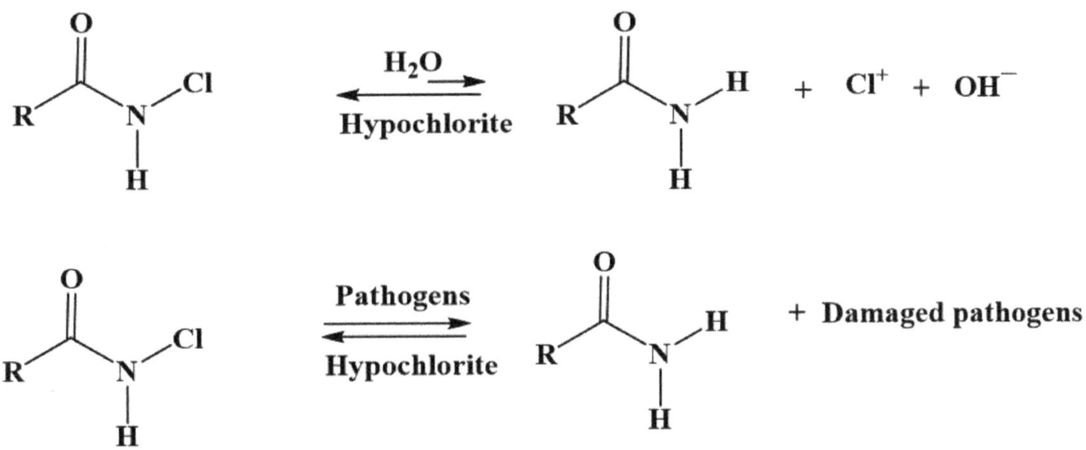

Figure 1. Oxidative transfer of chlorine from a chlorinated amide to (**1**) water and (**2**) pathogens.

Thymol is one of the main components in thyme, which is widely used in pharmacology [29]. This molecule has anti-bacterial, anti fungal, and some antiviral activities [29–33]. There are many examples of anti-bacterial and anti-fungal surfaces that contain thymol [34–37] albeit only a few reports on thymol as an antiviral agent [31,38,39]. Thymol can inactivate herpes from infecting cells, probably by damaging the virion envelope structure required for adsorption/entry [40]. Moreover, Castanas et al. showed that essential oils mixture (mainly menthol, 1,8-cienol and limonene) can reduce the release of the delta variant pseudo virus of SARS-CoV-2 by more than 80% [41,42]. Other studies have shown the antiviral activity of other mixtures of essential oils, including thymol, against SARS-CoV-2 [42–45]. It is worth mentioning that those studies report on the antiviral activity of essential oil mixtures and not on their ability to serve as antiviral coatings.

An in silico study on the transmembrane protease serine 2 (TMPRSS2) suggests that thymol can inhibit this enzyme's host, which plays a major role in the entry of SARS-CoV-2 [46]. Despite all of the above, we are not aware of any report on antiviral coatings based on essential oils. Here, we report a new antiviral coating on PC film composed of SiO_2-urea that may bind to activated Cl and/or thymol and allow their controlled release, both separately and together, upon contact with a virus. Thin SiO_2-urea thin coatings on

PC film were prepared by modified Stöber polymerization of 1-[3-(trimethoxysilyl)propyl] urea (TMSPU) in an ethanol/H_2O continuous phase, as shown in Figure 2.

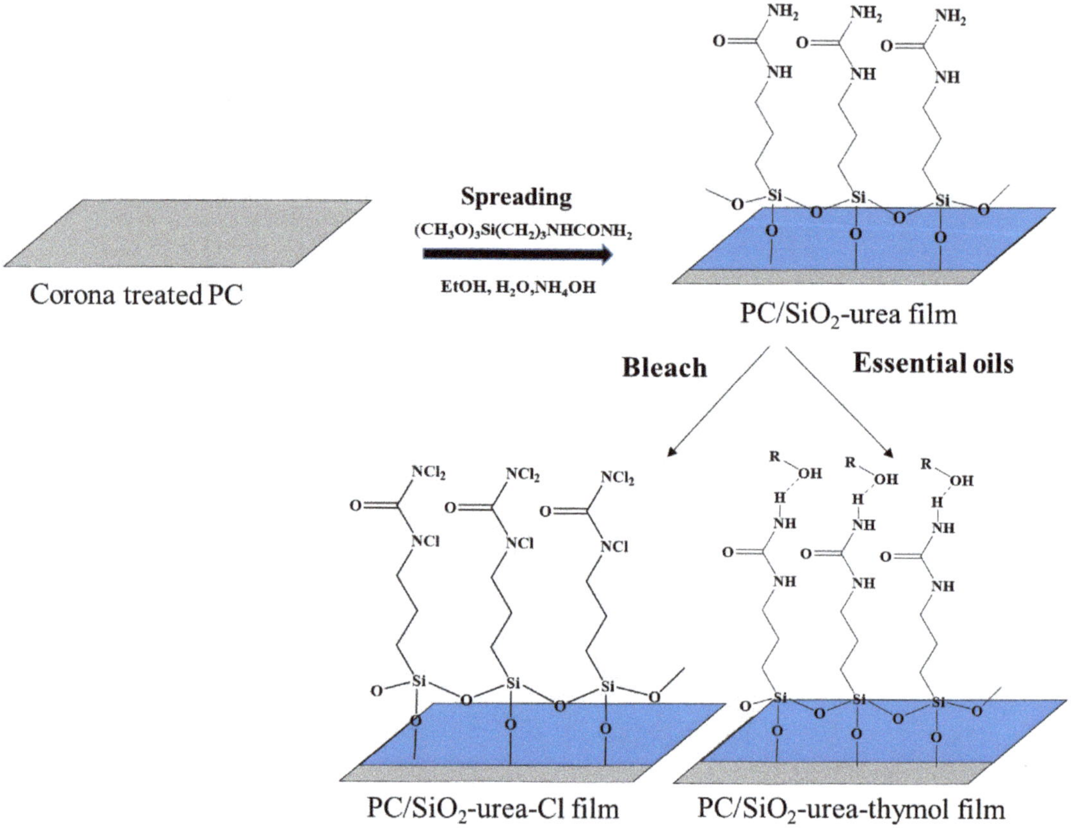

Figure 2. Production of SiO_2-urea-Cl and SiO_2-urea-thymol thin coatings on polycarbonate (PC) films.

Chlorine-releasing coatings were prepared by chlorination of the SiO_2-urea by NaOCl via the urea amide groups to obtain N-chloramine groups, as shown in Figure 2 [34]. Thymol-releasing coating was prepared by adding the thymol (Figure 3) to the silane-urea monomer (TMSPU, Figure 3), followed by Stöber polymerization in presence of the corona-treated PC film, as shown in Figure 2. The antiviral activity of SiO_2-urea-Cl, SiO_2-urea-thymol and SiO_2-urea-Cl-thymol coatings on the PC films was measured against two different viruses: T4 bacteriophage and canine coronavirus (CCV). This covering can be used as a stable coating for protecting partitions against pathogenic viruses, including SARS-CoV-2.

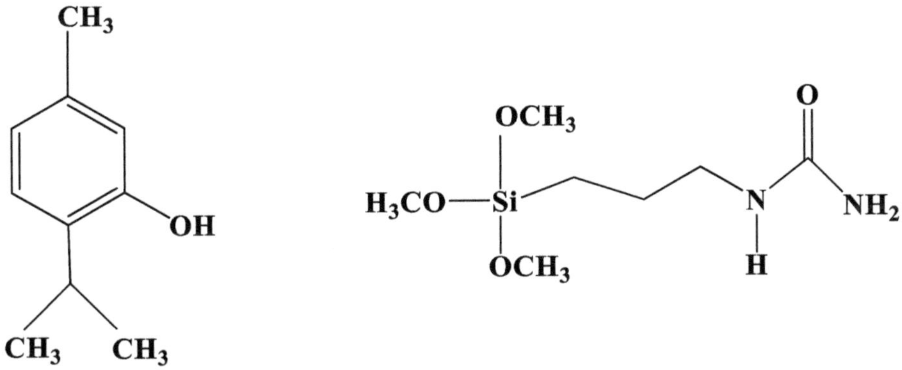

Figure 3. Chemical structures of thymol (**left**) and TMSPU (**right**).

2. Experimental

2.1. Materials

The following analytical-grade chemicals were purchased from Sigma-Aldrich (St. Louis, MO, USA) and used without further purification: ethanol (AR grade), ammonium hydroxide (NH_4OH, 28%), 1-[3-(trimethoxysilyl)propyl] urea (TMSPU), sodium hypochlorite (NaOCl) aqueous solution (5%) and thymol oil. Double-distilled water (DDW) and triple-distilled water (TDW) were obtained from a TREION™ purification system (Tel-Aviv, Israel). Corona-treated PC films were provided by Plaskolite (Gazit, Tel Aviv, Israel).

Escherichia coli strain B (Migula), Castellani and Chalmers (ATCC 11303) and *E. coli* bacteriophage T4 (ATCC 11303-B4) were obtained from the ATCC (Manassas, VA, USA). Lysogeny broth (LB) and agar were, respectively acquired from Merck (Rahway, NJ, USA) and Becton Dickinson (Franklin Lakes, NJ, USA). Crandell Rees feline kidney (CRFK) cells (ATCC CCL94), CCV (VR2068), L-alanyl-L-glutamine (200 mM), medium with essential minimum Earl salt (EMEM), penicillin−streptomycin−amphotericin (PSA), trypsin−EDTA, donor horse serum (DHS) and Dulbecco's phosphate-buffered saline (DPBS) were obtained from Biological Industries (Beit-HaEmek, Israel). Trypsin 1:250 was acquired from Bio-World (Dublin, OH, USA).

2.2. Methodology

2.2.1. Preparation of SiO_2-Urea and SiO_2-Urea-Thymol Coatings on PC Film

The SiO_2-urea coating on PC film was prepared by modified Stöber polymerization of TMSPU in continuous ethanol/water phase. In a typical experiment, surface oxidation of PC films was carried out by air corona treatment (450 W·min/m^2) [47]. Ethanol (18.75 mL), water (3 mL), ammonium hydroxide (0.675 mL) and TMSPU (1.5 mL) in the absence or presence of thymol (1% w/v, 0.25 g) were added to 50 mL tubes and mixed at room temperature (RT) for about half an hour. The corona-treated PC films were then coated by a Mayer rod with the formed coating dispersion.

2.2.2. Rod-Coating Technique

The Corona-treated PC film was coated with TMSPU by spreading the basic mixture of TMSPU in ethanol/water in the absence or presence of thymol using a Mayer rod obtained from RK Print Coat Instruments Ltd. (Litlington, UK) [16,17]; this is a metal bar wrapped in a wire, which draws a solution over the surface of the substrate (Figure 4). The film thickness is determined by the wire diameter. The substrate was dried for a few minutes in an oven at 110 °C. The formed PC/SiO_2-urea and PC/SiO_2-urea-thymol films were washed (with water) and dried in air.

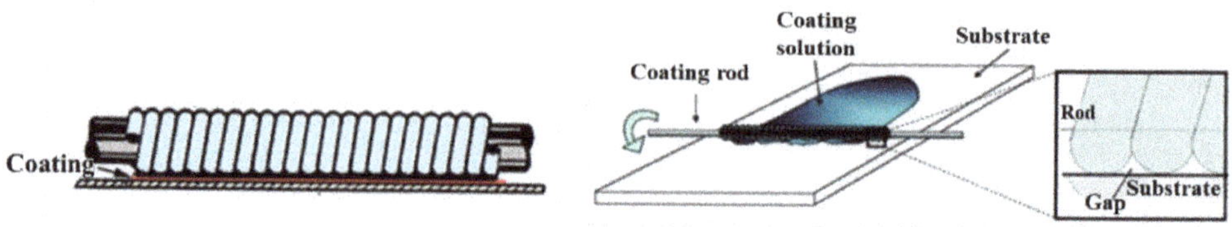

Figure 4. Mayer rod system.

Rods are available in a wide variety of wire sizes to provide a range of coating thickness. The concentration of TMSPU in the coating solution determines the dry thickness. In this study, uniform layers with a wet thickness of 400, 200 and 120 μm were prepared. For biological studies, only coatings with a wet thickness of 400 μm were used.

2.2.3. SiO_2-Urea-Cl Coating on PC Film

Chlorination of the SiO_2-Urea Coating

After completing the preparation of the SiO_2-urea coating on the PC film, activated chlorine binding to the amide groups of the urea was carried out by the following chlorination process. Sodium hypochlorite aqueous solution (0.5% w/v) was put in a box that contained the coated film. The box was then shaken at 40 °C for 12 h, followed by removal of excess sodium hypochlorite through extensive washing with water.

Active Chlorine Determination

Iodometric sodium thiosulfate titration was used to determine the content of chlorine that was bound [48,49]:

$$[Cl^+]\ (mM) = \frac{1000VN}{2}$$

where V is the volume and N is the normality of the titrated solution.

2.2.4. Characterization of the Coatings

Atomic Force Microscope

Atomic force microscopy (AFM) measurements of the films were carried out using a scanning probe Bio FastScan microscope (Bruker AXS), with quantitative nanomechanical mapping (QNM) Peak Force mode using a silicon probe (FastScan-C, Bruker, Billerica, MA, USA) with 0.45 N/m spring constant. An acoustic hood was used for minimal vibrational noise, with a scan rate of 1.7 Hz in the retrace direction. The resolution was 512 samples per line. Nanoscope Analysis software (v1.5) was used for image processing and thickness analysis, applying the "flatting" and "planefit" functions. Three areas of the film were averaged (three measurements per area).

Fourier Transform Infrared Spectroscopy

Fourier transform infrared spectroscopy (FTIR) measurements used the attenuated total reflectance (ATR) technique with ALPHA-FTIR (Bruker) QuickSnap sampling with a platinum diamond ATR module.

Contact Angles

Water contact angle (CA) sessile drop measurements were made with a goniometer OCA20 system from Data Physics Instruments Gmbh (Fielderstadt, Germany). Drops of DDW (5 μL) were placed on three areas of the film, and images were captured after a few seconds. Laplace–Young curve fitting was used to obtain static CA values. All measurements were performed under the same conditions.

X-ray Photoelectron Spectroscopy

X-ray photoelectron spectroscopy (XPS) measurements were made using a Nexsa (Bristol, UK) spectrometer with a monochromated micro-focused Al Kα source (with a low photon energy of 1486.6 eV). Survey/high-resolution spectra were acquired at pass energy of 200/50 eV. The usual source power was 72 W. Binding energies were recalibrated with CC/CH of C 1s set at 285 eV. UHV conditions were applied with a base pressure of 5×10^{-10} torr (up to 3×10^{-9} torr). Analysis and deconvolution were performed using software by Vision 2 (Kratos, UK).

2.2.5. Release Rates at Different Temperatures of Activated Cl from PC/SiO_2-Urea-Cl Films

PC/SiO_2-urea-Cl and PC/SiO_2-urea-Cl-thymol films (1 cm^2 pieces) were exposed at RT, 4 and −4 °C for 0, 1, 3, 4, 24, 48, 72, 96 and 500 h, during which time Cl was released from the coatings. After the set exposure time, the films were tested by iodometric/thiosulfate titration to quantify the concentration of Cl^+ still present on the coatings.

2.2.6. Determination of Thymol Content in PC/SiO_2-Urea Thymol-Containing Films

PC/SiO_2-urea-Cl and PC/SiO_2-urea-Cl-thymol films (4 cm^2 pieces) in 20 mL ethanol were centrifuged (6500 rpm, 35 min) until complete thymol extraction (further centrifugation indicated no remnants of trapped thymol). The content of thymol was obtained using a pre-calibrated curve (max absorption at λ = 276 nm) with a UV-visible spectrophotometer (CARY 1E, Varian, Palo Alto, CA, USA). The extracted thymol concentration was derived from the calibration curve (Figure S1). Measurements were made in triplicate.

2.2.7. Release Rates of Thymol from PC/SiO_2-Urea-Thymol and PC/SiO_2-Urea-Cl-Thymol Films

PC/SiO_2 urea-thymol and PC/SiO_2-urea-Cl-thymol coated films were cut to 4 cm^2 pieces and left exposed for 0, 4, 8, 22 and 264 h at RT, during which time thymol vapors were released. After each exposure time, the sample in ethanol was centrifuged (6500 rpm, 35 min) until complete thymol extraction. The release rate was determined by subtracting the amount of thymol in the ethanol extract from the initial amount. Measurements were made in triplicate.

2.2.8. Durability of SiO_2-Urea Coating on PC Film

The strength of the interaction between the SiO_2-urea coating and the film was tested. An adhesive tape was pressed firmly on the film, then slowly peeled off (Figure S2) [50]. The process was repeated 25 times, and FTIR spectra were measured periodically. The coating's stability in water was evaluated by soaking for six months and drying.

2.2.9. Bacteriophage Assay

The *E. coli* bacteriophage T4 stock solution was diluted to 1.0×10^6 PFU/mL. 16 μL drops were incubated at RT on the bare and coated PC films for 24 h with Parafilm on top of the drop. The size of the films and the Parafilm were 1 cm × 1 cm. The incubation was carried out in a humid environment by adding 2 mL of DDW near each sample. After incubation, soya casein digest lecithin polysorbate (SCDLP, 2 mL) was added and the samples were shaken (15 min, 120 rpm) to remove the phages from the surface to the solution. The solution was diluted with Luria–Bertani (LB, 8 mg/mL) solution to a 10-fold dilution series. The T4 and *E. coli* starter solution were mixed in 0.6% agarose solution and spread on plates with agar incubated at 37 °C overnight to form plaques. The plaques were manually quantified and normalized with respect to the bare polycarbonate.

2.2.10. CCV Assay

Some 16 μL of diluted T4 solution was incubated as described above using TDW instead of DDW. The viruses were removed from the surface by adding 2 mL of TDW. A

10-fold dilution series in EMEM was prepared with L-alanyl-L-glutamine (2 mM) PSA (1% solution), and the $TCID_{50}$/mL was measured. Six surfaces were included (three coated/uncoated). In a 96-well plate, CRFK cells (10,000/well) were incubated to 80–90% confluence and washed with DPBS (0.2 mL/well). The diluted CCV solution was added (0.1 mL/well) and the plate was incubated (37 °C, 5% CO_2, 5–6 repeats). After 2 h, EMEM (2 mM L-Glu, 1% PSA), DHS (1% final concentration), and trypsin (1:250, 1 µg/mL final concentration) were added (total volume 0.2 mL/well). The plates were incubated for 5–7 days until a cytopathic effect was observed using a Zeiss Primovert inverted microscope (Carl Zeiss AG, Jena, Germany). Based on the number of wells with an observed cytopathic effect, TCID50/mL values were calculated according to the improved Kärber method [21]. Micrographs were taken using Axiocam ICC3 (Carl Zeiss AG, Jena, Germany).

3. Results and Discussion

3.1. Characterization of the Coatings

3.1.1. Atomic Force Microscopy (AFM)

Air corona-treated PC films coated with SiO_2-urea, SiO_2-urea-Cl, SiO_2-urea-thymol and SiO_2-urea-Cl-thymol were measured for their surface roughness using AFM (Table 1, Figure 5). The treated films had much higher roughness: 11.7 ± 0.5 compared to only 0.40 ± 0.02 nm for the non-treated film, due to pores created by collisions with oxygen ions. After coating with SiO_2-urea and SiO_2-urea-thymol, Table 1 shows a pronounced decrease in roughness (0.85 ± 0.01 and 1.2 ± 0.1 nm, respectively) possibly due to filling of the pores achieved by the corona treatment. However, coating the PC/SiO_2-urea with activated chlorine increased the roughness considerably from 0.85 ± 0.01 to 6.7 ± 0.3 nm due to the use of a strong oxidizer (NaOCl) on the coated surface. Thymol increased the roughness further to 9.1 ± 0.7 nm due to combination with the active chlorine.

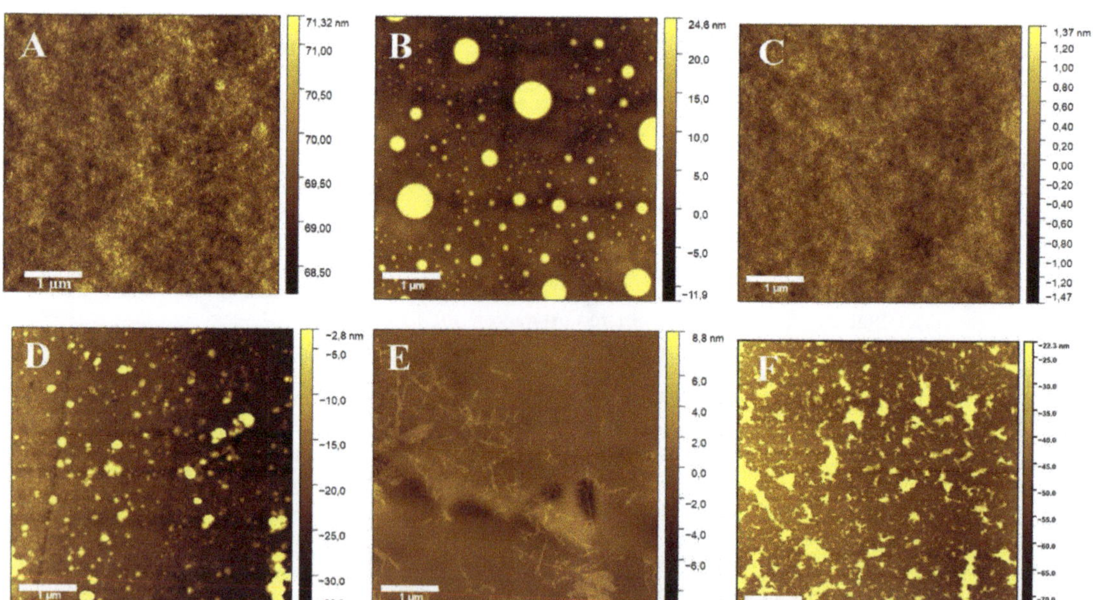

Figure 5. AFM topography analysis of PC (**A**), corona-treated PC (**B**), PC/SiO_2-urea (**C**), PC/SiO_2–urea-Cl (**D**), PC/SiO_2–urea-thymol (**E**) and PC/SiO_2-urea-Cl-thymol films (**F**).

Table 1. Surface roughness of the various films.

Film [a]	Roughness (nm) [b]
PC	0.40 ± 0.02
Corona-treated PC	11.7 ± 0.5
PC/SiO_2-urea	0.85 ± 0.01
PC/SiO_2-urea-Cl	6.7 ± 0.3
PC/SiO_2-urea-thymol	1.2 ± 0.1
PC/SiO_2-urea-Cl-thymol	9.1 ± 0.7

[a] PC films were corona-treated and then coated with TMSPU using a rod-coating technique, followed by chlorination or thymol interaction. [b] Average of three areas of the film (three measurements per area).

3.1.2. Fourier Transform IR/Attenuated Total Reflection

FTIR/ATR was used to verify TMSPU polymerization on the corona-treated PC films, followed by chlorination and/or thymol linking to the SiO_2-urea coating. Figure 6 presents spectra of the corona-treated PC and PC/SiO_2-urea films before and after chlorination (top). The principal peaks of the PC spectrum [51] are shown in Figure 6 (bottom): aromatic C–H deformations (~3000 cm^{-1}), C=O deformations (~1775 cm^{-1}), C=C vibrations (1506 cm^{-1}), asymmetric O–C–O carbonate deformations (1232–1164 cm^{-1}) and CH_3 vibrations (2800–3000 cm^{-1}). The PC/SiO_2-urea film spectrum (Figure 6, middle) includes Si-O-Si and SiOH deformations around 1132 and 1041 cm^{-1}, and deformations of the urea N-H vibrations (3348 cm^{-1}). The PC/SiO_2-urea-Cl spectrum (Figure 6, top) displays the same peaks as the PC/SiO_2-urea film, except for the urea peak at 3348 cm^{-1}, which decreases, and an additional peak (1673 cm^{-1}) corresponding to the new Cl-activated N–Cl bond [52].

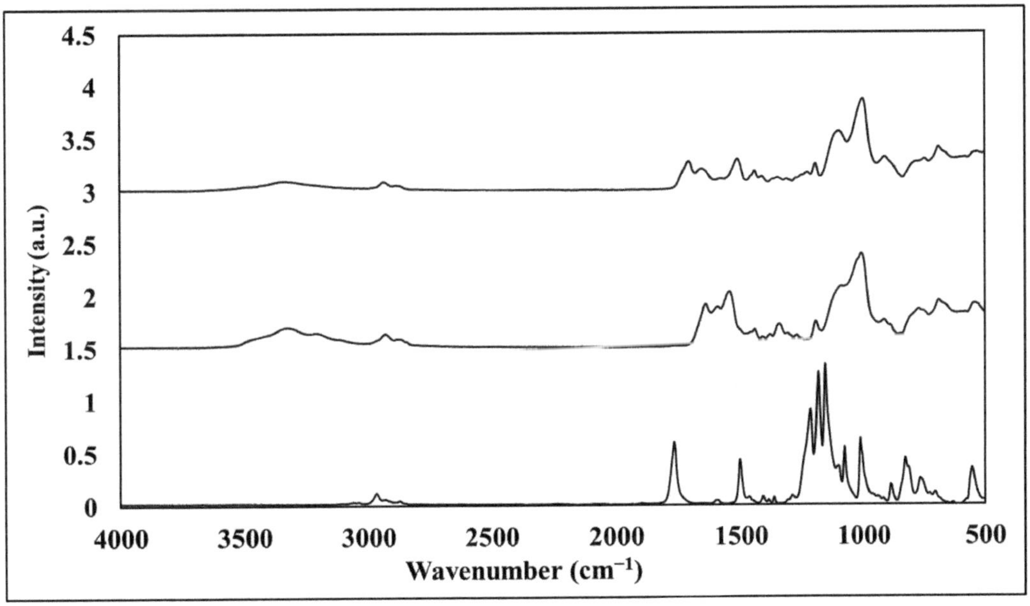

Figure 6. FTIR/ATR spectra of PC (bottom), PC/SiO_2-urea (middle) and PC/SiO_2-urea-Cl (top) films.

Figure 7 presents the spectra of thymol, PC/SiO_2-urea-thymol and PC/SiO_2-urea-Cl-thymol. The thymol spectrum [53] includes (bottom of Figure 7) peaks of hydroxyl (OH) around 3323 cm^{-1}, CH_3 stretching at 2957–2930 cm^{-1}, phenyl (1454 cm^{-1}), and phenol C-O stretching and in-plane OH bending (1344 and 1380 cm^{-1}, respectively). At 806 cm^{-1}, the peak is ascribed to the aromatic out-of-plane C-H and the ring at 947–806 cm^{-1}. The

PC/SiO_2-urea-thymol film spectrum is presented in the middle of Figure 7. This spectrum shows that thymol is present in the PC/silica-urea film, as indicated by CH_3 stretching (2930–2853 cm^{-1}), the phenyl ring (1454, 930–782 cm^{-1}), and aromatic out-of-plane C-H wagging vibrations (806 cm^{-1}). The PC/SiO_2-urea-Cl-thymol spectrum presented at the top of Figure 7 illustrates the presence of both thymol and activated chlorine.

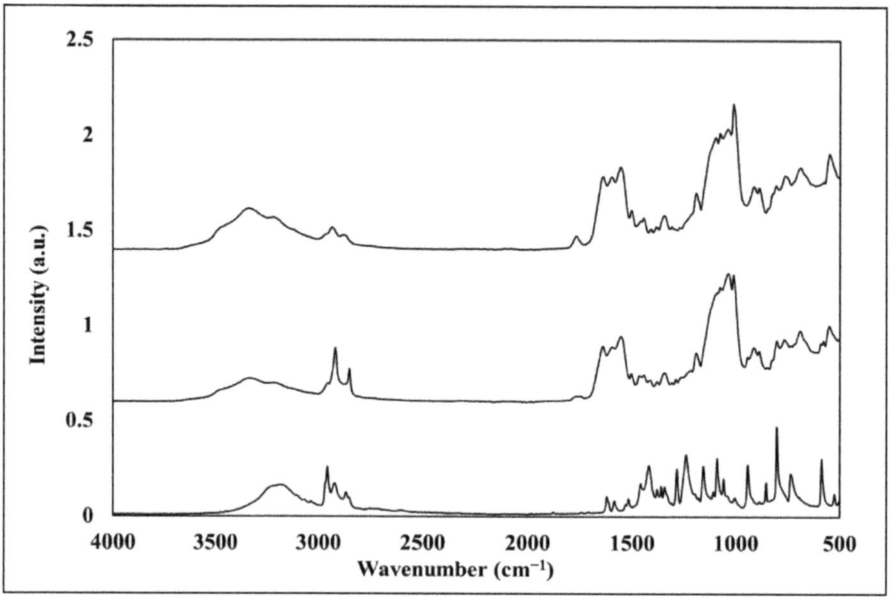

Figure 7. FTIR/ATR spectra of thymol (bottom), PC/SiO_2-urea-thymol (middle) and PC/SiO_2-urea-Cl-thymol (top) films.

3.1.3. Contact Angle (CA)

Water CAs were determined to evaluate the hydrophilicity/hydrophobicity of the different thin coatings. Table 2 shows, as expected, that corona treatment decreased the CA from $74 \pm 2°$ to $40 \pm 2°$, probably due to surface oxidation. Thin coating of the treated PC film with SiO_2-urea, followed by chlorination or linking of thymol, lead to significant increase to $93 \pm 4°$ and $110 \pm 6°$, respectively, due to the hydrophobic nature of Cl and thymol.

Table 2. Water CAs of PC and corona-treated PC and PC/SiO_2-urea films.

Film [a]	CA (°)
PC	74 ± 2
Corona-treated PC	40 ± 2
PC/SiO_2-urea	69 ± 3
PC/SiO_2-urea-Cl	93 ± 4
PC/SiO_2-urea-thymol	110 ± 6
PC/SiO_2-urea-Cl-thymol	102 ± 4

[a] Corona-treated PC films were coated according to the experimental procedure.

3.2. *Determination of Cl Content in PC/SiO_2-Urea-Cl Film*

Table 3 illustrates a direct connection between the thickness of the SiO_2-urea coating and the Cl concentration bonded to the SiO_2-urea coating. Higher coating thickness leads to increased bound Cl. Increasing the wet thickness (as determined from the Mayer rod) of the SiO_2-urea-Cl onto PC films from 120 to 200/400 μm increased the Cl content from 0.75 to 1.5/3.5 μmol/cm^2.

Table 3. Effect of SiO_2-urea-Cl coating thickness on the concentration of Cl bonded to the coating.

Wet Film Thickness [a] **(μm)**	**[Cl] (μmol/cm^2)**
120	0.75
200	1.5
400	3.5

[a] Corona-treated PC films were coated as described above and chlorinated with 0.5% bleach (12 h, 40 °C).

3.2.1. X-ray Photoelectron Spectroscopy

The surface composition of the films was measured using XPS (Figure 8). The characteristic peaks of N 1s (400.02 eV) and Si 2p (102.1 eV) in Figure 8B,C, respectively, indicate the presence of the Si-urea coating in both PC/SiO_2-urea (orange) and PC/SiO_2-urea-Cl (gray) films. Figure 8D exhibits the success of the chlorination of the SiO_2-urea coating by the Cl 2p peak at 200 eV. PC/SiO_2-urea-thymol films were not tested, as thymol can evaporate during XPS.

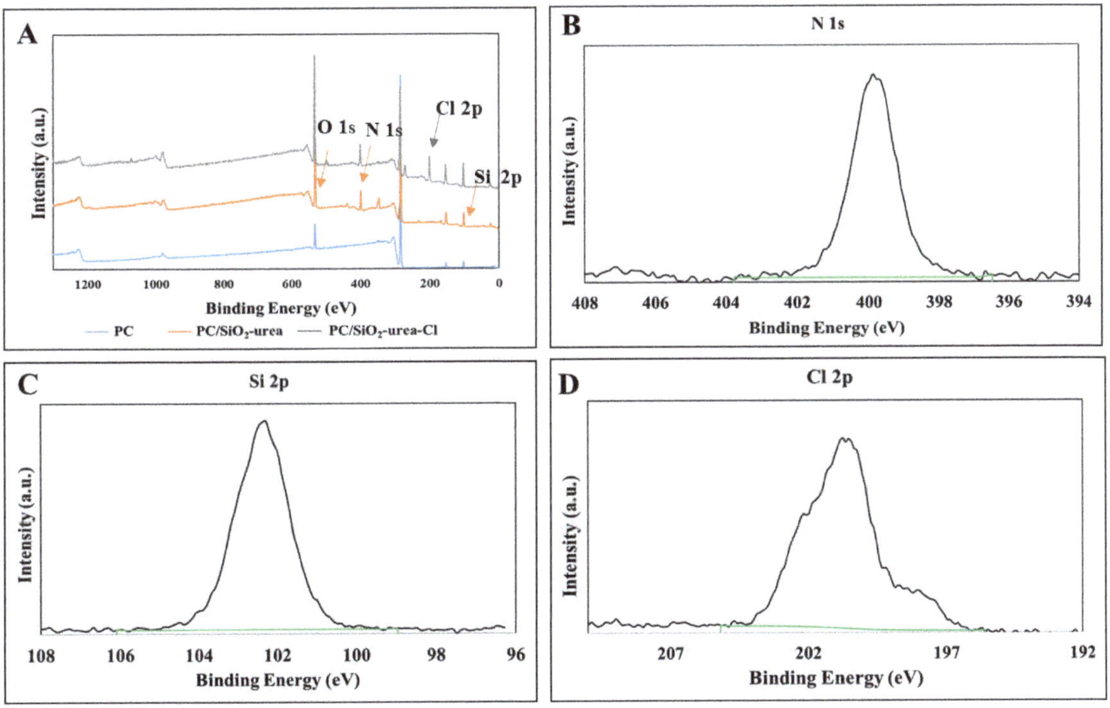

Figure 8. (**A**) XPS spectra of corona-treated PC (blue), PC/SiO_2-urea (orange) and PC/SiO_2-urea-Cl (gray) films. Magnification of the PC/SiO_2-urea film spectra depicts the peaks corresponding to (**B**) N 1s and (**C**) Si 2p, belonging to the SiO_2–urea coating. Magnification of the PC/SiO_2-urea-Cl film spectrum shows the peaks corresponding to (**D**) Cl 2p, belonging to SiO_2-urea-Cl coating.

3.2.2. Release Rates of Chlorine from PC/SiO_2-Urea-Cl and PC/SiO_2-Urea-Cl-Thymol Films

The Section 2 describes the activated chlorine analysis and the evaluation of Cl^+ release rates from the PC/SiO_2-urea-Cl and PC/SiO_2-urea-Cl-thymol films. Figure 9 exhibits release rates from PC/SiO_2-urea-Cl at RT, 4 °C and −4 °C. In the first 24 h, 20% of the Cl^+ content is released; after 48 h, 70% is released, which increases to 90% after a few weeks (500 h). Storing the coating at 4 °C (green line) significantly slows down the release rate, as shown in Figure 9. In addition, storing the coating at −4 °C does not release any chlorine even after 500 h. This is probably due to the significant decrease in exposure to microorganisms at these temperatures. Similar release rates of Cl^+ as a function of temperature were also observed for the PC/SiO_2-urea-Cl-thymol film.

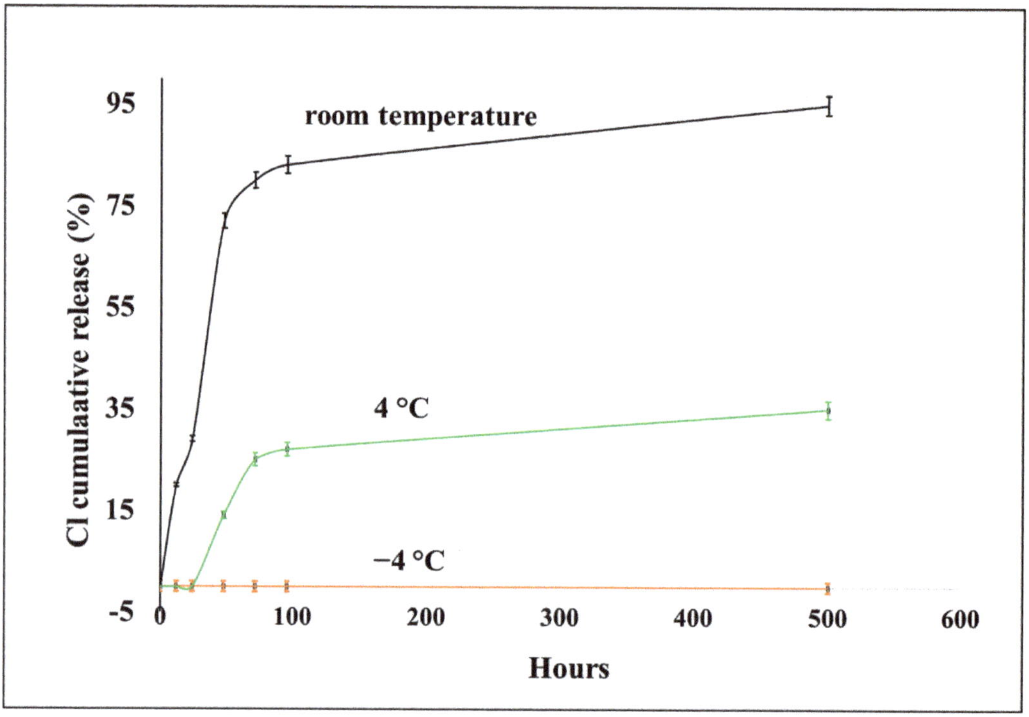

Figure 9. Release rates of activated chlorine from PC/SiO_2-urea-Cl film at room temperature, 4 °C and −4 °C.

3.3. *Determination of Thymol Content in PC/SiO_2-Urea Films*

Thymol bound to films with and without added Cl was determined as described in the Section 2. PC/SiO_2-urea-thymol and PC/SiO_2-urea-Cl-thymol films exhibited similar binding: 933 ± 81 μg and 924 ± 12 μg of thymol per 4 cm^2, respectively.

Release Rates of Thymol from PC/SiO_2-Urea-Thymol and PC/SiO_2-Urea-Cl-Thymol Films

Figure 10 exhibits the release rates of thymol at room and low temperatures (−4 °C). At RT, both coatings exhibit a massive release of thymol after 10 h, after which the rate decreases. After 22 h, about 60% of the thymol was released from PC/SiO_2-urea-thymol. However, the release from PC/SiO_2-urea-Cl-thymol is considerably slower (~25%). This is in accordance with hydrophobic interactions between the aromatic ring of thymol and the chlorine of the N-halamine, which are expected to decrease the release rate. Figure 10 also shows that storing both films at −4 °C completely prevents the release of thymol, a behavior similar to that illustrated in Figure 9 for the release of activated Cl from the PC/SiO_2-urea-Cl film.

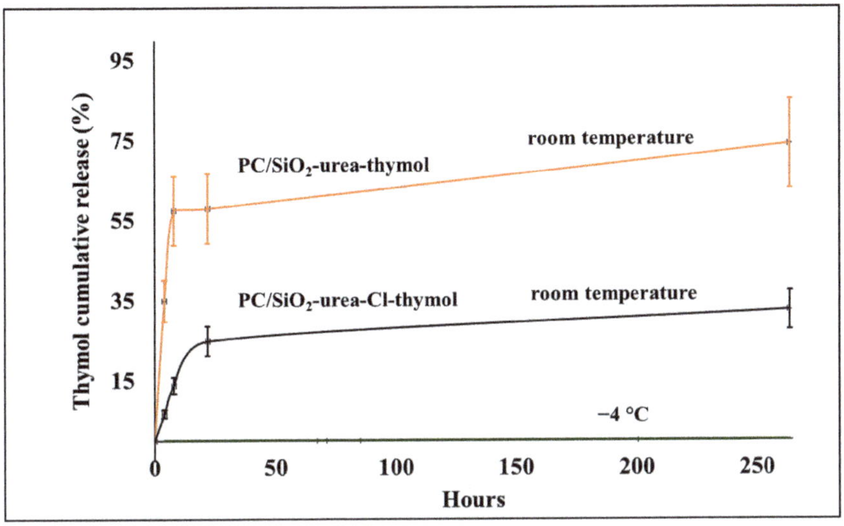

Figure 10. Release rates of thymol from PC/SiO_2-urea films at room temperature and −4 °C.

3.4. Durability of SiO_2-Urea Coating on PC Film

SiO_2-urea coating durability was examined as described in the Section 2. Similar results were obtained in FTIR after the tape and soaking tests (Figures S2 and S3, respectively). Hence, the PC/SiO_2-urea coatings are stable and durable in water for six months.

3.5. Antiviral Activity

The coatings were tested towards two different viruses: T4 bacteriophage and canine coronavirus (CCV). As the bacteriophage host, *E. coli*, replicates every ~30 min, the effect of the infection can be seen after 24 h from the time of incubation. This is why we chose T4 bacteriophage as a model virus. As the world is still handling the COVID-19 pandemic, and SARS-CoV-2 shows prolonged survival on surfaces, we wanted to show that our coating is effective. CCV was used as a safe surrogate.

The antiviral activity against CCV was examined using the well-known $TCID_{50}$/mL method. This technique is based on the cytopathic effect. If the cells appear round and detached from each other, they are infected by the virus. On the other hand, if the cells are oblong-shaped and layer the surface, there is no cytopathic effect, the viruses are inactive and have not infected the cells.

The PC/SiO_2-urea-Cl and PC/SiO_2-urea-Cl-thymol coatings reduced the $TCID_{50}$/mL of CCV below the detection level (at least ~1.5 log) compared to corona-treated PC surfaces used as a control. The $TCID_{50}$/mL value of the bare PC surfaces was 1.06×10^5, and the viruses inoculated on the coated surfaces could not be detected (Figure 11A). The PC/SiO_2-urea-thymol titer of the inoculated viruses was below the detection limit.

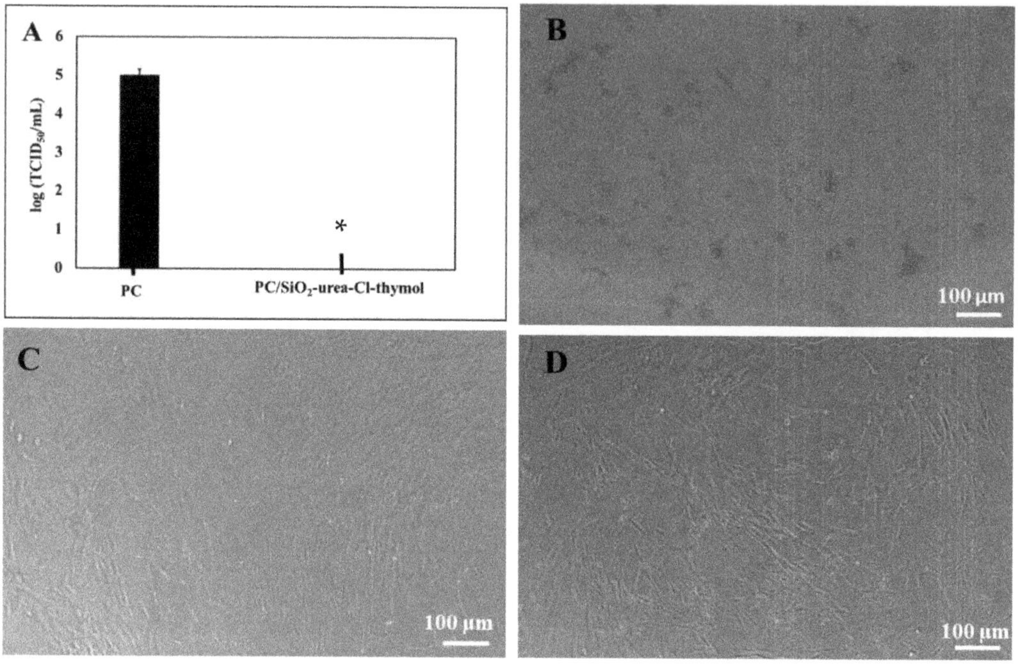

Figure 11. Antiviral activity against CCV (**A**) showing log($TCID_{50}$/mL) on the bare PC and SiO_2-urea-Cl-thymol coated surfaces. Microscope images ×100 of CRFK cells incubated with CCV on bare PC (**B**) and on PC coated with SiO_2-urea-Cl (**C**) or SiO_2-urea-Cl-thymol (**D**). The SD is based on one experiment in triplicate. * represents Student's *t*-test at a significance level of $p < 0.05$.

CCV was detected by examining the cytopathic effect of the CRFK cells. A cytopathic effect is well recognized for CRFK CCL94 cells inoculated with CCV for three hours on bare polycarbonate surfaces (Figure 11B). On both coated surfaces, there is no cytopathic effect, indicating that the coatings effectively inactivate the virus (Figure 11C,D).

To determine which component of the coating is antiviral, plaque assays were performed with T4 bacteriophage. The number of plaques corresponds to the number of active viruses. Three types of coatings were examined: PC/SiO_2-urea-Cl, PC/SiO_2-urea-thymol, and PC/SiO_2-urea-Cl-thymol. Our results indicate that PC/SiO_2-urea-thymol enhances viral persistence on the surface, while PC/SiO_2-urea-Cl reduces the number of bacteriophages by about 84% (Figure 12A). Surprisingly, the combination of thymol and Cl had an improved antiviral activity, as it reduced the number of viruses by four orders of magnitude. This suggests that thymol and Cl act synergistically as antiviral agents. Figure 12B,C show agar plates with a layer of *E. coli* cells infected with bacteriophages inoculated on the bare and coated surface, respectively. Each transparent spot represents a single bacteriophage. Transparent spots could not be detected on the coated surface, indicating inactivation of the bacteriophages that were inoculated on it.

Overall, these antiviral assays show that the coating that combines thymol and chlorine exhibits excellent activity toward two rather different viruses based on both RNA and DNA. The presence of thymol improves the activity, yet it is not crucial for the performance of this coating. However, the presence of activated chlorine is essential. If the coating does not contain Cl^+, the number of phages increases.

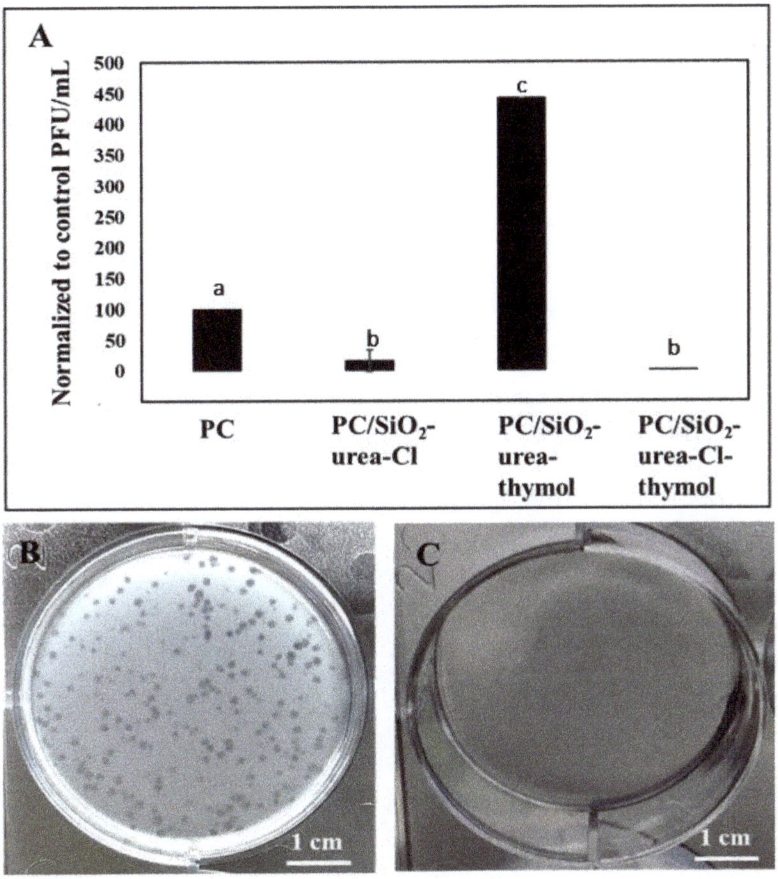

Figure 12. Antiviral activity against bacteriophage T4. (**A**) Normalized PFU/mL on bare PC and coated surfaces. Images show a layer of *E. coli* on wells with 1.75 cm radius after seeding bacteriophages from (**B**) bare PC and (**C**) PC/SiO_2-urea-Cl-thymol. The SD is based on three experiments performed in triplicate. Each letter represents an insignificant group, using an ANOVA test and Fisher's least significant difference (LSD) post hoc test, with a significance level of $p < 0.05$.

4. Summary and Discussion

SiO_2-urea, SiO_2-urea-Cl, SiO_2-urea-thymol and SiO_2-urea-Cl-thymol thin coatings on PC films have been successfully prepared, as evidenced by AFM, FTIR/ATR, CA and XPS. The durability of these coatings was demonstrated via an adhesive tape peeling test and their long-term stability in aqueous solution. Increasing the thickness of the SiO_2-urea-Cl coating on the PC film led to increased activated Cl^+ content.

The release rates of Cl^+ as well as thymol were found to be temperature-dependent. The rate is much slower at 4 °C than at room temperature. Storing the coated PC films at −4 °C completely prevented the release of both Cl^+ and thymol.

The antiviral activity of the coatings was tested against two different viruses, T4 bacteriophage (DNA) and canine coronavirus (RNA), safely mimicking the highly significant and widely spread SARS-CoV-2. The present study illustrates that PC/SiO_2-urea-thymol enhances viral persistence on the surface, while PC/SiO_2-urea-Cl reduces the number of bacteriophages by about 84%.

Surprisingly, the combination of thymol and Cl showed improved activity, as it reduced the number of viruses by four orders of magnitude. This suggests that thymol and Cl act synergistically as antiviral agents. Further work examining the effect of the above thin coatings on other types of viruses, as well as other microorganisms including bacteria and fungi, are ongoing in our laboratories.

Supplementary Materials: The following supporting information can be downloaded at: https://www.mdpi.com/article/10.3390/jfb14050270/s1, Figure S1: Thymol calibration curve; Figure S2: Adhesive test of coated film by tape. Figure S3: FTIR spectra of PC/SiO_2-urea before (blue) and after (orange) tape test. Figure S4: FTIR spectra of PC/SiO_2-urea before (blue) and after (orange) 6-month soaking test in water.

Author Contributions: Conceptualization, M.R. and S.M.; Methodology, O.A. and E.M.; Validation, E.S., O.A. and E.M.; Investigation, E.S., O.A. and E.M.; Writing—original draft, E.S. All authors have read and agreed to the published version of the manuscript.

Funding: This research received no external funding.

Data Availability Statement: Raw/processed data cannot be shared at this time due to technical and time limitations.

Acknowledgments: We acknowledge the professional writing assistance and editing by Yuval Elias for and the dedicated AFM analysis of surfaces by Eti Teiblum.

Conflicts of Interest: The authors declare no conflict of interest.

References

1. McIntosh, K.; Hirsch, M.S.; Bloom, A. Coronavirus disease (COVID-19). *UpToDate Hirsch MS Bloom* **2020**, *5*, 873.
2. Rawlinson, S.; Ciric, L.; Cloutman-Green, E. COVID-19 pandemic—Let's not forget surfaces. *J. Hosp. Infect.* **2020**, *105*, 790–791.
3. Patients, L.; Taylor, D.; Lindsay, A.C.; Halcox, J.P. Aerosol and surface stability of SARS-CoV-2 as compared with SARS-CoV-1. *N. Engl. J. Med.* **2020**, *382*, 1564–1567.
4. Kampf, G.; Todt, D.; Pfaender, S.; Steinmann, E. Persistence of coronaviruses on inanimate surfaces and their inactivation with biocidal agents. *J. Hosp. Infect.* **2020**, *104*, 246–251.
5. Imani, S.M.; Ladouceur, L.; Marshall, T.; Maclachlan, R.; Soleymani, L.; Didar, T.F. Antimicrobial nanomaterials and coatings: Current mechanisms and future perspectives to control the spread of viruses including SARS-CoV-2. *ACS Nano* **2020**, *14*, 12341–12369.
6. Chiari, W.; Damayanti, R.; Harapan, H.; Puspita, K.; Saiful, S.; Rahmi, R.; Rizki, D.R.; Iqhrammullah, M. Trend of polymer research related to COVID-19 pandemic: Bibliometric analysis. *Polymers* **2022**, *14*, 3297.
7. Sharun, K.; Timari, R.; Yatoo, M.I.; Natesan, S.; Megawati, D.; Singh, K.P.; Michalak, I.; Dhama, K. A comprehensive review on pharmacologic agents, immunotherapies and supportive therapeutics for COVID-19. *Narra J.* **2022**, *2*, e92.
8. Pirtle, E.C.; Beran, G.W. Virus survival in the environment. *Rev. Sci. Tech.* **1991**, *10*, 733–748.
9. Centers for Disease Control (US). *Recommendations for Prevention of HIV Transmission in Health-Care Settings*; US Department of Health and Human Services, Public Health Service, Centers for Disease Control: Atlanta, GA, USA, 1987; Volume 37.
10. Otter, J.A.; Donskey, C.; Yezli, S.; Douthwaite, S.; Goldenberg, S.; Weber, D.J. Transmission of SARS and MERS coronaviruses and influenza virus in healthcare settings: The possible role of dry surface contamination. *J. Hosp. Infect.* **2016**, *92*, 235–250.
11. Noyce, J.O.; Michels, H.; Keevil, C.W. Inactivation of influenza A virus on copper versus stainless steel surfaces. *Appl. Environ. Microbiol.* **2007**, *73*, 2748–2750.
12. Karlstrom, A.R.; Levine, R.L. Copper inhibits the protease from human immunodeficiency virus 1 by both cysteine-dependent and cysteine-independent mechanisms. *Proc. Natl. Acad. Sci. USA* **1991**, *88*, 5552–5556.
13. Ingle, A.P.; Duran, N.; Rai, M. Bioactivity, mechanism of action, and cytotoxicity of copper-based nanoparticles: A review. *Appl. Microbiol. Biotechnol.* **2014**, *98*, 1001–1009.
14. Rai, M.; Deshmukh, S.D.; Ingle, A.P.; Gupta, I.R.; Galdiero, M.; Galdiero, S. Metal nanoparticles: The protective nanoshield against virus infection. *Crit. Rev. Microbiol.* **2016**, *42*, 46–56.
15. Lara, H.H.; Ayala-Nuñez, N.V.; Ixtepan-Turrent, L.; Rodriguez-Padilla, C. Mode of antiviral action of silver nanoparticles against HIV-1. *J. Nanobiotechnol.* **2010**, *8*, 1.
16. Gupta, P.; Rapp, F. Effect of zinc ions on synthesis of herpes simplex virus type 2-induced polypeptides. *Proc. Soc. Exp. Biol. Med.* **1976**, *152*, 455–458.
17. Fridlender, B.; Chejanovsky, N.; Becker, Y. Selective inhibition of herpes simplex virus type 1 DNA polymerase by zinc ions. *Virology* **1978**, *84*, 551–554.
18. Arens, M.; Travis, S. Zinc salts inactivate clinical isolates of herpes simplex virus in vitro. *J. Clin. Microbiol.* **2000**, *38*, 1758–1762.

19. Gelman, F.; Lewis, K.; Klibanov, A.M. Drastically lowering the titer of waterborne bacteriophage PRD1 by exposure to immobilized hydrophobic polycations. *Biotechnol. Lett.* **2004**, *26*, 1695–1700.
20. Liu, H.; Kim, Y.; Mello, K.; Lovaasen, J.; Shah, A.; Rice, N.; Yim, J.H.; Pappas, D.; Klibanov, A.M. Aerosol-assisted plasma deposition of hydrophobic polycations makes surfaces highly antimicrobial. *Appl. Biochem. Biotechnol.* **2014**, *172*, 1254–1264.
21. Hu, T.; Agazani, O.; Nir, S.; Cohen, M.; Pan, S.; Reches, M. Antiviral activity of peptide-based assemblies. *ACS Appl. Mater. Interfaces* **2021**, *13*, 48469–48477.
22. Steinman, N.Y.; Hu, T.; Dombrovsky, A.; Reches, M.; Domb, A.J. Antiviral polymers based on N-halamine polyurea. *Biomacromolecules* **2021**, *10*, 4357–4364.
23. Padmanabhuni, R.V.; Luo, J.; Cao, Z.; Sun, Y. Preparation and characterization of N-halamiine-based antimicrobial fillers. *Ind. Eng. Chem. Res.* **2012**, *51*, 5148–5156.
24. Kocer, H.B.; Cerkez, I.; Worley, S.D.; Broughton, R.M.; Huang, T.S. N-halamine copolymers for use in antimicrobial paints. *ACS Appl. Mater. Interfaces* **2011**, *3*, 3189–3194.
25. Lauten, S.D.; Sarvis, H.; Wheatley, W.B.; Williams, D.E.; Mora, E.C.; Worley, S.D. Efficacies of novel N-halamine disinfectants against Salmonella and Pseudomonas species. *Appl. Environ. Microbiol.* **1992**, *58*, 1240–1243.
26. Eknoian, M.W.; Worley, S.D.; Bickert, J.; Williams, J.F. Novel antimicrobial N-halamine polymer coatings generated by emulsion polymerization. *Polymer* **1999**, *40*, 1367–1371.
27. Zhao, N.; Liu, S. Thermoplastic semi-IPN of polypropylene (PP) and polymeric N-halamine for efficient and durable antibacterial activity. *Eur. Polym. J.* **2011**, *47*, 1654–1663.
28. Liang, J.; Chen, Y.; Ren, X.; Wu, R.; Barnes, K.; Worley, S.D.; Broughton, R.M.; Cho, U.; Kocer, H.; Huang, T.S. Fabric treated with antimicrobial N-halamine epoxides. *Ind. Eng. Chem. Res.* **2007**, *46*, 6425–6429.
29. Escobar, A.; Pérez, M.; Romanelli, G.; Blustein, G. Thymol bioactivity: A review focusing on practical applications. *Arab. J. Chem.* **2020**, *13*, 9243–9269.
30. Kachur, K.; Suntres, Z. The antibacterial properties of phenolic isomers, carvacrol and thymol. *Crit. Rev. Food Sci. Nutr.* **2020**, *60*, 3042–3053.
31. Salehi, B.; Mishra, A.P.; Shukla, I.; Sharifi-Rad, M.; Contreras, M.D.M.; Segura-Carretero, A.; Fathi, H.; Nasrabadi, N.N.; Kobarfard, F.; Sharifi-Rad, J. Thymol, thyme, and other plant sources: Health and potential uses. *Phyther. Res.* **2018**, *32*, 1688–1706.
32. Marchese, A.; Orhan, I.E.; Daglia, M.; Barbieri, R.; Di Lorenzo, A.; Nabavi, S.F.; Gortzi, O.; Izadi, M.; Nabavi, S.M. Antibacterial and antifungal activities of thymol: A brief review of the literature. *Food Chem.* **2016**, *210*, 402–414.
33. Kulkarni, S.A.; Nagarajan, S.K.; Ramesh, V.; Palaniyandi, V.; Selvam, S.P.; Madhavan, T. Computational evaluation of major components from plant essential oils as potent inhibitors of SARS-CoV-2 spike protein. *J. Mol. Struct.* **2020**, *1221*, 128823.
34. Walczak, M.; Michalska-Sionkowska, M.; Kaczmarek, B.; Sionkowska, A. Surface and antibacterial properties of thin films based on collagen and thymol. *Mater. Today Commun.* **2020**, *22*, 100949.
35. Yue, Y.; Gong, X.; Jiao, W.; Li, Y.; Yin, X.; Si, Y.; Yu, J.; Ding, B. In-situ electrospinning of thymol-loaded polyurethane fibrous membranes for waterproof, breathable, and antibacterial wound dressing application. *J. Colloid Interface Sci.* **2021**, *592*, 310–318.
36. Robledo, N.; Vera, P.; López, L.; Yazdani-Pedram, M.; Tapia, C.; Abugoch, L. Thymol nanoemulsions incorporated in quinoa protein/chitosan edible films; antifungal effect in cherry tomatoes. *Food Chem.* **2018**, *246*, 211–219.
37. Ribes, S.; Ruiz-Rico, M.; Pérez-Estave, É.; Fuentes, E.; Barat, J.M. Eugenol and thymol immobilised on mesoporous silica-based material as an innovative antifungal system: Application in strawberry jam. *Food Control* **2017**, *81*, 181–188.
38. Lai, W.L.; Chuang, H.S.; Lee, M.H.; Wei, C.L.; Lin, C.F.; Tsai, Y.C. Inhibition of herpes simplex virus type 1 by thymol-related monoterpenoids. *Planta Med.* **2012**, *78*, 1636–1638.
39. Alagawany, M.; Farag, M.R.; Abdelnour, S.A.; Elnesr, S.S. A review on the beneficial effect of thymol on health and production of fish. *Rev. Aquac.* **2021**, *13*, 632–641.
40. Aranega, A.; Boulaiz, H. Susceptibility of herpes simplex virus type 1 to monoterpenes thymol, carvacrol, p-cymene and essential oils of Sinapis arvensis L., Lallemantia royleana Benth. and Pulicaria vulgaris Gaertn. *Cell. Mol. Biol.* **2005**, *51*, 1.
41. Iqhrammullah, M.; Rizki, D.R.; Purnama, A.; Duta, T.F.; Harapan, H.; Idroes, R.; Gintig, B. Antiviral molecular targets of essential oils against SARS-CoV-2: A systematic review. *Sci. Pharm.* **2023**, *91*, 15.
42. Lionis, C.; Karakasiliotis, I.; Petelos, E.; Linardakis, M.; Diamantakis, A.; Symvoulakis, E.; Panopoulou, M.; Kampa, M.; Pirintsos, S.A.; Sourvinos, G.; et al. A mixture of essential oils from three Cretan Aromatic Plants (thyme, Greek sage and Cretan dittany, CAPeo) inhibits SASR-CoV-2 proliferation: In vitro evidence and a Proof-of-Concept intervention study in mild ambulatory COVID-19-positive patients. *MedRxiv* **2021**. [CrossRef]
43. Torres-Neto, L.; Monteiro, M.L.; Fernández-Romero, J.; Teleshova, N.; Sailer, J.; Conte-Junior, C.A. Essential oils block cellular entry of SARS-CoV-2 delta variant. *Sci. Rep.* **2022**, *12*, 20639.
44. Esharkawy, E.R.; Almalki, F.; Hadda, T.B. In vitro potential antiviral SARS-CoV-19-activity of natural product thymohydroquinone and dithymoquinone from *Nigella sativa*. *Bioorg. Chem.* **2022**, *120*, 105587.
45. Ćavar-Zeljković, S.; Schadich, E.; Džubák, P.; Hajdúch, M.; Tarkowski, P. Antiviral activity of selected lamiaceae essential oils and their monoterpenes against SARS-Cov-2. *Front. Pharmacol.* **2022**, *13*, 1589.
46. Yadav, P.K.; Jaiswal, A.; Singh, R.K. In silico study on spice-derived antiviral phytochemicals against SARS-CoV-2 TMPRSS2 target. *J. Biomol. Struct. Dyn.* **2021**, *40*, 11874–11884.

47. Hayne, S.; Margel, S. In situ coatings of polymeric films with core polystyrene, core-shell polystyrene/SiO2, and hollow SiO2 micro/nanoparticles and potential applications. *ACS Omega* **2023**, *8*, 11406–11413.
48. Gutman, O.; Natan, M.; Banin, E.; Margel, S. Characterization and antibacterial properties of N-halamine-derivatized cross-linked polymethacrylamide nanoparticles. *Biomaterials* **2014**, *35*, 5079–5087.
49. Jonathan, N. The infrared and Raman spectra and structure of acrylamide. *J. Mol. Spectrosc.* **1961**, *6*, 205–214.
50. Rickerby, D.S. A review of the methods for the measurement of coating-substrate adhesion. *Surf. Coatings Technol.* **1988**, *36*, 541–557.
51. Li, J.; Chen, F.; Yang, L.; Jiang, L.; Dan, Y. FTIR analysis on aging characteristics of ABS/PC blend under UV-irradiation in air. *Spectrochim. Acta-Part A Mol. Biomol. Spectrosc.* **2017**, *184*, 361–367.
52. Demir, B.; Broughton, R.M.; Qiao, M.; Huang, T.S.; Worley, S.D. N-halamine biocidal materials with superior antimicrobial efficacies for wound dressings. *Molecules* **2017**, *22*, 1582.
53. Michalska-Sionkowska, M.; Walczak, M.; Sionkowska, A. Antimicrobial activity of collagen material with thymol addition for potential application as wound dressing. *Polym. Test.* **2017**, *63*, 360–366.

lubricants

MDPI

Article

Effect of CH_4 Flow Rate on the Tribological Behaviors of TiCN Films against Si_3N_4 Ceramic and Steel Ball

Yanhong Lyu [1,*], Jianyun Zheng [2], Huilian Sun [1], Xinrong Deng [1], Yang Liu [1] and Qiaoyu Zhang [1]

1 School of Physical and Chemistry, Hunan First Normal University, Changsha 410205, China
2 State Key Laboratory of Solid Lubrication, Lanzhou Institute of Chemical Physics, Chinese Academy of Sciences, Lanzhou 730000, China
* Correspondence: lyh@hnfnu.edu.cn

Abstract: Control of the structural, mechanical and tribological properties of TiCN films play an important role in its numerous applications, including the cutting tools, mechanical components, aeronautical and biomedical engineering industries. Direct-current magnetron sputtering (DCMS) system was applied to deposit TiCN films onto n-type silicon (100) at room temperature. The Ti-TiN interlayer was used to enhance the adhesive strength between the coating and the substrate. The composition and microstructure of the TiCN films were studied using X-ray photoelectron spectroscopy (XPS) and field-emitted scanning electron microscopy (FESEM). The mechanical properties of the films as a function of methane (CH_4) flow ratio were then characterized using nano-indentation measurements. The tribological behavior of TiCN films was investigated by UMT-2MT tribometer against a Si_3N_4 ceramic and AISI52100 steel ball. After the tribological tests, the wear rate of the films was obtained by the 3D surface profiler and the component content of wear debris was evaluated by energy dispersive X-ray spectroscopy (EDS). The results show that the tribological properties of TiCN films are a function of CH_4 flow rates. The film obtained at a 10 sccm CH_4 flow rate possesses a minimum average COF value of 0.1964 and reaches 72,000 cycles against a Si_3N_4 ball over the test duration. Furthermore, the wear rate was only 2.076×10^{-6} $mm^3/N{\cdot}m$. Furthermore, the TiCN films exhibited longer lifespan against the Si_3N_4 ball than against the steel ball under the normal load of 1 N, indicating that the TiCN films present better lubricative properties when against low-hardness counterparts than high-hardness counterparts.

Keywords: TiCN film; direct-current magnetron sputtering; CH_4 flow rate; tribological behavior

1. Introduction

Tool wear, including crater wear, nose wear, fracture wear and flank wear, are destructive and determine the tool's life [1]. Surface modification, such as hard film deposition, is a responsible way to enhance the tool's service life. In general, the wear resistance and optimized mechanical properties of hard films depend on grain size, chemical composition, hardness, surface stress and so on, which are highly related to preparation parameters, including deposition temperature, bias voltage, current density, gas flow rate, etc. Furthermore, the friction environment and friction medium also strongly affect the tribological properties of the films [2,3].

Early, hard films have been deposited by the CVD method at a high deposition temperature (1000 °C) [4]. In order to reduce the deposition temperature, various PVD techniques, including magnetron sputtering [5–7], laser ablation [8], cathodic arc [9] and plasma immersion ion implantation [10], have been developed. Among them, magnetron sputtering technology is the most attractive due to the high deposition rate, high ionization rate, stable sputtering reaction and low deposition temperatures [11–14]. Compared to other magnetron sputtering methods, direct-current magnetron sputtering (DCMS) has higher sputtering and film deposition rates and can be effectively control the stoichiometric ratio of film when the pure metal target and appropriate amount of reactive gas (such as O_2, N_2, etc.) added into the working gas (such as argon) are used for sputtering.

Several studies that focus on hard films have been reported. Among them, TiN and TiCN films have been used extensively as protective layers because of their high hardness, chemical stability and better wear resistance [15,16]. TiN is widely used in various tribological applications, but its performance is limited at high temperatures (≥600 °C) [17]. TiCN exhibits higher hardness and better resistance to oxidation at high temperatures compared to TiN because it combines the advantages of TiN and TiC [5]. In recent years, owing to its better oxidation resistance and high hardness, TiCN has gained wide attention by researchers. Cheng Y.H. et al. have studied the influence of CH_4 contents on the composition, structure and internal stress of TiCN films deposited by the large-area filtered-arc deposition (LAFAD) technique [4]. Compared to monolayer films, multilayer films exhibit much more excellent mechanical and tribological properties. Su Y.L. and Kao W.H. researched the tribological behavior and wear mechanisms of TiN-TiCN-TiN multilayer films [18]. Yang Y.L. et al. reported the phase constituents and mechanical properties TiCN/TiN films synthesized by laser in situ [15].

It is documented that PVD TiCN films are applied as protective films for cutting tools, automobile parts and biomedical instrumentation [19–22]. However, in order to obtain high-quality hard films and expand the application range, the relationship between the preparation parameters, the friction medium and the properties of the TiCN-based multilayer films should be further studied. In this paper, we deposited Ti/TiN/TiCN triple-layer films by DCMS and investigated the microstructure, mechanical properties and tribological behaviors of the top layer of TiCN films. It was found that the tribological behaviors of TiCN films against steel and Si_3N_4 balls are closely related to the microstructure and mechanical properties determined by CH_4 flow rate, and the optimal tribological behavior of films against steel and Si_3N_4 balls was obtained.

2. Materials and Methods

2.1. Film Deposition

n-type Si (100) wafers with a surface roughness of 20 nm and stainless steel (1Cr18Mn8Ni5N) were used as the substrate materials. The films were deposited by DC reactive-magnetron-sputtering titanium target (purity > 99.5 wt.%) at room temperature. The working gas and reactive gas were Argon (Ar) gas, nitrogen (N_2) gas and the mixed gas of nitrogen (N_2) and methane (CH_4). The purity of gases used were 99.99%. The schematic diagram of the deposition system is shown in Figure 1. All of the substrates were ultrasonically cleaned in ethyl alcohol and acetone for 10 min before deposition. The distance between the substrate and the target was controlled to 10 cm. The system was evacuated to a base pressure under 7×10^{-4} Pa before deposition. Prior to deposition, the substrates were cleaned by argon plasma bombardment for 10 min to remove any residual contaminants on the surfaces. After that, a Ti layer was deposited in a pure Ar atmosphere with 20 sccm. Then, the TiN layer was deposited under a 2.0 direct current for 25 min in mixed gas of Ar (20 sccm) and N_2 (6 sccm). Following this, the TiCN film was obtained in a gaseous mixture of CH_4 (3, 5, 8, 10, and 12 sccm) and N_2 (6 sccm) for 25 min. During deposition, a bias voltage of −100 V and duty cycle of 80% was applied to the substrates.

2.2. Film Characterization

Conventional Bragg–Brentano X-ray diffraction (XRD) using a Philips X'Pert-MRD type diffractometer with a Cu tube operated at 40 kV and 60 mA was performed to study the crystalline structure and calculate the grain size in the films. The morphologies of the fractured cross-section and surface of the films were observed using FESEM (JSM-6701F). The composition of the films was analyzed by XPS with monochromated Al Kα radiation at a pass energy of 29.4 eV. The microstructure of the films was also demonstrated by a HR800 Raman microscope with a 532 nm Ar ion laser and a resolution of 1 cm^{-1}. The hardness of the films was determined using a nano-indenter (Nanotest 600) with a Berkovich diamond tip. The indentation depth was about 10% of the film thickness in order to reduce the influence of substrates [9]. Five replicate indentations were made for each sample and the hardness was calculated from the loading–unloading curves.

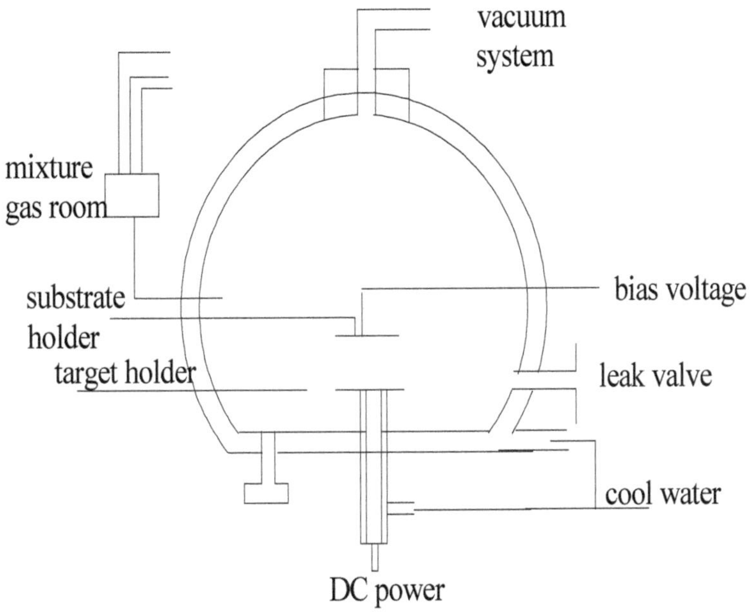

Figure 1. Top view of the DCMS deposition system.

2.3. Tribological Tests

The tribological behavior of the films was evaluated by UMT-2MT tribometer using the counterparts of the AISI52100 steel ball and the Si_3N_4 ball at a room temperature of about 25 °C and a humidity of 47%. The diameter and surface roughness of the AISI52100 steel and Si_3N_4 ball were 3 mm and 0.02 μm, and their hardness was 850 HV and 1600 HV, respectively. All the tests were conducted at a sliding speed of 600 rpm and a load of 1 N. The sliding stroke was about 5 mm. The wear volume was measured using Micro XAM-3D Surface Profile. The wear rate was calculated from the wear volume. The components of wear debris were analyzed by energy dispersive X-ray spectroscopy (EDS) in a scanning electron microscope (SEM).

3. Results

3.1. Composition and Microstructure

The XPS results of the deposited films under various CH_4 flow rates are given in Table 1. It can be found that the ratio of Ti and N decreased as the CH_4 flow rates increased, while C firstly enhanced and was then stationary. In general, by adding CH_4 gas, the concentration of C ions rises, which induces the dilution of the whole plasma. In addition, the reactivity between Ti and N ions decreased in the plasma because C ions participate in the reaction of the Ti ions, so the N ions reduced. In addition, the increase in CH_4 gas molecules also led to the Ti atom suffering from higher scattering in the collisions [8], which is the main reason for the decrease in Ti ions. However, the increase in C can form the TiC phase inlaid in the TiN plane or amorphous carbon located at the border of the TiN crystal. Furthermore, O abruptly dropped and was then an invariable trend. It should be noted that O was not deliberately introduced, and its presence was attributed to the exposure in the air, resulting in the spontaneous oxidation process of Ti [19]. When the TiCN films were deposited under CH_4 flow rates of 8 sccm, the C relative contents increased drastically and the Ti, N and O relative contents decreased obviously. This could be attributed to the films producing a dense microstructure. Therefore, the microstructure of TiCN films has been further confirmed by XRD analysis, Raman spectra and FESEM measurements.

Table 1. Composition of the films' surface measured by XPS.

CH_4 (sccm)	Ti (at.%)	C (at.%)	N (at.%)	O (at.%)
3	26.70	29.05	22.95	21.30
5	21.49	37.17	20.08	21.26
8	11.69	65.75	10.79	11.77
10	9.28	69.96	7.66	13.10
12	7.32	70.42	10.13	12.13

The XRD patterns of the TiCN films with various CH_4 flow rates from 5 to 12 sccm are given in Figure 2. The results indicate that the CH_4 flow rate has a significant influence on the grain size, phase structure and preferred orientation of the TiCN films. In Figure 2, the films deposited at high CH_4 flow rates possess a finer grain size because of the broader full width at half maximum (FWHM). As a result, the films provided with a denser microstructure increased the CH_4 flow rate. In the films deposited under 10 sccm CH_4 flow rates, the average size of crystal was about 10 nm, evaluated from the FWHM of the TiN (200) peak by the Scherrer formula. This result is in agreement with the findings of other authors [20]. It can be seen that the films presented a NaCl-type TiN polycrystalline structure. However, the films deposited at the 10 sccm CH_4 flow rates showed an obvious TiC (220) peak. It is suggested that the TiC phase is formed via local epitaxial growth on the TiN (220) plane. With increasing CH_4 flow rates, preferred orientation of the TiCN films changed from TiN (111) to (200). The grain size calculated by Scherrer's formula is shown in Figure 2b [23,24]. It is observed the grain size decreased from 19.3 nm to 4.1 nm as the CH_4 flow rate increased, which would have great influence on structural and mechanical properties.

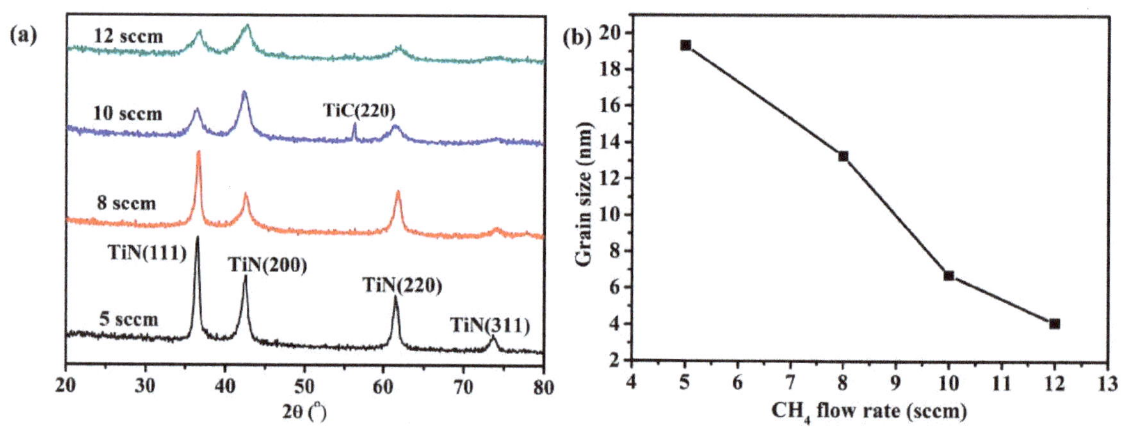

Figure 2. X-ray diffraction patterns (**a**)and grain size (**b**) of the films deposited in different CH_4 flow.

It is known that Raman can provide information about the carbon bonding configuration. Figure 3 shows the Raman spectra corresponding to the films grown at different CH_4 flow rates. The Raman shift of all spectrums ranges from 1000 to 2000 cm^{-1}. The films deposited under CH_4 flow rates from 5 to 12 sccm show the appearance of Raman signals at 1380 cm^{-1} and 1560 cm^{-1}, which is an indication of the presence of amorphous carbon [21]. This result is in agreement with the conclusion of XPS. It is also revealed that the films provide better lubricity, which is relative to the low or free C-containing films [22]. The tribological behaviors of the films are also included in Figure 4 and are be discussed later.

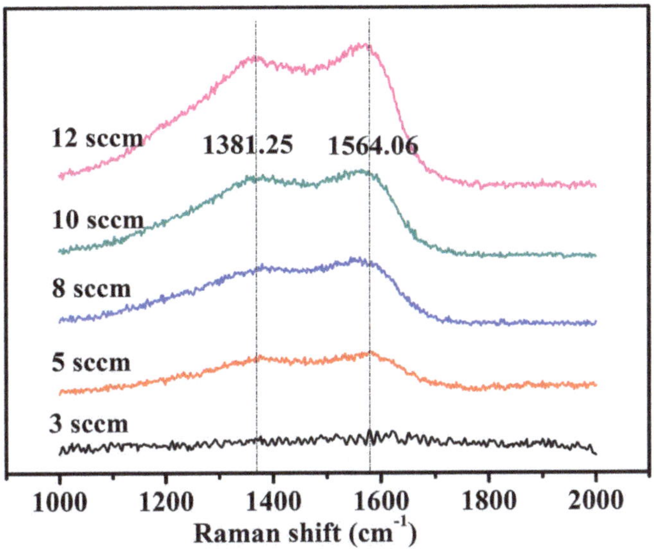

Figure 3. Raman spectra for the films deposited at different CH_4 flow rates.

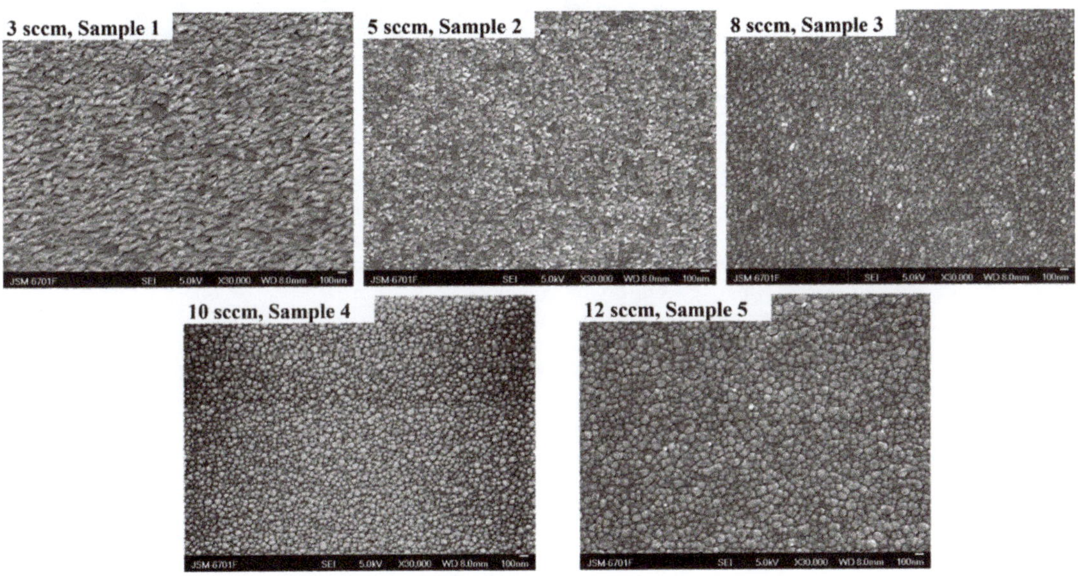

Figure 4. FESEM images of surface morphologies of the films deposited at different CH_4 flow.

The films deposited under CH_4 flow rates of 3, 5, 8, 10 and 12 sccm are referred to in sample 1 to 5, respectively. In order to acquire the microstructure information of the films, the FESEM technique was employed to observe the morphologies of the surfaces and fractured cross-section. Figure 4 shows the typical SEM images of the TiCN films' surface morphologies deposited at variable CH_4 flow rates. As described in Figure 4, an obvious evolution of microstructures from a loose and sharp pyramid-like column to a dense and globular nanostructure arises with increasing CH_4 flow rates from 3 to 12 sccm. In these photographs, the grain sizes are gradually fine with increasing CH_4 flow rates, except that of sample 5 because the C atom has a bigger radius than the N atom [4], which is in line with the XRD result. In Figure 4, it is apparently revealed that the surfaces of samples 1 and 2 consist of the pyramid-like column grains separated by pore or titanium oxide, but those of samples 3 to 5 appear to be extremely tight without titanium oxide or pores, which can be attributed to the dense microstructure at the surfaces.

Figure 5 depicts the morphologies of the fractured cross-section of the samples. It can be seen that sample 1 possesses an obvious columnar structure. However, the TiN layer from the TiCN layer can hardly be distinguished just by Figure 5a. The border of the columnar structure of pure crystalline TiN is found to severely adhere to the amorphous C phases in sample 2 observed from Figure 5b. These results further confirm the conclusion of the previous section.

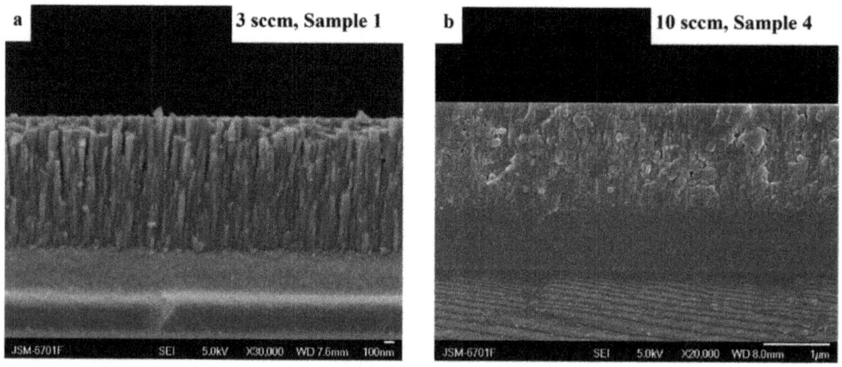

Figure 5. FESEM images of fractured cross-section of films deposited at CH_4 flow with 3 sccm (**a**) and 10 sccm (**b**).

3.2. Nanoindentation Tests

Due to the very low thickness of the single layers, we did not try to acquire values for each individual layer because very low loads lead to very high dispersions [25]. Therefore, the values of hardness in this work accord with the whole thickness of the films. Figure 6 shows the average hardness of the films deposited under different CH_4 flow rates. Generally, the hardness of films depends on many factors, such as phase composition, growth orientation, grain size, etc. Amorphous phase and low grain size are beneficial to the improvement of hardness. Regarding these aspects, sample 3 and sample 4 deposited on a CH_4 flow rate of 8 sccm and 10 sccm, respectively, own the maximum hardness value due to appropriate grain size and the formation of appropriate amount of amorphous carbon, hindering dislocation of the TiCN crystal [26].

The value of H^3/E^2 (shown in Figure 7a) that is significantly related to the wear resistance of films as a function of CH_4 flow rates is given in Figure 7. It can be found that the value of H^3/E^2 increases firstly and then decreases. It is worth noting that the ratio value of sample 4 is higher than that of other samples. The scratch test shown in Figure 7b–e exhibits that sample 4 has excellent adhesion force between the film and the substrate. These results show that sample 4 has good resistance to plastic deformation [17]. Therefore, we could conjecture that sample 4 deposited under a 10 sccm CH_4 flow rate should have the lowest wear rate. The detailed study is discussed in the subsequent section.

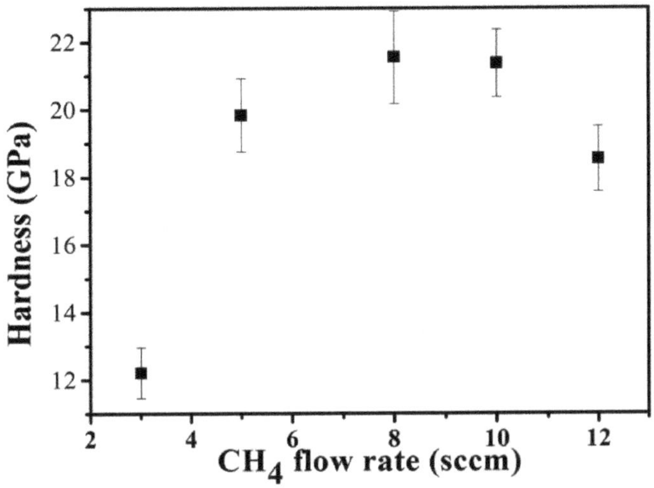

Figure 6. The average hardness of the films deposited under different CH_4 flow rates.

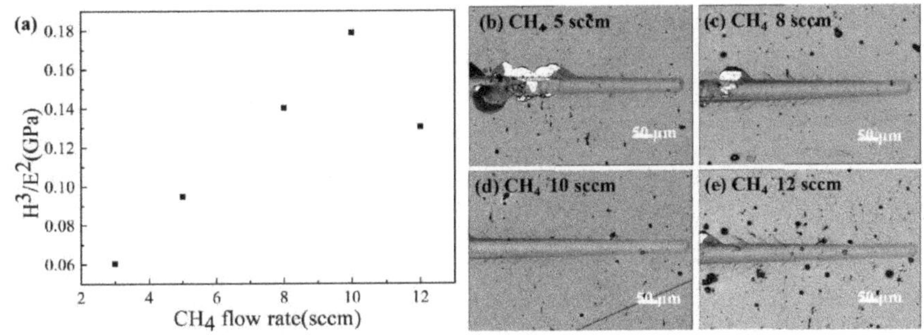

Figure 7. (**a**) The value of H^3/E^2 and (**b–e**) scratch test as a function of CH_4 flow rate.

3.3. *Tribological Behavior*

3.3.1. Against Steel Ball

The COF curves of sample 2 to 5 against the steel ball are given in Figure 8a. The average COF of the samples firstly reduces as the CH_4 flow rate increases during deposition and reaches the minimum average COF value of 0.1665 in sample 4, which is deposited under a CH_4 flow rate of 10 sccm. Then, the average COF increases slightly with a further increase in CH_4 flow rate and reaches 0.1792 when the CH_4 flow rate is at 12 sccm. The average COF value of sample 4 is lower than that reported by other authors [22,27,28]. The dense microstructure and existence of the amorphous C and TiC phase may lubricate the contact area during sliding against the steel ball and lower the COF. It is observed in Figure 8a that the time of the steady stage [29] of COF in sample 4 and 5 is hardly identical and is longer than in sample 2 and 3. However, the lubricative property of sample 5 is still maintained after the steady stage, and the COF value fluctuates from 0.18 to 0.28 in this period. The wear rates of sample 4 and 5 are 2.43×10^{-6} $mm^3/N{\cdot}m$ and 4.12×10^{-6} $mm^3/N{\cdot}m$ against the steel ball, respectively, as shown in Table 2. This result indicates that the wear rate is not in accordance with the wear life in sample 4 and 5, but in agreement with the result of the H^3/E^2 value. It is worth noticing that the wear debris of sample 4 contains about 10 at.% Fe and that of sample 5 only contains 8 at.% Fe, measured by EDS. It is obviously revealed that sample 4 is subjected to serious adhesion wear [30] when sliding against the steel ball. It explains why sample 4 exhibits a shorter life span and lower wear rate than sample 5.

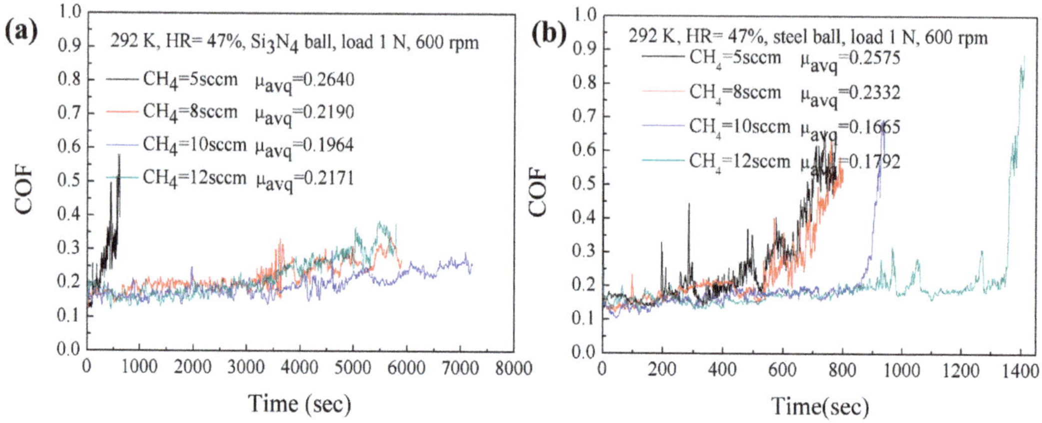

Figure 8. The COF curves of the films: (**a**) against steel ball, (**b**) against Si_3N_4 ball.

Table 2. The wear rate of the films deposited.

Samples	CH_4 (sccm)	Wear Rate (10^{-6} $mm^3/N{\cdot}m$)	
		Against Steel Ball	Against Si_3N_4 Ball
2	5	3.35	3.52
3	8	5.90	2.33
4	10	2.43	2.08
5	12	4.12	5.39

3.3.2. Against Si_3N_4 Ball

Figure 8b shows typical COF curves corresponding to the samples with CH_4 flow rates ranging from 5 to 12 sccm against with the Si_3N_4 ball. Sample 2 deposited under a CH_4 flow rate of 5 sccm presents poor tribological properties, including high COF and a short life span, because of the less amorphous carbon, loose microstructure and low hardness. The wear rate of sample 2 is 3.52×10^{-6} $mm^3/N{\cdot}m$ as shown in Table 2. The COF and life span for sample 3 and 5 are almost identical, approximately reaching 0.22 and 60,000 cycles, respectively. However, their wear rates are 2.33×10^{-6} $mm^3/N{\cdot}m$ and 5.39×10^{-6} $mm^3/N{\cdot}m$, respectively. However, sample 4 possessed the minimum average COF value of 0.1964 and reaches 72,000 cycles against the Si_3N_4 ball throughout the test duration. The wear rate of sample 4 is 2.08×10^{-6} $mm^3/N{\cdot}m$, which is lower than that of other samples. These results indicate that the sample deposited under a CH_4 flow rate of 10 sccm possesses optimal tribological behavior against the Si_3N_4 ball. Low or high C content in the samples shows certain wear problems. It is explained that the sample deposited under a CH_4 flow rate of 10 sccm possesses a dense microstructure, high hardness and good resistance to plastic deformation. This result is consistent with the conclusion of H^3/E^2 mentioned above.

3.3.3. Comparative Study

All the samples make the COF slightly lower against the steel ball than the Si_3N_4 ball except sample 3, as shown in Figure 8. It is suggested that TiCN films present better lubricative properties when against low-hardness counterparts than high-hardness counterparts. This result can be attributed to the fact that the translation layer formed during the sliding tests is softer by contrast with the Si_3N_4 ball. In addition, it is easily found that sample 4 deposited under a CH_4 flow rate of 10 sccm shows excellent friction properties, not only against the steel ball but also against the Si_3N_4 ball, resulting from the existence of a dense and globular microstructure, appropriate amount of amorphous carbon and TiC phase. Moreover, the life span of all the samples is longer against the Si_3N_4 ball than against the steel ball. It can be explained that the steel ball is easily oxidated by a tribo-chemical reaction [30,31] compared with the Si_3N_4 ball. The wear rate of the films is given in Table 2 against the Si_3N_4 ball and steel ball. The difference in wear rate for sample 2 between sliding against the steel ball and the Si_3N_4 ball is neglected due to the short time of the sliding test. The wear rate of other samples has an obvious difference against the Si_3N_4 ball and steel ball. It is worth noticing that wear rates for sample 3 and 4 are 5.905×10^{-6} $mm^3/N{\cdot}m$ and 2.43×10^{-6} $mm^3/N{\cdot}m$, respectively, against the steel ball. However, the wear rate of sample 5 is 4.12×10^{-6} $mm^3/N{\cdot}m$ against the steel ball, which is about 1.27×10^{-6} $mm^3/N{\cdot}m$ lower than against the Si_3N_4 ball. In Figures 6 and 7, it is easily found that the hardness and H^3/E^2 values of sample 3 and 4 are higher than that of sample 5. Therefore, the hard samples present poor wear resistance when sliding against the steel ball. On the contrary, the samples with relatively low hardness exhibit better wear resistance against the steel ball than against the Si_3N_4 ball. These results could be attributed to the wear rate, which is closely related to the difference in the hardness between the film and its counterpart. When the difference is small, the wear resistance is good.

4. Conclusions

TiCN films were deposited on stainless steel and silicon (100) substrates under different CH_4 flow rates by DCMS. The composition, microstructure and hardness of the films were studied, and the tribological behavior of the Ti-TiN-TiCN films sliding against a steel ball and a Si_3N_4 ball in air had already been investigated in detail. The main results can be summarized as:

1. The TiCN films displayed a gradual change in microstructure from loose and sharp pyramid-like to dense and globular nanostructure, with an increasing CH_4 flow rate. The TiC phase was found in the TiCN film deposited at 10 sccm CH_4 flow rates. Along with the rise in CH_4 flow rates, the content of amorphous carbon raised. The grain size calculated by Scherrer's formula decreased from 19.3 nm to 4.1 nm as the CH_4 flow rate increased.
2. The TiCN film deposited at 8 sccm CH_4 flow rates had an optimal value of hardness. However, the value of H^3/E^2 of the film deposited under a 10 sccm CH_4 flow rate was the uppermost value.
3. The COF of the films firstly decreased and then increased by adding a CH_4 flow rate while sliding against both the steel and the Si_3N_4 ball. The TiCN film obtained under a 10 sccm CH_4 flow rate exhibited excellent tribological behaviors against both the steel ball and the Si_3N_4 ball.
4. The films presented lower COF and shorter life span against the steel ball than the Si_3N_4 ball. However, the high hard films exhibited better wear resistance against the Si_3N_4 ball than the steel ball and vice versa.

Due to the films deposited at room temperature in this work, they may be limited to applications in high-temperature cutting tools. Therefore, the mechanical and tribological properties at high temperature of Ti/TiN/TiCN triple-layer films should be studied in the future.

Author Contributions: Conceptualization, methodology, software, validation, formal analysis, investigation, resources, data curation, writing—original draft preparation, writing—review, Y.L. (Yanhong Lyu) and J.Z.; editing, J.Z., X.D., H.S., Q.Z., Y.L. (Yang Liu) and Y.L. (Yanhong Lyu); supervision, Y.L. (Yanhong Lyu); project administration, Y.L. (Yanhong Lyu); funding acquisition, Y.L. (Yanhong Lyu). All authors have read and agreed to the published version of the manuscript.

Funding: This research was funded by Provincial Natural Science Foundation of Hunan, grant number 2021JJ40140 and the Scientific Research Fund of Hunan Provincial Education Department, grant number Grant No. 21B0812.

Data Availability Statement: Data is contained in the article.

Acknowledgments: This research also appreciates the help and guidance by Junying Hao.

Conflicts of Interest: The authors declare no conflict of interest.

References

1. Feng, W.R.; Zhou, H.; Yang, S.Z. Nano-indentation and wear-resistance behaviors of TiCN films by pulsed plasma on cemented carbide cutting tool. *Mater. Sci. Eng. A* **2010**, *527*, 4767–4770. [CrossRef]
2. Zhang, F.; He, J.; Chen, K.; Qin, Y.; Li, C.; Yin, F. Microstructure evolution and mechanical properties of TiCN-Cr nano/micro composite coatings prepared by reactive plasma spraying. *Appl. Surf. Sci.* **2018**, *427*, 905–914. [CrossRef]
3. Kumar, T.S.; Jebaraj, A.V.; Shankar, E.; Tamiloli, N.; Sivakumar, K. Metallurgical and mechanical characterization of TiCN/TiAlN and TiAlN/TiCN bilayer nitride coatings. *Surf. Inter.* **2019**, *15*, 256–264.
4. Cheng, Y.H.; Browne, T.; Heckerman, B. Influence of CH_4 fraction on the composition, structure, and internal stress of the TiCN coatings deposited by LAFAD technique. *Vacuum* **2010**, *85*, 89–94. [CrossRef]
5. Hsieh, J.; Li, C.; Wu, W.; Tan, A. Synthesis of TiC, N, O coatings by unbalanced magnetron sputtering. *J. Mater. Process. Technol.* **2003**, *140*, 662–667. [CrossRef]
6. Balaceanu, M.; Braic, M.; Braic, V.; Pavelescu, G. Properties of arc plasma deposited TiCN/ZrCN superlattice coatings. *Surf. Coat. Technol.* **2005**, *200*, 1084–1087. [CrossRef]
7. Podchernyaeva, I.A.; Lavrenko, V.A.; Umanskii, A.P.; Koror, A.A.; Timofeeva, I.I.; Yuga, A.I. Electric-spark alloying of structural alloys with a composite based on TiCN-AlN. *Powder Metall. Met. Ceram.* **2000**, *39*, 232–238. [CrossRef]
8. Lackner, J.M.; Waldhauser, W.; Ebner, R.; Bakker, R.J. Chemistry and microstructure of PLD Ti, AlCxN1-x coatings deposited at room temperature. *Appl. Phys. A* **2004**, *79*, 1469–1471. [CrossRef]
9. Bemporad, E.; Pecchio, C.; De Rossi, S.; Carassiti, F. Characterization and hardness modelling of alternate TiN/TiCN multilayer cathodic arc PVD coating on tool steel. *Surf. Coat. Technol.* **2001**, *146*, 363–370. [CrossRef]
10. Huber, P.; Manova, D.; Mandl, S.; Rauschenbach, B. Formation of TiN, TiC and TiCN by metal plasma immersion ion implantation and deposition. *Surf. Coat. Technol.* **2003**, *174*, 1243–1247. [CrossRef]
11. Zheng, J.Y.; Hao, J.Y.; Liu, X.Q.; Gong, Q.Y.; Liu, W.M. TiN/TiCN Multilayer Films Modified by Argon Plasma Treatment. *Appl. Surf. Sci.* **2013**, *280*, 764. [CrossRef]
12. Zheng, J.Y.; Ren, X.D.; Hao, H.Y.; Li, A.; Liu, W.M. Carbon nanohoops as attractive toughening and lubricant agents in TiN porous films. *Appl. Surf. Sci.* **2017**, *393*, 60–66. [CrossRef]
13. Wang, Q.; Jin, X.; Zhou, F. Comparison of mechanical and tribological properties of CrBN coatings modified by Ni or Cu incorporation. *Friction* **2022**, *10*, 516. [CrossRef]
14. Serro, A.P.; Completo, C.; Colaço, R.; dos Santos, F.; Lobato da Silva, C.; Cabral, J.M.S.; Araújo, H.; Pires, E.; Saramago, B. A comparative study of titanium nitrides, TiN, TiNbN and TiCN, as coatings for biomedical applications. *Surf. Coat. Technol.* **2009**, *203*, 3701–3707. [CrossRef]
15. Yang, Y.L.; Yao, W.M.; Zhang, H.Z. Phase constituents and mechanical properties of laser in-situ synthesized TiCN/TiN composite coating on Ti-6Al-4V. *Surf. Coat. Technol.* **2010**, *205*, 620–624. [CrossRef]
16. Canteli, J.A.; Cantero, J.L.; Marín, N.C.; Gómez, B.; Gordo, E.; Miguélez, M.H. Cutting performance of TiCN-HSS cermet in dry machining. *J. Mater. Process. Technol.* **2010**, *210*, 122–128. [CrossRef]
17. Guruvenket, S.; Li, D.; Klemberg-Sapieha, J.E.; Martinu, L.; Szpunar, J. Mechanical and tribological properties of duplex treated TiN, nc-TiN/a-SiNx and nc-TiCN/a-SiCN coatings deposited on 410 low alloy stainless steel. *Surf. Coat. Technol.* **2009**, *203*, 2905–2911. [CrossRef]
18. Su, Y.L.; Kao, W.H. Tribological behavior and wear mechanisms of TiN/TiCN/TiN multilayer coatings. *J. Mater. Eng. Perform.* **1998**, *7*, 601–612. [CrossRef]
19. Eizenberg, M.; Littau, L.; Ghanayem, S.; Liao, M.; Mosely, R.; Sinha, A.K. Chemical vapor deposited TiCN: A new barrier metallization for submicron via and contact applications. *J. Vac. Sci. Technol. A* **1995**, *13*, 590–595. [CrossRef]
20. Klemberg-Sapieha, J.E.; Jedrzejowski, P.; Martinu, L. Mechanical and optical characteristics of superhard nanocomposite TiN/a-Si_3N_4 and TiCN/a-SiCN coatings produced by PECVD. *J. Superhard Mater.* **2007**, *29*, 147–152. [CrossRef]
21. Escobar-Alarcón, L.; Camps, E.; Romero, S.; Muhl, S.; Camps, I.; Haro-Poniatowski, E. TiCN thin films grown by reactive crossed beam pulsed laser deposition. *Appl. Phys. A Mater. Sci. Process.* **2010**, *101*, 771–775. [CrossRef]
22. Deng, J.G.; Braun, M.; Gudowska, I. Properties of TiCN coatings prepared by magnetron sputtering. *J. Vac. Sci. Technol. A* **1994**, *12*, 733–736. [CrossRef]
23. Rabiei, M.; Palevicius, A.; Monshi, A.; Nasiri, S.; Vilkauskas, A.; Janusas, G. Comparing Methods for Calculating Nano Crystal Size of Natural Hydroxyapatite Using X-Ray Diffraction. *Nanomaterials* **2020**, *10*, 1627. [CrossRef] [PubMed]

24. Klug, H.P.; Alexander, L.E. *X-ray Diffraction Procedures for Polycrystalline and Amorphous Materials*; Wiley-Interscience: New York, NY, USA, 1974.
25. Morant, C.; Prieto, P.; Forn, A.; Picas, J.A.; Elizalde, E.; Sanz, J.M. Hardness enchancement by CN/TiCN/TiN multilayer films. *Surf. Coat. Technol.* **2004**, *180*, 512–518. [CrossRef]
26. Caicedo, J.C.; Amaya, C.; Yate, L.; Gómez, M.E.; Zambrano, G.; Alvarado-Rivera, J.; Muñoz-Saldaña, J.; Prieto, P. TiCN/TiNbCN multilayer coatings with enhanced mechanical properties. *Appl. Surf. Sci.* **2010**, *256*, 5898–5904. [CrossRef]
27. Zukerman, I.; Raveh, A.; Landau, Y.; Weiss, R.; Shneck, R.; Shneor, Y.; Kalman, H.; Klemberg-Saphieha, J.E.; Martinu, L. Tribological properties of duplex treated TiN/TiCN coatings on plasma nitrided PH15-5 steel. *Surf. Coat. Technol.* **2007**, *201*, 6171–6175. [CrossRef]
28. Polcar, T.; Novák, R.; Široký, P. The tribological characteristics of TiCN coating at elevated temperatures. *Wear* **2006**, *260*, 40–49. [CrossRef]
29. Mo, J.L.; Zhu, M.H. Sliding tribological behaviors of PVD CrN and AlCrN coatings against Si_3N_4 ceramic and pure titanium. *Wear* **2009**, *267*, 874–881. [CrossRef]
30. Mo, J.L.; Zhu, M.H.; Lei, B.; Leng, Y.X.; Huang, N. Comparison of tribological behaviours of AlCrN and TiAlN coatings deposited by physical vapor deposition. *Wear* **2007**, *263*, 1423–1429. [CrossRef]
31. Hager, C.H., Jr.; Sanders, J.; Sharma, S.; Voevodin, A. Gross slip fretting wear of CrCN, TiAlN, Ni, and CuNiIn coatings on Ti6Al4V interfaces. *Wear* **2007**, *263*, 430–443. [CrossRef]

Journal of
Manufacturing and Materials Processing

MDPI

Article

Process Map Definition for Laser Metal Deposition of VDM Alloy 780 on the 316L Substrate

Pascal Paulus [1,*], Yannick Ruppert [2], Michael Vielhaber [1] and Juergen Griebsch [2]

[1] Institute of Engineering Design, Saarland University, Campus E2 9, 66123 Saarbruecken, Germany
[2] School of Engineering, University of Applied Sciences Saarland, Goebenstrasse 40, 66117 Saarbruecken, Germany
* Correspondence: paulus@lkt.uni-saarland.de

Abstract: VDM Alloy 780 is a novel Ni-based superalloy that allows for approximately 50 °C higher operating temperatures, compared to Inconel 718, without a significant decrease in mechanical properties. The age hardenable NiCoCr Alloy combines increased temperature strength with oxidation resistance, as well as improved microstructural stability due to γ'-precipitation. These advantages make it suitable for wear- and corrosion-resistant coatings that can be used in high temperature applications. However, VDM Alloy 780 has not yet been sufficiently investigated for laser metal deposition applications. A design of experiments with single tracks on 316L specimens was carried out to evaluate the influence of the process parameters on clad quality. Subsequently, the quality of the clads was evaluated by means of destructive and non-destructive testing methods, in order to verify the suitability of VDM Alloy 780 for laser metal deposition applications. The single-track experiments provide a basis for coating or additive manufacturing applications. For conveying the results, scatter plots with regression lines are presented, which illustrate the influence of specific energy density on the resulting porosity, dilution, powder efficiency, aspect ratio, width and height. Finally, the clad quality, in terms of porosity, is visualized by two process maps with different mass per unit lengths.

Keywords: laser metal deposition; VDM Alloy 780; superalloy; 316L; process map; single track

1. Introduction

In order to meet today's environmental and climate challenges, both the production processes and the resulting products must have an increased sustainability aspect. Additive manufacturing processes, in this case laser metal deposition (LMD), can make a decisive contribution here, as it enables the production of near-net-shape geometries without significant material waste [1] or graded materials [2], as well as cladding operations [3]. The resulting parts can have advanced properties due to the build-up of parts from various materials.

LMD enables a wide range of applications due to a large product range of available powder materials, such as repair [4–6], additive manufacturing [7–9] and cladding [10–13]. Research in this field includes parameter optimizations [14,15], simulation [16–18], process monitoring and control [19–22] and material structure, as well as mechanical properties [23–25]. The process variants of laser cladding can be differentiated in terms of the material supplied. A differentiation is made between powder-based and wire-based laser metal deposition [26]. Powder-based LMD uses powder particles that are injected into the process by means of a carrier gas. Typical powder grain fractions are between 50 and 100 µm [27]. The feeding systems of additive materials in powder-based LMD processes can be divided into coaxial-continuous, coaxial-discontinuous and lateral systems [28].

Ni-based superalloys can be used for numerous applications, such as the manufacturing of static and rotating components in aircraft turbines [28,29], repair of gas turbines [30], as well as wear- and corrosion-resistant coatings [31–33]. Most research in the field of LMD features Inconel 718 [34–36] or Inconel 625 [36–39]. A novel material that has hardly been used in the LMD process so far is VDM Alloy 780. The age-hardenable NiCoCr superalloy features increased temperature and oxidation resistance, as well as improved microstructural stability due to γ'-precipitation [40,41].

The objective of this work is to create a process map for powder-based LMD of the VDM Alloy 780 on the 316L substrate, to provide a basis for advanced processing applications in cladding or additive manufacturing. Due to the high number of process parameters, a process map is helpful to enable user-friendly production by predicting the quality of the clads [42]. The result of the deposition process depends on various process parameters, such as laser power, feed rate, powder mass flow, nozzle and carrier gas flow, stand-off distance, preheating temperatures, cooling rates and material properties. However, the main process parameters, i.e., laser power, feed rate and powder mass flow, are crucial for a qualitative results [43]. The quality of the clad is evaluated by the degree of dilution [44,45], density [46], cracks [47,48] and the microstructure. Furthermore, economical aspects such as powder efficiency and build rate are considered. In addition, economic aspects, such as powder efficiency and build-up rate, as well as the relationships between process parameters and quality characteristics, are taken into account. Finally, the results consist of multiple scatter plots and two process maps. The scatter plots are provided to show the relationships between quality criteria and specific energy density, while the process maps help in a defect-free deposition of VDM Alloy 780 on 316L.

2. Materials and Methods

2.1. Material Preparation

As additive, the gas-atomized age-hardenable VDM Alloy 780 (VDM Metals, Germany) with a grain size of 53–100 µm is used. It is characterized by an austenitic microstructure, in which several phases can occur. The most important characteristic is the establishment of γ'-precipitation that replaces the γ''-hardening compared to known alloys, such as Inconel 718 [40]. While in Inconel 718 longtime exposure to temperatures above 650 °C leads to the transformation of the metastable γ''-phase and the loss of strengthening effects, VDM Alloy 780 features a higher temperature stability with excellent mechanical properties due to thermodynamically stable γ'-precipitation. The key difference of VDM Alloy 780 is the addition of Co, substituting Fe and Ni. This constrains the solvus temperature of the γ'-phase and consequently decelerates the precipitation [40]. The hardenability results from the alloy elements niobium, titanium and aluminum. Tillack [49] shows that increasing the contents of aluminum and titanium decreases the weldability, but increases the amount of strengthening γ'-precipitation. Studies on other age-hardenable Ni-based alloys with γ'-precipitation, such as Waspaloy [50] and IN 100 [51], have already been carried out. The chemical composition of the VDM Alloy 780 is listed in Table 1.

Table 1. Chemical composition (wt%) of the VDM Alloy 780.

Ni	Cr	Fe	C	Mn	Si	Cu	Mo	Co	Nb + Ta	Al	Ti	B	P	S
bal.	18	≤4	≤0.1	≤0.5	≤0.3	≤0.5	3	24	4–6	2	0.6	≤0.02	≤0.03	≤0.015

Figure 1 shows a comparison of the currently most used nickel-based superalloys in LMD. The dashed line indicates when an alloy is considered difficult to weld. As the figure shows, weldability is strongly influenced by the alloying elements aluminum and titanium. Due to its composition of 0.6 wt% Ti and 2 wt% Al, the VDM Alloy 780 exhibits significantly poorer weldability compared to Inconel 718 (IN 718). However, compared to U-700 and IN 100 alloys, it is still in a better weldability range.

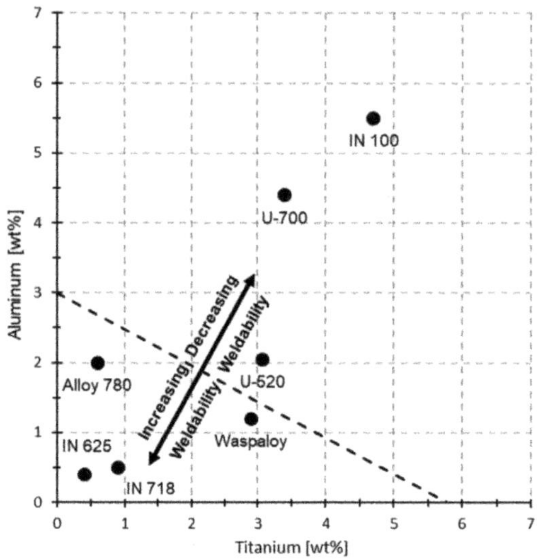

Figure 1. Weldability comparison of the most commonly used Ni-based superalloys in LMD [49].

As substrate material, an austenitic stainless steel (316L) with a thickness of 10 mm and a milled surface (Ra = 0.8 μm) is used. Due to the excellent corrosion resistance and well-defined mechanical properties, these steels have a wide range of applications. For experimental preparation, the surface of the specimens was cleaned and degreased with ethanol. The chemical composition of 316L is listed in Table 2. The 316L was chosen as a substrate material because of its face-centered cubic lattice structure, as well as the comparable amount of Cr with regards to the VDM Alloy 780.

Table 2. Chemical composition (wt%) of the 316L substrate.

Ni	Cr	Fe	C	Mn	Si	Mo	P	S
12.5	17.5	bal.	0.03	2	1	2.5	0.045	0.03

2.2. Experimental Setup

The experimental setup features a Trumpf TruDisk 4001 solid-state laser as a beam source, with a maximum output power of 4000 W and a wavelength of λ = 1030 nm. A 600 μm fiber optic cable is used to guide this radiation to the optics. For the optics and nozzle, a Trumpf BEO D70 processing optic and an ILT SO 16 three-jet nozzle are used. The imaging ratio of 2:1 and an adjustable collimation allow for focusing the diameters of the laser beam between 1.2 and 8 mm. The SO 16 powder nozzle has a processing distance of 16 mm and a powder focus diameter of 4 mm. Thus, single-track widths of 1.5–7 mm were generated. The transport of the gas and powder to the process is ensured by the GTV PF 2/2 powder conveyor. The conveyor features two powder containers with different conveying volumes. The maximum conveying volume of the main container is 80 g/min. Both containers are operated with up to 12 l/min helium 4.6 as the carrier gas. Argon 4.6 with a flow rate of up to 20 l/min is used as a shielding gas. A KUKA KR 16-2 is used as the handling system for the laser cladding process.

2.3. Experimental Procedure

- The parameters for the experimental study are defined on the basis of the literature data for comparable materials such as Inconel 718 [29] and 625 [39]. In advance, the laser power is determined by an approximation equation based on literature information [46]. The equation describes the required laser power as a function of machine- and material-specific factors. Radiation losses due to component reflection are not taken into account, since the assumption is made that these are nearly completely absorbed by the powder stream. For an economically optimized application, the following quality requirements are placed on the weld:

- To ensure that the powder is used optimally, the laser beam is adapted to match the constant powder focus diameter (4 mm), ensuring a nearly complete deposition of the additive.
- As quality characteristics, an aspect ratio of 4:1 [52] and a low degree of dilution (5–30%) [53] are desired.
- At the same time, high build-up rates should be realized in order to achieve an economical process.

Based on the limits of the experimental setup and the material properties of the VDM Alloy 780 and 316L, a laser power of approximately 3670 W at a track width of approximately 4 mm is required. As laser cladding is a multi-parameter problem and a new material pairing is used, a statistical experimental plan helps to efficiently implement the experiments. A design of experiments, based on the approximated initial parameters, is developed to investigate the influence of different energy densities at two constant massper unit lengths (27.9 g/m and 39.3 g/m). The first mass per unit length was based on previously performed tests with Inconel 718 and process parameters selected from literature data [29]. The second mass per unit length was selected due to the poorer weldability of the VDM Alloy 780 compared to Inconel 718.

Based on the conclusion of Tillack [49] that the lower feed rates favor a more stable process, lower feed rates were investigated in the second part of the design of experiments. To ensure the comparability of the results, the specific energy density is retained. This results in a higher mass per unit length and a longer interaction time between the laser and the material.

The process parameters for laser power, feed rate and powder mass flow rate are shown in Table 3. Carrier, as well as shielding gas flow rates are kept constant. The carrier gas flow rate used is 4 l/min, while the shielding gas flow rate is 16 l/min. Additionally, to eliminate external interferences and process errors induced by the system, a block formation with randomization is carried out.

Table 3. Design of experiments to investigate the influence of different specific energy densities at a constant mass per unit length: (a) feed rates for 27.9 g/m; (b) feed rates for 39.3 g/m.

Laser Power [W]	Feed Rate [mm/min] (a)	(b)	Powder Mass Flow Rate [g/min]	Specific Energy Density [kJ/g]
3000	1183	840	33	5.45
3250	1183	840	33	5.91
3500	1183	840	33	6.36
3750	1183	840	33	6.82
4000	1183	840	33	7.27
3000	1075	762	30	6.00
3250	1075	762	30	6.50
3500	1075	762	30	7.00
3750	1075	762	30	7.50
4000	1075	762	30	8.00
3000	968	690	27	6.67
3250	968	690	27	7.22
3500	968	690	27	7.78
3750	968	690	27	8.33
4000	968	690	27	8.89
3000	860	612	24	7.50
3250	860	612	24	8.13
3500	860	612	24	8.75
3750	860	612	24	9.38
4000	860	612	24	10.00

2.4. Metallographic Preparation and Evaluation of Quality Characteristics

All specimens were cut on a cut-off machine, using a cut-off wheel with 250 mm outer diameter. Subsequently, the samples were warm-bedded. The metallographic preparation involved a series of three grinding steps (220, 500 and 1000) for five minutes each. After each grinding process, the specimens are cleaned in an ultrasonic bath with ethanol for five minutes. The specimens are then polished with a 3 μm diamond polish for ten minutes. The polishing wheel is cleaned under water using detergent and a brush after each operation. The etching of the specimens was then carried out with V2A etchant for 45 s at 60 °C, which allows for both microstructural analysis and the investigation of flaws such as pores, bonding defects and cracks. The analysis of the cross-sections was carried out on a light microscope (Leica DM LM). In addition, the microstructure was examined with a scanning electron microscope (SEM) (JEOL 6460L), and energy-dispersive X-ray spectroscopy (EDX) analyses were performed. The quality characteristics considered are porosity, the degree of dilution, aspect ratio, powder efficiency, as well as the height and width of the seam.

The porosity ε_P provides information on how large the volume of the cavities is compared to the total volume of the clad. Pores and shrinkage cavities are formed during rapid solidification processes in which no outgassing of the molten material takes place and negatively affects the mechanical properties of the clad. The porosity of a welded clad can be calculated using the following Equation (1). A_P indicates the area of a pore in the cross section and A_T the total area of the clad [29].

$$\varepsilon_P = \frac{\sum A_P}{A_T} \tag{1}$$

In addition to porosity, the degree of dilution is an important parameter for assessing the quality of a clad. The degree of dilution indicates the percentage of the additive that does not directly build up the volume, but creates the bond between the substrate and the additive. The degree of dilution, A_D, of a clad can be calculated according to the following Equation (2). A_2 indicates the cross-sectional area which does not cause a direct volume build-up and has immersed into the substrate material. A_1 describes the cross-sectional area that is located above the substrate surface and achieves a volume build-up [54].

$$A_D = \frac{A_2}{A_1 + A_2} * 100\% \tag{2}$$

The aspect ratio, AR, provides information on the ratio of the width, W, to the height, H, of the clad and is calculated according to Equation (3). The ratio affects the tool path generation in coating and additive manufacturing applications, as the required overlap and layer height are affected.

$$AR = \frac{W}{H} \tag{3}$$

The powder efficiency indicates the percentage of the powder conveyed during the process that was effectively applied. The powder efficiency was calculated as the quotient of the applied mass m_A and the maximum applicable mass, $m_{A,max}$, according to Equation (4). The maximum applicable mass, $m_{A,max}$, can also be expressed as the product of the clad length, l, and powder mass flow rate, $\dot{m}$, divided by the feed rate, v.

$$\eta_P = \frac{m_A}{m_{A,max}} = \frac{m_A * v}{l * \dot{m}} \tag{4}$$

3. Results and Discussion

3.1. Metallographic Analysis

Figure 2 shows a cross section of a specimen with the lowest porosity of 0.091%, an aspect ratio of 3.8, 30% dilution, 1.38 mm height and 5.25 mm width. The specimen was fabricated using a laser power of 3500 W, a feed rate of 612 mm/min and a powder mass flow rate of 24 g/min. Spherical pores, but no cracks or bonding defects, were observed. The largest pore measured 17.2 μm in diameter and no local concentration of pores could be detected in the entire cross section. The etching with V2A etchant for 45 s at 60 °C reveals a columnar grain structure characterized by elongated, epitaxial grain growth. A solidification structure with elongated grains growing in a normal direction to the substrate material was observed across all specimens. Additional bright interdendritic phases can be detected. These were further investigated by SEM analyses. Figure 3 presents the SEM micrographs of the microstructure of an as-deposited VDM Alloy 780 clad. Light gray precipitations that were analyzed using EDX show high amounts of Nb (approx. 22%). This phase is similar to the Laves phase and usually precipitates in alloys containing high amounts of Fe. EDX analyses of the dark-colored regions (γ-Matrix) show high amounts of Fe (approx. 10 to 15%) due to the intermixing of the substrate and the additive.

Figure 2. VDM Alloy 780 clad with a porosity of 0.091%, aspect ratio of 3.8, 30% dilution, 1.38 mm height and 5.25 mm width, etched with a V2A etchant (45 s at 60 °C).

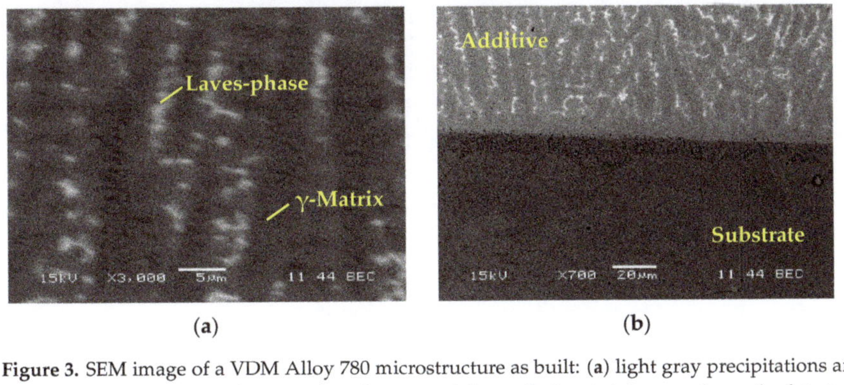

Figure 3. SEM image of a VDM Alloy 780 microstructure as built: (**a**) light gray precipitations are Nb-rich; (**b**) transition zone between the substrate and the applied material, no cracks can be detected.

3.2. EDX Linescan Analysis

Similar to Inconel 718, VDM Alloy 780 is a Nb-bearing, nickel-based superalloy. During LMD, the non-equilibrium solidification of the local molten pool leads to microsegregations of a high concentration of refractory elements such as Nb and eutectic Laves phases [55].

If different materials are used as substrates and additives, the mixing of the materials occurs in the dilution zone. Therefore, it is necessary to ensure that the mixing of the elements does not have a negative influence on the mechanical or chemical properties of the materials. A higher Fe content in Ni-based alloys, for example, promotes the formation of a Laves phase [56]. These brittle phases are crack-propagating, especially if they are precipitated in a chain-like manner [55].

Figure 4 shows that in the dilution zone, the amount of Fe changes. It can be observed that due to the LMD process, an increase in Fe within the additive has taken place. With a higher specific energy density, higher Fe contents can be detected in the additive. This value also strongly correlates with the dilution (Figure 5b). In order to achieve a low degree of dilution, lower specific energy densities are needed. This enables less mixing of the alloying elements, but results in higher porosity (Figure 5a). For the parameter sets applied, an increased amount of iron in the additive was detected in each specimen. The lowest Fe content (7.1%) was determined for a specific energy density of 5.45 kJ/g. The highest Fe content was observed at a specific energy density of 10 kJ/g, at 22.3%. EDX-based area determinations are averages, and local values may be higher. The Fe content in the LMD clad initially decreases steeply with the distance from the substrate, until a constant value is reached.

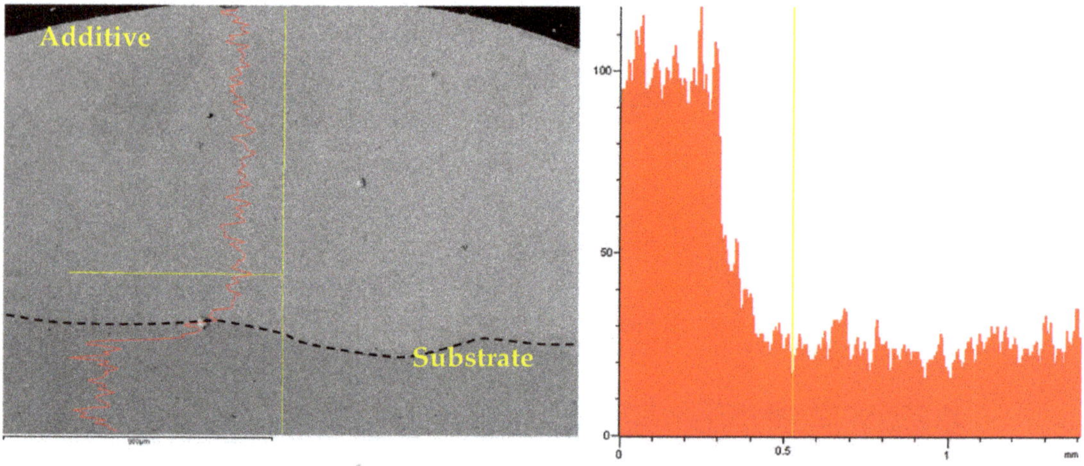

Figure 4. EDX linescan analysis of specimens with 6.67 kJ/g and 27.9 g/m, quantitative distribution of the iron content along the cross-section.

3.3. Statistical Analysis

The microscopic analyses of the clads show that there were no bonding defects, as well as cracking in the specimens. Figure 5 shows scatterplots of the results for the quality characteristics **porosity** (**a**), **dilution** (**b**), **powder efficiency** (**c**), **aspect ratio** (**d**), **width** (**e**) and **height** (**f**). While **a–e** are mainly influenced by specific energy densities, **f** seems to be mainly influenced by the used mass per unit length. The regression analyses for **c** to **f**, on the other hand, show a linear relationship. A higher specific energy density results in a lower **porosity** (**a**) for both mass per unit lengths, which asymptotically approaches a value of 0.1% in this experimental study.

The **degree of dilution** (**b**) increases with a higher specific energy density, but converges to 38.2% for 27.9 g/m and 37% for 39.3 g/m for >10 kJ/g. In coating applications, low mixing is desirable to avoid different chemical properties. In contrast, a higher degree of dilution has a positive effect on the bonding quality in additive manufacturing applications. High intermixing of the substrate and the additive leads to unstandardized alloys, in which phases and precipitates can form, which have a negative influence on, for example, the strength and corrosion resistance.

The **powder efficiency** (**c**), as a function of specific energy density, indicates the largest deviation in the regression analysis (Pearson correlation r = 0.763 for 27.9 g/m and r = 0.627 for 39.3 g/m). Due to the low material deposition in relation to the substrate weight, the deposited mass can only be determined inaccurately. Nevertheless, it is observed that as the specific energy density increases, the powder efficiency also increases. In **a**, **b** and **c**, the mass per unit length is significantly responsible for the steepness of the regression line.

The **aspect ratio** (**d**) is an important characteristic for coating applications and additive manufacturing [45]. In pure coating applications, it has a decisive influence on the path calculation of the surface to be coated (overlap ratio). In the production of volumes, important path planning parameters are based on the aspect ratio in order to create near-net-shape geometries and be able to design the post-processing efficiently. The aspect ratio can be predicted for both mass flows via the regression equation, with an r = 0.913 for 27.9 g/m and r = 0.893 for 39.3 g/m.

With a mass per unit length of 39.3 g/m, an ideal ratio of 4:1 is achieved with a specific energy density of 8.5–10 kJ/g. In addition, a low porosity and high powder efficiency can be observed.

If the **width** (**e**) and **height** (**f**) are considered as separate characteristics, the resulting characteristic can be predicted more accurately for lower mass per unit lengths. The regression lines in **d**, **e** and **f** are parallel to one another. Here, the mass per unit length mainly influences the *Y*-axis intercept, and defines the start and end values for the selected mass flows. Based on the regression analyses and the requirements for a high quality clad, a target optimization can be defined, which suggests the processing of the VDM Alloy 780 with $\geq$8.5 kJ/g. At the same time, a larger mass flow results in higher build-up rates.

However, the maximum processing parameters, such as laser power, feed rate and powder mass flow, are system-dependent and must therefore be adapted to the existing system with reference to the available specific energy density.

The determined parameters of the single-track tests are only suitable to a limited extent for adaptation in the generative build-up of thin-walled structures or volumes. The temperature gradient generated during the additive manufacturing process has a detrimental effect on the microstructure and the residual stresses. Furthermore, shape deviations, as a result of local heat accumulation, are possible. However, the single-track experiments provide geometrical and metallurgical properties of the clad, which save time and materials for further investigations.

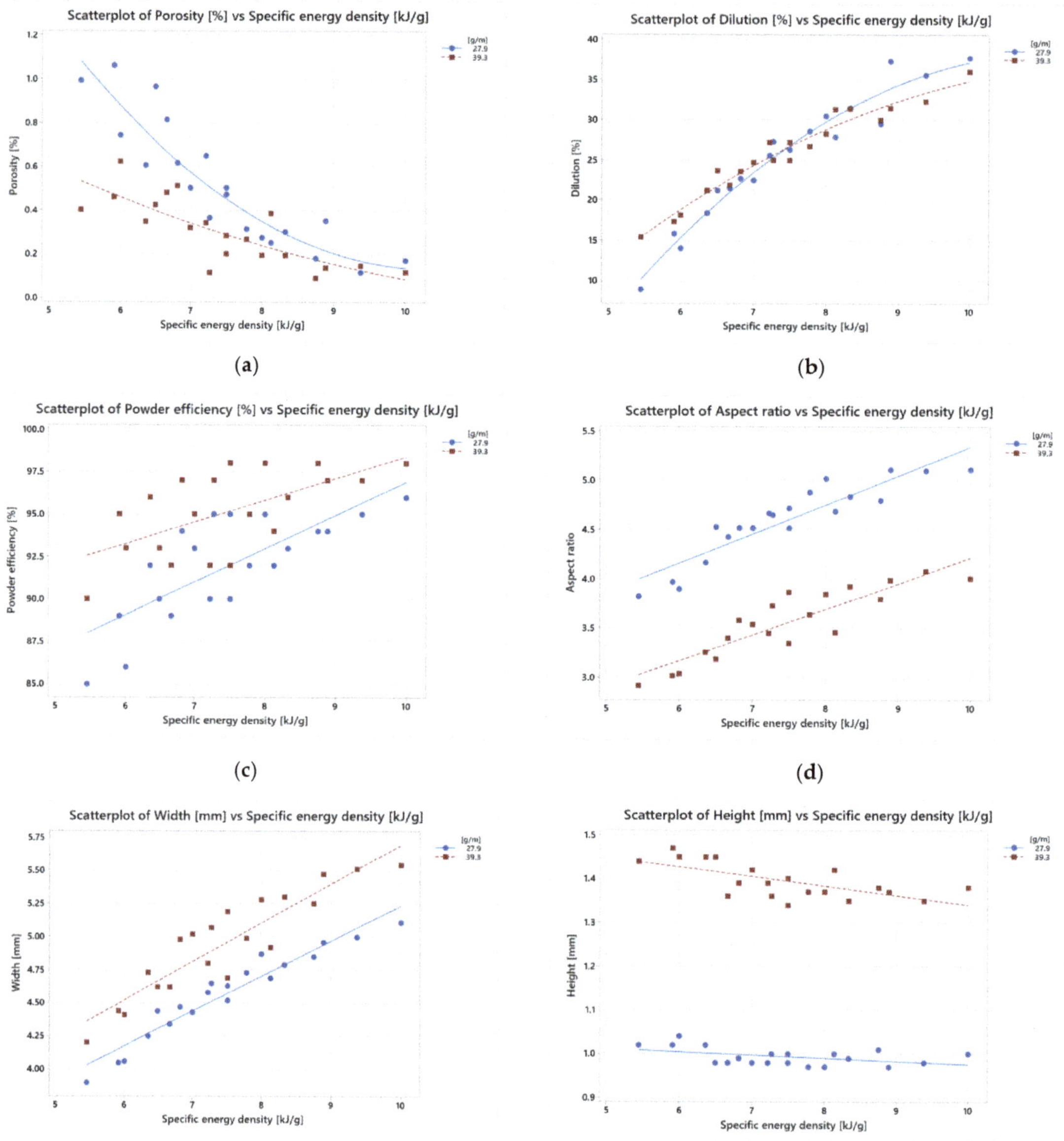

Figure 5. Scatterplots of quality characteristics vs. specific energy densities with 27.9 g/m (blue) and 39.3 g/m (red): (**a**) Porosity; (**b**) dilution; (**c**) powder efficiency; (**d**) aspect ratio; (**e**) width; (**f**) height.

3.4. Process Map Definiton

Figures 6 and 7 show two process maps for laser metal deposition of the VDM Alloy 780 using different mass per unit lengths. The investigations show that higher specific energy densities in combination with lower feed rates result in a lower porosity. Efficient deposition is possible with a higher mass per unit length, since lower porosity values can be achieved at the same build rate using a lower feed rate. The process window in Figure 7 already starts to open at a mass per unit length of 39.3 g/m from a level of 3400 W, allowing porosity values of less than 0.3% to be achieved. With a mass per unit length of 27.9 g/m, on the other hand, only a process window can be glimpsed in the upper left corner of Figure 6. Presumably, this window will open further by increasing the energy density; in other words, by increasing the laser power higher than P = 4000 W or reducing the feed rate. Contrary to this, a reduction of the feed rate is also undesirable, since this would result in economic losses. Considering the build rate of the process additionally, an increase in the mass per unit length seems reasonable, since this keeps the build rate constantly high and a further reduction of the porosity can presumably take place at lower energy densities.

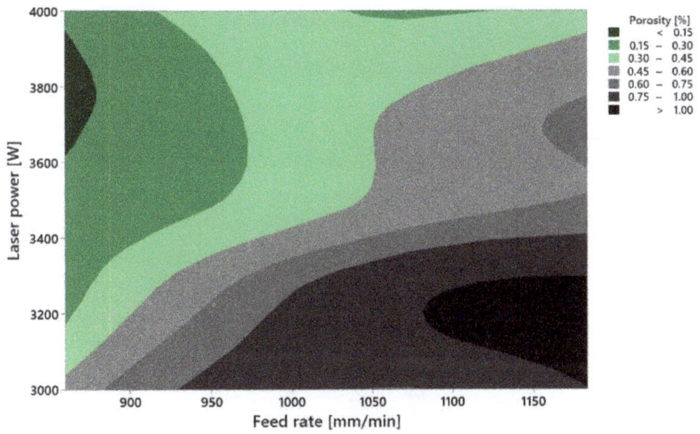

Figure 6. Process map of the VDM Alloy 780 regarding the feed rate and laser power, using a mass per unit length of 27.9 g/m.

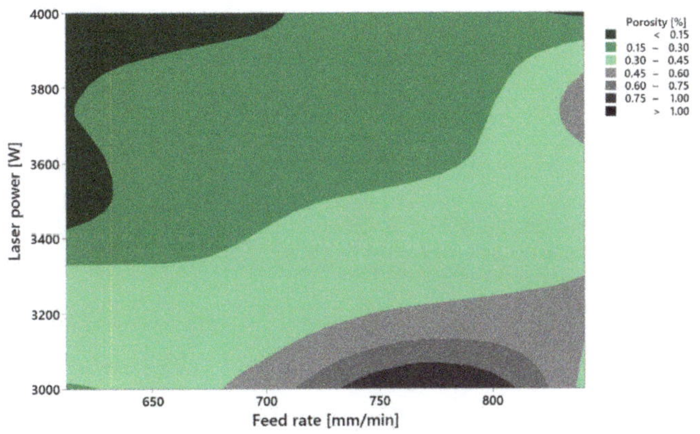

Figure 7. Process map of the VDM Alloy 780 regarding the feed rate and laser power, using a mass per unit length of 39.3 g/m.

4. Conclusions

The aim of this work was to create a process map for single-track clads of the VDM Alloy 780 on the 316L substrate. The main process parameters, such as laser power, feed rate and powder mass flow rate, are used in the design of the experiments to investigate the influence of different specific energy densities at constant mass per unit lengths. The results of the investigation show that a high-quality clad with <0.15% porosity can be achieved with 27.9, as well as 39.3 g/m. While the process window for economically efficient quality cladding is rather small at 27.9 g/m, higher mass per unit lengths seem to be advantageous for processing with lower specific energy densities. Optimization with masses per unit length > 39.3 g/m will further improve the economic efficiency of the process, as higher build rates can be realized. The most significant conclusions from the present study are as follows:

- In powder-based laser metal deposition of the VDM Alloy 780 on the 316L substrate, higher masses per unit length improve the overall porosity values.
- Higher mass per unit lengths are beneficial, since the degree of dilution is reduced at high specific energy densities with low porosity values [57]. In addition, the intermixing of the substrate and the additive must be kept low to avoid different chemical properties of the VDM Alloy 780 clads.
- A mass per unit length of 39.3 g/m with 8.75 kJ/g achieves high clad quality (0.091% porosity, aspect ratio of 3.8 and 30% dilution).
- The appearance of the Laves phase in deposited clads occurs due to the intermixing of the substrate and the additive.
- EDX analyses show that the Fe content strongly depends on the specific energy density of the process. The lowest Fe content of 7.1% was determined for a specific energy density of 5.45 kJ/g. The highest Fe content was observed at a specific energy density of 10 kJ/g, at 22.3%.
- In order to further reduce the porosity (<0.1%) and the Fe content in the clad, higher mass per unit lengths with lower feed rates should also be considered.

Author Contributions: P.P.: Conceptualization, Investigation, Methodology, Project administration, Formal analysis, Writing—original draft; Y.R.: Conceptualization, Investigation, Methodology, Visualization, Formal analysis, Writing—original draft; M.V.: Supervision, Resources, Writing—review and editing; J.G.: Supervision, Resources, Validation, Writing—review and editing. All authors have read and agreed to the published version of the manuscript.

Funding: This research was funded by European Regional Development Fund (ERDF), within the project "LPA2PECM".

Data Availability Statement: The data presented in this study are available on request from the corresponding author.

Conflicts of Interest: The authors declare no conflict of interest.

References

1. Heilemann, M.; Möller, M.; Emmelmann, C.; Burkhardt, I.; Riekehr, S.; Ventzke, V.; Kashaev, N.; Enz, J. Laser Metal Deposition of Ti-6Al-4V Structures: New building strategy for a decreased shape deviation and its influence on the microstructure and mechanical properties. In Proceedings of the Lasers in Manufacturing Conference, München, Germany, 26–29 June 2017.
2. Yan, J.; Battiato, I.; Fadel, G.M. Planning the process parameters for the direct metal deposition of functionally graded parts based on mathematical models. *J. Manuf. Process.* **2018**, *31*, 56–71. [CrossRef]
3. Güpner, M.; Patschger, A.; Bliedtner, J. Influence of Process Parameters on the Process Efficiency in Laser Metal Deposition Welding. *Phys. Procedia* **2016**, *83*, 657–666. [CrossRef]
4. Graf, B.; Gumenyuk, A.; Rethmeier, M. Laser Metal Deposition as Repair Technology for Stainless Steel and Titanium Alloys. *Phys. Procedia* **2012**, *39*, 376–381. [CrossRef]
5. Qi, H.; Azer, M.; Singh, P. Adaptive toolpath deposition method for laser net shape manufacturing and repair of turbine compressor airfoils. *Int. J. Adv. Manuf. Technol.* **2010**, *48*, 121–131. [CrossRef]
6. Graf, B.; Ammer, S.; Gumenyuk, A.; Rethmeier, M. Design of Experiments for Laser Metal Deposition in Maintenance, Repair and Overhaul Applications. *Procedia CIRP* **2013**, *11*, 245–248. [CrossRef]
7. Fu, Y.; Guo, N.; Wang, G.; Yu, M.; Cheng, Q.; Zhang, D. Underwater additive manufacturing of Ti-6Al-4V alloy by laser metal deposition: Formability, gran growth and microstructure evolution. *Mater. Des.* **2021**, *197*, 109196. [CrossRef]
8. Langebeck, A.; Bohlen, A.; Rentsch, R.; Vollertsen, F. Mechanical Properties of High Strength Aluminum Alloy EN AW-7075 Additively Manufactured by Directed Energy Deposition. *Metals* **2020**, *10*, 579. [CrossRef]
9. Gradl, P.; Tinker, D.C.; Park, A.; Mireles, O.R.; Garcia, M.; Wilkerson, R.; Mckinney, C. Robust Metal Additive Manufacturing Process Selection and Development for Aerospace Components. *J. Mater. Eng. Perform.* **2022**, *31*, 6013–6044. [CrossRef]
10. Amado, J.M.; Montero, J.; Tobar, M.J.; Yáñez, A. Laser Cladding of Ni-WC Layers with Graded WC Content. *Phys. Procedia* **2014**, *56*, 269–275. [CrossRef]

11. Janicki, D.; Górka, J.; Kotarska, A. Laser cladding of Inconel 625-based composite coatings. *Weld. Technol. Rev.* **2018**, *90*, 9. [CrossRef]
12. Łukasz, N.; Marta, W. Finishing Surface After Regeneration with Laser Cladding. *Procedia Eng.* **2017**, *192*, 1012–1015. [CrossRef]
13. Li, T.; Zhang, L.; Bultel, G.G.P.; Schopphoven, T.; Gasser, A.; Schleifenbaum, J.H.; Poprawe, R. Extreme High-Speed Laser Material Deposition (EHLA) of AISI 4340 Steel. *Coatings* **2019**, *9*, 778. [CrossRef]
14. Mahamood, R.M.; Akinlabi, E.T. Processing Parameters Optimization for Material Deposition Efficiency in Laser Metal Deposited Titanium Alloy. *Lasers Manuf. Mater. Process.* **2016**, *3*, 9–21. [CrossRef]
15. Caiazzo, F.; Alfieri, V.; Argenio, P.; Sergi, V. Additive manufacturing by means of laser-aided directed metal deposition of 2024 aluminium powder: Investigation and optimization. *Adv. Mech. Eng.* **2017**, *9*, 1687814017714982. [CrossRef]
16. Biegler, M.; Graf, B.; Rethmeier, M. In-situ distortions in LMD additive manufacturing walls can be measured with digital image correlation and predicted using numerical simulations. *Addit. Manuf.* **2018**, *20*, 101–110. [CrossRef]
17. Tamanna, N.; Crouch, R.; Naher, S. Progress in numerical simulation of the laser cladding process. *Opt. Lasers Eng.* **2019**, *122*, 151–163. [CrossRef]
18. Zeng, Q.; Tian, Y.; Xu, Z.; Qin, Y. Simulation of thermal behaviours and powder flow for direct laser metal deposition process. *MATEC Web Conf.* **2018**, *190*, 2001. [CrossRef]
19. Arrizubieta, J.I.; Martínez, S.; Lamikiz, A.; Ukar, E.; Arntz, K.; Klocke, F. Instantaneous powder flux regulation system for Laser Metal Deposition. *J. Manuf. Process.* **2017**, *29*, 242–251. [CrossRef]
20. Vandone, A.; Baraldo, S.; Valente, A.; Mazzucato, F. Vision-based melt pool monitoring system setup for additive manufacturing. *Procedia CIRP* **2019**, *81*, 747–752. [CrossRef]
21. Bi, G.; Gasser, A.; Wissenbach, K.; Drenker, A.; Poprawe, R. Identification and qualification of temperature signal for monitoring and control in laser cladding. *Opt. Lasers Eng.* **2006**, *44*, 1348–1359. [CrossRef]
22. Marko, A.; Bähring, S.; Raute, J.; Biegler, M.; Rethmeier, M. Quality Prediction in Directed Energy Deposition Using Artificial Neural Networks Based on Process Signals. *Appl. Sci.* **2022**, *12*, 3955. [CrossRef]
23. Jiang, B.; Zhenglong, L.; Xi, C.; Peng, L.; Nannan, L.; Yanbin, C. Microstructure and mechanical properties of TiB2-reinforced 7075 aluminum matrix composites fabricated by laser melting deposition. *Ceram. Int.* **2019**, *45*, 5680–5692. [CrossRef]
24. Li, L.; Wang, J.; Lin, P.; Liu, H. Microstructure and mechanical properties of functionally graded TiCp/Ti6Al4V composite fabricated by laser melting deposition. *Ceram. Int.* **2017**, *43*, 16638–16651. [CrossRef]
25. Loginova, I.S.; Khalil, A.M.; Churyumov, A.; Solonin, A.N.; Popov, N.A. Effect of Direct Laser Deposition on Microstructure and Mechanical Properties of 316L Stainless Steel. *KnE Eng.* **2019**, *1*, 82. [CrossRef]
26. Bambach, M.; Sizova, I.; Silze, F.; Schnick, M. Comparison of laser metal deposition of Inconel 718 from powder, hot and cold wire. *Procedia CIRP* **2018**, *74*, 206–209. [CrossRef]
27. Katayama, S. *Handbook of Laser Welding Technologies*; Woodhead Publsihing: Cambridge UK; Philadelphia, PA, USA, 2013.
28. Kelbassa, I. Qualifizieren des Laserstrahl-Auftragsschweißens von BLISKs aus Nickel- und Titanbasislegierungen. Ph.D. Thesis, RWTH Aachen University, Aachen, Germany, 2006.
29. Witzel, J.M.F. Qualifizierung des Laserstrahl-Auftragschweißens zur Generativen Fertigung von Luftfahrtkomponenten. Ph.D. Thesis, RWTH Aachen University, Aachen, Germany, 2014.
30. Petrat, T.; Graf, B.; Gumenyuk, A.; Rethmeier, M. Laser Metal Deposition as Repair Technology for a Gas Turbine Burner Made of Inconel 718. *Phys. Procedia* **2016**, *83*, 761–768. [CrossRef]
31. Rauter, R. Laserauftragschweißen von Wolframkarbidschichten in Nickelbasismatrizen zur Herstellung Verschleißfester Beschichtungen von Warmformwerkzeugen. Master's Thesis, Technischen Universität Graz, Graz, Austria, 2016.
32. Amado, J.M.; Tobar, M.J.; Yáñez, A.; Amigó, V.; Candel, J.J. Crack Free Tungsten Carbide Reinforced Ni(Cr) Layers obtained by Laser Cladding. *Phys. Procedia* **2011**, *12*, 338–344. [CrossRef]
33. Luo, J.; Liu, X.-B.; Xiang, Z.-F.; Shi, S.-H.; Chen, Y.; Shi, G.-L.; Wu, S.-H.; Wu, Y.-N. Synthesis of High-Temperature Self-lubricating Wear Resistant Composite Coating on Ti6Al4V Alloy by Laser Deposition. *J. Mater. Eng. Perform.* **2015**, *24*, 1881–1889. [CrossRef]
34. Bold, M.-N.; Linnenbrink, S.; Pirch, N.; Gasser, A.; Mund, J.; Schleifenbaum, J.H. Powder based laser material deposition on edges. *J. Laser Appl.* **2020**, *32*, 32001. [CrossRef]
35. Liu, F.; Lyu, F.; Liu, F.; Lin, X.; Huang, C. Laves phase control of inconel 718 superalloy fabricated by laser direct energy deposition via δ aging and solution treatment. *J. Mater. Res. Technol.* **2020**, *9*, 9753–9765. [CrossRef]
36. Ghiaasiaan, R.; Poudel, A.; Ahmad, N.; Gradl, P.R.; Shao, S.; Shamsaei, N. High Temperature Tensile and Fatigue Behaviors of Additively Manufactured IN625 and IN718. *Procedia Struct. Integr.* **2022**, *38*, 581–587. [CrossRef]
37. Artur, V.; Konstantin, B.; Anton, K.; Andrey, A.; Marina, G. Development of laser metal deposition process for a large IN625 part using small trial samples. *Procedia CIRP* **2020**, *94*, 310–313. [CrossRef]
38. Dutkiewicz, J.; Rogal, Ł.; Kalita, D.; Berent, K.; Antoszewski, B.; Danielewski, H.; Węglowski, M.S.; Łazińska, M.; Durejko, T.; Czujko, T. Microstructure and Properties of Inconel 625 Fabricated Using Two Types of Laser Metal Deposition Methods. *Materials* **2020**, *13*, 5050. [CrossRef]
39. Heilemann, M.; Prakash, V.J.; Beulting, L.; Emmelmann, C. Effect of heat accumulation on the single track formation during laser metal deposition and development of a framework for analyzing new process strategies. *J. Laser Appl.* **2021**, *33*, 12003. [CrossRef]
40. Rösler, J.; Hentrich, T.; Gehrmann, B. On the Development Concept for a New 718-Type Superalloy with Improved Temperature Capability. *Metals* **2019**, *9*, 1130. [CrossRef]

41. Solís, C.; Munke, J.; Bergner, M.; Kriele, A.; Mühlbauer, M.J.; Cheptiakov, D.V.; Gehrmann, B.; Rösler, J.; Gilles, R. In Situ Characterization at Elevated Temperatures of a New Ni-Based Superalloy VDM-780 Premium. *Met. Mater. Trans. A* **2018**, *49*, 4373–4381. [CrossRef]
42. Bax, B.; Rajput, R.; Kellet, R.; Reisacher, M. Systematic evaluation of process parameter maps for laser cladding and directed energy deposition. *Addit. Manuf.* **2018**, *21*, 487–494. [CrossRef]
43. Alvarez, P.; Montealegre, M.; Pulido-Jiménez, J.; Arrizubieta, J. Analysis of the Process Parameter Influence in Laser Cladding of 316L Stainless Steel. *J. Manuf. Mater. Process.* **2018**, *2*, 55. [CrossRef]
44. Song, B.; Hussain, T.; Voisey, K.T. Laser Cladding of Ni50Cr: A Parametric and Dilution Study. *Phys. Procedia* **2016**, *83*, 706–715. [CrossRef]
45. Xi, W.; Song, B.; Zhao, Y.; Yu, T.; Wang, J. Geometry and dilution rate analysis and prediction of laser cladding. *Int. J. Adv. Manuf. Technol.* **2019**, *103*, 4695–4702. [CrossRef]
46. Möller, M.L.B. *Prozessmanagement für das Laser-Pulver-Auftragschweißen*; Springer: Berlin/Heidelberg, Germany, 2021.
47. Alizadeh-Sh, M.; Marashi, S.; Ranjbarnodeh, E.; Shoja-Razavi, R.; Oliveira, J.P. Prediction of solidification cracking by an empirical-statistical analysis for laser cladding of Inconel 718 powder on a non-weldable substrate. *Opt. Laser Technol.* **2020**, *128*, 106244. [CrossRef]
48. Korsmik, R.; Klimova-Korsmik, O.; Valdaytseva, E.; Udin, I. Investigation of cracking causes during multi-pass laser cladding of heat-resistant single crystal nickel alloy. *Procedia CIRP* **2020**, *94*, 314–319. [CrossRef]
49. Tillack, D.J. Nickel alloys and stainless steels for elevated temperature service: Weldability considerations. In *Proceedings from Materials Solutions*; Nickel Development Institute: Toronto, ON, Canada, 1997.
50. Hussein, N.; Segal, J.; McCartney, D.G.; Pashby, I.R. Microstructure formation in Waspaloy multilayer builds following direct metal deposition with laser and wire. *Mater. Sci. Eng. A* **2008**, *497*, 260–269. [CrossRef]
51. Acharya, R.; Das, S. Additive Manufacturing of IN100 Superalloy Through Scanning Laser Epitaxy for Turbine Engine Hot-Section Component Repair: Process Development, Modeling, Microstructural Characterization, and Process Control. *Met. Mater. Trans. A* **2015**, *46*, 3864–3875. [CrossRef]
52. Jambor, T. Funktionalisierung von Bauteiloberflächen Durch Mikro-Laserauftragschweißen. Ph.D. Thesis, RWTH Aachen University, Aachen, Germany, 2012.
53. Hügel, H.; Graf, T. *Laser in der Fertigung: Strahlquellen, Systeme, Fertigungsverfahren*, 2nd ed.; neu bearb. Aufl; Vieweg + Teubner: Wiesbaden, Germany, 2009.
54. Guepner, M.; Rietzke, S.; Paetzold, R.; Bliedtner, J.; Bergman, J.P. Process characterization of powder based laser metal deposition on thin substrates. *J. Laser Appl.* **2018**, *30*, 32506. [CrossRef]
55. Liu, X.; Xiao, H.; Xiao, W.; Song, L. Microstructure and Crystallographic Texture of Laser Additive Manufactured Nickel-Based Superalloys with Different Scanning Strategies. *Crystals* **2021**, *11*, 591. [CrossRef]
56. Maier, H.J.; Niendorf, T.; Bürgel, R. *Handbuch Hochtemperatur-Werkstofftechnik: Grundlagen, Werkstoffbeanspruchungen, Hochtemperaturlegierungen und -Beschichtungen*, 6th ed.; überarbarbeitete und erweiterte Auflage; Springer Vieweg: Wiesbaden, Germany; Heidelberg, Germany, 2019.
57. Moussaoui, K.; Rubio, W.; Mousseigne, M.; Sultan, T.; Rezai, F. Effects of Selective Laser Melting additive manufacturing parameters of Inconel 718 on porosity, microstructure and mechanical properties. *Mater. Sci. Eng. A* **2018**, *735*, 182–190. [CrossRef]

machines

Article

Influence of Laser Surface Texture on the Anti-Friction Properties of 304 Stainless Steel

Xiashuang Li *, Guifeng Li, Yuesui Lei, Lei Gao, Lin Zhang and Kangkang Yang

College of Mechanical and Electrical Engineering, Shaanxi University of Science and Technology, Xi'an 710021, China
* Correspondence: lixiashuang@sust.edu.cn

Abstract: To enhance the anti-friction properties of 304 stainless steel, friction experiments were conducted on it after laser surface texturing. The influences of laser scanning speed, repetition frequency, processing times, laser beam line spacing, and lattice spacing on the friction properties of 304 stainless steel were investigated by contrast tests under annular filling mode. The results revealed that laser texturing improved the anti-friction properties of 304 stainless steel. The friction coefficient of the sample surface decreased first and then increased with the increase in scanning speed, repetition frequency, processing times, laser beam line spacing, and lattice spacing. Based on this, process optimization found that a stainless steel surface with good anti-friction properties could be obtained when the laser power was 0.3 W, the repetition frequency was 50 kHz, the scanning speed was 80 mm/s, the laser beam line spacing was 1 μm, the lattice spacing was 200 μm, and the number of processing times was two. Finally, scanning electron microscope (SEM) characterization of wear morphology on the sample surface showed that the laser textured surface could collect debris during effective friction, which reduced the occurrence of abrasive and adhesive wear. Meanwhile, the actual contact area of the friction pair was effectively reduced, thereby reducing friction force and wear. This study provided experimental data and a theoretical basis for improving the friction properties of the 304 stainless steel surface and laid the foundation for its reliable use under friction and wear conditions.

Keywords: 304 stainless steel; laser surface texturing; friction properties

1. Introduction

With its excellent corrosion resistance, 304 austenitic stainless steel has been widely employed in many areas of modern industrial production [1,2]. However, its poor wear resistance limits its applications in many engineering fields [3–5]. Surface deformation hardening [6] uses mechanical means, such as shot blasting [7], extrusion [8], roll pressing [9], and laser shock hardening [10,11], to produce strong plastic deformations on metal surfaces in each direction. Therefore, the grains on the surface of the material are refined to form a hardening layer of deformation, increasing the hardness of the surface layer. Furthermore, surface deformation strengthening causes grain breakage and high-density dislocation of metal surfaces. After sufficient time, the plastic flow of metal materials occurs on the workpiece surface. It produces plastic deformation of the overlapping holes, which causes compressive stress and stretches the surface structure during pit generation. Parts inside the workpiece that are not hammered obstruct this process, causing residual compressive stress to accumulate on and close to the workpiece surface, greatly enhancing the mechanical characteristics of the material. However, surface deformation strengthening may induce martensite phase transformation, the varying degrees of which affect the corrosion resistance of austenitic stainless steel differently. The application of surface strengthening in austenitic stainless steel is constrained by this, which is a significant factor. Nitriding as a surface strengthening treatment can enhance the mechanical characteristics of austenitic stainless steel [12,13]. Traditional gas nitriding requires a high temperature and a long time. Although it improves the wear resistance of surface and the hardness of the material, chromium nitrides are released from thermal decomposition, resulting in a decrease in the corrosion resistance of the material [14–17]. In recent years, various surface texturing methods have emerged, which can be roughly divided into mechanical processing, energy beam etching, and chemical etching. Mechanical processing methods mainly include scanning probe processing. Energy beam etching mostly uses laser and electrochemical etching to process textures. The main chemical etching methods include anodic oxidation and sol-gel [18]. Costa et al. [19] found that circular pit texture with a low area ratio under oil-rich conditions could not improve surface friction performance. However, for strip grooves with a high texture density on the surface (approximately 25%), the friction coefficient could be significantly reduced when the direction of the groove was perpendicular to the direction of motion. However, when the groove direction was parallel to the direction of motion, the friction coefficient increased, and the material exhibited poor wear resistance. Evidently, the texture distribution had an impact on tribological properties. Hua et al. [20] used laser processing technology to fabricate high-precision micro groove arrays on the surfaces of the stator and rotor of the ultrasonic motor. They found that textured surfaces can significantly improve the friction coefficient and lower the wear rate of contact surface, thus increasing the dry friction driving force between the motor stator and rotor. Moreover, the friction coefficient of the double-sided texture is higher than that of the single-sided texture. Xue et al. [21] prepared dimple textures with different area densities on Ti-20Zr-10Nb-4Ta alloy and investigated their tribological properties. It was found that when the texture diameter was fixed with the surface density of the texture from 10% to 40%, the COF and wear rate of the sample surface decreased significantly, and the wear mechanism on the sample surface was abrasive wear and fatigue wear. Siripuram et al. [22] studied the effect of the texture cross-section shape on hydrodynamic lubrication, finding that the friction coefficient did not change much when the depths were constant for microconvex and micropits with various cross-section shapes, but the effect of the cross-section shape on the cross-section area was significant, and both the shape of the cross-section and the size of the texture had significant effects on the leakage rate. Through theoretical investigation, Singh et al. [23] predicted the impact of spherical texture on the performance traits of a two-lobe journal bearing system. They discovered that spherical textures with different dimple aspect ratios can lower the liquid film friction coefficient and enhance bearing performance. Chen et al. [24] fabricated micropore and micrograting arrays on 316 L stainless steel surfaces with a nanosecond laser and tested their friction properties. The results revealed that under the same load and sliding speed, the wear marks on the smooth surfaces were deep and wide, while those on the textured surfaces were shallow and narrow. In addition, the variations in the friction resistance of the textured surface were smaller than those of the smooth surface. The denser the micropore and micrograting arrays, the smaller the friction resistance. Gajrani et al. [25] fabricated mechanical microtextures over the rake face of PN-HSS cutting tools through mechanical microtexturing method. Under the condition of little impact on tool hardness, the performance of the tool is improved by reducing the actual contact area between the tool and chip, and the proper lubrication of contact area through mechanical microstructures. Li et al. [26] used laser microprocessing to form groove textures on the smooth surface of 316 L stainless steel. The results showed that the contact load had little influence on the friction coefficient of the textured surface, while texturing significantly affected friction reduction and anti-wear performance under high loads. Panigrahi et al. [27] investigated the effect of positive deterministic textures with different cross-sectional shapes, sizes, and heights on the tribological properties of parallel sliding lubricated contacts, and found that weaves with fewer texture scales and larger texture heights provided better tribological and lubrication properties to the specimens. When Bharatish et al. [28] used a femtosecond laser to study the effects of scanning speed on the femtosecond laser texture characteristics of a tin bronze alloy containing 8% and 12% tin, they discovered that the wear rate of the textures produced at lower scanning speeds was lower at higher loads and higher frequencies. Researchers have conducted the studies on the laser preparation technology of micro-circular texture. Deng [29] placed four types of micro-circular textures on the upper surface of rails and performed lubricated reciprocal friction tests. The findings demonstrated that in comparison to the unwoven rail, all four varieties of micro-circular textures enhanced contact stress on the contact surface and decreased friction force during the stable friction phase. Circular cratered micro-circular textures demonstrated the most notable reduction in friction coefficient when compared to linear and lattice-line micro-circular textures. Borjalia et al. [30] found that micro-circular texture reduces material wear by laser ablation of polyethylene materials with different shapes and densities of micro-circular texture. Suitable densities of micro-circular texture structures were obtained by orthogonal experiments. Higher micro-circular texture densities were shown to be ineffective at reducing friction. By adding micro-circular textures to the surface of pure titanium drilling tools, Li [31] investigated the tribological characteristics of these materials when exposed to soil with various particle sizes. The findings demonstrate that when the abrasive's grain size is less than the micro-circular texture's diameter, the abrasive will penetrate the texture and lower the friction coefficient by sliding and rolling. The abrasive will create a sliding plow on the surface of the micro-circular texture when the particle size is greater than the diameter of the texture. While the two abrasives function differently, they can lessen the wear on drilling bits made of pure titanium.

A thorough literature survey reveals that in the above-mentioned research on the influence of surface texture on the properties of metal surfaces, some scholars use mechanical micromachining and photochemical processing to prepare the metal surface texture. When mechanical micromachining is used to prepare metal surface texture, the metal to be processed will suffer relatively large stress, which requires a high degree of hardness of the metal itself; and when using photochemical processing technology to build metal surface texture, the process of preparing the surface texture is more complicated, and it is difficult to control the etching time of the metal when using this method. Other research, meanwhile, using laser surface modification technology to construct texture on the metal surface, mainly focus on the study of the influence of texture shape, area density, and size on various characteristics of the metal material surface. In such studies, the laser processing parameters used are fixed and the types of metals studied vary with the properties of the metals affected by the surface weaving. There is little research on the tribological properties of 304 stainless steel by using laser surface texture technology to construct texture on the surface of 304 stainless steel, and there were still some deficiencies in the studies on laser parameters affecting the friction properties of material surface by affecting texture. The friction mechanism of 304 stainless steel surface after laser surface texture needs further research.

This article presents a thorough review of literature and focuses on the study of 304 stainless steel based on previous research. A nanosecond laser was used in an annular filling mode to investigate the impact of various laser parameters (such as scanning speed, repetition rate, processing times, and laser beam line spacing) on the surface friction performance of 304 stainless steel. The study also used a single-factor rotation method to evaluate the effect of surface texture created by different texture lattice spacing on surface friction performance. The surface morphology and depth of the texture were analyzed using an ultra-deep field microscope. Additionally, friction experiments were conducted on the surface of the 304 stainless steel before and after texturing using a friction and wear tester to examine the differences in surface friction performance. The surface of the textured 304 stainless steel was observed using a scanning electron microscope after friction experiments, and the friction mechanism of 304 stainless steel surface after laser surface texturing was further studied. The goal of this study is to enhance the anti-friction performance of 304 stainless steel surface and to provide a comprehensive and systematic laser processing solution.

2. Experiment

2.1. Testing Apparatus and Materials

The maximum laser power of the fiber laser marker used in the experiment was 20 W. Austenitic 304 stainless steel sheets were selected and cut by laser into round 40 mm samples with a thickness of 6 mm. The chemical composition is shown in Table 1.

Table 1. Chemical composition of 304 stainless steel austenitic [32].

Element	C	Si	Mn	Cr	Ni	S	P	Fe
Proportion (%)	0.07	1.00	2.00	17.15	8.02	0.03	0.035	71.965

2.2. Experiment Design

Using a nanosecond fiber laser marking machine (HGTECH-LG20, Wuhan Huagong Laser) with an emission wavelength of 1064 nm, a maximum average power of 20 W, a pulse duration of 200 ns, and a repetition frequency that can be continuously adjusted within the range of 30 kHz to 60 kHz. The laser beam was focused onto the substrate by means of a 110 mm focusing lens, resulting in a laser spot diameter of 50 μm after focusing. The machine was equipped with a galvanometer scanner, which was responsible for completing the texturing process.

The surface texture of the sample is processed by laser using the scanning strategy shown in Figure 1. Figure 1a shows the theoretical processing method, where overlapping of laser spots occurs when the distance between laser beam lines is smaller than the diameter of the laser spot, otherwise they will not. Under high magnification, some of the textures produced by the processing appear as polygons. This is because the galvanometer scanning system control software employs an annular filling method, as shown in Figure 1b. As a result, the laser path is automatically fitted from polygons to circles when processing the texture.

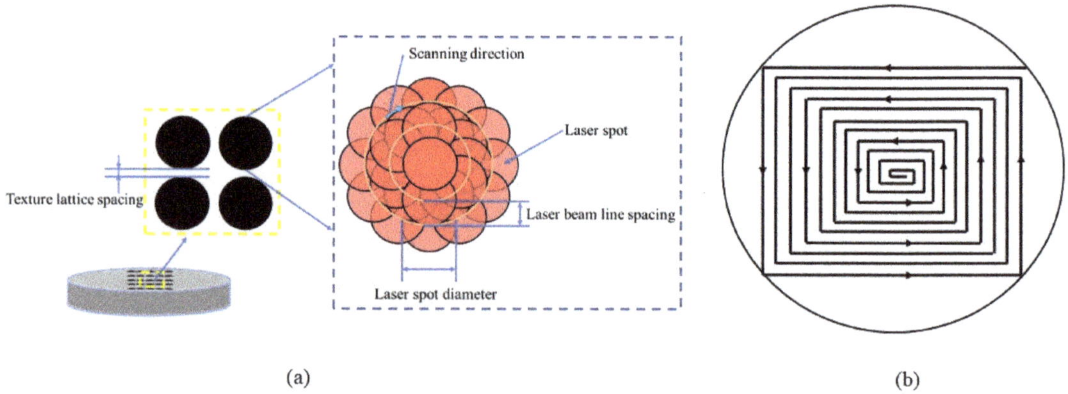

Figure 1. Scanning strategy for surface texture produced by laser: (**a**) theoretically, (**b**) actually.

During laser processing, a significant amount of slag remains on both the surface and inside the texture of the sample, affecting its roughness. These slags can have a profound impact on the depth of the texture's subsequent characterization. Therefore, using 2000# sandpaper to lightly polish the sample can remove the slag remaining on the surface of the sample without damaging the texture of the laser processed surface. Additionally, prior to conducting friction tests, it is essential to use ultrasonic cleaning to eliminate any slag or impurities inside the texture. This process not only enhances the texture's ability to collect debris but also prevents any polishing dust from entering the texture and affecting subsequent experiments. Moreover, ultrasonic cleaning makes the depth and morphology of the processed texture clearer.

To investigate the effect of surface texture on the friction properties of 304 stainless steel, the dry friction test was conducted on textured samples. The tests were performed using an electrochemical corrosion friction tester (MSR-2T) and a GGr15 steel ball with a diameter of 6 mm as the friction sub-mate material, as this type of steel is commonly used in practical applications with 304 stainless steel. The samples were fixed on a special fixture on the chassis, while the GGr15 steel ball was fixed on the upper end of the machine fixture. A loading force perpendicular to the sample surface was applied, and the friction test was carried out using a reciprocating linear motion with a 5 mm stroke and a 5 N applied load for 30 min. The surface morphology and internal structure of the substrates were examined using an ultra-deep field microscope (DSX-510) and a scanning electron microscope (SU3500), while the hardness of the samples before and after laser processing was measured using a digital micro-Vickers hardness tester (402SXV). For more information on the test process, please see Table 2.

Table 2. Test scheme for laser surface texturing.

No.	Laser Power/W	Scanning Speed/mm/s	Repetition Frequency/kHz	Processing Times	Laser Beam Line Spacing/μm	Lattice Spacing/μm
1	0.3	60	60	1	1	200
2	0.3	80	60	1	1	200
3	0.3	100	60	1	1	200
4	0.3	120	60	1	1	200
5	0.3	140	60	1	1	200
6	0.3	100	35	1	1	200
7	0.3	100	40	1	1	200
8	0.3	100	45	1	1	200
9	0.3	100	50	1	1	200
10	0.3	100	55	1	1	200
11	0.3	100	60	2	1	200
12	0.3	100	60	3	1	200
13	0.3	100	60	1	5	200
14	0.3	100	60	1	10	200
15	0.3	100	60	1	50	200
16	0.3	100	60	1	100	200
17	0.3	100	60	1	1	100
18	0.3	100	60	1	1	300

3. Results and Discussions

3.1. Effect of Scanning Speed on Surface Texturing and Anti-Friction Properties of 304 Stainless Steel

According to the experiment scheme, groups 1–5 used different scanning speeds to texture the samples. The texture morphologies observed under a super depth-of-field microscope are shown in Figure 2a–e. According to the analysis, as the scanning speed increased, the duration of energy acting at a certain point decreased, resulting in less material removal at that point and a lower average texture depth. This was consistent with the trend of the average texture depths measured experimentally (Figure 2f), showing that as the scanning speed increased, the average texture depth decreased. The trend of the friction coefficient of the material surface over time at various scanning speeds is shown in Figure 2g. The friction coefficient was very high at the beginning of the friction process. This is because at the beginning of grinding, there is still a small amount of residual molted material on the surface, resulting in excessive surface roughness. After 20 min, the friction coefficient gradually decreased and stabilized in the stable wear stage. This is because the texture acts as a trap for impurities such as abrasive dust, reducing the surface roughness of the substrate material while reducing abrasive wear on the surface of the substrate material. As displayed in Figure 2h, when the scanning speed reached 80 mm/s, the friction coefficient was the lowest at 0.5692. However, when the scanning speed increased further to 140 mm/s, the friction coefficient increased. A preliminary analysis indicated that the reason for this phenomenon was that when the scanning speed was too high, the time for the laser to act on one specific point on the material surface became shorter and there was poor continuity of the laser spot. Energy absorption was insufficient for effective removal of surface material. Instead, an oxidized layer was produced that could be peeled off more easily and has a higher roughness than the matrix material on the surface, resulting in a further increase in the friction coefficient. Therefore, within the appropriate scanning speed range, the friction coefficient of the laser-textured surface was lower than those of the nontextured surface, which proved that laser surface texturing could improve the anti-friction properties of materials. According to the analysis, deep textures on the material surface could mitigate abrasive and adhesive wear by collecting debris generated during the friction process, thus improving the anti-friction properties of the material surface. In other words, when the scanning speed was 80 mm/s, the prepared surface texture resulted in the lowest friction coefficient and the best anti-friction properties.

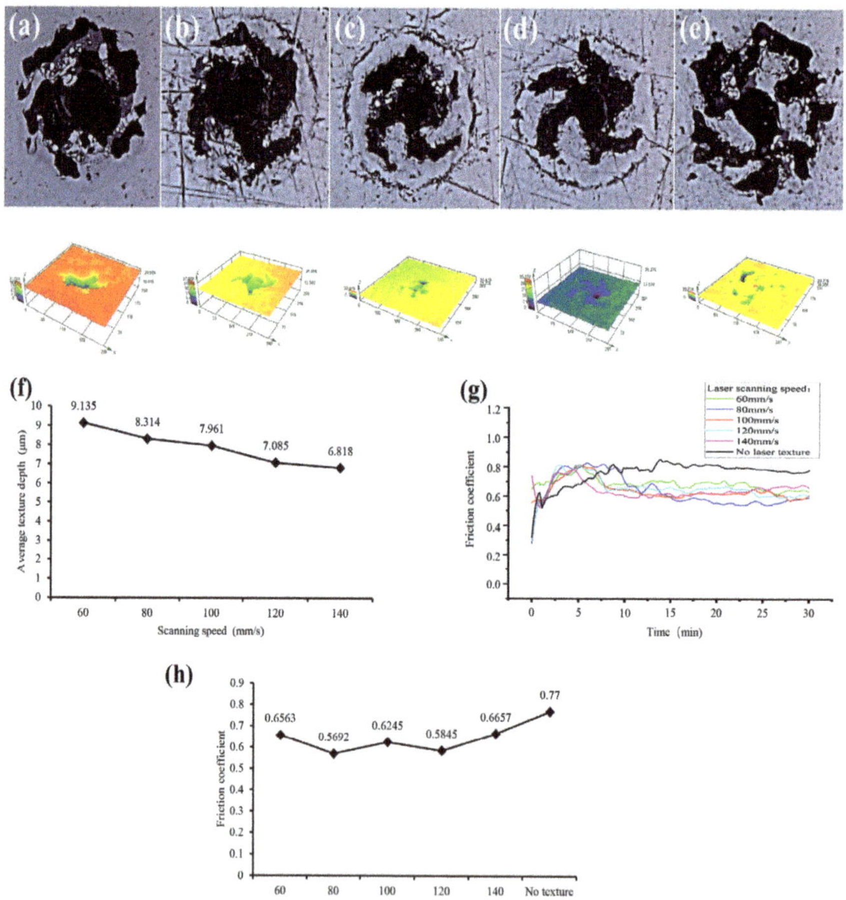

Figure 2. Texture morphology of sample surface at different scanning speeds. (**a**) 60 mm/s, (**b**) 80 mm/s, (**c**) 100 mm/s, (**d**) 120 mm/s, (**e**) 140 mm/s, (**f**) average depth of textures processed at different scanning speeds, (**g**) variation of friction coefficient with time at different scanning speeds, and (**h**) the friction coefficient of stable wear at different scanning speeds.

3.2. Effect of Repetition Frequency on Surface Texturing and Anti-Friction Properties of 304 Stainless Steel

Based on the first groups of investigations, groups 6–10 used a different repetition frequency to texture the samples. By observing the texture morphologies under the super depth of the field microscope as displayed in Figure 3a–f, it was found that with the increase in repetition frequency, the energy density acting at a certain point decreased, resulting in less material removal and a smaller texture depth. This was consistent with the average texture depths experimentally measured (Figure 3g). The overall average texture depth decreased with increasing repetition frequency. The trend of the friction coefficient for the material surface over time at various repetition frequencies is shown in Figure 3h, showing a very high friction coefficient at the beginning of the wear stage. This is because at the beginning of grinding, there is still a small amount of residual molted material on the surface, resulting in excessive surface roughness. After 20 min, the friction coefficient gradually decreased and stabilized when entering the stable wear stage. This is because the texture acts as a trap for impurities such as abrasive dust, reducing the surface roughness of the substrate material while reducing abrasive wear on the surface of the substrate material. As show in Figure 3i, when the repetition frequency was 50 kHz, the friction coefficient was 0.5273, which was smaller than those at other repetition frequencies. In the stable wear stage, with increasing repetition frequency, the friction coefficient gradually increased, the energy acting on the surface of the material at a single time decreased, and the texture depth decreased. However, the friction coefficient of the surface of the textured material was smaller than that of the nontextured material. The analysis revealed that the anti-friction properties of the material surface improved, which was consistent with the fact that the friction coefficient was smaller. Therefore, when the repetition frequency was 50 kHz, the prepared surface texture resulted in the lowest friction coefficient and the best anti-friction properties.

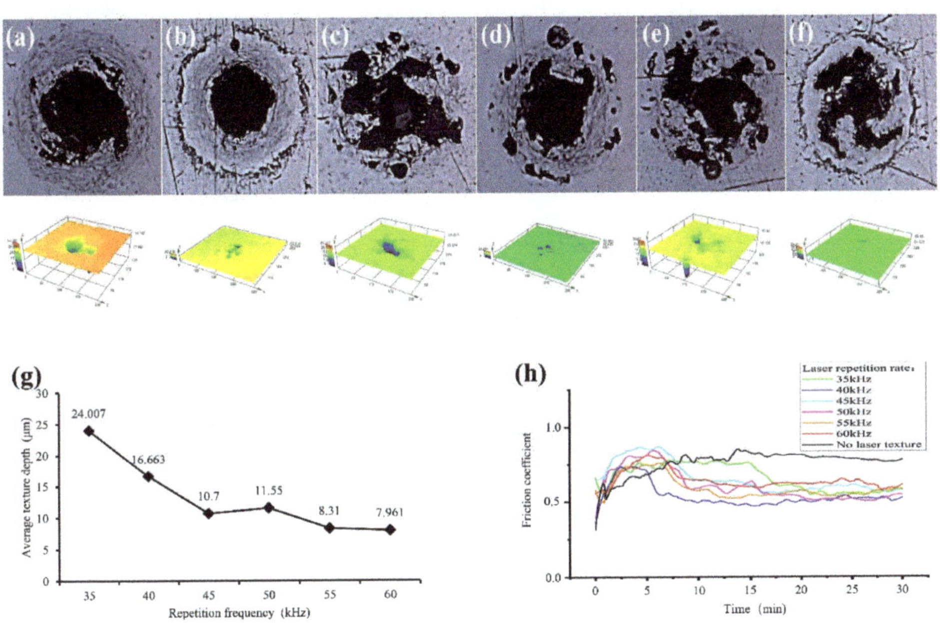

Figure 3. *Cont.*

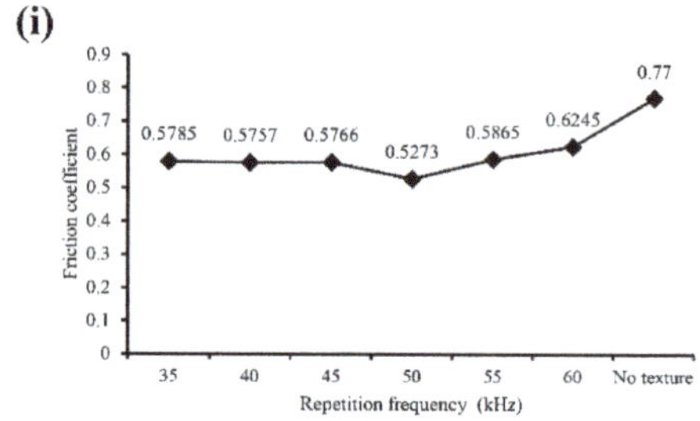

Figure 3. Texture morphology of sample surface at different repetition rates. (**a**) 35 kHz, (**b**) 40 kHz, (**c**) 45 kHz, (**d**) 50 kHz, (**e**) 55 kHz, (**f**) 60 kHz, (**g**) average depth of textures prepared at different repetition rates, (**h**) variation of friction coefficient with time at different repetition rates, (**i**) the friction coefficient of stable wear at different repetition rates.

3.3. Effect of Processing Times on Surface Texturing and Anti-Friction Properties of 304 Stainless Steel

Based on the first groups of investigations, groups 11–12 used different processing times to texture the samples. By observing the texture morphologies under the super depth of field microscope, as displayed in Figure 4a–c, it was found that as the processing times increased, the sum of laser energy acting on a certain point increased, the amount of removed material increased, and the average texture depth increased as well. This was consistent with the average texture depths experimentally measured and displayed in Figure 4d. The average texture depth increased with the number of processing times. The trend of the friction coefficient of the material surface over time at various processing times is shown in Figure 4e, showing that the friction coefficient was very high at the initial wear stage. This is because at the beginning of grinding, there is still a small amount of residual molted material on the surface, resulting in excessive surface roughness. After 15 min, the friction coefficient gradually decreased and stabilized when entering the stable wear stage. This is because the texture acts as a trap for impurities such as abrasive dust, reducing the surface roughness of the substrate material while reducing abrasive wear on the surface of the substrate material. As shown in Figure 4f, the friction coefficient measured during the stable wear stage was the smallest (0.5865) when processed twice. Increasing the amount of processing times could increase the texture depth and decrease the friction coefficient, but too many processing times could increase the thickness of the recast layer, increasing the friction coefficient. Therefore, when the processing times were 2, the texture of the prepared surface minimized the friction coefficient of the material surface and optimized the anti-friction properties of the material.

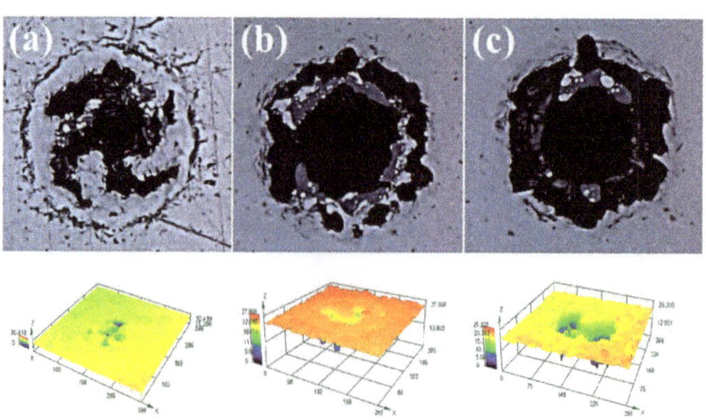

Figure 4. *Cont.*

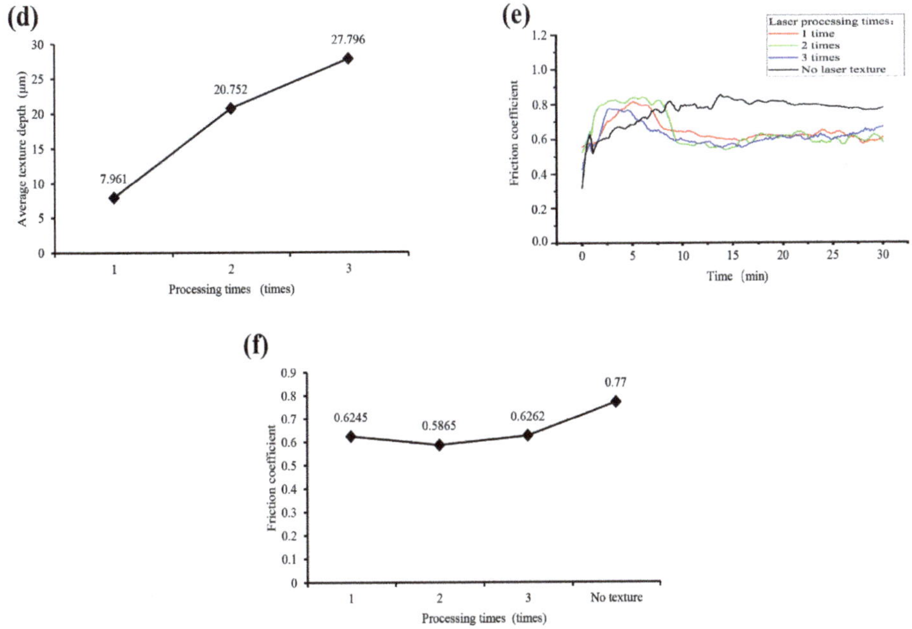

Figure 4. Texture morphology of sample surface under different processing times. (**a**) 1, (**b**) 2, (**c**) 3, (**d**) average depth of texture prepared by different processing times, (**e**) variation of friction coefficient with time under different machining times, (**f**) friction coefficient of stable wear under different machining times.

3.4. Effect of Laser Beam Line Spacing on Surface Texturing and Anti-Friction Properties of 304 Stainless Steel

Based on the first groups of investigations, groups 13–16 adopted different laser beam line spacing to texture the samples. By observing the texture morphologies under the super-depth of field microscope (Figure 5a–e), it was found that the overlapping area of the laser beam spot decreased with the increase in the laser beam line spacing. This resulted in a smaller energy density on the material surface, less material removal, and a smaller average texture depth. This was consistent with what is shown in Figure 5f, where the average depth of the texture decreased with the increasing spacing of the laser beam lines. The trend of the friction coefficient of the material surface over time under various laser beam line spacing conditions is shown in Figure 5g, which shows a very high coefficient of friction at the beginning of the wear stage. This is because at the beginning of grinding, there is still a small amount of residual molted material on the surface, resulting in excessive surface roughness. After 15 min, the friction coefficient gradually decreased and stabilized when entering the stable wear stage. This is because the texture acts as a trap for impurities such as abrasive dust, reducing the surface roughness of the substrate material while reducing abrasive wear on the surface of the substrate material. When the line spacing was much smaller than the laser spot diameter, a large energy density acted at the same point on the material surface, causing a large amount of material removal and deeper textures, which could effectively improve the anti-friction properties of the material by collecting debris. The friction coefficients measured during the stable wear stage (Figure 5h) showed that when the laser beam line spacing was 1 μm, the friction coefficient was the smallest at 0.6245. The friction coefficients of laser-textured material surfaces were smaller than those of the nontextured ones. However, when the line spacing was 5 μm, the friction coefficient of the material surface was the highest. Different line spacings resulted in varied texture morphologies. In Figure 5d,e, a line spacing larger than the minimum laser spot diameter was used, so the laser spots did not overlap on the processing path and materials were only removed on the laser path. The friction coefficient was taken as the main parameter to evaluate the friction properties of samples. When the laser beam line spacing was 1 μm, the texture of the prepared surface minimized the friction coefficient of the material surface.

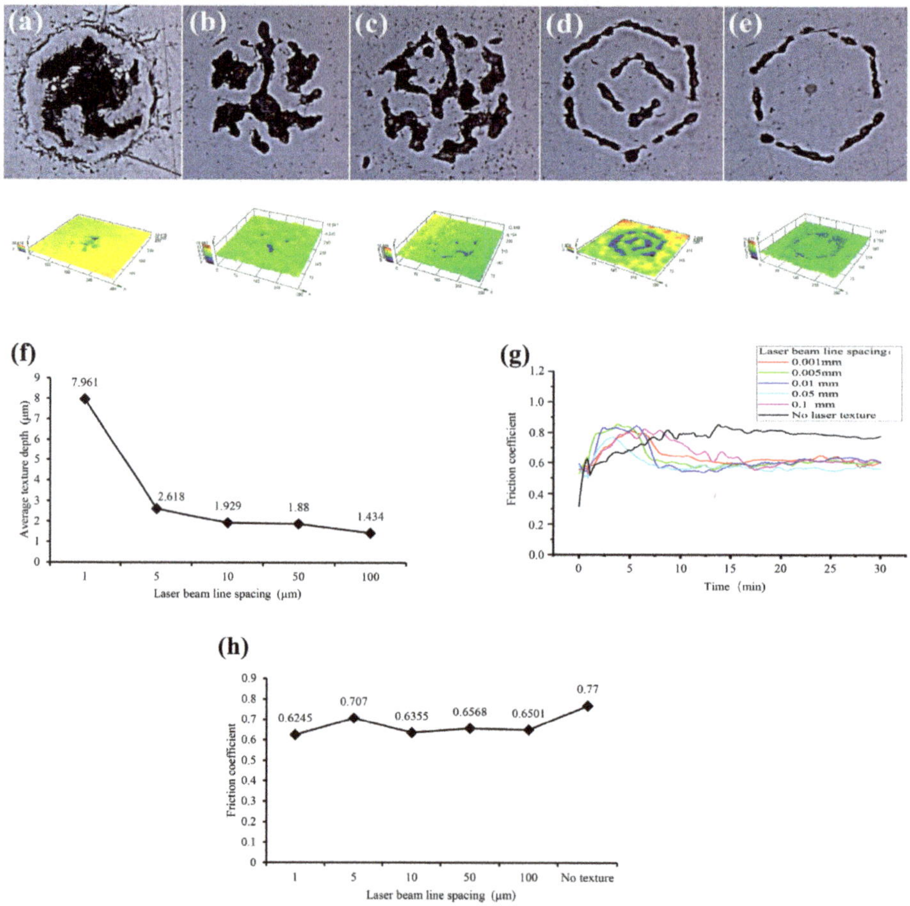

Figure 5. Texture morphology of sample surface under different line spacing. (**a**) 1 μm, (**b**) 5 μm, (**c**) 10 μm, (**d**) 50 μm, (**e**) 100 μm, (**f**) average depth of textures prepared with different line spacing, (**g**) change of friction coefficient with time under different laser beam line spacing conditions, (**h**) friction coefficient of stable wear at different line spacing.

3.5. Effect of Texture Lattice Spacing on Surface Texturing and Anti-Friction Properties of 304 Stainless Steel

Based on the first groups of investigations, groups 17–18 used different lattice spacing to texture the samples. The morphologies of the textures observed under the super depth of the field microscope are shown in Figure 6a–c, and the average depths of textures experimentally measured are displayed in Figure 6d. According to the analysis, since the laser parameters did not change, the textures under various lattice spacings (i.e., texture densities) exhibited little difference in texture morphology and average texture depth. The trend of the friction coefficient of the material surface over time under various lattice spacing conditions is shown in Figure 6e, showing a very high friction coefficient in the initial wear stage. This is because at the beginning of grinding, there is still a small amount of residual molted material on the surface, resulting in excessive surface roughness. After 20 min, the friction coefficient gradually decreased and stabilized when entering the stable wear stage. This is because the texture acts as a trap for impurities such as abrasive dust, reducing the surface roughness of the substrate material while reducing abrasive wear on the surface of the substrate material. The stable wear stage that the friction coefficient was the lowest at was 0.6245, as reflected in Figure 6f, when the lattice spacing was 200 μm. The texture density varied with the lattice spacing, but a suitable range existed. When the lattice spacing was moderate, the textures exhibited the best debris collection capability and the greatest improvement in the anti-friction properties of the material. The friction coefficient was taken as the main parameter to evaluate the friction properties of the samples. When the lattice spacing was 200 μm, the prepared surface texture made the friction properties of the material surface optimal.

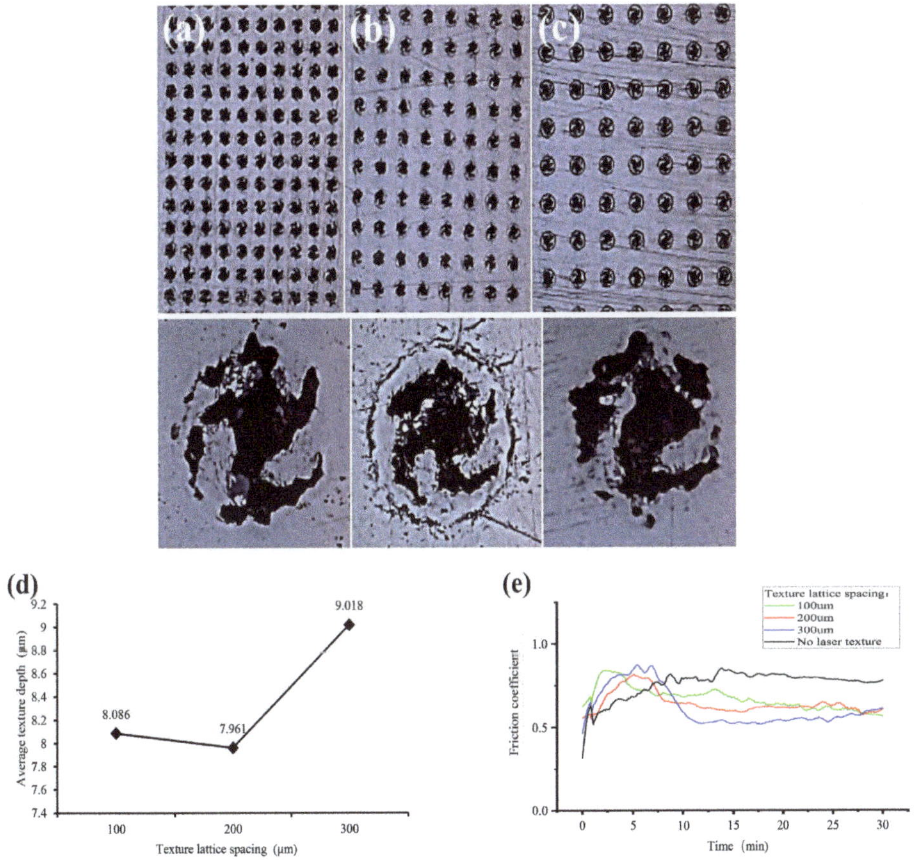

Figure 6. *Cont.*

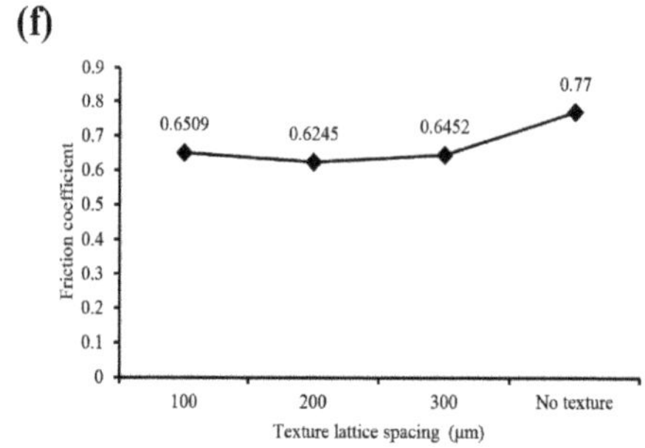

Figure 6. Surface morphology of samples with different textural lattice spacing. (**a**) 100 μm, (**b**) 200 μm, (**c**) 300 μm, (**d**) the average texture depth was prepared with different lattice spacing, (**e**) variation of friction coefficient with time under different lattice spacing, (**f**) friction coefficient of stable wear at different lattice spacing.

3.6. *Effect of Surface Texture on Material Surface Hardness*

In this test, the hardness of the material surface before and after processing the texture was measured and compared using a micro-Vickers hardness tester. The hardness of the material surface before processing the texture was 228.6 HV using the method of multiple measurements at multiple locations and taking the average value, and the hardness of the material surface after processing the texture was about 228.6 HV. Using the same method, the hardness of the different texture surfaces processed under various laser parameters was measured at around 229.70 HV, with a maximum value of 245.73 HV and a minimum value of 213.80 HV, as shown in Figure 7. The hardness of the material surface did not change significantly before and after the processing of the weave, indicating that the weave studied in this test did not cause significant changes in the hardness of the material, which in turn did not affect the frictional wear performance of the sample.

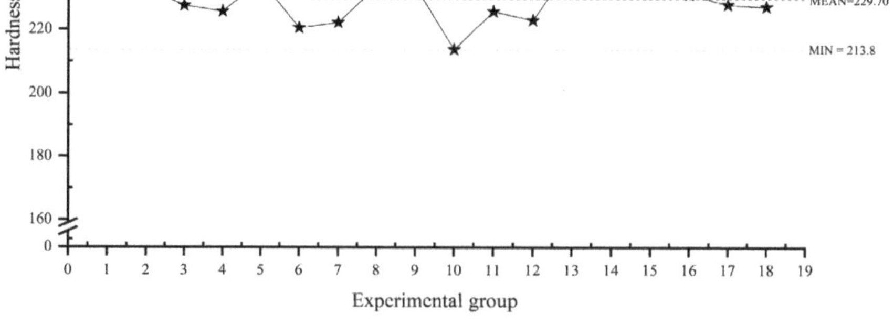

Figure 7. Average surface hardness of each test group after texturing treatment.

4. Wear Mechanism Analysis

The micromorphologies of the abrasion marks under various surface textures are displayed in Figure 8. Figure 8a shows the abrasion marks on the surface before laser texture. Figure 8b is the result after processing once with a laser power of 0.3 W, a repetition frequency of 60 kHz, a scanning speed of 100 mm/s, a laser beam line spacing of 1 μm, and a lattice spacing of 200 μm. The other abrasion marks in Figure 8 were obtained after changing a single variable according to Figure 8b. The abrasion morphologies on the textured surfaces processed under various parameters were similar to those without texturing, indicating that their wear mechanisms were the same. There were varying numbers of furrows, pits, and scaly fragments in each abrasion morphology in Figure 8. The debris generated during the wear process would become abrasive grains and produce clear grooves on the friction surface, called furrows, which was typical for abrasive wear. During the wear process, the friction pair moved relative to each other and their surfaces were subjected to cyclic loads, which were prone to deformations and cracks. As the cracks expanded, the materials would be peeled off to form pits, which was a typical characteristic of fatigue wear. As the friction progressed, the temperature of the friction surface increased and the wear debris adhered to the surface of the material, forming scaly fragments, which then caused the adhesive to wear [33].

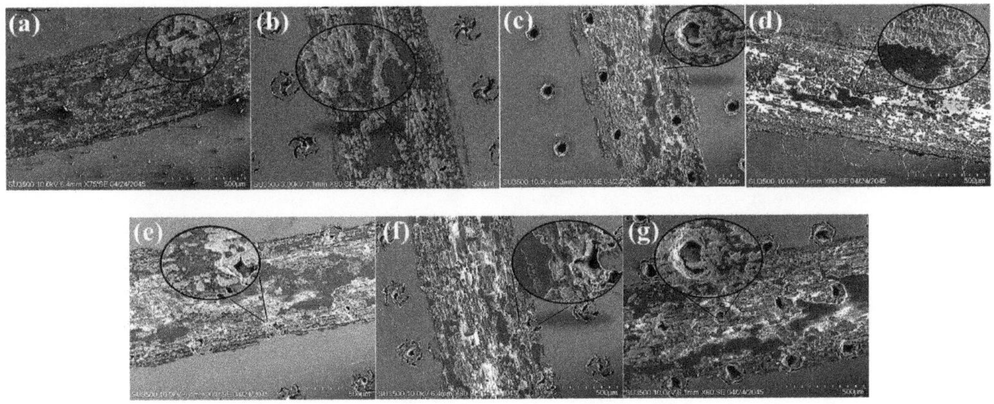

Figure 8. Micromorphologies of abrasion marks under surface texture. (**a**) Nontextured state, (**b**) Laser power 0.3 W, repetition frequency 60 kHz, scanning speed 100 mm/s, processed once, laser beam line spacing 1 μm, lattice spacing 200 μm, (**c**) repetition frequency 40 kHz, (**d**) laser beam line spacing 100 μm, (**e**) lattice spacing 300 μm, (**f**) scanning speed 120 mm/s, (**g**) processed twice.

Observation of each micromorphology revealed that the furrows on both sides of the abrasion mark were more obvious, whereas pits and scaly fragments were more frequent in the middle part. Analysis showed that abrasive wear occurred first as the main wear mode during the wear process and debris was generated during the movement of the friction pair. On the one hand, furrows were generated on the material surface; on the other hand, due to the increasing temperature, the debris adhered to the material surface, turning the type of wear to adhesive wear. During the whole process, cracks occurred on the material surface due to cyclic loading, which led to the material peeling and belonged to the fatigue wear stage. The peeled materials were then involved in abrasive and adhesive wear, such as abrasive grains and debris.

As shown in Figure 8a, adhesive wear and fatigue wear were the main types of wear for the nontextured surface. In Figure 8b,d,e, the overall depths of the textures were small, the texture surfaces were not very large, and there were more scaly fragments than pits, indicating that the adhesion wear was dominating. Figure 8c,f,g showed that the pits were in the majority, indicative of fatigue wear. Figure 8c,e show that the textures collected some debris. In summary, the added surface textures could collect debris during friction, thereby reducing the occurrence of abrasive and adhesive wear and thus the degree of wear.

Bowden proposed adhesive friction theory [34], revealing that when adhesive wear occurs, the relative sliding of the bond between the two friction surfaces will destroy the bonding point and produce a shear force, called the friction force. In this experiment, hard bearing steel and soft 304 stainless steel were the two surfaces subjected to friction, for which there was a hardness difference. Therefore, furrows formed on the surface of stainless steel with a smaller hardness. The force that produced the furrows was also part of the friction force.

Thus, in this experiment, the surface textures reduced friction and wear by lowering the actual contact area between the two friction surfaces.

5. Conclusions

In this work, a nanosecond laser marking machine was used to process the anti-friction textured surface of 304 stainless steel. The research results are summarized as follows:

(1) The selected texture filling method for this experiment is annular filling, which involves higher energy density more at the corners than in the straight-line regions. This leads to more melting and vaporization of material in the corner area, causing material to splash onto the surrounding region and accumulate. After undergoing slight polishing and ultrasonic cleaning, the laser processed texture displays deeper texture depth at the corners with annular filling. This results in a petal-shaped depression instead of the circular depression that was designed in theory.

(2) Hardness measurements have shown that there was no significant difference in the material's hardness on the surface of 304 stainless steel before and after laser processing was used to create a texture on it. This indicates that the surface texture does not cause any changes to the surface hardness of the material.

(3) Optimizing laser parameters can enhance the anti-friction performance of the 304 stainless steel surface by modifying the texture morphology. The experimental results indicate that decreasing the scanning speed, repetition frequency, and laser beam line spacing while maintaining an appropriate lattice spacing leads to a gradual increase in texture depth and better debris collection ability. The size and morphology of the structure are influenced by laser parameters. Thus, it can be concluded that using laser power of 0.3 W, repetition frequency of 50 kHz, scanning speed of 80 mm/s, laser beam line spacing of 1 μm, lattice spacing of 200 μm, and processing times of two yields per texture with distinct contours and efficient debris collection. This, in turn, reduces abrasive and adhesive wear.

(4) Through observation of surface scratches on the sample and analysis of the wear mechanism, it was discovered that in the absence of laser surface texturing, adhesive wear, abrasive wear, and fatigue wear all occur simultaneously. On the other hand, textured surfaces primarily experience fatigue wear. The designed texture can effectively collect the wear debris generated during the wear process, reducing the incidence of abrasive wear and adhesive wear. This, in turn, improves the material's frictional performance on the surface.

Author Contributions: X.L.: Conceptualization, Supervision, Methodology, Writing–original draft. G.L.: Investigation. Y.L.: Investigation. L.G.: Investigation. L.Z.: Investigation. K.Y.: Investigation. All authors have read and agreed to the published version of the manuscript.

Funding: The author(s) disclosed receipt of the following financial support for the research, authorship, and publication of this article: This work was supported by China Postdoctoral Science Foundation (2020M683406), Natural Science Foundation of Shaanxi province (2018JQ5201) and Science and technology Project of Education Department of Shaanxi Province (21JK0538).

Acknowledgments: The authors thank the members of our laboratory for their aid in sample preparation and analysis during the tests. The authors would also like to thank He Nairu and Chen Wei, members of the Center for New Technology and Service Behavior of Mechanical Surfaces in Shaanxi University of Science and Technology for their help in the analysis of friction and wear mechanisms.

Conflicts of Interest: The authors declare that they are not aware of competing economic interest or personal relationships that may affect the work reported in this paper.

References

1. Lo, K.H.; Shek, C.H.; Lai, J.K.L. Recent developments in stainless steels. *Mater. Sci. Eng. R. Rep.* **2009**, *65*, 39–104. [CrossRef]
2. Jiménez–Come, M.J.; Turias, I.J.; Trujillo, F.J. An automatic pitting corrosion detection approach for 316L stainless steel. *Mater. Des.* **2014**, *56*, 642–648. [CrossRef]
3. Sun, Y. Sliding wear behaviour of surface mechanical attrition treated AISI 304 stainless steel. *Tribol. Int.* **2013**, *57*, 67–75. [CrossRef]
4. Niu, W.; Li, Z.; Ernst, F.; Lillard, R. The passivity of low-temperature carburized austenitic stainless steel AISI-316L in a simulated boiling-water-reactor environment. *J. Nucl. Mater.* **2020**, *537*, 152197. [CrossRef]
5. Ceschini, L.; Chiavari, C.; Lanzoni, E.; Martini, C. Low-temperature carburised AISI 316L austenitic stainless steel: Wear and corrosion behaviour. *Mater. Des.* **2012**, *38*, 154–160. [CrossRef]
6. Xu, Z.; Chen, Z.; Huang, L. Research on the outline of strengthening technology of surface deformation. *Equip. Manuf. Technol.* **2007**, *04*, 69–71+75.
7. Efremov, D.; Gerasimova, A. Shotblasting process for surface hardening. *Mater. Today Proc.* **2021**, *38*, 1685–1688. [CrossRef]
8. Wang, Y.; Li, F.; Bian, N.; Du, H.Q.; Da Huo, P. Mechanism of plasticity enhancement of AZ31B magnesium alloy sheet by accumulative alternating back extrusion. *J. Magnes. Alloy.* **2021**. [CrossRef]
9. Du, Q.-L.; Li, C.; Cui, X.-H.; Kong, C.; Yu, H.-L. Fabrication of ultrafine-grained AA1060 sheets via accumulative roll bonding with subsequent cryorolling. *Trans. Nonferrous Met. Soc. China* **2021**, *31*, 3370–3379. [CrossRef]
10. Hu, X.; Yang, Y.; Zhao, J.; Lu, Y.; Wu, J.; Qiao, H. Surface integrity evolution of a Ni-based single crystal superalloy by laser shock peening. *Appl. Surf. Sci. Adv.* **2021**, *6*, 100183. [CrossRef]
11. Chi, J.; Cai, Z.; Zhang, H.; Guo, W.; Wan, Z.; Han, G.; Peng, P.; Zeng, Z. Combining manufacturing of titanium alloy through direct energy deposition and laser shock peening processes. *Mater. Des.* **2021**, *203*, 109626. [CrossRef]
12. Kikuchi, S.; Nakahara, Y.; Komotori, J. Fatigue properties of gas nitrided austenitic stainless steel pre-treated with fine particle peening. *Int. J. Fatigue* **2010**, *32*, 403–410. [CrossRef]
13. Ordoñez, M.F.C.; Amorim, C.L.G.; Krindges, I.; Aguzzoli, C.; Baumvol, I.J.; Figueroa, C.A.; Sinatora, A.; Souza, R.M.; Farias, M.C.M. Microstructure and micro-abrasive wear of sintered yttria-containing 316L stainless steel treated by plasma nitriding. *Surf. Coat. Technol.* **2019**, *374*, 700–712. [CrossRef]
14. Chen, Y.; Liu, C.; Yan, H.; Fan, Y.; Wang, J.; Cui, Y. Effect of gas nitriding on 316 L stainless steel lattice manufactured via selective laser melting. *Surf. Coat. Technol.* **2022**, *441*, 128559. [CrossRef]
15. Wang, Y.; Yin, Y.; Wu, G.; Li, L.; Yao, J.; Zhang, Q. The microstructure and cavitation erosion resistance of Ti6Al4V alloy treated by laser gas nitriding with scanning galvanometer. *Opt. Laser. Technol.* **2022**, *153*, 108270. [CrossRef]
16. Wang, B.; Sun, S.; Guo, M.; Jin, G.; Zhou, Z.; Fu, W. Study on pressurized gas nitriding characteristics for steel 38CrMoAlA. *Surf. Coat. Technol.* **2015**, *279*, 60–64. [CrossRef]
17. Mayr, W.; Lengauer, W.; Ettmayer, P.; Rafaja, D.; Baue, J.; Bohr, M. Phase equilibria and multiphase reaction diffusion in the Cr-C and Cr-N systems. *J. Equilib. Fluid* **1999**, *20*, 35–44. [CrossRef]
18. Wenbo, L. Tribological Experiment Study of Textured Hard Coating. Master's Thesis, Yanshan University, Qinhuangdao, China, 2019.
19. Costa, H.L.; Hutchings, I.M. Effects of die surface patterning on lubrication in strip drawing. *J. Mater. Process. Technol.* **2009**, *209*, 1175–1180. [CrossRef]
20. Hua, X.; Puoza, J.C.; Zhang, P. The influence of laser surface texture on the tribological properties of friction layer materials in ultrasound motors. *Proc. Inst. Mech. Eng. Part J J. Eng. Tribol.* **2022**, *236*, 1123–1132. [CrossRef]
21. Xue, X.; Lu, L.; Wang, Z.; Li, Y.; Guan, Y. Improving tribological behavior of laser textured Ti-20Zr-10Nb-4Ta alloy with dimple surface. *Mater. Lett.* **2021**, *305*, 130876. [CrossRef]
22. Siripuram, R.B.; Stephens, L.S. Effect of deterministic asperity geometry on hydrodynamic lubrication. *J. Trib.* **2004**, *126*, 527–534. [CrossRef]
23. Singh, N.; Awasthi, R. Theoretical investigation of surface texture effects on the performance characteristics of hydrodynamic two-lobe journal bearing. *Proc. Inst. Mech. Eng. Part J J. Eng. Tribol.* **2020**, *234*, 1712–1725. [CrossRef]
24. Chen, X.; Xu, Y.; Wang, Z. Study of micro-hole and micro-grating on antifriction performance of 316 L stainless steel surface. *Foundry Technol.* **2012**, *33*, 1283–1285.
25. Gajrani, K.K.; Reddy, R.P.K.; Sankar, M.R. Tribo-mechanical and surface morphological comparison of untextured, mechanical micro-textured (MμT), and coated-MμT cutting tools during machining. *Proc. Inst. Mech. Eng. Part J J. Eng. Tribol.* **2019**, *233*, 95–111. [CrossRef]

26. Li, G.; Shen, M.; Meng, X.; Li, J.; Li, X.; Peng, X. An experimental study on tribological properties of groove-textured surfaces of 316 L stainless steel. *J. Funct. Mater.* **2015**, *46*, 2033–2037.
27. Panigrahi, D.K.; Sarangi, M. Tribological performance of positive deterministic textured surfaces in parallel sliding lubricated contacts: Effect of texture size and height. *Proc. Inst. Mech. Eng. Part J J. Eng. Tribol.* **2020**, *234*, 1908–1925. [CrossRef]
28. Bharatish, A.; Harish, V.; Bathe, R.N.; Senthilselvan, J.; Soundarapandian, S. Effect of scanning speed and tin content on the tribological behavior of femtosecond laser textured tin-bronze alloy. *Opt. Laser Technol.* **2018**, *108*, 17–25. [CrossRef]
29. Yun, Z.; Jianxin, D.; Hongzhi, Y.; Xuemu, L. Effect of micro-texture on the surface friction properties of steel guide rails. *Manuf. Technol. Mach. Tool* **2019**, *11*, 17–23.
30. Borjali, A.; Langhorn, J.; Monson, K.; Raeymaekers, B. Using a patterned microtexture to reduce polyethylene wear in metal-on-polyethylene prosthetic bearing couples. *Wear* **2017**, *392–393*, 77–83. [CrossRef]
31. Xinliang, L.; Wen, Y.; Fei, H.; Jiajie, K.; Zhiqiang, F. The effect of abrasive particle size on the dry friction properties of pure titanium with micro circular texture. *J. Mech. Eng.* **2017**, *53*, 25–33.
32. Guangtian, H. Effect of Laser Shock on Corrosion Properties of 304 Austenitic Stainless Steel. Master's Thesis, Jiangsu University, Zhenjiang, China, 2013.
33. Li, Z. Combined Effects of Laser Shock Processing and Coating on the Mechanical Properties of Metallic Materials. Master's Thesis, Jiangsu University, Zhenjiang, China, 2018.
34. Bowden, F.P.; Bowden, F.P.; Tabor, D. *The Friction and Lubrication of Solids*; Oxford University Press: Oxford, UK, 2001.

machines

MDPI

Article

Effect of Coated Composite Micro–Texture Tool on Cutting Shape and Cutting Force during Aluminum Alloy Cutting

Qinghua Li, Chunlu Ma, Lintao Xie, Baizhong Wang and Shihong Zhang *

School of Mechanical and Vehicular Engineering, Changchun University, Changchun 130022, China
* Correspondence: zhangsh@ccu.edu.cn

Abstract: Aluminum alloy materials are very difficult to break in the processing process; chip accumulation leads to poor processing stability. In this paper, a coated composite micro–textured tool is presented. The influence of coated composite micro–textured tool on chip shape and cutting force under the same cutting parameters is studied. Firstly, the combination of cutting force and cutting temperature is optimized based on cutting test. Secondly, combined with the finite element simulation technology, three kinds of composite micro–texture tools are tested to determine the same cutting parameters, tools and workpiece materials, and optimize the composite micro–texture combination from the angle of chip shape, cutting temperature and cutting force. Finally, on the basis of optimizing the microstructure combination, the optimization of the cutting performance of aluminum alloy by coating material is studied. It was found that the chip was more easily broken in the composite microstructure when two tools were used. The cutting force decreased by 25% and the cutting temperature decreased by 9.09% compared with the non–micro–texture tools. The cutting temperature and cutting force decreased by 3% and 4.99% compared with those of uncoated composite microstructure.

Keywords: composite micro–texture; cutting force; chip; coating; cutting temperature

1. Introduction

With the innovation of modern manufacturing, processing technology is developing in the direction of high accuracy and speed [1,2]. This puts forward a higher requirement for the machining quality and production efficiency of parts. Due to its excellent mechanical and thermal properties, aluminum alloy materials are widely used in automotive, shipping, aerospace and other fields [3,4]. However, the aluminum alloy material has a low melting point, which produces plastic deformation under the action of high temperature and high pressure generated by the cutting process, and the chip continues to overflow with the processing, and softens under the action of high temperature, which is very easy to be captured by the tool and attached to the tool surface, seriously affecting the tool life, resulting in the reduction in the processing quality [5–7]. Therefore, in order to improve the cutting performance and prolong the service life of the tool, many scholars at home and abroad optimize the cutting performance of the tool from the aspects of tool structure and surface coating.

In recent years, with the development of bionics and tribology, it has been found that surfaces with microstructures have excellent wear resistance and abrasion resistance [8]. During the cutting process, the front surface of the tool is in close contact with the material of the workpiece to be cut, causing serious friction and causing tool wear [9,10]. Therefore, the application of micro–texture on the blade has attracted a lot of scholars' attention. Su, Y et al. studied the machining performance of microgroove tool in cutting titanium alloy and found that the friction coefficient between tool and chip was significantly increased under dry turning conditions [11]. Rathod et al. studied AL6063 dry cutting with microstructure tools and found that the cutting performance of microstructure tools was better than that of non-microstructure tools, and the cutting force of square microstructure tools decreased by 30% [12]. Patel, K et al. studied the effect of micro–texture parameters on the cutting performance of the tool when machining titanium alloy, and found that the width, depth and distance from the main cutting edge of the micro–texture have a significant impact on the cutting force [13]. Zhang Yan and others have studied the continuous wear and tear of microstructure tools in the cutting process, and found that the number of microstructure tools and the placement of rake face position have important influence on the wear and tear of tools. The surface temperature of tools is the key factor affecting the depth of tool wear [14]. Therefore, reasonable preparation of the micro–texture on tool surface is helpful to improve tool performance, but micro–texture is limited in the position of tool surface and the improvement of tool performance by single texture is also limited. Aiming at the advantages and disadvantages of a single micro–texture tool, a combined micro–texture tool based on a single micro–texture tool is proposed. Fu and others found that the combined micro–texture tool can greatly improve the stress distribution on the tool surface and reduce the cutting heat produced during the cutting process [15].

As the research deepened, researchers found that coating tool surfaces can effectively improve the shortcomings of traditional cutting tools [16]. Zheng, G and others carried out a high-speed cutting test under dry and coated conditions and found that coating can avoid direct contact and friction between tool body and workpiece, and can lubrication and friction reduction [17]. Li, Y et al. studied the surface roughness of coated tools for high-speed cutting AISI4340, and found that the surface roughness of the workpiece after machining with coated tools is 0.08~0.32 μm, which is helpful to improve the surface quality of machining difficult-to-machine materials [18].

After the above research, it is found that the cutting performance of composite microstructure tool in aluminum alloy cutting is seldom studied. In this paper, a variety of concave–convex composite micro–textured tools are developed for cutting test. The optimum cutting parameters were obtained by cutting AL7075 aluminum alloy. Optimal cutting parameters and composite micro–texture are used in lathe simulation tests. The chip shape and cutting force of the tool are analyzed, the influence of composite micro–texture on tool performance is discussed, and the most optimal composite micro–texture is obtained. In order to optimize the cutting tool performance, the coating test for the optimal composite micro–texture was carried out.

2. Cutting Test

2.1. Test Materials and Equipment

The CBN tool was used in the test. Tool material is a new kind of artificial material, which is widely used in the cutting industry because of its high hardness and good impact resistance [19,20]. Tool material performance parameters are shown in Table 1. The workpiece material is made of AL7075 aluminum alloy, 50 mm in diameter and 250 mm in length. The performance parameters of workpiece materials are shown in Table 2. The cutting force data were collected by CA6140A lathe and Kistler 2825A-02 high sensitivity piezoelectric three-way dynamometer. Thermal imagers are used to measure the cutting temperature produced during the cutting process. The test equipment is shown in Figure 1.

Table 1. CBN tool material performance parameters [21].

Material Properties	Young's Modulus (Gpa)	Thermal Conductivity (W/m·K)	Poisson Ratio	Density (g/cm^3)	Specific Heat (J/kg °C)
value	690	120	0.2	3.8	700

Table 2. 7075AL aluminum alloy material performance parameters [22].

Material Properties	Young's Modulus (Gpa)	Thermal Conductivity (W/m·K)	Poisson Ratio	Density (g/cm^3)	Specific Heat (J/kg·°C)
value	71.7	173	0.33	2.810	860

Figure 1. Test equipment diagram.

2.2. Cutting Test and Result

In the cutting process, the selection of cutting amount has an important influence on the generation of cutting heat and the stability of the cutting process [23]. In this paper, the cutting test scheme shown in Table 3 is designed according to the suitable machining parameter range of CBN tool and the guarantee of machining accuracy. Dry cutting is used in the cutting test.

Table 3. Cutting Test Scheme.

Group	Cutting Speed Vc (m/min)	Feed f (mm/r)	Cutting Depth a_p (mm)
1	190	0.2	0.3
2	150	0.2	0.3
3	125	0.2	0.3

The cutting force test results are shown in Figure 2 below. As the cutting speed decreases, the cutting force decreases gradually. When the cutting parameters are Vc = 125 m/min, f = 0.2 mm/r, a_p = 0.3 mm, the cutting force is the smallest. This is because with the decrease of cutting amount, the shape and size of cutting layer decrease, the shape deformation of workpiece decreases, leading to the decrease in cutting force. As Figure 3 shows, the chip entanglement around the aluminum alloy rod, even without spilling and stacking near the main edge of the cutter, increasing the length of the chip's contact with the tool and producing it prone to psoriasis.

Figure 4 shows data collected using a thermal imager, and Figure 5 shows the results of the cutting temperature test. It can be seen from the graph that the cutting temperature decreases with the cutting quantity, and the cutting deformation, friction and contact area decrease with the cutting quantity, which improves the heat dissipation conditions and promotes further heat dissipation of the tool.

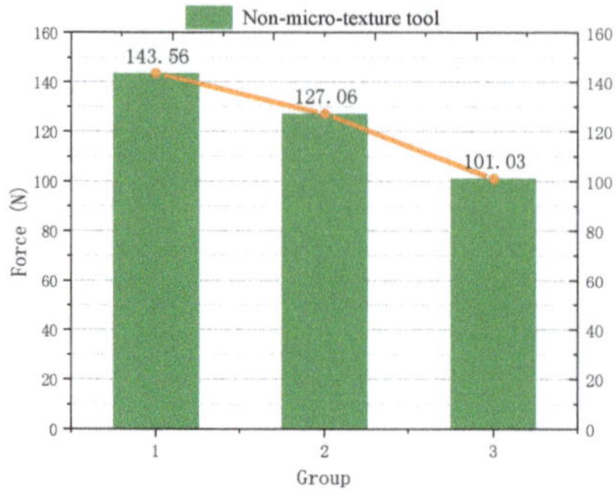

Figure 2. Cutting force of non-micro–textured tool.

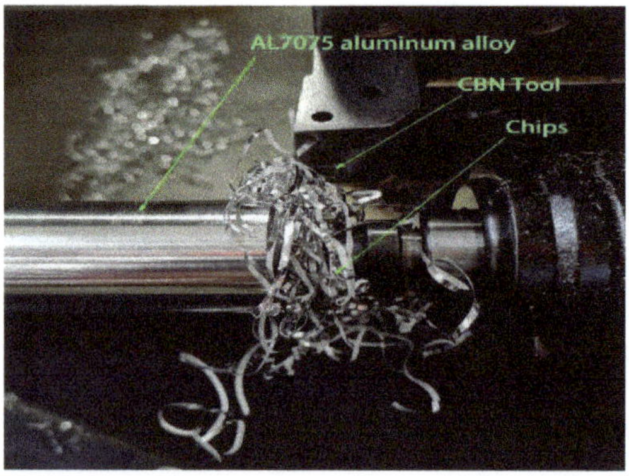

Figure 3. Chip winding.

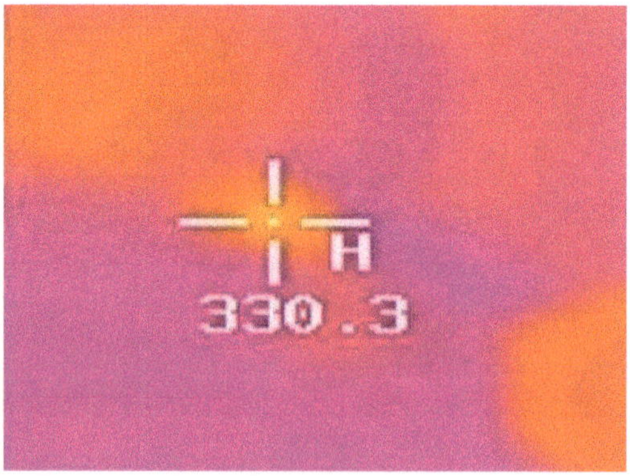

Figure 4. Cutting temperature thermal imaging.

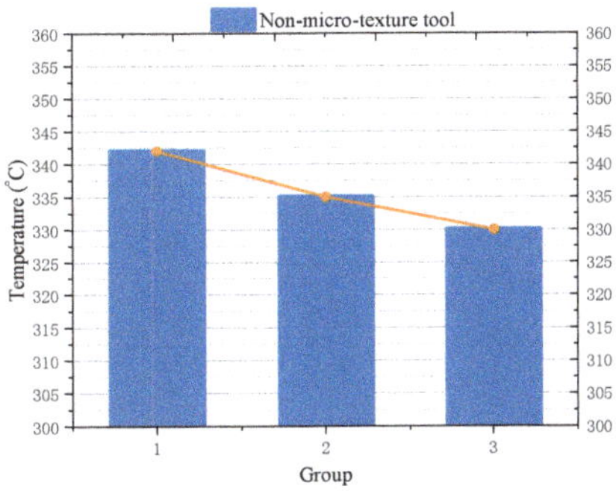

Figure 5. Cutting temperature of non-micro–texture tool.

3. Cutting Simulation Test of Composite Micro–Textured Tool

In view of the difficulty of aluminum alloy material processing, CBN tool is difficult to cut aluminum alloy material chips under conventional cutting conditions [24,25]. On the basis of using single micro–texture to improve tool performance, the concept of composite micro–texture tool is proposed. Using the synergy of concave micro–texture and convex micro–texture, the tool performance is further improved.

3.1. Finite Element Simulation

In order to study the effect of pitching and pitching microstructure on tool cutting performance, three kinds of composite micro–texture tools are designed in this paper, which are shown in Figure 6. Composite micro–texture two-tool (micro–texture combination with pit and then convex interlaced according to main cutting edge); composite micro–texture e III (a combination of three rows of micro–textures protruding from a pit first and then, based on the main cutting edge). Pit micro–texture depth 35 μm. Height of convex micro–textured tool 35 μm.

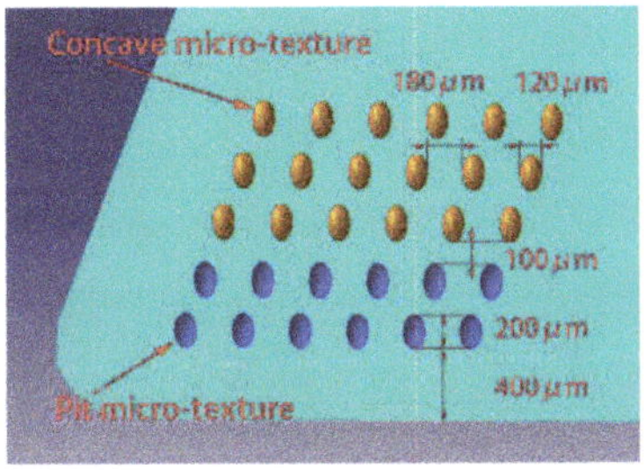

(**a**) Composite Micro–texture I

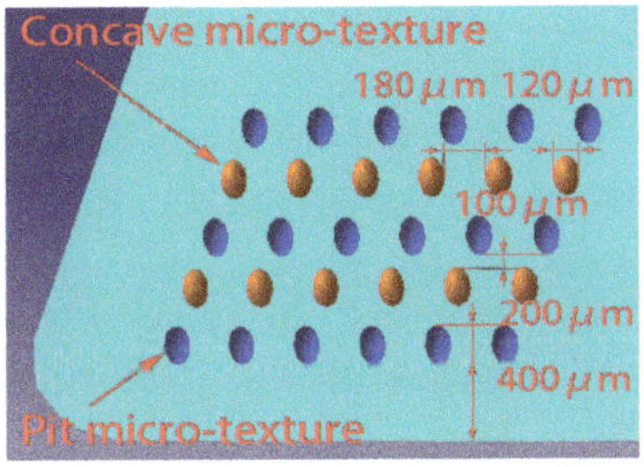

(**b**) Composite Micro–texture II

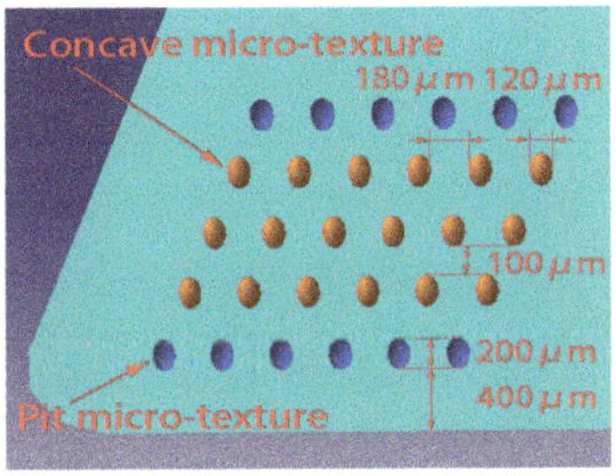

(**c**) Composite Micro–texture III

Figure 6. Micro–texture simulation mode.

The above three composite micro–texture models are imported into the finite element simulation software, and the tool and workpiece are meshed. The maximum mesh size of the tool is set to 1 mm, the minimum mesh size is set to 0.1 mm, the maximum mesh size of the workpiece is set to 3 mm, and the minimum mesh size is set to 0.1 mm. The finite element simulation mesh model of the tool workpiece is shown in Figure 7. The cutting amount scheme selects the minimum cutting amount combination of the above cutting forces, namely Vc = 125 m/min, f = 0.2 mm/r, a_p = 0.3 mm.

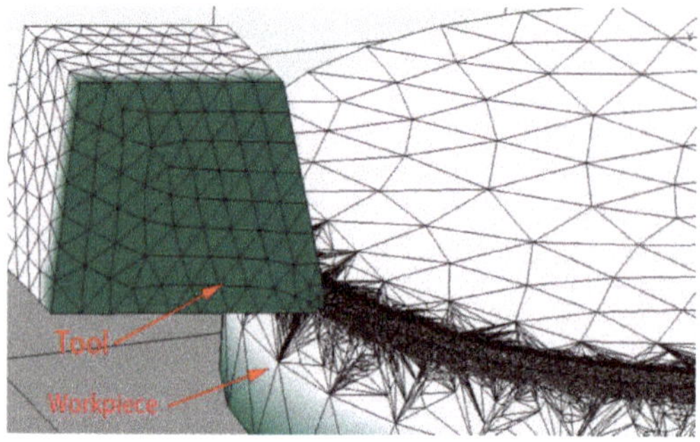

Figure 7. Finite element simulation model.

In the finite element analysis, the Johnson–Cook constitutive model is selected. The model shows that the strain, strain rate and temperature distribution are not uniform and vary greatly when the materials collide. It can accurately reflect the large strain and thermal softening effect of the materials in the cutting process. The parameters of the constitutive model are shown in Table 4 [26,27].

Table 4. Johnson-Cook constitutive model parameters of 7075AL aluminum alloy [28].

A (MPa)	B (MPa)	C	n	m	$t_0\cdot$(°C)	$t_m\cdot$(°C)
546	678	0.024	0.71	1.56	20	650

Empirical equation of Johnson-Cook constitutive model:

$$\sigma = \left[A + B(\varepsilon)^n\right] \bullet \left[1 + C\ln\left(\frac{\dot{\varepsilon}_p}{\dot{\varepsilon}_0}\right)\right] \bullet \left\{1 - \left(\frac{t - t_0}{t_m - t_0}\right)^m\right\} \tag{1}$$

where: σ—Plastic stress, A—yield strength, B—strain hardening constant; C—strain rate hardening coefficient, m—thermal softening parameter, n—strain hardening parameter; ε—Plastic strain, $\dot{\varepsilon}_p$—plastic strain rate, $\dot{\varepsilon}_0$—reference strain rate; t—material dynamic temperature, t_0—ambient reference temperature, t_m—material melting point.

3.2. *Analysis of Simulation Results*

3.2.1. Effect of Composite Microtextured Tool on Chip

Under the same cutting amount, the cutting simulation test of the three composite micro–texturing tools is carried out to obtain the chip morphology as shown in Figure 8 and the cutting temperature as shown in Figure 9. It is found in observation Figure 8 that when cutting with three tools, composite micro–texture I and composite micro–texture III, continuous strip chips are formed and curl out on the front of the blade. When working with a composite micro–texture two-tool, the chip is partially broken and spilled at the front of the cutter.

During cutting process, the cutting-edge part of the tool collides with the cutting material, the cutting layer of the tool is subjected to shear stress, which causes the cutting chips to form and flow out along the front of the tool and be rubbed by the front of the tool. Composite micro–texture two tools, because of the dented and crisscrossed composite micro–texture, when the chip flows through the part of the composite micro–texture, it causes extrusion and friction, which leads to a decrease in the curl radius of the chip, an increase in strain of the chip material and a fracture of the chip.

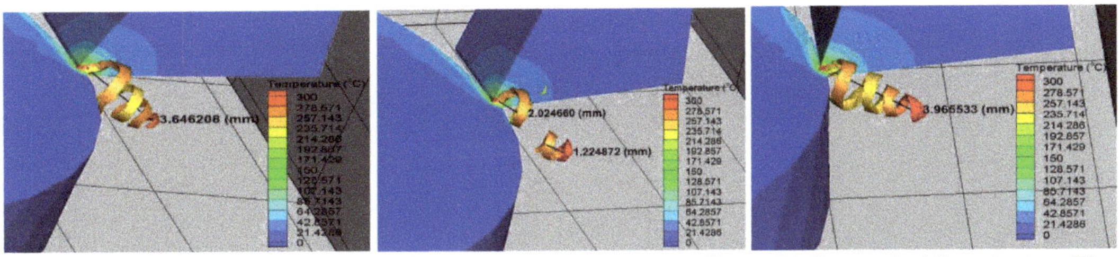

(**a**) Composite Micro–texture I (**b**) Composite Micro–texture II (**c**) Composite Micro–texture III

Figure 8. Chip morphology of three composite micro–textured tools.

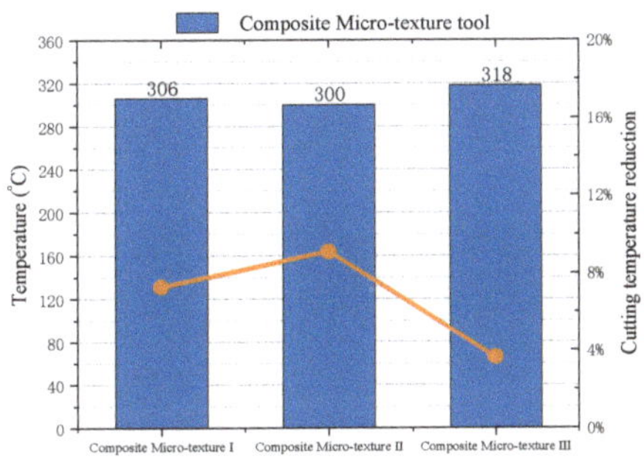

Figure 9. Cutting temperature of three kinds of composite micro–texture tools.

As can be seen from Figure 9, the lowest cutting temperature was obtained when using a composite micro–texture II tool. Compared to the same cutting conditions, the reduction in the micro–texture I, II and III tools was 7.27%, 9.09% and 3.64%. It is very easy to cause the workpiece material to be softened by heat, produce thermal deformation, attach to tool surface, and affect machining. The distribution of cutting temperature is mainly taken away by chip fracture and overflow, and the closer the chip meets the tool surface, the less heat can be dissipated. For the composite micro–texture of the two tools, most of the heat of chip breaking is taken away with chip breaking, and the contact area between knife and chip is reduced, the contact surface friction is reduced, and the heat dissipation is lost, so the cutting temperature is reduced.

3.2.2. Effect of Composite Micro–Textured Tool on Cutting Force

Collect the data of cutting force test results. To avoid errors due to the randomness of data collection, multiple data points were uniformly selected during the cutting process and their averages were calculated as cutting force test results, Figure 10 shows the mechanical fluctuation of the cutting force during simulation test and Figure 11 shows the result of cutting force test.

As can be seen from Figure 11, the cutting force of the three composite micro–textured tools decreased by 17.02%, 25% and 19.99%, respectively, compared to non-microstructural tools. On the one hand, because of the compound micro–texture on the rake tooth surface, the actual contact length with the chip is reduced, the friction coefficient is reduced and the friction force is reduced; on the other hand, the dent-convex micro–texture increases the extrusion degree the tool structure in the cutting layer of the workpiece material and reduces friction force. The composite micro–texture II tool has the greatest influence on the reduction in cutting force. This is because when cutting with the composite micro–texture II tool, the chips break and spill, and do not accumulate and clog at the tool tip, reducing secondary cutting and resulting in a reduction in cutting force.

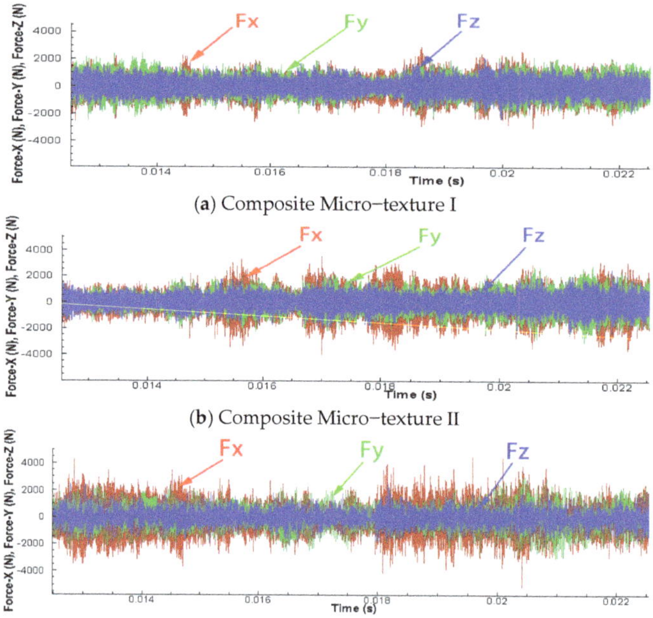

(**a**) Composite Micro–texture I

(**b**) Composite Micro–texture II

(**c**) Composite Micro–texture III

Figure 10. Mechanical wave diagram of cutting force of three kinds of composite micro–texture tools.

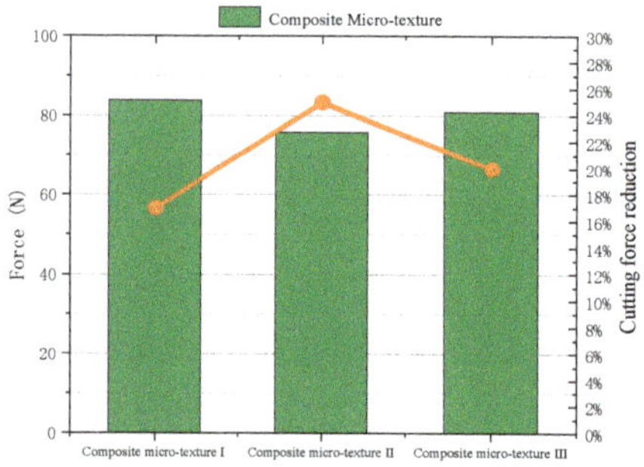

Figure 11. Cutting force of composite micro–texture tool.

Figure 12 shows a comparison of radial force with and without micro–textured tools. The first is the non-woven cutting test group, the second, third and fourth are the composite micro–textured tools one, II and III. As can be seen from Figure 8, the influence of three kinds of composite microstructure tools on the radial force is very significant. The smaller the radial force of the tool, the smaller the radial pulse, the more stable the machining state. It can be seen that composite microstructure tool can improve the stability of the cutting process and improve the quality of cutting products.

Figure 13 is the stress distribution diagram during the cutting process of the composite micro–textured tool. Figure (a,d,c) are the tool stress distribution at the beginning of cutting. (b,e,h) are the tool stress distribution maps for stable cutting; (c,f,i) are the tool stress profiles at the end of the cutting process. It can be seen from the figure that with the cutting process, the stress gradually spreads to the inside of the tool front face and gradually transfers to the micro–textured position. Comparing the stress nephogram of three kinds of composite micro–texture tools, it is found that the secondary stress of composite micro–texture is concentrated at the micro–texture position far away from the tool tip and the main cutting edge, and the stress value is the smallest. It can be concluded that the stress absorption effect of composite micro–texture II is significantly better than that of the other two micro–texture tools.

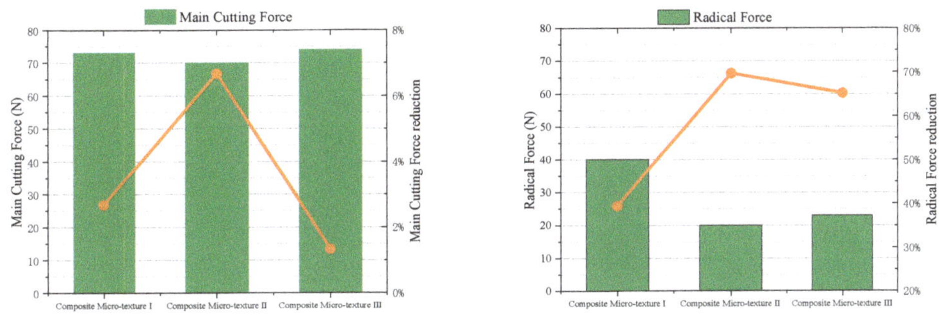

(**a**) Main cutting force image of composite micro–texture tool (**b**) Radial force image of composite micro–texture tool

Figure 12. Comparison of main cutting force and radial force.

Composite Micro–texture I:

Composite Micro–texture II:

Composite Micro–texture III:

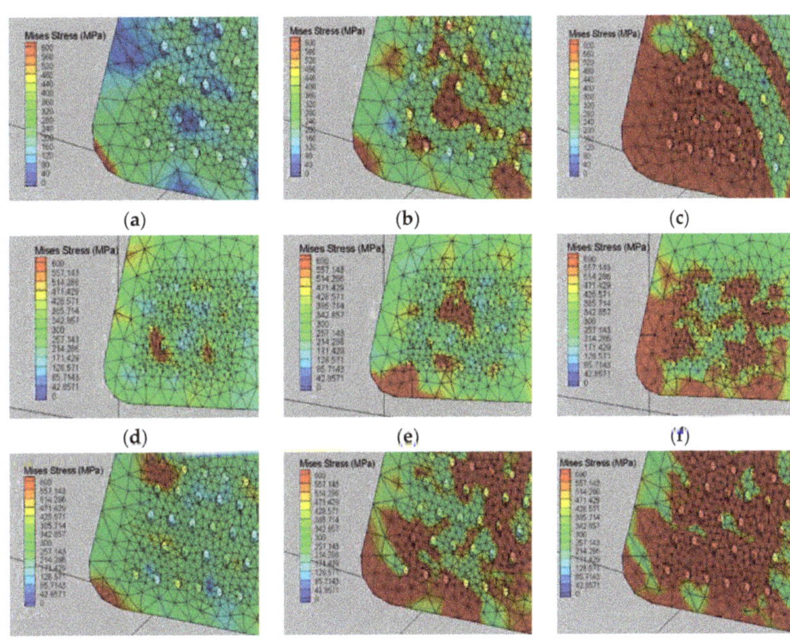

Figure 13. Stress distribution of three kinds of composite micro–texture tool.

It can be seen from the above three kinds of composite micro–texture simulation experiments that the chip, cutting force and stress distribution can be significantly improved by using two tools of composite micro–texture combination, namely, the pit-convex interwoven micro–texture combination.

4. Simulation Test of Coated Composite Micro–Textured Tool

During the cutting process, the cutting heat increases rapidly, and the aluminum alloy chips softens easily and adheres to the tool. In the cutting process of aluminum alloy materials, the main cause of tool wear is adhesive wear. In order to further improve the properties of aluminum alloy tool, the concept of coated tools is put forward. Compared with traditional tools, coated tools have higher abrasion resistance and thermal stability, and are now widely used in machining [29,30]. In this paper, the coating is combined with composite microstructure tool to optimize the cutting performance of aluminum alloy.

4.1. Effect of Coated Micro–Textured Tools on Cutting Temperature

TiN material micron was coated on the surface of the above selected composite micro–texture II tool 1.5μm and the cutting force is tested with minimum cutting dosage. Figure 14 shows a comparison of cutting temperatures between coated and uncoated micro–textured tools. It can be seen from the figure that the cutting temperature of coated tools is about 3% lower than that of uncoated tools. When using coated microstructure tool to process aluminum alloy material, TiN material clings to tool surface, TiN material coated on tool surface acts as thermal protective film, which reduces friction coefficient between tool and workpiece, has the effect of dry lubrication, promotes the smooth flow of chip and reduces the heat of cutting due to chip accumulation. Due to the chemical stability of TiN material, it is difficult to react with cutting material under high temperature and pressure, so the tool works well under high temperature.

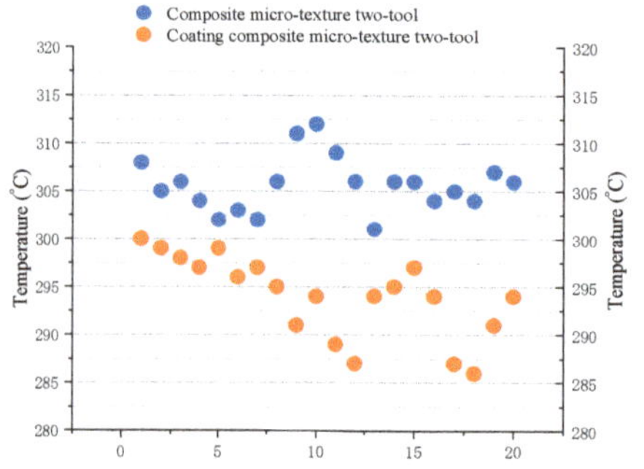

Figure 14. Comparison of cutting temperature of coated and non-micro–textured tools.

4.2. Effect of Coated Micro–Textured Tool on Cutting Force

Figure 15 shows a decrease of 28.75% in the cutting force of coated and uncoated micro–textured tools and 4.99% in the cutting force of coated and uncoated micro–textured tools. When TiN material works instead of tool body, TiN material acts as barrier, which makes it difficult to invade the chip material produced by external cutting. TiN material has higher hardness and is more abrasive than tool body material, which reduces internal stress and protects tool body structure.

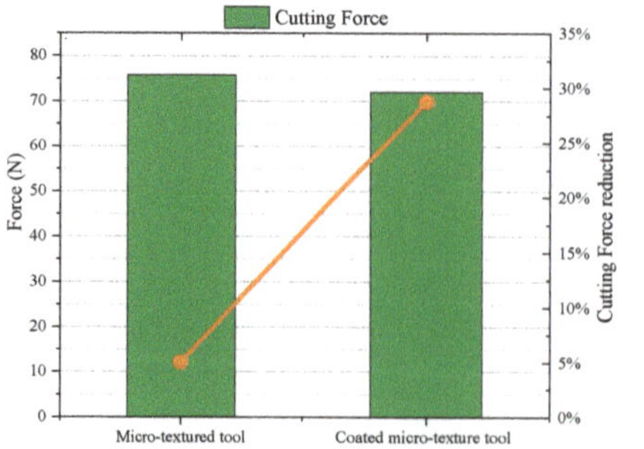

Figure 15. Comparison of cutting force of coated and uncoated tools.

Therefore, coating on the tool surface can alleviate the cutting force and heat produced during the cutting process, and effectively protect the tool. So, it is practical to apply coating technology to tool surface and optimize tool performance based on the optimization of composite microstructure tool.

5. Conclusions

In this paper, the optimal cutting dosage combination is selected by cutting test, taking cutting force and cutting temperature as the yardstick. The cutting performance of the composite micro–texture tool is simulated by means of finite element simulation software combined with the above optimal cutting quantity. By comparing the cutting performance of the different optimal composite micro–texture e tool, the cutting performance of different composite micro–texture tools is optimized, and coating cutting test is carried out on the preferred composite micro–texture tool. The results were as follows:

(1) Cutting tools with interlaced pit-convex micro–texture promote wafer curl and fracture. When the chip flows through the blade, the fracture overflows and the chip accumulation improves markedly;
(2) Changes in the distribution of dented tissue in the front of the tool will cause changes in cutting force and cutting temperature. pit-convex micro–texture combined with staggered micro–texture has good cutting performance in reducing cutting force and cutting temperature;
(3) Due to the high hardness and chemical stability of the coating material, it acts as a protective barrier in the event of sharp tool collisions and further optimizes the cutting performance of the tool;
(4) The combination of combined micro–texture and surface coating has solved the processing problem of aluminum alloy material chip winding with large fluctuations and provided a theoretical basis for aluminum alloy cutting.

Author Contributions: Q.L. research ideas, experimental design and data analysis are provided; C.M. writing manuscripts, charting, data collection.; L.X. data collection, data analysis, literature retrieval; B.W. document retrieval, data collection, research design.; S.Z. experimental design, data collection, data analysis. All authors have read and agreed to the published version of the manuscript.

Funding: Here, we need to thank the following organizations for their strong support: "Natural Science Foundation of Jilin Province-General Project, Study on the Machinability of Milling Titanium Alloy with Micro–texture Milling Cutter" (20220101227JC).

Data Availability Statement: The data that support the findings of this study are available from the corresponding author upon reasonable request.

Acknowledgments: Authors would like to thank the members of the project team for their dedications and efforts, and the teachers and schools for their help.

Conflicts of Interest: The authors declare no conflict of interest.

References

1. Debnath, S.; Reddy, M.M.; Yi, Q.S. Environmental friendly cutting fluids and cooling techniques in machining: A review. *J. Clean. Prod.* **2014**, *83*, 33–47. [CrossRef]
2. Zhou, Y.; He, Y.Y.; Yang, Z.T.; Yang, J.; Fu, Y.H. The performance of composite texture cutting tool for cutting aluminum alloy. *China Surf. Eng.* **2022**, *35*, 281–288.
3. Rambabu, P.P.N.K.V.; Eswara Prasad, N.; Kutumbarao, V.V.; Wanhill, R.J.H. Aluminium alloys for aerospace applications. In *Aerospace Materials and Material Technologies*; Springer: Singapore, 2017; Volume 1, Aerospace materials; pp. 29–52.
4. Georgantzia, E.; Gkantou, M.; Kamaris, G.S. Aluminium alloys as structural material: A review of research. *Eng. Struct.* **2021**, *227*, 111372. [CrossRef]
5. Dong, G.; Wang, X.; Gao, S. Molecular dynamics simulation and experiment research of cutting-tool wear mechanism for cutting aluminum alloy. *Int. J. Adv. Manuf. Technol.* **2018**, *96*, 1123–1137. [CrossRef]
6. Sharifi, M.; Akbari, M. Experimental investigation of the effect of process parameters on cutting region temperature and cutting edge quality in laser cutting of AL6061T6 alloy. *Optik* **2019**, *184*, 457–463. [CrossRef]

7. Sugihara, T.; Singh, P.; Enomoto, T. Development of novel cutting tools with dimple textured surfaces for dry machining of aluminum alloys. *Procedia Manuf.* **2017**, *14*, 111–117. [CrossRef]
8. Liu, W.; Liu, S.; Liang, G.Q.; Yuan, H.C. Simulation analysis of cutting performance and friction reduction effect of micro–textured tools. *Surf. Technol.* **2022**, *51*, 338–346.
9. Arulkirubakaran, D.; Senthilkumar, V.; Kumawat, V. Effect of micro–textured tools on machining of Ti–6Al–4V alloy: An experimental and numerical approach. *Int. J. Refract. Met. Hard Mater.* **2016**, *54*, 165–177. [CrossRef]
10. Duan, R.; Deng, J.; Lei, S.; Ge, D.; Liu, Y.; Li, X. Effect of derivative cutting on machining performance of micro textured tools. *J. Manuf. Process.* **2019**, *45*, 544–556. [CrossRef]
11. Su, Y.; Li, Z.; Li, L.; Wang, J.; Gao, H.; Wang, G. Cutting performance of micro–textured polycrystalline diamond tool in dry cutting. *J. Manuf. Process.* **2017**, *27*, 1–7. [CrossRef]
12. Rathod, P.; Aravindan, S. Performance evaluation of novel micro–textured tools in improving the machinability of aluminum alloy (Al 6063). *Procedia Technol.* **2016**, *23*, 296–303. [CrossRef]
13. Patel, K.; Liu, G.; Shah, S.R.; Özel, T. Effect of micro–textured tool parameters on forces, stresses, wear rate, and variable friction in titanium alloy machining. *J. Manuf. Sci. Eng.* **2020**, *142*, 021007. [CrossRef]
14. Zhang, Y.; Li, Q.H.; Yang, F.S. Modeling and simulation of continuous wear of micro–textured tools. *Tool Technol.* **2022**, *56*, 76–81.
15. Fu, Y.H.; Wang, H.B.; Gu, Y.L.; Li, Y.D. Microtexture design and cutting simulation analysis of tool surface. *Mach. Manuf.* **2016**, *54*, 5047–5055.
16. Morsy, M.K.; Sharoba, A.M.; Khalaf, H.H.; El-Tanahy, H.H.; Cutter, C.N. Efficacy of antimicrobial pullulan-based coating to improve internal quality and shelf-life of chicken eggs during storage. *J. Food Sci.* **2015**, *80*, M1066–M1074. [CrossRef]
17. Zheng, G.; Zhao, G.; Cheng, X.; Xu, R.; Zhao, J.; Zhang, H. Frictional and wear performance of TiAlN/TiN coated tool against high-strength steel. *Ceram. Int.* **2018**, *44*, 6878–6885. [CrossRef]
18. Li, Y.; Zheng, G.; Zhang, X.; Cheng, X.; Yang, X.; Xu, R. Cutting force, tool wear and surface roughness in high-speed milling of high-strength steel with coated tools. *J. Mech. Sci. Technol.* **2019**, *33*, 5393–5398. [CrossRef]
19. Rajasekaran, T.; Palanikumar, K.; Vinayagam, B.K. Application of fuzzy logic for modeling surface roughness in turning CFRP composites using CBN tool. *Prod. Eng.* **2011**, *5*, 191–199. [CrossRef]
20. Harto, B.; Umroh, B.; Darianto, D. Study on the CBN Tool Wear Mechanism on Dry High-Rate Turning Process for AISI 4140. *J. Mech. Eng. Manuf. Mater. Energy* **2018**, *2*, 20–26. [CrossRef]
21. Ślusarczyk, Ł.; Struzikiewicz, G. Hardened steel turning by means of modern CBN cutting tools. In *Key Engineering Materials*; Trans Tech Publications Ltd.: Baech, Switzerland, 2014; Volume 581, pp. 188–193.
22. Liu, P.; Hu, J.Y.; Li, H.X.; Sun, S.Y.; Zhang, Y.B. Effect of heat treatment on microstructure, hardness and corrosion resistance of 7075 Al alloys fabricated by SLM. *J. Manuf. Process.* **2020**, *60*, 578–585. [CrossRef]
23. Rao, C.J.; Rao, D.N.; Srihari, P. Influence of cutting parameters on cutting force and surface finish in turning operation. *Procedia Eng.* **2013**, *64*, 1405–1415. [CrossRef]
24. Qi, K.F.; Zhang, C.T.; Zhao, C.X. Research Status and Prospect of Microtexture Tools. *Mech. Eng.* **2021**, *82*, 77–78.
25. Viana, F.; Pinto, A.M.P.; Santos, H.M.C.; Lopes, A.B. Retrogression and re-ageing of 7075 aluminium alloy: Microstructural characterization. *J. Mater. Process. Technol.* **1999**, *92*, 54–59. [CrossRef]
26. Fan, L.; Deng, Z.; He, Y.; Zhu, X.L.; Gao, X.J.; Jin, Z. The effects of micro–texture shape on serrated chip geometry in the hardened steel AISI D2 cutting process. *Surf. Topogr. Metrol. Prop.* **2022**, *10*, 015031. [CrossRef]
27. Chen, W.L.; Yang, D.; Wu, M.; Ma, H.Q. Simulation Study on Right-Angle Cutting of 7050-T7451 Aluminum Alloy. *Mach. Manuf.* **2021**, *59*, 63–67. [CrossRef]
28. Peng, H.B.; Zhang, H.J. Research progress of constitutive model of metal materials. *Mech. Eng. Mater.* **2012**, *36*, 5–10+75.
29. Cai, F.; Wang, J.; Zhou, Q.; Zhang, S.; Zheng, J.; Wang, Q.; Kim, K.H. Reduced delamination and improved cutting performance of TiAlSiN multilayer coated cutter by tailoring the adhesion layers and intermediate layers. *Wear* **2022**, *488*, 204135. [CrossRef]
30. Sadiq, S.N.; Raguraman, T.R.; Kumar, D.T.; Rajasekaran, R.; Kannan, T.T.M. Optimization of milling parameters of OHNS steel using TiAlN coated cutter by design of experiment technique. *Int. J. Mech. Eng. Robot. Reearch* **2014**, *3*, 1.

machines

MDPI

Article

Conical Grinding Wheel Ultrasonic-Assisted Grinding Micro-Texture Surface Formation Mechanism

Jiaying Han [1,*], Yiqi Jiang [2,3], Xinrui Li [1] and Qing Li [4]

1 Maritime College, Tianjin University of Technology, Tianjin 300384, China
2 Tianjin Key Laboratory for Advanced Mechatronic System Design and Intelligent Control, School of Mechanical Engineering, Tianjin University of Technology, Tianjin 300384, China
3 National Demonstration Center for Experimental Mechanical and Electrical Engineering Education, Tianjin University of Technology, Tianjin 300384, China
4 School of Mechanical Engineering, Tianjin University, Tianjin 300350, China
* Correspondence: hanjiaying@email.tjut.edu.cn

Abstract: The rotating ultrasonic-assisted grinding (RUAG) experiment of the conical grinding wheel generated the intermittent pit-shaped micro-texture on the surface of the workpiece, reducing thermal damage and improving the lubrication characteristics compared with conventional grinding (CG). To further optimize the surface properties, this paper studied the formation mechanism of micro-texture. This study used as basis the theory that micro-debris volume equals the macroscopic material removal one to establish the mathematical equation of grinding depth. Thereafter, formulas of micro-texture feature parameters, including pit length, pit depth, and texture spacing were deduced. The solved microscopic grinding depth was alternatingly positive and negative, indicating that the alternating separation between the grinding grain and workpiece caused intermittent pits in the grinding. Through response surface analysis (RSA), this paper analyzed the relationships among macroscopic grinding depth, micro-texture feature parameters, and machining parameters (i.e., amplitude, feed rate, and rotational speed). Single-factor experiments of machining parameters, with finite element simulation and experiment methods, were performed to verify the theoretical micro-texture features. The simulated program formed three-dimensional surfaces with micro-textures. Their measurement results were consistent with the theoretical ones. Experimental results proved that the range of pit length covers the theoretical ones, further verifying the accuracy of the grinding depth model. For this grinding wheel, the 8–10 μm amplitude was optimal for better roughness, lubrication, and thermal damage. Roughness was improved when increasing the rotational speed or reducing the feed rate based on the experiment. If the rotational speed and feed rate exceed the limiting values, then continuous grinding will break down the abrasive grains and even damage the cubic boron nitride (CBN) coating. Experimental results likewise showed that the pit shape was closely related to the surface properties, which deserves further investigation.

Keywords: rotary ultrasonic assisted grinding; conical grinding wheel; grinding depth; micro-texture; simulation method

1. Introduction

The conical grinding wheel is adopted to grind spiral bevel and worm gears, among others, which are applied to the movement of intersecting axes, particularly in cases of high transmission efficiency, stable transmission, and low noise. Gear manufacturing technology is undergoing focus transformation from high dimensional precision [1,2] to improved surface properties, mainly including thermal damage, lubrication, and roughness. Therefore, the ultrasonic-assisted grinding (UAG) method was applied in the gear manufacture.

Ultrasonic assisted technology has been widely recognized for its influence on morphology. In 1998, Uhlmann et al. analyzed the kinematic model of slow feed grinding with ultrasonic vibration on the surface, proving that RUAGcan significantly improve material removal rate and will not cause surface damage [3]. From 1998 to 2013, most researchers made ultrasonic assisted devices to verify the advantages of RUAG metals through experimental measurements and analyses, which mainly include improving roughness, decreasing grinding forces, reducing thermal cracks of surfaces, and high material rates. Zhao et al. completed an experimental research on surface characteristics in ultrasonic ductile-regime honing ZrO_2 engineering ceramics with a new type of ultrasonic honing machine [4,5]. Tawakoli adopted ultrasonic assisted technology in dry grinding [6]. The results obtained showed that grinding forces decreased significantly. Mohsen Ghahramani Nik et al. proved the effect of ultrasonic vibration on the grinding of Ti6Al4V alloy [7]. W. M. Zeng et al. studied RUAG of advanced ceramics and discussed tool wear and cutting forces [8]. The Wu team [9–11] investigated elliptical ultrasonic-assisted grinding (EUAG) technologies. Most of the experiment results showed the feasibility of UAG in improving surface properties. Since 2014, the field of hard and brittle materials has found the application of UAG technologies [12,13] and carbon fiber-reinforced plastic [14], among others, other than special alloys. Research methods were transferred from qualitative to quantitative analyses. Past research has established mathematical models of cutting force [15,16], material removal rate [17,18], roughness [19], morphology [20], and cavitation bubble [21]; and further discussed the relationship between grinding parameters and ground surface.

Surface morphology is a critical influencing factor for surface performance. Recently, it has attracted considerable attention in the UAG field. Past research has investigated three types of ultrasonic vibration: longitudinal, tangential, and radial ultrasonic vibrations.

Early research has mainly focused on longitudinal UAG, in which ultrasonic vibration transfers to the tool head along the rotation axis. Wdowik et al. presented the results of measurements of surface micro-structure parameters after UAG and CG of ZrO_2-based ceramic material [22]. Guo et al. investigated the influences of ultrasonic vibration parameters and tilt angle on the ground quality of micro-structured surfaces [23]. Zheng et al. built surface micro-texture models based on multi-grains motion equations [24]. Wen et al. proposed an improved model of a rough surface profile to find the microscopic feature parameters, such as the curvature radius of the grain, which are suitable for contact analysis and calculation [25,26]. One-dimensional longitudinal vibration often adopts a slender horn with a small tool head. Longitudinal ultrasonic vibration produces sinusoidal micro-texture, which changes the linear micro-texture of CG. Interference action of multiple abrasive grains subdivides the surface further. Compared with CG, micro-texture reduces surface roughness and thermal damage and improves lubrication characteristics.

Tangential ultrasonic assistant vibration has the same direction as the tangential one of the grinding wheel. In general, ultrasonic vibration is along the workpiece axis. Its advantage is that the circumnutation of abrasive particles forms a separation between abrasive particles and workpiece surface, which helps to decrease the grinding temperature, further reducing thermal damage. Jiang et al. established the surface micro-structure formation mechanism model by considering the continuous cutting process of grains [27]. Wang et al. developed the surface model and realized the optimized microstructure and high-precision surface ground through theory calculation [28]. Zhao et al. proposed a novel processing of ultrasonic vibration-assisted forming grinding gear [29]. For this vibration, ultrasonic vibration can only be applied in a specified workpiece, the dimensions of which are limited by the ultrasonic generator's frequency.

Radial ultrasonic vibration is often combined with one of the other two vibrations to function. One case is the longitudinal radial ultrasonic vibration in the large disc grinding wheel. Radial amplitude is over 10% of the longitudinal one. The actual disc grinding wheel changes the longitudinal vibration of the transducer into a combination of longitudinal and radial. Compared with one-dimensional longitudinal ultrasonic, this case can form deeper micro-pits and also discontinuously separate abrasive particles and workpiece, thereby reducing thermal damage on the surface. Zhou et al. developed a comprehensive analytical model of MPD and micro-topography [30]. In this case, radial amplitude is determined by the dimension of the grinding wheel related to frequency; therefore, it is difficult to adjust. The other tangential radial ultrasonic vibration case was realized by elliptical vibration. Wang et al. [31] analyzed the topography-generation mechanism of a workpiece surface evolved using EUAG, with axial and radial ultrasonic vibration applied on the workpiece. Accordingly, they found the discontinuous motion by the radial ultrasonic vibration. Current theoretical research has focused on the trajectory interference mechanism of multiple abrasive grains, disregarding the influence of grinding depth and removal volume of abrasive grains.

The conical tool head studied in this paper decomposes the one-dimensional longitudinal vibration into two vibrations along the conical surface and vertical conical surface, which is equivalent to the longitudinal radial vibration. The longitudinal ultrasonic vibration and rotation motion of the grinding wheel can produce a sine wave trajectory. Moreover, radial ultrasonic vibration discontinuously hammers the workpiece surface, resulting in an intermittent grinding process and micro-pit textures. Compared with other radial vibrations, radial amplitude is more convenient to realize by the conical angle.

2. RUAG Kinematic Model

2.1. Experimental Setup

The RUAG device includes a CBN grinding wheel, amplitude horn, transducer, and electric slip ring, as shown in Figure 1. The machine tool spindle is bored to insert the device. The device is fixed and positioned by the cover and positioning journal, respectively. In the grinding process, the transducer generates the axial vibration and transmits it to the horn and grinding wheel. The grinding wheel simultaneously vibrates and rotates with the spindle, thereby forming the RUAG movement.

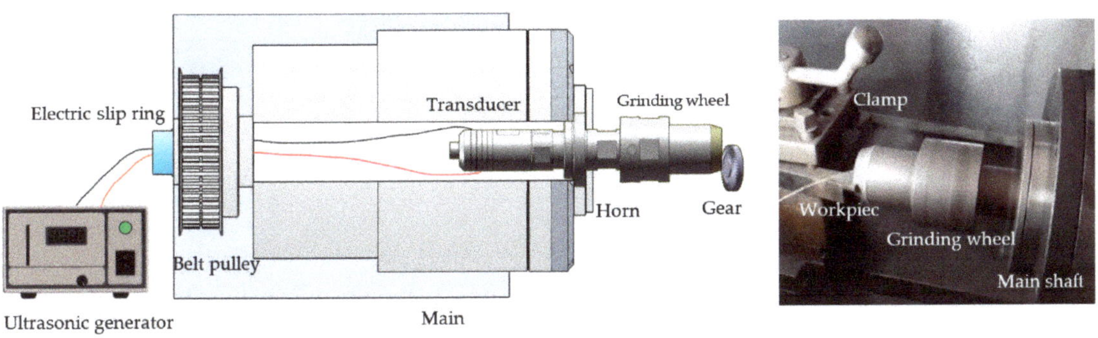

(a) (b)

Figure 1. RUAG setup and experiment: (**a**) RUAG setup and (**b**) grinding experiment.

2.2. Theoretical Kinematic Model

Figure 2a shows the cylindrical coordinate system $O(R, Z)$, where O is the central point of the grinding wheel top surface. The grinding wheel's dimensions are presented in Table 1. An abrasive grain is regarded as the frustum, located at a distance u from the top surface along the outside edge. Its motion comprises three components: (1) rotational speed n around the Z-axis with initial rotational angle φ_0; (2) ultrasonic vibration along the Z-axis with amplitude A, frequency f, and initial ultrasonic phase angle θ_0; and (3) feed speed V_s along the Z-axis. The subsequent theoretical and simulated calculations are based on the assumption that grains have the same shape and evenly distributed on the ground surface.

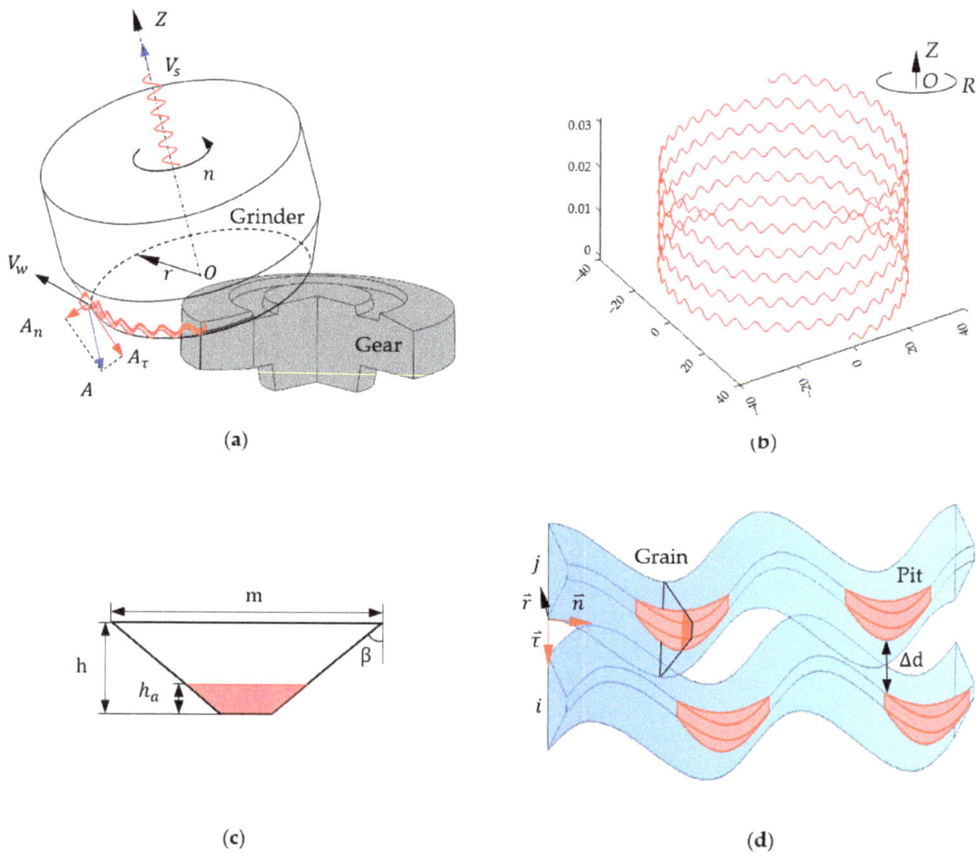

Figure 2. RUAG kinematic models: (**a**) grinding wheel model, (**b**) abrasive grain's trajectory, (**c**) abrasive grain model, and (**d**) micro-pit model.

Table 1. Grinding wheel's dimensions.

Parameters	Values
Wheel diameter (mm)	60.00
Outside pressure angle (°)	17.92
Inside pressure angle (°)	22.08
Point width (mm)	0.73
Outside edge radius (mm)	0.50
Inside edge radius (mm)	0.50
Mean radius (mm)	23.14

Axial amplitude of RUAG is composed of two components. Normal amplitude, which is perpendicular to the grinding edge, hammers on the workpiece surface once a cycle. Tangential amplitude, which is parallel to the grinding edge, is combined with rotational motion to form a wave path. The motion trajectory of the abrasive grain is presented as Equation (1), as shown in Figure 2a:

$$\begin{cases} R(u) = r + u\sin\alpha \\ \varphi = \frac{2\pi n}{60}t + \varphi_0 \\ Z(t) = V_s t + A\sin(\theta_0 + 2\pi f t) \end{cases} \tag{1}$$

where α_e is the outside pressure angle and can be changed into the negative inside pressure angle $-\alpha_i$. Velocity of the abrasive grain is deduced as Equation (2):

$$\begin{cases} V_w(t) = \frac{2\pi n}{60} r(u) \\ V_z(t) = V_s + 2\pi f A \cos(\theta_0 + 2\pi f t) \end{cases} \tag{2}$$

3. Micro-Texture Feature Models

3.1. Pit Model

The grinding path length of an abrasive grain l_p in an ultrasonic vibration cycle is as follows:

$$l_p = \int_0^{1/f} \sqrt{V_w^2 + V_z^2} \tag{3}$$

when A = 0, and the curve length of the grinding path for CG is as follows:

$$l_c = \int_0^{1/f} \sqrt{\frac{4\pi^2 n^2}{3600} (r + u\sin(\alpha_e))^2 + V_s^2}. \tag{4}$$

According to the traditional definition, macroscopic grinding depth h_p is the interference height between the grinding wheel and workpiece, often a positive constant. In RUAG, depth h_p is considered a distance between the grinding wheel and workpiece. Therefore, the microscopic grinding depth of grain h_a between the grain and workpiece changes with time or phase angle in the grinding, as shown in Figure 2c,d.

$$h_a = h_p + A_n \sin(2\pi f t + \theta_0) \tag{5}$$

The condition of intermittent grinding is $h_p < A_n$, in which depth h_a is alternating positive and negative. In one ultrasonic cycle, when height h_a is positive, grinding time t is in the range of $\left[-\frac{1}{2\beta f}(sin^{-1}\left(\frac{h_p}{A_n}\right) + \theta_0), \frac{1}{2f} + \frac{1}{2\beta f}(sin^{-1}(\frac{h_p}{A_n}\right) + \theta_0)]$ s. Therefore, the pit length is as follows:

$$l_s = 2\int_{-\frac{1}{2\beta f}(sin^{-1}(\frac{h_p}{A_n}) + \theta_0)}^{\frac{1}{4f}} \sqrt{V_w^2 + V_z^2} \tag{6}$$

The three-dimensional model of the pit is the interference geometry between the sweeping body of abrasive grain moving along the trajectory and workpiece body, as shown in Figure 2d. Microscopic grinding depth h_a is a variable. Therefore, the cross-sectional area of the interference part is described as follows:

$$S_p = (m - 2htan\beta + h_a tan\beta) h_a. \tag{7}$$

Micro-debris volume V of abrasive grain in one cycle of RUAG is as follows:

$$V = 2\int_{-\frac{1}{2\pi f}\sin^{-1}(\frac{h_p}{A_n})}^{\frac{1}{4f}} S_p \sqrt{V_w^2 + V_z^2} dt \tag{8}$$

3.2. Grinding Depth Equation

3.2.1. Effective Abrasive Gains Number

Abrasive grains are assumed to be distributed uniformly on the surface of the grinding wheel. The distance L between adjacent abrasive grains is as follows:

$$L = d_g \left(\sqrt{\frac{\pi}{4V_g}} - 1 \right), \tag{9}$$

where V_g is the volume of abrasive grain and d_g is the average diameter of abrasive grains. The proportion of abrasive grains P involved in grinding is as follows:

$$P = \frac{1}{\sqrt{2\pi}} \int_{p}^{+\infty} e^{\frac{-x^2}{2}}, \tag{10}$$

where p is the lower limit of integration and its expression is as follows:

$$P = \left(\frac{d_{max} - d_{min}}{2} - a_p \right) \times \frac{4.4}{\delta/2}, \tag{11}$$

where d_{max} and d_{min} are the maximum and minimum diameters, respectively, of abrasive grain.
In time Δt of the grinding wheel, the grinding wheel rotational angle $\Delta\varphi$ is as follows:

$$\Delta\varphi = \frac{2\pi n}{60} \cdot \Delta t. \tag{12}$$

The corresponding circumferential length of grinding wheel $S(u)$ is as follows:

$$S(u) = \Delta\varphi \cdot (r + u\sin(\alpha_e)). \tag{13}$$

Thereafter, dynamic effective abrasive gains number N of the grinding wheel is as follows:

$$N = P \int_{u_1}^{u_2} S(u) / \left(d_g + l\right)^2 du, \tag{14}$$

where l is the average distance between two adjacent grains.

3.2.2. Calculation of Grinding Depth

In RUAG, one part of the grinding wheel participates in the grinding, relative to the workpiece dimensions. Therefore, grinding time is less than the rotational time of the grinding wheel. If the rotational angle of grinding is $\Delta\varphi_b$, then the circumferential length of grinding wheel $S_b(u)$ is as follows:

$$S_b(u) = \Delta\varphi_b (r + u\sin(\alpha_e)), \tag{15}$$

and the effective vibration period number of abrasive grain n_t is as follows:

$$n_t = \frac{\Delta\varphi_b}{\frac{2\pi n}{60}} \cdot f \tag{16}$$

The micro-debris volume of dynamic effective abrasive grains is as follows:

$$V_r = V N n_t. \tag{17}$$

Macroscopically, the total grinding depth of the grinding wheel is as follows:

$$a_p = V_s \Delta t \sin\alpha. \tag{18}$$

The macroscopic volume of material removal is as follows:

$$V_R = a_p \int_{u_1}^{u_2} S_b(u) du. \tag{19}$$

The macroscopic material removal volume equals the microscopic grinding debris:

$$V_R = V_r, \tag{20}$$

in which macroscopic grinding depth h_p of the abrasive grain is solved.
Furthermore, depth of the micro-texture is as follows:

$$A_p = A_n + h_p. \tag{21}$$

3.3. Calculation of Texture Spacing

As shown in Figure 2d, two grinding paths generated by the adjacent abrasive grains i and j have the phase angles θ_i and θ_j, respectively:.

$$d_i = l + (h - h_p - A_n sin\theta_i)tan\beta + A_\tau sin\theta_i, \tag{22}$$

$$d_j = -(h - h_p - A_n sin\theta_j)tan\beta + A_\tau sin\theta_j. \tag{23}$$

Texture spacing between them is as follows:

$$\Delta d = d_i - d_j. \tag{24}$$

In intermittent grinding, if $\theta_i = -\sin^{-1}\left(\frac{h_p}{A_n}\right)$ and $\theta_j = \frac{\pi}{2}$, then texture spacing Δd has the minimum value Δd_{min}. When $\Delta d_{min} \leq 0$, an interference exists, but no interference occurs when $\Delta d_{min} > 0$.

In continuous or conventional grinding, if $\theta_i = \theta_j = \frac{\pi}{2}$, then Δd is minimum. When $A = 0$ in traditional grinding, Δd is constant as follows:

$$\Delta d = l + (h - h_p)tan\beta. \tag{25}$$

3.4. RSA of the Theoretical Model

Intermittent ultrasonic grinding facilitates the reduction of surface grinding temperature and avoidance of thermal damage. To achieve the proper parameters of intermittent grinding, RSA was used to analyze the influence of the machining parameters (i.e., feed speed, rotational speed, and amplitude) on grinding depth and micro-texture feature.

3.4.1. Experimental Design and Result

According to spiral bevel gear grinding conditions, the grinding wheel electroplates CBN of 400# grain dimension and 100% concentration. The workpiece material is 45# steel. Ultrasonic vibration frequency is about 19.82 KHz. The Box–Behnken module is used to carry out orthogonal experiments with the influencing factors: feed speed is 0–30 μm/s, rotational speed is 2000–3000 rpm, and amplitude is 6–12 μm. Response values include h_p, l_s,Δd, and A_p. The RSA results are shown in Figure 3.

3.4.2. Discussion of RSA

Macroscopic grinding depth h_p ranges from -0.348 μm to -4.800 μm. That is, intermittent grinding condition is satisfied in this range. Grinding height h_p is positively correlated with feed speed V_s and negatively correlated with other factors. Amplitude A has the most significant influence among the factors.

Texture spacing Δd is 0.023–0.032 mm, as shown in Figure 3d, indicating that no interference appears in the adjacent paths. Texture spacing Δd has a positive correlation to rotational speed n and a negative correlation with the others. The influence of feed speed is most remarkable.

Pit length l_s and depth A_p in Figure 3e,f are 0.018–0.120 mm and 0.004–1.95 μm, respectively. Length l_s has negative and positive relations to amplitude and the others, respectively. Pit depth A_p is negatively and positively correlated with rotational speed and the others, respectively. Influence of feed speed is most significant for both.

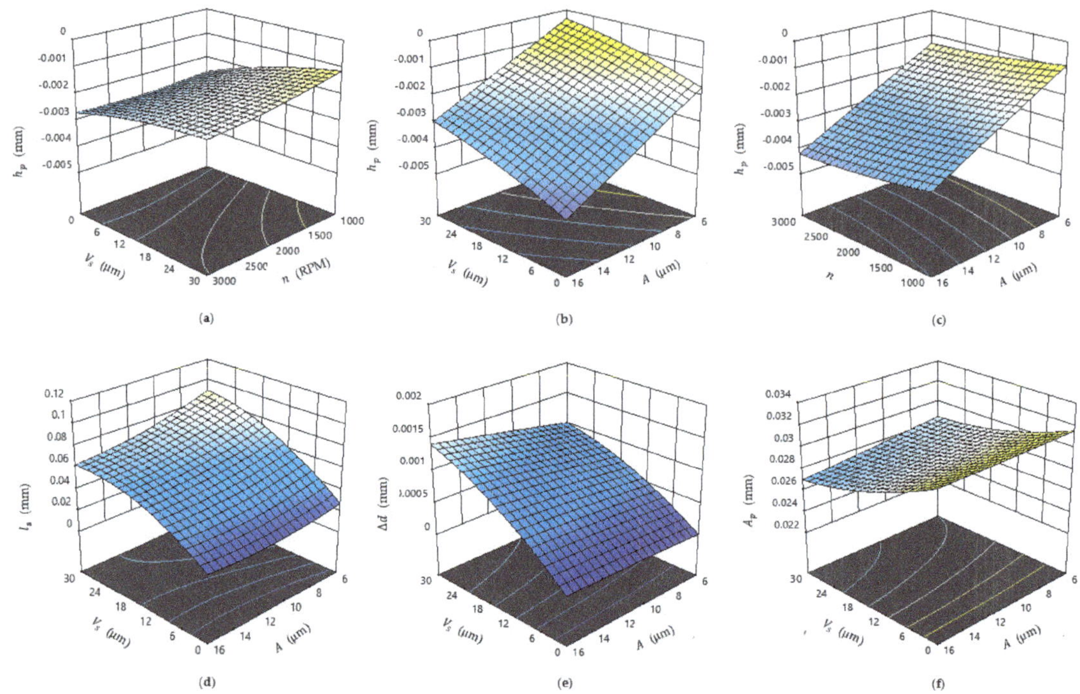

Figure 3. RSA of the relative parameters: (**a**) $V_s - n - h_p$, (**b**) $V_s - A - h_p$, (**c**) $n - A - h_p$, (**d**) $V_s - A - l_s$, (**e**) $V_s - A - \Delta d$, and (**f**) $V_s - A - A_p$.

4. RUAG Progress Simulation

4.1. Finite Element Simulation Preprocessing

Figure 4 shows the finite element models of the grinding wheel and workpiece. Grains on the outside surface of the grinding wheel grind the concave of the workpiece. In Abaqus2020 software, grinding wheel's meshes are tetrahedral, unit type is C3D4, and mesh size is 0.5 mm. The workpiece's meshes are hexahedral, unit type is C3D8R, and mesh size is 0.001 mm.

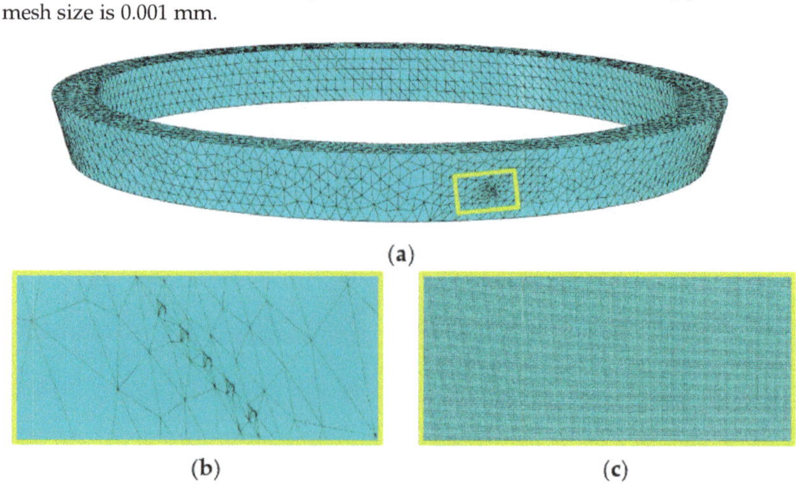

Figure 4. Geometries and Meshes: (**a**) grinding wheel, (**b**) abrasive gain, and (**c**) workpiece.

The simulation program selected the Johnson–Cook plastic constitutive model for the workpiece to calculate the constitutive relationship. The Johnson–Cook shear failure criterion was applied to describe the separation of grinding debris and the workpiece. The relative parameters are presented in Table 2.

Table 2. Material properties.

Material Properties	Workpiece (45# Steel)	Grain (CBN)
Density (kg/m^3)	7850	15,700
Elastic modulus (GPa)	210	705
Poisson's ratio	0.33	0.23
Specific heat capacity (J/(kg·°C))	526.3	178
Thermal conductivity (w/(m·°C))	6.7	24
Linear expansion coefficient (10^{-6}/°C)	9	5

The program adopted the hard contact and penalty functions to carry out the normal and tangential contact calculations between abrasive grain and workpiece, respectively, with the latter having a friction factor of 0.3. Rigid and coupling constraints were set between the grinding wheel and the coordinate system. The workpiece was set as fixed, while rotational speed and axial ultrasonic motion were imposed on the grinding wheel. Simulation time is 0.003 s.

4.2. Micro-Texture Characteristic Analysis

The single-factor experiments of machining parameters comprise three groups, as shown in Table 3. In the grinding simulation, macroscopic grinding depth h_p was set according to the theoretical calculation. The results are listed in Tables 4–6, where the depths of the micro-pit are all 1 μm because the depths are less than the mesh size.

Table 3. Machining parameters of the experiment.

Groups	Amplitude (μm)	Feed Speed (μm/s)	Rotational Speed (rpm)
1	0/8/10/12	1	3000
2	8	1/10/20/30	2000
3	8	1/10/20/30	3000

Table 4. Micro-pit dimension comparison of Group 1.

Amplitude A (μm)	Theoretical Length l_s(μm)	Theoretical Depth A_p (μm)	Theoretical Spacing Δd (μm)	Simulated Length l_s (μm)	Grinding Depth h_p (μm)
0	-	0.006	31.787	-	0.006
8	35.215	0.098	31.359	39.51	−2.363
10	32.663	0.106	31.325	36.153	−2.971
12	30.717	0.112	31.296	33.186	−3.580

Table 5. Micro-pit dimension comparison of Group 2.

Feed Speed V_s (μm/s)	Theoretical Length l_s(μm)	Theoretical Depth A_p (μm)	Theoretical Spacing Δd (μm)	Simulated Length l_s (μm)	Grinding Depth h_p (μm)
1	26.850	0.128	31.224	32.260	−2.333
10	57.038	0.559	29.275	62.500	−1.902
20	71.307	0.853	27.949	78.620	−1.609
30	81.210	1.084	26.904	85.680	−1.378

Table 6. Micro-pit dimension comparison of Group 3.

Feed Speed V_s (μm/s)	Theoretical Length l_s(μm)	Theoretical Depth A_p (μm)	Theoretical Spacing Δd (μm)	Simulated Length l_s (μm)	Grinding Depth h_p (μm)
1	35.215	0.098	46.660	46.660	−2.363
10	74.951	0.435	82.060	82.060	−2.027
20	93.371	0.668	100.790	100.790	−1.794
30	106.772	0.854	114.170	114.170	−1.608

Figure 5 shows the micro-texture of group 1. With the amplitude increasing, pit length decreases, depth increases, and texture spacing does not decrease significantly. Figure 6 shows the comparison of the micro-texture of groups 2 and 3. Figure 6a,c,e,g show the micro-texture of group 2. Note that with feed speed increasing, pit length and depth increase and texture spacing decreases. Compared with group 2, Figure 6b,d,f,h show that group 3 has an increased pit length and decreased depth. Texture spacing increment is not remarkable. The results of the three groups are consistent with the theoretical ones.

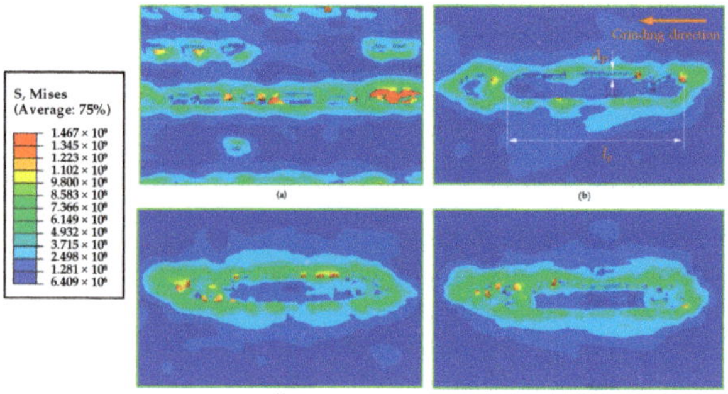

Figure 5. Simulated micro-pit of group 1: (**a**) $A = 0$, (**b**) $A = 8$ μm, (**c**) $A = 10$ μm, and (**d**) $A = 12$ μm.

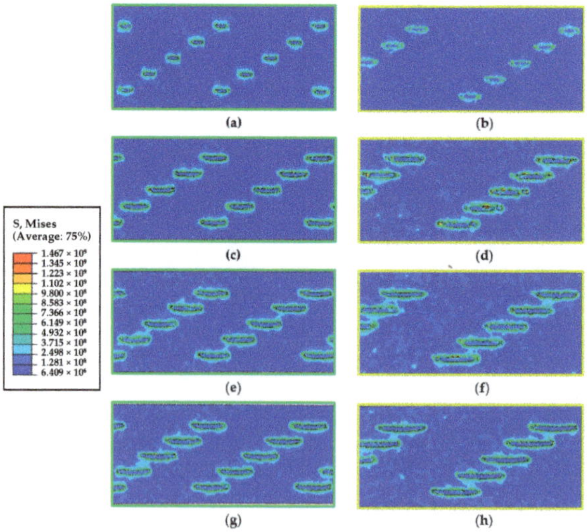

Figure 6. Simulation result comparisons of groups 2 and 3: (**a**) $n = 2000$ RPM, $V_s = 1$ μm/s, (**b**) $n = 3000$ RPM, $V_s = 1$ μm/s, (**c**) $n = 2000$ RPM, $V_s = 10$ μm/s, (**d**) $n = 3000$ RPM, $V_s = 10$ μm/s, (**e**) $n = 2000$ RPM, $V_s = 20$ μm/s, (**f**) $n = 3000$ RPM, $V_s = 20$ μm/s, (**g**) $n = 2000$ RPM, $V_s = 30$ μm/s and (**h**) $n = 3000$ RPM, $V_s = 30$ μm/s.

5. Experiment Results and Discussion

5.1. Influence of Amplitude

Figure 7 shows the surface texture comparisons in the different amplitudes, as measured using the Olympus DSX510 microscopic image analysis system. CG shown in Figure 7a has more thermal damage than the other surfaces. Compared to Figure 7b,c, Figure 7d has more thermal damage, which will increase with the amplitude increasing if it exceeds 8 µm. Therefore, the 8–10 µm amplitude is optimum to reduce thermal damage.

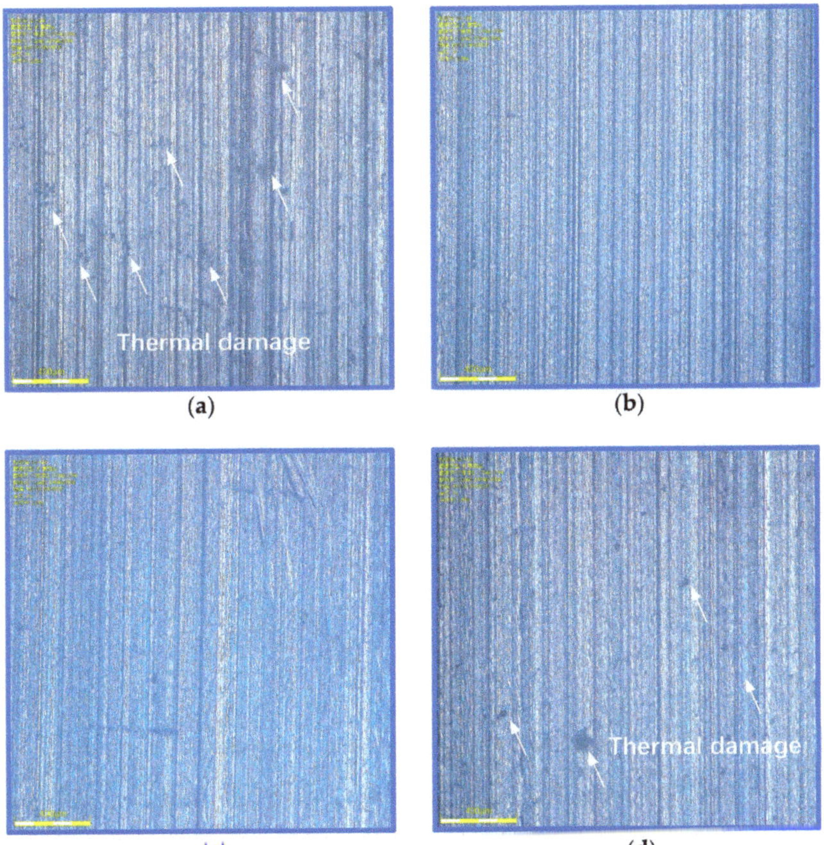

Figure 7. Surface texture comparison of group 1: (**a**) $A = 0$, (**b**) $A = 8$ µm, (**c**) $A = 10$ µm, and (**d**) $A = 12$ µm.

Contact angles of this group, as measured using the Kruss DSA30s contact angle measuring instrument, are 68.52°, 56.23°, 57.32°, and 57.39°, respectively. The 8 µm amplitude had optimal lubrication effect. When amplitude exceeds 8 µm, lubrication effect will descend with amplitude ascending.

Figure 8 shows the micro-pit comparison of a 50 um microscopic image with the length distribution results of micro-pits. In the last few seconds of practical grinding, when retracting the grinding wheel, the distance between abrasive grain and workpiece increases gradually, causing the length and depth of the micro-pit to decrease correspondingly. Therefore, longer lengths can reflect those of the micro-pit in the grinding process.

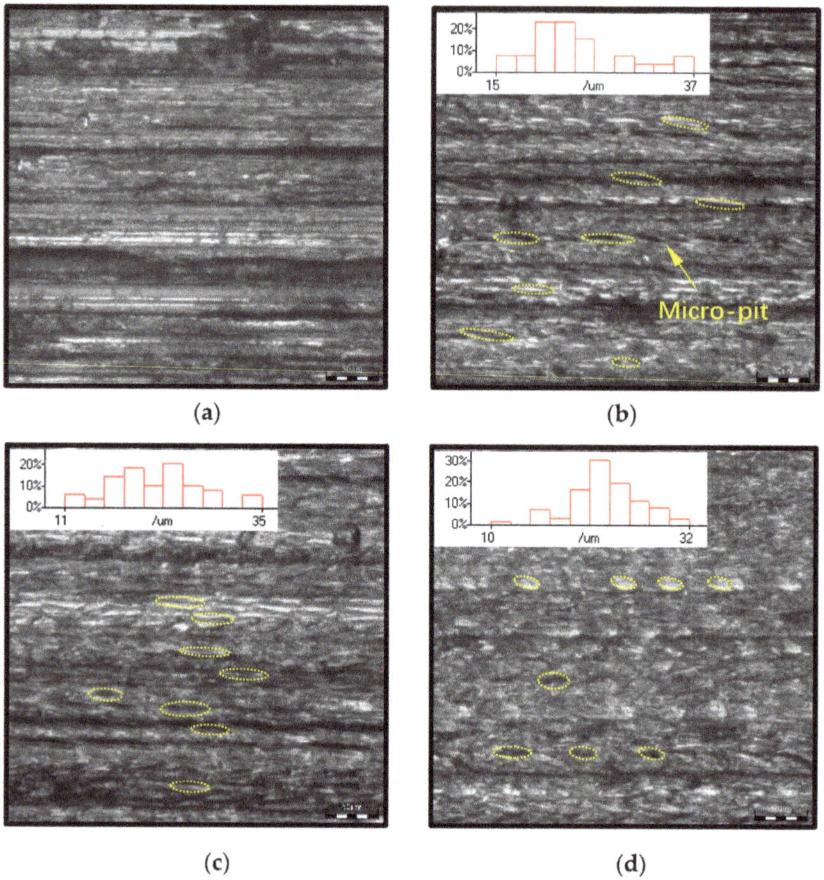

Figure 8. Micro-texture comparison of group 1: (**a**) $A = 0$, (**b**) $A = 8$ μm, (**c**) $A = 10$ μm, and (**d**) $A = 12$ μm.

Selected statistics are 10% of the longer lengths from five different positions on the surface, compared with the theoretical data calculated in the middle position of distance u on the grinding wheel and simulated data (see Figure 9a). Experimental pit length is negatively correlated with the amplitude, which is consistent with the theoretical and simulated results. Surface roughness initially decreases and increases eventually, as shown in Figure 9b. Theoretical depth of the pit is negatively related to roughness when amplitude exceeds the optimum. Therefore, the 8–10 μm amplitude range is the optimum solution for surface properties–thermal damage, lubrication, and roughness.

5.2. Influence of Feed and Rotational Speed

For group 2, pit length, depth, and roughness increase with feed speed increase, as shown in Figure 9c,d. The result is consistent with the theory and simulation trend. Practical length range measured in the uniform pit area covers the theoretical and simulated values.

Figure 10 shows the micro-texture comparison between groups 2 and 3, as measured using the Eclipse LV100ND metallographic microscope.

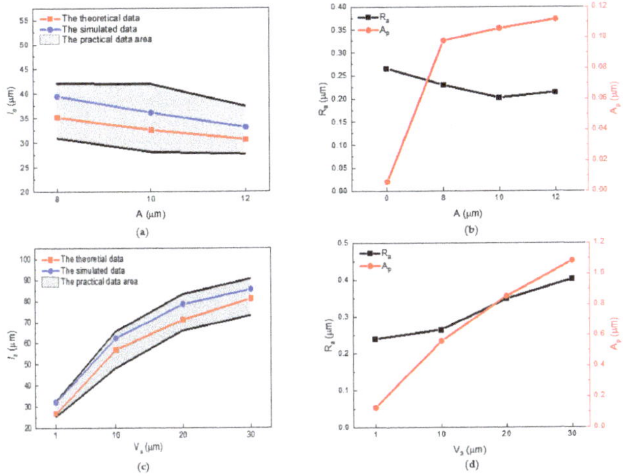

Figure 9. Micro-texture comparisons of groups 1 and 2: (**a**) $A - R_a - A_p$ of group 1, (**b**) $A - l_s$ of group 1, (**c**) $V_s - R_a - A_p$ of group 2, and (**d**) $V_s - l_s$ of group 2.

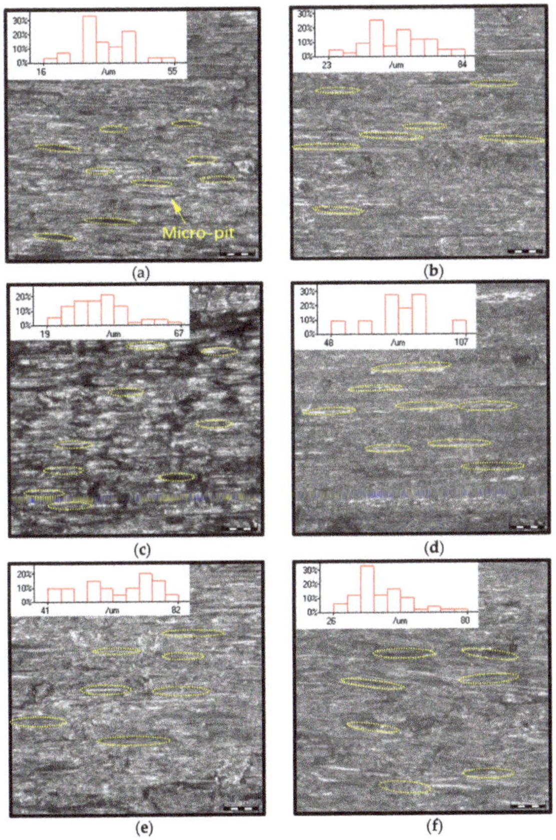

Figure 10. Experimental micro-pit comparisons: (**a**) n = 2000 RPM, V_s = 10 μm/s, (**b**) n = 2000 RPM, V_s = 20 μm/s, (**c**) n = 2000 RPM, V_s = 30 μm/s, (**d**) n = 3000 RPM, V_s = 10 μm/s, (**e**) n = 3000 RPM, V_s = 20 μm/s, and (**f**) n = 2000 RPM,V_s = 10 μm/s, convex.

For group 3, Figure 10b,d show that pit length increases significantly, depth decreases, and surface roughness decreases with increased rotational speed. When feed rate increases to 20 μm/s, the CBN coating is thermally damaged and causes direct friction between the grinding wheel body and workpiece. The presumption is that grinding a longer pit requires more continuous grinding time and further generates a large amount of heat, thereby damaging the CBN grains. Therefore, feed speed of 400# grinding grain cannot exceed 20 μm/s and 3000 rpm. In a grinding process with high feed speed, a large diameter grain should be selected, which will also result in a different surface performance.

When rotational speed is 2000 rpm, the convex and concave surfaces of the workpiece, generated by the inside and outside surfaces of the grinding wheel, are compared (see Figure 10e,f). According to theoretical calculations, pit length is negatively correlated with pressure angles and positively correlated with the radius. The experimental results showed that compared with the concave, the pit length of the convex surface descended and depth ascended because of the larger pressure angle and smaller radius. The results are consistent with the theoretical trend. The roughness of convex is worse than that of the concave.

6. Conclusions

This paper revealed the formation principle of intermittent grinding in RUAG of the conical grinding wheel. Radial vibration causes the microscopic grinding depth to change alternatingly. Therefore, there is an alternating separation between abrasive grain and workpiece. Grinding starts when grinding depth is positive and stops when grinding depth is negative to form the micro-pits. Furthermore, this study proposes the intermittent grinding condition of the conical grinding wheel.

The current paper also designed the equation to solve grinding depth on the basis of the theory that the volume of micro-debris is equal to that of the macroscopic removal material. Based on the grinding depth, the paper built the relationship between machining parameters and the micro-texture feature parameters. According to the RSA results, amplitude is the critical influencing factor for grinding depth. Moreover, grinding depth is reduced with increasing amplitude because increased amplitude will increase the material removal rate. The micro-pit's depth increases with increasing feed speed or amplitude. Lastly, the micro-pit's length increases with increasing feed speed or rotation speed.

This paper offered a 3D FEM method to simulate the micro-texture formation mechanism in RUAG. The simulation performed a single-factor experiment of machining parameters to verify the theoretical micro-texture feature parameters. The results were consistent with the theoretical ones. Moreover, the simulation of micro-texture provided a three-dimensional micro-pit structure for further research of surface properties.

The experiment proved that compared with CG, RUAG promoted surface properties, including thermal damage, lubrication, and roughness. The experimental results of the three groups showed that 8–10 um amplitude is optimal for roughness, thermal damage, and lubrication for 400# abrasive gain with decomposed radial amplitude that is 1/3 of the longitudinal one.

Compared with other radial vibrations, radial amplitude is easily optimized by adjusting the conical angle in the conical grinding wheel, which is adaptable for the different grinding wheels. This method is adaptable in grinding spiral bevel gear, worm gear, and other grinding surfaces.

Author Contributions: Conceptualization, J.H.; methodology, J.H.; validation, Y.J. and X.L.; formal analysis, X.L.; investigation, Y.J.; writing—original draft preparation, J.H.; writing—review and editing, Y.J. and J.H.; visualization, Y.J. and X.L.; supervision, Q.L.; project administration, J.H.; and funding acquisition, J.H. All authors have read and agreed to the published version of the manuscript.

Funding: This research was supported financially by the Natural Science Foundation of Tianjin, China (No. 19JCYBJC19200).

Data Availability Statement: All data are contained within the article, and fully available without restriction.

Conflicts of Interest: The authors declare no conflict of interest.

References

1. Litvin, F.L.; Wang, A.G.; Handschuh, R.F. Computerized generation and simulation of meshing and contact of spiral bevel gears with improved geometry. *Comput. Methods Appl. Mech. Eng.* **1998**, *158*, 35–64. [CrossRef]
2. Vimercati, M. Mathematical model for tooth surfaces representation of face-hobbed hypoid gears and its application to contact analysis and stress calculation. *Mech. Mach. Theory* **2007**, *42*, 668–690. [CrossRef]
3. Uhlmann, E.; Spur, G. Surface Formation in Creep Feed Grinding of Advanced Ceramics with and without Ultrasonic Assistance. *CIRP Ann.* **1998**, *47*, 249–252. [CrossRef]
4. Gao, G.F.; Zhao, B.; Xiang, D.H.; Kong, Q.H. Research on the surface characteristics in ultrasonic grinding nano-zirconia ceramics. *J. Mater. Process. Technol.* **2009**, *209*, 32–37. [CrossRef]
5. Zhao, B.; Liu, C.S.; Gao, G.F.; Jiao, F. Surface characteristics in the ultrasonic ductile honing of ZrO_2 ceramics using coarse grits. *J. Mater. Process. Technol.* **2002**, *123*, 54–60.
6. Tawakoli, T.; Azarhoushang, B. Influence of ultrasonic vibrations on dry grinding of soft steel. *Int. J. Mach. Tools Manuf.* **2008**, *48*, 1585–1591. [CrossRef]
7. Mohsen, G.N.; Mohammad, R.; Movahhedy, J.A. Ultrasonic-Assisted Grinding of Ti6Al4V Alloy. *Procedia CIRP* **2012**, *1*, 353–358.
8. Zeng, W.M.; Li, Z.C.; Pei, Z.J.; Treadwell, C. Experimental observation of tool wear in rotary ultrasonic machining of advanced ceramics. *Int. J. Mach. Tools Manuf.* **2005**, *45*, 1468–1473. [CrossRef]
9. Wu, Y.B.; Yokoyama, S.; Sato, T.; Lin, W.M.; Tachibana, T. Development of a new rotary ultrasonic spindle for precision ultrasonically assisted grinding. *Int. J. Mach. Tools Manuf.* **2009**, *49*, 933–938. [CrossRef]
10. Liang, Z.Q.; Wu, Y.B.; Wang, X.B.; Zhao, W.X. A new two-dimensional ultrasonic assisted grinding (2D-RUAG) method and its fundamental performance in monocrystal silicon machining. *Int. J. Mach. Tools Manuf.* **2010**, *50*, 728–736. [CrossRef]
11. Liang, Z.Q.; Wang, X.B.; Wu, Y.B.; Xie, L.J.; Jiao, L.; Zhao, W.X. Experimental study on brittle–ductile transition in elliptical ultrasonic assisted grinding (ERUAG) of monocrystal sapphire using single diamond abrasive grain. *Int. J. Mach. Tools Manuf.* **2013**, *71*, 41–51. [CrossRef]
12. Wang, Y.; Bin, L.; Cao, X.Y.; Wang, S.L. An experimental investigation of system matching in ultrasonic vibration assisted grinding for titanium. *J. Mater. Process. Technol.* **2014**, *214*, 1871–1878. [CrossRef]
13. Egashira, K.; Kumagai, R.; Okina, R.; Yamaguchi, K.; Ota, M. Drilling of microholes down to 10 μm in diameter using ultrasonic grinding. *Precis. Eng.* **2014**, *38*, 605–610. [CrossRef]
14. Ning, F.D.; Cong, W.L.; Pei, Z.J.; Treadwell, C. Rotary ultrasonic machining of CFRP: A comparison with grinding. *Ultrasonics* **2016**, *66*, 125–132. [CrossRef]
15. Wang, H.; Hu, Y.B.; Cong, W.L.; Hu, Z.L. A mechanistic model on feeding directional cutting force in surface grinding of CFRP composites using rotary ultrasonic machining with horizontal ultrasonic vibration. *Int. J. Mech. Sci.* **2019**, *155*, 450–460. [CrossRef]
16. Yang, Z.C.; Zhu, L.D.; Ni, C.B.; Ning, J.S. Investigation of surface topography formation mechanism based on abrasive workpiece contact rate model in tangential ultrasonic vibration assisted CBN grinding of ZrO_2 ceramics. *Int. J. Mech. Sci.* **2019**, *155*, 66–82. [CrossRef]
17. Xu, W.X.; Wu, Y.B. A novel approach to fabricate high aspect ratio micro-rod using ultrasonic vibration-assisted centreless grinding. *Int. J. Mech. Sci.* **2018**, *141*, 21–30. [CrossRef]
18. Li, C.; Zhang, F.H.; Meng, B.B.; Liu, L.F.; Rao, X.S. Material removal mechanism and grinding force modelling of ultrasonic vibration assisted grinding for SiC ceramics. *Ceram. Int.* **2017**, *43*, 2981–2993. [CrossRef]
19. Wei, S.L.; Zhao, H.; Jing, J.T. Investigation on three-dimensional surface roughness evaluation of engineering ceramic for rotary ultrasonic grinding machining. *Appl. Surf. Sci.* **2015**, *357*, 139–146. [CrossRef]
20. Wang, Q.Y.; Zhao, W.X.; Liang, Z.Q.; Wang, X.B.; Zhou, T.F.; Wu, Y.B.; Jiao, L. Investigation of diamond wheel topography in Elliptical Ultrasonic Assisted Grinding (ERUAG) of monocrystal sapphire using fractal analysis method. *Ultrasonics* **2018**, *84*, 87–95. [CrossRef]
21. Ye, L.Z.; Zhu, X.J.; Wang, L.J.; Guo, C. Study on characteristics of single cavitation bubble considering condensation and evaporation of kerosene steam under ultrasonic vibration honing. *Ultrason. Sonochem.* **2018**, *40*, 988–994. [CrossRef]
22. Wdowwik, R.; Porzycki, J.; Magdziak, M. Measurements of Surface Micro-structure Parameters after Ultrasonic Assisted and Conventional Grinding of ZrO_2 Based Ceramic Material Characterized by Different States of Sintering. *Procedia CIRP* **2017**, *62*, 293–298. [CrossRef]
23. Guo, B.; Zhao, Q.L. Ultrasonic vibration assisted grinding of hard and brittle linear micro-structured surfaces. *Precis. Eng.* **2017**, *48*, 98–106. [CrossRef]
24. Zheng, F.F.; Kang, R.K.; Dong, Z.G.; Guo, J.; Liu, J.T.; Zhang, J.T. A theoretical and experimental investigation on ultrasonic assisted grinding from the single-grain aspect. *Int. J. Mech. Sci.* **2018**, *148*, 667–675. [CrossRef]
25. Wen, Y.Q.; Tang, J.Y.; Zhou, W.; Zhu, C.C. Study on contact performance of ultrasonic-assisted grinding surface. *Ultrasonics* **2019**, *91*, 193–200. [CrossRef]
26. Qin, S.Q.; Zhu, L.; Hao, Y.P.; Shi, C.L.; Wang, S.F.; Yang, Z.C. Theoretical and experimental investigations of surface generation induced by ultrasonic assisted grinding. *Tribol. Int.* **2023**, *179*, 108120. [CrossRef]

27. Jiang, J.L.; Sun, S.F.; Wang, D.X.; Yang, Y.; Liu, X.F. Surface texture formation mechanism based on the ultrasonic vibration-assisted grinding process. *Int. J. Mach. Tools Manuf.* **2020**, *156*, 103595. [CrossRef]
28. Wang, Y.; Guangheng, D.Y.; Zhao, J.N.; Dong, Y.H.; Zhang, X.F.; Jiang, X.M.; Lin, B. Study on key factors influencing the surface generation in rotary ultrasonic grinding for hard and brittle materials. *J. Manuf. Process.* **2019**, *38*, 549–555. [CrossRef]
29. Zhao, B.; Guo, X.C.; Bie, W.B.; Chang, B.Q.; Zhao, C.Y. Thermo-mechanical coupling effect on surface residual stress during ultrasonic vibration-assisted forming grinding gear. *J. Manuf. Process.* **2020**, *59*, 19–32. [CrossRef]
30. Zhou, W.H.; Tang, J.Y.; Shao, W. Modelling of surface micro-structure and parameters matching considering the interaction of multiple rotation cycles in ultrasonic assisted grinding. *Int. J. Mech. Sci.* **2020**, *166*, 105246. [CrossRef]
31. Wang, Q.Y.; Liang, Z.Q.; Wang, X.B.; Bai, S.W.; Yeo, S.H.; Jia, S. Modelling and analysis of generation mechanism of micro-surface topography during elliptical ultrasonic assisted grinding. *J. Mater. Process. Technol.* **2020**, *279*, 116585. [CrossRef]

machines

Article

Investigation on White Layer Formation in Dry High-Speed Milling of Nickel-Based Superalloy GH4169

Jiamao Zhang [1,2], Jin Du [1,2,*], Binxun Li [1,2] and Guosheng Su [1,2]

[1] School of Mechanical Engineering, Qilu University of Technology (Shandong Academy of Sciences), Jinan 250300, China
[2] Shandong Institute of Mechanical Design and Research, Jinan 250300, China
* Correspondence: dj84105@126.com

Abstract: To investigate the formation mechanism of the white layer on the machined surface during high-speed milling of nickel-based superalloy GH4169, several cutting parameters were selected for milling experiments. Energy dispersive spectroscopy (EDS), X-ray diffraction (XRD), and electron backscattered diffraction (EBSD) were employed to characterize element distribution, phase transformation, and microstructure changes in the machined surface of the superalloy and then reveal the formation mechanism of the white layer on the machined surface. The results show that the white layer appears on the machined surface of GH4169, which is dense and has no obvious structural features. The total amount of elements in the white layer remains unchanged, but the distribution of elements such as C, N, O, Fe, and Ni changes due to phase change. The formation mechanism of the white layer is due to the dynamic recovery and dynamic recrystallization caused by the heat–force coupling effect, which leads to the grain refinement of the material and thus forms the white layer. This investigation can provide theoretical support to improve the service life of the parts in actual machining.

Keywords: high-speed milling; superalloy; machined surface; white layer; dynamic recrystallization

1. Introduction

Nickel-based superalloy GH4169 is widely used to manufacture the hottest parts of aero-engines and various gas turbines due to its high strength at high temperatures, good oxidation resistance, corrosion resistance, fatigue resistance, and other comprehensive properties [1]. However, nickel-based superalloys are typically difficult to machine because of their low thermal conductivity, high hardness, and severe work hardening. Therefore, when machining nickel-based superalloy GH4169, there are usually present high cutting forces, high cutting temperatures and severe tool wear, poor machined surface quality and the appearance of white layer on the machined surface, large residual stresses, and other defects in the machined surface [2]. High-speed cutting is currently an effective way to solve the problem of the difficult machining of superalloy. It can also be used to improve machining efficiency and surface quality [3,4]. However, in the process of high-speed cutting, there is a multi-field coupling effect of thermal–mechanical–chemical, resulting in high temperature, high stress, high strain, and high strain rates, which will lead to drastic changes in the microstructure of the workpiece machining surface, resulting in the appearance of a metamorphic layer on the machining surface—the most obvious effect is the white layer. Griffiths [5] first introduced the concept of the white layer in 1987; white layer is a generic term for a hard layer that exists under different conditions on the surface or subsurface of a metallic material, which has no characteristic morphology and appears white under a light microscope after etching with metallographic reagents. It has two characteristics: (1) higher hardness than the matrix; (2) no characteristic histological form. The physical and mechanical properties of the white layer are also quite different from the matrix material and are usually harder and more brittle than the matrix [2], and lead to the development and extension of fatigue cracks, thus reducing the fatigue life and wear resistance of the workpiece [6].

Based on the current research on the mechanism of white layer formation on the surface of high-speed cutting superalloys, it is generally believed that there are three formation mechanisms of white layer [7], namely: (a) the plastic deformation mechanism—through plastic deformation resulting in grain refinement, the machined surface produces an inhomogeneous structure or has a very fine grain structure; (b) rapid heating and quenching to produce transformation, i.e., the phase-change mechanism; (c) the mechanism of a chemical reaction between the cutting surface and elements such as carbon, nitrogen, and oxygen, etc. in the air. The (a) and (b) mechanisms are difficult to separate and may depend on each other, for example, the phase-transition temperature may be affected by plastic strain or strain rate. Among the different mechanisms, the austenite-to-martensite conversion due to rapid heating and quenching is seen as the main reason for the formation of white layer on the machined surfaces of hardened steels [8]. However, for the nickel-based superalloy GH4169, the austenitic structure of the matrix remains stable from room temperature to melting [9]; therefore, the austenite-martensite transformation mechanism is not suitable for the white layer formation mechanism of GH4169.

With the profound study of white layer on the machined surfaces of superalloys, the formation mechanism of the white layer is now generally considered to be caused by continuous dynamic recrystallization (DRX) at high strain rates, while DRX and grain refinement due to severe plastic deformation are considered to be the formation mechanism of white layer [10–12]. However, many views attribute the formation of the white layer to the heat concentration near the surface area and the low thermal conductivity of the nickel-based superalloy [13,14]. The main reason is that there are a large number of thermal, force, and chemical interactions in the formation of the white layer, which are difficult to distinguish, so there is a lot of debate about the formation mechanism of the white layer.

It was found that the white layer on the machined surface of superalloys is not a simple recrystallization layer, but a combination of a transition layer and a recrystallization layer consisting of low-angle grain boundaries and sub-grain. A transition layer with low-angular grain boundaries and sub-grains was formed below the recrystallization layer, while larger grains were mainly in the matrix [15]. During machining, high stresses and high cutting heat lead to severe plastic deformation of the workpiece, resulting in grain deformation and grain refinement of the material, especially when machining superalloys, which have high mechanical properties (i.e., high yield rate and ultimate tensile strength, hardness) and generate high cutting forces and tool wear rates during machining [16], which deforms the surface and subsurface layer [17]. In this case, the slip of dislocations within the crystal causes the deformation of the grain, while the grain boundaries are difficult to form new dislocations due to the strengthening effect of dislocations [18,19]. Thus, the high dislocation density inside the grain and the accumulation of dislocations at the grain boundaries not only lead to plastic deformation of the material but also induce the fusion of low-angle grain boundaries to form low-angle grain boundaries and recrystallization, which in turn leads to the generation of white layer. The change in cutting parameters has an important influence on the formation of white layer, among which the cutting speed and depth of cut have the greatest influence. Firstly, changes in cutting parameters can lead to changes in surface integrity, including surface roughness, residual stresses, and metamorphic layers. It was found that the white layer formation was greatly influenced by cutting parameters, mainly in the effect of changing cutting parameters on cutting forces and cutting temperatures [20]. Li et al. [21] found that the effect of cutting force on the thickness of the white layer was mainly due to the plastic deformation generated by the machining refining the grains on the machined surface and creating a high density of dislocations within it, thus affecting the phase transformation temperature of austenite and martensite in the tissue and promoting the generation of the white layer. Du et al. [22] analyzed the performance and mechanism of white layer formation on the surface of nickel-based superalloy FGH95 machined at different cutting speeds and found that when the cutting speed was less than 2000 m/min, the increase in cutting speed led to an increase in the thickness of the white layer. When other cutting parameters were kept constant, the thickness of the white layer decreased with the increase in cutting speed when the cutting speed was greater than 2000 m/min.

Variations in cutting temperature at different cutting parameters also have a significant effect on the formation of the white layer. For example, the use of cutting fluids during machining plays a greater role in reducing the thickness of the white layer [23]. Lo'pez de Lacalle et al. [24] found that the application of PAM (Plasma Assisted Milling) reduced the cutting temperature and increased productivity when machining 718. Saoubi et al. [25] found that, compared with traditional oil-based coolant, milling Inconel-718 with low-temperature machining significantly reduced the process temperature in the cutting zone, reduced tool wear, improved surface quality, and reduced the thickness of white layer. It has also been found [26] that the white layer is formed as a result of thermal concentration in the sub-surface portions of the material due to machining, and that the low thermal conductivity of nickel-based superalloys enhances this effect, while the increase in cutting temperature produces high strain rates. This combined effect leads to an increase in the thickness of the heat-affected layer and plastic deformation zone, resulting in work hardening of the superalloy [27,28].

As mentioned above, it is found that the current research on white layer formation during superalloy machining is mainly focused on the grain refinement mechanism. Most of the studies have shown the existence of dynamic recrystallization structure in the white layer, but the mechanism of dynamic recrystallization and grain refinement still needs further in-depth study. The research on the influence of cutting parameters on the white layer is also limited to the research on the influence of the thickness of the white layer, the research on the mechanism of the formation of the white layer is not quite perfect, and the mechanism of the formation of the white layer is still more controversial. Therefore, to study the formation mechanism of the white layer, to find out the changes in the material during the cutting process, to reduce the thickness of the white layer, to reduce the generation of the white layer, to study the role played by different cutting parameters during the cutting process, and to find out the different effects of different cutting parameters on the formation of the white layer, this paper analyzes the overall microstructure of the white layer on the surface of GH4169 after high-speed milling and investigates the different effects of cutting parameters during the cutting process on the formation of the white layer and the physical and mechanical properties of the white layer. The formation mechanism of white layer formation was analyzed through the change in material microstructure, which provides a theoretical basis for reducing the generation of white layer and improving the service life of parts.

2. Experimental Procedures

2.1. Material

The workpiece material for this experiment is GH4169, a nickel-based superalloy with a hardness of 43–45 HRC, whose main components are shown in Table 1. The heat treatment method is as follows: solution treatment at 960 °C for 1 h and then air-cooling to room temperature, followed by heating to 720 °C for 8 h; then cooling to 620 °C in the furnace at 50 °C/h, holding for 8 h, and cooling to room temperature. After heat treatment, GH4169 material is mainly composed of matrix γ, reinforced γ', and γ'' and precipitated δ. The γ'' is the main strengthening phase, which is dispersion co-lattice precipitation in the matrix, and the co-lattice strain between the γ'' and the γ forms a large mismatch to achieve the strengthening effect of the γ''. The γ' is a secondary strengthening phase which is spherical and diffusely precipitated in the matrix with nanometer size, and has a co-grid strain strengthening effect. The material size of the workpiece was chosen as 50 mm × 30 mm × 3 mm and mounted on a special fixture.

Table 1. Main chemical composition of superalloy GH4169 (wt.%).

C	Cr	Co	Mo	Mn	AL	Nb	Cu	Fe	Ni
2.80	17.60	0.75	2.95	0.22	0.44	4.96	0.32	16.28	Bal

The machine tool used for the cutting experiments is a CNC vertical machining center Doosan DNM-415 (Korea) with a maximum speed of 12,000 r/min and power of 15 KW. The cutting tool employed in this experiment was a face milling cutter supplied by KENNAMETAL INC, which can approximate orthogonal cutting. The cutting inserts were SNHX12L5PZTNGP KC725M with TiAlN-PVD-coated. Before each test cutting, the insert was changed to a fresh one in order to eliminate the influence of tool wear on the machined surface integrity.

2.2. Milling Experiments

The cutting speeds were 500, 600, 700, 800, 900, and 1000 m/min, the radial depths of cut were 0.1, 0.2, and 0.3 mm, and the feeds were kept constant at 0.02 mm/r, total of 18 sets of cutting tests. The cutting tests were conducted under dry cutting conditions. The milling experimental setup is shown in Figure 1a. The cutting force and cutting temperature were collected during the cutting process, the force-measuring device was Kistler-9129AA three-way dynamic-cutting-force-measuring instrument; the cutting-force-measuring direction is shown in Figure 1a. The cutting temperature testing device was a FlirA315 infrared thermal imager from the USA.

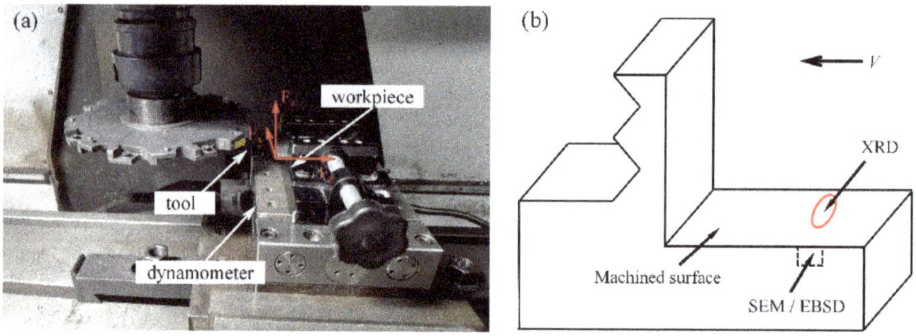

Figure 1. Experimental procedures: (**a**) Schematic diagram of the milling process; (**b**) schematic diagram of test areas of SEM XRD and EBSD.

2.3. Microstructure Characterization

The specimens were cut by EDM machine ZJK 7532A after each cutting process, and a 10 mm × 10 mm × 3 mm rectangular specimen was cut out of the part of the machined specimen containing the machined surface to obtain the cross-section containing the white layer, and the specimens were inlaid with metallographic thermal inlay material in the inlay machine to ensure a flat surface for subsequent grinding and polishing. After grinding and polishing the specimens to obtain a mirror image of the scratch-free surface, etching treatment was carried out with an etching solution of 2.5% copper chloride, 48.8% hydrochloric acid, and 48.7% ethanol, and the etching time was about 25 s. The surface of the specimens was then rinsed with running water and an alcohol solution.

The metallographic changes on the machined surface were observed using an ultra-deep field 3D viewing microscope and a Scanning Electron Microscope (SEM). The measurement points for both white layer and metamorphic layer thickness were selected at an angle of 45° to the tool feed direction. An EDS spectrometer was used to analyze the distribution and elements contained in the white layer. The phases in the white layer of the machined surface were analyzed by XRD using Cu-Ka radiation at 40 kV and 150 mA with a 2θ scan range of 30° to 90° and a scan increment of 0.02°. Grain size and orientation changes in the surface layer were measured using EBSD. EBSD measurements were performed at an accelerating voltage of 20 kV with a tilt angle of 70° and a step size of 0.2 mm, and the experimental results of EBSD were analyzed using Channel 5 software. Vickers microhardness tester was used to measure the change in hardness of the machined surface of the workpiece. The test area of the experiment is shown in Figure 1b.

3. Results and Discussion

3.1. Results

3.1.1. Microstructure Analysis

The images of the inlaid metallographic specimens after corrosion treatment are shown in Figure 2. From Figure 2a, the microstructure and grain boundaries of the matrix of GH4169 alloy block are visible, and the machined surface is covered with a thin layer of a significantly different color from the matrix, which is bright white under the optical microscope, it is the "white layer". The microstructure of the white layer under SEM is shown in Figure 2b, and this layer exhibits a significantly different microstructure from that of the substrate material, being dense and without obvious crystal features. From Figure 2a, it can be seen that there is a dark region at the transition to the substrate below the white layer, called the transition zone or plastic flow layer [29], where there is a strong plastic deformation flow along the cutting direction, and there are obvious slip lines on the machined surface and significant elongation of the surface grains, indicating that plastic flow occurs along the cutting direction during the cutting process, resulting in significant plastic deformation [8].

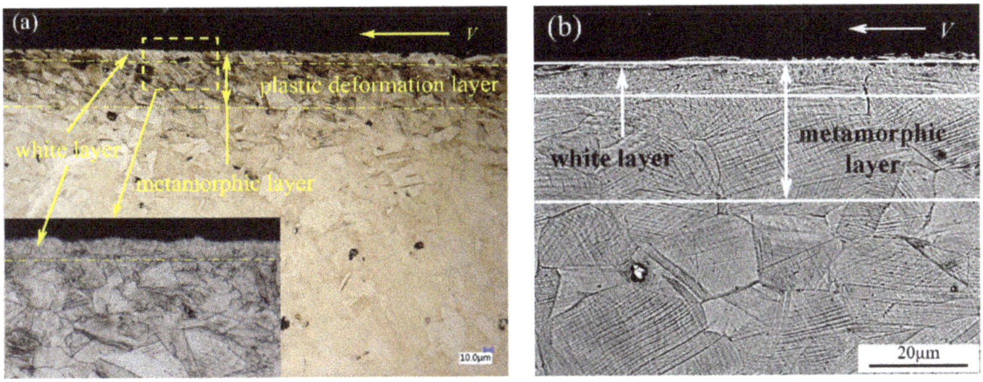

Figure 2. The etched transverse section under the cutting condition of V = 800 m/min, a_p = 0.2 mm: (**a**) white layer under a microscope; (**b**) white layer under SEM.

The presence of white layer can be observed in the workpiece samples after milling under all cutting parameters. Four representative groups of parameters were selected for analysis, and their measured cutting forces and cutting temperatures are shown in Table 2.

Table 2. Variation in cutting force and cutting temperature under different cutting parameters.

Cutting Speed (r/min)	Cutting Depth (mm)	Cutting Force (N)	Cutting Temperature (°C)
500	0.1	265.3	493.3
700	0.1	246.8	560.9
800	0.3	298.6	655.7
1000	0.3	287.8	798.5

It is obvious from Figure 3 that the white layer produced by machining under all parameters shows the characteristics of dense, no obvious crystal features; the plastic deformation area of the machined surface shows obvious grain tilting, pore elongation and compression, and slip lines; and the machined surface has shear flow, and the appearance of slip lines is due to the slip of crystals resulting in the surface of the specimen after polishing with varying heights. After corrosion, the presence of slip lines can be seen in the metallographic image, which also proves from the side that strong plastic deformation leads to grain slip.

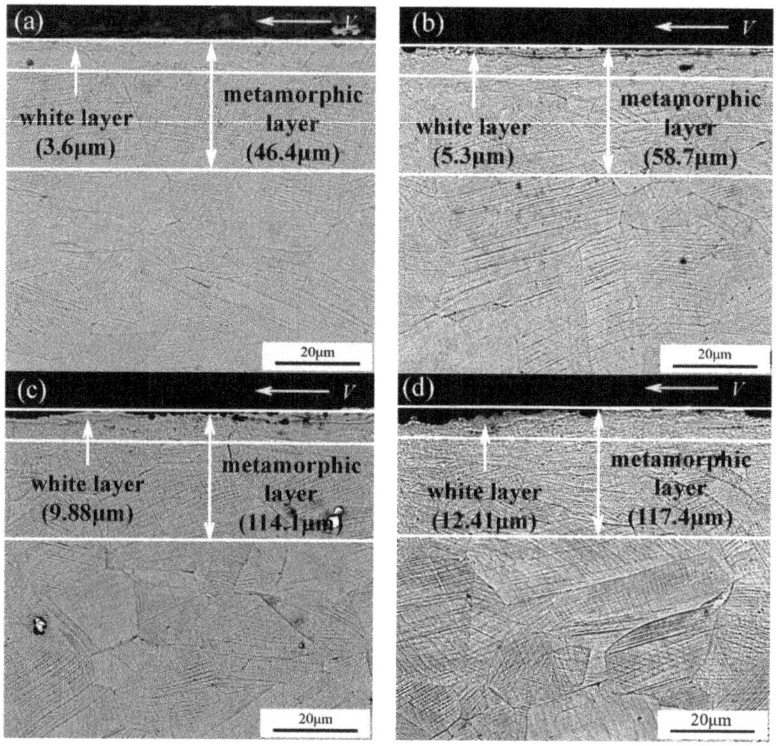

Figure 3. Microscopic morphology of the material under different processing parameters: (**a**) $V = 500$ m/min, $a_p = 0.1$ mm; (**b**) $V = 700$ m/min, $a_p = 0.1$ mm; (**c**) $V = 800$ m/min, $a_p = 0.3$ mm; (**d**) $V = 1000$ m/min, $a_p = 0.3$ mm.

3.1.2. EDS Analysis

The variation in elemental content and distribution in the white layer can provide a reliable basis for the analysis of the phase change in the workpiece. The changes in the content of the main elements in the white layer and the matrix material were analyzed by EDS and the results are shown in Table 3. It can be seen that the content of major elements such as Ni, Cr, and Fe in the white layer is significantly lower compared to the matrix, while the content of elements such as C and O at the boundary is significantly higher. At the same time, the elemental content of Mo, Ti, and Nb in the white layer is higher and that of Cr and Co is lower. In GH4169, Nb and Mo elements are mainly distributed in the γ', γ'', and δ phases. It is inferred that as the content of Ti and Nb elements increases after machining, the content of the strengthening phase γ' also increases, with an increase of 8–15%, while Cr and Co elements are mainly concentrated in γ, indicating that γ' appears to be diffusely precipitated during the machining process. In addition, the C and O content in the white layer increases more obviously, which can be inferred from the fact that the oxide and carbide content in the white layer increases due to the chemical reaction between the workpiece material and air under high temperature and pressure during the machining process.

Table 3. The main element's content in the matrix and white layer (wt.%).

	Ni	Cr	Fe	Nb	Mo	Ti	C	O
Matrix	Bal.	17.61	16.28	4.26	2.95	0.68	2.80	3.20
White layer	Bal.	13.53	13.08	4.94	3.01	0.78	12.68	9.09

The EDS analysis of the machined surface of GH4169 along the depth of cut direction is shown in Figure 4, with different color fluorescence maps indicating different elemental distributions. The EDS line analysis of the elements on the machined surface and the base material is shown in Figure 5.

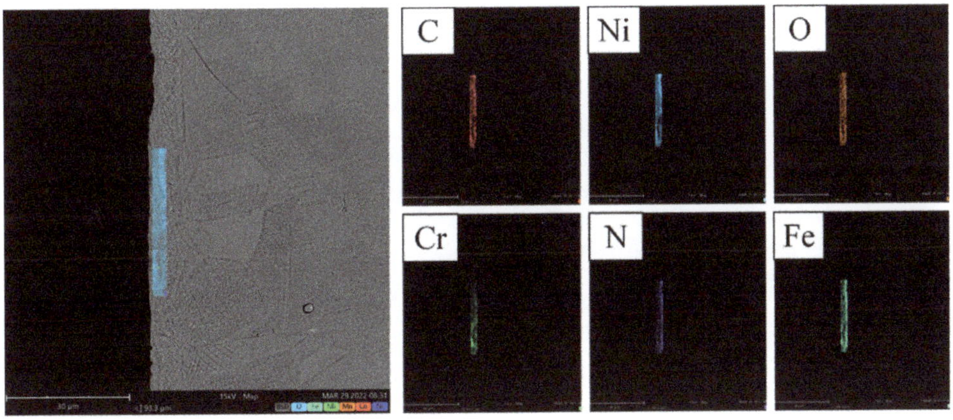

Figure 4. EDS surface scan of white layer.

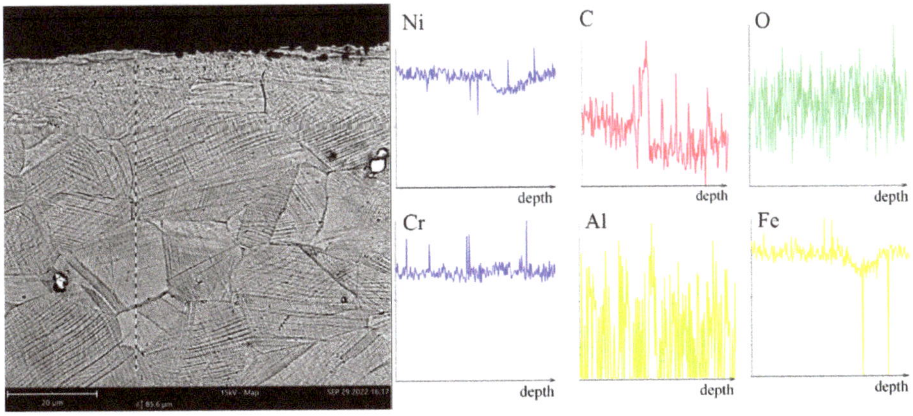

Figure 5. EDS line sweep.

According to the EDS elemental analysis, the composition and content of the elements in the metamorphic layer basically did not change, but the content of the elements in the white layer microstructure changed significantly; the content of carbon and oxygen elements especially increased significantly, according to the analysis of previous research, in the process of high-speed cutting due to strong plastic deformation and high cutting heat, which prompted the material oxidation reaction and carburizing, resulting in the aggregation of oxygen and carbon elements in the material. The clustering of C and O can also be seen in the fluorescence map in Figure 4, which indicates that the transfer of elements occurred during processing.

As can be seen in Figure 5, the total content of the elements is the same in the workpiece matrix and machined surface, i.e., in the white layer, the metamorphic layer, and the matrix, but if there are pores or cracks present, there are fluctuations or vacancies in the element content at the pores or cracks. Combined with the fluorogram analysis, C and O appear to be clustered, and the line sweep also indicates that a change in the elemental distribution has occurred. This proves that there is no chemical change in the material during processing—only a change in the location of the element distribution.

3.1.3. XRD Analysis

The XRD analysis of the matrix and machined surface of the GH4169 superalloy was carried out to analyze the effect of the cutting process on the microstructure of the workpiece. The diffraction spectra of the matrix and machined surface are shown in Figure 6.

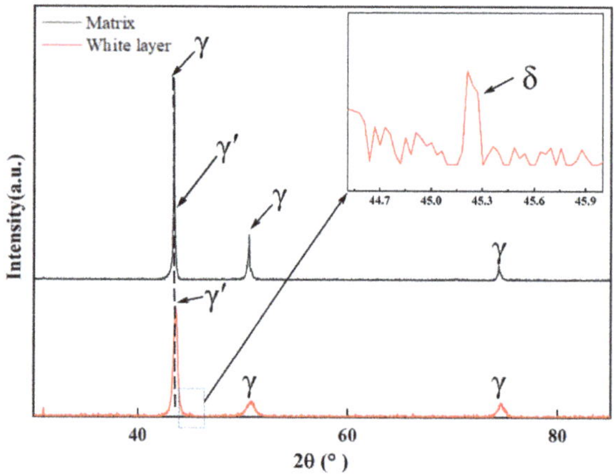

Figure 6. XRD analysis of white layer and matrix.

The XRD patterns of the GH4169 matrix and white layer show five peaks with diffraction angles 2θ of 43.80°, 45.75°, 46.37°, 50.99°, and 74.81°, respectively. By matching the standard PDF cards of the phases with MDI Jade software, the corresponding phases were measured as γ (Ni, Fe), γ' $Ni_3(Al, Ti)$, γ'' (Ni_3Nb), and δ (Ni_3Nb). From Figure 6, it can be observed that the XRD peaks of the processed surface white layer have lower peak heights and wider peak shapes than those of the base material. Fully crystallized crystals have larger grains, so the diffraction peaks are stronger, sharper, and symmetrical, while less crystalline crystals tend to have internal defects such as dislocations and finer grains, so the diffraction peaks are wider and less intense [22]. Therefore, it shows that the white layer is less crystalline and has smaller grains than the matrix material, which proves that grain refinement occurs in the white layer.

Combined with the EDS elemental analysis, the content of Ni, Fe, etc. decreased and the content of Ti and Nb increased, while the diffraction peak of the white layer was shifted, as seen in the enlarged figure; this proved that the content of γ in the white layer decreased while the content of γ' increased, thus indicating that a phase change occurred during the white layer formation.

3.1.4. EBSD Analysis

The specimens with V = 800 m/min and a_p = 0.2 mm were selected for EBSD analysis, and the inverse polar diagram (IPF), Kernel Average Misorientation (KAM), and grain orientation spread (GOS) plots were obtained as shown below, in which grain boundaries of different sizes were plotted with different color lines, with low-angle grain boundaries defined as 2°~10° and high-angle grain boundaries defined as more than 10°.

As shown in the IPF of Figure 7, compared to the inside of the deformed grains, low-angular grain boundaries formed by recrystallization appear at the surface grain boundaries of the machined surface in addition to the initially formed high angular grain boundaries, while there is no significant change in the grain size of the plastically deformed layer compared to that of the matrix. A clear refinement of the grains is observed in the most superficial layer, with elongated grains at the white layer and a large number of low-angle grain boundaries (LAGBs) appearing inside the high-angle grain boundaries (HAGBs). The sub-grain formation due to dynamic recovery can be clearly seen in Figure 7 [11]. The initial formation of HAGB fragments can be observed in the subsurface layer, which is surrounded by dense LAGBs with LAGBs also attached at the ends.

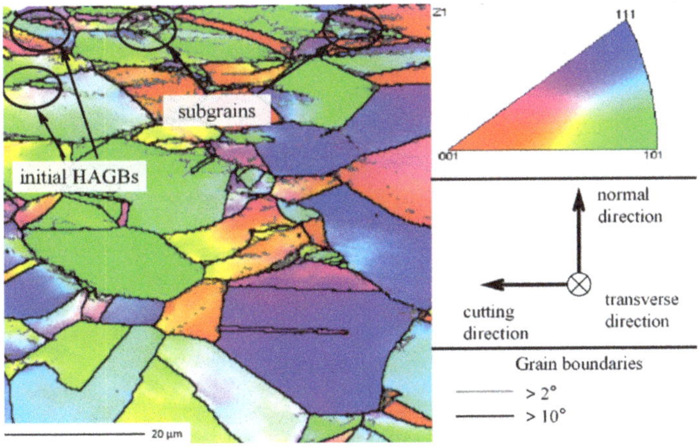

Figure 7. Inverse pole figure (IPF).

To specifically analyze the changes in recrystallization within the white layer, the EBSD orientation results were water-averaged along the depth direction by the sub-subset function of the channel 5 software to extract the proportion of LAGBs and HAGBs at different depths; the results are shown in Figure 8. It can be seen from Figure 8 that the proportion of LAGBs is higher the closer to the machined surface, and the proportion of LAGBs increases from 86.5% to 98.1% from the inside of the material to the plastic deformation zone, and the proportion of HAGBs gradually decreases from 13.5% to 1.9%. This result indicates that the grains in the transition zone are extruded, elongated, bent, etc. during the cutting process, resulting in a large number of dislocations inside the grains, which are driven by the combination of deformation and high temperature to form LAGBs, resulting in a significant increase in the proportion of LAGBs. At the white layer region, the percentage of LAGBs decreased to 97.2%, and the percentage of HAGBs increased from 1.9% to 2.8%, which proved that during the cutting process, LAGBs fused to form HAGBs, and the degree of grain recrystallization inside the white layer increased to form HAGBs, which made the percentage of HAGBs increase.

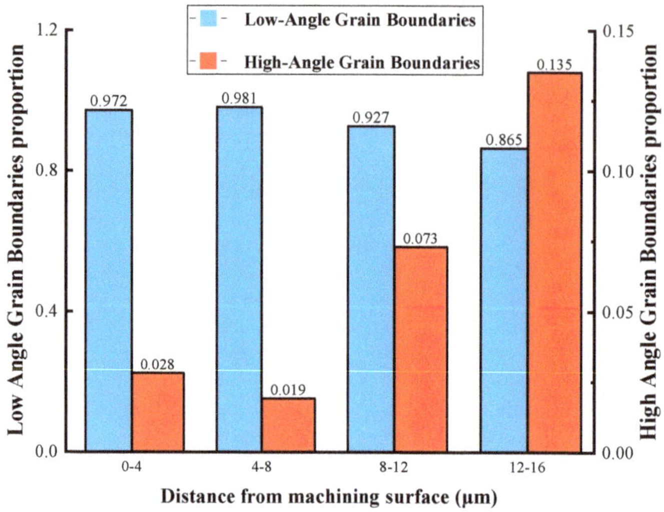

Figure 8. Distribution of high-/low-angle grain boundaries in the surface layer of the milling process.

To further analyze the variation in grains and the extent of dislocations in the metamorphic layer region, the analysis of the KAM was carried out, as shown in Figure 9. The KAM can represent the average orientation difference distribution at each point inside the grains. From Figure 9, it can be seen that the degree of grain deformation gradually increases from the inside of the material to the plastic deformation layer, and the dislocation density also gradually increases. The dislocation density at the grain junction is significantly higher than that inside the grain, indicating that the grain produces an uneven deformation, resulting in a higher degree of strain at the grain boundary than inside the grain. The high strain and high strain rate generated during the milling process lead to the formation of high-density dislocations, and the dislocation density increases further with the further increase in strain rate and strain. In contrast, the presence of a grain refinement layer on the surface and a significant decrease in the dislocation density inside the grains suggest that the grains on the surface may nucleate by accommodating dislocations from highly deformed grains under strain and high temperature, producing equiaxed grains with well-defined boundaries [30], referred to as DRX, resulting in a reduction in dislocations on the surface.

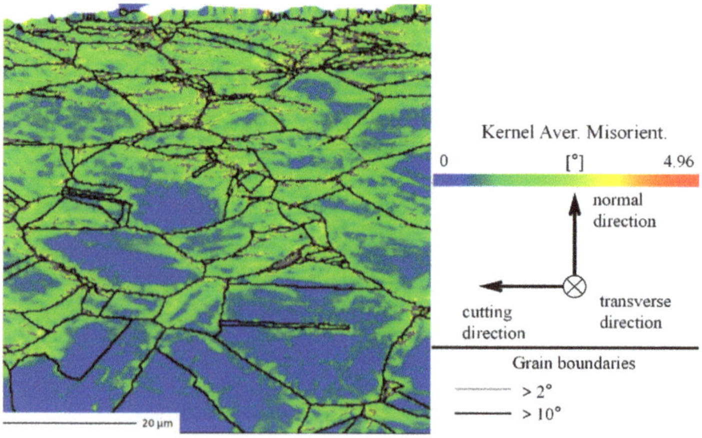

Figure 9. Kernel average misorientation (KAM).

Figure 10 shows the GOS, which is one of the methods to distinguish dynamically recrystallized grains from deformed substrates; a larger GOS value means a larger degree of lattice distortion and a higher dislocation density. The threshold value of GOS for recrystallized grains is generally chosen as 2°, i.e., a GOS value less than 2° is considered as a recrystallized grain. As can be seen in Figure 10, there is a small number of fine recrystallized grains in the white layer region, at the grain boundaries in the plastic deformation zone, and inside the grains.

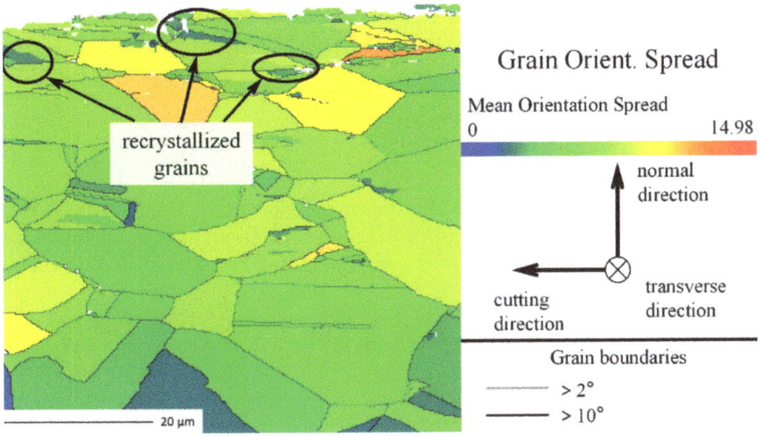

Figure 10. Grain orientation spread (GOS).

3.2. Discussion

Through analyzing EDS, XRD, and EBSD, it can be seen that there is obvious phase transformation and recrystallization in the sub-surface. The white layer zone is inferred from the decrease in Al elements and γ' phase content that the γ' phase dissolved in the white layer. The dissolution of γ' phase and the fragmentation and refinement of δ phase make the white layer more homogeneous and change the physical and mechanical properties of the material. Compared with the workpiece, the lattice mismatch of γ and γ'' phases on the processed surface is reduced [31]. This means that the strain and strain energy at the interface is reduced, which reduces the yield strength, increases the stability, and increases the hardness of the white layer.

With the assistance of EBSD analysis, it can be seen from the IPF diagram that the proportion of LAGB in the white layer is much larger than that in the matrix, and the proportion of LAGB gradually decreases from the white layer region to the plastic deformation region; meanwhile, combined with the KAM and GOS diagrams, it can be seen that LAGB is mainly formed in the region of higher dislocation density, dividing the large grains into many sub-grains, which proves that the material undergoes dynamic recovery (DRV).

DRX can be classified as discontinuous dynamic recrystallization (DDRX), continuous dynamic recrystallization (CDRX), and geometric dynamic recrystallization. K. Huang [32] pointed out that the identification of the DRX mechanism concerns whether the HAGBs formed after severe plastic deformation was transformed from LAGBs, micro shear, deformation bands, or pristine HAGBs. To determine whether DRX was generated, as shown in Figure 9, more LAGBs and sub-grains were formed near the grain boundaries relative to the interior of the deformed grains, and it can be observed that the initially formed HAGB fragments were surrounded by dense LAGBs with LAGBs also attached at the ends, indicating that the HAGBs were transformed from LAGBs, which is evidence that DRX occurred in the most superficial region.

At the same time, elongated grains are formed in the white layer region, and the internal crystal defects are reduced. DRV and DRX absorb the dislocations and reduce the defect density, resulting in a dense white layer [15]. In the plastic deformation zone, the dislocation density is reduced and mainly concentrated at the intersection of grain boundaries because the stress and strain in the subsurface layer gradually decrease and the transferred cutting heat also gradually decreases; thus, the dynamic recrystallization cannot be produced in the plastic deformation zone, therefore the proportion of HAGB gradually increases and the grains are mainly subjected to plastic deformation by extrusion, elongation, and bending. The effect of material anisotropy on cutting forces can be explained and quantified by the means of crystal and grain morphology effects with the help of pole diagrams [33]. Due to the decrease in temperature and strain, DRX occurs mainly in the outermost layer driven by dislocations, and only plastic deformation occurs in the subsurface layer, which also indicates that the white layer is produced by DRX due to high temperature and strong strain [32,34].

4. Conclusions

In this paper, the high-speed milling of nickel-based superalloy GH4169 was carried out via different machining parameters to investigate the formation mechanism of white layer during machining; additionally, the effects of cutting speeds and depths of cut on the physical and mechanical properties of the workpiece were analyzed. The main conclusions are as follows:

1. White layer is formed on the machined surface of GH4169 superalloy, with the microstructure characteristics greatly different from the matrix, and showing no obvious crystal features.

2. The total amount of elements in the white layer remains unchanged, but the distribution of elements such as C, N, O, Fe, and Ni changes due to phase change, and the crystallinity of the white layer is observed to be poorer and finer than that of the matrix material, which proves that grain refinement occurs in the white layer.

3. White layer shows no obvious crystal features due to the occurrence of dynamic recovery and dynamic recrystallization, absorption of dislocations, and reduction in surface defects. The mechanism of white layer formation is continuous dynamic recrystallization due to the thermal coupling effect, which makes the LAGBs increase to produce white layer.

Author Contributions: J.Z.: Conceptualization; Resources; Writing—Original Draft; Writing—Review and Editing; J.D.: Conceptualization; Resources; Writing—Original Draft; Writing—Review and Editing; Oversight and leadership; Project administration; B.L.: Conceptualization; Resources; Writing—Review and Editing; Oversight and leadership; Project administration; G.S.: Conceptualization; Resources; Writing—Review and Editing; Oversight and leadership; Project administration. All authors have read and agreed to the published version of the manuscript.

Funding: This research was funded by the National Natural Science Foundation of China (52275438, 51675289), the Natural Science Foundation of Shandong Province (ZR2020ME160, ZR2021QE230) and the Basic research project of pilot project of integration of science, education and production (Qi-lu University of Technology (Shandong Academy of Sciences)) (2022PY007).

Data Availability Statement: The datasets generated and/or analyzed during the current study are available from the corresponding author upon reasonable request.

Conflicts of Interest: The authors declare that they have no competing interests.

References

1. Akhtar, W.; Sun, J.F.; Chen, W.Y. Effect of Machining Parameters on Surface Integrity in High Speed Milling of Super Alloy GH4169/Inconel 718. Mater. *Manuf. Process.* **2016**, *31*, 620–627. [CrossRef]
2. Yin, Q.; Liu, Z.; Wang, B.; Song, Q.; Cai, Y. Recent progress of machinability and surface integrity for mechanical machining Inconel 718: A review. *J. Adv. Manuf. Technol.* **2020**, *109*, 215–245. [CrossRef]
3. Mohsan, A.U.H.; Liu, Z.Q.; Padhy, G.K. A review on the progress towards improvement in surface integrity of Inconel 718 under high pressure and flood cooling conditions. *Int. J. Adv. Manuf. Technol.* **2017**, *91*, 107–125. [CrossRef]
4. Segreto, T.; D'Addona, D.; Teti, R. Tool wear estimation in turning of Inconel 718 based on wavelet sensor signal analysis and machine learning paradigms. *Prod. Eng.* **2020**, *14*, 693–705. [CrossRef]

5. Griffiths, B.J. White layer formations at machined surfaces and their relationship to white layer formations at worn surfaces. *J. Tribol.* **1985**, *107*, 165–171. [CrossRef]
6. Liang, X.; Liu, Z.; Wang, B. State-of-the-art of surface integrity induced by tool wear effects in machining process of titanium and nickel alloys: A review. *Measurement* **2019**, *132*, 150–181. [CrossRef]
7. Zhang, F.Y.; Duan, C.Z.; Sun, W.; Ju, K. Influence of White Layer and Residual Stress Induced by Hard Cutting on Wear Resistance During Sliding Friction. *J. Mater. Eng. Perform.* **2019**, *28*, 7649–7662.
8. Zeng, H.H.; Yan, R.; Hu, T.T.; Du, P.L.; Wang, W.; Peng, F.Y. Analytical Modeling of White Layer Formation in Orthogonal Cutting of AerMet100 Steel Based on Phase Transformation Mechanism. *J. Manuf. Sci. Eng.* **2019**, *141*, 1087–1357. [CrossRef]
9. Cherdyntsev, V.V.; Pustov, L.Y.; Kaloshkin, S.D.; Tomilin, I.A.; Shelekhov, E.V.; Estrin, E.I. Phase transformations in powder iron-nickel alloys produced by mechanical alloying. *Phys. Met. Metallogr.* **2009**, *107*, 466–477. [CrossRef]
10. Herbert, C.; Axinte, D.; Hardy, M.; Brown, P.D. Investigation into the characteristics of white layers produced in a nickel-based superalloy from drilling operations. *Mach. Sci. Technol.* **2012**, *16*, 40–52. [CrossRef]
11. Agmell, M.; Ahadi, A.; Zhou, J.M.; Peng, R.L.; Bushlya, V.; Stahl, J.E. Modeling subsurface deformation induced by machining of Inconel 718. *Mach. Sci. Technol.* **2017**, *21*, 103–120. [CrossRef]
12. Imran, M.; Mativenga, P.T.; Gholinia, A.; Withers, P.J. Assessment of surface integrity of Ni superalloy after electrical discharge, laser, and mechanical micro-drilling processes. *Int. J. Adv. Manuf. Technol.* **2015**, *79*, 1303–1311. [CrossRef]
13. Herbert, C.R.J.; Kwong, J.; Kong, M.C.; Axinte, D.A.; Hardy, M.C.; Withers, P.J. An evaluation of the evolution of workpiece surface integrity in hole-making operations for a nickel-based superalloy. *J. Mater. Process. Technol.* **2012**, *212*, 1723–1730. [CrossRef]
14. Thakur, A.; Mohanty, A.; Gangopadhyay, S. Comparative study of surface integrity aspects of Incoloy 825 during machining with uncoated and CVD multilayer coated inserts. *Appl. Surf. Sci.* **2014**, *320*, 829–837. [CrossRef]
15. Liao, Z.R.; Polyakov, M.; Diaz, O.G.; Axinte, D.; Mohanty, G.; Maeder, X. Grain refinement mechanism of nickel-based superalloy by severe plastic deformation—Mechanical machining case. *Acta Mater.* **2019**, *180*, 2–14. [CrossRef]
16. Liao, Z.R.; Gao, D.; Lu, Y.; Lv, Z.K. Multi-scale hybrid HMM for tool wear condition monitoring. *Int. J. Adv. Manuf. Technol.* **2016**, *84*, 2437–2448. [CrossRef]
17. Ding, R.; Knaggs, C.; Li, H.; Li, Y.G.; Bowen, P. Characterization of plastic deformation induced by machining in a Ni-based superalloy. *Mater. Sci. Eng. A* **2020**, *778*, 139104. [CrossRef]
18. Jia, P.; Rong, Y.; Huang, Y. Condition monitoring of the feed drive system of a machine tool based on long-term operational modal analysis. *Int. J. Mach. Tools Manuf.* **2019**, *146*, 103454. [CrossRef]
19. Shaikh, V.A.; Scharf, T.W.; Boubekri, N. Microlubrication machining of 1018 steel: The effect of a biodegradable lubricant on the microstructural integrity. *Lubr. Sci.* **2017**, *29*, 357–376. [CrossRef]
20. Thirumalai, R.; Techato, K.; Chandrasekaran, M.; Venkatapathy, K.; Seenivasan, M. Experimental investigation during turning process of titanium material for surface roughness. *Mater. Today Proc.* **2021**, *45*, 1423–1426. [CrossRef]
21. Li, L.; Lai, D.; Ji, Q.; Huang, J.; Lin, Y.; Pan, M. Influence of tool characteristics on white layer produced by cutting hardened steel and prediction of white layer thickness. *Int. J. Adv. Manuf. Technol.* **2021**, *113*, 1215–1228. [CrossRef]
22. Du, J.; Liu, Z.; Lv, S. Deformation-phase transformation coupling mechanism of white layer formation in high-speed machining of FGH95 Ni-based superalloy. *Appl. Surf. Sci.* **2014**, *292*, 197–203. [CrossRef]
23. Paturi, U.M.R.; Reddy, N.S. Progress of machinability on the machining of Inconel 718: A comprehensive review on the perception of cleaner machining. *Clean. Technol. Environ. Policy* **2021**, *5*, 100323. [CrossRef]
24. De Lacalle, L.N.L., Sanchez, J.A.; Lamikiz, A.; Celaya, A. Plasma assisted milling of heat-resistant superalloys. *J. Manuf. Sci. Eng.* **2004**, *126*, 274–285. [CrossRef]
25. M'Saoubi, R.; Axinte, D.; Soo, S.L.; Nobel, C.; Attia, H.; Kappmeyer, G. High-performance cutting of advanced aerospace alloys and composite materials. *CIRP Ann.* **2015**, *64*, 557–580. [CrossRef]
26. Chaize, E.; Dumont, F.; Truffart, B.; Girinon, M.; Brosse, A.; Dorlin, T. Influence of lubrication mode onto residual stress generation in turning. *Procedia CIRP* **2022**, *108*, 390–393. [CrossRef]
27. Sharman, A.R.C.; Hughes, J.I.; Ridgway, K. Workpiece surface integrity and tool life issues when turning inconel 718™ nickel-based superalloy. *Mach. Sci. Technol.* **2004**, *8*, 399–414. [CrossRef]
28. Sharman, A.R.C.; Hughes, J.I.; Ridgway, K. Surface integrity and tool life when turning Inconel 718 using ultra-high pressure and flood coolant systems. *Proc. Inst. Mech. Eng. Part B-J. Eng. Manuf.* **2008**, *222*, 653–664. [CrossRef]
29. Jin, D.; Liu, Z. Damage of the machined surface and subsurface in orthogonal milling of FGH95 superalloy. *Int. J. Adv. Manuf. Technol.* **2013**, *68*, 1573–1581. [CrossRef]
30. La Monaca, A.; Murray, J.W.; Liao, Z.; Speidel, A.; Robles-Linares, J.A.; Axinte, D.A. Surface integrity in metal machining—Part II: Functional performance. *Int. J. Mach. Tools Manuf.* **2021**, *164*, 103718. [CrossRef]
31. Kruk, A.; Wusatowska-Sarnek, A.M.; Ziętara, M.; Jemielniak, K.; Siemiątkowski, Z.; Czyrska-Filemonowicz, A. Characterization on White Etching Layer Formed During Ceramic Milling of Inconel 718. *Met. Mater. Int.* **2018**, *24*, 1036–1045. [CrossRef]
32. Huang, K.; Logé, R.E. A review of dynamic recrystallization phenomena in metallic materials. *Mater. Des.* **2016**, *111*, 548–574. [CrossRef]

33. Pérez-Ruiz, J.D.; de Lacalle, L.N.L.; Urbikain, G.; Pereira, O.; Martínez, S.; Bris, J. On the relationship between cutting forces and anisotropy features in the milling of LPBF Inconel 718 for near net shape parts. *Int. J. Mach. Tools Manuf.* **2021**, *170*, 103801. [CrossRef]
34. Zhang, H.; Zhang, K.; Jiang, S.; Zhou, H.; Zhao, C.; Yang, X. Dynamic recrystallization behavior of a γ'-hardened nickel-based superalloy during hot deformation. *J. Alloys Compd.* **2015**, *623*, 374–385. [CrossRef]

Journal of
Functional Biomaterials

MDPI

Article

Effect of Interbody Implants on the Biomechanical Behavior of Lateral Lumbar Interbody Fusion: A Finite Element Study

Hangkai Shen [1], Jia Zhu [2], Chenhui Huang [1], Dingding Xiang [3,4,*] and Weiqiang Liu [2,4,*]

[1] China United Engineering Corporation, Hangzhou 310000, China
[2] Tsinghua Shenzhen International Graduate School, Tsinghua University, Shenzhen 518055, China
[3] School of Mechanical Engineering and Automation, Northeastern University, Shenyang 110057, China
[4] State Key Laboratory of Tribology in Advanced Equipment, Tsinghua University, Beijing 100084, China
* Correspondence: xiangdd@mail.neu.edu.cn (D.X.); weiqliu@hotmail.com (W.L.)

Abstract: Porous titanium interbody scaffolds are growing in popularity due to their appealing advantages for bone ingrowth. This study aimed to investigate the biomechanical effects of scaffold materials in both normal and osteoporotic lumbar spines using a finite element (FE) model. Four scaffold materials were compared: Ti6Al4V (Ti), PEEK, porous titanium of 65% porosity (P65), and porous titanium of 80% porosity (P80). In addition, the range of motion (ROM), endplate stress, scaffold stress, and pedicle screw stress were calculated and compared. The results showed that the ROM decreased by more than 96% after surgery, and the solid Ti scaffold provided the lowest ROM (1.2–3.4% of the intact case) at the surgical segment among all models. Compared to solid Ti, PEEK decreased the scaffold stress by 53–66 and the endplate stress by 0–33%, while porous Ti decreased the scaffold stress by 20–32% and the endplate stress by 0–32%. Further, compared with P65, P80 slightly increased the ROM (<0.03°) and pedicle screw stress (<4%) and decreased the endplate stress by 0–13% and scaffold stress by approximately 18%. Moreover, the osteoporotic lumbar spine provided higher ROMs, endplate stresses, scaffold stresses, and pedicle screw stresses in all motion modes. The porous Ti scaffolds may offer an alternative for lateral lumbar interbody fusion.

Keywords: finite element; porous scaffold; lumbar interbody fusion; osteoporosis; endplate stress

1. Introduction

Lumbar interbody fusion has been widely used in the treatment of lumbar degenerative diseases [1]. Due to the large footprint area, lateral lumbar interbody fusion had the advantage of significantly reducing the range of motion (ROM) of the index segment compared to other surgical techniques [2]. Traditional titanium scaffolds could reduce the ROM effectively and achieve a satisfactory fusion rate [1,3]. However, previous studies reported that titanium scaffolds may result in complications such as stress shielding, scaffold migration, and scaffold subsidence because of the high stiffness of the solid titanium [4,5]. Polyetheretherketone (PEEK) scaffolds have shown suitable elastic modulus and excellent clinical performance [6]. However, PEEK scaffolds showed a higher risk of instability (up to 32.8%) because of their poor cell adhesion effect [7].

Recently, porous titanium scaffolds manufactured by additive manufacturing (AM) have also been used in lumbar interbody fusion. By adjusting pore topology and porosity, the porous AM scaffold could achieve optimal mechanical properties [8,9]. In addition, the porous scaffold facilitated bone ingrowth, which was beneficial for long-term stability [10]. Moreover, the porous structure was similar to that of the cancellous bone, which could promote nutrient transportation [11]. The interlock between the porous structure and bony elements could also reduce the risk of scaffold migration [12]. Some studies have investigated the biomechanical performance of porous scaffolds using finite element (FE) and experimental methods. For example, Meena et al. [13] reported that porous scaffolds with pore sizes ranging from 400–600 μm could significantly decrease the stress shielding. Similarly, Tsuruga et al. [14] also reported that pore sizes of 300–400 μm were optimum for bone ingrowth. In the numerical and experimental study of Tsai et al. [15], they reported that a porosity of 69–80% achieved satisfactory biomechanical performances. Wang et al. [9] also optimized the design of porous scaffolds using methods of topology optimization.

In addition, the prevalence of osteoporosis has increased dramatically among the elderly in recent years, and an increasing number of patients requiring lumbar interbody fusion suffer from osteoporosis [16]. Previous studies reported that osteoporosis significantly altered the normal biomechanics of the lumbar spine [17] and showed a higher risk of fracture, internal fixation failure, and scaffold subsidence [18,19]. However, according to the existing literature, the biomechanics of lateral scaffolds with various materials were not fully understood. Further, few studies have investigated the impact of scaffold materials on the biomechanical performance of the osteoporotic lumbar spine. It was necessary to study the postoperative biomechanics of the osteoporotic lumbar spine.

Consequently, the aim of this study was to evaluate the biomechanical effects of scaffold materials in both the normal and osteoporotic lumbar spines using the finite element (FE) method. The main novelty of the present study is the application of porous scaffolds in both the normal and osteoporotic lumbar spines. Four scaffold materials were compared: PEEK, Ti6Al4V, 65% porous Ti, and 80% porous Ti. The material properties of porous Ti were measured by the compression test. A total of 10 FE models were developed and simulated. The results of ROM, endplate stress, scaffold stress, and pedicle screw stress were calculated and simulated.

2. Materials and Methods

2.1. Finite Element Model of the Lumbar Spine

A previously developed and validated FE model of the intact L1-5 human lumbar spine was used in this study (Figure 1) [20]. The geometry of the current model was obtained from 0.7-mm-thick computed tomography (CT) scans of a healthy volunteer (female, age 37 years, height 158 cm, weight 52 kg). Then a total of 492 CT images were transformed into a 3D geometric lumbar model, which was meshed in Hypermesh (Version 14.0, Altair Technologies, Inc., Fremont, CA, USA). The current model is composed of cortical bone (1 mm thick) [21], cancellous bone, posterior bone, intervertebral disc, and all seven kinds of ligaments: anterior longitudinal ligament, posterior longitudinal ligament, ligament flava, interspinous ligament, supraspinal ligament, intertransverse ligament, and capsular ligament. The intervertebral disc included a cartilaginous endplate (0.5 mm thick), an annulus fibrosus, and a nucleus pulposus [21]. The ligaments were simulated by tension-only truss elements. The frictionless surface-to-surface contact was applied between the facet joints [22]. Finally, the intact L1-5 FE model included 120,978 nodes and 555,063 elements, which could effectively ensure the accuracy of the calculation. Abaqus (Version 14.0, Simulia, Inc., Providence, RI, USA) was used to perform the FE simulation.

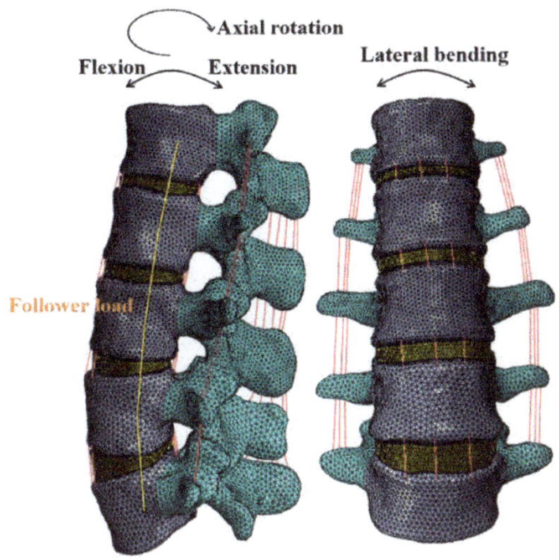

Figure 1. Finite element model of the L1-5 lumbar spine.

2.2. Finite Element Models of L3-4 Lateral Interbody Fusion Construct

In order to simulate the lateral interbody fusion (LIF) procedure, the total nucleus pulposus and partial annulus were removed at the L3-4 segment [23]. The scaffold was inserted at the L3-4 disc space with bilateral pedicle screw fixation (Figure 2). PEEK, titanium alloy (Ti6Al4V, Ti), and two titanium scaffolds with varying porosities (65% and 80%, P65, and P80) were simulated as lateral scaffold materials. [23,24]. All the scaffolds were filled with graft bone to promote bone ingrowth. The porous scaffolds were produced through 3D printing technology, with an average pore size of 350–400 μm [23,24]. All the scaffolds were 28-mm long, 9-mm wide, and 7-mm high. The pedicle screws (45 mm in length and 6 mm in diameter) were interconnected by longitudinal rods (6 mm in diameter). The bone-screw and bone-scaffold interfaces were simulated as tie constraints to represent the long-term effects after instrumentation [25]. The material of the pedicle screws was Ti6Al4V. The porous scaffolds and pedicle screws used in this study were created using Solidworks 2014 software (Dassault Systemes SolidWorks Corp., Waltham, MA, USA).

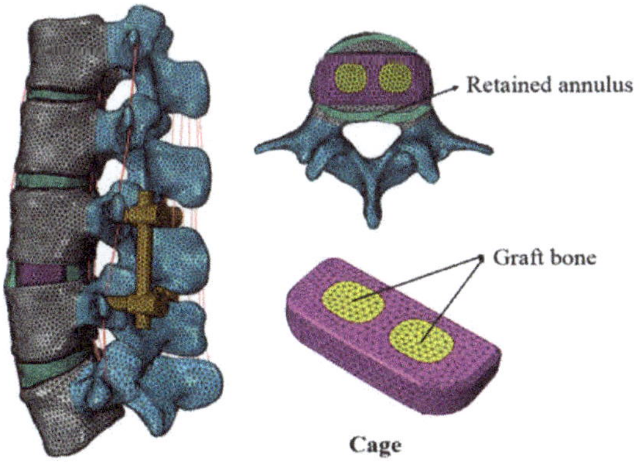

Figure 2. Finite element model of L3-4 lateral interbody fusion.

Figure 3a displays the mechanical test for the porous materials. A compression test was performed based on the ISO13314:2011 standard [23,24] by using a universal material testing system (Instron 8874, Instron Corporation, Canto, MA, USA). The compression test was carried out according to the displacement-controlled protocol [26]. The loading rate during the test was set at 1 mm/min. The load-displacement curves were recorded, and the Young's modulus of the porous titanium materials was obtained. The average Young's modulus was 8800 MPa for the P65 samples and 5000 MPa for the P80 samples.

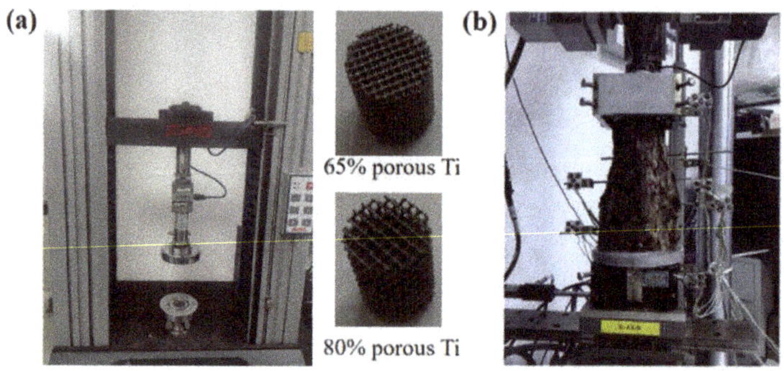

Figure 3. (**a**) Mechanical tests for the porous materials; (**b**) In vitro test with 6 human cadaver lumbar spines.

In order to further investigate the impact of the scaffold materials on the osteoporotic lumbar spine, the osteoporotic lumbar FE model was also developed. According to the previous studies [27,28], the Young's modulus of all bony structures was reduced, by 66% for the cancellous bone, and by 33% for the cortical bone endplates, and posterior elements. The thickness of the cortical bone was not significantly changed [29]. Table 1 summarizes the material properties used in the FE models.

Table 1. Material properties used in the FE models.

Component	Young's Modulus (MPa)	Poisson Ratio	Cross-Sectional Area (mm^2)	References
Cortical bone	12,000 (osteoporosis:8040)	0.3	-	[23]
Cancellous bone	100 (osteoporosis:34)	0.2	-	[23]
Posterior bone	3500 (osteoporosis:2345)	0.25	-	[21]
Endplate	24 (osteoporosis:16.1)	0.25	-	[30]
Nucleus pulposus	1	0.49	-	[23]
Annulus fibrosus	4.2	0.45	-	[23]
Anterior longitudinal	20	0.3	63.7	
Posterior longitudinal	20	0.3	20	
Ligament flava	19.5	0.3	40	
Interspinal	11.6	0.3	40	[21]
Supraspinal	15	0.3	30	
Intertransverse	58.7	0.3	3.6	
Capsular	32.9	0.3	60	
PEEK	3500	0.3	-	[23]
Ti6Al4V	110,000	0.3	-	[23]
65% porous Ti	8800	0.05	-	Experiment
80% porous Ti	5000	0.05	-	Experiment
Graft Bone	3500	0.25	-	[23]

Additionally, to validate the intact lumbar model, we followed the same protocol used in the previous study [26]. The inferior surface of L5 was constrained in all directions, and then four moments (8 Nm flexion, 6 Nm extension, 6 Nm lateral bending, and 4 Nm axial rotation) were applied to the L1 segment, respectively. The simulated results of the FE model were compared with the experimental results of Renner et al. [26].

Further, we used six human cadaver lumbar spines (L1-5) to further validate the FE model (Figure 3b). The lumbar spines were obtained from the Research Institute of Tsinghua University in Shenzhen, and the experiment followed an institutional medical ethics procedure. The L5 segment of the lumbar spine specimens were fixed by Wood's alloy.

A compressive follower load of 280 N, which represented partial body weight, was applied along the curvature of the lumbar spine, and then four moments (8 Nm flexion, 6 Nm extension, 6 Nm lateral bending, and 4 Nm axial rotation) were applied at the L1 segment through the MTS Bionix Spine Simulation Machine (MTS system, Inc., Minneapolis, MN, USA), respectively. The experiment was performed at a room temperature of 23 °C and a humidity of 35–80%. In addition, 0.9% saline was sprayed on the specimen periodically to keep it moist. Each segment of the lumbar spine was implanted with an NDI marker (NDI, Inc., Waterloo, ON, Canada), which consisted of four targets. The location of each segment was represented by the coordinates of the NDI markers. The rate at the moment was 0.1°/s. Each motion mode (flexion, extension, lateral bending, and axial rotation) was repeated three times, and the results of the ROM were measured and averaged for analysis. Finally, the results of the ROM of each segment were compared with the results of the intact FE model in order to validate the FE model.

2.3. Boundary and Loading Conditions

In this study, apart from the intact model, four lateral scaffold materials were simulated: PEEK, Ti6Al4V (Ti), 65% porous Ti (P65), and 80% porous Ti (P80). The inferior surface of the L5 vertebral body was fixed in all directions, and a follower load of 280 N was applied along the contour of the lumbar spine to represent the body weight [21]. Then, a moment of 7.5 Nm was applied at the L1 vertebral body to simulate flexion, extension, lateral bending, and axial rotation [21,24]. Both the normal and osteoporotic spine conditions were perform. In total, 40 simulation calculations for 10 models (five normal models and five osteoporotic models) and four motion modes were performed. The following results were simulated and measured: ROM at the surgical segment, endplate stress, scaffold stress, and pedicle screw stress.

3. Results

3.1. Model Validation

Under 280 N follower load and 7.5 Nm moment, the simulated results of L1-5 ROM were compared with the experimental results of Renner et al. [26]. These results were also compared with the results of our cadaveric specimen test, as shown in Figure 4. The total ROMs of the intact L1-5 lumbar spine were 34.4° in flexion-extension, 13.4° in axial rotation, and 30.8° in left-right lateral bending. The predicted ROM of each segment was within one standard deviation of the results derived from previous studies [26] and our cadaveric measurement. Therefore, the current FE model was validated.

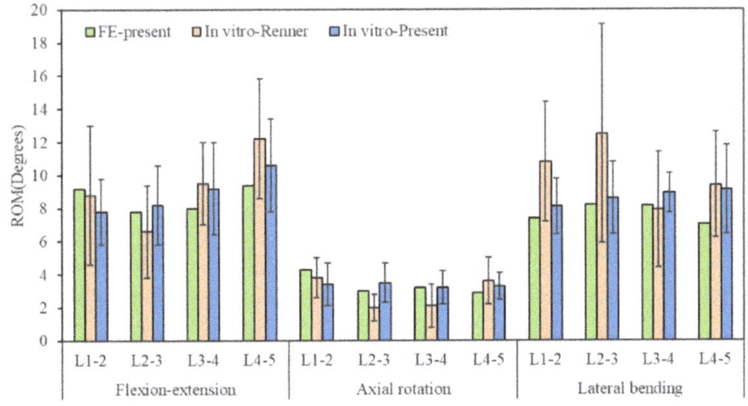

Figure 4. Validation of the finite element model.

3.2. ROM of the Surgical Segment

Under 280 N follower load and 7.5 Nm moment, the ROM of the surgical segment was displayed in Figure 5. The ROMs of the surgical models were normalized to the intact model [23]. For the normal spine condition (Figure 5a), the predicted ROMs for all surgical models decreased by more than 96% after the lateral scaffold was inserted. ROMs of the solid Ti scaffold were the lowest compared among all the scaffolds (1.2–3.4% of the intact model). ROMs of the P65 scaffold were slightly lower than those of the P80 scaffold (deviation within 0.03°). The PEEK scaffold and the P80 scaffold showed similar ROMs in all motion modes.

For the osteoporotic spine condition, the ROMs of the surgical models decreased by more than 94% compared with the intact model (Figure 5b). Compared with the normal spine, the ROMs were higher in the osteoporotic spine, and the trend of the ROMs was similar.

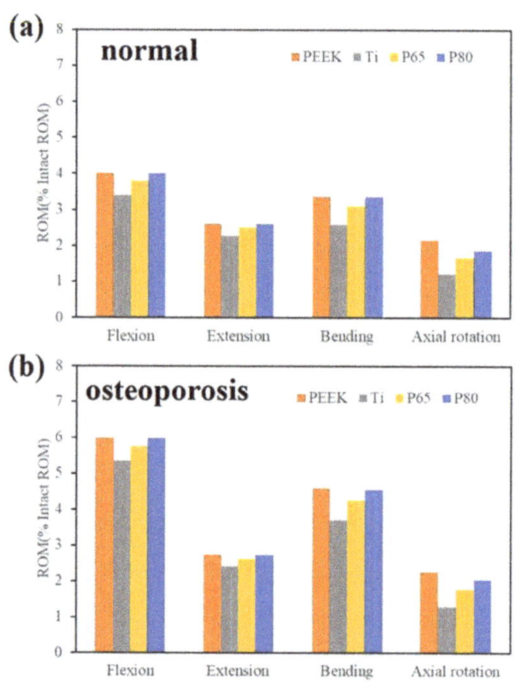

Figure 5. ROM of the surgical segment. (**a**) Normal spine condition; (**b**) Osteoporotic spine condition.

3.3. Endplate Stress

Figure 6 displays the endplate stress of the L3 inferior endplate. For the normal spine condition, the Ti scaffold provided the highest endplate stress among all surgical models (Figure 6a). The PEEK and P80 scaffolds appeared to provide similar endplate stresses in all motion modes. Compared with P80, the endplate stress of P65 was slightly higher in extension, lateral bending, and axial rotation. For the osteoporotic spine condition, the endplate stress increased by 3.4–15.9% compared with the normal spine condition (Figure 6b). The Ti scaffold provided the highest endplate stress in extension and axial rotation. However, in flexion and lateral bending, PEEK and P80 showed higher endplate stresses.

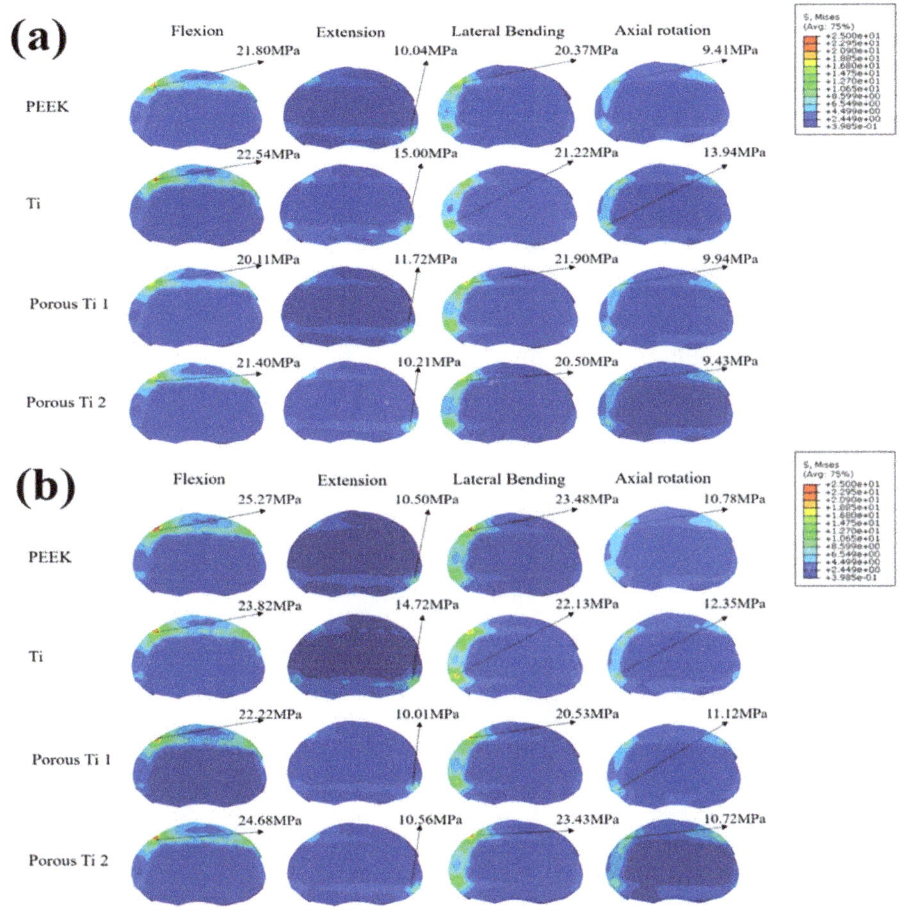

Figure 6. L3 inferior endplate stress. (**a**) Normal spine condition; (**b**) Osteoporotic spine condition.

Figure 7 displays the endplate stress of the L4 superior endplate. For both normal and osteoporotic spine conditions, the Ti scaffold provided the highest endplate stress in all motion modes. Compared with P80, P65 provided higher endplate stresses. Compared with the normal spine condition, the endplate stress increased by 3.0–25% for osteoporotic spines.

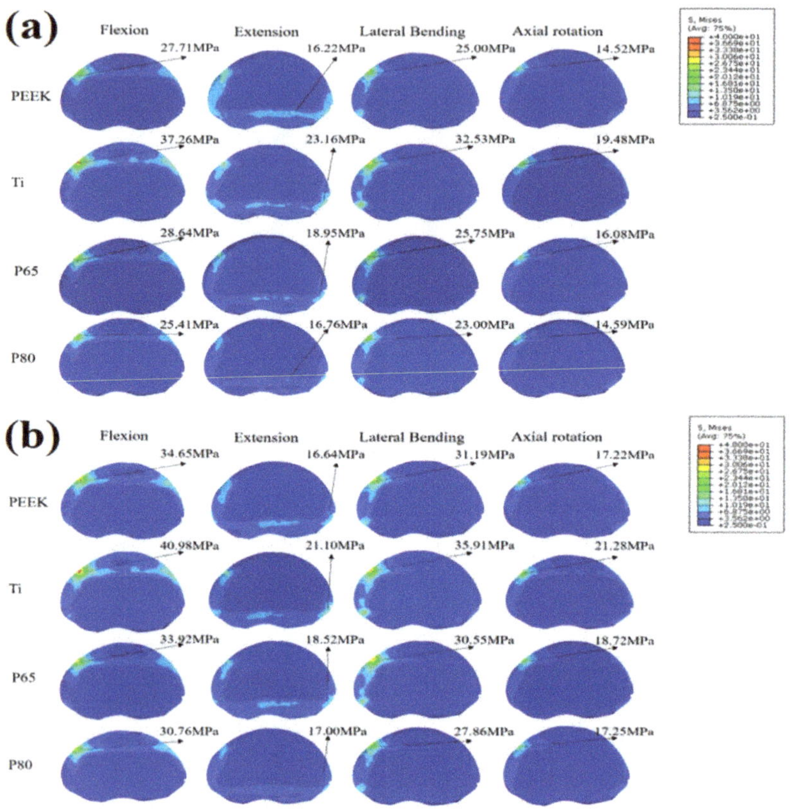

Figure 7. L4 superior endplate stress. (**a**) Normal spine condition; (**b**) Osteoporotic spine condition.

3.4. Scaffold Stress

Figure 8 showed the maximum scaffold stresses of the surgical models. For both normal and osteoporotic spines, the Ti scaffold provided the highest scaffold stress in all motion modes, followed by P65 and P80, while PEEK had the lowest scaffold stress (Figure 8a). Compared to solid Ti, PEEK decreased the scaffold stress by 53–66%, and porous Ti decreased the scaffold stress by 20–32%. In addition, P80 decreased the scaffold stress by approximately 18% compared with P65.

Compared with normal spines, osteoporotic spines increased the maximum scaffold stress by 13–30% in flexion, 13–22% in extension, 12–29% in lateral bending, and 20–28% in axial rotation. In addition, as shown in Figure 8b, the maximum scaffold stress always appeared at the edges of the scaffold.

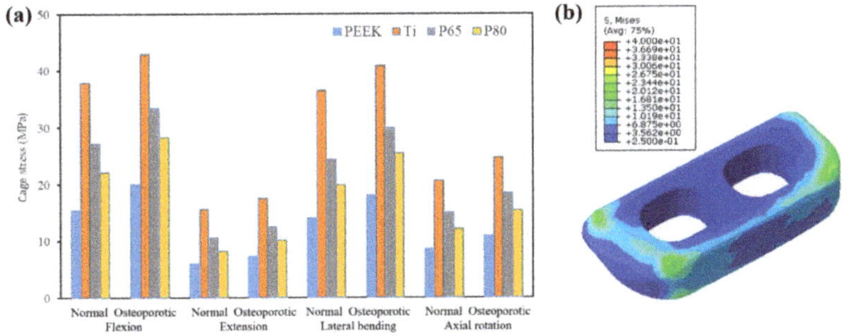

Figure 8. (**a**) The maximum scaffold stress of the surgical models; (**b**) The maximum scaffold stress appeared at the edges of the scaffold.

3.5. Pedicle Screw Stress

Figure 9 showed the maximum pedicle screw stresses of the surgical models. For the Ti scaffold, the pedicle screw stress was slightly lower than that of PEEK, P65, and P80, but the difference was not statistically significant. Compared with normal spines, osteoporotic spines increased the maximum pedicle screw stress by 6–13% in flexion, 6–7% in extension, 28–31% in lateral bending, and 14–17% in axial rotation. In addition, as shown in Figure 9b, the maximum pedicle screw stress appeared at the joint between the rod and screw.

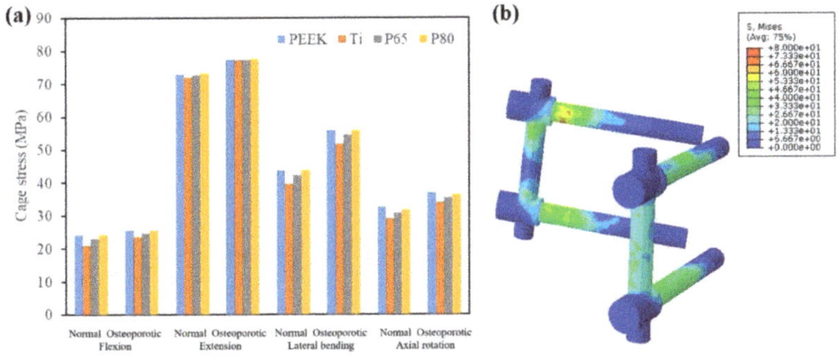

Figure 9. (**a**) The maximum pedicle screw stress of the surgical models; (**b**) The maximum pedicle screw stress appeared at the joint between rod and screw.

4. Discussion

The current study investigated the biomechanical effect of lateral scaffolds made of various materials in both normal and osteoporotic lumbar spines using the FE method. Four scaffold materials were evaluated: solid Ti, PEEK, P65, and P80, of which the material properties of the porous Ti were measured via mechanical testing. Such finite element research was not feasible in a human cadaveric experiment because of the difficulty in measuring soft tissue stresses. The predicted ROM of each segment was comparable to a previous study [26] and our cadaveric measurement. Additionally, the effects of the scaffold materials on endplate stress, scaffold stress, and pedicle screw stress were also investigated. We believe this study may be helpful in understanding the role of porous Ti scaffolds in lateral lumbar interbody fusion.

As displayed in Figure 5, the ROM of the surgical segment decreased by more than 90% in all motion modes compared with the intact model, and the Ti scaffold offered the least ROM. In addition, the ROM of P80 was slightly higher compared with P65. These findings were consistent with the FE research of Zhang et al. [23], who reported that the ROM increased as the porosity increased. Compared with the normal spine, ROMs were higher in the osteoporotic spine, but the differences were not significant. This finding indicated that osteoporosis did not significantly alter the ROM of the lumbar spine, similar to the findings of previous studies [29]. This was due to the fact that the intervertebral disc of the osteoporotic spine did not degenerate significantly, and its mobility was not markedly changed [31].

Scaffold subsidence was associated with the stresses at the bone-scaffold interface [24]. As shown in Figures 6 and 7, the porous Ti scaffolds showed advantages in lowering the endplate stresses. This was because the stress shielding was inversely proportional to the stiffness of the scaffold [4], and the porous Ti provided a Young's modulus more comparable to bone elements. Therefore, the porous scaffold may decrease the risk of scaffold subsidence. Due to its lower mechanical strength, the osteoporotic spine had a higher risk of scaffold subsidence [32]. The current study found that the endplate stresses increased significantly in osteoporotic spines, consistent with the findings in the research of Zhang et al. [33]. Furthermore, compared to solid Ti, PEEK, P65, and P80, they decreased the endplate stresses of the osteoporotic spine in extension, lateral bending, and axial rotation, indicating a lower risk of subsidence. In addition, PEEK and P80 showed similar endplate stresses this may be because the Young's moduli of the two materials were similar.

A previous study has reported that the increased scaffold stress might result in subsidence and scaffold failure [32]. Porous Ti had advantages in reducing cage stress compared to solid Ti, and P65 had a greater cage stress than P80. These findings were consistent with the biomechanical research of Zhang et al. [23]. In addition, compared with the normal spine, the scaffold stress increased by 12–30% in osteoporotic spines. Similarly, Liu et al. [29] also reported that the stresses on the scaffolds of patients with osteoporosis were slightly higher than those of a normal spine. These findings indicated that the osteoporotic bone had a higher risk of scaffold failure, and the PEEK and porous Ti scaffolds might decrease the risk of scaffold failure, especially in osteoporotic spines.

As shown in Figure 9, the solid Ti scaffold slightly decreased the pedicle screw stress compared with PEEK, P65, and P80. This was because the solid Ti scaffold provided the highest stiffness in flexion, extension, lateral bending, and axial rotation. In addition, we found that the maximum pedicle screw stress appeared at the joint between the rod and screw. This finding agreed with the results of Fan et al. [34], who reported that the stress was concentrated in the neck regions of the pedicle screws and the middle regions of the rods. In addition, we also found that the osteoporotic bone had a higher risk of pedicle screw breakage.

In summary, the porous Ti scaffolds showed many advantages over the solid Ti scaffolds and may be an alternative to PEEK scaffolds in lateral lumbar interbody fusion. In addition to its favorable biomechanical performances, porous Ti also showed advantages in facilitating bone ingrowth, promoting nutrient transportation, and reducing the risk of scaffold migration because of its micro-pore structure. A recent prospective study (40 patients with 53 segments) reported that the bone fusion rates for porous Ti scaffolds ranged from 86.7–94.3% [35]. Additionally, a recent study compared the clinical outcomes of porous Ti versus PEEK scaffolds for stand-alone lateral lumbar interbody fusion [36], and the results showed that patients in the porous Ti group had a significantly lower overall subsidence rate (20% vs. 58.8%, $P = 0.004$) and revision rate. Moreover, patients in the porous Ti group showed significantly greater pain reduction. In the biocompatibility evaluation (porcine model) of Tsai et al. [37], they found high porosities, especially 60 and 80%, facilitated more bone formation inside the implant, and bone formation was higher in the porous Ti group than in the PEEK group. This may be explained by the poor cell adhesion effect of the PEEK scaffolds [12]. Similarly, McGilvray et al. [38] also found that porous Ti significantly increased bone ingrowth over PEEK in their ovine lumbar fusion model. Recently, a number of porous Ti cages have been commercially available, but there have been limited long-term clinical trials supporting their efficacy in promoting early osseointegration and fusion [39]. Although increasing the porosity may be more beneficial to bone growth, the mechanical strength of the porous Ti should be taken into account [24]. Therefore, it required careful consideration to further increase the porosity.

Although the lateral scaffolds showed favorable clinical outcomes and advantages in stabilizing the surgical segment, the high incidence of adjacent segment degeneration (ASD) should be taken into consideration. According to a previous review of research, the incidence of radiographic ASD ranged from 5.2% to 100% after lumbar interbody fusion [40]. This may be because the stress and motion of adjacent segments have increased extremely after lumbar fusion [23,24]. In our future study, we will focus on the biomechanical performances of the adjacent segments, and optimize the structure of the lateral scaffolds to mitigate ASD through preserving spinal movement and improving load transmission in the lumbar spine.

In addition, the current research performed the FE simulation in static loading conditions (280 N follower load + four pure moments). However, patients may be inevitably exposed to dynamic loads in daily life, such as the whole-body vibration caused by vehicles [20]. Guo et al. [41] investigated the stress responses of the lumbar spine under dynamic loads and reported that, compared with static loads, dynamic loads significantly increased the intradiscal pressure by 242.4%. This finding indicated that dynamic loads may cause more damage to the lumbar spine compared to static loads. Therefore, the next task for our team is to investigate the dynamic biomechanical performances of porous Ti scaffolds.

The limitations in the current FE study should be pointed out. Firstly, because only one unique FE model was employed, the simulated results might not be representative of an average person. Secondly, the porous scaffolds were simplified as "solid-mimic" scaffolds rather than porous structures, and the mechanical properties were estimated from the mechanical test. However, many previous studies have adopted this simplification to improve computing efficiency [23,24]. In addition, the same follower load was applied for the lumbar spine, although there were some changes in muscle activities after lateral lumbar interbody fusion. However, the tendency of the predicted results would not be significantly changed by these simplifications.

5. Conclusions

The present research attempted to investigate the biomechanical effects of scaffold materials in both the normal and osteoporotic lumbar spines. The solid Ti scaffold showed advantages in stabilizing the surgical segment and decreasing the risk of pedicle screw breakage. However, solid Ti showed higher risks of scaffold breakage, scaffold subsidence, and stress shielding. PEEK and porous Ti showed significant advantages over solid Ti in decreasing endplate stresses and scaffold stresses, especially in osteoporotic spines. Compared with the normal spine, the osteoporotic lumbar spine provided higher ROMs, endplate stresses, scaffold stresses, and pedicle screw stresses in all motion modes. The porous Ti scaffolds may offer an alternative for lateral lumbar interbody fusion. Further clinical studies are necessary to validate the findings of this study.

Author Contributions: Conceptualization, H.S. and J.Z.; methodology, H.S.; software, H.S.; validation, H.S.; formal analysis, J.Z. and C.H.; investigation, H.S.; resources, C.H. and D.X.; data curation, C.H.; writing—original draft preparation, H.S.; writing—review and editing, J.Z. and D.X.; visualization, C.H.; supervision, W.L.; project administration, H.S.; funding acquisition, J.Z. and D.X. All authors have read and agreed to the published version of the manuscript.

Funding: This research was funded by the Guangdong Basic and Applied Basic Research Foundation (Grant No. 2020B1515120082), the China Postdoctoral Science Foundation (2022M710633), the Liaoning Province Natural Science Foundation (2022-BS-078) and the Tribology Science Fund of the State Key Laboratory of Tribology (SKLTKF21B03).

Institutional Review Board Statement: The study was conducted in accordance with the Declaration of Helsinki and approved by the Ethics Committee of the Biomechanics and Biotechnology Lab, Research Institute of Tsinghua University in Shenzhen.

Informed Consent Statement: Informed consent was obtained from all subjects involved in the study.

Conflicts of Interest: The authors declare no conflict of interest.

References

1. Tan, J.-H.; Cheong, C.K.; Hey, H.W.D. Titanium (Ti) scaffolds may be superior to polyetheretherketone (PEEK) scaffolds in lumbar interbody fusion: A systematic review and meta-analysis of clinical and radiological outcomes of spinal interbody fusions using Ti versus PEEK scaffolds. *Eur. Spine J.* **2021**, *30*, 1285–1295. [CrossRef] [PubMed]
2. Perez-Orribo, L.; Kalb, S.; Reyes, P.M.; Chang, S.W.; Crawford, N.R. Biomechanics of Lumbar Cortical Screw-Rod Fixation Versus Pedicle Screw-Rod Fixation with and Without Interbody Support. *Spine* **2013**, *38*, 635–641. [CrossRef] [PubMed]
3. Massaad, E.; Fatima, N.; Kiapour, A.; Hadzipasic, M.; Shankar, G.M.; Shin, J.H. Polyetheretherketone Versus Titanium Scaffolds for Posterior Lumbar Interbody Fusion: Meta-Analysis and Review of the Literature. *Neurospine* **2020**, *17*, 125–135. [CrossRef]
4. Lu, T.; Ren, J.; Sun, Z.; Zhang, J.; Xu, K.; Sun, L.; Yang, P.; Wang, D.; Lian, Y.; Zhai, J.; et al. Relationship between the elastic modulus of the scaffold material and the biomechanical properties of transforaminal lumbar interbody fusion: A logarithmic regression analysis based on parametric finite element simulations. *Comput. Methods Programs Biomed.* **2022**, *214*, 106570. [CrossRef] [PubMed]
5. Seaman, S.; Kerezoudis, P.; Bydon, M.; Torner, J.C.; Hitchon, P.W. Titanium vs. polyetheretherketone (PEEK) interbody fusion: Meta-analysis and review of the literature. *J. Clin. Neurosci.* **2017**, *44*, 23–29. [CrossRef]
6. Schnake, K.J.; Fleiter, N.; Hoffmann, C.; Pingel, A.; Scholz, M.; Langheinrich, A.; Kandziora, F. P LIF surgery with titanium-coated PEEK or uncoated PEEK scaffolds: A prospective randomised clinical and radiological study. *Eur. Spine J.* **2021**, *30*, 114–121. [CrossRef]
7. Olivares-Navarrete, R.; Gittens, R.A.; Schneider, J.M.; Hyzy, S.L.; Haithcock, D.A.; Ullrich, P.F.; Schwartz, Z.; Boyan, B.D. Osteoblasts exhibit a more differentiated phenotype and increased bone morphogenetic protein production on titanium alloy substrates than on poly-ether-ether-ketone. *Spine J.* **2012**, *12*, 265–272. [CrossRef]
8. Chen, Y.-N.; Chang, C.-W. Computational comparison of three different scaffold porosities in posterior lumbar interbody fusion with porous scaffold. *Comput. Biol. Med.* **2021**, *139*, 105036. [CrossRef]
9. Wang, H.; Wan, Y.; Li, Q.; Xia, Y.; Liu, X.; Liu, Z.; Li, X. Porous fusion scaffold design via integrated global-local topology optimization and biomechanical analysis of performance. *J. Mech. Behav. Biomed. Mater.* **2020**, *112*, 103982. [CrossRef]
10. Lee, Y.-H.; Chung, C.-J.; Wang, C.-W.; Peng, Y.-T.; Chang, C.-H.; Chen, C.-H.; Chen, Y.-N.; Li, C.-T. Computational comparison of three posterior lumbar interbody fusion techniques by using porous titanium interbody scaffolds with 50% porosity. *Comput. Biol. Med.* **2016**, *71*, 35–45. [CrossRef]
11. Zhang, N.-Z.; Xiong, Q.-S.; Yao, J.; Liu, B.-L.; Zhang, M.; Cheng, C.-K. Biomechanical changes at the adjacent segments induced by a lordotic porous interbody fusion scaffold. *Comput. Biol. Med.* **2022**, *143*, 105320. [CrossRef] [PubMed]
12. Makino, T.; Takaneka, S.; Sakai, Y.; Yoshikawa, H.; Kaito, T. Impact of mechanical stability on the progress of bone ongrowth on the frame surfaces of a titanium-coated PEEK scaffold and a 3D porous titanium alloy scaffold: In vivo analysis using CT color mapping. *Eur. Spine J.* **2021**, *30*, 1303–1313. [CrossRef] [PubMed]
13. Meena, V.K.; Kalra, P.; Sinha, R.K. Finite element study on the influence of pore size and structure on stress shielding effect of additive manufactured spinal scaffold. *Comput. Methods Biomech. Biomed. Eng.* **2022**, *25*, 566–577. [CrossRef]
14. Tsuruga, E.; Takita, H.; Itoh, H.; Wakisaka, Y.; Kuboki, Y. Pore size of porous hydroxyapatite as the cell-substratum controls BMP-induced osteogenesis. *J. Biochem.* **1997**, *121*, 317–324. [CrossRef] [PubMed]
15. Tsai, P.-I.; Hsu, C.-C.; Chen, S.-Y.; Wu, T.-H.; Huang, C.-C. Biomechanical investigation into the structural design of porous additive manufactured scaffolds using numerical and experimental approaches. *Comput. Biol. Med.* **2016**, *76*, 14–23. [CrossRef]
16. Song, S.; Guo, Y.; Yang, Y.; Fu, D. Advances in pathogenesis and therapeutic strategies for osteoporosis. *Pharmacol. Ther.* **2022**, *237*, 108168. [CrossRef]
17. Kang, S.; Park, C.-H.; Jung, H.; Lee, S.; Min, Y.-S.; Kim, C.-H.; Cho, M.; Jung, G.-H.; Kim, D.-H.; Kim, K.-T.; et al. Analysis of the physiological load on lumbar vertebrae in patients with osteoporosis: A finite-element study. *Sci. Rep.* **2022**, *12*, 11001. [CrossRef]
18. Wang, Z.; Ma, R.; Cai, Z.; Wang, Z.; Yang, S.; Ge, Z. Biomechanical Evaluation of Stand-Alone Oblique Lateral Lumbar Interbody Fusion Under 3 Different Bone Mineral Density Conditions: A Finite Element Analysis. *World Neurosurg.* **2021**, *155*, E285–E293. [CrossRef]
19. Cawthon, P.M.; Shahnazari, M.; Orwoll, E.S.; Lane, N.E. Osteoporosis in men: Findings from the Osteoporotic Fractures in Men Study (MrOS). *Ther. Adv. Musculoskelet. Dis.* **2016**, *8*, 15–27. [CrossRef]
20. Zhu, J.; Shen, H.; Cui, Y.; Fogel, G.R.; Liao, Z.; Liu, W. Biomechanical Evaluation of Transforaminal Lumbar Interbody Fusion with Coflex-F and Pedicle Screw Fixation: Finite Element Analysis of Static and Vibration Conditions. *Orthop. Surg.* **2022**, *14*, 2339–2349. [CrossRef]
21. Choi, J.; Shin, D.-A.; Kim, S. Biomechanical Effects of the Geometry of Ball-and-Socket Artificial Disc on Lumbar Spine. *Spine* **2017**, *42*, E332–E339. [CrossRef] [PubMed]
22. Fan, W.; Guo, L.-X.; Zhao, D. Posterior Lumbar Interbody Fusion Versus Transforaminal Lumbar Interbody Fusion: Finite Element Analysis of the Vibration Characteristics of Fused Lumbar Spine. *World Neurosurg.* **2021**, *150*, E81–E88. [CrossRef]

23. Zhang, Z.; Li, H.; Fogel, G.R.; Liao, Z.; Li, Y.; Liu, W. Biomechanical Analysis of Porous Additive Manufactured Scaffolds for Lateral Lumbar Interbody Fusion: A Finite Element Analysis. *World Neurosurg.* **2018**, *111*, E581–E591. [CrossRef] [PubMed]
24. Zhang, Z.; Li, H.; Fogel, G.R.; Xiang, D.; Liao, Z.; Liu, W. Finite element model predicts the biomechanical performance of transforaminal lumbar interbody fusion with various porous additive manufactured scaffolds. *Comput. Biol. Med.* **2018**, *95*, 167–174. [CrossRef]
25. Fan, W.; Guo, L.-X.; Zhang, M. Biomechanical analysis of lumbar interbody fusion supplemented with various posterior stabilization systems. *Eur. Spine J.* **2021**, *30*, 2342–2350. [CrossRef] [PubMed]
26. Renner, S.M.; Natarajan, R.N.; Patwardhan, A.G.; Havey, R.M.; Voronov, L.I.; Guo, B.Y.; Andersson, G.B.J.; An, H.S. Novel model to analyze the effect of a large compressive follower pre-load on range of motions in a lumbar spine. *J. Biomech.* **2007**, *40*, 1326–1332. [CrossRef] [PubMed]
27. Wang, Q.-D.; Guo, L.-X. Biomechanical role of cement augmentation in the vibration characteristics of the osteoporotic lumbar spine after lumbar interbody fusion. *J. Mater. Sci. -Mater. Med.* **2022**, *33*, 52. [CrossRef]
28. Li, J.; Xu, W.; Zhang, X.; Xi, Z.; Xie, L. Biomechanical role of osteoporosis affects the incidence of adjacent segment disease after percutaneous transforaminal endoscopic discectomy. *J. Orthop. Surg. Res.* **2019**, *14*, 131. [CrossRef]
29. Liu, Z.-X.; Gao, Z.-W.; Chen, C.; Liu, Z.-Y.; Cai, X.-Y.; Ren, Y.-N.; Sun, X.; Ma, X.-L.; Du, C.-F.; Yang, Q. Effects of osteoporosis on the biomechanics of various supplemental fixations co-applied with oblique lumbar interbody fusion (OLIF): A finite element analysis. *BMC Musculoskelet. Disord.* **2022**, *23*, 794. [CrossRef]
30. Liu, X.; Ma, J.; Park, P.; Huang, X.; Xie, N.; Ye, X. Biomechanical comparison of multilevel lateral interbody fusion with and without supplementary instrumentation: A three-dimensional finite element study. *BMC Musculoskelet. Disord.* **2017**, *18*, 63. [CrossRef]
31. Yang, Z.; Griffith, J.F.; Leung, P.C.; Lee, R. Effect of Osteoporosis on Morphology and Mobility of the Lumbar Spine. *Spine* **2009**, *34*, E115–E121. [CrossRef] [PubMed]
32. Rastegar, S.; Arnoux, P.-J.; Wang, X.; Aubin, C.-E. Biomechanical analysis of segmental lumbar lordosis and risk of scaffold subsidence with different scaffold heights and alternative placements in transforaminal lumbar interbody fusion. *Comput. Methods Biomech. Biomed. Eng.* **2020**, *23*, 456–466. [CrossRef] [PubMed]
33. Zhang, Q.; Zhang, Y.; Chon, T.E.; Baker, J.S.; Gu, Y. Analysis of stress and stabilization in adolescent with osteoporotic idiopathic scoliosis: Finite element method. *Comput. Methods Biomech. Biomed. Eng.* **2022**, *26*, 12–24. [CrossRef] [PubMed]
34. Fan, W.; Guo, L.-X.; Zhao, D. Stress analysis of the implants in transforaminal lumbar interbody fusion under static and vibration loadings: A comparison between pedicle screw fixation system with rigid and flexible rods. *J. Mater. Sci.-Mater. Med.* **2019**, *30*, 118. [CrossRef] [PubMed]
35. Chung, S.S.; Lee, K.J.; Kwon, Y.B.; Kang, K.C. Characteristics and Efficacy of a New 3-Dimensional Printed Mesh Structure Titanium Alloy Spacer for Posterior Lumbar Interbody Fusion. *Orthopedics* **2017**, *40*, 880–885. [CrossRef] [PubMed]
36. Amini, D.A.; Moser, M.; Oezel, L.; Zhu, J.; Okano, I.; Shue, J.; Sama, A.A.; Cammisa, F.P.; Girardi, F.P.; Hughes, A.P. Early Outcomes of Three-Dimensional–Printed Porous Titanium versus Polyetheretherketone Cage Implantation for Stand-Alone Lateral Lumbar Interbody Fusion in the Treatment of Symptomatic Adjacent Segment Degeneration. *World Neurosurg.* **2022**, *162*, 14–20. [CrossRef]
37. Tsai, P.I.; Wu, M.H.; Li, Y.Y.; Lin, T.H.; Tsai, J.S.C.; Huang, H.I.; Lai, H.J.; Lee, M.H.; Chen, C.Y. Additive-manufactured Ti-6Al-4V/Polyetheretherketone composite porous cage for Interbody fusion: Bone growth and biocompatibility evaluation in a porcine model. *BMC Musculoskelet. Disord.* **2021**, *22*, 171. [CrossRef]
38. McGilvray, K.C.; Easley, J.; Seim, H.B.; Regan, D.; Berven, S.H.; Hsu, W.K.; Mroz, T.E.; Puttlitz, C.M. Bony ingrowth potential of 3D-printed porous titanium alloy: A direct comparison of interbody cage materials in an in vivo ovine lumbar fusion model. *Spine J.* **2018**, *18*, 1250–1260. [CrossRef]
39. Toop, N.; Gifford, C.; Motiei Langroudi, R.; Farzadi, A.; Boulter, D.; Forghani, R.; Farhadi, H.F. Can activated titanium interbody cages accelerate or enhance spinal fusion? A review of the literature and a design for clinical trials. *J. Mater. Sci.-Mater. Med.* **2021**, *33*, 1. [CrossRef]
40. Park, P.; Garton, H.J.; Gala, V.C.; Hoff, J.T.; McGillicuddy, J.E. Adjacent segment disease after lumbar or lumbosacral fusion: Review of the literature. *Spine* **2004**, *29*, 1938–1944. [CrossRef]
41. Guo, L.X.; Fan, W. Dynamic Response of the Lumbar Spine to Whole-body Vibration Under a Compressive Follower Preload. *Spine* **2018**, *43*, 143–153. [CrossRef] [PubMed]

lubricants

MDPI

Article

Simulation Study on Bearing Lubrication Mechanism and Friction Characteristics of the Biomimetic Non-Smooth Surface of a Cross-Scale, Second-Order Compound Microstructure

Yingna Liang [1,2,*], Cunyuan Wang [1,2], Zongyi Zhang [1,2], Zhepeng Zhang [1,2], Wei Wang [1,2], Hao Xing [1,2], Tianyuan Guan [1,2] and Dianrong Gao [1,2,*]

1 School of Mechanical Engineering, Yanshan University, Qinhuangdao 066004, China
2 Heibei Provincial Key Laboratory of Heavy Machinery Fluid Power Transmission and Control, Yanshan University, Qinhuangdao 066004, China
* Correspondence: liangyingna@ysu.edu.cn (Y.L.); gaodr@ysu.edu.cn (D.G.)

Abstract: The reasonable design of biomimetic non-smooth surfaces is a novel and effective way to solve problems such as the poor lubricity and serious friction and wear of friction pairs of seawater axial piston pumps. Inspired by cross-scale, second-order compound microstructures on the surfaces of some living organisms, a hydrodynamic lubrication model of a slipper pair with a surface featuring spherical pits containing spherical convex hulls was built. This study analyzed the bearing lubrication mechanism and friction characteristics of cross-scale, second-order compound microstructure from the microflow perspective via the CFD method and optimized the working and geometric parameters using a hybrid orthogonal test scheme. The study's results show that the cross-scale, second-order compound microstructure can produce a superimposed hydrodynamic pressure effect to improve the bearing capacity of the lubrication film of a slipper pair, reducing the friction coefficient. The orders of factors (the working parameter and geometric parameters) under multiple indices (the total pressure-bearing capacity and the friction coefficient) were found. The optimal combination is a spherical pit with a first order diameter of 0.7 mm, a first order depth-to-diameter ratio of 0.1, an area rate of 20%, an arrangement angle of $\alpha/3$ and a spherical convex hull with a second order diameter of 0.13 mm, and a second order depth-to-diameter ratio of 0.3. Compared to a smooth surface and a first-order, non-smooth microstructure, the cross-scale, second-order compound microstructure has an 11.0% and 8.9% higher total pressure-bearing capacity, respectively, and the friction coefficient decreased by 9.5% and 5.4%, respectively.

Keywords: slipper pair; biomimetic non-smooth surface; cross-scale, second-order compound microstructure; bearing lubrication mechanism; friction characteristics

1. Introduction

The surfaces of many organisms' bodies have evolved into microscopic non-smooth morphologies such as pits, bumps, scales, bristles, grooves, or ripples under the long-term repeated stimulation and action of the outside world; such morphologies provide excellent desorption, drag reduction, and wear resistance [1–3]. Based on this rationale, research has been carried out to construct a single regular microstructure arrangement on the surfaces of different friction pairs to improve their tribological properties [4–10]. The drag reduction and wear resistance mechanisms of microscopic non-smooth surfaces are attributed to (a) the hydrodynamic effect generated by the convergent gap between the non-smooth structure and the relative motion between friction surfaces potentially producing the bearing capacity; (b) the lubricant entering the friction surfaces to avoid the direct contact of the two surfaces; and (c) the non-smooth structure diminishing wear during the friction process by storing the wear debris [11–13].

Uddin et al. [14] numerically simulated the lubrication-bearing characteristics of circular, elliptical, V-shaped, and triangular pits via the Reynolds equation and finite difference method. It was found that the triangular effect enhances the convergence effect of the micro-wedge and thus increases the bearing capacity of the lubrication film. On this basis, the design and optimization of a "star" composed of a series of triangles around the center point were proposed. The simulation results showed that six-point star pits at a texture density of 40% can reduce the friction coefficient to the greatest extent. Ni et al. [15,16] placed three kinds of micro-pits (conical, cylindrical, and square) on the bottom of a slipper to study the influence of the change in the micro-pits' morphology parameters on the bearing capacity and temperature increase of an oil film under high-speed and high-pressure conditions. The results showed that the oil film in a conical pit has the best bearing capacity, and the average pressure of the contact surface increases obviously with the increase in the depth-to-diameter ratio. A square pit shows the lowest temperature increase when the area ratio is small and the depth-to-diameter ratio is large. Sugihara et al. [17,18] studied the influence of micro-pits and micro-grooves of different sizes and arrangements on the surface abrasion resistance and cutting force of steel cutting tools. The micro-pits show greater improvement in the surface resistance and friction properties of cutting tools than micro-grooves, especially under bad lubrication conditions. Liu et al. [19] studied the vibrational and tribological effects of laser surface textures of stainless steel via a tribometer and a data acquisition and signal processing (DASP) system. Surface textures can reduce friction coefficients, enhance wear resistance, and improve the dynamical performance of frictional surfaces. Naduvinamani et al. [20] investigated the impact of surface roughness and a porous medium on the dynamic characteristics of the Rayleigh step bearing by considering a squeezing action.

Further studies found that the microscopic non-smooth body surface of most organisms actually has a cross-scale, second-order compound structure. At the millimeter scale, shark shield scales have a micron-scale distribution of grooves, and these small grooves along the flow direction can effectively reduce the friction resistance of the scale wall [21,22]. The micron-scale mastoids on lotus leaves are composed of nano-scale hairdrites, which makes the surfaces of lotus leaves have a large water contact angle and high hydrophobicity [23,24]. The outer edge of the dung beetle's chest section has pits with a central bulge so that the dung beetle does not adhere to the soil/dung [25,26]. However, few studies on the lubrication mechanism and drag reduction characteristics of these cross-scale, second-order compound microstructures have been reported.

Seawater hydraulic transmission technology is a kind of green transmission technology using seawater instead of mineral oil as the working medium. It has the advantages of obtaining materials locally, saving energy, protecting the environment, safety, and stability. The seawater axial piston pump, with its advantages of high-volume efficiency, high working pressure, and a compact structure, has become the preferred power component of seawater hydraulic transmission technology, and it plays a key role in important areas of ocean engineering, such as seawater desalination, ocean mineral oil and gas development, ocean ship transportation, and so on [27]. However, seawater has low viscosity, poor lubrication, strong corrosion, and high vaporization pressure and thus has a great impact on the tribological performance of the key friction pair of axial piston pumps operating in it [28,29], especially for slipper pairs under the most complex movement and force conditions. There are some key technical problems that seriously affect the quality, efficiency, working reliability, and service life of seawater axial piston pumps, such as increased clearance leakage, reduced water film bearing capacity, and intensified friction and wear. These problems currently restrict the development of seawater hydraulic transmission technology [30].

This study presents the novel and effective idea of replicating the microscopic, non-smooth morphology of a cross-scale, second-order compound structure of a creature's body surface in the slipper pair of seawater axial piston pumps to improve the lubrication ability, bearing characteristics, and friction and wear performance of the slipper pair. By designing cross-scale, second-order compound micro-pits or micro-grooves with a certain size and arrangement on the surface of the slipper pair, drag reduction and wear reduction can be achieved by utilizing the hydrodynamic lubrication effect. On the other hand, it is important to control the shape, size, and arrangement of micro-pits or micro-grooves well; otherwise, the drag reduction and wear reduction effects will not be obvious, even resulting in negative effects. In this study, firstly, the hydrodynamic lubrication model of the slipper pair with a biomimetic, non-smooth surface (spherical pits containing spherical convex hulls) was built. Secondly, the bearing lubrication mechanism and friction characteristics of the cross-scale, second-order compound microstructure were analyzed from the microflow perspective via the CFD method. Thirdly, the influence of the working parameter (speed), size (diameter and depth-to-diameter ratio), and distribution (area rate and arrangement angle) of the cross-scale, second-order compound microstructure on the total pressure-bearing capacity of the lubrication film and the friction coefficient was investigated using a hybrid orthogonal test scheme. Furthermore, an optimization analysis was carried out. Finally, the excellent bearing lubrication capacity and friction performance of the cross-scale, second-order compound microstructure were confirmed through comparative analysis with the smooth surface and a first-order, non-smooth microstructure. This study is expected to provide a theoretical reference for the practical engineering design of slipper pairs of seawater axial piston pumps.

2. Models and Methods

2.1. Model Description

For the slipper pair of seawater axial piston pumps (Figure 1a), the cross-scale compound biomimetic structures (spherical pits containing spherical convex hulls) were arranged on the surface of a swash plate, and the surface of the slippers remained smooth (Figure 1b). Assuming that the slipper pair is fully lubricated and setting the origin coordinate at the center of the front surface of the swash plate, the fluid field models of the water film and the cross-scale compound biomimetic structures were established. The small red circle on the right in Figure 1c is the location of one of the slippers.

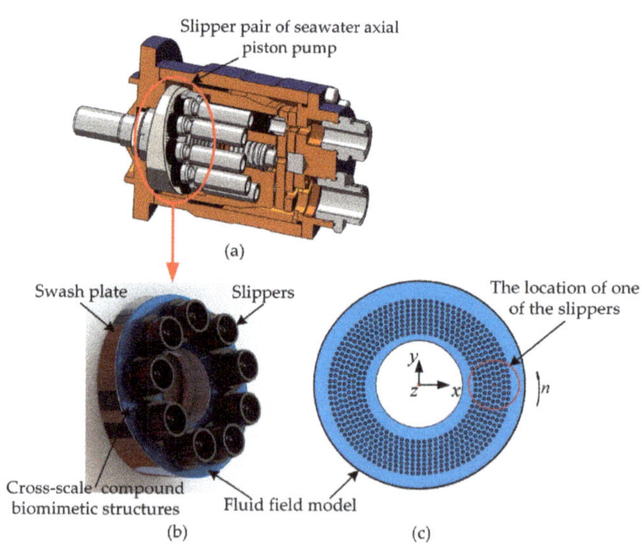

Figure 1. Slipper pair with biomimetic non-smooth surface and the fluid field model. (**a**) The seawater axial piston pump; (**b**) the slipper pair with cross-scale compound biomimetic structures (spherical pits containing spherical convex hulls); and (**c**) the fluid field models of water film and cross-scale compound biomimetic structures.

In order to study the lubrication mechanism and influence law of the cross-scale compound biomimetic structures, a fan-shaped region covering about 3 × 3 spherical pits containing spherical convex hulls under the center of one slipper was selected as the computed fluid field, as shown in Figure 2. Each spherical pit is located in the center of an imaginary fan-shaped control unit, and the ratio between the total area of spherical pits and the total area of the fan shape is defined as the area rate S. Then:

$$S = \frac{90Nd_1{}^2}{(R_2{}^2 - R_1{}^2)\theta} \tag{1}$$

where R_1 = 28.75 mm, R_2 = 34.75 mm, θ = 10.8°; r_1 = 29.75 mm, r_2 = 31.75 mm, and r_3 = 33.75 mm; N is the number of spherical pits containing spherical convex hulls; d_1 is the diameter of the spherical pit (first order diameter); and n represents the direction in which the slippers rotate.

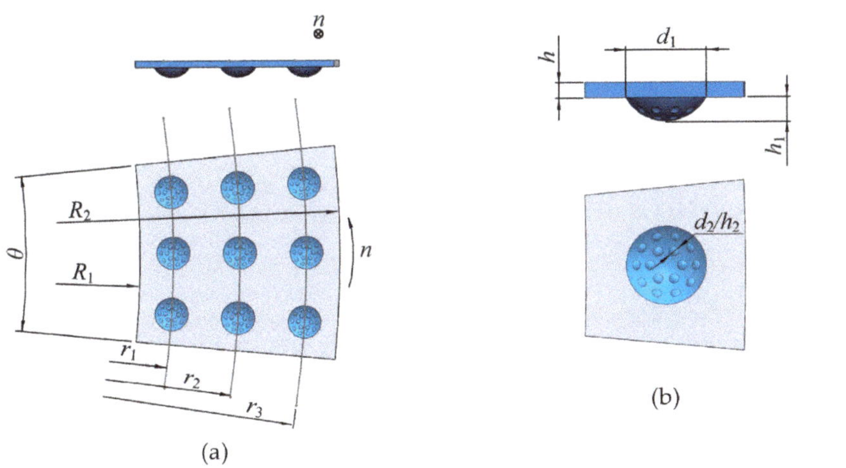

Figure 2. The fan-shaped computed fluid field. (**a**) The arrangement of spherical pits containing spherical convex hulls; (**b**) the structure of spherical pits containing spherical convex hulls.

As shown in Figure 2b, h is the thickness of the water film, h_1 is the depth of the spherical pit (first order depth), and the ratio of the depth of spherical pit h_1 to the diameter of spherical pit d_1 is defined as the first order depth-to-diameter ratio δ_1. h_2 is the depth of the spherical convex hull (second order depth), d_2 is the diameter of the spherical convex hull (second order diameter), and the ratio of the depth of spherical convex hull h_2 to the diameter of spherical convex hull d_2 is defined as the second order depth-to-diameter ratio δ_2. Then:

$$\delta_1 = \frac{h_1}{d_1} \tag{2}$$

$$\delta_2 = \frac{h_2}{d_2} \tag{3}$$

The diameter of a spherical pit or a spherical convex hull is controlled by the area rate, and the depth of a spherical pit or a spherical convex hull is controlled by the depth-to-diameter ratio.

As shown in Figure 3, α is the circumferential interval angle, which is the central angle formed by the centers of two adjacent spherical pits at the same distribution circle and the center of the swash plate; φ is the arrangement angle, which is the central angle formed the centers of two spherical pits at two adjacent distribution circles but not along the same radius and the center of the swash plate.

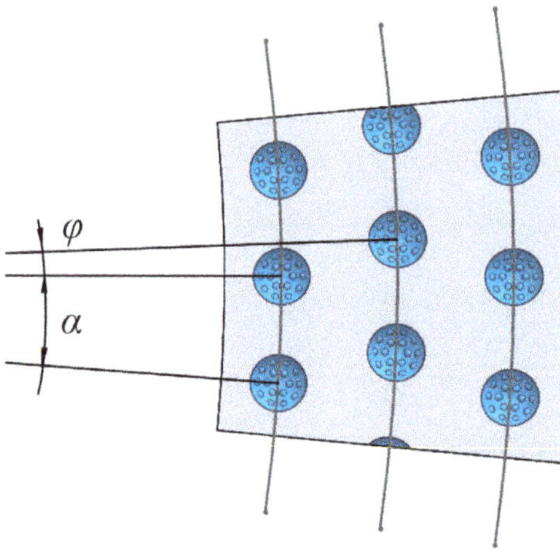

Figure 3. The arrangement angle of spherical pits containing spherical convex hulls.

In order to study the influence of different shapes, sizes, and distributions of the cross-scale compound biomimetic structures on dynamic bearing characteristics under different working conditions, an orthogonal test scheme for numerical simulation was designed under the mixed levels of seven factors: rotate speed A, first order diameter B, first order depth-to-diameter ratio C, second order diameter D, second order depth-to-diameter ratio E, area rate F, and arrangement angle G. The details are shown in Table 1.

Table 1. $L_{18}(6 \times 3^6)$ orthogonal test scheme for numerical simulation.

Number	Experimental Factor							Total Pressure-Bearing Capacity F_t (N)	Friction Coefficient μ
	A Rotate Speed n (r/min)	B First Order Diameter d_1 (mm)	C First Order Depth-to-Diameter Ratio δ_1	D Second Order Diameter d_2 (mm)	E Second Order Depth-to-Diameter Ratio δ_2	F Area Rate S (%)	G Arrangement Angle φ (°)		
1	1(1000)	1(0.7)	1(0.1)	1(0.07)	1(0.1)	1(10)	1(0)	0.1353	0.0760
2	1	2(1.0)	2(0.3)	2(0.1)	2(0.3)	2(20)	2(α/3)	0.1246	0.0999
3	1	3(1.3)	3(0.5)	3(0.13)	3(0.5)	3(30)	3(α/2)	0.1093	0.1450
4	2(1250)	1	1	2	2	3	3	0.2107	0.0611
5	2	2	2	3	3	1	1	0.1989	0.0731
6	2	3	3	1	1	2	2	0.1797	0.1012
7	3(1500)	1	2	1	3	2	3	0.2889	0.0589
8	3	2	3	2	1	3	1	0.2336	0.1075
9	3	3	1	3	2	1	2	0.2981	0.0554
10	4(1750)	1	3	3	2	2	1	0.3484	0.0723
11	4	2	1	1	3	3	2	0.3981	0.0485
12	4	3	2	2	1	1	3	0.3620	0.0653
13	5(2000)	1	2	3	1	3	2	0.4808	0.0533
14	5	2	3	1	2	1	3	0.4610	0.0613
15	5	3	1	2	3	2	1	0.5111	0.0453
16	6(2250)	1	3	2	3	1	2	0.5907	0.0519
17	6	2	1	3	1	2	3	0.6349	0.0423
18	6	3	2	1	2	3	1	0.5123	0.0762

2.2. Governing Equation

The working medium in this study is seawater, which is a Newtonian fluid. The typical film thickness of hydrodynamic lubrication is 1~100 μm [31]. The lubrication film thickness of the slipper pair in a water pump can only reach one-third of that in an oil pump, which is controlled within 3.3~10 μm. The flow of the seawater in the narrow space between the surfaces of a slipper pair should follow the Navier–Stokes equation. The continuity equation and the Navier–Stokes equation [32] are employed to describe the problem studied:

$$\frac{\partial u}{\partial x} + \frac{\partial v}{\partial y} + \frac{\partial w}{\partial z} = 0 \tag{4}$$

$$\begin{cases} \rho\left(u\frac{\partial u}{\partial x} + v\frac{\partial u}{\partial y} + w\frac{\partial u}{\partial z}\right) = -\frac{\partial p}{\partial x} + \eta\left(\frac{\partial^2 u}{\partial x^2} + \frac{\partial^2 u}{\partial y^2} + \frac{\partial^2 u}{\partial z^2}\right) \\ \rho\left(u\frac{\partial v}{\partial x} + v\frac{\partial v}{\partial y} + w\frac{\partial v}{\partial z}\right) = -\frac{\partial p}{\partial y} + \eta\left(\frac{\partial^2 v}{\partial x^2} + \frac{\partial^2 v}{\partial y^2} + \frac{\partial^2 v}{\partial z^2}\right) \\ \rho\left(u\frac{\partial w}{\partial x} + v\frac{\partial w}{\partial y} + w\frac{\partial w}{\partial z}\right) = -\frac{\partial p}{\partial z} + \eta\left(\frac{\partial^2 w}{\partial x^2} + \frac{\partial^2 w}{\partial y^2} + \frac{\partial^2 w}{\partial z^2}\right) \end{cases} \tag{5}$$

where u, v, and w are the velocity vectors in directions x, y, and z, respectively; ρ is the density of seawater (1.025 g/cm^3); η is the dynamic viscosity of seawater (6.15×10^{-4} Pa·s); and p is the pressure of the lubricating film.

2.3. Simulation Strategy

The fan-shaped computed fluid field, which contains about 3 × 3 spherical pits containing spherical convex hulls, was modeled and imported into the CFD pre-processing software ICEM for grid division. The grid division method, grid quantity, and grid quality are vital to ensure the reliability of the numerical simulation results [33]. In order to explore the appropriate grid quantity for accurate calculation results, a grid independence verification was carried out, and the results are shown in Figure 4. When the grid quantity is less than 600,000, the calculation results are quite different. When the grid quantity is higher than 900,000, the error of the calculation results can be controlled within 0.1%. Therefore, the seawater film was meshed by hexahedral grids with the size of 0.02, the first-order spherical pits were meshed by tetrahedral grids with the size of 0.02, and the second-order spherical convex hulls were meshed by tetrahedral grids with the size of 0.008. The total number of units in the grid is about 2~19 million.

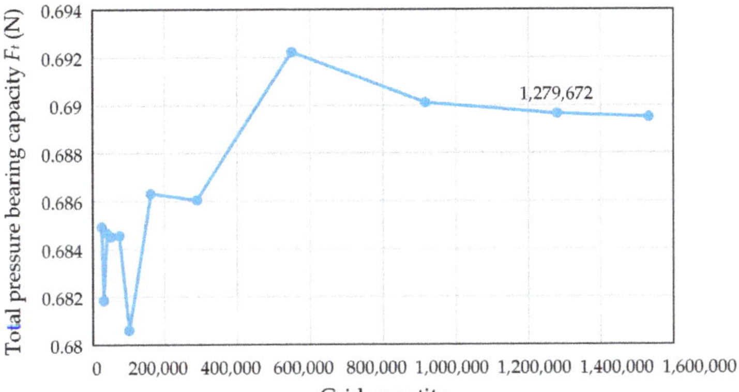

Figure 4. Grid independence verification.

The model mesh was imported into the CFD software Fluent for numerical simulation. The flow of seawater film is similar to a plate differential pressure flow and can be regarded as being in a laminar state. Due to the low viscosity of seawater, the flow inertia force is greater than the viscous force [34], which makes it easy to generate local eddy currents in spherical pits and around the spherical convex hulls. Therefore, the standard k-ε model was used for calculation.

As shown in Figure 5, the inner and outer cylinder surfaces of the fan-shaped seawater film were set to pressure-inlet and pressure-outlet boundary conditions, respectively. The left and right surfaces were both set to periodic boundary rotation conditions. The upper surface of the seawater film (i.e., the lower surface of the slipper) was set to the moving wall boundary condition, which rotates around the coordinate axis perpendicular to the lower surface of the seawater film (i.e., the center line of the swash plate) at the corresponding speed listed in Table 1. The lower surface of the seawater film and the upper surfaces of the spherical pits were both set to interface boundary conditions to carry out the data transfer of two parts. The lower surfaces of the spherical pits and the spherical convex hulls were both set to stationary wall boundary conditions.

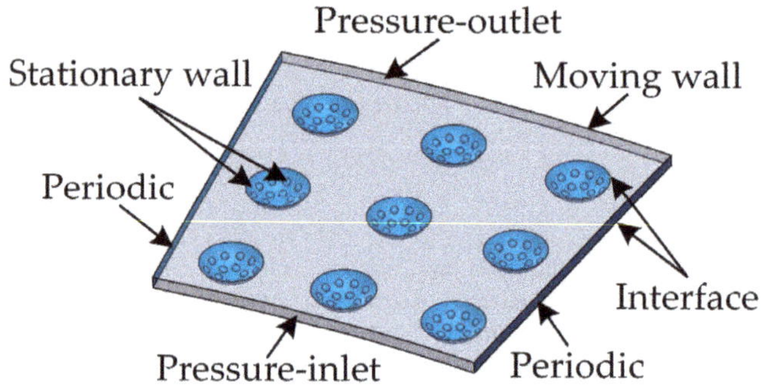

Figure 5. Boundary condition setting.

In order to verify the reliability of this solver, the hydrodynamic pressure lubrication problem of two-dimensional rectangular micro-pit texture solved by the Navier–Stokes equation in the reference literature [35] was firstly verified and calculated. The comparison of calculation results is shown in Figure 6. It can be seen that the shape of the pressure distribution curve is basically the same, and the differences of the extreme value of pressure are about 8%.

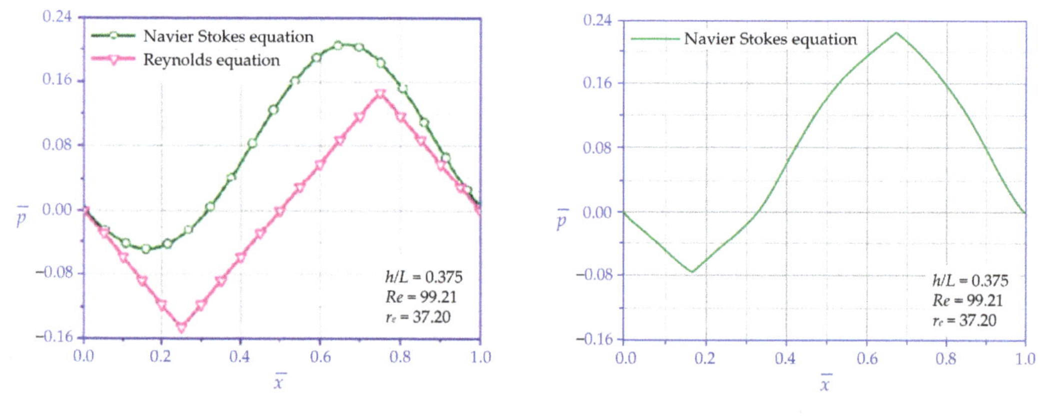

Figure 6. Validation of the current solver. (**a**) Pressure distribution in Ref. [35]; (**b**) pressure distribution with the current solver.

3. Results and Discussion

3.1. Analysis of Single Group Test Simulation Results

Taking the first calculation result in Table 1 as an example, there are 3 × 3 spherical pits containing spherical convex hulls uniformly distributed in the fan-shaped computed fluid field. Figures 7 and 8 show the 3D pressure peak and 2D pressure contour of the upper surface of the seawater film, respectively.

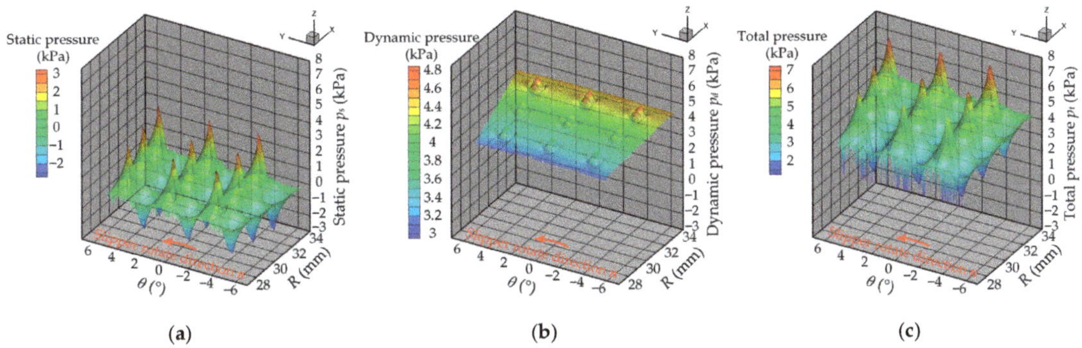

Figure 7. The 3D pressure peak of the upper surface of the seawater film. (**a**) The 3D static pressure peak; (**b**) the 3D dynamic pressure peak; (**c**) the 3D total pressure peak.

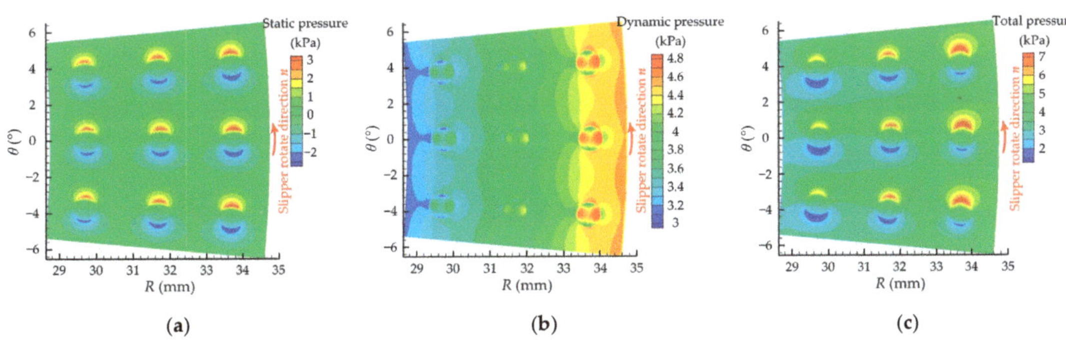

Figure 8. The 2D pressure contour of the upper surface of the seawater film. (**a**) The 2D static pressure contour; (**b**) the 2D dynamic pressure contour; (**c**) the 2D total pressure contour.

In the calculation results using Fluent, static pressure is defined as the difference from the operating pressure (i.e., atmospheric pressure). Dynamic pressure is the pressure related to the velocity, and its value is $\rho v^2/2$. In an incompressible fluid, the total pressure is the sum of static pressure and dynamic pressure, which corresponds to the pressure p mentioned in the right term of the Navier–Stokes equation. As shown in Figure 7, n represents the rotation direction of the slipper, which rotates counterclockwise around the axis of the swash plate (z-axis). The static pressure ranges from −2.5 to 3 kPa, the dynamic pressure ranges from 3 to 5 kPa, and the total pressure ranges from 1.5 to 7 kPa, indicating that the total pressure mainly comes from the contribution of dynamic pressure.

It can be seen from Figures 7a and 8a that the static pressure distribution of 3×3 spherical pits on the surface of the seawater film is positive in the front half and negative in the back half (along the speed direction of the slipper), and the static pressure distribution of the three pits on the same distribution circle is basically the same. With the increase in the radius of the distribution circle, the area and intensity of positive and negative pressure slightly increase.

Figures 7b and 8b show that the dynamic pressure distribution in the spherical pit is low at the two sides and high in the center along the circular direction of the distribution circle and high at the two sides and low in the center along the radial direction of the distribution circle. The fluid flow is the relative sliding between many extremely thin fluid layers. The viscosity of the fluid generates shear stress between layers sliding over each other, which slows down the layers with faster flows and accelerates the layers with slower flows; the movement is transmitted layer by layer [31]. Figure 9 shows the velocity vector of the central spherical pit longitudinally sliced by the distribution cylinder of r_2 = 31.75 mm. The maximum velocity appears on the upper surface of the seawater film which is in contact with the slipper sole. When the slipper drives the upper fluid layer to move above the spherical pit, the depth increases and the velocity needs to be transmitted down to more fluid layers, so the velocity loss of the upper surface of the seawater film here is larger than in places without pits. However, the seawater in the upper part of the pit first moves following the seawater at the lower surface of the seawater film. After hitting the front half of the pit wall, it turns downward to flow back along the bottom of the pit. After hitting the back half of the pit wall, it turns upward to move in the same direction as the seawater at the lower surface of the seawater film again; that is, a vortex ring is formed in the spherical pit. This flow vortex ring can also be found in the simulation study by Wang et al. [36]. The upper part of the vortex ring transmits velocity to the middle layer of the seawater film together with the upper layer. Therefore, the velocity loss of the upper surface of the seawater film at the center of the spherical pit is compensated, and the dynamic pressure is basically equal to areas without pits. The vortex ring changes direction at the front and back edges of the spherical pit, which increases the velocity loss. That is why the dynamic pressure decreases at the two sides of the pit along the circular direction of the distribution circle. Along the radial direction of the distribution circle, the dynamic pressure is higher at the two sides of the pit. Because the depth decreases gradually from the center to the edge, the velocity compensation effect of the vortex ring is gradually enhanced. As the radius of the distribution circle increases, the linear velocity increases obviously, and the dynamic pressure increases obviously as well.

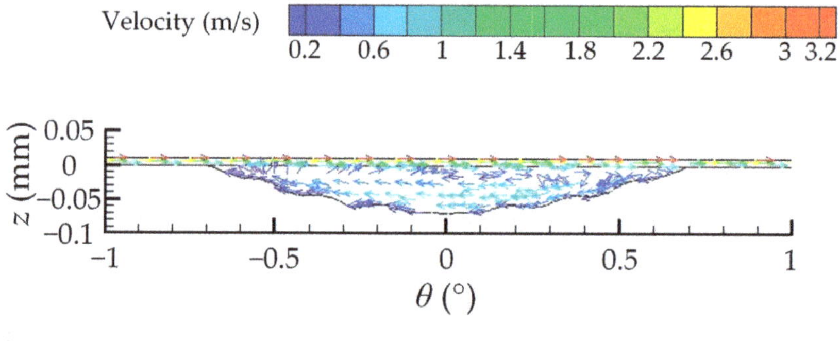

Figure 9. Velocity vector of the central spherical pit longitudinally sliced by the distribution cylinder of r_2 = 31.75 mm.

Figures 7c and 8c show that the total pressure distribution is similar to the static pressure, while the high-pressure area increases and the low-pressure area decreases with the increase in the distribution circle's radius. The total pressure of the upper surface of the seawater film is all positive, indicating that the spherical pits containing spherical convex hulls could produce a hydrodynamic pressure effect. The results are consistent with the previous study results by Tang et al. [37].

Figure 10 shows the 2D pressure contour of the central spherical pit longitudinally sliced by the distribution cylinder of r_2 = 31.75 mm. As shown in Figure 10a, the maximum positive static pressure appears at the front edge of the spherical pit, the minimum negative static pressure appears at the back edge, and there is a rainbow-shaped gradient distribution in the middle.

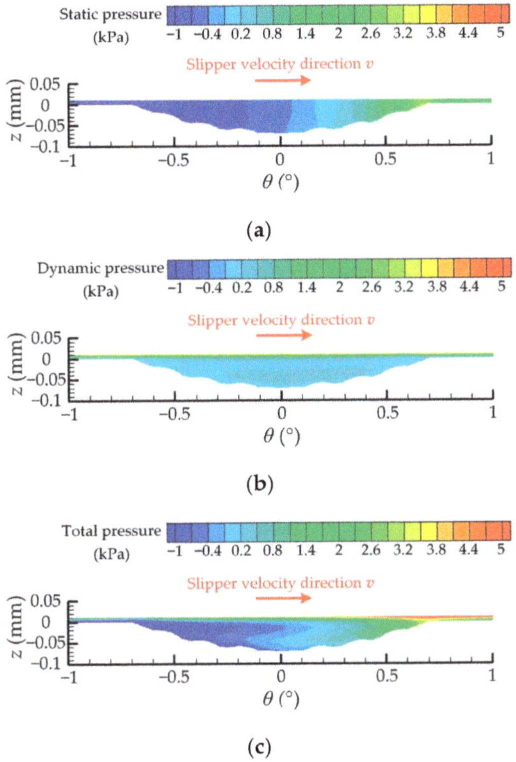

Figure 10. The 2D pressure contour of the central spherical pit longitudinally sliced by the distribution cylinder of r_2 = 31.75 mm. (**a**) The 2D static pressure contour; (**b**) the 2D dynamic pressure contour; (**c**) the 2D total pressure contour.

Dynamic pressure is the pressure change caused by the velocity change, which is proportional to the square of the velocity. As shown in Figure 10b, the dynamic pressure is positive. The maximum dynamic pressure appears on the upper surface of the seawater film and decreases along the thickness direction. The dynamic pressure value at the lower part of the vortex ring rebounds slightly.

The total pressure is the synthesis of static pressure and dynamic pressure, which is the macroscopic manifestation of the hydrodynamic pressure effect of the spherical pits containing spherical convex hulls. When the upper surface of the seawater film moves to the right at a certain speed, a convergence gap is formed along the direction of the movement, the seawater flows from the large section to the small section, and the flow rate per unit length caused by velocity flow gradually decreases. To keep the flow continuous, the pressure distribution shown in Figure 7c is generated. The pressure flow caused by this pressure distribution will reduce the inflow at the large section and increase the outflow at the small section to keep the flow equal across the sections. As shown in Figure 10c, the total pressure is almost all positive. Similar to static pressure distribution, the maximum total pressure appears at the front edge and the minimum total pressure appears at the back edge. However, the high-pressure contour lines of the rainbow-shaped gradient distribution significantly bend to the left near the bottom of the spherical pit. This is because the spherical convex hulls produce second-order hydrodynamic pressure effects and enlarge the range of the high-pressure area.

3.2. Range Analysis of Multi-Index Orthogonal Test Results

By integrating the static pressure distribution, dynamic pressure distribution, and total pressure distribution of the upper surface of the seawater film on the fan-shaped domain, the static pressure-bearing capacity F_S, dynamic pressure-bearing capacity F_d, and total pressure-bearing capacity F_t of the calculation domain could be obtained. Meanwhile, the shear stress of the upper surface of the seawater film was extracted, and the friction coefficient μ was calculated. The total pressure-bearing capacity and friction coefficient were added to the last two columns of Table 1. The histogram of the bearing capacity and the line chart of the friction coefficient for 18 groups of tests in the orthogonal simulation scheme are shown in Figure 11.

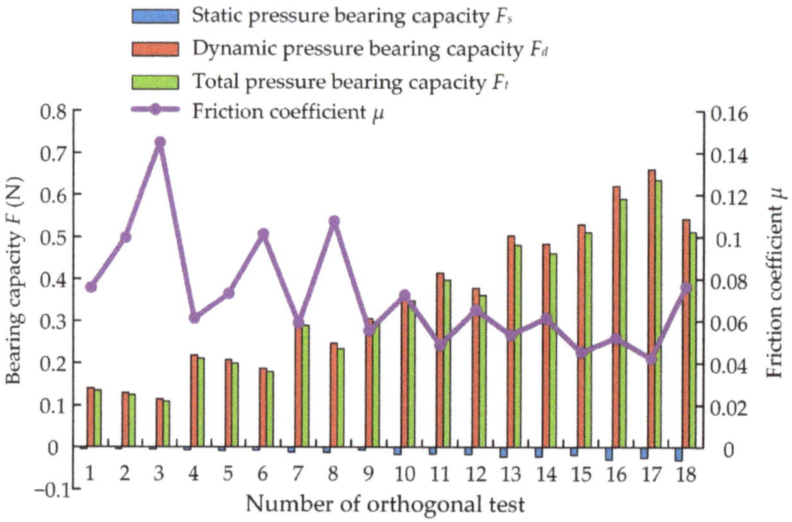

Figure 11. Histogram of bearing capacity and line chart of friction coefficient of 18 groups of tests in the orthogonal simulation scheme.

As shown in Figure 11, the static pressure-bearing capacity is negative and the value is low. The dynamic pressure-bearing capacity and total pressure-bearing capacity are both positive, and the heights of their respective columns are much higher than that of the static pressure-bearing capacity, indicating that the hydrodynamic pressure effect can produce load bearing and lubrication effects and that the total pressure-bearing capacity mainly comes from the contribution of dynamic pressure-bearing capacity. The line chart shows that the friction coefficient is inversely proportional to the bearing capacity. The bearing capacity increases and the friction coefficient decreases obviously with the increase in the rotation speed.

Table 2 shows the range analysis of $L_{18}(6 \times 3^6)$ multi-index orthogonal test results. The sum K_i and average k_i of the two test indices of total pressure-bearing capacity and friction coefficient corresponding to each level of each factor were calculated, and the range R and the adjusted range R' of each factor were calculated. According to the adjusted range R', the order of the factors under the total pressure-bearing capacity is A > C > F > E > G > D > B; the order of the factors under the friction coefficient is C > A > B > F > G > E > D. According to the average k_i of each level of each factor, the optimal combination is determined as $A_6B_1C_1D_3E_3F_2G_2$ under the total pressure-bearing capacity and $A_5B_1C_1D_1E_3F_1G_2$ under the friction coefficient.

Table 2. Range analysis of $L_{18}(6 \times 3^6)$ multi-index orthogonal test results.

		Experimental Factor						
		A **Rotate Speed** n **(r/min)**	**B** **First Order Diameter** d_1 **(mm)**	**C** **First Order Depth-to-Diameter Ratio** δ_1	**D** **Second Order Diameter** d_2 **(mm)**	**E** **Second Order Depth-to-Diameter Ratio** δ_2	**F** **Area Rate** S **(%)**	**G** **Arrangement Angle** φ **(°)**
Total pressure-bearing capacity F_t (N)	K_1	0.3692	2.0548	2.1881	1.9753	2.0263	2.0460	1.9396
	K_2	0.5893	2.0511	1.9676	2.0327	1.9551	2.0875	2.0718
	K_3	0.8206	1.9725	1.9227	2.0703	2.0970	1.9448	2.0669
	K_4	1.1085						
	K_5	1.4528						
	K_6	1.7379						
	k_1	0.1231	0.3425	0.3647	0.3292	0.3377	0.3410	0.3233
	k_2	0.1964	0.3418	0.3279	0.3388	0.3258	0.3479	0.3453
	k_3	0.2735	0.3288	0.3204	0.3451	0.3495	0.3241	0.3445
	k_4	0.3695						
	k_5	0.4843						
	k_6	0.5793						
	Range R	0.4562	0.0137	0.0442	0.0158	0.0237	0.0238	0.0220
	Adjusted range R'	0.2924	0.0175	0.0564	0.0202	0.0301	0.0303	0.0281
	Order				A > C > F > E > G > D > B			
	Optimal level	A_6	B_1	C_1	D_3	E_3	F_2	G_2
	Optimal combination				$A_6B_1C_1D_3E_3F_2G_2$			
Friction coefficient μ	K_1	0.3209	0.3736	0.3285	0.4221	0.4456	0.3828	0.4504
	K_2	0.2354	0.4326	0.4267	0.4310	0.4263	0.4199	0.4103
	K_3	0.2218	0.4883	0.5392	0.4414	0.4226	0.4918	0.4339
	K_4	0.1862						
	K_5	0.1599						
	K_6	0.1703						
	k_1	0.1070	0.0623	0.0548	0.0703	0.0743	0.0638	0.0751
	k_2	0.0785	0.0721	0.0711	0.0718	0.0710	0.0610	0.0684
	k_3	0.0739	0.0814	0.0899	0.0736	0.0704	0.0820	0.0723
	k_4	0.0621						
	k_5	0.0533						
	k_6	0.0568						
	Range R	0.0537	0.0191	0.0351	0.0032	0.0038	0.0182	0.0067
	Adjusted range R'	0.0344	0.0244	0.0447	0.0041	0.0049	0.0231	0.0085
	Order				C > A > B > F > G > E > D			
	Optimal level	A_5	B_1	C_1	D_1	E_3	F_1	G_2
	Optimal combination				$A_5B_1C_1D_1E_3F_1G_2$			

The comprehensive balance method is adopted to determine the optimal combination that makes each index as good as possible [38]. For factor A, the rank under the total pressure-bearing capacity is first and the optimal level is A_6, and the rank under the friction coefficient is second and the optimal level is A_5. If A_6 is taken, the total pressure-bearing capacity increases by 19.6% compared with A_5, while the friction coefficient only increases by 6.6%, so A_6 is preferable to A_5 after comprehensive comparison. Similarly analyzed, D is taken as D_3 and F is taken as F_2. The optimal combination is determined as $A_6B_1C_1D_3E_3F_2G_2$ through comprehensive balancing.

The structural combination of $B_1C_1D_3E_3$ would make the spherical convex hulls higher than the spherical pit plane, so the optimal combination was modified to $A_6B_1C_1D_3E_2F_2G_2$. The above strategy was used to simulate the optimal combination. The results show that the total pressure-bearing capacity can reach 0.6897 N, which is higher than the maximum of 0.6349 N across the 18 tests. The friction coefficient is reduced to 0.0401, which is lower than the minimum of 0.0423 across the 18 tests.

3.3. Comparative Analysis of Bearing Capacity and Lubrication Characteristics of Different Surfaces

Based on the structure of the optimal combination $A_6B_1C_1D_3E_2F_2G_2$ (second-order non-smooth structure), the spherical convex hulls were removed, and the spherical pit surface (first-order non-smooth structure) with the same size was modeled and simulated. The smooth surface was also simulated as a control. Figure 12 shows the velocity vector and the 2D total pressure contour of the profile of r_2 = 31.75 mm of different surfaces. Figure 13 shows the 2D total pressure contour of the upper surface of the seawater film of different surfaces.

As analyzed in Section 3.1, the maximum velocity of a smooth surface only decreases layer by layer from the upper surface of the seawater film, and no hydrodynamic pressure effect is generated. The total pressure on the upper surface of the seawater film is distributed in a circular zonal pattern, gradually increasing from inside to outside, and only reaches 10~20 kPa.

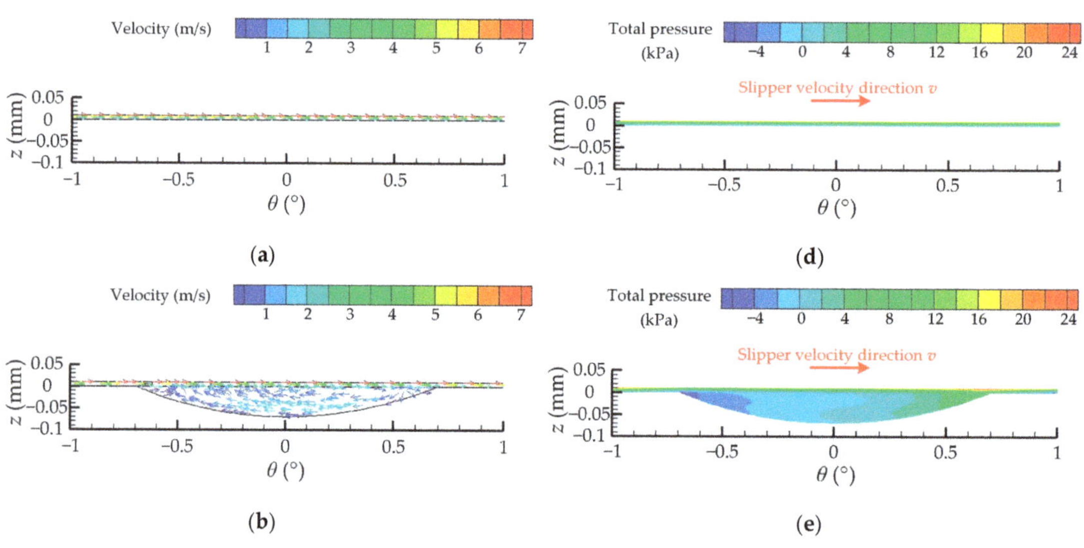

Figure 12. *Cont.*

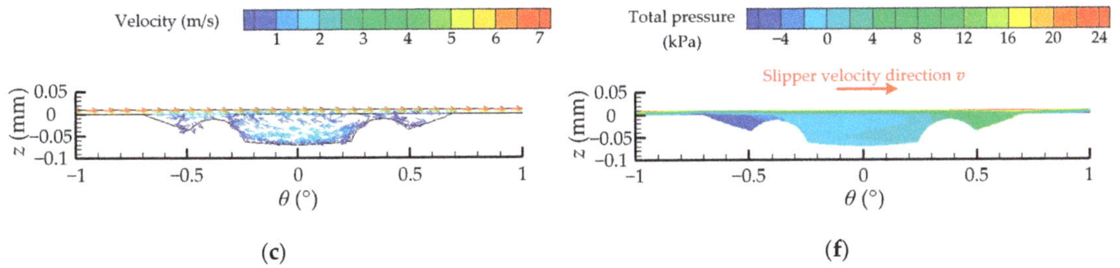

Figure 12. Velocity vector and 2D total pressure contour of the profile of r_2 = 31.75 mm of different surfaces. (**a**) Velocity vector of a smooth surface; (**b**) velocity vector of a spherical pit; (**c**) velocity vector of a spherical pit containing spherical convex hulls; (**d**) 2D total pressure contour of a smooth surface; (**e**) 2D total pressure contour of a spherical pit; and (**f**) 2D total pressure contour of a spherical pit containing spherical convex hulls.

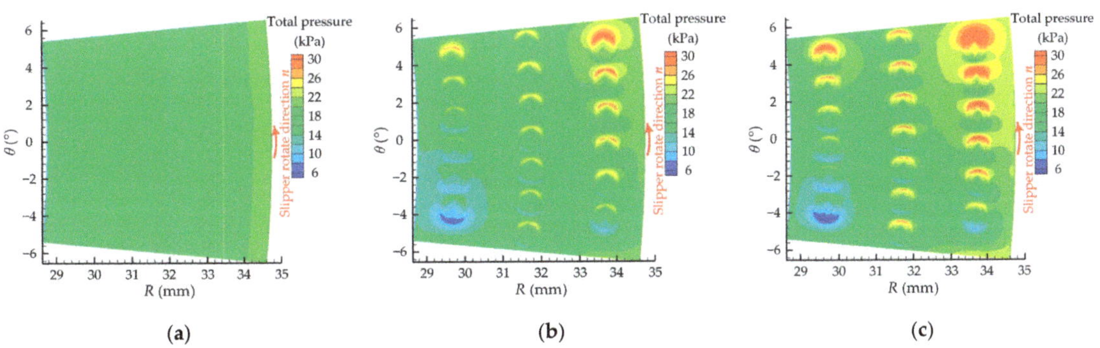

Figure 13. The 2D total pressure contour of the upper surface of the seawater film of different surfaces: (**a**) smooth surface; (**b**) spherical pit; (**c**) spherical pit containing spherical convex hulls.

The non-smooth surface produces a hydrodynamic pressure effect at the pit, and the total pressure on the surface of the seawater film can reach 30 kPa. Comparing Figure 12b,c,e,f, the second-order non-smooth structure generates not only a large vortex ring in the first order spherical pit, but also some small vortex rings near the second order spherical convex hulls. Therefore, the superimposed hydrodynamic pressure effect can be generated near the second order spherical convex hulls to expand the range of the high-pressure zone, thus producing a larger bearing capacity and a lower friction coefficient and obtaining better lubrication performance.

Table 3 shows the total pressure-bearing capacity and friction coefficient of different surfaces. It can be seen that the total pressure-bearing capacity of the second-order non-smooth structure increased by 11.0% and 8.9% over that of the smooth surface and the first-order non-smooth structure, respectively, and the friction coefficient decreased by 9.5% and 5.4% over that of the smooth surface and the first-order non-smooth structure, respectively.

Table 3. Total pressure-bearing capacity and friction coefficient of different surfaces.

	Smooth Surface	Spherical Pit	Spherical Pit Containing Spherical Convex Hulls
Total pressure-bearing capacity F_t (N)	0.6215	0.6335	0.6897
Friction coefficient μ	0.0443	0.0424	0.0401

4. Conclusions

Aiming to address problems of the poor lubricity and the serious friction and wear of friction pairs of seawater axial piston pumps, a hydrodynamic lubrication model of a slipper pair with a surface featuring spherical pits containing spherical convex hulls was built. Through the simulation analysis of the bearing lubrication mechanism and friction characteristics of the cross-scale, second-order compound microstructure from the microflow perspective, the following conclusions are drawn:

(1) The cross-scale, second-order compound microstructure could produce a hydrodynamic pressure effect to improve the bearing capacity of the lubrication film of a slipper pair. Along the velocity direction, seawater flows through the front part of the spherical pits (convergence wedge) to produce high pressure and through the back part of the spherical pits (divergence wedge) to produce low pressure, both of which have positive values. When seawater flows to spherical convex hulls, a hydrodynamic pressure effect is also generated to further expand the high-pressure zone.

(2) The total pressure is the synthesis of the static pressure and dynamic pressure, which is the macroscopic manifestation of the hydrodynamic pressure effect produced by the cross-scale, second-order compound microstructure. The total pressure-bearing capacity mainly comes from the contribution of the dynamic pressure to the bearing capacity that is proportional to the square of the velocity. Moreover, the friction coefficient is inversely proportional to the total pressure-bearing capacity. Thus, increasing the rotating speed can improve the bearing capacity of the lubrication film of the slipper pair, reduce the friction coefficient, and improve the lubrication performance.

(3) A hybrid orthogonal test scheme is used to find the order and optimal combination of factors under multiple indices. The order of the factors under the total pressure-bearing capacity is A (rotate speed) > C (first order depth-to-diameter ratio) > F (area rate) > E (second order depth-to-diameter ratio) > G (arrangement angle) > D (second order diameter) > B (first order diameter). The order under the friction coefficient is C > A > B > F > G > E > D. The optimal combination is $A_6B_1C_1D_3E_3F_2G_2$ under the total pressure-bearing capacity and $A_5B_1C_1D_1E_3F_1G_2$ under the friction coefficient. The final optimal combination is modified to $A_6B_1C_1D_3E_2F_2G_2$ through comprehensive balancing and analysis.

(4) The superimposed hydrodynamic pressure effect makes the cross-scale second-order compound microstructure obtain better bearing lubrication capacity and friction performance. Compared to the smooth surface and the first-order non-smooth microstructure, the total pressure-bearing capacity increased by 11.0% and 8.9%, respectively, and the friction coefficient decreased by 9.5% and 5.4%, respectively.

This study analyzed the bearing lubrication mechanism and friction characteristics of the cross-scale, second-order compound microstructure from the microflow perspective via the CFD method and optimized the working parameter and geometric parameters using a hybrid orthogonal test scheme. This study provides an effective solution and optimization reference for solving problems of the poor lubricity and the serious friction and wear of slipper pairs of seawater axial piston pumps. This solution can be extended to the design of friction pairs of other hydraulic components. Corresponding friction and wear tests need to be carried out in the future work to verify the reliability of the simulation analysis and to investigate the wear mechanism.

Author Contributions: Conceptualization, Y.L. and D.G.; methodology, Y.L.; software, C.W. and Z.Z. (Zongyi Zhang); validation, Z.Z. (Zhepeng Zhang) and W.W.; formal analysis, H.X. and T.G.; investigation, Y.L. and C.W.; resources, Y.L.; data curation, Z.Z. (Zongyi Zhang); writing—original draft preparation, Y.L., C.W. and Z.Z. (Zongyi Zhang); writing—review and editing, D.G., Z.Z. (Zhepeng Zhang), W.W., H.X. and T.G.; supervision, D.G.; project administration, D.G.; funding acquisition, Y.L. and D.G. All authors have read and agreed to the published version of the manuscript.

Funding: This research was supported by the National Natural Science Foundation of China (Grant No. 52005428), the Hebei Natural Science Foundation (Grant No. E2020203107 and Grant No. E2021203099), and the Open Fund Project of Shaanxi Provincial Key Laboratory of Hydraulic Technology (No. YYJS2022KF04).

Institutional Review Board Statement: Not applicable.

Informed Consent Statement: Not applicable.

Data Availability Statement: All data generated or analyzed during this study are included in this article.

Conflicts of Interest: The authors declare no conflict of interest.

References

1. Ren, L.Q.; Yang, Z.J.; Han, Z.W. Non-smooth wearable surface of living creatures and their bionic application. *Trans. Chin. Soc. Agric. Mach.* **2005**, *36*, 144–147.
2. Li, X.F.; Guo, Z.W.; Huang, Q.R.; Yuan, C.Q. Application of bionic tribology in water-lubricated bearing: A review. *J. Bionic Eng.* **2022**, *19*, 902–934. [CrossRef]
3. Tian, G.Z.; Zhang, Y.S.; Feng, X.M.; Hu, Y.S. Focus on bioinspired textured surfaces toward fluid drag reduction: Recent progresses and challenges. *Adv. Eng. Mater.* **2022**, *24*, 2100696. [CrossRef]
4. Gachot, C.; Rosenkranz, A.; Hsu, S.M.; Costa, H.L. A critical assessment of surface texturing for friction and wear improvement. *Wear* **2017**, *372–373*, 21–41. [CrossRef]
5. Zhang, J.; Chen, Y.; Xu, B.; Chao, Q.; Zhu, Y.; Huang, X.C. Effect of surface texture on wear reduction of the tilting cylinder and the valve plate for a high-speed electro-hydrostatic actuator pump. *Wear* **2018**, *414–415*, 68–78. [CrossRef]
6. Wang, Z.Q.; Fu, Q.; Wood, R.J.K.; Wu, J.; Wang, S.C. Influence of bionic non-smooth surface texture on tribological characteristics of carbon-fiber-reinforced polyetheretherketone under seawater lubrication. *Tribol. Int.* **2020**, *144*, 106100. [CrossRef]
7. Watanabe, S.; Kodama, E.; Sakakibara, K.; Sasaki, S.; Tsujii, Y. Effect of surface texturing on the durability of concentrated polymer brushes. *Tribol. Int.* **2021**, *155*, 106668. [CrossRef]
8. Song, F.; Yang, X.F.; Dong, W.L.; Zhu, Y.Q.; Wang, Z.Y.; Wu, M. Research and prospect of textured sliding bearing. *Int. J. Adv. Manuf. Technol.* **2022**, *121*, 1–25. [CrossRef]
9. Hou, Q.M.; Yang, X.F.; Cheng, J.; Wang, S.R.; Duan, D.R.; Xiao, J.P.; Li, W.Y. Optimization of performance parameters and mechanism of bionic texture on friction surface. *Coatings* **2020**, *10*, 171. [CrossRef]
10. Wei, Y.; Resendiz, J.; Tomkowski, R.; Liu, X. An experimental study of micro-dimpled texture in friction control under dry and lubricated conditions. *Micromachines* **2022**, *13*, 70. [CrossRef]
11. Liang, Y.N.; Gao, D.R.; Chen, B.; Zhao, J.H. Friction and wear study on friction pairs with a biomimetic non-smooth surface of 316L relative to CF/PEEK under a seawater lubricated condition. *Chin. J. Mech. Eng.* **2019**, *32*, 66. [CrossRef]
12. Liang, Y.N.; Gao, D.R.; Zhao, J.H. Tribological properties of friction pair between 316L stainless steel and CF/PEEK with nonsmooth surface under seawater lubrication. *Tribol. Trans.* **2020**, *63*, 658–671. [CrossRef]
13. Liang, Y.N.; Gao, J.X.; Gao, D.R.; Sun, Y.N.; Zhang, Z.Y.; Zhao, J.H. Effect of micro-textured surfaces and sliding speed on the lubrication mechanism and friction-wear characteristics of CF/PEEK rubbing against 316L stainless steel under seawater lubrication. *Appl. Sci.* **2021**, *11*, 9915. [CrossRef]
14. Uddin, M.S.; Liu, Y.W. Design and optimization of a new geometric texture shape for the enhancement of hydrodynamic lubrication performance of parallel slider surfaces. *Biosurface Biotribology* **2016**, *2*, 59–69. [CrossRef]
15. Ni, S.L.; Wu, H.C.; Zhao, L.M.; Yang, L. Effect of surface micro-pit on oil film performance of slipper pair in axial piston pump under high pressure. *Lubr. Eng.* **2018**, *43*, 30–34, 42.
16. Ni, S.L.; Wu, H.C.; Zhao, L.M.; Yang, L. Study on the effect of material matching to the performance of dry friction of the slipper pair in axial piston pump under high pressure. *Mach. Des. Manuf.* **2019**, *3*, 12–15.
17. Sugihara, T.; Enomoto, T. Performance of cutting tools with dimple textured surfaces: A comparative study of different texture patterns. *Precis. Eng.* **2017**, *49*, 52–60. [CrossRef]
18. Lin, X.K.; Kinoshita, S.; Sugihara, T.; Enomoto, T. Exploring the role of the interface adhesion phenomena focusing on surface expansion distribution. *Tribol. Int.* **2023**, *179*, 108160. [CrossRef]
19. Liu, S.C.; Sai, Q.Y.; Wang, S.W.; Williams, J. Effects of laser surface texturing and lubrication on the vibrational and tribological performance of sliding contact. *Lubricants* **2022**, *10*, 10. [CrossRef]
20. Naduvinamani, N.; Angadi, A. Static and dynamic characteristics of rough porous Rayleigh step bearing lubricated with couple stress fluid. *Lubricants* **2022**, *10*, 257. [CrossRef]
21. Liu, B. Studies on the Morphological Dimensions of the Riblet Surface of Fast-Swimming Sharks. Master Degree Thesis, Qingdao University of Science & Technology, Qingdao, China, 2008.
22. Lloyd, C.J.; Peakall, J.; Burns, A.D.; Keevil, G.M.; Dorrell, R.M.; Wignall, P.B.; Fletcher, T.M. Hydrodynamic efficiency in sharks: The combined role of riblets and denticles. *Bioinspiration Biomim.* **2021**, *16*, 046008. [CrossRef]

23. Barthlott, W.; Neinhuis, C. Purity of the sacred lotus, or escape from contamination in biological surfaces. *Planta* **1997**, *202*, 1–8. [CrossRef]
24. Lu, Y. Superior lubrication properties of biomimetic surfaces with hierarchical structure. *Tribol. Int.* **2018**, *119*, 131–142. [CrossRef]
25. Sun, J.R.; Cheng, H.; Cong, Q.; Li, J.Q.; Chen, B.C.; Ren, L.Q. Bionic study on the dung beetle (Copris ochus Motschulsky) for reduction of soil adhesion. *Acta Biophys. Sin.* **2001**, *17*, 785–793.
26. Cheng, H.; Sun, J.R.; Li, J.Q.; Ren, L.Q. Structure of the integumentary surface of the dung beetle *Copris ochus Motschulsky* and its relation to non-adherence of substrate particles. *Acta Entomol. Sin.* **2002**, *45*, 175–181.
27. Liu, Y.S.; Wu, D.F.; Li, D.L.; Deng, Y.P. Applications and research progress of hydraulic technology in deep sea. *J. Mech. Eng.* **2018**, *54*, 14–23. [CrossRef]
28. Yan, X.P.; Bai, X.Q.; Yuan, C.Q. Discussion on connotation and research scopes of ocean tribology and its research progress. *J. Mech. Eng.* **2013**, *49*, 95–103. [CrossRef]
29. Wang, W.; Wen, H.X.; Chen, W. Research status on tribological behaviors of materials under seawater environment. *Mater. Rep.* **2017**, *31*, 51–58.
30. Nie, S.L.; Yin, F.L. Progress and prospect of water hydraulic piston pump. *Chin. Hydraul. Pneum.* **2015**, *1*, 1–7.
31. Wen, S.Z.; Huang, P.; Tian, Y.; Ma, L.R. *Principles of Tribology*, 5th ed.; Tsinghua University Press: Beijing, China, 2018; p. 3.
32. Chen, Z.R.; Wang, H.J.; Liu, Q.Z.; Cai, W.H. *Engineering Fluid Mechanics*, 3rd ed.; Higher Education Press: Beijing, China, 2013.
33. Feng, J.A.; Tang, X.Q.; Wang, W.B.; Ying, R.; Zhang, T. Reliability verification method of numerical simulation based on grid independence and time independence. *J. Shihezi Univ. Nat. Sci.* **2017**, *35*, 52–56.
34. Lachner, H. Hydrostatic lagerungen in axialkolbenmachine. *Hydraul. Pneum.* **1974**, *18*, 605–611.
35. Han, J. Study on the Lubrication and Friction Properties of Macroscale and Microscale Surface Texture. Doctor Degree Thesis, China University of Mining & Technology, Xuzhou, China, 2013.
36. Wang, Y.J.; Jacobs, G.; König, F.; Zhang, S.; Goeldel, V.S. Investigation of microflow effects in textures on hydrodynamic performance of journal bearings using CFD simulations. *Lubricants* **2023**, *11*, 20. [CrossRef]
37. Tang, W.; Zhou, Y.K.; Zhu, H.; Yang, H.F. The effect of surface texturing on reducing the friction and wear of steel under lubricated sliding contact. *Appl. Surf. Sci.* **2013**, *273*, 199–204. [CrossRef]
38. He, W.; Xue, W.D.; Tang, B. *Optimize Experimental Design Method and Data Analysis*; Chemical Industry Press: Beijing, China, 2012.

machines

MDPI

Article

Novel Tool Path Generation Method for Pocket Machining Using Sound Field Synthesis Theory

Xuefeng Yang *, **Xulin Cai**, **Wenan Yang** and **Youpeng You**

College of Mechanical and Electrical Engineering, Nanjing University of Aeronautics and Astronautics, Nanjing 210016, China
* Correspondence: yangxuefeng0309@nuaa.edu.cn

Abstract: Contour parallel tool paths have been proved to be a preferred machining strategy for their advantage of less tool retractions and less sharp turns. The traditional geometrical algorithm-based tool path generation method often makes it hard to simply and simultaneously solve the problems of self-intersection, no residual, and smoothness at the same time due to their contradictions. To address this issue, a contoured parallel tool path generation method for pocket machining is developed in this study. It is based on sound field synthesis theory inspired by the phenomenon of sound wave propagation. Firstly, the simplified medial axis (SMA) tree of the pocket is extracted and the propagation direction of each SMA segment is calculated on account of the geometric characteristics of the pocket boundary. Secondly, the final tool path is obtained through the synthesis of the sound field. Finally, the novel method is verified on five different pockets to generate a contoured parallel milling tool path. After machining these pockets and measuring the machining time, roughness, and cutting force, the experimental results demonstrate that the tool path obtained by the novel method has advantages in improving machining quality and efficiency.

Keywords: tool path generation; pocket milling; sound field synthesis; computer-aided manufacturing; simplified medial axis

1. Introduction

Aviation structural parts are becoming integrated and lightweight as the development of aviation industry continues, which leads to the design of structures with complex geometric characteristics and puts forward higher requirements on machining quality and efficiency [1]. As one of the most important parts of machining technology, tool path has been widely discussed in the past decades, and many tool path generation strategies for pocket machining have been developed by experts and scholars, which can be roughly divided into direction parallel [2,3] and contour parallel [4,5] tool path generation strategies. Of the two strategies, contour parallel tool paths have been proved to be a preferred machining strategy for their advantage of less tool retractions and less sharp turns [6,7].

The core idea of contour parallel tool path generation is isometric offset based on a given pocket, and the generation strategies can be divided into the following categories according to different principles: the pairwise offset method [8–10], Voronoi diagram-based method [11–14], and pixel-based method [15–18]. The pairwise offset method usually includes the following steps: Calculate the offset segment of the boundary curve and add arcs on it to form the initial offset curve, detect the intersection points of the initial curve and eliminate both local loops and global loops of it. Choi et al. [8] designed an effective pairwise offset algorithm firstly to generate a tool path for a 2D closed pocket. Lin et al. [9] developed a pairwise offset method to round the inferior corners while generating contour parallel tool paths, which involves residual detection based on Boolean operation. Bahloul et al. [10] established an analytical model for contour parallel tool path generation based on curves of pocket boundary, including a corner residual elimination method and a center residual elimination method. Although the pairwise offset method can generate tool paths directly by initial boundary segment offsetting, it also constantly introduces a certain amount of local and global invalid loops that require further adjustments to rationalize the final tool path, and the generation of additional residuals is usually unavoidable. The Voronoi diagram-based method usually includes the following steps: Calculate two endpoints of the offset segment of the boundary curve, and then calculate the path formed by obtained endpoints based on Voronoi skeleton. Jeong et al. [11] developed a tool path generation method for free-form shaped pockets based on Voronoi diagrams, which can be utilized to generate tool paths in any pocket with quadratic differentiable curve boundary. Lai et al. [12] developed an incremental algorithm based on the Voronoi diagram to calculate the offset distance and minimum passage width in the tool path generation of a pocket. Hinduja et al. [14] developed a linking algorithm for contour-parallel tool paths generation by using the segments on a Voronoi diagram to overcome defects such as being not based on clear geometric principles, non-optimal, and not considering technical factors. The main defect of this method is that the smoothness of the generated tool path is not ideal and the calculation process is complicated. The main idea of the pixel-based method is to decompose the interior region defined by the curve of pocket boundary into pixels, and then generate tool paths through machining simulation. Dhanik et al. [17] developed a tool path generation method based on the boundary formulation of the pocket boundary and fast marching method, in which topological changes and slope discontinuities were dealt with by adding "entropy conditions" to the numerical implementation during the process of offset. Xu et al. [18] developed a tool path generation method based on the digital image method, in which the region within the pocket boundary is converted into a binary image and machining residuals and smoothness are optimized by adjusting the blurring and sharpness parameters of the image. The main advantage of the pixel-based method is that it does not need to calculate the offset segment, and the precision of the offset curve only depends on the mesh dispersion of the region inside the pocket, but this method can only be applied to a closed pocket in principle.

The biggest three problems in tool path generation are self-intersection, sharp corners, and uncut residual. Therefore, a large number of methods have been designed to solve these problems and effectively improve the machining efficiency under the premise of machining quality. Li et al. [19] developed a smooth tool path generation method for pocket machining by using linear interpolation, exponential regularization, and B-spline curve fitting. Xu et al. [20] developed a self-intersection identification and elimination algorithm based on mapping to process offset tool paths for milling operation. Zhou et al. [21] developed a residual regions identification method and a looping tool paths generation method to remove uncut areas left by pocket roughing while make the generated tool paths achieve G1 continuous and progressive radial cutting depths. Chen et al. [22] introduced the central axis (MA) transformation of the pocket and proposed a tool path generation optimization method for pocket machining in order to avoid interference and achieve the highest efficiency. Similarly, Huang et al. [23] developed an improved spiral toolpath generation method based on MA transformation to improve the machining performance. Although the problems of self-intersecting loops, sharp turning, and uncut residual have been well addressed by previous studies, it is important to note that most solutions are not only complicated but also not universal, which makes them potentially invalid on pockets of other type or geometry.

It is undeniable that most tool path generation methods are based on computational geometry, but people often fall into the dilemma of geometric algorithm to find a better solution. However, most geometry-based solutions are prone to result in trouble since the two problems of "smoothness" and "no residual" are usually contradictory and difficult to solve simultaneously. In view of the above questions, the diffusion process of sound waves with time is deeply studied. The research results show that if the sound source of the sound field is properly processed, the ideal steady-state sound field distribution can be obtained. The diffusion process of steady sound field is almost parallel, which is very similar to the geometric shape of the cutting tool path, and the diffusion line is generally smooth and non-intersecting. This can not only match the geometry of the pocket boundary well, but also effectively achieve the goal of smoothness, no self-intersection, and no residual. Therefore, based on the theory of sound field synthesis (SFS), a tool path generation method using the plane wave synthesis of a continuous linear sound source is developed in this study. In order to facilitate readers' understanding, the theory of SFS utilized in this study is introduced in Section 2; the tool path generation method based on SFS is described in Section 3, including the medial axis transformation of a pocket, the calculation method of step-over, and propagation direction for different geometry. In Section 4, the validity of the developed method is proved by three examples with different geometric characteristics, moreover, the method is experimentally verified in a closed pocket, open pocket, and pocket with an island. Conclusions are summarized in Section 5.

2. The Theory of SFS

In this study, sound field $s_p(\mathbf{x}, t)$ represents the sound pressure, that is, the local pressure deviation from the ambient pressure caused by a sound wave. In the time-frequency domain, $S_p(\mathbf{x}, \omega)$ denotes sound field or sound pressure, $S_{\text{pd}}(\mathbf{x}, \omega)$ is referred to as sound pressure deviation caused by a plane wave sound field, which can be given by

$$S_{\text{pd}}(\mathbf{x}, \omega) = \overline{S}_{\text{pd}}(\omega) e^{-i\mathbf{k}_{\text{pd}}^T \mathbf{x}} \tag{1}$$

where $\mathbf{x} = [x\ y\ z]$ represents Cartesian coordinates, $\mathbf{k}_{\text{pd}}^T = \left[k_{pd,x} k_{pd,y} k_{pd,z}\right]$ denotes propagation direction, $\overline{S}_{\text{pd}}(\omega)$ denotes time-frequency component.

According to the literature [24], the SFS means that the sound field emitted by individual elementary sound sources is superimposed on an extended domain to form a sound field with given a desired property, which happens to coincide with the generation process of the tool path. The employed elementary sound source is referred as a "secondary source" in the remainder of this study. For the sake of facilitating the mathematical treatment, it can be assumed that the ensemble of the secondary source considered is continuous. Therefore, it will be called the secondary sources distribution (SSD). Similarly, as the distribution considered can be also assumed to enclose the target region expected to synthesize the desired sound field, then the SFS problem can be expressed as

$$S_p(\mathbf{x}, \omega) = \oint_{\partial\Omega} D(\mathbf{x}_0, \omega) G(\mathbf{x}, \mathbf{x}_0, \omega) dF(\mathbf{x}_0) \tag{2}$$

where $\partial\Omega$ denotes the geometry of the distribution, ω is the radian frequency, $D(\mathbf{x}_0, \omega)$, $G(\mathbf{x}, \mathbf{x}_0, \omega)$, and $dF(\mathbf{x}_0)$ denote the driving function, spatial transfer function, and infinitesimal surface element of the secondary source located at $\mathbf{x}_0$, respectively.

$s_p(\mathbf{x}, t)$ can be physically possible as long as the scalar wave equation in the domain of interest is fulfilled. Therefore, considering source-free domains, the wave equation can be expressed as [24]

$$\Delta^2 S_P(\mathbf{x}, t) - \frac{1}{v^2} \frac{\partial^2 S_P(\mathbf{x}, t)}{\partial t^2} = 0 \tag{3}$$

where v represents the speed of sound, and the scalar differential operator Δ^2 can be obtained by using twice the gradient Δ in Cartesian coordinates, which can be described as follow.

$$\Delta = \frac{\partial}{\partial x} \mathbf{e}_x + \frac{\partial}{\partial y} \mathbf{e}_y + \frac{\partial}{\partial z} \mathbf{e}_z \tag{4}$$

where $\mathbf{e}_i (i = x, y, z)$ represents the unit vector of indexed direction.

Using the temporal Fourier transform, Equation (3) can be obtained as below

$$\Delta^2 S_p(\mathbf{x}, \omega) - k^2 S_p(\mathbf{x}, \omega) = 0 \tag{5}$$

where k is termed wavenumber and can be expressed as

$$k = \frac{\omega}{v} \tag{6}$$

Generally speaking, the solution of a continuous distribution boundary condition in a sound field can be roughly divided into the arbitrary shaped simple connected SSD, spherical SSD, circular SSD, plane SSD, linear SSD, and so on. In this study, by analyzing the possibility and complexity of SFS by using different source distribution schemes in tool path generation, linear SSD is finally selected to SFS.

Based on Equation (2), the SFS problem of linear source distribution can be expressed as

$$S_p(\mathbf{x},\omega)=\int_{-\infty}^{\infty}D(\mathbf{x}_0,\omega)G(\mathbf{x}-\mathbf{x}_0,\omega)d\mathbf{x}_0 \tag{7}$$

where $\mathbf{x}_0=\begin{bmatrix}x_0 & 0 & 0\end{bmatrix}^T$.

Just as literature [24], the explicit solution for linear SSD can be given as follows:

$$S_p(\mathbf{x},\omega)=\frac{e^{-ik_{\text{pd},y}y_{\text{ref}}}}{H_0^{(2)}\left(k_{\text{pd},y}y_{\text{ref}}\right)}e^{-ik_{\text{pd},x}x}H_0^{(2)}\left(k_{\text{pd},y}\sqrt{y^2+z^2}\right) \tag{8}$$

and it consists with the required sound field for $y=y_{\text{ref}}$ and $z=0$.

Substituting $H_0^{(2)}(z)=2/(\pi z)\exp -i(z-\pi/4)$ [25] into Equation (8), the approximated synthesized sound field can then be expressed as

$$S_{\text{appr}}(\mathbf{x},\omega)=\sqrt{\frac{y_{\text{ref}}}{\sqrt{y^2+z^2}}}e^{-ik_{\text{pd},x}x}e^{-ik_{\text{pd},y}\sqrt{y^2+z^2}} \tag{9}$$

which can be used as an option to reduce the amount of computation in the algorithm without a great loss of accuracy.

In order to actually apply the SSD, the SFS systems and milling tool path cannot be of infinite length. Therefore, incorporating a suitable window function $f(x_0)$ into Equation (7), the sound field $S_{tr}(\mathbf{x},\omega\ )$ of a truncated linear source distribution in x-dimension can be expressed as

$$S_{tr}(\mathbf{x},\omega)=\int_{-\infty}^{\infty}f(x_0)D(\mathbf{x}_0,\omega)G(\mathbf{x}-\mathbf{x}_0,\omega)d\mathbf{x}_0 \tag{10}$$

Using the convolution theorem, the synthesized sound field $\overline{S}_{tr}(k_x,y,z,\omega)$ can be given by

$$\begin{aligned}\overline{S}_{tr}(k_x,y,z,\omega)&=\tfrac{1}{2\pi}\left(\overline{\omega}(k_x)*_{k_x}\overline{D}(k_x,\omega)\right)\overline{G}(k_x,y,z,\omega)\\&=\tfrac{1}{2\pi}\overline{D}_{tr}(k_x,\omega)\overline{G}(k_x,y,z,\omega)\end{aligned} \tag{11}$$

where the symbol $*_{k_x}$ represents convolution with respect to k_x. $\overline{D}_{tr}(k_x,\omega)$ represents the truncated driving function.

The finite extent of an SSD with length L that started at $x=0$ and ended at $x=L$ can be given by a rectangular window function $f_R(x)$ as

$$f_R(x)=\text{rect}\left(\frac{x}{L}\right)=\begin{cases}1 & \text{if } 0\le x\le L\\ 0 & \text{else}\end{cases} \tag{12}$$

Using the Fourier transformation and Equation (12) can be expressed as

$$\overline{f}_R(k_x)=L\cdot\sin\text{c}\left(\frac{k_xL}{2}\right) \tag{13}$$

As can be seen from the above formula, the convolution of $\overline{D}(k_x,\omega)$ with $\overline{f}_R(k_x)$ is essentially a spatial low pass filtering operation smearing $\overline{D}(k_x,\omega)$ along the k_x axis. Therefore, the truncated SSD reveals distinctive complex radiation properties. According to Equation (10), the properties of $\overline{S}_{tr}(k_x,y,z,\omega)$ can be directly gained from the performances of $\overline{D}_{tr}(k_x,\omega)$.

3. Universal Tool Path Generation Method

Pocket machining has been widely applied in the aerospace and mold industry. Although almost all commercial CAM packages provide the implementation of traditional strategy, the generated tool path is only based on the idea of geometric bias, which cannot simply and directly guarantee the high efficiency of the generated path. However, machining time is the most concerned evaluation index of tool path in roughing.

In this study, inspired by the phenomenon of sound field fluctuation and diffusion, a rough machining tool path generation method based on SFS is developed. The developed method is not only of great universality, but also has high machining efficiency, which is very suitable for rough machining of pockets. The specific generation method and process will be described in this section.

3.1. Medial Axis Transformation of Pocket

The medial axis transform (MAT) of a pocket is a completed shape descriptor, which is of great importance in describing the boundary of complex pocket [22]. The MAT is based on medial axis (MA) and locally inscribed circles. The MA of the pocket is composed of all the locus of centers of the tangent circles, and it is also called the MA tree, as shown in Figure 1a. The locally inscribed circles are confined to the boundary and can be placed anywhere in the pocket. A detailed description of MA can be found in literature [22], including the definition of the locally inscribed circle, end point, bisector point, and branch point.

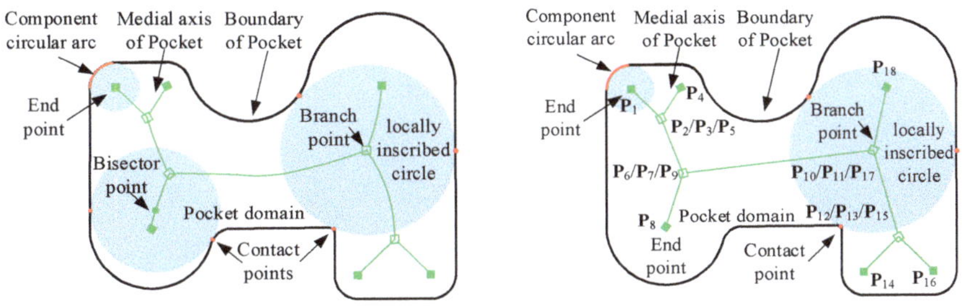

(**a**) MA tree of a pocket (**b**) SMA tree of a pocket

Figure 1. Illustration of the basic MA and SMA tree of a pocket.

It is essential to calculate locally inscribed circles of bisector points accurately and efficiently in the calculation of MAT, and the method used in [22] is directly used in this study. After that, the MAT should be stored in a tree data structure, and several specialized data structures have been developed in the past few years to represent the MA tree. In order to store the MA tree as simply as possible under the premise of application requirements, the strategy proposed in [23] is adopted in this study and some adjustments are made to meet the requirements of the tool path generation method developed in this study.

In this method, complex MA tree structure is transformed into simple closed loop called the MA cycle by a clockwise or counterclockwise traversal strategy. The MA cycle begins at the MA point with the minimum locally tangent circle and visits the branch point more than twice, the bisector points twice, and the end point once.

Even though the MA consists of a series of curves that can be approximated by a series of linear segments, such a complex MA is not needed in the method of tool path generation in this study. Instead, only linear continuous secondary sources combined with the acoustic propagation method is needed to achieve the fitting of the straight line and curve boundary, which will be described in Section 3.3. Therefore, after the calculation of MA, only branch points and end points are needed to construct a simplified MA (SMA), which can be marked as $P_i(i = 1, 2, \cdots, n)$. An example is shown in Figure 1b, linear segments connecting branch points and end points are the simplified MA, which and can be represented by $\overline{\mathbf{P}_1\mathbf{P}_2\mathbf{P}_3\cdots\mathbf{P}_{16}\mathbf{P}_{17}\mathbf{P}_{18}}$. The Euclidean distance from P_i to P_j can be expressed as $len(P_i, P_j)$.

3.2. Step-Over Calculation for Tool Path

The first problem to be solved is to analyze the mapping between the step-over of the tool path and sound field parameters. It is not difficult to see that the mathematical description of the sound field caused by a plane wave in the time–frequency domain is a complex wave, which seems difficult to relate to the step-over. However, only the real part of it is used as the sound field in practical application, and the real part of Equation (1) can be expressed as

$$\Re\left(S_{\text{pd}}(\mathbf{x},\omega)\right) = \overline{S}_{\text{pd}}(\omega)\cos\left(\mathbf{k}_{\text{pd}}^{T}\mathbf{x}\right) \tag{14}$$

In this study, 2.5-D SFS is utilized in the generation of the tool path, however, if the theory of SFS is used for calculation, it is not difficult conclude that the calculation result is a wave surface by analyzing Equation (14), as shown in Figure 2. It also shows that the diffusion process of the sound field is almost parallel and non-intersecting, which is very similar to the geometry of the cutting tool path.

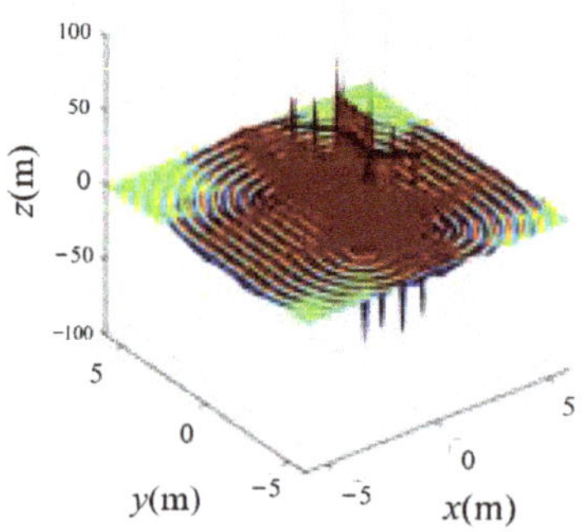

Figure 2. Illustration of wave surface.

To handle this problem, the intersection line of the surface and the x-y plane is used as the generated tool path, which means that the step-over between two adjacent tool paths is related to the wave cycle of the sound field and is half of the wave cycle in terms of spatial length. Therefore, the following relationship can be obtained by combining Equations (6) and (14)

$$stepover = \frac{\pi}{k} = \frac{\pi v}{\omega} = \frac{v}{2f} \tag{15}$$

which means the step-over of the tool path is determined by the time frequency and sound speed of the sound field, meaning that SFS can fully meet the actual needs of tool path generation.

3.3. *Propagation Direction Calculation Considering Pocket Boundary Geometry*

A SMA tree consisting of linear segments has been constructed in Section 3.1 to generate the tool path by using continuous linear secondary sources. However, the SMA tree cannot reflect the geometric characteristics of the pocket boundary well, but instead only the segmentation of the boundary. Therefore, in order to make the synthesized sound field effectively reflect the geometric characteristics of the pocket boundary, different propagation directions of acoustic waves are studied to realize the matching of the synthetic sound field to different boundary features in this section.

3.3.1. Propagation Direction for Linear Boundary

The sound field solution method of linear secondary sources situated along the x-axis and propagating to the y-axis has been described in Section 2, However, in order to use the linear sequences of SMA in Section 3.1 to synthesize the sound field reflecting the comprehensive geometric characteristics of the pocket, the position and attitude of the secondary source must be transformed firstly, including translation and rotation. As shown in Figure 3, the transformation of the secondary source position and attitude is described by taking the SMA $\mathbf{P}_1\mathbf{P}_2$ as an example, where $\mathbf{P}_g$ is the global coordinate of the pocket and $\mathbf{P}_l$ is the local coordinate of $\mathbf{P}_1\mathbf{P}_2$. The transformation relationship is as follows

$$\begin{pmatrix} \mathbf{P}_l \\ 1 \end{pmatrix} = \begin{pmatrix} {}^G\mathbf{R}_L & {}^G\mathbf{p}_L \\ 0 & 1 \end{pmatrix} \begin{pmatrix} \mathbf{P}_g \\ 1 \end{pmatrix} \tag{16}$$

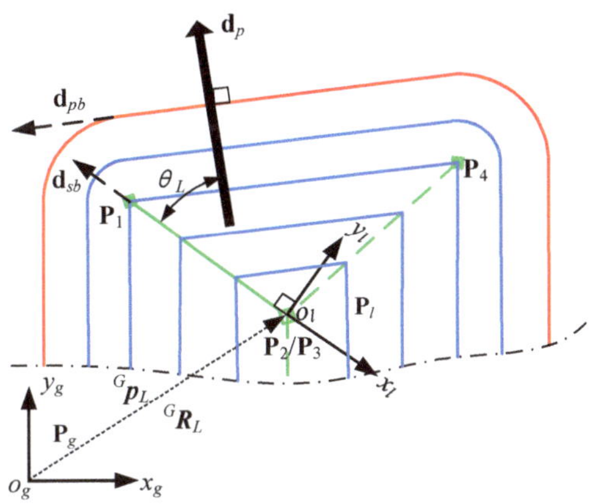

Figure 3. Illustration of propagation direction against linear boundary.

After the coordinate transformation, the propagation direction of each linear secondary source (segment of SMA) can be calculated in the local coordinates. Taking θ_L to represent the propagation direction of $\mathbf{P}_1\mathbf{P}_2$, and that the propagation direction is along the direction of y_l-axis if $\theta_L = \pi/2$. In order to meet the needs of different linear boundaries, $\mathbf{d}_{pd}$ and $\mathbf{d}_{sb}$ are utilized to represent the direction of linear boundary and the direction of the linear SMA, respectively, and then the propagation direction can be expressed as

$$\theta_L = \pi/2 - \left\langle \mathbf{d}_{pd}, \mathbf{d}_{sb} \right\rangle \tag{17}$$

where $\left\langle \mathbf{d}_{pd}, \mathbf{d}_{sb} \right\rangle$ denotes the angle between $\mathbf{d}_{pd}$ and $\mathbf{d}_{sb}$.

An example of tool path generation based on the abovementioned method is shown in Figure 4; it is not difficult to see the feasibility of the developed method and the advantage of smoothness at the corner.

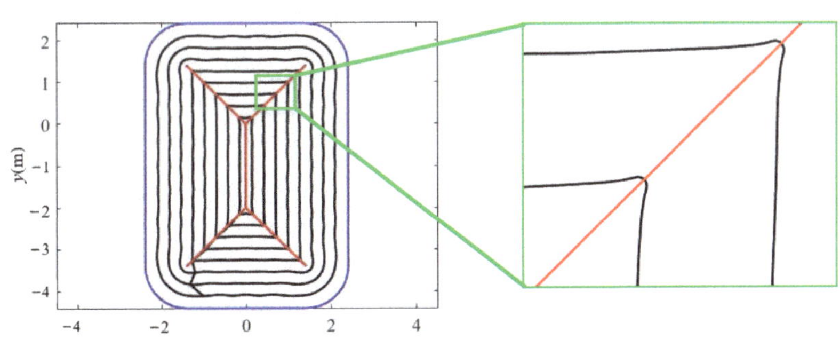

Figure 4. Tool path generation for linear boundary pocket.

3.3.2. Propagation Direction for Convex Arc Boundary

Analogous to the linear boundary, the propagation direction of each linear secondary source can be calculated in the local coordinates to approximate the convex arc boundary after the coordinate transformation, but the difference is that the approximation of a convex curve requires at least two linear secondary sources, as shown in Figure 5. Taking θ_L and θ_R to represent the propagation direction of $\mathbf{P}_1\mathbf{P}_2$ and $\mathbf{P}_3\mathbf{P}_4$, respectively, to meet the approximation of the convex arc boundary, $\mathbf{d}_{sb,L}$ and $\mathbf{d}_{sb,R}$ are utilized to represent the direction of linear SMA at the left and right, respectively. The normal vector and tangent vector of convex arc boundary at left and right ends of the curve can be expressed in turn as $\mathbf{n}_L$, $\mathbf{t}_L$, $\mathbf{n}_R$ and $\mathbf{t}_R$, respectively, and the calculation of them are not complicated and will not be repeated here. After defining and calculating the above variables, the propagation direction of $\mathbf{P}_1\mathbf{P}_2$ and $\mathbf{P}_3\mathbf{P}_4$ can be expressed as

$$\begin{cases} \theta_L = \langle \mathbf{t}_L, \mathbf{d}_{sb,L} \rangle - \pi/2 \\ \theta_R = \langle \mathbf{t}_R, \mathbf{d}_{sb,L} \rangle - \pi/2 \end{cases} \tag{18}$$

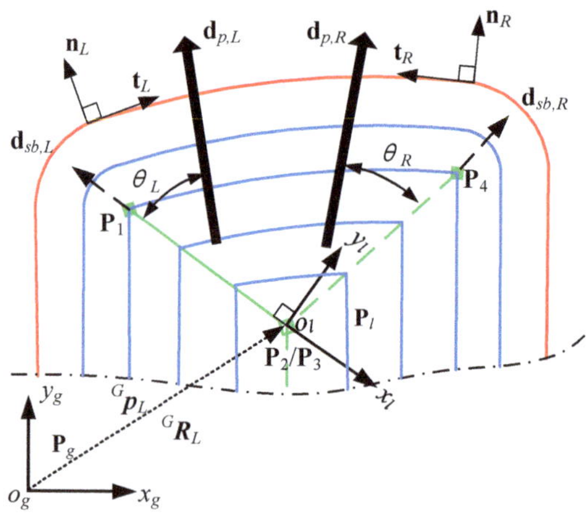

Figure 5. Illustration of propagation direction against convex arc boundary.

An example is also shown in Figure 6; it is not difficult to see the feasibility of the developed method and the advantage of smoothness at the corner.

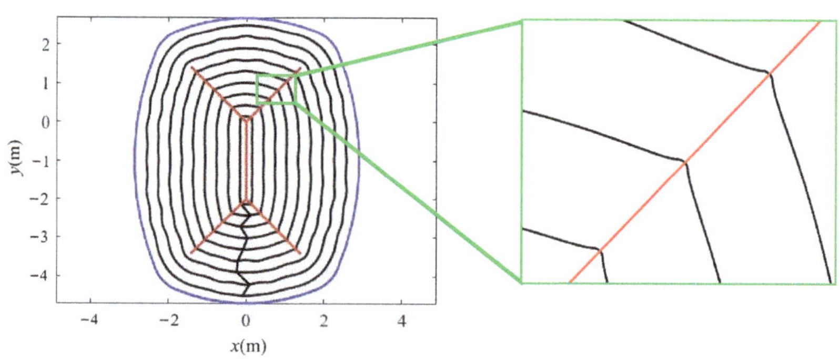

Figure 6. Tool path generation for convex arc boundary pocket.

3.3.3. Propagation Direction for Concave Arc Boundary

The same as for the convex arc boundary, the propagation direction of each linear secondary source can be calculated in the local coordinates to approximate concave arc boundary after the coordinate transformation, as shown in Figure 7. Taking θ_L and θ_R to represent the propagation direction of $\mathbf{P}_1\mathbf{P}_2$ and $\mathbf{P}_3\mathbf{P}_4$, respectively, to meet the approximation of convex arc boundary, $\mathbf{d}_{sb,L}$ and $\mathbf{d}_{sb,R}$ are utilized to represent the direction of linear SMA at the left and right, respectively. The normal vector and tangent vector of convex arc boundary at left and right ends of the curve can be expressed in turn as $\mathbf{n}_L$, $\mathbf{t}_L$, $\mathbf{n}_R$, and $\mathbf{t}_R$, and the propagation direction of $\mathbf{P}_1\mathbf{P}_2$ and $\mathbf{P}_3\mathbf{P}_4$ can be expressed as

$$\begin{cases} \theta_L = \langle \mathbf{t}_L, \mathbf{d}_{sb,L} \rangle - \pi/2 \\ \theta_R = \langle \mathbf{t}_R, \mathbf{d}_{sb,L} \rangle - \pi/2 \end{cases} \tag{19}$$

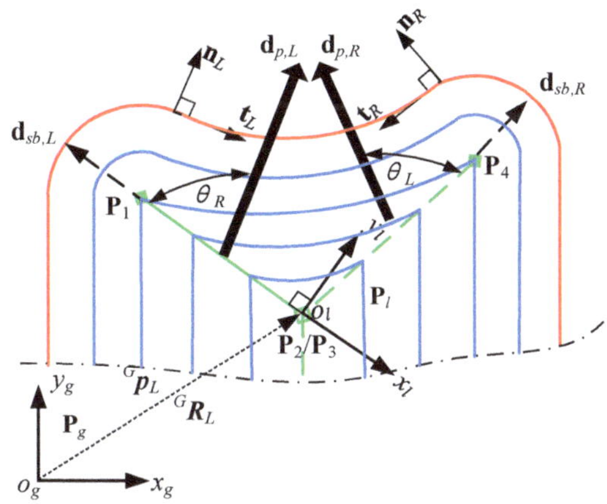

Figure 7. Illustration of propagation direction against concave arc boundary.

An example is also shown in Figure 8; it is not difficult to see the feasibility of the developed method and the advantage of smoothness at the corner.

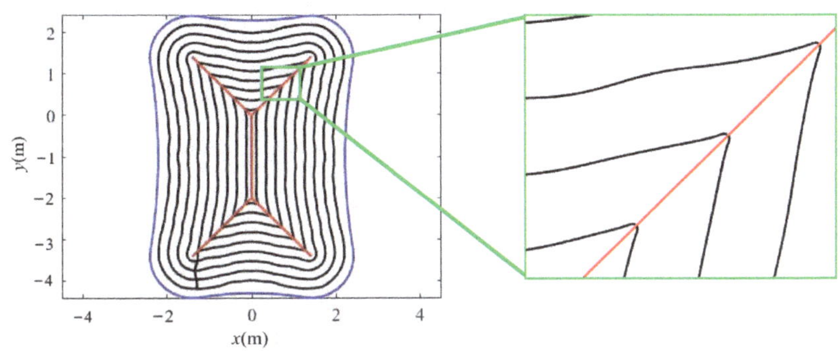

Figure 8. Tool path generation for concave arc boundary pocket.

4. Experimental Verification

4.1. Simulation Results

To verify the validity and universality of the developed SFS-based method, five pockets are selected as test cases for tool path generation, including an open pocket (pocket 1), three closed pockets (pockets 2, 3, and 4), and a pocket with an island (pocket 5). At the same time, for the purpose of comparison, tool paths of these five pockets are generated by using famous commercial CAM software NX® under the same machining parameters, and the generated tool paths are shown in Table 1.

Table 1. Simulation comparison of tool path generation methods.

Method / No.	SFS-Based	NX®-Based
1	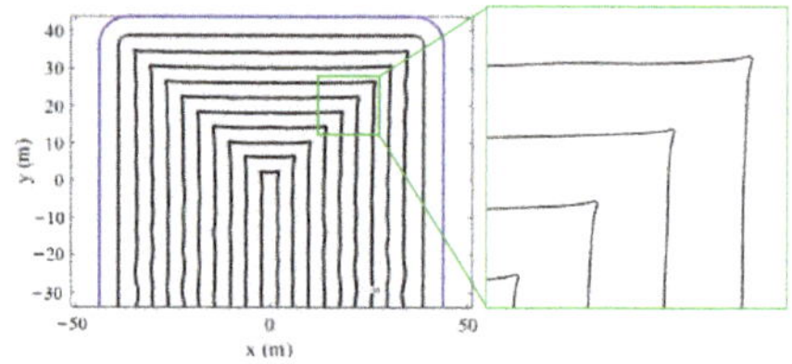	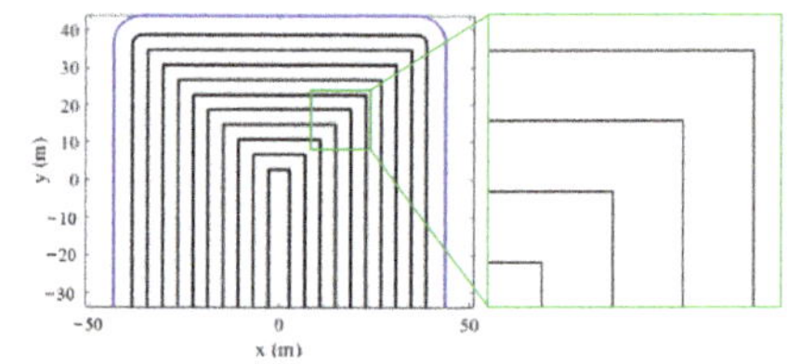
2	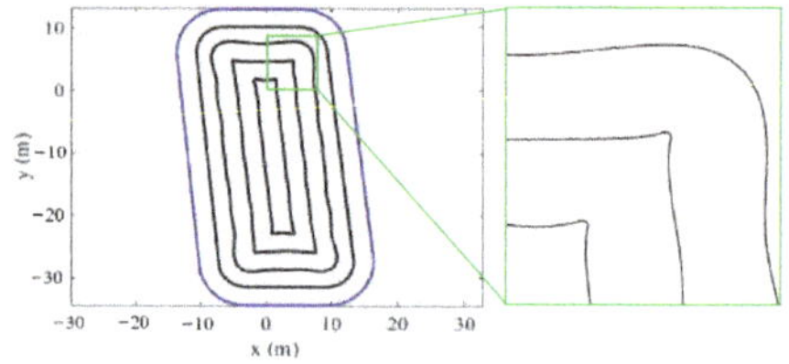	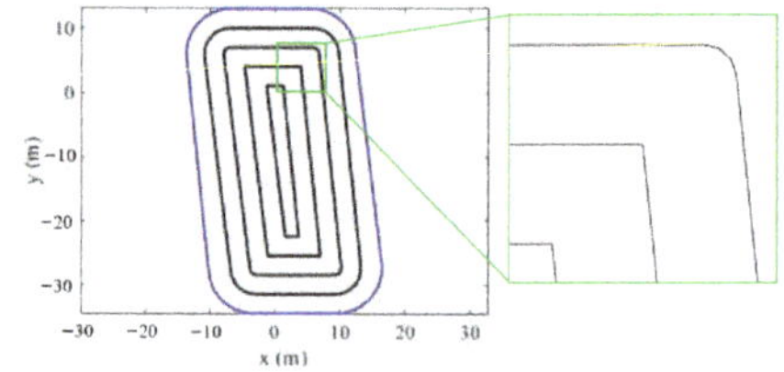
3	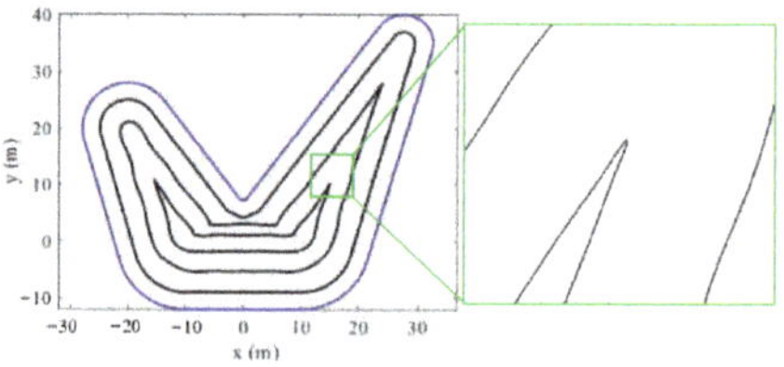	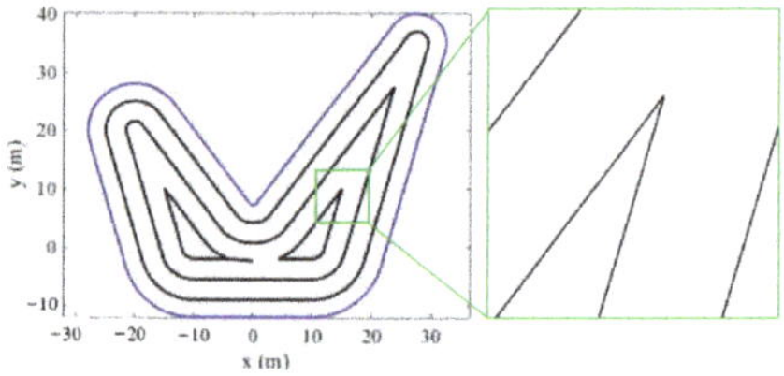
4	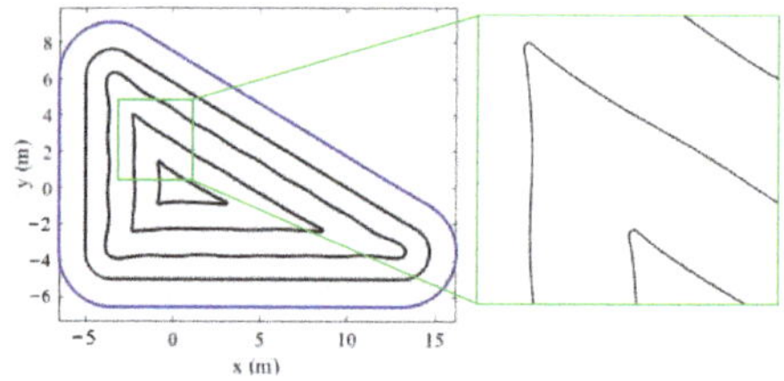	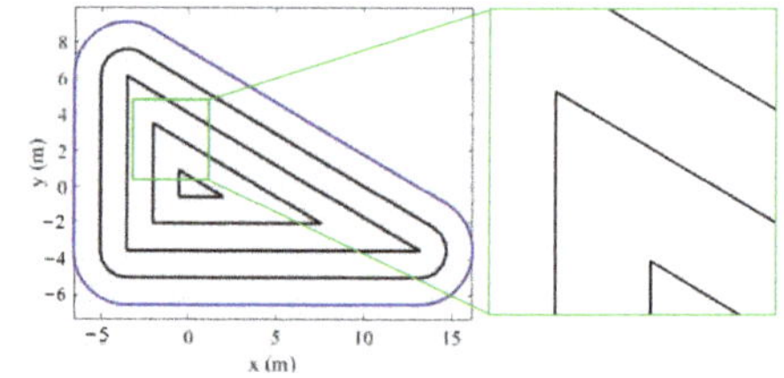
5	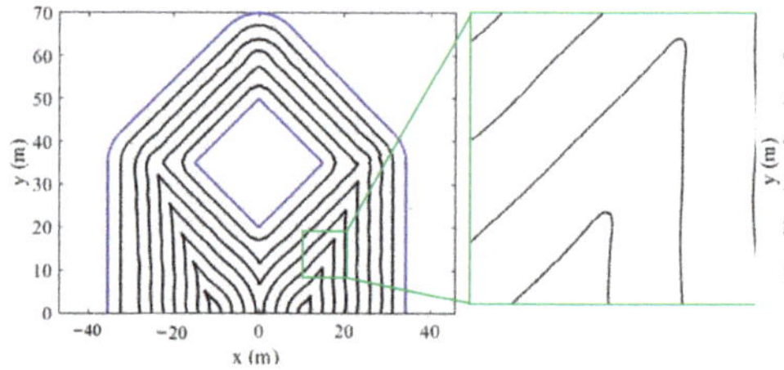	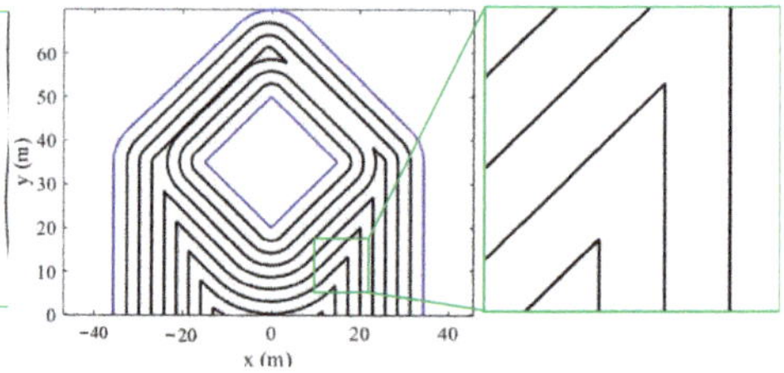

It can be seen from Table 1 that the developed tool path generation method can effectively generate tool paths for different pocket types and improve the smoothness of tool paths. In addition, the total length of tool paths obtained by these two methods are also compared in this simulation, and the test results are shown in Figure 9.

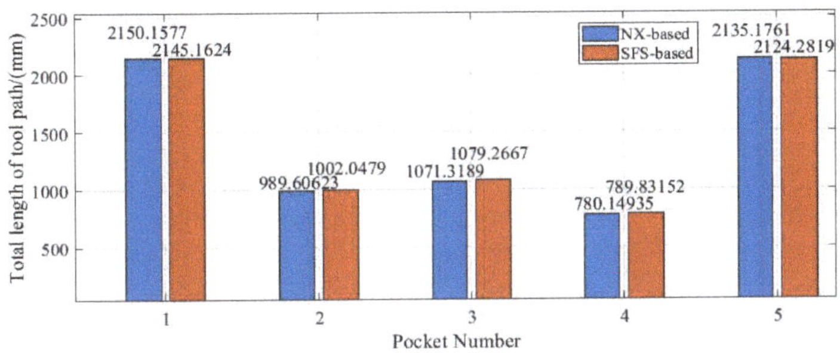

Figure 9. Comparison of tool path length.

As we can see from Figure 9, the length of the tool path generated by using the developed SFS-based method is slightly less than NX® in two out of five cases and slightly greater than NX® in the other case. However, this does not mean that the generated path by the developed SFS-based method has no advantage in machining efficiency; because the smooth tool path can effectively reduce the frequent acceleration and deceleration in the process of machining, it is not only beneficial to improve the efficiency, but also to improve the stability and dynamic accuracy of machining, so physical experiments should be carried out.

4.2. Physical Experimental Results

To further verify the above conclusions, physical experiments were carried out on a CNC machine tool, and all five examples in Table 1 were cut on the machine tool. The test system is shown in Figure 10. The machine tool utilized on this experiment is XK-L540, the force sensor is YD-F3220, the data acquisition card is uT3408M, and the data acquisition system is uTekAcqu. Of course, except for the tool path, the tool, fixture, and cutting parameters of the comparison experiment are strictly consistent to ensure the objectivity of the results. To verify the advantages of the developed method in improving the dynamic characteristics of tool paths, cutting forces were measured at the corner of the closed pocket 3 and the open pocket 5 in the machining process, and surface roughness was measured by surface roughness measuring instrument MarSurf PS 10 to verify the advantages of the developed method in improving the machining quality.

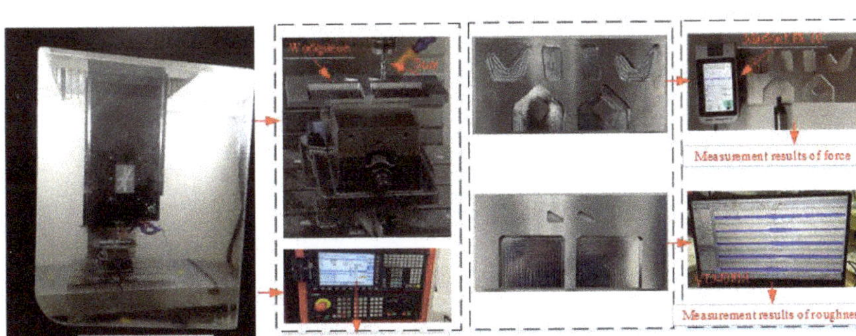

Figure 10. Physical experimental system.

Machining time directly reflects the machining efficiency of tool paths. As shown in Figure 11, the machining time of tool paths generated by NX® and the developed method was recorded in the process of machining and data which are obtained from the machine tool directly. The machining time of the tool path generated by using the developed SFS-based method is slightly less than NX® in two out of five cases and slightly greater than NX® in the other two cases, but it does not mean that our developed SFS-based method has no advantage over the tool path generated by NX® in machining efficiency, because all the tool paths are converted to small line segments in the post-processing stage for the consideration of fair comparison, resulting in the machining time being closely related to the length of the tool path, and the advantage of path smoothness in machining efficiency and quality is not reflected. Therefore, the developed SFS-based tool path generation method can achieve higher machining efficiency and accuracy by the b-spline post-processing strategy, in theory. Of course, from the perspective of statistical data, the total machining time of the developed SFS-based method on all the five pockets is improved by 2.10% compared with NX®. Therefore, a conclusion can be drawn that the developed SFS-based method has a certain improvement in machining efficiency compared with NX®.

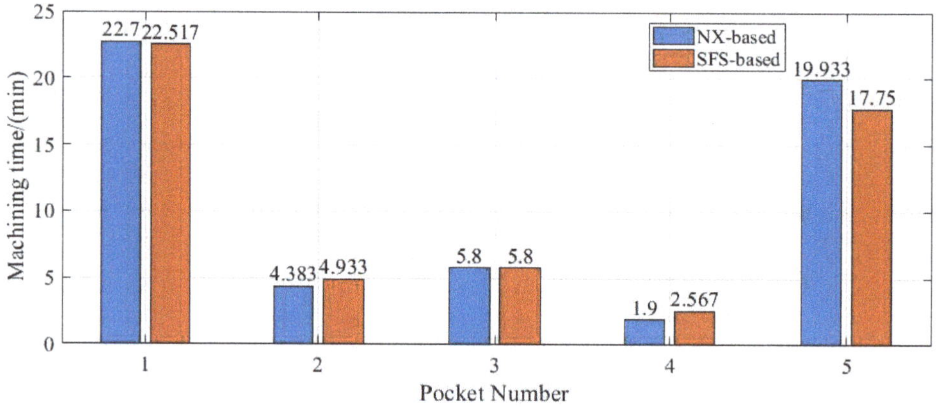

Figure 11. Comparison of machining time.

Not only are the machining times recorded in the cutting process, but the cutting force of closed pocket 3 and the open pocket 5 are also measured. As shown in Figure 12, the relationship between the cutting force and machining time were measured in the sampled corner, where one corner in the tool path of pocket 3 and two corners in the tool path of pocket 5 are sampled, as shown in Figure 12a. It can be seen in Figure 12b that the amplitude and change of cutting force are lower than that of the NX® due to the developed method produces a smooth transition at the corner, as we can see in Table 1, indicating that the tool path obtained by the developed SFS-based method has better dynamic characteristics compared to NX®.

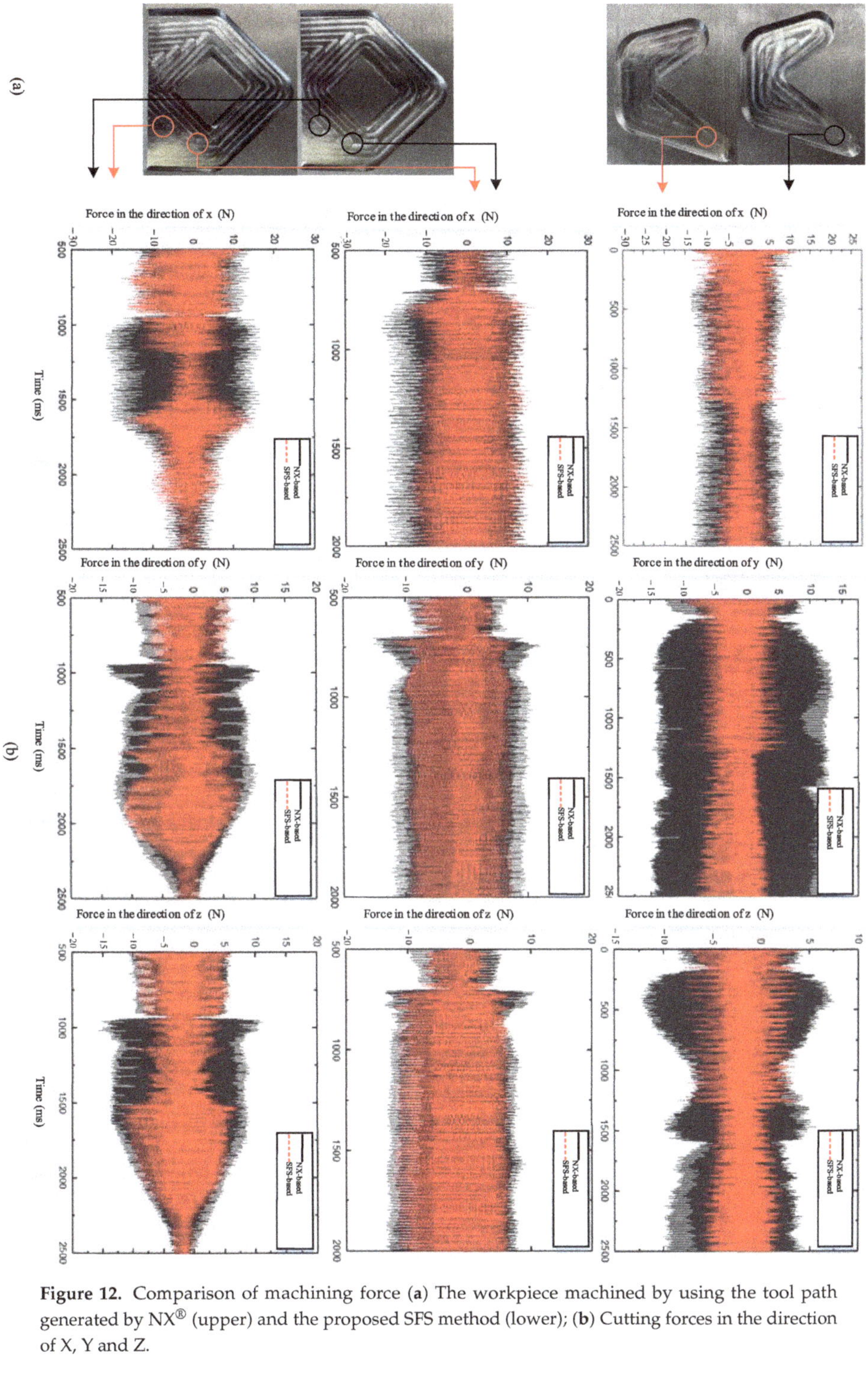

Figure 12. Comparison of machining force (**a**) The workpiece machined by using the tool path generated by NX® (upper) and the proposed SFS method (lower); (**b**) Cutting forces in the direction of X, Y and Z.

Better dynamic performance of tool paths generally means that the machining quality of workpiece is also better, but it cannot be directly concluded that the machining quality of the workpiece machined by using the developed SFS-based method is better. Therefore, the surface roughness of open pockets 1 and 5 are measured after machining to compare the average roughness (Ra), the mean roughness depth (Rz), and the maximum roughness depth (Rmax) of the machined surface. To reduce the measurement error, roughness values were measured at multiple points, and the distribution of measurement points is shown in Figure 13a; average values were measured at each point three times, and then the surface roughness values were averaged at each point, as shown in Figure 13b.

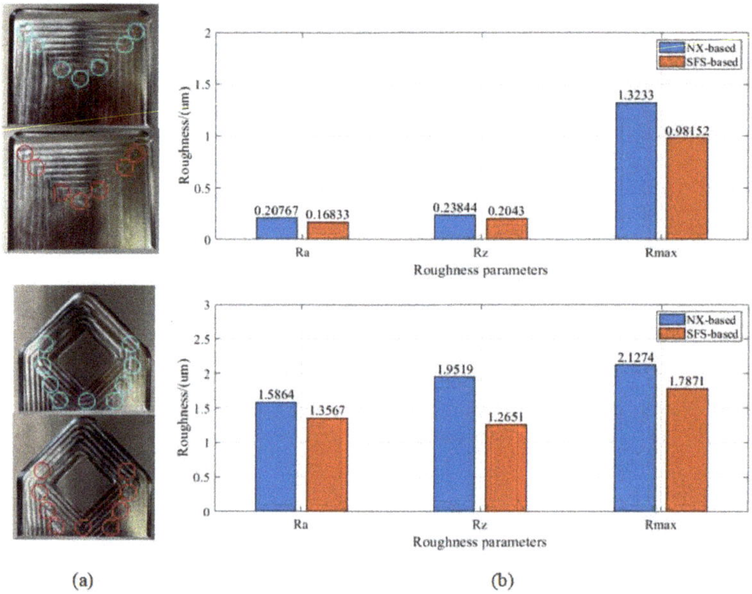

Figure 13. Comparison of roughness (**a**) The distribution of measurement points; (**b**) Measured surface roughness.

According to the measurement results, compared with the tool path generated by NX®, the roughness Ra, Rz, and Rmax obtained by using the developed SFS-based method were reduced by 18.94%, 14.32%, and 25.83%, respectively, on pocket 1. For pocket 5, the roughness Ra, Rz, and Rmax obtained by using the developed SFS-based method were reduced by 14.48%, 35.19%, and 16.00%, respectively, which indicated that the developed tool path generation method has obvious advantages in improving machining quality.

5. Conclusions

In this paper, a novel contoured parallel tool path generation method is developed. Compared to previous works, the SMA tree of the pocket is extracted in the proposed method and then the propagation direction of each SMA segment is calculated based on the geometric characteristics of the pocket boundary. Moreover, the final tool path is obtained through the synthesis of the sound field, which can not only match the geometry of the pocket boundary well, but also effectively achieves the goal of smoothness, no self-intersection, and no residual. Finally, the experimental results show that the tool path obtained by the developed SFS-based method has better advantages in smoothness, and the cutting force variation range during machining is much smaller, which reflected better dynamic performance. At the same time, the SFS-based method can effectively improve the machining quality and machining efficiency.

There is a research direction that deserves to be investigated in the future. Although this study is aimed at pocket milling, its idea can be extended to the milling of complex surfaces.

Author Contributions: Writing—original draft, X.Y. and X.C.; Writing—review and editing, X.Y. and W.Y.; Supervision and Validation, Y.Y. All authors have read and agreed to the published version of the manuscript.

Funding: This study was funded partially by the National Science Foundation of China (51775279) and National Key Research and Development Plan of China (2018YFB1309203).

Data Availability Statement: Data sharing is not applicable to this article.

Conflicts of Interest: The author declares that there is no conflict of interest.

References

1. Adamsk, W. Manufacturing development strategies in aviation industry. *Adv. Manuf. Sci. Technol.* **2010**, *34*, 73–84.
2. Kim, H.-C.; Lee, S.G.; Yang, M.-Y. An optimized contour parallel tool path for 2D milling with flat endmill. *Int. J. Adv. Manuf. Technol.* **2006**, *31*, 567–573. [CrossRef]
3. Abdullah, H.; Ramli, R.; Wahab, D.A. Tool path length optimisation of contour parallel milling based on modified ant colony optimisation. *Int. J. Adv. Manuf. Technol.* **2017**, *92*, 1263–1276. [CrossRef]
4. Yao, Z.; Gupta, S.K. Cutter path generation for 2.5 D milling by combining multiple different cutter path patterns. *Int. J. Prod. Res.* **2004**, *42*, 2141–2161. [CrossRef]
5. Kim, H.C. Optimum tool path generation for 2.5 D direction-parallel milling with incomplete mesh model. *J. Mech. Sci. Technol.* **2010**, *24*, 1019–1027. [CrossRef]
6. Kim, B.H.; Choi, B.K. Machining efficiency comparison direction-parallel tool path with contour-parallel tool path. *Comput.-Aided Des.* **2002**, *34*, 89–95. [CrossRef]
7. Lasemi, A.; Xue, D.; Gu, P. Recent development in CNC machining of freeform surfaces: A state-of-the-art review. *Comput.-Aided Des.* **2010**, *42*, 641–654. [CrossRef]
8. Choi, B.K.; Park, S.C. A pair-wise offset algorithm for 2D point-sequence curve. *Comput.-Aided Des.* **1999**, *31*, 735–745. [CrossRef]
9. Lin, Z.; Fu, J.; Shen, H.; Yao, X. Smooth contour-parallel tool path generation for high-speed machining through a dual offset procedure. *Int. J. Adv. Manuf. Technol.* **2015**, *81*, 1233–1245. [CrossRef]
10. Bahloul, E.; Brioua, M.; Rebiai, C. An efficient contour parallel tool path generation for arbitrary pocket shape without uncut regions. *Int. J. Precis. Eng. Manuf.* **2015**, *16*, 1157–1169. [CrossRef]
11. Jeong, J.; Kim, K. Generating tool paths for free-form pocket machining using z-buffer-based Voronoi diagrams. *Int. J. Adv. Manuf. Technol.* **1999**, *15*, 182–187. [CrossRef]
12. Lai, W.; Faddis, T.; Sorem, R. Incremental algorithms for finding the offset distance and minimum passage width in a pocket machining toolpath using the Voronoi technique. *J. Mater. Process. Technol.* **2000**, *100*, 30–35. [CrossRef]
13. Mansor, M.S.A.; Hinduja, S.; Owodunni, O.O. Voronoi diagram-based tool path compensations for removing uncut material in $2\frac{1}{2}$D pocket machining. *Comput.-Aided Des.* **2006**, *38*, 194–209. [CrossRef]
14. Hinduja, S.; Mansor, M.S.A.; Owodunni, O.O. Voronoi-diagram-based linking of contour-parallel tool paths for two-and-a-half-dimensional closed-pocket machining. *Proc. Inst. Mech. Eng. Part B J. Eng. Manuf.* **2010**, *224*, 1329–1350. [CrossRef]
15. Kim, H.C. Tool Path Modification for Optimized Pocket Milling. *Int. J. Prod. Res.* **2007**, *45*, 5715–5729. [CrossRef]
16. Molina-Carmona, R.; Jimeno, A.; Davia, M. Contour pocketing computation using mathematical morphology. *Int. J. Adv. Manuf. Technol.* **2008**, *36*, 334–342. [CrossRef]
17. Dhanik, S.; Xirouchakis, P. Contour parallel milling tool path generation for arbitrary pocket shape using a fast marching method. *Int. J. Adv. Manuf. Technol.* **2010**, *50*, 1101–1111. [CrossRef]
18. Xu, K.; Li, Y.; Xiang, B. Image processing-based contour parallel tool path optimization for arbitrary pocket shape. *Int. J. Adv. Manuf. Technol.* **2019**, *102*, 1091–1105. [CrossRef]
19. Li, X.; Liang, J.; Ni, P.; Wang, Y.; Song, Y.; Tong, L. Novel path generation algorithm for high-speed pocket milling. *Int. J. Prod. Res.* **2014**, *52*, 397–404. [CrossRef]
20. Xu, J.; Sun, Y.; Zhang, L. A mapping-based approach to eliminating self-intersection of offset paths on mesh surfaces for CNC machining. *Comput.-Aided Des.* **2015**, *62*, 131–142. [CrossRef]
21. Zhou, M.; Zheng, G.; Chen, S. Identification and looping tool path generation for removing residual areas left by pocket roughing. *Int. J. Adv. Manuf. Technol.* **2016**, *87*, 765–778. [CrossRef]

22. Chen, Z.C.; Fu, Q. An optimal approach to multiple tool selection and their numerical control path generation for aggressive rough machining of pockets with free-form boundaries. *Comput.-Aided Des.* **2011**, *43*, 651–663. [CrossRef]
23. Huang, N.; Jin, Y.; Lu, Y.; Yi, B.; Li, X.; Wu, S. Spiral toolpath generation method for pocket machining. *Comput. Ind. Eng.* **2020**, *139*, 106142. [CrossRef]
24. Ahrens, J. *Analytic Methods of Sound Field Synthesis*; Springer Science & Business Media: Berlin/Heidelberg, Germany, 2012.
25. Williams, E.G. *Fourier Acoustics: Sound Radiation and Nearfield Acoustical Holography*; Academic Press: Cambridge, MA, USA, 1999.

lubricants

Article

Effect of Microstructure of TiN /TiCN Layer on the Structural, Mechanical and Tribological Properties of the Ti/TiN/TiCN Films

Yanhong Lyu *, Qiaoyu Zhang, Yang Liu, Xinrong Deng, Huilian Sun and Min Mo

School of Physical and Chemistry, Hunan First Normal University, Changsha 410205, China
* Correspondence: lvyanhong603@163.com

Abstract: A direct current magnetron sputtering (DCMS) system at room temperature was applied to deposit the Ti/TiN/TiCN films. In order to research the effect of the microstructure of the TiN/TiCN layer on the Ti/TiN/TiCN films, the deposition time ratio of the TiN/TiCN layer ranged from 28.57 to 200%, and the whole deposition time of the films and the deposition time of Ti layer were constant. In this work, the relationship between structure and mechanical properties of films were investigated. The research results showed that the composition and structure of the films only slightly changed, while the crystalline orientation of the TiCN layer was of significant variation with the deposition time of the TiN layer. It is shown that the adhesion strength and internal stresses were 18.6 ± 1.5 N and 140 MPa when the deposition time ratio was 50%. By adjusting the deposition time ratio, the films displayed significant improvement in tribological behaviors. The coefficients of friction (COF) for the films deposited under deposition time ratio of 50% were about only 0.139 when the value of COF was stabilized. This work can provide a good wear-resisting film prepared approach at room temperature.

Keywords: Ti/TiN/TiCN films; DCMS; deposition time ratio; mechanical and tribological properties

1. Introduction

For about three decades, titanium carbonitride (TiCN)-based films have been receiving great attention due to their combined performance, such as high hardness, high toughness, low friction coefficient, excellent wear, and abrasion resistance [1–4]. Recently, methods, such as laser technology [5–7], magnetron sputtering [8,9], plasma enhanced chemical vapor deposition (PECVD) [10,11], and cathodic arc technique [12], have been developed for producing TiCN-based hard films with an improved wear resistance. The composition, structure, morphology, and properties of TiCN-based films are investigated in several studies [1,13–16]. It has been shown that TiCN-based films are a solid solution of TiN and TiC and can incorporate the advantages and characteristics of both. So, the hardness and abrasion resistance of TiCN-based films are superior to TiN, and the toughness and chemical stability of TiCN-based films are superior to TiC.

In the past 15 to 20 years, several researchers have been able to generate a single-layer [13,17] or graded TiCN film [18] and multi-layered TiCN film [19,20]. In these multi-layered TiCN films, the composition of films most frequently investigated are TiN/TiCN-based films. Morant et al. [21] have studied CN/TiCN/TiN multi-layer films by dual ion beam enhanced hardness. Yang et al. [7] have discussed the phase constituents, microstructure, microhardness, and wear resistance of TiCN/TiN films by laser in situ synthesized. Su et al. [22] have researched the tribological behavior and wear mechanisms for TiN/TiCN multi-layer films deposited by multi-arc PVD. Bemporad et al. [16] have investigated the wear resistance of TiN/TiCN multi-layer films by alternate layers of TiN and TiCN. It is reported that the Ti-based compound films show good tribological properties, which can be used in the industrial area as promising materials [23–26]. However, few people show the effects of the deposition time ratio of the TiN layer/TiCN layer on the properties of the Ti/TiN/TiCN films.

In this study, the Ti/TiN/TiCN films were deposited on n-type Si(100) by DCMS technique from Ti targets under an atmosphere of a mixture of Ar, N_2, and CH_4 gases. The deposition time of the Ti layer was 5 min and the whole deposition time of the films was set at 50 min. We focused on investigating the effects of the microstructure of the TiN/TiCN layer on the mechanical and tribological properties of the Ti-TiN-TiCN films.

2. Materials and Methods

2.1. Specimens Preparation

The Ti/TiN/TiCN films were deposited in a closed field unbalanced reactive DCMS system. Argon (Ar) (99.99%) was used as work gas, and the gas mixture of nitrogen (N_2) (99.99%) and methane (CH_4) (99.99%) was used as reaction gas.

The details about the films deposited by the system can be found elsewhere [27]. The n-type Si (100) wafers with surface roughness of 20 nm were used as the substrate material. Before loading into the chamber, all of the substrates were ultrasonically cleaned in the ethyl alcohol and acetone for 10 min, respectively. The base pressure of the chamber was inferior to 6.5×10^{-4} Pa. The distance between substrate and target was 10 cm. The silicon substrates were sputter-cleaned using argon plasma before depositing. During deposition, the direct current and argon working flow rate were 4 A and 20 standard cubic centimeter per minute (sccm), respectively. The deposition time of the TiN layer ranged from 10 to 30 min. Due to the fact that the whole deposition time was 50 min, the deposition time ratio of TiN/TiCN was set at 28.57%, 50%, 80%, 125%, and 200%, hereinafter referred to as a-28, b-50, c-80, d-125, and e-200, respectively. The Ti/TiN films were also prepared for comparation, which is named f-TN for short. After deposition, the average roughness of these films was about 27 nm measured by a 2206-type surface roughness tester.

2.2. Films Characterization

The composition of the films was analyzed with a PHI-5702 x-ray photoelectron spectroscope (XPS) with monochromated Al Kα radiation at pass energy of 29.4 eV. X-ray diffraction (XRD) analysis was used to study the crystalline structure of the films on Philips X'perts diffractometer using Cu Kα radiation. Raman spectroscopy was obtained on a HR800 Raman microscope instrument, with 532 nm Ar ion laser and a resolution of 1 cm^{-1}.

2.3. Properties

For adhesion evaluation, a scratch tester (Kaihua MFT-4000) was used. During scratch testing, the load was fixedly set at 30 N with a loading rate of 50 N/min. The critical load for adhesion failure was identified when a sudden change of the first order derivative of friction and audio frequency versus load curve was observed. The thin film stress distribution tester (BIOET BGS-6341) was employed to survey the internal stresses of the films by the radius of curvature technique. The internal stress was calculated by the Stoney equation:

$$\sigma = -\frac{1}{6}\frac{E_s}{1-v_s}\frac{h_s^2}{h_c}\left(\frac{1}{R_1}-\frac{1}{R_2}\right)$$

where E_s is Young's modulus, v_s is Poisson's ratio of the substrate, and h_s and h_c denote the thicknesses of the substrate and films, respectively. Moreover, $1/R_1$ and $1/R_2$ are the substrate curvature radii before and after film deposition. The thickness of films was measured by field emission scanning electron microscopy (FESEM).

The tribological behavior of the samples was evaluated by using the reciprocating ball-on-disc UMT-2MT tribometer with the counterparts of the AISI52100 steel ball at room temperature and about a relative humidity of 25%. The diameter, surface roughness, and hardness of the steel ball were 3 mm, 0.02 μm, and 850 HV, respectively. All the tests were conducted at a revolving velocity of 600 rpm, a sliding stroke of 5 mm, and a load of 1 N. The wear volume was measured using Micro XAM-3D Surface Profile. Then, the wear rate was acquired from the wear volume divided by the total sliding distance and applied load. The wear scars of the films were observed by scanning electron microscope (SEM) in order to observe the wear mechanism.

3. Results

3.1. Composition and Microstructure

Table 1 provides the relative contents of titanium, carbon, nitrogen, and oxygen of the surface of the films deposited under different deposition time ratios ranging from 28.57 to 200% (a–e). One can find that the relative contents of titanium, nitrogen, carbon, and oxygen are slightly fluctuating in all films. It should be noted that oxygen is not deliberately introduced and its presence is attributed to the films that are exposed to the air resulting from the spontaneous oxidation process of Ti on the surface [28]. In addition, Sun C.Q et al. [29] have discussed that N reaction induced tensile stress on the surface and the C reaction induced compressive stress. It is estimated that film b should possess lowest tensile stress or highest compressive stress as shown in Table 1. Further study will be carried out through the later section.

Table 1. Thickness and surface elemental composition of the films measured by FESEM and XPS.

Deposition Time Ratio	Thickness (μm)	Ti (at. %)	C (at. %)	N (at. %)	O (at. %)
28.57	2.1	24.21	41.85	13.95	19.99
50.00	2.2	25.55	45.32	11.78	17.35
80.00	2.4	22.79	44.44	11.28	21.49
125.00	3.1	24.39	44.77	12.68	18.16
200.00	3.3	25.59	38.93	15.54	19.94

XRD is used to investigate the crystalline structure of the Ti/TiN/TiCN films. Figure 1 depicts the XRD of the Ti/TiN/TiCN structure for films a-28, b-50, c-80, d-125, and e-200 and the Ti/TiNstructure for film f-TN. The whole deposition time and deposition time of the Ti layer were 50 min and 5 min in the Ti/TiN film. Five peaks corresponding to the (111), (200), (220), (311), and (222) plane of the cubic TiCN phase are observed in the films. A slight variation of intensity of the TiCN (311) peak with increasing deposition time ratio ranging from 28.57 to 200% is observed. After adding 21.43% of deposition time ratio from a to b, a significant increase in the intensity of the TiCN (220) peak and a reduction in the intensity of the TiCN (111) are seen, indicating a change in the preferred orientation from (111) to (220). However, for the films deposited with 80% and 200% of deposition time ratio, the relative intensity of the (111) and (220) peaks clearly drop, and the films exhibit an evident (200) preferred orientation. In addition, film d-125 merely shows a slight (220) preferred orientation. As seen in Figure 1, the XRD patterns of film f-TN display a significant shift of the peak positions to higher angles in relation to dash lines, indicating that the lattice spacing of the film was reduced. It is known that the lattice spacing of TiN and TiC is 0.4241 nm and 0.4329 nm, respectively. Therefore, the lattice is shrunk due to the formation of TiN phase instead of TiC phase [10]. In the films c-80 and e-125, the texture of the TiCN layer presents the preferential orientation (200) corresponding with the preferential orientation of the TiN layer. Moreover, the reflexes of films c-80 and e-200 belong to a face center cubic (FCC) lattice. This result suggests the possibility of a cube-on-cube epitaxial growth [8]. However, having an obvious preferential orientation (220), the TiCN layer of film b-50 restrains crystal growth of the TiN layer in the films. This result could hint that the films possess high hardness due to crystal lattice mismatching [30,31]. In Figure 1, some peaks of the films from a to e experienced a tiny shift of lower angles compared with the dash lines, due to the residual stress in the films [8].

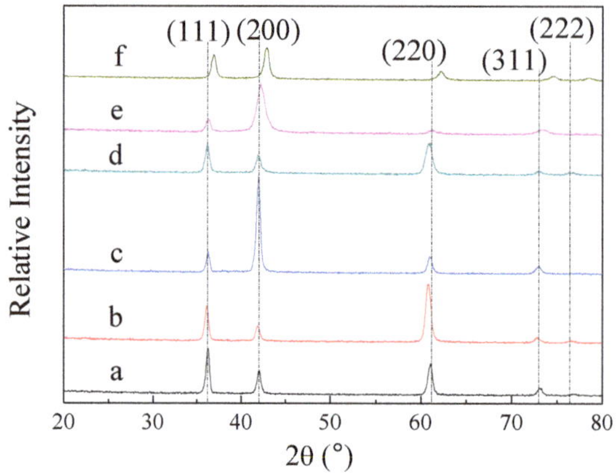

Figure 1. X-ray diffraction patterns of the films with deposition time ratio. (a) film a-28; (b) film b-50; (c) film c-80; (d) film d-125; (e) film e-200.

It is known that Raman spectroscopy can provide information about the bonding configuration. Figure 2 shows the Raman spectra corresponding to different deposition time ratios of the films. It is worth noting that the Raman spectra of film e-200 did not show significant D ("disorder", ~1360 cm^{-1}) and G ("graphitic", ~1560 cm^{-1}) bands, indicating the absence of an amorphous carbon phase. The reason could be that the deposition time of the TiCN layer of film e-200 was too short and unable to obtain more C element. The result is in agreement with the data of XPS. In Figure 2, all films possess acoustical and optical modes [32]. Due to a slight change of element contents in all films, the shift of the peaks scarcely ever varies. However, the intensity changes are observed. Obviously, films b-50, c-80, and d-125 exhibit higher defect concentration in the films than the samples a and e, due to higher total peak intensity produced [32]. It could be ascribed to the effect of the deposition time ratio on the crystalline growth of the TiCN layer mentioned in the previous paragraph.

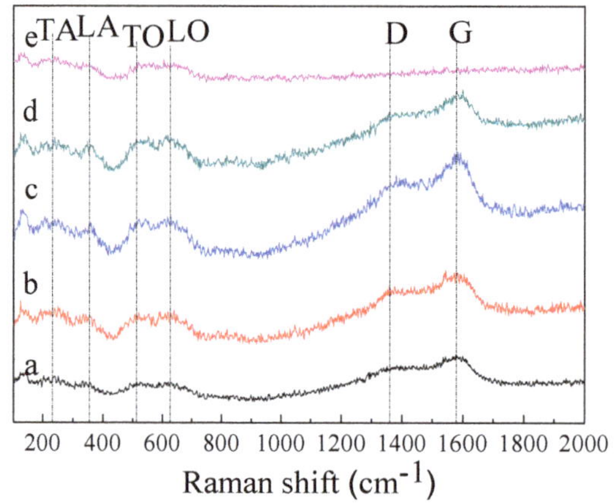

Figure 2. Raman spectra for the films with different deposition times. (a) film a-28; (b) film b-50; (c) film c-80; (d) film d-125; (e) film e-200.

3.2. Stress and Scratch Test

In the films, substrate curvature is typically used to provide a measure of the average residual stress. Although the stress may not be uniform, the average value is a useful metric for comparing trends in processing parameters [33]. Average residual stress as a function of deposition time ratio for the Ti/TiN/TiCN films is plotted in Figure 3. As we have seen, the average stress with different deposition time ratio has a significant change. Above all, the average stress of all films is the tensile stress. With increasing the deposition time ratio, the average stress of the films firstly declines, then increases, and finally declines. However, film b-50 deposited on deposition time ratio of 50% displays the lowest average stress on account of the maximum contents of carbon and the minimum contents of nitrogen in the film. On the other hand, film b-50 has an obvious preferential orientation (220) than other films. This results in a vast alteration in the stress aspect [34]. Due to the thinner TiCN layer, the loose surface of film e-200 has easy oxidation. As a result, the compression stress for film e-200 increases when the film is exposed to air [35,36]. Therefore, the average stress of film e-200 has a sudden decrease compared with that of film d-125. Further experimental explorations will be carried out in the later research.

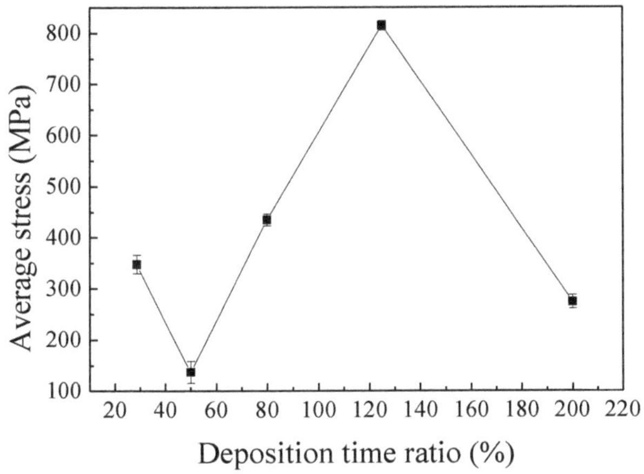

Figure 3. Average residual stress as a function of deposition time ratio for TiCN films.

Table 2 presents the results obtained from the scratch test performed on the films. The COF and the width of the scratch track corresponding to the critical load L_c are also included. However, it is worth to point out that the film thickness, friction coefficient at L_c, and residual stress are the main factors that have a large effect on L_c [37]. Since there is no appreciable difference for the COF of all the films in Table 2, the adhesion strength is primarily dependent on the residual stress. For the films, the critical load firstly increases and then decreases as the deposition time ratio increases. This result mainly agrees with the conclusion of average stress. However, the result of the scratch test for film e-200 is inconsistent with the previous conclusion. It can be attributed to the fact that the structure of the film is loose, and a mass of oxidation products leads to bad scratch behaviors.

Table 2. Scratch test results from films a-28 to e-200.

Film	Critical Load, L_c (N)	Friction Coefficient at L_c, μ_c	Scratch Track Width at L_c, w_c (mm)
a-28	17.9 ± 3.3	0.38	0.1795
b-50	18.6 ± 1.5	0.40	0.1632
c-80	17.0 ± 2.1	0.39	0.1702
d-125	15.8 ± 0.8	0.38	0.2123
e-200	12.0 ± 2.5	0.36	0.2145

Figure 4 reveals 2D micrograph mapping for the failures in each film. It is worth noting that the failures of all films are brittle failures, which propagated along or near the interface, causing spalling of the films inside and at the edges of the scratch track. For films c-80 and d-125, the main failure mode is recovery spallation [37], which gave rise to chipping at the track edge. Nevertheless, the adhesion failure mode of films a-28, b-50, and e-200 was buckling ahead of the stylus [37]. As seen in Figure 4a-e, the dark portion of the picture was the adhesion failures inside the track. These results suggest that films a-28, b-50, and e-200 present better homogeneous structure than other films.

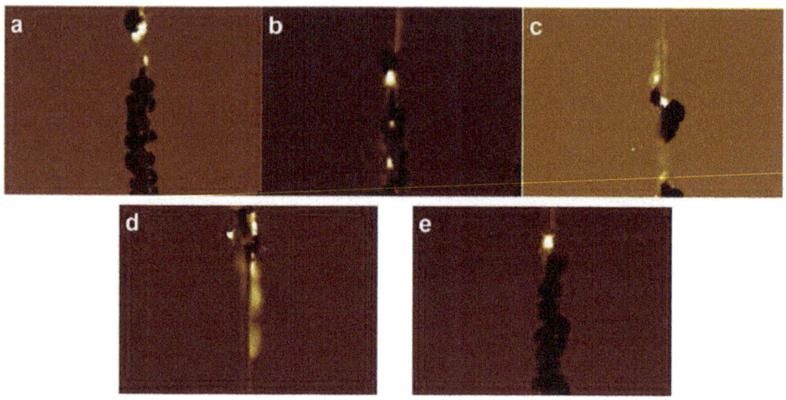

Figure 4. 2D micrograph of the film scratch tracks at critical load: (**a**) film a-28; (**b**) film b-50; (**c**) film c-80; (**d**) film d-125; (**e**) film e-200.

3.3. Tribological Properties

The COF of the film is continuously recorded during the wear tests and plots as a function of sliding time for films prepared under different deposition time ratios as shown in Figure 5. During the 30 min sliding test, three stages of the evolution processes of COF can be observed. Firstly, the COF for each film is extremely high in the initial stage due to forming the oxidation layer on the film surfaces [38]. However, the time of the initial stage of film b-50 was 50 sec, which was shorter than that of other samples. This is because film b-50 has a dense microstructure leading to it being difficult to be oxidated relative to other samples. Then, the lubricant stage corresponding to the value of COF is stabilized at a low value after the oxidation layer wears out. It should be noted that films a-28 and e-200 maintain extremely high COF in the whole sliding process unlike films b-50, c-80, and d-125 owning a lubricant stage. It can be attributed to the fact that the TiN layer for film a-28 is thinner than other films leading to it being unable to support the TiCN layer when the film is sliding against a steel ball. So, the lubrication for film a-28 becomes invalid fleetingly. However, film e-200 possessed high COF due to the thinner TiCN layer than other samples. Among films b-50, c-80, and d-125, the average COF of the lubricant stage is about 0.139. However, the COF curve of film b-50 hardly fluctuated compared with films c-80 and d-125. It is indicated that film b-50 has more stable tribological behavior. Finally, the invalid stage corresponding to the COF moves from a low value to a relatively higher value and the COF curve has significant fluctuation. As seen in Figure 5, the COF of films b-50 and d-125 begins to ascend slightly between 1600 and 1800 sec. During this period, the COF of film c-80 varies little. These results present that film b-50 deposited under deposition time ratio of 50% remains in the lubricant stage.

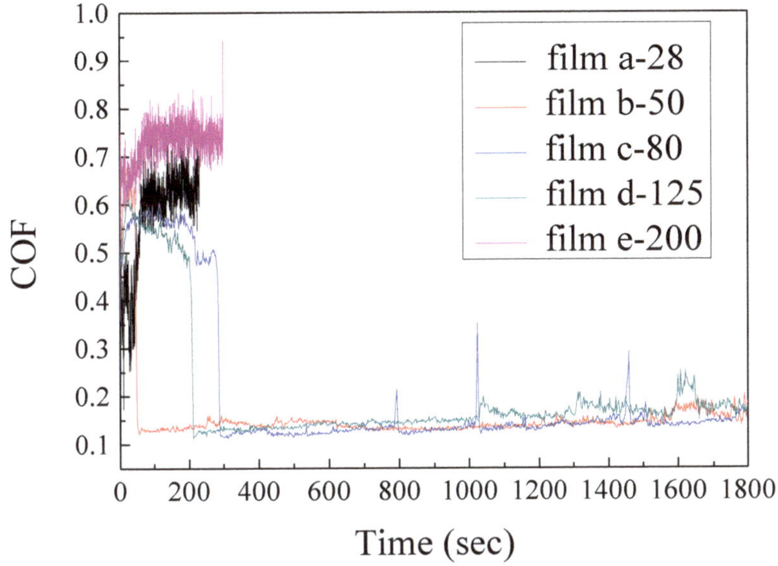

Figure 5. The coefficient of friction curves of the films against steel balls.

The wear volume of the films is measured with a 3D surface profiler. Table 3 summarizes the wear rate of all the films, and Figure 6 displays wear scar of the films sliding against steel balls. According to the figures in Table 3, the wear rate of film a-28 is highest, which is two hundred times higher than film b-50. However, it could be observed that film b-50 has good tribological performance with values for the wear rate as low as 1.163×10^{-6} $mm^3 \cdot m^{-1} \cdot N^{-1}$ sliding in the 30 min. As seen in Table 3, the deposition time ratio of film b-50 is 50% and represents 1.75 times that of film a-28. It is indicated that the film deposited under the low deposition time of TiN layer has lack of supporting for the TiCN layer. However, with increasing the deposition time ratio, the wear rate of the films appears as a slowly rising trend. Due to possessing relatively low adhesion strength and different crystalline orientation compared with film b-50, films c-80, d-125, and e-200 possess less resistant wear. It also could be explained that the deposition time of the TiCN layer is too short to form efficient lubricate. Generally, the deposition time ratio of the film is a determining factor in its resistant wear performance.

Table 3. Wear rate of films with deposition time ratio ranging from 28.57 to 200%.

Samples	Deposition Time Ratio (%)	Wear Rate against Steel Balls ($\times 10^{-6}$ $mm^3/N \cdot m$)
a-28	28.57	224.7
b-50	50.00	1.163
c-80	80.00	1.196
d-125	125.00	2.668
e-200	200.00	5.800

In Figure 6, it is worth noting that all films, with the exception of film a, have relatively smooth and neat wear scar. As illustrated in Figure 6a, the wear debris and particles are observed in the wear scar of the film and the edges of the wear scar have large flaking. It is suggested that adhesive wear shows in the contact surface and adheres to the film, which forms an overlapped transfer layer that induces severe wear [27]. Moreover, the delamination also appears in film a-28. As a result, film a-28 has extremely bad tribological behavior. In Figure 6, the width of the wear scar of film b-50 is shorter than the other samples. In addition, no concave-convex is surveyed on the edges of the wear trace of films b-50, c-80, and d-125. A similar wear mechanism could occur in films b-50, c-80, and d-125 due to similar COF, wear rate, and wear scar. Film e-200 possesses obvious wear debris and slight concave-convex.

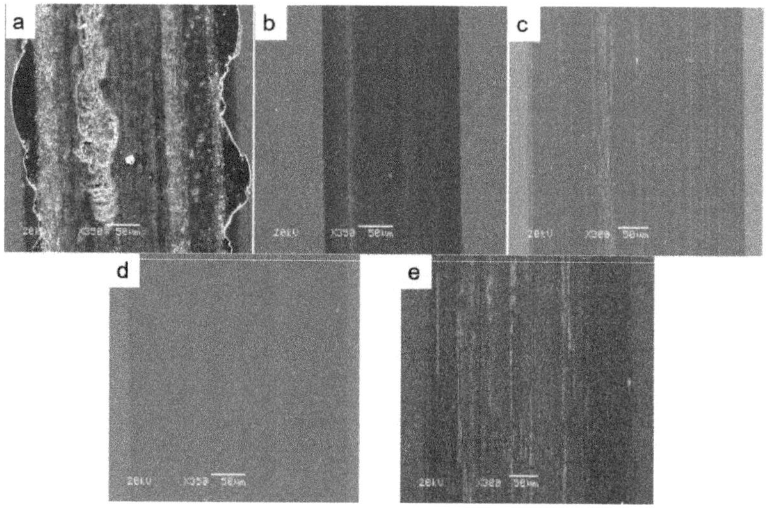

Figure 6. SEM micrograph of the films wear scar. (**a**) film a-28; (**b**) film b-50; (**c**) film c-80; (**d**) film d-125; (**e**) film e-200.

4. Conclusions

The Ti/TiN/TiCN triple-layer films were deposited under different deposition time ratios by direct current reactive sputtering titanium target. With varying the deposition time ratio, the composition, crystalline structure, and bonding structure of the films were researched, and the properties of the Ti/TiN/TiCN films were investigated in detail. The main results can be summarized as follows:

1. The composition and bonding structure of the films only slightly changed, while crystalline structure was of significant variation with the deposition time ratio. The film deposited under deposition time ratio of 50% had an obvious change in the preferred orientation from (111) to (220).
2. The adhesion strength and internal stresses were closely related to the deposition time ratio. The film deposited under deposition time ratio of 50% had an optimal value of adhesion strength. However, the stress curve firstly declined, then increased, and finally declined.
3. The COF of the films was divided into the three stages: initial stage, lubricant stage, and invalid stage. The COF curve included the stages when the films were deposited under deposition time ratios of 50% and 125%. Among the films, the COF was about 0.139 in the steady stage.
4. The films deposited under deposition time ratio of 50% was of lower wear rate than the other samples, which was 1.163×10^{-6} $mm^3 \cdot m^{-1} \cdot N^{-1}$. Moreover, the width of its wear scar was shorter.

Author Contributions: Conceptualization, methodology, software, validation, formal analysis, investigation, resources, data curation, writing—original draft preparation, and writing—review, Y.L. (Yanhong Lyu); editing, Q.Z., Y.L. (Yang Liu), X.D., H.S. and Y.L. (Yanhong Lyu); supervision, Y.L. (Yanhong Lyu); project administration, Y.L. (Yanhong Lyu) and M.M.; funding acquisition, Y.L. (Yanhong Lyu). All authors have read and agreed to the published version of the manuscript.

Funding: This research was funded by Provincial Natural Science Foundation of Hunan, China (Grant Nos. 2021JJ40140, 2021JJ30170), and the Scientific Research Fund of Hunan Provincial Education Department, Hunan, China (Grant Nos. 21B0812, 21A0580).

Conflicts of Interest: The authors declare no conflict of interest.

References

1. Mekgwe, G.N.; Akinribide, O.J.; Akinwamide, S.O.; Olubambi, P.A. Fabrication of graphite reinforced TiCxNy by spark plasma sintering technique: A comparative assessment of microstructural integrity and nanoindentation properties. *Vacuum* **2021**, *187*, 110144. [CrossRef]
2. Kumar, C.S.; Patel, S.K. Effect of duplex nanostructured TiAlSiN/TiSiN/TiAlN-TiAlN and TiAlN-TiAlSiN/TiSiN/TiAlN coatings on the hard turning performance of Al_2O_3-TiCN ceramic cutting tools. *Wear* **2019**, *418–419*, 226–240. [CrossRef]
3. Polcar, T.; Novak, R.; Siroky, P. The tribological characteristics of TiCN coating at elevated temperatures. *Wear* **2006**, *260*, 40–49. [CrossRef]
4. Cheng, Y.H.; Browne, T.; Heckerman, B. Influence of CH_4 fraction on the composition, structure, and internal stress of the TiCN coatings deposited by LAFAD technique. *Vacuum* **2010**, *85*, 89–94. [CrossRef]
5. Yang, Y.L.; Yan, W.; Zhang, D.; Song, G.Z.; Zheng, Y.R. In Situ-Fabrication of TiCN Ceramic Coating on Titanium Alloy by Laser Cladding Technology. *Key Eng. Mater.* **2010**, *434–435*, 485–488. [CrossRef]
6. Yang, Y.; Zhang, D.; Yan, W.; Zheng, Y. Microstructure and wear properties of TiCN/Ti coatings on titanium alloy by laser cladding. *Opt. Lasers Eng.* **2010**, *48*, 119–124. [CrossRef]
7. Yang, Y.; Yao, W.; Zhang, H. Phase constituents and mechanical properties of laser in-situ synthesized TiCN/TiN composite coating on Ti-6Al-4V. *Surf. Coat. Technol.* **2010**, *205*, 620–624. [CrossRef]
8. Caicedo, J.C.; Amaya, C.; Yate, L.; Gómez, M.E.; Zambrano, G.; Alvarado-Rivera, J.; Muñoz-Saldaña, J.; Prieto, P. TiCN/TiNbCN multilayer coatings with enhanced mechanical properties. *Appl. Surf. Sci.* **2010**, *256*, 5898–5904. [CrossRef]
9. Agudelo, L.C.; Ospina, R.; Castillo, H.A.; Devia, A. Synthesis of Ti/TiN/TiCN coatings grown in graded form by sputtering dc. *Phys. Scripta.* **2008**, *T131*, 014006. [CrossRef]
10. Guruvenket, S.; Li, D.; Klemberg-Sapieha, J.E.; Martinu, L.; Szpunar, J. Mechanical and tribological properties of duplex treated TiN, nc-TiN/a-SiN_x and nc-TiCN/a-SiCN coatings deposited on 410 low alloy stainless steel. *Surf. Coat. Technol.* **2009**, *203*, 2905–2911. [CrossRef]
11. Klemberg-Sapieha, J.E.; Jedrzejowski, P.; Martinu, L. Mechanical and optical characteristics of superhard nanocomposite TiN/a-Si_3N_4 and TiCN/a-SiCN coatings produced by PECVD. *J. Superhard Mater.* **2007**, *29*, 147–152. [CrossRef]
12. Balaceanu, M.; Braic, M.; Braic, V.; Pavelescu, G. Properties of arc plasma deposited TiCN/ZrCN superlattice coatings. *Surf. Coat. Technol.* **2005**, *200*, 1084–1087. [CrossRef]
13. Sun, H.; Billard, A.; Luo, H.; Zheng, W.; Zheng, X.; Dai, M.; Lin, S.; Shi, Q.; Sanchette, F. Influence of carbon content on the mechanical properties of TiCN–Cu nanocomposite coatings prepared by multi-arc ion plating. *Vacuum* **2021**, *187*, 110139. [CrossRef]
14. Wang, Q.; Jin, X.; Zhou, F. Comparison of mechanical and tribological properties of CrBN coatings modified by Ni or Cu incorporation. *Friction* **2022**, *10*, 516–529. [CrossRef]
15. Rebelodefigueiredo, M.; Neidhardt, J.; Kaindl, R.; Reiter, A.; Tessadri, R.; Mitterer, C. Formation mechanisms of low-friction tribo-layers on arc-evaporated $TiC_{1-x}N_x$ hard coatings. *Wear* **2008**, *265*, 525–532. [CrossRef]
16. Xian, G.; Xiong, J.; Zhao, H.; Xian, L.; Fan, H.; Li, Z.; Du, H. Study on the growth and wear behavior of the TiAlN-based composite coating deposited on TiCN-based cermets with different binder phase. *Wear* **2020**, *460–461*, 203460. [CrossRef]
17. Zhu, B.; Mardel, J.; Kelly, G.L. An investigation of tribological properties of CN and TiCN coatings. *J. Mater. Eng. Perform.* **2004**, *13*, 481–487. [CrossRef]
18. Wei, C.H.; Lin, J.F.; Jiang, T.H.; Ai, C.F. Tribological characteristics of titanium nitride and titanium carbonitride multilayer films: Part I. The effect of coating sequence on material and mechanical properties. *Thin Solid Films* **2001**, *381*, 94–103. [CrossRef]
19. Zheng, J.Y.; Hao, J.Y.; Liu, X.Q.; Gong, Q.Y.; Liu, W.M. TiN/TiCN Multilayer Films Modified by Argon Plasma Treatment. *Appl. Surf. Sci.* **2013**, *280*, 764–771. [CrossRef]
20. Meng, J.; Lu, J.; Wang, J.; Yang, S. Tribological behavior of TiCN-based cermets at elevated temperatures. *Materi. Sci. Eng. A* **2006**, *418*, 68–76. [CrossRef]
21. Morant, C.; Prieto, P.; Forn, A.; Picas, J.A.; Elizalde, E.; Sanz, J.M. Hardness enchancement by CN/TiCN/TiN multilayer films. *Surf. Coat. Technol.* **2004**, *180–181*, 512–518. [CrossRef]
22. Zheng, J.Y.; Ren, X.D.; Hao, H.Y.; Li, A.; Liu, W.M. Carbon nanohoops as attractive toughening and lubricant agents in TiN porous films. *Appl. Surf. Sci.* **2017**, *393*, 60–66. [CrossRef]
23. Alisir, S.H.; Evrensel, D. Investigation into Coating Structure and Wear Environment Effects on Tribological Properties of Pistn Ring Coated with Monolayer TiAlN and Multilayer TiN/TiAlN. *J. Mater. Eng. Perform.* **2022**, *31*, 1654–1666. [CrossRef]
24. Wróblewski, P.; Rogólski, R. Experimental Analysis of the Influence of the Application of TiN, TiAlN, CrN and DLC1 Coatings on the Friction Losses in an Aviation Internal Combustion Engine Intended for the Propulsion of Ultralight Aircraft. *Materials* **2021**, *14*, 6839. [CrossRef]

25. Ruan, H.T.; Wang, Z.Y.; Wang, L.; Sun, L.L.; Peng, H.; Ke, P.L.; Wang, A.Y. Designed Ti/TiN sub-layers suppressing the crack and erosion of TiAlN coatings. *Surf. Coat. Technol.* **2022**, *438*, 128419. [CrossRef]
26. Wróblewski, P.; Koszalka, G. An Experimental Study on Frictional Losses of Coated Piston Rings with Symmetric and Asymmetric Geometry. *SAE Int. J. Engines* **2021**, *14*, 85386. [CrossRef]
27. Su, Y.L.; Kao, W.H. Tribological behavior and wear mechanisms of Ti-C: H/TiC/TiCN/TiN/Ti coatings when sliding against steel, bronze and aluminum alloy rods. *J. Mater. Sci.* **2001**, *36*, 189–199. [CrossRef]
28. Talib, R.J.; Toff, M.R.M.; Ariff, H.M. Wear mechanism of TiN, TiAlN and TiCN coated drills during drilling of carbon steel. *J. Phys. Sci.* **2007**, *18*, 75–85.
29. Sun, C.Q.; Fu, Y.Q.; Yan, B.B.; Hsieh, J.H.; Lau, S.P.; Sun, X.W.; Tay, B.K. Improving diamond–metal adhesion with graded TiCN interlayers. *J. Appl. Phys.* **2002**, *91*, 2051–2054. [CrossRef]
30. Lackner, J.M.; Waldhauser, W.; Ebner, R.; Bakker, R.J. Chemistry and microstructure of PLD (Ti,Al)C_xN_{1-x} coatings deposited at room temperature. *Appl. Phys. A* **2004**, *79*, 1469–1471. [CrossRef]
31. Peters, A.M.; Nastasi, M. Effect of carrier gas on the deposition of titanium carbo-nitride coatings by a novel organo-metallic plasma immersion ion processing technique. *Vacuum* **2002**, *67*, 169–175. [CrossRef]
32. Dreiling, I.; Stiens, D.; Chassé, T. Raman spectroscopy investigations of $TiB_xC_yN_z$ coatings deposited by low pressure chemical vapor deposition. *Surf. Coat. Technol.* **2010**, *205*, 1339–1344. [CrossRef]
33. Detor, A.J.; Hodge, A.M.; Chason, E.; Wang, Y.M.; Xu, H.W.; Conyers, M.; Nikroo, A.; Hamza, A. Stress and microstructure evolution in thick sputtered films. *Acta Mater.* **2009**, *57*, 2055–2065. [CrossRef]
34. Abadias, G. Stress and preferred orientation in nitride-based PVD coatings. *Surf. Coat. Technol.* **2008**, *202*, 2223–2235. [CrossRef]
35. Shi, Y.L.; Peng, H.R.; Xie, Y.; Xie, G.W.; Zhao, C.; Li, S.Z. Plasma CVD of hard coatings Ti(CNO) using metallo-organic compound $Ti(OC_3H_7)_4$. *Surf. Coat. Technol.* **2000**, *132*, 26–30.
36. Stanishevsky, A.; Lappalainen, R. Tribological properties of composite Ti(N,O,C) coatings containing hard amorphous carbon layers. *Surf. Coat Technol.* **2000**, *123*, 101–105. [CrossRef]
37. Bull, S.J.; Bhat, D.G.; Staia, M.H. Properties and performance of commercial TiCN coatings. Part 2: Tribological performance. *Surf. Coat. Technol.* **2003**, *163–164*, 507–514. [CrossRef]
38. Mo, J.L.; Zhu, M.H. Sliding tribological behaviors of PVD CrN and AlCrN coatings against Si_3N_4 ceramic and pure titanium. *Wear* **2009**, *267*, 874–881. [CrossRef]

lubricants

MDPI

Article

Fretting Wear Characteristics of SLM-Formed 316L Stainless Steel in Seawater

Mingji Huang [1,2,*], Ping Chen [1] and Xiaoxi Qiao [1]

[1] School of Mechanical Engineering, University of Science and Technology Beijing, Beijing 100083, China
[2] Shunde Innovation School, University of Science and Technology Beijing, Foshan 528300, China
* Correspondence: huangmingji@ustb.edu.cn

Abstract: The fretting wear characteristics of two different energy density 316L stainless steels formed by selective laser melting (SLM) under different friction conditions are studied. The image method was used to study the porosity of two samples with different energy densities (46.88 J/mm^3, 98.96 J/mm^3) formed by SLM. The dynamic wear test, respectively, evaluates its wear morphology and wear depth under three conditions: dry friction, distilled water, and an 3.5% NaCl solution. The porosity of the samples with SLM forming an energy density of 46.88 J/mm^3 and 98.96 J/mm^3 are 7.66% and 1.00%, respectively. Under the three conditions, the friction coefficient and wear depth of the samples with high energy density are smaller than those of the samples with low energy density; the friction of the samples with two energy densities in aqueous solution is faster than dry friction in air and tends to be stable. The friction coefficient in 3.5% NaCl solution is the smallest; when the energy density is constant, the wear depth of the fretting wear is the largest under dry friction and the smallest in distilled water. Under dry-friction conditions, the wear mechanisms of fretting wear are mainly oxidative wear and adhesive wear. In the fretting wear in the distilled water and the 3.5% NaCl solution, both wear mechanisms are abrasive wear and fatigue wear.

Keywords: selective laser melting; 316L stainless steel; energy density; porosity; fretting wear

1. Introduction

Selective laser melting (selective laser melting, SLM) is a technology that uses laser heat to quickly melt and solidify metal powder and generate a physical model layer by layer. It can produce very complex and personalized metal parts and has been widely used in multiple engineering fields [1–3]. Porosity is a common defect in the forming process of SLM. It usually reduces the density of the material and affects other related properties. Porosity is used to quantify the pore defects of the sample, such as pores, cracks, holes, etc. When producing parts through SLM technology, the first goal is to obtain close to full-density components. High porosity is not conducive to mechanical properties. Olakanmi et al. [4] stated that SLM is a very complex process and believed that process parameters would affect the densification mechanism and microstructure characteristics of SLM processed materials. Cherry et al. [5] studied the influence of laser energy density on SLM forming and summarized the optimal energy density.

The literature [6] shows that changing the scanning distance and energy density will directly affect the surface roughness, porosity, and friction characteristics of the formed sample. These documents show that changing the energy density can directly affect the performance of the formed sample.

The material properties are influenced by the forming process [7,8]. 316L stainless steel has good mechanical properties, but it is easy to wear [9,10]. Sun et al. [11] showed that the pore defects produced in the process of forming 316L stainless steel by SLM will lead to accelerated wear in the process of non-lubricating friction. The literature [12] compared the effects of three different processing techniques (selective laser melting, hot press sintering, and conventional casting) on the microstructure and mechanical properties of 316L. The results show that SLM can improve its mechanical and wear properties in forming 316L. Huang et al. [13] studied the wear performance of SLM-formed 316L specimens and concluded that SLM-formed 316L has better wear performance than rolled 316L.

316L stainless steel is extensively used in marine engineering due to its superior corrosion resistance [14,15]. Seawater is a natural electrolyte and can damage the passive film of the metals, leading to corrosion, which will affect the tribological properties [16,17]. Meanwhile, small displacements can occur between close-fitting parts due to wind and wave effects, causing fretting wear. Several scholars have investigated the fretting wear characteristics of metallic materials in seawater [18,19], but less research exists on SLM-formed 316L.

In summary, the 3.5% NaCl solution is used to simulate seawater, and the fretting wear under dry friction and distilled water conditions is compared with the fretting wear under 3.5% NaCl conditions to study the performance of SLM-formed 316L stainless steel samples with different energy densities in seawater.

2. Materials and Methods

SLM molding uses 316L stainless steel powder. The particle size of the powder is shown in Figure 1, and its element composition is shown in Table 1. Selective laser melting adopts a metal 3D printer model EP-M100T (made in Eplus3D Corporation, Beijing, China) with a maximum power of 200 W. The machine is equipped with an IPGYt fiber laser (made in Eplus3D Corporation, Beijing, China) with a wavelength of 1030 nm and a spot diameter of 50 μm. All samples are in an atmosphere with an oxygen content of ≤1000 mg/L.

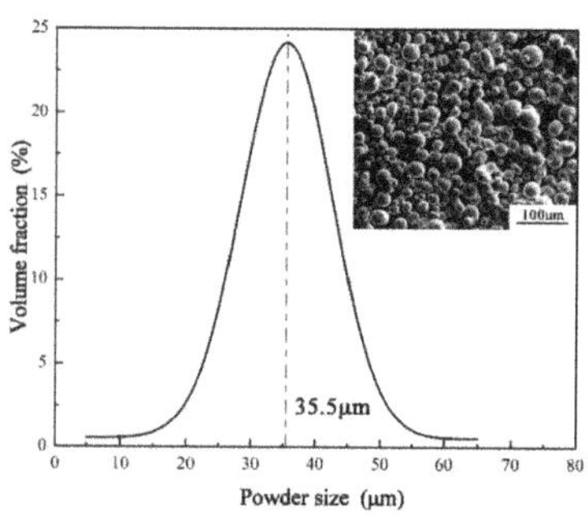

Figure 1. 316L powder micro-morphology and particle size distribution.

Table 1. Chemical composition of gas-atomized 316L powder (wt%).

C	Si	Mn	S	P	Cr	Ni	Mo	Fe
≤0.03	≤1.0	≤2.0	≤0.03	≤0.035	16.0–18.0	10.0–14.0	2.0–3.0	Bal.

The main process parameters in the SLM-forming process are laser power L (W), scanning speed v (mm/s), layer thickness t (mm), and scanning distance h (mm). According to formula (1) [20], these variables can be combined into energy density E (J/mm^3). The literature [6] studied the effect of energy density on the porosity of 316L. The study showed that in the range of energy density of 50 J/mm^3~200 J/mm^3, the porosity first decreased and then increased with the increase in energy density, and when the energy density was 125 J/mm^3, the porosity reached the lowest value (the lowest value was 0.63%). In this experiment, two samples with energy densities of 46.88 J/mm^3 and 98.96 J/mm^3 were selected to study the friction performance, and the SLM-forming parameters were selected:

scanning speed v = 800 mm/s; layer thickness t = 0.03 mm; hatch space h = 0.08 mm; and power L = 90 W, 190 W.

$$E = \frac{L}{v \cdot t \cdot h} \quad (1)$$

2.1. Porosity and Hardness Test

Polish the upper surface of the SLM-formed sample until there are no scratches visible to the naked eye. Take five images at different positions in the center of the sample, as shown in Figure 2. According to this scheme, the measurement range covers about one-third of the upper surface of the sample. Use Image J software (version 1.7.0) to measure the porosity of 5 images and finally obtain the corresponding porosity by calculating the average value of the porosity. Make 5 indentations in the porosity observation area and find the average value of hardness. The test-loading pressure is 1.96 N (200 gf), and the holding time is 15 s. When testing indentation, try to avoid large pore defects, so as not to affect the measurement results.

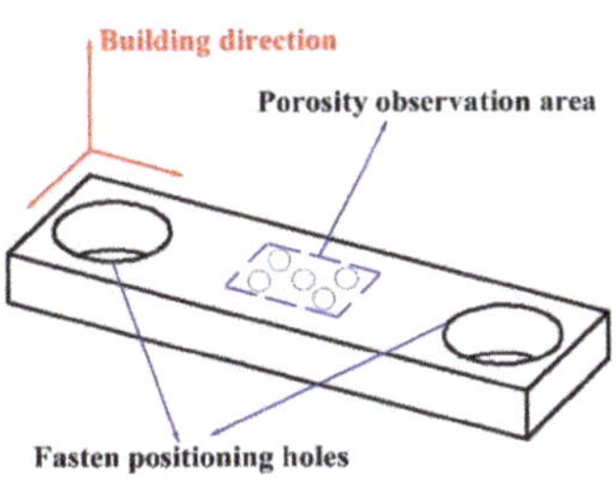

Figure 2. Schematic of specimen forming and measurement.

2.2. Fretting Wear Test

The fretting wear experiment is carried out with self-made experimental equipment, and the principle is shown in Figure 3. The slider 3 is driven in a reciprocating motion by a screw connected to the stepper motor 1. The number of pulses of the stepper motor enables the precise adjustment of the friction stroke, while the velocity of the movement is regulated by controlling the number of pulses per unit of time. The lower specimen is mounted on fixture 5 and moves in a linear reciprocating motion with workbench 4, while the upper specimen is fixed to a vertical rod on fixture 6 and remains stationary. Weight 8 applies the test load to the specimen. The horizontal elastic beam 9 is deformed by the friction force, which causes the strain gauges 10 to elongate or shorten, and friction force can be measured by means of strain-gauge force transducers.

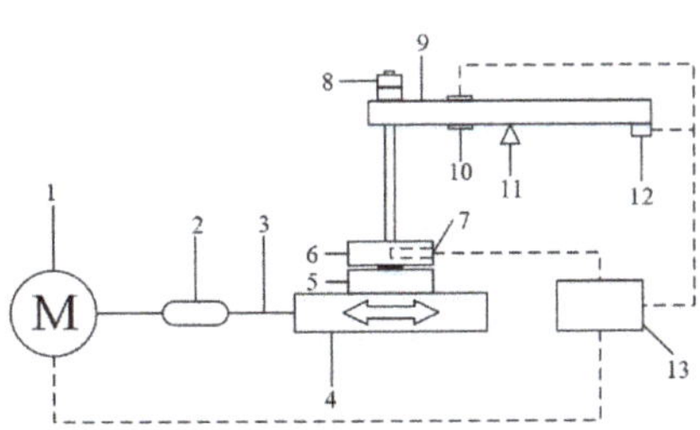

Figure 3. Structure diagram of testing machine. 1, Stepper motor; 2, Coupling; 3, Slider; 4, Workbench; 5, Lower specimen fixture; 6, Upper specimen fixture; 7, Temperature sensor; 8, Weight; 9, Horizontal elastic beam; 10, Resistance strain gauge; 11, Pivot; 12, Motion detector; 13, Electrical control system.

The testing machine is shown in Figure 4. In this research, the upper sample is quenched C45E4, the friction head is spherical, the spherical radius is 4.75 mm, the hardness is about 45HRC, and it is fixed. The lower sample is a sample formed by SLM, which is controlled by a motor to make a reciprocating linear movement and contact the fixed upper sample. The test load is 10 N, the speed is 4 mm/s, the single stroke is 0.6 mm, and the frequency is 2.5 Hz. The friction was carried out under three conditions (dry friction, distilled water, and 3.5% NaCl solution), each set of experiments was repeated at least 3 times, and the duration of each experiment was 60 min. When the water-based solution is used for lubrication and friction, it is necessary to ensure that the contact area is completely enclosed by the liquid. The experiment was carried out at room temperature, and the wear depth and friction coefficient change curve were recorded by computer.

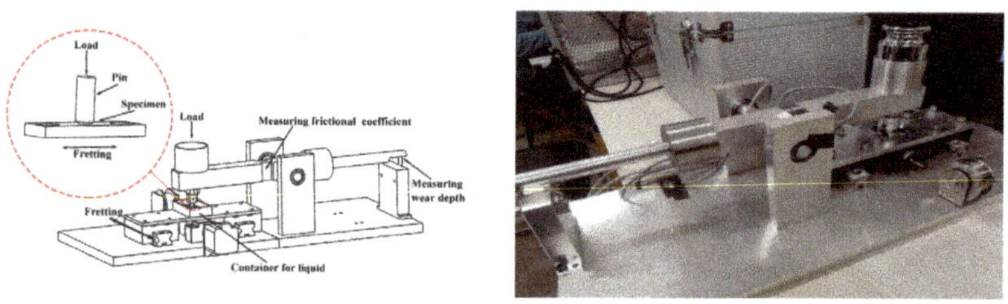

(a) (b)

Figure 4. Fretting wear testing machine: (**a**) schematic; (**b**) testing machine.

3. Results

3.1. Porosity Analysis

The pores in SLM are mostly caused by the process. When the energy density is too low, the 316L powder is not completely melted due to insufficient power, resulting in poor bonding between adjacent layers, the extremely irregular shape of the defects, and many irregular pores, such as in Figure 5a. When the energy density is too high, jetting may occur during powder melting. This is because the gas generated by the high energy density is trapped at the bottom of these molten pools to form pores. Therefore, too high energy density will cause the high melting temperature of the molten pool, resulting in high residual stress caused by rapid cooling of the molten pool, and the shape of hole defects is regular and close to circle Shape, as shown in Figure 5b. Figure 6 is the OM image of the sample porosity observation optical microscope. The experimentally measured porosities of the samples with energy densities of 46.88 J/mm^3 and 98.96 J/mm^3 are 7.66% and 1.00%, respectively. The sample with an energy density of 46.88 J/mm^3 has fewer defects than the sample with an energy density of 46.88 J/mm^3. The hardness of two samples measured by the microhardness tester is used to calculate the average value. The hardnesses of samples 46.88 J/mm^3 and 98.96 J/mm^3 are 206.6 HV and 250.6 HV, respectively. Combined with the analysis of porosity, the porosity of the sample with lower energy density is larger, and the corresponding hardness will be smaller.

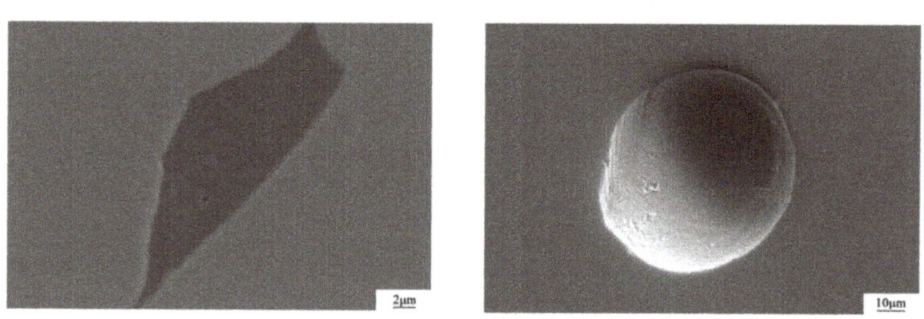

(a) (b)

Figure 5. Types of defects in SLM-formed 316L: (**a**) irregular defect; (**b**) regular defect.

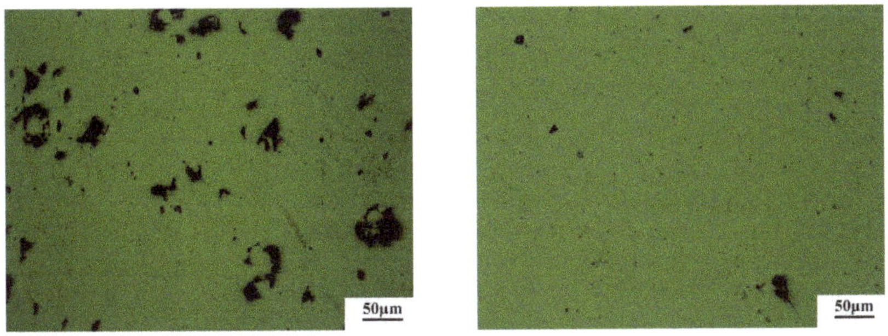

(**a**) (**b**)

Figure 6. OM diagram of porosity observation: (**a**) 46.88 J/mm^3; (**b**) 98.96 J/mm^3.

3.2. Friction Coefficient and Wear Depth

Figure 7a,b are the friction coefficients of samples with energy densities of 46.88 J/mm^3 and 98.96 J/mm^3 in air, distilled water, and 3.5% NaCl solution. It can be seen in Figure 6 that the friction coefficient of SLM-formed 316L changes with time and can be divided into two main stages. In the first stage, the friction coefficient increases rapidly. In the second stage, the friction coefficient tends to be stable and fluctuates more uniformly around a certain value. When the energy density is 46.88 J/mm^3 and 98.96 J/mm^3, the friction coefficient during dry friction is stable at about 1.007 and 0.898, respectively. When in the distilled water, the friction coefficient is stable at about 0.567 and 0.534, respectively, and when in the 3.5% NaCl solution, the friction coefficient is stable at about 0.506 and 0.476, respectively. The main reasons for these two stages are as follows: in the first stage, since the surfaces of 316L stainless steel and C45E4 steel start to come into to contact, the movement between the upper and lower pairs increases the frictional resistance and causes adhesion on the contact surface of the two, and the friction coefficient is further increased. In the second stage, the small vibration amplitude of the fretting friction makes it difficult for the wear debris between the two contact surfaces to be discharged, thereby forming a third body. The third body has protective and lubricating effects [21], which can inhibit adhesion. Therefore, when the generation and overflow of debris reach a balance, the friction coefficient changes relatively stable. Furthermore, sudden changes in the friction coefficient can be observed in both stages under dry friction. This is because the material is more prone to peeling and deformation under dry friction, and when large debris peeling occurs, the friction coefficient changes suddenly. As the friction continues, the debris is overflowed or flattened and the friction coefficient re-stabilizes.

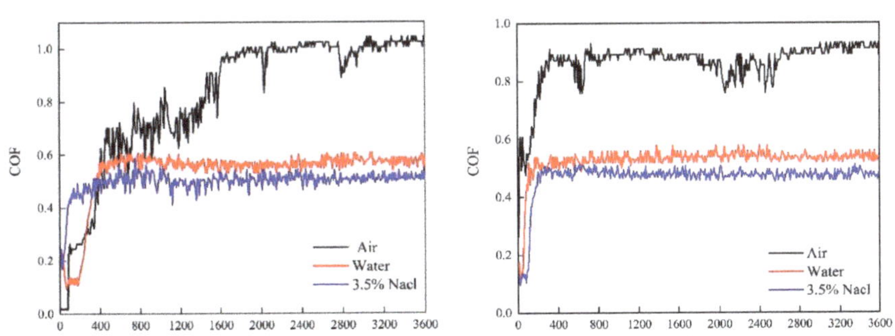

(**a**) (**b**)

Figure 7. The effect of the lubricant on the friction coefficient: (**a**) 46.88 J/mm^3; (**b**) 98.96 J/mm^3.

Under the two energy densities, the friction in the aqueous solution tends to be more stable and the friction coefficient is smaller than in the dry friction. This is because the liquid plays a role in lubricating and reducing friction [22]. The friction coefficient in the 3.5% NaCl solution is smaller than that in the distilled water. Active elements such as Cl are present in the 3.5% NaCl solution in the form of sodium chloride. Owing to the friction heat generated during the friction process, these elements will react with the element Fe, and an easy-shear friction film containing ferric chloride forms. This prevents direct contact between the friction pairs, and the two-body wear becomes three-body wear, which further reduces the friction coefficient [23–25]. Under the three friction conditions, the sample with an energy density of 98.96 J/mm^3 is denser, so it is less prone to wear than the sample with an energy density of 46.88 J/mm^3. The closeness of the friction coefficients of the two samples in the aqueous solution is due to the lubrication of the aqueous solution.

Figure 8a is a schematic diagram of the comparison of wear depth. The slope of the wear depth after 30 min is defined as the wear rate (Figure 8b). It can be seen from the figure that the wear depth of the samples is less than 10 μm, the maximum wear depth is generated when the energy density is 46.88 J/mm^3 sample dry friction, and the wear depth is about 9.225 μm. The minimum depth is produced in the sample distilled water medium with an energy density of 98.96 J/mm^3, and the wear depth is about 2.292 μm. The trend of the wear rate is similar to the wear depth.

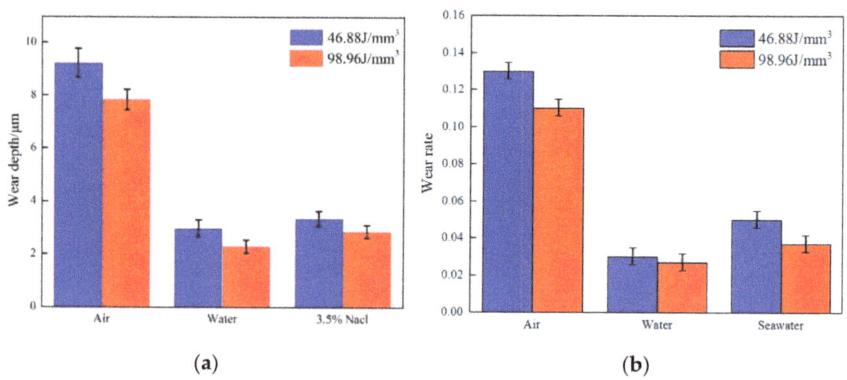

Figure 8. Wear depth and wear rate. (**a**) Wear depth, (**b**) Wear rate.

Under the three friction conditions, the wear depth and wear rate of the sample with an energy density of 46.88 J/mm^3 is greater than that of the sample with an energy density of 98.96 J/mm^3. This is due to the fact that the low-energy density sample has more pores and the structure is not dense. When the energy density of the sample is constant, the wear depth and wear rate in the aqueous solution are much smaller than during dry friction. This is mainly because the lubrication effect of the aqueous solution greatly reduces friction and wear. The wear depth and wear rate in the distilled water are lower than those in the 3.5% NaCl solution. In the 3.5% NaCl solution, wear is accompanied by corrosion, so the wear depth and wear rate in the 3.5% NaCl solution are slightly higher than those in the distilled water.

3.3. Wear Morphology

Figure 9a is the wear scar and oxygen Energy Dispersive Spectroscopy (EDS) during dry friction. The surface of the abrasion marks was covered with an oxide film, and the surface was severely cracked. In some areas, the oxide film was removed to form wear debris, leaving a dark and rough surface. The oxidation of the sliding wear surface is the result of friction heating in the ambient air. Additionally, in this process, plastic deformation and material peeling are produced, which cause the rough surface of the wear to produce many grooves and cracks. Therefore, the wear mechanism is mainly oxidative wear and adhesive wear. The wear scar morphology and oxygen EDS in the distilled water are shown in Figure 9b. The wear marks in the distilled water have clear contours, and furrows are formed in the wear marks. Under the cooling and lubricating action of distilled water, the adhesion and peeling of the material surface are inhibited. There are furrows and micro cracks along the fretting direction on the wear scar, and distilled water plays a role in blocking oxygen, so the surface oxidation is inhibited. This indicates that the wear mechanism of 316L stainless steel formed by SLM in the distilled water is mainly abrasive wear, accompanied by fatigue wear. From Figure 9c, it is found that the wear scars in the 3.5% NaCl solution are covered with furrows along the fretting direction and microcracks in parallel and perpendicular fretting directions. Compared with dry friction, 3.5% NaCl solution inhibits the occurrence of oxidation reaction. There is almost no adhesion and material peeling on the wear surface in the 3.5% NaCl solution. The wear surface is mainly composed of furrows and microcracks, so it is in sea water. The wear mechanism is the same as that in the distilled water—mainly abrasive wear accompanied by fatigue wear.

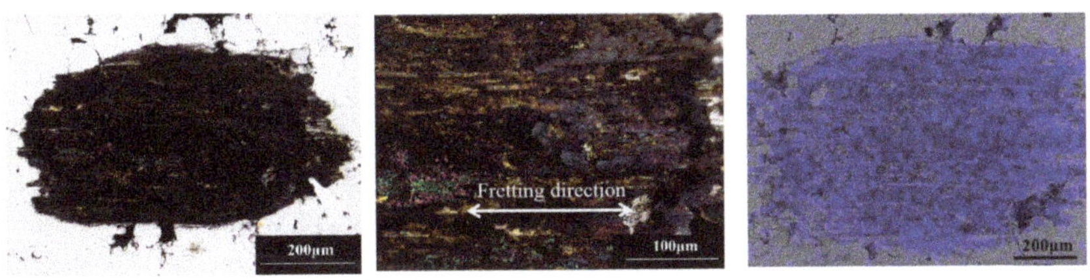

(a)

(b)

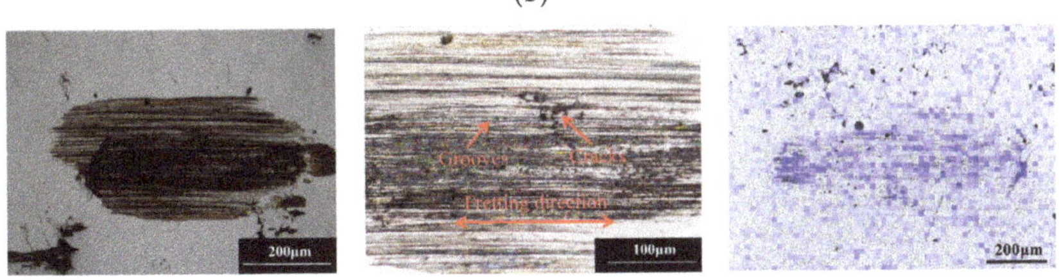

(c)

Figure 9. OM image of wear scar and EDS image of oxygen element: (**a**) air; (**b**) water; and (**c**) 3.5% NaCl.

4. Conclusions

Based on the results obtained, several conclusions can be drawn about the fretting wear characteristics of SLM-formed 316L stainless steel made by the 3D printer in seawater.

(1) The energy density of SLM will affect the porosity of the 316L sample, thereby affecting the compactness of the sample, and it will further affect the fretting wear characteristics of the material. Within a certain range of energy density, the smaller the energy density, the less dense the material, and the less wear resistant the material is.

(2) Under the three friction conditions, the friction coefficient of 316L stainless steel formed by SLM first increased and then gradually stabilized. The aqueous solution

will accelerate the stabilization and reduce sudden changes. When the energy density is the same, the friction coefficient is highest in air and lowest in seawater. Compared to distilled water, the easy-shear friction film produced in seawater can further reduce the friction coefficient.

(3) Under dry-friction conditions, the main wear mechanisms are oxidative wear and adhesive wear. The fretting wear mechanism in the distilled water and 3.5% NaCl solution are the same as abrasive wear accompanied by fatigue wear. The lack of lubrication results in higher wear depth and wear rate than those in aqueous solution. The friction coefficient in the 3.5% NaCl solution is generally smaller than that in the distilled water. However, due to friction and corrosion during this friction process, the wear depth and wear rate are slightly greater those that in the distilled water.

Author Contributions: Conceptualization, M.H., P.C. and X.Q.; methodology, M.H. and P.C.; software, M.H.; validation, M.H., P.C. and X.Q.; formal analysis, X.Q.; investigation, M.H.; resources, M.H.; data curation, M.H.; writing—original draft preparation, M.H.; writing—review and editing, P.C.; visualization, X.Q.; supervision, P.C.; project administration, X.Q.; and funding acquisition, X.Q. All authors have read and agreed to the published version of the manuscript.

Funding: This research was funded by the National Natural Science Foundation of China (No. 51905032; No. 51975042), and the Science and Technology Innovation Special Fund, Foshan (No. BK20BE024; No. BK21BE018).

Data Availability Statement: The processed data required to reproduce these findings cannot be shared at this time because the data also form part of ongoing research.

Conflicts of Interest: The authors declare no conflict of interest.

References

1. Blakey-Milner, B.; Gradl, P.; Snedden, G.; Brooks, M.; Pitot, J.; Lopez, E.; Leary, M.; Berto, F.; du Plessis, A. Metal additive manufacturing in aerospace: A review. *Mater. Des.* **2021**, *209*, 110008. [CrossRef]
2. Yu, W.H.; Sing, S.L.; Chua, C.K.; Kuo, C.N.; Tian, X.L. Particle-reinforced metal matrix nanocomposites fabricated by selective laser melting: A state of the art review. *Prog. Mater. Sci.* **2019**, *104*, 330–379. [CrossRef]
3. Ghorbani, J.; Li, J.; Srivastava, A.K. Application of optimized laser surface re-melting process on selective laser melted 316L stainless steel inclined parts. *J. Manuf. Process.* **2020**, *56*, 726–734. [CrossRef]
4. Olakanmi, E.O.; Cochrane, R.F.; Dalgarno, K.W. A review on selective laser sintering/melting (SLS/SLM) of aluminium alloy powders: Processing, microstructure, and properties. *Prog. Mater. Sci.* **2015**, *74*, 401–477. [CrossRef]
5. Cherry, J.A.; Davies, H.M.; Mehmood, S.; Lavery, N.P.; Brown, S.G.R.; Sienz, J. Investigation into the effect of process parameters on microstructural and physical properties of 316L stainless steel parts by selective laser melting. *Int. J. Adv. Manuf. Technol.* **2015**, *76*, 869–879. [CrossRef]
6. Huang, M.; Yang, Y.; Feng, S. Effect of 316L SLM forming process on sliding wear characteristics and hardness. *Surf. Technol.* **2020**, *49*, 221–227.
7. Yan, J.; Huang, J.; Li, K.; Zhou, P.; Wang, Y.; Qiu, J. Effect of Boron Content on Microstructure, Mechanical Properties, and Oxidation Resistance of Mo-Si-B Composites. *J. Mater. Eng. Perform.* **2018**, *27*, 6218–6226. [CrossRef]
8. Huang, M.; Hang, J.; Dong, X. Tribological properties of the SLM-316L filament under the grease lubrication condition. *Chin. J. Eng.* **2021**, *43*, 835–842.
9. Peng, P.; Wang, K.; Wang, W.; Han, P.; Zhang, T.; Liu, Q.; Zhang, S.; Wang, H.; Qiao, K.; Liu, J. Relationship between microstructure and mechanical properties of friction stir processed AISI 316L steel produced by selective laser melting. *Mater. Charact.* **2020**, *163*, 110283. [CrossRef]
10. Shin, W.-S.; Son, B.; Song, W.; Sohn, H.; Jang, H.; Kim, Y.-J.; Park, C. Heat treatment effect on the microstructure, mechanical properties, and wear behaviors of stainless steel 316L prepared via selective laser melting. *Mater. Sci. Eng. A* **2021**, *806*, 140805. [CrossRef]
11. Sun, Y.; Moroz, A.; Alrbaey, K. Sliding Wear Characteristics and Corrosion Behaviour of Selective Laser Melted 316L Stainless Steel. *J. Mater. Eng. Perform.* **2014**, *23*, 518–526. [CrossRef]
12. Bartolomeu, F.; Buciumeanu, M.; Pinto, E.; Alves, N.; Carvalho, O.; Silva, F.S.; Miranda, G. 316L stainless steel mechanical and tribological behavior—A comparison between selective laser melting, hot pressing and conventional casting. *Addit. Manuf.* **2017**, *16*, 81–89. [CrossRef]
13. Huang, Y.; Yang, S.; Gu, J.; Xiong, Q.; Duan, C.; Meng, X.; Fang, Y. Microstructure and wear properties of selective laser melting 316L. *Mater. Chem. Phys.* **2020**, *254*, 123487. [CrossRef]

14. Zhang, P.; Shan, L.; Su, X.; Huang, Z.; Luo, L.; Chen, J. Microstructure and tribological performance of CrTiSiCN coatings on 316L and TC4 in seawater. *Tribol. Int.* **2021**, *156*, 106832. [CrossRef]
15. Hoseinieh, S.M.; Shahrabi, T. Influence of ionic species on scaling and corrosion performance of AISI 316L rotating disk electrodes in artificial seawater. *Desalination* **2017**, *409*, 32–46. [CrossRef]
16. Alkan, S.; Gök, M.S. Effect of sliding wear and electrochemical potential on tribocorrosion behaviour of AISI 316 stainless steel in seawater. *Eng. Sci. Technol. Int. J.* **2021**, *24*, 524–532. [CrossRef]
17. Zhang, X.; Xia, Y.; Zhang, Y.; Tan, Q.; Zhu, Z.; Lin, L. Experimental study on wear behaviors of TBM disc cutter ring under drying, water and seawater conditions. *Wear* **2017**, *392–393*, 109–117. [CrossRef]
18. Fan, N.; Wang, Y.; Wang, Q.; Wang, M.; Yan, F. Effect of displacement amplitude on fretting wear of 304 stainless steel in air and sea water. *Lubr. Sci.* **2018**, *30*, 116–125. [CrossRef]
19. Hou, R.; Lu, S.; Chen, S.; Dou, W.; Liu, G. The corrosion of 316L stainless steel induced by methanocossus mariplaudis through indirect electron transfer in seawater. *Bioelectrochemistry* **2022**, *149*, 108310. [CrossRef]
20. Simchi, A. Direct laser sintering of metal powders: Mechanism, kinetics and microstructural features. *Mater. Sci. Eng. A* **2006**, *428*, 148–158. [CrossRef]
21. Varenberg, M.; Halperin, G.; Etsion, I. Different aspects of the role of wear debris in fretting wear. *Wear* **2002**, *252*, 902–910. [CrossRef]
22. Chen, J.; Wang, J.; Chen, B.; Yan, F. Tribocorrosion Behaviors of Inconel 625 Alloy Sliding against 316 Steel in Seawater. *Tribol. Trans.* **2011**, *54*, 514–522. [CrossRef]
23. Wang, M.; Wang, Y.; Fang, N.; Wang, Q.; Yan, F. Fretting Wear Properties of 2507 Super Duplex Stainless Steel in Artificial Seawater. *Lubr. Eng.* **2019**, *44*, 14–26.
24. Ding, H.Y.; Dai, Z.D.; Zhang, Y.; Zhou, G.H. Corrosive Fretting Wear Behavior of a Titanium Alloy TC11 in Artificial Seawater. In *Advanced Tribology*; Springer: Berlin/Heidelberg, Germany, 2009; pp. 322–326.
25. Ding, H.Y.; Zhou, G.H.; Hui, D. Friction and Wear Performance of An Aluminium Alloy in Artificial Seawater. *Proc. Inst. Mech. Eng. Part J.* **2011**, *225*, 43–49. [CrossRef]

machines

MDPI

Article

Effect of Process Parameters on Stress Field of Laser Additive Manufacturing

Yulin Liu, Qi Li, Zhaohui Ren *, Zeyu Jiang, Hengfa Luo and Xingwen Zhang

School of Mechanical Engineering and Automation, Northeastern University, Shenyang 110819, China
* Correspondence: zhhren@mail.neu.edu.cn

Abstract: In order to optimize the additive manufacturing process and find the process parameters affecting the mechanical properties of the parts, an additive manufacturing simulation model of Ti-6Al-4V titanium alloy was established, and the effects of ambient temperature, substrate thickness and wire temperature on the stress field and residual stress field were analyzed. The results show that the ambient temperature is inversely proportional to the residual stress of the cladding layer, while the substrate thickness and wire temperature are positively correlated to the residual stress of the cladding layer. When the ambient temperature increases from 0 °C to 600 °C, the maximum residual stress decreases by 36.0%, the maximum residual stress increases by 10.0% when the substrate thickness increases from 25 mm to 55 mm and the maximum residual stress increases by 7.48% when the temperature increases from 0 °C to 600 °C. The influence of the three parameters on the maximum residual stress is as follows: ambient temperature > substrate thickness > wire temperature. The research results can provide reference for stress control during actual manufacturing.

Keywords: additive manufacturing; process parameters; residual stress; Ti-6Al-4V titanium alloy

1. Introduction

Additive manufacturing (AM) technology is mainly the process of melting materials and stacking them layer by layer to form required parts by making use of various energies, also known as "the third industrial revolution" [1–3]. At present, there are dozens of kinds of AM technologies, involving many fields such as industry, education, medicine, etc. Meanwhile, the materials used in AM are various as well, such as steel, titanium alloy, aluminum alloy in industrial fields, various kinds of cells in the medical industry, etc. [4]. Compared with traditional manufacturing methods, AM has advantages of saving molds and additive materials, low cost, short cycle, wide range of applicable materials and applicability for steel materials that are hard to process by traditional methods. In addition, the parts used have the advantages of fine grain, dense structure, excellent performance in mechanical properties and uniform composition [5,6].

The AM parts made by AM also have many shortcomings, such as large residual stress, high surface roughness and cracks [7,8]. Therefore, it is very necessary to carry out relevant research on optimizing AM technologies. Sun et al. [9] studied the impacts of different scanning routines on sheet deformation during arc AM, and created the S-value discriminant method for sheet deformation degree, which was applied to the AM process of solid parts. By taking 5356 aluminum alloy as the welded object of AC argon tungsten arc welding, Shen et al. [10] studied the impact of different intervals on the cladding layer and microstructure. The results show that various intervals can produce better parts. Through numerical simulation and experiments, Ogino et al. [11] studied the impact of cooling and welding direction in arc welding, and the results show that under appropriate cooling conditions, the shape of the deposited layer becomes larger and thinner, and for the linear and cylindrical deposition, the welding direction has an obvious impact on forming quality. Through the mutual verification between experiments and simulation, Daniel et al. [12] analyzed the life, width and depth of a molten pool under different electron beam powers, scanning speeds and energies, and stimulated the temperature distribution during the selective melting process of the electron beam by using the finite element simulation tool which was developed by them independently. The results show that the life of the molten pool increases linearly along with linear energy, while the size of the molten pool is nonlinear, related to linear energy. Zinoviev et al. [13] established the evolutionary mathematical model of grain structure in each component used for prediction of the laser melting process, and simulated the selective laser melting process. In the simulation, the growth and structure formation of grains during laser additive manufacturing (LAM) were fully described. Chu [14] established the simulation model of laser cladding Al_2O_3 aluminum alloy, and carried out research on the roles of preheating and slow cooling on the laser cladding and the influence law of different process parameters on the temperature gradient. Loh et al. [15] created a selective laser melting simulation model taking the transformation from powder to solid into account, and carried out research on the relationship between laser power, scanning speed and melt size, melting and evaporation of powder and temperature change rate. By comprehensively taking free surface, temperature, component concentration and molten pool flow velocity into consideration, Dubrov et al. [16] established the dynamic equation for the laser melting process of metal powder, and provided its numerical implementation algorithm, and tested calculation results of the temperature field, flow structure of the molten pool, solute distribution and geometric characteristics of the object to be formed. Zhao et al. [17] analyzed the influence law of thermal cycle and cladding direction on AM part quality during AM of sheet parts. Miranda et al. [18] conducted research on the impact of the relative position of wire feeder and substrate and ratio of laser radius to wire radius on the forming quality. The results show that the former has impacts on the transfer mode of molten metal and the pressure of the wire tip on the molten pool, while the latter has an influence on procedure efficiency. The two parameters control the surface finish of the parts. Zhan et al. [19] discussed the effect of process parameters on residual stress. The results show that the residual stress is positively correlated with the laser power, and it is negatively correlated with the scanning speed and powder feeding. Fan et al. [20] analyzed the residual stress field of TC4/TC11 FGM fabricated via LAM using the finite element method. Samodurova et al. [21] developed a methodology for manufacturing titanium alloy products and studied the microstructure of the obtained details and measured the microhardness of the samples. Products made according to the developed technique do not have visible defects and pores.

There are many elements causing residual stress of AM parts. Three parameters were selected, which are studied rarely at present but very important in real processing. The research results can provide a significant theoretical basis for residual stress control in LAM. Due to advantages such as relatively better high-temperature resistance, oxidation resistance and corrosion resistance, at present, Ti-6Al-4V titanium alloy is mainly used in the aerospace and maritime fields [6]. Based on the simulation of the AM process, this research describes the change in internal properties of the cladding layer, manufactured by laser adding material with titanium alloy wire.

AM is a process of "rapid heating and rapid cooling", and the formed cladding layer is heated unevenly and the temperature changes dramatically, so the cladding layer will produce thermal stress due to expansion with heat and contraction with cold. When this thermal stress exceeds the yield and tensile strength of Ti-6Al-4V titanium alloy, it is easy to deform the cladding layer and substrate. Therefore, the mechanism and action law of thermal stress in AM are a difficult problem.

2. Modeling of LAM

2.1. Finite Element Model

In this paper, the simulation model of laser wire additive manufacturing is established by ABAQUS, and the function of MODEL CHANGE in ABAQUS is used to simulate the addition of Ti-6Al-4V titanium alloy material. In order to make the numerical simulation results more accurate, this paper determines the simulation process parameters according to the actual experimental data [22]. The parameters are as follows. Laser scanning speed is 6 mm/s. Laser power is 180 W. Laser diameter is 0.8 mm. Simplified finite element model of the substrate is set to 60 mm × 14 mm × 5 mm. In addition, the length, width and thickness of the cladding layer are 50 mm, 0.85 mm and 0.54 mm. The four corners of the substrate are completely fixed.

As displayed in Figure 1, the substrate is meshed, and the area near the cladding layer is densely meshed, with 40,680 nodes and 34,582 elements.

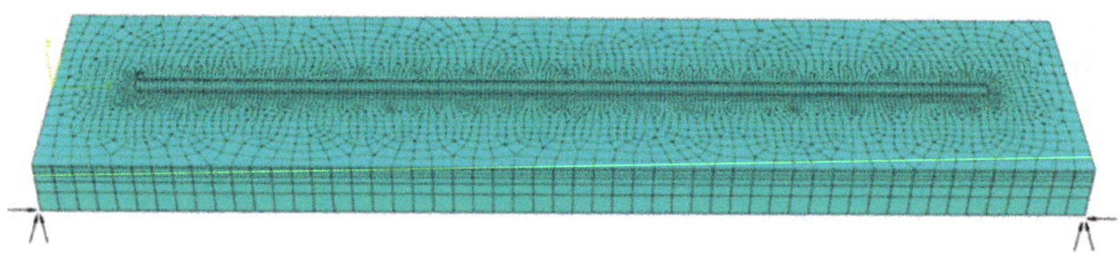

Figure 1. Grid diagram of substrate and cladding layer in AM.

In the finite element model of the cladding layer and the substrate, the surface heat transfer coefficient and thermal radiation coefficient are set according to Equations (1)–(3) to simulate the actual heat dissipation process. The initial temperature of the cladding layer and the substrate is set at 20 °C and the ambient temperature is set at 20 °C. In addition, a birth–death element strategy is used to simulate the material addition process.

The heat production equation is

$$\frac{\partial(\rho c_P T)}{\partial t} + \nabla(\rho c_P v T) - \nabla(K \nabla T) = Q \tag{1}$$

in which Q is heat generation of laser heat source; K is thermal conductivity of titanium alloy; c_p is specific heat capacity; ρ is density of titanium alloy; t is laser heat source action time; v is laser scanning speed.

The initial condition of the heat conduction equation of the laser heat source is

$$-K(\nabla T \boldsymbol{n})|_{\Omega} = \begin{cases} \alpha I(x,y,z,t) - h(T - T_0) - \varepsilon_t \sigma(T^4 - T_0^4) & \text{if } \Omega \in \Gamma \\ -h_c(T - T_0) - \varepsilon_t \sigma(T^4 - T_0^4) & \text{if } \Omega \notin \Gamma \end{cases} \tag{2}$$

in which $\boldsymbol{n}$ is the normal vector of the cladding layer and substrate surface; $I\ (x, y, z, t)$ is the energy distribution of the cladding layer and substrate surface; α is laser absorption coefficients of Ti-6Al-4V titanium alloys; h_c is convection coefficient of the cladding layer and substrate; ε_t is radiance of the cladding layer and substrate; σ is Stefan–Boltzmann constant; Ω is cladding layer and substrate surface; Γ is laser heating surface; T_0 is ambient temperature.

The boundary condition is

$$\begin{cases} T(x,y,z,0) = T_0 \\ T(x,y,z,\infty) = T_0 \end{cases} \tag{3}$$

2.2. Basic Assumption

In this paper, the following assumptions are made for the simulation process of LAM.

- The interaction between laser heat source and material follows the traditional heat transfer theory [23].
- The materials in this paper are considered as isotropic and the effect of temperature on density is not considered.
- The influence of molten pool flow and vaporization on temperature is neglected.

- Since the deformation of the material has little effect on the temperature, the effect of material deformation on temperature is neglected.
- The effect of substrate deformation on the relaxation of the residual stresses is neglected.

2.3. Heat Source Model

In the actual laser AM process, the energy of the heat source acting on the cladding layer gradually decreases from center to periphery. At the same time, due to the continuous movement of the laser heat source, the energy distribution is different before and after. Therefore, the double ellipsoid heat source is selected to simulate the loading process of the laser heat source, and the coordinate directions of the finite element model are defined by the coordinate directions of the heat source. The double ellipsoid heat source is shown in Figure 2 [24].

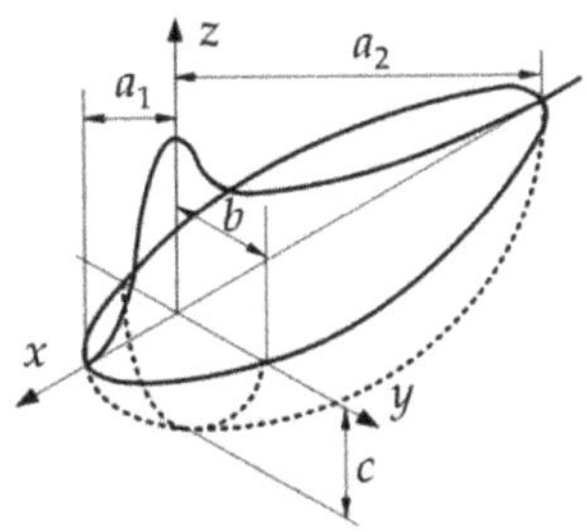

Figure 2. Sketch of the three-dimensional model of the double ellipsoid heat source.

The expression of the double ellipsoid heat source model is

$$\begin{cases} Q_1 = \frac{6\sqrt{3}p\omega}{\pi^{3/2}a_1bc}\exp\left\{-3\left[\left(\frac{x-x_0}{a_1}\right)^2+\left(\frac{y-y_0}{b}\right)^2+\left(\frac{z-z_0}{c}\right)^2\right]\right\} \\ Q_2 = \frac{6\sqrt{3}p\omega}{\pi^{3/2}a_2bc}\exp\left\{-3\left[\left(\frac{x-x_0}{a_2}\right)^2+\left(\frac{y-y_0}{b}\right)^2+\left(\frac{z-z_0}{c}\right)^2\right]\right\} \end{cases} \tag{4}$$

in which Q_1 and Q_2 are the quantity of heat generated by the laser heat source; p is laser power; ω is the energy absorption efficiency of Ti-6Al-4V titanium alloy; a_1, a_2, b and c are shape parameters of the double ellipsoid heat source; x_0, y_0 and z_0 are position coordinates of the starting point of the cladding layer.

2.4. Material Model

The physical parameters of titanium alloy change with the change in temperature, so the setting of thermal physical parameters of titanium alloy is a major factor affecting the accuracy of the results. The line chart of thermophysical performance parameters of titanium alloy is shown in Figure 3.

In the finite element simulation experiment, the materials of the substrate and wire are Ti-6Al-4V titanium alloy, and their chemical compositions are listed in Table 1.

Table 1. Chemical compositions of Ti-6Al-4V wire and substrate.

Composition	Al	V	Fe	C	Ti
wire	5.8	3.8	0.09	0.01	Bal
substrate	5.5	4.8	0.4	0.2	Bal

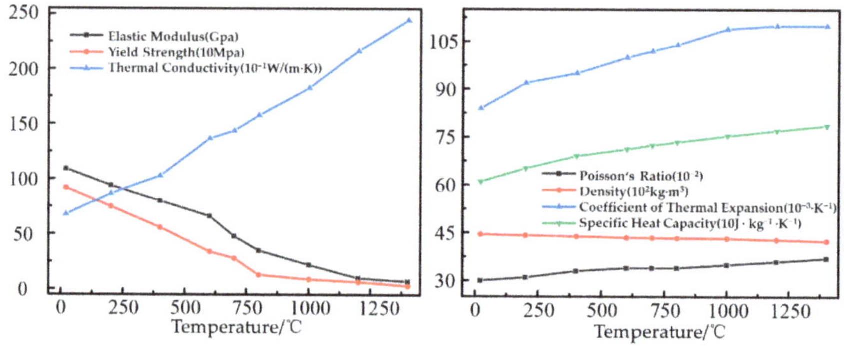

Figure 3. Thermophysical performance parameters of Ti-6Al-4V.

3. Analysis of Thermal Stress Field of LAM

3.1. Model Validation

In order to verify the authenticity of the established model, the material is set as the material used in the experiment of reference [22] for AM simulation without changing other parameters. The temperature curves at different points are obtained through selecting three points at equal intervals between cladding layers, which are compared with experimental results and shown as Figure 4. From the simulation results in Figure 4b, it can be seen that all temperatures at each point of the cladding layers have two processes of increasing and decreasing. When the laser heat source arrives, the temperature of the cladding layer increases rapidly, and when heat source leaves this point, the temperature of the cladding layer decreases rapidly and the cooling speed is slower than the speed of temperature increase. In addition, along with the consistent processing of the laser cladding, the heat dissipation capacity of the substrate and cladding layers gradually decreases, meaning the temperature peak values at all points are continuously increased. The above conclusions are consistent with the temperature data shown in Figure 4a. Among them, the error of maximum temperature at 1–3 points of the cladding layer is less than 10%, which proves the accuracy of the simulation model.

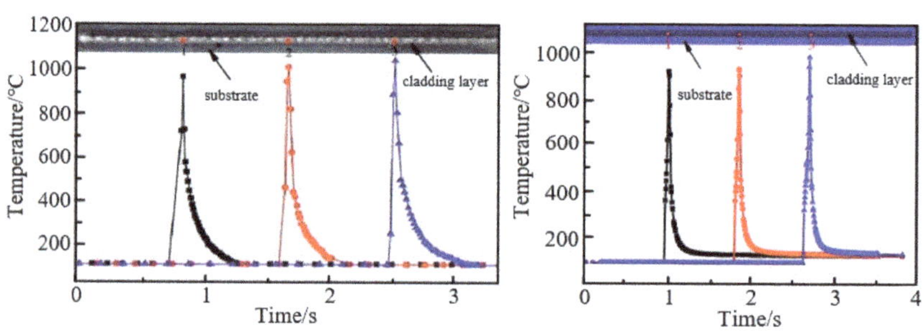

Figure 4. Temperature curve at different points of the cladding layer at equal intervals: (**a**) Temperature curve at different points of cladding layer in reference [22]; (**b**) finite element simulation of temperature curve at different points of cladding layer.

3.2. Results of Stress Field

The middle point of the cladding layer is selected for analysis and defined as point A. Figure 5 shows the stress cloud diagram and stress curve diagram of point A when the laser heat source moves to point A. It can be seen from Figure 5a that when the laser heat source passes through the substrate, the titanium alloy wire and the substrate melt to form a liquid molten pool, so the internal equivalent stress is the minimum, which is in the zero-stress state. At the front of the molten pool, due to the action of the laser heat source, the substrate and the wire are heated and expanded to produce a large area of thermal stress. Additionally, at the back end of the heat source, due to the heat dissipation and distance away from the heat source, the heat-affected area of the cladding layer gradually decreases. Therefore, the stress is concentrated near the cladding layer. It can be seen from Figure 5b that when the heat source gradually approaches point A, the stress increases, and when the molten pool is formed the stress decreases to zero. After the heat source passes through the substrate, the molten pool changes from liquid to solid, and then a large stress is formed.

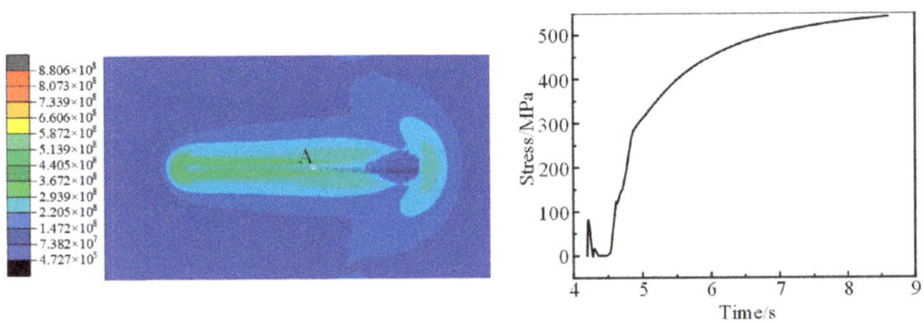

(a) **(b)**

Figure 5. Stress cloud diagram and stress curve diagram of point A: (**a**) Stress cloud diagram of LAM; (**b**) stress curve diagram of point A.

The residual stress distribution has a significant effect on the quality of the laser cladding layer and life of the workpiece. Figure 6 shows the residual stress distribution after the cladding layer is cooled for 900 s. It can be seen from Figure 6 that during the laser forming process, not only is residual stress generated in the cladding area, but also in the surrounding area without melting. However, the stress near the cladding layer is greater than that at other positions.

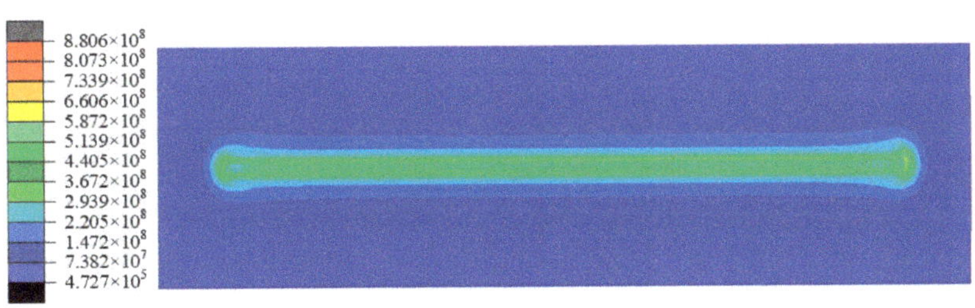

Figure 6. Residual stress cloud diagram of cladding layer.

Figure 7 shows the residual stress curve along the path of AM after the cladding layer is cooled for 900 s. It can be seen from Figure 7 that the residual stress in the middle of the cladding layer is much greater than that in the beginning and end positions. The main reason is that there is no constraint at the beginning and end positions, and the residual stress can be released well. However, in the middle of the cladding layer, due to the constraints of the substrate and the cladding layers, a large residual stress is generated, which increases the risk of cracks and other defects.

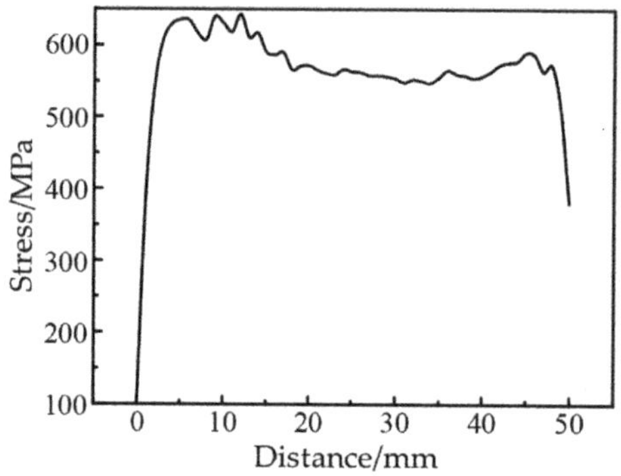

Figure 7. Residual stress curve along the path.

As shown in Figure 8, we firstly let X, Y and Z be the stress direction produced by LAM for the following discussion with ease.

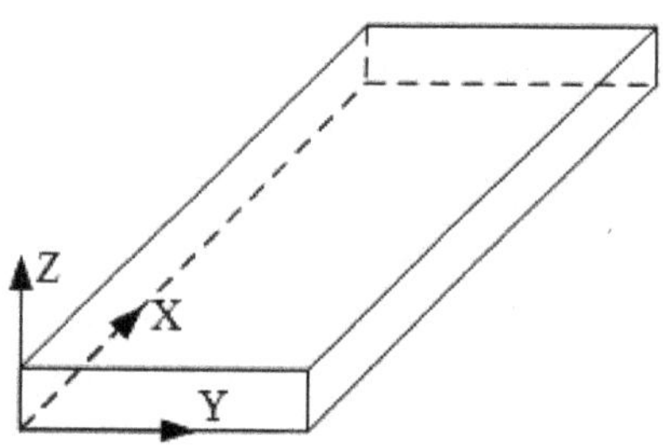

Figure 8. Specified stress revised direction.

4. Influence of Different Parameters on Stress Field of Formed Parts

4.1. Ambient Temperature

During the research on the influence of ambient temperature on the stress field of formed parts, the ambient temperatures of 0 °C, 200 °C, 400 °C and 600 °C are selected. Taking point A as the research object, the stress and stress component curve are obtained at different ambient temperatures, which is shown as Figure 9. From Figure 9a, it can be found that the variation trends of von Mises stress of point A at each ambient temperature are almost the same. However, along with the increment in ambient temperature, the peak value of stress is gradually decreased. It is decreased by 80 MPa and about 14.5% from 552 MPa at 0 °C to 472 MPa at 600 °C. From the stress curve of Figure 9b,d, it can be seen that the stress in the x direction and z direction is basically over time. The relationship between stress and ambient temperature is a negative correlation. From the curve of Figure 9c, the stress of the y direction basically has a downward trend, and lower temperature the higher formed pressure is. Therefore, since the stress in the x direction is the largest among all directions, cracks are more likely to occur in the vertical direction of this direction.

Figure 10a shows the curve of residual stress of the cladding layer at different ambient temperatures. Comparing the residual stress curves at 0 °C, 200 °C, 400 °C and 600 °C, it can be seen that the residual stress of the cladding layer decreases as the ambient temperature increases. Figure 10b shows the line chart of maximum residual stress of the cladding layer at different ambient temperatures. With the increase in ambient temperature, the maximum residual stress is decreased linearly. It is decreased by 254 MPa and about 36.0% from 706 MPa to 452 MPa. Therefore, ambient temperature has a great influence on the residual stress of the formed parts.

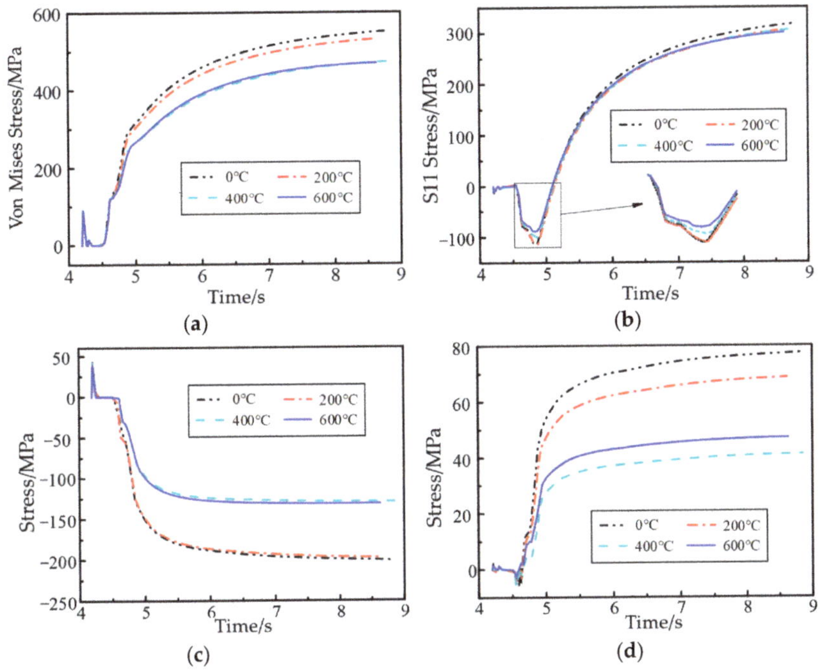

Figure 9. Stress curves under different ambient temperatures: (**a**) Von Mises stress curve; (**b**) stress curve in x direction; (**c**) stress curve in y direction; (**d**) stress curve in z direction.

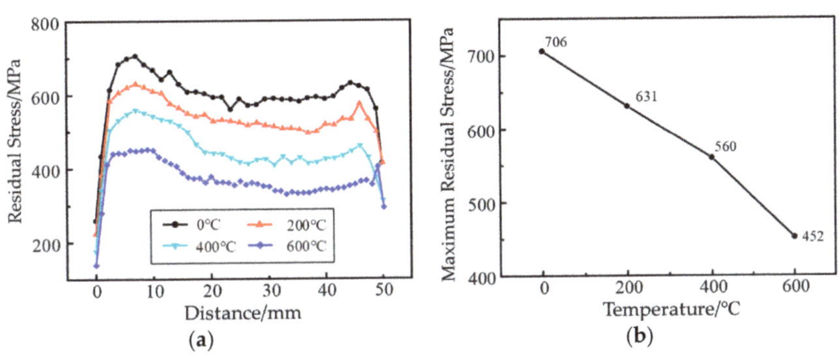

Figure 10. (**a**) Curve of residual stress at different ambient temperatures; (**b**) line chart of maximum residual stress at different ambient temperatures.

4.2. Substrate Thickness

During the research on the influence of substrate thickness on stress field, four variables of substrate thickness of 25 mm, 35 mm, 45 mm and 55 mm are selected. Figure 11 is the equivalent stress and stress component curve of point A. From Figure 11a, it can be seen that along with the increment in substrate thickness, the von Mises stress increases gradually, which is increased about 15.9% from 511 MPa at substrate thickness of 25 mm to 592 MPa at substrate thickness of 55 mm. However, in the actual process, a relatively obvious deformation would happen when substrate thickness is decreased to a certain thickness [25]. Figure 11b,d show that the stress in the x direction and z direction increases with the increase in substrate thickness. From the curve of Figure 11c, the pressure of the y direction is basically increased over time.

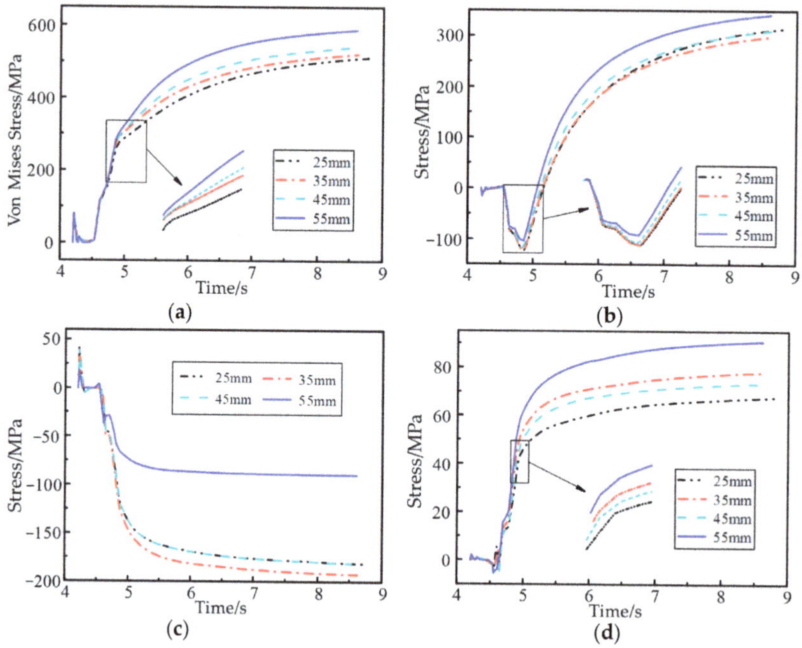

Figure 11. Stress curves under different substrate thicknesses: (**a**) Von Mises stress curve; (**b**) stress curve in x direction; (**c**) stress curve in y direction; (**d**) stress curve in z direction.

Figure 12a shows the curve of residual stress of the cladding layer under different substrate thicknesses. Comparing the residual stress curves at 25 mm, 35 mm, 45 mm and 55 mm, it can be seen that the residual stress of the cladding layer increases as the substrate thickness increases. Figure 12b shows the line chart of maximum residual stress of the cladding layer under different substrate thicknesses. With the increment in substrate thickness, the maximum residual stress is gradually increased. It is increased by 67 MPa and about 10.0% from 733 MPa to 666 MPa. Therefore, substrate thickness has a certain influence on the residual stress of the formed parts.

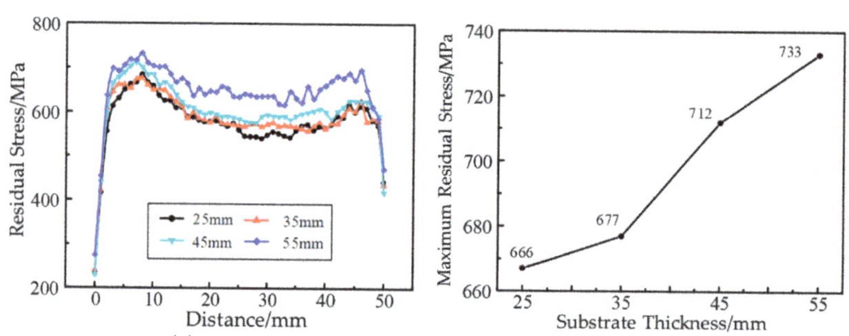

Figure 12. (**a**) Curve of residual stress under different substrate thicknesses; (**b**) line chart of maximum residual stress under different substrate thicknesses.

4.3. Wire Temperature

In the research of wire temperature, the wire temperatures of 0 °C, 200 °C, 400 °C and 600 °C are selected. Figure 13 is the equivalent stress and stress component curve of point A. From Figure 13a, it can be seen that along with the increment in wire temperature, the von Mises stress increases gradually, which is increased by 106 MPa and about 20.5% from 518 MPa at 0 °C to 624 MPa at 600 °C. Figure 13b,d show that the stress in the x direction and z direction increases with the increase in wire temperature. Figure 13c shows that the pressure of the y direction is basically increased over time.

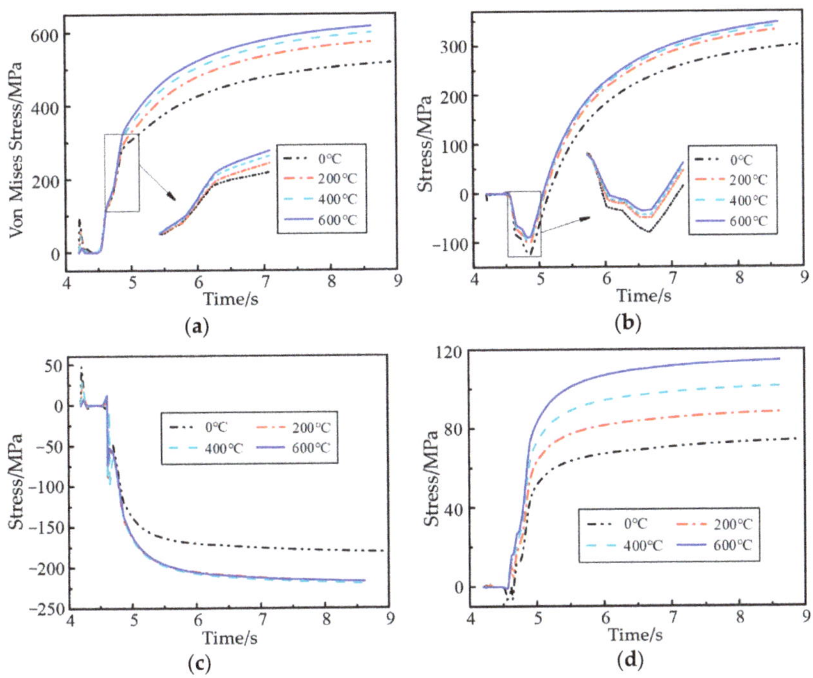

Figure 13. Stress curves under different wire temperatures: (**a**) Von Mises stress curve; (**b**) stress curve in x direction; (**c**) stress curve in y direction; (**d**) stress curve in z direction.

Figure 14a shows the curve of residual stress of the cladding layer under different wire temperatures. Comparing the residual stress curves at 0 °C, 200 °C, 400 °C and 600 °C, it can be seen that the residual stress of the cladding layer increases as the wire temperature increases. Figure 14b shows the line chart of maximum residual stress of the cladding layer under different wire temperatures. With the increase in wire temperature, the maximum residual stress is gradually increased. It is increased by 50 MPa and about 7.48% from 718 MPa to 668 MPa. Therefore, wire temperature has little effect on the residual stress of the formed parts.

By comprehensively comparing the effects of ambient temperature, substrate thickness and wire temperature on the stress field, it can be found that ambient temperature has the greatest effect on the residual stress of formed parts, and wire temperature has the least effect on the residual stress.

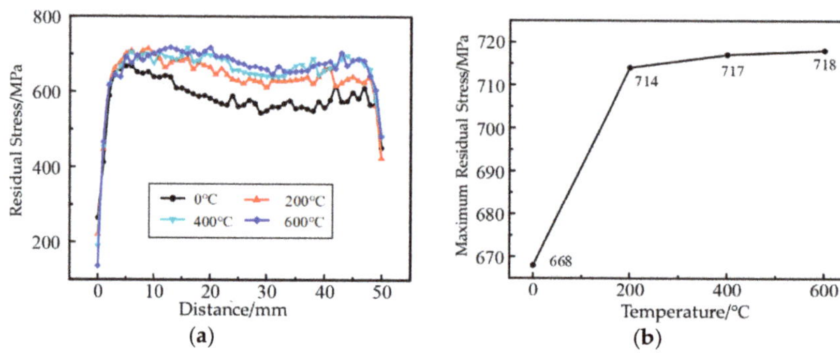

Figure 14. (**a**) Curve of residual stress under different wire temperatures; (**b**) line chart of maximum residual stress under different wire temperatures.

5. Conclusions

1. In the AM process, in a certain range of clamping conditions, temperature and substrate thickness, the residual stress basically increases first, then decreases and finally gradually increases and tends to be stable. The residual stress of the forming direction has the most influence on von Mises stress.
2. When the temperature increases from 0 °C to 600 °C, the maximum residual stress of the cladding layer basically decreases linearly. The maximum residual stress is decreased about 36.0% from 706 MPa to 452 MPa. When the substrate thickness is increased from 25 mm to 55 mm, the maximum residual stress of cladding layers is gradually increased by 10.0%. When the wire temperature increases from 0 °C to 600 °C, the maximum residual stress of the cladding layer is increased by 7.48%.
3. Through the comprehensive comparisons of residual stress of final formed parts affected by ambient temperature, substrate thickness and wire temperature, it can be found that the ambient temperature has the most influence, then the substrate thickness, and the wire temperature has the least influence.

Author Contributions: Y.L.: Writing—original draft. Q.L.: Investigation. Z.J.: Software. Z.R.: Writing—review & editing. H.L.: Methodology. X.Z.: Data curation, Investigation. All authors have read and agreed to the published version of the manuscript.

Funding: This research received no external funding.

Data Availability Statement: Not applicable.

Conflicts of Interest: The authors declare no conflict of interest.

References

1. Vrancken, B.; Thijs, L.; Kruth, J.P.; Van Humbeeck, J. Microstructure and Mechanical Properties of a Novel β Titanium Metallic Composite by Selective Laser Melting. *Acta Mater.* **2014**, *68*, 150–158. [CrossRef]
2. Huang, W.; Kovacevic, R. A Laser-Based Vision System for Weld Quality Inspection. *Sensors* **2011**, *11*, 506–521. [CrossRef] [PubMed]
3. Zhou, C.Y.; Luo, L.; Liu, Y.; Wu, J. Research Status of Additive Manufacturing Technology for Metal. *Hot Work. Technol.* **2018**, *47*, 9–14.
4. Song, B.; Dong, S.J.; Liao, H.L.; Coddet, C. Process Parameter Selection for Selective Laser Melting of Ti6Al4V Based on Temperature Distribution Simulation and Experimental Sintering. *Int. J. Adv. Manuf. Technol.* **2012**, *61*, 967–974. [CrossRef]
5. Baufeld, B.; Brandl, E.; Biest, O.V.D. Wire Based Additive Layer Manufacturing: Comparison of Microstructure and Mechanical Properties of Ti-6Al-4V Components Fabricated by Laser-Beam Deposition and Shaped Metal Deposition. *J. Mater. Process. Technol.* **2011**, *211*, 1146–1158. [CrossRef]
6. Mok, S.H.; Bi, G.; Folkes, J.; Pashby, L. Deposition of Ti-6Al-4V Using a High Power Diode Laser and Wire, Part I: Investigation on the Process Characteristics. *Surf. Coat. Technol.* **2008**, *202*, 3933–3939. [CrossRef]

7. Peng, Q.; Dong, S.Y.; Yan, S.X.; Men, P.; Wang, B. An Overview of Defects in Laser Melting Deposition Forming Products and the Corresponding Controlling Methods. *Mater. Rep.* **2018**, *32*, 2666–2671+2682.
8. Victor, C.; Philippe, Q.; Wilson, M.; Eric, C. Comparative Study of Fatigue Properties of Ti-6Al-4V Specimens Built by Electron Beam Melting (EBM) and Selective Laser Melting (SLM). *Mater. Charact.* **2018**, *143*, 76–81.
9. Sun, Q.J.; Sang, H.B.; Liu, Y.B.; Feng, J.C. Section Scanning Trajectory Planning Based on Arc Additive Manufacturing. *Trans. China Weld. Inst.* **2017**, *38*, 21–24+65.
10. Shen, J.Q.; Hu, S.S.; Liu, W.L.; Han, J.H. Effects of Time Interval in Rapid Prototyping of Al-alloy Based on Welding. *Trans. China Weld. Inst.* **2008**, *29*, 112–115+121.
11. Ogino, Y.; Asai, S.; Hirata, Y. Numerical Simulation of WAAM Process by a GMAW Weld Pool Model. *Weld. World* **2018**, *62*, 393–401. [CrossRef]
12. Daniel, R.; Thorsten, S.; Robert, F.S.; Paul, S.; Carolin, K.; Julia, M. Macroscopic Simulation and Experimental Measurement of Melt Pool Characteristics in Selective Electron Beam Melting of Ti-6Al-4V. *Int. J. Adv. Manuf. Technol.* **2017**, *88*, 1309–1317.
13. Zinoviev, A.; Zinovieva, O.; Ploshikhin, V.; Romanova, V.; Balokhonov, R. Evolution of Grain Structure During Laser Additive Manufacturing. Simulation by a Cellular Automata Method. *Mater. Des.* **2016**, *106*, 321–329. [CrossRef]
14. Chu, Y. Temperature Field Simulation of Dual-Beam Cladding Al_2O_3 and Design of Optical System. Master's Thesis, Dalian University of Technology, Dalian, China, 2014.
15. Loh, L.E.; Chua, C.K.; Yeong, W.Y.; Song, J.; Mapar, M.; Sing, S.L.; Liu, Z.H.; Zhang, D.Q. Numerical Investigation and an Effective Modelling on the Selective Laser Melting (SLM) Process with Aluminium Alloy 6061. *Int. J. Heat Mass Transf.* **2015**, *80*, 288–300. [CrossRef]
16. Dubrov, A.V.; Mirzade, F.K.; Dubrov, V.D. Mathematical Modeling of Thermal Behavior for Additive Manufacturing with Metal Powder Injection. *Procedia Eng.* **2017**, *201*, 478–488. [CrossRef]
17. Zhao, H.H.; Zhang, G.J.; Yin, Z.Q.; Liu, W. A 3D Dynamic Analysis of Thermal Behavior During Single-pass Multi-layer Weld-based Rapid Prototyping. *J. Mater. Process. Technol.* **2011**, *211*, 488–495. [CrossRef]
18. Miranda, R.M.; Lopes, G.; Quintino, L.; Rodrigues, J.P.; Williams, S. Rapid Prototyping with High Power Fiber Lasers. *Mater. Des.* **2008**, *29*, 2072–2075. [CrossRef]
19. Zhan, Y.; Liu, C.; Zhang, J.; Mo, G.; Liu, C. Measurement of Residual Stress in Laser Additive Manufacturing TC4 Titanium Alloy with the Laser Ultrasonic Technique. *Mater. Sci. Eng. A* **2019**, *762*, 138093. [CrossRef]
20. Fan, P.; Pan, J.; Ge, Y.; Zhan, Y. Finite Element Analysis of Residual Stress in TC4/TC11 Titanium Alloy Gradient Material Produced by Laser Additive Manufacturing. *Chin. J. Lasers* **2021**, *48*, 110–118.
21. Samodurova, M.; Logachev, I.; Shaburova, N.; Samoilova, O.; Radionova, L.; Zakirov, R.; Pashkeev, K.; Myasoedov, V.; Trofimov, E. A Study of the Structural Characteristics of Titanium Alloy Products Manufactured Using Additive Technologies by Combining the Selective Laser Melting and Direct Metal Deposition Methods. *Materials* **2019**, *12*, 3269. [CrossRef]
22. Li, H.M. Study on the Process Characteristics of Wire Based Laser Additive Manufacturing. Master's Thesis, Chongqing University, Chongqing, China, 2017.
23. Yin, B. Experimental Study on Laser Welding Technology of Invar Thick Plate Materials. Master's Thesis, Dalian University of Technology, Dalian, China, 2009.
24. Filippo, M.; Giuseppe, V.; Antonio, S.; Gianni, C. Finite Element Modelling of Wire-arc-additive-manufacturing Process. *Procedia CIRP* **2016**, *55*, 109–114.
25. Yuan, D.; Gao, H.B.; Sun, X.J.; Zhou, C.P.; Sun, L.B.; Chen, Y.J.; Guo, C.H.; Niu, Z.Y.; Jiang, F.C. Methods and Techniques for Improving Microstructure and Performance of Metal Additively Manufactured Materials. *Aeronaut. Manuf. Technol.* **2018**, *61*, 40–48.

MDPI AG
Grosspeteranlage 5
4052 Basel
Switzerland
Tel.: +41 61 683 77 34

MDPI Books Editorial Office
E-mail: books@mdpi.com
www.mdpi.com/books

www.ingramcontent.com/pod-product-compliance
Lightning Source LLC
Chambersburg PA
CBHW061446220926
51850CB00004B/26
* 9 7 8 3 7 2 5 8 4 6 5 1 1 *